VASCULAR PLANT FAMILIES AND GENERA

VASCULAR PLANT FAMILIES AND GENERA

A listing of the genera of vascular plants of the world according to their families, as recognised in the Kew Herbarium, with an analysis of relationships of the flowering plant families according to eight systems of classification

Compiled by

R.K. BRUMMITT

in collaboration with the Kew Herbarium Staff and with the assistance of botanists elsewhere

ROYAL BOTANIC GARDENS, KEW
1992

First published 1992

General Editor of Series J.M. Lock

Cover Design by Media Resources, Royal Botanic Gardens, Kew

ISBN 1842460862

Typeset at the Royal Botanic Gardens, Kew by Pam Arnold,
Christine Beard, Brenda Carey, Margaret Newman,
Pam Rosen and Helen Ward

CONTENTS

PREFACE

Kew has frequently been asked which authority it follows in deciding which plant families to adopt in its publications, particularly *Index Kewensis* and *Kew Record*. Editions 7 and 8 of J.C. Willis's *Dictionary of the Flowering Plants and Ferns* were prepared in the Kew Herbarium by the late H.K. Airy Shaw, published in 1966 and 1973 respectively, but the families recognised there differ in many instances from those recognised in other Kew publications. The answer to the question hitherto has been that Kew publications generally adopt the families used in the arrangement of the collections in the Kew Herbarium, which have never been defined in any publication and which may change with time anyway. The original main objective of the present volume was therefore to provide an answer to the question by listing all the genera currently recognised at Kew, indicating important synonyms, and assigning them to the families in which they are placed. This objective is addressed in Part 1 here.

In Part 2 the accepted genera are listed family by family. We hope that this will fill a major gap in botanical literature, for there is at present no modern publication to which one can turn to find all currently recognised genera listed for all the families of higher plants. Estimates of the numbers of genera in many families are often based on inadequate documentation.

A related second question has also been posed to Kew from time to time, going beyond mere delimitation of families and asking what sequence the accepted families should be arranged in. The last two decades have seen much activity in publication of new systems of classification of the flowering plants, and a number of competing and often very different modern systems are available. The herbarium collections at Kew, however, are basically still arranged according to the sequence of families in Bentham & Hooker's *Genera Plantarum* published last century, with relatively minor modifications. The logistic problems of rearranging 6 million specimens according to a new classification are sufficiently formidable to have prevented any serious consideration of the possibility of doing so. The present volume has therefore side-stepped the issue and has listed all families, and the genera within them, alphabetically in Part 2. At the same time, however, the reader is offered the possibility of considering how a number of systems of classification have treated the flowering plant families which Kew recognises. In Part 2, under each family heading a chart gives the position of that family in each of eight systems, six of them modern with two older ones added to give an historical perspective.

In Part 3 each of these eight systems is laid out separately with an enumeration of the families recognised under a hierarchy of higher groupings. A simple numerical cross reference for family names between Parts 2 and 3 is provided, which it is hoped will facilitate comparison of the different classifications. An indication is also given in both Parts 2 and 3 of how and where circumscriptions of families differ between Kew and each of the eight systems. In the fern allies and ferns the problems of generic concepts are probably greater than in the flowering plants, and here the number of genera recognised by different authorities, as well as their different family concepts, are given under the family headings.

The classification of the vascular plants is still very controversial in many areas at generic, family and higher levels. Changes will continue to be proposed for a long time yet. What Kew has tried to present here is admittedly only one institution's view at one point in time, though after wide consultation with other institutions and botanists around the world to whom our deep indebtedness is expressed in the extensive acknowledgements below. We have tried to make our lists of genera and families as comprehensive, accurate and up to date as possible, but are well aware that there are probably many omissions and errors. At the higher levels we have tried to present merely a non-committal perspective of some of the currently competing systems. We hope that others will find something of interest and value in this presentation of data on families and genera, and we will be pleased to receive corrections or critical comments with a view to improving our data-base for the future.

Production of the work has been very much a co-operative effort. My sincere personal thanks are due to my colleagues at Kew and to the many botanists elsewhere, individually acknowledged below, who have assisted in many ways. Their kind co-operation in so freely giving information and friendly advice has made my task of compilation a pleasure.

R.K. Brummitt
Royal Botanic Gardens, Kew
November 1991

INTRODUCTION

Notes on some general matters are given in these introductory comments. Explanations of the content and format of each of the three Parts are given separately at the head of each Part.

DEVELOPMENT OF THE DATA-BASE

The original compilation of a consolidated list of families and genera in the Kew Herbarium was started in 1974. The information was taken from the generic indexes kept in the herbarium cupboards for each family. Accepted names were marked in a copy of Airy Shaw's 8th edition of Willis's *Dictionary*, with notes of any differences in family placements. In 1981 these accepted names, together with their author or authors, were extracted from the copy of the *Dictionary* and keyed into the Wang computer held by the IUCN Threatened Plants Committee based at Kew, under the direction of A.H.M. Synge. Since then the list has been extensively used in checking names of genera and families in conservation projects and in other work at Kew, particularly in the living collections, in auxiliary parts of the Herbarium such as the spirit collection, and in compilation of the *Index Kewensis* and *Kew Record*.

The original rather crude data-base was gradually refined and was transferred to Kew's present PRIME mainframe computer in 1985. Synonymy given in the generic indexes in the Herbarium was added in, and genera thought to be taxonomically acceptable but not represented by any material at Kew were also included. For some families where work on a generic revision was being carried out at Kew, such as Gramineae and Palmae, complete synonymy was incorporated, but complete synonymy for all families has never been attempted. An indication of nomenclatural status was also introduced, so that names known to be invalid, or illegitimate either as superfluous names or later homonyms, or mere orthographic variants, could be indicated as such when the information was available, but no comprehensive analysis of all synonyms in the data-base has ever been undertaken. Much valuable time can be wasted by taxonomists in researching the identity or status of long-lost and useless synonyms.

In the last decade new generic names picked up by the compilers of *Index Kewensis* have been referred to the Herbarium staff for inclusion in the data-base as either accepted names or synonyms. In the last two or three years generic changes not involving newly described genera have been noted from current literature by the *Kew Record* compilers and similarly fed back to the Herbarium staff for consideration. In this way a hopefully comprehensive list of genera considered to be taxonomically accepted by the Herbarium staff, but not a comprehensive list of synonyms, has been built up. For the most part the specimens in the Herbarium are now arranged according to these accepted generic names, though the pressure of work has meant that curation of the collections has in a few cases fallen a little behind changes made in the data-base.

In the last three years before publication four major developments have taken place, all resulting in considerable changes in the data held. Firstly print-outs of accepted names and synonyms, arranged family by family, were sent to 22 other institutions with an invitation to comment, and very many constructive comments were received in return (see Acknowledgements below). Secondly the United States Department of Agriculture, through the kind offices of Dr C.R. Gunn and his colleagues, ran a check of our files against their own files of similar data (including records extracted from Airy Shaw's 1973 8th edition of Willis's *Dictionary* and D.J. Mabberley's 1987 *Plant Book*) and reported back on conflicting opinions, which brought to light many points needing correction. Thirdly, in connection with the IAPT 'Names in Current Use' project, all accepted names in our list were checked against *Index Nominum Genericorum* (see section on 'Generic Nomenclature' below for more detail). Fourthly, the names of authors of all the nearly 25000 names in the data-base were revised and cited according to the recommendations of *Authors of Plant Names* being produced simultaneously at Kew (again see below for details). Although these exercises have delayed the expected publication date of the present volume, this is considered well justified by the very considerable improvements achieved in the quality of the data.

The following table gives figures for names of genera and families included in the present publication:

	families recognised	names of accepted genera	all generic names
Fern Allies	5	10	48
Ferns	36	318	532
Gymnosperms	16	81	92
Dicotyledons	357	10712	18295
Monocotyledons	97	2767	5347
Flowering Plants total	454	13479	23642
Vascular Plants total	511	13888	24634

The following families have 100 genera or over:

Compositae	1509	Rutaceae	158
Orchidaceae	835	Sapindaceae	134
Leguminosae	677	Gesneriaceae	133
Gramineae	657	Boraginaceae	131
Rubiaceae	606	Aizoaceae	128
Umbelliferae	428	Myrtaceae	127
Cruciferae	381	Annonaceae	125
Euphorbiaceae	331	Cucurbitaceae	120
Asclepiadaceae	315	Malvaceae	119
Scrophulariaceae	292	Ericaceae	116
Acanthaceae	228	Chenopodiaceae	113
Labiatae	212	Bignoniaceae	111
Palmae	202	Araceae	105
Melastomataceae	194	Cyperaceae	102
Apocynaceae	168	Cactaceae	100

GENERIC NOMENCLATURE

At a time coinciding with editing of this list, the International Association for Plant Taxonomy (IAPT) was developing an initiative on 'Names in Current Use' designed to promote stability of plant names (see D.L. Hawksworth & W. Greuter in Taxon 38: 142–148 (1989)). As part of this operation, a list of generic names in current use will be published in 1992. This will cover all plant groups, including cryptogams and fossils, and first drafts for major groups were circulated widely in the taxonomic community in mid-1991. The IAPT draft list of generic names of vascular plants was supplied by Kew, and was a preliminary version of the list of accepted names published here, contributed in March 1991. This will be modified and augmented in the light of comments received, before incorporation into the final IAPT list which is to be published as a part of *Regnum Vegetabile*. Names included there will be proposed to the 1993 International Botanical Congress in Tokyo to be granted 'protected' status, which, if approved, would mean that these names would take precedence over any names not included in the list, despite any contrary indication on grounds of priority or legitimacy.

The IAPT list includes places of publication, printed as a subset from the data-base of *Index Nominum Genericorum* (ING). To enable this to be produced, all accepted names in the Kew list were checked against ING and scored for inclusion in the IAPT list, the work being done at Kew by Dr Lulu Rico in late 1990 and early 1991. Thanks are due to Dr Rico for her careful work involving nearly 14000 names. Several hundreds of names not in ING were recorded, mostly published in the period 1976–1990, and these were passed to the ING compiler in Utrecht, Gea Zijlstra, who proceeded to make the necessary additions to the ING data-base. At the same time, any differences between the Kew list and ING either in spelling of a name or in author citation were noted, producing another list of about 800 names. Original publications of all of these were checked by R.K. Brummitt. In a majority ING was found to be right and the Kew list was corrected, but in nearly 300 cases the version in ING was disputed. Sincere thanks are due to Gea Zijlstra for much painstaking correspondence during 1991 in reaching agreement on these cases. Many corrections have now been made both to the original Kew list and to ING, and the operation was certainly mutually beneficial. It is believed that significant improvements have been made in our respective data-bases of generic names of vascular plants.

AUTHORS OF NAMES

While this list of generic names has been in preparation in recent years, a parallel operation at Kew has been the revision and expansion of the *'Draft Index of Author Abbreviations'* originally compiled under the guidance of R.D. Meikle and published in 1980. A revised version has been produced after much consultation with a broad international working group, and, thanks to contributions from cryptogamists at other institutions, includes authors of names in cryptogamic groups as well as flowering plants. It is expected to be published at Kew in 1992 more or less simultaneously with the present volume, and will be entitled *Authors of Plant Names*, edited by R.K. Brummitt & C.E. Powell. The names of authors in the present list of generic names are cited in accordance with the recommendations of this publication. Often it has been necessary to check original publications to identify authors accurately, and the compiler of the present volume is indebted to Emma Powell for assistance in doing this.

FAMILY CONCEPTS

Until some twenty years ago the families recognised at Kew were more or less exactly those of Bentham & Hooker's *Genera Plantarum* (with some more recently discovered families added). In 1969, in connection with a reorganisation of work and the opening of a new wing in the Herbarium, a committee was set up under the chairmanship of R. Melville, including R.D. Meikle, C. Jeffrey and F.N. Hepper, together with N.K.B. Robson from the British Museum (Natural History), to modernise the family concepts, resulting in recognition of many more families. The new list was strictly adhered to until the mid 1980's.

In July 1984 a new committee, which became known as the Family Planning Committee, was set up to revise the list of families again. The chairman was L.L. Forman, and other members C. Jeffrey, R.K. Brummitt and J. Dransfield from Kew and N.K.B. Robson from the Natural History Museum (as it is now called); and for the Monocotyledons they were joined by B. Mathew (Kew Herbarium) and D.F. Cutler (Jodrell Laboratory, Kew). At the outset a bibliographic exercise was carried out in which all differences from the existing Kew list found in the publications of Cronquist, Takhtajan, Thorne, Dahlgren and Airy Shaw were recorded. The committee met for half a day at approximately two-weekly intervals until early 1986. Of course, much assistance was received from other members of the Herbarium staff. In considering the arguments for or against each case, the committee made a point of actually looking at specimens of the plants concerned every time, rather than merely consulting the literature. Perhaps the most striking aspect of the operation was the very high level of agreement between the members in nearly all cases. Between 1986 and 1991 the list was frequently up-dated to take account of new published data in the literature and even some newly described families.

From the figures quoted below (p.5) comparing the numbers of families recognised by other authors with those at Kew, it seems in retrospect that the committee's family concepts have been neither very broad nor very narrow but hopefully fairly middle-of-the-road. There was a general feeling that if the arguments for lumping or splitting were evenly balanced, then lumping was to be preferred. Marked adaptations associated with parasitism or saprophytism, such as the loss of chlorophyll or leaves, were usually considered to be insufficient characters on their own for recognising a family. Thus, for examples, Monotropaceae, Cuscutaceae and Geosiridaceae were not separated from Ericaceae, Convolvulaceae and Iridaceae respectively, since all other characters place them within these families. Similarly floral zygomorphy, usually considered an adaptation for specialised pollination, was not regarded as a family character on its own, and Fumariaceae and Lobeliaceae were not separated from Papaveraceae and Campanulaceae. In the case of Escalloniaceae, where rather striking but uncorrelated characters were found, and overall affinities seemed obvious, it was considered preferable to recognise one variable family rather than fragment it into up to eight families as others have done. However, each case was considered separately on its own merits.

When phylogenetic considerations are taken into account, questions may be raised against three of the larger families. There are strong indications that the Umbelliferae, Labiatae and Asclepiadaceae as here defined may not be monophyletic, having evolved from different groups within the Araliaceae, Verbenaceae and Apocynaceae respectively. A traditional concept has been retained here in each case pending further evaluation of the phylogenetic evidence.

In the petaloid Monocotyledons it was unanimously agreed, after consideration of floral, vegetative and anatomical characters, that splitting of the old concept of Liliales to a level comparable with that adopted by Dahlgren was essential to make the family concept more comparable with that adopted in other groups. Other options, where just a few of the more conspicuous families were separated from the Liliaceae, were found to be inconsistent and unsatisfactory. While a few slight adjustments may still need to be made, the committee was convinced that the great majority of the family concepts adopted in this group by Dahlgren, usually taken up by Takhtajan, and confirmed again here, will soon be very widely accepted.

There are other areas where family limits are still far from clear. The Saxifragales *sensu lato* are notoriously controversial. Although the Caryophyllidae sensu Cronquist and Takhtajan have been the subject of numerous studies recently, the optimal family delimitation is still far from clear, with the limits of Phytolaccaceae and Caryophyllaceae particularly in question. Splitting of Loganiaceae *sensu lato* is still open to debate, and the limits of Scrophulariaceae are also questionable. In many cases like these, further research on the phylogenetic relationships is still required, and the decisions represented here may be seen as merely the best the committee could produce with the time and knowledge available.

4

Kew wishes to record its thanks to the Natural History Museum for contributing Dr Robson's expertise to the discussions on family limits. His wide knowledge of the flowering plants was greatly appreciated by the other members of the committee. The decisions taken by the committee have now been implemented in the herbaria at Kew, at the Natural History Museum, and at the Royal Botanic Garden in Edinburgh.

SYSTEMS OF CLASSIFICATION OF FLOWERING PLANTS QUOTED IN PARTS 2 AND 3

For the flowering plant families recognised here, their position in eight different classifications is recorded in Part 2. Bibliographic details of these systems of classification are given in Part 3, where the families of each are enumerated in taxonomic sequence.

Five of these eight systems have been published in the last ten years, one was published nearly thirty years ago, and the other two are much older than that. Bentham & Hooker's system (1862-1883) is included because of its historical influence on Kew, and that of Dalla Torre & Harms (1900-1907) because it represents the influential Engler system and is still used for the arrangement of many major herbaria.

It might have been thought desirable to include a number of other published classifications, but space did not permit. A useful synopsis of major systems published prior to 1964 is given in the introduction to that of Melchior (*loc. cit.* in Part 3). Among more recent publications, the treatments by A. Goldberg of the Dicotyledons in *Smithsonian Contrib. Bot.* 58: 1-314 (1986) and of the Monocotyledons, *ibid.* 71: 1-74 (1989), might have merited a place here, but they do not recognise supra-ordinal taxa and give no indication of where very many families recognised by other authors are placed, and so are not easily comparable with other systems or adaptable to the format adopted here. Useful references to other modern classifications are given by Goldberg (*loc. cit.*).

It is of some interest to compare the numbers of flowering plant families recognised by each of the eight quoted systems with those recognised here by Kew. For explanation of abbreviations used see introduction to Part 2 and full bibliographic references in Part 3. In the following table Monocotyledons and Dicotyledons are recorded separately and then are combined in the right hand columns. While the decision on how to divide the Liliales *sensu lato* (Dahlgren, Takhtajan and Kew accepting many more families than others) obviously weights the figures heavily in the Monocotyledons, it is interesting to note that rather similar proportions are found each time in the Dicotyledons also.

	Dicots			Monocots			All flowering plants		
	a	b	c	a	b	c	a	b	c
B&H	157	4	165	62	1	34	219	5	199
DT&H	116	10	234	56	2	43	172	12	277
Melc	66	15	290	43	3	53	109	18	343
Thor	56	6	298	45	1	53	101	7	351
Dahl	32	40	367	6	11	102	38	51	469
Young	52	21	325	38	2	61	90	23	386
Takh	12	84	429	7	14	104	19	98	533
Cron	54	18	318	36	5	65	90	23	383
Kew	–	–	357	–	–	97	–	–	454

Column **a** – families recognised by Kew but not by system concerned (excluding families which had not been discovered at the time).
Column **b** – families recognised by the system concerned but not by Kew.
Column **c** – total number of families recognised.

To summarise, of the six more modern systems listed, those of Melchior, Thorne, Young and Cronquist tend to take a broader view of families than Kew does, while Dahlgren and Takhtajan tend to take a narrower view. The system with the fewest divergences from Kew is that of Dahlgren. But perhaps the main lesson emerging is that there is still very wide divergence of opinion between different authorities over how to circumscribe many families. Fortunately, however, with the notable exceptions of families grouped around Liliaceae and Saxifragaceae, and in the question of whether the Legumes are one family or three, most of the major families are relatively clearly circumscribed, and there is likely to be little dispute over the family to which the great majority of genera are assigned. Most of the disputes concern small families.

ACKNOWLEDGEMENTS — OUTSIDE KEW

Early copies of the complete list of families and genera were sent in 1988–89 to the following herbaria: Arnold Arboretum (A), Adelaide (AD), Berlin (B), Natural History Museum, London (BM), Meise, Brussels (BR), Brisbane (BRI), Copenhagen (C), Canberra (CANB), Christchurch (CHR), Edinburgh (E), Geneva (G), Leiden (L), Missouri Botanical Garden, St. Louis (MO), New York (NY), Sydney (NSW), Paris (P), Pretoria (PRE), Stockholm (S), Utrecht (U), Uppsala (UPS), Smithsonian Institution, Washington (US). Replies were received from all but two, with many very helpful comments. Our sincere thanks are due to the Directors and staff of these institutions for their expert advice freely given.

We are also very grateful to Dr C.R. Gunn and his colleagues at the United States Department of Agriculture, Beltsville, for checking our list of genera against their data-base and making the results of the comparison readily available to us. Many errors were eliminated through this operation.

Many botanists outside Kew have given advice, either in the replies from the institutions listed above or in other ways. Their comments range from a brief note on a single case to a complete generic revision of a family. The decision on whether their advice should be accepted or not has rested with the Kew Herbarium staff. In the list of acknowledgements below, mention of a family or other group after a person's name does not necessarily imply that he or she has revised (or even seen) the complete list of genera for that group, and certainly does not imply that the person necessarily accepts all the taxonomic decisions taken in that group. In many cases their advice may refer to only a single genus. However, in some instances where critical expertise has been lacking at Kew, the list of genera recognised in a family (or the family placement of a genus) has been based entirely on advice from somebody outside; such cases are indicated by an asterisk (*) after the group concerned. Whether their comments have been major or minor, we are grateful to the following for advice concerning the groups indicated: L.G. Adams (Canberra), Philydraceae, Compositae, Violaceae; I. Al-Shehbaz (St. Louis), Cruciferae; W.R. Anderson (Ann Arbor), Malpighiaceae*; G. Argent (Edinburgh), Ericaceae; S. Armbruster (Fairbanks), Euphorbiaceae; S.C. Arroyo (Buenos Aires and Berlin), Amaryllidaceae; P.S. Ashton (Cambridge, Mass.), Dipterocarpaceae*; P. Bamps (Meise, Brussels), Tiliaceae, Leguminosae; W.R. Barker (Adelaide), Scrophulariaceae; R.C. Barneby (New York), Leguminosae; L.C. Barnett (New York), Sterculiaceae, Theaceae; D.J. Bedford (Sydney), Xanthorrhoeaceae and related families (Calectasiaceae, Dasypogonaceae, Lomandraceae)*; C.C. Berg (Bergen), Moraceae*, Cecropiaceae*; B.M. Boom (New York), various comments; J. Bosser (Paris), various Madagascan families; L. Boulos (Kuwait), Chenopodiaceae; F.J. Breteler (Wageningen), Connaraceae*; Barbara G. Briggs (Sydney), Myrtaceae, Proteaceae, Restionaceae; H.-M. Burdet (Geneva), Cruciferae; B.L. Burtt (Edinburgh), Gesneriaceae; F. Butzin (Berlin), Misodendraceae, Orchidaceae; P. Cantino (Athens, Ohio), Labiatae, Verbenaceae; R. Carolin (Sydney), Portulacaceae, Goodeniaceae; D.F. Chamberlain (Edinburgh),

Berberidaceae; A.D. Chapman (Canberra), various Australian genera; R.J. Chinnock (Adelaide), Myoporaceae; B.J. Conn (Sydney), Labiatae; A. Cronquist (New York), numerous comments on family limits; J. Cullen (Edinburgh, later Cambridge), Papaveraceae; C. Cusset (Paris), Podostemaceae; T.F. Daniel (San Francisco), Acanthaceae; W.G. D'Arcy (St. Louis), Solanaceae; T. Deroin (Paris), Annonaceae; Ding Hou (Leiden), Anacardiaceae, Aristolochiaceae, Celastraceae, Rhizophoraceae, Thymelaeaceae; L.J. Dorr (New York), Bombacaceae, Sterculiaceae, Theaceae; H. Eichler (Canberra), Umbelliferae, Compositae, Zygophyllaceae; Joy Everett (Sydney), various Australian genera; A. Farjon (Utrecht), Gymnospermae; E. Farr (Washington), Melastomataceae and nomenclature of various genera; J.J. Floret (Paris), Rhizophoraceae, Anisophylleaceae; F. Friedmann (Paris), Sapindaceae, Araliaceae, Campanulaceae, Hypoxidaceae; I. Friis (Copenhagen), Urticaceae, Moraceae, Cecropiaceae, Amaryllidaceae; D.G. Frodin (Pennsylvania), Araliaceae; P.A. Fryxell (College Station, Texas), Malvaceae; P. Garnock-Jones (Christchurch), Hectorellaceae, Cruciferae and other New Zealand families; R. Geesink (Leiden), Leguminosae, Portulacaceae; A.H. Gentry (St. Louis), Bignoniaceae; M.G. Gilbert (London), Asclepiadaceae; L. Gillespie (Washington), Euphorbiaceae; D.R. Given (Christchurch), various New Zealand genera; P. Goetghebeur (Ghent), Cyperaceae; P. Goldblatt (St. Louis), Iridaceae; W. Greuter (Berlin), Med-Checklist genera; the late A.J.C. Grierson (Edinburgh), Compositae; N. El Hadidi (Cairo), Zygophyllaceae; L. Haegi (Adelaide), Compositae; N. Hallé (Paris), Celastraceae; B. Hammel (St. Louis), Clusiaceae, Cyclanthaceae; B. Hansen (Copenhagen), Acanthaceae, Cynomoriaceae; the late C. Hansen (Copenhagen), Melastomataceae; H.E.K. Hartmann (Hamburg), Aizoaceae subfams. Mesembryanthemoideae* and Ruschioideae*; O. Hedberg (Uppsala), Polygonaceae; I.C. Hedge (Edinburgh), Chenopodiaceae, Cruciferae, Labiatae, Umbelliferae; W.H.A. Hekking (Utrecht), Violaceae; E. Hennipman (Utrecht), Polypodiaceae; N.E.C.H. van Heusden (Utrecht), Annonaceae; P.Hiepko (Berlin), Ranunculaceae; O. Hilliard (Edinburgh), Scrophulariaceae; R.J. Hnatiuk (Canberra), various Australian genera; P.C. Hoch (St. Louis), Onagraceae*; N.H. Holmgren (New York), Scrophulariaceae; R.D. Hoogland (Paris), Cunoniaceae, Dilleniaceae, Escalloniaceae, Palmae, Rubiaceae and family nomenclature; M.J. Huft (Chicago), Euphorbiaceae; S.W.L. Jacobs (Sydney), Chenopodiaceae, Gramineae; M.J. Jansen-Jacobs (Utrecht), Verbenaceae; J.P. Jessop (Adelaide), various Australian genera; D.M. Johnson (New York, later Delaware, Ohio), ferns, Annonaceae, Malpighiaceae; L.A.S. Johnson (Sydney), many Australian families and genera; R.W. Johnson (Brisbane), various Australian genera; C. Jonghind (Wageningen), Connaraceae; B. Jonsell (Stockholm), Cruciferae; S.L. Jury (Reading), Umbelliferae; C. Kalkman (Leiden), Rosaceae; J. Kallunki (New York), Bombacaceae; P. Kessler (Leiden), Annonaceae; R.C. Kruijt (Utrecht), Euphorbiaceae; I. Kukkonen (Helsinki), Cyperaceae; H.W. Lack (Berlin), Compositae, Dipsacaceae; J.M. Lamond (Edinburgh), Umbelliferae; the late A. Lauener (Edinburgh), Ranunculaceae; J. Lawesson (Aarhus), Passifloraceae; G.J. Leach (Darwin), Eriocaulaceae; P.W. Leenhouts (Leiden), Buddlejaceae, Connaraceae, Loganiaceae, Sapindaceae; A.J.M. Leeuwenberg (Wageningen), Apocynaceae*, Loganiaceae; R.H.M.J. Lemmens (Wageningen), Connaraceae; J.-F. Leroy (Paris), Meliaceae, Rubiaceae, Winteraceae; M. Lescot (Paris), Meliaceae, Rubiaceae, Winteraceae; A. Le Thomas (Paris), Annonaceae; B.E. Leuenberger (Berlin), Bromeliaceae, Cactaceae; G. Levin (San Diego), Euphorbiaceae; D.G. Long (Edinburgh), Moraceae, Urticaceae etc.; Alicia Lourteig (Paris), Ranunculaceae, Primulaceae, Lythraceae, Celastraceae, Zygophyllaceae, Oxalidaceae and others; P.P. Lowry (St. Louis), Araliaceae; J.L. Luteyn (New York), Ericaceae and others; P.J.M. Maas (Utrecht), Annonaceae*, Costaceae, Gentianaceae, Triuridaceae; J. van der Maesen (Wageningen), Leguminosae; R. Makinson (Sydney), various Australian genera; U. Matthäs (Berlin), Med-Checklist genera; J. McNeill (Edinburgh, later Toronto), Caryophyllaceae, Portulacaceae; G. McPherson (St. Louis), Euphorbiaceae; R. van der Meijden (Leiden),

Polygalaceae; A.M.W. Mennega (Utrecht), Celastraceae; J.T. Mickel (New York), various comments; R.R. Mill (Edinburgh), Boraginaceae; P. Morat (Paris), Eriocaulaceae, Podocarpaceae, Sterculiaceae; S.A. Mori (New York), Vochysiaceae etc.; M.E. Newton (Liverpool), Cruciferae; D.H. Nicolson (Washington), Araceae; I. Nielsen (Aarhus) Leguminosae-Mimsoideae; H.P. Nooteboom (Leiden), Symplocaceae, Magnoliaceae, Linaceae, Simaroubaceae, Cyperaceae etc; I. Nordal (Oslo), various monocotyledonous families; B. Nordenstam et al. (Stockholm), Compositae; E.G.H. Oliver (Stellenbosch), Ericaceae; C.N. Page (Edinburgh), Gymnospermae; J.A.R. Paiva (Coimbra), Polygalaceae; Pham Hoang Ho, various families; M.G. Pimenov (Moscow), Umbelliferae; Jocelyn Powell (Sydney), Epacridaceae; P.H. Raven (St. Louis), Onagraceae*; J.L. Reveal (College Park, Maryland), Polygonaceae, family nomenclature; E. Robbrecht (Meise, Brussels), Rubiaceae; H. Robinson (Washington), Compositae; N.K.B. Robson (BM, London), Guttiferae*, Ochnaceae and many others; A.N. Rodd (Sydney), various Australian genera; R. Rollins (Cambridge, Mass.), Cruciferae; H. Scholz (Berlin), Cruciferae, Gramineae; M. Shaffer-Fehre (London), *Najas;* L.E. Skog (Washington), Gesneriaceae etc.; R.M. Smith (Edinburgh), Zingiberaceae; P.F. Stevens (Cambridge, Mass.), Guttiferae, Ericaceae; D.A. Sutton (BM, London), Scrophulariaceae; M. Tebbs (BM, London), Piperaceae; W.W. Thomas (New York), ferns, Cyperaceae, Ericaceae, Lecythidaceae, Simaroubaceae; M. Thulin (Uppsala), Campanulaceae, Leguminosae; C. Tirel (Paris), Monimiaceae, Euphorbiaceae; J.F. Veldkamp (Leiden), Elaeagnaceae, Oxalidaceae, Gramineae, Melastomataceae; J.-F. Villiers (Paris), Icacinaceae, Olacaceae, Pandanaceae, Capparaceae, Mimosoideae; W. Vink (Leiden), Hamamelidaceae, Sapotaceae, Winteraceae; E.F. de Vogel (Leiden), Orchidaceae; C.J. Webb (Christchurch), various New Zealand genera; G.L. Webster (Davis), Euphorbiaceae; H. van der Werff (St. Louis), Lauraceae; Judy West (Canberra), Caryophyllaceae, Illecebraceae; P. Weston (Sydney), Rutaceae; D.O. Wijnands (Wageningen), Hyacinthaceae; W.J.J.O. de Wilde (Leiden), Myristicaceae, Passifloraceae; Karen Wilson (Sydney), Asclepiadaceae, Polygonaceae, and Cyperaceae and various other monocots; Peter G. Wilson (Sydney), various Australian families including ferns and Myrtaceae; J. Wurdack (Washington), Melastomataceae; G. Zijlstra (Utrecht), many nomenclatural matters. If others have been inadvertently omitted, we sincerely apologise.

ACKNOWLEDGEMENTS — KEW

The generic concepts currently adopted in the Herbarium at Kew have been derived from curatorial practice over a period of 100 years or more, and it is impossible to mention all of the many people who have been responsible. Equally, however, it would seem impossible not to acknowledge certain members of the Herbarium staff who have made major contributions from their own specialist research experience; responsibility for taxonomic decisions in major groups has been as follows: Pteridophytes – B.S. Parris and R.J. Johns; Dicotyledons – Compositae, C. Jeffrey assisted by D.J.N. Hind; Leguminosae, R.M. Polhill and G.P. Lewis; Rubiaceae, D.M. Bridson; Euphorbiaceae, A. Radcliffe-Smith; Asclepiadaceae, D.J. Goyder; Acanthaceae, K. Vollesen and R.K. Brummitt; Labiatae, R.M. Harley assisted by A.J. Paton; Myrtaceae, E. NicLughadha; Cucurbitaceae, C. Jeffrey; Cactaceae, D.R. Hunt and N.P. Taylor; Loranthaceae, R.M. Polhill; Sapotaceae, T.D. Pennington; Menispermaceae, L.L. Forman; Amaranthaceae, C.C. Townsend; Chrysobalanaceae, G.T. Prance; Elaeocarpaceae, M.J.E. Coode; Monocotyledons – Orchidaceae, P.J. Cribb and J.J. Wood; Gramineae, W.D. Clayton, S. Renvoize and T.A. Cope; Palmae, J. Dransfield; petaloid monocots (Liliales sensu lato), B. Mathew; Cyperaceae, D.A. Simpson; Araceae, S.J. Mayo assisted by P. Boyce; Commelinaceae, D.R. Hunt. Among others who have advised on smaller families too numerous to list are B. Verdcourt, M. Cheek, B.L. Stannard, S. Zmarzty, M. Wilmot-Dear, S. Bidgood, P. Halliday, J. Cowley, S. Atkins and D.W. Kirkup.

Over the years numerous people have assisted in compilation of the data-base in different ways. The original work of extracting names from the indexes to the herbarium collections was done by the late Hilary Brummitt. First computerisation of the data on the IUCN Wang machine was supervised by A.H.M. Synge who also wrote the program and actively developed the project, while keyboarding of the data was done by Joan Taylor. Supervision of computing from 1985 was by W.N. Loader, programs for handling the data on the Kew PRIME machine were written and maintained by Mark Jackson and Sarah Edwards, and W.D. Clayton assisted in writing specifications. Much assistance in handling various sorts of data was given by Margaret Beyer and Laura Hastings. Rosemary Davies and Katherine Lloyd supplied data from *Index Kewensis,* and Suzy Dickerson and Lisa von Schlippe did the same from *Kew Record.* Emma Powell assisted in editing the standard forms of author names, Lulu Rico compared accepted names with *Index Nominum Genericorum,* and Ann McNeil assisted in checking other data. J.M. Lock has given general assistance in editing and production of final copy. The team of typesetters in the Typing Pool are acknowledged individually elsewhere. Sincere thanks are extended to all these.

PART 1

GENERIC NAMES ALPHABETICALLY

Accepted names and synonyms are here listed alphabetically and referred to families. The major group under which the family is listed in Part 2 is indicated in brackets at the end of each entry, as follows:

(FA) Fern Allies (p.465)
(F) Ferns (p.467)
(G) Gymnosperms (p.477)

(D) Dictyledons (p.481)
(M) Monocotyledons (p.695)

Synonymy is not intended to be comprehensive, and only names which happen to have so far been included in the data-base at Kew are listed. Originally this included names which appeared on the generic indexes to families in the Herbarium in 1974–75, which have been revised and augmented since. For a few major families which have been extensively worked on at Kew, such as Gramineae and Palmae, full synonymy has now been added, but there has been no attempt to do this for all families. Nonetheless, it is hoped that most names which have been used in botanical literature at least in this century have been accounted for.

Certain categories of synonyms which are considered not available for use under any taxonomic treatment are recognised, and are indicated when details have been checked. Names of uncertain application are also indicated. Codes for these categories are given after the author of the name, and are explained as follows:

SUH: Synonym, Unavailable for use, later Homonym
SUI: Synonym, Unavailable for use, Invalid
SUO: Synonym, Unavailable for use, Orthographic Variant
SUS: Synonym, Unavailable for use, Superfluous
UP: Unplaced, pending further research
UR: Unplaced despite efforts to identify the plant

Application of the above codes is not yet complete, and many listed synonyms with no such indication will eventually be found to fall into these categories.

For every synonym the accepted name to which it is referred is given. It should be noted that the accepted name given after a superfluous name is not necessarily the name which makes the synonym superfluous, just as after an orthographic variant it is not necessarily the accepted spelling of the name. For example, *Poarion* Rchb. is a superfluous name for *Aegialina* Schult. which is a synonym of *Rostraria* Trin.; the list below therefore gives *Poarion* (SUS) as a synonym of *Rostraria*, not of *Aegialina*. *Borrera* Spreng. is an unacceptable orthographic variant of *Borreria* G. Mey. which is a synonym of *Spermacoce* L.; the list below simply refers *Borrera* (SUO) to synonymy of *Spermacoce*, not to *Borreria*.

Aa Rchb.f. ORCHIDACEAE (M)
Aalius Lam. ex Kuntze (SUH) = Sauropus Blume EUPHORBIACEAE (D)
Aaronsohnia Warb. & Eig COMPOSITAE (D)
Abacopteris Fée = Pronephrium C.Presl THELYPTERIDACEAE (F)
Abarema Pittier LEGUMINOSAE–MIMOSOIDEAE (D)
Abatia Ruíz & Pav. FLACOURTIACEAE (D)
Abauria Becc. = Koompassia Maingay LEGUMINOSAE–CAESALPINIOIDEAE (D)
Abbevillea O.Berg = Campomanesia Ruíz & Pav. MYRTACEAE (D)
Abbottia F.Muell. = Timonius DC. RUBIACEAE (D)
Abdominea J.J.Sm. ORCHIDACEAE (M)
Abdra Greene = Draba L. CRUCIFERAE (D)
Abdulmajidia Whitmore LECYTHIDACEAE (D)
Abebaia Baehni = Manilkara Adans. SAPOTACEAE (D)
Abelia R.Br. CAPRIFOLIACEAE (D)
Abelicea Baill. = Zelkova Spach ULMACEAE (D)
Abeliophyllum Nakai OLEACEAE (D)

Abelmoschus Medik. MALVACEAE (D)
Aberemoa Aubl. (SUH) = Guatteria Ruíz & Pav. ANNONACEAE (D)
Aberia Hochst. = Dovyalis E.Mey. ex Arn. FLACOURTIACEAE (D)
Abies Mill. PINACEAE (G)
Abildgaardia Vahl = Fimbristylis Vahl CYPERACEAE (M)
Abobra Naudin CUCURBITACEAE (D)
Abola Lindl. = Caucaea Schltr. ORCHIDACEAE (M)
Abola Adans. (SUS) = Cinna L. GRAMINEAE (M)
Abolboda Bonpl. XYRIDACEAE (M)
Aboriella Bennet URTICACEAE (D)
Abortopetalum O.Deg. = Abutilon Mill. MALVACEAE (D)
Abramsia Gillespie = Airosperma K.Schum. & Lauterb. RUBIACEAE (D)
Abrodictyum C.Presl = Trichomanes L. HYMENOPHYLLACEAE (F)
Abroma Jacq. STERCULIACEAE (D)
Abromeitia Mez MYRSINACEAE (D)
Abromeitiella Mez BROMELIACEAE (M)
Abronia Juss. NYCTAGINACEAE (D)
Abrophyllum Hook.f. ex Benth. ESCALLONIACEAE (D)
Abrotanella Cass. COMPOSITAE (D)
Abrus Adans. LEGUMINOSAE–PAPILIONOIDEAE (D)
Absolmsia Kuntze ASCLEPIADACEAE (D)
Abuta Aubl. MENISPERMACEAE (D)
Abutilon Mill. MALVACEAE (D)
Abutilothamnus Ulbr. MALVACEAE (D)
Acacallis Lindl. = Aganisia Lindl. ORCHIDACEAE (M)
Acachmena H.P.Fuchs (UP) CRUCIFERAE (D)
Acacia Mill. LEGUMINOSAE–MIMOSOIDEAE (D)
Acaciella Britton & Rose = Acacia Mill. LEGUMINOSAE–MIMOSOIDEAE (D)
Acaciopsis Britton & Rose = Acacia Mill. LEGUMINOSAE–MIMOSOIDEAE (D)
Acaena Mutis ex L. ROSACEAE (D)
Acalypha L. EUPHORBIACEAE (D)
Acalyphes Hassk. = Acalypha L. EUPHORBIACEAE (D)
Acalyphopsis Pax & K.Hoffm. = Acalypha L. EUPHORBIACEAE (D)
Acampe Lindl. ORCHIDACEAE (M)
Acamptoclados Nash = Eragrostis Wolf GRAMINEAE (M)
Acamptopappus (A.Gray) A.Gray COMPOSITAE (D)
Acanthella Hook.f. MELASTOMATACEAE (D)
Acanthephippium Blume ORCHIDACEAE (M)
Acanthinophyllum Allemão = Clarisia Ruíz & Pav. MORACEAE (D)
Acanthocalycium Backeb. CACTACEAE (D)
Acanthocalyx (DC.) Tiegh. = Morina L. MORINACEAE (D)
Acanthocardamum Thell. CRUCIFERAE (D)
Acanthocarpus Lehm. LOMANDRACEAE (M)
Acanthocaulon Klotzsch ex Endl. = Platygyna Mercier EUPHORBIACEAE (D)
Acanthocephala Backeb. = Parodia Speg. CACTACEAE (D)
Acanthocephalus Kar. & Kir. COMPOSITAE (D)
Acanthocereus (Engelm. ex A.Berger) Britton & Rose CACTACEAE (D)
Acanthochiton Torr. = Amaranthus L. AMARANTHACEAE (D)
Acanthochlamys P.C.Kao VELLOZIACEAE (M)
Acanthocladium F.Muell. COMPOSITAE (D)
Acanthocladus Klotzsch ex Hassk. = Polygala L. POLYGALACEAE (D)
Acanthococos Barb.Rodr. = Acrocomia Mart. PALMAE (M)
Acanthodesmos C.D.Adams & duQuesnay COMPOSITAE (D)
Acanthodium Delile = Blepharis Juss. ACANTHACEAE (D)
Acanthogilia A.G.Day & Moran POLEMONIACEAE (D)

Acanthoglossum Blume = Pholidota Lindl. ex Hook. ORCHIDACEAE (M)
Acanthogonum Torr. = Chorizanthe R.Br. ex Benth. POLYGONACEAE (D)
Acantholepis Less. COMPOSITAE (D)
Acantholimon Boiss. PLUMBAGINACEAE (D)
Acantholippia Griseb. VERBENACEAE (D)
Acantholobivia Backeb. = Echinopsis Zucc. CACTACEAE (D)
Acantholoma Gaudich. ex Baill. = Pachystroma Muell.Arg. EUPHORBIACEAE (D)
Acanthomintha (A.Gray) A.Gray LABIATAE (D)
Acanthonema Hook.f. GESNERIACEAE (D)
Acanthopale C.B.Clarke ACANTHACEAE (D)
Acanthopanax (Decne. & Planch.) Witte = Eleutherococcus Maxim. ARALIACEAE (D)
Acanthopetalus Y.Ito = Echinopsis Zucc. CACTACEAE (D)
Acanthophaca Nevski = Astragalus L. LEGUMINOSAE–PAPILIONOIDEAE (D)
Acanthophippium Blume (SUO) = Acanthephippium Blume ORCHIDACEAE (M)
Acanthophoenix H.A.Wendl. PALMAE (M)
Acanthophora Merr. (SUH) = Aralia L. ARALIACEAE (D)
Acanthophyllum C.A.Mey. CARYOPHYLLACEAE (D)
Acanthopsis Harv. ACANTHACEAE (D)
Acanthopteron Britton = Mimosa L. LEGUMINOSAE–MIMOSOIDEAE (D)
Acanthopyxis Miq. ex Lanj. (SUI) = Caperonia A.St.–Hil. EUPHORBIACEAE (D)
Acanthorhipsalis (K.Schum.) Britton & Rose = Lepismium Pfeiff. CACTACEAE (D)
Acanthorhipsalis Kimnach (SUH) = Lepismium Pfeiff. CACTACEAE (D)
Acanthorrhinum Rothm. SCROPHULARIACEAE (D)
Acanthorrhiza H.A.Wendl. = Cryosophila Blume PALMAE (M)
Acanthosabal Prosch. = Acoelorrhaphe H.A.Wendl. PALMAE (M)
Acanthoscyphus Small = Oxytheca Nutt. POLYGONACEAE (D)
Acanthosicyos Welw. ex Hook.f. CUCURBITACEAE (D)
Acanthosicyus Post & Kuntze (SUO) = Acanthosicyos Welw. ex Hook.f.
    CUCURBITACEAE (D)
Acanthospermum Schrank COMPOSITAE (D)
Acanthosphaera Warb. (SUH) = Naucleopsis Miq. MORACEAE (D)
Acanthostachys Klotzsch BROMELIACEAE (M)
Acanthostelma Bidgood & Brummitt ACANTHACEAE (D)
Acanthostemma (Blume) Blume = Hoya R.Br. ASCLEPIADACEAE (D)
Acanthostyles R.M.King & H.Rob. COMPOSITAE (D)
Acanthosyris (Eichler) Griseb. SANTALACEAE (D)
Acanthothamnus Brandegee CANOTIACEAE (D)
Acanthotreculia Engl. = Treculia Decne. ex Trécul MORACEAE (D)
Acanthotrichilia (Urb.) O.F.Cook & G.N.Collins = Trichilia P.Browne MELIACEAE (D)
Acanthura Lindau ACANTHACEAE (D)
Acanthus L. ACANTHACEAE (D)
Acanthyllis Pomel = Anthyllis L. LEGUMINOSAE–PAPILIONOIDEAE (D)
Acareosperma Gagnep. VITACEAE (D)
Acarpha Griseb. CALYCERACEAE (D)
Acaulimalva Krapov. MALVACEAE (D)
Acaulon N.E.Br. (SUH) = Aloinopsis Schwantes AIZOACEAE (D)
Acca O.Berg MYRTACEAE (D)
Accara Landrum MYRTACEAE (D)
Accia A.St.–Hil. = Fragariopsis A.St.–Hil. EUPHORBIACEAE (D)
Acelica Rizzini = Justicia L. ACANTHACEAE (D)
Acentra Phil. VIOLACEAE (D)
Acer L. ACERACEAE (D)
Aceranthus C.Morren & Decne. BERBERIDACEAE (D)
Aceras R.Br. ORCHIDACEAE (M)
Acerates Elliott = Asclepias L. ASCLEPIADACEAE (D)

Aceratium DC. ELAEOCARPACEAE (D)
Aceratorchis Schltr. ORCHIDACEAE (M)
Aceriphyllum Engl. (SUH) = Mukdenia Koidz. SAXIFRAGACEAE (D)
Acetosa Mill. = Rumex L. POLYGONACEAE (D)
Acetosella (Meisn.) Fourr. = Rumex L. POLYGONACEAE (D)
Acetosella Kuntze (SUH) = Oxalis L. OXALIDACEAE (D)
Achaenipodium Brandegee = Verbesina L. COMPOSITAE (D)
Achaeta Fourn. = Calamagrostis Adans. GRAMINEAE (M)
Achaetogeron A.Gray COMPOSITAE (D)
Acharia Thunb. ACHARIACEAE (D)
Achasma Griff. = Etlingera Giseke ZINGIBERACEAE (M)
Achatocarpus Triana ACHATOCARPACEAE (D)
Achetaria Cham. & Schltdl. SCROPHULARIACEAE (D)
Achillea L. COMPOSITAE (D)
Achimenes P.Browne (SUH) = Columnea L. GESNERIACEAE (D)
Achimenes Pers. GESNERIACEAE (D)
Achlaena Griseb. = Arthropogon Nees GRAMINEAE (M)
Achlyphila Maguire & Wurdack XYRIDACEAE (M)
Achlys DC. BERBERIDACEAE (D)
Achnatherum P.Beauv. = Stipa L. GRAMINEAE (M)
Achneria Benth. (SUH) = Pentaschistis (Nees) Spach GRAMINEAE (M)
Achneria P.Beauv. = Eriachne R.Br. GRAMINEAE (M)
Achnodon Link (SUS) = Phleum L. GRAMINEAE (M)
Achnodonton P.Beauv. = Phleum L. GRAMINEAE (M)
Achnophora F.Muell. COMPOSITAE (D)
Achnopogon Maguire, Steyerm. & Wurdack COMPOSITAE (D)
Achoriphragma Soják CRUCIFERAE (D)
Achradelpha O.F.Cook = Pouteria Aubl. SAPOTACEAE (D)
Achradotypus Baill. = Pycnandra Benth. SAPOTACEAE (D)
Achras L. = Manilkara Adans. SAPOTACEAE (D)
Achroanthes Raf. = Malaxis Sol. ex Sw. ORCHIDACEAE (M)
Achrochloa B.D.Jacks. (SUO) = Koeleria Pers. GRAMINEAE (M)
Achroostachys Benth. (SUO) = Athroostachys Benth. GRAMINEAE (M)
Achrouteria Eyma = Chrysophyllum L. SAPOTACEAE (D)
Achuaria Gereau RUTACEAE (D)
Achudemia Blume URTICACEAE (D)
Achyrachaena Schauer COMPOSITAE (D)
Achyranthes L. AMARANTHACEAE (D)
Achyrobaccharis Sch.Bip. ex Walp. = Baccharis L. COMPOSITAE (D)
Achyrocalyx Benoist ACANTHACEAE (D)
Achyrocline (Less.) DC. COMPOSITAE (D)
Achyrodes Boehm. (SUS) = Lamarckia Moench GRAMINEAE (M)
Achyronychia Torr. & A.Gray ILLECEBRACEAE (D)
Achyropappus Kunth COMPOSITAE (D)
Achyropsis (Moq.) Hook.f. AMARANTHACEAE (D)
Achyrospermum Blume LABIATAE (D)
Achyrothalamus O.Hoffm. COMPOSITAE (D)
Aciachne Benth. GRAMINEAE (M)
Acianthera Scheidw. = Pleurothallis R.Br. ORCHIDACEAE (M)
Acianthus R.Br. ORCHIDACEAE (M)
Acicalyptus A.Gray = Cleistocalyx Blume MYRTACEAE (D)
Acicarpa Raddi (SUH) = Digitaria Haller GRAMINEAE (M)
Acicarpha Juss. CALYCERACEAE (D)
Acidanthera Hochst. = Gladiolus L. IRIDACEAE (M)
Acidocroton Griseb. EUPHORBIACEAE (D)

Acidocroton P.Browne (SUH) = Flueggea Willd. EUPHORBIACEAE (D)
Acidosasa C.D.Chu C.S.Chao GRAMINEAE (M)
Acidoton P.Browne (SUH) = Flueggea Willd. EUPHORBIACEAE (D)
Acidoton Sw. EUPHORBIACEAE (D)
Aciella Tiegh. = Amylotheca Tiegh. LORANTHACEAE (D)
Acilepidopsis H.Rob. = Vernonia Schreb. COMPOSITAE (D)
Acineta Lindl. ORCHIDACEAE (M)
Acinos Mill. LABIATAE (D)
Acioa Aubl. CHRYSOBALANACEAE (D)
Aciotis D.Don MELASTOMATACEAE (D)
Aciphylla J.R.Forst. & G.Forst. UMBELLIFERAE (D)
Acisanthera P.Browne MELASTOMATACEAE (D)
Ackama A.Cunn. = Caldcluvia D.Don CUNONIACEAE (D)
Acleisanthes A.Gray NYCTAGINACEAE (D)
Aclinia Griff. = Dendrobium Sw. ORCHIDACEAE (M)
Aclisia E.Mey. = Pollia Thunb. COMMELINACEAE (M)
Acmadenia Bartl. & H.L.Wendl. RUTACEAE (D)
Acmanthera Griseb. MALPIGHIACEAE (D)
Acmella L.C.Rich. ex Pers. COMPOSITAE (D)
Acmena DC. MYRTACEAE (D)
Acmenosperma Kausel MYRTACEAE (D)
Acmispon Raf. = Lotus L. LEGUMINOSAE–PAPILIONOIDEAE (D)
Acmopyle Pilg. PODOCARPACEAE (G)
Acmostemon Pilg. = Ipomoea L. CONVOLVULACEAE (D)
Acmostigma Raf. = Pavetta L. RUBIACEAE (D)
Acnida L. = Amaranthus L. AMARANTHACEAE (D)
Acnistus Schott SOLANACEAE (D)
Acoelorrhaphe H.A.Wendl. PALMAE (M)
Acoidium Lindl. = Trichocentrum Poepp. & Endl. ORCHIDACEAE (M)
Acokanthera G.Don APOCYNACEAE (D)
Acomastylis Greene = Geum L. ROSACEAE (D)
Acomis F.Muell. COMPOSITAE (D)
Aconceveibum Miq. = Mallotus Lour. EUPHORBIACEAE (D)
Aconiopteris C.Presl = Elaphoglossum Schott ex J.Sm. LOMARIOPSIDACEAE (F)
Aconitum L. RANUNCULACEAE (D)
Aconogonon (Meisn.) Rchb. POLYGONACEAE (D)
Aconogonum (Meisn.) Rchb. (SUO) = Aconogonon (Meisn.) Rchb. POLYGONACEAE (D)
Acopanea Steyerm. = Bonnetia Mart. GUTTIFERAE (D)
Acophorum Steud. (UR) GRAMINEAE (M)
Acorellus Palla = Cyperus L. CYPERACEAE (M)
Acoridium Nees & Meyen ORCHIDACEAE (M)
Acorus L. ACORACEAE (M)
Acosmium Schott LEGUMINOSAE–PAPILIONOIDEAE (D)
Acosmus Desv. = Aspicarpa Rich. MALPIGHIACEAE (D)
Acostaea Schltr. ORCHIDACEAE (M)
Acostia Swallen GRAMINEAE (M)
Acourtia D.Don COMPOSITAE (D)
Acrachne Wight & Arn. ex Chiov. GRAMINEAE (M)
Acradenia Kippist RUTACEAE (D)
Acraea Lindl. = Pterichis Lindl. ORCHIDACEAE (M)
Acrandra O.Berg = Campomanesia Ruíz & Pav. MYRTACEAE (D)
Acranthemum Tiegh. = Agelanthus Tiegh. LORANTHACEAE (D)
Acranthera Arn. ex Meisn. RUBIACEAE (D)
Acratherum Link = Arundinella Raddi GRAMINEAE (M)
Acreugenia Kausel = Myrcianthes O.Berg MYRTACEAE (D)

Acridocarpus Guill. & Perr. MALPIGHIACEAE (D)
Acrilia Griseb. = Trichilia P.Browne MELIACEAE (D)
Acriopsis Blume ORCHIDACEAE (M)
Acrisione B.Nord. COMPOSITAE (D)
Acrista O.F.Cook = Prestoea Hook.f. PALMAE (M)
Acritochaete Pilg. GRAMINEAE (M)
Acritopappus R.M.King & H.Rob. COMPOSITAE (D)
Acriulus Ridl. = Scleria Bergius CYPERACEAE (M)
Acroanthes Raf. = Malaxis Sol. ex Sw. ORCHIDACEAE (M)
Acrobotrys K.Schum. & K.Krause RUBIACEAE (D)
Acrocarpus Wight ex Arn. LEGUMINOSAE–CAESALPINIOIDEAE (D)
Acrocephalus Benth. LABIATAE (D)
Acroceras Stapf GRAMINEAE (M)
Acrochaene Lindl. = Monomeria Lindl. ORCHIDACEAE (M)
Acrochaete Peter (SUH) = Setaria P.Beauv. GRAMINEAE (M)
Acroclasia C.Presl = Mentzelia L. LOASACEAE (D)
Acroclinium A.Gray COMPOSITAE (D)
Acrocoelium Baill. = Leptaulus Benth. ICACINACEAE (D)
Acrocomia Mart. PALMAE (M)
Acrodiclidium Nees & Mart. = Licaria Aubl. LAURACEAE (D)
Acrodon N.E.Br. AIZOACEAE (D)
Acrodryon Spreng. = Cephalanthus L. RUBIACEAE (D)
Acroelytrum Steud. = Lophatherum Brongn. GRAMINEAE (M)
Acroglochin Schrad. ex Schult. CHENOPODIACEAE (D)
Acrolobus Klotzsch = Heisteria Jacq. OLACACEAE (D)
Acrolophia Pfitzer ORCHIDACEAE (M)
Acronema Falc. ex Edgew. UMBELLIFERAE (D)
Acronia C.Presl = Pleurothallis R.Br. ORCHIDACEAE (M)
Acronodia Blume = Elaeocarpus L. ELAEOCARPACEAE (D)
Acronychia J.R.Forst. & G.Forst. RUTACEAE (D)
Acropelta Nakai DRYOPTERIDACEAE (F)
Acropera Lindl. = Gongora Ruíz & Pav. ORCHIDACEAE (M)
Acrophorus C.Presl DRYOPTERIDACEAE (F)
Acrophyllum Benth. CUNONIACEAE (D)
Acropogon Schltr. STERCULIACEAE (D)
Acropteris Link = Asplenium L. ASPLENIACEAE (F)
Acropterygium (Diels) Nakai = Dicranopteris Bernh. GLEICHENIACEAE (F)
Acroptilon Cass. COMPOSITAE (D)
Acrorchis Dressler ORCHIDACEAE (M)
Acrorumohra (H.Ito) H.Ito = Arachniodes Blume DRYOPTERIDACEAE (F)
Acrosanthes Eckl. & Zeyh. AIZOACEAE (D)
Acroschizocarpus Gombocz = Christolea Cambess. ex Jacquem. CRUCIFERAE (D)
Acrosepalum Pierre = Ancistrocarpus Oliv. TILIACEAE (D)
Acrosorus Copel. GRAMMITIDACEAE (F)
Acrospelion Steud. (SUS) = Trisetum Pers. GRAMINEAE (M)
Acrostachys (Benth.) Tiegh. = Helixanthera Lour. LORANTHACEAE (D)
Acrostachys Herter (SUS) = Lycopodium L. LYCOPODIACEAE (FA)
Acrostemon Klotzsch ERICACEAE (D)
Acrostephanus Tiegh. = Tapinanthus (Blume) Blume LORANTHACEAE (D)
Acrostichum L. PTERIDACEAE (F)
Acrostigma O.F.Cook & Doyle = Catoblastus H.A.Wendl. PALMAE (M)
Acrostoma Didr. = Remijia DC. RUBIACEAE (D)
Acrostylia Frapp. ex Cordem. = Cynorkis Thouars ORCHIDACEAE (M)
Acrosynanthus Urb. = Remijia DC. RUBIACEAE (D)
Acrotome Benth. ex Endl. LABIATAE (D)

Acrotrema Jack DILLENIACEAE (D)
Acrotriche R.Br. EPACRIDACEAE (D)
Acroxis Steud. (SUI) = Muhlenbergia Schreb. GRAMINEAE (M)
Acrymia Prain LABIATAE (D)
Acsmithia Hoogland CUNONIACEAE (D)
Actaea L. RANUNCULACEAE (D)
Actegeton Blume = Azima Lam. SALVADORACEAE (D)
Actephila Blume EUPHORBIACEAE (D)
Actephilopsis Ridl. = Trigonostemon Blume EUPHORBIACEAE (D)
Actinanthella Balle LORANTHACEAE (D)
Actinanthus Ehrenb. UMBELLIFERAE (D)
Actinidia Lindl. ACTINIDIACEAE (D)
Actiniopteris Link ACTINIOPTERIDACEAE (F)
Actinobole Endl. COMPOSITAE (D)
Actinocarpus R.Br. = Damasonium Mill. ALISMATACEAE (M)
Actinocarya Benth. BORAGINACEAE (D)
Actinocheita F.A.Barkley ANACARDIACEAE (D)
Actinochloa Roem. & Schult. (SUS) = Chondrosum Desv. GRAMINEAE (M)
Actinochloris Steud. (SUI) = Chloris Sw. GRAMINEAE (M)
Actinocladum McClure ex Soderstr. GRAMINEAE (M)
Actinodaphne Nees LAURACEAE (D)
Actinodium Schauer MYRTACEAE (D)
Actinokentia Dammer PALMAE (M)
Actinolema Fenzl UMBELLIFERAE (D)
Actinomeris Nutt. = Verbesina L. COMPOSITAE (D)
Actinomorphe (Miq.) Miq. = Schefflera J.R.Forst. & G.Forst. ARALIACEAE (D)
Actinophlebia C.Presl = Cyathea Sm. CYATHEACEAE (F)
Actinophloeus (Becc.) Becc. = Ptychosperma Labill. PALMAE (M)
Actinophyllum Ruíz & Pav. = Schefflera J.R.Forst. & G.Forst. ARALIACEAE (D)
Actinorhytis H.A.Wendl. & Drude PALMAE (M)
Actinoschoenus Benth. CYPERACEAE (M)
Actinoscirpus (Ohwi) R.W.Haines & Lye CYPERACEAE (M)
Actinoseris (Endl.) Cabrera COMPOSITAE (D)
Actinospermum Elliott COMPOSITAE (D)
Actinostachys Wall. = Schizaea Sm. SCHIZAEACEAE (F)
Actinostemma Griff. CUCURBITACEAE (D)
Actinostemon Mart. ex Klotzsch EUPHORBIACEAE (D)
Actinostrobus Miq. CUPRESSACEAE (G)
Actinotus Labill. UMBELLIFERAE (D)
Actites Lander = Sonchus L. COMPOSITAE (D)
Acuan Medik. = Desmanthus Willd. LEGUMINOSAE–MIMOSOIDEAE (D)
Acunaeanthus Borhidi, Koml. & M.Moncada RUBIACEAE (D)
Acustelma Baill. = Cryptolepis R.Br. ASCLEPIADACEAE (D)
Acystopteris Nakai WOODSIACEAE (F)
Ada Lindl. ORCHIDACEAE (M)
Adactylus Rolfe = Apostasia Blume ORCHIDACEAE (M)
Adamia Wall. = Dichroa Lour. HYDRANGEACEAE (D)
Adamia Jacq.–Fél. (SUH) = Feliciadamia Bullock MELASTOMATACEAE (D)
Adansonia L. BOMBACACEAE (D)
Adectum Link = Dennstaedtia Bernh. DENNSTAEDTIACEAE (F)
Adelanthus Endl. = Pyrenacantha Wight ICACINACEAE (D)
Adelaster Lindl. ex Veitch = Fittonia Coem. ACANTHACEAE (D)
Adelbertia Meisn. = Meriania Sw. MELASTOMATACEAE (D)
Adelia L. EUPHORBIACEAE (D)
Adelia P.Browne (SUH) = Forestiera Poir. OLEACEAE (D)

Adeliopsis Benth. = Hypserpa Miers MENISPERMACEAE (D)
Adelmeria Ridl. = Alpinia Roxb. ZINGIBERACEAE (M)
Adelobotrys DC. MELASTOMATACEAE (D)
Adelocaryum Brand = Lindelofia Lehm. BORAGINACEAE (D)
Adelodypsis Becc. = Dypsis Noronha ex Mart. PALMAE (M)
Adelonema Schott = Homalomena Schott ARACEAE (M)
Adelonenga Hook.f. = Hydriastele H.A.Wendl. & Drude PALMAE (M)
Adelopetalum Fitzg. = Bulbophyllum Thouars ORCHIDACEAE (M)
Adelosa Blume VERBENACEAE (D)
Adelostemma Hook.f. ASCLEPIADACEAE (D)
Adelostigma Steetz COMPOSITAE (D)
Ademo Post & Kuntze = Euphorbia L. EUPHORBIACEAE (D)
Adenacanthus Nees = Strobilanthes Blume ACANTHACEAE (D)
Adenandra Willd. RUTACEAE (D)
Adenanthe Maguire, Steyerm. & Wurdack = Tyleria Gleason OCHNACEAE (D)
Adenanthellum B.Nord. COMPOSITAE (D)
Adenanthemum B.Nord. (SUH) = Adenanthellum B.Nord. COMPOSITAE (D)
Adenanthera L. LEGUMINOSAE–MIMOSOIDEAE (D)
Adenanthos Labill. PROTEACEAE (D)
Adenarake Maguire & Wurdack OCHNACEAE (D)
Adenaria Kunth LYTHRACEAE (D)
Adeneleuterophora Barb.Rodr. = Elleanthus C.Presl ORCHIDACEAE (M)
Adenia Forssk. PASSIFLORACEAE (D)
Adenium Roem. & Schult. APOCYNACEAE (D)
Adenoa Arbo TURNERACEAE (D)
Adenocalymma Mart. ex Meisn. BIGNONIACEAE (D)
Adenocarpus DC. LEGUMINOSAE–PAPILIONOIDEAE (D)
Adenocaulon Hook. COMPOSITAE (D)
Adenoceras Rchb.f. & Zoll. ex Baill. = Macaranga Thouars EUPHORBIACEAE (D)
Adenochilus Hook.f. ORCHIDACEAE (M)
Adenochlaena Boivin ex Baill. EUPHORBIACEAE (D)
Adenoclina Post & Kuntze (SUO) = Adenocline Turcz. EUPHORBIACEAE (D)
Adenocline Turcz. EUPHORBIACEAE (D)
Adenocrepis Blume = Baccaurea Lour. EUPHORBIACEAE (D)
Adenocritonia R.M.King & H.Rob. COMPOSITAE (D)
Adenoderris J.Sm. WOODSIACEAE (F)
Adenodolichos Harms LEGUMINOSAE–PAPILIONOIDEAE (D)
Adenoglossa B.Nord. COMPOSITAE (D)
Adenogramma Rchb. MOLLUGINACEAE (D)
Adenogramma Link ex Engl. (SUH) = Anogramma Link ADIANTACEAE (F)
Adenogyne Klotzsch = Sebastiania Spreng. EUPHORBIACEAE (D)
Adenogynum Rchb.f. & Zoll. (SUH) = Cladogynos Zipp. ex Spanoghe
    EUPHORBIACEAE (D)
Adenolisianthus Gilg = Irlbachia Mart. GENTIANACEAE (D)
Adenolobus (Harv. ex Benth.) Torre & Hillc. LEGUMINOSAE–
    CAESALPINIOIDEAE (D)
Adenoncos Blume ORCHIDACEAE (M)
Adenoon Dalzell COMPOSITAE (D)
Adenopappus Benth. COMPOSITAE (D)
Adenopeltis Bertero ex A.Juss. EUPHORBIACEAE (D)
Adenopetalum Klotzsch & Garcke = Euphorbia L. EUPHORBIACEAE (D)
Adenophaedra (Müll.Arg.) Müll.Arg. EUPHORBIACEAE (D)
Adenophora Fisch. CAMPANULACEAE (D)
Adenophorus Gaudich. GRAMMITIDACEAE (F)
Adenophyllum Pers. COMPOSITAE (D)

Adenoplea Radlk. = Buddleja L. BUDDLEJACEAE (D)
Adenoplusia Radlk. = Buddleja L. BUDDLEJACEAE (D)
Adenopodia C.Presl LEGUMINOSAE–MIMOSOIDEAE (D)
Adenoporces Small = Tetrapterys Cav. MALPIGHIACEAE (D)
Adenopus Benth. = Lagenaria Ser. CUCURBITACEAE (D)
Adenorhopium Rchb. = Jatropha L. EUPHORBIACEAE (D)
Adenorima Raf. = Euphorbia L. EUPHORBIACEAE (D)
Adenoropium Pohl (SUO) = Jatropha L. EUPHORBIACEAE (D)
Adenosacme Wall. = Mycetia Reinw. RUBIACEAE (D)
Adenosciadium H.Wolff UMBELLIFERAE (D)
Adenosma R.Br. SCROPHULARIACEAE (D)
Adenostachya Bremek. = Strobilanthes Blume ACANTHACEAE (D)
Adenostegia Benth. = Cordylanthus Nutt. ex Benth. SCROPHULARIACEAE (D)
Adenostemma J.R.Forst. & G.Forst. COMPOSITAE (D)
Adenostoma Hook. & Arn. ROSACEAE (D)
Adenostyles Cass. (SUS) = Cacalia L. COMPOSITAE (D)
Adenostylis Blume = Zeuxine Lindl. ORCHIDACEAE (M)
Adenothamnus D.Keck COMPOSITAE (D)
Adenothola Lem. = Manettia L. RUBIACEAE (D)
Adenotrias Jaub. & Spach = Hypericum L. GUTTIFERAE (D)
Adesmia DC. LEGUMINOSAE–PAPILIONOIDEAE (D)
Adhatoda Mill. = Justicia L. ACANTHACEAE (D)
Adiantopsis Fée ADIANTACEAE (F)
Adiantum L. ADIANTACEAE (F)
Adina Salisb. RUBIACEAE (D)
Adinandra Jack THEACEAE (D)
Adinandrella Exell = Ternstroemia Mutis ex L.f. THEACEAE (D)
Adinauclea Ridsdale RUBIACEAE (D)
Adinobotrys Dunn = Whitfordiodendron Elmer LEGUMINOSAE–PAPILIONOIDEAE (D)
Adisa Steud. = Sumbaviopsis J.J.Sm. EUPHORBIACEAE (D)
Adisca Blume = Sumbaviopsis J.J.Sm. EUPHORBIACEAE (D)
Adiscanthus Ducke RUTACEAE (D)
Adlumia Raf. ex DC. PAPAVERACEAE (D)
Adnula Raf. = Pelexia Poit. ex Lindl. ORCHIDACEAE (M)
Adolphia Meisn. RHAMNACEAE (D)
Adonidia Becc. = Veitchia H.A.Wendl. PALMAE (M)
Adonis L. RANUNCULACEAE (D)
Adoxa L. ADOXACEAE (D)
Adrastaea DC. = Hibbertia Andrews DILLENIACEAE (D)
Adriana Gaudich. EUPHORBIACEAE (D)
Adromischus Lem. CRASSULACEAE (D)
Adrorhizon Hook.f. ORCHIDACEAE (M)
Aechmandra Arn. = Kedrostis Medik. CUCURBITACEAE (D)
Aechmanthera Nees ACANTHACEAE (D)
Aechmea Ruíz & Pav. BROMELIACEAE (M)
Aechmolepis Decne. = Tacazzea Decne. ASCLEPIADACEAE (D)
Aechmophora Steud. (SUS) = Bromus L. GRAMINEAE (M)
Aedesia O.Hoffm. COMPOSITAE (D)
Aegialina Schult. = Rostraria Trin. GRAMINEAE (M)
Aegialitis Trin. (SUH) = Rostraria Trin. GRAMINEAE (M)
Aegialitis R.Br. PLUMBAGINACEAE (D)
Aegiceras Gaertn. MYRSINACEAE (D)
Aegicon Adans. (SUS) = Aegilops L. GRAMINEAE (M)
Aegilemma Á.Löve = Aegilops L. GRAMINEAE (M)
Aegilonearum Á.Löve = Aegilops L. GRAMINEAE (M)

Aegilopodes Á.Löve (SUS) = Aegilops L. GRAMINEAE (M)
Aegilops L. GRAMINEAE (M)
Aeginetia L. SCROPHULARIACEAE (D)
Aeginetia Cav. (SUH) = Bouvardia Salisb. RUBIACEAE (D)
Aegiphila Jacq. VERBENACEAE (D)
Aegle Corrêa RUTACEAE (D)
Aeglopsis Swingle RUTACEAE (D)
Aegochloa Benth. = Navarretia Ruíz & Pav. POLEMONIACEAE (D)
Aegokeras Raf. UMBELLIFERAE (D)
Aegonychon Gray = Lithospermum L. BORAGINACEAE (D)
Aegopicron Giseke (SUO) = Maprounea Aubl. EUPHORBIACEAE (D)
Aegopodium L. UMBELLIFERAE (D)
Aegopogon Humb. & Bonpl. ex Willd. GRAMINEAE (M)
Aegopordon Boiss. COMPOSITAE (D)
Aegopricon L.f. (SUO) = Maprounea Aubl. EUPHORBIACEAE (D)
Aegopricum L. (SUS) = Maprounea Aubl. EUPHORBIACEAE (D)
Aegotoxicum Endl. (SUO) = Aextoxicon Ruíz & Pav. AEXTOXICACEAE (D)
Aegtoxicon Molina (SUO) = Aextoxicon Ruíz & Pav. AEXTOXICACEAE (D)
Aelbroeckia DeMoor (SUS) = Aeluropus Trin. GRAMINEAE (M)
Aellenia Ulbr. = Halothamnus Jaub. & Spach CHENOPODIACEAE (D)
Aeluropus Trin. GRAMINEAE (M)
Aembilla Adans. = Scolopia Schreb. FLACOURTIACEAE (D)
Aenictophyton A.T.Lee LEGUMINOSAE–PAPILIONOIDEAE (D)
Aenigmopteris Holttum DRYOPTERIDACEAE (F)
Aeollanthus Mart. ex Spreng. LABIATAE (D)
Aeonia Lindl. = Oeonia Lindl. ORCHIDACEAE (M)
Aeoniopsis Rech.f. (SUS) = Bukiniczia Lincz. PLUMBAGINACEAE (D)
Aeonium Webb & Berthel. CRASSULACEAE (D)
Aequatorium B.Nord. COMPOSITAE (D)
Aera Asch. (SUO) = Aira L. GRAMINEAE (M)
Aerangis Rchb.f. ORCHIDACEAE (M)
Aeranthes Lindl. ORCHIDACEAE (M)
Aeria O.F.Cook = Gaussia H.A.Wendl. PALMAE (M)
Aerides Lour. ORCHIDACEAE (M)
Aeridium Salisb. = Aerides Lour. ORCHIDACEAE (M)
Aeridostachya (Hook.f.) Brieger = Eria Lindl. ORCHIDACEAE (M)
Aerisilvaea Radcl.–Sm. EUPHORBIACEAE (D)
Aerobion Kaempfer ex Spreng. = Angraecum Bory ORCHIDACEAE (M)
Aeropsis Asch. & Graebn. (SUS) = Airopsis Desv. GRAMINEAE (M)
Aerva Forssk. AMARANTHACEAE (D)
Aeschrion Vell. = Picrasma Blume SIMAROUBACEAE (D)
Aeschynanthus Jack GESNERIACEAE (D)
Aeschynomene L. LEGUMINOSAE–PAPILIONOIDEAE (D)
Aesculus L. HIPPOCASTANACEAE (D)
Aetanthus (Eichler) Engl. LORANTHACEAE (D)
Aetheolaena Cass. = Lasiocephalus Schltdl. COMPOSITAE (D)
Aetheolirion Forman COMMELINACEAE (M)
Aetheonema Rchb. = Gaertnera Lam. RUBIACEAE (D)
Aetheorhiza Cass. COMPOSITAE (D)
Aethephyllum N.E.Br. AIZOACEAE (D)
Aetheria Blume ex Endl. = Stenorrhynchos Rich. ex Spreng. ORCHIDACEAE (M)
Aethiocarpa Vollesen STERCULIACEAE (D)
Aethionema R.Br. CRUCIFERAE (D)
Aethonopogon Kuntze (SUI) = Polytrias Hack. GRAMINEAE (M)
Aethusa L. UMBELLIFERAE (D)

Aetia Adans. = Combretum Loefl. COMBRETACEAE (D)
Aetopteron Ehrh. ex House = Polystichum Roth DRYOPTERIDACEAE (F)
Aetoxylon (Airy Shaw) Airy Shaw THYMELAEACEAE (D)
Aextoxicon Ruíz & Pav. AEXTOXICACEAE (D)
Affonsea A.St.-Hil. LEGUMINOSAE–MIMOSOIDEAE (D)
Afgekia Craib LEGUMINOSAE–PAPILIONOIDEAE (D)
Afrachneria Sprague = Pentaschistis (Nees) Spach GRAMINEAE (M)
Afraegle (Swingle) Engl. RUTACEAE (D)
Afrafzelia Pierre = Afzelia Sm. LEGUMINOSAE–CAESALPINIOIDEAE (D)
Aframmi C.Norman UMBELLIFERAE (D)
Aframomum K.Schum. ZINGIBERACEAE (M)
Afrardisia Mez = Ardisia Sw. MYRSINACEAE (D)
Afraurantium A.Chev. (UP) RUTACEAE (D)
Afridia Duthie = Nepeta L. LABIATAE (D)
Afrobrunnichia Hutch. & Dalziel POLYGONACEAE (D)
Afrocalathea K.Schum. MARANTACEAE (M)
Afrocarpus (Buchholz & E.Gray) C.N.Page PODOCARPACEAE (G)
Afrocarum Rauschert UMBELLIFERAE (D)
Afrocrania (Harms) Hutch. CORNACEAE (D)
Afrofittonia Lindau ACANTHACEAE (D)
Afroguatteria Boutique ANNONACEAE (D)
Afrohamelia Wernham = Atractogyne Pierre RUBIACEAE (D)
Afroknoxia Verdc. = Knoxia L. RUBIACEAE (D)
Afrolicania Mildbr. CHRYSOBALANACEAE (D)
Afroligusticum C.Norman UMBELLIFERAE (D)
Afrolimon Lincz. = Limonium Mill. PLUMBAGINACEAE (D)
Afromendoncia Gilg ex Lindau = Mendoncia Vell. ex Vand. ACANTHACEAE (D)
Afropteris Alston PTERIDACEAE (F)
Afrorhaphidophora Engl. = Rhaphidophora Hassk. ARACEAE (M)
Afrormosia Harms = Pericopsis Thwaites LEGUMINOSAE–PAPILIONOIDEAE (D)
Afrosersalisia A.Chev. = Synsepalum (A.DC.) Daniell SAPOTACEAE (D)
Afrosison H.Wolff UMBELLIFERAE (D)
Afrostyrax Perkins & Gilg HUACEAE (D)
Afrothismia (Engl.) Schltr. BURMANNIACEAE (M)
Afrotrewia Pax & K.Hoffm. EUPHORBIACEAE (D)
Afrotrichloris Chiov. GRAMINEAE (M)
Afrotrilepis (Gilly) J.Raynal CYPERACEAE (M)
Afrotysonia Rauschert BORAGINACEAE (D)
Afrovivella A.Berger = Rosularia (DC.) Stapf CRASSULACEAE (D)
Afzelia Sm. LEGUMINOSAE–CAESALPINIOIDEAE (D)
Afzeliella Gilg = Guyonia Naudin MELASTOMATACEAE (D)
Agalinis Raf. SCROPHULARIACEAE (D)
Agallis Phil. CRUCIFERAE (D)
Agalma Miq. = Schefflera J.R.Forst. & G.Forst. ARALIACEAE (D)
Agalmanthus (Endl.) Hombr. & Jacquinot = Metrosideros Banks ex Gaertn.
     MYRTACEAE (D)
Agalmyla Blume GESNERIACEAE (D)
Agaloma Raf. = Euphorbia L. EUPHORBIACEAE (D)
Aganisia Lindl. ORCHIDACEAE (M)
Aganonerion Pierre ex Spire APOCYNACEAE (D)
Aganope Miq. LEGUMINOSAE–PAPILIONOIDEAE (D)
Aganosma (Blume) G.Don APOCYNACEAE (D)
Agaosizia Spach = Camissonia Link ONAGRACEAE (D)
Agapanthus L'Hér. ALLIACEAE (M)
Agapetes D.Don ex G.Don ERICACEAE (D)

Agardhia Spreng. (SUH) = Qualea Aubl. VOCHYSIACEAE (D)
Agarista D.Don ex G.Don ERICACEAE (D)
Agastache Gronov. LABIATAE (D)
Agastachys R.Br. PROTEACEAE (D)
Agasyllis Spreng. UMBELLIFERAE (D)
Agatea A.Gray VIOLACEAE (D)
Agathelpis Choisy SCROPHULARIACEAE (D)
Agathis Salisb. ARAUCARIACEAE (G)
Agathisanthemum Klotzsch RUBIACEAE (D)
Agathophora (Fenzl) Bunge CHENOPODIACEAE (D)
Agathosma Willd. RUTACEAE (D)
Agati Adans. = Sesbania Scop. LEGUMINOSAE–PAPILIONOIDEAE (D)
Agation Brongn. (SUS) = Agatea A.Gray VIOLACEAE (D)
Agauria (DC.) Hook.f. = Agarista D.Don ex G.Don ERICACEAE (D)
Agave L. AGAVACEAE (M)
Agdestis Moçino & Sessé ex DC. AGDESTIDACEAE (D)
Agelaea Sol. ex Planch. CONNARACEAE (D)
Agelandra Engl. & Pax (SUO) = Croton L. EUPHORBIACEAE (D)
Agelanthus Tiegh. LORANTHACEAE (D)
Agenium Nees GRAMINEAE (M)
Ageomoron Raf. = Caucalis L. UMBELLIFERAE (D)
Ageratella A.Gray ex S.Watson COMPOSITAE (D)
Ageratina Spach COMPOSITAE (D)
Ageratinastrum Mattf. COMPOSITAE (D)
Ageratum L. COMPOSITAE (D)
Aggeianthus Wight = Porpax Lindl. ORCHIDACEAE (M)
Agiabampoa Rose ex O.Hoffm. COMPOSITAE (D)
Agianthus Greene CRUCIFERAE (D)
Agirta Baill. = Tragia L. EUPHORBIACEAE (D)
Aglaia Lour. MELIACEAE (D)
Aglaodorum Schott ARACEAE (M)
Aglaomorpha Schott POLYPODIACEAE (F)
Aglaonema Schott ARACEAE (M)
Aglossorhyncha Schltr. ORCHIDACEAE (M)
Aglycia Steud. (SUI) = Eriochloa Kunth GRAMINEAE (M)
Agnirictus Schwantes = Stomatium Schwantes AIZOACEAE (D)
Agonandra Miers ex Hook.f. OPILIACEAE (D)
Agoneissos Zoll. ex Nied. = Tristellateia Thouars MALPIGHIACEAE (D)
Agonis (DC.) Sweet MYRTACEAE (D)
Agoseris Raf. COMPOSITAE (D)
Agraulus P.Beauv. = Agrostis L. GRAMINEAE (M)
Agrestis Bubani (SUS) = Agrostis L. GRAMINEAE (M)
Agrianthus Mart. ex DC. COMPOSITAE (D)
Agrimonia L. ROSACEAE (D)
Agriophyllum M.Bieb. CHENOPODIACEAE (D)
Agrocharis Hochst. UMBELLIFERAE (D)
Agropyron Gaertn. GRAMINEAE (M)
Agropyropsis (Batt. & Trab.) A.Camus GRAMINEAE (M)
Agrostemma L. CARYOPHYLLACEAE (D)
Agrosticula Raddi = Sporobolus R.Br. GRAMINEAE (M)
Agrostis L. GRAMINEAE (M)
Agrostistachys Dalzell EUPHORBIACEAE (D)
Agrostocrinum F.Muell. PHORMIACEAE (M)
Agrostomia Cerv. = Chloris Sw. GRAMINEAE (M)
Agrostophyllum Blume ORCHIDACEAE (M)

Aguava Raf. = Myrcia DC. ex Guill. MYRTACEAE (D)
Aguiaria Ducke BOMBACACEAE (D)
Agylophora Neck. ex Raf. (SUI) = Uncaria Schreb. RUBIACEAE (D)
Agyneia L. = Glochidion J.R.Forst. & G.Forst. EUPHORBIACEAE (D)
Agyneja Vent. (SUH) = Sauropus Blume EUPHORBIACEAE (D)
Agyneja L. = Glochidion J.R.Forst. & G.Forst. EUPHORBIACEAE (D)
Ahernia Merr. FLACOURTIACEAE (D)
Ahzolia Standl. & Steyerm. = Sechium P.Browne CUCURBITACEAE (D)
Aichryson Webb & Berthel. CRASSULACEAE (D)
Aidia Lour. RUBIACEAE (D)
Aidiopsis Tirveng. RUBIACEAE (D)
Aidomene Stopp ASCLEPIADACEAE (D)
Aikinia R.Br. (SUH) = Epithema Blume GESNERIACEAE (D)
Aikinia Wall. (SUI) = Ratzeburgia Kunth GRAMINEAE (M)
Ailanthus Desf. SIMAROUBACEAE (D)
Ailantopsis Gagnep. = Trichilia P.Browne MELIACEAE (D)
Ainea Ravenna IRIDACEAE (M)
Ainsliaea DC. COMPOSITAE (D)
Ainsworthia Boiss. UMBELLIFERAE (D)
Aiolotheca DC. = Zaluzania Pers. COMPOSITAE (D)
Aiouea Aubl. LAURACEAE (D)
Aiphanes Willd. PALMAE (M)
Aipyanthus Steven = Arnebia Forssk. BORAGINACEAE (D)
Aira L. GRAMINEAE (M)
Airampoa Frič = Opuntia Mill. CACTACEAE (D)
Airella (Dumort.) Dumort. (SUS) = Aira L. GRAMINEAE (M)
Airidium Steud. = Deschampsia P.Beauv. GRAMINEAE (M)
Airochloa Link = Koeleria Pers. GRAMINEAE (M)
Airopsis Desv. GRAMINEAE (M)
Airosperma K.Schum. & Lauterb. RUBIACEAE (D)
Airyantha Brummitt LEGUMINOSAE–PAPILIONOIDEAE (D)
Aisandra Pierre = Diploknema Pierre SAPOTACEAE (D)
Aistocaulon Poelln. (SUS) = Aloinopsis Schwantes AIZOACEAE (D)
Aistopetalum Schltr. CUNONIACEAE (D)
Aitchisonia Hemsl. ex Aitch. RUBIACEAE (D)
Aithales Webb & Berthel. = Sedum L. CRASSULACEAE (D)
Aitonia Thunb. (SUH) = Nymania Lindb. MELIACEAE (D)
Aizoanthemum Dinter ex Friedrich AIZOACEAE (D)
Aizoon L. AIZOACEAE (D)
Aizopsis Grulich = Sedum L. CRASSULACEAE (D)
Ajania Poljakov COMPOSITAE (D)
Ajaniopsis C.Shih COMPOSITAE (D)
Ajuga L. LABIATAE (D)
Ajugoides Makino LABIATAE (D)
Akania Hook.f. AKANIACEAE (D)
Akebia Decne. LARDIZABALACEAE (D)
Akentra L.Benj. = Utricularia L. LENTIBULARIACEAE (D)
Akersia Buining = Cleistocactus Lem. CACTACEAE (D)
Aklema Raf. = Euphorbia L. EUPHORBIACEAE (D)
Aladenia Pichon = Farquharia Stapf APOCYNACEAE (D)
Alafia Thouars APOCYNACEAE (D)
Alagophyla Raf. = Sinningia Nees GESNERIACEAE (D)
Alajja Ikonn. LABIATAE (D)
Alamania Lex. ORCHIDACEAE (M)
Alangium Lam. ALANGIACEAE (D)

Alania Endl. ANTHERICACEAE (M)
Alatoseta Compton COMPOSITAE (D)
Alberta E.Mey. RUBIACEAE (D)
Albertia Regel & Schmalh. (SUH) = Kozlovia Lipsky UMBELLIFERAE (D)
Albertinia Spreng. COMPOSITAE (D)
Albertisia Becc. MENISPERMACEAE (D)
Albertisiella Pierre ex Aubrév. = Pouteria Aubl. SAPOTACEAE (D)
Albidella Pichon = Echinodorus Rich. ex Engelm. ALISMATACEAE (M)
Albina Giseke = Alpinia Roxb. ZINGIBERACEAE (M)
Albizia Durazz. LEGUMINOSAE–MIMOSOIDEAE (D)
Albovia Schischk. = Pimpinella L. UMBELLIFERAE (D)
Albraunia Speta SCROPHULARIACEAE (D)
Albuca L. HYACINTHACEAE (M)
Alcantara Glaz. ex G.M.Barroso COMPOSITAE (D)
Alcantarea (E.Morren) Harms = Vriesea Lindl. BROMELIACEAE (M)
Alcea L. MALVACEAE (D)
Alchemilla L. ROSACEAE (D)
Alchornea Sw. EUPHORBIACEAE (D)
Alchorneopsis Müll.Arg. EUPHORBIACEAE (D)
Alcicornium Gaudich. ex Underw. = Platycerium Desv. POLYPODIACEAE (F)
Alcimandra Dandy = Magnolia L. MAGNOLIACEAE (D)
Alcineanthus Merr. = Neoscortechinia Pax EUPHORBIACEAE (D)
Alciope DC. ex Lindl. COMPOSITAE (D)
Alcmene Urb. = Duguetia A.St.–Hil. ANNONACEAE (D)
Alcoceratothrix Nied. = Byrsonima Rich. ex Kunth MALPIGHIACEAE (D)
Alcoceria Fernald = Dalembertia Baill. EUPHORBIACEAE (D)
Aldama La Llave COMPOSITAE (D)
Aldina Endl. LEGUMINOSAE–PAPILIONOIDEAE (D)
Aldinia Raf. = Croton L. EUPHORBIACEAE (D)
Aldrovanda L. DROSERACEAE (D)
Alectoridia A.Rich. = Arthraxon P.Beauv. GRAMINEAE (M)
Alectoroctonum Schltdl. = Euphorbia L. EUPHORBIACEAE (D)
Alectorurus Makino (SUH) = Comospermum Rauschert ANTHERICACEAE (M)
Alectra Thunb. SCROPHULARIACEAE (D)
Alectorolophus Zinn = Rhinanthus L. SCROPHULARIACEAE (D)
Alectryon Gaertn. SAPINDACEAE (D)
Aleisanthia Ridl. RUBIACEAE (D)
Alepidea F.Delaroche UMBELLIFERAE (D)
Alepidocalyx Piper = Phaseolus L. LEGUMINOSAE–PAPILIONOIDEAE (D)
Alepidocline S.F.Blake COMPOSITAE (D)
Alepis Tiegh. LORANTHACEAE (D)
Alepyrum R.Br. = Centrolepis Labill. CENTROLEPIDACEAE (M)
Aletes J.M.Coult. & Rose UMBELLIFERAE (D)
Aletris L. MELANTHIACEAE (M)
Aleurites J.R.Forst. & G.Forst. EUPHORBIACEAE (D)
Aleuritopteris Fée = Cheilanthes Sw. ADIANTACEAE (F)
Alevia Baill. = Bernardia Mill. EUPHORBIACEAE (D)
Alexa Moq. LEGUMINOSAE–PAPILIONOIDEAE (D)
Alexandra Bunge CHENOPODIACEAE (D)
Alexeya Pakhamova = Paraquilegia J.R.Drumm. & Hutch. RANUNCULACEAE (D)
Alexgeorgea Carlquist RESTIONACEAE (M)
Alexitoxicon St.–Lag. = Vincetoxicum Wolf ASCLEPIADACEAE (D)
Alfaroa Standl. JUGLANDACEAE (D)
Alfonsia Kunth = Elaeis Jacq. PALMAE (M)
Alfredia Cass. COMPOSITAE (D)

Algernonia Baill. EUPHORBIACEAE (D)
Alhagi Gagnebin LEGUMINOSAE–PAPILIONOIDEAE (D)
Alibertia A.Rich. ex DC. RUBIACEAE (D)
Alibrexia Miers = Nolana L.f. SOLANACEAE (D)
Alicabon Raf. = Withania Pauquy SOLANACEAE (D)
Alicastrum P.Browne = Brosimum Sw. MORACEAE (D)
Aliciella Brand = Gilia Ruíz & Pav. POLEMONIACEAE (D)
Aliella Qaiser & Lack COMPOSITAE (D)
Alifana Raf. = Brachyotum (DC.) Triana MELASTOMATACEAE (D)
Alifanus Adans. = Rhexia L. MELASTOMATACEAE (D)
Aligera Suksd. VALERIANACEAE (D)
Alinula J.Raynal CYPERACEAE (M)
Alipsea Hoffmanns. = Liparis Rich. ORCHIDACEAE (M)
Alisma L. ALISMATACEAE (M)
Alismorkis Thouars = Calanthe Ker–Gawl. ORCHIDACEAE (M)
Alistilus N.E.Br. LEGUMINOSAE–PAPILIONOIDEAE (D)
Aliteria Benoist = Clarisia Ruíz & Pav. MORACEAE (D)
Alkanna Tausch BORAGINACEAE (D)
Alkekengi Mill. = Physalis L. SOLANACEAE (D)
Allaeanthus Thwaites = Broussonetia L'Hér. ex Vent. MORACEAE (D)
Allaeophania Thwaites RUBIACEAE (D)
Allagopappus Cass. COMPOSITAE (D)
Allagoptera Nees PALMAE (M)
Allagosperma M.Roem. = Melothria L. CUCURBITACEAE (D)
Allagostachyum Steud. (SUI) = Tribolium Desv. GRAMINEAE (M)
Allamanda L. APOCYNACEAE (D)
Allanblackia Oliv. ex Benth. GUTTIFERAE (D)
Allantodia R.Br. = Diplazium Sw. WOODSIACEAE (F)
Allantoma Miers LECYTHIDACEAE (D)
Allantospermum Forman IXONANTHACEAE (D)
Allardia Decne. COMPOSITAE (D)
Alleizettea Dubard & Dop = Danais Comm. ex Vent. RUBIACEAE (D)
Alleizettella Pit. RUBIACEAE (D)
Allelotheca Steud. = Lophatherum Brongn. GRAMINEAE (M)
Allemanda L. (SUO) = Allamanda L. APOCYNACEAE (D)
Allenanthus Standl. RUBIACEAE (D)
Allenia E.Phillips = Radyera Bullock MALVACEAE (D)
Allenia Ewart (SUH) = Micrantheum Desf. EUPHORBIACEAE (D)
Allenrolfea Kuntze CHENOPODIACEAE (D)
Allexis Pierre VIOLACEAE (D)
Alliaria Kuntze (SUH) = Dysoxylum Blume MELIACEAE (D)
Alliaria Heist. ex Fabr. CRUCIFERAE (D)
Allionia L. NYCTAGINACEAE (D)
Allioniella Rydb. NYCTAGINACEAE (D)
Allium L. ALLIACEAE (M)
Allmania R.Br. ex Wight AMARANTHACEAE (D)
Allmaniopsis Suess. AMARANTHACEAE (D)
Allobia Raf. = Euphorbia L. EUPHORBIACEAE (D)
Allobium Miers = Phoradendron Nutt. VISCACEAE (D)
Alloburkillia Whitmore = Burkilliodendron Sastry LEGUMINOSAE–
    PAPILIONOIDEAE (D)
Allocalyx Cordem. = Monocardia Pennell SCROPHULARIACEAE (D)
Allocarya Greene = Plagiobothrys Fisch. & C.A.Mey. BORAGINACEAE (D)
Allocaryastrum Brand = Plagiobothrys Fisch. & C.A.Mey. BORAGINACEAE (D)
Allocassine N.Robson CELASTRACEAE (D)

Allocasuarina L.A.S.Johnson CASUARINACEAE (D)
Allocheilos W.T.Wang GESNERIACEAE (D)
Allochilus Gagnep. = Goodyera R.Br. ORCHIDACEAE (M)
Allochrusa Bunge ex Boiss. CARYOPHYLLACEAE (D)
Alloeochaete C.E.Hubb. GRAMINEAE (M)
Allohemia Raf. = Oryctanthus (Griseb.) Eichl. LORANTHACEAE (D)
Alloiatheros Raf. (SUS) = Gymnopogon P.Beauv. GRAMINEAE (M)
Alloispermum Willd. COMPOSITAE (D)
Allolepis Soderstr. & H.F.Decker GRAMINEAE (M)
Allomaieta Gleason MELASTOMATACEAE (D)
Allomarkgrafia Woodson APOCYNACEAE (D)
Allomorphia Blume MELASTOMATACEAE (D)
Alloneuron Pilg. MELASTOMATACEAE (D)
Allophyllum (Nutt.) A.D.Grant & V.E.Grant POLEMONIACEAE (D)
Allophylus L. SAPINDACEAE (D)
Alloplectus Mart. GESNERIACEAE (D)
Allopterigeron Dunlop COMPOSITAE (D)
Allosandra Raf. = Tragia L. EUPHORBIACEAE (D)
Allosanthus Radlk. SAPINDACEAE (D)
Alloschemone Schott ARACEAE (M)
Alloschmidia H.E.Moore PALMAE (M)
Allosidastrum (Hochr.) Krapov., Fryxell & D.M.Bates MALVACEAE (D)
Allosorus Bernh. = Cheilanthes Sw. ADIANTACEAE (F)
Allospondias (Pierre) Stapf = Spondias L. ANACARDIACEAE (D)
Allostelites Börner = Equisetum L. EQUISETACEAE (FA)
Allostigma W.T.Wang GESNERIACEAE (D)
Allostis Raf. = Baeckea L. MYRTACEAE (D)
Allosyncarpia S.T.Blake MYRTACEAE (D)
Alloteropsis C.Presl GRAMINEAE (M)
Allotropa Torr. & A.Gray ERICACEAE (D)
Allowissadula D.M.Bates MALVACEAE (D)
Allowoodsonia Markgr. APOCYNACEAE (D)
Alloxylon P.H.Weston & Crisp PROTEACEAE (D)
Allozygia Naudin = Oxyspora DC. MELASTOMATACEAE (D)
Alluaudia (Drake) Drake DIDIEREACEAE (D)
Alluaudiopsis Humbert & Choux DIDIEREACEAE (D)
Almaleea Crisp & P.H.Weston LEGUMINOSAE–PAPILIONOIDEAE (D)
Almana Raf. = Sinningia Nees GESNERIACEAE (D)
Almeidea A.St.–Hil. RUTACEAE (D)
Almutaster Á.Löve & D.Löve = Aster L. COMPOSITAE (D)
Alniphyllum Matsum. STYRACACEAE (D)
Alnus Mill. BETULACEAE (D)
Alocasia (Schott) G.Don ARACEAE (M)
Alococarpum Riedl & Kuber UMBELLIFERAE (D)
Aloe L. ALOACEAE (M)
Aloinopsis Schwantes AIZOACEAE (D)
Aloitis Raf. = Gentianella Moench GENTIANACEAE (D)
Alomia Kunth COMPOSITAE (D)
Alomiella R.M.King & H.Rob. COMPOSITAE (D)
Alona Lindl. = Nolana L.f. SOLANACEAE (D)
Alonsoa Ruíz & Pav. SCROPHULARIACEAE (D)
Alopecuropsis Opiz = Alopecurus L. GRAMINEAE (M)
Alopecurus L. GRAMINEAE (M)
Alophia Herb. IRIDACEAE (M)
Aloysia Juss. VERBENACEAE (D)

Alpaminia O.E.Schulz = Weberbauera Gilg & Muschler CRUCIFERAE (D)
Alphandia Baill. EUPHORBIACEAE (D)
Alphitonia Reissek ex Endl. RHAMNACEAE (D)
Alphonsea Hook.f. & Thomson ANNONACEAE (D)
Alphonseopsis E.G.Baker = Polyceratocarpus Engl. & Diels ANNONACEAE (D)
Alpinia Roxb. ZINGIBERACEAE (M)
Alposelinum Pimenov UMBELLIFERAE (D)
Alrawia (Wendelbo) K.Persson & Wendelbo HYACINTHACEAE (M)
Alseis Schott RUBIACEAE (D)
Alseodaphne Nees LAURACEAE (D)
Alseuosmia A.Cunn. ALSEUOSMIACEAE (D)
Alsinanthe (Fenzl) Rchb. = Minuartia L. CARYOPHYLLACEAE (D)
Alsine Gaertn. (SUH) = Minuartia L. CARYOPHYLLACEAE (D)
Alsine L. = Stellaria L. CARYOPHYLLACEAE (D)
Alsinidendron H.Mann CARYOPHYLLACEAE (D)
Alsmithia H.E.Moore PALMAE (M)
Alsobia Hanst. = Episcia Mart. GESNERIACEAE (D)
Alsocydia Mart. ex J.C.Gomes f. = Cuspidaria DC. BIGNONIACEAE (D)
Alsodeia Thouars = Rinorea Aubl. VIOLACEAE (D)
Alsodeidium Engl. = Alsodeiopsis Oliv. ICACINACEAE (D)
Alsodeiopsis Oliv. ICACINACEAE (D)
Alsomitra (Blume) M.Roem. CUCURBITACEAE (D)
Alsophila R.Br. = Cyathea Sm. CYATHEACEAE (F)
Alstonia R.Br. APOCYNACEAE (D)
Alstroemeria L. ALSTROEMERIACEAE (M)
Altamiranoa Rose = Villadia Rose CRASSULACEAE (D)
Altensteinia Kunth ORCHIDACEAE (M)
Alternanthera Forssk. AMARANTHACEAE (D)
Alternasemina Silva Manso (UR) CUCURBITACEAE (D)
Althaea L. MALVACEAE (D)
Althenia F.Petit ZANNICHELLIACEAE (M)
Althoffia K.Schum. = Trichospermum Blume TILIACEAE (D)
Altingia Noronha HAMAMELIDACEAE (D)
Altora Adans. = Clutia L. EUPHORBIACEAE (D)
Alvaradoa Liebm. SIMAROUBACEAE (D)
Alveolina Tiegh. = Psittacanthus Mart. LORANTHACEAE (D)
Alvesia Welw. LABIATAE (D)
Alvimia Calderón ex Soderstr. & Londoño GRAMINEAE (M)
Alvimiantha Grey–Wilson RHAMNACEAE (D)
Alvisia Lindl. = Eria Lindl. ORCHIDACEAE (M)
Alvordia Brandegee COMPOSITAE (D)
Alycia Steud. (SUO) = Eriochloa Kunth GRAMINEAE (M)
Alyogyne Alef. MALVACEAE (D)
Alysicarpus Desv. LEGUMINOSAE–PAPILIONOIDEAE (D)
Alyssoides Mill. CRUCIFERAE (D)
Alyssopsis Boiss. CRUCIFERAE (D)
Alyssum L. CRUCIFERAE (D)
Alyxia Banks ex R.Br. APOCYNACEAE (D)
Alzatea Ruíz & Pav. ALZATEACEAE (D)
Amagris Raf. (SUS) = Calamagrostis Adans. GRAMINEAE (M)
Amaioua Aubl. RUBIACEAE (D)
Amalia Rchb. = Laelia Lindl. ORCHIDACEAE (M)
Amalocalyx Pierre APOCYNACEAE (D)
Amamelis Lem. = Hamamelis L. HAMAMELIDACEAE (D)
Amana Honda = Tulipa L. LILIACEAE (M)

Amannia Blume = Ammannia L. LYTHRACEAE (D)
Amanoa Aubl. EUPHORBIACEAE (D)
Amaraboya Linden ex Mast. = Blakea P.Browne MELASTOMATACEAE (D)
Amaracarpus Blume RUBIACEAE (D)
Amaracus Gled. = Origanum L. LABIATAE (D)
Amaralia Welw. ex Hook.f. = Sherbournia G.Don RUBIACEAE (D)
Amaranthus L. AMARANTHACEAE (D)
Amarolea Small = Osmanthus Lour. OLEACEAE (D)
Amaroria A.Gray SIMAROUBACEAE (D)
Amaryllis L. AMARYLLIDACEAE (M)
Amasonia L.f. VERBENACEAE (D)
Amathea Raf. = Aphelandra R.Br. ACANTHACEAE (D)
Amatlania Lundell = Ardisia Sw. MYRSINACEAE (D)
Amatula Medik. = Lycopersicon Mill. SOLANACEAE (D)
Amauria Benth. COMPOSITAE (D)
Amauriella Rendle = Anubias Schott ARACEAE (M)
Amauropelta Kunze THELYPTERIDACEAE (F)
Amaxitis Adans. (SUS) = Dactylis L. GRAMINEAE (M)
Ambaiba Adans. (SUS) = Cecropia Loefl. CECROPIACEAE (D)
Ambavia Le Thomas ANNONACEAE (D)
Ambelania Aubl. APOCYNACEAE (D)
Amberboa (Pers.) Less. COMPOSITAE (D)
Ambinux Comm. ex Juss. = Vernicia Lour. EUPHORBIACEAE (D)
Amblia C.Presl = Cyrtomium C.Presl DRYOPTERIDACEAE (F)
Amblostoma Scheidw. ORCHIDACEAE (M)
Amblyachyrum Steud. = Apocopis Nees GRAMINEAE (M)
Amblyanthe Rauschert = Dendrobium Sw. ORCHIDACEAE (M)
Amblyanthera Müll.Arg. (SUH) = Mandevilla Lindl. APOCYNACEAE (D)
Amblyanthera Blume = Osbeckia L. MELASTOMATACEAE (D)
Amblyanthopsis Mez MYRSINACEAE (D)
Amblyanthus A.DC. MYRSINACEAE (D)
Amblychloa Link (SUI) = Sclerochloa P.Beauv. GRAMINEAE (M)
Amblyglottis Blume = Calanthe Ker-Gawl. ORCHIDACEAE (M)
Amblygonocarpus Harms LEGUMINOSAE-MIMOSOIDEAE (D)
Amblynotopsis J.F.Macbr. = Antiphytum DC. ex Meisn. BORAGINACEAE (D)
Amblynotus (A.DC.) I.M.Johnst. BORAGINACEAE (D)
Amblyocalyx Benth. = Alstonia R.Br. APOCYNACEAE (D)
Amblyocarpum Fisch. & C.A.Mey. COMPOSITAE (D)
Amblyoglossum Turcz. = Tylophora R.Br. ASCLEPIADACEAE (D)
Amblyolepis DC. COMPOSITAE (D)
Amblyopappus Hook. & Arn. COMPOSITAE (D)
Amblyopetalum (Griseb.) Malme = Oxypetalum R.Br. ASCLEPIADACEAE (D)
Amblyopyrum Eig = Aegilops L. GRAMINEAE (M)
Amblystigma Benth. ASCLEPIADACEAE (D)
Amblytes Dulac (SUS) = Molinia Schrank GRAMINEAE (M)
Amblytropis Kitag. = Gueldenstaedtia Fisch. LEGUMINOSAE-PAPILIONOIDEAE (D)
Ambongia Benoist ACANTHACEAE (D)
Amborella Baill. AMBORELLACEAE (D)
Amboroa Cabrera COMPOSITAE (D)
Ambraria Fabr. = Anthospermum L. RUBIACEAE (D)
Ambrella H.Perrier ORCHIDACEAE (M)
Ambroma L.f. = Abroma Jacq. STERCULIACEAE (D)
Ambrosia L. COMPOSITAE (D)
Ambrosina Bassi ARACEAE (M)
Amburana Schwacke & Taub. LEGUMINOSAE-PAPILIONOIDEAE (D)

Ameghinoa Speg. COMPOSITAE (D)
Amelanchier Medik. ROSACEAE (D)
Amellus L. COMPOSITAE (D)
Amentotaxus Pilg. TAXACEAE (G)
Amerorchis Hultén ORCHIDACEAE (M)
Amerosedum Á.Löve & D.Löve = Sedum L. CRASSULACEAE (D)
Amesia A.Nelson & J.F.Macbr. = Epipactis Zinn ORCHIDACEAE (M)
Amesiella Schltr. ex Garay ORCHIDACEAE (M)
Amesiodendron Hu SAPINDACEAE (D)
Amesium Newman = Asplenium L. ASPLENIACEAE (F)
Amethystanthus Nakai = Isodon (Schrad. ex Benth.) Spach LABIATAE (D)
Amethystea L. LABIATAE (D)
Amherstia Wall. LEGUMINOSAE–CAESALPINIOIDEAE (D)
Amianthium A.Gray MELANTHIACEAE (M)
Amicia Kunth LEGUMINOSAE–PAPILIONOIDEAE (D)
Amischophacelus R.S.Rao & R.V.Kammathy = Cyanotis D.Don COMMELINACEAE (M)
Amischotolype Hassk. COMMELINACEAE (M)
Amitostigma Schltr. ORCHIDACEAE (M)
Ammandra O.F.Cook PALMAE (M)
Ammanella Miq. = Ammannia L. LYTHRACEAE (D)
Ammannia L. LYTHRACEAE (D)
Ammi L. UMBELLIFERAE (D)
Ammianthus Spruce ex Benth. = Retiniphyllum Bonpl. RUBIACEAE (D)
Ammiopsis Boiss. UMBELLIFERAE (D)
Ammobium R.Br. ex Sims COMPOSITAE (D)
Ammobroma Torr. LENNOACEAE (D)
Ammocharis Herb. AMARYLLIDACEAE (M)
Ammochloa Boiss. GRAMINEAE (M)
Ammocodon Standl. NYCTAGINACEAE (D)
Ammodaucus Coss. & Durieu UMBELLIFERAE (D)
Ammodendron Fisch. ex DC. LEGUMINOSAE–PAPILIONOIDEAE (D)
Ammoides Adans. UMBELLIFERAE (D)
Ammophila Host GRAMINEAE (M)
Ammopiptanthus S.H.Cheng LEGUMINOSAE–PAPILIONOIDEAE (D)
Ammoselinum Torr. & A.Gray UMBELLIFERAE (D)
Ammosperma Hook.f. CRUCIFERAE (D)
Ammothamnus Bunge = Sophora L. LEGUMINOSAE–PAPILIONOIDEAE (D)
Amoebophyllum N.E.Br. = Phyllobolus N.E.Br. AIZOACEAE (D)
Amolinia R.M.King & H.Rob. COMPOSITAE (D)
Amomis O.Berg = Pimenta Lindl. MYRTACEAE (D)
Amomum Roxb. ZINGIBERACEAE (M)
Amomyrtella Kausel MYRTACEAE (D)
Amomyrtus (Burret) D.Legrand & Kausel MYRTACEAE (D)
Amoora Roxb. = Aglaia Lour. MELIACEAE (D)
Amoreuxia Moçino & Sessé ex DC. COCHLOSPERMACEAE (D)
Amorpha L. LEGUMINOSAE–PAPILIONOIDEAE (D)
Amorphophallus Blume ex Decne. ARACEAE (M)
Amorphospermum F.Muell. = Niemeyera F.Muell. SAPOTACEAE (D)
Ampalis Bojer ex Bur. = Streblus Lour. MORACEAE (D)
Amparoa Schltr. ORCHIDACEAE (M)
Ampelamus Raf. ASCLEPIADACEAE (D)
Ampelocalamus S.L.Chen, T.H.Wen & G.Y.Sheng = Sinarundinaria Nakai
    GRAMINEAE (M)
Ampelocera Klotzsch ULMACEAE (D)
Ampelocissus Planch. VITACEAE (D)

Ampelodaphne Meisn. = Endlicheria Nees LAURACEAE (D)
Ampelodesmos Link GRAMINEAE (M)
Ampelodonax Lojac. = Ampelodesmos Link GRAMINEAE (M)
Ampelopsis Michx. VITACEAE (D)
Ampelopteris Kunze THELYPTERIDACEAE (F)
Ampelosicyos Thouars CUCURBITACEAE (D)
Ampelothamnus Small = Pieris D.Don ERICACEAE (D)
Ampelozizyphus Ducke RHAMNACEAE (D)
Amperea A.Juss. EUPHORBIACEAE (D)
Amphania Banks ex DC. = Ternstroemia Mutis ex L.f. THEACEAE (D)
Amphiachyris (DC.) Nutt. COMPOSITAE (D)
Amphianthus Torr. = Bacopa Aubl. SCROPHULARIACEAE (D)
Amphiasma Bremek. RUBIACEAE (D)
Amphiblemma Naudin MELASTOMATACEAE (D)
Amphiblestra C.Presl DRYOPTERIDACEAE (F)
Amphibolia L.Bolus ex A.G.J.Herre AIZOACEAE (D)
Amphibolis C.Agardh CYMODOCEACEAE (M)
Amphibologyne Brand BORAGINACEAE (D)
Amphibromus Nees = Helictotrichon Besser GRAMINEAE (M)
Amphicarpaea Elliott ex Nutt. LEGUMINOSAE–PAPILIONOIDEAE (D)
Amphicarpum Kunth GRAMINEAE (M)
Amphicome Royle = Incarvillea Juss. BIGNONIACEAE (D)
Amphicosmia Gardner = Cyathea Sm. CYATHEACEAE (F)
Amphidasya Standl. RUBIACEAE (D)
Amphidesmium Schott ex Kunze = Metaxya C.Presl METAXYACEAE (F)
Amphidetes Fourn. ASCLEPIADACEAE (D)
Amphidonax Nees = Arundo L. GRAMINEAE (M)
Amphidoxa DC. = Gnaphalium L. COMPOSITAE (D)
Amphiestes S.Moore = Hypoestes Sol. ex R.Br. ACANTHACEAE (D)
Amphigena Rolfe ORCHIDACEAE (M)
Amphigenes Janka = Festuca L. GRAMINEAE (M)
Amphiglossa DC. COMPOSITAE (D)
Amphiglottis Salisb. = Epidendrum L. ORCHIDACEAE (M)
Amphilochia Mart. = Qualea Aubl. VOCHYSIACEAE (D)
Amphilophis Nash = Bothriochloa Kuntze GRAMINEAE (M)
Amphilophium Kunth BIGNONIACEAE (D)
Amphimas Pierre ex Harms LEGUMINOSAE–PAPILIONOIDEAE (D)
Amphineurion (A.DC.) Pichon = Aganosma (Blume) G.Don APOCYNACEAE (D)
Amphineuron Holttum THELYPTERIDACEAE (F)
Amphiodon Huber = Poecilanthe Benth. LEGUMINOSAE–PAPILIONOIDEAE (D)
Amphipappus Torr. & A.Gray COMPOSITAE (D)
Amphipetalum Bacigalupo PORTULACACEAE (D)
Amphiphyllum Gleason RAPATEACEAE (M)
Amphipleis Raf. = Nicotiana L. SOLANACEAE (D)
Amphipogon R.Br. GRAMINEAE (M)
Amphipterum (Copel.) Copel. = Hymenophyllum Sm. HYMENOPHYLLACEAE (F)
Amphipterygium Schiede ex Standl. ANACARDIACEAE (D)
Amphirrhox Spreng. VIOLACEAE (D)
Amphiscirpus Oteng-Yeb. CYPERACEAE (M)
Amphiscopia Nees = Justicia L. ACANTHACEAE (D)
Amphisiphon W.F.Barker HYACINTHACEAE (M)
Amphistelma Griseb. = Metastelma R.Br. ASCLEPIADACEAE (D)
Amphitecna Miers BIGNONIACEAE (D)
Amphithalea Eckl. & Zeyh. LEGUMINOSAE–PAPILIONOIDEAE (D)
Amphitoma Gleason = Miconia Ruíz & Pav. MELASTOMATACEAE (D)

Amphochaeta Andersson = Pennisetum Rich. GRAMINEAE (M)
Amphoradenium Desv. = Adenophorus Gaudich. GRAMMITIDACEAE (F)
Amphorchis Thouars = Cynorkis Thouars ORCHIDACEAE (M)
Amphorella Brandegee = Matelea Aubl. ASCLEPIADACEAE (D)
Amphoricarpos Vis. COMPOSITAE (D)
Amphoricarpus Spruce ex Miers = Cariniana Casar. LECYTHIDACEAE (D)
Amphorocalyx Baker MELASTOMATACEAE (D)
Amphorogyne Stauffer & Hürl. SANTALACEAE (D)
Amsinckia Lehm. BORAGINACEAE (D)
Amsonia Walter APOCYNACEAE (D)
Amydrium Schott ARACEAE (M)
Amyema Tiegh. LORANTHACEAE (D)
Amygdalopersica Daniel = Prunus L. ROSACEAE (D)
Amygdalophora M.Roem. = Prunus L. ROSACEAE (D)
Amygdalopsis M.Roem. (SUH) = Prunus L. ROSACEAE (D)
Amygdalus L. = Prunus L. ROSACEAE (D)
Amylocarpus Barb.Rodr. = Bactris Jacq. PALMAE (M)
Amylotheca Tiegh. LORANTHACEAE (D)
Amyrea Leandri EUPHORBIACEAE (D)
Amyris P.Browne RUTACEAE (D)
Amyrsia Raf. = Myrteola O.Berg MYRTACEAE (D)
Amyxa Tiegh. THYMELAEACEAE (D)
Anabaena A.Juss. (SUH) = Romanoa Trevis. EUPHORBIACEAE (D)
Anabaenella Pax & K.Hoffm. (SUS) = Romanoa Trevis. EUPHORBIACEAE (D)
Anabasis L. CHENOPODIACEAE (D)
Anacampseros Mill. (SUH) = Sedum L. CRASSULACEAE (D)
Anacampseros L. PORTULACACEAE (D)
Anacampta Miers = Tabernaemontana L. APOCYNACEAE (D)
Anacamptis Rich. ORCHIDACEAE (M)
Anacantha (Iljin) Soják COMPOSITAE (D)
Anacaona Alain CUCURBITACEAE (D)
Anacardium L. ANACARDIACEAE (D)
Anacharis Rich. = Elodea Michx. HYDROCHARITACEAE (M)
Anacheilium Hoffmanns. = Epidendrum L. ORCHIDACEAE (M)
Anachortus Jirásek & Chrtek = Corynephorus P.Beauv. GRAMINEAE (M)
Anachyris Nees = Paspalum L. GRAMINEAE (M)
Anachyrium Steud. (SUO) = Paspalum L. GRAMINEAE (M)
Anaclanthe N.E.Br. = Antholyza L. IRIDACEAE (M)
Anacolosa (Blume) Blume OLACACEAE (D)
Anacyclia Hoffmanns. = Billbergia Thunb. BROMELIACEAE (M)
Anacyclus L. COMPOSITAE (D)
Anadelphia Hack. GRAMINEAE (M)
Anadenanthera Speg. LEGUMINOSAE–MIMOSOIDEAE (D)
Anadendrum Schott ARACEAE (M)
Anaectocalyx Triana ex Hook.f. MELASTOMATACEAE (D)
Anagallidium Griseb. = Swertia L. GENTIANACEAE (D)
Anagallis L. PRIMULACEAE (D)
Anagosperma Wettst. = Euphrasia L. SCROPHULARIACEAE (D)
Anagyris L. LEGUMINOSAE–PAPILIONOIDEAE (D)
Anakasia Philipson ARALIACEAE (D)
Anamirta Colebr. MENISPERMACEAE (D)
Anamomis Griseb. = Myrcianthes O.Berg MYRTACEAE (D)
Ananas Mill. BROMELIACEAE (M)
Ananthacorus Underw. & Maxon VITTARIACEAE (F)
Anantherix Nutt. = Asclepias L. ASCLEPIADACEAE (D)

Anapalina N.E.Br. IRIDACEAE (M)
Anapausia C.Presl = Bolbitis Schott LOMARIOPSIDACEAE (F)
Anapeltis J.Sm. = Microgramma C.Presl POLYPODIACEAE (F)
Anaphalioides (Benth.) Kirp. COMPOSITAE (D)
Anaphalis DC. COMPOSITAE (D)
Anaphrenium E.Mey. ex Endl. = Heeria Meisn. ANACARDIACEAE (D)
Anaphyllopsis A.Hay ARACEAE (M)
Anaphyllum Schott ARACEAE (M)
Anarrhinum Desf. SCROPHULARIACEAE (D)
Anarthria R.Br. ANARTHRIACEAE (M)
Anarthrophyllum Benth. LEGUMINOSAE–PAPILIONOIDEAE (D)
Anarthropteris Copel. POLYPODIACEAE (F)
Anartia Miers = Tabernaemontana L. APOCYNACEAE (D)
Anaspis Rech.f. = Scutellaria L. LABIATAE (D)
Anastatica L. CRUCIFERAE (D)
Anastrabe E.Mey. ex Benth. SCROPHULARIACEAE (D)
Anastraphia D.Don COMPOSITAE (D)
Anastrophea Wedd. = Sphaerothylax Bisch. ex Krauss PODOSTEMACEAE (D)
Anastrophus Schltdl. = Axonopus P.Beauv. GRAMINEAE (M)
Anathallis Barb.Rodr. = Pleurothallis R.Br. ORCHIDACEAE (M)
Anatherum Nábĕlek (SUH) = Festuca L. GRAMINEAE (M)
Anatherum P.Beauv. = Andropogon L. GRAMINEAE (M)
Anatropanthus Schltr. ASCLEPIADACEAE (D)
Anatropostylia (Plitmann) Kupicha = Ormosia G.Jacks. LEGUMINOSAE–
    PAPILIONOIDEAE (D)
Anaua Miq. = Drypetes Vahl EUPHORBIACEAE (D)
Anaueria Kosterm. = Beilschmiedia Nees LAURACEAE (D)
Anauxanopetalum Teijsm. & Binn. = Swintonia Griff. ANACARDIACEAE (D)
Anax Ravenna = Stenomesson Herb. AMARYLLIDACEAE (M)
Anaxagorea A.St.–Hil. ANNONACEAE (D)
Anaxeton Gaertn. COMPOSITAE (D)
Anaxetum Schott = Niphidium J.Sm. POLYPODIACEAE (F)
Ancalanthus Balf.f. (SUO) = Angkalanthus Balf.f. ACANTHACEAE (D)
Ancana F.Muell. ANNONACEAE (D)
Ancathia DC. COMPOSITAE (D)
Anchietea A.St.–Hil. VIOLACEAE (D)
Anchistea C.Presl = Woodwardia Sm. BLECHNACEAE (F)
Anchomanes Schott ARACEAE (M)
Anchonium DC. CRUCIFERAE (D)
Anchusa L. BORAGINACEAE (D)
Ancistrachne S.T.Blake GRAMINEAE (M)
Ancistragrostis S.T.Blake GRAMINEAE (M)
Ancistranthus Lindau ACANTHACEAE (D)
Ancistrocactus (K.Schum.) Britton & Rose = Sclerocactus Britton & Rose CACTACEAE (D)
Ancistrocarphus A.Gray = Stylocline Nutt. COMPOSITAE (D)
Ancistrocarpus Oliv. TILIACEAE (D)
Ancistrocarya Maxim. BORAGINACEAE (D)
Ancistrochilus Rolfe ORCHIDACEAE (M)
Ancistrochloa Honda = Calamagrostis Adans. GRAMINEAE (M)
Ancistrocladus Wall. ANCISTROCLADACEAE (D)
Ancistrodesmus Naudin = Microlepis (DC.) Miq. MELASTOMATACEAE (D)
Ancistrophora A.Gray COMPOSITAE (D)
Ancistrophyllum (G.Mann & H.A.Wendl.) H.A.Wendl. = Laccosperma (G.Mann &
    H.A.Wendl.) Drude PALMAE (M)
Ancistrorhynchus Finet ORCHIDACEAE (M)

Ancistrostylis Yamaz. SCROPHULARIACEAE (D)
Ancistrothyrsus Harms PASSIFLORACEAE (D)
Ancrumia Harv. ex Baker ALLIACEAE (M)
Ancylacanthus Lindau = Ptyssiglottis T.Anderson ACANTHACEAE (D)
Ancylanthos Desf. RUBIACEAE (D)
Ancylanthus Steud. = Ancylanthos Desf. RUBIACEAE (D)
Ancylobothrys Pierre APOCYNACEAE (D)
Ancylobotrys Pierre (SUO) = Ancylobothrys Pierre APOCYNACEAE (D)
Ancylocladus Wall. = Willughbeia Roxb. APOCYNACEAE (D)
Ancylostemon Craib GESNERIACEAE (D)
Ancyrostemma Poepp. & Endl. = Sclerothrix C.Presl LOASACEAE (D)
Anda Adr.Juss. (SUS) = Joannesia Vell. EUPHORBIACEAE (D)
Andenea Frič (SUI) = Echinopsis Zucc. CACTACEAE (D)
Andersonia R.Br. EPACRIDACEAE (D)
Andersonia Willd. (SUI) = Gaertnera Lam. RUBIACEAE (D)
Anderssoniopiper Trel. = Piper L. PIPERACEAE (D)
Andesia Hauman = Oxychloe Phil. JUNCACEAE (M)
Andicus Vell. (SUS) = Joannesia Vell. EUPHORBIACEAE (D)
Andira Juss. LEGUMINOSAE–PAPILIONOIDEAE (D)
Andrachne L. EUPHORBIACEAE (D)
Andradea Allemão NYCTAGINACEAE (D)
Andradia T.R.Sim = Dialium L. LEGUMINOSAE–CAESALPINIOIDEAE (D)
Andrea Mez BROMELIACEAE (M)
Andreettaea Luer = Pleurothallis R.Br. ORCHIDACEAE (M)
Andresia Sleumer = Cheilotheca Hook.f. ERICACEAE (D)
Androcalymma Dwyer LEGUMINOSAE–CAESALPINIOIDEAE (D)
Androcentrum Lem. = Bravaisia DC. ACANTHACEAE (D)
Androcera Nutt. = Solanum L. SOLANACEAE (D)
Androchilus Liebm. ex Hartman ORCHIDACEAE (M)
Androcorys Schltr. ORCHIDACEAE (M)
Androcymbium Willd. COLCHICACEAE (M)
Andrographis Wall. ex Nees ACANTHACEAE (D)
Androgyne Griff. = Panisea (Lindl.) Lindl. ORCHIDACEAE (M)
Androlepis Brongn. ex Houllet BROMELIACEAE (M)
Andromeda L. ERICACEAE (D)
Andromycia A.Rich. = Asterostigma Fisch. & C.A.Mey. ARACEAE (M)
Androphoranthus Karst. = Caperonia A.St.–Hil. EUPHORBIACEAE (D)
Andropogon L. GRAMINEAE (M)
Andropterum Stapf GRAMINEAE (M)
Andropus Brand = Nama L. HYDROPHYLLACEAE (D)
Androsace L. PRIMULACEAE (D)
Androsaemum Duhamel = Hypericum L. GUTTIFERAE (D)
Androscepia Brongn. = Themeda Forssk. GRAMINEAE (M)
Androsemum link = Hypericum L. GUTTIFERAE (D)
Androsiphon Schltr. HYACINTHACEAE (M)
Androsiphonia Stapf PASSIFLORACEAE (D)
Androstachys Prain EUPHORBIACEAE (D)
Androstephium Torr. ALLIACEAE (M)
Androstylanthus Ducke = Helianthostylis Baill. MORACEAE (D)
Androstylium Miq. = Clusia L. GUTTIFERAE (D)
Androsyne Salisb. = Walleria J.Kirk TECOPHILAEACEAE (M)
Androtium Stapf ANACARDIACEAE (D)
Androtrichum (Brongn.) Brongn. CYPERACEAE (M)
Androtropis R.Br. = Acranthera Arn. ex Meisn. RUBIACEAE (D)
Androya H.Perrier BUDDLEJACEAE (D)

Andruris Schltr. TRIURIDACEAE (M)
Andryala L. COMPOSITAE (D)
Andrzeiowskya Rchb. CRUCIFERAE (D)
Anechites Griseb. APOCYNACEAE (D)
Anectochilus Blume (SUO) = Anoectochilus Blume ORCHIDACEAE (M)
Aneilema R.Br. COMMELINACEAE (M)
Aneimiaebotrys Fée = Anemia Sw. SCHIZAEACEAE (F)
Anelsonia J.F.Macbr. & Payson CRUCIFERAE (D)
Anelytrum Hack. = Avena L. GRAMINEAE (M)
Anemagrostis Trin. (SUS) = Apera Adans. GRAMINEAE (M)
Anemanthele Veldkamp = Stipa L. GRAMINEAE (M)
Anemarrhena Bunge ANTHERICACEAE (M)
Anemia Sw. SCHIZAEACEAE (F)
Anemidictyum J.Sm. ex Hook. = Anemia Sw. SCHIZAEACEAE (F)
Anemirhiza J.Sm. = Anemia Sw. SCHIZAEACEAE (F)
Anemoclema (Franch.) W.T.Wang RANUNCULACEAE (D)
Anemone L. RANUNCULACEAE (D)
Anemonella Spach RANUNCULACEAE (D)
Anemonidium (Spach) Á.Löve & D.Löve = Anemone L. RANUNCULACEAE (D)
Anemonopsis Siebold & Zucc. RANUNCULACEAE (D)
Anemopaegma Mart. ex Meisn. BIGNONIACEAE (D)
Anemopsis Hook. & Arn. SAURURACEAE (D)
Anepsias Schott = Rhodospatha Poepp. ARACEAE (M)
Anerincleistus Korth. MELASTOMATACEAE (D)
Anetanthus Hiern ex Benth. GESNERIACEAE (D)
Anethum L. UMBELLIFERAE (D)
Anetilla Galushko = Anemone L. RANUNCULACEAE (D)
Anetium Splitg. VITTARIACEAE (F)
Aneulophus Benth. ERYTHROXYLACEAE (D)
Aneurolepidium Nevski = Leymus Hochst. GRAMINEAE (M)
Angadenia Miers APOCYNACEAE (D)
Angasomyrtus Trudgen & Keighery MYRTACEAE (D)
Angelandra Endl. = Croton L. EUPHORBIACEAE (D)
Angelesia Korth. = Licania Aubl. CHRYSOBALANACEAE (D)
Angelica L. UMBELLIFERAE (D)
Angelocarpa Rupr. UMBELLIFERAE (D)
Angelonia Bonpl. SCROPHULARIACEAE (D)
Angelphytum G.M.Barroso COMPOSITAE (D)
Angianthus J.C.Wendl. COMPOSITAE (D)
Anginon Raf. UMBELLIFERAE (D)
Angiopteris Adans. (SUH) = Onoclea L. WOODSIACEAE (F)
Angiopteris Hoffm. MARATTIACEAE (F)
Angkalanthus Balf.f. ACANTHACEAE (D)
Angolaea Wedd. PODOSTEMACEAE (D)
Angolluma R.Munster = Pachycymbium Leach ASCLEPIADACEAE (D)
Angophora Cav. MYRTACEAE (D)
Angorchis Thouars = Angraecum Bory ORCHIDACEAE (M)
Angoseseli Chiov. UMBELLIFERAE (D)
Angostura Roem. & Schult. RUTACEAE (D)
Angostyles Benth. = Angostylis Benth. EUPHORBIACEAE (D)
Angostylidium (Müll.Arg.) Pax & K.Hoffm. = Tetracarpidium Pax EUPHORBIACEAE (D)
Angostylis Benth. EUPHORBIACEAE (D)
Angraecopsis Kraenzl. ORCHIDACEAE (M)
Angraecum Bory ORCHIDACEAE (M)
Anguillaria R.Br. = Wurmbea Thunb. COLCHICACEAE (M)

Anguillicarpus Burkill = Spirorhynchus Kar. & Kir. CRUCIFERAE (D)
Anguina Mill. = Trichosanthes L. CUCURBITACEAE (D)
Anguloa Ruíz & Pav. ORCHIDACEAE (M)
Anguria Mill. = Citrullus Schrad. ex Eckl. & Zeyh. CUCURBITACEAE (D)
Anguria Jacq. (SUH) = Psiguria Neck. ex Arn. CUCURBITACEAE (D)
Anguriopsis J.R.Johnst. = Doyerea Grosourdy CUCURBITACEAE (D)
Angylocalyx Taub. LEGUMINOSAE–PAPILIONOIDEAE (D)
Anhalonium Lem. = Ariocarpus Scheidw. CACTACEAE (D)
Ania Lindl. ORCHIDACEAE (M)
Aniba Aubl. LAURACEAE (D)
Anigosanthos Lemée (SUO) = Anigozanthos Labill. HAEMODORACEAE (M)
Anigozanthos Labill. HAEMODORACEAE (M)
Aningeria Aubrév. & Pellegr. = Pouteria Aubl. SAPOTACEAE (D)
Aniotum Sol. ex Parkinson = Inocarpus J.R.Forst. & G.Forst. LEGUMINOSAE–
    PAPILIONOIDEAE (D)
Anisacanthus Nees ACANTHACEAE (D)
Anisachne Keng = Calamagrostis Adans. GRAMINEAE (M)
Anisadenia Wall. ex Meisn. LINACEAE (D)
Anisantha K.Koch = Bromus L. GRAMINEAE (M)
Anisanthera Raf. = Caccinia Savi BORAGINACEAE (D)
Anisantherina Pennell SCROPHULARIACEAE (D)
Aniseia Choisy CONVOLVULACEAE (D)
Aniselytron Merr. = Calamagrostis Adans. GRAMINEAE (M)
Anisepta Raf. = Croton L. EUPHORBIACEAE (D)
Aniserica N.E.Br. ERICACEAE (D)
Anisocalyx L.Bolus (SUH) = Drosanthemopsis Rauschert AIZOACEAE (D)
Anisocampium C.Presl WOODSIACEAE (F)
Anisocentrum Turcz. = Acisanthera P.Browne MELASTOMATACEAE (D)
Anisocereus Backeb. = Escontria Rose CACTACEAE (D)
Anisochaeta DC. COMPOSITAE (D)
Anisochilus Wall. ex Benth. LABIATAE (D)
Anisocoma Torr. & A.Gray COMPOSITAE (D)
Anisocycla Baill. MENISPERMACEAE (D)
Anisodontea C.Presl MALVACEAE (D)
Anisodus Link ex Spreng. SOLANACEAE (D)
Anisogonium C.Presl = Diplazium Sw. WOODSIACEAE (F)
Anisolobus A.DC. = Odontadenia Benth. APOCYNACEAE (D)
Anisomallon Baill. = Apodytes E.Mey. ex Arn. ICACINACEAE (D)
Anisomeles R.Br. LABIATAE (D)
Anisomeria D.Don PHYTOLACCACEAE (D)
Anisomeris C.Presl = Chomelia Jacq. RUBIACEAE (D)
Anisonema A.Juss. = Phyllanthus L. EUPHORBIACEAE (D)
Anisopappus Hook. & Arn. COMPOSITAE (D)
Anisopetalum Hook. = Bulbophyllum Thouars ORCHIDACEAE (M)
Anisophyllea R.Br. ex Sabine ANISOPHYLLEACEAE (D)
Anisophyllum G.Don (SUO) = Anisophyllea R.Br. ex Sabine ANISOPHYLLEACEAE (D)
Anisophyllum Boivin ex Baill. (SUH) = Croton L. EUPHORBIACEAE (D)
Anisophyllum Haw. (SUH) = Euphorbia L. EUPHORBIACEAE (D)
Anisoplectus Oerst. = Drymonia Mart. GESNERIACEAE (D)
Anisopoda Baker UMBELLIFERAE (D)
Anisopogon R.Br. GRAMINEAE (M)
Anisoptera Korth. DIPTEROCARPACEAE (D)
Anisopus N.E.Br. ASCLEPIADACEAE (D)
Anisopyrum (Griseb.) Gren. & Duval (SUI) = Leymus Hochst. GRAMINEAE (M)
Anisosciadium DC. UMBELLIFERAE (D)

Anisosepalum E.Hossain ACANTHACEAE (D)
Anisosorus Trevis. ex Maxon = Lonchitis L. DENNSTAEDTIACEAE (F)
Anisosperma Silva Manso = Fevillea L. CUCURBITACEAE (D)
Anisostachya Nees = Justicia L. ACANTHACEAE (D)
Anisostemon Turcz. = Connarus L. CONNARACEAE (D)
Anisostichus Bureau = Bignonia L. BIGNONIACEAE (D)
Anisostigma Schinz = Tetragonia L. AIZOACEAE (D)
Anisotes Nees ACANTHACEAE (D)
Anisothrix O.Hoffm. ex Kuntze COMPOSITAE (D)
Anisotoma Fenzl ASCLEPIADACEAE (D)
Anisotome Hook.f. UMBELLIFERAE (D)
Anistelma Raf. = Hedyotis L. RUBIACEAE (D)
Anistylis Raf. = Liparis Rich. ORCHIDACEAE (M)
Ankylocheilos Summerh. = Taeniophyllum Blume ORCHIDACEAE (M)
Ankyropetalum Fenzl CARYOPHYLLACEAE (D)
Anna Pellegr. GESNERIACEAE (D)
Annaea Kolak. = Campanula L. CAMPANULACEAE (D)
Anneliesia Brieger & Lückel ORCHIDACEAE (M)
Annesijoa Pax & K.Hoffm. EUPHORBIACEAE (D)
Anneslea Wall. THEACEAE (D)
Anneslia Salisb. = Calliandra Benth. LEGUMINOSAE–MIMOSOIDEAE (D)
Annesorhiza Cham. & Schltdl. UMBELLIFERAE (D)
Annickia Setten & Maas ANNONACEAE (D)
Annona L. ANNONACEAE (D)
Annulodiscus Tardieu = Salacia L. CELASTRACEAE (D)
Anocheile Hoffmanns. ex Rchb. = Epidendrum L. ORCHIDACEAE (M)
Anochilus Rolfe ORCHIDACEAE (M)
Anoda Cav. MALVACEAE (D)
Anodendron A.DC. APOCYNACEAE (D)
Anodiscus Benth. GESNERIACEAE (D)
Anodopetalum A.Cunn. ex Endl. CUNONIACEAE (D)
Anoectocalyx Hook.f. (SUO) = Anaectocalyx Triana ex Hook.f.
    MELASTOMATACEAE (D)
Anoectochilus Blume ORCHIDACEAE (M)
Anogeissus (DC.) Wall. COMBRETACEAE (D)
Anogra Spach = Oenothera L. ONAGRACEAE (D)
Anogramma Link ADIANTACEAE (F)
Anoiganthus Baker = Cyrtanthus Aiton AMARYLLIDACEAE (M)
Anomacanthus R.D.Good ACANTHACEAE (D)
Anomalanthus Klotzsch ERICACEAE (D)
Anomalesia N.E.Br. = Gladiolus L. IRIDACEAE (M)
Anomalocalyx Ducke EUPHORBIACEAE (D)
Anomalopteris (DC.) G.Don (SUO) = Acridocarpus Guill. & Perr. MALPIGHIACEAE (D)
Anomalopterys (DC.) G.Don = Acridocarpus Guill. & Perr. MALPIGHIACEAE (D)
Anomalosicyos Gentry = Sicyos L. CUCURBITACEAE (D)
Anomalostylus R.C.Foster = Trimezia Salisb. ex Herb. IRIDACEAE (M)
Anomalotis Steud. = Agrostis L. GRAMINEAE (M)
Anomanthodia Hook.f. RUBIACEAE (D)
Anomatheca Ker Gawl. IRIDACEAE (M)
Anomianthus Zoll. ANNONACEAE (D)
Anomochloa Brongn. GRAMINEAE (M)
Anomoctenium Pichon = Pithecoctenium Mart. ex Meisn. BIGNONIACEAE (D)
Anomopanax Harms = Mackinlaya F.Muell. ARALIACEAE (D)
Anomosanthes Blume = Lepisanthes Blume SAPINDACEAE (D)
Anomospermum Miers MENISPERMACEAE (D)

Anomospermum Dalzell (SUH) = Actephila Blume EUPHORBIACEAE (D)
Anomostachys (Baill.) Hurus. = Excoecaria L. EUPHORBIACEAE (D)
Anomostephium DC. COMPOSITAE (D)
Anomotassa K.Schum. ASCLEPIADACEAE (D)
Anonidium Engl. & Diels ANNONACEAE (D)
Anoniodes Schltr. = Sloanea L. ELAEOCARPACEAE (D)
Anonocarpus Ducke = Batocarpus Karst. MORACEAE (D)
Anoplanthus Endl. = Phelypaea L. SCROPHULARIACEAE (D)
Anoplia Steud. (SUI) = Leptochloa P.Beauv. GRAMINEAE (M)
Anoplocaryum Ledeb. BORAGINACEAE (D)
Anoplon Rchb. = Phelypaea L. SCROPHULARIACEAE (D)
Anopteris Prantl ex Diels PTERIDACEAE (F)
Anopterus Labill. ESCALLONIACEAE (D)
Anopyxis (Pierre) Engl. RHIZOPHORACEAE (D)
Anosporum Nees = Cyperus L. CYPERACEAE (M)
Anota Schltr. = Rhynchostylis Blume ORCHIDACEAE (M)
Anotea (DC.) Kunth MALVACEAE (D)
Anothea O.F.Cook = Chamaedorea Willd. PALMAE (M)
Anotis DC. = Arcytophyllum Willd. ex Schult. & Schu RUBIACEAE (D)
Anplectrella Furtado = Creochiton Blume MELASTOMATACEAE (D)
Anplectrum A.Gray = Diplectria (Blume) Rchb. MELASTOMATACEAE (D)
Anredera Juss. BASELLACEAE (D)
Ansellia Lindl. ORCHIDACEAE (M)
Anstrutheria Gardner = Cassipourea Aubl. RHIZOPHORACEAE (D)
Antacanthus A.Rich. ex DC. = Scolosanthus Vahl RUBIACEAE (D)
Antagonia Griseb. = Cayaponia Silva Manso CUCURBITACEAE (D)
Antegibbaeum Schwantes ex C.Weber AIZOACEAE (D)
Antelaea Gaertn. = Melia L. MELIACEAE (D)
Antennaria Gaertn. COMPOSITAE (D)
Antenoron Raf. = Persicaria (L.) Mill. POLYGONACEAE (D)
Anteriorchis E.Klein & Strack = Orchis L. ORCHIDACEAE (M)
Anthacantha Lem. = Euphorbia L. EUPHORBIACEAE (D)
Anthacanthus Nees ACANTHACEAE (D)
Anthactinia Bory ex M.Roem. = Passiflora L. PASSIFLORACEAE (D)
Anthaenantiopsis Mex ex Pilger GRAMINEAE (M)
Anthagathis Harms = Jollydora Pierre ex Gilg CONNARACEAE (D)
Antheliacanthus Ridl. = Pseuderanthemum Radlk. ACANTHACEAE (D)
Anthemis L. COMPOSITAE (D)
Anthenantia P.Beauv. GRAMINEAE (M)
Anthephora Schreb. GRAMINEAE (M)
Anthericlis Raf. = Tipularia Nutt. ORCHIDACEAE (M)
Anthericopsis Engl. COMMELINACEAE (M)
Anthericum L. ANTHERICACEAE (M)
Antherolophus Gagnep. = Aspidistra Ker–Gawl. CONVALLARIACEAE (M)
Antheropeas Rydb. COMPOSITAE (D)
Antheroporum Gagnep. LEGUMINOSAE–PAPILIONOIDEAE (D)
Antherostele Bremek. RUBIACEAE (D)
Antherostylis C.A.Gardner = Velleia Sm. GOODENIACEAE (D)
Antherothamnus N.E.Br. SCROPHULARIACEAE (D)
Antherotoma (Naudin) Hook.f. MELASTOMATACEAE (D)
Antherura Lour. = Psychotria L. RUBIACEAE (D)
Antherylium Rohr = Ginoria Jacq. LYTHRACEAE (D)
Anthipsimus Raf. = Muhlenbergia Schreb. GRAMINEAE (M)
Anthistiria L.f. = Themeda Forssk. GRAMINEAE (M)
Anthobembix Perkins MONIMIACEAE (D)

Anthobolus R.Br. SANTALACEAE (D)
Anthobryum Phil. FRANKENIACEAE (D)
Anthocarapa Pierre MELIACEAE (D)
Anthocephalus A.Rich. = Breonia A.Rich. ex DC. RUBIACEAE (D)
Anthocercis Labill. SOLANACEAE (D)
Anthochlamys Fenzl CHENOPODIACEAE (D)
Anthochloa Nees & Meyen GRAMINEAE (M)
Anthochortus Nees RESTIONACEAE (M)
Anthocleista Afzel. ex R.Br. LOGANIACEAE (D)
Anthoclitandra (Pierre) Pichon = Landolphia P.Beauv. APOCYNACEAE (D)
Anthodiscus G.Mey. CARYOCARACEAE (D)
Anthodon Ruíz & Pav. CELASTRACEAE (D)
Anthogonium Wall. ex Lindl. ORCHIDACEAE (M)
Anthogyas Raf. = Bletia Ruíz & Pav. ORCHIDACEAE (M)
Antholoma Labill. = Sloanea L. ELAEOCARPACEAE (D)
Antholyza L. IRIDACEAE (M)
Anthonotha P.Beauv. LEGUMINOSAE–CAESALPINIOIDEAE (D)
Anthopogon Nutt. (SUS) = Gymnopogon P.Beauv. GRAMINEAE (M)
Anthopteropsis A.C.Sm. ERICACEAE (D)
Anthopterus Hook. ERICACEAE (D)
Anthorrhiza C.Huxley & Jebb RUBIACEAE (D)
Anthosachne Steud. = Elymus L. GRAMINEAE (M)
Anthosiphon Schltr. ORCHIDACEAE (M)
Anthospermum L. RUBIACEAE (D)
Anthostema Juss. EUPHORBIACEAE (D)
Anthostyrax Pierre = Styrax L. STYRACACEAE (D)
Anthotium R.Br. GOODENIACEAE (D)
Anthotroche Endl. SOLANACEAE (D)
Anthoxanthum L. GRAMINEAE (M)
Anthriscus Pers. UMBELLIFERAE (D)
Anthurium Schott ARACEAE (M)
Anthyllis L. LEGUMINOSAE–PAPILIONOIDEAE (D)
Antia O.F.Cook = Coccothrinax Sarg. PALMAE (M)
Antiaris Lesch. MORACEAE (D)
Antiaropsis K.Schum. MORACEAE (D)
Anticharis Endl. SCROPHULARIACEAE (D)
Anticheirostylis Fitzg. = Prasophyllum R.Br. ORCHIDACEAE (M)
Antichloa Steud. (SUI) = Chondrosum Desv. GRAMINEAE (M)
Antidaphne Poepp. & Endl. EREMOLEPIDACEAE (D)
Antidesma L. EUPHORBIACEAE (D)
Antigonon Endl. POLYGONACEAE (D)
Antigramma C.Presl ASPLENIACEAE (F)
Antillanorchis Garay ORCHIDACEAE (M)
Antillia R.M.King & H.Rob. COMPOSITAE (D)
Antimima N.E.Br. AIZOACEAE (D)
Antimion Raf. = Lycopersicon Mill. SOLANACEAE (D)
Antinisa (Tul.) Hutch. = Homalium Jacq. FLACOURTIACEAE (D)
Antinoria Parl. GRAMINEAE (M)
Antiotrema Hand.–Mazz. BORAGINACEAE (D)
Antiphiona Merxm. COMPOSITAE (D)
Antiphytum DC. ex Meisn. BORAGINACEAE (D)
Antirhea Comm. ex Juss. RUBIACEAE (D)
Antirhoea Comm. ex Juss. = Antirhea Comm. ex Juss. RUBIACEAE (D)
Antirrhinum L. SCROPHULARIACEAE (D)
Antirrhoea Endl. = Antirhea Comm. ex Juss. RUBIACEAE (D)

Antistrophe A.DC. MYRSINACEAE (D)
Antithrixia DC. COMPOSITAE (D)
Antitragus Gaertn. (SUS) = Crypsis Aiton GRAMINEAE (M)
Antizoma Miers MENISPERMACEAE (D)
Antochortus Nees (SUO) = Anthochortus Nees RESTIONACEAE (M)
Antonella Caro = Tridens Roem. & Schult. GRAMINEAE (M)
Antongilia Jum. = Neodypsis Baill. PALMAE (M)
Antonia Pohl LOGANIACEAE (D)
Antonia R.Br. (SUH) = Rhynchoglossum Blume GESNERIACEAE (D)
Antoniana Tussac = Faramea Aubl. RUBIACEAE (D)
Antonina Vved. = Calamintha Mill. LABIATAE (D)
Antopetitia A.Rich. LEGUMINOSAE–PAPILIONOIDEAE (D)
Antoschmidtia Boiss. (SUS) = Schmidtia Steud. ex J.A.Schmidt GRAMINEAE (M)
Antriba Raf. = Scurrula L. LORANTHACEAE (D)
Antrocaryon Pierre ANACARDIACEAE (D)
Antrophora I.M.Johnst. = Lepidocordia Ducke BORAGINACEAE (D)
Antrophyum Kaulf. VITTARIACEAE (F)
Anubias Schott ARACEAE (M)
Anulocaulis Standl. NYCTAGINACEAE (D)
Anura (Juz.) Tscherneva COMPOSITAE (D)
Anvillea DC. COMPOSITAE (D)
Anvilleina Maire = Anvillea DC. COMPOSITAE (D)
Anychia Michx. ILLECEBRACEAE (D)
Aonikena Speg. = Chiropetalum A.Juss. EUPHORBIACEAE (D)
Aopla Lindl. = Herminium L. ORCHIDACEAE (M)
Aoranthe Somers RUBIACEAE (D)
Aotus Sm. LEGUMINOSAE–PAPILIONOIDEAE (D)
Apacheria C.T.Mason CROSSOSOMATACEAE (D)
Apalanthe Planch. HYDROCHARITACEAE (M)
Apalochlamys (Cass.) Cass. COMPOSITAE (D)
Apalophlebia C.Presl = Pyrrosia Mirb. POLYPODIACEAE (F)
Apaloxylon Drake = Neoapaloxylon Rauschert LEGUMINOSAE–
    CAESALPINIOIDEAE (D)
Apama Lam. = Thottea Rottb. ARISTOLOCHIACEAE (D)
Apargidium Torr. & A.Gray COMPOSITAE (D)
Aparinanthus Fourr. = Galium L. RUBIACEAE (D)
Aparine Guett. = Galium L. RUBIACEAE (D)
Aparinella Fourr. = Galium L. RUBIACEAE (D)
Aparisthmium Endl. EUPHORBIACEAE (D)
Apassalus Kobuski ACANTHACEAE (D)
Apatemone Schott = Schismatoglottis Zoll. & Moritzi ARACEAE (M)
Apatesia N.E.Br. AIZOACEAE (D)
Apation Blume = Liparis Rich. ORCHIDACEAE (M)
Apatitia Desv. = Bellucia Neck. ex Raf. MELASTOMATACEAE (D)
Apatophyllum McGill. CELASTRACEAE (D)
Apatostelis Garay (SUH) = Stelis Sw. ORCHIDACEAE (M)
Apaturia Lindl. = Pachystoma Blume ORCHIDACEAE (M)
Apatzingania Dieterle CUCURBITACEAE (D)
Apeiba Aubl. TILIACEAE (D)
Apemon Raf. = Datura L. SOLANACEAE (D)
Apentostera Raf. = Penstemon Schmidel SCROPHULARIACEAE (D)
Apera Adans. GRAMINEAE (M)
Apetahia Baill. CAMPANULACEAE (D)
Apetalon Wight = Didymoplexis Griff. ORCHIDACEAE (M)
Aphaenandra Miq. RUBIACEAE (D)

Aphaerema Miers FLACOURTIACEAE (D)
Aphanactis Wedd. COMPOSITAE (D)
Aphanamixis Blume MELIACEAE (D)
Aphanandrium Lindau = Neriacanthus Benth. ACANTHACEAE (D)
Aphananthe Planch. ULMACEAE (D)
Aphandra Barfod PALMAE (M)
Aphanelytrum Hack. GRAMINEAE (M)
Aphanes L. ROSACEAE (D)
Aphania Blume = Lepisanthes Blume SAPINDACEAE (D)
Aphanisma Nutt. ex Moq. CHENOPODIACEAE (D)
Aphanocalyx Oliv. LEGUMINOSAE–CAESALPINIOIDEAE (D)
Aphanocarpus Steyerm. RUBIACEAE (D)
Aphanochilus Benth. = Elsholtzia Willd. LABIATAE (D)
Aphanococcus Radlk. = Lepisanthes Blume SAPINDACEAE (D)
Aphanomyrtus Miq. = Syzygium Gaertn. MYRTACEAE (D)
Aphanopetalum Endl. CUNONIACEAE (D)
Aphanopleura Boiss. UMBELLIFERAE (D)
Aphanosperma T.F.Daniel ACANTHACEAE (D)
Aphanostelma Malme = Melinia Decne. ASCLEPIADACEAE (D)
Aphanostemma St.–Hil. = Ranunculus L. RANUNCULACEAE (D)
Aphanostephus DC. COMPOSITAE (D)
Aphanostylis Pierre = Landolphia P.Beauv. APOCYNACEAE (D)
Aphantochaeta A.Gray = Chaetopappa DC. COMPOSITAE (D)
Aphelandra R.Br. ACANTHACEAE (D)
Aphelandrella Mildbr. ACANTHACEAE (D)
Aphelia R.Br. CENTROLEPIDACEAE (M)
Aphloia (DC.) Bennett FLACOURTIACEAE (D)
Aphora Nutt. = Ditaxis Vahl ex A.Juss. EUPHORBIACEAE (D)
Aphragmia Nees = Ruellia L. ACANTHACEAE (D)
Aphragmus Andrz. ex DC. CRUCIFERAE (D)
Aphyllanthes L. APHYLLANTHACEAE (M)
Aphyllarum S.Moore = Caladium Vent. ARACEAE (M)
Aphyllocalpa Lag., D.García & Clemente = Osmunda L. OSMUNDACEAE (F)
Aphyllocladus Wedd. COMPOSITAE (D)
Aphyllodium (DC.) Gagnep. = Hedysarum L. LEGUMINOSAE–PAPILIONOIDEAE (D)
Aphyllon Mitch. SCROPHULARIACEAE (D)
Aphyllorchis Blume ORCHIDACEAE (M)
Apiastrum Nutt. UMBELLIFERAE (D)
Apicra Willd. = Haworthia Duval ALOACEAE (M)
Apilia Raf. (SUO) = Fraxinus L. OLEACEAE (D)
Apinagia Tul. PODOSTEMACEAE (D)
Apiopetalum Baill. ARALIACEAE (D)
Apios Fabr. LEGUMINOSAE–PAPILIONOIDEAE (D)
Apista Blume = Podochilus Blume ORCHIDACEAE (M)
Apium L. UMBELLIFERAE (D)
Aplanodes Marais CRUCIFERAE (D)
Aplarina Raf. = Euphorbia L. EUPHORBIACEAE (D)
Aplectrum (Nutt.) Torr. ORCHIDACEAE (M)
Aplectrum Blume (SUH) = Diplectria (Blume) Rchb. MELASTOMATACEAE (D)
Apleura Phil. = Azorella Lam. UMBELLIFERAE (D)
Aplexia Raf. = Leersia Sw. GRAMINEAE (M)
Aplilia Raf. = Fraxinus L. OLEACEAE (D)
Aploca Neck. ex Kuntze = Periploca L. ASCLEPIADACEAE (D)
Aplocarya Lindl. = Nolana L.f. SOLANACEAE (D)
Aplocera Raf. (SUS) = Ctenium Panz. GRAMINEAE (M)

Aploleia Raf. = Callisia Loefl. COMMELINACEAE (M)
Apluda L. GRAMINEAE (M)
Apocaulon R.S.Cowan RUTACEAE (D)
Apochaete (C.E.Hubb.) J.B.Phipps = Tristachya Nees GRAMINEAE (M)
Apochiton C.E.Hubb. GRAMINEAE (M)
Apochoris Duby = Lysimachia L. PRIMULACEAE (D)
Apoclada McClure GRAMINEAE (M)
Apocopis Nees GRAMINEAE (M)
Apocynum L. APOCYNACEAE (D)
Apodandra Pax & K.Hoffm. EUPHORBIACEAE (D)
Apodanthera Arn. CUCURBITACEAE (D)
Apodanthes Poit. RAFFLESIACEAE (D)
Apodicarpum Makino UMBELLIFERAE (D)
Apodina Tiegh. = Psittacanthus Mart. LORANTHACEAE (D)
Apodiscus Hutch. EUPHORBIACEAE (D)
Apodocephala Baker COMPOSITAE (D)
Apodolirion Baker AMARYLLIDACEAE (M)
Apodostigma R.Wilczek CELASTRACEAE (D)
Apodytes E.Mey. ex Arn. ICACINACEAE (D)
Apogon Steud. (SUI) = Chloris Sw. GRAMINEAE (M)
Apogonia (Nutt.) E.Fourn. = Coelorachis Brongn. GRAMINEAE (M)
Apoia Merr. = Sarcosperma Hook.f. SAPOTACEAE (D)
Apolepsis (Blume) Hassk. = Lepidagathis Willd. ACANTHACEAE (D)
Apollonias Nees LAURACEAE (D)
Apomuria Bremek. = Psychotria L. RUBIACEAE (D)
Aponoa Raf. = Columnea L. GESNERIACEAE (D)
Aponogeton L.f. APONOGETONACEAE (M)
Apophyllum F.Muell. CAPPARACEAE (D)
Apoplanesia C.Presl LEGUMINOSAE–PAPILIONOIDEAE (D)
Aporocactus Lem. = Disocactus Lindl. CACTACEAE (D)
Aporocereus Frič & Kreuz. (SUO) = Disocactus Lindl. CACTACEAE (D)
Aporosa Blume (SUO) = Aporusa Blume EUPHORBIACEAE (D)
Aporosella Chodat = Phyllanthus L. EUPHORBIACEAE (D)
Aporostylis Rupp & Hatch ORCHIDACEAE (M)
Aporrhiza Radlk. SAPINDACEAE (D)
Aporum Blume = Dendrobium Sw. ORCHIDACEAE (M)
Aporusa Blume EUPHORBIACEAE (D)
Aposeris Neck. ex Cass. COMPOSITAE (D)
Apostasia Blume ORCHIDACEAE (M)
Apostates Lander COMPOSITAE (D)
Apoxyanthera Hochst. = Raphionacme Harv. ASCLEPIADACEAE (D)
Appendicula Blume ORCHIDACEAE (M)
Appendicularia DC. MELASTOMATACEAE (D)
Appertiella C.D.K.Cook & Triest HYDROCHARITACEAE (M)
Appunettia R.D.Good = Morinda L. RUBIACEAE (D)
Appunia Hook.f. = Morinda L. RUBIACEAE (D)
Aprevalia Baill. = Delonix Raf. LEGUMINOSAE–CAESALPINIOIDEAE (D)
Aptandra Miers OLACACEAE (D)
Aptandropsis Ducke = Heisteria Jacq. OLACACEAE (D)
Aptenia N.E.Br. AIZOACEAE (D)
Apterantha C.H.Wright = Lagrezia Moq. AMARANTHACEAE (D)
Apterigia Galushko (UP) CRUCIFERAE (D)
Apterokarpos Rizzini = Loxopterygium Hook.f. ANACARDIACEAE (D)
Apteron Kurz = Ventilago Gaertn. RHAMNACEAE (D)
Apteropteris (Copel.) Copel. = Sphaerocionium C.Presl HYMENOPHYLLACEAE (F)

Apterosperma H.T.Chang THEACEAE (D)
Apterygia Baehni = Sideroxylon L. SAPOTACEAE (D)
Apteuxis Griff. = Pternandra Jack MELASTOMATACEAE (D)
Aptosimum Burch. ex Benth. SCROPHULARIACEAE (D)
Aptotheca Miers = Forsteronia G.Mey. APOCYNACEAE (D)
Apuleia Mart. LEGUMINOSAE–CAESALPINIOIDEAE (D)
Apurimacia Harms LEGUMINOSAE–PAPILIONOIDEAE (D)
Aquartia Jacq. = Solanum L. SOLANACEAE (D)
Aquilaria Lam. THYMELAEACEAE (D)
Aquilegia L. RANUNCULACEAE (D)
Arabidella (F.Muell.) O.E.Schulz CRUCIFERAE (D)
Arabidium Spach = Arabis L. CRUCIFERAE (D)
Arabidopsis Heynh. = Arabis L. CRUCIFERAE (D)
Arabis L. CRUCIFERAE (D)
Arabisa Rchb. = Arabis L. CRUCIFERAE (D)
Aracamunia Carnevali & I.Ramírez ORCHIDACEAE (M)
Arachis L. LEGUMINOSAE–PAPILIONOIDEAE (D)
Arachnanthe Blume = Arachnis Blume ORCHIDACEAE (M)
Arachne (Endl.) Pojark. = Leptopus Decne. EUPHORBIACEAE (D)
Arachnimorpha Desv. = Rondeletia L. RUBIACEAE (D)
Arachniodes Blume DRYOPTERIDACEAE (F)
Arachnis Blume ORCHIDACEAE (M)
Arachnites F.W.Schmidt = Ophrys L. ORCHIDACEAE (M)
Arachnitis Phil. CORSIACEAE (M)
Arachnocalyx Compton ERICACEAE (D)
Arachnodes Gagnep. = Phyllanthus L. EUPHORBIACEAE (D)
Arachnothryx Planch. RUBIACEAE (D)
Araeoandra Lefor = Viviania Cav. GERANIACEAE (D)
Araeococcus Brongn. BROMELIACEAE (M)
Arafoe Pimenov & Lavrova UMBELLIFERAE (D)
Aragallus Neck. ex Greene = Oxytropis DC. LEGUMINOSAE–PAPILIONOIDEAE (D)
Aragoa Kunth SCROPHULARIACEAE (D)
Araiostegia Copel. DAVALLIACEAE (F)
Aralia L. ARALIACEAE (D)
Aralidium Miq. ARALIDIACEAE (D)
Araliopsis Engl. RUTACEAE (D)
Araliorhamnus H.Perrier = Berchemia Neck. ex DC. RHAMNACEAE (D)
Aranella Barnhart = Utricularia L. LENTIBULARIACEAE (D)
Arapatiella Rizzini & A.Mattos LEGUMINOSAE–CAESALPINIOIDEAE (D)
Arariba Mart. = Simira Aubl. RUBIACEAE (D)
Ararocarpus Scheff. = Meiogyne Miq. ANNONACEAE (D)
Aratitiyopea Steyerm. XYRIDACEAE (M)
Araucaria Juss. ARAUCARIACEAE (G)
Araujia Brot. ASCLEPIADACEAE (D)
Arausiaca Blume = Orania Zipp. PALMAE (M)
Arbelaezaster Cuatrec. COMPOSITAE (D)
Arberella Soderstr. & Calderón GRAMINEAE (M)
Arbulocarpus Tennant = Spermacoce L. RUBIACEAE (D)
Arbutus L. ERICACEAE (D)
Arcangelina Kuntze (SUS) = Tripogon Roem. & Schult. GRAMINEAE (M)
Arcangelisia Becc. MENISPERMACEAE (D)
Arceuthobium M.Bieb. VISCACEAE (D)
Arceuthos Antoine & Kotschy = Juniperus L. CUPRESSACEAE (G)
Archangiopteris Christ & Giesenh. = Angiopteris Hoffm. MARATTIACEAE (F)
Archboldia E.Beer & H.J.Lam VERBENACEAE (D)

Archboldiodendron Kobuski THEACEAE (D)
Archeria Hook.f. EPACRIDACEAE (D)
Archiatriplex G.L.Chu CHENOPODIACEAE (D)
Archibaccharis Heering COMPOSITAE (D)
Archiboehmeria C.J.Chen URTICACEAE (D)
Archiclematis (Tamura) Tamura = Clematis L. RANUNCULACEAE (D)
Archidendron F.Muell. LEGUMINOSAE–MIMOSOIDEAE (D)
Archidendropsis I.C.Nielsen LEGUMINOSAE–MIMOSOIDEAE (D)
Archimedea Leandro = Lophophytum Schott & Endl. BALANOPHORACEAE (D)
Archineottia S.C.Chen ORCHIDACEAE (M)
Archiphysalis Kuang SOLANACEAE (D)
Archirhodomyrtus (Nied.) Burret MYRTACEAE (D)
Archontophoenix H.A.Wendl. & Drude PALMAE (M)
Archytaea Mart. GUTTIFERAE (D)
Arcoa Urb. LEGUMINOSAE–CAESALPINIOIDEAE (D)
Arctagrostis Griseb. GRAMINEAE (M)
Arctanthemum (Tsvelev) Tsvelev COMPOSITAE (D)
Arcteranthis Greene = Ranunculus L. RANUNCULACEAE (D)
Arcterica Coville = Pieris D.Don ERICACEAE (D)
Arctium L. COMPOSITAE (D)
Arctocalyx Fenzl = Solenophora Benth. GESNERIACEAE (D)
Arctocrania Nakai = Cornus L. CORNACEAE (D)
Arctogentia A.Löve = Gentianella Moench GENTIANACEAE (D)
Arctogeron DC. COMPOSITAE (D)
Arctomecon Torr. & Frém. PAPAVERACEAE (D)
Arctophila (Rupr.) N.J.Andersson GRAMINEAE (M)
Arctopoa (Griseb.) Probatova = Poa L. GRAMINEAE (M)
Arctopus L. UMBELLIFERAE (D)
Arctostaphylos Adans. ERICACEAE (D)
Arctotheca J.C.Wendl. COMPOSITAE (D)
Arctotis L. COMPOSITAE (D)
Arctottonia Trel. PIPERACEAE (D)
Arctous Nied. = Arctostaphylos Adans. ERICACEAE (D)
Arcuatopterus Rupr. (UP) UMBELLIFERAE (D)
Arculus Tiegh. = Amylotheca Tiegh. LORANTHACEAE (D)
Arcynospermum Turcz. (UP) MALVACEAE (D)
Arcyosperma O.E.Schulz CRUCIFERAE (D)
Arcypteris Underw. = Pleocnemia C.Presl DRYOPTERIDACEAE (F)
Arcytophyllum Willd. ex Schult. & Schult.f. RUBIACEAE (D)
Ardinghalia Comm. ex Juss. = Phyllanthus L. EUPHORBIACEAE (D)
Ardisia Sw. MYRSINACEAE (D)
Ardisiandra Hook.f. PRIMULACEAE (D)
Areca L. PALMAE (M)
Arecastrum (Drude) Becc. = Syagrus Mart. PALMAE (M)
Arechavaletaia Speg. = Azara Ruíz & Pav. FLACOURTIACEAE (D)
Aregelia Kuntze (SUS) = Nidularium Lem. BROMELIACEAE (M)
Aremonia Neck. ex Nestl. ROSACEAE (D)
Arenaria L. CARYOPHYLLACEAE (D)
Arenga Labill. PALMAE (M)
Arenifera A.G.J.Herre AIZOACEAE (D)
Arequipa Britton & Rose = Oreocereus (A.Berger) Riccob. CACTACEAE (D)
Arequipiopsis Kreuz. & Buining = Oreocereus (A.Berger) Riccob. CACTACEAE (D)
Arethusa L. ORCHIDACEAE (M)
Aretia L. = Androsace L. PRIMULACEAE (D)
Aretia Link (SUH) = Primula L. PRIMULACEAE (D)

Aretiastrum (DC.) Spach VALERIANACEAE (D)
Arfeuillea Pierre ex Radlk. SAPINDACEAE (D)
Argania Roem. & Schult. SAPOTACEAE (D)
Argelia Decne. = Solenostemma Hayne ASCLEPIADACEAE (D)
Argemone L. PAPAVERACEAE (D)
Argentina Hill = Potentilla L. ROSACEAE (D)
Argeta N.E.Br. = Gibbaeum Haw. AIZOACEAE (D)
Argillochloa W.A.Weber = Festuca L. GRAMINEAE (M)
Argithamnia Sw. (SUO) = Argythamnia P.Browne EUPHORBIACEAE (D)
Argocoffea (Pierre ex De Wild.) Lebrun (SUI) = Argocoffeopsis Lebrun RUBIACEAE (D)
Argocoffeopsis Lebrun RUBIACEAE (D)
Argomuellera Pax EUPHORBIACEAE (D)
Argophilum Blanco = Aglaia Lour. MELIACEAE (D)
Argophyllum J.R.Forst. & G.Forst. ESCALLONIACEAE (D)
Argopogon Mimeur = Ischaemum L. GRAMINEAE (M)
Argostemma Wall. RUBIACEAE (D)
Argostemmella Ridl. = Argostemma Wall. RUBIACEAE (D)
Argothamnia Spreng. (SUO) = Argythamnia P.Browne EUPHORBIACEAE (D)
Argusia Boehm. BORAGINACEAE (D)
Argylia D.Don BIGNONIACEAE (D)
Argyra Noronha ex Baill. = Croton L. EUPHORBIACEAE (D)
Argyranthemum Webb COMPOSITAE (D)
Argyreia Lour. CONVOLVULACEAE (D)
Argyrella Naudin = Dissotis Benth. MELASTOMATACEAE (D)
Argyrocalymma K.Schum. & Lauterb. = Carpodetus J.R.Forst. & G.Forst.
    ESCALLONIACEAE (D)
Argyrochosma (J.Sm.) Windham = Cheilanthes Sw. ADIANTACEAE (F)
Argyrocytisus (Maire) Raynaud LEGUMINOSAE–PAPILIONOIDEAE (D)
Argyrodendron (Endl.) Klotzsch (SUH) = Croton L. EUPHORBIACEAE (D)
Argyrodendron F.Muell. = Heritiera Aiton STERCULIACEAE (D)
Argyroderma N.E.Br. AIZOACEAE (D)
Argyroglottis Turcz. COMPOSITAE (D)
Argyrolobium Eckl. & Zeyh. LEGUMINOSAE–PAPILIONOIDEAE (D)
Argyronerium Pit. = Epigynum Wight APOCYNACEAE (D)
Argyrophanes Schltdl. COMPOSITAE (D)
Argyrorchis Blume = Macodes (Blume) Lindl. ORCHIDACEAE (M)
Argyrothamnia Müll.Arg. (SUO) = Argythamnia P.Browne EUPHORBIACEAE (D)
Argyrovernonia MacLeish = Chresta Vell. ex DC. COMPOSITAE (D)
Argyroxiphium DC. COMPOSITAE (D)
Argytamnia Duchesne (SUO) = Argythamnia P.Browne EUPHORBIACEAE (D)
Argythamnia P.Browne EUPHORBIACEAE (D)
Arhynchium Lindl. & Paxton = Armodorum Breda ORCHIDACEAE (M)
Aria (Pers.) Host = Sorbus L. ROSACEAE (D)
Ariadne Urb. RUBIACEAE (D)
Aridaria N.E.Br. = Phyllobolus N.E.Br. AIZOACEAE (D)
Aridarum Ridl. ARACEAE (M)
Arietinum Beck = Cypripedium L. ORCHIDACEAE (M)
Arikury Becc. = Syagrus Mart. PALMAE (M)
Arikuryroba Barb.Rodr. = Syagrus Mart. PALMAE (M)
Arillastrum Panch. ex Baill. MYRTACEAE (D)
Ariocarpus Scheidw. CACTACEAE (D)
Ariopsis Nimmo ARACEAE (M)
Ariosorbus Koidz. = Sorbus L. ROSACEAE (D)
Arisaema Mart. ARACEAE (M)
Arisanorchis Hayata = Cheirostylis Blume ORCHIDACEAE (M)

Arisarum Mill. ARACEAE (M)
Arischrada Pobed. = Salvia L. LABIATAE (D)
Aristaria Jungh. = Themeda Forssk. GRAMINEAE (M)
Aristavena F.Albers & Butzin = Deschampsia P.Beauv. GRAMINEAE (M)
Aristea Sol. ex Aiton IRIDACEAE (M)
Aristeguietia R.M.King & H.Rob. COMPOSITAE (D)
Aristella Bertol. = Stipa L. GRAMINEAE (M)
Aristeyera H.E.Moore = Asterogyne H.A.Wendl. PALMAE (M)
Aristida L. GRAMINEAE (M)
Aristidium (Endl.) Lindl. = Bouteloua Lag. GRAMINEAE (M)
Aristocapsa Reveal & Hardham POLYGONACEAE (D)
Aristogeitonia Prain EUPHORBIACEAE (D)
Aristolelea Lour. = Spiranthes Rich. ORCHIDACEAE (M)
Aristolochia L. ARISTOLOCHIACEAE (D)
Aristopsis Catasus = Aristida L. GRAMINEAE (M)
Aristotelia L'Hér. ELAEOCARPACEAE (D)
Aristotelia Comm. ex Lam. (SUH) = Terminalia L. COMBRETACEAE (D)
Arjona Cav. SANTALACEAE (D)
Arkezostis Raf. = Cayaponia Silva Manso CUCURBITACEAE (D)
Armatocereus Backeb. CACTACEAE (D)
Armentaria Thouars ex Baill. = Uvaria L. ANNONACEAE (D)
Armeria Willd. PLUMBAGINACEAE (D)
Armodorum Breda ORCHIDACEAE (M)
Armola (Kirschl.) Montandon = Atriplex L. CHENOPODIACEAE (D)
Armoracia P.Gaertn., B.Mey. & Scherb. CRUCIFERAE (D)
Armourea Lewton = Thespesia Sol. ex Corrêa MALVACEAE (D)
Arnaldoa Cabrera COMPOSITAE (D)
Arnanthus Baehni = Pichonia Pierre SAPOTACEAE (D)
Arnebia Forssk. BORAGINACEAE (D)
Arnebiola Chiov. = Arnebia Forssk. BORAGINACEAE (D)
Arnedina Rchb.f. = Arundina Blume ORCHIDACEAE (M)
Arnhemia Airy Shaw THYMELAEACEAE (D)
Arnica L. COMPOSITAE (D)
Arnicastrum Greenm. COMPOSITAE (D)
Arnicratea N.Hallé CELASTRACEAE (D)
Arnocrinum Endl. & Lehm. ANTHERICACEAE (M)
Arnoglossum Raf. COMPOSITAE (D)
Arnoldia Blume = Weinmannia L. CUNONIACEAE (D)
Arnoseris Gaertn. COMPOSITAE (D)
Arnottia A.Rich. ORCHIDACEAE (M)
Arodendron Werth = Typhonodorum Schott ARACEAE (M)
Aromadendron Andrews ex Steud. (SUI) = Eucalyptus L'Hér. MYRTACEAE (D)
Aromadendron Blume = Magnolia L. MAGNOLIACEAE (D)
Aronia Medik. = Photinia Lindl. ROSACEAE (D)
Arophyton Jum. ARACEAE (M)
Aropsis Rojas Acosta = Spathicarpa Hook. ARACEAE (M)
Aropteris Alston = Afropteris Alston PTERIDACEAE (F)
Arpitium Neck. ex Sweet = Pachypleurum Ledeb. UMBELLIFERAE (D)
Arpophyllum Lex. ORCHIDACEAE (M)
Arrabidaea DC. BIGNONIACEAE (D)
Arracacia Bancr. UMBELLIFERAE (D)
Arrhenatherum P.Beauv. GRAMINEAE (M)
Arrhenechthites Mattf. COMPOSITAE (D)
Arrhostoxylum Nees = Ruellia L. ACANTHACEAE (D)
Arrojadoa Britton & Rose CACTACEAE (D)

Arrojadocharis Mattf. COMPOSITAE (D)
Arrowsmithia DC. COMPOSITAE (D)
Arrozia Kunth (SUS) = Luziola Juss. GRAMINEAE (M)
Arrudaria Macedo = Copernicia Mart. ex Endl. PALMAE (M)
Arsenjevia Starod. = Anemone L. RANUNCULACEAE (D)
Artabotrys R.Br. ANNONACEAE (D)
Artanema D.Don SCROPHULARIACEAE (D)
Artedia L. UMBELLIFERAE (D)
Artemisia L. COMPOSITAE (D)
Artemisiopsis S.Moore COMPOSITAE (D)
Arthraerua (Kuntze) Schinz AMARANTHACEAE (D)
Arthragrostis Lazarides GRAMINEAE (M)
Arthratherum P.Beauv. (SUS) = Aristida L. GRAMINEAE (M)
Arthraxella Nakai = Psittacanthus Mart. LORANTHACEAE (D)
Arthraxon Tiegh. (SUH) = Psittacanthus Mart. LORANTHACEAE (D)
Arthraxon P.Beauv. GRAMINEAE (M)
Arthrobotrya J.Sm. = Teratophyllum Mett. ex Kuhn LOMARIOPSIDACEAE (F)
Arthrobotrys (C.Presl) Lindl. = Dryopteris Adans. DRYOPTERIDACEAE (F)
Arthrocarpum Balf.f. LEGUMINOSAE–PAPILIONOIDEAE (D)
Arthrocereus A.Berger & F.Knuth CACTACEAE (D)
Arthrochilium Beck = Epipactis Zinn ORCHIDACEAE (M)
Arthrochilus F.Muell. ORCHIDACEAE (M)
Arthrochlaena Benth. (SUI) = Sclerodactylon Stapf GRAMINEAE (M)
Arthrochloa Lorch (SUH) = Acrachne Wight & Arn. ex Chiov. GRAMINEAE (M)
Arthrochloa R.Br. (SUS) = Holcus L. GRAMINEAE (M)
Arthrochortus Lowe = Lolium L. GRAMINEAE (M)
Arthroclianthus Baill. LEGUMINOSAE–PAPILIONOIDEAE (D)
Arthrocnemum Moq. CHENOPODIACEAE (D)
Arthrolophis (Trin.) Chiov. (SUO) = Andropogon L. GRAMINEAE (M)
Arthromeris (T.Moore) J.Sm. POLYPODIACEAE (F)
Arthrophyllum Blume ARALIACEAE (D)
Arthrophyllum Bojer ex DC. (SUH) = Phyllarthron DC. BIGNONIACEAE (D)
Arthrophytum Schrenk CHENOPODIACEAE (D)
Arthropodium R.Br. ANTHERICACEAE (M)
Arthropogon Nees GRAMINEAE (M)
Arthropteris J.Sm. OLEANDRACEAE (F)
Arthrosamanea Britton & Rose = Albizia Durazz. LEGUMINOSAE–MIMOSOIDEAE (D)
Arthrosolen C.A.Mey. = Gnidia L. THYMELAEACEAE (D)
Arthrostachya Link = Gaudinia P.Beauv. GRAMINEAE (M)
Arthrostachys Desv. = Andropogon L. GRAMINEAE (M)
Arthrostemma Pav. ex D.Don MELASTOMATACEAE (D)
Arthrostylidium Rupr. GRAMINEAE (M)
Arthrostylis R.Br. CYPERACEAE (M)
Arthrothamnus Klotzsch & Garcke = Euphorbia L. EUPHORBIACEAE (D)
Artia Guillaumin APOCYNACEAE (D)
Artocarpus J.R.Forst. & G.Forst. MORACEAE (D)
Artomeria Breda = Eria Lindl. ORCHIDACEAE (M)
Artorhiza Raf. = Solanum L. SOLANACEAE (D)
Artorima Dressler & G.E.Pollard ORCHIDACEAE (M)
Artrolobium Desv. = Ornithopus L. LEGUMINOSAE–PAPILIONOIDEAE (D)
Arum L. ARACEAE (M)
Aruncus L. ROSACEAE (D)
Arundarbor Kuntze (SUS) = Bambusa Schreb. GRAMINEAE (M)
Arundina Blume ORCHIDACEAE (M)
Arundinaria Michx. GRAMINEAE (M)

Arundinella Raddi GRAMINEAE (M)
Arundo L. GRAMINEAE (M)
Arundoclaytonia Davidse & R.P.Ellis GRAMINEAE (M)
Arytera Blume SAPINDACEAE (D)
Asaemia (Harv.) Benth. = Athanasia L. COMPOSITAE (D)
Asagraea Baill. = Psorothamnus Rydb. LEGUMINOSAE–PAPILIONOIDEAE (D)
Asanthus R.M.King & H.Rob. COMPOSITAE (D)
Asarca Lindl. = Chloraea Lindl. ORCHIDACEAE (M)
Asarina Mill. SCROPHULARIACEAE (D)
Asarum L. ARISTOLOCHIACEAE (D)
Ascarina J.R.Forst. & G.Forst. CHLORANTHACEAE (D)
Ascarinopsis Humbert & Capuron = Ascarina J.R.Forst. & G.Forst.
    CHLORANTHACEAE (D)
Aschenbornia S.Schauer = Calea L. COMPOSITAE (D)
Aschersoniodoxa Gilg & Muschler CRUCIFERAE (D)
Aschistanthera C.Hansen MELASTOMATACEAE (D)
Asciadium Griseb. UMBELLIFERAE (D)
Ascidieria Seidenf. ORCHIDACEAE (M)
Ascidiogyne Cuatrec. COMPOSITAE (D)
Asclepias L. ASCLEPIADACEAE (D)
Asclepiodella Small = Asclepias L. ASCLEPIADACEAE (D)
Asclepiodora A.Gray = Asclepias L. ASCLEPIADACEAE (D)
Ascocarydion G.Taylor = Plectranthus L'Hér. LABIATAE (D)
Ascocentrum Schltr. ex J.J.Sm. ORCHIDACEAE (M)
Ascochilopsis C.E.Carr ORCHIDACEAE (M)
Ascochilus Ridl. ORCHIDACEAE (M)
Ascoglossum Schltr. ORCHIDACEAE (M)
Ascolabium S.S.Ying = Ascocentrum Schltr. ex J.J.Sm. ORCHIDACEAE (M)
Ascolepis Nees ex Steud. CYPERACEAE (M)
Ascopholis C.E.C.Fisch. = Cyperus L. CYPERACEAE (M)
Ascotainia Ridl. = Tainia Blume ORCHIDACEAE (M)
Ascotheca Heine ACANTHACEAE (D)
Ascyrum L. = Hypericum L. GUTTIFERAE (D)
Ascyrum Mill. (SUH) = Hypericum L. GUTTIFERAE (D)
Asemanthia Ridl. = Mussaenda L. RUBIACEAE (D)
Asemnantha Hook.f. RUBIACEAE (D)
Asepalum Marais CYCLOCHEILACEAE (D)
Asephananthes Bory = Passiflora L. PASSIFLORACEAE (D)
Ashtonia Airy Shaw EUPHORBIACEAE (D)
Asimia Kunth (SUO) = Asimina Adans. ANNONACEAE (D)
Asimina Adans. ANNONACEAE (D)
Asiphonia Griff. = Thottea Rottb. ARISTOLOCHIACEAE (D)
Askellia W.A.Weber = Crepis L. COMPOSITAE (D)
Asketanthera Woodson APOCYNACEAE (D)
Askidiosperma Steud. RESTIONACEAE (M)
Askofake Raf. = Utricularia L. LENTIBULARIACEAE (D)
Aspalathoides (DC.) K.Koch = Anthyllis L. LEGUMINOSAE–PAPILIONOIDEAE (D)
Aspalathus L. LEGUMINOSAE–PAPILIONOIDEAE (D)
Aspalthium Medik. = Bituminaria Heist. ex Fabr. LEGUMINOSAE–
    PAPILIONOIDEAE (D)
Asparagus L. ASPARAGACEAE (M)
Aspasia Lindl. ORCHIDACEAE (M)
Aspazoma N.E.Br. AIZOACEAE (D)
Aspegrenia Poepp. & Endl. = Octomeria R.Br. ORCHIDACEAE (M)
Aspera Moench = Galium L. RUBIACEAE (D)

Asperella Humb. (SUH) = Hystrix Moench GRAMINEAE (M)
Asperuginoides Rauschert CRUCIFERAE (D)
Asperugo L. BORAGINACEAE (D)
Asperula L. RUBIACEAE (D)
Asphodeline Rchb. ASPHODELACEAE (M)
Asphodelus L. ASPHODELACEAE (M)
Aspicarpa Rich. MALPIGHIACEAE (D)
Aspidanthera Benth. = Ferdinandusa Pohl RUBIACEAE (D)
Aspidistra Ker Gawl. CONVALLARIACEAE (M)
Aspidium Sw. = Tectaria Cav. DRYOPTERIDACEAE (F)
Aspidixia Tiegh. = Viscum L. VISCACEAE (D)
Aspidocarya Hook.f. & Thomson MENISPERMACEAE (D)
Aspidogenia Burret = Myrcianthes O.Berg MYRTACEAE (D)
Aspidoglossum E.Mey. ASCLEPIADACEAE (D)
Aspidogyne Garay ORCHIDACEAE (M)
Aspidophyllum Ulbr. = Ranunculus L. RANUNCULACEAE (D)
Aspidopterys A.Juss. MALPIGHIACEAE (D)
Aspidosperma Mart. & Zucc. APOCYNACEAE (D)
Aspidostemon Rohwer & H.G.Richt. LAURACEAE (D)
Aspidotis (Nutt. ex Hook.) Copel. ADIANTACEAE (F)
Aspilia Thouars COMPOSITAE (D)
Aspiliopsis Greenm. = Podachaenium Benth. ex Oerst. COMPOSITAE (D)
Aspla Rchb. = Herminium L. ORCHIDACEAE (M)
Asplenidictyum J.Sm. = Asplenium L. ASPLENIACEAE (F)
Aspleniopsis Mett. ex Kuhn = Austrogramme Fourn. ADIANTACEAE (F)
Asplenium L. ASPLENIACEAE (F)
Asplundia Harling CYCLANTHACEAE (M)
Asplundianthus R.M.King & H.Rob. COMPOSITAE (D)
Asprella Host (SUH) = Psilurus Trin. GRAMINEAE (M)
Asprella Schreb. (SUS) = Leersia Sw. GRAMINEAE (M)
Aspris Adans. (SUS) = Aira L. GRAMINEAE (M)
Asraoa Joseph = Wallichia Roxb. PALMAE (M)
Assidora A.Chev. = Schumanniophyton Harms RUBIACEAE (D)
Asta Klotzsch ex O.E.Schulz CRUCIFERAE (D)
Astartea DC. MYRTACEAE (D)
Astelia Banks & Sol. ex R.Br. ASTELIACEAE (M)
Astelma Schltr. (SUS) = Papuastelma Bullock ASCLEPIADACEAE (D)
Astemma Less. = Monactis Kunth COMPOSITAE (D)
Astemon Regel = Lepechinia Willd. LABIATAE (D)
Astenolobium Nevski = Astragalus L. LEGUMINOSAE–PAPILIONOIDEAE (D)
Astephanus R.Br. ASCLEPIADACEAE (D)
Aster L. COMPOSITAE (D)
Asteracantha Nees = Hygrophila R.Br. ACANTHACEAE (D)
Asterandra Klotzsch = Phyllanthus L. EUPHORBIACEAE (D)
Asteranthe Engl. & Diels ANNONACEAE (D)
Asteranthera Hanst. GESNERIACEAE (D)
Asteranthopsis Kuntze = Asteranthe Engl. & Diels ANNONACEAE (D)
Asteranthos Desf. LECYTHIDACEAE (D)
Asteranthus Spreng. = Asteranthos Desf. LECYTHIDACEAE (D)
Asteriastigma Bedd. = Hydnocarpus Gaertn. FLACOURTIACEAE (D)
Asteridea Lindl. COMPOSITAE (D)
Asteriscium Cham. & Schltdl. UMBELLIFERAE (D)
Asteriscus Mill. COMPOSITAE (D)
Asterochaete Nees = Carpha Banks & Sol. ex R.Br. CYPERACEAE (M)
Asterochlaena Garcke = Pavonia Cav. MALVACEAE (D)

Asterocytisus (W.D.J.Koch) Schur ex Fuss = Genista L. LEGUMINOSAE–
PAPILIONOIDEAE (D)
Asterogyne H.A.Wendl. PALMAE (M)
Asterohyptis Epling LABIATAE (D)
Asterolasia F.Muell. RUTACEAE (D)
Asterolepidion Ducke = Dendrobangia Rusby ICACINACEAE (D)
Asterolinon Hoffmanns. & Link PRIMULACEAE (D)
Asteromoea Blume COMPOSITAE (D)
Asteromyrtus Schauer MYRTACEAE (D)
Asteropeia Thouars ASTEROPEIACEAE (D)
Asterophorum Sprague TILIACEAE (D)
Asterophyllum W.Schimp. & Spenn. = Asperula L. RUBIACEAE (D)
Asteropyrum J.R.Drumm. & Hutch. RANUNCULACEAE (D)
Asterosedum Grulich = Sedum L. CRASSULACEAE (D)
Asterostemma Decne. ASCLEPIADACEAE (D)
Asterostigma Fisch. & C.A.Mey. ARACEAE (M)
Asterostoma Blume = Osbeckia L. MELASTOMATACEAE (D)
Asterotricha V.V.Botschantzeva (SUH) = Pterygostemon V.V.Botschantzeva
CRUCIFERAE (D)
Asterotrichion Klotzsch MALVACEAE (D)
Asthenatherum Nevski (SUS) = Centropodia Rchb. GRAMINEAE (M)
Asthenochloa Büse GRAMINEAE (M)
Asthotheca Miers ex Planch. & Triana = Clusia L. GUTTIFERAE (D)
Astianthus D.Don BIGNONIACEAE (D)
Astiella Jovet RUBIACEAE (D)
Astilbe Buch.–Ham. ex G.Don SAXIFRAGACEAE (D)
Astilboides (Hemsl.) Engl. SAXIFRAGACEAE (D)
Astiria Lindl. STERCULIACEAE (D)
Astoma DC. (SUH) = Astomaea Rchb. UMBELLIFERAE (D)
Astomaea Rchb. UMBELLIFERAE (D)
Astomatopsis Korovin = Astomaea Rchb. UMBELLIFERAE (D)
Astracantha Podlech LEGUMINOSAE–PAPILIONOIDEAE (D)
Astraea Schauer (SUH) = Thryptomene Endl. MYRTACEAE (D)
Astraea Klotzsch (SUH) = Croton L. EUPHORBIACEAE (D)
Astragalus L. LEGUMINOSAE–PAPILIONOIDEAE (D)
Astranthium Nutt. COMPOSITAE (D)
Astrantia L. UMBELLIFERAE (D)
Astrebla F.Muell. GRAMINEAE (M)
Astrephia Dufr. VALERIANACEAE (D)
Astridia Dinter AIZOACEAE (D)
Astripomoea A.D.J.Meeuse CONVOLVULACEAE (D)
Astrocalyx Merr. MELASTOMATACEAE (D)
Astrocarpa Dumort. = Sesamoides Ortega RESEDACEAE (D)
Astrocarpus Duby (SUS) = Sesamoides Ortega RESEDACEAE (D)
Astrocaryum G.Mey. PALMAE (M)
Astrocasia B.L.Rob. & Millsp. EUPHORBIACEAE (D)
Astrochlaena Hallier f. (SUH) = Astripomoea A.D.J.Meeuse CONVOLVULACEAE (D)
Astrococcus Benth. EUPHORBIACEAE (D)
Astrocodon Fedorov = Campanula L. CAMPANULACEAE (D)
Astrocoma Neck. (SUI) = Staavia Dahl BRUNIACEAE (D)
Astrodaucus Drude UMBELLIFERAE (D)
Astroglossus Rchb. ex Benth. (SUI) = Stellilabium Schltr. ORCHIDACEAE (M)
Astrogyne Benth. = Croton L. EUPHORBIACEAE (D)
Astroloba Uitew. = Haworthia Duval ALOACEAE (M)
Astroloma R.Br. EPACRIDACEAE (D)

Astromerremia Pilg. = Merremia Dennst. ex Endl. CONVOLVULACEAE (D)
Astronia Blume MELASTOMATACEAE (D)
Astronidium A.Gray MELASTOMATACEAE (D)
Astronium Jacq. ANACARDIACEAE (D)
Astropanax Seem. = Schefflera J.R.Forst. & G.Forst. ARALIACEAE (D)
Astropetalum Griff. = Swintonia Griff. ANACARDIACEAE (D)
Astrophyllum Torr. & A.Gray = Choisya Kunth RUTACEAE (D)
Astrophyton Lawr. (SUO) = Astrophytum Lem. CACTACEAE (D)
Astrophytum Lem. CACTACEAE (D)
Astrostemma Benth. = Absolmsia Kuntze ASCLEPIADACEAE (D)
Astrothalamus C.B.Rob. URTICACEAE (D)
Astrotheca Vesque = Clusia L. GUTTIFERAE (D)
Astrotricha DC. ARALIACEAE (D)
Astrotrichilia (Harms) T.D.Penn. & Styles MELIACEAE (D)
Astydamia DC. UMBELLIFERAE (D)
Astylis Wight = Drypetes Vahl EUPHORBIACEAE (D)
Asyneuma Griseb. & Schenck CAMPANULACEAE (D)
Asystasia Blume ACANTHACEAE (D)
Asystasiella Lindau ACANTHACEAE (D)
Ataenidia Gagnep. MARANTACEAE (M)
Atalantia Corrêa RUTACEAE (D)
Atalaya Blume SAPINDACEAE (D)
Atalopteris Maxon & C.Chr. DRYOPTERIDACEAE (F)
Atamisquea Miers ex Hook. & Arn. CAPPARACEAE (D)
Ataxia R.Br. = Hierochloe R.Br. GRAMINEAE (M)
Ataxipteris Holttum DRYOPTERIDACEAE (F)
Ate Lindl. = Habenaria Willd. ORCHIDACEAE (M)
Ateixa Ravenna = Sarcodraba Gilg & Muschler CRUCIFERAE (D)
Atelanthera Hook.f. & Thomson CRUCIFERAE (D)
Ateleia (DC.) Benth. LEGUMINOSAE–PAPILIONOIDEAE (D)
Atelophragma Rydb. = Astragalus L. LEGUMINOSAE–PAPILIONOIDEAE (D)
Atemnosiphon Leandri = Gnidia L. THYMELAEACEAE (D)
Ateramnus P.Browne = Gymnanthes Sw. EUPHORBIACEAE (D)
Athamanta L. UMBELLIFERAE (D)
Athanasia L. COMPOSITAE (D)
Athenaea Sendtn. SOLANACEAE (D)
Atherandra Decne. ASCLEPIADACEAE (D)
Atheranthera Mast. = Gerrardanthus Harv. ex Hook.f. CUCURBITACEAE (D)
Athernotus Dulac (SUS) = Calamagrostis Adans. GRAMINEAE (M)
Atherolepis Hook.f. ASCLEPIADACEAE (D)
Atherophora Steud. (SUI) = Aegopogon Humb. & Bonpl. ex Willd. GRAMINEAE (M)
Atheropogon Willd. = Bouteloua Lag. GRAMINEAE (M)
Atherosperma Labill. MONIMIACEAE (D)
Atherostemon Blume ASCLEPIADACEAE (D)
Athrixia Ker Gawl. COMPOSITAE (D)
Athroandra (Hook.f.) Pax & K.Hoffm. = Erythrococca Benth. EUPHORBIACEAE (D)
Athroisma Griff. (SUH) = Trigonostemon Blume EUPHORBIACEAE (D)
Athroisma DC. COMPOSITAE (D)
Athrolophis (Trin.) Chiov. (SUS) = Andropogon L. GRAMINEAE (M)
Athroostachys Benth. GRAMINEAE (M)
Athrotaxis D.Don TAXODIACEAE (G)
Athyana (Griseb.) Radlk. SAPINDACEAE (D)
Athyriopsis Ching = Lunathyrium Koidz. WOODSIACEAE (F)
Athyrium Roth WOODSIACEAE (F)
Athyrocarpus Schltdl. ex Benth. = Commelina L. COMMELINACEAE (M)

Athysanus Greene CRUCIFERAE (D)
Atirbesia Raf. = Marrubium L. LABIATAE (D)
Atitara O.F.Cook = Desmoncus Mart. PALMAE (M)
Atkinsia R.A.Howard = Thespesia Sol. ex Corrêa MALVACEAE (D)
Atkinsonia F.Muell. LORANTHACEAE (D)
Atlanthemum Raynaud = Helianthemum Mill. CISTACEAE (D)
Atomostigma Kuntze (UP) ROSACEAE (D)
Atopocarpus Cuatrec. = Clonodia Griseb. MALPIGHIACEAE (D)
Atopostema Boutique = Monanthotaxis Baill. ANNONACEAE (D)
Atractantha McClure GRAMINEAE (M)
Atractocarpa Franch. = Puelia Franch. GRAMINEAE (M)
Atractocarpus Schltr. & K.Krause RUBIACEAE (D)
Atractogyne Pierre RUBIACEAE (D)
Atractylis L. COMPOSITAE (D)
Atractylodes DC. COMPOSITAE (D)
Atragene L. = Clematis L. RANUNCULACEAE (D)
Atraphaxis L. POLYGONACEAE (D)
Atrichantha Hilliard & B.L.Burtt COMPOSITAE (D)
Atrichodendron Gagnep. SOLANACEAE (D)
Atrichoseris A.Gray COMPOSITAE (D)
Atriplex L. CHENOPODIACEAE (D)
Atropa L. SOLANACEAE (D)
Atropanthe Pascher SOLANACEAE (D)
Atropatenia F.K.Mey. = Thlaspi L. CRUCIFERAE (D)
Atropis (Trin.) Griseb. (SUS) = Puccinellia Parl. GRAMINEAE (M)
Atroxima Stapf POLYGALACEAE (D)
Atrutegia Bedd. = Goniothalamus (Blume) Hook.f. & Thomson ANNONACEAE (D)
Attalea Kunth PALMAE (M)
Atuna Raf. CHRYSOBALANACEAE (D)
Atylosia Wight & Arn. = Cajanus DC. LEGUMINOSAE–PAPILIONOIDEAE (D)
Aubertia Chapel. ex Baill. = Croton L. EUPHORBIACEAE (D)
Aubregrinia Heine SAPOTACEAE (D)
Aubrevillea Pellegr. LEGUMINOSAE–MIMOSOIDEAE (D)
Aubrieta Adans. CRUCIFERAE (D)
Aucoumea Pierre BURSERACEAE (D)
Aucuba Thunb. AUCUBACEAE (D)
Aucubaephyllum Ahlb. = Psychotria L. RUBIACEAE (D)
Audibertia Benth. (1831) (SUH) = Salvia L. LABIATAE (D)
Audibertia Benth. (1829) = Mentha L. LABIATAE (D)
Audibertiella Briq. = Salvia L. LABIATAE (D)
Audouinia Brongn. BRUNIACEAE (D)
Auerodendrona Urb. RHAMNACEAE (D)
Augea Thunb. ZYGOPHYLLACEAE (D)
Augouardia Pellegr. LEGUMINOSAE–CAESALPINIOIDEAE (D)
Augusta Pohl RUBIACEAE (D)
Augustea DC. (SUO) = Augusta Pohl RUBIACEAE (D)
Augustinea Mart. = Bactris Jacq. PALMAE (M)
Aulacocalyx Hook.f. RUBIACEAE (D)
Aulacocarpus O.Berg = Mouriri Aubl. MELASTOMATACEAE (D)
Aulacodiscus Hook.f. = Pleiocarpidia K.Schum. RUBIACEAE (D)
Aulacolepis Hack. (SUH) = Calamagrostis Adans. GRAMINEAE (M)
Aulacospermum Ledeb. UMBELLIFERAE (D)
Aulacothele Monv. (SUI) = Coryphantha (Engelm.) Lem. CACTACEAE (D)
Aulandra H.J.Lam SAPOTACEAE (D)
Aulax Bergius PROTEACEAE (D)

Aulaxanthus Elliott = Anthenantia P.Beauv. GRAMINEAE (M)
Aulaxia Nutt. (SUS) = Anthenantia P.Beauv. GRAMINEAE (M)
Aulaya Harv. = Harveya Hook. SCROPHULARIACEAE (D)
Auliza Small = Epidendrum L. ORCHIDACEAE (M)
Aulojusticia Lindau = Justicia L. ACANTHACEAE (D)
Aulomyrcia O.Berg = Myrcia DC. ex Guill. MYRTACEAE (D)
Aulonemia Goudot GRAMINEAE (M)
Aulosepalum Garay ORCHIDACEAE (M)
Aulospermum J.M.Coult. & Rose = Cymopterus Raf. UMBELLIFERAE (D)
Aulostephanus Schltr. = Brachystelma R.Br. ASCLEPIADACEAE (D)
Aulostylis Schltr. = Calanthe Ker–Gawl. ORCHIDACEAE (M)
Aulotandra Gagnep. ZINGIBERACEAE (M)
Aureilobivia Frič (SUI) = Echinopsis Zucc. CACTACEAE (D)
Aureliana Sendtn. SOLANACEAE (D)
Aureolaria Raf. SCROPHULARIACEAE (D)
Auriculardisia Lundell = Ardisia Sw. MYRSINACEAE (D)
Aurinia Desv. CRUCIFERAE (D)
Australina Gaudich. URTICACEAE (D)
Australopyrum A.Löve = Agropyron Gaertn. GRAMINEAE (M)
Australorchis Brieger = Dendrobium Sw. ORCHIDACEAE (M)
Austroamericium Hendrych = Thesium L. SANTALACEAE (D)
Austrobaileya C.T.White AUSTROBAILEYACEAE (D)
Austrobrickellia R.M.King & H.Rob. COMPOSITAE (D)
Austrobuxus Miq. EUPHORBIACEAE (D)
Austrocactus Britton & Rose CACTACEAE (D)
Austrocedrus Florin & Boutelje CUPRESSACEAE (G)
Austrocephalocereus Backeb. = Micranthocereus Backeb. CACTACEAE (D)
Austrochloris Lazarides GRAMINEAE (M)
Austrocritonia R.M.King & H.Rob. COMPOSITAE (D)
Austrocylindropuntia Backeb. = Opuntia Mill. CACTACEAE (D)
Austrocynoglossum Popov ex R.R.Mill BORAGINACEAE (D)
Austrodolichos Verdc. LEGUMINOSAE–PAPILIONOIDEAE (D)
Austroeupatorium R.M.King & H.Rob. COMPOSITAE (D)
Austrofestuca (Tsvelev) Alexeev GRAMINEAE (M)
Austrogambeya Aubrév. & Pellegr. = Chrysophyllum L. SAPOTACEAE (D)
Austrogramme Fourn. ADIANTACEAE (F)
Austroliabum H.Rob. & Brettell COMPOSITAE (D)
Austrolycopodium Holub = Lycopodium LYCOPODIACEAE (FA)
Austromatthaea L.S.Sm. MONIMIACEAE (D)
Austromimusops A.D.J.Meeuse = Vitellariopsis Baill. ex Dubard SAPOTACEAE (D)
Austromuellera C.T.White PROTEACEAE (D)
Austromyrtus (Nied.) Burret MYRTACEAE (D)
Austropeucedanum Mathias & Constance UMBELLIFERAE (D)
Austrosteenisia Geesink LEGUMINOSAE–PAPILIONOIDEAE (D)
Austrosynotis C.Jeffrey COMPOSITAE (D)
Austrotaxus Compton TAXACEAE (G)
Auticoryne Turcz. = Baeckea L. MYRTACEAE (D)
Autrandra Pierre ex Prain = Erythrococca Benth. EUPHORBIACEAE (D)
Autranella A.Chev. SAPOTACEAE (D)
Autrania C.Winkl. & Barbey = Jurinea Cass. COMPOSITAE (D)
Autumnalia Pimenov UMBELLIFERAE (D)
Auxemma Miers BORAGINACEAE (D)
Auxopus Schltr. ORCHIDACEAE (M)
Auzuba Juss. = Sideroxylon L. SAPOTACEAE (D)
Aveledoa Pittier = Metteniusa Karst. ICACINACEAE (D)

Avellanita Phil. EUPHORBIACEAE (D)
Avellara Blanca & Díaz Guard. COMPOSITAE (D)
Avellinia Parl. = Trisetaria Forssk. GRAMINEAE (M)
Avena Scop. (SUS) = Lagurus L. GRAMINEAE (M)
Avena L. GRAMINEAE (M)
Avenaria Fabr. (SUS) = Bromus L. GRAMINEAE (M)
Avenastrum Opiz (SUS) = Helictotrichon Besser GRAMINEAE (M)
Avenella Parl. = Deschampsia P.Beauv. GRAMINEAE (M)
Avenochloa Holub (SUS) = Helictotrichon Besser GRAMINEAE (M)
Avenula (Dumort.) Dumort. = Helictotrichon Besser GRAMINEAE (M)
Averia Leonard = Justicia L. ACANTHACEAE (D)
Averrhoa L. OXALIDACEAE (D)
Averrhoidium Baill. SAPINDACEAE (D)
Avesicaria (Kamieński) Barnhart = Utricularia L. LENTIBULARIACEAE (D)
Avetra H.Perrier DIOSCOREACEAE (M)
Avicennia L. AVICENNIACEAE (D)
Aviceps Lindl. = Satyrium Sw. ORCHIDACEAE (M)
Avoira Giseke = Astrocaryum G.Mey. PALMAE (M)
Axanthes Blume = Urophyllum Wall. RUBIACEAE (D)
Axanthopsis Korth. = Urophyllum Wall. RUBIACEAE (D)
Axenfeldia Baill. = Mallotus Lour. EUPHORBIACEAE (D)
Axinaea Ruíz & Pav. MELASTOMATACEAE (D)
Axinandra Thwaites CRYPTERONIACEAE (D)
Axinanthera Karst. = Bellucia Neck. ex Raf. MELASTOMATACEAE (D)
Axinea Juss. (SUO) = Axinaea Ruíz & Pav. MELASTOMATACEAE (D)
Axiniphyllum Benth. COMPOSITAE (D)
Axolus Raf. = Cephalanthus L. RUBIACEAE (D)
Axonopus P.Beauv. GRAMINEAE (M)
Axyris L. CHENOPODIACEAE (D)
Ayapana Spach COMPOSITAE (D)
Ayapanopsis R.M.King & H.Rob. COMPOSITAE (D)
Aydendron Nees = Aniba Aubl. LAURACEAE (D)
Ayenia L. STERCULIACEAE (D)
Ayensua L.B.Sm. BROMELIACEAE (M)
Aylacophora Cabrera COMPOSITAE (D)
Aylostera Speg. = Rebutia K.Schum. CACTACEAE (D)
Aylthonia N.L.Menezes VELLOZIACEAE (M)
Aynia H.Rob. = Vernonia Schreb. COMPOSITAE (D)
Aytonia L.f. (SUO) = Nymania Lindb. MELIACEAE (D)
Azadehdelia Braem = Cribbia Senghas ORCHIDACEAE (M)
Azadirachta A.Juss. MELIACEAE (D)
Azalea L. = Rhododendron L. ERICACEAE (D)
Azanza Alef. MALVACEAE (D)
Azaola Blanco = Madhuca Ham. ex J.F.Gmel. SAPOTACEAE (D)
Azara Ruíz & Pav. FLACOURTIACEAE (D)
Azedarach Mill. = Melia L. MELIACEAE (D)
Azilia Hedge & Lamond UMBELLIFERAE (D)
Azima Lam. SALVADORACEAE (D)
Azolla Lam. AZOLLACEAE (F)
Azorella Lam. UMBELLIFERAE (D)
Azorina Feer CAMPANULACEAE (D)
Aztekium Boed. CACTACEAE (D)
Azukia Takah. ex Ohwi = Vigna Savi LEGUMINOSAE–PAPILIONOIDEAE (D)
Azureocereus Akers & H.Johnson = Browningia Britton & Rose CACTACEAE (D)

Babactes DC. = Chirita Buch.–Ham. ex D.Don GESNERIACEAE (D)
Babbagia F.Muell. CHENOPODIACEAE (D)
Babcockia Boulos = Sonchus L. COMPOSITAE (D)
Babiana Ker Gawl. ex Sims IRIDACEAE (M)
Babingtonia Lindl. = Baeckea L. MYRTACEAE (D)
Baccaurea Lour. EUPHORBIACEAE (D)
Baccaureopsis Pax = Thecacoris A.Juss. EUPHORBIACEAE (D)
Baccharidastrum Cabrera = Baccharis L. COMPOSITAE (D)
Baccharidiopsis G.M.Barroso = Aster L. COMPOSITAE (D)
Baccharis L. COMPOSITAE (D)
Baccharoides Moench = Vernonia Schreb. COMPOSITAE (D)
Bachmannia Pax CAPPARACEAE (D)
Backebergia H.Bravo (SUS) = Pachycereus (A.Berger) Britton & Rose CACTACEAE (D)
Backeria Bakh.f. = Diplectria (Blume) Rchb. MELASTOMATACEAE (D)
Backhousia Hook. & Harv. MYRTACEAE (D)
Baconia DC. = Pavetta L. RUBIACEAE (D)
Bacopa Aubl. SCROPHULARIACEAE (D)
Bactris Jacq. PALMAE (M)
Bacularia F.Muell. ex Hook.f. = Linospadix H.A.Wendl. PALMAE (M)
Badamia Gaertn. = Terminalia L. COMBRETACEAE (D)
Badiera DC. POLYGALACEAE (D)
Badilloa R.M.King & H.Rob. COMPOSITAE (D)
Badula Juss. MYRSINACEAE (D)
Badusa A.Gray RUBIACEAE (D)
Baeckea L. MYRTACEAE (D)
Baeolepis Decne. ex Moq. ASCLEPIADACEAE (D)
Baeometra Salisb. ex Endl. COLCHICACEAE (M)
Baeothryon A.Dietr. = Eleocharis R.Br. CYPERACEAE (M)
Baeriopsis J.T.Howell COMPOSITAE (D)
Bafodeya Prance ex F.White CHRYSOBALANACEAE (D)
Bafutia C.D.Adams COMPOSITAE (D)
Bagassa Aubl. MORACEAE (D)
Bagnisia Becc. = Thismia Griff. BURMANNIACEAE (M)
Bahamia Britton & Rose = Acacia Mill. LEGUMINOSAE–MIMOSOIDEAE (D)
Bahia Lag. COMPOSITAE (D)
Bahianthus R.M.King & H.Rob. COMPOSITAE (D)
Baikiaea Benth. LEGUMINOSAE–CAESALPINIOIDEAE (D)
Baileya Harv. & A.Gray COMPOSITAE (D)
Baileyoxylon C.T.White FLACOURTIACEAE (D)
Baillandea Roberty = Calycobolus Willd. ex J.A.Schult. CONVOLVULACEAE (D)
Baillonella Pierre SAPOTACEAE (D)
Baillonia Bocquillon VERBENACEAE (D)
Baissea A.DC. APOCYNACEAE (D)
Bakerantha L.B.Sm. = Hechtia Klotzsch BROMELIACEAE (M)
Bakerella Tiegh. LORANTHACEAE (D)
Bakeria E.F.Andre (SUH) = Hechtia Klotzsch BROMELIACEAE (M)
Bakeria Seem. = Schefflera J.R.Forst. & G.Forst. ARALIACEAE (D)
Bakeridesia Hochr. MALVACEAE (D)
Bakeriella Pierre ex Dubard = Synsepalum (A.DC.) Daniell SAPOTACEAE (D)
Bakerisideroxylon Engl. = Synsepalum (A.DC.) Daniell SAPOTACEAE (D)
Bakerophyton (J.Léonard) Hutch. = Aeschynomene L. LEGUMINOSAE–
    PAPILIONOIDEAE (D)
Bakeropteris Kuntze = Cassebeera Kaulf. ADIANTACEAE (F)
Balaka Becc. PALMAE (M)
Balanites Delile BALANITACEAE (D)

Balanocarpus Bedd. = Hopea Roxb. DIPTEROCARPACEAE (D)
Balanophora J.R.Forst. & G.Forst. BALANOPHORACEAE (D)
Balanops Baill. BALANOPACEAE (D)
Balanostreblus Kurz = Sorocea A.St.–Hil. MORACEAE (D)
Balansaea Boiss. & Reut. = Geocaryum Coss. UMBELLIFERAE (D)
Balansochloa Kuntze (SUS) = Germainia Balansa & Poitrasson GRAMINEAE (M)
Balantium Desv. ex Ham. (SUH) = Parinari Aubl. CHRYSOBALANACEAE (D)
Balantium Kaulf. = Dicksonia L'Hér. DICKSONIACEAE (F)
Balaustion Hook. MYRTACEAE (D)
Balbisia Cav. GERANIACEAE (D)
Balboa Planch. & Triana = Chrysochlamys Poepp. GUTTIFERAE (D)
Baldellia Parl. ALISMATACEAE (M)
Baldingera P.Gaertn., B.Mey. & Scherb. (SUS) = Phalaris L. GRAMINEAE (M)
Baldomiria Herter = Leptochloa P.Beauv. GRAMINEAE (M)
Balduina Nutt. COMPOSITAE (D)
Baldwinia Raf. = Passiflora L. PASSIFLORACEAE (D)
Balfourina Kuntze = Didymaea Hook.f. RUBIACEAE (D)
Balfourodendron Corr.Mello ex Oliv. RUTACEAE (D)
Balgoya Morat & Meijden POLYGALACEAE (D)
Baliospermum Blume EUPHORBIACEAE (D)
Balisaea Taub. = Aeschynomene L. LEGUMINOSAE–PAPILIONOIDEAE (D)
Ballantinia Hook.f. ex E.A.Shaw CRUCIFERAE (D)
Ballardia Montrouz. = Carpolepis (J.W.Dawson) J.W.Dawson MYRTACEAE (D)
Ballochia Balf.f. ACANTHACEAE (D)
Ballota L. LABIATAE (D)
Ballya Brenan = Aneilema R.Br. COMMELINACEAE (M)
Balmea Martínez RUBIACEAE (D)
Baloghia Endl. EUPHORBIACEAE (D)
Balonga Le Thomas ANNONACEAE (D)
Balsamita Mill. = Tanacetum L. COMPOSITAE (D)
Balsamocarpon Clos LEGUMINOSAE–CAESALPINIOIDEAE (D)
Balsamocitrus Stapf RUTACEAE (D)
Balsamodendrum Kunth = Commiphora Jacq. BURSERACEAE (D)
Balsamorhiza Hook. ex Nutt. COMPOSITAE (D)
Balthasaria Verdc. THEACEAE (D)
Baltimora L. COMPOSITAE (D)
Bambekea Cogn. CUCURBITACEAE (D)
Bamboga Baill. (SUO) = Mitragyna Korth. RUBIACEAE (D)
Bambos Retz. (SUS) = Bambusa Schreb. GRAMINEAE (M)
Bamburanta L.Linden = Trachyphrynium Benth. MARANTACEAE (M)
Bambusa Caldas (SUH) = Bambusa Schreb. GRAMINEAE (M)
Bambusa Schreb. GRAMINEAE (M)
Bamiania Lincz. PLUMBAGINACEAE (D)
Bamlera K.Schum. & Lauterb. = Astronidium A.Gray MELASTOMATACEAE (D)
Bampsia Lisowski & Mielcarek SCROPHULARIACEAE (D)
Banalia Moq. (SUH) = Indobanalia A.N.Henry & Roy AMARANTHACEAE (D)
Banalia Raf. = Croton L. EUPHORBIACEAE (D)
Banara Aubl. FLACOURTIACEAE (D)
Bancalus Kuntze (UP) RUBIACEAE (D)
Bandeiraea Benth. = Griffonia Baill. LEGUMINOSAE–CAESALPINIOIDEAE (D)
Banisteria L. (SUH) = Heteropterys Kunth MALPIGHIACEAE (D)
Banisterioides Dubard & Dop = Sphedamnocarpus Planch. ex Benth.
    MALPIGHIACEAE (D)
Banisteriopsis C.B.Rob. ex Small MALPIGHIACEAE (D)
Banksia L.f. PROTEACEAE (D)

Baolia H.W.Kung & G.L.Chu = Chenopodium L. CHENOPODIACEAE (D)
Baoulia A.Chev. = Murdannia Royle COMMELINACEAE (M)
Baphia Afzel. ex Lodd. LEGUMINOSAE–PAPILIONOIDEAE (D)
Baphiastrum Harms LEGUMINOSAE–PAPILIONOIDEAE (D)
Baphicacanthus Bremek. = Strobilanthes Blume ACANTHACEAE (D)
Baphiopsis Benth. ex Baker LEGUMINOSAE–PAPILIONOIDEAE (D)
Baphorhiza Link = Alkanna Tausch BORAGINACEAE (D)
Baprea Pierre ex Pax & K.Hoffm. = Cladogynos Zipp. ex Spanoghe
    EUPHORBIACEAE (D)
Baptisia Vent. LEGUMINOSAE–PAPILIONOIDEAE (D)
Baptistonia Barb.Rodr. ORCHIDACEAE (M)
Baptorhachis Clayton & Renvoize GRAMINEAE (M)
Barathranthus (Korth.) Miq. LORANTHACEAE (D)
Barbacenia Vand. VELLOZIACEAE (M)
Barbaceniopsis L.B.Sm. VELLOZIACEAE (M)
Barbarea R.Br. CRUCIFERAE (D)
Barberetta Harv. HAEMODORACEAE (M)
Barbeuia Thouars BARBEUIACEAE (D)
Barbeya Schweinf. BARBEYACEAE (D)
Barbeyastrum Cogn. = Dichaetanthera Endl. MELASTOMATACEAE (D)
Barbieria Spreng. = Clitoria L. LEGUMINOSAE–PAPILIONOIDEAE (D)
Barbilus P.Browne = Trichilia P.Browne MELIACEAE (D)
Barbosa Becc. = Syagrus Mart. PALMAE (M)
Barbosella Schltr. ORCHIDACEAE (M)
Barbrodria Luer ORCHIDACEAE (M)
Barcella (Trail) Trail ex Drude PALMAE (M)
Barcena Duges = Colubrina Rich. ex Brongn. RHAMNACEAE (D)
Barclaya Wall. NYMPHAEACEAE (D)
Bargemontia Gaudich. = Nolana L.f. SOLANACEAE (D)
Barhamia Klotzsch = Croton L. EUPHORBIACEAE (D)
Barjonia Decne. ASCLEPIADACEAE (D)
Barkeria Knowles & Westc. ORCHIDACEAE (M)
Barkerwebbia Becc. = Heterospathe Scheff. PALMAE (M)
Barkleyanthus H.Rob. & Brettell COMPOSITAE (D)
Barklya F.Muell. = Bauhinia L. LEGUMINOSAE–CAESALPINIOIDEAE (D)
Barlaea Rchb.f. = Cynorkis Thouars ORCHIDACEAE (M)
Barleria L. ACANTHACEAE (D)
Barleriacanthus Oerst. = Barleria L. ACANTHACEAE (D)
Barlerianthus Oerst. = Barleria L. ACANTHACEAE (D)
Barleriola Oerst. ACANTHACEAE (D)
Barleriosiphon Oerst. = Barleria L. ACANTHACEAE (D)
Barlerites Oerst. = Barleria L. ACANTHACEAE (D)
Barlia Parl. ORCHIDACEAE (M)
Barnadesia Mutis ex L.f. COMPOSITAE (D)
Barnardiella Goldblatt IRIDACEAE (M)
Barnebya W.R.Anderson & B.Gates MALPIGHIACEAE (D)
Barneoudia Gay RANUNCULACEAE (D)
Barnettia Santisuk (SUH) = Santisukia Brummitt BIGNONIACEAE (D)
Barnhartia Gleason POLYGALACEAE (D)
Barola Adans. (SUS) = Trichilia P.Browne MELIACEAE (D)
Barombia Schltr. ORCHIDACEAE (M)
Barongia Peter G.Wilson & B.Hyland MYRTACEAE (D)
Baroniella Costantin & Gallaud ASCLEPIADACEAE (D)
Barosma Willd. = Agathosma Willd. RUTACEAE (D)
Barraldeia Thouars = Carallia Roxb. RHIZOPHORACEAE (D)

Barrettia T.R.Sim = Ricinodendron Muell.Arg. EUPHORBIACEAE (D)
Barringtonia J.R.Forst. & G.Forst. LECYTHIDACEAE (D)
Barroetea A.Gray COMPOSITAE (D)
Barrosoa R.M.King & H.Rob. COMPOSITAE (D)
Barteria Hook.f. PASSIFLORACEAE (D)
Barthea Hook.f. MELASTOMATACEAE (D)
Bartholina R.Br. ORCHIDACEAE (M)
Barthollesia Silva Manso = Bertholletia Bonpl. LECYTHIDACEAE (D)
Bartholomaea Standl. & Steyerm. FLACOURTIACEAE (D)
Bartlettia A.Gray COMPOSITAE (D)
Bartlettina R.M.King & H.Rob. COMPOSITAE (D)
Bartlingia Rchb. = Plocama Aiton RUBIACEAE (D)
Bartonia Mühl. ex Willd. GENTIANACEAE (D)
Bartonia Pursh ex Sims (SUH) = Mentzelia L. LOASACEAE (D)
Bartramia Salisb. = Penstemon Schmidel SCROPHULARIACEAE (D)
Bartschella Britton & Rose = Mammillaria Haw. CACTACEAE (D)
Bartsia L. SCROPHULARIACEAE (D)
Barylucuma Ducke = Pouteria Aubl. SAPOTACEAE (D)
Basanacantha Hook.f. = Randia L. RUBIACEAE (D)
Basananthe Peyr. PASSIFLORACEAE (D)
Basedowia E.Pritz. COMPOSITAE (D)
Basella L. BASELLACEAE (D)
Baseonema Schltr. & Rendle ASCLEPIADACEAE (D)
Bashania Keng f. & Yi = Arundinaria Michx. GRAMINEAE (M)
Basicarpus Tiegh. = Antidaphne Poepp. & Endl. EREMOLEPIDACEAE (D)
Basigyne J.J.Sm. = Dendrochilum Blume ORCHIDACEAE (M)
Basileophyta F.Muell. = Fieldia A.Cunn. GESNERIACEAE (D)
Basilicum Moench LABIATAE (D)
Basiloxylon K.Schum. = Pterygota Schott & Endl. STERCULIACEAE (D)
Basiphyllaea Schltr. ORCHIDACEAE (M)
Basisperma C.T.White MYRTACEAE (D)
Basistelma Bartlett ASCLEPIADACEAE (D)
Basistemon Turcz. SCROPHULARIACEAE (D)
Baskervillea Lindl. ORCHIDACEAE (M)
Basselinia Vieill. PALMAE (M)
Bassia K.Koenig (SUH) = Madhuca Ham. ex J.F.Gmel. SAPOTACEAE (D)
Bassia All. CHENOPODIACEAE (D)
Bassovia Aubl. = Solanum L. SOLANACEAE (D)
Bastardia Kunth MALVACEAE (D)
Bastardiastrum (Rose) D.M.Bates MALVACEAE (D)
Bastardiopsis (K.Schum.) Hassl. MALVACEAE (D)
Basutica E.Phillips THYMELAEACEAE (D)
Batania Hatus. = Pycnarrhena Miers ex Hook.f. & Thomson MENISPERMACEAE (D)
Batanthes Raf. = Ipomopsis Michx. POLEMONIACEAE (D)
Bataprine Nieuwl. = Galium L. RUBIACEAE (D)
Batatas Choisy = Ipomoea L. CONVOLVULACEAE (D)
Batemannia Lindl. ORCHIDACEAE (M)
Batesanthus N.E.Br. ASCLEPIADACEAE (D)
Batesia Spruce ex Benth. LEGUMINOSAE–CAESALPINIOIDEAE (D)
Batesimalva Fryxell MALVACEAE (D)
Bathiaea Drake LEGUMINOSAE–CAESALPINIOIDEAE (D)
Bathiea Schltr. = Neobathiea Schltr. ORCHIDACEAE (M)
Bathiorhamnus Capuron RHAMNACEAE (D)
Bathmium C.Presl ex Link = Tectaria Cav. DRYOPTERIDACEAE (F)
Bathysa C.Presl RUBIACEAE (D)

Bathysograya Kuntze = Badusa A.Gray RUBIACEAE (D)
Batidaea (Dumort.) Greene = Rubus L. ROSACEAE (D)
Batidophaca Rydb. = Astragalus L. LEGUMINOSAE–PAPILIONOIDEAE (D)
Batis P.Browne BATACEAE (D)
Batocarpus Karst. MORACEAE (D)
Batopedina Verdc. RUBIACEAE (D)
Batrachium (DC.) Gray = Ranunculus L. RANUNCULACEAE (D)
Batratherum Nees = Arthraxon P.Beauv. GRAMINEAE (M)
Battandiera Maire = Ornithogalum L. HYACINTHACEAE (M)
Battata Hill = Solanum L. SOLANACEAE (D)
Bauchea Fourn. = Sporobolus R.Br. GRAMINEAE (M)
Baudinia Lesch. ex DC. = Calothamnus Labill. MYRTACEAE (D)
Baudouinia Baill. LEGUMINOSAE–CAESALPINIOIDEAE (D)
Bauera Banks ex Andr. CUNONIACEAE (D)
Bauerella Borzi = Acronychia J.R.Forst. & G.Forst. RUTACEAE (D)
Bauerella Schindl. (SUH) = Psoralea L. LEGUMINOSAE–PAPILIONOIDEAE (D)
Baueropsis Hutch. = Cullen Medik. LEGUMINOSAE–PAPILIONOIDEAE (D)
Bauhinia L. LEGUMINOSAE–CAESALPINIOIDEAE (D)
Baukea Vatke LEGUMINOSAE–PAPILIONOIDEAE (D)
Baumannia K.Schum. (SUH) = Knoxia L. RUBIACEAE (D)
Baumannia DC. = Damnacanthus C.F.Gaertn. RUBIACEAE (D)
Baumannia Spach (SUH) = Oenothera L. ONAGRACEAE (D)
Baumea Gaudich. = Machaerina Vahl CYPERACEAE (M)
Baumia Engl. & Gilg SCROPHULARIACEAE (D)
Baumiella H.Wolff (SUH) = Afrocarum Rauschert UMBELLIFERAE (D)
Baxtera Rchb. = Loniceroides Bullock ASCLEPIADACEAE (D)
Baxteria R.Br. ex Hook. LOMANDRACEAE (M)
Bayonia Dugand = Mansoa DC. BIGNONIACEAE (D)
Bdallophyton Eichler RAFFLESIACEAE (D)
Beadlea Small = Cyclopogon C.Presl ORCHIDACEAE (M)
Bealia Scribn. = Muhlenbergia Schreb. GRAMINEAE (M)
Beata O.F.Cook = Coccothrinax Sargent PALMAE (M)
Beatonia Herb. = Tigridia Juss. IRIDACEAE (M)
Beatsonia Roxb. FRANKENIACEAE (D)
Beaucarnea Lem. DRACAENACEAE (M)
Beaufortia R.Br. MYRTACEAE (D)
Beaumontia Wall. APOCYNACEAE (D)
Beauprea Brongn. & Gris PROTEACEAE (D)
Beaupreopsis Virot PROTEACEAE (D)
Beautempsia Gaudich. CAPPARACEAE (D)
Beauverdia Herter = Leucocoryne Lindl. ALLIACEAE (M)
Beauvisagea Pierre = Pouteria Aubl. SAPOTACEAE (D)
Bebbia Greene COMPOSITAE (D)
Beccabunga Hill = Veronica L. SCROPHULARIACEAE (D)
Beccarianthus Cogn. = Astronidium A.Gray MELASTOMATACEAE (D)
Beccariella Pierre = Pouteria Aubl. SAPOTACEAE (D)
Beccarimnia Pierre ex Koord. = Pouteria Aubl. SAPOTACEAE (D)
Beccarina Tiegh. (SUS) = Trithecanthera Tiegh. LORANTHACEAE (D)
Beccarinda Kuntze GESNERIACEAE (D)
Beccariodendron Warb. = Goniothalamus (Blume) Hook.f. & Thomson
ANNONACEAE (D)
Beccariophoenix Jum. & H.Perrier PALMAE (M)
Becheria Ridl. (UP) RUBIACEAE (D)
Becium Lindl. LABIATAE (D)
Beckera Fres. (SUH) = Snowdenia C.E.Hubb. GRAMINEAE (M)

Beckeria Bernh. = Melica L. GRAMINEAE (M)
Beckeropsis Figari & De Not. = Pennisetum Rich. GRAMINEAE (M)
Beckmannia Host GRAMINEAE (M)
Beckwithia Jeps. = Ranunculus L. RANUNCULACEAE (D)
Beclardia A.Rich. ORCHIDACEAE (M)
Becquerelia Brongn. CYPERACEAE (M)
Beddomea Hook.f. = Aglaia Lour. MELIACEAE (D)
Bedfordia DC. COMPOSITAE (D)
Beesha Munro (SUI) = Ochlandra Thwaites GRAMINEAE (M)
Beesha Kunth (SUI) = Melocanna Trin. GRAMINEAE (M)
Beesia Balf.f. & W.W.Sm. RANUNCULACEAE (D)
Beethovenia Engel = Ceroxylon Bonpl. ex DC. PALMAE (M)
Befaria Mutis ex L. (SUO) = Bejaria Mutis ex L. ERICACEAE (D)
Begonia L. BEGONIACEAE (D)
Begoniella Oliv. = Begonia L. BEGONIACEAE (D)
Beguea Capuron SAPINDACEAE (D)
Behaimia Griseb. LEGUMINOSAE–PAPILIONOIDEAE (D)
Behnia Didr. PHILESIACEAE (M)
Behria Greene = Bessera Schult. ALLIACEAE (M)
Behuria Cham. MELASTOMATACEAE (D)
Beilschmiedia Nees LAURACEAE (D)
Beirnaertia Louis ex Troupin MENISPERMACEAE (D)
Beiselia Forman BURSERACEAE (D)
Bejaranoa R.M.King & H.Rob. COMPOSITAE (D)
Bejaria Mutis ex L. ERICACEAE (D)
Bejaudia Gagnep. = Myrialepis Becc. PALMAE (M)
Belairia A.Rich. LEGUMINOSAE–PAPILIONOIDEAE (D)
Belamcanda Adans. IRIDACEAE (M)
Belandra S.F.Blake = Prestonia R.Br. APOCYNACEAE (D)
Belangera Cambess. = Lamanonia Vell. CUNONIACEAE (D)
Belemia Pires NYCTAGINACEAE (D)
Belencita Karst. CAPPARACEAE (D)
Belenia Decne. = Physochlaina G.Don SOLANACEAE (D)
Beliceodendron Lundell = Lecointea Ducke LEGUMINOSAE–PAPILIONOIDEAE (D)
Belicia Lundell = Morinda L. RUBIACEAE (D)
Belilla Adans. = Mussaenda L. RUBIACEAE (D)
Belladona Duhamel = Atropa L. SOLANACEAE (D)
Bellardia Schreb. (SUH) = Coccocypselum P.Browne RUBIACEAE (D)
Bellardia All. = Bartsia L. SCROPHULARIACEAE (D)
Bellardiochloa Chiov. = Poa L. GRAMINEAE (M)
Bellendena R.Br. PROTEACEAE (D)
Bellevalia Lapeyr. HYACINTHACEAE (M)
Bellevalia Roem. & Schult. (SUH) = Richeria Vahl EUPHORBIACEAE (D)
Bellida Ewart COMPOSITAE (D)
Bellingia Pierre = Zollingeria Kurz SAPINDACEAE (D)
Bellinia Roem. & Schult. = Saracha Ruíz & Pav. SOLANACEAE (D)
Belliolum Tiegh. = Zygogynum Baill. WINTERACEAE (D)
Bellis L. COMPOSITAE (D)
Bellium L. COMPOSITAE (D)
Belloa J.Remy COMPOSITAE (D)
Bellonia L. GESNERIACEAE (D)
Bellota Gay = Ocotea Aubl. LAURACEAE (D)
Bellucia Neck. ex Raf. MELASTOMATACEAE (D)
Bellynkxia Müll.Arg. = Morinda L. RUBIACEAE (D)
Belmontia E.Mey. = Sebaea Sol. ex R.Br. GENTIANACEAE (D)

Beloglottis Schltr. = Spiranthes Rich. ORCHIDACEAE (M)
Belonanthus Graebn. VALERIANACEAE (D)
Belonophora Hook.f. RUBIACEAE (D)
Beloperone Nees = Justicia L. ACANTHACEAE (D)
Belostemma Wall. ex Wight = Tylophora R.Br. ASCLEPIADACEAE (D)
Belosynapsis Hassk. COMMELINACEAE (M)
Belotia A.Rich. = Trichospermum Blume TILIACEAE (D)
Beltrania Miranda = Enriquebeltrania Rzed. EUPHORBIACEAE (D)
Belvisia Mirb. POLYPODIACEAE (F)
Bemarivea Choux = Tinopsis Radlk. SAPINDACEAE (D)
Bembicia Oliv. FLACOURTIACEAE (D)
Bembicidium Rydb. LEGUMINOSAE–PAPILIONOIDEAE (D)
Bemsetia Raf. = Ixora L. RUBIACEAE (D)
Bencomia Webb & Berthel. ROSACEAE (D)
Benedictaea Toledo = Ottelia Pers. HYDROCHARITACEAE (M)
Benedictella Maire = Lotus L. LEGUMINOSAE–PAPILIONOIDEAE (D)
Benevidesia Saldanha & Cogn. MELASTOMATACEAE (D)
Benguellia G.Taylor LABIATAE (D)
Benincasa Savi CUCURBITACEAE (D)
Benitoa D.Keck COMPOSITAE (D)
Benjaminia Mart. ex Benj. SCROPHULARIACEAE (D)
Benkara Adans. RUBIACEAE (D)
Bennetia Raf. (SUH) = Sporobolus R.Br. GRAMINEAE (M)
Bennettia R.Br. (SUH) = Galearia Zoll. & Moritzi PANDACEAE (D)
Bennettiodendron Merr. FLACOURTIACEAE (D)
Benoicanthus Heine & A.Raynal ACANTHACEAE (D)
Benoistia H.Perrier & Leandri EUPHORBIACEAE (D)
Bensonia Abrams & Bacigalupi (SUH) = Bensoniella C.Morton SAXIFRAGACEAE (D)
Bensoniella C.Morton SAXIFRAGACEAE (D)
Benteca Adans. = Hymenodictyon Wall. RUBIACEAE (D)
Benthamantha Alef. = Coursetia DC. LEGUMINOSAE–PAPILIONOIDEAE (D)
Benthamia Lindl. (SUH) = Cornus L. CORNACEAE (D)
Benthamia A.Rich. ORCHIDACEAE (M)
Benthamidia Spach = Cornus L. CORNACEAE (D)
Benthamiella Speg. SOLANACEAE (D)
Benthamina Tiegh. LORANTHACEAE (D)
Benthamistella Kuntze = Buchnera L. SCROPHULARIACEAE (D)
Bentia Rolfe = Justicia L. ACANTHACEAE (D)
Bentinckia Berry ex Roxb. PALMAE (M)
Bentinckiopsis Becc. = Clinostigma H.A.Wendl. PALMAE (M)
Bentleya E.M.Benn. PITTOSPORACEAE (D)
Benzingia Dodson ORCHIDACEAE (M)
Benzonia Schumach. RUBIACEAE (D)
Bequaertia R.Wilczek CELASTRACEAE (D)
Bequaertiodendron De Wild. = Englerophytum K.Krause SAPOTACEAE (D)
Berardia Vill. COMPOSITAE (D)
Berberidopsis Hook.f. FLACOURTIACEAE (D)
Berberis L. BERBERIDACEAE (D)
Berchemia Neck. ex DC. RHAMNACEAE (D)
Berchemiella Nakai RHAMNACEAE (D)
Berchtoldia C.Presl = Chaetium Nees GRAMINEAE (M)
Berendtia A.Gray (SUH) = Berendtiella Wettst. & Harms SCROPHULARIACEAE (D)
Berendtiella Wettst. & Harms SCROPHULARIACEAE (D)
Berenice Tul. CAMPANULACEAE (D)
Bergena Adans. = Lecythis Loefl. LECYTHIDACEAE (D)

Bergenia Moench SAXIFRAGACEAE (D)
Bergeranthus Schwantes AIZOACEAE (D)
Bergerocactus Britton & Rose CACTACEAE (D)
Bergerocereus Frič & Kreuz. (SUO) = Bergerocactus Britton & Rose CACTACEAE (D)
Bergeronia Micheli LEGUMINOSAE–PAPILIONOIDEAE (D)
Berghausia Endl. = Garnotia Brongn. GRAMINEAE (M)
Berghesia Nees RUBIACEAE (D)
Berghias Juss. = Gardenia Ellis RUBIACEAE (D)
Bergia L. ELATINACEAE (D)
Berginia Harv. = Holographis Nees ACANTHACEAE (D)
Bergkias Sonn. = Gardenia Ellis RUBIACEAE (D)
Bergsmia Blume = Ryparosa Blume FLACOURTIACEAE (D)
Berhautia Balle LORANTHACEAE (D)
Berkheya Ehrh. COMPOSITAE (D)
Berlandiera DC. COMPOSITAE (D)
Berlinia Sol. ex Hook.f. LEGUMINOSAE–CAESALPINIOIDEAE (D)
Berlinianche (Harms) Vattimo RAFFLESIACEAE (D)
Bernarda Adans. (SUO) = Bernardia Mill. EUPHORBIACEAE (D)
Bernardia Mill. EUPHORBIACEAE (D)
Bernardinia Planch. = Rourea Aubl. CONNARACEAE (D)
Berneuxia Decne. DIAPENSIACEAE (D)
Bernhardia Post & Kuntze (SUO) = Bernardia Mill. EUPHORBIACEAE (D)
Bernhardia Willd. ex Bernh. = Psilotum Sw. PSILOTACEAE (FA)
Bernieria Baill. = Beilschmiedia Nees LAURACEAE (D)
Bernoullia Oliv. BOMBACACEAE (D)
Berrisfordia L.Bolus AIZOACEAE (D)
Berroa Beauverd COMPOSITAE (D)
Berrya Roxb. TILIACEAE (D)
Bersama Fres. MELIANTHACEAE (D)
Berteroa DC. CRUCIFERAE (D)
Berteroella O.E.Schulz CRUCIFERAE (D)
Berthiera Vent. (SUO) = Bertiera Aubl. RUBIACEAE (D)
Bertholletia Bonpl. LECYTHIDACEAE (D)
Bertiera Aubl. RUBIACEAE (D)
Bertolonia Raddi MELASTOMATACEAE (D)
Bertuchia Dennst. = Gardenia Ellis RUBIACEAE (D)
Bertya Planch. EUPHORBIACEAE (D)
Berula Besser & W.D.J.Koch UMBELLIFERAE (D)
Beruniella Zak. & Nabiev = Heliotropium L. BORAGINACEAE (D)
Berzelia Brongn. BRUNIACEAE (D)
Beschorneria Kunth AGAVACEAE (M)
Besleria L. GESNERIACEAE (D)
Bessera Schult. ALLIACEAE (M)
Bessera Spreng. (SUH) = Flueggea Willd. EUPHORBIACEAE (D)
Besseya Rydb. SCROPHULARIACEAE (D)
Bestram Adans. = Antidesma L. EUPHORBIACEAE (D)
Beta L. CHENOPODIACEAE (D)
Betchea Schltr. = Caldcluvia D.Don CUNONIACEAE (D)
Betonica L. = Stachys L. LABIATAE (D)
Betula L. BETULACEAE (D)
Beurreria Jacq. (SUO) = Bourreria P.Browne BORAGINACEAE (D)
Bewsia Goossens GRAMINEAE (M)
Beyeria Miq. EUPHORBIACEAE (D)
Beyeriopsis Müll.Arg. = Beyeria Miq. EUPHORBIACEAE (D)
Beyrichia Cham. & Schltdl. = Achetaria Cham. & Schltdl. SCROPHULARIACEAE (D)

Bhesa Buch.–Ham. ex Arn. CELASTRACEAE (D)
Bhidea Stapf ex Bor GRAMINEAE (M)
Bia Klotzsch = Tragia L. EUPHORBIACEAE (D)
Biancaea Tod. = Caesalpinia L. LEGUMINOSAE–CAESALPINIOIDEAE (D)
Biarum Schott ARACEAE (M)
Biaslia Vand. = Mayaca Aubl. MAYACACEAE (M)
Biasolettia W.D.J.Koch (SUH) = Geocaryum Coss. UMBELLIFERAE (D)
Biatherium Desv. = Gymnopogon P.Beauv. GRAMINEAE (M)
Bicchia Parl. = Habenaria Willd. ORCHIDACEAE (M)
Bicornella Lindl. = Cynorkis Thouars ORCHIDACEAE (M)
Bicuspidaria Rydb. = Mentzelia L. LOASACEAE (D)
Bidens L. COMPOSITAE (D)
Biebersteinia Stephan GERANIACEAE (D)
Bieneria Rchb.f. = Chloraea Lindl. ORCHIDACEAE (M)
Bienertia Bunge ex Boiss. CHENOPODIACEAE (D)
Biermannia King & Pantl. ORCHIDACEAE (M)
Bifaria (Hack.) Kuntze (SUH) = Mesosetum Steud. GRAMINEAE (M)
Bifaria Tiegh. = Korthalsella Tiegh. VISCACEAE (D)
Bifolium Nieuwl. = Listera R.Br. ORCHIDACEAE (M)
Bifora Hoffm. UMBELLIFERAE (D)
Bifrenaria Lindl. ORCHIDACEAE (M)
Bigelovia Sm. = Forestiera Poir. OLEACEAE (D)
Bigelovia Spreng. (SUH) = Spermacoce L. RUBIACEAE (D)
Bigelowia DC. (SUH) = Spermacoce L. RUBIACEAE (D)
Bigelowia DC. COMPOSITAE (D)
Biglandularia Seem. = Sinningia Nees GESNERIACEAE (D)
Bignonia L. BIGNONIACEAE (D)
Bijlia N.E.Br. AIZOACEAE (D)
Bikkia Reinw. RUBIACEAE (D)
Bikkiopsis Brongn. & Gris = Bikkia Reinw. RUBIACEAE (D)
Bilabium Miq. = Chirita Buch.–Ham. ex D.Don GESNERIACEAE (D)
Bilabrella Lindl. = Habenaria Willd. ORCHIDACEAE (M)
Bilacunaria Pimenov & V.N.Tikhom. UMBELLIFERAE (D)
Bilderdykia Dumort. = Fallopia Adans. POLYGONACEAE (D)
Bilegnum Brand = Rindera Pall. BORAGINACEAE (D)
Billardiera Vahl (SUH) = Coussarea Aubl. RUBIACEAE (D)
Billardiera Sm. PITTOSPORACEAE (D)
Billbergia Thunb. BROMELIACEAE (M)
Billia Peyr. HIPPOCASTANACEAE (D)
Billieturnera Fryxell MALVACEAE (D)
Billiotia G.Don (SUO) = Melanopsidium Colla RUBIACEAE (D)
Billiottia DC. = Melanopsidium Colla RUBIACEAE (D)
Billotia R.Br. ex G.Don = Agonis (DC.) Sweet MYRTACEAE (D)
Billottia Colla = Calothamnus Labill. MYRTACEAE (D)
Billya Cass. (SUH) = Petalacte D.Don COMPOSITAE (D)
Binectaria Forssk. = Mimusops L. SAPOTACEAE (D)
Bingeria A.Chev. = Turraeanthus Baill. MELIACEAE (D)
Binghamia Britton & Rose (SUH) = Espostoa Britton & Rose CACTACEAE (D)
Binia Noronha ex Thouars (SUI) = Noronhia Stadman ex Thouars OLEACEAE (D)
Binotia Rolfe ORCHIDACEAE (M)
Biondia Schltr. (UP) ASCLEPIADACEAE (D)
Biophytum DC. OXALIDACEAE (D)
Biota (D.Don) Endl. = Thuja L. CUPRESSACEAE (G)
Biovularia Kamieński = Utricularia L. LENTIBULARIACEAE (D)
Bipinnula Comm. ex Juss. ORCHIDACEAE (M)

Bipontia S.F.Blake = Soaresia Sch.Bip. COMPOSITAE (D)
Birchea A.Rich. = Luisia Gaudich. ORCHIDACEAE (M)
Biropteris Kümmerle = Asplenium L. ASPLENIACEAE (F)
Bisboeckelera Kuntze CYPERACEAE (M)
Bischofia Blume EUPHORBIACEAE (D)
Biscutella L. CRUCIFERAE (D)
Bisedmondia Hutch. = Calycophysum Karst. & Triana CUCURBITACEAE (D)
Biserrula L. LEGUMINOSAE–PAPILIONOIDEAE (D)
Bisglaziovia Cogn. MELASTOMATACEAE (D)
Bisgoeppertia Kuntze GENTIANACEAE (D)
Bishopalea H.Rob. COMPOSITAE (D)
Bishopanthus H.Rob. COMPOSITAE (D)
Bishopiella R.M.King & H.Rob. COMPOSITAE (D)
Bishovia R.M.King & H.Rob. COMPOSITAE (D)
Bismarckia Hildebrandt & H.A.Wendl. PALMAE (M)
Bisnaga Orcutt = Ferocactus Britton & Rose CACTACEAE (D)
Bisnicholsonia Kuntze = Neonicholsonia Dammer PALMAE (M)
Bisquamaria Pichon = Laxoplumeria Markgr. APOCYNACEAE (D)
Bisrautanenia Kuntze = Neorautanenia Schinz LEGUMINOSAE–PAPILIONOIDEAE (D)
Bissea V.R.Fuentes (SUS) = Henoonia Griseb. GOETZEACEAE (D)
Bistella Adans. = Vahlia Thunb. VAHLIACEAE (D)
Bistorta (L.) Adans. POLYGONACEAE (D)
Biswarea Cogn. CUCURBITACEAE (D)
Bituminaria Heist. ex Fabr. LEGUMINOSAE–PAPILIONOIDEAE (D)
Biventraria Small = Asclepias L. ASCLEPIADACEAE (D)
Bivinia Jaub. ex Tul. FLACOURTIACEAE (D)
Bivonaea DC. CRUCIFERAE (D)
Bivonea Raf. (SUH) = Cnidoscolus Pohl EUPHORBIACEAE (D)
Bivonia Spreng. (SUH) = Bernardia Mill. EUPHORBIACEAE (D)
Bixa L. BIXACEAE (D)
Bizonula Pellegr. SAPINDACEAE (D)
Blabea Baehni = Pouteria Aubl. SAPOTACEAE (D)
Blaberopus A.DC. = Alstonia R.Br. APOCYNACEAE (D)
Blachia Baill. EUPHORBIACEAE (D)
Blackallia C.A.Gardner RHAMNACEAE (D)
Blackiella Aellen = Atriplex L. CHENOPODIACEAE (D)
Blackstonia Huds. GENTIANACEAE (D)
Blaeria L. = Erica L. ERICACEAE (D)
Blainvillea Cass. COMPOSITAE (D)
Blakea P.Browne MELASTOMATACEAE (D)
Blakeanthus R.M.King & H.Rob. COMPOSITAE (D)
Blakeochloa Veldkamp (SUS) = Plinthanthesis Steud. GRAMINEAE (M)
Blakiella Cuatrec. COMPOSITAE (D)
Blanchetia DC. COMPOSITAE (D)
Blanchetiastrum Hassl. MALVACEAE (D)
Blancoa Blume (SUH) = Arenga Labill. PALMAE (M)
Blancoa Lindl. HAEMODORACEAE (M)
Blandfordia Sm. BLANDFORDIACEAE (M)
Blandibractea Wernham RUBIACEAE (D)
Blandowia Willd. = Apinagia Tul. PODOSTEMACEAE (D)
Blastania Kotschy & Peyr. = Ctenolepis Hook.f. CUCURBITACEAE (D)
Blastemanthus Planch. OCHNACEAE (D)
Blastocaulon Ruhland ERIOCAULACEAE (M)
Blastus Lour. MELASTOMATACEAE (D)
Bleasdalea F.Muell. ex Domin PROTEACEAE (D)

Blechnidium T.Moore = Blechnum L. BLECHNACEAE (F)
Blechnopsis C.Presl = Blechnum L. BLECHNACEAE (F)
Blechnum L. BLECHNACEAE (F)
Blechum P.Browne ACANTHACEAE (D)
Bleekeria Miq. (SUH) = Alchornea Sw. EUPHORBIACEAE (D)
Bleekeria Hassk. = Ochrosia Juss. APOCYNACEAE (D)
Bleekrodea Blume MORACEAE (D)
Blennoderma Spach = Oenothera L. ONAGRACEAE (D)
Blennodia R.Br. CRUCIFERAE (D)
Blennosperma Less. COMPOSITAE (D)
Blennospora A.Gray COMPOSITAE (D)
Blenocoes Raf. (SUS) = Nicotiana L. SOLANACEAE (D)
Blepharacanthus Nees ex Lindl. = Blepharis Juss. ACANTHACEAE (D)
Blepharandra Griseb. MALPIGHIACEAE (D)
Blepharanthera Schltr. = Brachystelma R.Br. ASCLEPIADACEAE (D)
Blepharanthes Sm. (SUI) = Adenia Forssk. PASSIFLORACEAE (D)
Blepharidachne Hack. GRAMINEAE (M)
Blepharidium Standl. RUBIACEAE (D)
Blephariglottis Raf. (SUH) = Platanthera Rich. ORCHIDACEAE (M)
Blepharipappus Hook. COMPOSITAE (D)
Blepharis Juss. ACANTHACEAE (D)
Blepharispermum DC. COMPOSITAE (D)
Blepharistemma Wall. ex Benth. RHIZOPHORACEAE (D)
Blepharitheca Pichon = Cuspidaria DC. BIGNONIACEAE (D)
Blepharizonia (A.Gray) Greene COMPOSITAE (D)
Blepharocalyx O.Berg MYRTACEAE (D)
Blepharocarya F.Muell. ANACARDIACEAE (D)
Blepharochloa Endl. = Leersia Sw. GRAMINEAE (M)
Blepharodon Decne. ASCLEPIADACEAE (D)
Blepharolepis Nees = Scirpus L. CYPERACEAE (M)
Blepharoneuron Nash GRAMINEAE (M)
Blepharostemma Fourr. = Asperula L. RUBIACEAE (D)
Blephilia Raf. LABIATAE (D)
Bletia Ruíz & Pav. ORCHIDACEAE (M)
Bletiana Raf. = Bletia Ruíz & Pav. ORCHIDACEAE (M)
Bletilla Rchb.f. ORCHIDACEAE (M)
Blighia K.König SAPINDACEAE (D)
Blighiopsis Veken SAPINDACEAE (D)
Blinkworthia Choisy CONVOLVULACEAE (D)
Blomia Miranda SAPINDACEAE (D)
Bloomeria Kellogg ALLIACEAE (M)
Blossfeldia Werderm. CACTACEAE (D)
Blotia Leandri EUPHORBIACEAE (D)
Blotiella R.M.Tryon DENNSTAEDTIACEAE (F)
Bluffia Nees = Alloteropsis C.Presl GRAMINEAE (M)
Blumea DC. COMPOSITAE (D)
Blumella Tiegh. = Elytranthe (Blume) Blume LORANTHACEAE (D)
Blumenbachia Koeler (SUH) = Sorghum Moench GRAMINEAE (M)
Blumenbachia Schrad. LOASACEAE (D)
Blumeodendron (Müll.Arg.) Kurz EUPHORBIACEAE (D)
Blumeopsis Gagnep. COMPOSITAE (D)
Blutaparon Raf. AMARANTHACEAE (D)
Blysmocarex N.Ivanova = Kobresia Willd. CYPERACEAE (M)
Blysmopsis Oteng-Yeb. = Blysmus Panz. ex Schult. CYPERACEAE (M)
Blysmus Panz. ex Schult. CYPERACEAE (M)

Blyttia Fr. (SUH) = Cinna L. GRAMINEAE (M)
Blyttia Arn. ASCLEPIADACEAE (D)
Blyxa Noronha ex Thouars HYDROCHARITACEAE (M)
Boaria A.DC. = Maytenus Molina CELASTRACEAE (D)
Bobartia L. IRIDACEAE (M)
Bobea Gaudich. RUBIACEAE (D)
Bocagea A.St.-Hil. ANNONACEAE (D)
Bocageopsis R.E.Fr. ANNONACEAE (D)
Bocconia L. PAPAVERACEAE (D)
Bocoa Aubl. LEGUMINOSAE–PAPILIONOIDEAE (D)
Bocquillonia Baill. EUPHORBIACEAE (D)
Boea Comm. ex Lam. GESNERIACEAE (D)
Boebera Willd. = Dyssodia Cav. COMPOSITAE (D)
Boeberastrum (A.Gray) Rydb. COMPOSITAE (D)
Boeberoides (DC.) Strother COMPOSITAE (D)
Boechera Á.Löve & D.Löve = Arabis L. CRUCIFERAE (D)
Boeckeleria T.Durand = Tetraria P.Beauv. CYPERACEAE (M)
Boehmeria Jacq. URTICACEAE (D)
Boehmeriopsis Kom. = Fatoua Gaudich. MORACEAE (D)
Boeica T.Anderson ex C.B.Clarke GESNERIACEAE (D)
Boeicopsis H.W.Li = Boeica T.Anderson ex C.B.Clarke GESNERIACEAE (D)
Boenninghausenia Rchb. ex Meisn. RUTACEAE (D)
Boerhavia L. NYCTAGINACEAE (D)
Boerlagea Cogn. MELASTOMATACEAE (D)
Boerlagella Pierre ex Cogn. (UP) SAPOTACEAE (D)
Boerlagiodendron Harms = Osmoxylon Miq. ARALIACEAE (D)
Boesenbergia Kuntze ZINGIBERACEAE (M)
Bogenhardia Rchb. = Herissantia Medik. MALVACEAE (D)
Bognera Mayo & Nicolson ARACEAE (M)
Bogoria J.J.Sm. ORCHIDACEAE (M)
Boholia Merr. RUBIACEAE (D)
Boisduvalia Spach ONAGRACEAE (D)
Boissiera Hochst. ex Steud. GRAMINEAE (M)
Boivinella A.Camus = Cyphochlaena Hack. GRAMINEAE (M)
Boivinella Pierre ex Aubrév. & Pellegr. (SUH) = Englerophytum K.Krause
    SAPOTACEAE (D)
Bojeria DC. = Inula L. COMPOSITAE (D)
Bojeria Raf. (SUH) = Euphorbia L. EUPHORBIACEAE (D)
Bokkeveldia D.Müll.-Doblies & U.Müll.-Doblies AMARYLLIDACEAE (M)
Bolandra A.Gray SAXIFRAGACEAE (D)
Bolanosa A.Gray COMPOSITAE (D)
Bolanthus (Ser.) Rchb. CARYOPHYLLACEAE (D)
Bolax Comm. ex Juss. UMBELLIFERAE (D)
Bolbidium Lindl. = Dendrobium Sw. ORCHIDACEAE (M)
Bolbitis Schott LOMARIOPSIDACEAE (F)
Bolborchis Lindl. = Coelogyne Lindl. ORCHIDACEAE (M)
Bolboschoenus (Asch.) Palla CYPERACEAE (M)
Bolbostemma Franquet CUCURBITACEAE (D)
Bolboxalis Small = Oxalis L. OXALIDACEAE (D)
Boldoa Cav. ex Lag. NYCTAGINACEAE (D)
Bolelia Raf. = Downingia Torr. CAMPANULACEAE (D)
Boleum Desv. CRUCIFERAE (D)
Bolivaria Cham. & Schltdl. = Menodora Bonpl. OLEACEAE (D)
Bolivicereus Cárdenas = Cleistocactus Lem. CACTACEAE (D)
Bollea Rchb.f. ORCHIDACEAE (M)

Bollwilleria Zabel (SUI) = Sorbopyrus C.K.Schneid. ROSACEAE (D)
Bolocephalus Hand.-Mazz. = Dolomiaea DC. COMPOSITAE (D)
Bolophyta Nutt. COMPOSITAE (D)
Boltonia L'Hér. COMPOSITAE (D)
Bolusafra Kuntze LEGUMINOSAE–PAPILIONOIDEAE (D)
Bolusanthemum Schwantes = Bijlia N.E.Br. AIZOACEAE (D)
Bolusanthus Harms LEGUMINOSAE–PAPILIONOIDEAE (D)
Bolusia Benth. LEGUMINOSAE–PAPILIONOIDEAE (D)
Bolusiella Schltr. ORCHIDACEAE (M)
Bomarea Mirb. ALSTROEMERIACEAE (M)
Bombacopsis Pittier BOMBACACEAE (D)
Bombax L. BOMBACACEAE (D)
Bombycidendron Zoll. & Moritzi MALVACEAE (D)
Bombycilaena (DC.) Smoljan. COMPOSITAE (D)
Bommeria Fourn. ADIANTACEAE (F)
Bonafousia A.DC. = Tabernaemontana L. APOCYNACEAE (D)
Bonamia Thouars CONVOLVULACEAE (D)
Bonamica Vell. = Chionanthus L. OLEACEAE (D)
Bonania A.Rich. EUPHORBIACEAE (D)
Bonannia Guss. UMBELLIFERAE (D)
Bonanox Raf. = Ipomoea L. CONVOLVULACEAE (D)
Bonatea Willd. ORCHIDACEAE (M)
Bonatia Schltr. & K.Krause (SUH) = Tarenna Gaertn. RUBIACEAE (D)
Bonaveria Scop. = Coronilla L. LEGUMINOSAE–PAPILIONOIDEAE (D)
Bonellia Bertero ex Colla = Jacquinia L. THEOPHRASTACEAE (D)
Bonetiella Rzed. ANACARDIACEAE (D)
Bongardia C.A.Mey. BERBERIDACEAE (D)
Bonia Balansa = Bambusa Schreb. GRAMINEAE (M)
Bonifacia Silva Manso ex Steud. = Augusta Pohl RUBIACEAE (D)
Bonifazia Standl. & Steyerm. = Disocactus Lindl. CACTACEAE (D)
Boninia Planch. RUTACEAE (D)
Boniniella Hayata = Asplenium L. ASPLENIACEAE (F)
Boninofatsia Nakai = Fatsia Decne. & Planch. ARALIACEAE (D)
Boniodendron Gagnep. SAPINDACEAE (D)
Bonnaya Link & Otto = Lindernia All. SCROPHULARIACEAE (D)
Bonnayodes Blatter & Hallberg = Limnophila R.Br. SCROPHULARIACEAE (D)
Bonnetia Mart. GUTTIFERAE (D)
Bonniera Cordem. ORCHIDACEAE (M)
Bonnierella R.Vig. = Polyscias J.R.Forst. & G.Forst. ARALIACEAE (D)
Bonplandia Cav. POLEMONIACEAE (D)
Bontia L. MYOPORACEAE (D)
Bonyunia Schomb. ex Progel LOGANIACEAE (D)
Boophone Herb. AMARYLLIDACEAE (M)
Boopis Juss. CALYCERACEAE (D)
Boottia Wall. = Ottelia Pers. HYDROCHARITACEAE (M)
Bopusia C.Presl = Graderia Benth. SCROPHULARIACEAE (D)
Boquila Decne. LARDIZABALACEAE (D)
Boraginella Kuntze = Trichodesma R.Br. BORAGINACEAE (D)
Borago L. BORAGINACEAE (D)
Borassodendron Becc. PALMAE (M)
Borassus L. PALMAE (M)
Borbonia L. = Aspalathus L. LEGUMINOSAE–PAPILIONOIDEAE (D)
Borderea Miègev. DIOSCOREACEAE (M)
Boreava Jaub. & Spach CRUCIFERAE (D)
Boriskelleru Terekhov = Eragrostis Wolf GRAMINEAE (M)

Borismene Barneby MENISPERMACEAE (D)
Borkonstia Ignatov = Aster L. COMPOSITAE (D)
Borneacanthus Bremek. ACANTHACEAE (D)
Borneodendron Airy Shaw EUPHORBIACEAE (D)
Bornmuellera Hausskn. CRUCIFERAE (D)
Borodinia N.Busch CRUCIFERAE (D)
Borojoa Cuatrec. RUBIACEAE (D)
Boronella Baill. RUTACEAE (D)
Boronia Sm. RUTACEAE (D)
Borrachinea Lavy = Borago L. BORAGINACEAE (D)
Borraginoides Boehm. = Trichodesma R.Br. BORAGINACEAE (D)
Borrera Spreng. (SUO) = Spermacoce L. RUBIACEAE (D)
Borreria G.Mey. = Spermacoce L. RUBIACEAE (D)
Borrichia Adans. COMPOSITAE (D)
Borsczowia Bunge CHENOPODIACEAE (D)
Borthwickia W.W.Sm. CAPPARACEAE (D)
Borya Willd. (SUS) = Forestiera Poir. OLEACEAE (D)
Borya Labill. ANTHERICACEAE (M)
Borzicactella F.Ritter = Cleistocactus Lem. CACTACEAE (D)
Borzicactus Riccob. = Cleistocactus Lem. CACTACEAE (D)
Borzicereus Frič & Kreuz. (SUO) = Cleistocactus Lem. CACTACEAE (D)
Boschia Korth. BOMBACACEAE (D)
Boschniakia C.A.Mey. ex Bong. SCROPHULARIACEAE (D)
Boscia Lam. CAPPARACEAE (D)
Bosea L. AMARANTHACEAE (D)
Bosistoa F.Muell. RUTACEAE (D)
Bosleria A.Nelson = Solanum L. SOLANACEAE (D)
Bosqueia Thouars ex Baill. = Trilepisium Thouars MORACEAE (D)
Bosqueiopsis De Wild. & T.Durand MORACEAE (D)
Bosscheria Vriese & Teijsm. = Ficus L. MORACEAE (D)
Bossera Leandri = Alchornea Sw. EUPHORBIACEAE (D)
Bossiaea Vent. LEGUMINOSAE–PAPILIONOIDEAE (D)
Bostrychanthera Benth. LABIATAE (D)
Bostrychode Miq. ex O.Berg = Syzygium Gaertn. MYRTACEAE (D)
Boswellia Roxb. ex Colebr. BURSERACEAE (D)
Botelua Lag. (SUO) = Bouteloua Lag. GRAMINEAE (M)
Bothriochilus Lem. ORCHIDACEAE (M)
Bothriochloa Kuntze GRAMINEAE (M)
Bothriocline Oliv. ex Benth. COMPOSITAE (D)
Bothriopodium Rizzini = Urbanolophium Melch. BIGNONIACEAE (D)
Bothriospermum Bunge BORAGINACEAE (D)
Bothriospora Hook.f. RUBIACEAE (D)
Botryanthe Klotzsch = Fragariopsis A.St.–Hil. EUPHORBIACEAE (D)
Botryanthus Kunth = Muscari Mill. HYACINTHACEAE (M)
Botryarrhena Ducke RUBIACEAE (D)
Botrycarpum A.Rich. = Ribes L. GROSSULARIACEAE (D)
Botryceras Willd. = Laurophyllus Thunb. ANACARDIACEAE (D)
Botrychium Sw. OPHIOGLOSSACEAE (F)
Botryocarpium Spach = Ribes L. GROSSULARIACEAE (D)
Botryodendrum Endl. = Meryta J.R.Forst. & G.Forst. ARALIACEAE (D)
Botryogramme Fée = Llavea Lag. ADIANTACEAE (F)
Botryoloranthus (Engl. & K.Krause) Balle = Oedina Tiegh. LORANTHACEAE (D)
Botryomeryta R.Vig. = Meryta J.R.Forst. & G.Forst. ARALIACEAE (D)
Botryopanax Miq. = Polyscias J.R.Forst. & G.Forst. ARALIACEAE (D)
Botryophora Hook.f. EUPHORBIACEAE (D)

Botryopleuron Hemsl. = Veronicastrum Heist. ex Fabr. SCROPHULARIACEAE (D)
Botryopteris C.Presl = Helminthostachys Kaulf. OPHIOGLOSSACEAE (F)
Botryospora B.D.Jacks. (SUO) = Botryophora Hook.f. EUPHORBIACEAE (D)
Botryostege Stapf ERICACEAE (D)
Botryphora Post & Kuntze (SUO) = Botryophora Hook.f. EUPHORBIACEAE (D)
Botrypus Michx. = Botrychium Sw. OPHIOGLOSSACEAE (F)
Botschantzevia Nabiev CRUCIFERAE (D)
Bottegoa Chiov. SAPINDACEAE (D)
Bottionea Colla = Trichopetalum Lindl. ANTHERICACEAE (M)
Boucerosia Wight & Arn. = Caralluma R.Br. ASCLEPIADACEAE (D)
Bouchardatia Baill. RUTACEAE (D)
Bouchea Cham. VERBENACEAE (D)
Bouchetia Dunal SOLANACEAE (D)
Bouea Meisn. ANACARDIACEAE (D)
Bouetia A.Chev. = Hemizygia (Benth.) Briq. LABIATAE (D)
Bougainvillea Comm. ex Juss. NYCTAGINACEAE (D)
Bougueria Decne. PLANTAGINACEAE (D)
Bourdaria A.Chev. = Cincinnobotrys Gilg MELASTOMATACEAE (D)
Bourdonia Greene = Chaetopappa DC. COMPOSITAE (D)
Bourjotia Pomel = Heliotropium L. BORAGINACEAE (D)
Bournea Oliv. GESNERIACEAE (D)
Bourreria P.Browne BORAGINACEAE (D)
Bousigonia Pierre APOCYNACEAE (D)
Boussingaultia Kunth BASELLACEAE (D)
Bouteloua Lag. GRAMINEAE (M)
Boutiquea Le Thomas ANNONACEAE (D)
Boutonia DC. ACANTHACEAE (D)
Boutonia Bojer ex Baill. (SUH) = Cordemoya Baill. EUPHORBIACEAE (D)
Bouvardia Salisb. RUBIACEAE (D)
Bouzetia Montrouz. (UP) RUTACEAE (D)
Bovonia Chiov. LABIATAE (D)
Bowdichia Kunth LEGUMINOSAE–PAPILIONOIDEAE (D)
Bowenia Hook. BOWENIACEAE (G)
Bowiea Harv. ex Hook.f. HYACINTHACEAE (M)
Bowkeria Harv. SCROPHULARIACEAE (D)
Bowlesia Ruíz & Pav. UMBELLIFERAE (D)
Bowringia Hook. (SUH) = Brainea J.Sm. BLECHNACEAE (F)
Bowringia Champ. ex Benth. LEGUMINOSAE–PAPILIONOIDEAE (D)
Boyania Wurdack MELASTOMATACEAE (D)
Boykinia Nutt. SAXIFRAGACEAE (D)
Braasiella Braem, Lückel & Russmann = Oncidium Sw. ORCHIDACEAE (M)
Brabejum L. PROTEACEAE (D)
Bracea Britton (SUH) = Neobracea Britton APOCYNACEAE (D)
Bracea King = Sarcosperma Hook.f. SAPOTACEAE (D)
Brachanthemum DC. COMPOSITAE (D)
Brachatera Desv. (SUS) = Danthonia DC. GRAMINEAE (M)
Brachiaria (Trin.) Griseb. GRAMINEAE (M)
Brachionidium Lindl. ORCHIDACEAE (M)
Brachionostylum Mattf. COMPOSITAE (D)
Brachiostemon Hand.–Mazz. = Ornithoboea Parish ex C.B.Clarke GESNERIACEAE (D)
Brachistus Miers SOLANACEAE (D)
Brachtia Rchb.f. ORCHIDACEAE (M)
Brachyachne (Benth.) Stapf GRAMINEAE (M)
Brachyandra Phil. = Helogyne Nutt. COMPOSITAE (D)
Brachyapium (Baill.) Maire – Stoibrax Raf. UMBELLIFERAE (D)

Brachyathera Kuntze (SUS) = Danthonia DC. GRAMINEAE (M)
Brachybotrys Maxim. ex Oliv. BORAGINACEAE (D)
Brachycalycium Backeb. = Gymnocalycium Pfeiff. CACTACEAE (D)
Brachycarpaea DC. CRUCIFERAE (D)
Brachycaulos Dixit & Panigrahi = Chamaerhodos Bunge ROSACEAE (D)
Brachycentrum Meisn. = Centronia D.Don MELASTOMATACEAE (D)
Brachycereus Britton & Rose CACTACEAE (D)
Brachychaeta Torr. & A.Gray = Solidago L. COMPOSITAE (D)
Brachychilum (R.Br. ex Wall.) O.Petersen = Hedychium J.König ZINGIBERACEAE (M)
Brachychiton Schott & Endl. STERCULIACEAE (D)
Brachychloa S.M.Phillips GRAMINEAE (M)
Brachyclados Gillies ex D.Don COMPOSITAE (D)
Brachycodon Fedorov = Campanula L. CAMPANULACEAE (D)
Brachycome Cass. (SUO) = Brachyscome Cass. COMPOSITAE (D)
Brachycorythis Lindl. ORCHIDACEAE (M)
Brachycylix (Harms) R.S.Cowan LEGUMINOSAE–CAESALPINIOIDEAE (D)
Brachycyrtis Koidz. = Tricyrtis Wall. CONVALLARIACEAE (M)
Brachyelytrum P.Beauv. GRAMINEAE (M)
Brachyglottis J.R.Forst. & G.Forst. COMPOSITAE (D)
Brachyhelus (Benth.) Post & Kuntze = Schwenckia L. SOLANACEAE (D)
Brachylaena R.Br. COMPOSITAE (D)
Brachylepis Hook. & Arn. (SUH) = Melinia Decne. ASCLEPIADACEAE (D)
Brachylepis Wight & Arn. (SUH) = Baeolepis Decne. ex Moq. ASCLEPIADACEAE (D)
Brachylepis C.A.Mey. CHENOPODIACEAE (D)
Brachyloma Hanst. (SUH) = Kohleria Regel GESNERIACEAE (D)
Brachyloma Sond. EPACRIDACEAE (D)
Brachylophon Oliv. MALPIGHIACEAE (D)
Brachymeris DC. = Phymaspermum Less. COMPOSITAE (D)
Brachynema F.Muell. (SUH) = Abrophyllum Hook.f. ex Benth. ESCALLONIACEAE (D)
Brachynema Benth. OLACACEAE (D)
Brachyonostylum Mattf. (SUO) = Brachionostylum Mattf. COMPOSITAE (D)
Brachyotum (DC.) Triana MELASTOMATACEAE (D)
Brachypeza Garay ORCHIDACEAE (M)
Brachyphragma Rydb. = Astragalus L. LEGUMINOSAE–PAPILIONOIDEAE (D)
Brachypodium P.Beauv. GRAMINEAE (M)
Brachypremna Gleason = Ernestia DC. MELASTOMATACEAE (D)
Brachypterys A.Juss. = Stigmaphyllon A.Juss. MALPIGHIACEAE (D)
Brachyscome Cass. COMPOSITAE (D)
Brachyscypha Baker = Lachenalia J.F.Jacq. ex Murray HYACINTHACEAE (M)
Brachysema R.Br. LEGUMINOSAE–PAPILIONOIDEAE (D)
Brachysiphon A.Juss. PENAEACEAE (D)
Brachysorus C.Presl = Diplazium Sw. WOODSIACEAE (F)
Brachystachys Klotzsch = Croton L. EUPHORBIACEAE (D)
Brachystachyum Keng = Semiarundinaria Nakai GRAMINEAE (M)
Brachystegia Benth. LEGUMINOSAE–CAESALPINIOIDEAE (D)
Brachystele Schltr. ORCHIDACEAE (M)
Brachystelma R.Br. ASCLEPIADACEAE (D)
Brachystelmaria Schltr. = Brachystelma R.Br. ASCLEPIADACEAE (D)
Brachystemma D.Don CARYOPHYLLACEAE (D)
Brachystephanus Nees ACANTHACEAE (D)
Brachystepis Pritz. = Oeonia Lindl. ORCHIDACEAE (M)
Brachystigma Pennell = Agalinis Raf. SCROPHULARIACEAE (D)
Brachystylus Dulac (SUS) = Koeleria Pers. GRAMINEAE (M)
Brachythalamus Gilg = Gyrinops Gaertn. THYMELAEACEAE (D)
Brachythrix Wild & G.V.Pope COMPOSITAE (D)

Brachytome Hook.f. RUBIACEAE (D)
Bracisepalum J.J.Sm. ORCHIDACEAE (M)
Brackenridgea A.Gray OCHNACEAE (D)
Braconotia Godr. (SUS) = Elymus L. GRAMINEAE (M)
Bracteantha Anderb. COMPOSITAE (D)
Bracteanthus Ducke MONIMIACEAE (D)
Bracteola Swallen (SUI) = Chrysochloa Swallen GRAMINEAE (M)
Bracteolanthus de Wit = Bauhinia L. LEGUMINOSAE–CAESALPINIOIDEAE (D)
Bradburia Torr. & A.Gray COMPOSITAE (D)
Bradburya Raf. = Centrosema (DC.) Benth. LEGUMINOSAE–PAPILIONOIDEAE (D)
Bradea Standl. ex Brade RUBIACEAE (D)
Bradleia Cav. (SUO) = Glochidion J.R.Forst. & G.Forst. EUPHORBIACEAE (D)
Bradleja Banks ex Gaertn. = Glochidion J.R.Forst. & G.Forst. EUPHORBIACEAE (D)
Braemia Jenny = Houlletia Brongn. ORCHIDACEAE (M)
Bragantia Lour. (SUH) = Thottea Rottb. ARISTOLOCHIACEAE (D)
Bragantia Vand. = Gomphrena L. AMARANTHACEAE (D)
Brahea Mart. ex Endl. PALMAE (M)
Brainea J.Sm. BLECHNACEAE (F)
Brami Adans. = Bacopa Aubl. SCROPHULARIACEAE (D)
Bramia Lam. = Bacopa Aubl. SCROPHULARIACEAE (D)
Brandegea Cogn. CUCURBITACEAE (D)
Brandella R.R.Mill BORAGINACEAE (D)
Brandesia Mart. = Alternanthera Forssk. AMARANTHACEAE (D)
Brandisia Hook.f. & Thomson SCROPHULARIACEAE (D)
Brandtia Kunth = Arundinella Raddi GRAMINEAE (M)
Brasenia Schreb. CABOMBACEAE (D)
Brasilia G.M.Barroso = Calea L. COMPOSITAE (D)
Brasilicactus Backeb. (SUS) = Parodia Speg. CACTACEAE (D)
Brasilicereus Backeb. CACTACEAE (D)
Brasiliopuntia (K.Schum.) A.Berger = Opuntia Mill. CACTACEAE (D)
Brasiliparodia F.Ritter = Parodia Speg. CACTACEAE (D)
Brasilocactus Frič (SUI) = Parodia Speg. CACTACEAE (D)
Brasilocalamus Nakai = Merostachys Spreng. GRAMINEAE (M)
Brassaia Endl. = Schefflera J.R.Forst. & G.Forst. ARALIACEAE (D)
Brassaiopsis Decne. & Planch. ARALIACEAE (D)
Brassavola R.Br. ORCHIDACEAE (M)
Brassia R.Br. ORCHIDACEAE (M)
Brassiantha A.C.Sm. CELASTRACEAE (D)
Brassica L. CRUCIFERAE (D)
Brassicaria Pomel = Brassica L. CRUCIFERAE (D)
Brassicella Fourr. ex O.E.Schulz = Coincya Rouy CRUCIFERAE (D)
Brassiodendron C.K.Allen = Endiandra R.Br. LAURACEAE (D)
Brassiophoenix Burret PALMAE (M)
Braunblanquetia Eskuche SCROPHULARIACEAE (D)
Braunsia Schwantes AIZOACEAE (D)
Bravaisia DC. ACANTHACEAE (D)
Bravoa Lex. AGAVACEAE (M)
Braxireon Raf. AMARYLLIDACEAE (M)
Braya Sternb. & Hoppe CRUCIFERAE (D)
Brayera Kunth = Hagenia J.F.Gmel. ROSACEAE (D)
Brayopsis Gilg & Muschler CRUCIFERAE (D)
Brayulinea Small = Guilleminea Kunth AMARANTHACEAE (D)
Brazoria Engelm. & A.Gray LABIATAE (D)
Brazzeia Baill. SCYTOPETALACEAE (D)
Brebissonia Spach = Fuchsia L. ONAGRACEAE (D)

Bredemeyera Willd. POLYGALACEAE (D)
Bredia Blume MELASTOMATACEAE (D)
Breitungia Á.Löve & D.Löve = Sedum L. CRASSULACEAE (D)
Bremekampia Sreem. (SUS) = Haplanthodes Kuntze ACANTHACEAE (D)
Bremontiera DC. = Indigofera L. LEGUMINOSAE–PAPILIONOIDEAE (D)
Brenania Keay RUBIACEAE (D)
Brenesia Schltr. = Pleurothallis R.Br. ORCHIDACEAE (M)
Brenierea Humbert LEGUMINOSAE–CAESALPINIOIDEAE (D)
Breonadia Ridsdale RUBIACEAE (D)
Breonia A.Rich. ex DC. RUBIACEAE (D)
Bretschneidera Hemsl. BRETSCHNEIDERACEAE (D)
Breueria R.Br. = Bonamia Thouars CONVOLVULACEAE (D)
Breueriopsis Roberty = Bonamia Thouars CONVOLVULACEAE (D)
Breuieropsis Roberty (SUO) = Bonamia Thouars CONVOLVULACEAE (D)
Breviea Aubrév. & Pellegr. SAPOTACEAE (D)
Brevipodium Á.Löve & D.Löve = Brachypodium P.Beauv. GRAMINEAE (M)
Brevoortia A.Wood = Dichelostemma Kunth ALLIACEAE (M)
Brewcaria L.B.Sm., Steyerm. & H.E.Rob. BROMELIACEAE (M)
Brewerina A.Gray CARYOPHYLLACEAE (D)
Brexia Noronha ex Thouars ESCALLONIACEAE (D)
Brexiella H.Perrier CELASTRACEAE (D)
Brexiopsis H.Perrier = Drypetes Vahl EUPHORBIACEAE (D)
Breynia J.R.Forst. & G.Forst. EUPHORBIACEAE (D)
Breyniopsis Beille = Sauropus Blume EUPHORBIACEAE (D)
Brezia Moq. = Suaeda Forssk. ex Scop. CHENOPODIACEAE (D)
Brickellia Elliott COMPOSITAE (D)
Brickelliastrum R.M.King & H.Rob. COMPOSITAE (D)
Bridelia Willd. EUPHORBIACEAE (D)
Bridgesia Bertero ex Cambess. SAPINDACEAE (D)
Bridgesia Backeb. (SUH) = Rebutia K.Schum. CACTACEAE (D)
Briegeria Senghas ORCHIDACEAE (M)
Brieya De Wild. = Piptostigma Oliv. ANNONACEAE (D)
Briggsia Craib GESNERIACEAE (D)
Briggsiopsis K.Y.Pan GESNERIACEAE (D)
Brighamia A.Gray CAMPANULACEAE (D)
Brignolia DC. = Isertia Schreb. RUBIACEAE (D)
Brillantaisia P.Beauv. ACANTHACEAE (D)
Brimeura Salisb. HYACINTHACEAE (M)
Brintonia Greene = Solidago L. COMPOSITAE (D)
Briquetastrum Robyns & Lebrun = Leocus A.Chev. LABIATAE (D)
Briquetia Hochr. MALVACEAE (D)
Briquetina J.F.Macbr. = Citronella D.Don ICACINACEAE (D)
Britoa O.Berg = Campomanesia Ruíz & Pav. MYRTACEAE (D)
Brittenia Cogn. MELASTOMATACEAE (D)
Brittonastrum Briq. = Agastache Gronov. LABIATAE (D)
Brittonella Rusby = Mionandra Griseb. MALPIGHIACEAE (D)
Brittonia C.A.Armstr. (SUI) = Ferocactus Britton & Rose CACTACEAE (D)
Brittonrosea Speg. (SUS) = Echinocactus Link & Otto CACTACEAE (D)
Briza L. GRAMINEAE (M)
Brizochloa Jirásek & Chrtek = Briza L. GRAMINEAE (M)
Brizopyrum Stapf (SUH) = Tribolium Desv. GRAMINEAE (M)
Brizopyrum Link (SUS) = Desmazeria Dumort. GRAMINEAE (M)
Brizula Hieron. CENTROLEPIDACEAE (M)
Brocchia Mauri ex Ten. (SUH) = Simmondsia Nutt. SIMMONDSIACEAE (D)
Brocchinia Schult.f. BROMELIACEAE (M)

Brochoneura Warb. MYRISTICACEAE (D)
Brochosiphon Nees = Dicliptera Juss. ACANTHACEAE (D)
Brockmania W.Fitzg. = Hibiscus L. MALVACEAE (D)
Brodiaea Sm. ALLIACEAE (M)
Brodriguesia R.S.Cowan LEGUMINOSAE–CAESALPINIOIDEAE (D)
Brombya F.Muell. RUTACEAE (D)
Bromelia L. BROMELIACEAE (M)
Bromelica (Thurb.) Farw. = Melica L. GRAMINEAE (M)
Bromheadia Lindl. ORCHIDACEAE (M)
Bromidium Nees & Meyen = Agrostis L. GRAMINEAE (M)
Bromopsis (Dumort.) Fourr. = Bromus L. GRAMINEAE (M)
Bromuniola Stapf & C.E.Hubb. GRAMINEAE (M)
Bromus Scop. (SUI) = Triticum L. GRAMINEAE (M)
Bromus L. GRAMINEAE (M)
Brongniartia Kunth LEGUMINOSAE–PAPILIONOIDEAE (D)
Brongniartikentia Becc. PALMAE (M)
Brookea Benth. SCROPHULARIACEAE (D)
Brosimopsis S.Moore = Brosimum Sw. MORACEAE (D)
Brosimum Sw. MORACEAE (D)
Brossaea L. (SUO) = Gaultheria Kalm ex L. ERICACEAE (D)
Brossardia Boiss. CRUCIFERAE (D)
Brossea Kuntze (SUO) = Gaultheria Kalm ex L. ERICACEAE (D)
Brotera Spreng. (SUH) = Hyptis Jacq. LABIATAE (D)
Broughtonia R.Br. ORCHIDACEAE (M)
Brousemichea Balansa = Zoysia Willd. GRAMINEAE (M)
Broussaisia Gaudich. HYDRANGEACEAE (D)
Broussonetia L'Hér. ex Vent. MORACEAE (D)
Browallia L. SOLANACEAE (D)
Brownanthus Schwantes AIZOACEAE (D)
Brownea Jacq. LEGUMINOSAE–CAESALPINIOIDEAE (D)
Browneopsis Huber LEGUMINOSAE–CAESALPINIOIDEAE (D)
Browningia Britton & Rose CACTACEAE (D)
Brownleea Harv. ex Lindl. ORCHIDACEAE (M)
Brownlowia Roxb. TILIACEAE (D)
Brucea J.F.Mill. SIMAROUBACEAE (D)
Bruchmannia Nutt. (SUS) = Beckmannia Host GRAMINEAE (M)
Bruckenthalia Rchb. ERICACEAE (D)
Bruea Gaudich. (UP) EUPHORBIACEAE (D)
Brugmansia Pers. SOLANACEAE (D)
Brugmansia Blume (SUH) = Rhizanthes Dumort. RAFFLESIACEAE (D)
Bruguiera Sav. RHIZOPHORACEAE (D)
Bruguiera Thouars (SUH) = Lumnitzera Willd. COMBRETACEAE (D)
Bruinsmania Miq. = Isertia Schreb. RUBIACEAE (D)
Bruinsmia Boerl. & Koord. STYRACACEAE (D)
Brunella Mill. = Prunella L. LABIATAE (D)
Brunellia Ruíz & Pav. BRUNELLIACEAE (D)
Brunfelsia L. SOLANACEAE (D)
Brunfelsiopsis (Urb.) Kuntze = Brunfelsia L. SOLANACEAE (D)
Brunia Lam. BRUNIACEAE (D)
Brunnera Steven BORAGINACEAE (D)
Brunnichia Banks ex Gaertn. POLYGONACEAE (D)
Brunonia Sm. GOODENIACEAE (D)
Brunoniella Bremek. ACANTHACEAE (D)
Brunsvigia Heist. AMARYLLIDACEAE (M)
Brya Vell. (SUH) = Hirtella L. CHRYSOBALANACEAE (D)

Brya P.Browne LEGUMINOSAE–PAPILIONOIDEAE (D)
Bryantea Raf. = Neolitsea (Benth.) Merr. LAURACEAE (D)
Bryanthus J.G.Gmel. ERICACEAE (D)
Bryaspis P.A.Duvign. LEGUMINOSAE–PAPILIONOIDEAE (D)
Brylkinia F.Schmidt GRAMINEAE (M)
Bryobium Lindl. = Eria Lindl. ORCHIDACEAE (M)
Bryocarpum Hook.f. & Thomson PRIMULACEAE (D)
Bryodes Benth. SCROPHULARIACEAE (D)
Bryomorphe Harv. COMPOSITAE (D)
Bryonia L. CUCURBITACEAE (D)
Bryonopsis Arn. = Kedrostis Medik. CUCURBITACEAE (D)
Bryophyllum Salisb. CRASSULACEAE (D)
Bubalina Raf. = Burchellia R.Br. RUBIACEAE (D)
Bubania Girard PLUMBAGINACEAE (D)
Bubbia Tiegh. = Zygogynum Baill. WINTERACEAE (D)
Bubon L. = Athamanta L. UMBELLIFERAE (D)
Bubonium Hill (SUS) = Asteriscus Mill. COMPOSITAE (D)
Bucanion Steven = Heliotropium L. BORAGINACEAE (D)
Bucculina Lindl. = Holothrix Rich. ex Lindl. ORCHIDACEAE (M)
Bucephalandra Schott ARACEAE (M)
Bucephalon L. = Trophis P.Browne MORACEAE (D)
Bucephalora Pau = Rumex L. POLYGONACEAE (D)
Bucera P.Browne = Bucida L. COMBRETACEAE (D)
Buceragenia Greenm. ACANTHACEAE (D)
Bucetum Parn. (SUS) = Festuca L. GRAMINEAE (M)
Buchanania Spreng. ANACARDIACEAE (D)
Buchenavia Eichler COMBRETACEAE (D)
Buchenroedera Eckl. & Zeyh. = Lotononis (DC.) Eckl. & Zeyh. LEGUMINOSAE–
    PAPILIONOIDEAE (D)
Bucheria Heynh. = Thryptomene Endl. MYRTACEAE (D)
Buchholzia Engl. CAPPARACEAE (D)
Buchia Kunth = Perama Aubl. RUBIACEAE (D)
Buchingera Boiss. & Hohen. = Asperuginoides Rauschert CRUCIFERAE (D)
Buchloe Engelm. GRAMINEAE (M)
Buchlomimus J.R.Reeder, C.G.Reeder & Rzed. GRAMINEAE (M)
Buchnera L. SCROPHULARIACEAE (D)
Buchnerodendron Gürke FLACOURTIACEAE (D)
Buchozia L'Hér. ex Juss. = Serissa Comm. ex Juss. RUBIACEAE (D)
Buchtienia Schltr. ORCHIDACEAE (M)
Bucida L. COMBRETACEAE (D)
Bucinella Wiehler (SUH) = Columnea L. GESNERIACEAE (D)
Bucinellina Wiehler = Columnea L. GESNERIACEAE (D)
Buckinghamia F.Muell. PROTEACEAE (D)
Bucklandia R.Br. ex Griff. (SUH) = Exbucklandia R.W.Br. HAMAMELIDACEAE (D)
Buckleya Torr. SANTALACEAE (D)
Bucquetia DC. MELASTOMATACEAE (D)
Bucranion Raf. = Utricularia L. LENTIBULARIACEAE (D)
Buddleja L. BUDDLEJACEAE (D)
Buekia Giseke = Alpinia Roxb. ZINGIBERACEAE (M)
Buena Cav. (SUH) = Gonzalagunia Ruíz & Pav. RUBIACEAE (D)
Buena Pohl = Cosmibuena Ruíz & Pav. RUBIACEAE (D)
Buergeria Miq. = Maackia Rupr. LEGUMINOSAE–PAPILIONOIDEAE (D)
Buergersiochloa Pilg. GRAMINEAE (M)
Buesia (C.Morton) Copel. = Hymenophyllum Sm. HYMENOPHYLLACEAE (F)
Buettneria Murray = Byttneria Loefl. STERCULIACEAE (D)

Bufonia L. CARYOPHYLLACEAE (D)
Buforrestia C.B.Clarke COMMELINACEAE (M)
Buglossoides Moench BORAGINACEAE (D)
Bugranopsis Pomel = Ononis L. LEGUMINOSAE–PAPILIONOIDEAE (D)
Buhsia Bunge CAPPARACEAE (D)
Buiningia Buxb. = Coleocephalocereus Backeb. CACTACEAE (D)
Bukiniczia Lincz. PLUMBAGINACEAE (D)
Bulbilis Kuntze (SUS) = Buchloe Engelm. GRAMINEAE (M)
Bulbine Wolf ASPHODELACEAE (M)
Bulbinella Kunth ASPHODELACEAE (M)
Bulbinopsis Borzi = Bulbine Wolf ASPHODELACEAE (M)
Bulbocodium L. COLCHICACEAE (M)
Bulbophyllaria Rchb.f. = Bulbophyllum Thouars ORCHIDACEAE (M)
Bulbophyllopsis Rchb.f. = Bulbophyllum Thouars ORCHIDACEAE (M)
Bulbophyllum Thouars ORCHIDACEAE (M)
Bulbostylis Kunth CYPERACEAE (M)
Bulbulus Swallen (SUI) = Rehia Fijten GRAMINEAE (M)
Bulleyia Schltr. ORCHIDACEAE (M)
Bulliarda DC. (SUH) = Crassula L. CRASSULACEAE (D)
Bulnesia Gay ZYGOPHYLLACEAE (D)
Bulweria F.Muell. = Deplanchea Vieill. BIGNONIACEAE (D)
Bumelia Sw. = Sideroxylon L. SAPOTACEAE (D)
Bunburya Meisn. ex Hochst. = Tricalysia A.Rich. ex DC. RUBIACEAE (D)
Bunchosia Rich. ex Kunth MALPIGHIACEAE (D)
Bungea C.A.Mey. SCROPHULARIACEAE (D)
Bunias L. CRUCIFERAE (D)
Buniella Schischk. = Bunium L. UMBELLIFERAE (D)
Buniotrinia Stapf & Wettst. = Ferula L. UMBELLIFERAE (D)
Bunium L. UMBELLIFERAE (D)
Bunophila Willd. ex Roem. & Schult. = Machaonia Bonpl. RUBIACEAE (D)
Buphthalmum L. COMPOSITAE (D)
Bupleuroides Moench = Phyllis L. RUBIACEAE (D)
Bupleurum L. UMBELLIFERAE (D)
Buraeavia Baill. = Austrobuxus Miq. EUPHORBIACEAE (D)
Burasaia Thouars MENISPERMACEAE (D)
Burbidgea Hook.f. ZINGIBERACEAE (M)
Burcardia Neck. ex Raf. = Campomanesia Ruíz & Pav. MYRTACEAE (D)
Burchardia R.Br. COLCHICACEAE (M)
Burchellia R.Br. RUBIACEAE (D)
Burckella Pierre SAPOTACEAE (D)
Burdachia Mart. ex A.Juss. MALPIGHIACEAE (D)
Bureaua Kuntze = Austrobuxus Miq. EUPHORBIACEAE (D)
Bureava Baill. = Combretum Loefl. COMBRETACEAE (D)
Bureavella Pierre = Pouteria Aubl. SAPOTACEAE (D)
Burgesia F.Muell. LEGUMINOSAE–PAPILIONOIDEAE (D)
Burkartia Crisci COMPOSITAE (D)
Burkea Hook. LEGUMINOSAE–CAESALPINIOIDEAE (D)
Burkillanthus Swingle RUTACEAE (D)
Burkillia Ridl. = Burkilliodendron Sastry LEGUMINOSAE–PAPILIONOIDEAE (D)
Burkilliodendron Sastry LEGUMINOSAE–PAPILIONOIDEAE (D)
Burlemarxia N.L.Menezes & Semir VELLOZIACEAE (M)
Burmabambus Keng f. = Sinarundinaria Nakai GRAMINEAE (M)
Burmannia L. BURMANNIACEAE (M)
Burmeistera Karst. & Triana CAMPANULACEAE (D)
Burnatastrum Briq. = Plectranthus L'Hér. LABIATAE (D)

Burnatia Micheli ALISMATACEAE (M)
Burnettia Lindl. ORCHIDACEAE (M)
Burneya Cham. & Schltdl. = Timonius DC. RUBIACEAE (D)
Burragea Donn.Sm. & Rose = Gongylocarpus Schltdl. & Cham. ONAGRACEAE (D)
Burretiodendron Rehder TILIACEAE (D)
Burretiokentia Pic.Serm. PALMAE (M)
Burroughsia Moldenke = Lippia L. VERBENACEAE (D)
Bursaria Cav. PITTOSPORACEAE (D)
Bursera Jacq. ex L. BURSERACEAE (D)
Burtonia R.Br. LEGUMINOSAE–PAPILIONOIDEAE (D)
Burttdavya Hoyle RUBIACEAE (D)
Burttia Baker f. & Exell CONNARACEAE (D)
Busbeckia Hecart (SUI) = Syringa L. OLEACEAE (D)
Buseria T.Durand = Coffea L. RUBIACEAE (D)
Bushia Nieuwl. = Kochia Roth CHENOPODIACEAE (D)
Bussea Harms LEGUMINOSAE–CAESALPINIOIDEAE (D)
Bustelma Fourn. = Oxystelma R.Br. ASCLEPIADACEAE (D)
Bustillosia Clos = Asteriscium Cham. & Schltdl. UMBELLIFERAE (D)
Butania Keng f. = Arundinaria Michx. GRAMINEAE (M)
Butayea De Wild. = Sclerochiton Harv. ACANTHACEAE (D)
Butea Roxb. ex Willd. LEGUMINOSAE–PAPILIONOIDEAE (D)
Buteraea Nees = Strobilanthes Blume ACANTHACEAE (D)
Butia (Becc.) Becc. PALMAE (M)
Butinia Boiss. = Conopodium W.D.J.Koch UMBELLIFERAE (D)
Butomopsis Kunth LIMNOCHARITACEAE (M)
Butomus L. BUTOMACEAE (M)
Buttonia McKen ex Benth. SCROPHULARIACEAE (D)
Butumia G.Taylor PODOSTEMACEAE (D)
Butyrospermum Kotschy = Vitellaria C.F.Gaertn. SAPOTACEAE (D)
Buxiphyllum W.T.Wang & C.Z.Gao = Paraboea (C.B.Clarke) Ridl. GESNERIACEAE (D)
Buxus L. BUXACEAE (D)
Byblis Salisb. BYBLIDACEAE (D)
Byrnesia Rose = Graptopetalum Rose CRASSULACEAE (D)
Byrsanthes C.Presl = Siphocampylus Pohl CAMPANULACEAE (D)
Byrsanthus Guill. FLACOURTIACEAE (D)
Byrsocarpus Schumach. & Thonn. = Rourea Aubl. CONNARACEAE (D)
Byrsonima Rich. ex Kunth MALPIGHIACEAE (D)
Byrsophyllum Hook.f. RUBIACEAE (D)
Byrsopteris C.Morton = Arachniodes Blume DRYOPTERIDACEAE (F)
Bystropogon L'Hér. LABIATAE (D)
Bythophyton Hook.f. SCROPHULARIACEAE (D)
Byttneria Loefl. STERCULIACEAE (D)
Byurlingtonia Lindl. = Rodriguezia Ruíz & Pav. ORCHIDACEAE (M)

Cabi Ducke = Callaeum Small MALPIGHIACEAE (D)
Cabomba Aubl. CABOMBACEAE (D)
Cabralea A.Juss. MELIACEAE (D)
Cabrera Lag. = Axonopus P.Beauv. GRAMINEAE (M)
Cabreriella Cuatrec. COMPOSITAE (D)
Cabucala Pichon APOCYNACEAE (D)
Cacabus Bernh. = Exodeconus Raf. SOLANACEAE (D)
Cacalia L. COMPOSITAE (D)
Cacaliopsis A.Gray COMPOSITAE (D)
Caccinia Savi BORAGINACEAE (D)
Cachrys L. UMBELLIFERAE (D)

Cacosmanthus De Vriese (SUO) = Madhuca Ham. ex J.F.Gmel. SAPOTACEAE (D)
Cacosmia Kunth COMPOSITAE (D)
Cacoucia Aubl. = Combretum Loefl. COMBRETACEAE (D)
Cactodendron Bigelow (SUI) = Opuntia Mill. CACTACEAE (D)
Cactus Lem. (SUH) = Opuntia Mill. CACTACEAE (D)
Cactus Britton & Rose (SUH) = Melocactus Link & Otto CACTACEAE (D)
Cactus L. (SUS) = Mammillaria Haw. CACTACEAE (D)
Cadaba Forssk. CAPPARACEAE (D)
Cadalvena Fenzl = Costus L. COSTACEAE (M)
Cadamba Sonn. = Guettarda L. RUBIACEAE (D)
Cadellia F.Muell. SURIANACEAE (D)
Cadetia Gaudich. ORCHIDACEAE (M)
Cadia Forssk. LEGUMINOSAE–PAPILIONOIDEAE (D)
Cadiscus E.Mey. ex DC. COMPOSITAE (D)
Caelodepas Benth. (SUO) = Koilodepas Hassk. EUPHORBIACEAE (D)
Caelospermum Blume (SUO) = Coelospermum Blume RUBIACEAE (D)
Caenopteris Bergius = Asplenium L. ASPLENIACEAE (F)
Caesalpinia L. LEGUMINOSAE–CAESALPINIOIDEAE (D)
Caesarea Cambess. = Viviania Cav. GERANIACEAE (D)
Caesia R.Br. ANTHERICACEAE (M)
Caesulia Roxb. COMPOSITAE (D)
Cafe Adans. (SUS) = Coffea L. RUBIACEAE (D)
Cailliea Guill. & Perr. = Dichrostachys (A.DC.) Wight & Arn. LEGUMINOSAE–MIMOSOIDEAE (D)
Cailliella Jacq.-Fél. MELASTOMATACEAE (D)
Cainito Adans. = Chrysophyllum L. SAPOTACEAE (D)
Caiophora C.Presl LOASACEAE (D)
Cajalbania Urb. = Gliricidia Kunth LEGUMINOSAE–PAPILIONOIDEAE (D)
Cajanus DC. LEGUMINOSAE–PAPILIONOIDEAE (D)
Cajophora C.Presl (SUO) = Caiophora C.Presl LOASACEAE (D)
Cajuputi Adans. = Melaleuca L. MYRTACEAE (D)
Cakile Mill. CRUCIFERAE (D)
Calacanthus T.Anderson ex Benth. ACANTHACEAE (D)
Caladenia R.Br. ORCHIDACEAE (M)
Caladiopsis Engl. = Chlorospatha Engl. ARACEAE (M)
Caladium Vent. ARACEAE (M)
Calamagrostis Adans. GRAMINEAE (M)
Calamaria Boehm. = Isoetes L. ISOETACEAE (FA)
Calamina P.Beauv. (SUS) = Apluda L. GRAMINEAE (M)
Calamintha Mill. LABIATAE (D)
Calamistrum L. ex Kuntze = Pilularia L. MARSILEACEAE (F)
Calamochloa Fourn. (SUH) = Sohnsia Airy Shaw GRAMINEAE (M)
Calamochloe Rchb. = Arundinella Raddi GRAMINEAE (M)
Calamophyllum Schwantes = Cylindrophyllum Schwantes AIZOACEAE (D)
Calamosagus Griff. = Korthalsia Blume PALMAE (M)
Calamovilfa (A.Gray) Hack. GRAMINEAE (M)
Calampelis D.Don = Eccremocarpus Ruíz & Pav. BIGNONIACEAE (D)
Calamus L. PALMAE (M)
Calanda K.Schum. RUBIACEAE (D)
Calandrinia Kunth PORTULACACEAE (D)
Calandriniopsis Franz = Calandrinia Kunth PORTULACACEAE (D)
Calanthe Ker Gawl. ORCHIDACEAE (M)
Calanthera Hook. (SUI) = Buchloe Engelm. GRAMINEAE (M)
Calanthidium Pfitzer = Calanthe Ker-Gawl. ORCHIDACEAE (M)
Calanthus Oerst. ex Hanst. = Drymonia Mart. GESNERIACEAE (D)

Calantica Jaub. ex Tul. FLACOURTIACEAE (D)
Calappa Steck = Cocos L. PALMAE (M)
Calathea G.Mey. MARANTACEAE (M)
Calathiana Delarbre = Gentiana L. GENTIANACEAE (D)
Calathodes Hook.f. & Thomson RANUNCULACEAE (D)
Calathostelma Fourn. ASCLEPIADACEAE (D)
Calatola Standl. ICACINACEAE (D)
Calcareoboea C.Y.Wu = Platyadenia B.L.Burtt GESNERIACEAE (D)
Calcearia Blume = Corybas Salisb. ORCHIDACEAE (M)
Calceolaria Fabr. (SUH) = Cypripedium L. ORCHIDACEAE (M)
Calceolaria L. SCROPHULARIACEAE (D)
Calceolus Nieuwl. = Cypripedium L. ORCHIDACEAE (M)
Caldasia Lag. (SUH) = Oreomyrrhis Endl. UMBELLIFERAE (D)
Caldasia Willd. = Bonplandia Cav. POLEMONIACEAE (D)
Caldcluvia D.Don CUNONIACEAE (D)
Calderonella Soderstr. & H.F.Decker GRAMINEAE (M)
Calderonia Standl. = Simira Aubl. RUBIACEAE (D)
Caldesia Parl. ALISMATACEAE (M)
Calea L. COMPOSITAE (D)
Caleana R.Br. ORCHIDACEAE (M)
Calectasia R.Br. CALECTASIACEAE (M)
Calendula L. COMPOSITAE (D)
Caleopsis Fedde = Goldmanella Greenm. COMPOSITAE (D)
Calepina Adans. CRUCIFERAE (D)
Caletia Baill. = Micrantheum Desf. EUPHORBIACEAE (D)
Caleya R.Br. = Caleana R.Br. ORCHIDACEAE (M)
Calia Berland. = Sophora L. LEGUMINOSAE–PAPILIONOIDEAE (D)
Calibanus Rose DRACAENACEAE (M)
Calibrachoa Cerv. SOLANACEAE (D)
Calicorema Hook.f. AMARANTHACEAE (D)
Calicotome Link LEGUMINOSAE–PAPILIONOIDEAE (D)
Caliphruria Herb. AMARYLLIDACEAE (M)
Calispepla Vved. LEGUMINOSAE–PAPILIONOIDEAE (D)
Calius Blanco = Streblus Lour. MORACEAE (D)
Calla L. ARACEAE (M)
Callaeolepium Karst. = Fimbristemma Turcz. ASCLEPIADACEAE (D)
Callaeum Small MALPIGHIACEAE (D)
Callianassa Webb & Berthel. = Isoplexis (Lindl.) J.C.Loudon SCROPHULARIACEAE (D)
Calliandra Benth. LEGUMINOSAE–MIMOSOIDEAE (D)
Calliandropsis H.M.Hern. & Guinet LEGUMINOSAE–MIMOSOIDEAE (D)
Callianthemum C.A.Mey. RANUNCULACEAE (D)
Calliaspidia Bremek. = Justicia L. ACANTHACEAE (D)
Callicarpa L. VERBENACEAE (D)
Callicephalus C.A.Mey. COMPOSITAE (D)
Callichilia Stapf APOCYNACEAE (D)
Callichlamys Miq. BIGNONIACEAE (D)
Callichloea Steud. (SUI) = Elionurus Humb. & Bonpl. ex Willd. GRAMINEAE (M)
Callicocca Schreb. = Psychotria L. RUBIACEAE (D)
Callicoma Andrews CUNONIACEAE (D)
Calligonum L. POLYGONACEAE (D)
Callilepis DC. COMPOSITAE (D)
Callipeltis Steven RUBIACEAE (D)
Calliphyllon Bubani = Epipactis Zinn ORCHIDACEAE (M)
Callipsyche Herb. = Eucrosia Ker–Gawl. AMARYLLIDACEAE (M)
Callipteris Bory = Diplazium Sw. WOODSIACEAE (F)

Callirhoe Nutt. MALVACEAE (D)
Callisace Fisch. = Angelica L. UMBELLIFERAE (D)
Callisia Loefl. COMMELINACEAE (M)
Callista Lour. = Dendrobium Sw. ORCHIDACEAE (M)
Callistachys Vent. LEGUMINOSAE–PAPILIONOIDEAE (D)
Callistemon R.Br. MYRTACEAE (D)
Callistephus Cass. COMPOSITAE (D)
Callisteris Greene = Ipomopsis Michx. POLEMONIACEAE (D)
Callisthene Mart. VOCHYSIACEAE (D)
Callistigma Dinter & Schwantes = Mesembryanthemum L. AIZOACEAE (D)
Callistopteris Copel. = Cephalomanes C.Presl HYMENOPHYLLACEAE (F)
Callistylon Pittier = Coursetia DC. LEGUMINOSAE–PAPILIONOIDEAE (D)
Callithauma Herb. = Stenomesson Herb. AMARYLLIDACEAE (M)
Callithronum Ehrh. = Cephalanthera Rich. ORCHIDACEAE (M)
Callitriche L. CALLITRICHACEAE (D)
Callitris Vent. CUPRESSACEAE (G)
Callogramme Fée = Syngramma J.Sm. ADIANTACEAE (F)
Callopsis Engl. ARACEAE (M)
Callostylis Blume = Eria Lindl. ORCHIDACEAE (M)
Callothlaspi F.K.Mey. = Thlaspi L. CRUCIFERAE (D)
Calluna Salisb. ERICACEAE (D)
Calobotrya Spach = Ribes L. GROSSULARIACEAE (D)
Calocarpum Pierre = Pouteria Aubl. SAPOTACEAE (D)
Calocedrus Kurz CUPRESSACEAE (G)
Calocephalus R.Br. COMPOSITAE (D)
Calochilus R.Br. ORCHIDACEAE (M)
Calochlaena (Maxon) R.A.White & M.D.Turner DICKSONIACEAE (F)
Calochloa Kunze (SUO) = Elionurus Humb. & Bonpl. ex Willd. GRAMINEAE (M)
Calochone Keay RUBIACEAE (D)
Calochortus Pursh LILIACEAE (M)
Calococcus Kurz ex Teijsm. = Margaritaria L.f. EUPHORBIACEAE (D)
Calocrater K.Schum. APOCYNACEAE (D)
Calodecaryia J.–F.Leroy MELIACEAE (D)
Calodendrum Thunb. RUTACEAE (D)
Calodryum Desv. = Turraea L. MELIACEAE (D)
Caloglossum Schltr. = Cymbidiella Rolfe ORCHIDACEAE (M)
Calogyne R.Br. = Goodenia Sm. GOODENIACEAE (D)
Calolisianthus Gilg = Irlbachia Mart. GENTIANACEAE (D)
Calomelanos (C.Presl) Lindl. = Pityrogramma Link ADIANTACEAE (F)
Calomeria Vent. COMPOSITAE (D)
Caloncoba Gilg FLACOURTIACEAE (D)
Calonyction Choisy = Ipomoea L. CONVOLVULACEAE (D)
Calopappus Meyen COMPOSITAE (D)
Calophaca Fisch. ex DC. LEGUMINOSAE–PAPILIONOIDEAE (D)
Calophanes D.Don = Dyschoriste Nees ACANTHACEAE (D)
Calophanoides (C.B.Clarke) Ridl. = Justicia L. ACANTHACEAE (D)
Calophyllum L. GUTTIFERAE (D)
Calophyllum L. (SUH) = Epipactis Zinn ORCHIDACEAE (M)
Calophysa DC. = Clidemia D.Don MELASTOMATACEAE (D)
Caloplectus Oerst. = Drymonia Mart. GESNERIACEAE (D)
Calopogon R.Br. ORCHIDACEAE (M)
Calopogonium Desv. LEGUMINOSAE–PAPILIONOIDEAE (D)
Calopsis P.Beauv. ex Desv. RESTIONACEAE (M)
Calopteryx A.C.Sm. ERICACEAE (D)
Calopyxis Tul. COMBRETACEAE (D)

Calorchis Barb.Rodr. = Ponthieva R.Br. ORCHIDACEAE (M)
Calorhabdos Benth. = Veronicastrum Heist. ex Fabr. SCROPHULARIACEAE (D)
Calorophus Labill. RESTIONACEAE (M)
Calosacme Wall. = Chirita Buch.–Ham. ex D.Don GESNERIACEAE (D)
Calosanthes Blume = Oroxylum Vent. BIGNONIACEAE (D)
Caloscordum Herb. ALLIACEAE (M)
Calospatha Becc. PALMAE (M)
Calospermum Pierre = Pouteria Aubl. SAPOTACEAE (D)
Calosteca Desv. = Briza L. GRAMINEAE (M)
Calostemma R.Br. AMARYLLIDACEAE (M)
Calostephane Benth. COMPOSITAE (D)
Calostigma Decne. ASCLEPIADACEAE (D)
Calotesta Karis COMPOSITAE (D)
Calothamnus Labill. MYRTACEAE (D)
Calotheca P.Beauv. (SUO) = Briza L. GRAMINEAE (M)
Calotheria Steud. (SUI) = Enneapogon Desv. ex P.Beauv. GRAMINEAE (M)
Calotis R.Br. COMPOSITAE (D)
Calotropis R.Br. ASCLEPIADACEAE (D)
Calpicarpum G.Don = Ochrosia Juss. APOCYNACEAE (D)
Calpidia Thouars = Pisonia L. NYCTAGINACEAE (D)
Calpidisca Barnhart = Utricularia L. LENTIBULARIACEAE (D)
Calpidochlamys Diels = Trophis P.Browne MORACEAE (D)
Calpidosicyos Harms = Momordica L. CUCURBITACEAE (D)
Calpigyne Blume = Koilodepas Hassk. EUPHORBIACEAE (D)
Calpocalyx Harms LEGUMINOSAE–MIMOSOIDEAE (D)
Calpurnia E.Mey. LEGUMINOSAE–PAPILIONOIDEAE (D)
Caltha L. RANUNCULACEAE (D)
Caluera Dodson & Determann ORCHIDACEAE (M)
Calvaria Comm. ex C.F.Gaertn. = Sideroxylon L. SAPOTACEAE (D)
Calvelia Moq. = Suaeda Forssk. ex Scop. CHENOPODIACEAE (D)
Calvoa Hook.f. MELASTOMATACEAE (D)
Calycacanthus K.Schum. ACANTHACEAE (D)
Calycadenia DC. COMPOSITAE (D)
Calycampe O.Berg = Myrcia DC. ex Guill. MYRTACEAE (D)
Calycantherum Klotzsch = Ipomoea L. CONVOLVULACEAE (D)
Calycanthus L. CALYCANTHACEAE (D)
Calycera Cav. CALYCERACEAE (D)
Calycobolus Willd. ex J.A.Schult. CONVOLVULACEAE (D)
Calycocarpum Nutt. ex Spach MENISPERMACEAE (D)
Calycocorsus F.W.Schmidt COMPOSITAE (D)
Calycodendron A.C.Sm. = Psychotria L. RUBIACEAE (D)
Calycodon Nutt. = Muhlenbergia Schreb. GRAMINEAE (M)
Calycogonium DC. MELASTOMATACEAE (D)
Calycolpus O.Berg MYRTACEAE (D)
Calycomelia Kostel. = Fraxinus L. OLEACEAE (D)
Calycomis R.Br. ex Nees & Sinning (SUS) = Callicoma Andrews CUNONIACEAE (D)
Calycomis D.Don (SUH) = Acrophyllum Benth. CUNONIACEAE (D)
Calycopeplus Planch. EUPHORBIACEAE (D)
Calycophyllum DC. RUBIACEAE (D)
Calycophysum Karst. & Triana CUCURBITACEAE (D)
Calycopteris Lam. = Getonia Roxb. COMBRETACEAE (D)
Calycorectes O.Berg MYRTACEAE (D)
Calycoseris A.Gray COMPOSITAE (D)
Calycosia A.Gray RUBIACEAE (D)
Calycosiphonia Pierre ex Robbr. RUBIACEAE (D)

Calycostemma Hanst. = Kohleria Regel GESNERIACEAE (D)
Calycothrix Meisn. = Calytrix Labill. MYRTACEAE (D)
Calycotropis Turcz. = Polycarpaea Lam. CARYOPHYLLACEAE (D)
Calyculogygas Krapov. MALVACEAE (D)
Calydermos Ruíz & Pav. = Nicandra Adans. SOLANACEAE (D)
Calydorea Herb. IRIDACEAE (M)
Calylophis Spach (SUO) = Calylophus Spach ONAGRACEAE (D)
Calylophus Spach ONAGRACEAE (D)
Calymella C.Presl = Gleichenia Sm. GLEICHENIACEAE (F)
Calymmanthera Schltr. ORCHIDACEAE (M)
Calymmanthium F.Ritter CACTACEAE (D)
Calymmatium O.E.Schulz CRUCIFERAE (D)
Calymmodon C.Presl GRAMMITIDACEAE (F)
Calymmostachya Bremek. = Justicia L. ACANTHACEAE (D)
Calyntranthele Nied. = Byrsonima Rich. ex Kunth MALPIGHIACEAE (D)
Calyplectus Ruíz & Pav. = Lafoensia Vand. LYTHRACEAE (D)
Calypso Salisb. ORCHIDACEAE (M)
Calypsodium Link = Calypso Salisb. ORCHIDACEAE (M)
Calypteriopetalon Hassk. = Croton L. EUPHORBIACEAE (D)
Calypterium Bernh. = Onoclea L. WOODSIACEAE (F)
Calyptocarpus Less. COMPOSITAE (D)
Calyptochloa C.E.Hubb. GRAMINEAE (M)
Calyptosepalum S.Moore = Drypetes Vahl EUPHORBIACEAE (D)
Calyptostylis Arènes MALPIGHIACEAE (D)
Calyptraemalva Krapov. MALVACEAE (D)
Calyptranthes Sw. MYRTACEAE (D)
Calyptranthus Blume = Syzygium Gaertn. MYRTACEAE (D)
Calyptraria Naudin = Centronia D.Don MELASTOMATACEAE (D)
Calyptrella Naudin = Graffenrieda DC. MELASTOMATACEAE (D)
Calyptridium Nutt. PORTULACACEAE (D)
Calyptrocalyx Blume PALMAE (M)
Calyptrocarya Nees CYPERACEAE (M)
Calyptrochilum Kraenzl. ORCHIDACEAE (M)
Calyptrocoryne Schott = Theriophonum Blume ARACEAE (M)
Calyptrogenia Burret MYRTACEAE (D)
Calyptrogyne H.A.Wendl. PALMAE (M)
Calyptromyrcia O.Berg = Myrcia DC. ex Guill. MYRTACEAE (D)
Calyptronoma Griseb. PALMAE (M)
Calyptroon Miq. = Baccaurea Lour. EUPHORBIACEAE (D)
Calyptropsidium O.Berg = Psidium L. MYRTACEAE (D)
Calyptrosciadium Rech.f. & Kuber UMBELLIFERAE (D)
Calyptrosicyos Keraudren = Corallocarpus Welw. ex Hook.f. CUCURBITACEAE (D)
Calyptrospatha Klotzsch ex Baill. = Acalypha L. EUPHORBIACEAE (D)
Calyptrospermum A.Dietr. (SUS) = Menodora Bonpl. OLEACEAE (D)
Calyptrostigma Klotzsch = Beyeria Miq. EUPHORBIACEAE (D)
Calyptrotheca Gilg PORTULACACEAE (D)
Calystegia R.Br. CONVOLVULACEAE (D)
Calythrix Labill. = Calytrix Labill. MYRTACEAE (D)
Calythropsis C.A.Gardner MYRTACEAE (D)
Calytrix Labill. MYRTACEAE (D)
Camarea A.St.–Hil. MALPIGHIACEAE (D)
Camaridium Lindl. = Maxillaria Ruíz & Pav. ORCHIDACEAE (M)
Camarotea Scott Elliot ACANTHACEAE (D)
Camarotis Lindl. = Micropera Lindl. ORCHIDACEAE (M)
Camassia Lindl. HYACINTHACEAE (M)

Cambania Comm. ex M.Roem. = Dysoxylum Blume MELIACEAE (D)
Cambessedesia DC. MELASTOMATACEAE (D)
Cambogia L. = Garcinia L. GUTTIFERAE (D)
Camchaya Gagnep. COMPOSITAE (D)
Camelina Crantz CRUCIFERAE (D)
Camelinopsis A.G.Mill. CRUCIFERAE (D)
Camellia L. THEACEAE (D)
Camelostalix Pfitzer = Pholidota Lindl. ex Hook. ORCHIDACEAE (M)
Cameraria L. APOCYNACEAE (D)
Camerunia (Pichon) Boiteau = Tabernaemontana L. APOCYNACEAE (D)
Camilleugenia Frapp. ex Cordem. = Cynorkis Thouars ORCHIDACEAE (M)
Camirium Gaertn. = Aleurites J.R.Forst. & G.Forst. EUPHORBIACEAE (D)
Camissonia Link ONAGRACEAE (D)
Camoensia Welw. ex Benth. LEGUMINOSAE–PAPILIONOIDEAE (D)
Camolenga Post & Kuntze = Benincasa Savi CUCURBITACEAE (D)
Campana Post & Kuntze = Tecomanthe Baill. BIGNONIACEAE (D)
Campanea Decne. (SUO) = Capanea Decne. GESNERIACEAE (D)
Campanocalyx Valeton = Keenania Hook.f. RUBIACEAE (D)
Campanolea Gilg & Schellenb. = Chionanthus L. OLEACEAE (D)
Campanula L. CAMPANULACEAE (D)
Campanulastrum Small = Campanula L. CAMPANULACEAE (D)
Campanulorchis Brieger = Eria Lindl. ORCHIDACEAE (M)
Campanumoea Blume = Codonopsis Wall. CAMPANULACEAE (D)
Campbellia Wight SCROPHULARIACEAE (D)
Campderia Benth. = Coccoloba P.Browne POLYGONACEAE (D)
Campecarpus H.A.Wendl. ex Becc. PALMAE (M)
Campeiostachys Drobov = Elymus L. GRAMINEAE (M)
Campelepis Falc. = Periploca L. ASCLEPIADACEAE (D)
Campelia Kunth (SUO) = Deschampsia P.Beauv. GRAMINEAE (M)
Campelia Rich. = Tradescantia L. COMMELINACEAE (M)
Campella Link (SUS) = Deschampsia P.Beauv. GRAMINEAE (M)
Campestigma Pierre ex Costantin ASCLEPIADACEAE (D)
Camphoromyrtus Schauer = Baeckea L. MYRTACEAE (D)
Camphorosma L. CHENOPODIACEAE (D)
Campimia Ridl. MELASTOMATACEAE (D)
Campium C.Presl = Bolbitis Schott LOMARIOPSIDACEAE (F)
Campnosperma Thwaites ANACARDIACEAE (D)
Campomanesia Ruíz & Pav. MYRTACEAE (D)
Campovassouria R.M.King & H.Rob. COMPOSITAE (D)
Campsiandra Benth. LEGUMINOSAE–CAESALPINIOIDEAE (D)
Campsidium Seem. BIGNONIACEAE (D)
Campsis Lour. BIGNONIACEAE (D)
Camptacra N.T.Burb. COMPOSITAE (D)
Camptandra Ridl. ZINGIBERACEAE (M)
Campteria C.Presl = Pteris L. PTERIDACEAE (F)
Camptocarpus Decne. ASCLEPIADACEAE (D)
Camptocarpus K.Koch (SUH) = Alkanna Tausch BORAGINACEAE (D)
Camptodium Fée DRYOPTERIDACEAE (F)
Camptolepis Radlk. SAPINDACEAE (D)
Camptoloma Benth. SCROPHULARIACEAE (D)
Camptophytum Pierre ex A.Chev. (SUI) = Tarenna Gaertn. RUBIACEAE (D)
Camptopus Hook.f. = Psychotria L. RUBIACEAE (D)
Camptorrhiza Hutch. COLCHICACEAE (M)
Camptosema Hook. & Arn. LEGUMINOSAE–PAPILIONOIDEAE (D)
Camptosorus Link ASPLENIACEAE (F)

Camptostemon Mast. BOMBACACEAE (D)
Camptostylus Gilg FLACOURTIACEAE (D)
Camptotheca Decne. CORNACEAE (D)
Campuloa Desv. (SUS) = Ctenium Panz. GRAMINEAE (M)
Campuloclinium DC. COMPOSITAE (D)
Campulosus Desv. = Ctenium Panz. GRAMINEAE (M)
Campylandra Baker = Tupistra Ker–Gawl. CONVALLARIACEAE (M)
Campylanthera Schott & Endl. = Ceiba Mill. BOMBACACEAE (D)
Campylanthus Roth SCROPHULARIACEAE (D)
Campylobotrys Lem. = Hoffmannia Sw. RUBIACEAE (D)
Campylocaryum DC. ex Meisn. = Alkanna Tausch BORAGINACEAE (D)
Campylocentrum Benth. ORCHIDACEAE (M)
Campylochiton Welw. ex Hiern = Combretum Loefl. COMBRETACEAE (D)
Campylogramma Alderw. = Microsorum Link POLYPODIACEAE (F)
Campylogyne Welw. ex Hemsl. = Quisqualis L. COMBRETACEAE (D)
Campyloneurum C.Presl POLYPODIACEAE (F)
Campylopetalum Forman PODOACEAE (D)
Campyloptera Boiss. = Aethionema R.Br. CRUCIFERAE (D)
Campylosiphon Benth. BURMANNIACEAE (M)
Campylospermum Tiegh. = Gomphia Schreb. OCHNACEAE (D)
Campylostachys Kunth STILBACEAE (D)
Campylostemon Welw. CELASTRACEAE (D)
Campylotheca Cass. = Bidens L. COMPOSITAE (D)
Campylotropis Bunge LEGUMINOSAE–PAPILIONOIDEAE (D)
Campynema Labill. MELANTHIACEAE (M)
Campynemanthe Baill. MELANTHIACEAE (M)
Camunium Roxb. (SUH) = Aglaia Lour. MELIACEAE (D)
Camusia Lorch = Acrachne Wight & Arn. ex Chiov. GRAMINEAE (M)
Camusiella Bosser = Setaria P.Beauv. GRAMINEAE (M)
Canaca Guillaumin = Austrobuxus Miq. EUPHORBIACEAE (D)
Canacomyrica Guillaumin MYRICACEAE (D)
Canacorchis Guillaumin = Bulbophyllum Thouars ORCHIDACEAE (M)
Cananga Aubl. (SUH) = Guatteria Ruíz & Pav. ANNONACEAE (D)
Cananga (DC.) Hook.f. & Thomson ANNONACEAE (D)
Canangium Baill. (SUS) = Cananga (DC.) Hook.f. & Thomson ANNONACEAE (D)
Canariastrum Engl. (SUI) = Uapaca Baill. EUPHORBIACEAE (D)
Canariellum Engl. = Canarium L. BURSERACEAE (D)
Canarina L. CAMPANULACEAE (D)
Canarium L. BURSERACEAE (D)
Canavalia DC. LEGUMINOSAE–PAPILIONOIDEAE (D)
Canbya Parry ex A.Gray PAPAVERACEAE (D)
Cancellaria (DC.) Mattei = Pavonia Cav. MALVACEAE (D)
Cancrinia Kar. & Kir. COMPOSITAE (D)
Cancriniella Tzvelev COMPOSITAE (D)
Candelabria Hochst. = Bridelia Willd. EUPHORBIACEAE (D)
Candollea Labill. (SUH) = Hibbertia Andrews DILLENIACEAE (D)
Candollea Steud. (SUS) = Agrostis L. GRAMINEAE (M)
Candollea Mirb. = Pyrrosia Mirb. POLYPODIACEAE (F)
Candolleodendron R.S.Cowan LEGUMINOSAE–PAPILIONOIDEAE (D)
Candollina Tiegh. = Amyema Tiegh. LORANTHACEAE (D)
Canella P.Browne CANELLACEAE (D)
Canephora Juss. RUBIACEAE (D)
Canistrum E.Morren BROMELIACEAE (M)
Canizaresia Britton = Piscidia L. LEGUMINOSAE–PAPILIONOIDEAE (D)
Canna L. CANNACEAE (M)

Cannabis L. CANNABACEAE (D)
Cannomois P.Beauv. ex Desv. RESTIONACEAE (M)
Canotia Torr. CANOTIACEAE (D)
Canschi Adans. = Trewia L. EUPHORBIACEAE (D)
Canscora Lam. GENTIANACEAE (D)
Cansjera Juss. OPILIACEAE (D)
Cantalea Raf. = Lycium L. SOLANACEAE (D)
Cantharospermum Wight & Arn. = Cajanus DC. LEGUMINOSAE–PAPILIONOIDEAE (D)
Canthiopsis Seem. = Tarenna Gaertn. RUBIACEAE (D)
Canthium Lam. RUBIACEAE (D)
Canthopsis Miq. = Catunaregam Wolf RUBIACEAE (D)
Cantleya Ridl. ICACINACEAE (D)
Cantua Juss. ex Lam. POLEMONIACEAE (D)
Cantuffa J.F.Gmel. = Pterolobium R.Br. ex Wight & Arn. LEGUMINOSAE–
    CAESALPINIOIDEAE (D)
Caoutchoua J.F.Gmel. = Hevea Aubl. EUPHORBIACEAE (D)
Capanea Decne. GESNERIACEAE (D)
Capanemia Barb.Rodr. ORCHIDACEAE (M)
Capassa Klotzsch = Lonchocarpus Kunth LEGUMINOSAE–PAPILIONOIDEAE (D)
Capellenia Teijsm. & Binn. = Endospermum Benth. EUPHORBIACEAE (D)
Caperonia A.St.-Hil. EUPHORBIACEAE (D)
Capethia Britton = Oreithales Schltdl. RANUNCULACEAE (D)
Capillipedium Stapf GRAMINEAE (M)
Capirona Spruce RUBIACEAE (D)
Capitanopsis S.Moore LABIATAE (D)
Capitanya Schweinf. ex Gürke LABIATAE (D)
Capitellaria Naudin = Clidemia D.Don MELASTOMATACEAE (D)
Capitularia J.V.Suringar = Capitularina Kern CYPERACEAE (M)
Capitularina Kern CYPERACEAE (M)
Capnoides Mill. PAPAVERACEAE (D)
Capnophyllum Gaertn. UMBELLIFERAE (D)
Capparidastrum Hutch. = Capparis L. CAPPARACEAE (D)
Capparis L. CAPPARACEAE (D)
Capraria L. SCROPHULARIACEAE (D)
Caprificus Gasp. = Ficus L. MORACEAE (D)
Capriola Adans. (SUS) = Cynodon Rich. GRAMINEAE (M)
Caprosma G.Don (SUO) = Coprosma J.R.Forst. & G.Forst. RUBIACEAE (D)
Capsella Medik. CRUCIFERAE (D)
Capsicodendron Hoehne CANELLACEAE (D)
Capsicum L. SOLANACEAE (D)
Captaincookia N.Hallé RUBIACEAE (D)
Capurodendron Aubrév. SAPOTACEAE (D)
Capuronetta Markgr. = Tabernaemontana L. APOCYNACEAE (D)
Capuronia Lourteig LYTHRACEAE (D)
Capuronianthus J.-F.Leroy MELIACEAE (D)
Capusia Lecomte = Siphonodon Griff. CELASTRACEAE (D)
Caquepira J.F.Gmel. = Gardenia Ellis RUBIACEAE (D)
Caracasia Szyszyl. MARCGRAVIACEAE (D)
Caragana Fabr. LEGUMINOSAE–PAPILIONOIDEAE (D)
Caraguata Lindl. (SUH) = Guzmania Ruíz & Pav. BROMELIACEAE (M)
Caraipa Aubl. GUTTIFERAE (D)
Carajaea (Tul.) Wedd. = Castelnavia Tul. & Wedd. PODOSTEMACEAE (D)
Carallia Roxb. RHIZOPHORACEAE (D)
Caralluma R.Br. ASCLEPIADACEAE (D)
Caramuri Aubrév. & Pellegr. = Pouteria Aubl. SAPOTACEAE (D)

Carandra Gaertn. = Psydrax Gaertn. RUBIACEAE (D)
Carania Chiov. = Basananthe Peyr. PASSIFLORACEAE (D)
Carapa Aubl. MELIACEAE (D)
Carapichea Aubl. = Psychotria L. RUBIACEAE (D)
Carcia Raeusch. = Garcia Rohr EUPHORBIACEAE (D)
Cardamine L. CRUCIFERAE (D)
Cardaminopsis (C.A.Mey.) Hayek = Arabis L. CRUCIFERAE (D)
Cardanthera Buch.–Ham. ex Benth. (SUS) = Hygrophila R.Br. ACANTHACEAE (D)
Cardaria Desv. = Lepidium L. CRUCIFERAE (D)
Cardenanthus R.C.Foster IRIDACEAE (M)
Cardenasia Rusby = Bauhinia L. LEGUMINOSAE–CAESALPINIOIDEAE (D)
Cardenasiodendron F.A.Barkley ANACARDIACEAE (D)
Cardiacanthus Nees & Schauer = Carlowrightia A.Gray ACANTHACEAE (D)
Cardiandra Siebold & Zucc. HYDRANGEACEAE (D)
Cardiochilos P.J.Cribb ORCHIDACEAE (M)
Cardiochlaena Fée = Tectaria Cav. DRYOPTERIDACEAE (F)
Cardiochlamys Oliv. CONVOLVULACEAE (D)
Cardiocrinum (Endl.) Lindl. LILIACEAE (M)
Cardiogyne Bureau = Maclura Nutt. MORACEAE (D)
Cardiomanes C.Presl HYMENOPHYLLACEAE (F)
Cardionema DC. ILLECEBRACEAE (D)
Cardiopetalum Schltdl. ANNONACEAE (D)
Cardiophyllarium Choux = Doratoxylon Thouars ex Hook.f. SAPINDACEAE (D)
Cardiophyllum Ehrh. = Listera R.Br. ORCHIDACEAE (M)
Cardiopteris Wall. ex Royle CARDIOPTERIDACEAE (D)
Cardiospermum L. SAPINDACEAE (D)
Cardiostigma Baker = Calydorea Herb. IRIDACEAE (M)
Cardioteucris C.Y.Wu = Caryopteris Bunge VERBENACEAE (D)
Cardonaea Aristeg., Maguire & Steyerm. = Gongylolepis R.H.Schomb. COMPOSITAE (D)
Cardopatium Juss. COMPOSITAE (D)
Carduncellus Adans. COMPOSITAE (D)
Carduus L. COMPOSITAE (D)
Cardwellia F.Muell. PROTEACEAE (D)
Carenophila Ridl. = Geostachys (Baker) Ridl. ZINGIBERACEAE (M)
Carex L. CYPERACEAE (M)
Careya Roxb. LECYTHIDACEAE (D)
Caribea Alain NYCTAGINACEAE (D)
Carica L. CARICACEAE (D)
Carinavalva Ising CRUCIFERAE (D)
Cariniana Casar. LECYTHIDACEAE (D)
Carinivalva Airy Shaw (SUO) = Carinavalva Ising CRUCIFERAE (D)
Carinta W.Wight = Geophila D.Don RUBIACEAE (D)
Carionia Naudin MELASTOMATACEAE (D)
Carissa L. APOCYNACEAE (D)
Carissophyllum Pichon = Tachiadenus Griseb. GENTIANACEAE (D)
Carlemannia Benth. CARLEMANNIACEAE (D)
Carlephyton Jum. ARACEAE (M)
Carlesia Dunn UMBELLIFERAE (D)
Carlina L. COMPOSITAE (D)
Carlostephania Bubani = Circaea L. ONAGRACEAE (D)
Carlowrightia A.Gray ACANTHACEAE (D)
Carludovica Ruíz & Pav. CYCLANTHACEAE (M)
Carmenocania Wernham = Pogonopus Klotzsch RUBIACEAE (D)
Carmichaelia R.Br. LEGUMINOSAE–PAPILIONOIDEAE (D)
Carminatia Moçino ex DC. COMPOSITAE (D)

Carmona Cav. BORAGINACEAE (D)
Carnarvonia F.Muell. PROTEACEAE (D)
Carnegiea Perkins (SUH) = Carnegieodoxa Perkins MONIMIACEAE (D)
Carnegiea Britton & Rose CACTACEAE (D)
Carnegieodoxa Perkins MONIMIACEAE (D)
Carolofritschia Engl. = Acanthonema Hook.f. GESNERIACEAE (D)
Caropsis (Rouy & Camus) Rauschert UMBELLIFERAE (D)
Caroxylon Thunb. = Salsola L. CHENOPODIACEAE (D)
Carpacoce Sond. RUBIACEAE (D)
Carpanthea N.E.Br. AIZOACEAE (D)
Carpentaria Becc. PALMAE (M)
Carpenteria Torr. HYDRANGEACEAE (D)
Carpentia Ewart = Cressa L. CONVOLVULACEAE (D)
Carpesium L. COMPOSITAE (D)
Carpha Banks & Sol. ex R.Br. CYPERACEAE (M)
Carphalea Juss. RUBIACEAE (D)
Carphephorus Cass. COMPOSITAE (D)
Carphochaete A.Gray COMPOSITAE (D)
Carpinus L. CORYLACEAE (D)
Carpobrotus N.E.Br. AIZOACEAE (D)
Carpodetus J.R.Forst. & G.Forst. ESCALLONIACEAE (D)
Carpodinopsis Pichon = Pleiocarpa Benth. APOCYNACEAE (D)
Carpodinus R.Br. ex G.Don = Landolphia P.Beauv. APOCYNACEAE (D)
Carpodiptera Griseb. TILIACEAE (D)
Carpogymnia (H.P.Fuchs ex Janchen) D.Löve & Á.Löve = Gymnocarpium Newm.
   WOODSIACEAE (F)
Carpolepis (J.W.Dawson) J.W.Dawson MYRTACEAE (D)
Carpolobia G.Don POLYGALACEAE (D)
Carpolyza Salisb. AMARYLLIDACEAE (M)
Carponema Eckl. & Zeyh. = Heliophila Burm.f. ex L. CRUCIFERAE (D)
Carpothalis E.Mey. = Tricalysia A.Rich. ex DC. RUBIACEAE (D)
Carpotroche Endl. FLACOURTIACEAE (D)
Carpoxis Raf. = Forestiera Poir. OLEACEAE (D)
Carpoxylon H.A.Wendl. & Drude PALMAE (M)
Carptotepala Moldenke = Syngonanthus Ruhland ERIOCAULACEAE (M)
Carramboa Cuatrec. COMPOSITAE (D)
Carrichtera DC. CRUCIFERAE (D)
Carrierea Franch. FLACOURTIACEAE (D)
Carrissoa Baker f. LEGUMINOSAE–PAPILIONOIDEAE (D)
Carroa C.Presl = Marina Liebm. LEGUMINOSAE–PAPILIONOIDEAE (D)
Carronia F.Muell. MENISPERMACEAE (D)
Carruanthus (Schwantes) Schwantes AIZOACEAE (D)
Carruthersia Seem. APOCYNACEAE (D)
Carterella Terrell RUBIACEAE (D)
Carteretia A.Rich. = Cleisostoma Blume ORCHIDACEAE (M)
Carterothamnus R.M.King COMPOSITAE (D)
Carthamus L. COMPOSITAE (D)
Cartiera Greene CRUCIFERAE (D)
Cartonema R.Br. COMMELINACEAE (M)
Cartrema Raf. = Osmanthus Lour. OLEACEAE (D)
Caruelina Kuntze = Chomelia Jacq. RUBIACEAE (D)
Carum L. UMBELLIFERAE (D)
Carumbium Kurz (SUH) = Sapium P.Browne EUPHORBIACEAE (D)
Carumbium Reinw. = Homalanthus A.Juss. EUPHORBIACEAE (D)
Carvalhoa K.Schum. APOCYNACEAE (D)

Carvia Bremek. = Strobilanthes Blume ACANTHACEAE (D)
Carya Nutt. JUGLANDACEAE (D)
Caryocar L. CARYOCARACEAE (D)
Caryochloa Spreng. (SUH) = Piptochaetium J.Presl GRAMINEAE (M)
Caryochloa Trin. = Luziola Juss. GRAMINEAE (M)
Caryococca Willd. ex Roem. & Schult. (SUI) = Gonzalagunia Ruíz & Pav. RUBIACEAE (D)
Caryodaphne Blume ex Nees = Cryptocarya R.Br. LAURACEAE (D)
Caryodaphnopsis Airy Shaw LAURACEAE (D)
Caryodendron Karst. EUPHORBIACEAE (D)
Caryolobis Gaertn. = Shorea Roxb. ex C.F.Gaertn. DIPTEROCARPACEAE (D)
Caryolopha Fisch. ex Trautv. (SUS) = Pentaglottis Tausch BORAGINACEAE (D)
Caryomene Barneby & Krukoff MENISPERMACEAE (D)
Caryophyllea Opiz (SUS) = Aira L. GRAMINEAE (M)
Caryophyllus L. = Syzygium Gaertn. MYRTACEAE (D)
Caryopteris Bunge VERBENACEAE (D)
Caryospermum Blume = Perrottetia Kunth CELASTRACEAE (D)
Caryota L. PALMAE (M)
Caryotophora Leistner AIZOACEAE (D)
Casabitoa Alain (UP) EUPHORBIACEAE (D)
Casalea St.-Hil. = Ranunculus L. RANUNCULACEAE (D)
Casasia A.Rich. RUBIACEAE (D)
Cascarilla Wedd. (SUH) = Ladenbergia Klotzsch RUBIACEAE (D)
Cascarilla Adans. (SUH) = Croton L. EUPHORBIACEAE (D)
Cascaronia Griseb. LEGUMINOSAE–PAPILIONOIDEAE (D)
Casearia Jacq. FLACOURTIACEAE (D)
Cashalia Standl. = Dussia Krug & Urb. ex Taub. LEGUMINOSAE–PAPILIONOIDEAE (D)
Casimirella Hassl. ICACINACEAE (D)
Casimiroa La Llave RUTACEAE (D)
Casiostega Gal. (SUI) = Opizia C.Presl GRAMINEAE (M)
Caspareopsis Britton & Rose = Bauhinia L. LEGUMINOSAE–CAESALPINIOIDEAE (D)
Caspia Galushko = Salsola L. CHENOPODIACEAE (D)
Cassandra D.Don = Chamaedaphne Moench ERICACEAE (D)
Cassebeera Kaulf. ADIANTACEAE (F)
Cassebeeria Dennst. = Sonerila Roxb. MELASTOMATACEAE (D)
Casselia Nees & Mart. VERBENACEAE (D)
Cassia L. LEGUMINOSAE–CAESALPINIOIDEAE (D)
Cassidispermum Hemsl. = Burckella Pierre SAPOTACEAE (D)
Cassine L. CELASTRACEAE (D)
Cassinia R.Br. COMPOSITAE (D)
Cassinopsis Sond. ICACINACEAE (D)
Cassiope D.Don ERICACEAE (D)
Cassipourea Aubl. RHIZOPHORACEAE (D)
Cassupa Humb. & Bonpl. = Isertia Schreb. RUBIACEAE (D)
Cassyta J.Mill. (SUH) = Rhipsalis Gaertn. CACTACEAE (D)
Cassytha L. LAURACEAE (D)
Cassytha Mill. (SUH) = Rhipsalis Gaertn. CACTACEAE (D)
Castalis Cass. COMPOSITAE (D)
Castanea Mill. FAGACEAE (D)
Castanola Llanos = Agelaea Sol. ex Planch. CONNARACEAE (D)
Castanopsis (D.Don) Spach FAGACEAE (D)
Castanospermum A.Cunn. ex Hook. LEGUMINOSAE–PAPILIONOIDEAE (D)
Castanospora F.Muell. SAPINDACEAE (D)
Castela Turpin SIMAROUBACEAE (D)
Castelia Cav. VERBENACEAE (D)
Castellanoa Traub AMARYLLIDACEAE (M)

Castellanosia Cárdenas = Browningia Britton & Rose CACTACEAE (D)
Castellia Tineo GRAMINEAE (M)
Castelnavia Tul. & Wedd. PODOSTEMACEAE (D)
Castenedia R.M.King & H.Rob. COMPOSITAE (D)
Castiglionia Ruíz & Pav. = Jatropha L. EUPHORBIACEAE (D)
Castilla Cerv. MORACEAE (D)
Castilleja Mutis ex L.f. SCROPHULARIACEAE (D)
Castilloa Endl. = Castilla Cerv. MORACEAE (D)
Castratella Naudin MELASTOMATACEAE (D)
Casuarina L. CASUARINACEAE (D)
Catabrosa P.Beauv. GRAMINEAE (M)
Catabrosella (Tsvelev) Tsvelev = Colpodium Trin. GRAMINEAE (M)
Catachaetum Hoffmanns. = Catasetum Rich. ex Kunth ORCHIDACEAE (M)
Catadysia O.E.Schulz CRUCIFERAE (D)
Catagyna Hutch. & Dalziel = Scleria Bergius CYPERACEAE (M)
Catalepis Stapf & Stent GRAMINEAE (M)
Catalpa Scop. BIGNONIACEAE (D)
Catamixis Thomson COMPOSITAE (D)
Catananche L. COMPOSITAE (D)
Catanthera F.Muell. MELASTOMATACEAE (D)
Catapodium Link GRAMINEAE (M)
Catappa Gaertn. = Terminalia L. COMBRETACEAE (D)
Catasetum Rich. ex Kunth ORCHIDACEAE (M)
Catatherophora Steud. = Pennisetum Rich. GRAMINEAE (M)
Catatia Humbert COMPOSITAE (D)
Catenaria Benth. = Desmodium Desv. LEGUMINOSAE–PAPILIONOIDEAE (D)
Catenularia Botsch. (SUH) = Catenulina Soják CRUCIFERAE (D)
Catenulina Soják CRUCIFERAE (D)
Catesbaea L. RUBIACEAE (D)
Catesbya Cothen. (SUO) = Catesbaea L. RUBIACEAE (D)
Catha Forssk. ex Schreb. CELASTRACEAE (D)
Catharanthus G.Don APOCYNACEAE (D)
Cathastrum Turcz. = Pleurostylia Wight & Arn. CELASTRACEAE (D)
Cathaya Chun & Kuang PINACEAE (G)
Cathayanthe Chun GESNERIACEAE (D)
Cathayeia Ohwi = Idesia Maxim. FLACOURTIACEAE (D)
Cathcartia Hook.f. = Meconopsis R.Vig. PAPAVERACEAE (D)
Cathea Salisb. = Calopogon R.Br. ORCHIDACEAE (M)
Cathedra Miers OLACACEAE (D)
Cathestecum C.Presl GRAMINEAE (M)
Cathetostemma Blume = Hoya R.Br. ASCLEPIADACEAE (D)
Cathetus Lour. = Phyllanthus L. EUPHORBIACEAE (D)
Cathormion (Benth.) Hassk. = Albizia Durazz. LEGUMINOSAE–MIMOSOIDEAE (D)
Catila Ravenna = Calydorea Herb. IRIDACEAE (M)
Catimbium Juss. = Alpinia Roxb. ZINGIBERACEAE (M)
Catinga Aubl. = Calycorectes O.Berg MYRTACEAE (D)
Catis O.F.Cook = Euterpe Mart. PALMAE (M)
Catoblastus H.A.Wendl. PALMAE (M)
Catocoryne Hook.f. MELASTOMATACEAE (D)
Catoferia (Benth.) Benth. LABIATAE (D)
Catonia P.Browne = Miconia Ruíz & Pav. MELASTOMATACEAE (D)
Catopheria Benth. (SUO) = Catoferia (Benth.) Benth. LABIATAE (D)
Catophractes D.Don BIGNONIACEAE (D)
Catopsis Griseb. BROMELIACEAE (M)
Catosperma Benth. (SUO) = Goodenia Sm. GOODENIACEAE (D)

Catospermum Benth. = Goodenia Sm. GOODENIACEAE (D)
Catostemma Benth. BOMBACACEAE (D)
Catostigma O.F.Cook & Doyle = Catoblastus H.A.Wendl. PALMAE (M)
Cattleya Lindl. ORCHIDACEAE (M)
Cattleyopsis Lem. ORCHIDACEAE (M)
Cattutella Rchb. = Wendlandia Bartl. ex DC. RUBIACEAE (D)
Catunaregam Wolf RUBIACEAE (D)
Caturus L. = Acalypha L. EUPHORBIACEAE (D)
Caucaea Schltr. ORCHIDACEAE (M)
Caucaliopsis H.Wolff = Agrocharis Hochst. UMBELLIFERAE (D)
Caucalis L. UMBELLIFERAE (D)
Caucanthus Forssk. MALPIGHIACEAE (D)
Caudoleucaena Britton & Rose = Leucaena Benth. LEGUMINOSAE–MIMOSOIDEAE (D)
Caulanthus S.Watson CRUCIFERAE (D)
Caularthron Raf. ORCHIDACEAE (M)
Caulocarpus Baker = Tephrosia Pers. LEGUMINOSAE–PAPILIONOIDEAE (D)
Caulokaempferia K.Larsen ZINGIBERACEAE (M)
Caulophyllum Michx. BERBERIDACEAE (D)
Caulostramina Rollins CRUCIFERAE (D)
Causea Scop. = Hirtella L. CHRYSOBALANACEAE (D)
Caustis R.Br. CYPERACEAE (M)
Cautleya (Benth.) Hook.f. ZINGIBERACEAE (M)
Cavacoa J.Léonard EUPHORBIACEAE (D)
Cavalcantia R.M.King & H.Rob. COMPOSITAE (D)
Cavanilla Vell. (SUH) = Caperonia A.St.-Hil. EUPHORBIACEAE (D)
Cavanilla Thunb. = Pyrenacantha Wight ICACINACEAE (D)
Cavanillesia Ruíz & Pav. BOMBACACEAE (D)
Cavaraea Speg. = Tamarindus L. LEGUMINOSAE–CAESALPINIOIDEAE (D)
Cavea W.W.Sm. & Small COMPOSITAE (D)
Cavendishia Lindl. ERICACEAE (D)
Cayaponia Silva Manso CUCURBITACEAE (D)
Caylusea A.St.-Hil. RESEDACEAE (D)
Cayratia Juss. VITACEAE (D)
Ceanothus L. RHAMNACEAE (D)
Ceballosia G.Kunkel (SUI) = Tournefortia L. BORAGINACEAE (D)
Cecarria Barlow LORANTHACEAE (D)
Cecchia Chiov. = Oldfieldia Benth. & Hook.f. EUPHORBIACEAE (D)
Cecropia Loefl. CECROPIACEAE (D)
Cedrela P.Browne MELIACEAE (D)
Cedrelinga Ducke LEGUMINOSAE–MIMOSOIDEAE (D)
Cedrelopsis Baill. PTAEROXYLACEAE (D)
Cedronella Moench LABIATAE (D)
Cedronia Cuatrec. = Picrolemma Hook.f. SIMAROUBACEAE (D)
Cedrostis Post & Kuntze (SUO) = Kedrostis Medik. CUCURBITACEAE (D)
Cedrus Trew PINACEAE (G)
Cedrus Mill. (SUH) = Cedrela P.Browne MELIACEAE (D)
Ceiba Mill. BOMBACACEAE (D)
Celaenodendron Standl. EUPHORBIACEAE (D)
Celastrus L. CELASTRACEAE (D)
Celerina Benoist ACANTHACEAE (D)
Celianella Jabl. EUPHORBIACEAE (D)
Celiantha Maguire GENTIANACEAE (D)
Celmisia Cass. COMPOSITAE (D)
Celosia L. AMARANTHACEAE (D)
Celsia L. = Verbascum L. SCROPHULARIACEAE (D)

Celtis L. ULMACEAE (D)
Cenarrhenes Labill. PROTEACEAE (D)
Cenchropsis Nash = Cenchrus L. GRAMINEAE (M)
Cenchrus L. GRAMINEAE (M)
Cenesmon Gagnep. = Cnesmone Blume EUPHORBIACEAE (D)
Cenocentrum Gagnep. MALVACEAE (D)
Cenolophium W.D.J.Koch ex DC. UMBELLIFERAE (D)
Cenolophon Blume = Alpinia Roxb. ZINGIBERACEAE (M)
Cenostigma Tul. LEGUMINOSAE–CAESALPINIOIDEAE (D)
Censcus Gaertn. = Catunaregam Wolf RUBIACEAE (D)
Centaurea L. COMPOSITAE (D)
Centaurium Hill GENTIANACEAE (D)
Centaurodendron Johow COMPOSITAE (D)
Centauropsis Bojer ex DC. COMPOSITAE (D)
Centaurothamnus Wagenitz & Dittrich COMPOSITAE (D)
Centella L. UMBELLIFERAE (D)
Centema Hook.f. AMARANTHACEAE (D)
Centemopsis Schinz AMARANTHACEAE (D)
Centipeda Lour. COMPOSITAE (D)
Centosteca Desv. (SUO) = Centotheca Desv. GRAMINEAE (M)
Centotheca Desv. GRAMINEAE (M)
Centradenia G.Don MELASTOMATACEAE (D)
Centradeniastrum Cogn. MELASTOMATACEAE (D)
Centrandra Karst. = Julocroton Mart. EUPHORBIACEAE (D)
Centranthera R.Br. SCROPHULARIACEAE (D)
Centranthera Scheidw. (SUH) = Pleurothallis R.Br. ORCHIDACEAE (M)
Centrantheropsis Bonati = Phtheirospermum Bunge ex Fisch. & C.A.Mey. SCROPHULARIACEAE (D)
Centranthus Lam. & DC. VALERIANACEAE (D)
Centrapalus Cass. = Vernonia Schreb. COMPOSITAE (D)
Centratherum Cass. COMPOSITAE (D)
Centrilla Lindau ACANTHACEAE (D)
Centrochilus Schauer = Habenaria Willd. ORCHIDACEAE (M)
Centrochloa Swallen GRAMINEAE (M)
Centrodiscus Müll.Arg. = Caryodendron Karst. EUPHORBIACEAE (D)
Centrogenium Schltr. = Stenorrhynchos Rich. ex Spreng. ORCHIDACEAE (M)
Centroglossa Barb.Rodr. ORCHIDACEAE (M)
Centrolepis Labill. CENTROLEPIDACEAE (M)
Centrolobium Mart. ex Benth. LEGUMINOSAE–PAPILIONOIDEAE (D)
Centronia D.Don MELASTOMATACEAE (D)
Centropetalum Lindl. = Fernandezia Ruíz & Pav. ORCHIDACEAE (M)
Centrophorum Trin. = Chrysopogon Trin. GRAMINEAE (M)
Centroplacus Pierre PANDACEAE (D)
Centropodia Rchb. GRAMINEAE (M)
Centropogon C.Presl CAMPANULACEAE (D)
Centrosema (DC.) Benth. LEGUMINOSAE–PAPILIONOIDEAE (D)
Centrosia A.Rich. = Calanthe Ker–Gawl. ORCHIDACEAE (M)
Centrosis Thouars = Calanthe Ker–Gawl. ORCHIDACEAE (M)
Centrosolenia Benth. = Episcia Mart. GESNERIACEAE (D)
Centrostachys Wall. AMARANTHACEAE (D)
Centrostegia A.Gray POLYGONACEAE (D)
Centrostemma Decne. = Hoya R.Br. ASCLEPIADACEAE (D)
Centrostigma Schltr. ORCHIDACEAE (M)
Centrostylis Baill. = Adenochlaena Boivin ex Baill. EUPHORBIACEAE (D)
Centunculus L. = Anagallis L. PRIMULACEAE (D)

Ceodes J.R.Forst. & G.Forst. = Pisonia L. NYCTAGINACEAE (D)
Cepaea Fabr. = Sedum L. CRASSULACEAE (D)
Cephaelis Sw. = Psychotria L. RUBIACEAE (D)
Cephalacanthus Lindau ACANTHACEAE (D)
Cephalandra Schrad. ex Eckl. & Zeyh. = Coccinia Wight & Arn. CUCURBITACEAE (D)
Cephalangraecum Schltr. = Ancistrorhynchus Finet ORCHIDACEAE (M)
Cephalanthera Rich. ORCHIDACEAE (M)
Cephalantheropsis Guillaumin ORCHIDACEAE (M)
Cephalanthus L. RUBIACEAE (D)
Cephalaralia Harms ARALIACEAE (D)
Cephalaria Schrad. ex Roem. & Schult. DIPSACACEAE (D)
Cephaleis Vahl (SUO) = Psychotria L. RUBIACEAE (D)
Cephalidium A.Rich. ex DC. = Breonia A.Rich. ex DC. RUBIACEAE (D)
Cephalina Thonn. = Sarcocephalus Afzel. ex Sabine RUBIACEAE (D)
Cephalipterum A.Gray COMPOSITAE (D)
Cephalobembix Rydb. = Schkuhria Roth COMPOSITAE (D)
Cephalocarpus Nees CYPERACEAE (M)
Cephaloceraton Gennari = Isoetes L. ISOETACEAE (FA)
Cephalocereus Pfeiff. CACTACEAE (D)
Cephalochloa Coss. & Durieu (SUI) = Ammochloa Boiss. GRAMINEAE (M)
Cephalocleistocactus F.Ritter = Cleistocactus Lem. CACTACEAE (D)
Cephalocroton Hochst. EUPHORBIACEAE (D)
Cephalocrotonopsis Pax EUPHORBIACEAE (D)
Cephalodendron Steyerm. RUBIACEAE (D)
Cephalohibiscus Ulbr. MALVACEAE (D)
Cephalomammillaria Frič = Epithelantha F.A.C.Weber ex Britton & Rose CACTACEAE (D)
Cephalomanes C.Presl HYMENOPHYLLACEAE (F)
Cephalomappa Baill. EUPHORBIACEAE (D)
Cephalomedinilla Merr. = Medinilla Gaudich. MELASTOMATACEAE (D)
Cephalonema K.Schum. = Clappertonia Meisn. TILIACEAE (D)
Cephalopappus Nees & Mart. COMPOSITAE (D)
Cephalopentandra Chiov. CUCURBITACEAE (D)
Cephalophora Cav. = Helenium L. COMPOSITAE (D)
Cephalophyllum (Haw.) N.E.Br. AIZOACEAE (D)
Cephalopodium Korovin UMBELLIFERAE (D)
Cephalorhizum Popov & Korovin PLUMBAGINACEAE (D)
Cephalorhyncus Boiss. (SUO) = Cephalorrhynchus Boiss. COMPOSITAE (D)
Cephalorrhynchus Boiss. COMPOSITAE (D)
Cephaloschefflera (Harms) Merr. = Schefflera J.R.Forst. & G.Forst. ARALIACEAE (D)
Cephalosorus A.Gray COMPOSITAE (D)
Cephalosphaera Warb. MYRISTICACEAE (D)
Cephalostachyum Munro = Schizostachyum Nees GRAMINEAE (M)
Cephalostemon R.H.Schomb. RAPATEACEAE (M)
Cephalostigma A.DC. = Wahlenbergia Schrad. ex Roth CAMPANULACEAE (D)
Cephalostigmaton (Yakovlev) Yakovlev = Sophora L. LEGUMINOSAE–
    PAPILIONOIDEAE (D)
Cephalotaxus Siebold & Zucc. CEPHALOTAXACEAE (G)
Cephalotomandra Karst. & Triana NYCTAGINACEAE (D)
Cephalotrophis Blume = Trophis P.Browne MORACEAE (D)
Cephalotus Labill. CEPHALOTACEAE (D)
Ceradenia L.E.Bishop GRAMMITIDACEAE (F)
Ceraia Lour. = Dendrobium Sw. ORCHIDACEAE (M)
Ceramanthus Hassk. = Phyllanthus L. EUPHORBIACEAE (D)
Ceramanthus (Kunze) Malme (SUH) = Funastrum Fourn. ASCLEPIADACEAE (D)
Ceramicalyx Blume = Osbeckia L. MELASTOMATACEAE (D)

Ceramium Blume = Thottea Rottb. ARISTOLOCHIACEAE (D)
Ceranthera Raf. (SUH) = Solanum L. SOLANACEAE (D)
Ceranthera Elliott (SUH) = Dicerandra Benth. LABIATAE (D)
Ceranthus Schreb. = Chionanthus L. OLEACEAE (D)
Cerapadus Buia = Prunus L. ROSACEAE (D)
Ceraria H.Pearson & Stephens PORTULACACEAE (D)
Ceraseidos Siebold & Zucc. = Prunus L. ROSACEAE (D)
Ceraselma Wittst. = Euphorbia L. EUPHORBIACEAE (D)
Cerasiocarpum Hook.f. = Kedrostis Medik. CUCURBITACEAE (D)
Cerastium L. CARYOPHYLLACEAE (D)
Cerasus Mill. = Prunus L. ROSACEAE (D)
Ceratandra Eckl. ex F.A.Bauer ORCHIDACEAE (M)
Ceratanthus F.Muell. ex G.Taylor LABIATAE (D)
Ceratephorus De Vriese (SUO) = Payena A.DC. SAPOTACEAE (D)
Ceratiola Michx. EMPETRACEAE (D)
Ceratiosicyos Nees ACHARIACEAE (D)
Ceratistes Labour. (SUI) = Eriosyce Phil. CACTACEAE (D)
Ceratites Miers = Rudgea Salisb. RUBIACEAE (D)
Ceratium Blume = Eria Lindl. ORCHIDACEAE (M)
Ceratocapnos Durieu PAPAVERACEAE (D)
Ceratocarpus L. CHENOPODIACEAE (D)
Ceratocaryum Nees RESTIONACEAE (M)
Ceratocaulos (Bernh.) Reichenb. (SUS) = Datura L. SOLANACEAE (D)
Ceratocentron Senghas ORCHIDACEAE (M)
Ceratocephala Moench RANUNCULACEAE (D)
Ceratocephalus Pers. (SUO) = Ceratocephala Moench RANUNCULACEAE (D)
Ceratochaete Lunell (SUS) = Zizania L. GRAMINEAE (M)
Ceratochilus Blume ORCHIDACEAE (M)
Ceratochloa P.Beauv. = Bromus L. GRAMINEAE (M)
Ceratocnemum Coss. & Balansa CRUCIFERAE (D)
Ceratococcus Meisn. = Pterococcus Hassk. EUPHORBIACEAE (D)
Ceratodactylis J.Sm. = Llavea Lag. ADIANTACEAE (F)
Ceratogyne Turcz. COMPOSITAE (D)
Ceratogynum Wight = Sauropus Blume EUPHORBIACEAE (D)
Ceratoides Gagnebin (SUS) = Ceratocarpus L. CHENOPODIACEAE (D)
Ceratolacis (Tul.) Wedd. PODOSTEMACEAE (D)
Ceratolobus Blume PALMAE (M)
Ceratominthe Briq. = Satureja L. LABIATAE (D)
Ceratonia L. LEGUMINOSAE–CAESALPINIOIDEAE (D)
Ceratopetalum Sm. CUNONIACEAE (D)
Ceratophorus Sond. = Suregada Roxb. ex Rottler EUPHORBIACEAE (D)
Ceratophyllum L. CERATOPHYLLACEAE (D)
Ceratophytum Pittier BIGNONIACEAE (D)
Ceratopsis Lindl. = Epipogium Borkh. ORCHIDACEAE (M)
Ceratopteris Brongn. PARKERIACEAE (F)
Ceratopyxis Hook.f. RUBIACEAE (D)
Ceratosanthes Burm. ex Adans. CUCURBITACEAE (D)
Ceratoscyphus Chun = Ornithoboea Parish ex C.B.Clarke GESNERIACEAE (D)
Ceratosepalum Oerst. = Passiflora L. PASSIFLORACEAE (D)
Ceratosepalum Oliv. (SUH) = Triumfetta L. TILIACEAE (D)
Ceratostema Juss. ERICACEAE (D)
Ceratostigma Bunge PLUMBAGINACEAE (D)
Ceratostylis Blume ORCHIDACEAE (M)
Ceratotheca Endl. PEDALIACEAE (D)
Ceratozamia Brongn. ZAMIACEAE (G)

Cerbera L. APOCYNACEAE (D)
Cerberiopsis Vieill. ex Pancher & Sebert APOCYNACEAE (D)
Cercestis Schott ARACEAE (M)
Cercidiopsis Britton & Rose = Parkinsonia L. LEGUMINOSAE–CAESALPINIOIDEAE (D)
Cercidiphyllum Siebold & Zucc. CERCIDIPHYLLACEAE (D)
Cercidium Tul. = Parkinsonia L. LEGUMINOSAE–CAESALPINIOIDEAE (D)
Cercis L. LEGUMINOSAE–CAESALPINIOIDEAE (D)
Cercocarpus Kunth ROSACEAE (D)
Cercocoma Wall. ex G.Don = Strophanthus DC. APOCYNACEAE (D)
Cercophora Miers = Lecythis Loefl. LECYTHIDACEAE (D)
Cerdia Moçino & Sessé CARYOPHYLLACEAE (D)
Cerea Schltdl. (SUS) = Paspalum L. GRAMINEAE (M)
Ceresia Pers. = Paspalum L. GRAMINEAE (M)
Cereus Mill. CACTACEAE (D)
Cerinthe L. BORAGINACEAE (D)
Ceriops Arn. RHIZOPHORACEAE (D)
Ceriosperma (O.E.Schulz) Greuter & Burdet CRUCIFERAE (D)
Ceriscoides (Hook.f.) Tirveng. RUBIACEAE (D)
Ceriscus Gaertn. ex Nees (SUI) = Catunaregam Wolf RUBIACEAE (D)
Cerocarpus Colebr. ex Hassk. = Syzygium Gaertn. MYRTACEAE (D)
Cerochilus Lindl. = Hetaeria Blume ORCHIDACEAE (M)
Cerochlamys N.E.Br. AIZOACEAE (D)
Cerolepis Pierre = Camptostylus Gilg FLACOURTIACEAE (D)
Ceropegia L. ASCLEPIADACEAE (D)
Cerophyllum Spach = Ribes L. GROSSULARIACEAE (D)
Ceropteris Link = Pityrogramma Link ADIANTACEAE (F)
Cerosora (Baker) Domin ADIANTACEAE (F)
Ceroxylon Bonpl. ex DC. PALMAE (M)
Cerqueiria O.Berg = Gomidesia O.Berg MYRTACEAE (D)
Ceruana Forssk. COMPOSITAE (D)
Cervantesia Ruíz & Pav. SANTALACEAE (D)
Cervia Rodr. ex Lag. = Rochelia Rchb. BORAGINACEAE (D)
Cervicina Delile = Wahlenbergia Schrad. ex Roth CAMPANULACEAE (D)
Cespedesia Goudot OCHNACEAE (D)
Cestichis Thouars = Liparis Rich. ORCHIDACEAE (M)
Cestrum L. SOLANACEAE (D)
Ceterac Adans. = Asplenium L. ASPLENIACEAE (F)
Ceterach Willd. ASPLENIACEAE (F)
Ceterachopsis (J.Sm.) Ching = Ceterach Willd. ASPLENIACEAE (F)
Ceuthocarpus Aiello RUBIACEAE (D)
Ceuthostoma L.A.S.Johnson CASUARINACEAE (D)
Cevallia Lag. LOASACEAE (D)
Ceytosis Munro (SUI) = Crypsis Aiton GRAMINEAE (M)
Chaboissaea Fourn. = Muhlenbergia Schreb. GRAMINEAE (M)
Chacaya Escal. = Discaria Hook. RHAMNACEAE (D)
Chacoa R.M.King & H.Rob. COMPOSITAE (D)
Chadsia Bojer LEGUMINOSAE–PAPILIONOIDEAE (D)
Chaenactis DC. COMPOSITAE (D)
Chaenanthe Lindl. ORCHIDACEAE (M)
Chaenesthes Miers = Iochroma Benth. SOLANACEAE (D)
Chaenocarpus Juss. = Spermacoce L. RUBIACEAE (D)
Chaenocephalus Griseb. = Verbesina L. COMPOSITAE (D)
Chaenomeles Lindl. ROSACEAE (D)
Chaenorhinum (DC.) Rchb. SCROPHULARIACEAE (D)

Chaenorrhinum (DC.) Rchb. (SUO) = Chaenorhinum (DC.) Rchb.
   SCROPHULARIACEAE (D)
Chaenostoma Benth. = Sutera Roth SCROPHULARIACEAE (D)
Chaenotheca Urb. (SUH) = Chascotheca Urb. EUPHORBIACEAE (D)
Chaerophyllopsis Boissieu UMBELLIFERAE (D)
Chaerophyllum L. UMBELLIFERAE (D)
Chaetacanthus Nees ACANTHACEAE (D)
Chaetachme Planch. ULMACEAE (D)
Chaetacme Planch. (SUO) = Chaetachme Planch. ULMACEAE (D)
Chaetadelpha A.Gray ex S.Watson COMPOSITAE (D)
Chaetaea Jacq. = Byttneria Loefl. STERCULIACEAE (D)
Chaetanthera Ruíz & Pav. COMPOSITAE (D)
Chaetanthus R.Br. RESTIONACEAE (M)
Chaetaria P.Beauv. (SUS) = Aristida L. GRAMINEAE (M)
Chaetium Nees GRAMINEAE (M)
Chaetobromus Nees GRAMINEAE (M)
Chaetocalyx DC. LEGUMINOSAE–PAPILIONOIDEAE (D)
Chaetocarpus Thwaites EUPHORBIACEAE (D)
Chaetocarpus Schreb. (SUH) = Pouteria Aubl. SAPOTACEAE (D)
Chaetocephala Barb.Rodr. = Myoxanthus Poepp. & Endl. ORCHIDACEAE (M)
Chaetochlamys Lindau = Justicia L. ACANTHACEAE (D)
Chaetochloa Scribn. (SUS) = Setaria P.Beauv. GRAMINEAE (M)
Chaetocyperus Nees = Eleocharis R.Br. CYPERACEAE (M)
Chaetogastra DC. = Tibouchina Aubl. MELASTOMATACEAE (D)
Chaetolepis (DC.) Miq. MELASTOMATACEAE (D)
Chaetolimon (Bunge) Lincz. PLUMBAGINACEAE (D)
Chaetonychia (DC.) Sweet ILLECEBRACEAE (D)
Chaetopappa DC. COMPOSITAE (D)
Chaetopoa C.E.Hubb. GRAMINEAE (M)
Chaetopogon Janchen GRAMINEAE (M)
Chaetosciadium Boiss. UMBELLIFERAE (D)
Chaetospira S.F.Blake COMPOSITAE (D)
Chaetospora R.Br. = Schoenus L. CYPERACEAE (M)
Chaetostachydium Airy Shaw RUBIACEAE (D)
Chaetostachys Valeton = Chaetostachydium Airy Shaw RUBIACEAE (D)
Chaetostichium C.E.Hubb. = Oropetium Trin. GRAMINEAE (M)
Chaetostoma DC. MELASTOMATACEAE (D)
Chaetosus Benth. = Parsonsia R.Br. APOCYNACEAE (D)
Chaetothylax Nees = Justicia L. ACANTHACEAE (D)
Chaetothylopsis Oerst. = Justicia L. ACANTHACEAE (D)
Chaetotropis Kunth = Polypogon Desf. GRAMINEAE (M)
Chaeturus Link (SUH) = Chaetopogon Janchen GRAMINEAE (M)
Chaetymenia Hook. & Arn. COMPOSITAE (D)
Chaffeyopuntia Frič & Schelle = Opuntia Mill. CACTACEAE (D)
Chaiturus Ehrh. ex Willd. LABIATAE (D)
Chaixia Lapeyr. = Ramonda Rich. GESNERIACEAE (D)
Chalarothyrsus Lindau ACANTHACEAE (D)
Chalazocarpus Hiern = Schumanniophyton Harms RUBIACEAE (D)
Chalcanthus Boiss. CRUCIFERAE (D)
Chalcas L. = Murraya L. RUTACEAE (D)
Chalcoelytrum Lunell (SUS) = Chrysopogon Trin. GRAMINEAE (M)
Chalema Dieterle CUCURBITACEAE (D)
Chalepoa Hook.f. = Tribeles Phil. ESCALLONIACEAE (D)
Chalepophyllum Hook.f. RUBIACEAE (D)
Chalynochlamys Franch. (SUI) = Arundinella Raddi GRAMINEAE (M)

Chamabainia Wight URTICACEAE (D)
Chamaeacanthus Chiov. = Campylanthus Roth SCROPHULARIACEAE (D)
Chamaealoe A.Berger = Aloe L. ALOACEAE (M)
Chamaeangis Schltr. ORCHIDACEAE (M)
Chamaeanthus Schltr. ex J.J.Sm. ORCHIDACEAE (M)
Chamaeanthus Ule (SUH) = Geogenanthus Ule COMMELINACEAE (M)
Chamaebatia Benth. ROSACEAE (D)
Chamaebatiaria (Porter) Maxim. ROSACEAE (D)
Chamaecalamus Meyen = Calamagrostis Adans. GRAMINEAE (M)
Chamaecereus Britton & Rose = Echinopsis Zucc. CACTACEAE (D)
Chamaechaenactis Rydb. COMPOSITAE (D)
Chamaecladon Miq. = Homalomena Schott ARACEAE (M)
Chamaeclitandra (Stapf) Pichon APOCYNACEAE (D)
Chamaecrista Moench LEGUMINOSAE–CAESALPINIOIDEAE (D)
Chamaecyparis Spach CUPRESSACEAE (G)
Chamaecytisus Link LEGUMINOSAE–PAPILIONOIDEAE (D)
Chamaedactylis T.Nees = Aeluropus Trin. GRAMINEAE (M)
Chamaedaphne Mitch. (SUH) = Mitchella L. RUBIACEAE (D)
Chamaedaphne Moench ERICACEAE (D)
Chamaedorea Willd. PALMAE (M)
Chamaefilix Hill ex Farw. = Asplenium L. ASPLENIACEAE (F)
Chamaegastrodia Makino & F.Maek. ORCHIDACEAE (M)
Chamaegeron Schrenk COMPOSITAE (D)
Chamaegigas Dinter ex Heil = Lindernia All. SCROPHULARIACEAE (D)
Chamaegyne Suess. = Eleocharis R.Br. CYPERACEAE (M)
Chamaele Miq. UMBELLIFERAE (D)
Chamaeleon Cass. COMPOSITAE (D)
Chamaelirium Willd. MELANTHIACEAE (M)
Chamaelobivia Y.Ito (SUI) = Echinopsis Zucc. CACTACEAE (D)
Chamaemeles Lindl. ROSACEAE (D)
Chamaemelum Mill. COMPOSITAE (D)
Chamaenerion Spach = Epilobium L. ONAGRACEAE (D)
Chamaepentas Bremek. RUBIACEAE (D)
Chamaepericlymenum Hill = Cornus L. CORNACEAE (D)
Chamaephoenix H.A.Wendl. ex Curtiss = Pseudophoenix H.A.Wendl. ex Sarg.
    PALMAE (M)
Chamaepitys Hill = Ajuga L. LABIATAE (D)
Chamaepus Wagenitz COMPOSITAE (D)
Chamaeranthemum Nees ACANTHACEAE (D)
Chamaeraphis R.Br. GRAMINEAE (M)
Chamaerepes Spreng. = Herminium L. ORCHIDACEAE (M)
Chamaerhodiola Nakai = Rhodiola L. CRASSULACEAE (D)
Chamaerhodos Bunge ROSACEAE (D)
Chamaeriphe Steck = Chamaerops L. PALMAE (M)
Chamaeriphes Ponted. ex Gaertn. = Chamaerops L. PALMAE (M)
Chamaerops L. PALMAE (M)
Chamaesaracha (A.Gray) Benth. SOLANACEAE (D)
Chamaesciadium C.A.Mey. UMBELLIFERAE (D)
Chamaescilla F.Muell. ex Benth. ANTHERICACEAE (M)
Chamaesenna (DC.) Raf. ex Pittier = Senna Mill. LEGUMINOSAE–
    CAESALPINIOIDEAE (D)
Chamaesium H.Wolff UMBELLIFERAE (D)
Chamaespartium Adans. = Genista L. LEGUMINOSAE–PAPILIONOIDEAE (D)
Chamaesphacos Schrenk ex Fisch. & C.A.Mey. LABIATAE (D)
Chamaesyce Gray = Euphorbia L. EUPHORBIACEAE (D)

Chamaexeros Benth. LOMANDRACEAE (M)
Chamagrostis Borkh. (SUS) = Mibora Adans. GRAMINEAE (M)
Chamarea Eckl. & Zeyh. UMBELLIFERAE (D)
Chamartemisia Rydb. = Artemisia L. COMPOSITAE (D)
Chambeyronia Vieill. PALMAE (M)
Chamelaucium Desf. MYRTACEAE (D)
Chamelophyton Garay ORCHIDACEAE (M)
Chamelum Phil. = Olsynium Raf. IRIDACEAE (M)
Chamerion (Raf.) Raf. = Epilobium L. ONAGRACEAE (D)
Chamguava Landrum MYRTACEAE (D)
Chamira Thunb. CRUCIFERAE (D)
Chamisme Raf. = Houstonia L. RUBIACEAE (D)
Chamissoa Kunth AMARANTHACEAE (D)
Chamissoniophila Brand BORAGINACEAE (D)
Chamorchis Rich. ORCHIDACEAE (M)
Champereia Griff. OPILIACEAE (D)
Championella Bremek. = Strobilanthes Blume ACANTHACEAE (D)
Championia Gardner GESNERIACEAE (D)
Chandrasekharania V.J.Nair, V.S.Ramach. & Sreek. GRAMINEAE (M)
Chanekia Lundell = Licaria Aubl. LAURACEAE (D)
Changium H.Wolff UMBELLIFERAE (D)
Changnienia S.S.Chien ORCHIDACEAE (M)
Chapelieria A.Rich. ex DC. RUBIACEAE (D)
Chapmannia Torr. & A.Gray LEGUMINOSAE–PAPILIONOIDEAE (D)
Chapmanolirion Dinter AMARYLLIDACEAE (M)
Chaptalia Vent. COMPOSITAE (D)
Characias Gray = Euphorbia L. EUPHORBIACEAE (D)
Charadrophila Marloth SCROPHULARIACEAE (D)
Chardinia Desf. COMPOSITAE (D)
Charesia E.A.Busch = Silene L. CARYOPHYLLACEAE (D)
Charia C.E.C.Fisch. = Ekebergia Sparrm. MELIACEAE (D)
Charianthus D.Don MELASTOMATACEAE (D)
Charidia Baill. = Savia Willd. EUPHORBIACEAE (D)
Charidion Bong. = Luxemburgia A.St.–Hil. OCHNACEAE (D)
Chariessa Miq. = Citronella D.Don ICACINACEAE (D)
Charpentiera Gaudich. AMARANTHACEAE (D)
Charpentiera Vieill. (SUH) = Ixora L. RUBIACEAE (D)
Chartocalyx Regel = Otostegia Benth. LABIATAE (D)
Chartolepis Cass. = Centaurea L. COMPOSITAE (D)
Chartoloma Bunge CRUCIFERAE (D)
Chascanum E.Mey. VERBENACEAE (D)
Chascolytrum Desv. = Briza L. GRAMINEAE (M)
Chascotheca Urb. EUPHORBIACEAE (D)
Chasea Nieuwl. (SUI) = Panicum L. GRAMINEAE (M)
Chasechloa A.Camus = Echinolaena Desv. GRAMINEAE (M)
Chaseella Summerh. ORCHIDACEAE (M)
Chasmanthe N.E.Br. IRIDACEAE (M)
Chasmanthera Hochst. MENISPERMACEAE (D)
Chasmanthium Link GRAMINEAE (M)
Chasmatocallis R.C.Foster = Lapeirousia Pourr. IRIDACEAE (M)
Chasmatophyllum (Schwantes) Dinter & Schwantes AIZOACEAE (D)
Chasmia Schott ex Spreng. = Arrabidaea DC. BIGNONIACEAE (D)
Chasmopodium Stapf GRAMINEAE (M)
Chassalia Comm. ex Poir. RUBIACEAE (D)
Chastenaea DC. = Axinaea Ruíz & Pav. MELASTOMATACEAE (D)

Chatinia Tiegh. = Psittacanthus Mart. LORANTHACEAE (D)
Chaubardia Rchb.f. ORCHIDACEAE (M)
Chaubardiella Garay ORCHIDACEAE (M)
Chauliodon Summerh. ORCHIDACEAE (M)
Chaunanthus O.E.Schulz = Iodanthus (Torr. & A.Gray) Steud. CRUCIFERAE (D)
Chaunochiton Benth. OLACACEAE (D)
Chaunostoma Donn.Sm. LABIATAE (D)
Chauvinia Steud. (SUH) = Spartina Schreb. GRAMINEAE (M)
Chavannesia A.DC. = Urceola Roxb. APOCYNACEAE (D)
Chaydaia Pit. RHAMNACEAE (D)
Chayota Jacq. = Sechium P.Browne CUCURBITACEAE (D)
Chazaliella E.Petit & Verdc. RUBIACEAE (D)
Cheesemania O.E.Schulz CRUCIFERAE (D)
Cheilanthes Sw. ADIANTACEAE (F)
Cheilanthopsis Hieron. WOODSIACEAE (F)
Cheiloclinium Miers CELASTRACEAE (D)
Cheiloepton Fée = Lomagramma J.Sm. LOMARIOPSIDACEAE (F)
Cheilogramme (Blume) Underwood = Neurodium Fée POLYPODIACEAE (F)
Cheiloplecton Fée ADIANTACEAE (F)
Cheilopsis Moq. = Acanthus L. ACANTHACEAE (D)
Cheilosa Blume EUPHORBIACEAE (D)
Cheilosandra Griff. ex Lindl. = Rhynchotechum Blume GESNERIACEAE (D)
Cheilosoria Trevis. = Pellaea Link ADIANTACEAE (F)
Cheilotheca Hook.f. ERICACEAE (D)
Cheiradenia Lindl. ORCHIDACEAE (M)
Cheiranthera A.Cunn. ex Brongn. PITTOSPORACEAE (D)
Cheiranthus L. = Erysimum L. CRUCIFERAE (D)
Cheiridopsis N.E.Br. AIZOACEAE (D)
Cheirodendron Nutt. ex Seem. ARALIACEAE (D)
Cheiroglossa C.Presl = Ophioglossum L. OPHIOGLOSSACEAE (F)
Cheirolaena Benth. STERCULIACEAE (D)
Cheirolophus Cass. COMPOSITAE (D)
Cheiropleuria C.Presl CHEIROPLEURIACEAE (F)
Cheiropteris Christ = Neocheiropteris Christ POLYPODIACEAE (F)
Cheiropterocephalus Barb.Rodr. = Malaxis Sol. ex Sw. ORCHIDACEAE (M)
Cheirorchis C.E.Carr = Cordiglottis J.J.Sm. ORCHIDACEAE (M)
Cheirostemon Humb. & Bonpl. = Chiranthodendron Larreat. STERCULIACEAE (D)
Cheirostylis Blume ORCHIDACEAE (M)
Chelidonium L. PAPAVERACEAE (D)
Chelonanthera Blume = Pholidota Lindl. ex Hook. ORCHIDACEAE (M)
Chelonanthus (Griseb.) Gilg = Irlbachia Mart. GENTIANACEAE (D)
Chelone L. SCROPHULARIACEAE (D)
Chelonecarya Pierre = Rhaphiostylis Planch. ex Benth. ICACINACEAE (D)
Chelonespermum Hemsl. = Burckella Pierre SAPOTACEAE (D)
Chelonistele Pfitzer ORCHIDACEAE (M)
Chelonopsis Miq. LABIATAE (D)
Chelyocarpus Dammer PALMAE (M)
Chennapyrum A.Löve = Aegilops L. GRAMINEAE (M)
Chenolea Thunb. = Bassia All. CHENOPODIACEAE (D)
Chenoleoides Botsch. = Bassia All. CHENOPODIACEAE (D)
Chenopodiopsis Hilliard & B.L.Burtt SCROPHULARIACEAE (D)
Chenopodium L. CHENOPODIACEAE (D)
Chersodoma Phil. COMPOSITAE (D)
Chesnea Scop. = Psychotria L. RUBIACEAE (D)
Chesneya Lindl. ex Endl. LEGUMINOSAE-PAPILIONOIDEAE (D)

Chetyson Raf. = Sedum L. CRASSULACEAE (D)
Chevaliera Gaudich. ex Beer = Aechmea Ruíz & Pav. BROMELIACEAE (M)
Chevalierella A.Camus GRAMINEAE (M)
Chevalierodendron J.–F.Leroy = Streblus Lour. MORACEAE (D)
Chevreulia Cass. COMPOSITAE (D)
Cheynia J.Drumm. ex Harv. = Balaustion Hook. MYRTACEAE (D)
Chiangiodendron Wendt FLACOURTIACEAE (D)
Chiapasia Britton & Rose = Disocactus Lindl. CACTACEAE (D)
Chiarinia Chiov. = Lecaniodiscus Planch. ex Benth. SAPINDACEAE (D)
Chiastophyllum (Ledeb.) A.Berger CRASSULACEAE (D)
Chiazospermum Bernh. = Hypecoum L. PAPAVERACEAE (D)
Chicharronia A.Rich. = Terminalia L. COMBRETACEAE (D)
Chichipia Backeb. (SUI) = Polaskia Backeb. CACTACEAE (D)
Chickassia Wight & Arn. = Chukrasia A.Juss. MELIACEAE (D)
Chiclea Lundell = Manilkara Adans. SAPOTACEAE (D)
Chidlowia Hoyle LEGUMINOSAE–CAESALPINIOIDEAE (D)
Chienia W.T.Wang = Delphinium L. RANUNCULACEAE (D)
Chieniodendron Tsiang & P.T.Li ANNONACEAE (D)
Chieniopteris Ching = Woodwardia Sm. BLECHNACEAE (F)
Chigua D.W.Stev. ZAMIACEAE (G)
Chikusichloa Koidz. GRAMINEAE (M)
Chilenia Backeb. [1939] = Neoporteria Britton & Rose CACTACEAE (D)
Chilenia Backeb. [1938] = Neoporteria Britton & Rose CACTACEAE (D)
Chilenia Backeb. [1935] (SUI) = Neoporteria Britton & Rose CACTACEAE (D)
Chileniopsis Backeb. = Neoporteria Britton & Rose CACTACEAE (D)
Chileocactus Frič (SUI) = Neoporteria Britton & Rose CACTACEAE (D)
Chileorebutia Frič (SUI) = Neoporteria Britton & Rose CACTACEAE (D)
Chileranthemum Oerst. ACANTHACEAE (D)
Chiliadenus Cass. COMPOSITAE (D)
Chiliandra Griff. = Rhynchotechum Blume GESNERIACEAE (D)
Chilianthus Burch. = Buddleja L. BUDDLEJACEAE (D)
Chiliocephalum Benth. COMPOSITAE (D)
Chiliophyllum Phil. COMPOSITAE (D)
Chiliorebutia Frič (SUO) = Neoporteria Britton & Rose CACTACEAE (D)
Chiliotrichiopsis Cabrera COMPOSITAE (D)
Chiliotrichum Cass. COMPOSITAE (D)
Chilita Orcutt = Mammillaria Haw. CACTACEAE (D)
Chillania Roiv. CYPERACEAE (M)
Chilocalyx Klotzsch = Cleome L. CAPPARACEAE (D)
Chilocardamum O.E.Schulz CRUCIFERAE (D)
Chilocarpus Blume APOCYNACEAE (D)
Chilochloa P.Beauv. = Phleum L. GRAMINEAE (M)
Chiloglossa Oerst. = Justicia L. ACANTHACEAE (D)
Chiloglottis R.Br. ORCHIDACEAE (M)
Chilopogon Schltr. = Appendicula Blume ORCHIDACEAE (M)
Chilopsis D.Don BIGNONIACEAE (D)
Chilopteris (C.Presl) Lindl. = Grammitis Sw. GRAMMITIDACEAE (F)
Chiloschista Lindl. ORCHIDACEAE (M)
Chimantaea Maguire, Steyerm. & Wurdack COMPOSITAE (D)
Chimaphila Pursh ERICACEAE (D)
Chimarrhis Jacq. RUBIACEAE (D)
Chimonanthus Lindl. CALYCANTHACEAE (D)
Chimonobambusa Makino GRAMINEAE (M)
Chimonocalamus Hsueh & Yi = Sinarundinaria Nakai GRAMINEAE (M)
Chingia Holttum THELYPTERIDACEAE (F)

Chingiacanthus Hand.–Mazz. = Isoglossa Oerst. ACANTHACEAE (D)
Chiococca P.Browne RUBIACEAE (D)
Chiogenes Salisb. ex Torr. = Gaultheria Kalm ex L. ERICACEAE (D)
Chionachne R.Br. GRAMINEAE (M)
Chionanthus L. OLEACEAE (D)
Chione DC. RUBIACEAE (D)
Chionocharis I.M.Johnst. BORAGINACEAE (D)
Chionochloa Zotov GRAMINEAE (M)
Chionodoxa Boiss. HYACINTHACEAE (M)
Chionographis Maxim. MELANTHIACEAE (M)
Chionohebe B.G.Briggs & Ehrend. SCROPHULARIACEAE (D)
Chionolaena DC. COMPOSITAE (D)
Chionopappus Benth. COMPOSITAE (D)
Chionophila Benth. SCROPHULARIACEAE (D)
Chionothrix Hook.f. AMARANTHACEAE (D)
Chiovendaea Speg. = Coursetia DC. LEGUMINOSAE–PAPILIONOIDEAE (D)
Chiranthodendron Larreat. STERCULIACEAE (D)
Chiridium Tiegh. = Helixanthera Lour. LORANTHACEAE (D)
Chirita Buch.–Ham. ex D.Don GESNERIACEAE (D)
Chiritopsis W.T.Wang GESNERIACEAE (D)
Chirocalyx Meisn. = Erythrina L. LEGUMINOSAE–PAPILIONOIDEAE (D)
Chironia L. GENTIANACEAE (D)
Chiropetalum A.Juss. EUPHORBIACEAE (D)
Chirripoa Suess. = Guzmania Ruíz & Pav. BROMELIACEAE (M)
Chisocheton Blume MELIACEAE (D)
Chitonanthera Schltr. ORCHIDACEAE (M)
Chitonia Moçino & Sessé = Morkillia Rose & Painter ZYGOPHYLLACEAE (D)
Chitonochilus Schltr. ORCHIDACEAE (M)
Chlaenandra Miq. MENISPERMACEAE (D)
Chlaenosciadium C.Norman UMBELLIFERAE (D)
Chlainanthus Briq. = Lagochilus Bunge LABIATAE (D)
Chlamydacanthus Lindau ACANTHACEAE (D)
Chlamydanthus C.A.Mey. = Thymelaea Mill. THYMELAEACEAE (D)
Chlamydites J.R.Drumm. = Aster L. COMPOSITAE (D)
Chlamydoboea Stapf = Paraboea (C.B.Clarke) Ridl. GESNERIACEAE (D)
Chlamydocardia Lindau ACANTHACEAE (D)
Chlamydocarya Baill. ICACINACEAE (D)
Chlamydocola (K.Schum.) Bodard = Cola Schott & Endl. STERCULIACEAE (D)
Chlamydogramme Holttum DRYOPTERIDACEAE (F)
Chlamydojatropha Pax & K.Hoffm. EUPHORBIACEAE (D)
Chlamydophora Ehrenb. ex Less. COMPOSITAE (D)
Chlamydophytum Mildbr. BALANOPHORACEAE (D)
Chlamydostachya Mildbr. ACANTHACEAE (D)
Chlamydostylus Baker = Nemastylis Nutt. IRIDACEAE (M)
Chlidanthus Herb. AMARYLLIDACEAE (D)
Chloachne Stapf = Poecilostachys Hack. GRAMINEAE (M)
Chlovummia Raf. = Vulpia C.C.Gmelin GRAMINEAE (M)
Chloanthes R.Br. VERBENACEAE (D)
Chloothamnus Büse = Nastus Juss. GRAMINEAE (M)
Chlora Adans. = Blackstonia Huds. GENTIANACEAE (D)
Chloracantha G.L.Nesom, Y.B.Suh, D.R.Morgan, S.D.Sundb. & B.B.Simpson
    COMPOSITAE (D)
Chloradenia Baill. = Cladogynos Zipp. ex Spanoghe EUPHORBIACEAE (D)
Chloraea Lindl. ORCHIDACEAE (M)
Chloranthus Sw. CHLORANTHACEAE (D)

Chloridion Stapf (SUH) = Stereochlaena Hack. GRAMINEAE (M)
Chloridopsis Hack. (SUI) = Trichloris E.Fourn. ex Benth. GRAMINEAE (M)
Chloris Sw. GRAMINEAE (M)
Chlorocalymma Clayton GRAMINEAE (M)
Chlorocardium Rohwer, H.G.Richt. & van der Werff LAURACEAE (D)
Chlorocarpa Alston FLACOURTIACEAE (D)
Chlorochorion Puff & Robbr. = Pentanisia Harv. RUBIACEAE (D)
Chlorocodon Hook.f. = Mondia Skeels ASCLEPIADACEAE (D)
Chlorocrambe Rydb. CRUCIFERAE (D)
Chlorocyathus Oliv. ASCLEPIADACEAE (D)
Chlorocyperus Rikli = Cyperus L. CYPERACEAE (M)
Chlorogalum (Lindl.) Kunth HYACINTHACEAE (M)
Chloroides Regel (SUI) = Eustachys Desv. GRAMINEAE (M)
Chloroleucon (Benth.) Record LEGUMINOSAE–MIMOSOIDEAE (D)
Chloroleucum (Benth.) Record (SUO) = Chloroleucon (Benth.) Record LEGUMINOSAE–
    MIMOSOIDEAE (D)
Chloroluma Baill. = Chrysophyllum L. SAPOTACEAE (D)
Chloromyrtus Pierre = Eugenia L. MYRTACEAE (D)
Chloropatane Engl. = Erythrococca Benth. EUPHORBIACEAE (D)
Chlorophora Gaudich. = Maclura Nutt. MORACEAE (D)
Chlorophytum Ker Gawl. ANTHERICACEAE (M)
Chloropsis Kuntze (SUS) = Trichloris E.Fourn. ex Benth. GRAMINEAE (M)
Chlorosa Blume ORCHIDACEAE (M)
Chlorospatha Engl. ARACEAE (M)
Chlorostemma (Lange) Fourr. = Asperula L. RUBIACEAE (D)
Chlorostis Raf. (SUS) = Chloris Sw. GRAMINEAE (M)
Chloroxylon DC. RUTACEAE (D)
Chnoophora Kaulf. = Cyathea Sm. CYATHEACEAE (F)
Choananthus Rendle = Scadoxus Raf. AMARYLLIDACEAE (M)
Chocho Adans. = Sechium P.Browne CUCURBITACEAE (D)
Chodanthus Hassl. = Mansoa DC. BIGNONIACEAE (D)
Chodsha–Kasiana Rauschert (SUS) = Catenulina Soják CRUCIFERAE (D)
Choeradoplectron Schauer = Habenaria Willd. ORCHIDACEAE (M)
Choerospondias B.L.Burtt & A.W.Hill ANACARDIACEAE (D)
Choisya Kunth RUTACEAE (D)
Chomelia Jacq. RUBIACEAE (D)
Chondrachne R.Br. = Lepironia Rich. CYPERACEAE (M)
Chondrachyrum Nees = Briza L. GRAMINEAE (M)
Chondradenia Maxim. ex Makino = Orchis L. ORCHIDACEAE (M)
Chondrilla L. COMPOSITAE (D)
Chondrochlaena Kuntze (SUO) = Prionanthium Desv. GRAMINEAE (M)
Chondrococcus Steyerm. = Coccochondra Rauschert RUBIACEAE (D)
Chondrodendron Ruíz & Pav. MENISPERMACEAE (D)
Chondrolaena Nees (SUS) = Prionanthium Desv. GRAMINEAE (M)
Chondropetalum Rottb. RESTIONACEAE (M)
Chondrophylla A.Nelson = Gentiana L. GENTIANACEAE (D)
Chondropyxis D.A.Cooke COMPOSITAE (D)
Chondrorhyncha Lindl. ORCHIDACEAE (M)
Chondrospermum Wall. ex G.Don = Myxopyrum Blume OLEACEAE (D)
Chondrostylis Boerl. EUPHORBIACEAE (D)
Chondrosum Desv. GRAMINEAE (M)
Chonemorpha G.Don APOCYNACEAE (D)
Chonocentrum Pierre ex Pax & K.Hoffm. EUPHORBIACEAE (D)
Chonopetalum Radlk. SAPINDACEAE (D)
Chontalesia Lundell = Ardisia Sw. MYRSINACEAE (D)

Chordospartium Cheeseman LEGUMINOSAE–PAPILIONOIDEAE (D)
Choretrum R.Br. SANTALACEAE (D)
Choriantha Riedl BORAGINACEAE (D)
Choricarpia Domin MYRTACEAE (D)
Choriceras Baill. EUPHORBIACEAE (D)
Chorigyne R.Erikss. CYCLANTHACEAE (M)
Chorilaena Endl. RUTACEAE (D)
Chorilepidella Tiegh. = Lepidaria Tiegh. LORANTHACEAE (D)
Chorilepis Tiegh. = Lepidaria Tiegh. LORANTHACEAE (D)
Chorioluma Baill. = Pycnandra Benth. SAPOTACEAE (D)
Choriophyllum Benth. = Austrobuxus Miq. EUPHORBIACEAE (D)
Choriptera Botsch. CHENOPODIACEAE (D)
Chorisandra Wight = Phyllanthus L. EUPHORBIACEAE (D)
Chorisandrachne Airy Shaw EUPHORBIACEAE (D)
Chorisepalum Gleason & Wodehouse GENTIANACEAE (D)
Chorisia Kunth BOMBACACEAE (D)
Chorisis DC. COMPOSITAE (D)
Chorisiva (A.Gray) Rydb. = Iva L. COMPOSITAE (D)
Chorispermum R.Br. (SUO) = Chorispora R.Br. ex DC. CRUCIFERAE (D)
Chorispora R.Br. ex DC. CRUCIFERAE (D)
Choristegeres Tiegh. = Lepeostegeres Blume LORANTHACEAE (D)
Choristegia Tiegh. = Lepeostegeres Blume LORANTHACEAE (D)
Choristemon H.B.Williamson EPACRIDACEAE (D)
Choristes Benth. = Deppea Cham. & Schltdl. RUBIACEAE (D)
Choristosoria Mett. ex Kuhn = Pellaea Link ADIANTACEAE (F)
Choristylis Harv. ESCALLONIACEAE (D)
Choritaenia Benth. UMBELLIFERAE (D)
Chorizandra R.Br. CYPERACEAE (M)
Chorizanthe R.Br. ex Benth. POLYGONACEAE (D)
Chorizema Labill. LEGUMINOSAE–PAPILIONOIDEAE (D)
Chorizotheca Müll.Arg. = Pseudanthus Sieber ex A.Spreng. EUPHORBIACEAE (D)
Chortolirion A.Berger = Haworthia Duval ALOACEAE (M)
Chosenia Nakai = Salix L. SALICACEAE (D)
Choulettia Pomel RUBIACEAE (D)
Chouxia Capuron SAPINDACEAE (D)
Chresta Vell. ex DC. COMPOSITAE (D)
Chrestienia Montrouz. (SUI) = Pseuderanthemum Radlk. ACANTHACEAE (D)
Chretomeris Nutt. ex J.G.Sm. (SUI) = Sitanion Raf. GRAMINEAE (M)
Christella H.Lév. THELYPTERIDACEAE (F)
Christensenia Maxon MARATTIACEAE (F)
Christia Moench LEGUMINOSAE–PAPILIONOIDEAE (D)
Christiana DC. TILIACEAE (D)
Christisonia Gardner SCROPHULARIACEAE (D)
Christolea Cambess. ex Jacquem. CRUCIFERAE (D)
Christopteris Copel. POLYPODIACEAE (F)
Christya Ward & Harv. = Strophanthus DC. APOCYNACEAE (D)
Chroesthes Benoist ACANTHACEAE (D)
Chromolaena DC. COMPOSITAE (D)
Chromolepis Benth. COMPOSITAE (D)
Chromolucuma Ducke SAPOTACEAE (D)
Chronanthus K.Koch = Cytisus Desf. LEGUMINOSAE–PAPILIONOIDEAE (D)
Chroniochilus J.J.Sm. ORCHIDACEAE (M)
Chronopappus DC. COMPOSITAE (D)
Chrozophora Neck. ex Juss. EUPHORBIACEAE (D)
Chryostoma Lilja = Mentzelia L. LOASACEAE (D)

Chrysactinia A.Gray COMPOSITAE (D)
Chrysactinium (Kunth) Wedd. COMPOSITAE (D)
Chrysalidocarpus H.A.Wendl. PALMAE (M)
Chrysallidosperma H.E.Moore = Syagrus Mart. PALMAE (M)
Chrysamphora Greene = Darlingtonia Torr. SARRACENIACEAE (D)
Chrysanthellum Rich. COMPOSITAE (D)
Chrysanthemoides Fabr. COMPOSITAE (D)
Chrysanthemum L. COMPOSITAE (D)
Chrysitrix L. CYPERACEAE (M)
Chrysobalanus L. CHRYSOBALANACEAE (D)
Chrysobaphus Wall. = Anoectochilus Blume ORCHIDACEAE (M)
Chrysobotrya Spach = Ribes L. GROSSULARIACEAE (D)
Chrysobraya Hara CRUCIFERAE (D)
Chrysocactus Y.Ito (SUI) = Parodia Speg. CACTACEAE (D)
Chrysocephalum Walp. COMPOSITAE (D)
Chrysochamela (Fenzl) Boiss. CRUCIFERAE (D)
Chrysochlamys Poepp. GUTTIFERAE (D)
Chrysochloa Swallen GRAMINEAE (M)
Chrysochosma (J.Sm.) Kümmerle = Cheilanthes Sw. ADIANTACEAE (F)
Chrysocoma L. COMPOSITAE (D)
Chrysocoryne Zoellner (SUH) = Leucocoryne Lindl. ALLIACEAE (M)
Chrysocoryne Endl. = Gnephosis Cass. COMPOSITAE (D)
Chrysocycnis Linden & Rchb.f. ORCHIDACEAE (M)
Chrysodium Fée = Acrostichum L. PTERIDACEAE (F)
Chrysoglossum Blume ORCHIDACEAE (M)
Chrysogonum L. COMPOSITAE (D)
Chrysolaena H.Rob. = Vernonia Schreb. COMPOSITAE (D)
Chrysolarix H.E.Moore (SUS) = Pseudolarix Gordon PINACEAE (G)
Chrysolepis Hjelmq. FAGACEAE (D)
Chrysopappus Takht. = Centaurea L. COMPOSITAE (D)
Chrysophae Kozo–Pol. = Chaerophyllum L. UMBELLIFERAE (D)
Chrysophthalmum Sch.Bip. ex Walp. COMPOSITAE (D)
Chrysophyllum L. SAPOTACEAE (D)
Chrysopogon Trin. GRAMINEAE (M)
Chrysopsis (Nutt.) Elliott COMPOSITAE (D)
Chrysopteris Link = Phlebodium (R.Br.) J.Sm. POLYPODIACEAE (F)
Chrysorhoe Lindl. = Verticordia DC. MYRTACEAE (D)
Chrysoscias E.Mey. LEGUMINOSAE–PAPILIONOIDEAE (D)
Chrysosplenium L. SAXIFRAGACEAE (D)
Chrysostachys Pohl = Combretum Loefl. COMBRETACEAE (D)
Chrysostemon Klotzsch = Pseudanthus Sieber ex A.Spreng. EUPHORBIACEAE (D)
Chrysothamnus Nutt. COMPOSITAE (D)
Chrysothemis Decne. GESNERIACEAE (D)
Chrysoxylon Wedd. (SUH) = Pogonopus Klotzsch RUBIACEAE (D)
Chrystolia Montrouz. = Glycine Willd. LEGUMINOSAE–PAPILIONOIDEAE (D)
Chrysurus Pers. (SUS) = Lamarckia Moench GRAMINEAE (M)
Chthamalia Decne. = Lachnostoma Kunth ASCLEPIADACEAE (D)
Chthonocephalus Steetz COMPOSITAE (D)
Chuanminshen M.L.Sheh & R.H.Shan UMBELLIFERAE (D)
Chucoa Cabrera COMPOSITAE (D)
Chukrasia A.Juss. MELIACEAE (D)
Chumsriella Bor = Germainia Balansa & Poitrasson GRAMINEAE (M)
Chuncoa Pav. ex Juss. = Terminalia L. COMBRETACEAE (D)
Chunechites Tsiang APOCYNACEAE (D)
Chunia H.T.Chang HAMAMELIDACEAE (D)

Chuniodendron Hu = Aphanamixis Blume MELIACEAE (D)
Chuniophoenix Burret PALMAE (M)
Chuquiraga Juss. COMPOSITAE (D)
Chusquea Kunth GRAMINEAE (M)
Chusua Nevski ORCHIDACEAE (M)
Chydenanthus Miers LECYTHIDACEAE (D)
Chylismia Nutt. = Camissonia Link ONAGRACEAE (D)
Chylogala Fourr. = Euphorbia L. EUPHORBIACEAE (D)
Chymococca Meisn. = Passerina L. THYMELAEACEAE (D)
Chymocormus Harv. = Fockea Endl. ASCLEPIADACEAE (D)
Chymsydia Albov UMBELLIFERAE (D)
Chysis Lindl. ORCHIDACEAE (M)
Chytraculia P.Browne = Calyptranthes Sw. MYRTACEAE (D)
Chytralia Adans. = Calyptranthes Sw. MYRTACEAE (D)
Chytranthus Hook.f. SAPINDACEAE (D)
Chytroglossa Rchb.f. ORCHIDACEAE (M)
Chytroma Miers = Lecythis Loefl. LECYTHIDACEAE (D)
Chytropsia Bremek. = Psychotria L. RUBIACEAE (D)
Cibirhiza Bruyns ASCLEPIADACEAE (D)
Cibotarium O.E.Schulz = Sphaerocardamum Nees & Schauer CRUCIFERAE (D)
Cibotium Kaulf. DICKSONIACEAE (F)
Cicca L. = Phyllanthus L. EUPHORBIACEAE (D)
Cicendia Adans. GENTIANACEAE (D)
Cicer L. LEGUMINOSAE–PAPILIONOIDEAE (D)
Cicerbita Wallr. COMPOSITAE (D)
Ciceronia Urb. COMPOSITAE (D)
Cichlanthus (Endl.) Tiegh. = Scurrula L. LORANTHACEAE (D)
Cichorium L. COMPOSITAE (D)
Ciclospermum Lag. (SUO) = Cyclospermum Lag. UMBELLIFERAE (D)
Cicuta L. UMBELLIFERAE (D)
Cieca Medik. (SUH) = Passiflora L. PASSIFLORACEAE (D)
Cieca Adans. = Julocroton Mart. EUPHORBIACEAE (D)
Cienfuegosia Cav. MALVACEAE (D)
Cienkowskia Schweinf. (SUI) = Siphonochilus J.M.Wood & Franks ZINGIBERACEAE (M)
Cienkowskiella Y.K.Kam (SUS) = Siphonochilus J.M.Wood & Franks ZINGIBERACEAE (M)
Cigarrilla Aiello RUBIACEAE (D)
Cimicifuga Wernisch. RANUNCULACEAE (D)
Ciminalis Adans. = Gentiana L. GENTIANACEAE (D)
Cinchona L. RUBIACEAE (D)
Cincinalis Gled. ex Desv. (SUH) = Cheilanthes Sw. ADIANTACEAE (F)
Cincinalis Gled. = Pteridium Gled. ex Scop. DENNSTAEDTIACEAE (F)
Cincinnobotrys Gilg MELASTOMATACEAE (D)
Cineraria L. COMPOSITAE (D)
Cinna L. GRAMINEAE (M)
Cinnabarinea Frič ex F.Ritter = Echinopsis Zucc. CACTACEAE (D)
Cinnadenia Kosterm. LAURACEAE (D)
Cinnagrostis Griseb. = Calamagrostis Adans. GRAMINEAE (M)
Cinnamodendron Endl. CANELLACEAE (D)
Cinnamomum Schaeffer LAURACEAE (D)
Cinnamosma Baill. CANELLACEAE (D)
Cinnastrum Fourn. = Cinna L. GRAMINEAE (M)
Cionandra Griseb. = Cayaponia Silva Manso CUCURBITACEAE (D)
Cionidium T.Moore = Tectaria Cav. DRYOPTERIDACEAE (F)
Cionisaccus Breda = Goodyera R.Br. ORCHIDACEAE (M)
Cionomene Krukoff MENISPERMACEAE (D)

Cionosicyos Griseb. CUCURBITACEAE (D)
Cionosicys Griseb. (SUO) = Cionosicyos Griseb. CUCURBITACEAE (D)
Cionura Griseb. ASCLEPIADACEAE (D)
Cipadessa Blume MELIACEAE (D)
Cipocereus F.Ritter CACTACEAE (D)
Cipura Aubl. IRIDACEAE (M)
Cipuropsis Ule = Vriesea Lindl. BROMELIACEAE (M)
Circaea L. ONAGRACEAE (D)
Circaeaster Maxim. CIRCAEASTERACEAE (D)
Circaeocarpus C.Y.Wu SAURURACEAE (D)
Circandra N.E.Br. AIZOACEAE (D)
Ciripedium Zumagl. = Cypripedium L. ORCHIDACEAE (M)
Cirrhaea Lindl. ORCHIDACEAE (M)
Cirrhopetalum Lindl. = Bulbophyllum Thouars ORCHIDACEAE (M)
Cirsium Mill. COMPOSITAE (D)
Cischweinfia Dressler & N.H.Williams ORCHIDACEAE (M)
Cissampelopsis (DC.) Miq. COMPOSITAE (D)
Cissampelos L. MENISPERMACEAE (D)
Cissarobryon Kunze ex Poepp. = Viviania Cav. GERANIACEAE (D)
Cissus L. VITACEAE (D)
Cistanche Hoffmanns. & Link SCROPHULARIACEAE (D)
Cistanthera K.Schum. = Nesogordonia Baill. STERCULIACEAE (D)
Cistella Blume = Geodorum G.Jacks. ORCHIDACEAE (M)
Cistus L. CISTACEAE (D)
Cithareloma Bunge CRUCIFERAE (D)
Citharexylum Mill. VERBENACEAE (D)
Citriobatus A.Cunn. ex Putterl. PITTOSPORACEAE (D)
Citronella D.Don ICACINACEAE (D)
Citropsis (Engl.) Swingle & M.Kellerm. RUTACEAE (D)
Citrullus Schrad. ex Eckl. & Zeyh. CUCURBITACEAE (D)
Citrus L. RUTACEAE (D)
Cladandra O.F.Cook = Chamaedorea Willd. PALMAE (M)
Cladanthus Cass. COMPOSITAE (D)
Claderia Hook.f. ORCHIDACEAE (M)
Cladium P.Browne CYPERACEAE (M)
Cladobium Lindl. = Scaphyglottis Poepp. & Endl. ORCHIDACEAE (M)
Cladobium Schltr. (SUH) = Stenorrhynchos Rich. ex Spreng. ORCHIDACEAE (M)
Cladocarpa (St.John) St.John = Sicyos L. CUCURBITACEAE (D)
Cladoceras Bremek. RUBIACEAE (D)
Cladochaeta DC. COMPOSITAE (D)
Cladocolea Tiegh. LORANTHACEAE (D)
Cladodes Lour. = Alchornea Sw. EUPHORBIACEAE (D)
Cladogelonium Leandri EUPHORBIACEAE (D)
Cladogynos Zipp. ex Spanoghe EUPHORBIACEAE (D)
Cladomyza Danser SANTALACEAE (D)
Cladopus H.Moeller PODOSTEMACEAE (D)
Cladoraphis Franch. GRAMINEAE (M)
Cladosicyos Hook.f. = Cucumeropsis Naudin CUCURBITACEAE (D)
Cladosperma Griff. = Pinanga Blume PALMAE (M)
Cladostachys D.Don = Deeringia R.Br. AMARANTHACEAE (D)
Cladostemon A.Braun & Vatke CAPPARACEAE (D)
Cladostigma Radlk. CONVOLVULACEAE (D)
Cladothamnus Bong. = Elliottia Mühl. ex Elliott ERICACEAE (D)
Cladothrix (Moq.) Hook.f. = Tidestromia Standl. AMARANTHACEAE (D)
Cladrastis Raf. LEGUMINOSAE–PAPILIONOIDEAE (D)

Clambus Miers = Phyllanthus L. EUPHORBIACEAE (D)
Clandestinaria Spach = Rorippa Scop. CRUCIFERAE (D)
Claoxylon A.Juss. EUPHORBIACEAE (D)
Claoxylopsis Leandri EUPHORBIACEAE (D)
Clappertonia Meisn. TILIACEAE (D)
Clappia A.Gray COMPOSITAE (D)
Clara Kunth = Herreria Ruíz & Pav. HERRERIACEAE (M)
Clarckia Pursh = Clarkia Pursh ONAGRACEAE (D)
Clarisia Ruíz & Pav. MORACEAE (D)
Clarkella Hook.f. RUBIACEAE (D)
Clarkia Pursh ONAGRACEAE (D)
Clarorivinia Pax & K.Hoffm. = Ptychopyxis Miq. EUPHORBIACEAE (D)
Clastopus Bunge ex Boiss. CRUCIFERAE (D)
Clathrospermum Planch. ex Benth. = Monanthotaxis Baill. ANNONACEAE (D)
Clathrotropis (Benth.) Harms LEGUMINOSAE–PAPILIONOIDEAE (D)
Claudia Opiz (SUS) = Melica L. GRAMINEAE (M)
Clausena Burm.f. RUTACEAE (D)
Clausenellia Á.Löve & D.Löve = Sedum L. CRASSULACEAE (D)
Clausenopsis (Engl.) Engl. = Fagaropsis Mildbr. ex Siebenl. RUTACEAE (D)
Clausia Trotzky ex Hayek CRUCIFERAE (D)
Clausospicula Lazarides GRAMINEAE (M)
Clavapetalum Pulle = Dendrobangia Rusby ICACINACEAE (D)
Clavarioidia Kreuz. (SUI) = Opuntia Mill. CACTACEAE (D)
Clavenna Neck. ex Standl. (SUO) = Lucya DC. RUBIACEAE (D)
Clavennaea Neck. ex Post & Kuntze (SUS) = Lucya DC. RUBIACEAE (D)
Clavija Ruíz & Pav. THEOPHRASTACEAE (D)
Clavimyrtus Blume = Syzygium Gaertn. MYRTACEAE (D)
Clavinodum T.H.Wen = Arundinaria Michx. GRAMINEAE (M)
Clavipodium Desv. ex Grüning = Beyeria Miq. EUPHORBIACEAE (D)
Clavistylus J.J.Sm. = Megistostigma Hook.f. EUPHORBIACEAE (D)
Claytonia L. PORTULACACEAE (D)
Claytoniella Yurtsev = Montia L. PORTULACACEAE (D)
Cleachne Adans. (SUI) = Paspalum L. GRAMINEAE (M)
Cleanthe Salisb. = Aristea Sol. ex Aiton IRIDACEAE (M)
Cleghornia Wight APOCYNACEAE (D)
Cleidiocarpon Airy Shaw EUPHORBIACEAE (D)
Cleidion Blume EUPHORBIACEAE (D)
Cleisocentron Brühl ORCHIDACEAE (M)
Cleisocratera Korth. = Saprosma Blume RUBIACEAE (D)
Cleisomeria Lindl. ex G.Don ORCHIDACEAE (M)
Cleisostoma Blume ORCHIDACEAE (M)
Cleistachne Benth. GRAMINEAE (M)
Cleistanthopsis Capuron = Allantospermum Forman IXONANTHACEAE (D)
Cleistanthus Hook.f. ex Planch. EUPHORBIACEAE (D)
Cleistes Rich. ex Lindl. ORCHIDACEAE (M)
Cleistocactus Lem. CACTACEAE (D)
Cleistocalyx Blume MYRTACEAE (D)
Cleistocereus Frič & Kreuz. (SUO) = Cleistocactus Lem. CACTACEAE (D)
Cleistochlamys Oliv. ANNONACEAE (D)
Cleistochloa C.E.Hubb. GRAMINEAE (M)
Cleistogenes Keng (SUI) = Kengia Packer GRAMINEAE (M)
Cleistoloranthus Merr. = Amyema Tiegh. LORANTHACEAE (D)
Cleistopholis Pierre ex Engl. ANNONACEAE (D)
Clelandia J.M.Black = Hybanthus Jacq. VIOLACEAE (D)
Clemanthus Klotzsch = Adenia Forssk. PASSIFLORACEAE (D)

Clematepistephium N.Hallé ORCHIDACEAE (M)
Clematicissus Planch. VITACEAE (D)
Clematis L. RANUNCULACEAE (D)
Clematitaria Bureau = Pleonotoma Miers BIGNONIACEAE (D)
Clematoclethra (Franch.) Maxim. ACTINIDIACEAE (D)
Clematopsis Bojer ex Hutch. = Clematis L. RANUNCULACEAE (D)
Clemensia Merr. = Chisocheton Blume MELIACEAE (D)
Clemensia Schltr. (SUH) = Clemensiella Schltr. ASCLEPIADACEAE (D)
Clemensiella Schltr. ASCLEPIADACEAE (D)
Clementea Cav. = Angiopteris Hoffm. MARATTIACEAE (F)
Clementsia Rose = Rhodiola L. CRASSULACEAE (D)
Cleobulia Mart. ex Benth. LEGUMINOSAE–PAPILIONOIDEAE (D)
Cleochroma Miers = Iochroma Benth. SOLANACEAE (D)
Cleodora Klotzsch = Croton L. EUPHORBIACEAE (D)
Cleome L. CAPPARACEAE (D)
Cleomella DC. CAPPARACEAE (D)
Cleomena Roem. & Schult. (SUO) = Muhlenbergia Schreb. GRAMINEAE (M)
Cleonia L. LABIATAE (D)
Cleophora Gaertn. = Latania Comm. ex Juss. PALMAE (M)
Cleretum N.E.Br. AIZOACEAE (D)
Clermontia Gaudich. CAMPANULACEAE (D)
Clerodendranthus Kudo = Orthosiphon Benth. LABIATAE (D)
Clerodendrum L. VERBENACEAE (D)
Clethra L. CLETHRACEAE (D)
Clevelandia Greene SCROPHULARIACEAE (D)
Cleyera Thunb. THEACEAE (D)
Clianthus Sol. ex Lindl. LEGUMINOSAE–PAPILIONOIDEAE (D)
Clibadium L. COMPOSITAE (D)
Clidemia D.Don MELASTOMATACEAE (D)
Cliffortia L. ROSACEAE (D)
Cliftonia Banks ex Gaertn.f. CYRILLACEAE (D)
Climacoptera Botsch. CHENOPODIACEAE (D)
Climacorachis Hemsl. & Rose = Aeschynomene L. LEGUMINOSAE–
   PAPILIONOIDEAE (D)
Clinacanthus Nees ACANTHACEAE (D)
Clinelymus (Griseb.) Nevski (SUS) = Elymus L. GRAMINEAE (M)
Clinogyne K.Schum. (SUH) = Marantochloa Brongn. ex Gris MARANTACEAE (M)
Clinopodium L. LABIATAE (D)
Clinosperma Becc. PALMAE (M)
Clinostemon Kuhlm. & Samp. = Licaria Aubl. LAURACEAE (D)
Clinostigma H.A.Wendl. PALMAE (M)
Clinostigmopsis Becc. = Clinostigma H.A.Wendl. PALMAE (M)
Clintonia Dougl. ex Lindl. (SUH) = Downingia Torr. CAMPANULACEAE (D)
Clintonia Raf. CONVALLARIACEAE (M)
Cliocarpus Miers = Solanum L. SOLANACEAE (D)
Cliococca Bab. LINACEAE (D)
Clistanthocereus Backeb. = Cleistocactus Lem. CACTACEAE (D)
Clistanthus Post & Kuntze (SUO) = Cleistanthus Hook.f. ex Planch.
   EUPHORBIACEAE (D)
Clistax Mart. ACANTHACEAE (D)
Clistoyucca (Engelm.) Trel. AGAVACEAE (M)
Clistranthus Poit. ex Baill. = Pera Mutis EUPHORBIACEAE (D)
Clitandra Benth. APOCYNACEAE (D)
Clitandropsis S.Moore = Melodinus J.R.Forst. & G.Forst. APOCYNACEAE (D)
Clitoria L. LEGUMINOSAE–PAPILIONOIDEAE (D)

Clitoriopsis R.Wilczek LEGUMINOSAE–PAPILIONOIDEAE (D)
Clivia Lindl. AMARYLLIDACEAE (M)
Cloezia Brongn. & Gris MYRTACEAE (D)
Clomena P.Beauv. = Muhlenbergia Schreb. GRAMINEAE (M)
Clonodia Griseb. MALPIGHIACEAE (D)
Clonostachys Klotzsch = Sebastiania Spreng. EUPHORBIACEAE (D)
Clonostylis S.Moore = Spathiostemon Blume EUPHORBIACEAE (D)
Closaschima Korth. = Laplacea Kunth THEACEAE (D)
Clowesia Lindl. ORCHIDACEAE (M)
Cluacena Raf. = Myrteola O.Berg MYRTACEAE (D)
Clusia L. GUTTIFERAE (D)
Clusianthemum Vieill. = Garcinia L. GUTTIFERAE (D)
Clusiella Planch. & Triana GUTTIFERAE (D)
Clusiophyllea Baill. = Canthium Lam. RUBIACEAE (D)
Clusiophyllum Müll.Arg. = Micrandra Benth. EUPHORBIACEAE (D)
Clutia L. EUPHORBIACEAE (D)
Cluytia Aiton (SUO) = Clutia L. EUPHORBIACEAE (D)
Cluytiandra Müll.Arg. (SUS) = Meineckia Baill. EUPHORBIACEAE (D)
Clybatis Phil. = Leucheria Lag. COMPOSITAE (D)
Clymenia Swingle RUTACEAE (D)
Clyostomanthus Pichon = Clytostoma Miers ex Bur. BIGNONIACEAE (D)
Clypeola L. CRUCIFERAE (D)
Clytia Stokes (SUO) = Clutia L. EUPHORBIACEAE (D)
Clytostoma Miers ex Bur. BIGNONIACEAE (D)
Cnemidaria C.Presl = Cyathea Sm. CYATHEACEAE (F)
Cnemidia Lindl. = Tropidia Lindl. ORCHIDACEAE (M)
Cnemidiscus Pierre = Glenniea Hook.f. SAPINDACEAE (D)
Cnemidophacos Rydb. = Astragalus L. LEGUMINOSAE–PAPILIONOIDEAE (D)
Cnemidostachys Mart. = Sebastiania Spreng. EUPHORBIACEAE (D)
Cneoridium Hook.f. RUTACEAE (D)
Cneorum L. CNEORACEAE (D)
Cnesmone Blume EUPHORBIACEAE (D)
Cnesmosa Blume = Cnesmone Blume EUPHORBIACEAE (D)
Cnestidium Planch. CONNARACEAE (D)
Cnestis Juss. CONNARACEAE (D)
Cnicothamnus Griseb. COMPOSITAE (D)
Cnicus L. COMPOSITAE (D)
Cnidiocarpa Pimenov UMBELLIFERAE (D)
Cnidium Cusson ex Juss. UMBELLIFERAE (D)
Cnidoscolus Pohl EUPHORBIACEAE (D)
Coaxana J.M.Coult. & Rose UMBELLIFERAE (D)
Cobaea Cav. COBAEACEAE (D)
Cobana Ravenna IRIDACEAE (M)
Cobananthus Wiehler = Alloplectus Mart. GESNERIACEAE (D)
Coccanthera K.Koch & Hanst. = Codonanthe (Mart.) Hanst. GESNERIACEAE (D)
Coccineorchis Schltr. = Stenorrhynchos Rich. ex Spreng. ORCHIDACEAE (M)
Coccinia Wight & Arn. CUCURBITACEAE (D)
Coccoceras Miq. = Mallotus Lour. EUPHORBIACEAE (D)
Coccochondra Rauschert RUBIACEAE (D)
Coccocypselum P.Browne RUBIACEAE (D)
Coccoglochidion K.Schum. = Glochidion J.R.Forst. & G.Forst. EUPHORBIACEAE (D)
Coccoloba P.Browne POLYGONACEAE (D)
Coccomelia Reinw. = Baccaurea Lour. EUPHORBIACEAE (D)
Coccomelia Ridl. (SUH) = Licania Aubl. CHRYSOBALANACEAE (D)
Cocconerion Baill. EUPHORBIACEAE (D)

Coccosperma Klotzsch ERICACEAE (D)
Coccothrinax Sarg. PALMAE (M)
Cocculus DC. MENISPERMACEAE (D)
Coccus Mill. = Cocos L. PALMAE (M)
Cochemiea (K.Brandegee) Walton = Mammillaria Haw. CACTACEAE (D)
Cochiseia Earle = Escobaria Britton & Rose CACTACEAE (D)
Cochlanthera Choisy = Clusia L. GUTTIFERAE (D)
Cochlanthus Balf.f. (SUH) = Socotranthus Kuntze ASCLEPIADACEAE (D)
Cochleanthes Raf. ORCHIDACEAE (M)
Cochlearia L. CRUCIFERAE (D)
Cochleariella Y.H.Zhang & R.Vogt CRUCIFERAE (D)
Cochleariopsis Á.Löve & D.Löve = Cochlearia L. CRUCIFERAE (D)
Cochleariopsis Zhang (SUH) = Cochleariella Y.H.Zhang & R.Vogt CRUCIFERAE (D)
Cochlia Blume = Bulbophyllum Thouars ORCHIDACEAE (M)
Cochlianthus Benth. LEGUMINOSAE–PAPILIONOIDEAE (D)
Cochlidiosperma (Rchb.) Rchb. SCROPHULARIACEAE (D)
Cochlidium Kaulf. GRAMMITIDACEAE (F)
Cochlioda Lindl. ORCHIDACEAE (M)
Cochliostema Lem. COMMELINACEAE (M)
Cochlospermum Kunth COCHLOSPERMACEAE (D)
Cochranea Miers = Heliotropium L. BORAGINACEAE (D)
Cockaynea Zotov = Hystrix Moench GRAMINEAE (M)
Cockburnia Balf.f. = Poskea Vatke GLOBULARIACEAE (D)
Cockerellia Á.Löve & D.Löve = Sedum L. CRASSULACEAE (D)
Cocops O.F.Cook = Calyptronoma Griseb. PALMAE (M)
Cocos L. PALMAE (M)
Codaria Kuntze = Lerchea L. RUBIACEAE (D)
Codariocalyx Hassk. LEGUMINOSAE–PAPILIONOIDEAE (D)
Codda–Pana Adans. = Corypha L. PALMAE (M)
Coddia Verdc. RUBIACEAE (D)
Codia J.R.Forst. & G.Forst. CUNONIACEAE (D)
Codiaeum A.Juss. EUPHORBIACEAE (D)
Codiocarpus R.A.Howard ICACINACEAE (D)
Codochonia Dunal = Acnistus Schott SOLANACEAE (D)
Codon Royen ex L. HYDROPHYLLACEAE (D)
Codonacanthus Nees ACANTHACEAE (D)
Codonachne Steud. (SUI) = Tetrapogon Desf. GRAMINEAE (M)
Codonandra Karst. = Calliandra Benth. LEGUMINOSAE–MIMOSOIDEAE (D)
Codonanthe (Mart.) Hanst. GESNERIACEAE (D)
Codonanthopsis Mansf. GESNERIACEAE (D)
Codonechites Markgr. = Odontadenia Benth. APOCYNACEAE (D)
Codonoboea Ridl. GESNERIACEAE (D)
Codonocalyx Klotzsch ex Baill. = Croton L. EUPHORBIACEAE (D)
Codonocarpus A.Cunn. ex Endl. GYROSTEMONACEAE (D)
Codonocephalum Fenzl = Inula L. COMPOSITAE (D)
Codonochlamys Ulbr. MALVACEAE (D)
Codonopsis Wall. CAMPANULACEAE (D)
Codonorchis Lindl. ORCHIDACEAE (M)
Codonosiphon Schltr. ORCHIDACEAE (M)
Codonostigma Klotzsch ex Benth. ERICACEAE (D)
Codonura K.Schum. = Baissea A.DC. APOCYNACEAE (D)
Coelachne R.Br. GRAMINEAE (M)
Coelachyropsis Bor = Coelachyrum Hochst. & Nees GRAMINEAE (M)
Coelachyrum Hochst. & Nees GRAMINEAE (M)
Coelandria Fitzg. = Dendrobium Sw. ORCHIDACEAE (M)

Coelanthum E.Mey. ex Fenzl MOLLUGINACEAE (D)
Coelarthron Hook.f. = Microstegium Nees GRAMINEAE (M)
Coelebogyne J.Sm. (SUO) = Alchornea Sw. EUPHORBIACEAE (D)
Coelia Lindl. ORCHIDACEAE (M)
Coelidium Vogel ex Walp. LEGUMINOSAE–PAPILIONOIDEAE (D)
Coeliopsis Rchb.f. ORCHIDACEAE (M)
Coelobogyne J.Sm. = Alchornea Sw. EUPHORBIACEAE (D)
Coelocarpum Balf.f. VERBENACEAE (D)
Coelocaryon Warb. MYRISTICACEAE (D)
Coelochloa Steud. (SUI) = Coelachyrum Hochst. & Nees GRAMINEAE (M)
Coelocline A.DC. = Xylopia L. ANNONACEAE (D)
Coelococcus H.A.Wendl. = Metroxylon Rottb. PALMAE (M)
Coelodepas Hassk. (SUO) = Koilodepas Hassk. EUPHORBIACEAE (D)
Coelodiscus Baill. = Mallotus Lour. EUPHORBIACEAE (D)
Coeloglossum Hartm. ORCHIDACEAE (M)
Coelogyne Lindl. ORCHIDACEAE (M)
Coelonema Maxim. CRUCIFERAE (D)
Coeloneurum Radlk. GOETZEACEAE (D)
Coelophragmus O.E.Schulz CRUCIFERAE (D)
Coelopleurum Ledeb. = Angelica L. UMBELLIFERAE (D)
Coelopyrena Valeton RUBIACEAE (D)
Coelorachis Brongn. GRAMINEAE (M)
Coelospermum Blume RUBIACEAE (D)
Coelostegia Benth. BOMBACACEAE (D)
Coelostelma Fourn. (UP) ASCLEPIADACEAE (D)
Coelostylis (Juss.) Kuntze = Echinopterys A.Juss. MALPIGHIACEAE (D)
Coemansia Marchal (SUH) = Pentapanax Seem. ARALIACEAE (D)
Coespeletia Cuatrec. COMPOSITAE (D)
Cofeanthus A.Chev. = Psilanthus Hook.f. RUBIACEAE (D)
Coffea L. RUBIACEAE (D)
Cogniauxia Baill. CUCURBITACEAE (D)
Cogniauxiella Baill. = Cogniauxia Baill. CUCURBITACEAE (D)
Cogniauxiocharis (Schltr.) Hoehne = Stenorrhynchos Rich. ex Spreng.
    ORCHIDACEAE (M)
Cogswellia Spreng. = Lomatium Raf. UMBELLIFERAE (D)
Cohnia Rchb.f. (SUH) = Cohniella Pfitzer ORCHIDACEAE (M)
Cohnia Kunth = Cordyline Comm. ex R.Br. AGAVACEAE (M)
Cohniella Pfitzer ORCHIDACEAE (M)
Coilocarpus F.Muell. ex Domin = Sclerolaena R.Br. CHENOPODIACEAE (D)
Coilochilus Schltr. ORCHIDACEAE (M)
Coilostigma Klotzsch ERICACEAE (D)
Coilostylis Raf. = Epidendrum L. ORCHIDACEAE (M)
Coincya Rouy CRUCIFERAE (D)
Coinochlamys T.Anderson ex Benth. = Mostuea Didr. LOGANIACEAE (D)
Coix L. GRAMINEAE (M)
Cojoba Britton & Rose LEGUMINOSAE–MIMOSOIDEAE (D)
Cola Schott & Endl. STERCULIACEAE (D)
Colania Gagnep. = Aspidistra Ker–Gawl. CONVALLARIACEAE (M)
Colanthelia McClure & E.W.Sm. GRAMINEAE (M)
Colax Lindl. (SUH) = Pabstia Garay ORCHIDACEAE (M)
Colchicum L. COLCHICACEAE (M)
Coldenia L. BORAGINACEAE (D)
Colea Bojer ex Meisn. BIGNONIACEAE (D)
Coleactina N.Hallé RUBIACEAE (D)
Coleanthera Stschegl. EPACRIDACEAE (D)

Coleanthus J.Seidel GRAMINEAE (M)
Coleataenia Griseb. = Panicum L. GRAMINEAE (M)
Colebrookea Sm. LABIATAE (D)
Colensoa Hook.f. = Pratia Gaudich. CAMPANULACEAE (D)
Coleobotrys Tiegh. = Helixanthera Lour. LORANTHACEAE (D)
Coleocarya S.T.Blake RESTIONACEAE (M)
Coleocephalocereus Backeb. CACTACEAE (D)
Coleochloa Gilly CYPERACEAE (M)
Coleocoma F.Muell. COMPOSITAE (D)
Coleogyne Torr. ROSACEAE (D)
Coleonema Bartl. & H.L.Wendl. RUTACEAE (D)
Coleospadix Becc. = Drymophloeus Zipp. PALMAE (M)
Coleostachys A.Juss. MALPIGHIACEAE (D)
Coleostephus Cass. COMPOSITAE (D)
Coleotrype C.B.Clarke COMMELINACEAE (M)
Coletia Vell. = Mayaca Aubl. MAYACACEAE (M)
Coleus Lour. = Plectranthus L'Hér. LABIATAE (D)
Colicodendron Mart. = Capparis L. CAPPARACEAE (D)
Colignonia Endl. NYCTAGINACEAE (D)
Colina Greene = Mohria Sw. SCHIZAEACEAE (F)
Colinil Adans. = Tephrosia Pers. LEGUMINOSAE–PAPILIONOIDEAE (D)
Coliquea Bibra (SUI) = Chusquea Kunth GRAMINEAE (M)
Collabiopsis S.S.Ying = Collabium Blume ORCHIDACEAE (M)
Collabium Blume ORCHIDACEAE (M)
Colladoa Cav. = Ischaemum L. GRAMINEAE (M)
Colladonia Spreng. = Palicourea Aubl. RUBIACEAE (D)
Colladonia DC. (SUH) = Heptaptera Margot & Reut. UMBELLIFERAE (D)
Collaea DC. LEGUMINOSAE–PAPILIONOIDEAE (D)
Collandra Lem. = Columnea L. GESNERIACEAE (D)
Collania Schult. & Schult.f. = Urceolina Rchb. AMARYLLIDACEAE (M)
Collea Lindl. = Pelexia Poit. ex Lindl. ORCHIDACEAE (M)
Collenucia Chiov. = Jatropha L. EUPHORBIACEAE (D)
Colletia Comm. ex Juss. RHAMNACEAE (D)
Colletoecema E.Petit RUBIACEAE (D)
Colletogyne Buchet ARACEAE (M)
Colliguaja Molina EUPHORBIACEAE (D)
Colliguonia Endl. (SUO) = Colignonia Endl. NYCTAGINACEAE (D)
Collinia (Liebm.) Liebm. ex Oerst = Chamaedorea Willd. PALMAE (M)
Collinsia Nutt. SCROPHULARIACEAE (D)
Collinsonia L. LABIATAE (D)
Collomia Nutt. POLEMONIACEAE (D)
Collospermum Skottsb. ASTELIACEAE (M)
Colmeiroa Reut. = Flueggea Willd. EUPHORBIACEAE (D)
Colobachne P.Beauv. = Alopecurus L. GRAMINEAE (M)
Colobanthera Humbert COMPOSITAE (D)
Colobanthium (Rchb.) G.Taylor = Sphenopholis Scribn. GRAMINEAE (M)
Colobanthus (Trin.) Spach (SUH) = Sphenopholis Scribn. GRAMINEAE (M)
Colobanthus Bartl. CARYOPHYLLACEAE (D)
Colobogyne Gagnep. = Acmella L.C.Rich. ex Pers. COMPOSITAE (D)
Colocasia Schott ARACEAE (M)
Colocynthis Mill. = Citrullus Schrad. ex Eckl. & Zeyh. CUCURBITACEAE (D)
Cologania Kunth LEGUMINOSAE–PAPILIONOIDEAE (D)
Colombiana Ospina = Pleurothallis R.Br. ORCHIDACEAE (M)
Colombobalanus Nixon & Crepet = Trigonobalanus Forman FAGACEAE (D)
Colona Cav. TILIACEAE (D)

Colophospermum J.Léonard LEGUMINOSAE–CAESALPINIOIDEAE (D)
Coloptera J.M.Coult. & Rose = Cymopterus Raf. UMBELLIFERAE (D)
Coloradoa Boissev. & Davidson = Sclerocactus Britton & Rose CACTACEAE (D)
Colpias E.Mey. ex Benth. SCROPHULARIACEAE (D)
Colpodium Trin. GRAMINEAE (M)
Colpogyne B.L.Burtt GESNERIACEAE (D)
Colpoon Bergius SANTALACEAE (D)
Colpothrinax Griseb. & H.A.Wendl. PALMAE (M)
Colquhounia Wall. LABIATAE (D)
Colsmannia Lehm. = Onosma L. BORAGINACEAE (D)
Colubrina Rich. ex Brongn. RHAMNACEAE (D)
Columbia Pers. = Colona Cav. TILIACEAE (D)
Columella Lour. = Cayratia Juss. VITACEAE (D)
Columellia Ruíz & Pav. COLUMELLIACEAE (D)
Columnea L. GESNERIACEAE (D)
Coluria R.Br. ROSACEAE (D)
Colutea L. LEGUMINOSAE–PAPILIONOIDEAE (D)
Coluteocarpus Boiss. CRUCIFERAE (D)
Colvillea Bojer ex Hook. LEGUMINOSAE–CAESALPINIOIDEAE (D)
Colysis C.Presl POLYPODIACEAE (F)
Comaclinium Scheidw. & Planch. COMPOSITAE (D)
Comandra Nutt. SANTALACEAE (D)
Comanthera L.B.Sm. = Syngonanthus Ruhland ERIOCAULACEAE (M)
Comanthosphace S.Moore LABIATAE (D)
Comarella Rydb. = Potentilla L. ROSACEAE (D)
Comarobatia Greene = Rubus L. ROSACEAE (D)
Comaropsis Rich. = Waldsteinia Willd. ROSACEAE (D)
Comarostaphylis Zucc. ERICACEAE (D)
Comarum L. = Potentilla L. ROSACEAE (D)
Comastoma Toyok. GENTIANACEAE (D)
Comatocroton Karst. = Croton L. EUPHORBIACEAE (D)
Combera Sandwith SOLANACEAE (D)
Combesia A.Rich. = Crassula L. CRASSULACEAE (D)
Combretocarpus Hook.f. ANISOPHYLLEACEAE (D)
Combretodendron A.Chev. = Petersianthus Merr. LECYTHIDACEAE (D)
Combretum Loefl. COMBRETACEAE (D)
Comesperma Labill. POLYGALACEAE (D)
Cometes L. ILLECEBRACEAE (D)
Cometia Thouars ex Baill. EUPHORBIACEAE (D)
Comeurya Baill. = Dracontomelon Blume ANACARDIACEAE (D)
Cominsia Hemsl. MARANTACEAE (M)
Comiphyton Floret RHIZOPHORACEAE (D)
Commarum Schrank = Potentilla L. ROSACEAE (D)
Commelina L. COMMELINACEAE (M)
Commelinantia Tharp = Tinantia Scheidw. COMMELINACEAE (M)
Commelinidium Stapf = Acroceras Stapf GRAMINEAE (M)
Commelinopsis Pichon = Commelina L. COMMELINACEAE (M)
Commersonia J.R.Forst. & G.Forst. STERCULIACEAE (D)
Commia Lour. = Excoecaria L. EUPHORBIACEAE (D)
Commianthus Benth. = Retiniphyllum Bonpl. RUBIACEAE (D)
Commicarpus Standl. NYCTAGINACEAE (D)
Commidendrum DC. COMPOSITAE (D)
Commiphora Jacq. BURSERACEAE (D)
Commitheca Bremek. RUBIACEAE (D)
Comocladia P.Browne ANACARDIACEAE (D)

Comolia DC. MELASTOMATACEAE (D)
Comoliopsis Wurdack MELASTOMATACEAE (D)
Comomyrsine Hook.f. = Cybianthus Mart. MYRSINACEAE (D)
Comopyena Kuntze = Pycnocoma Benth. EUPHORBIACEAE (D)
Comopyrum A.Löve = Aegilops L. GRAMINEAE (M)
Comoranthus Knobl. OLEACEAE (D)
Comoroa Oliv. = Teclea Delile RUTACEAE (D)
Comospermum Rauschert ANTHERICACEAE (M)
Comparettia Poepp. & Endl. ORCHIDACEAE (M)
Comperia K.Koch ORCHIDACEAE (M)
Complaya Strother COMPOSITAE (D)
Compsoneura (DC.) Warb. MYRISTICACEAE (D)
Comptonanthus B.Nord. = Ifloga Cass. COMPOSITAE (D)
Comptonella Baker f. RUTACEAE (D)
Comptonia L'Hér. ex Aiton MYRICACEAE (D)
Comularia Pichon = Hunteria Roxb. APOCYNACEAE (D)
Conami Aubl. = Phyllanthus L. EUPHORBIACEAE (D)
Conamomum Ridl. = Amomum Roxb. ZINGIBERACEAE (M)
Conandrium (K.Schum.) Mez MYRSINACEAE (D)
Conandron Siebold & Zucc. GESNERIACEAE (D)
Conanthera Ruíz & Pav. TECOPHILAEACEAE (M)
Conanthus S.Watson = Nama L. HYDROPHYLLACEAE (D)
Conceveiba Aubl. EUPHORBIACEAE (D)
Conceveibastrum (Müll.Arg.) Pax & K.Hoffm. = Conceveiba Aubl. EUPHORBIACEAE (D)
Conchidium Griff. = Eria Lindl. ORCHIDACEAE (M)
Conchochilus Hassk. = Appendicula Blume ORCHIDACEAE (M)
Conchopetalum Radlk. SAPINDACEAE (D)
Conchophyllum Blume = Dischidia R.Br. ASCLEPIADACEAE (D)
Condalia Cav. RHAMNACEAE (D)
Condalia Ruíz & Pav. (SUH) = Coccocypselum P.Browne RUBIACEAE (D)
Condaliopsis (Weberb.) Suess. = Condalia Cav. RHAMNACEAE (D)
Condaminea DC. RUBIACEAE (D)
Condea Adans. = Hyptis Jacq. LABIATAE (D)
Condylago Luer ORCHIDACEAE (M)
Condylidium R.M.King & H.Rob. COMPOSITAE (D)
Condylocarpon Desf. APOCYNACEAE (D)
Condylopodium R.M.King & H.Rob. COMPOSITAE (D)
Condylostylis Piper = Vigna Savi LEGUMINOSAE–PAPILIONOIDEAE (D)
Congdonia Müll.Arg. = Declieuxia Kunth RUBIACEAE (D)
Congdonia Jeps. (SUH) = Sedum L. CRASSULACEAE (D)
Congea Roxb. VERBENACEAE (D)
Congolanthus A.Raynal GENTIANACEAE (D)
Coniandra Schrad. ex Eckl. & Zeyh. = Kedrostis Medik. CUCURBITACEAE (D)
Conicosia N.E.Br. AIZOACEAE (D)
Conimitella Rydb. SAXIFRAGACEAE (D)
Coniogeton Blume = Buchanania Spreng. ANACARDIACEAE (D)
Coniogramme Fée ADIANTACEAE (F)
Conioselinum Fisch. ex Hoffm. UMBELLIFERAE (D)
Conium L. UMBELLIFERAE (D)
Connarus L. CONNARACEAE (D)
Connellia N.E.Br. BROMELIACEAE (M)
Conobea Aubl. SCROPHULARIACEAE (D)
Conocalyx Benoist (UP) ACANTHACEAE (D)
Conocarpus L. COMBRETACEAE (D)
Conocephalus Blume = Poikilospermum Zipp. ex Miq. CECROPIACEAE (D)

Conocliniopsis R.M.King & H.Rob. COMPOSITAE (D)
Conoclinium DC. COMPOSITAE (D)
Conomitra Fenzl ASCLEPIADACEAE (D)
Conomorpha A.DC. = Cybianthus Mart. MYRSINACEAE (D)
Conopharyngia G.Don = Tabernaemontana L. APOCYNACEAE (D)
Conopholis Wallr. SCROPHULARIACEAE (D)
Conophyllum Schwantes = Mitrophyllum Schwantes AIZOACEAE (D)
Conophytum N.E.Br. AIZOACEAE (D)
Conopodium W.D.J.Koch UMBELLIFERAE (D)
Conosapium Müll.Arg. = Sapium P.Browne EUPHORBIACEAE (D)
Conosiphon Poepp. = Sphinctanthus Benth. RUBIACEAE (D)
Conospermum Sm. PROTEACEAE (D)
Conostalix (Kraenzl.) Brieger = Eria Lindl. ORCHIDACEAE (M)
Conostegia D.Don MELASTOMATACEAE (D)
Conostephium Benth. EPACRIDACEAE (D)
Conostomium (Stapf) Cufod. RUBIACEAE (D)
Conostylis R.Br. HAEMODORACEAE (M)
Conothamnus Lindl. MYRTACEAE (D)
Conotrichia A.Rich. = Manettia L. RUBIACEAE (D)
Conradia Mart. = Gesneria L. GESNERIACEAE (D)
Conradina A.Gray LABIATAE (D)
Conringia Heist. ex Fabr. CRUCIFERAE (D)
Consolea Lem. = Opuntia Mill. CACTACEAE (D)
Consolida Gray RANUNCULACEAE (D)
Constantia Barb.Rodr. ORCHIDACEAE (M)
Contortuplicata Medik. = Astragalus L. LEGUMINOSAE–PAPILIONOIDEAE (D)
Conuleum A.Rich. = Siparuna Aubl. MONIMIACEAE (D)
Convallaria L. CONVALLARIACEAE (M)
Convolvulus L. CONVOLVULACEAE (D)
Conyza Less. COMPOSITAE (D)
Conyzanthus Tamamsch. = Aster L. COMPOSITAE (D)
Conzattia Rose LEGUMINOSAE–CAESALPINIOIDEAE (D)
Coombea P.Royen RUTACEAE (D)
Cooperia Herb. = Zephyranthes Herb. AMARYLLIDACEAE (M)
Coopernookia Carolin GOODENIACEAE (D)
Copaifera L. LEGUMINOSAE–CAESALPINIOIDEAE (D)
Copedesma Gleason = Miconia Ruíz & Pav. MELASTOMATACEAE (D)
Copelandiopteris B.C.Stone = Pteris L. PTERIDACEAE (F)
Copernicia Mart. ex Endl. PALMAE (M)
Copiapoa Britton & Rose CACTACEAE (D)
Copioglossa Miers = Ruellia L. ACANTHACEAE (D)
Copodium Raf. (SUS) = Lycopodium L. LYCOPODIACEAE (FA)
Coprosma J.R.Forst. & G.Forst. RUBIACEAE (D)
Coptidipteris Nakai & Momose = Dennstaedtia Bernh. DENNSTAEDTIACEAE (F)
Coptis Salisb. RANUNCULACEAE (D)
Coptocheile Hoffmanns. GESNERIACEAE (D)
Coptophyllum Korth. RUBIACEAE (D)
Coptophyllum Gardner (SUH) = Anemia Sw. SCHIZAEACEAE (F)
Coptosapelta Korth. RUBIACEAE (D)
Coptospelta K.Schum. (SUO) = Coptosapelta Korth. RUBIACEAE (D)
Coptosperma Hook.f. = Tarenna Gaertn. RUBIACEAE (D)
Corallobotrys Hook.f. = Agapetes D.Don ex G.Don ERICACEAE (D)
Corallocarpus Welw. ex Hook.f. CUCURBITACEAE (D)
Corallodendron Kuntze = Erythrina L. LEGUMINOSAE–PAPILIONOIDEAE (D)
Corallodiscus Batalin GESNERIACEAE (D)

Corallorhiza Châtel. (SUO) = Corallorrhiza Gagnebin ORCHIDACEAE (M)
Corallorrhiza Gagnebin ORCHIDACEAE (M)
Corallospartium J.B.Armstr. LEGUMINOSAE–PAPILIONOIDEAE (D)
Corbassona Aubrév. = Niemeyera F.Muell. SAPOTACEAE (D)
Corbichonia Scop. MOLLUGINACEAE (D)
Corchoropsis Siebold & Zucc. TILIACEAE (D)
Corchorus L. TILIACEAE (D)
Cordaea Spreng. = Cyamopsis DC. LEGUMINOSAE–PAPILIONOIDEAE (D)
Cordeauxia Hemsl. LEGUMINOSAE–CAESALPINIOIDEAE (D)
Cordemoya Baill. EUPHORBIACEAE (D)
Cordia L. BORAGINACEAE (D)
Cordiada Vell. = Cordia L. BORAGINACEAE (D)
Cordiera A.Rich. = Alibertia A.Rich. ex DC. RUBIACEAE (D)
Cordiglottis J.J.Sm. ORCHIDACEAE (M)
Cordiopsis Desv. = Cordia L. BORAGINACEAE (D)
Cordisepalum Verdc. CONVOLVULACEAE (D)
Cordobia Nied. MALPIGHIACEAE (D)
Cordula Raf. = Paphiopedilum Pfitzer ORCHIDACEAE (M)
Cordyla Blume (SUH) = Nervilia Comm. ex Gaud. ORCHIDACEAE (M)
Cordyla Lour. LEGUMINOSAE–PAPILIONOIDEAE (D)
Cordylanthus Nutt. ex Benth. SCROPHULARIACEAE (D)
Cordylestylis Falc. = Goodyera R.Br. ORCHIDACEAE (M)
Cordyline Comm. ex R.Br. AGAVACEAE (M)
Cordyloblaste Hensch. ex Moritzi = Symplocos Jacq. SYMPLOCACEAE (D)
Cordylocarpus Desf. CRUCIFERAE (D)
Cordylogyne E.Mey. ASCLEPIADACEAE (D)
Cordylophorum Rydb. = Epilobium L. ONAGRACEAE (D)
Coreanomecon Nakai = Chelidonium L. PAPAVERACEAE (D)
Corema D.Don EMPETRACEAE (D)
Coreocarpus Benth. COMPOSITAE (D)
Coreopsis L. COMPOSITAE (D)
Coreosma Spach = Ribes L. GROSSULARIACEAE (D)
Corethamnium R.M.King & H.Rob. COMPOSITAE (D)
Corethrodendron Fisch. & Basiner = Hedysarum L. LEGUMINOSAE–
    PAPILIONOIDEAE (D)
Corethrogyne DC. COMPOSITAE (D)
Corethrum Vahl (UR) GRAMINEAE (M)
Coriandropsis H.Wolff = Coriandrum L. UMBELLIFERAE (D)
Coriandrum L. UMBELLIFERAE (D)
Coriaria L. CORIARIACEAE (D)
Coridochloa Nees = Alloteropsis C.Presl GRAMINEAE (M)
Coridothymus Rchb.f. LABIATAE (D)
Coriflora Weber (SUI) = Clematis L. RANUNCULACEAE (D)
Coriophyllus Rydb. = Cymopterus Raf. UMBELLIFERAE (D)
Coris L. PRIMULACEAE (D)
Corispermum L. CHENOPODIACEAE (D)
Cormiophyllum Newman = Cyathea Sm. CYATHEACEAE (F)
Cormonema Reissek ex Endl. = Colubrina Rich. ex Brongn. RHAMNACEAE (D)
Cormus Spach = Sorbus L. ROSACEAE (D)
Cornelia Ard. = Ammannia L. LYTHRACEAE (D)
Cornera Furtado = Calamus L. PALMAE (M)
Cornicina Boiss. = Anthyllis L. LEGUMINOSAE–PAPILIONOIDEAE (D)
Cornidia Ruíz & Pav. = Hydrangea L. HYDRANGEACEAE (D)
Cornopteris Nakai WOODSIACEAE (F)
Cornucopiae L. GRAMINEAE (M)

Cornuella Pierre = Chrysophyllum L. SAPOTACEAE (D)
Cornulaca Delile CHENOPODIACEAE (D)
Cornus L. CORNACEAE (D)
Cornutia L. VERBENACEAE (D)
Corokia A.Cunn. ESCALLONIACEAE (D)
Corollonema Schltr. ASCLEPIADACEAE (D)
Coronanthera Vieill. ex C.B.Clarke GESNERIACEAE (D)
Coronaria Guett. = Lychnis L. CARYOPHYLLACEAE (D)
Coronilla L. LEGUMINOSAE–PAPILIONOIDEAE (D)
Coronopus Zinn CRUCIFERAE (D)
Coroya Pierre LEGUMINOSAE–PAPILIONOIDEAE (D)
Corozo Jacq. ex Giseke = Elaeis Jacq. PALMAE (M)
Corpuscularia Schwantes AIZOACEAE (D)
Correa Andrews RUTACEAE (D)
Correllia A.M.Powell COMPOSITAE (D)
Correlliana D'Arcy = Cybianthus Mart. MYRSINACEAE (D)
Corrigiola L. MOLLUGINACEAE (D)
Corryocactus Britton & Rose CACTACEAE (D)
Corryocereus Frič & Kreuz. (SUO) = Corryocactus Britton & Rose CACTACEAE (D)
Corsia Becc. CORSIACEAE (M)
Cortaderia Stapf GRAMINEAE (M)
Cortesia Cav. BORAGINACEAE (D)
Cortia DC. UMBELLIFERAE (D)
Cortiella C.Norman UMBELLIFERAE (D)
Cortusa L. PRIMULACEAE (D)
Coryanthes Hook. ORCHIDACEAE (M)
Corybas Salisb. ORCHIDACEAE (M)
Corycarpus Spreng. (SUO) = Diarrhena P.Beauv. GRAMINEAE (M)
Corycium Sw. ORCHIDACEAE (M)
Corydalis DC. PAPAVERACEAE (D)
Corylopsis Siebold & Zucc. HAMAMELIDACEAE (D)
Corylus L. CORYLACEAE (D)
Corymbis Thouars = Corymborkis Thouars ORCHIDACEAE (M)
Corymbium L. COMPOSITAE (D)
Corymborchis Thouars = Corymborkis Thouars ORCHIDACEAE (M)
Corymborkis Thouars ORCHIDACEAE (M)
Corymbostachys Lindau = Justicia L. ACANTHACEAE (D)
Corynabutilon (K.Schum.) Kearney MALVACEAE (D)
Corynaea Hook.f. BALANOPHORACEAE (D)
Corynanthe Welw. RUBIACEAE (D)
Corynanthera J.W.Green MYRTACEAE (D)
Corynella DC. LEGUMINOSAE–PAPILIONOIDEAE (D)
Corynemyrtus (Kiaersk.) Mattos = Myrtus L. MYRTACEAE (D)
Corynephorus P.Beauv. GRAMINEAE (M)
Corynephyllum Rose = Sedum L. CRASSULACEAE (D)
Corynocarpus J.R.Forst. & G.Forst. CORYNOCARPACEAE (D)
Corynopuntia F.Knuth = Opuntia Mill. CACTACEAE (D)
Corynosicyos F.Muell. (SUI) = Cucumeropsis Naudin CUCURBITACEAE (D)
Corynostigma C.Presl = Ludwigia L. ONAGRACEAE (D)
Corynostylis Mart. VIOLACEAE (D)
Corynotheca F.Muell. ex Benth. ANTHERICACEAE (M)
Corynula Hook.f. = Leptostigma Arn. RUBIACEAE (D)
Corypha L. PALMAE (M)
Coryphadenia Morley = Votomita Aubl. MELASTOMATACEAE (D)
Coryphantha (Engelm.) Lem. CACTACEAE (D)

Coryphomia Rojas Acosta = Copernicia Mart. ex Endl. PALMAE (M)
Coryphopteris Holttum THELYPTERIDACEAE (F)
Coryphothamnus Steyerm. RUBIACEAE (D)
Corysadenia Griff. = Illigera Blume HERNANDIACEAE (D)
Corysanthera Wall. ex Endl. = Rhynchotechum Blume GESNERIACEAE (D)
Corysanthes R.Br. = Corybas Salisb. ORCHIDACEAE (M)
Corythea S.Watson = Acalypha L. EUPHORBIACEAE (D)
Corytholoma (Benth.) Decne. = Sinningia Nees GESNERIACEAE (D)
Corythophora R.Knuth LECYTHIDACEAE (D)
Corytoplectus Oerst. GESNERIACEAE (D)
Coryzadenia Griff. = Illigera Blume HERNANDIACEAE (D)
Coscinium Colebr. MENISPERMACEAE (D)
Cosentinia Tod. = Cheilanthes Sw. ADIANTACEAE (F)
Cosmelia R.Br. EPACRIDACEAE (D)
Cosmianthemum Bremek. ACANTHACEAE (D)
Cosmibuena Ruíz & Pav. RUBIACEAE (D)
Cosmiusa Alef. = Parochetus Buch.–Ham. ex D.Don LEGUMINOSAE–
    PAPILIONOIDEAE (D)
Cosmiza Raf. = Utricularia L. LENTIBULARIACEAE (D)
Cosmocalyx Standl. RUBIACEAE (D)
Cosmos Cav. COMPOSITAE (D)
Cosmostigma Wight ASCLEPIADACEAE (D)
Cossignia Comm. ex Lam. (SUO) = Cossinia Comm. ex Lam. SAPINDACEAE (D)
Cossinia Comm. ex Lam. SAPINDACEAE (D)
Cossonia Durieu = Raffenaldia Godr. CRUCIFERAE (D)
Costaea A.Rich. = Purdiaea Planch. CYRILLACEAE (D)
Costantina Bullock ASCLEPIADACEAE (D)
Costarica L.D.Gómez = Sicyos L. CUCURBITACEAE (D)
Costaricaea Schltr. = Hexisea Lindl. ORCHIDACEAE (M)
Costaricia Christ = Dennstaedtia Bernh. DENNSTAEDTIACEAE (F)
Costera J.J.Sm. ERICACEAE (D)
Costia Willk. (SUS) = Agropyron Gaertn. GRAMINEAE (M)
Costularia C.B.Clarke CYPERACEAE (M)
Costus L. COSTACEAE (M)
Cotema Britton & P.Wilson = Spirotecoma Baill. ex Dalla Torre & Harms
    BIGNONIACEAE (D)
Cotinus Mill. ANACARDIACEAE (D)
Cotoneaster Medik. ROSACEAE (D)
Cotopaxia Mathias & Constance UMBELLIFERAE (D)
Cottea Kunth GRAMINEAE (M)
Cottendorfia Schult.f. BROMELIACEAE (M)
Cottonia Wight ORCHIDACEAE (M)
Cottsia Dubard & Dop = Janusia A.Juss. MALPIGHIACEAE (D)
Cotula L. COMPOSITAE (D)
Cotylanthera Blume GENTIANACEAE (D)
Cotyledon L. CRASSULACEAE (D)
Cotylelobium Pierre DIPTEROCARPACEAE (D)
Cotyliphyllum Link = Cotyledon L. CRASSULACEAE (D)
Cotylodiscus Radlk. = Plagioscyphus Radlk. SAPINDACEAE (D)
Cotylolabium Garay = Stenorrhynchos Rich. ex Spreng. ORCHIDACEAE (M)
Cotylonychia Stapf STERCULIACEAE (D)
Coudenbergia Marchal = Pentapanax Seem. ARALIACEAE (D)
Couepia Aubl. CHRYSOBALANACEAE (D)
Coula Baill. OLACACEAE (D)
Coulejia Dennst. = Antidesma L. EUPHORBIACEAE (D)

Coulterella Vasey & Rose COMPOSITAE (D)
Coulterophytum B.L.Rob. UMBELLIFERAE (D)
Couma Aubl. APOCYNACEAE (D)
Coumarouna Aubl. = Dipteryx Schreb. LEGUMINOSAE–PAPILIONOIDEAE (D)
Coupoui Aubl. = Duroia L.f. RUBIACEAE (D)
Coupuia Raf. (SUO) = Duroia L.f. RUBIACEAE (D)
Coupuya Raf. (SUO) = Duroia L.f. RUBIACEAE (D)
Couralia Splitg. = Tabebuia Gomes ex DC. BIGNONIACEAE (D)
Courantia Lem. = Echeveria DC. CRASSULACEAE (D)
Couratari Aubl. LECYTHIDACEAE (D)
Courbonia Brongn. = Maerua Forssk. CAPPARACEAE (D)
Couroupita Aubl. LECYTHIDACEAE (D)
Coursetia DC. LEGUMINOSAE–PAPILIONOIDEAE (D)
Coursiana Homolle = Schismatoclada Baker RUBIACEAE (D)
Courtoisia Nees (SUH) = Courtoisina Soják CYPERACEAE (M)
Courtoisia Rchb. (SUH) = Collomia Nutt. POLEMONIACEAE (D)
Courtoisina Soják CYPERACEAE (M)
Cousinia Cass. COMPOSITAE (D)
Cousiniopsis Nevski COMPOSITAE (D)
Coussapoa Aubl. CECROPIACEAE (D)
Coussarea Aubl. RUBIACEAE (D)
Coutaportla Urb. RUBIACEAE (D)
Coutarea Aubl. RUBIACEAE (D)
Couthovia A.Gray = Neuburgia Blume LOGANIACEAE (D)
Coutinia Vell. = Aspidosperma Mart. & Zucc. APOCYNACEAE (D)
Coutoubea Aubl. GENTIANACEAE (D)
Covellia Gasp. = Ficus L. MORACEAE (D)
Coveniella Tindale DRYOPTERIDACEAE (F)
Covillea Vail = Larrea Cav. ZYGOPHYLLACEAE (D)
Cowania D.Don ROSACEAE (D)
Cowellocassia Britton = Crudia Schreb. LEGUMINOSAE–CAESALPINIOIDEAE (D)
Cowiea Wernham RUBIACEAE (D)
Coxella Cheeseman & Hemsl. UMBELLIFERAE (D)
Crabbea Harv. ACANTHACEAE (D)
Cracca Benth. = Coursetia DC. LEGUMINOSAE–PAPILIONOIDEAE (D)
Cracosna Gagnep. GENTIANACEAE (D)
Craepalia Schrank = Lolium L. GRAMINEAE (M)
Craspedoneuron Van den Bosch = Crepidomanes (C.Presl) C.Presl
    HYMENOPHYLLACEAE (F)
Craspedophyllum (C.Presl) Copel. = Hymenophyllum Sm. HYMENOPHYLLACEAE (F)
Craspedorhachis Benth. GRAMINEAE (M)
Craspedostoma Domke THYMELAEACEAE (D)
Craspidospermum Bojer ex A.DC. APOCYNACEAE (D)
Crassipes Swallen = Sclerochloa P.Beauv. GRAMINEAE (M)
Crassocephalum Moench COMPOSITAE (D)
Crassula L. CRASSULACEAE (D)
Crataegus L. ROSACEAE (D)
Crateranthus Baker f. LECYTHIDACEAE (D)
Cratericarpium Spach = Boisduvalia Spach ONAGRACEAE (D)
Crateriphytum Scheff. ex Koord. = Neuburgia Blume LOGANIACEAE (D)
Craterispermum Benth. RUBIACEAE (D)
Craterocapsa Hilliard & B.L.Burtt CAMPANULACEAE (D)
Craterocoma Mart. ex DC. = Lundia DC. BIGNONIACEAE (D)
Craterogyne Lanj. = Dorstenia L. MORACEAE (D)
Craterosiphon Engl. & Gilg THYMELAEACEAE (D)

Craterostemma K.Schum. = Brachystelma R.Br. ASCLEPIADACEAE (D)
Craterostigma Hochst. SCROPHULARIACEAE (D)
Crateva L. CAPPARACEAE (D)
Cratoxylon Blume (SUO) = Cratoxylum Blume GUTTIFERAE (D)
Cratoxylum Blume GUTTIFERAE (D)
Cratylia Mart. ex Benth. LEGUMINOSAE–PAPILIONOIDEAE (D)
Cratystylis S.Moore COMPOSITAE (D)
Crawfurdia Wall. GENTIANACEAE (D)
Creaghia Scort. = Mussaendopsis Baill. RUBIACEAE (D)
Creaghiella Stapf = Anerincleistus Korth. MELASTOMATACEAE (D)
Creatantha Standl. = Isertia Schreb. RUBIACEAE (D)
Cremanthodium Benth. COMPOSITAE (D)
Cremaspora Benth. RUBIACEAE (D)
Cremastopus Paul G.Wilson CUCURBITACEAE (D)
Cremastosciadium Rech.f. = Eriocycla Lindl. UMBELLIFERAE (D)
Cremastosperma R.E.Fr. ANNONACEAE (D)
Cremastra Lindl. ORCHIDACEAE (M)
Cremastus Miers = Arrabidaea DC. BIGNONIACEAE (D)
Cremnobates Ridl. = Schizomeria D.Don CUNONIACEAE (D)
Cremnophila Rose CRASSULACEAE (D)
Cremnophyton Brullo & Pavone CHENOPODIACEAE (D)
Cremocarpon Baill. RUBIACEAE (D)
Cremochilus Turcz. = Siphocampylus Pohl CAMPANULACEAE (D)
Cremolobus DC. CRUCIFERAE (D)
Cremophyllum Scheidw. = Dalechampia L. EUPHORBIACEAE (D)
Cremopyrum Schur (SUO) = Eremopyrum (Ledeb.) Jaub. & Spach GRAMINEAE (M)
Cremosperma Benth. GESNERIACEAE (D)
Cremostachys Tul. = Galearia Zoll. & Moritzi PANDACEAE (D)
Crena Scop. = Crenea Aubl. LYTHRACEAE (D)
Crenaea Schreb. = Crenea Aubl. LYTHRACEAE (D)
Crenea Aubl. LYTHRACEAE (D)
Crenias Spreng. = Mniopsis Mart. & Zucc. PODOSTEMACEAE (D)
Crenidium Haegi SOLANACEAE (D)
Crenosciadium Boiss. & Heldr. = Opopanax W.D.J.Koch UMBELLIFERAE (D)
Crenularia Boiss. = Aethionema R.Br. CRUCIFERAE (D)
Creochiton Blume MELASTOMATACEAE (D)
Crepidaria Haw. = Pedilanthus Neck. ex Poit. EUPHORBIACEAE (D)
Crepidiastrum Nakai COMPOSITAE (D)
Crepidium C.Presl = Crepidomanes (C.Presl) C.Presl HYMENOPHYLLACEAE (F)
Crepidium Blume (SUH) = Malaxis Sol. ex Sw. ORCHIDACEAE (M)
Crepidomanes (C.Presl) C.Presl HYMENOPHYLLACEAE (F)
Crepidophyllum C.F.Reed = Crepidomanes (C.Presl) C.Presl HYMENOPHYLLACEAE (F)
Crepidopsis Arv.–Touv. = Hieracium L. COMPOSITAE (D)
Crepidopteris Copel. = Crepidomanes (C.Presl) C.Presl HYMENOPHYLLACEAE (F)
Crepidorhopalon Eb.Fisch. = Torenia L. SCROPHULARIACEAE (D)
Crepidospermum Hook.f. BURSERACEAE (D)
Crepinella Marchal = Schefflera J.R.Forst. & G.Forst. ARALIACEAE (D)
Crepinodendron Pierre = Micropholis (Griseb.) Pierre SAPOTACEAE (D)
Crepis L. COMPOSITAE (D)
Crescentia L. BIGNONIACEAE (D)
Creslobus Lilja = Mentzelia L. LOASACEAE (D)
Cressa L. CONVOLVULACEAE (D)
Cribbia Senghas ORCHIDACEAE (M)
Criciuma Soderstr. & Londoño GRAMINEAE (M)
Crimaea Vassilcz. = Medicago L. LEGUMINOSAE–PAPILIONOIDEAE (D)

Crinipes Hochst. GRAMINEAE (M)
Crinita Houtt. = Pavetta L. RUBIACEAE (D)
Crinodendron Molina ELAEOCARPACEAE (D)
Crinonia Blume = Pholidota Lindl. ex Hook. ORCHIDACEAE (M)
Crinum L. AMARYLLIDACEAE (M)
Crioceras Pierre APOCYNACEAE (D)
Criogenes Salisb. = Cypripedium L. ORCHIDACEAE (M)
Crispiloba Steenis ALSEUOSMIACEAE (D)
Cristaria Cav. MALVACEAE (D)
Cristaria Sonn. (SUH) = Combretum Loefl. COMBRETACEAE (D)
Cristatella Nutt. CAPPARACEAE (D)
Critesion Raf. = Hordeum L. GRAMINEAE (M)
Crithmum L. UMBELLIFERAE (D)
Critho E.Mey. = Hordeum L. GRAMINEAE (M)
Crithodium Link = Triticum L. GRAMINEAE (M)
Crithopsis Jaub. & Spach GRAMINEAE (M)
Crithopyrum Steud. (SUI) = Elymus L. GRAMINEAE (M)
Critonia P.Browne COMPOSITAE (D)
Critoniadelphus R.M.King & H.Rob. COMPOSITAE (D)
Critoniella R.M.King & H.Rob. COMPOSITAE (D)
Critoniopsis Sch.Bip. COMPOSITAE (D)
Crobylanthe Bremek. RUBIACEAE (D)
Crocanthemum Spach = Halimium (Dunal) Spach CISTACEAE (D)
Crocanthus L.Bolus = Malephora N.E.Br. AIZOACEAE (D)
Crocidium Hook. COMPOSITAE (D)
Crociris Schur = Crocus L. IRIDACEAE (M)
Crocodeilanthe Rchb.f. = Pleurothallis R.Br. ORCHIDACEAE (M)
Crocopsis Pax = Stenomesson Herb. AMARYLLIDACEAE (M)
Crocosmia Planch. IRIDACEAE (M)
Crocoxylon Eckl. & Zeyh. CELASTRACEAE (D)
Crocus L. IRIDACEAE (M)
Crocyllis E.Mey. ex Hook.f. RUBIACEAE (D)
Croftia Small = Carlowrightia A.Gray ACANTHACEAE (D)
Croixia Pierre = Palaquium Blanco SAPOTACEAE (D)
Croizatia Steyerm. EUPHORBIACEAE (D)
Cromidon Compton SCROPHULARIACEAE (D)
Cronquistia R.M.King COMPOSITAE (D)
Cronquistianthus R.M.King & H.Rob. COMPOSITAE (D)
Croomia Torr. STEMONACEAE (M)
Croptilon Raf. COMPOSITAE (D)
Crossandra Salisb. ACANTHACEAE (D)
Crossandrella C.B.Clarke ACANTHACEAE (D)
Crossangis Schltr. = Diaphananthe Schltr. ORCHIDACEAE (M)
Crosslandia W.Fitzg. CYPERACEAE (M)
Crossonephelis Baill. = Glenniea Hook.f. SAPINDACEAE (D)
Crossopetalum P.Browne CELASTRACEAE (D)
Crossophora Link = Chrozophora Neck. ex Juss. EUPHORBIACEAE (D)
Crossopteryx Fenzl RUBIACEAE (D)
Crossosoma Nutt. CROSSOSOMATACEAE (D)
Crossostemma Planch. ex Benth. PASSIFLORACEAE (D)
Crossostephium Less. COMPOSITAE (D)
Crossostigma Spach = Epilobium L. ONAGRACEAE (D)
Crossostylis J.R.Forst. & G.Forst. RHIZOPHORACEAE (D)
Crossothamnus R.M.King & H.Rob. COMPOSITAE (D)
Crossotropis Stapf = Trichoneura Andersson GRAMINEAE (M)

Crotalaria L. LEGUMINOSAE–PAPILIONOIDEAE (D)
Croton L. EUPHORBIACEAE (D)
Crotonanthus Klotzsch ex Schltdl. = Croton L. EUPHORBIACEAE (D)
Crotonogyne Müll.Arg. EUPHORBIACEAE (D)
Crotonogynopsis Pax EUPHORBIACEAE (D)
Crotonopsis Michx. EUPHORBIACEAE (D)
Crowea Sm. RUTACEAE (D)
Crozophora A.Juss. (SUO) = Chrozophora Neck. ex Juss. EUPHORBIACEAE (D)
Crozophyla Raf. = Codiaeum A.Juss. EUPHORBIACEAE (D)
Crucianella L. RUBIACEAE (D)
Cruciata Mill. RUBIACEAE (D)
Crucicaryum Brand BORAGINACEAE (D)
Cruckshanksia Hook. & Arn. RUBIACEAE (D)
Cruddasia Prain LEGUMINOSAE–PAPILIONOIDEAE (D)
Crudia Schreb. LEGUMINOSAE–CAESALPINIOIDEAE (D)
Crudya Batsch (SUO) = Crudia Schreb. LEGUMINOSAE–CAESALPINIOIDEAE (D)
Crumenaria Mart. RHAMNACEAE (D)
Crunocallis Rydb. = Montia L. PORTULACACEAE (D)
Crupina (Pers.) DC. COMPOSITAE (D)
Crusea Cham. & Schltdl. RUBIACEAE (D)
Crusea A.Rich. (SUH) = Chione DC. RUBIACEAE (D)
Cruzia Phil. = Scutellaria L. LABIATAE (D)
Crybe Lindl. ORCHIDACEAE (M)
Cryophytum N.E.Br. = Mesembryanthemum L. AIZOACEAE (D)
Cryosophila Blume PALMAE (M)
Cryphia R.Br. = Prostanthera Labill. LABIATAE (D)
Crypsinna Fourn. = Muhlenbergia Schreb. GRAMINEAE (M)
Crypsinopsis Pic.Serm. = Crypsinus C.Presl POLYPODIACEAE (F)
Crypsinus C.Presl POLYPODIACEAE (F)
Crypsis Aiton GRAMINEAE (M)
Cryptadenia Meisn. THYMELAEACEAE (D)
Cryptandra Sm. RHAMNACEAE (D)
Cryptangium Schrad. ex Nees = Lagenocarpus Nees CYPERACEAE (M)
Cryptanopsis Ule = Orthophytum Beer BROMELIACEAE (M)
Cryptantha G.Don BORAGINACEAE (D)
Cryptanthela Gagnep. = Argyreia Lour. CONVOLVULACEAE (D)
Cryptanthemis Rupp = Rhizanthella R.S.Rogers ORCHIDACEAE (M)
Cryptanthus Otto & A.Dietr. BROMELIACEAE (M)
Cryptarrhena R.Br. ORCHIDACEAE (M)
Crypteronia Blume CRYPTERONIACEAE (D)
Cryptobasis Nevski = Iris L. IRIDACEAE (M)
Cryptocapnos Rech.f. PAPAVERACEAE (D)
Cryptocarpus Kunth NYCTAGINACEAE (D)
Cryptocarya R.Br. LAURACEAE (D)
Cryptocentrum Benth. ORCHIDACEAE (M)
Cryptocereus Alexander = Selenicereus (A.Berger) Britton & Rose CACTACEAE (D)
Cryptochilus Wall. ORCHIDACEAE (M)
Cryptochloa Swallen GRAMINEAE (M)
Cryptochloris Benth. = Tetrapogon Desf. GRAMINEAE (M)
Cryptocodon Fedorov CAMPANULACEAE (D)
Cryptocoryne Fisch. ex Wydl. ARACEAE (M)
Cryptodiscus Schrenk ex Fisch. & C.A.Mey. (SUH) = Prangos Lindl. UMBELLIFERAE (D)
Cryptoglottis Blume = Podochilus Blume ORCHIDACEAE (M)
Cryptogramma R.Br. ADIANTACEAE (F)
Cryptogyne Hook.f. = Sideroxylon L. SAPOTACEAE (D)

Cryptolepis R.Br. ASCLEPIADACEAE (D)
Cryptoloma Hanst. = Kohleria Regel GESNERIACEAE (D)
Cryptomeria D.Don TAXODIACEAE (G)
Cryptopetalum Hook. & Arn. = Lepuropetalon Elliott PARNASSIACEAE (D)
Cryptophaeseolus Kuntze = Canavalia DC. LEGUMINOSAE–PAPILIONOIDEAE (D)
Cryptophoranthus Barb.Rodr. ORCHIDACEAE (M)
Cryptophragmium Nees = Gymnostachyum Nees ACANTHACEAE (D)
Cryptophyllum Schltdl. (SUO) = Anemia Sw. SCHIZAEACEAE (F)
Cryptophysa Standl. & J.F.Macbr. = Conostegia D.Don MELASTOMATACEAE (D)
Cryptopodium Schrad. ex Nees = Scleria Bergius CYPERACEAE (M)
Cryptopus Lindl. ORCHIDACEAE (M)
Cryptopylos Garay ORCHIDACEAE (M)
Cryptopyrum Heynh. (SUI) = Elymus L. GRAMINEAE (M)
Cryptorhiza Urb. = Pimenta Lindl. MYRTACEAE (D)
Cryptorrhynchus Nevski = Astragalus L. LEGUMINOSAE–PAPILIONOIDEAE (D)
Cryptosanus Scheidw. = Leochilus Knowles & Westc. ORCHIDACEAE (M)
Cryptosepalum Benth. LEGUMINOSAE–CAESALPINIOIDEAE (D)
Cryptosorus Fée = Ctenopteris Blume ex Kunze GRAMMITIDACEAE (F)
Cryptospermum Young ex Pers. = Opercularia Gaertn. RUBIACEAE (D)
Cryptospora Kar. & Kir. CRUCIFERAE (D)
Cryptostachys Steud. = Sporobolus R.Br. GRAMINEAE (M)
Cryptostegia R.Br. ASCLEPIADACEAE (D)
Cryptostemma R.Br. ex W.T.Aiton = Arctotheca J.C.Wendl. COMPOSITAE (D)
Cryptostemon F.Muell. ex Miq. = Darwinia Rudge MYRTACEAE (D)
Cryptostephanus Welw. ex Baker AMARYLLIDACEAE (M)
Cryptostylis R.Br. ORCHIDACEAE (M)
Cryptotaenia DC. UMBELLIFERAE (D)
Cryptotaeniopsis Dunn = Pternopetalum Franch. UMBELLIFERAE (D)
Cryptothladia (Bunge) M.J.Cannon = Morina L. MORINACEAE (D)
Crypturus Trin. = Lolium L. GRAMINEAE (M)
Crystallopollen Steetz = Vernonia Schreb. COMPOSITAE (D)
Cteisium Michx. = Lygodium Sw. SCHIZAEACEAE (F)
Ctenadena Prokh. = Euphorbia L. EUPHORBIACEAE (D)
Ctenanthe Eichler MARANTACEAE (M)
Ctenardisia Ducke MYRSINACEAE (D)
Ctenitis (C.Chr.) C.Chr. DRYOPTERIDACEAE (F)
Ctenitopsis Ching ex Tardieu & C.Chr. = Heterogonium C.Presl DRYOPTERIDACEAE (F)
Ctenium Panz. GRAMINEAE (M)
Ctenocladium Airy Shaw = Dorstenia L. MORACEAE (D)
Ctenolepis Hook.f. CUCURBITACEAE (D)
Ctenolophon Oliv. CTENOLOPHONACEAE (D)
Ctenomeria Harv. EUPHORBIACEAE (D)
Ctenopaepale Bremek. = Strobilanthes Blume ACANTHACEAE (D)
Ctenophrynium K.Schum. = Saranthe (Regel & Koern.) Eichl. MARANTACEAE (M)
Ctenophyllum Rydb. = Astragalus L. LEGUMINOSAE–PAPILIONOIDEAE (D)
Ctenopsis Naudin (SUH) = Ctenolepis Hook.f. CUCURBITACEAE (D)
Ctenopsis De Not. = Vulpia C.C.Gmelin GRAMINEAE (M)
Ctenopteris Newman (SUH) = Polypodium L. POLYPODIACEAE (F)
Ctenopteris Blume ex Kunze GRAMMITIDACEAE (F)
Ctenorchis K.Schum. = Angraecum Bory ORCHIDACEAE (M)
Cuatrecasanthus H.Rob. = Vernonia Schreb. COMPOSITAE (D)
Cuatrecasasia Standl. = Cuatrecasasiodendron Standl. & Steyerm. RUBIACEAE (D)
Cuatrecasasiella H.Rob. COMPOSITAE (D)
Cuatrecasasiodendron Standl. & Steyerm. RUBIACEAE (D)
Cuatrecasea Dugand = Iriartella H.A.Wendl. PALMAE (M)

Cuatresia Hunz. SOLANACEAE (D)
Cubacroton Alain EUPHORBIACEAE (D)
Cubanola Aiello RUBIACEAE (D)
Cubanthus (Boiss.) Millsp. EUPHORBIACEAE (D)
Cubelium Raf. ex Britton & A.Br. = Hybanthus Jacq. VIOLACEAE (D)
Cubilia Blume SAPINDACEAE (D)
Cubitanthus Barringer GESNERIACEAE (D)
Cubospermum Lour. = Ludwigia L. ONAGRACEAE (D)
Cuchumatanea Seid. & Beaman COMPOSITAE (D)
Cucifera Delile = Hyphaene Gaertn. PALMAE (M)
Cucubalus L. CARYOPHYLLACEAE (D)
Cuculina Raf. = Catasetum Rich. ex Kunth ORCHIDACEAE (M)
Cucullaria Schreb. (SUS) = Vochysia Aubl. VOCHYSIACEAE (D)
Cucumella Chiov. CUCURBITACEAE (D)
Cucumeroides Gaertn. = Trichosanthes L. CUCURBITACEAE (D)
Cucumeropsis Naudin CUCURBITACEAE (D)
Cucumis L. CUCURBITACEAE (D)
Cucurbita L. CUCURBITACEAE (D)
Cucurbitella Walp. CUCURBITACEAE (D)
Cucurbitula (M.Roem.) Post & Kuntze = Zehneria Endl. CUCURBITACEAE (D)
Cudrania Trécul = Maclura Nutt. MORACEAE (D)
Cuervea Triana ex Miers CELASTRACEAE (D)
Cufodontia Woodson = Aspidosperma Mart. & Zucc. APOCYNACEAE (D)
Cuitlauzina La Llave & Lex. ORCHIDACEAE (M)
Culcasia P.Beauv. ARACEAE (M)
Culcita C.Presl DICKSONIACEAE (F)
Culcitium Bonpl. = Senecio L. COMPOSITAE (D)
Cullen Medik. LEGUMINOSAE–PAPILIONOIDEAE (D)
Cullenia Wight = Durio Adans. BOMBACACEAE (D)
Cullmannia Distef. = Peniocereus (A.Berger) Britton & Rose CACTACEAE (D)
Cullumia R.Br. COMPOSITAE (D)
Cumarinia Buxb. = Coryphantha (Engelm.) Lem. CACTACEAE (D)
Cumetea Raf. = Myrcia DC. ex Guill. MYRTACEAE (D)
Cumingia Vidal = Camptostemon Mast. BOMBACACEAE (D)
Cumingia Kunth (SUI) = Conanthera Ruíz & Pav. TECOPHILAEACEAE (M)
Cuminia Colla LABIATAE (D)
Cuminum L. UMBELLIFERAE (D)
Cummingia D.Don = Conanthera Ruíz & Pav. TECOPHILAEACEAE (M)
Cumminsia King ex Prain (SUI) = Meconopsis R.Vig. PAPAVERACEAE (D)
Cumulopuntia F.Ritter = Opuntia Mill. CACTACEAE (D)
Cuncea Buch.–Ham. ex D.Don = Knoxia L. RUBIACEAE (D)
Cunila Royen ex L. LABIATAE (D)
Cunina Gay = Nertera Banks & Sol. ex Gaertn. RUBIACEAE (D)
Cunninghamia R.Br. TAXODIACEAE (G)
Cunninghamia Schreb. (SUH) = Malanea Aubl. RUBIACEAE (D)
Cunonia L. CUNONIACEAE (D)
Cunuria Baill. = Micrandra Benth. EUPHORBIACEAE (D)
Cupameni Adans. = Acalypha L. EUPHORBIACEAE (D)
Cupania L. SAPINDACEAE (D)
Cupaniopsis Radlk. SAPINDACEAE (D)
Cuphaea Moench (SUO) = Cuphea P.Browne LYTHRACEAE (D)
Cuphea P.Browne LYTHRACEAE (D)
Cupheanthus Seem. MYRTACEAE (D)
Cuphocarpus Decne. & Planch. ARALIACEAE (D)
Cuphonotus O.E.Schulz CRUCIFERAE (D)

Cupi Adans. = Tarenna Gaertn. RUBIACEAE (D)
Cupia (Schult.) DC. = Anomanthodia Hook.f. RUBIACEAE (D)
Cupirana Miers = Duroia L.f. RUBIACEAE (D)
Cupressus L. CUPRESSACEAE (G)
Cupuia Raf. = Duroia L.f. RUBIACEAE (D)
Cupulanthus Hutch. LEGUMINOSAE–PAPILIONOIDEAE (D)
Cupulissa Raf. = Anemopaegma Mart. ex Meisn. BIGNONIACEAE (D)
Curanga Juss. = Picria Lour. SCROPHULARIACEAE (D)
Curarea Barneby & Krukoff MENISPERMACEAE (D)
Curatari J.F.Gmel. (SUO) = Couratari Aubl. LECYTHIDACEAE (D)
Curataria Spreng. (SUO) = Couratari Aubl. LECYTHIDACEAE (D)
Curatella Loefl. DILLENIACEAE (D)
Curcas Adans. = Jatropha L. EUPHORBIACEAE (D)
Curculigo Gaertn. HYPOXIDACEAE (M)
Curcuma L. ZINGIBERACEAE (M)
Curcumorpha A.S.Rao & D.M.Verma ZINGIBERACEAE (M)
Curima O.F.Cook = Aiphanes Willd. PALMAE (M)
Curinila Schult. ASCLEPIADACEAE (D)
Currania Copel. = Gymnocarpium Newm. WOODSIACEAE (F)
Curraniodendron Merr. = Quintinia A.DC. ESCALLONIACEAE (D)
Curroria Planch. ex Benth. ASCLEPIADACEAE (D)
Curtia Cham. & Schltdl. GENTIANACEAE (D)
Curtisia Aiton CORNACEAE (D)
Curtonus N.E.Br. = Crocosmia Planch. IRIDACEAE (M)
Curtopogon P.Beauv. = Aristida L. GRAMINEAE (M)
Curupira G.A.Black OLACACEAE (D)
Cuscatlania Standl. NYCTAGINACEAE (D)
Cuscuta L. CONVOLVULACEAE (D)
Cusickia M.E.Jones = Lomatium Raf. UMBELLIFERAE (D)
Cusickiella Rollins CRUCIFERAE (D)
Cusparia Humb. ex R.Br. = Angostura Roem. & Schult. RUTACEAE (D)
Cuspidaria Fée (SUH) = Dicranoglossum J.Sm. POLYPODIACEAE (F)
Cuspidaria DC. BIGNONIACEAE (D)
Cuspidia Gaertn. COMPOSITAE (D)
Cussonia Thunb. ARALIACEAE (D)
Cutandia Willk. GRAMINEAE (M)
Cuthbertia Small = Callisia Loefl. COMMELINACEAE (M)
Cutsis Burns–Bal., E.W.Greenwood & Gonzales = Dichromanthus Garay
      ORCHIDACEAE (M)
Cuttsia F.Muell. ESCALLONIACEAE (D)
Cuviera Koeler (SUH) = Hordelymus (Jess.) Jess. ex Harz GRAMINEAE (M)
Cuviera DC. RUBIACEAE (D)
Cwangayana Rauschert = Aralia L. ARALIACEAE (D)
Cyamopsis DC. LEGUMINOSAE–PAPILIONOIDEAE (D)
Cyanaeorchis Barb.Rodr. ORCHIDACEAE (M)
Cyanandrium Stapf MELASTOMATACEAE (D)
Cyananthus Griff. (SUH) = Stauranthera Benth. GESNERIACEAE (D)
Cyananthus Wall. ex Benth. CAMPANULACEAE (D)
Cyanastrum Oliv. CYANASTRACEAE (M)
Cyanea Gaudich. CAMPANULACEAE (D)
Cyanella Royen ex L. TECOPHILAEACEAE (M)
Cyanitis Reinw. = Dichroa Lour. HYDRANGEACEAE (D)
Cyanocarpus F.M.Bailey = Helicia Lour. PROTEACEAE (D)
Cyanodaphne Blume = Dehaasia Blume LAURACEAE (D)
Cyanopsis Cass. = Volutaria Cass. COMPOSITAE (D)

Cyanorchis Thouars = Phaius Lour. ORCHIDACEAE (M)
Cyanostegia Turcz. VERBENACEAE (D)
Cyanostremma Benth. ex Hook. & Arn. = Calopogonium Desv. LEGUMINOSAE–
    PAPILIONOIDEAE (D)
Cyanothamnus Lindl. RUTACEAE (D)
Cyanothyrsus Harms = Daniellia Benn. LEGUMINOSAE–CAESALPINIOIDEAE (D)
Cyanotis D.Don COMMELINACEAE (M)
Cyanthillium Blume = Vernonia Schreb. COMPOSITAE (D)
Cyanthanthus Engl. = Scyphosyce Baill. MORACEAE (D)
Cyathea Sm. CYATHEACEAE (F)
Cyathiscus Tiegh. = Barathranthus (Korth.) Miq. LORANTHACEAE (D)
Cyathobasis Aellen CHENOPODIACEAE (D)
Cyathocalyx Champ. ex Hook.f. & Thomson ANNONACEAE (D)
Cyathochaeta Nees CYPERACEAE (M)
Cyathocline Cass. COMPOSITAE (D)
Cyathocoma Nees = Tetraria P.Beauv. CYPERACEAE (M)
Cyathodes Labill. EPACRIDACEAE (D)
Cyathoglottis Poepp. & Endl. = Sobralia Ruíz & Pav. ORCHIDACEAE (M)
Cyathogyne Müll.Arg. = Thecacoris A.Juss. EUPHORBIACEAE (D)
Cyathomone S.F.Blake COMPOSITAE (D)
Cyathophora Raf. = Euphorbia L. EUPHORBIACEAE (D)
Cyathophylla Bocquet & Strid CARYOPHYLLACEAE (D)
Cyathopsis Brongn. & Gris EPACRIDACEAE (D)
Cyathopus Stapf GRAMINEAE (M)
Cyathorhachis Steud. = Polytoca R.Br. GRAMINEAE (M)
Cyathoselinum Benth. UMBELLIFERAE (D)
Cyathostegia (Benth.) Schery LEGUMINOSAE–PAPILIONOIDEAE (D)
Cyathostelma Fourn. ASCLEPIADACEAE (D)
Cyathostemma Griff. ANNONACEAE (D)
Cyathostemon Turcz. = Baeckea L. MYRTACEAE (D)
Cyathostyles Schott ex Meisn. = Cyphomandra Mart. ex Sendtn. SOLANACEAE (D)
Cyathula Blume AMARANTHACEAE (D)
Cybebus Garay ORCHIDACEAE (M)
Cybelion Spreng. = Ionopsis Kunth ORCHIDACEAE (M)
Cybianthopsis (Mez) Lundell = Cybianthus Mart. MYRSINACEAE (D)
Cybianthus Mart. MYRSINACEAE (D)
Cybistax Mart. ex Meisn. BIGNONIACEAE (D)
Cybistetes Milne–Redh. & Schweick. AMARYLLIDACEAE (M)
Cycas L. CYCADACEAE (G)
Cyclacanthus S.Moore ACANTHACEAE (D)
Cyclachaena Fres. ex Schltdl. COMPOSITAE (D)
Cycladenia Benth. APOCYNACEAE (D)
Cyclamen L. PRIMULACEAE (D)
Cyclandrophora Hassk. = Atuna Raf. CHRYSOBALANACEAE (D)
Cyclanthera Schrad. CUCURBITACEAE (D)
Cyclantheropsis Harms CUCURBITACEAE (D)
Cyclanthus Poit. CYCLANTHACEAE (M)
Cyclea Arn. ex Wight MENISPERMACEAE (D)
Cyclobalanopsis Oerst. = Quercus L. FAGACEAE (D)
Cyclocarpa Afzel. ex Baker LEGUMINOSAE–PAPILIONOIDEAE (D)
Cyclocarya Iljinsk. JUGLANDACEAE (D)
Cyclocheilon Oliv. CYCLOCHEILACEAE (D)
Cyclocodon Griff. = Codonopsis Wall. CAMPANULACEAE (D)
Cyclocotyla Stapf APOCYNACEAE (D)
Cyclodiscus Klotzsch = Thottea Rottb. ARISTOLOCHIACEAE (D)

Cyclodium C.Presl DRYOPTERIDACEAE (F)
Cyclogramma Tagawa THELYPTERIDACEAE (F)
Cyclolepis Gillies ex D.Don COMPOSITAE (D)
Cyclolobium Benth. LEGUMINOSAE–PAPILIONOIDEAE (D)
Cycloloma Moq. CHENOPODIACEAE (D)
Cyclopeltis J.Sm. DRYOPTERIDACEAE (F)
Cyclophorus Desv. = Pyrrosia Mirb. POLYPODIACEAE (F)
Cyclophyllum Hook.f. RUBIACEAE (D)
Cyclopia Vent. LEGUMINOSAE–PAPILIONOIDEAE (D)
Cyclopogon C.Presl ORCHIDACEAE (M)
Cyclopteris Schrad. ex Gray = Cystopteris Bernh. WOODSIACEAE (F)
Cycloptychis E.Mey. ex Sond. CRUCIFERAE (D)
Cyclorhiza M.L.Sheh & R.H.Shan UMBELLIFERAE (D)
Cyclosorus Link THELYPTERIDACEAE (F)
Cyclospathe O.F.Cook = Pseudophoenix H.A.Wendl. ex Sarg. PALMAE (M)
Cyclospermum Lag. UMBELLIFERAE (D)
Cyclostachya J.R.Reeder & C.G.Reeder GRAMINEAE (M)
Cyclostegia Benth. = Elsholtzia Willd. LABIATAE (D)
Cyclostemon Blume = Drypetes Vahl EUPHORBIACEAE (D)
Cyclostigma Klotzsch ex Seem. (SUH) = Croton L. EUPHORBIACEAE (D)
Cyclostigma Phil. (SUH) = Leptoglossis Benth. SOLANACEAE (D)
Cycloteria Stapf (SUI) = Coelorachis Brongn. GRAMINEAE (M)
Cyclotrichium (Boiss.) Manden. & Scheng. LABIATAE (D)
Cycniopsis Engl. SCROPHULARIACEAE (D)
Cycnium E.Mey. ex Benth. SCROPHULARIACEAE (D)
Cycnoches Lindl. ORCHIDACEAE (M)
Cycnopodium Naudin = Graffenrieda DC. MELASTOMATACEAE (D)
Cydista Miers BIGNONIACEAE (D)
Cydonia Mill. ROSACEAE (D)
Cylichnium Dulac (SUS) = Gaudinia P.Beauv. GRAMINEAE (M)
Cylicodiscus Harms LEGUMINOSAE–MIMOSOIDEAE (D)
Cylicomorpha Urb. CARICACEAE (D)
Cylindria Lour. = Chionanthus L. OLEACEAE (D)
Cylindrocarpa Regel CAMPANULACEAE (D)
Cylindrocline Cass. COMPOSITAE (D)
Cylindrokelupha Kosterm. = Archidendron F.Muell. LEGUMINOSAE–
    MIMOSOIDEAE (D)
Cylindrolobus (Blume) Brieger = Eria Lindl. ORCHIDACEAE (M)
Cylindrophyllum Schwantes AIZOACEAE (D)
Cylindropsis Pierre APOCYNACEAE (D)
Cylindropuntia (Engelm.) F.Knuth = Opuntia Mill. CACTACEAE (D)
Cylindropyrum (Jaub. & Spach) Á.Löve = Aegilops L. GRAMINEAE (M)
Cylindrorebutia Frič & Kreuz. = Rebutia K.Schum. CACTACEAE (D)
Cylindrosolenium Lindau ACANTHACEAE (D)
Cylindrosperma Ducke = Microplumeria Baill. APOCYNACEAE (D)
Cylista Aiton = Paracalyx Ali LEGUMINOSAE–PAPILIONOIDEAE (D)
Cylixylon Llanos = Gymnanthera R.Br. ASCLEPIADACEAE (D)
Cymaria Benth. LABIATAE (D)
Cymatocarpus O.E.Schulz CRUCIFERAE (D)
Cymatochloa Schltdl. = Paspalum L. GRAMINEAE (M)
Cymbachne Retz. (UR) GRAMINEAE (M)
Cymbalaria Hill SCROPHULARIACEAE (D)
Cymbanthelia Andersson (SUI) = Cymbopogon Spreng. GRAMINEAE (M)
Cymbaria L. SCROPHULARIACEAE (D)
Cymbidiella Rolfe ORCHIDACEAE (M)

Cymbidium Sw. ORCHIDACEAE (M)
Cymbiglossum Halbinger = Lemboglossum Halbinger ORCHIDACEAE (M)
Cymbispatha Pichon = Tradescantia L. COMMELINACEAE (M)
Cymbocarpa Miers BURMANNIACEAE (M)
Cymbocarpum DC. ex C.A.Mey. UMBELLIFERAE (D)
Cymboglossum (J.J.Sm.) Brieger = Eria Lindl. ORCHIDACEAE (M)
Cymbolaena Smoljan. COMPOSITAE (D)
Cymbonotus Cass. COMPOSITAE (D)
Cymbopappus B.Nord. COMPOSITAE (D)
Cymbopetalum Benth. ANNONACEAE (D)
Cymbopogon Spreng. GRAMINEAE (M)
Cymbosema Benth. LEGUMINOSAE–PAPILIONOIDEAE (D)
Cymbosepalum Baker = Haematoxylum L. LEGUMINOSAE–CAESALPINIOIDEAE (D)
Cymbosetaria Schweick. = Setaria P.Beauv. GRAMINEAE (M)
Cymelonema C.Presl = Urophyllum Wall. RUBIACEAE (D)
Cymodocea K.König CYMODOCEACEAE (M)
Cymophora B.L.Rob. COMPOSITAE (D)
Cymophyllus Mack. ex Britton & A.Br. CYPERACEAE (M)
Cymopterus Raf. UMBELLIFERAE (D)
Cynanchica Fourr. = Asperula L. RUBIACEAE (D)
Cynanchum L. ASCLEPIADACEAE (D)
Cynara L. COMPOSITAE (D)
Cynaropsis Kuntze = Cynara L. COMPOSITAE (D)
Cyne Danser LORANTHACEAE (D)
Cynocrambe Hill = Mercurialis L. EUPHORBIACEAE (D)
Cynoctonum J.F.Gmel. = Mitreola L. LOGANIACEAE (D)
Cynodendron Baehni = Chrysophyllum L. SAPOTACEAE (D)
Cynodon Rich. GRAMINEAE (M)
Cynoglossopsis Brand BORAGINACEAE (D)
Cynoglossum L. BORAGINACEAE (D)
Cynoglottis (Guşul.) Vural & Kit Tan BORAGINACEAE (D)
Cynomarathrum Nutt. = Lomatium Raf. UMBELLIFERAE (D)
Cynometra L. LEGUMINOSAE–CAESALPINIOIDEAE (D)
Cynomorium L. CYNOMORIACEAE (D)
Cynomyrtus Scriv. = Rhodomyrtus (DC.) Rchb. MYRTACEAE (D)
Cynorchis Thouars (SUO) = Cynorkis Thouars ORCHIDACEAE (M)
Cynorkis Thouars ORCHIDACEAE (M)
Cynorrhiza Eckl. & Zeyh. = Peucedanum L. UMBELLIFERAE (D)
Cynosciadium DC. UMBELLIFERAE (D)
Cynosorchis Thouars (SUO) = Cynorkis Thouars ORCHIDACEAE (M)
Cynosurus L. GRAMINEAE (M)
Cynoxylon Raf. = Cornus L. CORNACEAE (D)
Cypella Herb. IRIDACEAE (M)
Cyperochis Blume = Cymbidium Sw. ORCHIDACEAE (M)
Cyperochloa Lazarides & L.Watson GRAMINEAE (M)
Cyperus L. CYPERACEAE (M)
Cyphacanthus Leonard ACANTHACEAE (D)
Cyphanthera Miers SOLANACEAE (D)
Cyphia Bergius CAMPANULACEAE (D)
Cyphisia Rizzini = Justicia L. ACANTHACEAE (D)
Cyphocalyx Gagnep. (SUH) = Trungboa Rauschert SCROPHULARIACEAE (D)
Cyphocardamum Hedge CRUCIFERAE (D)
Cyphocarpus Miers CAMPANULACEAE (D)
Cyphochilus Schltr. ORCHIDACEAE (M)
Cyphochlaena Hack. GRAMINEAE (M)

Cyphokentia Brongn. PALMAE (M)
Cypholepis Chiov. = Coelachyrum Hochst. & Nees GRAMINEAE (M)
Cypholophus Wedd. URTICACEAE (D)
Cypholoron Dodson & Dressler ORCHIDACEAE (M)
Cyphomandra Mart. ex Sendtn. SOLANACEAE (D)
Cyphomattia Boiss. = Rindera Pall. BORAGINACEAE (D)
Cyphomeris Standl. NYCTAGINACEAE (D)
Cyphophoenix H.A.Wendl. ex Hook.f. PALMAE (M)
Cyphosperma H.A.Wendl. ex Hook.f. PALMAE (M)
Cyphostemma (Planch.) Alston VITACEAE (D)
Cyphostigma Benth. ZINGIBERACEAE (M)
Cyphostyla Gleason MELASTOMATACEAE (D)
Cyphotheca Diels MELASTOMATACEAE (D)
Cyprinia Browicz ASCLEPIADACEAE (D)
Cypripedium L. ORCHIDACEAE (M)
Cypselea Turpin AIZOACEAE (D)
Cypselocarpus F.Muell. GYROSTEMONACEAE (D)
Cyrenea F.Allam. (UR) GRAMINEAE (M)
Cyrilla Garden ex L. CYRILLACEAE (D)
Cyrillopsis Kuhlm. IXONANTHACEAE (D)
Cyrilwhitea Ising = Bassia All. CHENOPODIACEAE (D)
Cyrtandra J.R.Forst. & G.Forst. GESNERIACEAE (D)
Cyrtandroidea F.B.H.Br. = Cyrtandra J.R.Forst. & G.Forst. GESNERIACEAE (D)
Cyrtandromoea Zoll. SCROPHULARIACEAE (D)
Cyrtandropsis C.B.Clarke ex DC. = Cyrtandra J.R.Forst. & G.Forst. GESNERIACEAE (D)
Cyrtanthemum Oerst. = Besleria L. GESNERIACEAE (D)
Cyrtanthera Nees = Justicia L. ACANTHACEAE (D)
Cyrtantherella Oerst. = Justicia L. ACANTHACEAE (D)
Cyrtanthus Aiton AMARYLLIDACEAE (M)
Cyrtanthus Schreb. (SUH) = Posoqueria Aubl. RUBIACEAE (D)
Cyrtidiorchis Rauschert ORCHIDACEAE (M)
Cyrtidium Schltr. (SUH) = Cyrtidiorchis Rauschert ORCHIDACEAE (M)
Cyrtocarpa Kunth ANACARDIACEAE (D)
Cyrtoceras Benn. = Hoya R.Br. ASCLEPIADACEAE (D)
Cyrtochilum Kunth = Oncidium Sw. ORCHIDACEAE (M)
Cyrtococcum Stapf GRAMINEAE (M)
Cyrtocymura H. Rob. = Vernonia Schreb. COMPOSITAE (D)
Cyrtodeira Hanst. = Episcia Mart. GESNERIACEAE (D)
Cyrtoglottis Schltr. = Podochilus Blume ORCHIDACEAE (M)
Cyrtogonellum Ching = Cyrtomium C.Presl DRYOPTERIDACEAE (F)
Cyrtogonium J.Sm. = Bolbitis Schott LOMARIOPSIDACEAE (F)
Cyrtogonone Prain EUPHORBIACEAE (D)
Cyrtomidictyum Ching = Cyrtomium C.Presl DRYOPTERIDACEAE (F)
Cyrtomium C.Presl DRYOPTERIDACEAE (F)
Cyrtonema Schrad. ex Eckl. & Zeyh. = Kedrostis Medik. CUCURBITACEAE (D)
Cyrtopera Lindl. = Eulophia R.Br. ex Lindl. ORCHIDACEAE (M)
Cyrtophlebium (R.Br.) J.Sm. = Campyloneurum C.Presl POLYPODIACEAE (F)
Cyrtophyllum Reinw. = Fagraea Thunb. LOGANIACEAE (D)
Cyrtopodium R.Br. ORCHIDACEAE (M)
Cyrtopogon Spreng. (SUO) = Aristida L. GRAMINEAE (M)
Cyrtorchis Schltr. ORCHIDACEAE (M)
Cyrtorhyncha Nutt. = Ranunculus L. RANUNCULACEAE (D)
Cyrtosia Blume ORCHIDACEAE (M)
Cyrtosiphonia Miq. = Rauvolfia L. APOCYNACEAE (D)
Cyrtosperma Griff. ARACEAE (M)

Cyrtospermum Benth. = Campnosperma Thwaites ANACARDIACEAE (D)
Cyrtostachys Blume PALMAE (M)
Cyrtostylis R.Br. ORCHIDACEAE (M)
Cyrtoxiphus Harms = Cylicodiscus Harms LEGUMINOSAE–MIMOSOIDEAE (D)
Cystacanthus T.Anderson ACANTHACEAE (D)
Cyste Dulac = Cystopteris Bernh. WOODSIACEAE (F)
Cystea Sm. = Cystopteris Bernh. WOODSIACEAE (F)
Cysticapnos Mill. PAPAVERACEAE (D)
Cysticorydalis Fedde ex Ikonn. = Corydalis DC. PAPAVERACEAE (D)
Cystidospermum Prokh. = Euphorbia L. EUPHORBIACEAE (D)
Cystistemon Post & Kuntze (SUO) = Cystostemon Balf.f. BORAGINACEAE (D)
Cystium (Steven) Steven = Astragalus L. LEGUMINOSAE–PAPILIONOIDEAE (D)
Cystoathyrium Ching WOODSIACEAE (F)
Cystochilum Barb.Rodr. = Cranichis Sw. ORCHIDACEAE (M)
Cystodiopteris Rauschert = Cystodium J.Sm. DICKSONIACEAE (F)
Cystodium J.Sm. DICKSONIACEAE (F)
Cystogyne Gasp. = Ficus L. MORACEAE (D)
Cystopteris Bernh. WOODSIACEAE (F)
Cystopus Blume = Pristiglottis Cretz. & J.J.Sm. ORCHIDACEAE (M)
Cystorchis Blume ORCHIDACEAE (M)
Cystostemma E.Fourn. = Oxypetalum R.Br. ASCLEPIADACEAE (D)
Cystostemon Balf.f. BORAGINACEAE (D)
Cytherea Salisb. = Calypso Salisb. ORCHIDACEAE (M)
Cytheris Lindl. = Calanthe Ker–Gawl. ORCHIDACEAE (M)
Cytinus L. RAFFLESIACEAE (D)
Cytisophyllum O.Lang LEGUMINOSAE–PAPILIONOIDEAE (D)
Cytisopsis Jaub. & Spach LEGUMINOSAE–PAPILIONOIDEAE (D)
Cytisus Desf. LEGUMINOSAE–PAPILIONOIDEAE (D)
Cyttaranthus J.Léonard EUPHORBIACEAE (D)
Czerniaevia Ledeb. (SUI) = Deschampsia P.Beauv. GRAMINEAE (M)
Czernya C.Presl = Phragmites Adans. GRAMINEAE (M)

Daboecia D.Don ERICACEAE (D)
Dachel Adans. = Phoenix L. PALMAE (M)
Dacrycarpus (Endl.) de Laub. PODOCARPACEAE (G)
Dacrydium Lamb. PODOCARPACEAE (G)
Dacryodes Vahl BURSERACEAE (D)
Dacryotrichia Wild COMPOSITAE (D)
Dactilon Vill. = Cynodon Rich. GRAMINEAE (M)
Dactimala Raf. = Chrysophyllum L. SAPOTACEAE (D)
Dactyladenia Welw. = Acioa Aubl. CHRYSOBALANACEAE (D)
Dactylaea Fedde ex H.Wolff UMBELLIFERAE (D)
Dactylaena Schrad. ex Schult.f. CAPPARACEAE (D)
Dactylanthes Haw. = Euphorbia L. EUPHORBIACEAE (D)
Dactylanthocactus Y.Ito = Parodia Speg. CACTACEAE (D)
Dactylanthus Hook.f. BALANOPHORACEAE (D)
Dactyliandra (Hook.f.) Hook.f. CUCURBITACEAE (D)
Dactylicapnos Wall. PAPAVERACEAE (D)
Dactyliocapnos Spreng. (SUO) = Dicentra Borkh. ex Bernh. PAPAVERACEAE (D)
Dactyliophora Tiegh. LORANTHACEAE (D)
Dactyliota (Blume) Blume = Hypenanthe (Blume) Blume MELASTOMATACEAE (D)
Dactylis L. GRAMINEAE (M)
Dactylocardamum Al–Shehbaz CRUCIFERAE (D)
Dactylocladus Oliv. CRYPTERONIACEAE (D)
Dactyloctenium Willd. GRAMINEAE (M)

Dactylicapnos Wall. PAPAVERACEAE (D)
Dactyliocapnos Spreng. (SUO) = Dicentra Borkh. ex Bernh. PAPAVERACEAE (D)
Dactyliophora Tiegh. LORANTHACEAE (D)
Dactyliota (Blume) Blume = Hypenanthe (Blume) Blume MELASTOMATACEAE (D)
Dactylis L. GRAMINEAE (M)
Dactylocardamum Al–Shehbaz CRUCIFERAE (D)
Dactylocladus Oliv. CRYPTERONIACEAE (D)
Dactyloctenium Willd. GRAMINEAE (M)
Dactylodes Kuntze (SUS) = Tripsacum L. GRAMINEAE (M)
Dactylogramma Link = Muhlenbergia Schreb. GRAMINEAE (M)
Dactylopetalum Benth. = Cassipourea Aubl. RHIZOPHORACEAE (D)
Dactylophyllum Spach (SUH) = Linanthus Benth. POLEMONIACEAE (D)
Dactylopsis N.E.Br. AIZOACEAE (D)
Dactylorchis (Klinge) Vermeulen = Dactylorhiza Neck. ex Nevski ORCHIDACEAE (M)
Dactylorhiza Neck. ex Nevski ORCHIDACEAE (M)
Dactylorhynchus Schltr. ORCHIDACEAE (M)
Dactylostalix Rchb.f. ORCHIDACEAE (M)
Dactylostegium Nees = Dicliptera Juss. ACANTHACEAE (D)
Dactylostelma Schltr. (UP) ASCLEPIADACEAE (D)
Dactylostemon Klotzsch = Actinostemon Mart. ex Klotzsch EUPHORBIACEAE (D)
Dactylostigma D.F.Austin = Hildebrandtia Vatke CONVOLVULACEAE (D)
Dactylostylis Scheidw. = Zygostates Lindl. ORCHIDACEAE (M)
Dactylus Asch. (SUO) = Cynodon Rich. GRAMINEAE (M)
Daedalacanthus T.Anderson = Eranthemum L. ACANTHACEAE (D)
Daemia Poir. = Pergularia L. ASCLEPIADACEAE (D)
Daemonorops Blume PALMAE (M)
Daenikera Hürl. & Stauffer SANTALACEAE (D)
Dahlgrenia Steyerm. = Dictyocaryum H.A.Wendl. PALMAE (M)
Dahlgrenodendron J.J.M.van der Merwe & A.E.van Wyk = Cryptocarya R.Br.
    LAURACEAE (D)
Dahlia Cav. COMPOSITAE (D)
Dahlstedtia Malme LEGUMINOSAE–PAPILIONOIDEAE (D)
Dahuronia Scop. = Licania Aubl. CHRYSOBALANACEAE (D)
Dais L. THYMELAEACEAE (D)
Daiswa Raf. TRILLIACEAE (M)
Daknopholis Clayton GRAMINEAE (M)
Dalanum J.Dostál = Galeopsis L. LABIATAE (D)
Dalbergaria Tussac = Columnea L. GESNERIACEAE (D)
Dalbergia L.f. LEGUMINOSAE–PAPILIONOIDEAE (D)
Dalbergiella Baker f. LEGUMINOSAE–PAPILIONOIDEAE (D)
Dalea L. LEGUMINOSAE–PAPILIONOIDEAE (D)
Dalea Mill. (SUH) = Browallia L. SOLANACEAE (D)
Dalechampia L. EUPHORBIACEAE (D)
Dalechampsia Post & Kuntze = Dalechampia L. EUPHORBIACEAE (D)
Dalembertia Baill. EUPHORBIACEAE (D)
Dalenia Korth. MELASTOMATACEAE (D)
Dalhousiea Wall. ex Benth. LEGUMINOSAE–PAPILIONOIDEAE (D)
Dallachya F.Muell. = Rhamnella Miq. RHAMNACEAE (D)
Dalucum Adans. (SUS) = Melica L. GRAMINEAE (M)
Dalzellia Wight PODOSTEMACEAE (D)
Dalzielia Turrill ASCLEPIADACEAE (D)
Damasonium Mill. ALISMATACEAE (M)
Dammaropsis Warb. = Ficus L. MORACEAE (D)
Dammera Lauterb. & K.Schum. = Licuala Thunb. PALMAE (M)
Damnacanthus C.F.Gaertn. RUBIACEAE (D)

Damnamenia Given = Celmisia Cass. COMPOSITAE (D)
Damnxanthodium Strother COMPOSITAE (D)
Dampiera R.Br. GOODENIACEAE (D)
Damrongia Kerr ex Craib = Chirita Buch.–Ham. ex D.Don GESNERIACEAE (D)
Danaa All. = Physospermum Cusson ex Juss. UMBELLIFERAE (D)
Danae Medik. RUSCACEAE (M)
Danaea Sm. MARATTIACEAE (F)
Danaeopsis C.Presl = Danaea Sm. MARATTIACEAE (F)
Danais Comm. ex Vent. RUBIACEAE (D)
Dandya H.E.Moore ALLIACEAE (M)
Danguya Benoist ACANTHACEAE (D)
Danguyodrypetes Leandri EUPHORBIACEAE (D)
Danielia Lem. = Crassula L. CRASSULACEAE (D)
Daniella Corr.Mello = Mansoa DC. BIGNONIACEAE (D)
Daniellia Benn. LEGUMINOSAE–CAESALPINIOIDEAE (D)
Dankia Gagnep. THEACEAE (D)
Dansera Steenis = Dialium L. LEGUMINOSAE–CAESALPINIOIDEAE (D)
Danserella Balle (SUI) = Oncocalyx Tiegh. LORANTHACEAE (D)
Dansiea Byrnes COMBRETACEAE (D)
Danthia Steud. = Ludwigia L. ONAGRACEAE (D)
Danthonia DC. GRAMINEAE (M)
Danthoniastrum (Holub) Holub = Metcalfia Conert GRAMINEAE (M)
Danthonidium C.E.Hubb. GRAMINEAE (M)
Danthoniopsis Stapf GRAMINEAE (M)
Danthorhiza Ten. (SUI) = Helictotrichon Besser GRAMINEAE (M)
Dantia Boehm. = Ludwigia L. ONAGRACEAE (D)
Dapania Korth. OXALIDACEAE (D)
Daphnandra Benth. MONIMIACEAE (D)
Daphne L. THYMELAEACEAE (D)
Daphnidium Nees = Lindera Thunb. LAURACEAE (D)
Daphniluma Baill. = Pouteria Aubl. SAPOTACEAE (D)
Daphnimorpha Nakai THYMELAEACEAE (D)
Daphniphyllum Blume DAPHINIPHYLLACEAE (D)
Daphnopsis Mart. THYMELAEACEAE (D)
Darbya A.Gray = Nestronia Raf. SANTALACEAE (D)
Darea Juss. = Asplenium L. ASPLENIACEAE (F)
Darlingia F.Muell. PROTEACEAE (D)
Darlingtonia Torr. SARRACENIACEAE (D)
Darmera Voss SAXIFRAGACEAE (D)
Darniella Maire & Weiller = Salsola L. CHENOPODIACEAE (D)
Darwinia Rudge MYRTACEAE (D)
Darwiniella Braas & Lückel = Trichoceros Kunth ORCHIDACEAE (M)
Darwiniera Braas & Lückel ORCHIDACEAE (M)
Darwiniothamnus Harling COMPOSITAE (D)
Dasillipe Dubard = Madhuca Ham. ex J.F.Gmel. SAPOTACEAE (D)
Dasiola Raf. = Vulpia C.C.Gmelin GRAMINEAE (M)
Dasiphora Raf. = Potentilla L. ROSACEAE (D)
Dasispermum Raf. UMBELLIFERAE (D)
Dasistoma Raf. SCROPHULARIACEAE (D)
Dasoclema J.Sinclair ANNONACEAE (D)
Dasus Lour. = Lasianthus Jack RUBIACEAE (D)
Dasyaulus Thwaites = Madhuca Ham. ex J.F.Gmel. SAPOTACEAE (D)
Dasycarya Liebm. = Cyrtocarpa Kunth ANACARDIACEAE (D)
Dasycephala (DC.) Hook.f. = Diodia L. RUBIACEAE (D)
Dasycoleum Turcz. = Chisocheton Blume MELIACEAE (D)

Dasycondylus R.M.King & H.Rob. COMPOSITAE (D)

Dasydesmus Craib = Oreocharis Benth. GESNERIACEAE (D)

Dasylepis Oliv. FLACOURTIACEAE (D)

Dasylirion Zucc. DRACAENACEAE (M)

Dasymaschalon (Hook.f. & Thomson) Dalla Torre & Harms ANNONACEAE (D)

Dasynotus I.M.Johnst. BORAGINACEAE (D)

Dasyochloa Rydb. = Erioneuron Nash GRAMINEAE (M)

Dasypetalum Pierre ex A.Chev. = Scottellia Oliv. FLACOURTIACEAE (D)

Dasyphyllum Kunth COMPOSITAE (D)

Dasypoa Pilg. = Poa L. GRAMINEAE (M)

Dasypogon R.Br. DASYPOGONACEAE (M)

Dasypyrum (Coss. & Durieu) T.Durand GRAMINEAE (M)

Dasys Lem. = Lasianthus Jack RUBIACEAE (D)

Dasysphaera Volkens ex Gilg AMARANTHACEAE (D)

Dasystachys Oerst. = Chamaedorea Willd. PALMAE (M)

Dasystachys Baker (SUH) = Chlorophytum Ker–Gawl. ANTHERICACEAE (M)

Dasystephana Adans. = Gentiana L. GENTIANACEAE (D)

Dasystoma Raf. = Agalinis Raf. SCROPHULARIACEAE (D)

Dasytropis Urb. ACANTHACEAE (D)

Datisca L. DATISCACEAE (D)

Datura L. SOLANACEAE (D)

Daturicarpa Stapf = Tabernanthe Baill. APOCYNACEAE (D)

Daubentonia DC. = Sesbania Scop. LEGUMINOSAE–PAPILIONOIDEAE (D)

Daubentoniopsis Rydb. = Sesbania Scop. LEGUMINOSAE–PAPILIONOIDEAE (D)

Daubenya Lindl. HYACINTHACEAE (M)

Daucosma Engelm. & A.Gray UMBELLIFERAE (D)

Daucus L. UMBELLIFERAE (D)

Daumalia Arènes = Urospermum Scop. COMPOSITAE (D)

Daun–Contu Adans. = Paederia L. RUBIACEAE (D)

Dauphinea Hedge LABIATAE (D)

Davallia Sm. DAVALLIACEAE (F)

Davalliopsis Van den Bosch = Cephalomanes C.Presl HYMENOPHYLLACEAE (F)

Davallodes (Copel.) Copel. DAVALLIACEAE (F)

Daveaua Willk. ex Mariz COMPOSITAE (D)

Davidia Baill. CORNACEAE (D)

Davidsea Soderstr. & R.P.Ellis = Schizostachyum Nees GRAMINEAE (M)

Davidsonia F.Muell. DAVIDSONIACEAE (D)

Daviesia Sm. LEGUMINOSAE–PAPILIONOIDEAE (D)

Davilla Vand. DILLENIACEAE (D)

Davya DC. = Adelobotrys DC. MELASTOMATACEAE (D)

Davyella Hack. (SUS) = Neostapfia Burtt Davy GRAMINEAE (M)

Dayaoshania W.T.Wang GESNERIACEAE (D)

Deamia Britton & Rose = Selenicereus (A.Berger) Britton & Rose CACTACEAE (D)

Deanea J.M.Coult. & Rose = Rhodosciadium S.Watson UMBELLIFERAE (D)

Debesia Kuntze = Anthericum L. ANTHERICACEAE (M)

Debraea Roem. & Schult. (SUS) = Erisma Rudge VOCHYSIACEAE (D)

Debregeasia Gaudich. URTICACEAE (D)

Decabelone Decne. ASCLEPIADACEAE (D)

Decachaeta DC. COMPOSITAE (D)

Decagonocarpus Engl. RUTACEAE (D)

Decaisnea Lindl. (SUH) = Tropidia Lindl. ORCHIDACEAE (M)

Decaisnea Brongn. (SUH) = Prescottia Lindl. ORCHIDACEAE (M)

Decaisnea Hook.f. & Thomson LARDIZABALACEAE (D)

Decaisnina Tiegh. LORANTHACEAE (D)

Decalepidanthus Riedl BORAGINACEAE (D)

Decalepis Wight & Arn. ASCLEPIADACEAE (D)
Decaloba M.Roem. = Passiflora L. PASSIFLORACEAE (D)
Decalobanthus Ooststr. CONVOLVULACEAE (D)
Decalophium Turcz. = Chamelaucium Desf. MYRTACEAE (D)
Decameria Welw. = Gardenia Ellis RUBIACEAE (D)
Decamerium Nutt. = Gaylussacia Kunth ERICACEAE (D)
Decandolia Batard (SUS) = Agrostis L. GRAMINEAE (M)
Decanema Decne. ASCLEPIADACEAE (D)
Decanemopsis Costantin & Gallaud = Sarcostemma R.Br. ASCLEPIADACEAE (D)
Decapenta Raf. = Diodia L. RUBIACEAE (D)
Decaphalangium Melch. = Clusia L. GUTTIFERAE (D)
Decaptera Turcz. CRUCIFERAE (D)
Decarinum Raf. = Croton L. EUPHORBIACEAE (D)
Decarya Choux DIDIEREACEAE (D)
Decarydendron Danguy MONIMIACEAE (D)
Decaryella A.Camus GRAMINEAE (M)
Decaryochloa A.Camus GRAMINEAE (M)
Decaschistia Wight & Arn. MALVACEAE (D)
Decaspermum J.R.Forst. & G.Forst. MYRTACEAE (D)
Decastelma Schltr. ASCLEPIADACEAE (D)
Decastylocarpus Humbert COMPOSITAE (D)
Decateles Raf. = Sideroxylon L. SAPOTACEAE (D)
Decatoca F.Muell. EPACRIDACEAE (D)
Decatropis Hook.f. RUTACEAE (D)
Decazesia F.Muell. COMPOSITAE (D)
Decazyx Pittier & S.F.Blake RUTACEAE (D)
Deccania Tirveng. RUBIACEAE (D)
Decemium Raf. = Hydrophyllum L. HYDROPHYLLACEAE (D)
Decemium Raf. (UP) HYDROPHYLLACEAE (D)
Deckenia H.A.Wendl. ex Seem. PALMAE (M)
Deckeria Karst. = Iriartea Ruíz & Pav. PALMAE (M)
Declieuxia Kunth RUBIACEAE (D)
Decodon J.F.Gmel. LYTHRACEAE (D)
Decorsea R.Vig. LEGUMINOSAE–PAPILIONOIDEAE (D)
Decorsella A.Chev. VIOLACEAE (D)
Decumaria L. HYDRANGEACEAE (D)
Decussocarpus de Laub. (SUS) = Nageia Gaertn. PODOCARPACEAE (G)
Dedea Baill. = Quintinia A.DC. ESCALLONIACEAE (D)
Dedeckera Reveal & J.T.Howell POLYGONACEAE (D)
Deeringia R.Br. AMARANTHACEAE (D)
Deeringothamnus Small ANNONACEAE (D)
Defforgia Lam. (SUS) = Forgesia Comm. ex Juss. ESCALLONIACEAE (D)
Deflersia Schweinf. ex Penzig = Erythrococca Benth. EUPHORBIACEAE (D)
Degeneria I.W.Bailey & A.C.Sm. DEGENERIACEAE (D)
Degenia Hayek CRUCIFERAE (D)
Degranvillea Determann ORCHIDACEAE (M)
Deguelia Aubl. = Derris Lour. LEGUMINOSAE–PAPILIONOIDEAE (D)
Dehaasia Blume LAURACEAE (D)
Deherainia Decne. THEOPHRASTACEAE (D)
Deianira Cham. & Schltdl. GENTIANACEAE (D)
Deidamia Noronha ex Thouars PASSIFLORACEAE (D)
Deilanthe N.E.Br. = Aloinopsis Schwantes AIZOACEAE (D)
Deina Alef. = Triticum L. GRAMINEAE (M)
Deinacanthon Mez = Bromelia L. BROMELIACEAE (M)
Deinandra Greene = Hemizonia DC. COMPOSITAE (D)

Deinanthe Maxim. HYDRANGEACEAE (D)
Deinbollia Schumach. & Thonn. SAPINDACEAE (D)
Deinostema Yamaz. SCROPHULARIACEAE (D)
Deiregyne Schltr. ORCHIDACEAE (M)
Deiregynopsis Rauschert = Aulosepalum Garay ORCHIDACEAE (M)
Dekindtia Gilg = Chionanthus L. OLEACEAE (D)
Dekinia M.Martens & Gal. = Agastache Gronov. LABIATAE (D)
Delaetia Backeb. (SUI) = Neoporteria Britton & Rose CACTACEAE (D)
Delairea Lem. COMPOSITAE (D)
Delamerea S.Moore COMPOSITAE (D)
Delaportea Thorel ex Gagnep. = Acacia Mill. LEGUMINOSAE–MIMOSOIDEAE (D)
Delarbrea Vieill. ARALIACEAE (D)
Delastrea A.DC. = Labramia A.DC. SAPOTACEAE (D)
Delavaya Franch. SAPINDACEAE (D)
Delilia Spreng. COMPOSITAE (D)
Delima L. = Tetracera L. DILLENIACEAE (D)
Delissea Gaudich. CAMPANULACEAE (D)
Delognaea Cogn. = Ampelosicyos Thouars CUCURBITACEAE (D)
Delonix Raf. LEGUMINOSAE–CAESALPINIOIDEAE (D)
Delopyrum Small = Polygonella Michx. POLYGONACEAE (D)
Delosperma N.E.Br. AIZOACEAE (D)
Delostoma D.Don BIGNONIACEAE (D)
Delpechia Montrouz. = Psychotria L. RUBIACEAE (D)
Delphinacanthus Benoist = Pseudodicliptera Benoist ACANTHACEAE (D)
Delphinium L. RANUNCULACEAE (D)
Delphyodon K.Schum. APOCYNACEAE (D)
Delpinophytum Speg. CRUCIFERAE (D)
Delpya Pierre ex Radlk. (SUS) = Sisyrolepis Radlk. SAPINDACEAE (D)
Delpydora Pierre SAPOTACEAE (D)
Deltaria Steenis THYMELAEACEAE (D)
Deltocheilos W.T.Wang = Chirita Buch.–Ham. ex D.Don GESNERIACEAE (D)
Dematra Raf. = Euphorbia L. EUPHORBIACEAE (D)
Demavendia Pimenov UMBELLIFERAE (D)
Demazeria Dumort. (SUO) = Desmazeria Dumort. GRAMINEAE (M)
Demidium DC. = Gnaphalium L. COMPOSITAE (D)
Demnosa Frič = Cleistocactus Lem. CACTACEAE (D)
Democritea DC. = Serissa Comm. ex Juss. RUBIACEAE (D)
Demosthenesia A.C.Sm. ERICACEAE (D)
Dendragrostis B.D.Jacks. (SUI) = Chusquea Kunth GRAMINEAE (M)
Dendranthema (DC.) Des Moulins COMPOSITAE (D)
Dendriopoterium Svent. = Sanguisorba L. ROSACEAE (D)
Dendrobangia Rusby ICACINACEAE (D)
Dendrobenthamia Hutch. = Cornus L. CORNACEAE (D)
Dendrobium Sw. ORCHIDACEAE (M)
Dendrobrychis (DC.) Galushko = Onobrychis Mill. LEGUMINOSAE–
    PAPILIONOIDEAE (D)
Dendrobryon Klotzsch ex Pax & K.Hoffm. = Tetraplandra Baill. EUPHORBIACEAE (D)
Dendrocacalia (Nakai) Tuyama COMPOSITAE (D)
Dendrocalamopsis (L.C.Chia & H.L.Fung) Keng f. = Bambusa Schreb. GRAMINEAE (M)
Dendrocalamus Nees GRAMINEAE (M)
Dendrocereus Britton & Rose = Acanthocereus (Engelm. ex A.Berger) Britton & Rose
    CACTACEAE (D)
Dendrochilum Blume ORCHIDACEAE (M)
Dendrochloa C.E.Parkinson = Schizostachyum Nees GRAMINEAE (M)
Dendrocnide Miq. URTICACEAE (D)

Dendroconche Copel. = Microsorum Link POLYPODIACEAE (F)
Dendrocoryne (Lindl.) Brieger = Dendrobium Sw. ORCHIDACEAE (M)
Dendrocousinsia Millsp. EUPHORBIACEAE (D)
Dendroglossa C.Presl = Leptochilus Kaulf. POLYPODIACEAE (F)
Dendrokingstonia Rauschert ANNONACEAE (D)
Dendroleandria Arènes = Helmiopsiella Arènes STERCULIACEAE (D)
Dendrolirium Blume = Eria Lindl. ORCHIDACEAE (M)
Dendrolobium (Wight & Arn.) Benth. LEGUMINOSAE–PAPILIONOIDEAE (D)
Dendromecon Benth. PAPAVERACEAE (D)
Dendromyza Danser SANTALACEAE (D)
Dendropanax Decne. & Planch. ARALIACEAE (D)
Dendropemon (Blume) Rchb. LORANTHACEAE (D)
Dendrophthoe Mart. LORANTHACEAE (D)
Dendrophthora Eichler VISCACEAE (D)
Dendrophylax Rchb.f. ORCHIDACEAE (M)
Dendrophyllanthus S.Moore = Phyllanthus L. EUPHORBIACEAE (D)
Dendrorkis Thouars = Aerides Lour. ORCHIDACEAE (M)
Dendrosenecio (Hauman ex Humbert) B.Nord. COMPOSITAE (D)
Dendroseris D.Don COMPOSITAE (D)
Dendrosicus Raf. = Amphitecna Miers BIGNONIACEAE (D)
Dendrosicyos Balf.f. CUCURBITACEAE (D)
Dendrosida Fryxell MALVACEAE (D)
Dendrosipanea Ducke RUBIACEAE (D)
Dendrosma Pancher & Sebert RUTACEAE (D)
Dendrostellera (C.A.Mey.) Tiegh. THYMELAEACEAE (D)
Dendrostigma Gleason = Mayna Aubl. FLACOURTIACEAE (D)
Dendrostylis Karst. & Triana = Mayna Aubl. FLACOURTIACEAE (D)
Dendrotrophe Miq. SANTALACEAE (D)
Denea O.F.Cook = Howea Becc. PALMAE (M)
Denekia Thunb. COMPOSITAE (D)
Denhamia Meisn. CELASTRACEAE (D)
Denisonia F.Muell. = Pityrodia R.Br. VERBENACEAE (D)
Denisophytum R.Vig. = Caesalpinia L. LEGUMINOSAE–CAESALPINIOIDEAE (D)
Denmoza Britton & Rose CACTACEAE (D)
Dennettia Baker f. ANNONACEAE (D)
Dennstaedtia Bernh. DENNSTAEDTIACEAE (F)
Denslovia Rydb. = Habenaria Willd. ORCHIDACEAE (M)
Dentaria L. = Cardamine L. CRUCIFERAE (D)
Dentella J.R.Forst. & G.Forst. RUBIACEAE (D)
Dentillaria Kuntze = Knoxia L. RUBIACEAE (D)
Dentimetula Tiegh. = Agelanthus Tiegh. LORANTHACEAE (D)
Dentoceras Small = Polygonella Michx. POLYGONACEAE (D)
Deonia Pierre ex Pax = Blachia Baill. EUPHORBIACEAE (D)
Depacarpus N.E.Br. = Meyerophytum Schwantes AIZOACEAE (D)
Depanthus S.Moore GESNERIACEAE (D)
Deparia Hook. & Grev. = Athyrium Roth WOODSIACEAE (F)
Deplanchea Vieill. BIGNONIACEAE (D)
Deppea Cham. & Schltdl. RUBIACEAE (D)
Deppia Raf. = Lycaste Lindl. ORCHIDACEAE (M)
Deprea Raf. SOLANACEAE (D)
Derenbergia Schwantes = Conophytum N.E.Br. AIZOACEAE (D)
Derenbergiella Schwantes = Mesembryanthemum L. AIZOACEAE (D)
Dermatobotrys Bolus SCROPHULARIACEAE (D)
Dermatocalyx Oerst. = Schlegelia Miq. SCROPHULARIACEAE (D)
Dermatophlebium C.Presl = Sphaerocionium C.Presl HYMENOPHYLLACEAE (F)

Dermophylla Silva Manso = Cayaponia Silva Manso CUCURBITACEAE (D)
Deroemera Rchb.f. = Holothrix Rich. ex Lindl. ORCHIDACEAE (M)
Derris Lour. LEGUMINOSAE–PAPILIONOIDEAE (D)
Derwentia Raf. = Parahebe W.R.B.Oliv. SCROPHULARIACEAE (D)
Desbordesia Pierre ex Tiegh. IRVINGIACEAE (D)
Descantaria Schltdl. = Tripogandra Raf. COMMELINACEAE (M)
Deschampsia P.Beauv. GRAMINEAE (M)
Descurainia Webb & Berthel. CRUCIFERAE (D)
Desdemona S.Moore = Basistemon Turcz. SCROPHULARIACEAE (D)
Desfontaena Vell. = Chiropetalum A.Juss. EUPHORBIACEAE (D)
Desfontaina Steud. (SUO) = Chiropetalum A.Juss. EUPHORBIACEAE (D)
Desfontainea Rchb. (SUO) = Chiropetalum A.Juss. EUPHORBIACEAE (D)
Desfontainesia Hoffmanns. (SUO) = Fontanesia Labill. OLEACEAE (D)
Desfontainia Ruíz & Pav. LOGANIACEAE (D)
Desforgia Steud. (SUI) = Forgesia Comm. ex Juss. ESCALLONIACEAE (D)
Desideria Pamp. = Christolea Cambess. ex Jacquem. CRUCIFERAE (D)
Desmanthodium Benth. COMPOSITAE (D)
Desmanthus Willd. LEGUMINOSAE–MIMOSOIDEAE (D)
Desmaria Tiegh. LORANTHACEAE (D)
Desmazeria Dumort. GRAMINEAE (M)
Desmidorchis Ehrenb. = Caralluma R.Br. ASCLEPIADACEAE (D)
Desmodiastrum (Prain) A.Pramanik & Thoth. LEGUMINOSAE–PAPILIONOIDEAE (D)
Desmodiocassia Britton & Rose = Senna Mill. LEGUMINOSAE–CAESALPINIOIDEAE (D)
Desmodium Desv. LEGUMINOSAE–PAPILIONOIDEAE (D)
Desmofischera Holth. = Desmodium Desv. LEGUMINOSAE–PAPILIONOIDEAE (D)
Desmogyne King & Prain = Agapetes D.Don ex G.Don ERICACEAE (D)
Desmoncus Mart. PALMAE (M)
Desmonema Miers = Tinospora Miers MENISPERMACEAE (D)
Desmonema Raf. (SUH) = Euphorbia L. EUPHORBIACEAE (D)
Desmopsis Saff. ANNONACEAE (D)
Desmos Lour. ANNONACEAE (D)
Desmoscelis Naudin MELASTOMATACEAE (D)
Desmoschoenus Hook.f. CYPERACEAE (M)
Desmostachya (Stapf) Stapf GRAMINEAE (M)
Desmostachys Miers ICACINACEAE (D)
Desmostemon Thwaites = Fahrenheitia Rchb.f. & Zoll. EUPHORBIACEAE (D)
Desmothamnus Small = Lyonia Nutt. ERICACEAE (D)
Desmotrichum Blume = Flickingeria A.D.Hawkes ORCHIDACEAE (M)
Despeleza Nieuwl. = Lespedeza Michx. LEGUMINOSAE–PAPILIONOIDEAE (D)
Desplatsia Bocquillon TILIACEAE (D)
Despretzia Kunth = Zeugites P.Browne GRAMINEAE (M)
Desrousseauxia Tiegh. = Aetanthus (Eichler) Engl. LORANTHACEAE (D)
Detarium Juss. LEGUMINOSAE–CAESALPINIOIDEAE (D)
Dethawia Endl. UMBELLIFERAE (D)
Detzneria Schltr. ex Diels SCROPHULARIACEAE (D)
Deuterocohnia Mez BROMELIACEAE (M)
Deuteromallotus Pax & K.Hoffm. EUPHORBIACEAE (D)
Deutzia Thunb. HYDRANGEACEAE (D)
Deutzianthus Gagnep. EUPHORBIACEAE (D)
Devauxia Kunth (SUI) = Glyceria R.Br. GRAMINEAE (M)
Devauxia R.Br. = Centrolepis Labill. CENTROLEPIDACEAE (M)
Deverra DC. = Pituranthos Viv. UMBELLIFERAE (D)
Devia Goldblatt & J.C.Manning IRIDACEAE (M)
Devillea Tul. & Wedd. PODOSTEMACEAE (D)
Dewevrea Micheli LEGUMINOSAE–PAPILIONOIDEAE (D)

Dewevrella De Wild. APOCYNACEAE (D)
Deweya Torr. & A.Gray = Tauschia Schltdl. UMBELLIFERAE (D)
Dewildemania O.Hoffm. COMPOSITAE (D)
Dewindtia De Wild. = Cryptosepalum Benth. LEGUMINOSAE–CAESALPINIOIDEAE (D)
Deyeuxia Clarion ex P.Beauv. = Calamagrostis Adans. GRAMINEAE (M)
Dhofaria A.G.Mill. CAPPARACEAE (D)
Diacalpe Blume DRYOPTERIDACEAE (F)
Diacarpa T.R.Sim = Atalaya Blume SAPINDACEAE (D)
Diachroa Nutt. = Leptochloa P.Beauv. GRAMINEAE (M)
Diachyrium Griseb. = Sporobolus R.Br. GRAMINEAE (M)
Diacidia Griseb. MALPIGHIACEAE (D)
Diacisperma Kuntze (SUS) = Leptochloa P.Beauv. GRAMINEAE (M)
Diacranthera R.M.King & H.Rob. COMPOSITAE (D)
Diacrium Benth. = Caularthron Raf. ORCHIDACEAE (M)
Diacrodon Sprague RUBIACEAE (D)
Diadenaria Klotzsch & Garcke = Pedilanthus Neck. ex Poit. EUPHORBIACEAE (D)
Diadenium Poepp. & Endl. ORCHIDACEAE (M)
Dialissa Lindl. = Stelis Sw. ORCHIDACEAE (M)
Dialium L. LEGUMINOSAE–CAESALPINIOIDEAE (D)
Dialyanthera Warb. = Otoba (DC.) Karst. MYRISTICACEAE (D)
Dialyceras Capuron SPHAEROSEPALACEAE (D)
Dialypetalanthus Kuhlm. DIALYPETALANTHACEAE (D)
Dialypetalum Benth. CAMPANULACEAE (D)
Dialytheca Exell & Mendonça MENISPERMACEAE (D)
Diamena Ravenna = Anthericum L. ANTHERICACEAE (M)
Diamonon Raf. = Solanum L. SOLANACEAE (D)
Diamorpha Nutt. CRASSULACEAE (D)
Diandranthus Liou (SUI) = Miscanthus Andersson GRAMINEAE (M)
Diandriella Engl. = Homalomena Schott ARACEAE (M)
Diandrochloa De Winter = Eragrostis Wolf GRAMINEAE (M)
Diandrolyra Stapf GRAMINEAE (M)
Diandrostachya (C.E.Hubb.) Jacq.–Fél. = Loudetiopsis Conert GRAMINEAE (M)
Dianella Lam. PHORMIACEAE (M)
Diania Noronha ex Tul. = Dicoryphe Thouars HAMAMELIDACEAE (D)
Dianthella Clauson ex Pomel = Petrorhagia (Ser.) Link CARYOPHYLLACEAE (D)
Dianthera L. = Justicia L. ACANTHACEAE (D)
Dianthera Klotzsch (SUH) = Cleome L. CAPPARACEAE (D)
Dianthoseris Sch.Bip. COMPOSITAE (D)
Dianthoveus Hammel & Wilder CYCLANTHACEAE (M)
Dianthus L. CARYOPHYLLACEAE (D)
Diapensia L. DIAPENSIACEAE (D)
Diaperia Nutt. COMPOSITAE (D)
Diaphananthe Schltr. ORCHIDACEAE (M)
Diaphanoptera Rech.f. CARYOPHYLLACEAE (D)
Diaphractanthus Humbert COMPOSITAE (D)
Diaphycarpus Calest. = Bunium L. UMBELLIFERAE (D)
Diarina Raf. (SUS) = Diarrhena P.Beauv. GRAMINEAE (M)
Diarrhena P.Beauv. GRAMINEAE (M)
Diarthron Turcz. THYMELAEACEAE (D)
Diascia Link & Otto SCROPHULARIACEAE (D)
Diaspananthus Miq. = Ainsliaea DC. COMPOSITAE (D)
Diaspasis R.Br. GOODENIACEAE (D)
Diasperus Kuntze = Glochidion J.R.Forst. & G.Forst. EUPHORBIACEAE (D)
Diaspis Nied. = Caucanthus Forssk. MALPIGHIACEAE (D)
Diastatea Scheidw. CAMPANULACEAE (D)

Diastella Salisb. ex Knight PROTEACEAE (D)
Diastema Benth. GESNERIACEAE (D)
Diastemella Oerst. = Diastema Benth. GESNERIACEAE (D)
Diastemenanthe Desv. = Stenotaphrum Trin. GRAMINEAE (M)
Diateinacanthus Lindau = Odontonema Nees ACANTHACEAE (D)
Diatenopteryx Radlk. SAPINDACEAE (D)
Diatoma Lour. = Carallia Roxb. RHIZOPHORACEAE (D)
Diatremis Raf. = Ipomoea L. CONVOLVULACEAE (D)
Diberara Baill. = Nebelia Neck. ex Sweet BRUNIACEAE (D)
Diblemma J.Sm. POLYPODIACEAE (F)
Dibrachion Regel = Homalanthus A.Juss. EUPHORBIACEAE (D)
Dibrachionostylus Bremek. RUBIACEAE (D)
Dibrachium Walp. = Homalanthus A.Juss. EUPHORBIACEAE (D)
Dicaelosperma E.G.O.Muell. & Pax (SUO) = Dicoelospermum C.B.Clarke
    CUCURBITACEAE (D)
Dicaelospermum C.B.Clarke (SUO) = Dicoelospermum C.B.Clarke CUCURBITACEAE (D)
Dicarpellum (Loes.) A.C.Sm. = Salacia L. CELASTRACEAE (D)
Dicarpidium F.Muell. STERCULIACEAE (D)
Dicarpophora Speg. ASCLEPIADACEAE (D)
Dicella Griseb. MALPIGHIACEAE (D)
Dicellandra Hook.f. MELASTOMATACEAE (D)
Dicellostyles Benth. MALVACEAE (D)
Dicentra Borkh. ex Bernh. PAPAVERACEAE (D)
Dicentranthera T.Anderson = Asystasia Blume ACANTHACEAE (D)
Dicera J.R.Forst. & G.Forst. = Elaeocarpus L. ELAEOCARPACEAE (D)
Dicerandra Benth. LABIATAE (D)
Diceratella Boiss. CRUCIFERAE (D)
Diceratostele Summerh. ORCHIDACEAE (M)
Dicercoclados C.Jeffrey & Y.L.Chen COMPOSITAE (D)
Dicerma DC. LEGUMINOSAE–PAPILIONOIDEAE (D)
Dicerocaryum Bojer PEDALIACEAE (D)
Dicerolepis Blume = Gymnanthera R.Br. ASCLEPIADACEAE (D)
Dicerospermum Bakh.f. = Poikilogyne Baker f. MELASTOMATACEAE (D)
Dicerostylis Blume ORCHIDACEAE (M)
Dichaea Lindl. ORCHIDACEAE (M)
Dichaelia Harv. = Brachystelma R.Br. ASCLEPIADACEAE (D)
Dichaeopsis Pfitzer = Dichaea Lindl. ORCHIDACEAE (M)
Dichaetanthera Endl. MELASTOMATACEAE (D)
Dichaetaria Nees ex Steud. GRAMINEAE (M)
Dichaetophora A.Gray COMPOSITAE (D)
Dichanthelium (Hitchc. & Chase) Gould = Panicum L. GRAMINEAE (M)
Dichanthium Willemet GRAMINEAE (M)
Dichapetalum Thouars DICHAPETALACEAE (D)
Dichasianthus Ovcz. & Yanusov CRUCIFERAE (D)
Dichasium (A.Braun) Fée = Dryopteris Adans. DRYOPTERIDACEAE (F)
Dichazothece Lindau ACANTHACEAE (D)
Dichelachne Endl. GRAMINEAE (M)
Dichelostemma Kunth ALLIACEAE (M)
Dicheranthus Webb ILLECEBRACEAE (D)
Dichilanthe Thwaites RUBIACEAE (D)
Dichiloboea Stapf = Trisepalum C.B.Clarke GESNERIACEAE (D)
Dichilus DC. LEGUMINOSAE–PAPILIONOIDEAE (D)
Dichocarpum W.T.Wang & Hsiao RANUNCULACEAE (D)
Dicholactina Hance = Phyllanthus L. EUPHORBIACEAE (D)
Dichondra J.R.Forst. & G.Forst. CONVOLVULACEAE (D)

Dichondropsis Brandegee = Dichondra J.R.Forst. & G.Forst. CONVOLVULACEAE (D)
Dichone Salisb. = Ixia L. IRIDACEAE (M)
Dichopetalum F.Muell. (SUH) = Dichosciadium Domin UMBELLIFERAE (D)
Dichopogon Kunth ANTHERICACEAE (M)
Dichopsis Thwaites = Palaquium Blanco SAPOTACEAE (D)
Dichopus Blume = Dendrobium Sw. ORCHIDACEAE (M)
Dichorexia C.Presl = Cyathea Sm. CYATHEACEAE (F)
Dichorisandra J.C.Mikan COMMELINACEAE (M)
Dichosciadium Domin UMBELLIFERAE (D)
Dichostemma Pierre EUPHORBIACEAE (D)
Dichostylis P.Beauv. = Fimbristylis Vahl CYPERACEAE (M)
Dichotomanthes Kurz ROSACEAE (D)
Dichroa Lour. HYDRANGEACEAE (D)
Dichrocephala L'Hér. ex DC. COMPOSITAE (D)
Dichromanthus Garay ORCHIDACEAE (M)
Dichromena Michx. = Rhynchospora Vahl CYPERACEAE (M)
Dichromochlamys Dunlop COMPOSITAE (D)
Dichromus Schltdl. (SUI) = Paspalum L. GRAMINEAE (M)
Dichrophyllum Klotzsch & Garcke = Euphorbia L. EUPHORBIACEAE (D)
Dichrospermum Bremek. = Spermacoce L. RUBIACEAE (D)
Dichrostachys (A.DC.) Wight & Arn. LEGUMINOSAE–MIMOSOIDEAE (D)
Dichrotrichum Reinw. ex Vriese = Agalmyla Blume GESNERIACEAE (D)
Dichylium Britton = Euphorbia L. EUPHORBIACEAE (D)
Dichynchosia K.Muell. (SUO) = Caldcluvia D.Don CUNONIACEAE (D)
Dickasonia L.O.Williams ORCHIDACEAE (M)
Dickinsia Franch. UMBELLIFERAE (D)
Dicksonia L'Hér. DICKSONIACEAE (F)
Dicladanthera F.Muell. ACANTHACEAE (D)
Diclemia Naudin = Ossaea DC. MELASTOMATACEAE (D)
Diclidanthera Mart. POLYGALACEAE (D)
Diclidopteris Brack. = Vaginularia Fée VITTARIACEAE (F)
Diclidostigma Kunze = Melothria L. CUCURBITACEAE (D)
Diclinanona Diels ANNONACEAE (D)
Dicliptera Juss. ACANTHACEAE (D)
Diclis Benth. SCROPHULARIACEAE (D)
Diclisodon T.Moore = Dryopteris Adans. DRYOPTERIDACEAE (F)
Dicocca Thouars = Dicoryphe Thouars HAMAMELIDACEAE (D)
Dicoelia Benth. EUPHORBIACEAE (D)
Dicoelospermum C.B.Clarke CUCURBITACEAE (D)
Dicolus Phil. = Zephyra D.Don TECOPHILAEACEAE (M)
Dicoma Cass. COMPOSITAE (D)
Diconangia Adans. = Itea L. ESCALLONIACEAE (D)
Dicophe Roem. = Dicoryphe Thouars HAMAMELIDACEAE (D)
Dicoria Torr. & A.Gray COMPOSITAE (D)
Dicorynia Benth. LEGUMINOSAE–CAESALPINIOIDEAE (D)
Dicorypha R.A.Hedw. = Dicoryphe Thouars HAMAMELIDACEAE (D)
Dicoryphe Thouars HAMAMELIDACEAE (D)
Dicraeanthus Engl. PODOSTEMACEAE (D)
Dicraeia Thouars PODOSTEMACEAE (D)
Dicraeopetalum Harms LEGUMINOSAE–PAPILIONOIDEAE (D)
Dicranacanthus Oerst. = Barleria L. ACANTHACEAE (D)
Dicranocarpus A.Gray COMPOSITAE (D)
Dicranoglossum J.Sm. POLYPODIACEAE (F)
Dicranolepis Planch. THYMELAEACEAE (D)
Dicranophlebia (Mart.) Lindl. = Cyathea Sm. CYATHEACEAE (F)

Dicranopteris Bernh. GLEICHENIACEAE (F)
Dicranopygium Harling CYCLANTHACEAE (M)
Dicranostachys trecul = Myrianthus P.Beauv. CECROPIACEAE (D)
Dicranostigma Hook.f. & Thomson PAPAVERACEAE (D)
Dicranostyles Benth. CONVOLVULACEAE (D)
Dicranotaenia Finet = Microcoelia Lindl. ORCHIDACEAE (M)
Dicraspidia Standl. TILIACEAE (D)
Dicrastylis J.L.Drumm. ex Harv. VERBENACEAE (D)
Dicraurus Hook.f. = Iresine P.Browne AMARANTHACEAE (D)
Dicrobotryum Willd. ex Roem. & Schult. = Guettarda L. RUBIACEAE (D)
Dicrocaulon N.E.Br. AIZOACEAE (D)
Dicrosperma H.A.Wendl. & Drude ex W.Watson = Dictyosperma H.A.Wendl. & Drude
    PALMAE (M)
Dicrypta Lindl. = Maxillaria Ruíz & Pav. ORCHIDACEAE (M)
Dictamnus L. RUTACEAE (D)
Dictyandra Welw. ex Hook.f. RUBIACEAE (D)
Dictyanthus Decne. ASCLEPIADACEAE (D)
Dictymia J.Sm. POLYPODIACEAE (F)
Dictyocalyx Hook.f. = Exodeconus Raf. SOLANACEAE (D)
Dictyocaryum H.A.Wendl. PALMAE (M)
Dictyochloa (Murb.) E.G.Camus = Ammochloa Boiss. GRAMINEAE (M)
Dictyocline T.Moore = Stegnogramma Blume THELYPTERIDACEAE (F)
Dictyodroma Ching = Diplazium Sw. WOODSIACEAE (F)
Dictyoglossum J.Sm. = Elaphoglossum Schott ex J.Sm. LOMARIOPSIDACEAE (F)
Dictyogramme Fée = Coniogramme Fée ADIANTACEAE (F)
Dictyolimon Rech.f. PLUMBAGINACEAE (D)
Dictyoloma A.Juss. RUTACEAE (D)
Dictyoneura Blume SAPINDACEAE (D)
Dictyophleba Pierre APOCYNACEAE (D)
Dictyophragmus O.E.Schulz CRUCIFERAE (D)
Dictyophyllaria Garay ORCHIDACEAE (M)
Dictyopteris C.Presl = Pleocnemia C.Presl DRYOPTERIDACEAE (F)
Dictyosperma H.A.Wendl. & Drude PALMAE (M)
Dictyospermum Wight COMMELINACEAE (M)
Dictyospora Hook.f. = Hedyotis L. RUBIACEAE (D)
Dictyostega Miers BURMANNIACEAE (M)
Dictyoxiphium Hook. = Tectaria Cav. DRYOPTERIDACEAE (F)
Dicyclophora Boiss. UMBELLIFERAE (D)
Dicymanthes Danser = Amyema Tiegh. LORANTHACEAE (D)
Dicymbe Spruce ex Benth. LEGUMINOSAE–CAESALPINIOIDEAE (D)
Dicymbopsis Ducke = Dicymbe Spruce ex Benth. LEGUMINOSAE–
    CAESALPINIOIDEAE (D)
Dicypellium Nees & Mart. LAURACEAE (D)
Dicyrta Regel = Achimenes Pers. GESNERIACEAE (D)
Didactyle Lindl. = Bulbophyllum Thouars ORCHIDACEAE (M)
Didactylon Zoll. & Moritzi = Dimeria R.Br. GRAMINEAE (M)
Didelotia Baill. LEGUMINOSAE–CAESALPINIOIDEAE (D)
Didelta L'Hér. COMPOSITAE (D)
Diderotia Baill. = Lautembergia Baill. EUPHORBIACEAE (D)
Didesmandra Stapf DILLENIACEAE (D)
Didesmus Desv. CRUCIFERAE (D)
Didiciea King & Prain ORCHIDACEAE (M)
Didiclis P.Beauv. ex Mirbel = Selaginella P.Beauv. SELAGINELLACEAE (FA)
Didierea Baill. DIDIEREACEAE (D)
Didiplis Raf. = Lythrum L. LYTHRACEAE (D)

Didissandra C.B.Clarke GESNERIACEAE (D)
Didonica Luteyn & Wilbur ERICACEAE (D)
Didothion Raf. = Epidendrum L. ORCHIDACEAE (M)
Didymaea Hook.f. RUBIACEAE (D)
Didymanthus Endl. CHENOPODIACEAE (D)
Didymaotus N.E.Br. AIZOACEAE (D)
Didymeles Thouars DIDYMELACEAE (D)
Didymiandrum Gilly CYPERACEAE (M)
Didymocarpus Wall. GESNERIACEAE (D)
Didymochaeta Steud. = Agrostis L. GRAMINEAE (M)
Didymocheton Blume = Dysoxylum Blume MELIACEAE (D)
Didymochlaena Desv. DRYOPTERIDACEAE (F)
Didymochlamys Hook.f. RUBIACEAE (D)
Didymocistus Kuhlm. EUPHORBIACEAE (D)
Didymocolpus S.C.Chen = Acanthochlamys P.C.Kao VELLOZIACEAE (M)
Didymodoxa E.Mey. ex Wedd. URTICACEAE (D)
Didymoecium Bremek. RUBIACEAE (D)
Didymoglossum Desv. = Trichomanes L. HYMENOPHYLLACEAE (F)
Didymopanax Decne. & Planch. = Schefflera J.R.Forst. & G.Forst. ARALIACEAE (D)
Didymopelta Regel & Schmalh. = Astragalus L. LEGUMINOSAE–PAPILIONOIDEAE (D)
Didymophysa Boiss. CRUCIFERAE (D)
Didymoplexiella Garay ORCHIDACEAE (M)
Didymoplexis Griff. ORCHIDACEAE (M)
Didymopogon Bremek. RUBIACEAE (D)
Didymosalpinx Keay RUBIACEAE (D)
Didymosperma H.A.Wendl. & Drude ex Hook.f. = Arenga Labill. PALMAE (M)
Didymostigma W.T.Wang (UP) GESNERIACEAE (D)
Didymotheca Hook.f. = Gyrostemon Desf. GYROSTEMONACEAE (D)
Didyplosandra Wight ex Bremek. = Strobilanthes Blume ACANTHACEAE (D)
Diectomis P.Beauv. (SUH) = Anadelphia Hack. GRAMINEAE (M)
Diectomis Kunth = Andropogon L. GRAMINEAE (M)
Diedropetala Galushko = Delphinium L. RANUNCULACEAE (D)
Dieffenbachia Schott ARACEAE (M)
Diegodendron Capuron DIEGODENDRACEAE (D)
Dielitzia P.S.Short COMPOSITAE (D)
Diellia Brack. ASPLENIACEAE (F)
Dielsantha E.Wimm. CAMPANULACEAE (D)
Dielsia Gilg RESTIONACEAE (M)
Dielsia Kudo (SUH) = Plectranthus L'Hér. LABIATAE (D)
Dielsina Kuntze = Polyceratocarpus Engl. & Diels ANNONACEAE (D)
Dielsiocharis O.E.Schulz CRUCIFERAE (D)
Dielsiochloa Pilg. GRAMINEAE (M)
Dielsiothamnus R.E.Fr. ANNONACEAE (D)
Diemenia Korth. = Licania Aubl. CHRYSOBALANACEAE (D)
Dienia Lindl. = Malaxis Sol. ex Sw. ORCHIDACEAE (M)
Dierama K.Koch IRIDACEAE (M)
Dierbachia Spreng. = Dunalia Kunth SOLANACEAE (D)
Diervilla Mill. CAPRIFOLIACEAE (D)
Dieterlea Lott CUCURBITACEAE (D)
Dietes Salisb. ex Klatt IRIDACEAE (M)
Dieudonnaea Cogn. = Gurania (Schltdl.) Cogn. CUCURBITACEAE (D)
Diflugossa Bremek. = Strobilanthes Blume ACANTHACEAE (D)
Digastrium (Hack.) A.Camus = Ischaemum L. GRAMINEAE (M)
Digera Forssk. AMARANTHACEAE (D)
Digitacalia Pippen COMPOSITAE (D)

Digitalis L. SCROPHULARIACEAE (D)
Digitaria Fabr. (SUH) = Paspalum L. GRAMINEAE (M)
Digitaria Adans. (SUS) = Tripsacum L. GRAMINEAE (M)
Digitaria Haller GRAMINEAE (M)
Digitariella De Winter = Digitaria Haller GRAMINEAE (M)
Digitariopsis C.E.Hubb. = Digitaria Haller GRAMINEAE (M)
Digitorebutia Frič & Kreuz. = Rebutia K.Schum. CACTACEAE (D)
Diglossophyllum H.A.Wendl. ex Salomon = Serenoa Hook.f. PALMAE (M)
Diglyphis Blume (SUO) = Diglyphosa Blume ORCHIDACEAE (M)
Diglyphosa Blume ORCHIDACEAE (M)
Dignathe Lindl. ORCHIDACEAE (M)
Dignathia Stapf GRAMINEAE (M)
Digomphia Benth. BIGNONIACEAE (D)
Digomphotis Raf. = Habenaria Willd. ORCHIDACEAE (M)
Digoniopterys Arènes MALPIGHIACEAE (D)
Digrammaria C.Presl = Diplazium Sw. WOODSIACEAE (F)
Digraphis Trin. (SUS) = Phalaris L. GRAMINEAE (M)
Diheteropogon (Hack.) Stapf GRAMINEAE (M)
Diholcos Rydb. = Astragalus L. LEGUMINOSAE–PAPILIONOIDEAE (D)
Dikylikostigma Kraenzl. = Discyphus Schltr. ORCHIDACEAE (M)
Dilatris Bergius HAEMODORACEAE (M)
Dilepyrum Raf. (SUS) = Oryzopsis Michx. GRAMINEAE (M)
Dilepyrum Michx. = Muhlenbergia Schreb. GRAMINEAE (M)
Dileucaden (Raf.) Steud. (SUI) = Panicum L. GRAMINEAE (M)
Dilkea Mast. PASSIFLORACEAE (D)
Dillenia L. DILLENIACEAE (D)
Dillenia Heist. ex Fabr. (SUH) = Sherardia L. RUBIACEAE (D)
Dillonia Sacleux = Catha Forssk. ex Schreb. CELASTRACEAE (D)
Dillwynia Sm. LEGUMINOSAE–PAPILIONOIDEAE (D)
Dilobeia Thouars PROTEACEAE (D)
Dilochia Lindl. ORCHIDACEAE (M)
Dilochiopsis (Hook.f.) Brieger = Eria Lindl. ORCHIDACEAE (M)
Dilodendron Radlk. SAPINDACEAE (D)
Dilomilis Raf. ORCHIDACEAE (M)
Dilophia Thomson CRUCIFERAE (D)
Dilophotriche (C.E.Hubb.) Jacq.-Fél. GRAMINEAE (M)
Dimanisa Raf. = Justicia L. ACANTHACEAE (D)
Dimeiostemon Raf. = Andropogon L. GRAMINEAE (M)
Dimerandra Schltr. ORCHIDACEAE (M)
Dimeresia A.Gray COMPOSITAE (D)
Dimeria R.Br. GRAMINEAE (M)
Dimerocarpus Gagnep. = Streblus Lour. MORACEAE (D)
Dimerocostus Kuntze COSTACEAE (M)
Dimerodisus Gagnep. = Ipomoea L. CONVOLVULACEAE (D)
Dimerostemma Cass. COMPOSITAE (D)
Dimesia Raf. (SUS) = Hierochloe R.Br. GRAMINEAE (M)
Dimetia (wight & Arn.) Meisn. = Hedyotis L. RUBIACEAE (D)
Dimetra Kerr VERBENACEAE (D)
Dimitria Ravenna = Sisymbrium L. CRUCIFERAE (D)
Dimocarpus Lour. SAPINDACEAE (D)
Dimorphandra Schott LEGUMINOSAE–CAESALPINIOIDEAE (D)
Dimorphanthera (Drude) J.J.Sm. ERICACEAE (D)
Dimorphocalyx Thwaites EUPHORBIACEAE (D)
Dimorphocarpa Rollins CRUCIFERAE (D)
Dimorphochloa S.T.Blake = Cleistochloa C.E.Hubb. GRAMINEAE (M)

Dimorphocladium Britton = Phyllanthus L. EUPHORBIACEAE (D)
Dimorphoclamys Hook.f. = Momordica L. CUCURBITACEAE (D)
Dimorphocoma F.Muell. & Tate COMPOSITAE (D)
Dimorphopteris Tagawa & K.Iwats. = Pronephrium C.Presl THELYPTERIDACEAE (F)
Dimorphorchis Rolfe ORCHIDACEAE (M)
Dimorphosciadium Pimenov UMBELLIFERAE (D)
Dimorphostachys Fourn. = Paspalum L. GRAMINEAE (M)
Dimorphostemon Kitag. CRUCIFERAE (D)
Dimorphotheca Moench COMPOSITAE (D)
Dinacria Haw. = Crassula L. CRASSULACEAE (D)
Dinebra Jacq. GRAMINEAE (M)
Dinema Lindl. ORCHIDACEAE (M)
Dinemagonum A.Juss. MALPIGHIACEAE (D)
Dinemandra A.Juss. MALPIGHIACEAE (D)
Dinetopsis Roberty = Porana Burm.f. CONVOLVULACEAE (D)
Dinizia Ducke LEGUMINOSAE–MIMOSOIDEAE (D)
Dinklagea Gilg = Manotes Sol. ex Planch. CONNARACEAE (D)
Dinklageella Mansf. ORCHIDACEAE (M)
Dinklageodoxa Heine & Sandwith BIGNONIACEAE (D)
Dinocanthium Bremek. = Pyrostria Comm. ex Juss. RUBIACEAE (D)
Dinochloa Büse GRAMINEAE (M)
Dinophora Benth. MELASTOMATACEAE (D)
Dinoseris Griseb. COMPOSITAE (D)
Dintera Stapf SCROPHULARIACEAE (D)
Dinteracanthus C.B.Clarke ex Schinz = Ruellia L. ACANTHACEAE (D)
Dinteranthus Schwantes AIZOACEAE (D)
Dioclea Kunth LEGUMINOSAE–PAPILIONOIDEAE (D)
Diodella Small = Diodia L. RUBIACEAE (D)
Diodia L. RUBIACEAE (D)
Diodioides Loefl. = Spermacoce L. RUBIACEAE (D)
Diodois Pohl (SUI) = Psyllocarpus Mart. Zucc. RUBIACEAE (D)
Diodontium F.Muell. COMPOSITAE (D)
Diodosperma H.A.Wendl. = Trithrinax Mart. PALMAE (M)
Dioecrescis Tirveng. RUBIACEAE (D)
Diogenesia Sleumer ERICACEAE (D)
Diogoa Exell & Mendonça OLACACEAE (D)
Dioicodendron Steyerm. RUBIACEAE (D)
Diolena Naudin = Triolena Naudin MELASTOMATACEAE (D)
Diomma Engl. ex Harms = Spathelia L. RUTACEAE (D)
Dionaea Ellis DROSERACEAE (D)
Dioncophyllum Baill. DIONCOPHYLLACEAE (D)
Dioneiodon Raf. = Diodia L. RUBIACEAE (D)
Dionycha Naudin MELASTOMATACEAE (D)
Dionychastrum A.Fern. & R.Fern. MELASTOMATACEAE (D)
Dionychia Hook.f. (SUO) = Dionycha Naudin MELASTOMATACEAE (D)
Dionysia Fenzl PRIMULACEAE (D)
Dioon Lindl. ZAMIACEAE (G)
Diopogon A.Jord. & Fourr. = Sempervivum L. CRASSULACEAE (D)
Diora Ravenna ANTHERICACEAE (M)
Dioscorea L. DIOSCOREACEAE (M)
Dioscoreophyllum Engl. MENISPERMACEAE (D)
Diosma L. RUTACEAE (D)
Diosphaera Buser CAMPANULACEAE (D)
Diospyros L. EBENACEAE (D)
Diostea Miers VERBENACEAE (D)

Diotacanthus Benth. ACANTHACEAE (D)
Diothonea Lindl. ORCHIDACEAE (M)
Dioticarpus Dunn = Hopea Roxb. DIPTEROCARPACEAE (D)
Diotocarpus Hochst. = Pentanisia Harv. RUBIACEAE (D)
Diotocranus Bremek. = Mitrasacmopsis Jovet RUBIACEAE (D)
Dipanax Seem. = Tetraplasandra A.Gray ARALIACEAE (D)
Dipcadi Medik. HYACINTHACEAE (M)
Dipelta Maxim. CAPRIFOLIACEAE (D)
Dipentodon Dunn DIPENTODONTACEAE (D)
Dipera Spreng. = Disperis Sw. ORCHIDACEAE (M)
Diperium Desv. = Mnesithea Kunth GRAMINEAE (M)
Dipetalanthus A.Chev. = Hymenostegia (Benth.) Harms LEGUMINOSAE–
   CAESALPINIOIDEAE (D)
Diphaca Lour. = Ormocarpum P.Beauv. LEGUMINOSAE–PAPILIONOIDEAE (D)
Diphasia Pierre RUTACEAE (D)
Diphasiastrum Holub = Lycopodium L. LYCOPODIACEAE (FA)
Diphasiopsis Mendonça RUTACEAE (D)
Diphasium C.Presl ex Rothm. = Lycopodium L. LYCOPODIACEAE (FA)
Dipholis A.DC. = Sideroxylon L. SAPOTACEAE (D)
Diphragmus C.Presl = Spermacoce L. RUBIACEAE (D)
Diphryllum Raf. = Listera R.Br. ORCHIDACEAE (M)
Diphyes Blume = Bulbophyllum Thouars ORCHIDACEAE (M)
Diphylax Hook.f. ORCHIDACEAE (M)
Diphyllarium Gagnep. LEGUMINOSAE–PAPILIONOIDEAE (D)
Diphylleia Michx. BERBERIDACEAE (D)
Diphysa Jacq. LEGUMINOSAE–PAPILIONOIDEAE (D)
Dipidax Salisb. ex Benth. = Onixotis Raf. COLCHICACEAE (M)
Diplachne P.Beauv. = Leptochloa P.Beauv. GRAMINEAE (M)
Diplachyrium Nees (UR) GRAMINEAE (M)
Diplacrum R.Br. CYPERACEAE (M)
Diplacus Nutt. = Mimulus L. SCROPHULARIACEAE (D)
Dipladenia A.DC. = Mandevilla Lindl. APOCYNACEAE (D)
Diplandra Raf. (SUH) = Ludwigia L. ONAGRACEAE (D)
Diplandra Hook. & Arn. (SUH) = Lopezia Cav. ONAGRACEAE (D)
Diplandrorchis S.C.Chen ORCHIDACEAE (M)
Diplanthemum K.Schum. = Duboscia Bocquet TILIACEAE (D)
Diplanthera Banks & Sol. = Deplanchea Vieill. BIGNONIACEAE (D)
Diplanthera Raf. (SUH) = Platanthera Rich. ORCHIDACEAE (M)
Diplarche Hook.f. & Thomson DIAPENSIACEAE (D)
Diplarpea Triana MELASTOMATACEAE (D)
Diplarrena Labill. (SUO) = Diplarrhena Labill. IRIDACEAE (M)
Diplarrhena Labill. IRIDACEAE (M)
Diplasanthum Desv. = Dichanthium Willemet GRAMINEAE (M)
Diplasia Pers. CYPERACEAE (M)
Diplaspis Hook.f. UMBELLIFERAE (D)
Diplatia Tiegh. LORANTHACEAE (D)
Diplax Benn. = Ehrharta Thunb. GRAMINEAE (M)
Diplaziopsis C.Chr. WOODSIACEAE (F)
Diplazium Sw. WOODSIACEAE (F)
Diplazoptilon Y.Ling COMPOSITAE (D)
Diplectraden Raf. = Habenaria Willd. ORCHIDACEAE (M)
Diplectria (Blume) Rchb. MELASTOMATACEAE (D)
Diplectrum Pers. = Satyrium Sw. ORCHIDACEAE (M)
Diploblechnum Hayata = Blechnum L. BLECHNACEAE (F)
Diplobryum C.Cusset PODOSTEMACEAE (D)

Diplocarex Hayata = Carex L. CYPERACEAE (M)
Diplocaulobium (Rchb.f.) Kraenzl. ORCHIDACEAE (M)
Diplocea Raf. = Triplasis P.Beauv. GRAMINEAE (M)
Diplocentrum Lindl. ORCHIDACEAE (M)
Diplochilus Lindl. = Diplomeris D.Don ORCHIDACEAE (M)
Diplochlamys Müll.Arg. = Mallotus Lour. EUPHORBIACEAE (D)
Diploclisia Miers MENISPERMACEAE (D)
Diploconchium Schauer = Agrostophyllum Blume ORCHIDACEAE (M)
Diplocos Bureau = Streblus Lour. MORACEAE (D)
Diplocrater Hook.f. = Tricalysia A.Rich. ex DC. RUBIACEAE (D)
Diplocyatha N.E.Br. = Orbea Haw. ASCLEPIADACEAE (D)
Diplocyathium Heinr.Schmidt = Euphorbia L. EUPHORBIACEAE (D)
Diplocyclos (Endl.) Post & Kuntze CUCURBITACEAE (D)
Diplocyclus Post & Kuntze (SUO) = Diplocyclos (Endl.) Post & Kuntze
    CUCURBITACEAE (D)
Diplodiscus Turcz. TILIACEAE (D)
Diplodium Sw. = Pterostylis R.Br. ORCHIDACEAE (M)
Diplodon Spreng. = Diplusodon Pohl LYTHRACEAE (D)
Diplodonta Karst. = Clidemia D.Don MELASTOMATACEAE (D)
Diplofatsia Nakai = Fatsia Decne. & Planch. ARALIACEAE (D)
Diploglottis Hook.f. SAPINDACEAE (D)
Diplogon Poir. (SUO) = Diplopogon R.Br. GRAMINEAE (M)
Diplokeleba N.E.Br. SAPINDACEAE (D)
Diploknema Pierre SAPOTACEAE (D)
Diplolabellum F.Maek. ORCHIDACEAE (M)
Diplolaena R.Br. RUTACEAE (D)
Diplolegnon Rusby = Corytoplectus Oerst. GESNERIACEAE (D)
Diplolepis R.Br. ASCLEPIADACEAE (D)
Diplolophium Turcz. UMBELLIFERAE (D)
Diplomeris D.Don ORCHIDACEAE (M)
Diplomorpha Griff. (SUH) = Sauropus Blume EUPHORBIACEAE (D)
Diploon Cronquist SAPOTACEAE (D)
Diploophyllum Van den Bosch = Hymenophyllum Sm. HYMENOPHYLLACEAE (F)
Diplopanax Hand.–Mazz. CORNACEAE (D)
Diplopeltis Endl. SAPINDACEAE (D)
Diploperianthium F.Ritter (SUI) = Calymmanthium F.Ritter CACTACEAE (D)
Diplophractum Desf. = Colona Cav. TILIACEAE (D)
Diplophragma (Wight & Arn.) Meisn. = Hedyotis L. RUBIACEAE (D)
Diplopilosa Dvořák = Hesperis L. CRUCIFERAE (D)
Diplopogon R.Br. GRAMINEAE (M)
Diploprora Hook.f. ORCHIDACEAE (M)
Diploptera C.A.Gardner = Strangea Meisn. PROTEACEAE (D)
Diplopterygium (Diels) Nakai GLEICHENIACEAE (F)
Diplopterys A.Juss. MALPIGHIACEAE (D)
Diplora Baker ASPLENIACEAE (F)
Diplorhynchus Welw. ex Ficalho & Hiern APOCYNACEAE (D)
Diplosoma Schwantes AIZOACEAE (D)
Diplospora DC. RUBIACEAE (D)
Diplosporopsis Wernham = Belonophora Hook.f. RUBIACEAE (D)
Diplostachyum P.Beauv. = Selaginella P.Beauv. SELAGINELLACEAE (FA)
Diplostephium Kunth COMPOSITAE (D)
Diplostigma K.Schum. ASCLEPIADACEAE (D)
Diplostylis Sond. = Adenocline Turcz. EUPHORBIACEAE (D)
Diplotaenia Boiss. UMBELLIFERAE (D)
Diplotaxis DC. CRUCIFERAE (D)

Diplothemium Mart. = Allagoptera Nees PALMAE (M)
Diplothorax Gagnep. = Streblus Lour. MORACEAE (D)
Diplotropis Benth. LEGUMINOSAE–PAPILIONOIDEAE (D)
Diplousodon Meisn. (SUO) = Diplusodon Pohl LYTHRACEAE (D)
Diplukion Raf. = Iochroma Benth. SOLANACEAE (D)
Diplusodon Pohl LYTHRACEAE (D)
Diplycosia Blume ERICACEAE (D)
Dipodium R.Br. ORCHIDACEAE (M)
Dipodophyllum Tiegh. = Psittacanthus Mart. LORANTHACEAE (D)
Dipogon Liebm. LEGUMINOSAE–PAPILIONOIDEAE (D)
Dipogon Steud. (SUI) = Sorghastrum Nash GRAMINEAE (M)
Dipogonia P.Beauv. (SUS) = Diplopogon R.Br. GRAMINEAE (M)
Dipoma Franch. CRUCIFERAE (D)
Diporidium H.L.Wendl. = Ochna L. OCHNACEAE (D)
Diposis DC. UMBELLIFERAE (D)
Dipsacus L. DIPSACACEAE (D)
Dipteracanthus Nees = Ruellia L. ACANTHACEAE (D)
Dipteranthemum F.Muell. = Ptilotus R.Br. AMARANTHACEAE (D)
Dipteranthus Barb.Rodr. ORCHIDACEAE (M)
Dipteris Reinw. DIPTERIDACEAE (F)
Dipterocarpus C.F.Gaertn. DIPTEROCARPACEAE (D)
Dipterocome Fisch. & C.A.Mey. COMPOSITAE (D)
Dipterocypsela S.F.Blake COMPOSITAE (D)
Dipterodendron Radlk. = Dilodendron Radlk. SAPINDACEAE (D)
Dipteronia Oliv. ACERACEAE (D)
Dipteropeltis Hallier f. CONVOLVULACEAE (D)
Dipterosiphon Huber = Campylosiphon Benth. BURMANNIACEAE (M)
Dipterosperma Hassk. = Stereospermum Cham. BIGNONIACEAE (D)
Dipterostele Schltr. ORCHIDACEAE (M)
Dipterostemon Rydb. = Dichelostemma Kunth ALLIACEAE (M)
Dipterygium Decne. CAPPARACEAE (D)
Dipteryx Schreb. LEGUMINOSAE–PAPILIONOIDEAE (D)
Diptychandra Tul. LEGUMINOSAE–CAESALPINIOIDEAE (D)
Diptychocarpus Trautv. CRUCIFERAE (D)
Diptychum Dulac (SUS) = Sesleria Scop. GRAMINEAE (M)
Dipyrena Hook. VERBENACEAE (D)
Dirachma Schweinf. ex Balf.f. GERANIACEAE (D)
Dirca L. THYMELAEACEAE (D)
Dircaea Decne. = Sinningia Nees GESNERIACEAE (D)
Dirhamphis Krapov. MALVACEAE (D)
Dirhynchosia Blume = Caldcluvia D.Don CUNONIACEAE (D)
Dirichletia Klotzsch = Carphalea Juss. RUBIACEAE (D)
Disa Bergius ORCHIDACEAE (M)
Disaccanthus Greene CRUCIFERAE (D)
Disakisperma Steud. = Leptochloa P.Beauv. GRAMINEAE (M)
Disanthus Maxim. HAMAMELIDACEAE (D)
Disarrenum Labill. = Hierochloe R.Br. GRAMINEAE (M)
Disaster Gilli = Trymalium Fenzl RHAMNACEAE (D)
Discalyxia Markgr. = Alyxia Banks ex R.Br. APOCYNACEAE (D)
Discanthera Torr. & A.Gray = Cyclanthera Schrad. CUCURBITACEAE (D)
Discaria Hook. RHAMNACEAE (D)
Dischidanthus Tsiang ASCLEPIADACEAE (D)
Dischidia R.Br. ASCLEPIADACEAE (D)
Dischidiopsis Schltr. ASCLEPIADACEAE (D)
Dischisma Choisy SCROPHULARIACEAE (D)

Dischistocalyx T.Anderson ex Benth. ACANTHACEAE (D)
Disciphania Eichler MENISPERMACEAE (D)
Discipiper Trelease & Stehle = Piper L. PIPERACEAE (D)
Discocactus Pfeiff. CACTACEAE (D)
Discocalyx (A.DC.) Mez MYRSINACEAE (D)
Discocapnos Cham. & Schltdl. PAPAVERACEAE (D)
Discocarpus Klotzsch EUPHORBIACEAE (D)
Discoclaoxylon (Müll.Arg.) Pax & K.Hoffm. EUPHORBIACEAE (D)
Discocleidion (Müll.Arg.) Pax & K.Hoffm. EUPHORBIACEAE (D)
Discocnide Chew URTICACEAE (D)
Discocoffea A.Chev. = Tricalysia A.Rich. ex DC. RUBIACEAE (D)
Discoglypremna Prain EUPHORBIACEAE (D)
Discolobium Benth. LEGUMINOSAE–PAPILIONOIDEAE (D)
Discoluma Baill. = Pouteria Aubl. SAPOTACEAE (D)
Discoma O.F.Cook = Chamaedorea Willd. PALMAE (M)
Discophis Raf. = Drypetes Vahl EUPHORBIACEAE (D)
Discophora Miers ICACINACEAE (D)
Discopleura DC. = Ptilimnium Raf. UMBELLIFERAE (D)
Discoplis Raf. = Mercurialis L. EUPHORBIACEAE (D)
Discopodium Hochst. SOLANACEAE (D)
Discospermum Dalzell = Diplospora DC. RUBIACEAE (D)
Discostegia C.Presl = Marattia Sw. MARATTIACEAE (F)
Discovium Raf. CRUCIFERAE (D)
Discyphus Schltr. ORCHIDACEAE (M)
Diselma Hook.f. CUPRESSACEAE (G)
Disemma Labill. = Passiflora L. PASSIFLORACEAE (D)
Disepalum Hook.f. ANNONACEAE (D)
Diseris Wight = Disperis Sw. ORCHIDACEAE (M)
Disisocactus Kunze (SUS) = Disocactus Lindl. CACTACEAE (D)
Diskion Raf. = Saracha Ruíz & Pav. SOLANACEAE (D)
Disocactus Lindl. CACTACEAE (D)
Disocereus Frič & Kreuz. (SUO) = Disocactus Lindl. CACTACEAE (D)
Disodea Pers. = Paederia L. RUBIACEAE (D)
Disparago Gaertn. COMPOSITAE (D)
Disperis Sw. ORCHIDACEAE (M)
Disperma J.F.Gmel. = Mitchella L. RUBIACEAE (D)
Disperma J.F.Gmel. (SUH) = Duosperma Dayton ACANTHACEAE (D)
Disphenia C.Presl = Cyathea Sm. CYATHEACEAE (F)
Disphyma N.E.Br. AIZOACEAE (D)
Disporopsis Hance CONVALLARIACEAE (M)
Disporum Salisb. ex D.Don CONVALLARIACEAE (M)
Dissanthelium Trin. GRAMINEAE (M)
Dissiliaria F.Muell. EUPHORBIACEAE (D)
Dissocarpus F.Muell. CHENOPODIACEAE (D)
Dissochaeta Blume MELASTOMATACEAE (D)
Dissochondrus (W.F.Hillebr.) Kuntze GRAMINEAE (M)
Dissomeria Hook.f. ex Benth. FLACOURTIACEAE (D)
Dissorhynchium Schauer = Habenaria Willd. ORCHIDACEAE (M)
Dissosperma Soják = Corydalis DC. PAPAVERACEAE (D)
Dissothrix A.Gray COMPOSITAE (D)
Dissotis Benth. MELASTOMATACEAE (D)
Distaxia C.Presl = Blechnum L. BLECHNACEAE (F)
Disteganthus Lem. BROMELIACEAE (M)
Distemon Wedd. (SUH) = Neodistemon Babu & A.N.Henry URTICACEAE (D)
Distemonanthus Benth. LEGUMINOSAE–CAESALPINIOIDEAE (D)

Distephania Gagnep. = Indosinia J.E.Vidal OCHNACEAE (D)
Distephanus (Cass.) Cass. COMPOSITAE (D)
Disterigma (Klotzsch) Nied. ERICACEAE (D)
Distichella Tiegh. = Dendrophthora Eichler VISCACEAE (D)
Distichia Nees & Meyen JUNCACEAE (M)
Distichis Thouars = Liparis Rich. ORCHIDACEAE (M)
Distichlis Raf. GRAMINEAE (M)
Distichocalyx Benth. (SUO) = Dischistocalyx T.Anderson ex Benth. ACANTHACEAE (D)
Distichoselinum Garc.Mart. & Silvestre UMBELLIFERAE (D)
Distichostemon F.Muell. SAPINDACEAE (D)
Distictella Kuntze BIGNONIACEAE (D)
Distictis Bureau (SUH) = Distictella Kuntze BIGNONIACEAE (D)
Distictis Mart. ex Meisn. BIGNONIACEAE (D)
Distoecha Phil. = Hypochaeris L. COMPOSITAE (D)
Distomaea Spenn. = Listera R.Br. ORCHIDACEAE (M)
Distomocarpus O.E.Schulz = Rytidocarpus Coss. CRUCIFERAE (D)
Distomomischus Dulac (SUS) = Vulpia C.C.Gmel. GRAMINEAE (M)
Distrepta Miers = Tecophilaea Bertero ex Colla TECOPHILAEACEAE (M)
Distrianthes Danser LORANTHACEAE (D)
Distyliopsis P.K.Endress HAMAMELIDACEAE (D)
Distylium Siebold & Zucc. HAMAMELIDACEAE (D)
Distylodon Summerh. ORCHIDACEAE (M)
Disynaphia Hook. & Arn. ex DC. COMPOSITAE (D)
Disynstemon R.Vig. LEGUMINOSAE–PAPILIONOIDEAE (D)
Ditassa R.Br. ASCLEPIADACEAE (D)
Ditaxis Vahl ex A.Juss. EUPHORBIACEAE (D)
Diteilis Raf. = Liparis Rich. ORCHIDACEAE (M)
Ditepalanthus Fagerl. BALANOPHORACEAE (D)
Dithecina Tiegh. = Helixanthera Lour. LORANTHACEAE (D)
Dithecoluma Baill. = Pouteria Aubl. SAPOTACEAE (D)
Dithrix Schltr. ORCHIDACEAE (M)
Dithyrea Harv. CRUCIFERAE (D)
Dithyridanthus Garay = Schiedeella Schltr. ORCHIDACEAE (M)
Dithyrostegia A.Gray COMPOSITAE (D)
Ditmaria Spreng. (SUH) = Erisma Rudge VOCHYSIACEAE (D)
Ditrichospermum Bremek. = Strobilanthes Blume ACANTHACEAE (D)
Ditritra Raf. = Euphorbia L. EUPHORBIACEAE (D)
Ditrysinia Raf. = Sebastiania Spreng. EUPHORBIACEAE (D)
Ditta Griseb. EUPHORBIACEAE (D)
Dittoceras Hook.f. ASCLEPIADACEAE (D)
Dittostigma Phil. SOLANACEAE (D)
Dittrichia Greuter COMPOSITAE (D)
Ditulima Raf. = Dendrobium Sw. ORCHIDACEAE (M)
Diuranthera Hemsl. = Chlorophytum Ker–Gawl. ANTHERICACEAE (M)
Diuris Sm. ORCHIDACEAE (M)
Diurospermum Edgew. = Utricularia L. LENTIBULARIACEAE (D)
Diyaminauclea Ridsdale RUBIACEAE (D)
Dizygostemon (Benth.) Radlk. ex Wettst. SCROPHULARIACEAE (D)
Dizygotheca N.E.Br. = Schefflera J.R.Forst. & G.Forst. ARALIACEAE (D)
Djaloniella P.Taylor GENTIANACEAE (D)
Djinga C.Cusset PODOSTEMACEAE (D)
Dobera Juss. SALVADORACEAE (D)
Dobinea Buch.–Ham. ex D.Don PODOACEAE (D)
Docanthe O.F.Cook (SUI) = Chamaedorea Willd. PALMAE (M)
Dockrillia Brieger = Dendrobium Sw. ORCHIDACEAE (M)

Docynia Decne. ROSACEAE (D)
Dodartia L. SCROPHULARIACEAE (D)
Dodecadenia Nees LAURACEAE (D)
Dodecahema Reveal & Hardham POLYGONACEAE (D)
Dodecas L.f. = Crenea Aubl. LYTHRACEAE (D)
Dodecastemon Hassk. = Drypetes Vahl EUPHORBIACEAE (D)
Dodecastigma Ducke EUPHORBIACEAE (D)
Dodecatheon L. PRIMULACEAE (D)
Dodonaea Mill. SAPINDACEAE (D)
Dodsonia Ackerman ORCHIDACEAE (M)
Doellochloa Kuntze (SUS) = Gymnopogon P.Beauv. GRAMINEAE (M)
Doerpfeldia Urb. RHAMNACEAE (D)
Doga (Baill.) Baill. ex Nakai = Storckiella Seem. LEGUMINOSAE–
    CAESALPINIOIDEAE (D)
Dolia Lindl. = Nolana L.f. SOLANACEAE (D)
Dolianthus C.H.Wright = Amaracarpus Blume RUBIACEAE (D)
Dolichandra Cham. BIGNONIACEAE (D)
Dolichandrone (Fenzl) Seem. BIGNONIACEAE (D)
Dolichanthera Schltr. & K.Krause = Morierina Vieill. RUBIACEAE (D)
Dolichlasium Lag. COMPOSITAE (D)
Dolichocentrum (Schltr.) Brieger = Dendrobium Sw. ORCHIDACEAE (M)
Dolichochaete (C.E.Hubb.) J.B.Phipps = Tristachya Nees GRAMINEAE (M)
Dolichodelphys K.Schum. & K.Krause RUBIACEAE (D)
Dolichodiera Hanst. = Sinningia Nees GESNERIACEAE (D)
Dolichoglottis B.Nord. COMPOSITAE (D)
Dolichokentia Becc. = Cyphokentia Brongn. PALMAE (M)
Dolicholobium A.Gray RUBIACEAE (D)
Dolicholoma D.Fang & W.T.Wang GESNERIACEAE (D)
Dolicholus Medik. = Rhynchosia Lour. LEGUMINOSAE–PAPILIONOIDEAE (D)
Dolichometra K.Schum. RUBIACEAE (D)
Dolichopsis Hassl. LEGUMINOSAE–PAPILIONOIDEAE (D)
Dolichopterys Kosterm. = Lophopterys A.Juss. MALPIGHIACEAE (D)
Dolichorhynchus Hedge & Kit Tan CRUCIFERAE (D)
Dolichorrhiza (Pojark.) Galushko COMPOSITAE (D)
Dolichos L. LEGUMINOSAE–PAPILIONOIDEAE (D)
Dolichosiphon Phil. = Jaborosa Juss. SOLANACEAE (D)
Dolichostachys Benoist (UP) ACANTHACEAE (D)
Dolichostegia Schltr. ASCLEPIADACEAE (D)
Dolichostigma Miers = Jaborosa Juss. SOLANACEAE (D)
Dolichostylis Turcz. = Draba L. CRUCIFERAE (D)
Dolichothele (K.Schum.) Britton & Rose = Mammillaria Haw. CACTACEAE (D)
Dolichothrix Hilliard & B.L.Burtt COMPOSITAE (D)
Dolichoura Brade MELASTOMATACEAE (D)
Dolichovigna Hayata = Vigna Savi LEGUMINOSAE–PAPILIONOIDEAE (D)
Doliocarpus Rol. DILLENIACEAE (D)
Dollinera Endl. = Desmodium Desv. LEGUMINOSAE–PAPILIONOIDEAE (D)
Dolomiaea DC. COMPOSITAE (D)
Dolosanthus Klatt = Vernonia Schreb. COMPOSITAE (D)
Doma Poir. = Hyphaene Gaertn. PALMAE (M)
Dombeya L'Hér. (SUH) = Tourrettia Foug. BIGNONIACEAE (D)
Dombeya Cav. STERCULIACEAE (D)
Domeykoa Phil. UMBELLIFERAE (D)
Dominella E.Wimm. CAMPANULACEAE (D)
Domingoa Schltr. ORCHIDACEAE (M)
Dominia Fedde = Uldinia J.M.Black UMBELLIFERAE (D)

Domkeocarpa Markgr. = Tabernaemontana L. APOCYNACEAE (D)
Domohinea Leandri EUPHORBIACEAE (D)
Donacium Fr. (SUS) = Arundo L. GRAMINEAE (M)
Donatia J.R.Forst. & G.Forst. STYLIDIACEAE (D)
Donax P.Beauv. (SUS) = Arundo L. GRAMINEAE (M)
Donax Lour. MARANTACEAE (M)
Dondia Spreng. (SUH) = Hacquetia Neck. ex DC. UMBELLIFERAE (D)
Dondia Adans. = Suaeda Forssk. ex Scop. CHENOPODIACEAE (D)
Dondisia DC. = Canthium Lam. RUBIACEAE (D)
Donella Pierre ex Baill. = Chrysophyllum L. SAPOTACEAE (D)
Doniophyton Wedd. COMPOSITAE (D)
Donkelaaria Lem. = Guettarda L. RUBIACEAE (D)
Donnellia C.B.Clarke ex Donn.Sm. (SUH) = Tripogandra Raf. COMMELINACEAE (M)
Donnellsmithia J.M.Coult. & Rose UMBELLIFERAE (D)
Dontostemon Andrz. ex C.A.Mey. CRUCIFERAE (D)
Doodia R.Br. BLECHNACEAE (F)
Doona Thwaites = Shorea Roxb. ex C.F.Gaertn. DIPTEROCARPACEAE (D)
Dopatrium Buch.–Ham. SCROPHULARIACEAE (D)
Doratolepis Schltdl. = Leptorhynchos Less. COMPOSITAE (D)
Doratoxylon Thouars ex Hook.f. SAPINDACEAE (D)
Dorcapteris C.Presl = Olfersia Raddi DRYOPTERIDACEAE (F)
Dorcoceras Bunge = Boea Comm. ex Lam. GESNERIACEAE (D)
Dorema D.Don UMBELLIFERAE (D)
Doricera Verdc. RUBIACEAE (D)
Dorisia Gillespie = Mastixiodendron Melch. RUBIACEAE (D)
Doritis Lindl. ORCHIDACEAE (M)
Dorobaea Cass. = Senecio L. COMPOSITAE (D)
Doronicum L. COMPOSITAE (D)
Dorothea Wernham = Aulacocalyx Hook.f. RUBIACEAE (D)
Dorotheanthus Schwantes AIZOACEAE (D)
Dorstenia L. MORACEAE (D)
Dortania A.Chev. = Gladiolus L. IRIDACEAE (M)
Doryanthes Corrêa DORYANTHACEAE (M)
Dorycheile Rchb. = Cephalanthera Rich. ORCHIDACEAE (M)
Dorycnium Mill. = Lotus L. LEGUMINOSAE–PAPILIONOIDEAE (D)
Dorycnopsis Boiss. = Anthyllis L. LEGUMINOSAE–PAPILIONOIDEAE (D)
Doryopteris J.Sm. ADIANTACEAE (F)
Doryphora Endl. MONIMIACEAE (D)
Dorystaechas Boiss. & Heldr. ex Benth. LABIATAE (D)
Dorystephania Warb. ASCLEPIADACEAE (D)
Doryxylon Zoll. EUPHORBIACEAE (D)
Dossifluga Bremek. = Strobilanthes Blume ACANTHACEAE (D)
Dossinia C.Morren ORCHIDACEAE (M)
Dothilis Raf. = Spiranthes Rich. ORCHIDACEAE (M)
Dothilophis Raf. = Epidendrum L. ORCHIDACEAE (M)
Douarrea Montrouz. = Psychotria L. RUBIACEAE (D)
Douepea Cambess. CRUCIFERAE (D)
Douglasia Lindl. = Androsace L. PRIMULACEAE (D)
Douma Poir. = Hyphaene Gaertn. PALMAE (M)
Douradoa Sleumer OLACACEAE (D)
Dovea Kunth RESTIONACEAE (M)
Dovyalis E.Mey. ex Arn. FLACOURTIACEAE (D)
Downingia Torr. CAMPANULACEAE (D)
Doxantha Miers = Macfadyena A.DC. BIGNONIACEAE (D)
Doxosma Raf. = Epidendrum L. ORCHIDACEAE (M)

Doyerea Grosourdy CUCURBITACEAE (D)
Draba L. CRUCIFERAE (D)
Drabastrum (F.Muell.) O.E.Schulz CRUCIFERAE (D)
Drabella Nábělek (SUH) = Draba L. CRUCIFERAE (D)
Drabopsis K.Koch CRUCIFERAE (D)
Dracaena Vand. ex L. DRACAENACEAE (M)
Dracamine Nieuwl. = Cardamine L. CRUCIFERAE (D)
Dracocactus Y.Ito (SUI) = Neoporteria Britton & Rose CACTACEAE (D)
Dracocephalum L. LABIATAE (D)
Dracontioides Engl. ARACEAE (M)
Dracontium L. ARACEAE (M)
Dracontomelon Blume ANACARDIACEAE (D)
Dracophilus (Schwantes) Dinter & Schwantes AIZOACEAE (D)
Dracophyllum Labill. EPACRIDACEAE (D)
Dracopsis Cass. = Rudbeckia L. COMPOSITAE (D)
Dracosciadium Hilliard & B.L.Burtt UMBELLIFERAE (D)
Dracula Luer ORCHIDACEAE (M)
Dracunculus Mill. ARACEAE (M)
Drakaea Lindl. ORCHIDACEAE (M)
Drake–Brockmania Stapf GRAMINEAE (M)
Drakebrockmania A.C.White & B.Sloane = White–Sloanea Chiov. ASCLEPIADACEAE (D)
Draparnaudia Montrouz. = Xanthostemon F.Muell. MYRTACEAE (D)
Draperia Torr. HYDROPHYLLACEAE (D)
Drapetes Banks ex Lam. THYMELAEACEAE (D)
Drebbelia Zoll. = Olax L. OLACACEAE (D)
Dregea E.Mey. ASCLEPIADACEAE (D)
Dregeochloa Conert GRAMINEAE (M)
Drejera Nees (UP) ACANTHACEAE (D)
Drejerella Lindau = Justicia L. ACANTHACEAE (D)
Drepadenium Raf. = Croton L. EUPHORBIACEAE (D)
Drepananthus Maingay ex Hook.f. = Cyathocalyx Champ. ex Hook.f. & Thomson
    ANNONACEAE (D)
Drepanocarpus G.Mey. = Machaerium Pers. LEGUMINOSAE–PAPILIONOIDEAE (D)
Drepanocaryum Pojark. LABIATAE (D)
Drepanostachyum Keng f. = Sinarundinaria Nakai GRAMINEAE (M)
Drepanostemma Jum. & H.Perrier ASCLEPIADACEAE (D)
Dresslerella Luer ORCHIDACEAE (M)
Dressleria Dodson ORCHIDACEAE (M)
Dressleriella Brieger (SUI) = Jacquiniella Schltr. ORCHIDACEAE (M)
Dressleriopsis Dwyer = Lasianthus Jack RUBIACEAE (D)
Dresslerothamnus H.Rob. COMPOSITAE (D)
Driessenia Korth. MELASTOMATACEAE (D)
Drimia Jacq. ex Willd. HYACINTHACEAE (M)
Drimiopsis Lindl. & Paxton HYACINTHACEAE (M)
Drimycarpus Hook.f. ANACARDIACEAE (D)
Drimys J.R.Forst. & G.Forst. WINTERACEAE (D)
Droceloncia J.Léonard EUPHORBIACEAE (D)
Droguetia Gaudich. URTICACEAE (D)
Droogmansia De Wild. LEGUMINOSAE–PAPILIONOIDEAE (D)
Drosanthemopsis Rauschert AIZOACEAE (D)
Drosanthemum Schwantes AIZOACEAE (D)
Drosera L. DROSERACEAE (D)
Drosodendron Roem. = Baeckea L. MYRTACEAE (D)
Drosophyllum Link DROSERACEAE (D)
Drudeophytum J.M.Coult. & Rose = Oreonana Jeps. UMBELLIFERAE (D)

Drummondia DC. = Mitella L. SAXIFRAGACEAE (D)
Drummondita Harv. RUTACEAE (D)
Druparia Silva Manso = Cayaponia Silva Manso CUCURBITACEAE (D)
Drusa DC. UMBELLIFERAE (D)
Dryadanthe Endl. = Sibbaldia L. ROSACEAE (D)
Dryadella Luer ORCHIDACEAE (M)
Dryadodaphne S.Moore MONIMIACEAE (D)
Dryadorchis Schltr. ORCHIDACEAE (M)
Dryandra R.Br. PROTEACEAE (D)
Dryandra Thunb. (SUH) = Vernicia Lour. EUPHORBIACEAE (D)
Dryas L. ROSACEAE (D)
Drymaria Willd. ex Schult. CARYOPHYLLACEAE (D)
Drymoanthus Nicholls ORCHIDACEAE (M)
Drymocallis Fourr. ex Rydb. ROSACEAE (D)
Drymochloa Holub = Festuca L. GRAMINEAE (M)
Drymoda Lindl. ORCHIDACEAE (M)
Drymoglossum C.Presl = Pyrrosia Mirb. POLYPODIACEAE (F)
Drymonaetes Fourr. (SUI) = Festuca L. GRAMINEAE (M)
Drymonia Mart. GESNERIACEAE (D)
Drymophila R.Br. CONVALLARIACEAE (M)
Drymophloeus Zipp. PALMAE (M)
Drymotaenium Makino POLYPODIACEAE (F)
Drynaria (Bory) J.Sm. POLYPODIACEAE (F)
Drynariopsis (Copel.) Ching = Aglaomorpha Schott POLYPODIACEAE (F)
Dryoathyrium Ching = Lunathyrium Koldz. WOODSIACEAE (F)
Dryobalanops C.F.Gaertn. DIPTEROCARPACEAE (D)
Dryomenis Fée ex J.Sm. = Tectaria Cav. DRYOPTERIDACEAE (F)
Dryopeia Thouars = Disperis Sw. ORCHIDACEAE (M)
Dryopetalon A.Gray CRUCIFERAE (D)
Dryopoa Vickery GRAMINEAE (M)
Dryopolystichum Copel. DRYOPTERIDACEAE (F)
Dryopsis Holttum & P.J.Edwards DRYOPTERIDACEAE (F)
Dryopteris Adans. DRYOPTERIDACEAE (F)
Dryorchis Thouars = Disperis Sw. ORCHIDACEAE (M)
Dryostachyum J.Sm. = Aglaomorpha Schott POLYPODIACEAE (F)
Dryparia Post & Kuntze (SUO) = Cayaponia Silva Manso CUCURBITACEAE (D)
Drypetes Vahl EUPHORBIACEAE (D)
Drypis L. CARYOPHYLLACEAE (D)
Duabanga Buch.-Ham. LYTHRACEAE (D)
Duania Noronha = Homalanthus A.Juss. EUPHORBIACEAE (D)
Dubardella H.J.Lam = Pyrenaria Blume THEACEAE (D)
Dubautia Gaudich. COMPOSITAE (D)
Dubois–Reymondia Karst. = Myoxanthus Poepp. & Endl. ORCHIDACEAE (M)
Duboisia Karst. (SUH) = Myoxanthus Poepp. & Endl. ORCHIDACEAE (M)
Duboisia R.Br. SOLANACEAE (D)
Duboscia Bocquet TILIACEAE (D)
Dubouzetia Pancher ex Brongn. & Gris ELAEOCARPACEAE (D)
Dubyaea DC. COMPOSITAE (D)
Duchassaingia Walp. = Erythrina L. LEGUMINOSAE–PAPILIONOIDEAE (D)
Duchesnea Sm. ROSACEAE (D)
Duchola Adans. = Omphalea L. EUPHORBIACEAE (D)
Duckea Maguire RAPATEACEAE (M)
Duckeanthus R.E.Fr. ANNONACEAE (D)
Duckeella Porto ORCHIDACEAE (M)
Duckeodendron Kuhlm. DUCKEODENDRACEAE (D)

Duckera F.A.Barkley (SUS) = Rhus L. ANACARDIACEAE (D)
Duckesia Cuatrec. HUMIRIACEAE (D)
Ducrosia Boiss. UMBELLIFERAE (D)
Dudleya Britton & Rose CRASSULACEAE (D)
Dufourea Bory ex Willd. = Tristicha Thouars PODOSTEMACEAE (D)
Dufrenoya Chatin SANTALACEAE (D)
Dugaldia (Cass.) Cass. COMPOSITAE (D)
Dugandiodendron Loz.–Contr. = Magnolia L. MAGNOLIACEAE (D)
Dugesia A.Gray COMPOSITAE (D)
Duggena Vahl (SUI) = Gonzalagunia Ruíz & Pav. RUBIACEAE (D)
Dugortia Scop. = Parinari Aubl. CHRYSOBALANACEAE (D)
Duguetia A.St.–Hil. ANNONACEAE (D)
Duhaldea DC. COMPOSITAE (D)
Duhamelia Pers. = Hamelia Jacq. RUBIACEAE (D)
Duidaea S.F.Blake COMPOSITAE (D)
Duidania Standl. RUBIACEAE (D)
Dukea Dwyer = Raritebe Wernham RUBIACEAE (D)
Dulacia Vell. OLACACEAE (D)
Dulcamara Moench = Solanum L. SOLANACEAE (D)
Dulichium Pers. CYPERACEAE (M)
Dulongia Kunth = Phyllonoma Willd. ex Schult. ESCALLONIACEAE (D)
Dumartroya Gaudich. = Trophis P.Browne MORACEAE (D)
Dumasia DC. LEGUMINOSAE–PAPILIONOIDEAE (D)
Dumoria A.Chev. = Tieghemella Pierre SAPOTACEAE (D)
Dunalia Kunth SOLANACEAE (D)
Dunalia Spreng. (SUH) = Lucya DC. RUBIACEAE (D)
Dunbaria Wight & Arn. LEGUMINOSAE–PAPILIONOIDEAE (D)
Dunnia Tutcher RUBIACEAE (D)
Dunniella Rauschert (SUS) = Aboriella Bennet URTICACEAE (D)
Dunstervillea Garay ORCHIDACEAE (M)
Duosperma Dayton ACANTHACEAE (D)
Duparquetia Baill. LEGUMINOSAE–CAESALPINIOIDEAE (D)
Duperrea Pierre ex Pit. RUBIACEAE (D)
Dupineta Raf. = Dissotis Benth. MELASTOMATACEAE (D)
Dupinia Scop. = Ternstroemia Mutis ex L.f. THEACEAE (D)
Dupontia R.Br. GRAMINEAE (M)
Durandea Planch. LINACEAE (D)
Durandeeldia Kuntze = Acidoton Sw. EUPHORBIACEAE (D)
Duranta L. VERBENACEAE (D)
Duravia (S.Watson) Greene = Polygonum L. POLYGONACEAE (D)
Duriala (R.H.Anderson) Ulbr. = Maireana Moq. CHENOPODIACEAE (D)
Durio Adans. BOMBACACEAE (D)
Duroia L.f. RUBIACEAE (D)
Durringtonia R.J.F.Hend. & Guymer RUBIACEAE (D)
Duseniella K.Schum. COMPOSITAE (D)
Dussia Krug & Urb. ex Taub. LEGUMINOSAE–PAPILIONOIDEAE (D)
Dutaillyea Baill. RUTACEAE (D)
Duthiastrum M.P.de Vos IRIDACEAE (M)
Duthiea Hack. GRAMINEAE (M)
Duthiella M.P.de Vos (SUH) = Duthiastrum M.P.de Vos IRIDACEAE (M)
Duvalia Bonpl. (SUH) = Hypocalyptus Thunb. LEGUMINOSAE–PAPILIONOIDEAE (D)
Duvalia Haw. ASCLEPIADACEAE (D)
Duvaliandra M.G.Gilbert ASCLEPIADACEAE (D)
Duvaliella F.Heim = Dipterocarpus C.F.Gaertn. DIPTEROCARPACEAE (D)
Duvaua Kunth = Schinus L. ANACARDIACEAE (D)

Duvaucellia Bowdich = Kohautia Cham. & Schltdl. RUBIACEAE (D)
Duvernoia E.Mey. ex Nees = Justicia L. ACANTHACEAE (D)
Duvigneaudia J.Léonard EUPHORBIACEAE (D)
Dyakia Christenson ORCHIDACEAE (M)
Dybowskia Stapf = Hyparrhenia Fourn. GRAMINEAE (M)
Dychotria Raf. = Psychotria L. RUBIACEAE (D)
Dyckia Schult.f. BROMELIACEAE (M)
Dyctiogramme C.Presl = Coniogramme Fée ADIANTACEAE (F)
Dyctiospora Reinw. ex Korth. = Hedyotis L. RUBIACEAE (D)
Dyera Hook.f. APOCYNACEAE (D)
Dyerophytum Kuntze PLUMBAGINACEAE (D)
Dymondia Compton COMPOSITAE (D)
Dypsidium Baill. = Neophloga Baill. PALMAE (M)
Dypsis Noronha ex Mart. PALMAE (M)
Dyschoriste Nees ACANTHACEAE (D)
Dyscritogyne R.M.King & H.Rob. COMPOSITAE (D)
Dyscritothamnus B.L.Rob. COMPOSITAE (D)
Dysoda Lour. = Serissa Comm. ex Juss. RUBIACEAE (D)
Dysodidendron Gardner = Saprosma Blume RUBIACEAE (D)
Dysodiopsis (A.Gray) Rydb. COMPOSITAE (D)
Dysolobium (Benth.) Prain LEGUMINOSAE–PAPILIONOIDEAE (D)
Dysophylla Blume = Pogostemon Desf. LABIATAE (D)
Dysopsis Baill. EUPHORBIACEAE (D)
Dysosma Woodson BERBERIDACEAE (D)
Dysosmia Miq. = Saprosma Blume RUBIACEAE (D)
Dysoxylum Blume MELIACEAE (D)
Dyspemptemorion Bremek. = Justicia L. ACANTHACEAE (D)
Dysphania R.Br. CHENOPODIACEAE (D)
Dyssochroma Miers = Markea Rich. SOLANACEAE (D)
Dyssodia Cav. COMPOSITAE (D)
Dystaenia Kitag. UMBELLIFERAE (D)
Dystovomita (Engl.) D'Arcy GUTTIFERAE (D)

Eadesia F.Muell. = Anthocercis Labill. SOLANACEAE (D)
Earina Lindl. ORCHIDACEAE (M)
Earleocassia Britton = Senna Mill. LEGUMINOSAE–CAESALPINIOIDEAE (D)
Eastwoodia Brandegee COMPOSITAE (D)
Eatonella A.Gray COMPOSITAE (D)
Eatonia Raf. = Panicum L. GRAMINEAE (M)
Eatoniopteris Bommer = Cyathea Sm. CYATHEACEAE (F)
Ebandoua Pellegr. = Jollydora Pierre ex Gilg CONNARACEAE (D)
Ebelia Rchb. = Diodia L. RUBIACEAE (D)
Ebenopsis Britton & Rose = Havardia Small LEGUMINOSAE–MIMOSOIDEAE (D)
Ebenus L. LEGUMINOSAE–PAPILIONOIDEAE (D)
Eberhardtia Lecomte SAPOTACEAE (D)
Eberlanzia Schwantes AIZOACEAE (D)
Ebermaiera Nees = Staurogyne Wall. ACANTHACEAE (D)
Ebingeria Chrtek & Křísa = Luzula DC. JUNCACEAE (M)
Ebnerella Buxb. (SUS) = Mammillaria Haw. CACTACEAE (D)
Ebracteola Dinter & Schwantes AIZOACEAE (D)
Eburopetalum Becc. = Anaxagorea A.St.-Hil. ANNONACEAE (D)
Eburophyton A.Heller = Cephalanthera Rich. ORCHIDACEAE (M)
Ecastaphyllum P.Browne = Dalbergia L.f. LEGUMINOSAE–PAPILIONOIDEAE (D)
Ecballium A.Rich. CUCURBITACEAE (D)
Ecbolium Kurz ACANTHACEAE (D)

Ecclinusa Mart. SAPOTACEAE (D)

Eccoilopus Steud. = Spodiopogon Trin. GRAMINEAE (M)

Eccoptocarpha Launert GRAMINEAE (M)

Eccremis Baker PHORMIACEAE (M)

Eccremocactus Britton & Rose = Weberocereus Britton & Rose CACTACEAE (D)

Eccremocarpus Ruíz & Pav. BIGNONIACEAE (D)

Eccremocereus Frič & Kreuz. (SUO) = Weberocereus Britton & Rose CACTACEAE (D)

Ecdeiocolea F.Muell. ECDEIOCOLEACEAE (M)

Ecdysanthera Hook. & Arn. APOCYNACEAE (D)

Echeandia Ortega ANTHERICACEAE (M)

Echeveria DC. CRASSULACEAE (D)

Echidnium Schott = Dracontium L. ARACEAE (M)

Echidnopsis Hook.f. ASCLEPIADACEAE (D)

Echinacanthus Nees ACANTHACEAE (D)

Echinacea Moench COMPOSITAE (D)

Echinalysium Trin. (SUS) = Elytrophorus P.Beauv. GRAMINEAE (M)

Echinanthus Cerv. = Tragus Haller GRAMINEAE (M)

Echinaria Fabr. (SUS) = Cenchrus L. GRAMINEAE (M)

Echinaria Desf. GRAMINEAE (M)

Echinocactus Fabr. (SUH) = Melocactus Link & Otto CACTACEAE (D)

Echinocactus Link & Otto CACTACEAE (D)

Echinocalyx Benth. = Sindora Miq. LEGUMINOSAE–CAESALPINIOIDEAE (D)

Echinocarpus Blume = Sloanea L. ELAEOCARPACEAE (D)

Echinocassia Britton & Rose = Senna Mill. LEGUMINOSAE–CAESALPINIOIDEAE (D)

Echinocereus Engelm. CACTACEAE (D)

Echinochlaena Spreng. (SUO) = Echinolaena Desv. GRAMINEAE (M)

Echinochloa P.Beauv. GRAMINEAE (M)

Echinocitrus Tanaka RUTACEAE (D)

Echinocodon D.Y.Hong CAMPANULACEAE (D)

Echinocoryne H.Rob. = Vernonia Schreb. COMPOSITAE (D)

Echinocroton F.Muell. = Mallotus Lour. EUPHORBIACEAE (D)

Echinocystis Torr. & A.Gray CUCURBITACEAE (D)

Echinodendrum A.Rich. = Scolosanthus Vahl RUBIACEAE (D)

Echinodorus Rich. ex Engelm. ALISMATACEAE (M)

Echinofossulocactus Lawr. = Echinocactus Link & Otto CACTACEAE (D)

Echinofossulocactus Britton & Rose (SUH) = Stenocactus (K.Schum.) A.W.Hill
   CACTACEAE (D)

Echinoglochin (A.Gray) Brand = Plagiobothrys Fisch. & C.A.Mey. BORAGINACEAE (D)

Echinoglossum Blume = Cleisostoma Blume ORCHIDACEAE (M)

Echinolaena Desv. GRAMINEAE (M)

Echinolobivia Y.Ito (SUI) = Echinopsis Zucc. CACTACEAE (D)

Echinomastus Britton & Rose = Sclerocactus Britton & Rose CACTACEAE (D)

Echinonyctanthus Lem. (SUS) = Echinopsis Zucc. CACTACEAE (D)

Echinopaepale Bremek. = Strobilanthes Blume ACANTHACEAE (D)

Echinopanax Decne. & Planch. = Oplopanax (Torr. & A.Gray) Miq. ARALIACEAE (D)

Echinopepon Naudin CUCURBITACEAE (D)

Echinophora L. UMBELLIFERAE (D)

Echinopogon P.Beauv. GRAMINEAE (M)

Echinops L. COMPOSITAE (D)

Echinopsilon Moq. = Bassia All. CHENOPODIACEAE (D)

Echinopsis Zucc. CACTACEAE (D)

Echinopterys A.Juss. MALPIGHIACEAE (D)

Echinorebutia Frič (SUI) = Rebutia K.Schum. CACTACEAE (D)

Echinosophora Nakai = Sophora L. LEGUMINOSAE–PAPILIONOIDEAE (D)

Echinospartum (Spach) Fourr. LEGUMINOSAE–PAPILIONOIDEAE (D)

Echinospermum Sw. ex Lehm. = Lappula Moench BORAGINACEAE (D)
Echinosphaera Sieb. ex Steud. = Ricinocarpos Desf. EUPHORBIACEAE (D)
Echinostephia (Diels) Domin MENISPERMACEAE (D)
Echinothamnus Engl. = Adenia Forssk. PASSIFLORACEAE (D)
Echinus Lour. = Mallotus Lour. EUPHORBIACEAE (D)
Echinus L.Bolus (SUH) = Braunsia Schwantes AIZOACEAE (D)
Echiochilon Desf. BORAGINACEAE (D)
Echiochilopsis Caball. BORAGINACEAE (D)
Echioides Ortega (SUH) = Arnebia Forssk. BORAGINACEAE (D)
Echiopsis Rchb. = Lobostemon Lehm. BORAGINACEAE (D)
Echiostachys Levyns = Lobostemon Lehm. BORAGINACEAE (D)
Echites P.Browne APOCYNACEAE (D)
Echium L. BORAGINACEAE (D)
Eckartia Rchb.f. = Peristeria Hook. ORCHIDACEAE (M)
Ecklonea Steud. = Trianoptiles Fenzl ex Endl. CYPERACEAE (M)
Eclipta L. COMPOSITAE (D)
Ecliptostelma Brandegee ASCLEPIADACEAE (D)
Ecpoma K.Schum. RUBIACEAE (D)
Ectadiopsis Benth. ASCLEPIADACEAE (D)
Ectadium E.Mey. ASCLEPIADACEAE (D)
Ecteinanthus T.Anderson = Isoglossa Oerst. ACANTHACEAE (D)
Ectinocladus Benth. = Alafia Thouars APOCYNACEAE (D)
Ectopopterys W.R.Anderson MALPIGHIACEAE (D)
Ectosperma Swallen (SUH) = Swallenia Soderstr. & H.F.Decker GRAMINEAE (M)
Ectozoma Miers = Markea Rich. SOLANACEAE (D)
Ectrosia R.Br. GRAMINEAE (M)
Ectrosiopsis (Ohwi) Jansen GRAMINEAE (M)
Edanyoa Copel. = Bolbitis Schott LOMARIOPSIDACEAE (F)
Edbakeria R.Vig. = Pearsonia Dümmer LEGUMINOSAE–PAPILIONOIDEAE (D)
Eddya Torr. & A.Gray = Tiquilia Pers. BORAGINACEAE (D)
Edechi Loefl. = Guettarda L. RUBIACEAE (D)
Edgaria C.B.Clarke CUCURBITACEAE (D)
Edgeworthia Meisn. THYMELAEACEAE (D)
Edgeworthia Falc. (SUH) = Sideroxylon L. SAPOTACEAE (D)
Edisonia Small = Matelea Aubl. ASCLEPIADACEAE (D)
Edithcolea N.E.Br. ASCLEPIADACEAE (D)
Edithea Standl. = Omiltemia Standl. RUBIACEAE (D)
Edmondia Cogn. (SUH) = Calycophysum Karst. & Triana CUCURBITACEAE (D)
Edmondia Cass. COMPOSITAE (D)
Edraianthus (A.DC.) DC. CAMPANULACEAE (D)
Edrastima Raf. = Oldenlandia L. RUBIACEAE (D)
Edwardsia Salisb. = Sophora L. LEGUMINOSAE–PAPILIONOIDEAE (D)
Edwinia A.Heller = Jamesia Torr. & A.Gray HYDRANGEACEAE (D)
Efossus Orcutt (SUO) = Stenocactus (K.Schum.) A.W.Hill CACTACEAE (D)
Efulensia C.H.Wright PASSIFLORACEAE (D)
Eganthus Tiegh. = Minquartia Aubl. OLACACEAE (D)
Egenolfia Schott = Bolbitis Schott LOMARIOPSIDACEAE (F)
Egeria Planch. HYDROCHARITACEAE (M)
Eggelingia Summerh. ORCHIDACEAE (M)
Egleria G.Eiten CYPERACEAE (M)
Eglerodendron Aubrév. & Pellegr. = Pouteria Aubl. SAPOTACEAE (D)
Egletes Cass. COMPOSITAE (D)
Ehrenbergia Mart. (SUH) = Kallstroemia Scop. ZYGOPHYLLACEAE (D)
Ehrenbergia Spreng. = Amaioua Aubl. RUBIACEAE (D)
Ehretia P.Browne BORAGINACEAE (D)

Ehrharta Thunb. GRAMINEAE (M)
Ehrhartia Weber (SUS) = Leersia Sw. GRAMINEAE (M)
Eichhornia Kunth PONTEDERIACEAE (M)
Eichlerago Carrick = Prostanthera Labill. LABIATAE (D)
Eichleria Hartog (SUH) = Manilkara Adans. SAPOTACEAE (D)
Eichleria Progel = Rourea Aubl. CONNARACEAE (D)
Eichlerina Tiegh. = Struthanthus Mart. LORANTHACEAE (D)
Eichlerodendron Briq. = Xylosma G.Forst. FLACOURTIACEAE (D)
Eigia Soják CRUCIFERAE (D)
Einadia Raf. CHENOPODIACEAE (D)
Einsteinia Ducke = Kutchubaea Fisch. ex DC. RUBIACEAE (D)
Eionitis Bremek. = Oldenlandia L. RUBIACEAE (D)
Eirmocephala H.Rob. = Vernonia Schreb. COMPOSITAE (D)
Eisocreochiton Quisumb. & Merr. = Creochiton Blume MELASTOMATACEAE (D)
Eitenia R.M.King & H.Rob. COMPOSITAE (D)
Eizia Standl. RUBIACEAE (D)
Ekebergia Sparrm. MELIACEAE (D)
Ekmania Gleason COMPOSITAE (D)
Ekmanianthe Urb. BIGNONIACEAE (D)
Ekmaniocharis Urb. MELASTOMATACEAE (D)
Ekmanochloa Hitchc. GRAMINEAE (M)
Elachanthemum Y.Ling & Y.R.Ling = Stilpnolepis I.M.Kraschen. COMPOSITAE (D)
Elachanthera F.Muell. = Myrsiphyllum Willd. ASPARAGACEAE (M)
Elachanthus F.Muell. COMPOSITAE (D)
Elachocroton F.Muell. = Sebastiania Spreng. EUPHORBIACEAE (D)
Elacholoma F.Muell. & Tate SCROPHULARIACEAE (D)
Elachyptera A.C.Sm. CELASTRACEAE (D)
Elaeagia Wedd. RUBIACEAE (D)
Elaeagnus L. ELAEAGNACEAE (D)
Elaeis Jacq. PALMAE (M)
Elaeocarpus L. ELAEOCARPACEAE (D)
Elaeococca Comm. ex Juss. = Vernicia Lour. EUPHORBIACEAE (D)
Elaeodendron J.F.Jacq. CELASTRACEAE (D)
Elaeoluma Baill. SAPOTACEAE (D)
Elaeophora Ducke EUPHORBIACEAE (D)
Elaeophorbia Stapf EUPHORBIACEAE (D)
Elaeopleurum Korovin = Seseli L. UMBELLIFERAE (D)
Elaeoselinum W.D.J.Koch ex DC. UMBELLIFERAE (D)
Elaeosticta Fenzl UMBELLIFERAE (D)
Elaphandra Strother COMPOSITAE (D)
Elaphanthera N.Hallé SANTALACEAE (D)
Elaphoglossum Schott ex J.Sm. LOMARIOPSIDACEAE (F)
Elaphrium Jacq. = Bursera Jacq. ex L. BURSERACEAE (D)
Elasis D.R.Hunt COMMELINACEAE (M)
Elasmatium Dulac = Goodyera R.Br. ORCHIDACEAE (M)
Elaterioides Kuntze = Elateriospermum Blume EUPHORBIACEAE (D)
Elateriopsis Ernst CUCURBITACEAE (D)
Elateriospermum Blume EUPHORBIACEAE (D)
Elaterispermum Rchb. = Elateriospermum Blume EUPHORBIACEAE (D)
Elaterium Mill. = Ecballium A.Rich. CUCURBITACEAE (D)
Elaterium Jacq. (SUH) = Rytidostylis Hook. & Arn. CUCURBITACEAE (D)
Elatine L. ELATINACEAE (D)
Elatinoides (Chavannes) Wettst. SCROPHULARIACEAE (D)
Elatostema J.R.Forst. & G.Forst. URTICACEAE (D)
Elatostematoides B.L.Rob. = Elatostema J.R.Forst. & G.Forst. URTICACEAE (D)

Elattospermum Soler. = Breonia A.Rich. ex DC. RUBIACEAE (D)
Elattostachys Radlk. SAPINDACEAE (D)
Elburzia Hedge CRUCIFERAE (D)
Elcaja Forssk. = Trichilia P.Browne MELIACEAE (D)
Elcomarhiza Barb.Rodr. = Marsdenia R.Br. ASCLEPIADACEAE (D)
Electra Panz. = Schismus P.Beauv. GRAMINEAE (M)
Elegia L. RESTIONACEAE (M)
Eleiodoxa (Becc.) Burret PALMAE (M)
Eleiosina Raf. = Spiraea L. ROSACEAE (D)
Eleiotis DC. LEGUMINOSAE–PAPILIONOIDEAE (D)
Elengi Adans. = Mimusops L. SAPOTACEAE (D)
Eleocharis R.Br. CYPERACEAE (M)
Eleogiton Link (SUH) = Isolepis R.Br. CYPERACEAE (M)
Eleorchis F.Maek. ORCHIDACEAE (M)
Elephantomene Barneby & Krukoff MENISPERMACEAE (D)
Elephantopus L. COMPOSITAE (D)
Elephantorrhiza Benth. LEGUMINOSAE–MIMOSOIDEAE (D)
Elephantusia Willd. = Phytelephas Ruíz & Pav. PALMAE (M)
Elettaria Maton ZINGIBERACEAE (M)
Elettariopsis Baker ZINGIBERACEAE (M)
Eleusine Gaertn. GRAMINEAE (M)
Eleutharrhena Forman MENISPERMACEAE (D)
Eleutherandra Slooten FLACOURTIACEAE (D)
Eleutheranthera Poit. ex Bosc COMPOSITAE (D)
Eleutheranthus K.Schum. (SUO) = Eleuthranthes F.Muell. RUBIACEAE (D)
Eleutherine Herb. IRIDACEAE (M)
Eleutherococcus Maxim. ARALIACEAE (D)
Eleutheropetalum H.A.Wendl. = Chamaedorea Willd. PALMAE (M)
Eleutherospermum K.Koch UMBELLIFERAE (D)
Eleutherostemon Herz. = Diogenesia Sleumer ERICACEAE (D)
Eleutherostigma Pax & K.Hoffm. EUPHORBIACEAE (D)
Eleutherostylis Burret TILIACEAE (D)
Eleuthranthes F.Muell. RUBIACEAE (D)
Eliaea Cambess. (SUO) = Eliea Cambess. GUTTIFERAE (D)
Eliea Cambess. GUTTIFERAE (D)
Eligmocarpus Capuron LEGUMINOSAE–CAESALPINIOIDEAE (D)
Elingamita G.T.S.Baylis MYRSINACEAE (D)
Elionurus Humb. & Bonpl. ex Willd. GRAMINEAE (M)
Elisena Herb. AMARYLLIDACEAE (M)
Elisia Milano = Brugmansia Pers. SOLANACEAE (D)
Elissarrhena Miers = Anomospermum Miers MENISPERMACEAE (D)
Elizabetha Schomb. ex Benth. LEGUMINOSAE–CAESALPINIOIDEAE (D)
Elizaldia Willk. BORAGINACEAE (D)
Elleanthus C.Presl ORCHIDACEAE (M)
Elleimataenia Kozo–Pol. = Osmorhiza Raf. UMBELLIFERAE (D)
Ellenbergia Cuatrec. COMPOSITAE (D)
Ellertonia Wight = Kamettia Kostel. APOCYNACEAE (D)
Elliottia Mühl. ex Elliott ERICACEAE (D)
Ellipanthus Hook.f. CONNARACEAE (D)
Ellipeia Hook.f. & Thomson ANNONACEAE (D)
Ellipeiopsis R.E.Fr. ANNONACEAE (D)
Ellisia L. HYDROPHYLLACEAE (D)
Ellisiophyllum Maxim. SCROPHULARIACEAE (D)
Ellobium Lilja = Fuchsia L. ONAGRACEAE (D)
Ellobocarpus Kaulf. = Ceratopteris Brongn. PARKERIACEAE (F)

Ellobum Blume = Didissandra C.B.Clarke GESNERIACEAE (D)
Elmera Rydb. SAXIFRAGACEAE (D)
Elmeria Ridl. = Alpinia Roxb. ZINGIBERACEAE (M)
Elmerrillia Dandy MAGNOLIACEAE (D)
Elodea Michx. HYDROCHARITACEAE (M)
Eloyella Ortiz Vald. = Phymatidium Lindl. ORCHIDACEAE (M)
Elsholtzia Willd. LABIATAE (D)
Elsiea F.M.Leighton = Ornithogalum L. HYACINTHACEAE (M)
Eltroplectris Raf. = Stenorrhynchos Rich. ex Spreng. ORCHIDACEAE (M)
Eluteria Steud. = Croton L. EUPHORBIACEAE (D)
Elutheria M.Roem. (SUH) = Swietenia Jacq. MELIACEAE (D)
Elutheria P.Browne = Guarea L. MELIACEAE (D)
Elvasia DC. OCHNACEAE (D)
Elvira Cass. = Delilia Spreng. COMPOSITAE (D)
Elymandra Stapf GRAMINEAE (M)
Elymus Mitch. (SUH) = Zizania L. GRAMINEAE (M)
Elymus L. GRAMINEAE (M)
Elyna Schrad. = Kobresia Willd. CYPERACEAE (M)
Elynanthus Nees = Tetraria P.Beauv. CYPERACEAE (M)
Elyonurus Willd. (SUO) = Elionurus Humb. & Bonpl. ex Willd. GRAMINEAE (M)
Elythranthera (Endl.) A.S.George ORCHIDACEAE (M)
Elytranthe (Blume) Blume LORANTHACEAE (D)
Elytraria Michx. ACANTHACEAE (D)
Elytrigia Desv. = Elymus L. GRAMINEAE (M)
Elytroblepharum (Steud.) Schltdl. = Digitaria Haller GRAMINEAE (M)
Elytropappus Cass. COMPOSITAE (D)
Elytrophorus P.Beauv. GRAMINEAE (M)
Elytropus Müll.Arg. APOCYNACEAE (D)
Elytrostachys McClure GRAMINEAE (M)
Embadium J.M.Black BORAGINACEAE (D)
Embelia Burm.f. MYRSINACEAE (D)
Embergeria Boulos = Sonchus L. COMPOSITAE (D)
Emblemantha B.C.Stone MYRSINACEAE (D)
Emblica Gaertn. = Phyllanthus L. EUPHORBIACEAE (D)
Emblingia F.Muell. EMBLINGIACEAE (D)
Embolanthera Merr. HAMAMELIDACEAE (D)
Embothrium J.R.Forst. & G.Forst. PROTEACEAE (D)
Embreea Dodson ORCHIDACEAE (M)
Embryogonia Blume = Combretum Loefl. COMBRETACEAE (D)
Emelianthe Danser LORANTHACEAE (D)
Emex Neck. ex Campdera POLYGONACEAE (D)
Emicocarpus K.Schum. & Schltr. ASCLEPIADACEAE (D)
Emilia (Cass.) Cass. COMPOSITAE (D)
Emiliella S.Moore COMPOSITAE (D)
Emiliomarcetia T.Durand & H.Durand = Trichoscypha Hook.f. ANACARDIACEAE (D)
Eminia Taub. LEGUMINOSAE–PAPILIONOIDEAE (D)
Eminium (Blume) Schott ARACEAE (M)
Emmenanthe Benth. HYDROPHYLLACEAE (D)
Emmenopterys Oliv. RUBIACEAE (D)
Emmenopteryx Dalla Torre & Harms (SUO) = Emmenopterys Oliv. RUBIACEAE (D)
Emmenosperma F.Muell. RHAMNACEAE (D)
Emmeorhiza Pohl ex Endl. RUBIACEAE (D)
Emmotum Desv. ex Ham. ICACINACEAE (D)
Emodiopteris Ching & S.K.Wu = Dennstaedtia Bernh. DENNSTAEDTIACEAE (F)
Emorya Torr. BUDDLEJACEAE (D)

Empedoclesia Sleumer = Orthaea Klotzsch ERICACEAE (D)
Empetrum L. EMPETRACEAE (D)
Emplectanthus N.E.Br. ASCLEPIADACEAE (D)
Emplectocladus Torr. = Prunus L. ROSACEAE (D)
Empleuridium Sond. & Harv. RUTACEAE (D)
Empleurum Aiton RUTACEAE (D)
Empodisma L.A.S.Johnson & D.F.Cutler RESTIONACEAE (M)
Empodium Salisb. HYPOXIDACEAE (M)
Empogona Hook.f. = Tricalysia A.Rich. ex DC. RUBIACEAE (D)
Empusa Lindl. = Liparis Rich. ORCHIDACEAE (M)
Empusaria Rchb.f. = Liparis Rich. ORCHIDACEAE (M)
Emularia Raf. = Justicia L. ACANTHACEAE (D)
Emurtia Raf. = Eugenia L. MYRTACEAE (D)
Enaimon Raf. = Olea L. OLEACEAE (D)
Enallagma (Miers) Baill. = Amphitecna Miers BIGNONIACEAE (D)
Enantia Oliv. (SUH) = Annickia Setten & Maas ANNONACEAE (D)
Enantiophylla J.M.Coult. & Rose UMBELLIFERAE (D)
Enantiosparton K.Koch = Genista L. LEGUMINOSAE–PAPILIONOIDEAE (D)
Enarganthe N.E.Br. AIZOACEAE (D)
Enarthrocarpus Labill. CRUCIFERAE (D)
Enaulophyton Steenis MELASTOMATACEAE (D)
Encelia Adans. COMPOSITAE (D)
Enceliopsis (A.Gray) A.Nelson COMPOSITAE (D)
Encephalartos Lehm. ZAMIACEAE (G)
Encephalocarpus A.Berger = Pelecyphora Ehrenb. CACTACEAE (D)
Encephalosphaera Lindau ACANTHACEAE (D)
Encheiridion Summerh. = Microcoelia Lindl. ORCHIDACEAE (M)
Enchidium Jack = Trigonostemon Blume EUPHORBIACEAE (D)
Encholirium Mart. ex Schult.f. BROMELIACEAE (M)
Enchosanthera King & Stapf ex Guillaumin = Creochiton Blume
   MELASTOMATACEAE (D)
Enchylaena R.Br. CHENOPODIACEAE (D)
Enchysia C.Presl = Laurentia Adans. CAMPANULACEAE (D)
Encliandra Zucc. = Fuchsia L. ONAGRACEAE (D)
Encopa Griseb. (SUH) = Encopella Pennell SCROPHULARIACEAE (D)
Encopea C.Presl = Faramea Aubl. RUBIACEAE (D)
Encopella Pennell SCROPHULARIACEAE (D)
Encyclia Hook. ORCHIDACEAE (M)
Endadenium L.C.Leach EUPHORBIACEAE (D)
Endallex Raf. (SUS) = Phalaris L. GRAMINEAE (M)
Endeisa Raf. = Dendrobium Sw. ORCHIDACEAE (M)
Endemal Pritz. (SUO) = Eudema Humb. & Bonpl. CRUCIFERAE (D)
Endertia Steenis & de Wit LEGUMINOSAE–CAESALPINIOIDEAE (D)
Endesmia R.Br. = Eucalyptus L'Hér. MYRTACEAE (D)
Endiandra R.Br. LAURACEAE (D)
Endlichera C.Presl = Emmeorhiza Pohl ex Endl. RUBIACEAE (D)
Endlicheria Nees LAURACEAE (D)
Endocaulos C.Cusset PODOSTEMACEAE (D)
Endocellion Turcz. ex Herder COMPOSITAE (D)
Endocomia W.J.de Wilde MYRISTICACEAE (D)
Endodesmia Benth. GUTTIFERAE (D)
Endodia Raf. = Leersia Sw. GRAMINEAE (M)
Endogonia Turcz. = Trigonotis Steven BORAGINACEAE (D)
Endoisila Raf. = Euphorbia L. EUPHORBIACEAE (D)
Endolasia Turcz. = Manettia L. RUBIACEAE (D)

Endolithodes Bartl. (SUI) = Retiniphyllum Bonpl. RUBIACEAE (D)
Endoloma Raf. = Amphilophium Kunth BIGNONIACEAE (D)
Endomallus Gagnep. = Cajanus DC. LEGUMINOSAE–PAPILIONOIDEAE (D)
Endonema A.Juss. PENAEACEAE (D)
Endopappus Sch.Bip. COMPOSITAE (D)
Endopleura Cuatrec. HUMIRIACEAE (D)
Endopogon Raf. (SUH) = Diodia L. RUBIACEAE (D)
Endopogon Nees (SUH) = Strobilanthes Blume ACANTHACEAE (D)
Endopogon Nees = Mecardonia Ruíz & Pav. SCROPHULARIACEAE (D)
Endosamara Geesink LEGUMINOSAE–PAPILIONOIDEAE (D)
Endosiphon T.Anderson ex Benth. = Ruellia L. ACANTHACEAE (D)
Endospermum Benth. EUPHORBIACEAE (D)
Endostemon N.E.Br. LABIATAE (D)
Endotricha Aubrév. & Pellegr. (SUH) = Aubregrinia Heine SAPOTACEAE (D)
Endresiella Schltr. ORCHIDACEAE (M)
Endressia J.Gay UMBELLIFERAE (D)
Endusa Miers = Minquartia Aubl. OLACACEAE (D)
Endymion Dumort. = Hyacinthoides Heist. ex Fabr. HYACINTHACEAE (M)
Enemion Raf. RANUNCULACEAE (D)
Enetophyton Nieuwl. = Utricularia L. LENTIBULARIACEAE (D)
Engelhardtia Lesch. ex Blume JUGLANDACEAE (D)
Engelmannia Klotzsch (SUH) = Croton L. EUPHORBIACEAE (D)
Engelmannia A.Gray ex Nutt. COMPOSITAE (D)
Englerastrum Briq. LABIATAE (D)
Englerella Pierre = Pouteria Aubl. SAPOTACEAE (D)
Engleria O.Hoffm. COMPOSITAE (D)
Englerina Tiegh. LORANTHACEAE (D)
Englerocharis Muschler CRUCIFERAE (D)
Englerodaphne Gilg THYMELAEACEAE (D)
Englerodendron Harms LEGUMINOSAE–CAESALPINIOIDEAE (D)
Englerodoxa Hoerold = Ceratostema Juss. ERICACEAE (D)
Englerophoenix Kuntze = Maximiliana Mart. PALMAE (M)
Englerophytum K.Krause SAPOTACEAE (D)
Engysiphon G.J.Lewis = Geissorhiza Ker–Gawl. IRIDACEAE (M)
Enhalus Rich. HYDROCHARITACEAE (M)
Enhydrias Ridl. HYDROCHARITACEAE (M)
Enicosanthellum Ban = Disepalum Hook.f. ANNONACEAE (D)
Enicosanthum Becc. ANNONACEAE (D)
Enicostema Blume GENTIANACEAE (D)
Enkianthus Lour. ERICACEAE (D)
Enkleia Griff. THYMELAEACEAE (D)
Enkylia Griff. = Gynostemma Blume CUCURBITACEAE (D)
Enkylista Hook.f. (SUO) = Calycophyllum DC. RUBIACEAE (D)
Ennealophus N.E.Br. IRIDACEAE (M)
Enneapogon Desv. ex P.Beauv. GRAMINEAE (M)
Enneastemon Exell = Monanthotaxis Baill. ANNONACEAE (D)
Enneatypus Herz. POLYGONACEAE (D)
Enodium Pers. ex Gaudin (SUS) = Molinia Schrank GRAMINEAE (M)
Enosanthes Cunn. ex Schauer (SUI) = Homoranthus A.Cunn. ex Schauer
    MYRTACEAE (D)
Enothrea Raf. = Octomeria R.Br. ORCHIDACEAE (M)
Enriquebeltrania Rzed. EUPHORBIACEAE (D)
Ensete Horan. MUSACEAE (M)
Enskide Raf. = Utricularia L. LENTIBULARIACEAE (D)
Enslenia Nutt. = Ampelamus Raf. ASCLEPIADACEAE (D)

Entada Adans. LEGUMINOSAE–MIMOSOIDEAE (D)
Entadopsis Britton = Entada Adans. LEGUMINOSAE–MIMOSOIDEAE (D)
Entandrophragma C.E.C.Fisch. MELIACEAE (D)
Entaticus Gray = Habenaria Willd. ORCHIDACEAE (M)
Entelea R.Br. TILIACEAE (D)
Enterolobium Mart. LEGUMINOSAE–MIMOSOIDEAE (D)
Enteropogon Nees GRAMINEAE (M)
Enterosora Baker = Glyphotaenium J.Sm. GRAMMITIDACEAE (F)
Enterospermum Hiern = Tarenna Gaertn. RUBIACEAE (D)
Entolasia Stapf GRAMINEAE (M)
Entomophobia de Vogel ORCHIDACEAE (M)
Entoplocamia Stapf GRAMINEAE (M)
Entrecasteauxia Montrouz. = Duboisia R.Br. SOLANACEAE (D)
Enydra Lour. COMPOSITAE (D)
Eomatucana F.Ritter = Oreocereus (A.Berger) Riccob. CACTACEAE (D)
Eomecon Hance PAPAVERACEAE (D)
Eopepon Naudin = Trichosanthes L. CUCURBITACEAE (D)
Eora O.F.Cook = Rhopalostylis H.A.Wendl. & Drude PALMAE (M)
Eosanthe Urb. RUBIACEAE (D)
Epacris Cav. EPACRIDACEAE (D)
Epaltes Cass. COMPOSITAE (D)
Eparmatostigma Garay ORCHIDACEAE (M)
Eperua Aubl. LEGUMINOSAE–CAESALPINIOIDEAE (D)
Ephaiola Raf. = Acnistus Schott SOLANACEAE (D)
Ephebopogon Steud. (SUI) = Microstegium Nees GRAMINEAE (M)
Ephedra L. EPHEDRACEAE (G)
Ephedranthus S.Moore ANNONACEAE (D)
Ephemerantha Summerh. = Flickingeria A.D.Hawkes ORCHIDACEAE (M)
Ephippiandra Decne. MONIMIACEAE (D)
Ephippianthus Rchb.f. ORCHIDACEAE (M)
Ephippiocarpa Markgr. = Callichilia Stapf APOCYNACEAE (D)
Ephippium Blume = Bulbophyllum Thouars ORCHIDACEAE (M)
Epiblastus Schltr. ORCHIDACEAE (M)
Epiblema R.Br. ORCHIDACEAE (M)
Epicampes C.Presl = Muhlenbergia Schreb. GRAMINEAE (M)
Epicarpurus Blume = Streblus Lour. MORACEAE (D)
Epicharis Blume = Dysoxylum Blume MELIACEAE (D)
Epicion Small = Metastelma R.Br. ASCLEPIADACEAE (D)
Epicladium Small = Epidendrum L. ORCHIDACEAE (M)
Epiclastopelma Lindau ACANTHACEAE (D)
Epicoila Raf. = Tristerix Mart. LORANTHACEAE (D)
Epicranthes Blume = Bulbophyllum Thouars ORCHIDACEAE (M)
Epicrianthus Blume = Bulbophyllum Thouars ORCHIDACEAE (M)
Epidanthus L.O.Williams ORCHIDACEAE (M)
Epidendroides Sol. = Myrmecodia Jack RUBIACEAE (D)
Epidendropsis Garay & Dunst. ORCHIDACEAE (M)
Epidendrum L. ORCHIDACEAE (M)
Epidryos Maguire RAPATEACEAE (M)
Epifagus Nutt. SCROPHULARIACEAE (D)
Epigaea L. ERICACEAE (D)
Epigeneium Gagnep. ORCHIDACEAE (M)
Epigynum Wight APOCYNACEAE (D)
Epilasia (Bunge) Benth. COMPOSITAE (D)
Epilobium L. ONAGRACEAE (D)
Epiluma Baill. = Pichonia Pierre SAPOTACEAE (D)

Epilyna Schltr. = Elleanthus C.Presl ORCHIDACEAE (M)

Epimedium L. BERBERIDACEAE (D)

Epimeredi Adans. = Anisomeles R.Br. LABIATAE (D)

Epinetrum Hiern = Albertisia Becc. MENISPERMACEAE (D)

Epipactis Zinn ORCHIDACEAE (M)

Epipetrum Phil. DIOSCOREACEAE (M)

Epiphora Lindl. = Polystachya Hook. ORCHIDACEAE (M)

Epiphyllanthus A.Berger = Schlumbergera Lem. CACTACEAE (D)

Epiphyllopsis Backeb. & F.M.Knuth = Hatiora Britton & Rose CACTACEAE (D)

Epiphyllum Haw. CACTACEAE (D)

Epiphyllum Pfeiff. (SUH) = Schlumbergera Lem. CACTACEAE (D)

Epiphyton Maguire (SUH) = Epidryos Maguire RAPATEACEAE (M)

Epipogium Borkh. ORCHIDACEAE (M)

Epipogon S.G.Gmel. = Epipogium Borkh. ORCHIDACEAE (M)

Epipremnopsis Engl. = Amydrium Schott ARACEAE (M)

Epipremnum Schott ARACEAE (M)

Epiprinus Griff. EUPHORBIACEAE (D)

Epirhizanthes Blume POLYGALACEAE (D)

Epischoenus C.B.Clarke CYPERACEAE (M)

Episcia Mart. GESNERIACEAE (D)

Episcothamnus H.Rob. COMPOSITAE (D)

Episteira Raf. = Glochidion J.R.Forst. & G.Forst. EUPHORBIACEAE (D)

Epistemma D.V.Field & J.B.Hall ASCLEPIADACEAE (D)

Epistephium Kunth ORCHIDACEAE (M)

Epistylium Sw. = Phyllanthus L. EUPHORBIACEAE (D)

Episyzygium Suess. & A.Ludw. = Eugenia L. MYRTACEAE (D)

Epitaberna K.Schum. (SUH) = Heinsia DC. RUBIACEAE (D)

Epithelantha F.A.C.Weber ex Britton & Rose CACTACEAE (D)

Epithema Blume GESNERIACEAE (D)

Epithinia Jack = Scyphiphora C.F.Gaertn. RUBIACEAE (D)

Epitriche Turcz. COMPOSITAE (D)

Epixiphium (A.Gray) Munz SCROPHULARIACEAE (D)

Eplateia Raf. = Acnistus Schott SOLANACEAE (D)

Epleienda Raf. = Eugenia L. MYRTACEAE (D)

Eplejenda Post & Kuntze = Eugenia L. MYRTACEAE (D)

Eplingia L.O.Williams = Trichostema L. LABIATAE (D)

Epurga Fourr. = Euphorbia L. EUPHORBIACEAE (D)

Equisetum L. EQUISETACEAE (FA)

Eraeliss Forssk. = Andrachne L. EUPHORBIACEAE (D)

Eragrostiella Bor GRAMINEAE (M)

Eragrostis Wolf GRAMINEAE (M)

Eranthemum L. ACANTHACEAE (D)

Eranthis Salisb. RANUNCULACEAE (D)

Erasma R.Br. = Lonchostoma Wikstr. BRUNIACEAE (D)

Erato DC. COMPOSITAE (D)

Erblichia Seem. TURNERACEAE (D)

Ercilla A.Juss. PHYTOLACCACEAE (D)

Erdisia Britton & Rose = Corryocactus Britton & Rose CACTACEAE (D)

Erechtites Raf. COMPOSITAE (D)

Ereicotis (DC.) Kuntze = Arcytophyllum Willd. ex Schult. & Schult.f. RUBIACEAE (D)

Eremaea Lindl. MYRTACEAE (D)

Eremaeopsis Kuntze = Eremaea Lindl. MYRTACEAE (D)

Eremalche Greene MALVACEAE (D)

Eremanthus Less. COMPOSITAE (D)

Eremia D.Don ERICACEAE (D)

Eremiastrum A.Gray = Monoptilon Torr. & A.Gray COMPOSITAE (D)
Eremiella Compton ERICACEAE (D)
Eremiopsis N.E.Br. = Eremia D.Don ERICACEAE (D)
Eremitis Döll GRAMINEAE (M)
Eremobium Boiss. CRUCIFERAE (D)
Eremoblastus Botsch. CRUCIFERAE (D)
Eremocarpus Benth. EUPHORBIACEAE (D)
Eremocarya Greene = Cryptantha G.Don BORAGINACEAE (D)
Eremocaulon Soderstr. & Londoño GRAMINEAE (M)
Eremocharis Phil. UMBELLIFERAE (D)
Eremochion Gilli = Salsola L. CHENOPODIACEAE (D)
Eremochlamys Peter = Tricholaena Schrad. ex Schult. & Schult.f. GRAMINEAE (M)
Eremochloa Büse GRAMINEAE (M)
Eremochloe S.Watson (SUH) = Blepharidachne Hack. GRAMINEAE (M)
Eremocitrus Swingle RUTACEAE (D)
Eremocrinum M.E.Jones ANTHERICACEAE (M)
Eremodaucus Bunge UMBELLIFERAE (D)
Eremodraba O.E.Schulz CRUCIFERAE (D)
Eremogeton Standl. & L.O.Williams SCROPHULARIACEAE (D)
Eremogone Fenzl = Arenaria L. CARYOPHYLLACEAE (D)
Eremolaena Baill. SARCOLAENACEAE (D)
Eremolepis Griseb. = Antidaphne Poepp. & Endl. EREMOLEPIDACEAE (D)
Eremolimon Lincz. = Limonium Mill. PLUMBAGINACEAE (D)
Eremoluma Baill. = Pouteria Aubl. SAPOTACEAE (D)
Eremomastax Lindau ACANTHACEAE (D)
Eremopanax Baill. = Arthrophyllum Blume ARALIACEAE (D)
Eremophea Paul G.Wilson CHENOPODIACEAE (D)
Eremophila R.Br. MYOPORACEAE (D)
Eremophyton Bég. CRUCIFERAE (D)
Eremopoa Roshev. GRAMINEAE (M)
Eremopodium Trevis. = Asplenium L. ASPLENIACEAE (F)
Eremopogon Stapf = Dichanthium Willemet GRAMINEAE (M)
Eremopyrum (Ledeb.) Jaub. & Spach GRAMINEAE (M)
Eremopyxis Baill. = Baeckea L. MYRTACEAE (D)
Eremosemium Greene = Grayia Hook. & Arn. CHENOPODIACEAE (D)
Eremosis (DC.) Gleason COMPOSITAE (D)
Eremosparton Fisch. & C.A.Mey. LEGUMINOSAE–PAPILIONOIDEAE (D)
Eremospatha (G.Mann & H.A.Wendl.) H.A.Wendl. PALMAE (M)
Eremosperma Chiov. = Hewittia Wight & Arn. CONVOLVULACEAE (D)
Eremostachys Bunge LABIATAE (D)
Eremosyne Endl. EREMOSYNACEAE (D)
Eremothamnus O.Hoffm. COMPOSITAE (D)
Eremotropa Andres = Monotropastrum Andres ERICACEAE (D)
Eremurus M.Bieb. ASPHODELACEAE (M)
Erepsia N.E.Br. AIZOACEAE (D)
Eresimus Raf. = Cephalanthus L. RUBIACEAE (D)
Ergocarpon C.C.Towns. UMBELLIFERAE (D)
Eria Lindl. ORCHIDACEAE (M)
Eriachaenium Sch.Bip. COMPOSITAE (D)
Eriachne Phil. (SUH) = Digitaria Haller GRAMINEAE (M)
Eriachne R.Br. GRAMINEAE (M)
Eriadenia Miers = Mandevilla Lindl. APOCYNACEAE (D)
Eriandra P.Royen & Steenis POLYGALACEAE (D)
Eriandrostachys Baill. = Macphersonia Blume SAPINDACEAE (D)
Erianthecium Parodi GRAMINEAE (M)

Erianthemum Tiegh. LORANTHACEAE (D)
Erianthera Benth. (SUH) = Alajja Ikonn. LABIATAE (D)
Erianthus Michx. = Saccharum L. GRAMINEAE (M)
Eriastrum Woot. & Standl. POLEMONIACEAE (D)
Eriaxis Rchb.f. ORCHIDACEAE (M)
Erica L. ERICACEAE (D)
Ericameria Nutt. COMPOSITAE (D)
Ericentrodea S.F.Blake & Sherff COMPOSITAE (D)
Erichsenia Hemsl. LEGUMINOSAE–PAPILIONOIDEAE (D)
Ericinella Klotzsch = Erica L. ERICACEAE (D)
Ericomyrtus Turcz. = Baeckea L. MYRTACEAE (D)
Erigenia Nutt. UMBELLIFERAE (D)
Erigeron L. COMPOSITAE (D)
Erinacea Adans. LEGUMINOSAE–PAPILIONOIDEAE (D)
Erinna Phil. ALLIACEAE (M)
Erinocarpus Nimmo ex J.Graham TILIACEAE (D)
Erinus L. SCROPHULARIACEAE (D)
Erioblastus Honda = Deschampsia P.Beauv. GRAMINEAE (M)
Eriobotrya Lindl. ROSACEAE (D)
Eriocactus Backeb. (SUS) = Parodia Speg. CACTACEAE (D)
Eriocapitella Nakai = Anemone L. RANUNCULACEAE (D)
Eriocaucanthus Chiov. = Caucanthus Forssk. MALPIGHIACEAE (D)
Eriocaulon L. ERIOCAULACEAE (M)
Eriocephala Backeb. = Parodia Speg. CACTACEAE (D)
Eriocephalus L. COMPOSITAE (D)
Eriocereus Riccob. = Harrisia Britton CACTACEAE (D)
Eriochaeta Figari & De Not. = Pennisetum Rich. GRAMINEAE (M)
Eriochilum Ritgen = Eriochilus R.Br. ORCHIDACEAE (M)
Eriochilus R.Br. ORCHIDACEAE (M)
Eriochiton (R.Anderson) A.J.Scott CHENOPODIACEAE (D)
Eriochlamys Sond. & F.Muell. COMPOSITAE (D)
Eriochloa Kunth GRAMINEAE (M)
Eriochrysis P.Beauv. GRAMINEAE (M)
Eriochylus Steud. = Eriochilus R.Br. ORCHIDACEAE (M)
Eriocnema Naudin MELASTOMATACEAE (D)
Eriococcus Hassk. = Phyllanthus L. EUPHORBIACEAE (D)
Eriocoelum Hook.f. SAPINDACEAE (D)
Eriocoma Nutt. = Oryzopsis Michx. GRAMINEAE (M)
Eriocycla Lindl. UMBELLIFERAE (D)
Eriodendron DC. = Ceiba Mill. BOMBACACEAE (D)
Eriodes Rolfe ORCHIDACEAE (M)
Eriodictyon Benth. HYDROPHYLLACEAE (D)
Erioglossum Blume = Lepisanthes Blume SAPINDACEAE (D)
Eriogonella Goodman = Chorizanthe R.Br. ex Benth. POLYGONACEAE (D)
Eriogonum Michx. POLYGONACEAE (D)
Eriogynia Hook. = Luetkea Bong. ROSACEAE (D)
Eriolaena DC. STERCULIACEAE (D)
Eriolobus (DC.) M.Roem. = Malus Mill. ROSACEAE (D)
Eriolopha Ridl. = Alpinia Roxb. ZINGIBERACEAE (M)
Eriolytrum Kunth (SUI) = Panicum L. GRAMINEAE (M)
Erione Schott & Endl. == Ceiba Mill. BOMBACACEAE (D)
Erioneuron Nash GRAMINEAE (M)
Eriope Kunth ex Benth. LABIATAE (D)
Eriopexis (Schltr.) Brieger = Dendrobium Sw. ORCHIDACEAE (M)
Eriophorella Holub = Trichophorum Pers. CYPERACEAE (M)

Eriophoropsis Palla = Eriophorum L. CYPERACEAE (M)
Eriophorum L. CYPERACEAE (M)
Eriophyllum Lag. COMPOSITAE (D)
Eriophyton Benth. LABIATAE (D)
Eriopidion Harley LABIATAE (D)
Eriopodium Hochst. (SUI) = Andropogon L. GRAMINEAE (M)
Eriopsis Lindl. ORCHIDACEAE (M)
Erioscirpus Palla = Eriophorum L. CYPERACEAE (M)
Eriosema (DC.) Rchb. LEGUMINOSAE–PAPILIONOIDEAE (D)
Eriosemopsis Robyns RUBIACEAE (D)
Eriosolena Blume THYMELAEACEAE (D)
Eriosorus Fée ADIANTACEAE (F)
Eriospermum Jacq. ex Willd. ERIOSPERMACEAE (M)
Eriosphaera Less. (SUH) = Galeomma Rauschert COMPOSITAE (D)
Eriospora Hochst. ex A.Rich. = Coleochloa Gilly CYPERACEAE (M)
Eriostemon Sm. RUTACEAE (D)
Eriostoma Boivin ex Baill. (SUI) = Tricalysia A.Rich. ex DC. RUBIACEAE (D)
Eriostrobilus Bremek. = Strobilanthes Blume ACANTHACEAE (D)
Eriostylos C.C.Towns. AMARANTHACEAE (D)
Eriosyce Phil. CACTACEAE (D)
Eriosynaphe DC. UMBELLIFERAE (D)
Eriotheca Schott & Endl. BOMBACACEAE (D)
Eriothymus (Benth.) Rchb. LABIATAE (D)
Eriotrix Cass. COMPOSITAE (D)
Erioxantha Raf. = Eria Lindl. ORCHIDACEAE (M)
Erioxylum Rose & Standl. MALVACEAE (D)
Eriphia P.Browne = Besleria L. GESNERIACEAE (D)
Erisma Rudge VOCHYSIACEAE (D)
Erismadelphus Mildbr. VOCHYSIACEAE (D)
Erismanthus Wall. ex Müll.Arg. EUPHORBIACEAE (D)
Erithalis G.Forst. (SUH) = Timonius DC. RUBIACEAE (D)
Erithalis P.Browne RUBIACEAE (D)
Eritrichium Schrad. ex Gaudin BORAGINACEAE (D)
Erlangea Sch.Bip. COMPOSITAE (D)
Ermanea Cham. = Christolea Cambess. ex Jacquem. CRUCIFERAE (D)
Ermaniopsis Hara CRUCIFERAE (D)
Ernestia DC. MELASTOMATACEAE (D)
Ernestimeyera Kuntze (SUS) = Alberta E.Mey. RUBIACEAE (D)
Ernodea Sw. RUBIACEAE (D)
Erochloe Raf. = Eragrostis Wolf GRAMINEAE (M)
Erodiophyllum F.Muell. COMPOSITAE (D)
Erodium L'Hér. ex Aiton GERANIACEAE (D)
Erophila DC. CRUCIFERAE (D)
Erosion Lunell (SUS) = Eragrostis Wolf GRAMINEAE (M)
Erosmia A.Juss. = Hoffmannia Sw. RUBIACEAE (D)
Erotium Blanco = Trichospermum Blume TILIACEAE (D)
Erpetina Naudin = Medinilla Gaudich. MELASTOMATACEAE (D)
Erpetion Sweet = Viola L. VIOLACEAE (D)
Errazurizia Phil. LEGUMINOSAE–PAPILIONOIDEAE (D)
Ertela Adans. = Monnieria Loefl. RUTACEAE (D)
Eruca Mill. CRUCIFERAE (D)
Erucago Mill. = Bunias L. CRUCIFERAE (D)
Erucaria Gaertn. CRUCIFERAE (D)
Erucaria Cerv. (SUH) = Chondrosum Desv. GRAMINEAE (M)
Erucastrum (DC.) C.Presl CRUCIFERAE (D)

Ervatamia (A.DC.) Stapf = Tabernaemontana L. APOCYNACEAE (D)
Ervum L. = Vicia L. LEGUMINOSAE–PAPILIONOIDEAE (D)
Erycibe Roxb. CONVOLVULACEAE (D)
Erycina Lindl. ORCHIDACEAE (M)
Erymophyllum Paul G.Wilson COMPOSITAE (D)
Eryngiophyllum Greenm. COMPOSITAE (D)
Eryngium L. UMBELLIFERAE (D)
Erysimum L. CRUCIFERAE (D)
Erythea S.Watson = Brahea Mart. ex Endl. PALMAE (M)
Erythracanthus Nees = Staurogyne Wall. ACANTHACEAE (D)
Erythradenia (B.L.Rob.) R.M.King & H.Rob. COMPOSITAE (D)
Erythraea L. = Centaurium Hill GENTIANACEAE (D)
Erythranthera Zotov = Rytidosperma Steud. GRAMINEAE (M)
Erythranthus Oerst. ex Hanst. = Drymonia Mart. GESNERIACEAE (D)
Erythrina L. LEGUMINOSAE–PAPILIONOIDEAE (D)
Erythrocarpus M.Roem. (SUH) = Adenia Forssk. PASSIFLORACEAE (D)
Erythrocarpus Blume = Suregada Roxb. ex Rottler EUPHORBIACEAE (D)
Erythrocephalum Benth. COMPOSITAE (D)
Erythrocereus Houghton (UR) CACTACEAE (D)
Erythrochiton Nees & Mart. RUTACEAE (D)
Erythrochiton Griff. (SUH) = Ternstroemia Mutis ex L.f. THEACEAE (D)
Erythrochlamys Gürke LABIATAE (D)
Erythrochylus Reinw. ex Blume = Claoxylon A.Juss. EUPHORBIACEAE (D)
Erythrococca Benth. EUPHORBIACEAE (D)
Erythrocoma Greene = Geum L. ROSACEAE (D)
Erythrodanum Thouars = Nertera Banks & Sol. ex Gaertn. RUBIACEAE (D)
Erythrodes Blume ORCHIDACEAE (M)
Erythrogyne Gasp. = Ficus L. MORACEAE (D)
Erythronium L. LILIACEAE (M)
Erythropalum Blume OLACACEAE (D)
Erythrophleum Afzel. ex G.Don LEGUMINOSAE–CAESALPINIOIDEAE (D)
Erythrophysa E.Mey. ex Arn. SAPINDACEAE (D)
Erythrophysopsis Verdc. = Erythrophysa E.Mey. ex Arn. SAPINDACEAE (D)
Erythropsis Lindl. ex Schott & Endl. = Firmiana Marsili STERCULIACEAE (D)
Erythropyxis Engl. = Brazzeia Baill. SCYTOPETALACEAE (D)
Erythrorchis Blume ORCHIDACEAE (M)
Erythrorhipsalis A.Berger = Rhipsalis Gaertn. CACTACEAE (D)
Erythroselinum Chiov. UMBELLIFERAE (D)
Erythrospermum Lam. FLACOURTIACEAE (D)
Erythrostigma Hassk. = Connarus L. CONNARACEAE (D)
Erythroxylum P.Browne ERYTHROXYLACEAE (D)
Erytropyxis Pierre = Brazzeia Baill. SCYTOPETALACEAE (D)
Escallonia Mutis ex L.f. ESCALLONIACEAE (D)
Eschatogramme Trevis. ex C.Chr. = Dicranoglossum J.Sm. POLYPODIACEAE (F)
Eschholtzia Rchb. (SUO) = Eschscholzia Cham. PAPAVERACEAE (D)
Escholtzia Dumort. (SUO) = Eschscholzia Cham. PAPAVERACEAE (D)
Eschscholtzia Bernh. (SUO) = Eschscholzia Cham. PAPAVERACEAE (D)
Eschscholzia Cham. PAPAVERACEAE (D)
Eschweilera Mart. ex DC. LECYTHIDACEAE (D)
Eschweileria Zipp. ex Boerl. = Osmoxylon Miq. ARALIACEAE (D)
Escobaria Britton & Rose CACTACEAE (D)
Escobedia Ruíz & Pav. SCROPHULARIACEAE (D)
Escobesseya Hester = Escobaria Britton & Rose CACTACEAE (D)
Escontria Rose CACTACEAE (D)
Esenbeckia Kunth RUTACEAE (D)

Esfandiaria Charif & Aellen CHENOPODIACEAE (D)
Eskemukerjea Malick & Sengupta = Fagopyrum Mill. POLYGONACEAE (D)
Esmeralda Rchb.f. ORCHIDACEAE (M)
Esmeraldia Fourn. = Asclepias L. ASCLEPIADACEAE (D)
Espadaea A.Rich. GOETZEACEAE (D)
Espejoa DC. COMPOSITAE (D)
Espeletia Mutis ex Humb. & Bonpl. COMPOSITAE (D)
Espeletiopsis Cuatrec. COMPOSITAE (D)
Espostoa Britton & Rose CACTACEAE (D)
Espostoopsis Buxb. CACTACEAE (D)
Esquirolia H.Lév. = Ligustrum L. OLEACEAE (D)
Esquiroliella H.Lév. = Neomartinella Pilg. CRUCIFERAE (D)
Esterhazya J.C.Mikan SCROPHULARIACEAE (D)
Esterhuysenia L.Bolus AIZOACEAE (D)
Esula (Pers.) Haw. = Euphorbia L. EUPHORBIACEAE (D)
Etaballia Benth. LEGUMINOSAE–PAPILIONOIDEAE (D)
Etaeria Blume (SUO) = Hetaeria Blume ORCHIDACEAE (M)
Etericius Desv. (UP) RUBIACEAE (D)
Ethesia Raf. (SUH) = Justicia L. ACANTHACEAE (D)
Ethulia L.f. COMPOSITAE (D)
Etiosedum Á.Löve & D.Löve = Sedum L. CRASSULACEAE (D)
Etlingera Giseke ZINGIBERACEAE (M)
Etorloba Raf. = Jacaranda Juss. BIGNONIACEAE (D)
Etubila Raf. = Dendrophthoe Mart. LORANTHACEAE (D)
Euadenia Oliv. CAPPARACEAE (D)
Euanthe Schltr. ORCHIDACEAE (M)
Euaraliopsis Hutch. = Brassaiopsis Decne. & Planch. ARALIACEAE (D)
Euathronia Nutt. ex A.Gray = Coprosma J.R.Forst. & G.Forst. RUBIACEAE (D)
Eubasis Salisb. = Aucuba Thunb. AUCUBACEAE (D)
Eubrachion Hook.f. EREMOLEPIDACEAE (D)
Eucalypton St.–Lag. = Eucalyptus L'Hér. MYRTACEAE (D)
Eucalyptopsis C.T.White MYRTACEAE (D)
Eucalyptus L'Hér. MYRTACEAE (D)
Eucapnia Raf. = Nicotiana L. SOLANACEAE (D)
Eucarya T.L.Mitch. = Santalum L. SANTALACEAE (D)
Euceraea Mart. FLACOURTIACEAE (D)
Euchaetis Bartl. & H.L.Wendl. RUTACEAE (D)
Eucharidium Fisch. & C.A.Mey. = Clarkia Pursh ONAGRACEAE (D)
Eucharis Planch. & Linden AMARYLLIDACEAE (M)
Eucheila O.F.Cook (SUI) = Chamaedorea Willd. PALMAE (M)
Euchilopsis F.Muell. LEGUMINOSAE–PAPILIONOIDEAE (D)
Euchiton Cass. COMPOSITAE (D)
Euchlaena Schrad. = Zea L. GRAMINEAE (M)
Euchlora Eckl. & Zeyh. = Lotononis (DC.) Eckl. & Zeyh. LEGUMINOSAE–
   PAPILIONOIDEAE (D)
Euchorium Ekman & Radlk. SAPINDACEAE (D)
Euchresta Benn. LEGUMINOSAE–PAPILIONOIDEAE (D)
Euclasta Franch. GRAMINEAE (M)
Euclea L. EBENACEAE (D)
Euclidium R.Br. CRUCIFERAE (D)
Euclinia Salisb. RUBIACEAE (D)
Eucnemia Rchb. (SUO) = Govenia Lindl. ORCHIDACEAE (M)
Eucnemis Lindl. = Govenia Lindl. ORCHIDACEAE (M)
Eucnide Zucc. LOASACEAE (D)
Eucodonia Hanst. GESNERIACEAE (D)

Eucolum Salisb. = Gloxinia L'Hér. GESNERIACEAE (D)
Eucomis L'Hér. HYACINTHACEAE (M)
Eucommia Oliv. EUCOMMIACEAE (D)
Eucorymbia Stapf APOCYNACEAE (D)
Eucosia Blume ORCHIDACEAE (M)
Eucrosia Ker Gawl. AMARYLLIDACEAE (M)
Eucryphia Cav. EUCRYPHIACEAE (D)
Eucrypta Nutt. HYDROPHYLLACEAE (D)
Eudema Humb. & Bonpl. CRUCIFERAE (D)
Eudisanthema Neck. ex Post & Kuntze = Brassavola R.Br. ORCHIDACEAE (M)
Eudonax Fr. (SUS) = Arundo L. GRAMINEAE (M)
Euforbia Ten. (SUO) = Euphorbia L. EUPHORBIACEAE (D)
Eufournia J.R.Reeder (SUS) = Sohnsia Airy Shaw GRAMINEAE (M)
Eugeissona Griff. PALMAE (M)
Eugenia L. MYRTACEAE (D)
Eugeniopsis O.Berg = Marlierea Cambess. MYRTACEAE (D)
Euglypha Chodat & Hassl. ARISTOLOCHIACEAE (D)
Euhemus Raf. = Lycopus L. LABIATAE (D)
Euhesperida Brullo & Furnari LABIATAE (D)
Euklastaxon Steud. = Andropogon L. GRAMINEAE (M)
Euklisia Rydb. ex Small CRUCIFERAE (D)
Eukylista Benth. = Calycophyllum DC. RUBIACEAE (D)
Eulalia Kunth GRAMINEAE (M)
Eulaliopsis Honda GRAMINEAE (M)
Eulenburgia Pax = Momordica L. CUCURBITACEAE (D)
Euleria Urb. ANACARDIACEAE (D)
Eulobus Nutt. = Camissonia Link ONAGRACEAE (D)
Eulophia R.Br. ex Lindl. ORCHIDACEAE (M)
Eulophidium Pfitzer = Oeceoclades Lindl. ORCHIDACEAE (M)
Eulophiella Rolfe ORCHIDACEAE (M)
Eulophiopsis Pfitzer = Graphorkis Thouars ORCHIDACEAE (M)
Eulophus Nutt. ex DC. = Perideridia Rchb. UMBELLIFERAE (D)
Eulychnia Phil. CACTACEAE (D)
Eulychnocactus Backeb. (SUI) = Corryocactus Britton & Rose CACTACEAE (D)
Eumachia DC. = Psychotria L. RUBIACEAE (D)
Eumecanthus Klotzsch & Garcke = Euphorbia L. EUPHORBIACEAE (D)
Eumolpe Decne. ex Jacques & Hérincq = Achimenes Pers. GESNERIACEAE (D)
Eumorphanthus A.C.Sm. = Psychotria L. RUBIACEAE (D)
Eumorphia DC. COMPOSITAE (D)
Eunomia DC. = Aethionema R.Br. CRUCIFERAE (D)
Euodia J.R.Forst. & G.Forst. RUTACEAE (D)
Euonymus L. CELASTRACEAE (D)
Euosmia Kunth = Hoffmannia Sw. RUBIACEAE (D)
Euothonaea Rchb.f. = Hexisea Lindl. ORCHIDACEAE (M)
Eupatoriadelphus R.M.King & H.Rob. = Eupatorium L. COMPOSITAE (D)
Eupatoriastrum Greenm. COMPOSITAE (D)
Eupatorina R.M.King & H.Rob. COMPOSITAE (D)
Eupatoriopsis Hieron. COMPOSITAE (D)
Eupatorium L. COMPOSITAE (D)
Euphlebium Brieger = Dendrobium Sw. ORCHIDACEAE (M)
Euphora Griff. = Aglaia Lour. MELIACEAE (D)
Euphorbia L. EUPHORBIACEAE (D)
Euphorbiastrum Klotzsch & Garcke = Euphorbia L. EUPHORBIACEAE (D)
Euphorbiodendron Millsp. = Euphorbia L. EUPHORBIACEAE (D)
Euphorbiopsis H.Lév. = Euphorbia L. EUPHORBIACEAE (D)

Euphorbium Hill = Euphorbia L. EUPHORBIACEAE (D)
Euphoria Comm. ex Juss. = Litchi Sonn. SAPINDACEAE (D)
Euphorianthus Radlk. SAPINDACEAE (D)
Euphrasia L. SCROPHULARIACEAE (D)
Euphronia Mart. & Zucc. EUPHRONIACEAE (D)
Euphrosyne DC. COMPOSITAE (D)
Euplassa Salisb. ex Knight PROTEACEAE (D)
Euploca Nutt. = Heliotropium L. BORAGINACEAE (D)
Eupodium J.Sm. = Marattia Sw. MARATTIACEAE (F)
Eupogon Desv. (SUI) = Andropogon L. GRAMINEAE (M)
Eupomatia R.Br. EUPOMATIACEAE (D)
Euporteria Kreuz. & Buining = Neoporteria Britton & Rose CACTACEAE (D)
Eupritchardia Kuntze = Pritchardia Seem. & H.A.Wendl. PALMAE (M)
Euproboscis Griff. = Thelasis Blume ORCHIDACEAE (M)
Euptelea Siebold & Zucc. EUPTELEACEAE (D)
Eupteris Newman = Pteridium Gled. ex Scop. DENNSTAEDTIACEAE (F)
Eupteron Miq. = Polyscias J.R.Forst. & G.Forst. ARALIACEAE (D)
Eupyrena Wight & Arn. = Timonius DC. RUBIACEAE (D)
Euraphis (Trin.) Lindl. (SUS) = Boissiera Hochst. ex Steud. GRAMINEAE (M)
Eurebutia Frič (SUI) = Rebutia K.Schum. CACTACEAE (D)
Eureiandra Hook.f. CUCURBITACEAE (D)
Eurhotia Neck. (SUI) = Psychotria L. RUBIACEAE (D)
Euroschinus Hook.f. ANACARDIACEAE (D)
Eurotia Adans. (SUS) = Axyris L. CHENOPODIACEAE (D)
Eurya Thunb. THEACEAE (D)
Euryale Salisb. NYMPHAEACEAE (D)
Euryandra Hook.f. (SUO) = Eureiandra Hook.f. CUCURBITACEAE (D)
Euryangium Kauffman = Ferula L. UMBELLIFERAE (D)
Eurybiopsis DC. = Minuria DC. COMPOSITAE (D)
Eurycarpus Botsch. CRUCIFERAE (D)
Eurycentrum Schltr. ORCHIDACEAE (M)
Eurychone Schltr. ORCHIDACEAE (M)
Eurycles Salisb. ex Schult. & Schult.f. (SUS) = Proiphys Herb. AMARYLLIDACEAE (M)
Eurycoma Jack SIMAROUBACEAE (D)
Eurycorymbus Hand.–Mazz. SAPINDACEAE (D)
Eurydochus Maguire & Wurdack COMPOSITAE (D)
Eurygania Klotzsch = Thibaudia Ruíz & Pav. ERICACEAE (D)
Eurylobium Hochst. STILBACEAE (D)
Eurynome DC. = Coprosma J.R.Forst. & G.Forst. RUBIACEAE (D)
Eurynotia R.C.Foster IRIDACEAE (M)
Euryomyrtus Schauer = Baeckea L. MYRTACEAE (D)
Euryops (Cass.) Cass. COMPOSITAE (D)
Eurypetalum Harms LEGUMINOSAE–CAESALPINIOIDEAE (D)
Eurysolen Prain LABIATAE (D)
Eurystemon Alexander PONTEDERIACEAE (M)
Eurystigma L.Bolus (SUI) = Mesembryanthemum L. AIZOACEAE (D)
Eurystyles Wawra ORCHIDACEAE (M)
Eurytaenia Torr. & A.Gray UMBELLIFERAE (D)
Euscaphis Siebold & Zucc. STAPHYLEACEAE (D)
Eusideroxylon Teijsm. & Binn. LAURACEAE (D)
Eusiphon Benoist ACANTHACEAE (D)
Eustachys Desv. GRAMINEAE (M)
Eustegia R.Br. ASCLEPIADACEAE (D)
Eustegia Raf. (SUH) = Conostegia D.Don MELASTOMATACEAE (D)
Eustephia Cav. AMARYLLIDACEAE (M)

Eustephiopsis R.E.Fr. = Hieronymiella Pax AMARYLLIDACEAE (M)
Eusteralis Raf. = Pogostemon Desf. LABIATAE (D)
Eustigma Gardner & Champ. HAMAMELIDACEAE (D)
Eustoma Salisb. GENTIANACEAE (D)
Eustrephus R.Br. PHILESIACEAE (M)
Eustylis Engelm. & A.Gray = Alophia Herb. IRIDACEAE (M)
Eusynetra Raf. = Columnea L. GESNERIACEAE (D)
Eutaxia R.Br. LEGUMINOSAE–PAPILIONOIDEAE (D)
Euterpe Mart. PALMAE (M)
Eutetras A.Gray COMPOSITAE (D)
Euthamia (Nutt.) Elliott COMPOSITAE (D)
Euthamnus Schltr. = Aeschynanthus Jack GESNERIACEAE (D)
Euthemis Jack OCHNACEAE (D)
Eutheta Standl. = Melasma Bergius SCROPHULARIACEAE (D)
Euthryptochloa Cope GRAMINEAE (M)
Euthystachys A.DC. STILBACEAE (D)
Eutrema R.Br. CRUCIFERAE (D)
Eutriana Trin. = Bouteloua Lag. GRAMINEAE (M)
Eutrophia Klotzsch = Croton L. EUPHORBIACEAE (D)
Euxena Calest. = Arabis L. CRUCIFERAE (D)
Euxylophora Huber RUTACEAE (D)
Euzomodendron Coss. CRUCIFERAE (D)
Evacidium Pomel COMPOSITAE (D)
Evandra R.Br. CYPERACEAE (M)
Evanesca Raf. = Pimenta Lindl. MYRTACEAE (D)
Evax Gaertn. = Filago L. COMPOSITAE (D)
Evea Aubl. = Psychotria L. RUBIACEAE (D)
Evelyna Poepp. & Endl. = Elleanthus C.Presl ORCHIDACEAE (M)
Everardia Ridl. CYPERACEAE (M)
Everettia Merr. = Astronidium A.Gray MELASTOMATACEAE (D)
Everettiodendron Merr. = Baccaurea Lour. EUPHORBIACEAE (D)
Eversmannia Bunge LEGUMINOSAE–PAPILIONOIDEAE (D)
Evodia Lam. (SUO) = Euodia J.R.Forst. & G.Forst. RUTACEAE (D)
Evodianthus Oerst. CYCLANTHACEAE (M)
Evodiella Linden RUTACEAE (D)
Evodiopanax (Harms) Nakai = Gamblea C.B.Clarke ARALIACEAE (D)
Evoista Raf. = Lycium L. SOLANACEAE (D)
Evolvulus L. CONVOLVULACEAE (D)
Evonymopsis H.Perrier CELASTRACEAE (D)
Evosmia Humb. & Bonpl. = Hoffmannia Sw. RUBIACEAE (D)
Evrardia Gagnep. (SUH) = Evrardianthe Rauschert ORCHIDACEAE (M)
Evrardiana Averyanov (SUS) = Evrardianthe Rauschert ORCHIDACEAE (M)
Evrardianthe Rauschert ORCHIDACEAE (M)
Ewartia Beauverd COMPOSITAE (D)
Ewartiothamnus Anderb. COMPOSITAE (D)
Ewyckia Blume = Pternandra Jack MELASTOMATACEAE (D)
Exaculum Caruel GENTIANACEAE (D)
Exacum L. GENTIANACEAE (D)
Exagrostis Steud. (SUI) = Eragrostis Wolf GRAMINEAE (M)
Exallage Bremek. = Hedyotis L. RUBIACEAE (D)
Exandra Standl. = Simira Aubl. RUBIACEAE (D)
Exbucklandia R.W.Br. HAMAMELIDACEAE (D)
Excavatia Markgr. = Ochrosia Juss. APOCYNACEAE (D)
Excentrodendron H.T.Chang & R.H.Miau = Burretiodendron Rehder TILIACEAE (D)
Excoecaria L. EUPHORBIACEAE (D)

Excoecariopsis Pax = Spirostachys Sond. EUPHORBIACEAE (D)
Excremis Willd. (SUO) = Eccremis Baker PHORMIACEAE (M)
Exechostylus K.Schum. = Pavetta L. RUBIACEAE (D)
Exellia Boutique ANNONACEAE (D)
Exellodendron Prance CHRYSOBALANACEAE (D)
Exeria Raf. = Eria Lindl. ORCHIDACEAE (M)
Exitelia Blume = Maranthes Blume CHRYSOBALANACEAE (D)
Exoacantha Labill. UMBELLIFERAE (D)
Exocarpos Labill. SANTALACEAE (D)
Exocarya Benth. CYPERACEAE (M)
Exochaenium Griseb. = Sebaea Sol. ex R.Br. GENTIANACEAE (D)
Exochogyne C.B.Clarke CYPERACEAE (M)
Exochorda Lindl. ROSACEAE (D)
Exodeconus Raf. SOLANACEAE (D)
Exogonium Choisy = Ipomoea L. CONVOLVULACEAE (D)
Exohebea R.C.Foster = Tritoniopsis L.Bolus IRIDACEAE (M)
Exolobus Fourn. ASCLEPIADACEAE (D)
Exomiocarpon Lawalrée COMPOSITAE (D)
Exomis Fenzl ex Moq. CHENOPODIACEAE (D)
Exophya Raf. = Epidendrum L. ORCHIDACEAE (M)
Exorhopala Steenis BALANOPHORACEAE (D)
Exorrhiza Becc. = Clinostigma H.A.Wendl. PALMAE (M)
Exospermum Tiegh. = Zygogynum Baill. WINTERACEAE (D)
Exostema (Pers.) Humb. & Bonpl. RUBIACEAE (D)
Exostemma DC. (SUO) = Exostema (Pers.) Humb. & Bonpl. RUBIACEAE (D)
Exostemon Post & Kuntze (SUO) = Exostema (Pers.) Humb. & Bonpl. RUBIACEAE (D)
Exostyles Schott LEGUMINOSAE-PAPILIONOIDEAE (D)
Exothea Macfad. SAPINDACEAE (D)
Exotheca Andersson GRAMINEAE (M)
Exsertanthera Pichon = Lundia DC. BIGNONIACEAE (D)
Exydra Endl. = Glyceria R.Br. GRAMINEAE (M)
Eylesia S.Moore = Buchnera L. SCROPHULARIACEAE (D)
Eysenhardtia Kunth LEGUMINOSAE-PAPILIONOIDEAE (D)
Eystathes Lour. = Xanthophyllum Roxb. POLYGALACEAE (D)

Faberia Hemsl. COMPOSITAE (D)
Fabiana Ruíz & Pav. SOLANACEAE (D)
Fabricia Gaertn. (SUH) = Neofabricia Joy Thompson MYRTACEAE (D)
Fabrisinapis C.C.Towns. = Hemicrambe Webb CRUCIFERAE (D)
Facelis Cass. COMPOSITAE (D)
Facheiroa Britton & Rose CACTACEAE (D)
Factorovskya Eig = Medicago L. LEGUMINOSAE-PAPILIONOIDEAE (D)
Fadenia Aellen & C.C.Towns. CHENOPODIACEAE (D)
Fadogia Schweinf. RUBIACEAE (D)
Fadogiella Robyns RUBIACEAE (D)
Fadyenia Hook. DRYOPTERIDACEAE (F)
Fagara L. = Zanthoxylum L. RUTACEAE (D)
Fagaropsis Mildbr. ex Siebenl. RUTACEAE (D)
Fagelia Neck. ex DC. = Bolusafra Kuntze LEGUMINOSAE-PAPILIONOIDEAE (D)
Fagerlindia Tirveng. RUBIACEAE (D)
Fagonia L. ZYGOPHYLLACEAE (D)
Fagopyrum Mill. POLYGONACEAE (D)
Fagraea Thunb. LOGANIACEAE (D)
Faguetia Marchand ANACARDIACEAE (D)
Fagus L. FAGACEAE (D)

Fahrenheitia Rchb.f. & Zoll. EUPHORBIACEAE (D)
Faidherbia A.Chev. LEGUMINOSAE–MIMOSOIDEAE (D)
Faika Philipson MONIMIACEAE (D)
Fairchildia Britton & Rose = Swartzia Schreb. LEGUMINOSAE–PAPILIONOIDEAE (D)
Falcaria Fabr. UMBELLIFERAE (D)
Falcatifolium de Laub. PODOCARPACEAE (G)
Falconera Royle (SUH) = Sapium P.Browne EUPHORBIACEAE (D)
Falconeria Hook.f. (SUH) = Wulfenia Jacq. SCROPHULARIACEAE (D)
Falimiria Rchb. (SUS) = Gaudinia P.Beauv. GRAMINEAE (M)
Falkia L.f. CONVOLVULACEAE (D)
Fallopia Adans. POLYGONACEAE (D)
Fallugia Endl. ROSACEAE (D)
Falona Adans. = Cynosurus L. GRAMINEAE (M)
Famarea Vitman (SUO) = Faramea Aubl. RUBIACEAE (D)
Famatina Ravenna AMARYLLIDACEAE (M)
Fanninia Harv. ASCLEPIADACEAE (D)
Faradaya F.Muell. VERBENACEAE (D)
Faramea Aubl. RUBIACEAE (D)
Farfugium Lindl. COMPOSITAE (D)
Fargesia Franch. = Thamnocalamus Munro GRAMINEAE (M)
Farinopsis Chrtek & Soják = Potentilla L. ROSACEAE (D)
Farmeria Willis ex Trimen PODOSTEMACEAE (D)
Faroa Welw. GENTIANACEAE (D)
Farquharia Stapf APOCYNACEAE (D)
Farrago Clayton GRAMINEAE (M)
Farreria Balf.f. & W.W.Sm. = Wikstroemia Endl. THYMELAEACEAE (D)
Farringtonia Gleason = Siphanthera Pohl MELASTOMATACEAE (D)
Farsetia Turra CRUCIFERAE (D)
Fartis Adans. (SUS) = Zizania L. GRAMINEAE (M)
Fascicularia Mez BROMELIACEAE (M)
Fatoua Gaudich. MORACEAE (D)
Fatrea Juss. = Terminalia L. COMBRETACEAE (D)
Fatsia Decne. & Planch. ARALIACEAE (D)
Faucaria Schwantes AIZOACEAE (D)
Faucherea Lecomte SAPOTACEAE (D)
Faujasia Cass. COMPOSITAE (D)
Faulia Raf. = Ligustrum L. OLEACEAE (D)
Faurea Harv. PROTEACEAE (D)
Fauria Franch. = Nephrophyllidium Gilg MENYANTHACEAE (D)
Faustia Font Quer & Rothm. = Saccocalyx Coss. & Durieu LABIATAE (D)
Favargera Á.Löve & D.Löve = Gentiana L. GENTIANACEAE (D)
Fawcettia F.Muell. = Tinospora Miers MENISPERMACEAE (D)
Faxonanthus Greenm. SCROPHULARIACEAE (D)
Faxonia Brandegee COMPOSITAE (D)
Feddea Urb. COMPOSITAE (D)
Fedia Gaertn. VALERIANACEAE (D)
Fedorouia Yakovlev = Ormosia G.Jacks. LEGUMINOSAE–PAPILIONOIDEAE (D)
Fedorovia Kolak. = Campanula L. CAMPANULACEAE (D)
Fedtschenkiella Kudr. = Dracocephalum L. LABIATAE (D)
Fedtschenkoa Regel & Schmalh. = Leptaleum DC. CRUCIFERAE (D)
Feea Bory = Trichomanes L. HYMENOPHYLLACEAE (F)
Feeria Buser CAMPANULACEAE (D)
Fegimanra Pierre ANACARDIACEAE (D)
Feijoa O.Berg = Acca O.Berg MYRTACEAE (D)
Feldstonia P.S.Short COMPOSITAE (D)

Felicia Cass. COMPOSITAE (D)
Feliciadamia Bullock MELASTOMATACEAE (D)
Feliciana Benth. (SUO) = Myrrhinium Schott MYRTACEAE (D)
Felicianea Cambess. = Myrrhinium Schott MYRTACEAE (D)
Felipponia Hicken = Mangonia Schott ARACEAE (M)
Felipponiella Hicken = Mangonia Schott ARACEAE (M)
Femeniasia Susanna de la Serna = Centaurea L. COMPOSITAE (D)
Fendlera Engelm. & A.Gray HYDRANGEACEAE (D)
Fendlerella A.Heller HYDRANGEACEAE (D)
Fendleria Steud. = Oryzopsis Michx. GRAMINEAE (M)
Fenervia Diels = Polyalthia Blume ANNONACEAE (D)
Fenestraria N.E.Br. AIZOACEAE (D)
Fenixanthes Raf. = Salvia L. LABIATAE (D)
Fenixia Merr. COMPOSITAE (D)
Fenzlia Benth. = Linanthus Benth. POLEMONIACEAE (D)
Fenzlia Endl. (SUH) = Myrtella F.Muell. MYRTACEAE (D)
Ferdinanda Benth. (SUO) = Fernandoa Welw. ex Seem. BIGNONIACEAE (D)
Ferdinandea Pohl = Ferdinandusa Pohl RUBIACEAE (D)
Ferdinandia Welw. ex Seem. (SUO) = Fernandoa Welw. ex Seem. BIGNONIACEAE (D)
Ferdinandoa Seem. (SUO) = Fernandoa Welw. ex Seem. BIGNONIACEAE (D)
Ferdinandusa Pohl RUBIACEAE (D)
Fereiria Vell. ex Vand. = Hillia Jacq. RUBIACEAE (D)
Feretia Delile RUBIACEAE (D)
Fergania Pimenov UMBELLIFERAE (D)
Fergusonia Hook.f. RUBIACEAE (D)
Fernaldia Woodson APOCYNACEAE (D)
Fernandezia Ruíz & Pav. ORCHIDACEAE (M)
Fernandia Baill. (SUO) = Fernandoa Welw. ex Seem. BIGNONIACEAE (D)
Fernandoa Welw. ex Seem. BIGNONIACEAE (D)
Fernelia Comm. ex Lam. RUBIACEAE (D)
Fernseea Baker BROMELIACEAE (M)
Ferocactus Britton & Rose CACTACEAE (D)
Feronia Corrêa = Limonia L. RUTACEAE (D)
Feroniella Swingle RUTACEAE (D)
Ferraria Burm. ex Mill. IRIDACEAE (M)
Ferreirea Allemão = Sweetia Spreng. LEGUMINOSAE–PAPILIONOIDEAE (D)
Ferreyranthus H.Rob. & Brettell COMPOSITAE (D)
Ferreyrella S.F.Blake COMPOSITAE (D)
Ferrocalamus C.J.Hsueh & Keng f. = Indocalamus Nakai GRAMINEAE (M)
Ferula L. UMBELLIFERAE (D)
Ferulago W.D.J.Koch UMBELLIFERAE (D)
Ferulopsis Kitag. UMBELLIFERAE (D)
Festuca L. GRAMINEAE (M)
Festucaria Link (SUH) = Vulpia C.C.Gmel. GRAMINEAE (M)
Festucaria Fabr. (SUS) = Festuca L. GRAMINEAE (M)
Festucella E.B.Alexeev = Austrofestuca (Tzvelev) Alexeev GRAMINEAE (M)
Festucopsis (C.E.Hubb.) Meld. = Elymus L. GRAMINEAE (M)
Feuillaea Gled. (SUO) = Fevillea L. CUCURBITACEAE (D)
Feuillea Kuntze (SUO) = Fevillea L. CUCURBITACEAE (D)
Fevillea L. CUCURBITACEAE (D)
Fezia Pit. CRUCIFERAE (D)
Fibichia Koeler = Cynodon Rich. GRAMINEAE (M)
Fibigia Medik. CRUCIFERAE (D)
Fibocentrum Pierre ex Glaziou (SUI) = Chrysophyllum L. SAPOTACEAE (D)
Fibraurea Lour. MENISPERMACEAE (D)

Ficalhoa Hiern THEACEAE (D)
Ficaria Schaeffer = Ranunculus L. RANUNCULACEAE (D)
Ficindica St.-Lag. = Opuntia Mill. CACTACEAE (D)
Ficinia Schrad. CYPERACEAE (M)
Ficus L. MORACEAE (D)
Fiebrigia Fritsch = Gloxinia L'Hér. GESNERIACEAE (D)
Fiebrigiella Harms LEGUMINOSAE–PAPILIONOIDEAE (D)
Fieldia A.Cunn. GESNERIACEAE (D)
Fieldia Gaudich. (SUH) = Vandopsis Pfitzer ORCHIDACEAE (M)
Figuierea Montrouz. = Coelospermum Blume RUBIACEAE (D)
Filaginella Opiz (SUS) = Gnaphalium L. COMPOSITAE (D)
Filago L. COMPOSITAE (D)
Filarum Nicolson ARACEAE (M)
Filetia Miq. ACANTHACEAE (D)
Filicium Thwaites ex Benth. SAPINDACEAE (D)
Filicula Séguier = Cystopteris Bernh. WOODSIACEAE (F)
Filifolium Kitam. COMPOSITAE (D)
Filipedum Raizada & S.K.Jain = Capillipedium Stapf GRAMINEAE (M)
Filipendula Mill. ROSACEAE (D)
Filix Adans. (SUH) = Cystopteris Bernh. WOODSIACEAE (F)
Filix Séguier = Dryopteris Adans. DRYOPTERIDACEAE (F)
Filix Ludw. (SUH) = Pteridium Gled. ex Scop. DENNSTAEDTIACEAE (F)
Filix–Foemina Hill ex Farw. = Pteridium Gled. ex Scop. DENNSTAEDTIACEAE (F)
Filix–Mas Hill ex Farw. = Dryopteris Adans. DRYOPTERIDACEAE (F)
Fillaeopsis Harms LEGUMINOSAE–MIMOSOIDEAE (D)
Fimbriella Farw. ex Butzin = Platanthera Rich. ORCHIDACEAE (M)
Fimbripetalum (Turcz.) Ikonn. = Stellaria L. CARYOPHYLLACEAE (D)
Fimbristemma Turcz. ASCLEPIADACEAE (D)
Fimbristylis Vahl CYPERACEAE (M)
Fimbrolina Raf. = Sinningia Nees GESNERIACEAE (D)
Findlaya Hook.f. (SUH) = Orthaea Klotzsch ERICACEAE (D)
Finetia Schltr. (SUH) = Neofinetia Hu ORCHIDACEAE (M)
Finetia Gagnep. (SUS) = Anogeissus (DC.) Wall. COMBRETACEAE (D)
Fingerhuthia Nees GRAMINEAE (M)
Finlaysonia Wall. ASCLEPIADACEAE (D)
Finschia Warb. PROTEACEAE (D)
Fintelmannia Kunth = Trilepis Nees CYPERACEAE (M)
Fioria Mattei MALVACEAE (D)
Fiorinia Parl. = Aira L. GRAMINEAE (M)
Firmiana Marsili STERCULIACEAE (D)
Fischeria DC. ASCLEPIADACEAE (D)
Fishlockia Britton & Rose = Acacia Mill. LEGUMINOSAE–MIMOSOIDEAE (D)
Fissendocarpa (Haines) Bennet = Ludwigia L. ONAGRACEAE (D)
Fissenia Endl. (SUO) = Kissenia R.Br. ex Endl. LOASACEAE (D)
Fissicalyx Benth. LEGUMINOSAE–PAPILIONOIDEAE (D)
Fissiglossum Rusby (URI) ASCLEPIADACEAE (D)
Fissilia Comm. ex Juss. = Olax L. OLACACEAE (D)
Fissipes Small = Cypripedium L. ORCHIDACEAE (M)
Fissipetalum Merr. = Erycibe Roxb. CONVOLVULACEAE (D)
Fissistigma Griff. ANNONACEAE (D)
Fitchia Hook.f. COMPOSITAE (D)
Fittingia Mez MYRSINACEAE (D)
Fittonia Coem. ACANTHACEAE (D)
Fitzalania F.Muell. ANNONACEAE (D)
Fitzgeraldia F.Muell. = Cananga (DC.) Hook.f. & Thomson ANNONACEAE (D)

Fitzgeraldia F.Muell. (SUH) = Lyperanthus R.Br. ORCHIDACEAE (M)
Fitzroya Hook.f. ex Lindl. CUPRESSACEAE (G)
Fitzwillia P.S.Short COMPOSITAE (D)
Flabellaria Cav. MALPIGHIACEAE (D)
Flabellariopsis R.Wilczek MALPIGHIACEAE (D)
Flacourtia Comm. ex L'Hér. FLACOURTIACEAE (D)
Flagellaria L. FLAGELLARIACEAE (M)
Flagellarisaema Nakai = Arisaema Mart. ARACEAE (M)
Flagenium Baill. RUBIACEAE (D)
Flanagania Schltr. = Cynanchum L. ASCLEPIADACEAE (D)
Flaveria Juss. COMPOSITAE (D)
Flavia Fabr. (SUS) = Anthoxanthum L. GRAMINEAE (M)
Fleischmannia Sch.Bip. COMPOSITAE (D)
Fleischmanniana Sch.Bip. (SUO) = Fleischmannia Sch.Bip. COMPOSITAE (D)
Fleischmanniopsis R.M.King & H.Rob. COMPOSITAE (D)
Flemingia Hunter ex Ridl. (SUH) = Tarenna Gaertn. RUBIACEAE (D)
Flemingia Roxb. ex W.T.Aiton LEGUMINOSAE–PAPILIONOIDEAE (D)
Fleurya Gaudich. = Laportea Gaudich. URTICACEAE (D)
Fleurydora A.Chev. OCHNACEAE (D)
Flexanthera Rusby RUBIACEAE (D)
Flexularia Raf. (UR) GRAMINEAE (M)
Flickingeria A.D.Hawkes ORCHIDACEAE (M)
Flindersia R.Br. RUTACEAE (D)
Floerkea Willd. LIMNANTHACEAE (D)
Floresia Krainz & Ritter ex Backe (SUI) = Haageocereus Backeb. CACTACEAE (D)
Florestina Cass. COMPOSITAE (D)
Floribunda F.Ritter = Cipocereus F.Ritter CACTACEAE (D)
Floscaldasia Cuatrec. COMPOSITAE (D)
Floscopa Lour. COMMELINACEAE (M)
Flosmutisia Cuatrec. COMPOSITAE (D)
Flourensia DC. COMPOSITAE (D)
Flueckigera Kuntze = Ledenbergia Klotzsch ex Moq. PHYTOLACCACEAE (D)
Flueggea Willd. EUPHORBIACEAE (D)
Fluggeopsis K.Schum. = Phyllanthus L. EUPHORBIACEAE (D)
Fluminia Fr. = Scolochloa Link GRAMINEAE (M)
Flyriella R.M.King & H.Rob. COMPOSITAE (D)
Fobea Frič (SUI) = Escobaria Britton & Rose CACTACEAE (D)
Fockea Endl. ASCLEPIADACEAE (D)
Foeniculum Mill. UMBELLIFERAE (D)
Foenodorum E.H.L.Krause (SUS) = Anthoxanthum L. GRAMINEAE (M)
Foersteria Scop. = Breynia J.R.Forst. & G.Forst. EUPHORBIACEAE (D)
Foetidia Comm. ex Lam. LECYTHIDACEAE (D)
Fokienia A.Henry & H.H.Thomas CUPRESSACEAE (G)
Foleyola Maire CRUCIFERAE (D)
Folliculigera Pasq. = Trigonella L. LEGUMINOSAE–PAPILIONOIDEAE (D)
Folotsia Costantin & Bois ASCLEPIADACEAE (D)
Fonna Adans. = Phlox L. POLEMONIACEAE (D)
Fontainea Heckel EUPHORBIACEAE (D)
Fontainesia Post & Kuntze (SUO) = Fontanesia Labill. OLEACEAE (D)
Fontanesia Labill. OLEACEAE (D)
Fontbrunea Pierre = Pouteria Aubl. SAPOTACEAE (D)
Fontquera Maire = Perralderia Coss. COMPOSITAE (D)
Fontqueriella Rothm. = Triguera Cav. SOLANACEAE (D)
Forasaccus Bubani (SUS) = Bromus L. GRAMINEAE (M)
Forbesina Ridl. = Eria Lindl. ORCHIDACEAE (M)

Forchhammeria Liebm. CAPPARACEAE (D)
Forcipella Baill. ACANTHACEAE (D)
Fordia Hemsl. LEGUMINOSAE–PAPILIONOIDEAE (D)
Fordiophyton Stapf MELASTOMATACEAE (D)
Forestiera Poir. OLEACEAE (D)
Forficaria Lindl. ORCHIDACEAE (M)
Forgesia Comm. ex Juss. ESCALLONIACEAE (D)
Formania W.W.Sm. & Small COMPOSITAE (D)
Formanodendron Nixon & Crepet = Trigonobalanus Forman FAGACEAE (D)
Formosia Pichon = Anodendron A.DC. APOCYNACEAE (D)
Forrestia A.Rich. (SUH) = Amischotolype Hassk. COMMELINACEAE (M)
Forsellesia Greene (SUS) = Glossopetalon A.Gray CROSSOSOMATACEAE (D)
Forsgardia Vell. = Combretum Loefl. COMBRETACEAE (D)
Forsskaolea L. URTICACEAE (D)
Forstera L.f. STYLIDIACEAE (D)
Forsteria Steud. = Breynia J.R.Forst. & G.Forst. EUPHORBIACEAE (D)
Forsteronia G.Mey. APOCYNACEAE (D)
Forsythia Vahl OLEACEAE (D)
Forsythiopsis Baker = Oplonia Raf. ACANTHACEAE (D)
Fortunatia J.F.Macbr. HYACINTHACEAE (M)
Fortunearia Rehder & E.H.Wilson HAMAMELIDACEAE (D)
Fortunella Swingle RUTACEAE (D)
Fortuynia Shuttlew. ex Boiss. CRUCIFERAE (D)
Foscarenia Vell. ex Vand. = Randia L. RUBIACEAE (D)
Fosterelia Airy Shaw (SUI) = Fosterella L.B.Sm. BROMELIACEAE (M)
Fosterella L.B.Sm. BROMELIACEAE (M)
Fosteria Molseed IRIDACEAE (M)
Fothergilla L. HAMAMELIDACEAE (D)
Fouquieria Kunth FOUQUIERIACEAE (D)
Fourniera Bommer ex Fourn. = Cyathea Sm. CYATHEACEAE (F)
Fourniera Scribn. (SUH) = Soderstromia C.Morton GRAMINEAE (M)
Fourraea Greuter & Burdet = Arabis L. CRUCIFERAE (D)
Foveolaria Ruíz & Pav. = Styrax L. STYRACACEAE (D)
Foveolina Källersjö COMPOSITAE (D)
Fractiunguis Schltr. = Reichenbachanthus Barb.Rodr. ORCHIDACEAE (M)
Fragaria L. ROSACEAE (D)
Fragariopsis A.St.–Hil. EUPHORBIACEAE (D)
Frailea Britton & Rose CACTACEAE (D)
Franchetella Pierre = Pouteria Aubl. SAPOTACEAE (D)
Franchetia Baill. = Breonia A.Rich. ex DC. RUBIACEAE (D)
Franciella Guillaumin (SUH) = Neofranciella Guillaumin RUBIACEAE (D)
Franciscea Pohl = Brunfelsia L. SOLANACEAE (D)
Franciscodendron B.Hyland & Steenis STERCULIACEAE (D)
Francisia Endl. = Darwinia Rudge MYRTACEAE (D)
Francoa Cav. SAXIFRAGACEAE (D)
Francoeuria Cass. = Pulicaria Gaertn. COMPOSITAE (D)
Frangula Mill. = Rhamnus L. RHAMI ¡ACEAE (D)
Frankenia L. FRANKENIACEAE (D)
Franklandia R.Br. PROTEACEAE (D)
Franklinia W.Bartr. ex Marshall THEACEAE (D)
Frankoa Rchb. (SUO) = Francoa Cav. SAXIFRAGACEAE (D)
Franseria Cav. = Ambrosia L. COMPOSITAE (D)
Frantzia Pittier = Sechium P.Browne CUCURBITACEAE (D)
Frasera Walter GENTIANACEAE (D)
Fraunhofera Mart. CELASTRACEAE (D)

Fraxinoides Medik. = Fraxinus L. OLEACEAE (D)
Fraxinus L. OLEACEAE (D)
Fredolia (Coss. & Durieu ex Bunge) Ulbr. CHENOPODIACEAE (D)
Freeria Merr. = Pyrenacantha Wight ICACINACEAE (D)
Freesia Eckl. ex Klatt IRIDACEAE (M)
Fregea Rchb.f. ORCHIDACEAE (M)
Fregirardia Dunal ex Raff.–Del. = Cestrum L. SOLANACEAE (D)
Freireodendron Müll.Arg. = Drypetes Vahl EUPHORBIACEAE (D)
Fremontia Torr. = Sarcobatus Nees CHENOPODIACEAE (D)
Fremontia Torr. (SUH) = Fremontodendron Coville STERCULIACEAE (D)
Fremontodendron Coville STERCULIACEAE (D)
Fremya Brongn. & Gris = Xanthostemon F.Muell. MYRTACEAE (D)
Frerea Dalzell ASCLEPIADACEAE (D)
Freya Badillo (UP) COMPOSITAE (D)
Freycinetia Gaudich. PANDANACEAE (M)
Freyera Rchb. = Geocaryum Coss. UMBELLIFERAE (D)
Freyeria Scop. (SUS) = Chionanthus L. OLEACEAE (D)
Freylinia Colla SCROPHULARIACEAE (D)
Freziera Willd. THEACEAE (D)
Fridericia Mart. BIGNONIACEAE (D)
Friedlandia Cham. & Schltdl. = Diplusodon Pohl LYTHRACEAE (D)
Friedrichsthalia Fenzl = Trichodesma R.Br. BORAGINACEAE (D)
Friesia Frič (SUI) = Parodia Speg. CACTACEAE (D)
Friesia Spreng. = Crotonopsis Michx. EUPHORBIACEAE (D)
Friesodielsia Steenis ANNONACEAE (D)
Frithia N.E.Br. AIZOACEAE (D)
Fritillaria L. LILIACEAE (M)
Fritschiantha Kuntze = Gloxinia L'Hér. GESNERIACEAE (D)
Fritzschia Cham. MELASTOMATACEAE (D)
Froehlichia D.Dietr. (SUO) = Coussarea Aubl. RUBIACEAE (D)
Froelichia Vahl (SUH) = Coussarea Aubl. RUBIACEAE (D)
Froelichia Moench AMARANTHACEAE (D)
Froelichiella R.E.Fr. AMARANTHACEAE (D)
Froesia Pires QUIINACEAE (D)
Froesiochloa G.A.Black GRAMINEAE (M)
Froesiodendron R.E.Fr. ANNONACEAE (D)
Frommia H.Wolff UMBELLIFERAE (D)
Frondaria Luer ORCHIDACEAE (M)
Froriepia K.Koch UMBELLIFERAE (D)
Froscula Raf. = Dendrobium Sw. ORCHIDACEAE (M)
Fructesca DC. ex Meisn. = Gaertnera Lam. RUBIACEAE (D)
Frumentum E.H.L.Krause (SUS) = Triticum L. GRAMINEAE (M)
Fryxellia D.M.Bates MALVACEAE (D)
Fuchsia Sw. (SUH) = Schradera Vahl RUBIACEAE (D)
Fuchsia L. ONAGRACEAE (D)
Fuernrohria K.Koch UMBELLIFERAE (D)
Fuerstia T.C.E.Fr. LABIATAE (D)
Fuertesia Urb. LOASACEAE (D)
Fuertesiella Schltr. ORCHIDACEAE (M)
Fugosia Juss. = Cienfuegosia Cav. MALVACEAE (D)
Fuirena Rottb. CYPERACEAE (M)
Fulcaldea Poir. COMPOSITAE (D)
Fulchironia Lesch. = Phoenix L. PALMAE (M)
Fumana (Dunal) Spach CISTACEAE (D)
Fumaria L. PAPAVERACEAE (D)

Funastrum Fourn. ASCLEPIADACEAE (D)
Funckia Dennst. = Lumnitzera Willd. COMBRETACEAE (D)
Funifera Leandro ex C.A.Mey. THYMELAEACEAE (D)
Funkiella Schltr. = Schiedeella Schltr. ORCHIDACEAE (M)
Funtumia Stapf APOCYNACEAE (D)
Furarium Rizzini = Oryctanthus (Griseb.) Eichl. LORANTHACEAE (D)
Furcaria Boivin ex Baill. = Croton L. EUPHORBIACEAE (D)
Furcaria Desv. (SUH) = Ceratopteris Brongn. PARKERIACEAE (F)
Furcatella Baum.–Bod. (SUI) = Psychotria L. RUBIACEAE (D)
Furcilla Tiegh. (SUH) = Muellerina Tiegh. LORANTHACEAE (D)
Furcraea Vent. AGAVACEAE (M)
Furiolobivia Y.Ito (SUI) = Echinopsis Zucc. CACTACEAE (D)
Furtadoa M.Hotta ARACEAE (M)
Fusaea (Baill.) Saff. ANNONACEAE (D)
Fusanus R.Br. = Santalum L. SANTALACEAE (D)
Fusanus L. (SUH) = Colpoon Bergius SANTALACEAE (D)
Fusispermum Cuatrec. VIOLACEAE (D)
Fussia Schur (SUS) = Aira L. GRAMINEAE (M)
Fusticus Raf. = Maclura Nutt. MORACEAE (D)
Fuziifilix Nakai & Momose = Microlepia C.Presl DENNSTAEDTIACEAE (F)

Gabertia Gaudich. = Grammatophyllum Blume ORCHIDACEAE (M)
Gabunia K.Schum. ex Stapf = Tabernaemontana L. APOCYNACEAE (D)
Gadellia Shulkina = Campanula L. CAMPANULACEAE (D)
Gaedawakka Kuntze = Chaetocarpus Thwaites EUPHORBIACEAE (D)
Gaertnera Retz. (SUH) = Sphenoclea Gaertn. CAMPANULACEAE (D)
Gaertnera Schreb. (SUH) = Hiptage Gaertn. MALPIGHIACEAE (D)
Gaertnera Lam. RUBIACEAE (D)
Gagea Salisb. LILIACEAE (M)
Gagnebina Neck. ex DC. LEGUMINOSAE–MIMOSOIDEAE (D)
Gagnepainia K.Schum. ZINGIBERACEAE (M)
Gagria M.Král = Pachyphragma (DC.) Rchb. CRUCIFERAE (D)
Gahnia J.R.Forst. & G.Forst. CYPERACEAE (M)
Gaiadendron G.Don LORANTHACEAE (D)
Gaillardia Foug. COMPOSITAE (D)
Gaillionia Endl. (SUO) = Neogaillonia Lincz. RUBIACEAE (D)
Gaillonia A.Rich. ex DC. (SUH) = Neogaillonia Lincz. RUBIACEAE (D)
Gaimardia Gaudich. CENTROLEPIDACEAE (M)
Galactia P.Browne LEGUMINOSAE–PAPILIONOIDEAE (D)
Galactites Moench COMPOSITAE (D)
Galactodendron Kunth = Brosimum Sw. MORACEAE (D)
Galactophora Woodson APOCYNACEAE (D)
Galactoxylon Pierre = Palaquium Blanco SAPOTACEAE (D)
Galagania Lipsky UMBELLIFERAE (D)
Galanthus L. AMARYLLIDACEAE (M)
Galapagoa Hook.f. = Tiquilia Pers. BORAGINACEAE (D)
Galarhoeus Haw. = Euphorbia L. EUPHORBIACEAE (D)
Galatella (Cass.) Cass. = Aster L. COMPOSITAE (D)
Galax Sims DIAPENSIACEAE (D)
Galaxia Thunb. IRIDACEAE (M)
Galbulimima F.M.Bailey HIMANTANDRACEAE (D)
Galeana La Llave COMPOSITAE (D)
Galeandra Lindl. ORCHIDACEAE (M)
Galearia Zoll. & Moritzi PANDACEAE (D)
Galearis Raf. ORCHIDACEAE (M)

Galega L. LEGUMINOSAE–PAPILIONOIDEAE (D)
Galenia L. AIZOACEAE (D)
Galeobdolon Huds. = Lamium L. LABIATAE (D)
Galeoglossa C.Presl = Pyrrosia Mirb. POLYPODIACEAE (F)
Galeoglossum A.Rich. & Galeotti = Prescottia Lindl. ORCHIDACEAE (M)
Galeola Lour. ORCHIDACEAE (M)
Galeomma Rauschert COMPOSITAE (D)
Galeopsis L. LABIATAE (D)
Galeorchis Rydb. = Orchis L. ORCHIDACEAE (M)
Galeottia A.Rich. = Mendoncella A.D.Hawkes ORCHIDACEAE (M)
Galeottia M.Martens & Gal. (SUI) = Zeugites P.Browne GRAMINEAE (M)
Galeottiella Schltr. = Brachystele Schltr. ORCHIDACEAE (M)
Galera Blume = Epipogium Borkh. ORCHIDACEAE (M)
Galianthe Griseb. = Spermacoce L. RUBIACEAE (D)
Galilea Parl. = Cyperus L. CYPERACEAE (M)
Galiniera Delile RUBIACEAE (D)
Galinsoga Ruíz & Pav. COMPOSITAE (D)
Galipea Aubl. RUTACEAE (D)
Galitzkya V.V.Botschantzeva CRUCIFERAE (D)
Galium L. RUBIACEAE (D)
Gallardoa Hicken MALPIGHIACEAE (D)
Gallesia Casar. PHYTOLACCACEAE (D)
Gallienia Dubard & Dop RUBIACEAE (D)
Gallion Pohl (SUO) = Galium L. RUBIACEAE (D)
Gallium Mill. (SUO) = Galium L. RUBIACEAE (D)
Galoglychia Gasp. = Ficus L. MORACEAE (D)
Galopina Thunb. RUBIACEAE (D)
Galphimia Cav. MALPIGHIACEAE (D)
Galpinia N.E.Br. LYTHRACEAE (D)
Galpinsia Britton = Calylophus Spach ONAGRACEAE (D)
Galtonia Decne. HYACINTHACEAE (M)
Galurus Spreng. = Acalypha L. EUPHORBIACEAE (D)
Galvania Vand. = Psychotria L. RUBIACEAE (D)
Galvezia Dombey ex Juss. SCROPHULARIACEAE (D)
Gamanthus Bunge = Halanthium K.Koch CHENOPODIACEAE (D)
Gamaria Raf. = Disa Bergius ORCHIDACEAE (M)
Gamatopea Bremek. = Psychotria L. RUBIACEAE (D)
Gambelia Nutt. SCROPHULARIACEAE (D)
Gambeya Pierre = Chrysophyllum L. SAPOTACEAE (D)
Gambeyobotrys Aubrév. = Chrysophyllum L. SAPOTACEAE (D)
Gamblea C.B.Clarke ARALIACEAE (D)
Gamelythrum Nees = Amphipogon R.Br. GRAMINEAE (M)
Gamocarpha DC. CALYCERACEAE (D)
Gamochaeta Wedd. COMPOSITAE (D)
Gamochaetopsis Anderb. & Freire COMPOSITAE (D)
Gamogyne N.E.Br. = Piptospatha N.E.Br. ARACEAE (M)
Gamoplexis Falc. = Gastrodia R.Br. ORCHIDACEAE (M)
Gamopoda Baker = Rhaptonema Miers MENISPERMACEAE (D)
Gamosepalum Schltr. = Aulosepalum Garay ORCHIDACEAE (M)
Gamosepalum Hausskn. = Alyssum L. CRUCIFERAE (D)
Gampsoceras Steven = Ranunculus L. RANUNCULACEAE (D)
Gamwellia Baker = Gleditsia L. LEGUMINOSAE–CAESALPINIOIDEAE (D)
Ganguebina Vell. (SUO) = Manettia L. RUBIACEAE (D)
Ganitrus Gaertn. = Elaeocarpus L. ELAEOCARPACEAE (D)
Ganophyllum Blume SAPINDACEAE (D)

Gantelbua Bremek. = Hemigraphis Nees ACANTHACEAE (D)
Ganua Pierre ex Dubard = Madhuca Ham. ex J.F.Gmel. SAPOTACEAE (D)
Garapatica Karst. = Alibertia A.Rich. ex DC. RUBIACEAE (D)
Garaventia Looser ALLIACEAE (M)
Garayella Brieger (SUI) = Chamelophyton Garay ORCHIDACEAE (M)
Garberia A.Gray COMPOSITAE (D)
Garcia Rohr EUPHORBIACEAE (D)
Garcibarrigoa Cuatrec. COMPOSITAE (D)
Garcilassa Poepp. COMPOSITAE (D)
Garcinia L. GUTTIFERAE (D)
Gardena Adans. (SUO) = Gardenia Ellis RUBIACEAE (D)
Gardenia Ellis RUBIACEAE (D)
Gardeniola Cham. = Alibertia A.Rich. ex DC. RUBIACEAE (D)
Gardeniopsis Miq. RUBIACEAE (D)
Gardneria Wall. LOGANIACEAE (D)
Gardnerina R.M.King & H.Rob. COMPOSITAE (D)
Gardnerodoxa Sandwith BIGNONIACEAE (D)
Gardoquia Ruíz & Pav. LABIATAE (D)
Garhadiolus Jaub. & Spach = Rhagadiolus Scop. COMPOSITAE (D)
Garidella L. = Nigella L. RANUNCULACEAE (D)
Garnieria Brongn. & Gris PROTEACEAE (D)
Garnotia Brongn. GRAMINEAE (M)
Garnotiella Stapf = Asthenochloa Büse GRAMINEAE (M)
Garretia Welw. = Khaya A.Juss. MELIACEAE (D)
Garrettia H.R.Fletcher VERBENACEAE (D)
Garrya Douglas ex Lindl. GARRYACEAE (D)
Garuga Roxb. BURSERACEAE (D)
Garuleum Cass. COMPOSITAE (D)
Gasoul Adans. = Mesembryanthemum L. AIZOACEAE (D)
Gasteranthopsis Oerst. = Besleria L. GESNERIACEAE (D)
Gasteranthus Benth. GESNERIACEAE (D)
Gasteria Duval ALOACEAE (M)
Gastonia Comm. ex Lam. ARALIACEAE (D)
Gastorchis Thouars = Phaius Lour. ORCHIDACEAE (M)
Gastranthus Moritz ex Benth. & Hook. = Stenostephanus Nees ACANTHACEAE (D)
Gastranthus F.Muell. (SUI) = Parsonsia R.Br. APOCYNACEAE (D)
Gastridium P.Beauv. GRAMINEAE (M)
Gastrocalyx Schischk. (SUH) = Schischkiniella Steenis CARYOPHYLLACEAE (D)
Gastrochilus Wall. (SUH) = Boesenbergia Kuntze ZINGIBERACEAE (M)
Gastrochilus D.Don ORCHIDACEAE (M)
Gastrococos Morales PALMAE (M)
Gastrocotyle Bunge BORAGINACEAE (D)
Gastrodia R.Br. ORCHIDACEAE (M)
Gastroglottis Blume = Liparis Rich. ORCHIDACEAE (M)
Gastrolepis Tiegh. ICACINACEAE (D)
Gastrolobium R.Br. LEGUMINOSAE–PAPILIONOIDEAE (D)
Gastromeria D.Don = Melasma Bergius SCROPHULARIACEAE (D)
Gastropyrum (Jaub. & Spach) Á.Löve = Aegilops L. GRAMINEAE (M)
Gastrorchis Schltr. = Phaius Lour. ORCHIDACEAE (M)
Gatesia A.Gray (SUH) = Yeatesia Small ACANTHACEAE (D)
Gatnaia Gagnep. = Baccaurea Lour. EUPHORBIACEAE (D)
Gaudichaudia Kunth MALPIGHIACEAE (D)
Gaudinia P.Beauv. GRAMINEAE (M)
Gaudinopsis (Boiss.) Eig = Ventenata Koeler GRAMINEAE (M)
Gaultheria Kalm ex L. ERICACEAE (D)

Gaumerocassia Britton = Senna Mill. LEGUMINOSAE–CAESALPINIOIDEAE (D)
Gaura L. ONAGRACEAE (D)
Gaurella Small = Oenothera L. ONAGRACEAE (D)
Gauridium Spach = Gaura L. ONAGRACEAE (D)
Gauropsis (Torr. & Frém.) Cockerell (SUH) = Oenothera L. ONAGRACEAE (D)
Gauropsis C.Presl = Clarkia Pursh ONAGRACEAE (D)
Gaussia H.A.Wendl. PALMAE (M)
Gavarretia Baill. EUPHORBIACEAE (D)
Gavilea Poepp. ORCHIDACEAE (M)
Gaya Kunth MALVACEAE (D)
Gaya Gaudin (SUH) = Pachypleurum Ledeb. UMBELLIFERAE (D)
Gayella Pierre = Pouteria Aubl. SAPOTACEAE (D)
Gaylussacia Kunth ERICACEAE (D)
Gayoides (Endl.) Small = Herissantia Medik. MALVACEAE (D)
Gayophytum A.Juss. ONAGRACEAE (D)
Gazachloa J.B.Phipps = Danthoniopsis Stapf GRAMINEAE (M)
Gazania Gaertn. COMPOSITAE (D)
Gazaniopsis C.Huber (UP) COMPOSITAE (D)
Geanthemum (R.E.Fr.) Saff. = Duguetia A.St.-Hil. ANNONACEAE (D)
Geanthus Reinw. = Etlingera Giseke ZINGIBERACEAE (M)
Gearum N.E.Br. ARACEAE (M)
Geblera Fisch. & C.A.Mey. = Flueggea Willd. EUPHORBIACEAE (D)
Geesinkorchis de Vogel ORCHIDACEAE (M)
Geigeria Griess. COMPOSITAE (D)
Geijera Schott RUTACEAE (D)
Geisarina Raf. = Forestiera Poir. OLEACEAE (D)
Geiseleria Klotzsch = Croton L. EUPHORBIACEAE (D)
Geissanthera Schltr. = Microtatorchis Schltr. ORCHIDACEAE (M)
Geissanthus Hook.f. MYRSINACEAE (D)
Geissaspis Wight & Arn. LEGUMINOSAE–PAPILIONOIDEAE (D)
Geissois Labill. CUNONIACEAE (D)
Geissolepis B.L.Rob. COMPOSITAE (D)
Geissoloma Lindl. ex Kunth GEISSOLOMATACEAE (D)
Geissomeria Lindl. ACANTHACEAE (D)
Geissopappus Benth. COMPOSITAE (D)
Geissorhiza Ker Gawl. IRIDACEAE (M)
Geissospermum Allemão APOCYNACEAE (D)
Geitonoplesium A.Cunn. ex R.Br. PHILESIACEAE (M)
Gelasine Herb. IRIDACEAE (M)
Geleznowia Turcz. RUTACEAE (D)
Gelibia Hutch. = Polyscias J.R.Forst. & G.Forst. ARALIACEAE (D)
Gelidocalamus T.H.Wen = Indocalamus Nakai GRAMINEAE (M)
Gelonium Roxb. ex Willd. (SUH) = Suregada Roxb. ex Rottler EUPHORBIACEAE (D)
Gelpkea Blume = Syzygium Gaertn. MYRTACEAE (D)
Gelsemium Juss. LOGANIACEAE (D)
Gembanga Blume = Corypha L. PALMAE (M)
Geminaria Raf. = Savia Willd. EUPHORBIACEAE (D)
Gemmaria Salisb. AMARYLLIDACEAE (M)
Gendarussa Nees = Justicia L. ACANTHACEAE (D)
Genea (Dumort.) Dumort. (SUH) = Bromus L. GRAMINEAE (M)
Genesiphylla L'Hér. = Phyllanthus L. EUPHORBIACEAE (D)
Genetyllis DC. = Darwinia Rudge MYRTACEAE (D)
Genianthus Hook.f. ASCLEPIADACEAE (D)
Geniosporum Wall. ex Benth. LABIATAE (D)
Geniostemon Engelm. & A.Gray GENTIANACEAE (D)

Geniostoma J.R.Forst. & G.Forst. LOGANIACEAE (D)
Genipa L. RUBIACEAE (D)
Genipella A.Rich. ex DC. (SUI) = Alibertia A.Rich. ex DC. RUBIACEAE (D)
Genista L. LEGUMINOSAE–PAPILIONOIDEAE (D)
Genistella Ortega = Genista L. LEGUMINOSAE–PAPILIONOIDEAE (D)
Genistidium I.M.Johnst. LEGUMINOSAE–PAPILIONOIDEAE (D)
Genlisea A.St.-Hil. LENTIBULARIACEAE (D)
Gennaria Parl. ORCHIDACEAE (M)
Genoplesium R.Br. ORCHIDACEAE (M)
Gentiana L. GENTIANACEAE (D)
Gentianella Moench GENTIANACEAE (D)
Gentianodes Á.Löve & D.Löve = Gentiana L. GENTIANACEAE (D)
Gentianopsis Ma GENTIANACEAE (D)
Gentianothamnus Humbert GENTIANACEAE (D)
Gentilia Chev. & Beille = Bridelia Willd. EUPHORBIACEAE (D)
Gentingia J.T.Johanss. & K.M.Wong RUBIACEAE (D)
Gentlea Lundell MYRSINACEAE (D)
Gentrya Breedlove & Heckard SCROPHULARIACEAE (D)
Genyorchis Schltr. ORCHIDACEAE (M)
Geobalanus Small = Licania Aubl. CHRYSOBALANACEAE (D)
Geobina Raf. = Goodyera R.Br. ORCHIDACEAE (M)
Geoblasta Barb.Rodr. ORCHIDACEAE (M)
Geocalpa Brieger (SUI) = Pleurothallis R.Br. ORCHIDACEAE (M)
Geocardia Standl. = Geophila D.Don RUBIACEAE (D)
Geocarpon Mack. ILLECEBRACEAE (D)
Geocaryum Coss. UMBELLIFERAE (D)
Geocaulon Fernald SANTALACEAE (D)
Geocharis (K.Schum.) Ridl. ZINGIBERACEAE (M)
Geochorda Cham. & Schltdl. SCROPHULARIACEAE (D)
Geococcus J.L.Drumm. ex Harv. CRUCIFERAE (D)
Geodorum G.Jacks. ORCHIDACEAE (M)
Geoffraya Bonati = Lindernia All. SCROPHULARIACEAE (D)
Geoffroea Jacq. LEGUMINOSAE–PAPILIONOIDEAE (D)
Geogenanthus Ule COMMELINACEAE (M)
Geoherpum Willd. ex Schult. = Mitchella L. RUBIACEAE (D)
Geomitra Becc. BURMANNIACEAE (M)
Geonoma Willd. PALMAE (M)
Geopanax Hemsl. = Schefflera J.R.Forst. & G.Forst. ARALIACEAE (D)
Geophila D.Don RUBIACEAE (D)
Geopogon Steud. (SUI) = Chloris Sw. GRAMINEAE (M)
Geoprumnon Rydb. = Astragalus L. LEGUMINOSAE–PAPILIONOIDEAE (D)
Georchis Lindl. = Goodyera R.Br. ORCHIDACEAE (M)
Geosiris Baill. IRIDACEAE (M)
Geostachys (Baker) Ridl. ZINGIBERACEAE (M)
Geraea Torr. & A.Gray COMPOSITAE (D)
Geraniopsis Chrtek = Geranium L. GERANIACEAE (D)
Geranium L. GERANIACEAE (D)
Gerardia Benth. = Stenandrium Nees ACANTHACEAE (D)
Gerardia Benth. (SUH) = Agalinis Raf. SCROPHULARIACEAE (D)
Gerardiina Engl. SCROPHULARIACEAE (D)
Gerardiopsis Engl. = Anticharis Endl. SCROPHULARIACEAE (D)
Gerascanthus B.Browne = Cordia L. BORAGINACEAE (D)
Gerbera L. COMPOSITAE (D)
Germainia Balansa & Poitrasson GRAMINEAE (M)
Gerocephalus F.Ritter (SUS) = Espostoopsis Buxb. CACTACEAE (D)

Gerontogea Cham. & Schltdl. = Oldenlandia L. RUBIACEAE (D)
Geropogon L. COMPOSITAE (D)
Gerrardanthus Harv. ex Hook.f. CUCURBITACEAE (D)
Gerrardina Oliv. FLACOURTIACEAE (D)
Gersinia Néraud = Bulbophyllum Thouars ORCHIDACEAE (M)
Gertrudia K.Schum. = Ryparosa Blume FLACOURTIACEAE (D)
Gesnera Mart. = Sinningia Nees GESNERIACEAE (D)
Gesneria L. GESNERIACEAE (D)
Gesnouinia Gaudich. URTICACEAE (D)
Gestroa Becc. = Erythrospermum Lam. FLACOURTIACEAE (D)
Gethyllis L. AMARYLLIDACEAE (M)
Gethyum Phil. ALLIACEAE (M)
Getonia Roxb. COMBRETACEAE (D)
Geum L. ROSACEAE (D)
Geunsia Blume = Callicarpa L. VERBENACEAE (D)
Gevuina Molina PROTEACEAE (D)
Ghaznianthus Lincz. PLUMBAGINACEAE (D)
Ghiesbreghtia A.Rich. & Galeotti = Calanthe Ker-Gawl. ORCHIDACEAE (M)
Ghiesbreghtia A.Gray (SUH) = Eremogeton Standl. & L.O.Williams
    SCROPHULARIACEAE (D)
Ghikaea Volkens & Schweinf. SCROPHULARIACEAE (D)
Ghinia Schreb. = Tamonea Aubl. VERBENACEAE (D)
Giadotrum Pichon = Cleghornia Wight APOCYNACEAE (D)
Gibasis Raf. COMMELINACEAE (M)
Gibasoides D.R.Hunt COMMELINACEAE (M)
Gibbaeum Haw. AIZOACEAE (D)
Gibbaria Cass. COMPOSITAE (D)
Gibbsia Rendle URTICACEAE (D)
Gibsoniothamnus L.O.Williams SCROPHULARIACEAE (D)
Giesleria Regel = Kohleria Regel GESNERIACEAE (D)
Gifola Cass. = Filago L. COMPOSITAE (D)
Gigachilon Seidl (SUI) = Triticum L. GRAMINEAE (M)
Gigalobium P.Browne = Entada Adans. LEGUMINOSAE-MIMOSOIDEAE (D)
Gigantochloa Kurz ex Munro GRAMINEAE (M)
Gigasiphon Drake = Bauhinia L. LEGUMINOSAE-CAESALPINIOIDEAE (D)
Gigliolia Barb.Rodr. (SUH) = Octomeria R.Br. ORCHIDACEAE (M)
Gigliolia Becc. = Areca L. PALMAE (M)
Gijefa (M.Roem.) Post & Kuntze = Kedrostis Medik. CUCURBITACEAE (D)
Gilberta Turcz. COMPOSITAE (D)
Gilbertiella Boutique ANNONACEAE (D)
Gilbertiodendron J.Léonard LEGUMINOSAE-CAESALPINIOIDEAE (D)
Gilesia F.Muell. STERCULIACEAE (D)
Gilgiochloa Pilg. GRAMINEAE (M)
Gilgiodaphne Domke = Synandrodaphne Gilg THYMELAEACEAE (D)
Gilia Ruíz & Pav. POLEMONIACEAE (D)
Giliastrum Rydb. = Gilia Ruíz & Pav. POLEMONIACEAE (D)
Gilibertia Ruíz & Pav. (SUH) = Dendropanax Decne. & Planch. ARALIACEAE (D)
Gilibertia J.F.Gmel. = Turraea L. MELIACEAE (D)
Gilipus Raf. (UP) RUBIACEAE (D)
Gillbeea F.Muell. CUNONIACEAE (D)
Gillenia Moench ROSACEAE (D)
Gillespiea A.C.Sm. RUBIACEAE (D)
Gilletiella De Wild. & T.Durand (SUH) = Anomacanthus R.D.Good ACANTHACEAE (D)
Gilletiodendron Vermoesen LEGUMINOSAE-CAESALPINIOIDEAE (D)
Gillettia Rendle = Anthericopsis Engl. COMMELINACEAE (M)

Gilliesia Lindl. ALLIACEAE (M)
Gilmania Coville POLYGONACEAE (D)
Gilruthia Ewart COMPOSITAE (D)
Gimbernatea Ruíz & Pav. = Terminalia L. COMBRETACEAE (D)
Ginalloa Korth. VISCACEAE (D)
Ginannia Bubani (SUS) = Holcus L. GRAMINEAE (M)
Gingidia J.W.Dawson UMBELLIFERAE (D)
Gingidium J.R.Forst. & G.Forst. (SUH) = Gingidia J.W.Dawson UMBELLIFERAE (D)
Ginkgo L. GINKGOACEAE (G)
Ginnania M.Roem. (SUH) = Turraea L. MELIACEAE (D)
Ginora L. = Ginoria Jacq. LYTHRACEAE (D)
Ginoria Jacq. LYTHRACEAE (D)
Giorgiella De Wild. = Efulensia C.H.Wright PASSIFLORACEAE (D)
Giraldiella Dammer = Lloydia Salisb. ex Rchb. LILIACEAE (M)
Girardinia Gaudich. URTICACEAE (D)
Girgensohnia Bunge CHENOPODIACEAE (D)
Gironniera Gaudich. ULMACEAE (D)
Gisekia L. GISEKIACEAE (D)
Gisopteris Bernh. = Lygodium Sw. SCHIZAEACEAE (F)
Gitara Pax & K.Hoffm. EUPHORBIACEAE (D)
Githopsis Nutt. CAMPANULACEAE (D)
Giulianettia Rolfe = Glossorhyncha Ridl. ORCHIDACEAE (M)
Givotia Griff. EUPHORBIACEAE (D)
Gjellerupia Lauterb. OPILIACEAE (D)
Gladiolimon Mob. PLUMBAGINACEAE (D)
Gladiolus L. IRIDACEAE (M)
Gladiopappus Humbert COMPOSITAE (D)
Glandiloba (Raf.) Steud. (SUI) = Eriochloa Kunth GRAMINEAE (M)
Glandonia Griseb. MALPIGHIACEAE (D)
Glandularia J.F.Gmel. VERBENACEAE (D)
Glandulicactus Backeb. = Sclerocactus Britton & Rose CACTACEAE (D)
Glandulifera Frič (SUH) = Coryphantha (Engelm.) Lem. CACTACEAE (D)
Glaphyria Jack = Leptospermum J.F.Forst. & G.Forst. MYRTACEAE (D)
Glaphyropteridopsis Ching THELYPTERIDACEAE (F)
Glaphyropteris (Fée) C.Presl ex Fée THELYPTERIDACEAE (F)
Glaribraya Hara CRUCIFERAE (D)
Glastaria Boiss. CRUCIFERAE (D)
Glaucidium Siebold & Zucc. GLAUCIDIACEAE (D)
Glaucium Mill. PAPAVERACEAE (D)
Glaucocarpum Rollins CRUCIFERAE (D)
Glaucocochlearia (O.E.Schulz) Pobed. = Cochlearia L. CRUCIFERAE (D)
Glaucosciadium B.L.Burtt & P.H.Davis UMBELLIFERAE (D)
Glaucothea O.F.Cook = Brahea Mart. ex Endl. PALMAE (M)
Glaux L. PRIMULACEAE (D)
Glaziocharis Taub. ex Warm. = Thismia Griff. BURMANNIACEAE (M)
Glaziophyton Franch. GRAMINEAE (M)
Glaziostelma E.Fourn. = Tassadia Decne. ASCLEPIADACEAE (D)
Glaziova Mart. ex Drude (SUH) = Lytocaryum Toledo PALMAE (M)
Glaziova Bureau BIGNONIACEAE (D)
Glaziovianthus G.M.Barroso = Chresta Vell. ex DC. COMPOSITAE (D)
Gleadovia Gamble & Prain SCROPHULARIACEAE (D)
Gleasonia Standl. RUBIACEAE (D)
Glechoma L. LABIATAE (D)
Glechon Spreng. LABIATAE (D)
Gleditschia Scop. = Gleditsia L. LEGUMINOSAE–CAESALPINIOIDEAE (D)

Gleditsia L. LEGUMINOSAE–CAESALPINIOIDEAE (D)
Glehnia F.Schmidt ex Miq. UMBELLIFERAE (D)
Gleichenella Ching = Dicranopteris Bernh. GLEICHENIACEAE (F)
Gleichenia Sm. GLEICHENIACEAE (F)
Gleicheniastrum C.Presl = Gleichenia Sm. GLEICHENIACEAE (F)
Glekia Hilliard SCROPHULARIACEAE (D)
Glenniea Hook.f. SAPINDACEAE (D)
Glia Sond. UMBELLIFERAE (D)
Glinus L. MOLLUGINACEAE (D)
Glionnetia Tirveng. RUBIACEAE (D)
Gliopsis Rauschert (SUS) = Rutheopsis A.Hansen & Kunkel UMBELLIFERAE (D)
Gliricidia Kunth LEGUMINOSAE–PAPILIONOIDEAE (D)
Glischrocaryon Endl. HALORAGACEAE (D)
Glischrocolla (Endl.) A.DC. PENAEACEAE (D)
Glischrothamnus Pilg. MOLLUGINACEAE (D)
Globba L. ZINGIBERACEAE (M)
Globimetula Tiegh. LORANTHACEAE (D)
Globularia L. GLOBULARIACEAE (D)
Globulariopsis Compton SCROPHULARIACEAE (D)
Globulostylis Wernham = Cuviera DC. RUBIACEAE (D)
Glochidion J.R.Forst. & G.Forst. EUPHORBIACEAE (D)
Glochidionopsis Blume = Glochidion J.R.Forst. & G.Forst. EUPHORBIACEAE (D)
Glochidotheca Fenzl UMBELLIFERAE (D)
Glochisandra Wight = Glochidion J.R.Forst. & G.Forst. EUPHORBIACEAE (D)
Glockeria Nees (SUH) = Habracanthus Nees ACANTHACEAE (D)
Gloeocarpus Radlk. SAPINDACEAE (D)
Gloeospermum Triana & Planch. VIOLACEAE (D)
Glomera Blume ORCHIDACEAE (M)
Glomeropitcairnia Mez BROMELIACEAE (M)
Gloneria E.F.Andre = Psychotria L. RUBIACEAE (D)
Gloriosa L. COLCHICACEAE (M)
Glosarithys Rizzini = Justicia L. ACANTHACEAE (D)
Glosocomia D.Don (SUO) = Codonopsis Wall. CAMPANULACEAE (D)
Glossanthus Klein ex Benth. = Rhynchoglossum Blume GESNERIACEAE (D)
Glossapis Spreng. = Habenaria Willd. ORCHIDACEAE (M)
Glossarion Maguire & Wurdack COMPOSITAE (D)
Glossidea Tiegh. = Psittacanthus Mart. LORANTHACEAE (D)
Glossocalyx Benth. MONIMIACEAE (D)
Glossocardia Cass. COMPOSITAE (D)
Glossocarya Wall. ex Griff. VERBENACEAE (D)
Glossochilus Nees ACANTHACEAE (D)
Glossocomia Rchb. (SUO) = Codonopsis Wall. CAMPANULACEAE (D)
Glossodia R.Br. ORCHIDACEAE (M)
Glossogyne Cass. = Glossocardia Cass. COMPOSITAE (D)
Glossolepis Gilg = Chytranthus Hook.f. SAPINDACEAE (D)
Glossoloma Hanst. = Alloplectus Mart. GESNERIACEAE (D)
Glossoma Schreb. = Votomita Aubl. MELASTOMATACEAE (D)
Glossonema Decne. ASCLEPIADACEAE (D)
Glossopappus Kunze COMPOSITAE (D)
Glossopetalon A.Gray CROSSOSOMATACEAE (D)
Glossopholis Pierre = Tiliacora Colebr. MENISPERMACEAE (D)
Glossopteris Raf. (SUH) = Asplenium L. ASPLENIACEAE (F)
Glossorhyncha Ridl. ORCHIDACEAE (M)
Glossostelma Schltr. ASCLEPIADACEAE (D)
Glossostemon Desf. STERCULIACEAE (D)

Glossostephanus E.Mey. = Oncinema Arn. ASCLEPIADACEAE (D)
Glossostigma Wight & Arn. SCROPHULARIACEAE (D)
Glossostipula Lorence RUBIACEAE (D)
Glossula Lindl. = Habenaria Willd. ORCHIDACEAE (M)
Glottidium Desv. LEGUMINOSAE–PAPILIONOIDEAE (D)
Glottiphyllum Haw. AIZOACEAE (D)
Gloxinia L'Hér. GESNERIACEAE (D)
Gluema Aubrév. & Pellegr. SAPOTACEAE (D)
Glumicalyx Hiern SCROPHULARIACEAE (D)
Gluta L. ANACARDIACEAE (D)
Glutago Comm. ex Poir. = Oryctanthus (Griseb.) Eichl. LORANTHACEAE (D)
Glycanthes Raf. = Columnea L. GESNERIACEAE (D)
Glyceria R.Br. GRAMINEAE (M)
Glycideras DC. = Psiadia Jacq. COMPOSITAE (D)
Glycine Willd. LEGUMINOSAE–PAPILIONOIDEAE (D)
Glycosmis Corrêa RUTACEAE (D)
Glycoxylon Ducke = Pradosia Liais SAPOTACEAE (D)
Glycydendron Ducke EUPHORBIACEAE (D)
Glycyrrhiza L. LEGUMINOSAE–PAPILIONOIDEAE (D)
Glyphaea Hook.f. TILIACEAE (D)
Glyphochloa Clayton GRAMINEAE (M)
Glyphosperma S.Watson = Asphodelus L. ASPHODELACEAE (M)
Glyphostylus Gagnep. EUPHORBIACEAE (D)
Glyphotaenium J.Sm. GRAMMITIDACEAE (F)
Glyptocarpa Hu = Pyrenaria Blume THEACEAE (D)
Glyptocaryopsis Brand = Plagiobothrys Fisch. & C.A.Mey. BORAGINACEAE (D)
Glyptopetalum Thwaites CELASTRACEAE (D)
Glyptopleura Eaton COMPOSITAE (D)
Glyptostrobus Endl. TAXODIACEAE (G)
Gmelina L. VERBENACEAE (D)
Gnaphaliothamnus Kirp. COMPOSITAE (D)
Gnaphalium L. COMPOSITAE (D)
Gnaphalodes Mill. = Actinobole Endl. COMPOSITAE (D)
Gnephosis Cass. COMPOSITAE (D)
Gnetum L. GNETACEAE (G)
Gnidia L. THYMELAEACEAE (D)
Gnomonia Lunell (SUS) = Festuca L. GRAMINEAE (M)
Gnoteris Raf. (SUS) = Hyptis Jacq. LABIATAE (D)
Goadbyella R.S.Rogers = Microtis R.Br. ORCHIDACEAE (M)
Gochnatia Kunth COMPOSITAE (D)
Godefroya Gagnep. = Cleistanthus Hook.f. ex Planch. EUPHORBIACEAE (D)
Godetia Spach = Clarkia Pursh ONAGRACEAE (D)
Godmania Hemsl. BIGNONIACEAE (D)
Godoya Ruíz & Pav. OCHNACEAE (D)
Goebelia Bunge ex Boiss. = Sophora L. LEGUMINOSAE–PAPILIONOIDEAE (D)
Goeldinia Huber = Allantoma Miers LECYTHIDACEAE (D)
Goeppertia Griseb. (SUH) = Bisgoeppertia Kuntze GENTIANACEAE (D)
Goeppertia Nees = Endlicheria Nees LAURACEAE (D)
Goerziella Urb. CHENOPODIACEAE (D)
Goethalsia Pittier TILIACEAE (D)
Goethartia Herz. = Pouzolzia Gaudich. URTICACEAE (D)
Goethea Nees MALVACEAE (D)
Goetzea Wydl. GOETZEACEAE (D)
Golaea Chiov. ACANTHACEAE (D)
Goldbachia DC. CRUCIFERAE (D)

Goldbachia Trin. (SUH) = Arundinella Raddi GRAMINEAE (M)
Goldfussia Nees = Strobilanthes Blume ACANTHACEAE (D)
Goldmanella Greenm. COMPOSITAE (D)
Goldmania Rose ex Micheli LEGUMINOSAE–MIMOSOIDEAE (D)
Goldschmidtia Dammer = Dendrobium Sw. ORCHIDACEAE (M)
Gomara Adans. = Crassula L. CRASSULACEAE (D)
Gomara Ruíz & Pav. (SUH) = Sanango Bunting & Duke BUDDLEJACEAE (D)
Gomaranthus Rauschert = Sanango Bunting & Duke BUDDLEJACEAE (D)
Gomaria Spreng. (SUO) = Sanango Bunting & Duke BUDDLEJACEAE (D)
Gomesa R.Br. ORCHIDACEAE (M)
Gomezia Mutis (SUO) = Nertera Banks & Sol. ex Gaertn. RUBIACEAE (D)
Gomidesia O.Berg MYRTACEAE (D)
Gomortega Ruíz & Pav. GOMORTEGACEAE (D)
Gomosia Lam. (SUO) = Nertera Banks & Sol. ex Gaertn. RUBIACEAE (D)
Gomoza Cothen. (SUO) = Nertera Banks & Sol. ex Gaertn. RUBIACEAE (D)
Gomozia Mutis ex L.f. = Nertera Banks & Sol. ex Gaertn. RUBIACEAE (D)
Gomphandra Wall. ex Lindl. ICACINACEAE (D)
Gomphia Schreb. OCHNACEAE (D)
Gomphichis Lindl. ORCHIDACEAE (M)
Gomphiluma Baill. = Pouteria Aubl. SAPOTACEAE (D)
Gomphocalyx Baker RUBIACEAE (D)
Gomphocarpus R.Br. ASCLEPIADACEAE (D)
Gomphogyne Griff. CUCURBITACEAE (D)
Gompholobium Sm. LEGUMINOSAE–PAPILIONOIDEAE (D)
Gomphopetalum Turcz. = Angelica L. UMBELLIFERAE (D)
Gomphosia Wedd. = Ferdinandusa Pohl RUBIACEAE (D)
Gomphostemma Wall. ex Benth. LABIATAE (D)
Gomphostigma Turcz. BUDDLEJACEAE (D)
Gomphotis Raf. = Thryptomene Endl. MYRTACEAE (D)
Gomphrena L. AMARANTHACEAE (D)
Gomutus Corrêa = Arenga Labill. PALMAE (M)
Gonatanthus Klotzsch ARACEAE (M)
Gonatogyne Klotzsch ex Muell.Arg. = Savia Willd. EUPHORBIACEAE (D)
Gonatopus Hook.f. ex Engl. ARACEAE (M)
Gonatostylis Schltr. ORCHIDACEAE (M)
Gongora Ruíz & Pav. ORCHIDACEAE (M)
Gongrodiscus Radlk. SAPINDACEAE (D)
Gongronema (Endl.) Decne. ASCLEPIADACEAE (D)
Gongrospermum Radlk. SAPINDACEAE (D)
Gongrostylus R.M.King & H.Rob. COMPOSITAE (D)
Gongrothamnus Steetz = Distephanus (Cass.) Cass. COMPOSITAE (D)
Gongylocarpus Schltdl. & Cham. ONAGRACEAE (D)
Gongylolepis R.H.Schomb. COMPOSITAE (D)
Gongylosciadium Rech.f. UMBELLIFERAE (D)
Gongylosperma King & Gamble ASCLEPIADACEAE (D)
Gonianthes A.Rich. (SUH) = Portlandia P.Browne RUBIACEAE (D)
Gonioanthela Malme ASCLEPIADACEAE (D)
Goniocaulon Cass. COMPOSITAE (D)
Goniochilus M.W.Chase ORCHIDACEAE (M)
Goniocladus Burret PALMAE (M)
Goniodiscus Kuhlm. CELASTRACEAE (D)
Goniogyna DC. = Crotalaria L. LEGUMINOSAE–PAPILIONOIDEAE (D)
Goniolimon Boiss. PLUMBAGINACEAE (D)
Gonioma E.Mey. APOCYNACEAE (D)
Goniophlebium C.Presl POLYPODIACEAE (F)

Goniopteris C.Presl THELYPTERIDACEAE (F)
Goniorrhachis Taub. LEGUMINOSAE–CAESALPINIOIDEAE (D)
Gonioscypha Baker CONVALLARIACEAE (M)
Goniosperma Burret = Physokentia Becc. PALMAE (M)
Goniostemma Wight ASCLEPIADACEAE (D)
Goniothalamus (Blume) Hook.f. & Thomson ANNONACEAE (D)
Gonocalyx Planch. & Linden ERICACEAE (D)
Gonocarpus Thunb. HALORAGACEAE (D)
Gonocarpus Ham. (SUH) = Combretum Loefl. COMBRETACEAE (D)
Gonocaryum Miq. ICACINACEAE (D)
Gonocormus Van den Bosch = Crepidomanes (C.Presl) C.Presl
    HYMENOPHYLLACEAE (F)
Gonocrypta Baill. ASCLEPIADACEAE (D)
Gonocytisus Spach LEGUMINOSAE–PAPILIONOIDEAE (D)
Gonogona Link = Goodyera R.Br. ORCHIDACEAE (M)
Gonolobus Michx. ASCLEPIADACEAE (D)
Gonoptera Turcz. = Bulnesia Gay ZYGOPHYLLACEAE (D)
Gonopyrum Fisch. & C.A.Mey. = Polygonella Michx. POLYGONACEAE (D)
Gonospermum Less. COMPOSITAE (D)
Gonostegia Turcz. = Hyrtanandra Miq. URTICACEAE (D)
Gonosuke Raf. = Ficus L. MORACEAE (D)
Gonotheca Blume ex DC. = Oldenlandia L. RUBIACEAE (D)
Gonsii Adans. (SUS) = Adenanthera L. LEGUMINOSAE–MIMOSOIDEAE (D)
Gontscharovia Borisova LABIATAE (D)
Gonyanera Korth. = Acranthera Arn. ex Meisn. RUBIACEAE (D)
Gonypetalum Ule = Tapura Aubl. DICHAPETALACEAE (D)
Gonystylus Teijsm. & Binn. THYMELAEACEAE (D)
Gonzalagunia Ruíz & Pav. RUBIACEAE (D)
Gonzalea Pers. = Gonzalagunia Ruíz & Pav. RUBIACEAE (D)
Goodallia Benth. THYMELAEACEAE (D)
Goodenia Sm. GOODENIACEAE (D)
Goodia Salisb. LEGUMINOSAE–PAPILIONOIDEAE (D)
Goodmania Reveal & Ertter POLYGONACEAE (D)
Goodyera R.Br. ORCHIDACEAE (M)
Gooringia F.N.Williams = Arenaria L. CARYOPHYLLACEAE (D)
Gorceixia Baker COMPOSITAE (D)
Gordonia Ellis THEACEAE (D)
Gorgoglossum F.Lehm. = Sievekingia Rchb.f. ORCHIDACEAE (M)
Gorgonidium Schott ARACEAE (M)
Gormania Britton = Sedum L. CRASSULACEAE (D)
Gorodkovia Botsch. & Karav. CRUCIFERAE (D)
Gorteria L. COMPOSITAE (D)
Gosela Choisy SCROPHULARIACEAE (D)
Gossweilera S.Moore COMPOSITAE (D)
Gossweilerochloa Renvoize = Tridens Roem. & Schult. GRAMINEAE (M)
Gossweilerodendron Harms LEGUMINOSAE–CAESALPINIOIDEAE (D)
Gossypianthus Hook. = Guilleminea Kunth AMARANTHACEAE (D)
Gossypioides Skovst. ex J.B.Hutch. MALVACEAE (D)
Gossypiospermum (Griseb.) Urb. = Casearia Jacq. FLACOURTIACEAE (D)
Gossypium L. MALVACEAE (D)
Gothofreda Vent. = Oxypetalum R.Br. ASCLEPIADACEAE (D)
Gouania Jacq. RHAMNACEAE (D)
Gouffeia Robill. & Castagne ex Lam. & DC. = Arenaria L. CARYOPHYLLACEAE (D)
Goughia Wight = Daphniphyllum Blume DAPHNIPHYLLACEAE (D)
Gouinia E.Fourn. GRAMINEAE (M)

Goulardia Husn. = Elymus L. GRAMINEAE (M)
Gouldia A.Gray RUBIACEAE (D)
Gouldochloa J.Valdés, Morden & S.L.Hatch = Chasmanthium Link GRAMINEAE (M)
Goupia Aubl. GOUPIACEAE (D)
Gourliea Gillies ex Hook. & Arn. = Geoffroea Jacq. LEGUMINOSAE–
  PAPILIONOIDEAE (D)
Govenia Lindl. ORCHIDACEAE (M)
Govindooia Wight = Tropidia Lindl. ORCHIDACEAE (M)
Goyazia Taub. GESNERIACEAE (D)
Goyazianthus R.M.King & H.Rob. COMPOSITAE (D)
Grabowskia Schltdl. SOLANACEAE (D)
Graciela Rzed. = Strotheria B.L.Turner COMPOSITAE (D)
Gracilea Hook.f. = Melanocenchris Nees GRAMINEAE (M)
Graderia Benth. SCROPHULARIACEAE (D)
Graeffea Seem. = Trichospermum Blume TILIACEAE (D)
Graeffenrieda D.Dietr. (SUO) = Graffenrieda DC. MELASTOMATACEAE (D)
Graellsia Boiss. CRUCIFERAE (D)
Graffenrieda DC. MELASTOMATACEAE (D)
Grafia Rchb. UMBELLIFERAE (D)
Grafia A.D.Hawkes (SUH) = Phalaenopsis Blume ORCHIDACEAE (M)
Grahamia Gillies ex Hook. & Arn. PORTULACACEAE (D)
Grajalesia Miranda NYCTAGINACEAE (D)
Gramen Krause (SUI) = Festuca L. GRAMINEAE (M)
Gramen Séguier (SUS) = Secale L. GRAMINEAE (M)
Gramen W.Young (URI) GRAMINEAE (M)
Gramerium Desv. = Digitaria Haller GRAMINEAE (M)
Graminastrum E.H.L.Krause = Dissanthelium Trin. GRAMINEAE (M)
Grammadenia Benth. = Cybianthus Mart. MYRSINACEAE (D)
Grammangis Rchb.f. ORCHIDACEAE (M)
Grammanthes DC. = Crassula L. CRASSULACEAE (D)
Grammatocarpus C.Presl = Scyphanthus Sweet LOASACEAE (D)
Grammatophyllum Blume ORCHIDACEAE (M)
Grammatopteridium Alderw. POLYPODIACEAE (F)
Grammatopteris Alderw. = Grammatopteridium Alderw. POLYPODIACEAE (F)
Grammatosorus Regel = Tectaria Cav. DRYOPTERIDACEAE (F)
Grammatotheca C.Presl CAMPANULACEAE (D)
Grammica Lour. = Cuscuta L. CONVOLVULACEAE (D)
Grammitis Sw. GRAMMITIDACEAE (F)
Grammosciadium DC. UMBELLIFERAE (D)
Grammosolen Haegi SOLANACEAE (D)
Grammosperma O.E.Schulz CRUCIFERAE (D)
Granadilla Mill. = Passiflora L. PASSIFLORACEAE (D)
Grandidiera Jaub. FLACOURTIACEAE (D)
Grangea Adans. COMPOSITAE (D)
Grangeopsis Humbert COMPOSITAE (D)
Grangeria Comm. ex Juss. CHRYSOBALANACEAE (D)
Grantia Boiss. (SUH) = Iphiona Cass. COMPOSITAE (D)
Graphandra J.B.Imlay ACANTHACEAE (D)
Graphardisia (Mez) Lundell MYRSINACEAE (D)
Graphephorum Desv. GRAMINEAE (M)
Graphistemma (Champ. ex Benth.) Champ. ex Benth. ASCLEPIADACEAE (D)
Graphistylis B.Nord. COMPOSITAE (D)
Graphorkis Thouars ORCHIDACEAE (M)
Graptopetalum Rose CRASSULACEAE (D)
Graptophyllum Nees ACANTHACEAE (D)

Grastidium Blume = Dendrobium Sw. ORCHIDACEAE (M)
Gratiola L. SCROPHULARIACEAE (D)
Gratwickia F.Muell. COMPOSITAE (D)
Grauanthus Fayed COMPOSITAE (D)
Gravenshorstia Nees = Lonchostoma Wikstr. BRUNIACEAE (D)
Gravesia Naudin MELASTOMATACEAE (D)
Gravesiella A.Fern. & R.Fern. = Cincinnobotrys Gilg MELASTOMATACEAE (D)
Gravisia Mez = Aechmea Ruíz & Pav. BROMELIACEAE (M)
Graya Steud. (SUH) = Sphaerocaryum Nees ex Hook.f. GRAMINEAE (M)
Grayia Hook. & Arn. CHENOPODIACEAE (D)
Grazielia R.M.King & H.Rob. COMPOSITAE (D)
Grazielodendron H.C.Lima LEGUMINOSAE–PAPILIONOIDEAE (D)
Greenea Wight & Arn. RUBIACEAE (D)
Greenella A.Gray = Gutierrezia Lag. COMPOSITAE (D)
Greeneocharis Gürke & Harms = Cryptantha G.Don BORAGINACEAE (D)
Greenia Nutt. (SUH) = Limnodea L.H.Dewey GRAMINEAE (M)
Greeniopsis Merr. RUBIACEAE (D)
Greenmaniella W.M.Sharp COMPOSITAE (D)
Greenovia Webb & Berthel. CRASSULACEAE (D)
Greenwayodendron Verdc. ANNONACEAE (D)
Greenwoodia Burns–Bal. ORCHIDACEAE (M)
Greggia A.Gray (SUH) = Nerisyrenia Greene CRUCIFERAE (D)
Greggia Gaertn. (SUH) = Eugenia L. MYRTACEAE (D)
Gregoria Duby = Androsace L. PRIMULACEAE (D)
Greigia Regel BROMELIACEAE (M)
Grenacheria Mez MYRSINACEAE (D)
Greslania Balansa GRAMINEAE (M)
Grevea Baill. MONTINIACEAE (D)
Grevellina Baill. = Turraea L. MELIACEAE (D)
Grevillea R.Br. ex Knight PROTEACEAE (D)
Grewia L. TILIACEAE (D)
Grewiopsis De Wild. & T.Durand = Desplatsia Bocquillon TILIACEAE (D)
Greyia Hook. & Harv. GREYIACEAE (D)
Grias L. LECYTHIDACEAE (D)
Grielum L. NEURADACEAE (D)
Griffinia Ker Gawl. AMARYLLIDACEAE (M)
Griffithella (Tul.) Warm. PODOSTEMACEAE (D)
Griffithia Maingay ex King (SUH) = Enicosanthum Becc. ANNONACEAE (D)
Griffithia Wight & Arn. = Benkara Adans. RUBIACEAE (D)
Griffithianthus Merr. = Enicosanthum Becc. ANNONACEAE (D)
Griffithsochloa G.J.Pierce GRAMINEAE (M)
Griffonia Baill. LEGUMINOSAE–CAESALPINIOIDEAE (D)
Griffonia Hook.f. (SUH) = Acioa Aubl. CHRYSOBALANACEAE (D)
Grimaldia Schrank = Chamaecrista Moench LEGUMINOSAE–CAESALPINIOIDEAE (D)
Grimmeodendron Urb. EUPHORBIACEAE (D)
Grindelia Willd. COMPOSITAE (D)
Gripidea Miers = Caiophora C.Presl LOASACEAE (D)
Grischowia Karst. = Monochaetum (DC.) Naudin MELASTOMATACEAE (D)
Grisebachia Drude & H.A.Wendl. (SUH) = Howea Becc. PALMAE (M)
Grisebachia Klotzsch ERICACEAE (D)
Grisebachianthus R.M.King & H.Rob. COMPOSITAE (D)
Grisebachiella Lorentz ASCLEPIADACEAE (D)
Griselinia J.R.Forst. & G.Forst. GRISELINIACEAE (D)
Grisia Brongn. = Bikkia Reinw. RUBIACEAE (D)
Grislea Loefl. (SUI) = Pehria Sprague LYTHRACEAE (D)

Grislea L. = Combretum Loefl. COMBRETACEAE (D)
Grisollea Baill. ICACINACEAE (D)
Grisseea Bakh.f. APOCYNACEAE (D)
Grobya Lindl. ORCHIDACEAE (M)
Groenlandia J.Gay POTAMOGETONACEAE (M)
Gronophyllum Scheff. PALMAE (M)
Gronovia L. LOASACEAE (D)
Gronovia Blanco (SUH) = Illigera Blume HERNANDIACEAE (D)
Grosourdya Rchb.f. ORCHIDACEAE (M)
Grossera Pax EUPHORBIACEAE (D)
Grossularia Mill. = Ribes L. GROSSULARIACEAE (D)
Grosvenoria R.M.King & H.Rob. COMPOSITAE (D)
Grotefendia Seem. = Polyscias J.R.Forst. & G.Forst. ARALIACEAE (D)
Grubbia Bergius GRUBBIACEAE (D)
Gruhlmannia Neck. ex Raf. = Spermacoce L. RUBIACEAE (D)
Grumilea Gaertn. = Psychotria L. RUBIACEAE (D)
Grundlea Steud. (SUO) = Psychotria L. RUBIACEAE (D)
Grunilea Poir. (SUO) = Psychotria L. RUBIACEAE (D)
Grusonia F.Rchb. ex Britton & Rose = Opuntia Mill. CACTACEAE (D)
Guacamaya Maguire RAPATEACEAE (M)
Guadella Franch. (SUO) = Guaduella Franch. GRAMINEAE (M)
Guadua Kunth = Bambusa Schreb. GRAMINEAE (M)
Guaduella Franch. GRAMINEAE (M)
Guagnebina Vell. = Manettia L. RUBIACEAE (D)
Guaiacum L. ZYGOPHYLLACEAE (D)
Guaicaia Maguire = Glossarion Maguire & Wurdack COMPOSITAE (D)
Guajacum L. (SUO) = Guiacum L. ZYGOPHYLLACEAE (D)
Guajava Mill. = Psidium L. MYRTACEAE (D)
Guamatela Donn.Sm. ROSACEAE (D)
Guamia Merr. ANNONACEAE (D)
Guanabanus Mill. = Annona L. ANNONACEAE (D)
Guapeba Gomes = Pouteria Aubl. SAPOTACEAE (D)
Guapira Aubl. NYCTAGINACEAE (D)
Guarania Wedd. ex Baill. = Richeria Vahl EUPHORBIACEAE (D)
Guardiola Cerv. ex Humb. & Bonpl. COMPOSITAE (D)
Guarea L. MELIACEAE (D)
Guatteria Ruíz & Pav. ANNONACEAE (D)
Guatteriella R.E.Fr. ANNONACEAE (D)
Guatteriopsis R.E.Fr. ANNONACEAE (D)
Guayabilla Sessé & Moçino = Samyda Jacq. FLACOURTIACEAE (D)
Guayania R.M.King & H.Rob. COMPOSITAE (D)
Guaymasia Britton & Rose = Caesalpinia L. LEGUMINOSAE–CAESALPINIOIDEAE (D)
Guazuma Mill. STERCULIACEAE (D)
Gubleria Gaudich. = Nolana L.f. SOLANACEAE (D)
Gueldenstaedtia Fisch. LEGUMINOSAE–PAPILIONOIDEAE (D)
Guerinia J.Sm. = Lindsaea Dryand. ex Sm. DENNSTAEDTIACEAE (F)
Guerkea K.Schum. = Baissea A.DC. APOCYNACEAE (D)
Guerreroia Merr. = Glossocardia Cass. COMPOSITAE (D)
Guersentia Raf. = Chrysophyllum L. SAPOTACEAE (D)
Guettarda L. RUBIACEAE (D)
Guettardella Benth. = Antirhea Comm. ex Juss. RUBIACEAE (D)
Guevaria R.M.King & H.Rob. COMPOSITAE (D)
Guibourtia Benn. LEGUMINOSAE–CAESALPINIOIDEAE (D)
Guichenotia J.Gay STERCULIACEAE (D)
Guidonia (DC.) Griseb. (SUH) = Samyda Jacq. FLACOURTIACEAE (D)

Guidonia Mill. = Samyda Jacq. FLACOURTIACEAE (D)
Guidonia P.Browne (SUH) = Laetia L. FLACOURTIACEAE (D)
Guiera Adans. ex Juss. COMBRETACEAE (D)
Guihaia J.Dransf., S.K.Lee & F.N.Wei PALMAE (M)
Guilandina L. = Caesalpinia L. LEGUMINOSAE–CAESALPINIOIDEAE (D)
Guilfoylia F.Muell. SURIANACEAE (D)
Guilielma Mart. = Bactris Jacq. PALMAE (M)
Guillainia Ridl. = Alpinia Roxb. ZINGIBERACEAE (M)
Guilleminea Kunth AMARANTHACEAE (D)
Guillenia Greene CRUCIFERAE (D)
Guillonea Coss. UMBELLIFERAE (D)
Guindilia Gillies ex Hook. & Arn. SAPINDACEAE (D)
Guioa Cav. SAPINDACEAE (D)
Guiraoa Coss. CRUCIFERAE (D)
Guizotia Cass. COMPOSITAE (D)
Gularia Garay = Schiedeella Schltr. ORCHIDACEAE (M)
Gulubia Becc. PALMAE (M)
Gulubiopsis Becc. = Gulubia Becc. PALMAE (M)
Gumillea Ruíz & Pav. CUNONIACEAE (D)
Gundelia L. COMPOSITAE (D)
Gundlachia A.Gray COMPOSITAE (D)
Gunillaea Thulin CAMPANULACEAE (D)
Gunnarella Senghas ORCHIDACEAE (M)
Gunnarorchis Brieger = Dendrobium Sw. ORCHIDACEAE (M)
Gunnera L. GUNNERACEAE (D)
Gunnessia P.I.Forst. ASCLEPIADACEAE (D)
Gunnia Lindl. (SUH) = Sarcochilus R.Br. ORCHIDACEAE (M)
Gunnia F.Muell. = Neogunnia Pax & K.Hoffm. AIZOACEAE (D)
Gunniopsis Pax AIZOACEAE (D)
Gurania (Schltdl.) Cogn. CUCURBITACEAE (D)
Guraniopsis Cogn. CUCURBITACEAE (D)
Guru Buch.–Ham. ex Wight = Finlaysonia Wall. ASCLEPIADACEAE (D)
Gussonea A.Rich. = Solenangis Schltr. ORCHIDACEAE (M)
Gussonia Spreng. = Sebastiania Spreng. EUPHORBIACEAE (D)
Gustavia L. LECYTHIDACEAE (D)
Gutenbergia Sch.Bip. COMPOSITAE (D)
Gutenbergia Walp. (SUO) = Morinda L. RUBIACEAE (D)
Guthnickia Regel = Achimenes Pers. GESNERIACEAE (D)
Guthriea Bolus ACHARIACEAE (D)
Gutierrezia Lag. COMPOSITAE (D)
Guttenbergia Zoll. & Moritzi = Morinda L. RUBIACEAE (D)
Gutzlaffia Hance = Strobilanthes Blume ACANTHACEAE (D)
Guya Frapp. = Drypetes Vahl EUPHORBIACEAE (D)
Guyonia Naudin MELASTOMATACEAE (D)
Guzmania Ruíz & Pav. BROMELIACEAE (M)
Gyaladenia Schltr. = Brachycorythis Lindl. ORCHIDACEAE (M)
Gyas Salisb. = Bletia Ruíz & Pav. ORCHIDACEAE (M)
Gymapsis Bremek. = Strobilanthes Blume ACANTHACEAE (D)
Gymelaea (Endl.) Spach = Nestegis Raf. OLEACEAE (D)
Gyminda Sarg. CELASTRACEAE (D)
Gymnacanthus Oerst. = Ruellia L. ACANTHACEAE (D)
Gymnachne Parodi = Rhombolytrum Link GRAMINEAE (M)
Gymnacranthera Warb. MYRISTICACEAE (D)
Gymnadenia R.Br. ORCHIDACEAE (M)
Gymnadeniopsis Rydb. = Platanthera Rich. ORCHIDACEAE (M)

Gymnagathis Stapf (SUH) = Stapfiophyton H.L.Li MELASTOMATACEAE (D)
Gymnagathis Schauer = Melaleuca L. MYRTACEAE (D)
Gymnalypha Griseb. = Acalypha L. EUPHORBIACEAE (D)
Gymnandropogon (Nees) Duthie = Bothriochloa Kuntze GRAMINEAE (M)
Gymnantha Y.Ito = Rebutia K.Schum. CACTACEAE (D)
Gymnanthelia Schweinf. (SUI) = Cymbopogon Spreng. GRAMINEAE (M)
Gymnanthemum Cass. = Vernonia Schreb. COMPOSITAE (D)
Gymnanthera R.Br. ASCLEPIADACEAE (D)
Gymnanthes Sw. EUPHORBIACEAE (D)
Gymnanthocereus Backeb. = Cleistocactus Lem. CACTACEAE (D)
Gymnarrhena Desf. COMPOSITAE (D)
Gymnartocarpus Boerl. = Parartocarpus Baill. MORACEAE (D)
Gymnema R.Br. ASCLEPIADACEAE (D)
Gymnemopsis Costantin ASCLEPIADACEAE (D)
Gymnobothrys Wall. ex Baill. = Sapium P.Browne EUPHORBIACEAE (D)
Gymnocactus John & Riha (SUH) = Turbinicarpus (Backeb.) Buxb. & Backeb.
    CACTACEAE (D)
Gymnocactus Backeb. = Turbinicarpus (Backeb.) Buxb. & Backeb. CACTACEAE (D)
Gymnocalycium Pfeiff. CACTACEAE (D)
Gymnocarpium Newman WOODSIACEAE (F)
Gymnocarpos Forssk. ILLECEBRACEAE (D)
Gymnocereus Rauh & Backeb. = Browningia Britton & Rose CACTACEAE (D)
Gymnochilus Blume ORCHIDACEAE (M)
Gymnocladus Lam. LEGUMINOSAE–CAESALPINIOIDEAE (D)
Gymnocondylus R.M.King & H.Rob. COMPOSITAE (D)
Gymnocoronis DC. COMPOSITAE (D)
Gymnodiscus Less. COMPOSITAE (D)
Gymnogonum Parry = Goodmania Reveal & Ertter POLYGONACEAE (D)
Gymnogramma Desv. = Gymnopteris Bernh. ADIANTACEAE (F)
Gymnogrammitis Griff. DAVALLIACEAE (F)
Gymnogynum P.Beauv. = Selaginella P.Beauv. SELAGINELLACEAE (FA)
Gymnolaema Benth. (SUH) = Sacleuxia Baill. ASCLEPIADACEAE (D)
Gymnolaena (DC.) Rydb. COMPOSITAE (D)
Gymnoleima Decne. = Moltkia Lehm. BORAGINACEAE (D)
Gymnolomia Kunth = Aspilia Thouars COMPOSITAE (D)
Gymnoluma Baill. = Elaeoluma Baill. SAPOTACEAE (D)
Gymnomesium Schott = Arum L. ARACEAE (M)
Gymnopentzia Benth. COMPOSITAE (D)
Gymnopetalum Arn. CUCURBITACEAE (D)
Gymnophyton Clos UMBELLIFERAE (D)
Gymnopodium Rolfe POLYGONACEAE (D)
Gymnopogon P.Beauv. GRAMINEAE (M)
Gymnopremnon Lindig = Cyathea Sm. CYATHEACEAE (F)
Gymnopteris Bernh. ADIANTACEAE (F)
Gymnopteris C.Presl (SUH) = Leptochilus Kaulf. POLYPODIACEAE (F)
Gymnorinorea Keay = Decorsella A.Chev. VIOLACEAE (D)
Gymnoschoenus Nees CYPERACEAE (M)
Gymnosciadium Hochst. = Pimpinella L. UMBELLIFERAE (D)
Gymnosiphon Blume BURMANNIACEAE (M)
Gymnosperma Less. COMPOSITAE (D)
Gymnospermium Spach BERBERIDACEAE (D)
Gymnosphaera Blume = Cyathea Sm. CYATHEACEAE (F)
Gymnosporia (Wight & Arn.) Hook.f. = Maytenus Molina CELASTRACEAE (D)
Gymnostachys R.Br. ARACEAE (M)
Gymnostachyum Nees ACANTHACEAE (D)

Gymnostemon Aubrév. & Pellegr. SIMAROUBACEAE (D)
Gymnostephium Less. COMPOSITAE (D)
Gymnosteris Greene POLEMONIACEAE (D)
Gymnostichum Schreb. (SUS) = Hystrix Moench GRAMINEAE (M)
Gymnostillingia Müll.Arg. = Stillingia Garden ex L. EUPHORBIACEAE (D)
Gymnostoma L.A.S.Johnson CASUARINACEAE (D)
Gymnostyles Juss. = Soliva Ruíz & Pav. COMPOSITAE (D)
Gymnotheca C.Presl (SUH) = Marattia Sw. MARATTIACEAE (F)
Gymnotheca Decne. SAURURACEAE (D)
Gymnotrix P.Beauv. = Pennisetum Rich. GRAMINEAE (M)
Gynaecopachys Hassk. (SUO) = Gynopachis Blume RUBIACEAE (D)
Gynamblosis Torr. = Croton L. EUPHORBIACEAE (D)
Gynandriris Parl. IRIDACEAE (M)
Gynandropsis DC. = Cleome L. CAPPARACEAE (D)
Gynatrix Alef. MALVACEAE (D)
Gynerium P.Beauv. GRAMINEAE (M)
Gynestum Poit. = Geonoma Willd. PALMAE (M)
Gynizodon Raf. = Oncidium Sw. ORCHIDACEAE (M)
Gynocardia R.Br. FLACOURTIACEAE (D)
Gynochthodes Blume RUBIACEAE (D)
Gynocraterium Bremek. ACANTHACEAE (D)
Gynoglottis J.J.Sm. ORCHIDACEAE (M)
Gynoon Juss. = Glochidion J.R.Forst. & G.Forst. EUPHORBIACEAE (D)
Gynopachis Blume RUBIACEAE (D)
Gynophoraria Rydb. = Astragalus L. LEGUMINOSAE–PAPILIONOIDEAE (D)
Gynophorea Gilli CRUCIFERAE (D)
Gynophyge Gilli = Agrocharis Hochst. UMBELLIFERAE (D)
Gynopleura Cav. = Malesherbia Ruíz & Pav. MALESHERBIACEAE (D)
Gynostemma Blume CUCURBITACEAE (D)
Gynotroches Blume RHIZOPHORACEAE (D)
Gynoxys Cass. COMPOSITAE (D)
Gynura Cass. COMPOSITAE (D)
Gypothamnium Phil. COMPOSITAE (D)
Gypsacanthus Lott, Jaramillo & Rzed. ACANTHACEAE (D)
Gypsophila L. CARYOPHYLLACEAE (D)
Gyptidium R.M.King & H.Rob. COMPOSITAE (D)
Gyptis (Cass.) Cass. COMPOSITAE (D)
Gyrandra Wall. (SUI) = Daphniphyllum Blume DAPHNIPHYLLACEAE (D)
Gyranthera Pittier BOMBACACEAE (D)
Gyrinops Gaertn. THYMELAEACEAE (D)
Gyrinopsis Decne. = Aquilaria Lam. THYMELAEACEAE (D)
Gyrocarpus Jacq. HERNANDIACEAE (D)
Gyrocaryum B.Valdés BORAGINACEAE (D)
Gyrocheilos W.T.Wang (UP) GESNERIACEAE (D)
Gyrodoma Wild COMPOSITAE (D)
Gyrogyne W.T.Wang (UP) GESNERIACEAE (D)
Gyroptera Botsch. CHENOPODIACEAE (D)
Gyrosorium C.Presl = Pyrrosia Mirb. POLYPODIACEAE (F)
Gyrostachis Blume = Spiranthes Rich. ORCHIDACEAE (M)
Gyrostachys Pers. ex Blume = Spiranthes Rich. ORCHIDACEAE (M)
Gyrostelma E.Fourn. = Matelea Aubl. ASCLEPIADACEAE (D)
Gyrostemon Desf. GYROSTEMONACEAE (D)
Gyrostipula J.–F.Leroy RUBIACEAE (D)
Gyrotaenia Griseb. URTICACEAE (D)

Haagea Frič (SUH) = Mammillaria Haw. CACTACEAE (D)
Haageocactus Backeb. (SUI) = Haageocereus Backeb. CACTACEAE (D)
Haageocereus Backeb. CACTACEAE (D)
Haarera Hutch. & E.A.Bruce = Erlangea Sch.Bip. COMPOSITAE (D)
Haastia Hook.f. COMPOSITAE (D)
Habenaria Nimmo (SUH) = Habenaria Willd. ORCHIDACEAE (M)
Habenaria Willd. ORCHIDACEAE (M)
Habenella Small = Habenaria Willd. ORCHIDACEAE (M)
Habenorkis Thouars = Habenaria Willd. ORCHIDACEAE (M)
Haberlea Friv. GESNERIACEAE (D)
Hablitzia M.Bieb. CHENOPODIACEAE (D)
Habracanthus Nees ACANTHACEAE (D)
Habranthus Herb. AMARYLLIDACEAE (M)
Habrochloa C.E.Hubb. GRAMINEAE (M)
Habroneuron Standl. RUBIACEAE (D)
Habropetalum Airy Shaw DIONCOPHYLLACEAE (D)
Habrosia Fenzl CARYOPHYLLACEAE (D)
Habrothamnus Endl. = Cestrum L. SOLANACEAE (D)
Habrurus Hochst. (SUI) = Elionurus Humb. & Bonpl. ex Willd. GRAMINEAE (M)
Habsia Steud. = Guettarda L. RUBIACEAE (D)
Habzelia A.DC. = Xylopia L. ANNONACEAE (D)
Hachettea Baill. BALANOPHORACEAE (D)
Hackelia Vasey (SUI) = Gouinia E.Fourn. GRAMINEAE (M)
Hackelia Opiz BORAGINACEAE (D)
Hackelochloa Kuntze GRAMINEAE (M)
Hacquetia Neck. ex DC. UMBELLIFERAE (D)
Hadrodemas H.E.Moore = Callisia Loefl. COMMELINACEAE (M)
Haeckeria F.Muell. COMPOSITAE (D)
Haegiela P.S.Short COMPOSITAE (D)
Haemacanthus S.Moore = Satanocrater Schweinf. ACANTHACEAE (D)
Haemanthus L. AMARYLLIDACEAE (M)
Haemaria Lindl. = Ludisia A.Rich. ORCHIDACEAE (M)
Haematocarpus Miers MENISPERMACEAE (D)
Haematodendron Capuron MYRISTICACEAE (D)
Haematorchis Blume = Galeola Lour. ORCHIDACEAE (M)
Haematospermum Wall. = Homonoia Lour. EUPHORBIACEAE (D)
Haematostaphis Hook.f. ANACARDIACEAE (D)
Haematostemon (Müll.Arg.) Pax & K.Hoffm. EUPHORBIACEAE (D)
Haematoxylum L. LEGUMINOSAE–CAESALPINIOIDEAE (D)
Haemax E.Mey. = Microloma R.Br. ASCLEPIADACEAE (D)
Haemocharis Salisb. ex Mart. & Zucc. = Laplacea Kunth THEACEAE (D)
Haemodorum Sm. HAEMODORACEAE (M)
Haenianthus Griseb. OLEACEAE (D)
Hafunia Chiov. = Sphaerocoma T.Anderson ILLECEBRACEAE (D)
Hagenbachia Nees & Mart. ANTHERICACEAE (M)
Hagenia J.F.Gmel. ROSACEAE (D)
Hagsatera Tamayo ORCHIDACEAE (M)
Hainania Merr. TILIACEAE (D)
Hainardia Greuter GRAMINEAE (M)
Haitia Urb. LYTHRACEAE (D)
Haitiella L.H.Bailey = Coccothrinax Sargent PALMAE (M)
Haitimimosa Britton = Mimosa L. LEGUMINOSAE–MIMOSOIDEAE (D)
Hakea Schrad. PROTEACEAE (D)
Hakoneaste F.Maek. = Ephippianthus Rchb.f. ORCHIDACEAE (M)
Hakonechloa Makino ex Honda GRAMINEAE (M)

Halacsya Dörfl. BORAGINACEAE (D)
Halacsyella Janchen = Edraianthus (A.DC.) DC. CAMPANULACEAE (D)
Halanthium K.Koch CHENOPODIACEAE (D)
Halarchon Bunge CHENOPODIACEAE (D)
Halconia Merr. = Trichospermum Blume TILIACEAE (D)
Haldina Ridsdale RUBIACEAE (D)
Halecus Raf. = Croton L. EUPHORBIACEAE (D)
Halenbergia Dinter = Mesembryanthemum AIZOACEAE (D)
Halenia Borkh. GENTIANACEAE (D)
Halerpestes Greene RANUNCULACEAE (D)
Halesia Ellis ex L. STYRACACEAE (D)
Halesia P.Browne (SUH) = Guettarda L. RUBIACEAE (D)
Halfordia F.Muell. RUTACEAE (D)
Halgania Gaudich. BORAGINACEAE (D)
Halimione Aellen CHENOPODIACEAE (D)
Halimium (Dunal) Spach CISTACEAE (D)
Halimocnemis C.A.Mey. CHENOPODIACEAE (D)
Halimodendron Fisch. ex DC. LEGUMINOSAE–PAPILIONOIDEAE (D)
Halimolobos Tausch CRUCIFERAE (D)
Halimolobus Tausch (SUO) = Halimolobos Tausch CRUCIFERAE (D)
Halimus Wallr. = Atriplex L. CHENOPODIACEAE (D)
Hallea J.-F.Leroy RUBIACEAE (D)
Halleria L. SCROPHULARIACEAE (D)
Hallia Thunb. LEGUMINOSAE–PAPILIONOIDEAE (D)
Hallianthus H.E.K.Hartmann AIZOACEAE (D)
Hallieracantha Stapf = Ptyssiglottis T.Anderson ACANTHACEAE (D)
Halliophytum I.M.Johnst. = Tetracoccus Engelm. ex Parry EUPHORBIACEAE (D)
Halmoorea J.Dransf. & N.W.Uhl PALMAE (M)
Halocarpus Quinn PODOCARPACEAE (G)
Halocharis Moq. CHENOPODIACEAE (D)
Halochloa Griseb. (SUH) = Monanthochloe Engelm. GRAMINEAE (M)
Halocnemum M.Bieb. CHENOPODIACEAE (D)
Halodule Endl. CYMODOCEACEAE (M)
Halogeton C.A.Mey. CHENOPODIACEAE (D)
Halopegia K.Schum. MARANTACEAE (M)
Halopeplis Bunge ex Ung.-Sternb. CHENOPODIACEAE (D)
Halophila Thouars HYDROCHARITACEAE (M)
Halophlebia (Mart.) Lindl. = Cyathea Sm. CYATHEACEAE (F)
Halophytum Speg. HALOPHYTACEAE (D)
Halopyrum Stapf GRAMINEAE (M)
Haloragis J.R.Forst. & G.Forst. HALORAGACEAE (D)
Haloragodendron Orchard HALORAGACEAE (D)
Halosarcia Paul G.Wilson CHENOPODIACEAE (D)
Halosciastrum Koidz. UMBELLIFERAE (D)
Halosicyos Mart.Crov. CUCURBITACEAE (D)
Halostachys C.A.Mey. ex Schrenk CHENOPODIACEAE (D)
Halothamnus F.Muell. (SUH) = Lawrencia Hook. MALVACEAE (D)
Halothamnus Jaub. & Spach CHENOPODIACEAE (D)
Halotis Bunge = Halimocnemis C.A.Mey. CHENOPODIACEAE (D)
Haloxanthium Ulbr. = Atriplex L. CHENOPODIACEAE (D)
Haloxylon Bunge CHENOPODIACEAE (D)
Halphophyllum Mansf. = Gasteranthus Benth. GESNERIACEAE (D)
Hamadryas Comm. ex Juss. RANUNCULACEAE (D)
Hamamelis L. HAMAMELIDACEAE (D)
Hamatocactus Britton & Rose = Thelocactus (K.Schum.) Britton & Rose CACTACEAE (D)

Hambergera Scop. = Combretum Loefl. COMBRETACEAE (D)
Hamelia Jacq. RUBIACEAE (D)
Hamilcoa Prain EUPHORBIACEAE (D)
Hamiltonia Roxb. = Spermadictyon Roxb. RUBIACEAE (D)
Hammada Iljin CHENOPODIACEAE (D)
Hammarbya Kuntze ORCHIDACEAE (M)
Hammatolobium Fenzl LEGUMINOSAE–PAPILIONOIDEAE (D)
Hamosa Medik. = Astragalus L. LEGUMINOSAE–PAPILIONOIDEAE (D)
Hampea Schltdl. MALVACEAE (D)
Hamulia Raf. = Utricularia L. LENTIBULARIACEAE (D)
Hanabusaya Nakai CAMPANULACEAE (D)
Hanburia Seem. CUCURBITACEAE (D)
Hanburyophyton Bureau = Mansoa DC. BIGNONIACEAE (D)
Hancea Hemsl. (SUH) = Hanceola Kudo LABIATAE (D)
Hancea Seem. (SUH) = Mallotus Lour. EUPHORBIACEAE (D)
Hanceola Kudo LABIATAE (D)
Hancockia Rolfe ORCHIDACEAE (M)
Hancornia Gomes APOCYNACEAE (D)
Handelia Heimerl COMPOSITAE (D)
Handeliodendron Rehder = Sideroxylon L. SAPOTACEAE (D)
Handroanthus Mattos = Tabebuia Gomes ex DC. BIGNONIACEAE (D)
Hanghomia Gagnep. & Thénint APOCYNACEAE (D)
Hanguana Blume HANGUANACEAE (M)
Haniffia Holttum ZINGIBERACEAE (M)
Hannafordia F.Muell. STERCULIACEAE (D)
Hannoa Planch. SIMAROUBACEAE (D)
Hannonia Braun-Blanq. & Maire AMARYLLIDACEAE (M)
Hansemannia K.Schum. = Archidendron F.Muell. LEGUMINOSAE–MIMOSOIDEAE (D)
Hanslia Schindl. = Desmodium Desv. LEGUMINOSAE–PAPILIONOIDEAE (D)
Hansteinia Oerst. = Habracanthus Nees ACANTHACEAE (D)
Hapaline Schott ARACEAE (M)
Hapaloceras Hassk. = Payena A.DC. SAPOTACEAE (D)
Hapalochilus (Schltr.) Senghas = Bulbophyllum Thouars ORCHIDACEAE (M)
Hapalorchis Schltr. ORCHIDACEAE (M)
Haplachne C.Presl = Dimeria R.Br. GRAMINEAE (M)
Haplanthera Hochst. = Ruttya Harv. ACANTHACEAE (D)
Haplanthodes Kuntze ACANTHACEAE (D)
Haplanthoides H.W.Li (SUH) = Andrographis Wall. ex Nees ACANTHACEAE (D)
Haplanthus Nees = Andrographis Wall. ex Nees ACANTHACEAE (D)
Haplocalymma S.F.Blake COMPOSITAE (D)
Haplocarpha Less. COMPOSITAE (D)
Haplochilus Endl. = Zeuxine Lindl. ORCHIDACEAE (M)
Haplochorema K.Schum. ZINGIBERACEAE (M)
Haploclathra Benth. GUTTIFERAE (D)
Haplocoelum Radlk. SAPINDACEAE (D)
Haplodesmium Naudin = Chaetolepis (DC.) Miq. MELASTOMATACEAE (D)
Haplodictyum C.Presl = Pronephrium C.Presl THELYPTERIDACEAE (F)
Haplodypsis Baill. = Neophloga Baill. PALMAE (M)
Haploesthes A.Gray COMPOSITAE (D)
Haplolobus H.J.Lam BURSERACEAE (D)
Haplolophium Cham. BIGNONIACEAE (D)
Haplopappus Cass. COMPOSITAE (D)
Haplopetalon A.Gray = Crossostylis J.R.Forst. & G.Forst. RHIZOPHORACEAE (D)
Haplopetalum Miq. (SUO) = Crossostylis J.R.Forst. & G.Forst. RHIZOPHORACEAE (D)
Haplophandra Pichon = Odontadenia Benth. APOCYNACEAE (D)

Haplophloga Baill. = Neophloga Baill. PALMAE (M)

Haplophragma Dop = Fernandoa Welw. ex Seem. BIGNONIACEAE (D)

Haplophyllophorus (Brenan) A.Fern. & R.Fern. = Cincinnobotrys Gilg MELASTOMATACEAE (D)

Haplophyllum A.Juss. RUTACEAE (D)

Haplophyton A.DC. APOCYNACEAE (D)

Haplopteris C.Presl = Vittaria Sm. VITTARIACEAE (F)

Haplorhus Engl. ANACARDIACEAE (D)

Haplormosia Harms LEGUMINOSAE–PAPILIONOIDEAE (D)

Haplosciadium Hochst. UMBELLIFERAE (D)

Haploseseli H.Wolff & Hand.–Mazz. = Physospermopsis H.Wolff UMBELLIFERAE (D)

Haplosphaera Hand.–Mazz. UMBELLIFERAE (D)

Haplostachys (A.Gray) W.F.Hillebr. LABIATAE (D)

Haplostellis Endl. = Nervilia Comm. ex Gaud. ORCHIDACEAE (M)

Haplostephium Mart. ex DC. COMPOSITAE (D)

Haplostichanthus F.Muell. ANNONACEAE (D)

Haplothismia Airy Shaw BURMANNIACEAE (M)

Haptocarpum Ule CAPPARACEAE (D)

Haradjania Rech.f. = Myopordon Boiss. COMPOSITAE (D)

Haraella Kudo ORCHIDACEAE (M)

Harbouria J.M.Coult. & Rose UMBELLIFERAE (D)

Hardenbergia Benth. LEGUMINOSAE–PAPILIONOIDEAE (D)

Hardwickia Roxb. LEGUMINOSAE–CAESALPINIOIDEAE (D)

Harfordia Greene & Parry POLYGONACEAE (D)

Harina Buch.–Ham. = Wallichia Roxb. PALMAE (M)

Hariota DC. (SUH) = Hatiora Britton & Rose CACTACEAE (D)

Hariota Adans. (SUS) = Rhipsalis Gaertn. CACTACEAE (D)

Harlandia Hance = Solena Lour. CUCURBITACEAE (D)

Harlanlewisia Epling = Scutellaria L. LABIATAE (D)

Harleya S.F.Blake COMPOSITAE (D)

Harleyodendron R.S.Cowan LEGUMINOSAE–PAPILIONOIDEAE (D)

Harmandia Pierre ex Baill. OLACACEAE (D)

Harmandiella Costantin ASCLEPIADACEAE (D)

Harmogia Schauer = Baeckea L. MYRTACEAE (D)

Harmsia K.Schum. STERCULIACEAE (D)

Harmsiella Briq. = Otostegia Benth. LABIATAE (D)

Harmsiodoxa O.E.Schulz CRUCIFERAE (D)

Harmsiopanax Warb. ARALIACEAE (D)

Harnackia Urb. COMPOSITAE (D)

Harnieria Solms–Laub. = Justicia L. ACANTHACEAE (D)

Haronga Thouars = Harungana Lam. GUTTIFERAE (D)

Harpachne Hochst. ex A.Rich. GRAMINEAE (M)

Harpagocarpus Hutch. & Dandy = Fagopyrum Mill. POLYGONACEAE (D)

Harpagonella A.Gray BORAGINACEAE (D)

Harpagophytum DC. ex Meisn. PEDALIACEAE (D)

Harpalyce Moçino & Sessé ex DC. LEGUMINOSAE–PAPILIONOIDEAE (D)

Harpanema Decne. ASCLEPIADACEAE (D)

Harpephyllum Bernh. ex Krauss ANACARDIACEAE (D)

Harperella Rose = Ptilimnium Raf. UMBELLIFERAE (D)

Harperia Rose (SUH) = Ptilimnium Raf. UMBELLIFERAE (D)

Harperia W.Fitzg. RESTIONACEAE (M)

Harperocallis McDaniel MELANTHIACEAE (M)

Harpochilus Nees ACANTHACEAE (D)

Harpochloa Kunth GRAMINEAE (M)

Harpullia Roxb. SAPINDACEAE (D)

Harrachia Jacq. = Crossandra Salisb. ACANTHACEAE (D)
Harrimanella Coville ERICACEAE (D)
Harrisella Fawc. & Rendle ORCHIDACEAE (M)
Harrisia Britton CACTACEAE (D)
Harrisonia Hook. (SUH) = Loniceroides Bullock ASCLEPIADACEAE (D)
Harrisonia R.Br. ex Adr.Juss. SIMAROUBACEAE (D)
Harrysmithia H.Wolff UMBELLIFERAE (D)
Harthamnus H.Rob. = Plazia Ruíz & Pav. COMPOSITAE (D)
Hartia Dunn = Stuartia L. THEACEAE (D)
Hartigshea A.Juss. = Dysoxylum Blume MELIACEAE (D)
Hartleya Sleumer ICACINACEAE (D)
Hartmannia Spach = Oenothera L. ONAGRACEAE (D)
Hartogia Thunb. ex L.f. = Hartogiella Codd CELASTRACEAE (D)
Hartogiella Codd CELASTRACEAE (D)
Hartogiopsis H.Perrier CELASTRACEAE (D)
Hartwegia Lindl. (SUH) = Nageliella L.O.Williams ORCHIDACEAE (M)
Hartwegiella O.E.Schulz = Mancoa Wedd. CRUCIFERAE (D)
Hartwrightia A.Gray ex S.Watson COMPOSITAE (D)
Harungana Lam. GUTTIFERAE (D)
Harveya Hook. SCROPHULARIACEAE (D)
Haselhoffia Lindau = Physacanthus Benth. ACANTHACEAE (D)
Haseltonia Backeb. = Cephalocereus Pfeiff. CACTACEAE (D)
Hasseanthus Rose = Dudleya Britton & Rose CRASSULACEAE (D)
Hasseltia Kunth TILIACEAE (D)
Hasseltiopsis Sleumer TILIACEAE (D)
Hasskarlia Baill. (SUH) = Tetrorchidium Poepp. EUPHORBIACEAE (D)
Hasslerella Chod. = Polypremum L. BUDDLEJACEAE (D)
Hassleropsis Chodat = Basistemon Turcz. SCROPHULARIACEAE (D)
Hasteola Raf. COMPOSITAE (D)
Hastingsia S.Watson HYACINTHACEAE (M)
Hatiora Britton & Rose CACTACEAE (D)
Hatschbachia L.B.Sm. = Napeanthus Gardner GESNERIACEAE (D)
Hatschbachiella R.M.King & H.Rob. COMPOSITAE (D)
Haumania J.Léonard MARANTACEAE (M)
Haumaniastrum P.A.Duvign. & Plancke LABIATAE (D)
Haussknechtia Boiss. UMBELLIFERAE (D)
Haussmannia F.Muell. = Neosepicaea Diels BIGNONIACEAE (D)
Haussmannianthes Steenis = Neosepicaea Diels BIGNONIACEAE (D)
Hauya Moçino & Sessé ex DC. ONAGRACEAE (D)
Havardia Small LEGUMINOSAE-MIMOSOIDEAE (D)
Havetia Kunth GUTTIFERAE (D)
Havetiopsis Planch. & Triana GUTTIFERAE (D)
Havilandia Stapf = Trigonotis Steven BORAGINACEAE (D)
Hawkesiophyton Hunz. SOLANACEAE (D)
Haworthia Duval ALOACEAE (M)
Haya Balf.f. ILLECEBRACEAE (D)
Hayataella Masam. RUBIACEAE (D)
Haydonia R.Wilczek = Vigna Savi LEGUMINOSAE-PAPILIONOIDEAE (D)
Haylockia Herb. AMARYLLIDACEAE (M)
Haynaldia Kanitz = Lobelia L. CAMPANULACEAE (D)
Haynaldia Schur (SUH) = Dasypyrum (Coss. & Durieu) T.Durand GRAMINEAE (M)
Haynea Schumach. & Thonn. = Laportea Gaudich. URTICACEAE (D)
Hazardia Greene COMPOSITAE (D)
Hazomalamia Capuron = Hernandia L. HERNANDIACEAE (D)
Hazunta Pichon = Tabernaemontana L. APOCYNACEAE (D)

Hearnia F.Muell. = Aglaia Lour. MELIACEAE (D)
Hebantha Mart. = Pfaffia Mart. AMARANTHACEAE (D)
Hebe Comm. ex Juss. SCROPHULARIACEAE (D)
Hebea L.Bolus (SUH) = Tritoniopsis L.Bolus IRIDACEAE (M)
Hebea (Pers.) R.A.Hedw. = Gladiolus L. IRIDACEAE (M)
Hebecladus Miers SOLANACEAE (D)
Hebeclinium DC. COMPOSITAE (D)
Hebecocca Beurl. = Omphalea L. EUPHORBIACEAE (D)
Hebecoccus Radlk. = Lepisanthes Blume SAPINDACEAE (D)
Hebenstreitia L. (SUO) = Hebenstretia L. SCROPHULARIACEAE (D)
Hebenstretia L. SCROPHULARIACEAE (D)
Hebepetalum Benth. LINACEAE (D)
Heberdenia Banks ex DC. MYRSINACEAE (D)
Hebestigma Urb. LEGUMINOSAE–PAPILIONOIDEAE (D)
Hebonga Radlk. = Ailanthus Desf. SIMAROUBACEAE (D)
Hecabe Raf. = Phaius Lour. ORCHIDACEAE (M)
Hecastocleis A.Gray COMPOSITAE (D)
Hecatea Thouars = Omphalea L. EUPHORBIACEAE (D)
Hecatostemon S.F.Blake FLACOURTIACEAE (D)
Hechtia Klotzsch BROMELIACEAE (M)
Hecistocarpus Post & Kuntze (SUO) = Hekistocarpa Hook.f. RUBIACEAE (D)
Hecistopteris J.Sm. VITTARIACEAE (F)
Heckeldora Pierre MELIACEAE (D)
Hectorella Hook.f. HECTORELLACEAE (D)
Hedaroma Lindl. = Darwinia Rudge MYRTACEAE (D)
Hedbergia Molau SCROPHULARIACEAE (D)
Hedeoma Pers. LABIATAE (D)
Hedeomoides Briq. = Pogogyne Benth. LABIATAE (D)
Hedera L. ARALIACEAE (D)
Hederella Stapf = Catanthera F.Muell. MELASTOMATACEAE (D)
Hederopsis C.B.Clarke = Macropanax Miq. ARALIACEAE (D)
Hederorkis Thouars ORCHIDACEAE (M)
Hedinia Ostenf. CRUCIFERAE (D)
Hediniopsis Botsch. & Petrovsky CRUCIFERAE (D)
Hedraeanthus Griseb. (SUO) = Edraianthus (A.DC.) DC. CAMPANULACEAE (D)
Hedraianthera F.Muell. CELASTRACEAE (D)
Hedraiostylus Hassk. = Pterococcus Hassk. EUPHORBIACEAE (D)
Hedranthera (Stapf) Pichon = Callichilia Stapf APOCYNACEAE (D)
Hedstromia A.C.Sm. RUBIACEAE (D)
Hedusa Raf. = Dissotis Benth. MELASTOMATACEAE (D)
Hedwigia Sw. = Tetragastris Gaertn. BURSERACEAE (D)
Hedyachras Radlk. = Glenniea Hook.f. SAPINDACEAE (D)
Hedycarpus Jack = Baccaurea Lour. EUPHORBIACEAE (D)
Hedycarya J.R.Forst. & G.Forst. MONIMIACEAE (D)
Hedycaryopsis Danguy MONIMIACEAE (D)
Hedychium J.König ZINGIBERACEAE (M)
Hedycrea Schreb. = Licania Aubl. CHRYSOBALANACEAE (D)
Hedyosmos Mitch. = Cunila Royen ex L. LABIATAE (D)
Hedyosmum Sw. CHLORANTHACEAE (D)
Hedyotis L. RUBIACEAE (D)
Hedyphylla Steven = Astragalus L. LEGUMINOSAE–PAPILIONOIDEAE (D)
Hedypnois Mill. COMPOSITAE (D)
Hedysarum L. LEGUMINOSAE–PAPILIONOIDEAE (D)
Hedyscepe H.A.Wendl. & Drude PALMAE (M)
Hedythyrsus Bremek. RUBIACEAE (D)

Heeria Meisn. ANACARDIACEAE (D)
Heeria Schltdl. (SUH) = Heterocentron Hook. & Arn. MELASTOMATACEAE (D)
Hegnera Schindl. = Desmodium Desv. LEGUMINOSAE–PAPILIONOIDEAE (D)
Heimerlia Skottsb. = Pisonia L. NYCTAGINACEAE (D)
Heimerliodendron Skottsb. = Pisonia L. NYCTAGINACEAE (D)
Heimia Link LYTHRACEAE (D)
Heimodendron Sillans = Entandrophragma C.E.C.Fisch. MELIACEAE (D)
Heinekenia Webb ex Christ = Lotus L. LEGUMINOSAE–PAPILIONOIDEAE (D)
Heinsenia K.Schum. RUBIACEAE (D)
Heinsia DC. RUBIACEAE (D)
Heintzia Karst. = Alloplectus Mart. GESNERIACEAE (D)
Heinzelia Nees = Justicia L. ACANTHACEAE (D)
Heisteria Jacq. OLACACEAE (D)
Hekaterosachne Steud. = Oplismenus P.Beauv. GRAMINEAE (M)
Hekistocarpa Hook.f. RUBIACEAE (D)
Heladena A.Juss. MALPIGHIACEAE (D)
Helanthium (Benth.) Engelm. ex Britton = Echinodorus Rich. ex Engelm.
    ALISMATACEAE (M)
Helcia Lindl. ORCHIDACEAE (M)
Heldreichia Boiss. CRUCIFERAE (D)
Helenium L. COMPOSITAE (D)
Heleocharis Lestib. = Eleocharis R.Br. CYPERACEAE (M)
Heleochloa Fr. (SUI) = Glyceria R.Br. GRAMINEAE (M)
Heleochloa P.Beauv. (SUH) = Phleum L. GRAMINEAE (M)
Heleochloa Roem. = Crypsis Aiton GRAMINEAE (M)
Helia Mart. = Irlbachia Mart. GENTIANACEAE (D)
Heliabravoa Backeb. = Polaskia Backeb. CACTACEAE (D)
Heliamphora Benth. SARRACENIACEAE (D)
Helianthella Torr. & A.Gray COMPOSITAE (D)
Helianthemum Mill. CISTACEAE (D)
Helianthocereus Backeb. = Echinopsis Zucc. CACTACEAE (D)
Helianthopsis H.Rob. COMPOSITAE (D)
Helianthostylis Baill. MORACEAE (D)
Helianthus L. COMPOSITAE (D)
Helicanthera Roem. & Schult. (SUO) = Helixanthera Lour. LORANTHACEAE (D)
Helicanthes Danser LORANTHACEAE (D)
Helichrysopsis Kirp. COMPOSITAE (D)
Helichrysum Mill. COMPOSITAE (D)
Helicia Pers. (SUH) = Helixanthera Lour. LORANTHACEAE (D)
Helicia Lour. PROTEACEAE (D)
Helicilla Moq. CHENOPODIACEAE (D)
Heliciopsis Sleumer PROTEACEAE (D)
Helicodiceros Schott ex K.Koch ARACEAE (M)
Heliconia L. HELICONIACEAE (M)
Helicophyllum Schott = Eminium (Blume) Schott ARACEAE (M)
Helicostylis Trécul MORACEAE (D)
Helicteres L. STERCULIACEAE (D)
Helicteropsis Hochr. MALVACEAE (D)
Helictonema Pierre CELASTRACEAE (D)
Helictonia Ehrh. = Spiranthes Rich. ORCHIDACEAE (M)
Helictotrichon Besser GRAMINEAE (M)
Helietta Tul. RUTACEAE (D)
Helinus E.Mey. ex Endl. RHAMNACEAE (D)
Heliocarpus L. TILIACEAE (D)
Heliocarya Bunge BORAGINACEAE (D)

Heliocauta Humphries COMPOSITAE (D)
Heliocereus (A.Berger) Britton & Rose = Disocactus Lindl. CACTACEAE (D)
Heliomeris Nutt. COMPOSITAE (D)
Heliophila Burm.f. ex L. CRUCIFERAE (D)
Heliopsis Pers. COMPOSITAE (D)
Heliosperma Rchb. = Silene L. CARYOPHYLLACEAE (D)
Heliospora Hook.f. (SUO) = Timonius DC. RUBIACEAE (D)
Heliostemma Woodson ASCLEPIADACEAE (D)
Heliotropium L. BORAGINACEAE (D)
Helipterum DC. ex Lindl. (SUH) = Syncarpha DC. COMPOSITAE (D)
Helixanthera Lour. LORANTHACEAE (D)
Helixyra Salisb. ex N.E.Br. = Gynandriris Parl. IRIDACEAE (M)
Helladia M.Kral = Sedum L. CRASSULACEAE (D)
Helleborine Kuntze = Calopogon R.Br. ORCHIDACEAE (M)
Helleborus L. RANUNCULACEAE (D)
Hellenia Willd. = Alpinia Roxb. ZINGIBERACEAE (M)
Hellenocarum H.Wolff UMBELLIFERAE (D)
Hellera Döll (SUI) = Raddia Bertol. GRAMINEAE (M)
Helleria Fourn. (SUH) = Festuca L. GRAMINEAE (M)
Helleriella A.D.Hawkes ORCHIDACEAE (M)
Hellerochloa Rauschert = Festuca L. GRAMINEAE (M)
Hellerorchis A.D.Hawkes = Rodrigueziella Kuntze ORCHIDACEAE (M)
Hellmuthia Steud. CYPERACEAE (M)
Hellwigia Warb. = Alpinia Roxb. ZINGIBERACEAE (M)
Helmholtzia F.Muell. PHILYDRACEAE (M)
Helminthia Juss. = Picris L. COMPOSITAE (D)
Helminthocarpon A.Rich. = Lotus L. LEGUMINOSAE–PAPILIONOIDEAE (D)
Helminthospermum Thwaites = Gironniera Gaudich. ULMACEAE (D)
Helminthostachys Kaulf. OPHIOGLOSSACEAE (F)
Helmiopsiella Arènes STERCULIACEAE (D)
Helmiopsis H.Perrier STERCULIACEAE (D)
Helmontia Cogn. CUCURBITACEAE (D)
Helogyne Nutt. COMPOSITAE (D)
Helonema Suess. = Eleocharis R.Br. CYPERACEAE (M)
Helonias L. MELANTHIACEAE (M)
Heloniopsis A.Gray MELANTHIACEAE (M)
Helonoma Garay ORCHIDACEAE (M)
Helopus Trin. = Eriochloa Kunth GRAMINEAE (M)
Helorchis Schltr. = Cynorkis Thouars ORCHIDACEAE (M)
Helosciadium W.D.J.Koch = Apium L. UMBELLIFERAE (D)
Helosis Rich. BALANOPHORACEAE (D)
Helospora Jack = Timonius DC. RUBIACEAE (D)
Helwingia Willd. HELWINGIACEAE (D)
Helxine Bubani = Soleirolia Gaudich. URTICACEAE (D)
Hemandradenia Stapf CONNARACEAE (D)
Hemarthria R.Br. GRAMINEAE (M)
Hemerocallis L. HEMEROCALLIDACEAE (M)
Hemesteum H.Lév. = Polystichum Roth DRYOPTERIDACEAE (F)
Hemestheum Newman = Thelypteris Schmidel THELYPTERIDACEAE (F)
Hemiadelphis Nees = Hygrophila L. ACANTHACEAE (D)
Hemiandra R.Br. LABIATAE (D)
Hemiandrina Hook.f. = Agelaea Sol. ex Planch. CONNARACEAE (D)
Hemianemia (Prantl) C.F.Reed = Anemia Sw. SCHIZAEACEAE (F)
Hemiangium A.C.Sm. = Hippocratea L. CELASTRACEAE (D)
Hemianthus Nutt. = Micranthemum Michx. SCROPHULARIACEAE (D)

Hemiarrhena Benth. = Lindernia All. SCROPHULARIACEAE (D)
Hemiarthron Tiegh. = Psittacanthus Mart. LORANTHACEAE (D)
Hemiboea C.B.Clarke GESNERIACEAE (D)
Hemiboeopsis W.T.Wang (UP) GESNERIACEAE (D)
Hemibromus Steud. = Glyceria R.Br. GRAMINEAE (M)
Hemicardion Fée = Cyclopeltis J.Sm. DRYOPTERIDACEAE (F)
Hemicarex Benth. = Kobresia Willd. CYPERACEAE (M)
Hemicarpha Nees = Lipocarpha R.Br. CYPERACEAE (M)
Hemichaena Benth. SCROPHULARIACEAE (D)
Hemichlaena Schrad. = Ficinia Schrad. CYPERACEAE (M)
Hemichoriste Nees = Justicia L. ACANTHACEAE (D)
Hemichroa R.Br. CHENOPODIACEAE (D)
Hemicicca Baill. = Phyllanthus L. EUPHORBIACEAE (D)
Hemicrambe Webb CRUCIFERAE (D)
Hemicrepidospermum Swart = Crepidospermum Hook.f. BURSERACEAE (D)
Hemicyatheon (Domin) Copel. = Hymenophyllum Sm. HYMENOPHYLLACEAE (F)
Hemicyclia Wight & Arn. = Drypetes Vahl EUPHORBIACEAE (D)
Hemidesma Raf. = Neptunia Lour. LEGUMINOSAE–MIMOSOIDEAE (D)
Hemidesmus R.Br. ASCLEPIADACEAE (D)
Hemidictyum C.Presl WOODSIACEAE (F)
Hemidiodia K.Schum. = Diodia L. RUBIACEAE (D)
Hemieva Raf. = Suksdorfia A.Gray SAXIFRAGACEAE (D)
Hemifuchsia Herrera (UP) ONAGRACEAE (D)
Hemigenia R.Br. LABIATAE (D)
Hemiglochidion (Müll.Arg.) K.Schum. = Phyllanthus L. EUPHORBIACEAE (D)
Hemigramma Christ = Tectaria Cav. DRYOPTERIDACEAE (F)
Hemigraphis Nees ACANTHACEAE (D)
Hemigymnia Stapf (SUH) = Ottochloa Dandy GRAMINEAE (M)
Hemiheisteria Tiegh. = Heisteria Jacq. OLACACEAE (D)
Hemilobium Welw. = Apodytes E.Mey. ex Arn. ICACINACEAE (D)
Hemilophia Franch. CRUCIFERAE (D)
Hemimeris L.f. SCROPHULARIACEAE (D)
Hemimunroa Parodi = Munroa Torr. GRAMINEAE (M)
Hemionitis L. ADIANTACEAE (F)
Hemiorchis Kurz ZINGIBERACEAE (M)
Hemiperis Frapp. ex Cordem. = Habenaria Willd. ORCHIDACEAE (M)
Hemiphlebium C.Presl = Trichomanes L. HYMENOPHYLLACEAE (F)
Hemiphora (F.Muell.) F.Muell. VERBENACEAE (D)
Hemiphragma Wall. SCROPHULARIACEAE (D)
Hemiphylacus S.Watson ASPHODELACEAE (M)
Hemipilia Lindl. ORCHIDACEAE (M)
Hemipogon Decne. ASCLEPIADACEAE (D)
Hemiptelea Planch. ULMACEAE (D)
Hemipteris Rosenst. = Pteris L. PTERIDACEAE (F)
Hemisacris Steud. = Schismus P.Beauv. GRAMINEAE (M)
Hemisandra Scheidw. = Aphelandra R.Br. ACANTHACEAE (D)
Hemisantiria H.J.Lam = Dacryodes Vahl BURSERACEAE (D)
Hemiscleria Lindl. ORCHIDACEAE (M)
Hemiscolopia Slooten FLACOURTIACEAE (D)
Hemisiphonia Urb. = Micranthemum Michx. SCROPHULARIACEAE (D)
Hemisorghum C.E.Hubb. ex Bor GRAMINEAE (M)
Hemistachyum (Copel.) Ching = Aglaomorpha Schott POLYPODIACEAE (F)
Hemistegia C.Presl = Cyathea Sm. CYATHEACEAE (F)
Hemisteptia Fisch. & C.A.Mey. COMPOSITAE (D)
Hemistylus Benth. URTICACEAE (D)

Hemitelia R.Br. = Cyathea Sm. CYATHEACEAE (F)
Hemithrinax Hook.f. = Thrinax Sw. PALMAE (M)
Hemitomes A.Gray ERICACEAE (D)
Hemitria Raf. = Phthirusa Mart. LORANTHACEAE (D)
Hemizonia DC. COMPOSITAE (D)
Hemizygia (Benth.) Briq. LABIATAE (D)
Hemprichia Ehrenb. = Commiphora Jacq. BURSERACEAE (D)
Hemsleia Kudo = Ceratanthus F.Muell. ex G.Taylor LABIATAE (D)
Hemsleya Cogn. ex F.B.Forbes & Hemsl. CUCURBITACEAE (D)
Hemsleyna Kuntze = Thryallis Mart. MALPIGHIACEAE (D)
Henckelia Spreng. = Didymocarpus Wall. GESNERIACEAE (D)
Hendecandras Eschsch. = Croton L. EUPHORBIACEAE (D)
Henicosanthum Dalla Torre & Harms (SUO) = Enicosanthum Becc. ANNONACEAE (D)
Henisia Walp. (SUO) = Heinsia DC. RUBIACEAE (D)
Henlea Karst. (UP) RUBIACEAE (D)
Henlea Griseb. = Thryallis Mart. MALPIGHIACEAE (D)
Henleophytum Karst. MALPIGHIACEAE (D)
Hennecartia Poiss. MONIMIACEAE (D)
Henonia Moq. AMARANTHACEAE (D)
Henoonia Griseb. GOETZEACEAE (D)
Henophyton Coss. & Durieu CRUCIFERAE (D)
Henosis Hook.f. = Bulbophyllum Thouars ORCHIDACEAE (M)
Henrardia C.E.Hubb. GRAMINEAE (M)
Henribaillonia Kuntze = Thecacoris A.Juss. EUPHORBIACEAE (D)
Henricia Cass. = Psiadia Jacq. COMPOSITAE (D)
Henricksonia B.L.Turner COMPOSITAE (D)
Henrietia Rchb. = Henriettea DC. MELASTOMATACEAE (D)
Henrietta Macfad. = Henriettea DC. MELASTOMATACEAE (D)
Henriettea DC. MELASTOMATACEAE (D)
Henriettella Naudin MELASTOMATACEAE (D)
Henriquezia Spruce ex Benth. RUBIACEAE (D)
Henrya Hemsl. (SUH) = Tylophora R.Br. ASCLEPIADACEAE (D)
Henrya Nees ACANTHACEAE (D)
Henryettana Brand = Antiotrema Hand.–Mazz. BORAGINACEAE (D)
Henschelia C.Presl = Illigera Blume HERNANDIACEAE (D)
Henslowia Blume (SUH) = Dendrotrophe Miq. SANTALACEAE (D)
Henslowia Lowe ex DC. (SUI) = Picconia DC. OLEACEAE (D)
Hensmania W.Fitzg. ANTHERICACEAE (M)
Hepatica Mill. RANUNCULACEAE (D)
Hephestionia Naudin = Tibouchina Aubl. MELASTOMATACEAE (D)
Heppiella Regel GESNERIACEAE (D)
Heptacodium Rehder CAPRIFOLIACEAE (D)
Heptallon Raf. = Croton L. EUPHORBIACEAE (D)
Heptanthus Griseb. COMPOSITAE (D)
Heptapleurum Gaertn. = Schefflera J.R.Forst. & G.Forst. ARALIACEAE (D)
Heptaptera Margot & Reut. UMBELLIFERAE (D)
Heptaseta Koidz. (SUI) = Agrostis L. GRAMINEAE (M)
Heracleum L. UMBELLIFERAE (D)
Herbertia Sweet IRIDACEAE (M)
Herbstia Sohmer AMARANTHACEAE (D)
Herderia Cass. COMPOSITAE (D)
Hereroa (Schwantes) Dinter & Schwantes AIZOACEAE (D)
Herissantia Medik. MALVACEAE (D)
Heritiera Aiton STERCULIACEAE (D)
Hermannia L. STERCULIACEAE (D)

Hermas L. UMBELLIFERAE (D)
Hermbstaedtia Rchb. AMARANTHACEAE (D)
Hermesia Humb. & Bonpl. = Alchornea Sw. EUPHORBIACEAE (D)
Hermidium S.Watson = Mirabilis L. NYCTAGINACEAE (D)
Herminiera Guill. & Perr. = Aeschynomene L. LEGUMINOSAE-PAPILIONOIDEAE (D)
Herminium L. ORCHIDACEAE (M)
Hermodactylus Mill. IRIDACEAE (M)
Hernandia L. HERNANDIACEAE (D)
Hernandiopsis Meisn. = Hernandia L. HERNANDIACEAE (D)
Herniaria L. ILLECEBRACEAE (D)
Herodotia Urb. & Ekman COMPOSITAE (D)
Herpestis Gaertn. SCROPHULARIACEAE (D)
Herpetacanthus Nees ACANTHACEAE (D)
Herpetica (DC.) Raf. = Senna Mill. LEGUMINOSAE-CAESALPINIOIDEAE (D)
Herpetophytum (Schltr.) Brieger = Dendrobium Sw. ORCHIDACEAE (M)
Herpetospermum Wall. ex Hook.f. CUCURBITACEAE (D)
Herpolirion Hook.f. ANTHERICACEAE (M)
Herpysma Lindl. ORCHIDACEAE (M)
Herpyza Sauvalle LEGUMINOSAE-PAPILIONOIDEAE (D)
Herrania Goudot STERCULIACEAE (D)
Herrea Schwantes = Conicosia N.E.Br. AIZOACEAE (D)
Herreanthus Schwantes AIZOACEAE (D)
Herrera Adans. = Erithalis P.Browne RUBIACEAE (D)
Herreria Ruíz & Pav. HERRERIACEAE (M)
Herreriopsis H.Perrier HERRERIACEAE (M)
Herrickia Woot. & Standl. = Aster L. COMPOSITAE (D)
Herschelia Lindl. (SUH) = Herschelianthe Rauschert ORCHIDACEAE (M)
Herschelia T.E.Bowdich = Physalis L. SOLANACEAE (D)
Herschelianthe Rauschert ORCHIDACEAE (M)
Hertelia Neck. (SUI) = Hernandia L. HERNANDIACEAE (D)
Hertia Less. = Othonna L. COMPOSITAE (D)
Hertrichocereus Backeb. = Stenocereus (A.Berger) Riccob. CACTACEAE (D)
Herya Cordem. = Pleurostylia Wight & Arn. CELASTRACEAE (D)
Hesioda Vell. = Heisteria Jacq. OLACACEAE (D)
Hesperaloe Engelm. AGAVACEAE (M)
Hesperantha Ker-Gawl. IRIDACEAE (M)
Hesperastragalus A.Heller = Astragalus L. LEGUMINOSAE-PAPILIONOIDEAE (D)
Hesperelaea A.Gray OLEACEAE (D)
Hesperethusa M.Roem. = Naringi Adans. RUTACEAE (D)
Hesperidanthus (B.L.Rob.) Rydb. = Schoenocrambe Greene CRUCIFERAE (D)
Hesperis L. CRUCIFERAE (D)
Hesperocallis A.Gray HYACINTHACEAE (M)
Hesperochiron S.Watson HYDROPHYLLACEAE (D)
Hesperochloa (Piper) Rydb. = Festuca L. GRAMINEAE (M)
Hesperocnide Torr. URTICACEAE (D)
Hesperogenia J.M.Coult. & Rose = Tauschia Schltdl. UMBELLIFERAE (D)
Hesperogreigia Skottsb. = Greigia Regel BROMELIACEAE (M)
Hesperolaburnum Maire LEGUMINOSAE-PAPILIONOIDEAE (D)
Hesperolinon (A.Gray) Small LINACEAE (D)
Hesperomannia A.Gray COMPOSITAE (D)
Hesperomecon Greene PAPAVERACEAE (D)
Hesperomeles Lindl. ROSACEAE (D)
Hesperonia Standl. NYCTAGINACEAE (D)
Hesperonix Rydb. = Astragalus L. LEGUMINOSAE-PAPILIONOIDEAE (D)
Hesperopeuce (Engelm.) Lemmon PINACEAE (G)

Hesperoseris Skottsb. = Dendroseris D.Don COMPOSITAE (D)
Hesperothamnus Brandegee LEGUMINOSAE–PAPILIONOIDEAE (D)
Hesperoxalis Small = Oxalis L. OXALIDACEAE (D)
Hesperoxiphion Baker IRIDACEAE (M)
Hesperoyucca Baker = Yucca L. AGAVACEAE (M)
Hesperozygis Epling LABIATAE (D)
Hessea Herb. AMARYLLIDACEAE (M)
Hetaeria Blume ORCHIDACEAE (M)
Heterachne Benth. GRAMINEAE (M)
Heteracia Fisch. & C.A.Mey. COMPOSITAE (D)
Heteradelphia Lindau ACANTHACEAE (D)
Heteranthelium Jaub. & Spach GRAMINEAE (M)
Heteranthemis Schott COMPOSITAE (D)
Heteranthera Ruíz & Pav. PONTEDERIACEAE (M)
Heteranthia Nees & Mart. SCROPHULARIACEAE (D)
Heteranthoecia Stapf GRAMINEAE (M)
Heteranthus Borkh. = Ventenata Koeler GRAMINEAE (M)
Heteraspidia Rizzini = Justicia L. ACANTHACEAE (D)
Heterelytron Jungh. = Themeda Forssk. GRAMINEAE (M)
Heterixia Tiegh. = Korthalsella Tiegh. VISCACEAE (D)
Heteroaridarum M.Hotta ARACEAE (M)
Heteroarisaema Nakai = Arisaema Mart. ARACEAE (M)
Heterocalymnantha Domin = Sauropus Blume EUPHORBIACEAE (D)
Heterocalycium Rauschert = Cuspidaria DC. BIGNONIACEAE (D)
Heterocalyx Gagnep. = Agrostistachys Dalzell EUPHORBIACEAE (D)
Heterocarpha Stapf & C.E.Hubb. = Drake–Brockmania Stapf GRAMINEAE (M)
Heterocarpus Phil. (SUH) = Cardamine L. CRUCIFERAE (D)
Heterocaryum A.DC. = Lappula Moench BORAGINACEAE (D)
Heterocentron Hook. & Arn. MELASTOMATACEAE (D)
Heterochaenia A.DC. CAMPANULACEAE (D)
Heterochaeta Schult. (SUS) = Ventenata Koeler GRAMINEAE (M)
Heterochlamys Turcz. = Julocroton Mart. EUPHORBIACEAE (D)
Heterochloa Desv. = Andropogon L. GRAMINEAE (M)
Heterocodon Nutt. CAMPANULACEAE (D)
Heterocoma DC. COMPOSITAE (D)
Heterocondylus R.M.King & H.Rob. COMPOSITAE (D)
Heterocroton S.Moore = Croton L. EUPHORBIACEAE (D)
Heterocypsela H.Rob. COMPOSITAE (D)
Heterodanaea C.Presl = Danaea Sm. MARATTIACEAE (F)
Heterodendrum Desf. = Alectryon Gaertn. SAPINDACEAE (D)
Heteroderis (Bunge) Boiss. COMPOSITAE (D)
Heterodon Meisn. = Nebelia Neck. ex Sweet BRUNIACEAE (D)
Heterodraba Greene CRUCIFERAE (D)
Heterogaura Rothr. ONAGRACEAE (D)
Heterogonium C.Presl DRYOPTERIDACEAE (F)
Heterolamium C.Y.Wu LABIATAE (D)
Heterolepis Boiss. (SUI) = Chloris Sw. GRAMINEAE (M)
Heterolepis Cass. COMPOSITAE (D)
Heterolobium Peter = Gonatopus Hook.f. ex Engl. ARACEAE (M)
Heterolobivia Y.Ito (SUI) = Echinopsis Zucc. CACTACEAE (D)
Heteromeles M.Roem. = Photinia Lindl. ROSACEAE (D)
Heteromera Montrouz. ex Beauvis. (SUI) = Leptostylis Benth. SAPOTACEAE (D)
Heteromera Pomel COMPOSITAE (D)
Heteromma Benth. COMPOSITAE (D)
Heteromorpha Cham. & Schltdl. UMBELLIFERAE (D)

Heteromyrtus Blume (SUI) = Blepharocalyx O.Berg MYRTACEAE (D)
Heteroneuron Hook.f. = Loreya DC. MELASTOMATACEAE (D)
Heteroneurum C.Presl = Bolbitis Schott LOMARIOPSIDACEAE (F)
Heteronoma DC. = Arthrostemma Pav. ex D.Don MELASTOMATACEAE (D)
Heteropanax Seem. ARALIACEAE (D)
Heteropappus Less. COMPOSITAE (D)
Heteropetalum Benth. ANNONACEAE (D)
Heterophlebium Fée = Pteris L. PTERIDACEAE (F)
Heteropholis C.E.Hubb. GRAMINEAE (M)
Heterophragma DC. BIGNONIACEAE (D)
Heterophyllaea Hook.f. RUBIACEAE (D)
Heterophyllium Hieron. ex Börner = Selaginella P.Beauv. SELAGINELLACEAE (FA)
Heteroplexis C.C.Chang COMPOSITAE (D)
Heteropogon Pers. GRAMINEAE (M)
Heteropsis Kunth ARACEAE (M)
Heteropteris Kunth (SUO) = Heteropterys Kunth MALPIGHIACEAE (D)
Heteropteris Fée (SUH) = Doryopteris J.Sm. ADIANTACEAE (F)
Heteropterys Kunth MALPIGHIACEAE (D)
Heteroptilis E.Mey. ex Meisn. = Dasispermum Raf. UMBELLIFERAE (D)
Heteropyxis Harv. MYRTACEAE (D)
Heterorhachis Sch.Bip. ex Walp. COMPOSITAE (D)
Heterosciadium Lange (UP) UMBELLIFERAE (D)
Heterosicyos (S.Watson) Cockerell (SUH) = Cremastopus Paul G.Wilson
  CUCURBITACEAE (D)
Heterosicyos Welw. ex Hook.f. = Trochomeria Hook.f. CUCURBITACEAE (D)
Heterosicyus Post & Kuntze (SUO) = Trochomeria Hook.f. CUCURBITACEAE (D)
Heterosmilax Kunth SMILACACEAE (M)
Heterospathe Scheff. PALMAE (M)
Heterosperma Cav. COMPOSITAE (D)
Heterostachys Ung.–Sternb. CHENOPODIACEAE (D)
Heterosteca Desv. = Bouteloua Lag. GRAMINEAE (M)
Heterostemma Wight & Arn. ASCLEPIADACEAE (D)
Heterostemon Desf. LEGUMINOSAE–CAESALPINIOIDEAE (D)
Heterotaxis Lindl. = Maxillaria Ruíz & Pav. ORCHIDACEAE (M)
Heterothalamus Less. COMPOSITAE (D)
Heterotheca Cass. COMPOSITAE (D)
Heterothrix (B.L.Rob.) Rydb. = Pennellia Nieuwl. CRUCIFERAE (D)
Heterotis Benth. = Dissotis Benth. MELASTOMATACEAE (D)
Heterotoma Zucc. CAMPANULACEAE (D)
Heterotrichum DC. MELASTOMATACEAE (D)
Heterotristicha Tobl. PODOSTEMACEAE (D)
Heterotropa C.Morren & Decne. = Asarum L. ARISTOLOCHIACEAE (D)
Heterozeuxine Hashim. = Zeuxine Lindl. ORCHIDACEAE (M)
Heterozostera (Setch.) Hartog ZOSTERACEAE (M)
Heuchera L. SAXIFRAGACEAE (D)
Heuffelia Schur (SUS) = Helictotrichon Besser GRAMINEAE (M)
Hevea Aubl. EUPHORBIACEAE (D)
Hewardia J.Sm. = Adiantum L. ADIANTACEAE (F)
Hewardia Hook.f. (SUH) = Isophysis T.Moore IRIDACEAE (M)
Hewittia Wight & Arn. CONVOLVULACEAE (D)
Hexachlamys O.Berg MYRTACEAE (D)
Hexactina Willd. ex Schltdl. = Amaioua Aubl. RUBIACEAE (D)
Hexacyrtis Dinter COLCHICACEAE (M)
Hexadena Raf. = Phyllanthus L. EUPHORBIACEAE (D)
Hexadenia Klotzsch & Garcke = Pedilanthus Neck. ex Poit. EUPHORBIACEAE (D)

Hexadesmia Brongn. ORCHIDACEAE (M)
Hexaglottis Vent. IRIDACEAE (M)
Hexakestra Hook.f. (SUO) = Leptopus Decne. EUPHORBIACEAE (D)
Hexakistra Hook.f. = Leptopus Decne. EUPHORBIACEAE (D)
Hexalectris Raf. ORCHIDACEAE (M)
Hexalobus A.DC. ANNONACEAE (D)
Hexameria R.Br. (SUH) = Podochilus Blume ORCHIDACEAE (M)
Hexameria Torr. & A.Gray = Echinocystis Torr. & A.Gray CUCURBITACEAE (D)
Hexaneurocarpon Dop = Fernandoa Welw. ex Seem. BIGNONIACEAE (D)
Hexapora Hook.f. = Micropora Hook.f. LAURACEAE (D)
Hexaptera Hook. = Menonvillea R.Br. ex DC. CRUCIFERAE (D)
Hexapterella Urb. BURMANNIACEAE (M)
Hexarrhena C.Presl = Hilaria Kunth GRAMINEAE (M)
Hexasepalum Bartl. ex DC. = Diodia L. RUBIACEAE (D)
Hexaspermum Domin EUPHORBIACEAE (D)
Hexaspora C.T.White CELASTRACEAE (D)
Hexastemon Klotzsch ERICACEAE (D)
Hexastylis Raf. = Asarum L. ARISTOLOCHIACEAE (D)
Hexatheca C.B.Clarke GESNERIACEAE (D)
Hexepta Raf. = Coffea L. RUBIACEAE (D)
Hexisea Lindl. ORCHIDACEAE (M)
Hexodontocarpus Dulac = Sherardia L. RUBIACEAE (D)
Hexopetion Burret = Astrocaryum G.Mey. PALMAE (M)
Hexopia Bateman ex Lindl. = Scaphyglottis Poepp. & Endl. ORCHIDACEAE (M)
Hexuris Miers (SUS) = Peltophyllum Gardner TRIURIDACEAE (M)
Heyderia K.Koch (SUH) = Calocedrus Kurz CUPRESSACEAE (G)
Heylandia DC. = Crotalaria L. LEGUMINOSAE–PAPILIONOIDEAE (D)
Heymassoli Aubl. = Ximenia L. OLACACEAE (D)
Heymia Dennst. = Dentella J.R.Forst. & G.Forst. RUBIACEAE (D)
Heynea Roxb. ex Sims = Trichilia P.Browne MELIACEAE (D)
Heynella Backer ASCLEPIADACEAE (D)
Heywoodia T.R.Sim EUPHORBIACEAE (D)
Heywoodiella Svent. & Bramwell = Hypochaeris L. COMPOSITAE (D)
Hibbertia Andrews DILLENIACEAE (D)
Hibiscadelphus Rock MALVACEAE (D)
Hibiscus L. MALVACEAE (D)
Hickelia A.Camus GRAMINEAE (M)
Hickenia Britton & Rose (SUH) = Parodia Speg. CACTACEAE (D)
Hickenia Lillo = Oxypetalum R.Br. ASCLEPIADACEAE (D)
Hicksbeachia F.Muell. PROTEACEAE (D)
Hicorius Raf. = Carya Nutt. JUGLANDACEAE (D)
Hicriopteris C.Presl = Dicranopteris Bernh. GLEICHENIACEAE (F)
Hidalgoa La Llave COMPOSITAE (D)
Hieracium L. COMPOSITAE (D)
Hieranthes Raf. = Stereospermum Cham. BIGNONIACEAE (D)
Hieris Steenis BIGNONIACEAE (D)
Hiernia S.Moore SCROPHULARIACEAE (D)
Hierobotana Briq. VERBENACEAE (D)
Hierochloe R.Br. GRAMINEAE (M)
Hieronima Allemão EUPHORBIACEAE (D)
Hieronyma Baill. = Hieronima Allemão EUPHORBIACEAE (D)
Hieronymiella Pax AMARYLLIDACEAE (M)
Hieronymusia Engl. = Suksdorfia A.Gray SAXIFRAGACEAE (D)
Higginsia Pers. = Hoffmannia Sw. RUBIACEAE (D)
Hilaria Kunth GRAMINEAE (M)

Hilariophyton Pichon (SUS) = Paragonia Bureau BIGNONIACEAE (D)
Hildebrandtia Vatke CONVOLVULACEAE (D)
Hildegardia Schott & Endl. STERCULIACEAE (D)
Hildewintera F.Ritter = Cleistocactus Lem. CACTACEAE (D)
Hildmannia Kreuz. & Buining (SUS) = Neoporteria Britton & Rose CACTACEAE (D)
Hillebrandia Oliv. BEGONIACEAE (D)
Hilleria Vell. PHYTOLACCACEAE (D)
Hillia Jacq. RUBIACEAE (D)
Hilliardia B.Nord. COMPOSITAE (D)
Hilliella (O.E.Schulz) Zhang & Li = Cochlearia L. CRUCIFERAE (D)
Himalayacalamus Keng f. = Thamnocalamus Munro GRAMINEAE (M)
Himalrandia Yamaz. RUBIACEAE (D)
Himantandra F.Muell. ex Diels = Galbulimima F.M.Bailey HIMANTANDRACEAE (D)
Himantochilus T.Anderson ex Benth. = Anisotes Nees ACANTHACEAE (D)
Himantoglossum K.Koch ORCHIDACEAE (M)
Himantostemma A.Gray ASCLEPIADACEAE (D)
Himatanthus Willd. ex Schult. APOCYNACEAE (D)
Hindsia Benth. ex Lindl. RUBIACEAE (D)
Hinterhubera Sch.Bip.ex Wedd. COMPOSITAE (D)
Hintonella Ames ORCHIDACEAE (M)
Hintonia Bullock RUBIACEAE (D)
Hionanthera A.Fern. & Diniz LYTHRACEAE (D)
Hippagrostis Kuntze (SUS) = Oplismenus P.Beauv. GRAMINEAE (M)
Hippeastrum Herb. AMARYLLIDACEAE (M)
Hippeophyllum Schltr. ORCHIDACEAE (M)
Hippia L. COMPOSITAE (D)
Hippobroma G.Don = Laurentia Adans. CAMPANULACEAE (D)
Hippobromus Eckl. & Zeyh. SAPINDACEAE (D)
Hippochaete J.Milde = Equisetum L. EQUISETACEAE (FA)
Hippocratea L. CELASTRACEAE (D)
Hippocrepandra Müll.Arg. = Monotaxis Brongn. EUPHORBIACEAE (D)
Hippocrepis L. LEGUMINOSAE–PAPILIONOIDEAE (D)
Hippodamia Decne. = Solenophora Benth. GESNERIACEAE (D)
Hippoglossum Breda = Bulbophyllum Thouars ORCHIDACEAE (M)
Hippolytia Poljakov COMPOSITAE (D)
Hippomane L. EUPHORBIACEAE (D)
Hippophae L. ELAEAGNACEAE (D)
Hippopodium Harv. = Ceratandra Eckl. ex F.A.Bauer ORCHIDACEAE (M)
Hipporkis Thouars = Satyrium Sw. ORCHIDACEAE (M)
Hippotis Ruíz & Pav. RUBIACEAE (D)
Hippoxylon Raf. = Oroxylum Vent. BIGNONIACEAE (D)
Hippuris L. HIPPURIDACEAE (D)
Hiptage Gaertn. MALPIGHIACEAE (D)
Hiraea Jacq. MALPIGHIACEAE (D)
Hirpicium Cass. COMPOSITAE (D)
Hirschfeldia Moench CRUCIFERAE (D)
Hirschia Baker = Iphiona Cass. COMPOSITAE (D)
Hirtella L. CHRYSOBALANACEAE (D)
Hirtzia Dodson ORCHIDACEAE (M)
Hispaniella Braem ORCHIDACEAE (M)
Hispidella Barnadez ex Lam. COMPOSITAE (D)
Histiopteris (J.G.Agardh) J.Sm. DENNSTAEDTIACEAE (F)
Hitchcockella A.Camus GRAMINEAE (M)
Hitchenia Wall. ZINGIBERACEAE (M)
Hitcheniopsis (Baker) Ridl. = Scaphochlamys Baker ZINGIBERACEAE (M)

Hitoa Nadeaud = Ixora L. RUBIACEAE (D)
Hjaltalinia Á.Löve & D.Löve = Sedum L. CRASSULACEAE (D)
Hladnikia Rchb. UMBELLIFERAE (D)
Hladnikia W.D.J.Koch (SUH) = Grafia Rchb. UMBELLIFERAE (D)
Hochreutinera Krapov. MALVACEAE (D)
Hochstetteria DC. = Dicoma Cass. COMPOSITAE (D)
Hockinia Gardner GENTIANACEAE (D)
Hodgkinsonia F.Muell. RUBIACEAE (D)
Hodgsonia Hook.f. & Thomson CUCURBITACEAE (D)
Hodgsoniola F.Muell. ANTHERICACEAE (M)
Hoehnea Epling LABIATAE (D)
Hoehneella Ruschi ORCHIDACEAE (M)
Hoehnelia Schweinf. = Ethulia L.f. COMPOSITAE (D)
Hoehnephytum Cabrera COMPOSITAE (D)
Hoepfneria Vatke = Abrus Adans. LEGUMINOSAE–PAPILIONOIDEAE (D)
Hoferia Scop. = Ternstroemia Mutis ex L.f. THEACEAE (D)
Hoffmannia Sw. RUBIACEAE (D)
Hoffmannia Willd. (SUH) = Psilotum Sw. PSILOTACEAE (FA)
Hoffmanniella Schltr. ex Lawalrée COMPOSITAE (D)
Hoffmannseggia Cav. LEGUMINOSAE–CAESALPINIOIDEAE (D)
Hofmeistera Rchb.f. = Hofmeisterella Rchb.f. ORCHIDACEAE (M)
Hofmeisterella Rchb.f. ORCHIDACEAE (M)
Hofmeisteria Walp. COMPOSITAE (D)
Hohenackeria Fisch. & C.A.Mey. UMBELLIFERAE (D)
Hohenbergia Schult.f. BROMELIACEAE (M)
Hohenbergiopsis L.B.Sm. & Read BROMELIACEAE (M)
Hoheria A.Cunn. MALVACEAE (D)
Hoita Rydb. LEGUMINOSAE–PAPILIONOIDEAE (D)
Hoitzia Juss. = Loeselia L. POLEMONIACEAE (D)
Holacantha A.Gray = Castela Turpin SIMAROUBACEAE (D)
Holalafia Stapf = Alafia Thouars APOCYNACEAE (D)
Holarrhena R.Br. APOCYNACEAE (D)
Holboellia Hook. (SUH) = Lopholepis Decne. GRAMINEAE (M)
Holboellia Wall. LARDIZABALACEAE (D)
Holcoglossum Schltr. ORCHIDACEAE (M)
Holcolemma Stapf & C.E.Hubb. GRAMINEAE (M)
Holcophacos Rydb. = Astragalus L. LEGUMINOSAE–PAPILIONOIDEAE (D)
Holcosorus T.Moore POLYPODIACEAE (F)
Holcus L. GRAMINEAE (M)
Holigarna Buch.–Ham. ex Roxb. ANACARDIACEAE (D)
Hollandaea F.Muell. PROTEACEAE (D)
Hollermayera O.E.Schulz CRUCIFERAE (D)
Hollisteria S.Watson POLYGONACEAE (D)
Hollrungia K.Schum. PASSIFLORACEAE (D)
Holmbergia Hicken CHENOPODIACEAE (D)
Holmesia P.J.Cribb (SUH) = Angraecopsis Kraenzl. ORCHIDACEAE (M)
Holmgrenanthe Elisens SCROPHULARIACEAE (D)
Holmskioldia Retz. VERBENACEAE (D)
Holocalyx Micheli LEGUMINOSAE–PAPILIONOIDEAE (D)
Holocarpa Baker = Pentanisia Harv. RUBIACEAE (D)
Holocarpha Greene COMPOSITAE (D)
Holocarya T.Durand (SUO) = Pentanisia Harv. RUBIACEAE (D)
Holocheila (Kudo) S.Chow LABIATAE (D)
Holocheilus Cass. COMPOSITAE (D)
Holochlamys Engl. ARACEAE (M)

Holodictyum Maxon ASPLENIACEAE (F)
Holodiscus (K.Koch) Maxim. ROSACEAE (D)
Hologamium Nees = Sehima Forssk. GRAMINEAE (M)
Holographis Nees ACANTHACEAE (D)
Hologyne Pfitzer = Coelogyne Lindl. ORCHIDACEAE (M)
Hololachna Ehrenb. TAMARICACEAE (D)
Hololeion Kitam. COMPOSITAE (D)
Hololepis DC. = Vernonia Schreb. COMPOSITAE (D)
Holopogon Kom. & Nevski ORCHIDACEAE (M)
Holoptelea Planch. ULMACEAE (D)
Holopyxidium Ducke = Lecythis Loefl. LECYTHIDACEAE (D)
Holoschoenus Link (SUH) = Scirpoides Séguier CYPERACEAE (M)
Holosetum Steud. = Alloteropsis C.Presl GRAMINEAE (M)
Holostachyum (Copel.) Ching = Aglaomorpha Schott POLYPODIACEAE (F)
Holostemma R.Br. ASCLEPIADACEAE (D)
Holosteum L. CARYOPHYLLACEAE (D)
Holostigma Spach = Camissonia Link ONAGRACEAE (D)
Holostyla Endl. (SUO) = Coelospermum Blume RUBIACEAE (D)
Holostylis Duch. ARISTOLOCHIACEAE (D)
Holostylon Robyns & Lebrun LABIATAE (D)
Holothrix Rich. ex Lindl. ORCHIDACEAE (M)
Holozonia Greene COMPOSITAE (D)
Holstia Pax (SUH) = Neoholstia Rauschert EUPHORBIACEAE (D)
Holstianthus Steyerm. RUBIACEAE (D)
Holtonia Standl. = Simira Aubl. RUBIACEAE (D)
Holttumia Copel. = Taenitis Willd. ex Schkuhr ADIANTACEAE (F)
Holttumiella Copel. = Taenitis Willd. ex Schkuhr ADIANTACEAE (F)
Holtzea Schindl. = Desmodium Desv. LEGUMINOSAE–PAPILIONOIDEAE (D)
Holubia Oliv. PEDALIACEAE (D)
Holubogentia Á.Löve & D.Löve = Gentiana L. GENTIANACEAE (D)
Holzneria Speta SCROPHULARIACEAE (D)
Homalachna Kuntze (SUS) = Holcus L. GRAMINEAE (M)
Homalanthus A.Juss. EUPHORBIACEAE (D)
Homalium Jacq. FLACOURTIACEAE (D)
Homalobus Nutt. = Astragalus L. LEGUMINOSAE–PAPILIONOIDEAE (D)
Homalocalyx F.Muell. MYRTACEAE (D)
Homalocarpus Hook. & Arn. UMBELLIFERAE (D)
Homalocenchrus Miègev. (SUS) = Leersia Sw. GRAMINEAE (M)
Homalocephala Britton & Rose = Echinocactus Link & Otto CACTACEAE (D)
Homalocheilos J.K.Morton = Isodon (Schrad. ex Benth.) Spach LABIATAE (D)
Homalocladium (F.Muell.) L.H.Bailey POLYGONACEAE (D)
Homaloclados Hook.f. = Faramea Aubl. RUBIACEAE (D)
Homalodiscus Bunge ex Boiss. RESEDACEAE (D)
Homalomena Schott ARACEAE (M)
Homalopetalum Rolfe ORCHIDACEAE (M)
Homalosciadium Domin UMBELLIFERAE (D)
Homalosorus Small ex Pic. Serm. = Diplaziopsis C.Chr. WOODSIACEAE (F)
Homalospermum Schauer MYRTACEAE (D)
Homeria Vent. IRIDACEAE (M)
Homocentria Naudin = Oxyspora DC. MELASTOMATACEAE (D)
Homocodon D.Y.Hong = Heterocodon Nutt. CAMPANULACEAE (D)
Homoeatherum Nees = Andropogon L. GRAMINEAE (M)
Homoeotes Bory = Trichomanes L. HYMENOPHYLLACEAE (F)
Homoglossum Salisb. = Gladiolus L. IRIDACEAE (M)
Homognaphalium Kirp. (SUS) = Gnaphalium L. COMPOSITAE (D)

Homogyne Cass. COMPOSITAE (D)
Homoiachne Pilg. = Deschampsia P.Beauv. GRAMINEAE (M)
Homolepis Chase GRAMINEAE (M)
Homollea Arènes RUBIACEAE (D)
Homolliella Arènes RUBIACEAE (D)
Homonoia Lour. EUPHORBIACEAE (D)
Homonoma Bello = Nepsera Naudin MELASTOMATACEAE (D)
Homopholis C.E.Hubb. GRAMINEAE (M)
Homophyllum Merino = Blechnum L. BLECHNACEAE (F)
Homoplitis Trin. (SUS) = Pogonatherum P.Beauv. GRAMINEAE (M)
Homopogon Stapf = Trachypogon Nees GRAMINEAE (M)
Homoranthus A.Cunn. ex Schauer MYRTACEAE (D)
Homozeugos Stapf GRAMINEAE (M)
Honckenya Ehrh. CARYOPHYLLACEAE (D)
Honckenya Willd. (SUO) = Clappertonia Meisn. TILIACEAE (D)
Hondbessen Adans. (UP) RUBIACEAE (D)
Honkenya Ehrh. (SUO) = Honckenya Ehrh. CARYOPHYLLACEAE (D)
Honkenya Cothen. (SUH) = Clappertonia Meisn. TILIACEAE (D)
Hoodia Sweet ex Decne. ASCLEPIADACEAE (D)
Hookerella Tiegh. = Muellerina Tiegh. LORANTHACEAE (D)
Hookerochloa E.B.Alexeev = Austrofestuca (Tzvelev) Alexeev GRAMINEAE (M)
Hopea Roxb. DIPTEROCARPACEAE (D)
Hopkinsia W.Fitzg. RESTIONACEAE (M)
Hoplestigma Pierre HOPLESTIGMATACEAE (D)
Hoplismenus Hassk. (SUO) = Oplismenus P.Beauv. GRAMINEAE (M)
Hoplophyllum DC. COMPOSITAE (D)
Hoppea Willd. GENTIANACEAE (D)
Hoppia Nees = Bisboeckelera Kuntze CYPERACEAE (M)
Horaninovia Fisch. & C.A.Mey. CHENOPODIACEAE (D)
Hordelymus (Jess.) Jess. ex Harz GRAMINEAE (M)
Hordeum L. GRAMINEAE (M)
Horichia Jenny ORCHIDACEAE (M)
Horkelia Cham. & Schltdl. ROSACEAE (D)
Horkeliella (Rydb.) Rydb. ROSACEAE (D)
Hormathophylla Cullen & T.R.Dudley CRUCIFERAE (D)
Hormidium (Lindl.) Heynh. ORCHIDACEAE (M)
Horminum L. LABIATAE (D)
Hormocalyx Gleason = Myrmidone Mart. MELASTOMATACEAE (D)
Hormogyne A.DC. = Pouteria Aubl. SAPOTACEAE (D)
Hormuzakia Guşul. = Anchusa L. BORAGINACEAE (D)
Hornea Baker SAPINDACEAE (D)
Hornemannia Vahl = Symphysia C.Presl ERICACEAE (D)
Hornschuchia Nees ANNONACEAE (D)
Hornstedtia Retz. ZINGIBERACEAE (M)
Hornungia Rchb. CRUCIFERAE (D)
Horridocactus Backeb. = Neoporteria Britton & Rose CACTACEAE (D)
Horsfieldia Blume ex DC. (SUH) = Harmsiopanax Warb. ARALIACEAE (D)
Horsfieldia Chifflot (SUH) = Monophyllaea R.Br. GESNERIACEAE (D)
Horsfieldia Willd. MYRISTICACEAE (D)
Horsfordia A.Gray MALVACEAE (D)
Horstrissea Greuter, Gerstb. & Egli UMBELLIFERAE (D)
Horta Vell. = Clavija Ruíz & Pav. THEOPHRASTACEAE (D)
Hortensia Comm. = Hydrangea L. HYDRANGEACEAE (D)
Hortia Vand. RUTACEAE (D)
Hortonia Wight MONIMIACEAE (D)

Horvatia Garay ORCHIDACEAE (M)
Horwoodia Turrill CRUCIFERAE (D)
Hosackia Benth. ex Lindl. = Lotus L. LEGUMINOSAE–PAPILIONOIDEAE (D)
Hosea Ridl. VERBENACEAE (D)
Hoseanthus Merr. (SUS) = Hosea Ridl. VERBENACEAE (D)
Hoshiarpuria Hajra, P.Daniel & Philcox = Rotala L. LYTHRACEAE (D)
Hosiea Hemsl. & E.H.Wilson ICACINACEAE (D)
Hoslundia Vahl LABIATAE (D)
Hosta Tratt. HOSTACEAE (M)
Hoteia C.Morren & Decne. = Astilbe Buch.–Ham. ex G.Don SAXIFRAGACEAE (D)
Hotnima Chev. = Manihot Mill. EUPHORBIACEAE (D)
Hottarum Bogner & Nicolson ARACEAE (M)
Hottea Urb. MYRTACEAE (D)
Hottonia L. PRIMULACEAE (D)
Houlletia Brongn. ORCHIDACEAE (M)
Hounea Baill. = Paropsia Noronha ex Thouars PASSIFLORACEAE (D)
Houssayanthus Hunz. SAPINDACEAE (D)
Houstonia L. RUBIACEAE (D)
Houttea Heynh. = Vanhouttea Lem. GESNERIACEAE (D)
Houttuynia Thunb. SAURURACEAE (D)
Houzeaubambus Mattei = Oxytenanthera Munro GRAMINEAE (M)
Hovea R.Br. ex W.T.Aiton LEGUMINOSAE–PAPILIONOIDEAE (D)
Hovenia Thunb. RHAMNACEAE (D)
Hoverdenia Nees ACANTHACEAE (D)
Howardia Wedd. = Pogonopus Klotzsch RUBIACEAE (D)
Howea Becc. PALMAE (M)
Howellia A.Gray CAMPANULACEAE (D)
Howelliella Rothm. SCROPHULARIACEAE (D)
Howethoa Rauschert = Lepisanthes Blume SAPINDACEAE (D)
Howittia F.Muell. MALVACEAE (D)
Hoya R.Br. ASCLEPIADACEAE (D)
Hoyella Ridl. ASCLEPIADACEAE (D)
Hua Pierre ex De Wild. HUACEAE (D)
Hualania Phil. = Bredemeyera Willd. POLYGALACEAE (D)
Huanaca Cav. UMBELLIFERAE (D)
Huanaca Raf. (SUH) = Dunalia Kunth SOLANACEAE (D)
Huarpea Cabrera COMPOSITAE (D)
Hubbardia Bor GRAMINEAE (M)
Hubbardochloa Auquier GRAMINEAE (M)
Huberia DC. MELASTOMATACEAE (D)
Huberodaphne Ducke = Endlicheria Nees LAURACEAE (D)
Huberodendron Ducke BOMBACACEAE (D)
Hubertia Bory COMPOSITAE (D)
Hudsonia L. CISTACEAE (D)
Hudsonia A.Rob. ex Lunan (SUH) = Terminalia L. COMBRETACEAE (D)
Hueblia Speta = Chaenorhinum (DC.) Rchb. SCROPHULARIACEAE (D)
Huebneria Schltr. = Orleanesia Barb.Rodr. ORCHIDACEAE (M)
Huernia R.Br. ASCLEPIADACEAE (D)
Huerniopsis N.E.Br. ASCLEPIADACEAE (D)
Huertea Ruíz & Pav. STAPHYLEACEAE (D)
Hufelandia Nees = Beilschmiedia Nees LAURACEAE (D)
Hugelia Benth. (SUH) = Eriastrum Woot. & Standl. POLEMONIACEAE (D)
Hughesia R.M.King & H.Rob. COMPOSITAE (D)
Hugonia Cav. ex Roem. = Lygodium Sw. SCHIZAEACEAE (F)
Hugonia L. LINACEAE (D)

Hugueninia Rchb. CRUCIFERAE (D)
Huidobria Gay = Loasa Adans. LOASACEAE (D)
Huilaea Wurdack MELASTOMATACEAE (D)
Hulemacanthus S.Moore ACANTHACEAE (D)
Hullettia King ex Hook.f. MORACEAE (D)
Hulsea Torr. & A.Gray COMPOSITAE (D)
Hulteniella Tzvelev = Dendranthema (DC.) Des Moulins COMPOSITAE (D)
Hulthemia Dumort. = Rosa L. ROSACEAE (D)
Hulthemosa Juz. = Rosa L. ROSACEAE (D)
Humata Cav. DAVALLIACEAE (F)
Humbertia Comm. ex Lam. CONVOLVULACEAE (D)
Humbertianthus Hochr. MALVACEAE (D)
Humbertiella Hochr. MALVACEAE (D)
Humbertina Buchet = Arophyton Jum. ARACEAE (M)
Humbertiodendron Leandri TRIGONIACEAE (D)
Humbertioturraea J.–F.Leroy MELIACEAE (D)
Humbertochloa A.Camus & Stapf GRAMINEAE (M)
Humbertodendron Leandri (SUO) = Humbertiodendron Leandri TRIGONIACEAE (D)
Humblotia Baill. = Drypetes Vahl EUPHORBIACEAE (D)
Humblotiella Tardieu = Lindsaea Dryand. ex Sm. DENNSTAEDTIACEAE (F)
Humblotiodendron Engl. = Vepris Comm. ex A.Juss. RUTACEAE (D)
Humboldtia Vahl LEGUMINOSAE–CAESALPINIOIDEAE (D)
Humboldtia Ruíz & Pav. (SUH) = Stelis Sw. ORCHIDACEAE (M)
Humboldtiella Harms = Coursetia DC. LEGUMINOSAE–PAPILIONOIDEAE (D)
Humeocline Anderb. COMPOSITAE (D)
Humiria Aubl. HUMIRIACEAE (D)
Humirianthera Huber ICACINACEAE (D)
Humiriastrum (Urb.) Cuatrec. HUMIRIACEAE (D)
Humularia P.A.Duvign. LEGUMINOSAE–PAPILIONOIDEAE (D)
Humulopsis Grudz. = Humulus L. CANNABACEAE (D)
Humulus L. CANNABACEAE (D)
Hunaniopanax Qi & Cao ARALIACEAE (D)
Hunga Pancher ex Prance CHRYSOBALANACEAE (D)
Hunnemannia Sweet PAPAVERACEAE (D)
Hunteria Roxb. APOCYNACEAE (D)
Huntleya Bateman ex Lindl. ORCHIDACEAE (M)
Hunzikeria D'Arcy SOLANACEAE (D)
Huodendron Rehder STYRACACEAE (D)
Huperzia Bernh. LYCOPODIACEAE (FA)
Hura L. EUPHORBIACEAE (D)
Husnotia Fourn. = Ditassa R.Br. ASCLEPIADACEAE (D)
Hussonia Boiss. = Erucaria Gaertn. CRUCIFERAE (D)
Hutchinsia R.Br. (SUS) = Thlaspi L. CRUCIFERAE (D)
Hutchinsiella O.E.Schulz CRUCIFERAE (D)
Hutchinsonia Robyns RUBIACEAE (D)
Hutera Porta = Coincya Rouy CRUCIFERAE (D)
Huthamnus Tsiang (UP) ASCLEPIADACEAE (D)
Huthia Brand POLEMONIACEAE (D)
Huttonaea Harv. ORCHIDACEAE (M)
Huttonella Kirk = Carmichaelia R.Br. LEGUMINOSAE–PAPILIONOIDEAE (D)
Huxleya Ewart VERBENACEAE (D)
Huynhia Greuter = Arnebia Forssk. BORAGINACEAE (D)
Hyacinthella Schur HYACINTHACEAE (M)
Hyacinthoides Heist. ex Fabr. HYACINTHACEAE (M)
Hyacinthorchis Blume = Cremastra Lindl. ORCHIDACEAE (M)

Hyacinthus L. HYACINTHACEAE (M)
Hyaenanche Lamb. EUPHORBIACEAE (D)
Hyalis D.Don ex Hook. & Arn. COMPOSITAE (D)
Hyalisma Champ. = Sciaphila Blume TRIURIDACEAE (M)
Hyalocalyx Rolfe TURNERACEAE (D)
Hyalochaete Dittrich & Rech.f. COMPOSITAE (D)
Hyalochlamys A.Gray COMPOSITAE (D)
Hyalocystis Hallier f. CONVOLVULACEAE (D)
Hyalolaena Bunge UMBELLIFERAE (D)
Hyalolepis Kunze = Belvisia Mirb. POLYPODIACEAE (F)
Hyalopoa (Tzvelev) Tzvelev = Colpodium Trin. GRAMINEAE (M)
Hyalosema Rolfe = Bulbophyllum Thouars ORCHIDACEAE (M)
Hyalosepalum Troupin = Tinospora Miers MENISPERMACEAE (D)
Hyaloseris Griseb. COMPOSITAE (D)
Hyalosperma Steetz COMPOSITAE (D)
Hyalostemma Wall. ex Meisn. = Miliusa Lesch. ex A.DC. ANNONACEAE (D)
Hyalotricha Copel. = Hyalotrichopteris W.H.Wagner POLYPODIACEAE (F)
Hyalotrichopteris W.H.Wagner POLYPODIACEAE (F)
Hybanthus Jacq. VIOLACEAE (D)
Hybochilus Schltr. ORCHIDACEAE (M)
Hybophrynium K.Schum. (SUS) = Trachyphrynium Benth. MARANTACEAE (M)
Hybosema Harms LEGUMINOSAE–PAPILIONOIDEAE (D)
Hybosperma Urb. RHAMNACEAE (D)
Hybridella Cass. COMPOSITAE (D)
Hydatella Diels HYDATELLACEAE (M)
Hydnocarpus Gaertn. FLACOURTIACEAE (D)
Hydnophytum Jack RUBIACEAE (D)
Hydnora Thunb. HYDNORACEAE (D)
Hydrangea L. HYDRANGEACEAE (D)
Hydranthelium Kunth SCROPHULARIACEAE (D)
Hydranthus Kuhl & Hasselt ex Rchb.f. = Dipodium R.Br. ORCHIDACEAE (M)
Hydrastis L. RANUNCULACEAE (D)
Hydriastele H.A.Wendl. & Drude PALMAE (M)
Hydrilla Rich. HYDROCHARITACEAE (M)
Hydrobryopsis Engl. = Hydrobryum Endl. PODOSTEMACEAE (D)
Hydrobryum Endl. PODOSTEMACEAE (D)
Hydrocera Blume BALSAMINACEAE (D)
Hydrocharis L. HYDROCHARITACEAE (M)
Hydrochloa Hartm. (SUS) = Glyceria R.Br. GRAMINEAE (M)
Hydrochloa P.Beauv. = Luziola Juss. GRAMINEAE (M)
Hydrocleys Rich. LIMNOCHARITACEAE (M)
Hydrocotyle L. UMBELLIFERAE (D)
Hydrodea N.E.Br. = Mesembryanthemum L. AIZOACEAE (D)
Hydrodyssodia B.L.Turner COMPOSITAE (D)
Hydrogaster Kuhlm. TILIACEAE (D)
Hydroglossum Willd. = Lygodium Sw. SCHIZAEACEAE (F)
Hydroidea Karis COMPOSITAE (D)
Hydrolea L. HYDROPHYLLACEAE (D)
Hydrolythrum Hook.f. = Rotala L. LYTHRACEAE (D)
Hydromestus Scheidw. = Aphelandra R.Br. ACANTHACEAE (D)
Hydromystria G.Mey. = Limnobium Rich. HYDROCHARITACEAE (M)
Hydropectis Rydb. COMPOSITAE (D)
Hydrophilus Linder RESTIONACEAE (M)
Hydrophylax L.f. RUBIACEAE (D)
Hydrophyllum L. HYDROPHYLLACEAE (D)

Hydropoa (Dumort.) Dumort. (SUS) = Glyceria R.Br. GRAMINEAE (M)
Hydropyrum Link = Zizania L. GRAMINEAE (M)
Hydrosme Schott = Amorphophallus Blume ex Decne. ARACEAE (M)
Hydrostachys Thouars HYDROSTACHYACEAE (D)
Hydrostemma Wall. = Barclaya Wall. NYMPHAEACEAE (D)
Hydrotaenia Lindl. = Tigridia Juss. IRIDACEAE (M)
Hydrothauma C.E.Hubb. GRAMINEAE (M)
Hydrothrix Hook.f. PONTEDERIACEAE (M)
Hydrotriche Zucc. SCROPHULARIACEAE (D)
Hyerochilus Pfitzer = Vandopsis Pfitzer ORCHIDACEAE (M)
Hyeronima Allemão = Hieronima Allemão EUPHORBIACEAE (D)
Hygea Hanst. GESNERIACEAE (D)
Hygrocharis Hochst. = Nephrophyllum A.Rich. CONVOLVULACEAE (D)
Hygrochilus Pfitzer ORCHIDACEAE (M)
Hygrochloa Lazarides GRAMINEAE (M)
Hygrophila R.Br. ACANTHACEAE (D)
Hygrorhiza Benth. (SUO) = Hygroryza Nees GRAMINEAE (M)
Hygroryza Nees GRAMINEAE (M)
Hylaeanthe Jonk.–Verh. & Jonk. MARANTACEAE (M)
Hylandia Airy Shaw EUPHORBIACEAE (D)
Hylandra Á.Löve = Arabis L. CRUCIFERAE (D)
Hylebates Chippindall GRAMINEAE (M)
Hylenaea Miers CELASTRACEAE (D)
Hyline Herb. AMARYLLIDACEAE (M)
Hylocarpa Cuatrec. HUMIRIACEAE (D)
Hylocereus (A.Berger) Britton & Rose CACTACEAE (D)
Hylocharis Miq. = Oxyspora DC. MELASTOMATACEAE (D)
Hylodendron Taub. LEGUMINOSAE–CAESALPINIOIDEAE (D)
Hylomecon Maxim. PAPAVERACEAE (D)
Hylophila Lindl. ORCHIDACEAE (M)
Hylotelephium H.Ohba CRASSULACEAE (D)
Hymanthoglossum Tod. = Himantoglossum K.Koch ORCHIDACEAE (M)
Hymenachne P.Beauv. GRAMINEAE (M)
Hymenaea L. LEGUMINOSAE–CAESALPINIOIDEAE (D)
Hymenandra (DC.) Spach MYRSINACEAE (D)
Hymenanthera R.Br. = Melicytus J.R.Forst. & G.Forst. VIOLACEAE (D)
Hymenasplenium Hayata = Asplenium L. ASPLENIACEAE (F)
Hymenatherum Cass. = Dyssodia Cav. COMPOSITAE (D)
Hymendocarpum Pierre ex Pit. = Nostolachma T.Durand RUBIACEAE (D)
Hymenella Moçino & Sessé = Minuartia L. CARYOPHYLLACEAE (D)
Hymenocallis Salisb. AMARYLLIDACEAE (M)
Hymenocardia Wall. ex Lindl. EUPHORBIACEAE (D)
Hymenocarpos Savi LEGUMINOSAE–PAPILIONOIDEAE (D)
Hymenocephalus Jaub. & Spach COMPOSITAE (D)
Hymenochaeta P.Beauv. (SUH) = Actinoscirpus (Ohwi) R.W.Haines & Lye
    CYPERACEAE (M)
Hymenocharis Salisb. == Ischnosiphon Koern. MARANTACEAE (M)
Hymenochlaena Bremek. = Strobilanthes Blume ACANTHACEAE (D)
Hymenoclea Torr. & A.Gray COMPOSITAE (D)
Hymenocnemis Hook.f. RUBIACEAE (D)
Hymenocoleus Robbr. RUBIACEAE (D)
Hymenocrater Fisch. & C.A.Mey. LABIATAE (D)
Hymenocyclus Dinter & Schwantes = Malephora N.E.Br. AIZOACEAE (D)
Hymenocystis C.A.Mey. = Woodsia R.Br. WOODSIACEAE (F)
Hymenodictyon Wall. RUBIACEAE (D)

Hymenodium Fée = Elaphoglossum Schott ex J.Sm. LOMARIOPSIDACEAE (F)
Hymenoglossum C.Presl HYMENOPHYLLACEAE (F)
Hymenogyne Haw. AIZOACEAE (D)
Hymenolaena DC. UMBELLIFERAE (D)
Hymenolepis Cass. COMPOSITAE (D)
Hymenolepis Kaulf. (SUH) = Belvisia Mirb. POLYPODIACEAE (F)
Hymenolobium Benth. LEGUMINOSAE–PAPILIONOIDEAE (D)
Hymenolobus Nutt. CRUCIFERAE (D)
Hymenolophus Boerl. (UP) APOCYNACEAE (D)
Hymenolyma Korovin = Hyalolaena Bunge UMBELLIFERAE (D)
Hymenonema Cass. COMPOSITAE (D)
Hymenopappus L'Hér. COMPOSITAE (D)
Hymenophyllopsis K.I.Goebel HYMENOPHYLLOPSIDACEAE (F)
Hymenophyllum Sm. HYMENOPHYLLACEAE (F)
Hymenophysa C.A.Mey. = Lepidium L. CRUCIFERAE (D)
Hymenopogon Wall. (SUH) = Neohymenopogon Bennet RUBIACEAE (D)
Hymenopyramis Wall. ex Griff. VERBENACEAE (D)
Hymenorchis Schltr. ORCHIDACEAE (M)
Hymenorebulobivia Frič (SUI) = Echinopsis Zucc. CACTACEAE (D)
Hymenorebutia Frič ex Buining = Echinopsis Zucc. CACTACEAE (D)
Hymenosicyos Chiov. = Oreosyce Hook.f. CUCURBITACEAE (D)
Hymenosporum R.Br. ex F.Muell. PITTOSPORACEAE (D)
Hymenostachys Bory = Trichomanes L. HYMENOPHYLLACEAE (F)
Hymenostegia (Benth.) Harms LEGUMINOSAE–CAESALPINIOIDEAE (D)
Hymenostemma Kunze ex Willk. COMPOSITAE (D)
Hymenostephium Benth. COMPOSITAE (D)
Hymenothecium Lag. = Aegopogon Humb. & Bonpl. ex Willd. GRAMINEAE (M)
Hymenothrix A.Gray COMPOSITAE (D)
Hymenoxys Cass. COMPOSITAE (D)
Hyobanche L. SCROPHULARIACEAE (D)
Hyophorbe Gaertn. PALMAE (M)
Hyoscyamus L. SOLANACEAE (D)
Hyoseris L. COMPOSITAE (D)
Hyospathe Mart. PALMAE (M)
Hypacanthium Juz. COMPOSITAE (D)
Hypagophytum A.Berger CRASSULACEAE (D)
Hypanthera Silva Manso = Fevillea L. CUCURBITACEAE (D)
Hyparrhenia Fourn. GRAMINEAE (M)
Hypecoum L. PAPAVERACEAE (D)
Hypelate P.Browne SAPINDACEAE (D)
Hypelichrysum Kirp. COMPOSITAE (D)
Hypenanthe (Blume) Blume MELASTOMATACEAE (D)
Hypenia (Mart. ex Benth.) Harley LABIATAE (D)
Hyperacanthus E.Mey. ex Bridson RUBIACEAE (D)
Hyperaspis Briq. = Ocimum L. LABIATAE (D)
Hyperbaena Miers ex Benth. MENISPERMACEAE (D)
Hypericophyllum Steetz COMPOSITAE (D)
Hypericopsis Boiss. FRANKENIACEAE (D)
Hypericum L. GUTTIFERAE (D)
Hypertelis E.Mey. ex Fenzl MOLLUGINACEAE (D)
Hyperthelia Clayton GRAMINEAE (M)
Hyphaene Gaertn. PALMAE (M)
Hypobathrum Blume RUBIACEAE (D)
Hypocalymma (Endl.) Endl. MYRTACEAE (D)
Hypocalyptus Thunb. LEGUMINOSAE–PAPILIONOIDEAE (D)

Hypochaeris L. COMPOSITAE (D)
Hypochlamys Fée = Diplazium Sw. WOODSIACEAE (F)
Hypochoeris L. (SUO) = Hypochaeris L. COMPOSITAE (D)
Hypocoton Urb. = Bonania A.Rich. EUPHORBIACEAE (D)
Hypocylix Woloszczak = Salsola L. CHENOPODIACEAE (D)
Hypocyrta Mart. = Nematanthus Schrad. GESNERIACEAE (D)
Hypodaphnis Stapf LAURACEAE (D)
Hypodema Rchb. = Cypripedium L. ORCHIDACEAE (M)
Hypodematium A.Rich. (SUH) = Eulophia R.Br. ex Lindl. ORCHIDACEAE (M)
Hypodematium A.Rich. (SUH) = Spermacoce L. RUBIACEAE (D)
Hypodematium Kunze WOODSIACEAE (F)
Hypoderris R.Br. ex Hook. DRYOPTERIDACEAE (F)
Hypodiscus Nees RESTIONACEAE (M)
Hypoestes Sol. ex R.Br. ACANTHACEAE (D)
Hypogomphia Bunge LABIATAE (D)
Hypogon Raf. = Collinsonia L. LABIATAE (D)
Hypogynium Nees = Andropogon L. GRAMINEAE (M)
Hypolaena R.Br. RESTIONACEAE (M)
Hypolepis Bernh. DENNSTAEDTIACEAE (F)
Hypolobus Fourn. ASCLEPIADACEAE (D)
Hypolytrum Rich. ex Pers. CYPERACEAE (M)
Hypopeltis Michx. = Polystichum Roth DRYOPTERIDACEAE (F)
Hypophyllanthus Regel = Helicteres L. STERCULIACEAE (D)
Hypopitys Hill (SUS) = Monotropa L. ERICACEAE (D)
Hypopterygiopsis Sakurai = Selaginella P.Beauv. SELAGINELLACEAE (FA)
Hypothronia Schrank = Hyptis Jacq. LABIATAE (D)
Hypoxidia Friedmann HYPOXIDACEAE (M)
Hypoxis L. HYPOXIDACEAE (M)
Hypsela C.Presl CAMPANULACEAE (D)
Hypselandra Pax & K.Hoffm. (SUI) = Boscia Lam. CAPPARACEAE (D)
Hypselodelphys (K.Schum.) Milne–Redh. MARANTACEAE (M)
Hypseloderma Radlk. SAPINDACEAE (D)
Hypseocharis J.Remy OXALIDACEAE (D)
Hypseochloa C.E.Hubb. GRAMINEAE (M)
Hypserpa Miers MENISPERMACEAE (D)
Hypsophila F.Muell. CELASTRACEAE (D)
Hyptiandra Hook.f. (SUH) = Quassia L. SIMAROUBACEAE (D)
Hyptianthera Wight & Arn. RUBIACEAE (D)
Hyptidendron Harley LABIATAE (D)
Hyptiodaphne Urb. = Daphnopsis Mart. THYMELAEACEAE (D)
Hyptis Jacq. LABIATAE (D)
Hypudaerus A.Braun (SUI) = Anthephora Schreb. GRAMINEAE (M)
Hyrtanandra Miq. URTICACEAE (D)
Hyssaria Kolak. = Campanula L. CAMPANULACEAE (D)
Hyssopus L. LABIATAE (D)
Hysteria Reinw. = Corymborkis Thouars ORCHIDACEAE (M)
Hystericina Steud. = Echinopogon P.Beauv. GRAMINEAE (M)
Hysterionica Willd. COMPOSITAE (D)
Hystrichophora Mattf. COMPOSITAE (D)
Hystringium Steud. (SUI) = Tribolium Desv. GRAMINEAE (M)
Hystrix Moench GRAMINEAE (M)
Hytophrynium K.Schum. = Trachyphrynium Benth. MARANTACEAE (M)

Iacranda Pers. = Jacaranda Juss. BIGNONIACEAE (D)
Iantha Hook. = Ionopsis Kunth ORCHIDACEAE (M)

Ianthe Salisb. = Hypoxis L. HYPOXIDACEAE (M)
Ibadja A.Chev. = Loesenera Harms LEGUMINOSAE–CAESALPINIOIDEAE (D)
Ibarraea Lundell = Ardisia Sw. MYRSINACEAE (D)
Ibatia Decne. ASCLEPIADACEAE (D)
Iberidella Boiss. = Aethionema R.Br. CRUCIFERAE (D)
Iberis L. CRUCIFERAE (D)
Ibervillea Greene CUCURBITACEAE (D)
Ibetralia Bremek. = Alibertia A.Rich. ex DC. RUBIACEAE (D)
Ibicella Van Eselt. PEDALIACEAE (D)
Ibidium Salisb. ex Small = Spiranthes Rich. ORCHIDACEAE (M)
Ibina Noronha = Sauropus Blume EUPHORBIACEAE (D)
Iboza N.E.Br. = Tetradenia Benth. LABIATAE (D)
Icacina A.Juss. ICACINACEAE (D)
Icacinopsis Roberty = Dichapetalum Thouars DICHAPETALACEAE (D)
Icacorea Aubl. = Ardisia Sw. MYRSINACEAE (D)
Icaria J.F.Macbr. = Miconia Ruíz & Pav. MELASTOMATACEAE (D)
Ichnanthus P.Beauv. GRAMINEAE (M)
Ichnocarpus R.Br. APOCYNACEAE (D)
Ichthyophora Baehni = Pouteria Aubl. SAPOTACEAE (D)
Ichthyothere Mart. COMPOSITAE (D)
Ichtyomethia Kuntze = Piscidia L. LEGUMINOSAE–PAPILIONOIDEAE (D)
Icianthus Greene CRUCIFERAE (D)
Icica Aubl. = Protium Burm.f. BURSERACEAE (D)
Icicaster Ridl. = Santiria Blume BURSERACEAE (D)
Icomum Hua = Aeollanthus Mart. ex Spreng. LABIATAE (D)
Icosandra Phil. = Cryptocarya R.Br. LAURACEAE (D)
Idahoa A.Nelson & J.F.Macbr. CRUCIFERAE (D)
Ideleria Kunth = Tetraria P.Beauv. CYPERACEAE (M)
Idertia Farron = Gomphia Schreb. OCHNACEAE (D)
Idesia Scop. (SUH) = Diospyros L. EBENACEAE (D)
Idesia Maxim. FLACOURTIACEAE (D)
Idesia Scop. (SUH) = Diospyros L. EBENACEAE (D)
Idiopteris T.G.Walker = Pteris L. PTERIDACEAE (F)
Idiospermum S.T.Blake IDIOSPERMACEAE (D)
Idiothamnus R.M.King & H.Rob. COMPOSITAE (D)
Idria Kellogg = Fouquieria Kunth FOUQUIERIACEAE (D)
Iebine Raf. = Liparis Rich. ORCHIDACEAE (M)
Ifloga Cass. COMPOSITAE (D)
Ighermia Wiklund COMPOSITAE (D)
Iguanura Blume PALMAE (M)
Ikonnikovia Lincz. PLUMBAGINACEAE (D)
Ildefonsia Gardner = Bacopa Aubl. SCROPHULARIACEAE (D)
Ileostylus Tiegh. LORANTHACEAE (D)
Ilex L. AQUIFOLIACEAE (D)
Iliamna Greene = Sphaeralcea A.St.–Hil. MALVACEAE (D)
Iljinia Korovin = Haloxylon Bunge CHENOPODIACEAE (D)
Illairea Lenné & K.Koch = Caiophora C.Presl LOASACEAE (D)
Illecebrum L. ILLECEBRACEAE (D)
Illicium L. ILLICIACEAE (D)
Illigera Blume HERNANDIACEAE (D)
Illipe Koenig ex Gras = Madhuca Ham. ex J.F.Gmel. SAPOTACEAE (D)
Ilocania Merr. = Diplocyclos (Endl.) Post & Kuntze CUCURBITACEAE (D)
Iltisia S.F.Blake COMPOSITAE (D)
Ilysanthes Raf. = Lindernia All. SCROPHULARIACEAE (D)
Imantina Hook.f. = Morinda L. RUBIACEAE (D)

Imbralyx Geesink = Fordia Hemsl. LEGUMINOSAE–PAPILIONOIDEAE (D)
Imbricaria Comm. ex Juss. = Mimusops L. SAPOTACEAE (D)
Imbricaria Sm. (SUH) = Baeckea L. MYRTACEAE (D)
Imeria R.M.King & H.Rob. COMPOSITAE (D)
Imerinaea Schltr. ORCHIDACEAE (M)
Imitaria N.E.Br. AIZOACEAE (D)
Impatiens L. BALSAMINACEAE (D)
Impatientella H.Perrier = Impatiens L. BALSAMINACEAE (D)
Imperata Cyr. GRAMINEAE (M)
Imperatoria L. UMBELLIFERAE (D)
Incarvillea Juss. BIGNONIACEAE (D)
Indigastrum Jaub. & Spach = Indigofera L. LEGUMINOSAE–PAPILIONOIDEAE (D)
Indigofera L. LEGUMINOSAE–PAPILIONOIDEAE (D)
Indobanalia A.N.Henry & B.Roy AMARANTHACEAE (D)
Indocalamus Nakai GRAMINEAE (M)
Indochloa Bor = Euclasta Franch. GRAMINEAE (M)
Indocourtoisia Bennet & Raizada (SUS) = Courtoisina Soják CYPERACEAE (M)
Indofevillea Chatterjee CUCURBITACEAE (D)
Indokingia Hemsl. = Gastonia Comm. ex Lam. ARALIACEAE (D)
Indoneesiella Sreem. ACANTHACEAE (D)
Indopiptadenia Brenan LEGUMINOSAE–MIMOSOIDEAE (D)
Indopoa Bor GRAMINEAE (M)
Indopolysolenia Bennet RUBIACEAE (D)
Indorouchera Hallier f. LINACEAE (D)
Indoryza A.N.Henry & B.Roy (SUS) = Porteresia Tateoka GRAMINEAE (M)
Indosasa McClure GRAMINEAE (M)
Indosinia J.E.Vidal OCHNACEAE (D)
Indotristicha P.Royen PODOSTEMACEAE (D)
Indovethia Boerl. = Sauvagesia L. OCHNACEAE (D)
Inezia E.Phillips COMPOSITAE (D)
Inga Mill. LEGUMINOSAE–MIMOSOIDEAE (D)
Inhambanella (Engl.) Dubard SAPOTACEAE (D)
Inobulbon (Schltr.) Schltr. = Dendrobium Sw. ORCHIDACEAE (M)
Inocarpus J.R.Forst. & G.Forst. LEGUMINOSAE–PAPILIONOIDEAE (D)
Inodes O.F.Cook = Sabal Adans. PALMAE (M)
Inophloeum Pittier = Poulsenia Eggers MORACEAE (D)
Intsia Thouars LEGUMINOSAE–CAESALPINIOIDEAE (D)
Inula L. COMPOSITAE (D)
Inulanthera Källersjö COMPOSITAE (D)
Involucraria Ser. = Trichosanthes L. CUCURBITACEAE (D)
Inyonia M.E.Jones = Psathyrotes A.Gray COMPOSITAE (D)
Ioackima Ten. (SUO) = Beckmannia Host GRAMINEAE (M)
Iocenes B.Nord. COMPOSITAE (D)
Iochroma Benth. SOLANACEAE (D)
Iodanthus (Torr. & A.Gray) Steud. CRUCIFERAE (D)
Iodes Blume ICACINACEAE (D)
Iodina Hook. & Arn. ex Meisn. SANTALACEAE (D)
Iodocephalus Thorel ex Gagnep. COMPOSITAE (D)
Iogeton Strother COMPOSITAE (D)
Ionacanthus Benoist ACANTHACEAE (D)
Iondra Raf. = Aethionema R.Br. CRUCIFERAE (D)
Iondraba Rchb. = Biscutella L. CRUCIFERAE (D)
Ione Lindl. = Sunipia Buch.–Ham. ex Lindl. ORCHIDACEAE (M)
Ionidium Vent. = Hybanthus Jacq. VIOLACEAE (D)
Ionopsidium Rchb. CRUCIFERAE (D)

Ionopsis Kunth ORCHIDACEAE (M)
Ionorchis Beck = Limodorum Boehm. ORCHIDACEAE (M)
Ionoxalis Small = Oxalis L. OXALIDACEAE (D)
Iostephane Benth. COMPOSITAE (D)
Ioxylon Raf. = Maclura Nutt. MORACEAE (D)
Ipecacuahna Arruda = Psychotria L. RUBIACEAE (D)
Ipheion Raf. ALLIACEAE (M)
Iphigenia Kunth COLCHICACEAE (M)
Iphiona Cass. COMPOSITAE (D)
Iphionopsis Anderb. COMPOSITAE (D)
Ipnum Phil. = Leptochloa P.Beauv. GRAMINEAE (M)
Ipomoea L. CONVOLVULACEAE (D)
Ipomopsis Michx. POLEMONIACEAE (D)
Ipsea Lindl. ORCHIDACEAE (M)
Iranecio B.Nord. COMPOSITAE (D)
Irenella Suess. AMARANTHACEAE (D)
Irenepharsus Hewson CRUCIFERAE (D)
Iresine P.Browne AMARANTHACEAE (D)
Iriartea Ruíz & Pav. PALMAE (M)
Iriartella H.A.Wendl. PALMAE (M)
Iridodictyum Rodion. = Iris L. IRIDACEAE (M)
Iridorchis Blume = Cymbidium Sw. ORCHIDACEAE (M)
Iridorkis Thouars = Oberonia Lindl. ORCHIDACEAE (M)
Iris L. IRIDACEAE (M)
Irlbachia Mart. GENTIANACEAE (D)
Irmischia Schltdl. = Metastelma R.Br. ASCLEPIADACEAE (D)
Irulia Bedd. (SUI) = Ochlandra Thwaites GRAMINEAE (M)
Irvingbaileya R.A.Howard ICACINACEAE (D)
Irvingia F.Muell. (SUH) = Polyscias J.R.Forst. & G.Forst. ARALIACEAE (D)
Irvingia Hook.f. IRVINGIACEAE (D)
Irwinia Barroso COMPOSITAE (D)
Iryanthera Warb. MYRISTICACEAE (D)
Isabelia Barb.Rodr. ORCHIDACEAE (M)
Isacanthus Nees = Sclerochiton Harv. ACANTHACEAE (D)
Isachne R.Br. GRAMINEAE (M)
Isaloa Humbert = Barleria L. ACANTHACEAE (D)
Isalus J.B.Phipps = Tristachya Nees GRAMINEAE (M)
Isandra F.Muell. (SUH) = Symonanthus Haegi SOLANACEAE (D)
Isandraea Rauschert (SUS) = Symonanthus Haegi SOLANACEAE (D)
Isanthera Nees = Rhynchotechum Blume GESNERIACEAE (D)
Isanthus Michx. = Trichostema L. LABIATAE (D)
Isartia Dumort. (SUO) = Isertia Schreb. RUBIACEAE (D)
Isatis L. CRUCIFERAE (D)
Ischaemopogon Griseb. = Ischaemum L. GRAMINEAE (M)
Ischaemum L. GRAMINEAE (M)
Ischnanthus Roem. & Schult. (SUS) = Ichnanthus P.Beauv. GRAMINEAE (M)
Ischnanthus (Engl.) Tiegh. (SUH) = Englerina Tiegh. LORANTHACEAE (D)
Ischnea F.Muell. COMPOSITAE (D)
Ischnocarpus O.E.Schulz CRUCIFERAE (D)
Ischnocentrum Schltr. ORCHIDACEAE (M)
Ischnochloa Hook.f. = Microstegium Nees GRAMINEAE (M)
Ischnogyne Schltr. ORCHIDACEAE (M)
Ischnolepis Jum. & H.Perrier ASCLEPIADACEAE (D)
Ischnosiphon Koern. MARANTACEAE (M)
Ischnostemma King & Gamble ASCLEPIADACEAE (D)

Ischnurus Balf.f. = Lepturus R.Br. GRAMINEAE (M)
Ischurochloa Büse = Bambusa Schreb. GRAMINEAE (M)
Ischyrolepis Steud. RESTIONACEAE (M)
Iseia O'Don. CONVOLVULACEAE (D)
Iseilema Andersson GRAMINEAE (M)
Isertia Schreb. RUBIACEAE (D)
Isidorea A.Rich. ex DC. RUBIACEAE (D)
Isinia Rech.f. = Lavandula L. LABIATAE (D)
Iskandera N.Busch CRUCIFERAE (D)
Islaya Backeb. = Neoporteria Britton & Rose CACTACEAE (D)
Ismene Salisb. ex Herb. = Hymenocallis Salisb. AMARYLLIDACEAE (M)
Isnardia L. = Ludwigia L. ONAGRACEAE (D)
Isoberlinia Craib & Stapf ex Holland LEGUMINOSAE–CAESALPINIOIDEAE (D)
Isocarpha R.Br. COMPOSITAE (D)
Isocaulon Tiegh. = Psittacanthus Mart. LORANTHACEAE (D)
Isochilus R.Br. ORCHIDACEAE (M)
Isochoriste Miq. = Asystasia Blume ACANTHACEAE (D)
Isocoma Nutt. COMPOSITAE (D)
Isodendrion A.Gray VIOLACEAE (D)
Isodesmia Gardner = Chaetocalyx DC. LEGUMINOSAE–PAPILIONOIDEAE (D)
Isodictyophorus Briq. LABIATAE (D)
Isodon (Schrad. ex Benth.) Spach LABIATAE (D)
Isoetella Gennari = Isoetes L. ISOETACEAE (FA)
Isoetes L. ISOETACEAE (FA)
Isoetopsis Turcz. COMPOSITAE (D)
Isoglossa Oerst. ACANTHACEAE (D)
Isolatocereus (Backeb.) Backeb. = Stenocereus (A.Berger) Riccob. CACTACEAE (D)
Isolepis R.Br. CYPERACEAE (M)
Isoleucas O.Schwartz LABIATAE (D)
Isolobus A.DC. = Lobelia L. CAMPANULACEAE (D)
Isoloma Benth. ex Decne. (SUH) = Kohleria Regel GESNERIACEAE (D)
Isoloma J.Sm. = Lindsaea Dryand. ex Sm. DENNSTAEDTIACEAE (F)
Isolona Engl. ANNONACEAE (D)
Isomacrolobium Aubrév. & Pellegr. = Anthonotha P.Beauv. LEGUMINOSAE–
    CAESALPINIOIDEAE (D)
Isomeris Nutt. CAPPARACEAE (D)
Isomerocarpa A.C.Sm. = Dryadodaphne S.Moore MONIMIACEAE (D)
Isometrum Craib GESNERIACEAE (D)
Isonandra Wight SAPOTACEAE (D)
Isonema R.Br. APOCYNACEAE (D)
Isopappus Torr. & A.Gray COMPOSITAE (D)
Isophysis T.Moore IRIDACEAE (M)
Isoplexis (Lindl.) J.C.Loudon SCROPHULARIACEAE (D)
Isopogon R.Br. ex Knight PROTEACEAE (D)
Isoptera Scheff. ex Burck = Shorea Roxb. ex C.F.Gaertn. DIPTEROCARPACEAE (D)
Isopyrum L. RANUNCULACEAE (D)
Isorium Raf. = Lobostemon Lehm. BORAGINACEAE (D)
Isostigma Less. COMPOSITAE (D)
Isotheca Turrill ACANTHACEAE (D)
Isotoma (R.Br.) Lindl. = Laurentia Adans. CAMPANULACEAE (D)
Isotrema Raf. ARISTOLOCHIACEAE (D)
Isotria Raf. ORCHIDACEAE (M)
Isotropis Benth. LEGUMINOSAE–PAPILIONOIDEAE (D)
Itacania Raf. = Elytranthe (Blume) Blume LORANTHACEAE (D)
Itaculumia Hoehne = Habenaria Willd. ORCHIDACEAE (M)

Itaobimia Rizzini = Riedeliella Harms LEGUMINOSAE–PAPILIONOIDEAE (D)
Itasina Raf. UMBELLIFERAE (D)
Itatiaia Ule = Tibouchina Aubl. MELASTOMATACEAE (D)
Itaya H.E.Moore PALMAE (M)
Itea L. ESCALLONIACEAE (D)
Iteadaphne Blume = Lindera Thunb. LAURACEAE (D)
Iteiluma Baill. = Pouteria Aubl. SAPOTACEAE (D)
Ithycaulon Copel. = Saccoloma Kaulf. DENNSTAEDTIACEAE (F)
Iti Garn.–Jones & P.N.John CRUCIFERAE (D)
Itoa Hemsl. FLACOURTIACEAE (D)
Ituridendron De Wild. = Omphalocarpum P.Beauv. SAPOTACEAE (D)
Itysa Ravenna = Calydorea Herb. IRIDACEAE (M)
Itzaea Standl. & Steyerm. CONVOLVULACEAE (D)
Iva L. COMPOSITAE (D)
Ivania O.E.Schulz CRUCIFERAE (D)
Ivanjohnstonia Kazmi BORAGINACEAE (D)
Ivesia Torr. & A.Gray ROSACEAE (D)
Ivodea Capuron RUTACEAE (D)
Ixalum G.Forst. (SUI) = Spinifex L. GRAMINEAE (M)
Ixanthus Griseb. GENTIANACEAE (D)
Ixerba A.Cunn. ESCALLONIACEAE (D)
Ixeridium (A.Gray) Tzvelev COMPOSITAE (D)
Ixeris (Cass.) Cass. COMPOSITAE (D)
Ixia L. IRIDACEAE (M)
Ixianthes Benth. SCROPHULARIACEAE (D)
Ixidium Eichler = Antidaphne Poepp. & Endl. EREMOLEPIDACEAE (D)
Ixiochlamys F.Muell. & Sond. COMPOSITAE (D)
Ixiolaena Benth. COMPOSITAE (D)
Ixiolirion Herb. IXIOLIRIACEAE (M)
Ixocactus Rizzini LORANTHACEAE (D)
Ixodia R.Br. COMPOSITAE (D)
Ixodonerium Pit. APOCYNACEAE (D)
Ixonanthes Jack IXONANTHACEAE (D)
Ixophorus Schltdl. GRAMINEAE (M)
Ixora L. RUBIACEAE (D)
Ixtlania M.E.Jones = Justicia L. ACANTHACEAE (D)
Izabalaea Lundell NYCTAGINACEAE (D)

Jablonskia G.L.Webster EUPHORBIACEAE (D)
Jaborosa Juss. SOLANACEAE (D)
Jacaima Rendle ASCLEPIADACEAE (D)
Jacaranda Juss. BIGNONIACEAE (D)
Jacaratia A.DC. CARICACEAE (D)
Jackia Wall. (SUH) = Jackiopsis Ridsdale RUBIACEAE (D)
Jackiopsis Ridsdale RUBIACEAE (D)
Jacksonia R.Br. ex Sm. LEGUMINOSAE–PAPILIONOIDEAE (D)
Jacksonia Hort. ex Schltdl. (SUI) = Jasminum L. OLEACEAE (D)
Jacmaia B.Nord. COMPOSITAE (D)
Jacobinia Nees ex Moricand = Justicia L. ACANTHACEAE (D)
Jacobsenia L.Bolus & Schwantes AIZOACEAE (D)
Jacquemontia Choisy CONVOLVULACEAE (D)
Jacquesfelixia J.B.Phipps = Danthoniopsis Stapf GRAMINEAE (M)
Jacqueshuberia Ducke LEGUMINOSAE–CAESALPINIOIDEAE (D)
Jacquinia L. THEOPHRASTACEAE (D)
Jacquiniella Schltr. ORCHIDACEAE (M)

Jadunia Lindau ACANTHACEAE (D)
Jaegeria Kunth COMPOSITAE (D)
Jaeggia Schinz = Adenia Forssk. PASSIFLORACEAE (D)
Jaeschkea Kurz GENTIANACEAE (D)
Jagera Blume SAPINDACEAE (D)
Jahnia Pittier & S.F.Blake = Turpinia Vent. STAPHYLEACEAE (D)
Jaimenostia Guinea & Gómez Moreno = Sauromatum Schott ARACEAE (M)
Jainia Balakr. = Coptophyllum Korth. RUBIACEAE (D)
Jalcophila Dillon & Sagast. COMPOSITAE (D)
Jaliscoa S.Watson COMPOSITAE (D)
Jaltomata Schltdl. SOLANACEAE (D)
Jamaiciella Braem = Oncidium Sw. ORCHIDACEAE (M)
Jambos Adans. = Syzygium Gaertn. MYRTACEAE (D)
Jambosa Adans. = Syzygium Gaertn. MYRTACEAE (D)
Jamesbrittenia Kuntze SCROPHULARIACEAE (D)
Jamesia Torr. & A.Gray HYDRANGEACEAE (D)
Jamesianthus S.F.Blake & Sherff COMPOSITAE (D)
Jamesonia Hook. & Grev. ADIANTACEAE (F)
Janakia Joseph & Chandras. ASCLEPIADACEAE (D)
Jancaea Boiss. GESNERIACEAE (D)
Jangaraca Raf. = Hamelia Jacq. RUBIACEAE (D)
Janipha Kunth = Manihot Mill. EUPHORBIACEAE (D)
Jankaea Boiss. (SUO) = Jancaea Boiss. GESNERIACEAE (D)
Janotia J.-F.Leroy RUBIACEAE (D)
Jansenella Bor GRAMINEAE (M)
Jansenia Barb.Rodr. = Plectrophora H.Focke ORCHIDACEAE (M)
Jansonia Kippist LEGUMINOSAE–PAPILIONOIDEAE (D)
Janusia A.Juss. MALPIGHIACEAE (D)
Japanobotrychium Masam. = Botrychium Sw. OPHIOGLOSSACEAE (F)
Japarandiba Adans. = Gustavia L. LECYTHIDACEAE (D)
Japonolirion Nakai MELANTHIACEAE (M)
Jaramilloa R.M.King & H.Rob. COMPOSITAE (D)
Jarandersonia Kosterm. TILIACEAE (D)
Jarava Ruíz & Pav. = Stipa L. GRAMINEAE (M)
Jardinea Steud. = Phacelurus Griseb. GRAMINEAE (M)
Jarilla Rusby CARICACEAE (D)
Jarilla I.M.Johnst. (SUH) = Jarilla Rusby CARICACEAE (D)
Jasarum Bunting ARACEAE (M)
Jasione L. CAMPANULACEAE (D)
Jasionella Stoj. & Stef. = Jasione L. CAMPANULACEAE (D)
Jasminium Dumort. (SUO) = Jasminum L. OLEACEAE (D)
Jasminocereus Britton & Rose CACTACEAE (D)
Jasminochyla (Stapf) Pichon = Landolphia P.Beauv. APOCYNACEAE (D)
Jasminoides Duhamel = Lycium L. SOLANACEAE (D)
Jasminum L. OLEACEAE (D)
Jasonia (Cass.) Cass. COMPOSITAE (D)
Jateorhiza Miers MENISPERMACEAE (D)
Jatropa Scop. (SUO) = Jatropha L. EUPHORBIACEAE (D)
Jatropha L. EUPHORBIACEAE (D)
Jaubertia Guill. RUBIACEAE (D)
Jaumea Pers. COMPOSITAE (D)
Jaundea Gilg = Rourea Aubl. CONNARACEAE (D)
Javorkaea Borhidi & Koml. RUBIACEAE (D)
Jedda J.R.Clarkson THYMELAEACEAE (D)
Jefea Strother COMPOSITAE (D)

Jeffersonia Barton BERBERIDACEAE (D)
Jeffreya Wild COMPOSITAE (D)
Jehlia Rose = Lopezia Cav. ONAGRACEAE (D)
Jejosephia A.N.Rao & Mani ORCHIDACEAE (M)
Jenkinsia Hook. = Bolbitis Schott LOMARIOPSIDACEAE (F)
Jenkinsia Wall. ex Voigt (SUI) = Myriopteron Griff. ASCLEPIADACEAE (D)
Jenmania Rolfe = Palmorchis Barb.Rodr. ORCHIDACEAE (M)
Jenmaniella Engl. PODOSTEMACEAE (D)
Jensenobotrya A.G.J.Herre AIZOACEAE (D)
Jensoa Raf. = Cymbidium Sw. ORCHIDACEAE (M)
Jepsonia Small SAXIFRAGACEAE (D)
Jerdonia Wight SCROPHULARIACEAE (D)
Jessenia Karst. = Oenocarpus Mart. PALMAE (M)
Jimensia Raf. = Bletilla Rchb.f. ORCHIDACEAE (M)
Joachima Ten. = Beckmannia Host GRAMINEAE (M)
Joannegria Chiov. = Lintonia Stapf GRAMINEAE (M)
Joannesia Vell. EUPHORBIACEAE (D)
Jobinia Fourn. ASCLEPIADACEAE (D)
Jocayena Raf. (SUO) = Tocoyena Aubl. RUBIACEAE (D)
Jodrellia Baijnath ASPHODELACEAE (M)
Johannesteijsmannia H.E.Moore PALMAE (M)
Johnia Roxb. = Salacia L. CELASTRACEAE (D)
Johnsonia R.Br. ANTHERICACEAE (M)
Johnsonia Adans. (SUH) = Cedrela P.Browne MELIACEAE (D)
Johnstonella Brand = Cryptantha G.Don BORAGINACEAE (D)
Johowia Epling & Looser = Cuminia Colla LABIATAE (D)
Johrenia DC. UMBELLIFERAE (D)
Johreniopsis Pimenov UMBELLIFERAE (D)
Joinvillea Gaudich. ex Brongn. & Gris JOINVILLEACEAE (M)
Joliffia Bojer ex Delile = Telfairia Hook. CUCURBITACEAE (D)
Jollya Pierre = Pycnandra Benth. SAPOTACEAE (D)
Jollydora Pierre ex Gilg CONNARACEAE (D)
Jondraba Medik. = Biscutella L. CRUCIFERAE (D)
Jonesia Roxb. = Saraca L. LEGUMINOSAE–CAESALPINIOIDEAE (D)
Jonesiella Rydb. = Astragalus L. LEGUMINOSAE–PAPILIONOIDEAE (D)
Jontanea Raf. = Coccocypselum P.Browne RUBIACEAE (D)
Joosia Karst. RUBIACEAE (D)
Jordaaniella H.E.K.Hartmann AIZOACEAE (D)
Joseanthus H.Rob. = Vernonia Schreb. COMPOSITAE (D)
Josephia Wight = Sirhookera Kuntze ORCHIDACEAE (M)
Josephinia Vent. PEDALIACEAE (D)
Jossinia Comm. ex DC. = Eugenia L. MYRTACEAE (D)
Jouvea Fourn. GRAMINEAE (M)
Jovellana Ruíz & Pav. SCROPHULARIACEAE (D)
Jovetia Guédès RUBIACEAE (D)
Jovibarba (DC.) Opiz = Sempervivum L. CRASSULACEAE (D)
Juania Drude PALMAE (M)
Juanulloa Ruíz & Pav. SOLANACEAE (D)
Jubaea Kunth PALMAE (M)
Jubaeopsis Becc. PALMAE (M)
Jubelina A.Juss. MALPIGHIACEAE (D)
Jubistylis Rusby = Banisteriopsis C.B.Rob. ex Small MALPIGHIACEAE (D)
Juchia M.Roem. = Solena Lour. CUCURBITACEAE (D)
Juellia Aspl. = Ombrophytum Poepp. ex Endl. BALANOPHORACEAE (D)
Jugastrum Miers = Eschweilera Mart. ex DC. LECYTHIDACEAE (D)

Juglans L. JUGLANDACEAE (D)
Julbernardia Pellegr. LEGUMINOSAE–CAESALPINIOIDEAE (D)
Juliania Schltdl. (SUH) = Amphipterygium Schiede ex Standl. ANACARDIACEAE (D)
Julibaria Mez = Loheria Merr. MYRSINACEAE (D)
Julocroton Mart. EUPHORBIACEAE (D)
Julostylis Thwaites MALVACEAE (D)
Jumellea Schltr. ORCHIDACEAE (M)
Jumelleanthus Hochr. MALVACEAE (D)
Juncellus C.B.Clarke = Cyperus L. CYPERACEAE (M)
Juncus L. JUNCACEAE (M)
Junellia Moldenke VERBENACEAE (D)
Junghuhnia Miq. (SUH) = Codiaeum A.Juss. EUPHORBIACEAE (D)
Jungia Gaertn. (SUH) = Baeckea L. MYRTACEAE (D)
Jungia L.f. COMPOSITAE (D)
Juniperus L. CUPRESSACEAE (G)
Juno Tratt. = Iris L. IRIDACEAE (M)
Junopsis Wern.Schulze = Iris L. IRIDACEAE (M)
Juppia Merr. = Zanonia L. CUCURBITACEAE (D)
Jupunba Britton & Rose = Abarema Pittier LEGUMINOSAE–MIMOSOIDEAE (D)
Jurgensia Raf. = Spermacoce L. RUBIACEAE (D)
Jurinea Cass. COMPOSITAE (D)
Jurinella Jaub. & Spach = Jurinea Cass. COMPOSITAE (D)
Jurtsevia Á.Löve & D.Löve = Anemone L. RANUNCULACEAE (D)
Juruasia Lindau ACANTHACEAE (D)
Jussiaea L. = Ludwigia L. ONAGRACEAE (D)
Jussieuia Houstoun = Cnidoscolus Pohl EUPHORBIACEAE (D)
Justago Kuntze = Cleome L. CAPPARACEAE (D)
Justenia Hiern = Bertiera Aubl. RUBIACEAE (D)
Justicia L. ACANTHACEAE (D)
Juttadinteria Schwantes AIZOACEAE (D)

Kabulia Bor & C.Fisch. ILLECEBRACEAE (D)
Kadali Adans. = Osbeckia L. MELASTOMATACEAE (D)
Kadalia Raf. = Dissotis Benth. MELASTOMATACEAE (D)
Kadenia Lavrova & V.N.Tikhom. UMBELLIFERAE (D)
Kadsura Juss. SCHISANDRACEAE (D)
Kadua Cham. & Schltdl. = Hedyotis L. RUBIACEAE (D)
Kaempferia L. ZINGIBERACEAE (M)
Kafirnigania Kamelin & Kinzik. UMBELLIFERAE (D)
Kageneckia Ruíz & Pav. ROSACEAE (D)
Kaieteuria Dwyer = Ouratea Aubl. OCHNACEAE (D)
Kailarsenia Tirveng. RUBIACEAE (D)
Kairoa Philipson MONIMIACEAE (D)
Kairothamnus Airy Shaw EUPHORBIACEAE (D)
Kajewskia Guillaumin = Veitchia H.A.Wendl. PALMAE (M)
Kajewskiella Merr. & Perry RUBIACEAE (D)
Kakosmanthus Hassk. = Madhuca Ham. ex J.F.Gmel. SAPOTACEAE (D)
Kalaharia Baill. = Clerodendrum L. VERBENACEAE (D)
Kalakia Alava UMBELLIFERAE (D)
Kalanchoe Adans. CRASSULACEAE (D)
Kalappia Kosterm. LEGUMINOSAE–CAESALPINIOIDEAE (D)
Kalbreyera Burret = Geonoma Willd. PALMAE (M)
Kalbreyeracanthus Wassh. = Habracanthus Nees ACANTHACEAE (D)
Kalbreyeriella Lindau ACANTHACEAE (D)
Kalidiopsis Aellen CHENOPODIACEAE (D)

Kalidium Moq. CHENOPODIACEAE (D)
Kalimeris (Cass.) Cass. COMPOSITAE (D)
Kalimpongia Pradhan = Dickasonia L.O.Williams ORCHIDACEAE (M)
Kaliphora Hook.f. MELANOPHYLLACEAE (D)
Kallstroemia Scop. ZYGOPHYLLACEAE (D)
Kalmia L. ERICACEAE (D)
Kalmiella Small = Kalmia L. ERICACEAE (D)
Kalmiopsis Rehder ERICACEAE (D)
Kalopanax Miq. ARALIACEAE (D)
Kalopternix Garay & Dunst. ORCHIDACEAE (M)
Kaluhaburunghos Kuntze = Cleistanthus Hook.f. ex Planch. EUPHORBIACEAE (D)
Kamettia Kostel. APOCYNACEAE (D)
Kamiella Vassilcz. = Medicago L. LEGUMINOSAE–PAPILIONOIDEAE (D)
Kampmannia Steud. (UR) GRAMINEAE (M)
Kampochloa Clayton GRAMINEAE (M)
Kamptzia Nees = Syncarpia Ten. MYRTACEAE (D)
Kanahia R.Br. ASCLEPIADACEAE (D)
Kandaharia Alava UMBELLIFERAE (D)
Kandelia (DC.) Wight & Arn. RHIZOPHORACEAE (D)
Kania Schltr. MYRTACEAE (D)
Kanilia Blume = Bruguiera Sav. RHIZOPHORACEAE (D)
Kanimia Gardner = Mikania Willd. COMPOSITAE (D)
Kanjarum Ramam. = Strobilanthes Blume ACANTHACEAE (D)
Kanopikon Raf. = Euphorbia L. EUPHORBIACEAE (D)
Kantou Aubrév. & Pellegr. = Inhambanella (Engl.) Dubard SAPOTACEAE (D)
Kaokochloa De Winter GRAMINEAE (M)
Kaoue Pellegr. = Stachyothyrsus Harms LEGUMINOSAE–CAESALPINIOIDEAE (D)
Karamyschewia Fisch. & C.A.Mey. = Oldenlandia L. RUBIACEAE (D)
Karatas Mill. = Bromelia L. BROMELIACEAE (M)
Karatavia Pimenov & Lavrova UMBELLIFERAE (D)
Kardanoglyphos Schltdl. CRUCIFERAE (D)
Karekandel Wolf = Carallia Roxb. RHIZOPHORACEAE (D)
Karelinia Less. COMPOSITAE (D)
Karimbolea Descoings ASCLEPIADACEAE (D)
Karina Boutique GENTIANACEAE (D)
Karivia Arn. = Solena Lour. CUCURBITACEAE (D)
Karlea Pierre = Maesopsis Engl. RHAMNACEAE (D)
Karnataka P.K.Mukh. & Constance UMBELLIFERAE (D)
Karomia Dop VERBENACEAE (D)
Karroochloa Conert & Turpe = Rytidosperma Steud. GRAMINEAE (M)
Karvandarina Rech.f. COMPOSITAE (D)
Karwinskia Zucc. RHAMNACEAE (D)
Kaschgaria Poljakov COMPOSITAE (D)
Kashmiria D.Y.Hong SCROPHULARIACEAE (D)
Katafa Costantin & J.Poiss. CELASTRACEAE (D)
Katherinea A.D.Hawkes = Epigeneium Gagnep. ORCHIDACEAE (M)
Katoutheka Adans. = Wendlandia Bartl. ex DC. RUBIACEAE (D)
Kaufmannia Regel PRIMULACEAE (D)
Kaukenia Kuntze = Mimusops L. SAPOTACEAE (D)
Kaulfussia C.Presl = Christensenia Maxon MARATTIACEAE (F)
Kaulinia B.K.Nayar = Microsorum Link POLYPODIACEAE (F)
Kaunia R.M.King & H.Rob. COMPOSITAE (D)
Kayea Wall. GUTTIFERAE (D)
Kearnemalvastrum D.M.Bates MALVACEAE (D)
Keayodendron Leandri EUPHORBIACEAE (D)

Keckia Straw (SUH) = Keckiella Straw SCROPHULARIACEAE (D)
Keckiella Straw SCROPHULARIACEAE (D)
Kedarnatha P.K.Mukh. & Constance UMBELLIFERAE (D)
Kedrostis Medik. CUCURBITACEAE (D)
Keenania Hook.f. RUBIACEAE (D)
Keetia E.Phillips RUBIACEAE (D)
Kefersteinia Rchb.f. ORCHIDACEAE (M)
Kegelia Rchb.f. = Kegeliella Mansf. ORCHIDACEAE (M)
Kegeliella Mansf. ORCHIDACEAE (M)
Keiria S.Bowdich (UR) OLEACEAE (D)
Keiskea Miq. LABIATAE (D)
Keithia Benth. = Rhabdocaulon (Benth.) Epling LABIATAE (D)
Keitia Regel = Eleutherine Herb. IRIDACEAE (M)
Kelissa Ravenna IRIDACEAE (M)
Kelleria Endl. THYMELAEACEAE (D)
Kelleronia Schinz ZYGOPHYLLACEAE (D)
Kelloggia Torr. ex Benth. RUBIACEAE (D)
Kelseya Rydb. ROSACEAE (D)
Kendrickia Hook.f. MELASTOMATACEAE (D)
Kengia Packer GRAMINEAE (M)
Kengyilia C.Yen & J.L.Yang = Elymus L. GRAMINEAE (M)
Keniochloa Meld. = Colpodium Trin. GRAMINEAE (M)
Kennedia Vent. LEGUMINOSAE–PAPILIONOIDEAE (D)
Kenopleurum Candargy = Thapsia L. UMBELLIFERAE (D)
Kensitia = Erepsia N.E.Br. Fedde AIZOACEAE (D)
Kentia Blume = Gronophyllum Scheff. PALMAE (M)
Kentia Blume (SUH) = Mitrella Miq. ANNONACEAE (D)
Kentiopsis Brongn. PALMAE (M)
Kentrochrosia K.Schum. & Lauterb. = Kopsia Blume APOCYNACEAE (D)
Kentrophyta Nutt. = Astragalus L. LEGUMINOSAE–PAPILIONOIDEAE (D)
Kentrosiphon N.E.Br. = Gladiolus L. IRIDACEAE (M)
Kentrosphaera Volkens = Volkensinia Schinz AMARANTHACEAE (D)
Kentrothamnus Suess. & Overk. RHAMNACEAE (D)
Keppleria Meisn. (SUH) = Oncosperma Blume PALMAE (M)
Keppleria Mart. ex Endl. = Bentinckia Berry ex Roxb. PALMAE (M)
Keracia (Coss.) Calest. = Hohenackeria Fisch. & C.A.Mey. UMBELLIFERAE (D)
Keramanthus Hook.f. = Adenia Forssk. PASSIFLORACEAE (D)
Keramocarpus Fenzl = Coriandrum L. UMBELLIFERAE (D)
Keratephorus Hassk. = Payena A.DC. SAPOTACEAE (D)
Keratolepis Rose ex Fröd. = Sedum L. CRASSULACEAE (D)
Keraudrenia J.Gay STERCULIACEAE (D)
Keraymonia Farille UMBELLIFERAE (D)
Kerbera E.Fourn. = Melinia Decne. ASCLEPIADACEAE (D)
Kerianthera J.H.Kirkbr. RUBIACEAE (D)
Kerigomnia P.Royen = Chitonanthera Schltr. ORCHIDACEAE (M)
Kerinozoma Steud. = Xerochloa R.Br. GRAMINEAE (M)
Kermadecia Brongn. & Gris PROTEACEAE (D)
Kernera Medik. CRUCIFERAE (D)
Kerrdora Gagnep. = Enkleia Griff. THYMELAEACEAE (D)
Kerria DC. ROSACEAE (D)
Kerriochloa C.E.Hubb. GRAMINEAE (M)
Kerriodoxa J.Dransf. PALMAE (M)
Kerriothyrsus C.Hansen MELASTOMATACEAE (D)
Kerstania Rech.f. = Lotus L. LEGUMINOSAE–PAPILIONOIDEAE (D)
Kerstingia K.Schum. = Belonophora Hook.f. RUBIACEAE (D)

Kerstingiella Harms = Macrotyloma (Wight & Arn.) Verdc. LEGUMINOSAE–
  PAPILIONOIDEAE (D)
Keteleeria Carrière PINACEAE (G)
Keyserlingia Bunge ex Boiss. = Sophora L. LEGUMINOSAE–PAPILIONOIDEAE (D)
Keysseria Lauterb. COMPOSITAE (D)
Khadia N.E.Br. AIZOACEAE (D)
Khasiaclunea Ridsdale RUBIACEAE (D)
Khaya A.Juss. MELIACEAE (D)
Kiapasia Woronow ex Grossh. = Astragalus L. LEGUMINOSAE–PAPILIONOIDEAE (D)
Kibara Endl. MONIMIACEAE (D)
Kibaropsis Vieill. ex Jérémie MONIMIACEAE (D)
Kibatalia G.Don APOCYNACEAE (D)
Kibbesia Walp. (SUO) = Pternandra Jack MELASTOMATACEAE (D)
Kibessia DC. = Pternandra Jack MELASTOMATACEAE (D)
Kickxia Dumort. SCROPHULARIACEAE (D)
Kielboul Adans. (SUS) = Aristida L. GRAMINEAE (M)
Kielmeyera Mart. GUTTIFERAE (D)
Kierschlegeria Spach = Fuchsia L. ONAGRACEAE (D)
Kigelia DC. BIGNONIACEAE (D)
Kigelianthe Baill. = Fernandoa Welw. ex Seem. BIGNONIACEAE (D)
Kigelkeia Raf. = Kigelia DC. BIGNONIACEAE (D)
Kiggelaria L. FLACOURTIACEAE (D)
Kiharapyrum Á.Löve = Aegilops L. GRAMINEAE (M)
Killipia Gleason MELASTOMATACEAE (D)
Killipiella A.C.Sm. = Disterigma (Klotzsch) Nied. ERICACEAE (D)
Killipiodendron Kobuski THEACEAE (D)
Kinepetalum Schltr. ASCLEPIADACEAE (D)
Kinetochilus (Schltr.) Brieger = Dendrobium Sw. ORCHIDACEAE (M)
Kinetostigma Dammer = Chamaedorea Willd. PALMAE (M)
Kingdon–Wardia Marquand = Swertia L. GENTIANACEAE (D)
Kingdonia Balf.f. & W.W.Sm. RANUNCULACEAE (D)
Kingella Tiegh. LORANTHACEAE (D)
Kinghamia C.Jeffrey COMPOSITAE (D)
Kingia R.Br. DASYPOGONACEAE (M)
Kingianthus H.Rob. COMPOSITAE (D)
Kingidium P.F.Hunt ORCHIDACEAE (M)
Kingiella Rolfe = Kingidium P.F.Hunt ORCHIDACEAE (M)
Kinginda Kuntze = Mitrephora (Blume) Hook.f. & Thomson ANNONACEAE (D)
Kingiodendron Harms LEGUMINOSAE–CAESALPINIOIDEAE (D)
Kingstonia Hook.f. & Thomson = Dendrokingstonia Rauschert ANNONACEAE (D)
Kinkina Adans. = Cinchona L. RUBIACEAE (D)
Kinostemon Kudo = Teucrium L. LABIATAE (D)
Kinugasa Tatew. & Suto TRILLIACEAE (M)
Kionophyton Garay = Stenorrhynchos Rich. ex Spreng. ORCHIDACEAE (M)
Kippistia F.Muell. COMPOSITAE (D)
Kirengeshoma Yatabe HYDRANGEACEAE (D)
Kirganelia Juss. – Phyllanthus L. EUPHORBIACEAE (D)
Kirilowia Bunge CHENOPODIACEAE (D)
Kirkbridea Wurdack MELASTOMATACEAE (D)
Kirkia Oliv. SIMAROUBACEAE (D)
Kirkianella Allan COMPOSITAE (D)
Kirkophytum (Harms) Allan = Stilbocarpa (Hook.f.) Decne. & Planch. ARALIACEAE (D)
Kirpicznikovia Á.Löve & D.Löve = Rhodiola L. CRASSULACEAE (D)
Kirschlegera Rchb. (SUO) = Fuchsia L. ONAGRACEAE (D)
Kissenia R.Br. ex Endl. LOASACEAE (D)

Kissodendron Seem. = Polyscias J.R.Forst. & G.Forst. ARALIACEAE (D)
Kita A.Chev. = Hygrophila R.Br. ACANTHACEAE (D)
Kitagawia Pimenov UMBELLIFERAE (D)
Kitaibela Willd. MALVACEAE (D)
Kitamuraea Rauschert (SUS) = Aster L. COMPOSITAE (D)
Kitamuraster Soják (SUS) = Aster L. COMPOSITAE (D)
Kitchingia Baker = Kalanchoe Adans. CRASSULACEAE (D)
Kittelia Rchb. = Cyanea Gaudich. CAMPANULACEAE (D)
Kjellbergia Bremek. = Strobilanthes Blume ACANTHACEAE (D)
Kjellbergiodendron Burret MYRTACEAE (D)
Klaineanthus Pierre ex Prain EUPHORBIACEAE (D)
Klaineastrum Pierre ex A.Chev. = Memecylon L. MELASTOMATACEAE (D)
Klainedoxa Pierre ex Engl. IRVINGIACEAE (D)
Klanderia F.Muell. = Prostanthera Labill. LABIATAE (D)
Klaprothia Kunth LOASACEAE (D)
Klattia Baker IRIDACEAE (M)
Kleinhovia L. STERCULIACEAE (D)
Kleinia Crantz (SUH) = Quisqualis L. COMBRETACEAE (D)
Kleinia Mill. COMPOSITAE (D)
Kleinodendron L.B.Sm. & Downs EUPHORBIACEAE (D)
Klemachloa R.Parker = Dendrocalamus Nees GRAMINEAE (M)
Klingia Schönl. = Gethyllis L. AMARYLLIDACEAE (M)
Klopstockia Karst. = Ceroxylon Bonpl. ex DC. PALMAE (M)
Klossia Ridl. RUBIACEAE (D)
Klotzschia Cham. UMBELLIFERAE (D)
Klotzschiphytum Baill. = Croton L. EUPHORBIACEAE (D)
Klugia Schltdl. = Rhynchoglossum Blume GESNERIACEAE (D)
Klugiodendron Britton & Killip = Abarema Pittier LEGUMINOSAE–MIMOSOIDEAE (D)
Kmeria (Pierre) Dandy MAGNOLIACEAE (D)
Knappia Sm. (SUS) = Mibora Adans. GRAMINEAE (M)
Knautia L. DIPSACACEAE (D)
Kneiffia Spach = Oenothera L. ONAGRACEAE (D)
Knema Lour. MYRISTICACEAE (D)
Knightia R.Br. PROTEACEAE (D)
Kniphofia Moench ASPHODELACEAE (M)
Kniphofia Scop. (SUH) = Terminalia L. COMBRETACEAE (D)
Knorringia (Czukav.) Tsvelev = Aconogonon (Meisn.) Rchb. POLYGONACEAE (D)
Knowltonia Salisb. RANUNCULACEAE (D)
Knoxia L. RUBIACEAE (D)
Koanophyllon Arruda COMPOSITAE (D)
Koanophyllum Arruda (SUO) = Koanophyllon Arruda COMPOSITAE (D)
Kobiosis Raf. = Euphorbia L. EUPHORBIACEAE (D)
Kobresia Willd. CYPERACEAE (M)
Kochia Roth CHENOPODIACEAE (D)
Kochiophyton Schltr. ex Cogn. = Aganisia Lindl. ORCHIDACEAE (M)
Kochummenia K.M.Wong RUBIACEAE (D)
Kodalyodendron Borhidi & Acuña RUTACEAE (D)
Koeberlinia Zucc. CAPPARACEAE (D)
Koechlea Endl. = Ptilostemon Cass. COMPOSITAE (D)
Koehneago Kuntze = Hoffmannia Sw. RUBIACEAE (D)
Koehneola Urb. COMPOSITAE (D)
Koehneria S.A.Graham, Tobe & Baas LYTHRACEAE (D)
Koeiea Rech.f. = Prionotrichon Botsch. & Vved. CRUCIFERAE (D)
Koeleria Pers. GRAMINEAE (M)
Koellensteinia Rchb.f. ORCHIDACEAE (M)

Koellia Moench = Pycnanthemum Michx. LABIATAE (D)
Koellikeria Regel GESNERIACEAE (D)
Koelpinia Pall. COMPOSITAE (D)
Koelreuteria Laxm. SAPINDACEAE (D)
Koelzella Hiroe = Prangos Lindl. UMBELLIFERAE (D)
Koelzia Rech.f. = Christolea Cambess. ex Jacquem. CRUCIFERAE (D)
Koenigia L. POLYGONACEAE (D)
Koenigia Post & Kuntze (SUH) = Lobularia Desv. CRUCIFERAE (D)
Koernickanthe L.Andersson MARANTACEAE (M)
Kohautia Cham. & Schltdl. RUBIACEAE (D)
Kohleria Regel GESNERIACEAE (D)
Kohlerianthus Fritsch = Columnea L. GESNERIACEAE (D)
Kohlrauschia Kunth = Petrorhagia (Ser.) Link CARYOPHYLLACEAE (D)
Koilodepas Hassk. EUPHORBIACEAE (D)
Kokabus Raf. = Hebecladus Miers SOLANACEAE (D)
Kokia Lewton MALVACEAE (D)
Kokoona Thwaites CELASTRACEAE (D)
Kokoschkinia Turcz. = Tecoma Juss. BIGNONIACEAE (D)
Kolbia P.Beauv. = Adenia Forssk. PASSIFLORACEAE (D)
Kolkwitzia Graebn. CAPRIFOLIACEAE (D)
Kolobochilus Lindau ACANTHACEAE (D)
Kolobopetalum Engl. MENISPERMACEAE (D)
Kolowratia C.Presl = Alpinia Roxb. ZINGIBERACEAE (M)
Komaroffia Kuntze = Nigella L. RANUNCULACEAE (D)
Komarovia Korovin UMBELLIFERAE (D)
Kompitsia Costantin & Gallaud = Gonocrypta Baill. ASCLEPIADACEAE (D)
Konantzia Dodson & N.H.Williams ORCHIDACEAE (M)
Konig Adans. = Lobularia Desv. CRUCIFERAE (D)
Konxikas Raf. = Lathyrus L. LEGUMINOSAE–PAPILIONOIDEAE (D)
Koompassia Maingay LEGUMINOSAE–CAESALPINIOIDEAE (D)
Koordersiochloa Merr. = Streblochaete Pilg. GRAMINEAE (M)
Koordersiodendron Engl. ANACARDIACEAE (D)
Kopsia Blume APOCYNACEAE (D)
Kopsiopsis (Beck) Beck SCROPHULARIACEAE (D)
Kordelestris Arruda = Jacaranda Juss. BIGNONIACEAE (D)
Korolkowia Regel LILIACEAE (M)
Korovinia Nevski & Vved. = Galagania Lipsky UMBELLIFERAE (D)
Korshinskia Lipsky UMBELLIFERAE (D)
Korthalsella Tiegh. VISCACEAE (D)
Korthalsia Blume PALMAE (M)
Korycarpus Lag. (SUS) = Diarrhena P.Beauv. GRAMINEAE (M)
Kosaria Forssk. = Dorstenia L. MORACEAE (D)
Kosmosiphon Lindau ACANTHACEAE (D)
Kosopoljanskia Korovin UMBELLIFERAE (D)
Kosteletzkya C.Presl MALVACEAE (D)
Kostermansia Soegeng BOMBACACEAE (D)
Kostermanthus Prance CHRYSOBALANACEAE (D)
Kostyczewa Korsh. = Chesneya Lindl. ex Endl. LEGUMINOSAE–PAPILIONOIDEAE (D)
Kotchubaea Hook.f. (SUO) = Kutchubaea Fisch. ex DC. RUBIACEAE (D)
Kotschya Endl. LEGUMINOSAE–PAPILIONOIDEAE (D)
Kotschyella F.K.Mey. = Thlaspi L. CRUCIFERAE (D)
Koyamaea W.W.Thomas & Davidse CYPERACEAE (M)
Kozlovia Lipsky UMBELLIFERAE (D)
Kraenzlinella Kuntze = Pleurothallis R.Br. ORCHIDACEAE (M)
Krainzia Backeb. = Mammillaria Haw. CACTACEAE (D)

Kralikia Coss. & Durieu = Tripogon Roem. & Schult. GRAMINEAE (M)
Kralikiella Batt. & Trab. (SUS) = Tripogon Roem. & Schult. GRAMINEAE (M)
Krameria L. ex Loefl. KRAMERIACEAE (D)
Krapfia DC. = Ranunculus L. RANUNCULACEAE (D)
Krapovickasia Fryxell MALVACEAE (D)
Krascheninnikovia Gueldenst. CHENOPODIACEAE (D)
Krascheninnikovia Turcz. ex Fenzl (SUH) = Pseudostellaria Pax CARYOPHYLLACEAE (D)
Krasnovia Popov ex Schischk. UMBELLIFERAE (D)
Krassera O.Schwartz = Anerincleistus Korth. MELASTOMATACEAE (D)
Kratzmannia Opiz (SUI) = Agropyron Gaertn. GRAMINEAE (M)
Kraunhia Raf. = Wisteria Nutt. LEGUMINOSAE–PAPILIONOIDEAE (D)
Krausella H.J.Lam = Pouteria Aubl. SAPOTACEAE (D)
Krauseola Pax & K.Hoffm. CARYOPHYLLACEAE (D)
Kraussia Harv. RUBIACEAE (D)
Krebsia Harv. = Stenostelma Schltr. ASCLEPIADACEAE (D)
Kremeria Durieu = Leucanthemum Mill. COMPOSITAE (D)
Kremeriella Maire CRUCIFERAE (D)
Kreodanthus Garay ORCHIDACEAE (M)
Kreysigia Rchb. = Schelhammera R.Br. CONVALLARIACEAE (M)
Krigia Schreb. COMPOSITAE (D)
Krokia Urb. = Pimenta Lindl. MYRTACEAE (D)
Krombholzia Fourn. = Zeugites P.Browne GRAMINEAE (M)
Krubera Hoffm. UMBELLIFERAE (D)
Krugella Pierre = Pouteria Aubl. SAPOTACEAE (D)
Krugia Urb. = Marlierea Cambess. MYRTACEAE (D)
Krugiodendron Urb. RHAMNACEAE (D)
Krukoviella A.C.Sm. OCHNACEAE (D)
Kryptostoma (Summerh.) Geerinck = Habenaria Willd. ORCHIDACEAE (M)
Ktenosachne Steud. = Rostraria Trin. GRAMINEAE (M)
Kubitzkia van der Werff LAURACEAE (D)
Kudoacanthus Hosok. ACANTHACEAE (D)
Kudrjaschevia Pojark. LABIATAE (D)
Kuepferella M.Laínz = Gentiana L. GENTIANACEAE (D)
Kuestera Regel = Justicia L. ACANTHACEAE (D)
Kuhitangia Ovcz. (UP) CARYOPHYLLACEAE (D)
Kuhlhasseltia J.J.Sm. ORCHIDACEAE (M)
Kuhlia Kunth = Banara Aubl. FLACOURTIACEAE (D)
Kuhlmannia J.C.Gomes = Pleonotoma Miers BIGNONIACEAE (D)
Kuhlmanniella Barroso = Dicranostyles Benth. CONVOLVULACEAE (D)
Kuhnia L. (SUH) = Brickellia Elliott COMPOSITAE (D)
Kuhnistera Lam. = Dalea L. LEGUMINOSAE–PAPILIONOIDEAE (D)
Kukolis Raf. = Hebecladus Miers SOLANACEAE (D)
Kumbaya Endl. ex Steud. = Gardenia Ellis RUBIACEAE (D)
Kumlienia Greene = Ranunculus L. RANUNCULACEAE (D)
Kummeria Mart. = Discophora Miers ICACINACEAE (D)
Kummerowia Schindl. LEGUMINOSAE–PAPILIONOIDEAE (D)
Kundmannia Scop. UMBELLIFERAE (D)
Kunhardtia Maguire RAPATEACEAE (M)
Kuniwatsukia Pic.Serm. WOODSIACEAE (F)
Kunkeliella Stearn SANTALACEAE (D)
Kunstleria Prain LEGUMINOSAE–PAPILIONOIDEAE (D)
Kunstlerodendron Ridl. = Chondrostylis Boerl. EUPHORBIACEAE (D)
Kunthea Humb. & Bonpl. = Chamaedorea Willd. PALMAE (M)
Kuntheria Conran & Clifford CONVALLARIACEAE (M)
Kunzea Rchb. MYRTACEAE (D)

Kurkas Raf. = Croton L. EUPHORBIACEAE (D)
Kurria Hochst. ex Steud. = Hymenodictyon Wall. RUBIACEAE (D)
Kurrimia Wall. ex Meisn. (SUH) = Itea L. ESCALLONIACEAE (D)
Kurrimia Wall. ex Thwaites = Bhesa Buch.-Ham. ex Arn. CELASTRACEAE (D)
Kurzamra Kuntze LABIATAE (D)
Kurziodendron N.P.Balakr. = Trigonostemon Blume EUPHORBIACEAE (D)
Kuschakewiczia Regel & Smirn. = Solenanthus Ledeb. BORAGINACEAE (D)
Kutchubaea Fisch. ex DC. RUBIACEAE (D)
Kydia Roxb. MALVACEAE (D)
Kyllinga Rottb. CYPERACEAE (M)
Kyllingiella R.W.Haines & Lye CYPERACEAE (M)
Kyphocarpa (Fenzl ex Endl.) Lopr. AMARANTHACEAE (D)
Kyrsteniopsis R.M.King & H.Rob. COMPOSITAE (D)
Kyrtanthus J.F.Gmel. = Posoqueria Aubl. RUBIACEAE (D)

Labatia Sw. = Pouteria Aubl. SAPOTACEAE (D)
Labichea Gaudich. ex DC. LEGUMINOSAE–CAESALPINIOIDEAE (D)
Labidostelma Schltr. ASCLEPIADACEAE (D)
Labisia Lindl. MYRSINACEAE (D)
Lablab Adans. LEGUMINOSAE–PAPILIONOIDEAE (D)
Labordia Gaudich. = Geniostoma J.R.Forst. & G.Forst. LOGANIACEAE (D)
Labourdonnaisia Bojer SAPOTACEAE (D)
Labramia A.DC. SAPOTACEAE (D)
Laburnum Fabr. LEGUMINOSAE–PAPILIONOIDEAE (D)
Lacaena Lindl. ORCHIDACEAE (M)
Lacaitaea Brand BORAGINACEAE (D)
Lacandonia E.Martínez & Ramos = Triuris Miers TRIURIDACEAE (M)
Lacanthis Raf. = Euphorbia L. EUPHORBIACEAE (D)
Laccodiscus Radlk. SAPINDACEAE (D)
Laccopetalum Ulbr. RANUNCULACEAE (D)
Laccospadix Drude & H.A.Wendl. PALMAE (M)
Laccosperma (G.Mann & H.A.Wendl.) Drude PALMAE (M)
Lacerdaea O.Berg = Campomanesia Ruíz & Pav. MYRTACEAE (D)
Lachanodes DC. COMPOSITAE (D)
Lachemilla (Kocke) Rydb. = Alchemilla L. ROSACEAE (D)
Lachenalia J.F.Jacq. ex Murray HYACINTHACEAE (M)
Lachnaea L. THYMELAEACEAE (D)
Lachnagrostis Trin. = Agrostis L. GRAMINEAE (M)
Lachnanthes Elliott HAEMODORACEAE (M)
Lachnastoma Korth. (SUH) = Nostolachma T.Durand RUBIACEAE (D)
Lachnocapsa Balf.f. CRUCIFERAE (D)
Lachnocaulon Kunth ERIOCAULACEAE (M)
Lachnochloa Steud. (UR) GRAMINEAE (M)
Lachnoloma Bunge CRUCIFERAE (D)
Lachnophyllum Bunge COMPOSITAE (D)
Lachnopodium Blume = Otanthera Blume MELASTOMATACEAE (D)
Lachnopylis Hochst. = Nuxia Comm. ex Lam. BUDDLEJACEAE (D)
Lachnorhiza A.Rich. COMPOSITAE (D)
Lachnosiphonium Hochst. = Catunaregam Wolf RUBIACEAE (D)
Lachnospermum Willd. COMPOSITAE (D)
Lachnostachys Hook. VERBENACEAE (D)
Lachnostoma Kunth ASCLEPIADACEAE (D)
Lachnostoma Hassk. (SUO) = Nostolachma T.Durand RUBIACEAE (D)
Lachnostylis Turcz. EUPHORBIACEAE (D)
Lachryma-jobi Ortega (SUS) = Coix L. GRAMINEAE (M)

Lachrymaria Fabr. (SUS) = Coix L. GRAMINEAE (M)
Laciala Kuntze = Schizoptera Turcz. COMPOSITAE (D)
Lacis Lindl. = Tulasneantha P.Royen PODOSTEMACEAE (D)
Lacistema Sw. LACISTEMATACEAE (D)
Lacmellea Karst. APOCYNACEAE (D)
Lacostea Van den Bosch = Trichomanes L. HYMENOPHYLLACEAE (F)
Lacosteopsis (Prantl) Nakaike = Crepidomanes (C.Presl) C.Presl
    HYMENOPHYLLACEAE (F)
Lacryma Medik. (SUS) = Coix L. GRAMINEAE (M)
Lactomammillaria Frič (SUI) = Mammillaria Haw. CACTACEAE (D)
Lactoris Phil. LACTORIDACEAE (D)
Lactuca L. COMPOSITAE (D)
Lactucosonchus (Sch.Bip.) Svent. COMPOSITAE (D)
Lacunaria Ducke QUIINACEAE (D)
Ladenbergia Klotzsch RUBIACEAE (D)
Ladyginia Lipsky UMBELLIFERAE (D)
Laelia Lindl. ORCHIDACEAE (M)
Laeliopsis Lindl. & Paxton ORCHIDACEAE (M)
Laennecia Cass. COMPOSITAE (D)
Laertia Gromov (SUS) = Leersia Sw. GRAMINEAE (M)
Laestadia Kunth ex Less. COMPOSITAE (D)
Laetia L. FLACOURTIACEAE (D)
Lafoensia Vand. LYTHRACEAE (D)
Lafuentea Lag. SCROPHULARIACEAE (D)
Lafuentia Lag. (SUO) = Lafuentea Lag. SCROPHULARIACEAE (D)
Lagarinthus E.Mey. = Schizoglossum E.Mey. ASCLEPIADACEAE (D)
Lagarosiphon Harv. HYDROCHARITACEAE (M)
Lagarosolen W.T.Wang (UP) GESNERIACEAE (D)
Lagarostrobos Quinn PODOCARPACEAE (G)
Lagascea Cav. COMPOSITAE (D)
Lagedium Soják = Mulgedium Cass. COMPOSITAE (D)
Lagenandra Dalzell ARACEAE (M)
Lagenantha Chiov. CHENOPODIACEAE (D)
Lagenanthus Gilg = Lehmanniella Gilg GENTIANACEAE (D)
Lagenaria Ser. CUCURBITACEAE (D)
Lagenia E.Fourn. = Araujia Brot. ASCLEPIADACEAE (D)
Lagenias E.Mey. = Sebaea Sol. ex R.Br. GENTIANACEAE (D)
Lagenifera Cass. (SUO) = Lagenophora Cass. COMPOSITAE (D)
Lagenocarpus Klotzsch (SUH) = Nagelocarpus Bullock ERICACEAE (D)
Lagenocarpus Nees CYPERACEAE (M)
Lagenophora Cass. COMPOSITAE (D)
Lageropyxis Miq. = Radermachera Zoll. & Moritzi BIGNONIACEAE (D)
Lagerstroemia L. LYTHRACEAE (D)
Lagetta Juss. THYMELAEACEAE (D)
Laggera Sch.Bip. ex Benth. COMPOSITAE (D)
Lagoa T.Durand ASCLEPIADACEAE (D)
Lagochilium Nees = Aphelandra R.Br. ACANTHACEAE (D)
Lagochilopsis Knorring = Lagochilus Bunge LABIATAE (D)
Lagochilus Bunge LABIATAE (D)
Lagoecia L. UMBELLIFERAE (D)
Lagonychium M.Bieb. = Prosopis L. LEGUMINOSAE–MIMOSOIDEAE (D)
Lagophylla Nutt. COMPOSITAE (D)
Lagopsis Bunge = Marrubium L. LABIATAE (D)
Lagoseris M.Bieb. = Crepis L. COMPOSITAE (D)
Lagotis Gaertn. SCROPHULARIACEAE (D)

Lagotis E.Mey. (SUH) = Carpacoce Sond. RUBIACEAE (D)
Lagrezia Moq. AMARANTHACEAE (D)
Laguna Cav. (UP) MALVACEAE (D)
Lagunaria (DC.) Rchb. MALVACEAE (D)
Laguncularia C.F.Gaertn. COMBRETACEAE (D)
Lagurus L. GRAMINEAE (M)
Lagynias E.Mey. ex Robyns RUBIACEAE (D)
Lahia Hassk. BOMBACACEAE (D)
Lalldhwojia Farille UMBELLIFERAE (D)
Lallemantia Fisch. & C.A.Mey. LABIATAE (D)
Lamanonia Vell. CUNONIACEAE (D)
Lamarchea Gaudich. MYRTACEAE (D)
Lamarckia Moench GRAMINEAE (M)
Lamarkea Rchb. (SUO) = Lamarchea Gaudich. MYRTACEAE (D)
Lamarkia Moench (SUO) = Lamarckia Moench GRAMINEAE (M)
Lambertia Sm. PROTEACEAE (D)
Lambertya F.Muell. = Bertya Planch. EUPHORBIACEAE (D)
Lamechites Markgr. = Microchites Miq. APOCYNACEAE (D)
Lamellisepalum Engl. = Sageretia Brongn. RHAMNACEAE (D)
Lamiacanthus Kuntze = Strobilanthes Blume ACANTHACEAE (D)
Lamiastrum Heist. ex Fabr. = Lamium L. LABIATAE (D)
Lamiodendron Steenis BIGNONIACEAE (D)
Lamiofrutex Lauterb. = Vavaea Benth. MELIACEAE (D)
Lamiophlomis Kudo = Phlomis L. LABIATAE (D)
Lamium L. LABIATAE (D)
Lamourouxia Kunth SCROPHULARIACEAE (D)
Lampas Danser LORANTHACEAE (D)
Lampayo Phil. VERBENACEAE (D)
Lampocarya R.Br. = Gahnia J.R.Forst. & G.Forst. CYPERACEAE (M)
Lamprachaenium Benth. COMPOSITAE (D)
Lampranthus N.E.Br. AIZOACEAE (D)
Lamprocaulos Mast. = Elegia L. RESTIONACEAE (M)
Lamprocephalus B.Nord. COMPOSITAE (D)
Lamprococcus Beer = Aechmea Ruíz & Pav. BROMELIACEAE (M)
Lamprolobium Benth. LEGUMINOSAE–PAPILIONOIDEAE (D)
Lamprophragma O.E.Schulz = Pennellia Nieuwl. CRUCIFERAE (D)
Lamprothamnus Hiern RUBIACEAE (D)
Lamprothyrsus Pilg. GRAMINEAE (M)
Lamyropappus Knorring & Tamamsch. COMPOSITAE (D)
Lamyropsis (Charadze) Dittrich COMPOSITAE (D)
Lanaria Aiton LANARIACEAE (M)
Lancea Hook.f. & Thomson SCROPHULARIACEAE (D)
Landersia Macfad. = Melothria L. CUCURBITACEAE (D)
Landia Comm. ex Juss. = Mussaenda L. RUBIACEAE (D)
Landolphia P.Beauv. APOCYNACEAE (D)
Landtia Less. = Haplocarpha Less. COMPOSITAE (D)
Landukia Planch. = Parthenocissus Planch. VITACEAE (D)
Laneasagum Bedd. = Drypetes Vahl EUPHORBIACEAE (D)
Lanessania Baill. = Trymatococcus Poepp. & Endl. MORACEAE (D)
Langebergia Anderb. COMPOSITAE (D)
Langlassea H.Wolff = Prionosciadium S.Watson UMBELLIFERAE (D)
Langloisia Greene POLEMONIACEAE (D)
Langsdorffia Mart. BALANOPHORACEAE (D)
Langsdorffia Regel (SUI) = Eustachys Desv. GRAMINEAE (M)
Langsdorffia Raddi (SUH) = Syagrus Mart. PALMAE (M)

Langsdorfia Raf. = Nicotiana L. SOLANACEAE (D)
Languas Koenig ex Small = Alpinia Roxb. ZINGIBERACEAE (M)
Lanium (Lindl.) Benth. ORCHIDACEAE (M)
Lankesterella Ames = Stenorrhynchos Rich. ex Spreng. ORCHIDACEAE (M)
Lankesteria Lindl. ACANTHACEAE (D)
Lannea A.Rich. ANACARDIACEAE (D)
Lanneoma Delile = Lannea A.Rich. ANACARDIACEAE (D)
Lansium Corrêa MELIACEAE (D)
Lantana L. VERBENACEAE (D)
Lantanopsis C.Wright COMPOSITAE (D)
Lanthorus C.Presl = Helixanthera Lour. LORANTHACEAE (D)
Lanugia N.E.Br. = Mascarenhasia A.DC. APOCYNACEAE (D)
Lapageria Ruíz & Pav. PHILESIACEAE (M)
Lapeirousia Pourr. IRIDACEAE (M)
Laphamia A.Gray = Perityle Benth. COMPOSITAE (D)
Lapidaria (Dinter & Schwantes) N.E.Br. AIZOACEAE (D)
Lapiedra Lag. AMARYLLIDACEAE (M)
Lapithea Griseb. GENTIANACEAE (D)
Laplacea Kunth THEACEAE (D)
Laportea Gaudich. URTICACEAE (D)
Lappago Schreb. (SUS) = Tragus Haller GRAMINEAE (M)
Lappagopsis Steud. = Axonopus P.Beauv. GRAMINEAE (M)
Lappula Moench BORAGINACEAE (D)
Lapsana L. COMPOSITAE (D)
Lardizabala Ruíz & Pav. LARDIZABALACEAE (D)
Larentia Klatt IRIDACEAE (M)
Laretia Gillies & Hook. UMBELLIFERAE (D)
Larix Mill. PINACEAE (G)
Larnandra Raf. = Epidendrum L. ORCHIDACEAE (M)
Larrea Cav. ZYGOPHYLLACEAE (D)
Larsenia Bremek. = Strobilanthes Blume ACANTHACEAE (D)
Lasallea Greene = Aster L. COMPOSITAE (D)
Laseguea A.DC. = Mandevilla Lindl. APOCYNACEAE (D)
Laser P.Gaertn., B.Mey. & Scherb. UMBELLIFERAE (D)
Laserpitium L. UMBELLIFERAE (D)
Lasia Lour. ARACEAE (M)
Lasiacis (Griseb.) Hitchc. GRAMINEAE (M)
Lasiadenia Benth. THYMELAEACEAE (D)
Lasiagrostis Link (SUS) = Stipa L. GRAMINEAE (M)
Lasiandra DC. = Tibouchina Aubl. MELASTOMATACEAE (D)
Lasianthaea DC. COMPOSITAE (D)
Lasianthera P.Beauv. ICACINACEAE (D)
Lasianthus Jack RUBIACEAE (D)
Lasiarrhenum I.M.Johnst. BORAGINACEAE (D)
Lasimorpha Schott ARACEAE (M)
Lasiobema (Korth.) Miq. = Bauhinia L. LEGUMINOSAE–CAESALPINIOIDEAE (D)
Lasiocarpus Liebm. MALPIGHIACEAE (D)
Lasiocaryum I.M.Johnst. BORAGINACEAE (D)
Lasiocephalus Schltdl. COMPOSITAE (D)
Lasiocereus F.Ritter = Haageocereus Backeb. CACTACEAE (D)
Lasiochlamys Pax & K.Hoffm. FLACOURTIACEAE (D)
Lasiochloa Kunth = Tribolium Desv. GRAMINEAE (M)
Lasiocladus Bojer ex Nees ACANTHACEAE (D)
Lasiococca Hook.f. EUPHORBIACEAE (D)
Lasiocoma Bolus = Euryops (Cass.) Cass. COMPOSITAE (D)

Lasiocorys Benth. = Leucas R.Br. LABIATAE (D)
Lasiocroton Griseb. EUPHORBIACEAE (D)
Lasiodiscus Hook.f. RHAMNACEAE (D)
Lasiogyne Klotzsch = Croton L. EUPHORBIACEAE (D)
Lasiolaena R.M.King & H.Rob. COMPOSITAE (D)
Lasiolytrum Steud. = Arthraxon P.Beauv. GRAMINEAE (M)
Lasionema D.Don = Macrocnemum P.Browne RUBIACEAE (D)
Lasiopetalum Sm. STERCULIACEAE (D)
Lasiopogon Cass. COMPOSITAE (D)
Lasiorhachis (Hack.) Stapf = Saccharum L. GRAMINEAE (M)
Lasiosiphon Fres. = Gnidia L. THYMELAEACEAE (D)
Lasiospermum Lag. COMPOSITAE (D)
Lasiostega Benth. (SUI) = Buchloe Engelm. GRAMINEAE (M)
Lasiostelma Benth. = Brachystelma R.Br. ASCLEPIADACEAE (D)
Lasiostoma Benth. = Hydnophytum Jack RUBIACEAE (D)
Lasiostyles C.Presl = Cleidion Blume EUPHORBIACEAE (D)
Lasiotrichos Lehm. (SUI) = Fingerhuthia Nees GRAMINEAE (M)
Lasipana Raf. = Mallotus Lour. EUPHORBIACEAE (D)
Lasiurus Boiss. GRAMINEAE (M)
Lassia Baill. = Tragia L. EUPHORBIACEAE (D)
Lastarriaea J.Remy POLYGONACEAE (D)
Lasthenia Cass. COMPOSITAE (D)
Laston C.Pau = Festuca L. GRAMINEAE (M)
Lastrea Bory = Oreopteris Holub THELYPTERIDACEAE (F)
Lastrella (H.Ito) Nakai = Phegopteris Fée THELYPTERIDACEAE (F)
Lastreopsis Ching DRYOPTERIDACEAE (F)
Latace Phil. ALLIACEAE (M)
Latania Comm. ex Juss. PALMAE (M)
Lateristachys Holub = Lycopodiella Holub LYCOPODIACEAE (FA)
Lateropora A.C.Sm. ERICACEAE (D)
Lathraea L. SCROPHULARIACEAE (D)
Lathraeocarpa Bremek. RUBIACEAE (D)
Lathriogyne Eckl. & Zeyh. = Amphithalea Eckl. & Zeyh. LEGUMINOSAE–
    PAPILIONOIDEAE (D)
Lathrisia Sw. = Bartholina R.Br. ORCHIDACEAE (M)
Lathrophytum Eichler BALANOPHORACEAE (D)
Lathyris Trew = Euphorbia L. EUPHORBIACEAE (D)
Lathyropteris Christ = Pteris L. PTERIDACEAE (F)
Lathyrus L. LEGUMINOSAE–PAPILIONOIDEAE (D)
Latipes Kunth = Leptothrium Kunth GRAMINEAE (M)
Latnax Miers = Athenaea Sendtn. SOLANACEAE (D)
Latouchea Franch. GENTIANACEAE (D)
Latourea Blume = Dendrobium Sw. ORCHIDACEAE (M)
Latourorchis Brieger = Dendrobium Sw. ORCHIDACEAE (M)
Latraeophila Leandro (SUI) = Helosis Rich. BALANOPHORACEAE (D)
Latrobea Meisn. LEGUMINOSAE–PAPILIONOIDEAE (D)
Latua Phil. SOLANACEAE (D)
Laubertia A.DC. APOCYNACEAE (D)
Laugeria Vahl ex Hook.f. = Neolaugeria Nicolson RUBIACEAE (D)
Laugeria L. (SUO) = Guettarda L. RUBIACEAE (D)
Laumoniera Nooteb. SIMAROUBACEAE (D)
Launaea Cass. COMPOSITAE (D)
Laurea Gaudich. = Bagassa Aubl. MORACEAE (D)
Laurelia Juss. MONIMIACEAE (D)
Laureliopsis Schodde = Laurelia Juss. MONIMIACEAE (D)

Laurembergia Bergius HALORAGACEAE (D)
Laurentia Adans. CAMPANULACEAE (D)
Laureria Schltdl. = Juanulloa Ruíz & Pav. SOLANACEAE (D)
Lauridia Eckl. & Zeyh. = Elaeodendron J.F.Jacq. CELASTRACEAE (D)
Lauro–Cerasus Duhamel = Prunus L. ROSACEAE (D)
Lauromerrillia C.K.Allen = Beilschmiedia Nees LAURACEAE (D)
Laurophyllus Thunb. ANACARDIACEAE (D)
Laurus L. LAURACEAE (D)
Lautea F.B.H.Br. = Corokia A.Cunn. ESCALLONIACEAE (D)
Lautembergia Baill. EUPHORBIACEAE (D)
Lauterbachia Perkins MONIMIACEAE (D)
Lavallea Baill. = Strombosia Blume OLACACEAE (D)
Lavalleopsis Tiegh. = Strombosia Blume OLACACEAE (D)
Lavana Raf. = Vestia Willd. SOLANACEAE (D)
Lavandula L. LABIATAE (D)
Lavatera L. MALVACEAE (D)
Lavauxia Spach = Oenothera L. ONAGRACEAE (D)
Lavigeria Pierre ICACINACEAE (D)
Lavoisiera DC. MELASTOMATACEAE (D)
Lavoisieria Spreng. (SUO) = Lavoisiera DC. MELASTOMATACEAE (D)
Lavoixia H.E.Moore PALMAE (M)
Lavradia Vell. ex Vand. = Sauvagesia L. OCHNACEAE (D)
Lavrania Plowes ASCLEPIADACEAE (D)
Lawia Griff. ex Tul. (SUH) = Dalzellia Wight PODOSTEMACEAE (D)
Lawia Wight (SUH) = Mycetia Reinw. RUBIACEAE (D)
Lawrencella Lindl. COMPOSITAE (D)
Lawrencia Hook. MALVACEAE (D)
Lawsonia L. LYTHRACEAE (D)
Laxmannia R.Br. ANTHERICACEAE (M)
Laxmannia S.G.Gmel. ex Trin. (SUH) = Phuopsis (Griseb.) Hook.f. RUBIACEAE (D)
Laxoplumeria Markgr. APOCYNACEAE (D)
Layia Hook. & Arn. (SUH) = Ormosia G.Jacks. LEGUMINOSAE–PAPILIONOIDEAE (D)
Layia Hook. & Arn. ex DC. COMPOSITAE (D)
Le–Monniera Lecomte (SUH) = Neolemonniera Heine SAPOTACEAE (D)
Leandra Raddi MELASTOMATACEAE (D)
Leandriella Benoist ACANTHACEAE (D)
Leaoa Schltr. & C.Porto = Hexadesmia Brongn. ORCHIDACEAE (M)
Leavenworthia Torr. CRUCIFERAE (D)
Lebeckia Thunb. LEGUMINOSAE–PAPILIONOIDEAE (D)
Lebetanthus Endl. EPACRIDACEAE (D)
Lebetina Cass. = Adenophylium Pers. COMPOSITAE (D)
Lebidiera Baill. = Cleistanthus Hook.f. ex Planch. EUPHORBIACEAE (D)
Lebidieropsis Müll.Arg. = Cleistanthus Hook.f. ex Planch. EUPHORBIACEAE (D)
Lebronnecia Fosberg MALVACEAE (D)
Lebrunia Staner GUTTIFERAE (D)
Lebruniodendron J.Léonard LEGUMINOSAE–CAESALPINIOIDEAE (D)
Lecananthus Jack RUBIACEAE (D)
Lecaniodiscus Planch. ex Benth. SAPINDACEAE (D)
Lecanium C.Presl (SUH) = Trichomanes L. HYMENOPHYLLACEAE (F)
Lecanium Reinw. = Lecanopteris Reinw. POLYPODIACEAE (F)
Lecanocarpus Nees = Acroglochin Schrad. ex Schult. CHENOPODIACEAE (D)
Lecanolepis Pic.Serm. = Trichomanes L. HYMENOPHYLLACEAE (F)
Lecanopteris Reinw. POLYPODIACEAE (F)
Lecanorchis Blume ORCHIDACEAE (M)
Lecanosperma Rusby = Heterophyllaea Hook.f. RUBIACEAE (D)

Lecanthus Griseb. (SUH) = Lisianthius P.Browne GENTIANACEAE (D)
Lecanthus Wedd. URTICACEAE (D)
Lecardia Poiss. ex Guillaumin = Salaciopsis Baker f. CELASTRACEAE (D)
Lecariocalyx Bremek. RUBIACEAE (D)
Lechea L. CISTACEAE (D)
Lechenaultia R.Br. (SUO) = Leschenaultia R.Br. GOODENIACEAE (D)
Lechlera Steud. (SUI) = Relchela Steud. GRAMINEAE (M)
Lecocarpus Decne. COMPOSITAE (D)
Lecointea Ducke LEGUMINOSAE–PAPILIONOIDEAE (D)
Lecokia DC. UMBELLIFERAE (D)
Lecomtea Koidz. = Cladopus H.Moeller PODOSTEMACEAE (D)
Lecomtedoxa (Pierre ex Engl.) Dubard SAPOTACEAE (D)
Lecomtella A.Camus GRAMINEAE (M)
Lecontea A.Rich. = Paederia L. RUBIACEAE (D)
Lectandra J.J.Sm. = Poaephyllum Ridl. ORCHIDACEAE (M)
Lecticula Barnhart = Utricularia L. LENTIBULARIACEAE (D)
Lecythis Loefl. LECYTHIDACEAE (D)
Lecythopsis Schrank = Couratari Aubl. LECYTHIDACEAE (D)
Leda C.B.Clarke = Isoglossa Oerst. ACANTHACEAE (D)
Ledebouria Roth HYACINTHACEAE (M)
Ledebouriella H.Wolff UMBELLIFERAE (D)
Ledenbergia Klotzsch ex Moq. PHYTOLACCACEAE (D)
Ledermannia Mildbr. & Burret = Desplatsia Bocquillon TILIACEAE (D)
Ledermanniella Engl. PODOSTEMACEAE (D)
Ledgeria F.Muell. = Galeola Lour. ORCHIDACEAE (M)
Ledocarpon Desf. = Balbisia Cav. GERANIACEAE (D)
Ledothamnus Meisn. ERICACEAE (D)
Ledum L. = Rhododendron L. ERICACEAE (D)
Leea L. LEEACEAE (D)
Leersia Sw. GRAMINEAE (M)
Leeuwenbergia Letouzey & N.Hallé EUPHORBIACEAE (D)
Lefebvrea A.Rich. UMBELLIFERAE (D)
Legendrea Webb & Berthel. = Turbina Raf. CONVOLVULACEAE (D)
Legenere McVaugh CAMPANULACEAE (D)
Legnea O.F.Cook (SUI) = Chamaedorea Willd. PALMAE (M)
Legnephora Miers MENISPERMACEAE (D)
Legnotis Sw. = Cassipourea Aubl. RHIZOPHORACEAE (D)
Legousia Durande CAMPANULACEAE (D)
Legrandia Kausel MYRTACEAE (D)
Lehmannia Spreng. = Nicotiana L. SOLANACEAE (D)
Lehmanniella Gilg GENTIANACEAE (D)
Leiandra Raf. = Callisia Loefl. COMMELINACEAE (M)
Leianthus Griseb. = Lisianthius P.Browne GENTIANACEAE (D)
Leibergia J.M.Coult. & Rose UMBELLIFERAE (D)
Leibnitzia Cass. COMPOSITAE (D)
Leiboldia Schltdl. ex Gleason COMPOSITAE (D)
Leichardtia R.Br. ASCLEPIADACEAE (D)
Leichhardtia F.Muell. = Phyllanthus L. EUPHORBIACEAE (D)
Leidesia Müll.Arg. EUPHORBIACEAE (D)
Leiocarpus Blume = Aporusa Blume EUPHORBIACEAE (D)
Leiochilus Hook.f. = Coffea L. RUBIACEAE (D)
Leiogyne K.Schum. = Pithecoctenium Mart. ex Meisn. BIGNONIACEAE (D)
Leioluma Baill. = Pouteria Aubl. SAPOTACEAE (D)
Leiophaca Lindau = Whitfieldia Hook. ACANTHACEAE (D)
Leiophyllum (Pers.) R.G.Hedw. ERICACEAE (D)

Leiopoa Ohwi = Festuca L. GRAMINEAE (M)
Leioptyx Pierre ex De Wild. = Entandrophragma C.E.C.Fisch. MELIACEAE (D)
Leiopyxis Miq. = Cleistanthus Hook.f. ex Planch. EUPHORBIACEAE (D)
Leiospermum D.Don = Weinmannia L. CUNONIACEAE (D)
Leiospora (C.A.Mey.) Vassilieva CRUCIFERAE (D)
Leiostemon Raf. = Penstemon Schmidel SCROPHULARIACEAE (D)
Leiothrix Ruhland ERIOCAULACEAE (M)
Leiothylax Warm. PODOSTEMACEAE (D)
Leiphaimos Cham. & Schltdl. = Voyria Aubl. GENTIANACEAE (D)
Leipoldtia L.Bolus AIZOACEAE (D)
Leitgebia Eichler = Sauvagesia L. OCHNACEAE (D)
Leitneria Chapm. LEITNERIACEAE (D)
Leleba Nakai = Bambusa Schreb. GRAMINEAE (M)
Lellingeria A.R.Sm. & R.C.Moran GRAMMITIDACEAE (F)
Leloutrea Gaudich. = Nolana L.f. SOLANACEAE (D)
Lelya Bremek. RUBIACEAE (D)
Lemaireocereus Britton & Rose = Pachycereus (A.Berger) Britton & Rose CACTACEAE (D)
Lemapteris Raf. = Pteris L. PTERIDACEAE (F)
Lembertia Greene COMPOSITAE (D)
Lembocarpus Leeuwenb. GESNERIACEAE (D)
Lemboglossum Halbinger ORCHIDACEAE (M)
Lembotropis Griseb. = Cytisus Desf. LEGUMINOSAE–PAPILIONOIDEAE (D)
Lemma Juss. ex Adans. = Marsilea L. MARSILEACEAE (F)
Lemmaphyllum C.Presl POLYPODIACEAE (F)
Lemmonia A.Gray HYDROPHYLLACEAE (D)
Lemna L. LEMNACEAE (M)
Lemooria P.S.Short COMPOSITAE (D)
Lemphoria O.E.Schulz = Arabidella (F.Muell.) O.E.Schulz CRUCIFERAE (D)
Lemuranthe Schltr. = Cynorkis Thouars ORCHIDACEAE (M)
Lemurella Schltr. ORCHIDACEAE (M)
Lemurodendron Villiers & Guinet LEGUMINOSAE–MIMOSOIDEAE (D)
Lemurophoenix J.Dransf. PALMAE (M)
Lemuropisum H.Perrier LEGUMINOSAE–CAESALPINIOIDEAE (D)
Lemurorchis Kraenzl. ORCHIDACEAE (M)
Lemurosicyos Keraudren CUCURBITACEAE (D)
Lemyrea (A.Chev.) A.Chev. & Beille RUBIACEAE (D)
Lenbrassia G.W.Gillett GESNERIACEAE (D)
Lenda Koidz. = Tectaria Cav. DRYOPTERIDACEAE (F)
Lendneria Minod = Poarium Desv. SCROPHULARIACEAE (D)
Lennea Klotzsch LEGUMINOSAE–PAPILIONOIDEAE (D)
Lennoa Lex. LENNOACEAE (D)
Lenophyllum Rose CRASSULACEAE (D)
Lenormandia Steud. (SUI) = Vetiveria Bory GRAMINEAE (M)
Lens Mill. LEGUMINOSAE–PAPILIONOIDEAE (D)
Lentibularia Séguier = Utricularia L. LENTIBULARIACEAE (D)
Lenzia Phil. PORTULACACEAE (D)
Leobordea Delile = Lotononis (DC.) Eckl. & Zeyh. LEGUMINOSAE–
    PAPILIONOIDEAE (D)
Leocereus Britton & Rose CACTACEAE (D)
Leochilus Knowles & Westc. ORCHIDACEAE (M)
Leocus A.Chev. LABIATAE (D)
Leonardendron Aubrév. = Anthonotha P.Beauv. LEGUMINOSAE–
    CAESALPINIOIDEAE (D)
Leonardoxa Aubrév. LEGUMINOSAE–CAESALPINIOIDEAE (D)
Leonia Ruíz & Pav. VIOLACEAE (D)

Leonocassia Britton = Senna Mill. LEGUMINOSAE–CAESALPINIOIDEAE (D)
Leonohebe Heads = Hebe Comm. ex Juss. SCROPHULARIACEAE (D)
Leonotis (Pers.) R.Br. LABIATAE (D)
Leontia Rchb. = Croton L. EUPHORBIACEAE (D)
Leontice L. BERBERIDACEAE (D)
Leontochir Phil. ALSTROEMERIACEAE (M)
Leontodon L. COMPOSITAE (D)
Leontonyx Cass. = Helichrysum Mill. COMPOSITAE (D)
Leontopodium (Pers.) R.Br. ex Cass. COMPOSITAE (D)
Leonuroides Rauschert (SUS) = Panzerina Soják LABIATAE (D)
Leonurus L. LABIATAE (D)
Leopardanthus Blume = Dipodium R.Br. ORCHIDACEAE (M)
Leopoldia Parl. HYACINTHACEAE (M)
Leopoldinia Mart. PALMAE (M)
Lepadanthus Ridl. = Ornithoboea Parish ex C.B.Clarke GESNERIACEAE (D)
Lepadena Raf. = Euphorbia L. EUPHORBIACEAE (D)
Lepanthes Sw. ORCHIDACEAE (M)
Lepanthopsis Ames ORCHIDACEAE (M)
Lepargochloa Launert = Loxodera Launert GRAMINEAE (M)
Lepechinia Willd. LABIATAE (D)
Lepechiniella Popov BORAGINACEAE (D)
Lepeocercis Trin. = Dichanthium Willemet GRAMINEAE (M)
Lepeostegeres Blume LORANTHACEAE (D)
Lepervenchea Cordem. = Angraecum Bory ORCHIDACEAE (M)
Lepiactis Raf. = Utricularia L. LENTIBULARIACEAE (D)
Lepiaglaia Pierre = Aglaia Lour. MELIACEAE (D)
Lepianthes Raf. = Piper L. PIPERACEAE (D)
Lepicochlea Rojas Acosta = Coronopus Zinn CRUCIFERAE (D)
Lepicystis (J.Sm.) J.Sm. = Polypodium L. POLYPODIACEAE (F)
Lepidacanthus C.Presl = Aphelandra R.Br. ACANTHACEAE (D)
Lepidagathis Willd. ACANTHACEAE (D)
Lepidanthemum Klotzsch = Dissotis Benth. MELASTOMATACEAE (D)
Lepidaploa (Cass.) Cass. = Vernonia Schreb. COMPOSITAE (D)
Lepidaria Tiegh. LORANTHACEAE (D)
Lepideilema Trin. = Streptochaeta Schrad. ex Nees GRAMINEAE (M)
Lepidella Tiegh. = Lepidaria Tiegh. LORANTHACEAE (D)
Lepiderema Radlk. SAPINDACEAE (D)
Lepidesmia Klatt COMPOSITAE (D)
Lepidium L. CRUCIFERAE (D)
Lepidobolus Nees RESTIONACEAE (M)
Lepidobotrys Engl. LEPIDOBOTRYACEAE (D)
Lepidocarpa Korth. = Parinari Aubl. CHRYSOBALANACEAE (D)
Lepidocaryum Mart. PALMAE (M)
Lepidocaulon Copel. = Histiopteris (J.G.Agardh) J.Sm. DENNSTAEDTIACEAE (F)
Lepidoceras Hook.f. EREMOLEPIDACEAE (D)
Lepidococcus H.A.Wendl. & Drude = Mauritiella Burret PALMAE (M)
Lepidococea Turcz. = Caperonia A.St.-Hil. EUPHORBIACEAE (D)
Lepidocordia Ducke BORAGINACEAE (D)
Lepidocoryphantha Backeb. = Coryphantha (Engelm.) Lem. CACTACEAE (D)
Lepidogrammitis Ching = Pleopeltis Humb. & Bonpl. ex Willd. POLYPODIACEAE (F)
Lepidogyne Blume ORCHIDACEAE (M)
Lepidolopha C.Winkl. COMPOSITAE (D)
Lepidolopsis Poljakov COMPOSITAE (D)
Lepidomicrosorium Ching & K.H.Shing = Microsorum Link POLYPODIACEAE (F)
Lepidoneuron Fée = Nephrolepis Schott OLEANDRACEAE (F)

Lepidonia S.F.Blake COMPOSITAE (D)
Lepidopelma Klotzsch = Sarcococca Lindl. BUXACEAE (D)
Lepidopetalum Blume SAPINDACEAE (D)
Lepidophorum Neck. ex DC. COMPOSITAE (D)
Lepidophyllum Cass. COMPOSITAE (D)
Lepidopironia A.Rich. = Tetrapogon Desf. GRAMINEAE (M)
Lepidorrhachis (H.A.Wendl. & Drude) O.F.Cook PALMAE (M)
Lepidospartum (A.Gray) A.Gray COMPOSITAE (D)
Lepidosperma Labill. CYPERACEAE (M)
Lepidostachys Wall. = Aporusa Blume EUPHORBIACEAE (D)
Lepidostemon Hook.f. & Thomson CRUCIFERAE (D)
Lepidostemon Hassk. (SUH) = Lepistemon Blume CONVOLVULACEAE (D)
Lepidostephium Oliv. COMPOSITAE (D)
Lepidostoma Bremek. RUBIACEAE (D)
Lepidothamnus Phil. PODOCARPACEAE (G)
Lepidotis P.Beauv. ex Mirb. = Lycopodium L. LYCOPODIACEAE (FA)
Lepidotrichilia (Harms) J.–F.Leroy MELIACEAE (D)
Lepidotrichum Velen. & Bornm. CRUCIFERAE (D)
Lepidoturus Baill. = Alchornea Sw. EUPHORBIACEAE (D)
Lepidozamia Regel ZAMIACEAE (G)
Lepidurus Janchen (SUI) = Parapholis C.E.Hubb. GRAMINEAE (M)
Lepilaena J.L.Drumm. ex Harv. ZANNICHELLIACEAE (M)
Lepinia Decne. APOCYNACEAE (D)
Lepiniopsis Valeton APOCYNACEAE (D)
Lepionurus Blume OPILIACEAE (D)
Lepipogon G.Bertol = Catunaregam Wolf RUBIACEAE (D)
Lepironia Rich. CYPERACEAE (M)
Lepisanthes Blume SAPINDACEAE (D)
Lepismium Pfeiff. CACTACEAE (D)
Lepisorus (J.Sm.) Ching POLYPODIACEAE (F)
Lepistemon Blume CONVOLVULACEAE (D)
Lepistemonopsis Dammer CONVOLVULACEAE (D)
Lepistoma Blume = Cryptolepis R.Br. ASCLEPIADACEAE (D)
Lepitoma Steud. (SUI) = Pleuropogon R.Br. GRAMINEAE (M)
Lepiurus Dumort. (SUS) = Lepturus R.Br. GRAMINEAE (M)
Leplaea Vermoesen = Guarea L. MELIACEAE (D)
Leporella A.S.George ORCHIDACEAE (M)
Leposma Blume = Cryptolepis R.Br. ASCLEPIADACEAE (D)
Leptacanthus Nees = Strobilanthes Blume ACANTHACEAE (D)
Leptactina Hook.f. RUBIACEAE (D)
Leptadenia R.Br. ASCLEPIADACEAE (D)
Leptagrostis C.E.Hubb. GRAMINEAE (M)
Leptaleum DC. CRUCIFERAE (D)
Leptalix Raf. = Fraxinus L. OLEACEAE (D)
Leptandra Nutt. = Veronicastrum Heist. ex Fabr. SCROPHULARIACEAE (D)
Leptanthe Klotzsch = Arnebia Forssk. BORAGINACEAE (D)
Leptarrhena R.Br. SAXIFRAGACEAE (D)
Leptaspis R.Br. GRAMINEAE (M)
Leptatherum Nees = Microstegium Nees GRAMINEAE (M)
Leptaulus Benth. ICACINACEAE (D)
Leptaxis Raf. = Tolmiea Torr. & A.Gray SAXIFRAGACEAE (D)
Lepteirs Raf. = Penstemon Schmidel SCROPHULARIACEAE (D)
Leptemon Raf. = Crotonopsis Michx. EUPHORBIACEAE (D)
Lepterica N.E.Br. ERICACEAE (D)
Leptinella Cass. COMPOSITAE (D)

Leptoboea Benth. GESNERIACEAE (D)
Leptobotrys Baill. = Tragia L. EUPHORBIACEAE (D)
Leptocallisia (Benth.) Pichon = Callisia Loefl. COMMELINACEAE (M)
Leptocanna L.C.Chia & H.L.Fung = Schizostachyum Nees GRAMINEAE (M)
Leptocarpha DC. COMPOSITAE (D)
Leptocarpus R.Br. RESTIONACEAE (M)
Leptocarydion Stapf GRAMINEAE (M)
Leptocaulis Nutt. ex DC. = Spermolepis Raf. UMBELLIFERAE (D)
Leptocentrum Schltr. = Rangaeris (Schltr.) Summerh. ORCHIDACEAE (M)
Leptoceras (R.Br.) Lindl. = Caladenia R.Br. ORCHIDACEAE (M)
Leptocercus Raf. (SUS) = Lepturus R.Br. GRAMINEAE (M)
Leptocereus (A.Berger) Britton & Rose CACTACEAE (D)
Leptochilus Kaulf. POLYPODIACEAE (F)
Leptochiton Sealy = Hymenocallis Salisb. AMARYLLIDACEAE (M)
Leptochloa P.Beauv. GRAMINEAE (M)
Leptochloopsis H.O.Yates = Uniola L. GRAMINEAE (M)
Leptochloris Kuntze (SUI) = Trichloris E.Fourn. ex Benth. GRAMINEAE (M)
Leptocionium C.Presl = Hymenophyllum Sm. HYMENOPHYLLACEAE (F)
Leptocladia Buxb. (SUH) = Mammillaria Haw. CACTACEAE (D)
Leptocladodia Buxb. = Mammillaria Haw. CACTACEAE (D)
Leptocladus Oliv. = Mostuea Didr. LOGANIACEAE (D)
Leptoclinium (Nutt.) Benth. COMPOSITAE (D)
Leptocodon Sond. (SUH) = Treichelia Vatke CAMPANULACEAE (D)
Leptocodon (Hook.f.) Lem. CAMPANULACEAE (D)
Leptocoryphium Nees GRAMINEAE (M)
Leptodactylon Hook. & Arn. POLEMONIACEAE (D)
Leptodermis Wall. RUBIACEAE (D)
Leptoderris Dunn LEGUMINOSAE–PAPILIONOIDEAE (D)
Leptodesmia (Benth.) Benth. LEGUMINOSAE–PAPILIONOIDEAE (D)
Leptofeddea Diels = Leptoglossis Benth. SOLANACEAE (D)
Leptoglossis Benth. SOLANACEAE (D)
Leptoglottis DC. = Schrankia Willd. LEGUMINOSAE–MIMOSOIDEAE (D)
Leptogonum Benth. POLYGONACEAE (D)
Leptogramma J.Sm. = Stegnogramma Blume THELYPTERIDACEAE (F)
Leptolaena Thouars SARCOLAENACEAE (D)
Leptolepia Prantl DENNSTAEDTIACEAE (F)
Leptolepidium Hsing & S.K.Wu = Cheilanthes Sw. ADIANTACEAE (F)
Leptoloma Chase = Digitaria Haller GRAMINEAE (M)
Leptomeria R.Br. SANTALACEAE (D)
Leptomeria Sieb. (SUH) = Amperea A.Juss. EUPHORBIACEAE (D)
Leptomischus Drake RUBIACEAE (D)
Leptomyrtus Miq. ex O.Berg = Syzygium Gaertn. MYRTACEAE (D)
Leptonema A.Juss. EUPHORBIACEAE (D)
Leptonema Hook. (SUH) = Draba L. CRUCIFERAE (D)
Leptonium Griff. = Lepionurus Blume OPILIACEAE (D)
Leptonychia Turcz. STERCULIACEAE (D)
Leptonychiopsis Ridl. STERCULIACEAE (D)
Leptopaetia Harv. = Tacazzea Decne. ASCLEPIADACEAE (D)
Leptopetalum Hook. & Arn. = Hedyotis L. RUBIACEAE (D)
Leptopharyngia (Stapf) Boiteau = Tabernaemontana L. APOCYNACEAE (D)
Leptopharynx Rydb. = Perityle Benth. COMPOSITAE (D)
Leptophoenix Becc. = Gronophyllum Scheff. PALMAE (M)
Leptophragma Benth. ex Dunal = Calibrachoa La Llave & Lex. SOLANACEAE (D)
Leptophyllochloa Calderón = Koeleria Pers. GRAMINEAE (M)
Leptoplax O.E.Schulz = Peltaria Jacq. CRUCIFERAE (D)

Leptopleuron C.Presl = Nephrolepis Schott OLEANDRACEAE (F)
Leptopogon Roberty = Andropogon L. GRAMINEAE (M)
Leptopteris C.Presl OSMUNDACEAE (F)
Leptopus Klotzsch & Garcke (SUH) = Euphorbia L. EUPHORBIACEAE (D)
Leptopus Decne. EUPHORBIACEAE (D)
Leptopyrum Raf. (URI) GRAMINEAE (M)
Leptopyrum Rchb. RANUNCULACEAE (D)
Leptorhabdos Schrenk SCROPHULARIACEAE (D)
Leptorhachis Klotzsch = Tragia L. EUPHORBIACEAE (D)
Leptorhoeo C.B.Clarke = Callisia Loefl. COMMELINACEAE (M)
Leptorhynchos Less. COMPOSITAE (D)
Leptorkis Thouars = Liparis Rich. ORCHIDACEAE (M)
Leptorumohra (H.Ito) H.Ito DRYOPTERIDACEAE (F)
Leptosaccharum (Hack.) A.Camus = Eriochrysis P.Beauv. GRAMINEAE (M)
Leptoscela Hook.f. RUBIACEAE (D)
Leptosema Benth. LEGUMINOSAE–PAPILIONOIDEAE (D)
Leptosiphon Benth. = Linanthus Benth. POLEMONIACEAE (D)
Leptosiphonium F.Muell. ACANTHACEAE (D)
Leptosolena C.Presl ZINGIBERACEAE (M)
Leptospermopsis S.Moore = Leptospermum J.R.Forst. & G.Forst. MYRTACEAE (D)
Leptospermum J.R.Forst. & G.Forst. MYRTACEAE (D)
Leptostachya Nees ACANTHACEAE (D)
Leptostachys G.Mey. (SUS) = Leptochloa P.Beauv. GRAMINEAE (M)
Leptostegia D.Don = Onychium Kaulf. ADIANTACEAE (F)
Leptostigma Arn. RUBIACEAE (D)
Leptostylis Benth. SAPOTACEAE (D)
Leptotaenia Nutt. = Lomatium Raf. UMBELLIFERAE (D)
Leptoterantha Louis ex Troupin MENISPERMACEAE (D)
Leptotes Lindl. ORCHIDACEAE (M)
Leptothrium Kunth GRAMINEAE (M)
Leptothrix (Dumort.) Dumort. (SUH) = Hordelymus (Jess.) Jess. ex Harz GRAMINEAE (M)
Leptothyrsa Hook.f. RUTACEAE (D)
Leptotriche Turcz. COMPOSITAE (D)
Leptunis Steven RUBIACEAE (D)
Lepturella Stapf = Oropetium Trin. GRAMINEAE (M)
Lepturidium Hitchc. & Ekman GRAMINEAE (M)
Lepturopetium Morat GRAMINEAE (M)
Lepturopsis Steud. = Rhytachne Desv. ex W.Ham. GRAMINEAE (M)
Lepturus R.Br. GRAMINEAE (M)
Lepurandra Graham = Antiaris Lesch. MORACEAE (D)
Lepuropetalon Elliott PARNASSIACEAE (D)
Lepyrodia R.Br. RESTIONACEAE (M)
Lepyrodiclis Fenzl CARYOPHYLLACEAE (D)
Lepyroxis Fourn. (SUI) = Muhlenbergia Schreb. GRAMINEAE (M)
Lequeetia Bubani = Limodorum Boehm. ORCHIDACEAE (M)
Lerchea L. RUBIACEAE (D)
Lerchenfeldia Schur (SUS) = Deschampsia P.Beauv. GRAMINEAE (M)
Lerchia Zinn = Suaeda Forssk. ex Scop. CHENOPODIACEAE (D)
Lereschia Boiss. = Cryptotaenia DC. UMBELLIFERAE (D)
Leretia Vell. = Mappia Jacq. ICACINACEAE (D)
Leroya Cavaco RUBIACEAE (D)
Leroyia Cavaco (SUO) = Leroya Cavaco RUBIACEAE (D)
Lescaillea Griseb. COMPOSITAE (D)
Leschenaultia R.Br. GOODENIACEAE (D)
Lesliea Seidenf. ORCHIDACEAE (M)

Lesourdia Fourn. = Scleropogon Phil. GRAMINEAE (M)
Lespedeza Michx. LEGUMINOSAE–PAPILIONOIDEAE (D)
Lesquerella S.Watson CRUCIFERAE (D)
Lesquereuxia Boiss. & Reut. SCROPHULARIACEAE (D)
Lessertia DC. LEGUMINOSAE–PAPILIONOIDEAE (D)
Lessingia Cham. COMPOSITAE (D)
Lessingianthus H.Rob. = Vernonia Schreb. COMPOSITAE (D)
Letestua Lecomte SAPOTACEAE (D)
Letestudoxa Pellegr. ANNONACEAE (D)
Letestuella G.Taylor PODOSTEMACEAE (D)
Lethedon Spreng. THYMELAEACEAE (D)
Lethia Ravenna IRIDACEAE (M)
Leto Phil. = Helogyne Nutt. COMPOSITAE (D)
Lettowianthus Diels ANNONACEAE (D)
Lettsomia Roxb. = Argyreia Lour. CONVOLVULACEAE (D)
Leucactinia Rydb. COMPOSITAE (D)
Leucadendron R.Br. PROTEACEAE (D)
Leucadenia Klotzsch ex Baill. = Croton L. EUPHORBIACEAE (D)
Leucaena Benth. LEGUMINOSAE–MIMOSOIDEAE (D)
Leucampyx A.Gray ex Benth. = Hymenopappus L'Hér. COMPOSITAE (D)
Leucandra Klotzsch = Tragia L. EUPHORBIACEAE (D)
Leucanthea Scheele = Salpiglossis Ruíz & Pav. SOLANACEAE (D)
Leucanthemella Tzvelev COMPOSITAE (D)
Leucanthemopsis (Giroux) Heywood COMPOSITAE (D)
Leucanthemum Mill. COMPOSITAE (D)
Leucas R.Br. LABIATAE (D)
Leucaster Choisy NYCTAGINACEAE (D)
Leucelene Greene = Chaetopappa DC. COMPOSITAE (D)
Leuceres Calest. = Endressia J.Gay UMBELLIFERAE (D)
Leucheria Lag. COMPOSITAE (D)
Leuchtenbergia Hook. CACTACEAE (D)
Leuciva Rydb. = Iva L. COMPOSITAE (D)
Leucobarleria Lindau = Neuracanthus Nees ACANTHACEAE (D)
Leucoblepharis Arn. COMPOSITAE (D)
Leucobotrys Tiegh. = Helixanthera Lour. LORANTHACEAE (D)
Leucocalantha Barb.Rodr. BIGNONIACEAE (D)
Leucocarpus D.Don SCROPHULARIACEAE (D)
Leucocasia Schott = Colocasia Schott ARACEAE (M)
Leucocera Turcz. = Calycera Cav. CALYCERACEAE (D)
Leucocodon Gardner RUBIACEAE (D)
Leucocoryne Lindl. ALLIACEAE (M)
Leucocrinum Nutt. ex A.Gray ANTHERICACEAE (M)
Leucocroton Griseb. EUPHORBIACEAE (D)
Leucocyclus Boiss. COMPOSITAE (D)
Leucogenes Beauverd COMPOSITAE (D)
Leucohyle Klotzsch ORCHIDACEAE (M)
Leucojum L. AMARYLLIDACEAE (M)
Leucolaena Ridl. = Didymoplexis Griff. ORCHIDACEAE (M)
Leucolophus Bremek. RUBIACEAE (D)
Leucomanes C.Presl = Crepidomanes (C.Presl) C.Presl HYMENOPHYLLACEAE (F)
Leucomeris D.Don = Gochnatia Kunth COMPOSITAE (D)
Leucomphalos Benth. ex Planch. LEGUMINOSAE–PAPILIONOIDEAE (D)
Leuconotis Jack APOCYNACEAE (D)
Leucophae Webb & Berthel. = Sideritis L. LABIATAE (D)
Leucopholis Gardner = Chionolaena DC. COMPOSITAE (D)

*Leucophrys*Rendle = Brachiaria (Trin.)Griseb. GRAMINEAE (M)
Leucophyllum Bonpl. SCROPHULARIACEAE (D)
Leucophysalis Rydb. SOLANACEAE (D)
Leucopoa Griseb. = Festuca L. GRAMINEAE (M)
Leucopogon R.Br. EPACRIDACEAE (D)
Leucopsis (DC.) Baker COMPOSITAE (D)
Leucoptera B.Nord. COMPOSITAE (D)
Leucoraphis Nees = Brillantaisia P.Beauv. ACANTHACEAE (D)
Leucorchis E.Mey. = Pseudorchis Séguier ORCHIDACEAE (M)
Leucorchis Blume = Didymoplexis Griff. ORCHIDACEAE (M)
Leucosalpa Scott Elliot SCROPHULARIACEAE (D)
Leucosceptrum Sm. LABIATAE (D)
Leucosedum Fourr. = Sedum L. CRASSULACEAE (D)
Leucosidea Eckl. & Zeyh. ROSACEAE (D)
Leucosmis Benth. = Phaleria Jack THYMELAEACEAE (D)
Leucospermum R.Br. PROTEACEAE (D)
Leucosphaera Gilg AMARANTHACEAE (D)
Leucospora Nutt. = Schistophragma Benth. ex Endl. SCROPHULARIACEAE (D)
Leucostachys Hoffmanns. = Goodyera R.Br. ORCHIDACEAE (M)
Leucostegane Prain LEGUMINOSAE–CAESALPINIOIDEAE (D)
Leucostegia C.Presl DAVALLIACEAE (F)
Leucostele Backeb. = Echinopsis Zucc. CACTACEAE (D)
Leucosyke Zoll. & Moritzi URTICACEAE (D)
Leucosyris Greene = Aster L. COMPOSITAE (D)
Leucothoe D.Don ERICACEAE (D)
Leucoxylon Raf. = Tabebuia Gomes ex DC. BIGNONIACEAE (D)
Leunisia Phil. COMPOSITAE (D)
Leuranthus Knobl. = Olea L. OLEACEAE (D)
Leurocline S.Moore = Echiochilon Desf. BORAGINACEAE (D)
Leutea Pimenov UMBELLIFERAE (D)
Leuzea DC. COMPOSITAE (D)
Leveillea Vaniot = Blumea DC. COMPOSITAE (D)
Levenhookia R.Br. STYLIDIACEAE (D)
Levieria Becc. MONIMIACEAE (D)
Levisanus Schreb. = Staavia Dahl BRUNIACEAE (D)
Levisticum Hill UMBELLIFERAE (D)
Levya Bureau ex Baill. = Cydista Miers BIGNONIACEAE (D)
Lewisia Pursh PORTULACACEAE (D)
Leycephyllum Piper = Rhynchosia Lour. LEGUMINOSAE–PAPILIONOIDEAE (D)
Leycesteria Wall. CAPRIFOLIACEAE (D)
Leymus Hochst. GRAMINEAE (M)
Leysera L. COMPOSITAE (D)
Lhotskya Schauer = Calytrix Labill. MYRTACEAE (D)
Lhotzkyella Rauschert ASCLEPIADACEAE (D)
Liabellum Rydb. = Sinclairia Hook. & Arn. COMPOSITAE (D)
Liabum Adans. COMPOSITAE (D)
Liatris Gaertn. ex Schreb. COMPOSITAE (D)
Libanothamnus Ernst COMPOSITAE (D)
Libanotis Haller ex Zinn. = Seseli L. UMBELLIFERAE (D)
Liberatia Rizzini = Lophostachys Pohl ACANTHACEAE (D)
Liberbaileya Furtado = Maxburretia Furtado PALMAE (M)
Libertia Lej. (SUS) = Bromus L. GRAMINEAE (M)
Libertia Spreng. IRIDACEAE (M)
Libocedrus Endl. CUPRESSACEAE (G)
Libonia K.Koch = Justicia L. ACANTHACEAE (D)

Librevillea Hoyle LEGUMINOSAE–CAESALPINIOIDEAE (D)
Libyella Pamp. GRAMINEAE (M)
Licania Aubl. CHRYSOBALANACEAE (D)
Licaria Aubl. LAURACEAE (D)
Lichinora Wight = Porpax Lindl. ORCHIDACEAE (M)
Lichtensteinia J.C.Wendl. (SUI) = Tapinanthus (Blume) Blume LORANTHACEAE (D)
Lichtensteinia Cham. & Schltdl. UMBELLIFERAE (D)
Licuala Thunb. PALMAE (M)
Lidbeckia Bergius COMPOSITAE (D)
Lidia Á.Löve & D.Löve = Minuartia L. CARYOPHYLLACEAE (D)
Liebichia Opiz = Ribes L. GROSSULARIACEAE (D)
Liebrechtsia De Wild. = Vigna Savi LEGUMINOSAE–PAPILIONOIDEAE (D)
Lietzia Regel GESNERIACEAE (D)
Lifago Schweinf. & Muschl. COMPOSITAE (D)
Ligaria Tiegh. LORANTHACEAE (D)
Ligea Poit. ex Tul. = Apinagia Tul. PODOSTEMACEAE (D)
Ligeophila Garay ORCHIDACEAE (M)
Ligeria Decne. = Sinningia Nees GESNERIACEAE (D)
Lightfootia L'Hér. = Wahlenbergia Schrad. ex Roth CAMPANULACEAE (D)
Lightfootia Schreb. (SUH) = Rondeletia L. RUBIACEAE (D)
Lightia R.H.Schomb. (SUH) = Euphronia Mart. & Zucc. EUPHRONIACEAE (D)
Lightiodendron Rauschert = Euphronia Mart. & Zucc. EUPHRONIACEAE (D)
Lignariella Baehni CRUCIFERAE (D)
Lignieria A.Chev. = Dissotis Benth. MELASTOMATACEAE (D)
Lignocarpa J.W.Dawson UMBELLIFERAE (D)
Ligularia Cass. COMPOSITAE (D)
Ligusticella J.M.Coult. & Rose UMBELLIFERAE (D)
Ligusticopsis Leute UMBELLIFERAE (D)
Ligusticum L. UMBELLIFERAE (D)
Ligustridium Spach = Ligustrum L. OLEACEAE (D)
Ligustrina Rupr. = Syringa L. OLEACEAE (D)
Ligustrum L. OLEACEAE (D)
Lijndenia Zoll. & Moritzi MELASTOMATACEAE (D)
Lilac Mill. = Syringa L. OLEACEAE (D)
Lilaca Raf. (SUS) = Syringa L. OLEACEAE (D)
Lilaea Bonpl. LILAEACEAE (M)
Lilaeopsis Greene UMBELLIFERAE (D)
Liliacum Renault (SUI) = Syringa L. OLEACEAE (D)
Lilium L. LILIACEAE (M)
Lilloa Speg. = Synandrospadix Engl. ARACEAE (M)
Limacia Lour. MENISPERMACEAE (D)
Limaciopsis Engl. MENISPERMACEAE (D)
Limatodes Blume = Calanthe Ker–Gawl. ORCHIDACEAE (M)
Limbarda Adans. COMPOSITAE (D)
Limeum L. MOLLUGINACEAE (D)
Limnalsine Rydb. = Montia L. PORTULACACEAE (D)
Limnanthemum S.G.Gmel. = Nymphoides Séguier MENYANTHACEAE (D)
Limnanthes R.Br. LIMNANTHACEAE (D)
Limnas Trin. GRAMINEAE (M)
Limnetis Rich. (SUS) = Spartina Schreb. GRAMINEAE (M)
Limnia Haw. = Claytonia L. PORTULACACEAE (D)
Limniboza R.E.Fr. LABIATAE (D)
Limnobium Rich. HYDROCHARITACEAE (M)
Limnocharis Bonpl. LIMNOCHARITACEAE (M)
Limnocitrus Swingle RUTACEAE (D)

Limnodea L.H.Dewey GRAMINEAE (M)
Limnophila R.Br. SCROPHULARIACEAE (D)
Limnophyton Miq. ALISMATACEAE (M)
Limnopoa C.E.Hubb. GRAMINEAE (M)
Limnorchis Rydb. = Platanthera Rich. ORCHIDACEAE (M)
Limnosciadium Mathias & Constance UMBELLIFERAE (D)
Limnosipanea Hook.f. RUBIACEAE (D)
Limodorum Boehm. ORCHIDACEAE (M)
Limonia L. RUTACEAE (D)
Limoniastrum Fabr. PLUMBAGINACEAE (D)
Limoniopsis Lincz. PLUMBAGINACEAE (D)
Limonium Mill. PLUMBAGINACEAE (D)
Limosella L. SCROPHULARIACEAE (D)
Linanthastrum Ewan = Linanthus Benth. POLEMONIACEAE (D)
Linanthus Benth. POLEMONIACEAE (D)
Linaria Mill. SCROPHULARIACEAE (D)
Linariantha B.L.Burtt & Ros.M.Sm. ACANTHACEAE (D)
Linariopsis Welw. PEDALIACEAE (D)
Linconia L. BRUNIACEAE (D)
Lindackeria C.Presl FLACOURTIACEAE (D)
Lindauea Rendle ACANTHACEAE (D)
Lindbergella Bor GRAMINEAE (M)
Lindbergia Bor (SUH) = Lindbergella Bor GRAMINEAE (M)
Lindblomia Fr. = Coeloglossum Hartm. ORCHIDACEAE (M)
Lindelofia Lehm. BORAGINACEAE (D)
Lindenbergia Lehm. SCROPHULARIACEAE (D)
Lindenia Benth. RUBIACEAE (D)
Lindeniopiper Trel. = Piper L. PIPERACEAE (D)
Lindera Thunb. LAURACEAE (D)
Lindernia All. SCROPHULARIACEAE (D)
Lindheimera A.Gray & Engelm. COMPOSITAE (D)
Lindleya Nees (SUH) = Laplacea Kunth THEACEAE (D)
Lindleya Kunth ROSACEAE (D)
Lindleyella Rydb. = Lindleya Kunth ROSACEAE (D)
Lindleyella Schltr. (SUH) = Rudolfiella Hoehne ORCHIDACEAE (M)
Lindmania Mez BROMELIACEAE (M)
Lindsaea Dryand. ex Sm. DENNSTAEDTIACEAE (F)
Lindsaenium Fée = Lindsaea Dryand. ex Sm. DENNSTAEDTIACEAE (F)
Lindsaeopsis Kuhn = Odontosoria Fée DENNSTAEDTIACEAE (F)
Lindsayella Ames & C.Schweinf. ORCHIDACEAE (M)
Lindsayoides Nakai = Nephrolepis Schott OLEANDRACEAE (F)
Lindsayomyrtus B.Hyland & Steenis MYRTACEAE (D)
Lingelsheimia Pax EUPHORBIACEAE (D)
Lingnania McClure = Bambusa Schreb. GRAMINEAE (M)
Linkagrostis Rom. García, Blanca & C.Morales = Agrostis L. GRAMINEAE (M)
Linnaea L. CAPRIFOLIACEAE (D)
Linnaeobreynia Hutch. = Capparis L. CAPPARACEAE (D)
Linnaeopsis Engl. GESNERIACEAE (D)
Linocalyx Lindau = Justicia L. ACANTHACEAE (D)
Linociera Sw. ex Schreb. = Chionanthus L. OLEACEAE (D)
Linodendron Griseb. THYMELAEACEAE (D)
Linoma O.F.Cook = Dictyosperma H.A.Wendl. & Drude PALMAE (M)
Linospadix Becc. ex Hook.f. (SUH) = Calyptrocalyx Blume PALMAE (M)
Linospadix H.A.Wendl. PALMAE (M)
Linosparton Adans. (SUS) = Lygeum L. GRAMINEAE (M)

Linostachys Klotzsch ex Schltdl. = Acalypha L. EUPHORBIACEAE (D)
Linostoma Wall. ex Endl. THYMELAEACEAE (D)
Lintonia Stapf GRAMINEAE (M)
Linum L. LINACEAE (D)
Liodendron H.Keng = Drypetes Vahl EUPHORBIACEAE (D)
Liparena Poit. ex Leman = Drypetes Vahl EUPHORBIACEAE (D)
Liparene Baill. = Drypetes Vahl EUPHORBIACEAE (D)
Liparia L. LEGUMINOSAE–PAPILIONOIDEAE (D)
Liparis Rich. ORCHIDACEAE (M)
Liparophyllum Hook.f. MENYANTHACEAE (D)
Lipocarpha R.Br. CYPERACEAE (M)
Lipochaeta DC. COMPOSITAE (D)
Lipophragma Schott & Kotschy ex Boiss. = Aethionema R.Br. CRUCIFERAE (D)
Lipostoma D.Don = Coccocypselum P.Browne RUBIACEAE (D)
Lippaya Endl. = Dentella J.R.Forst. & G.Forst. RUBIACEAE (D)
Lippia L. VERBENACEAE (D)
Lipskya (Kozo–Pol.) Nevski UMBELLIFERAE (D)
Lipskyella Juz. COMPOSITAE (D)
Liquidambar L. HAMAMELIDACEAE (D)
Lirayea Pierre = Mendoncia Vell. ex Vand. ACANTHACEAE (D)
Liriodendron L. MAGNOLIACEAE (D)
Liriope Lour. CONVALLARIACEAE (M)
Liriosma Poepp. & Endl. = Dulacia Vell. OLACACEAE (D)
Liriothamnus Schltr. = Trachyandra Kunth ASPHODELACEAE (M)
Lisaea Boiss. UMBELLIFERAE (D)
Lisianthius P.Browne GENTIANACEAE (D)
Lisianthus P.Browne (SUO) = Lisianthius P.Browne GENTIANACEAE (D)
Lissanthe R.Br. EPACRIDACEAE (D)
Lissocarpa Benth. LISSOCARPACEAE (D)
Lissochilus R.Br. = Eulophia R.Br. ex Lindl. ORCHIDACEAE (M)
Lissospermum Bremek. = Strobilanthes Blume ACANTHACEAE (D)
Listera R.Br. ORCHIDACEAE (M)
Listeria Neck. ex Raf. = Oldenlandia L. RUBIACEAE (D)
Listia E.Mey. = Lotononis (DC.) Eckl. & Zeyh. LEGUMINOSAE–PAPILIONOIDEAE (D)
Listrobanthes Bremek. = Strobilanthes Blume ACANTHACEAE (D)
Listrostachys Rchb. ORCHIDACEAE (M)
Litanthus Harv. HYACINTHACEAE (M)
Litchi Sonn. SAPINDACEAE (D)
Lithachne P.Beauv. GRAMINEAE (M)
Lithagrostis Gaertn. (SUS) = Coix L. GRAMINEAE (M)
Lithobium Bong. MELASTOMATACEAE (D)
Lithocardium Kuntze = Cordia L. BORAGINACEAE (D)
Lithocarpos Targ.–Toz. = Attalea Kunth PALMAE (M)
Lithocarpus Blume FAGACEAE (D)
Lithocaulon Bally = Pseudolithos Bally ASCLEPIADACEAE (D)
Lithococca Small ex Rydb. = Heliotropium L. BORAGINACEAE (D)
Lithodora Griseb. BORAGINACEAE (D)
Lithodraba Boelcke = Xerodraba Skottsb. CRUCIFERAE (D)
Lithophila Sw. AMARANTHACEAE (D)
Lithophragma (Nutt.) Torr. & A.Gray SAXIFRAGACEAE (D)
Lithophytum Brandegee = Plocosperma Benth. LOGANIACEAE (D)
Lithops N.E.Br. AIZOACEAE (D)
Lithospermum L. BORAGINACEAE (D)
Lithostegia Ching DRYOPTERIDACEAE (F)
Lithoxylon Endl. = Actephila Blume EUPHORBIACEAE (D)

Lithraea Miers ex Hook. & Arn. ANACARDIACEAE (D)
Litobrochia C.Presl = Pteris L. PTERIDACEAE (F)
Litocarpus L.Bolus = Aptenia N.E.Br. AIZOACEAE (D)
Litolobium Newman = Dennstaedtia Bernh. DENNSTAEDTIACEAE (F)
Litosanthes Blume RUBIACEAE (D)
Litosiphon Pierre ex Harms = Lovoa Harms MELIACEAE (D)
Litothamnus R.M.King & H.Rob. COMPOSITAE (D)
Litrisa Small COMPOSITAE (D)
Litsea Lam. LAURACEAE (D)
Littledalea Hemsl. GRAMINEAE (M)
Littonia Hook. COLCHICACEAE (M)
Littorella Bergius PLANTAGINACEAE (D)
Litwinowia Woronow = Euclidium R.Br. CRUCIFERAE (D)
Livistona R.Br. PALMAE (M)
Llagunoa Ruíz & Pav. SAPINDACEAE (D)
Llanosia Blanco = Ternstroemia Mutis ex L.f. THEACEAE (D)
Llavea Liebm. (SUH) = Neopringlea S.Watson FLACOURTIACEAE (D)
Llavea Lag. ADIANTACEAE (F)
Llerasia Triana COMPOSITAE (D)
Llewelynia Pittier MELASTOMATACEAE (D)
Lloydia Salisb. ex Rchb. LILIACEAE (M)
Loasa Adans. LOASACEAE (D)
Loasella Baill. = Eucnide Zucc. LOASACEAE (D)
Lobanilia Radcl.–Sm. EUPHORBIACEAE (D)
Lobbia Planch. = Thottea Rottb. ARISTOLOCHIACEAE (D)
Lobeira Alexander = Disocactus Lindl. CACTACEAE (D)
Lobelia L. CAMPANULACEAE (D)
Lobia O.F.Cook (SUI) = Chamaedorea Willd. PALMAE (M)
Lobirebutia Frič (SUI) = Echinopsis Zucc. CACTACEAE (D)
Lobivia Britton & Rose = Echinopsis Zucc. CACTACEAE (D)
Lobiviopsis Frič (SUI) = Echinopsis Zucc. CACTACEAE (D)
Lobocarpus Wight & Arn. = Glochidion J.R.Forst. & G.Forst. EUPHORBIACEAE (D)
Lobogyne Schltr. = Appendicula Blume ORCHIDACEAE (M)
Lobostema Spreng. = Lobostemon Lehm. BORAGINACEAE (D)
Lobostemon Lehm. BORAGINACEAE (D)
Lobostephanus N.E.Br. = Emicocarpus K.Schum. & Schltr. ASCLEPIADACEAE (D)
Lobularia Desv. CRUCIFERAE (D)
Locella Tiegh. = Taxillus Tiegh. LORANTHACEAE (D)
Locheria Regel = Achimenes Pers. GESNERIACEAE (D)
Lochia Balf.f. ILLECEBRACEAE (D)
Lochmocydia Mart. ex DC. = Cuspidaria DC. BIGNONIACEAE (D)
Lochnera Endl. = Catharanthus G.Don APOCYNACEAE (D)
Lockhartia Hook. ORCHIDACEAE (M)
Loddigesia Sims = Hypocalyptus Thunb. LEGUMINOSAE–PAPILIONOIDEAE (D)
Lodicularia P.Beauv. = Hemarthria R.Br. GRAMINEAE (M)
Lodoicea Comm. ex DC. PALMAE (M)
Loefgrenianthus Hoehne ORCHIDACEAE (M)
Loeflingia L. CARYOPHYLLACEAE (D)
Loerzingia Airy Shaw EUPHORBIACEAE (D)
Loeselia L. POLEMONIACEAE (D)
Loeseliastrum (Brand) Timbrook POLEMONIACEAE (D)
Loesenera Harms LEGUMINOSAE–CAESALPINIOIDEAE (D)
Loeseneriella A.C.Sm. CELASTRACEAE (D)
Loevigia Karst. & Triana = Monochaetum (DC.) Naudin MELASTOMATACEAE (D)
Loewia Urb. TURNERACEAE (D)

Logania R.Br. LOGANIACEAE (D)
Logfia Cass. COMPOSITAE (D)
Loheria Merr. MYRSINACEAE (D)
Loiseleuria Desv. ERICACEAE (D)
Lojaconoa Gand. = Festuca L. GRAMINEAE (M)
Loliolum Krecz. & Bobrov GRAMINEAE (M)
Lolium L. GRAMINEAE (M)
Lomagramma J.Sm. LOMARIOPSIDACEAE (F)
Lomandra Labill. LOMANDRACEAE (M)
Lomanodia Raf. = Astronidium A.Gray MELASTOMATACEAE (D)
Lomanthes Raf. = Phyllanthus L. EUPHORBIACEAE (D)
Lomaphlebia J.Sm. = Grammitis Sw. GRAMMITIDACEAE (F)
Lomaria Willd. = Blechnum L. BLECHNACEAE (F)
Lomaridium C.Presl = Blechnum L. BLECHNACEAE (F)
Lomariobotrys Fée = Stenochlaena J.Sm. BLECHNACEAE (F)
Lomariopsis Fée LOMARIOPSIDACEAE (F)
Lomastelma Raf. = Acmena DC. MYRTACEAE (D)
Lomatia R.Br. PROTEACEAE (D)
Lomatium Raf. UMBELLIFERAE (D)
Lomatocarpa Pimenov UMBELLIFERAE (D)
Lomatogoniopsis T.N.Ho & S.W.Liu GENTIANACEAE (D)
Lomatogonium A.Braun GENTIANACEAE (D)
Lomatophyllum Willd. ALOACEAE (M)
Lomatopodium Fisch. & C.A.Mey. = Seseli L. UMBELLIFERAE (D)
Lomatozona Baker COMPOSITAE (D)
Lombardochloa Roseng. & B.R.Arill. = Briza L. GRAMINEAE (M)
Lomelosia Raf. = Scabiosa L. DIPSACACEAE (D)
Lomilis Raf. = Hamamelis L. HAMAMELIDACEAE (D)
Lomoplis Raf. = Mimosa L. LEGUMINOSAE-MIMOSOIDEAE (D)
Lonas Adans. COMPOSITAE (D)
Lonchestigma Dunal = Jaborosa Juss. SOLANACEAE (D)
Lonchitis Bubani (SUH) = Serapias L. ORCHIDACEAE (M)
Lonchitis L. DENNSTAEDTIACEAE (F)
Lonchitis–Aspera Hill ex Farw. = Blechnum L. BLECHNACEAE (F)
Lonchocarpus Kunth LEGUMINOSAE-PAPILIONOIDEAE (D)
Lonchomera Hook.f. & Thomson = Mezzettia Becc. ANNONACEAE (D)
Lonchophaca Rydb. = Astragalus L. LEGUMINOSAE-PAPILIONOIDEAE (D)
Lonchophora Durieu = Matthiola R.Br. CRUCIFERAE (D)
Lonchostephus Tul. PODOSTEMACEAE (D)
Lonchostoma Wikstr. BRUNIACEAE (D)
Londesia Fisch. & C.A.Mey. = Kirilowia Bunge CHENOPODIACEAE (D)
Longetia Baill. = Austrobuxus Miq. EUPHORBIACEAE (D)
Lonicera L. CAPRIFOLIACEAE (D)
Lonicera Gaertn. (SUH) = Dendrophthoe Mart. LORANTHACEAE (D)
Loniceroides Bullock ASCLEPIADACEAE (D)
Lontarus Adans. = Borassus L. PALMAE (M)
Lopezia Cav. ONAGRACEAE (D)
Lophacme Stapf GRAMINEAE (M)
Lophalix Raf. = Alloplectus Mart. GESNERIACEAE (D)
Lophanthera A.Juss. MALPIGHIACEAE (D)
Lophanthus Adans. LABIATAE (D)
Lophatherum Brongn. GRAMINEAE (M)
Lophia Desv. (SUS) = Alloplectus Mart. GESNERIACEAE (D)
Lophiaris Raf. = Oncidium Sw. ORCHIDACEAE (M)
Lophidium Rich. = Schizaea Sm. SCHIZAEACEAE (F)

Lophiocarpus Turcz. PHYTOLACCACEAE (D)
Lophiocarpus (Kunth) Miq. (SUH) = Sagittaria L. ALISMATACEAE (M)
Lophiola Ker Gawl. MELANTHIACEAE (M)
Lophira Banks ex Gaertn.f. OCHNACEAE (D)
Lophobios Raf. = Euphorbia L. EUPHORBIACEAE (D)
Lophocarpinia Burkart LEGUMINOSAE–CAESALPINIOIDEAE (D)
Lophocarpus Boeck. = Schoenus L. CYPERACEAE (M)
Lophocereus (A.Berger) Britton & Rose = Pachycereus (A.Berger) Britton & Rose
  CACTACEAE (D)
Lophochlaena Nees = Pleuropogon R.Br. GRAMINEAE (M)
Lophochloa Rchb. = Rostraria Trin. GRAMINEAE (M)
Lophodium Newman = Dryopteris Adans. DRYOPTERIDACEAE (F)
Lophoglottis Raf. = Sophronitis Lindl. ORCHIDACEAE (M)
Lophogyne Tul. PODOSTEMACEAE (D)
Lopholaena DC. COMPOSITAE (D)
Lopholepis Decne. GRAMINEAE (M)
Lopholepis (J.Sm.) J.Sm. (SUH) = Microgramma C.Presl POLYPODIACEAE (F)
Lophomyrtus Burret MYRTACEAE (D)
Lophopappus Rusby COMPOSITAE (D)
Lophopetalum Wight ex Arn. CELASTRACEAE (D)
Lophophora J.M.Coult. CACTACEAE (D)
Lophophytum Schott & Endl. BALANOPHORACEAE (D)
Lophopogon Hack. GRAMINEAE (M)
Lophopterys A.Juss. MALPIGHIACEAE (D)
Lophopyrum Á.Löve = Elymus L. GRAMINEAE (M)
Lophopyxis Hook.f. LOPHOPYXIDACEAE (D)
Lophoschoenus Stapf = Costularia C.B.Clarke CYPERACEAE (M)
Lophosciadium DC. = Ferulago W.D.J.Koch UMBELLIFERAE (D)
Lophosoria C.Presl DICKSONIACEAE (F)
Lophospatha Burret = Salacca Reinw. PALMAE (M)
Lophospermum D.Don SCROPHULARIACEAE (D)
Lophostachys Pohl ACANTHACEAE (D)
Lophostemon Schott MYRTACEAE (D)
Lophostigma Radlk. SAPINDACEAE (D)
Lophostoma (Meisn.) Meisn. THYMELAEACEAE (D)
Lophothecium Rizzini = Justicia L. ACANTHACEAE (D)
Lophothele O.F.Cook (SUI) = Chamaedorea Willd. PALMAE (M)
Lophotocarpus T.Durand = Sagittaria L. ALISMATACEAE (M)
Lopimia Mart. MALVACEAE (D)
Lopriorea Schinz AMARANTHACEAE (D)
Loranthus Jacq. LORANTHACEAE (D)
Lordhowea B.Nord. COMPOSITAE (D)
Lorentzianthus R.M.King & H.Rob. COMPOSITAE (D)
Lorenzochloa J.R.Reeder & C.G.Reeder = Ortachne Nees ex Steud. GRAMINEAE (M)
Loretia Duval–Jouve = Vulpia C.C.Gmelin GRAMINEAE (M)
Loretoa Standl. = Capirona Spruce RUBIACEAE (D)
Loreya DC. MELASTOMATACEAE (D)
Loricalepis Brade MELASTOMATACEAE (D)
Loricaria Wedd. COMPOSITAE (D)
Lorinseria C.Presl = Woodwardia Sm. BLECHNACEAE (F)
Loroglossum Rich. = Himantoglossum K.Koch ORCHIDACEAE (M)
Loroma O.F.Cook = Archontophoenix H.A.Wendl. & Drude PALMAE (M)
Loropetalum R.Br. ex Rchb. HAMAMELIDACEAE (D)
Lorostelma Fourn. ASCLEPIADACEAE (D)
Lorostemon Ducke GUTTIFERAE (D)

Lortia Rendle = Monadenium Pax EUPHORBIACEAE (D)
Lothiania Kraenzl. = Porroglossum Schltr. ORCHIDACEAE (M)
Lothoniana Kraenzl. = Porroglossum Schltr. ORCHIDACEAE (M)
Lotononis (DC.) Eckl. & Zeyh. LEGUMINOSAE–PAPILIONOIDEAE (D)
Lotoxalis Small = Oxalis L. OXALIDACEAE (D)
Lotus L. LEGUMINOSAE–PAPILIONOIDEAE (D)
Lotzea Klotzsch & Karst. = Diplazium Sw. WOODSIACEAE (F)
Loudetia A.Braun (SUS) = Tristachya Nees GRAMINEAE (M)
Loudetia Hochst. ex Steud. GRAMINEAE (M)
Loudetiopsis Conert GRAMINEAE (M)
Loudonia Lindl. = Glischrocaryon Endl. HALORAGACEAE (D)
Louisiella C.E.Hubb. & J.Léonard GRAMINEAE (M)
Lourea Desv. = Christia Moench LEGUMINOSAE–PAPILIONOIDEAE (D)
Loureira Cav. = Jatropha L. EUPHORBIACEAE (D)
Lourteigia R.M.King & H.Rob. COMPOSITAE (D)
Lourtella S.A.Graham, Baas & Tobe LYTHRACEAE (D)
Louteridium S.Watson ACANTHACEAE (D)
Louvelia Jum. & H.Perrier PALMAE (M)
Lovanafia M.Peltier = Dicraeopetalum Harms LEGUMINOSAE–PAPILIONOIDEAE (D)
Lovoa Harms MELIACEAE (D)
Lowia Scort. = Orchidantha N.E.Br. LOWIACEAE (M)
Loxania Tiegh. = Cladocolea Tiegh. LORANTHACEAE (D)
Loxanthera (Blume) Blume LORANTHACEAE (D)
Loxanthocereus Backeb. = Cleistocactus Lem. CACTACEAE (D)
Loxidium Vent. = Swainsona Salisb. LEGUMINOSAE–PAPILIONOIDEAE (D)
Loxocalyx Hemsl. LABIATAE (D)
Loxocarpus R.Br. GESNERIACEAE (D)
Loxocarya R.Br. RESTIONACEAE (M)
Loxococcus H.A.Wendl. & Drude PALMAE (M)
Loxodera Launert GRAMINEAE (M)
Loxodiscus Hook.f. SAPINDACEAE (D)
Loxogramme (Blume) C.Presl POLYPODIACEAE (F)
Loxomorchis Rauschert (SUS) = Smithsonia C.J.Saldanha ORCHIDACEAE (M)
Loxonia Jack GESNERIACEAE (D)
Loxophyllum Blume = Loxonia Jack GESNERIACEAE (D)
Loxoptera O.E.Schulz CRUCIFERAE (D)
Loxopterygium Hook.f. ANACARDIACEAE (D)
Loxoscaphe T.Moore = Asplenium L. ASPLENIACEAE (F)
Loxostachys Peter (SUS) = Pseudechinolaena Stapf GRAMINEAE (M)
Loxostemon Hook.f. & Thomson CRUCIFERAE (D)
Loxostigma C.B.Clarke GESNERIACEAE (D)
Loxostylis A.Spreng. ex Rchb. ANACARDIACEAE (D)
Loxothysanus B.L.Rob. COMPOSITAE (D)
Loxotis (R.Br.) Benth. = Rhynchoglossum Blume GESNERIACEAE (D)
Loxsoma R.Br. ex A.Cunn. LOXSOMATACEAE (F)
Loxsomopsis Christ LOXSOMATACEAE (F)
Loydia Delile (SUH) = Pennisetum Rich. GRAMINEAE (M)
Lozanella Greenm. ULMACEAE (D)
Lozania S.Mutis ex Caldas LACISTEMATACEAE (D)
Lubaria Pittier RUTACEAE (D)
Lucaea Kunth = Arthraxon P.Beauv. GRAMINEAE (M)
Lucilia Cass. COMPOSITAE (D)
Luciliocline Anderb. & Freire COMPOSITAE (D)
Luciliopsis Wedd. = Chaetanthera Ruíz & Pav. COMPOSITAE (D)
Lucinaea DC. RUBIACEAE (D)

Luckhoffia A.C.White & B.Sloane = × Hoodiapelia G.D.Rowley ASCLEPIADACEAE (D)
Luculia Sweet RUBIACEAE (D)
Lucuma Molina = Pouteria Aubl. SAPOTACEAE (D)
Lucya DC. RUBIACEAE (D)
Ludekia Ridsdale RUBIACEAE (D)
Ludia Comm. ex Juss. FLACOURTIACEAE (D)
Ludisia A.Rich. ORCHIDACEAE (M)
Ludolphia Willd. (SUS) = Arundinaria Michx. GRAMINEAE (M)
Ludovia Brongn. CYCLANTHACEAE (M)
Ludwigia L. ONAGRACEAE (D)
Ludwigiantha (Torr. & A.Gray) Small = Ludwigia L. ONAGRACEAE (D)
Lueddemannia Rchb.f. ORCHIDACEAE (M)
Luehea Willd. TILIACEAE (D)
Lueheopsis Burret TILIACEAE (D)
Luerella Brass = Masdevallia Ruíz & Pav. ORCHIDACEAE (M)
Luerssenia Kuhn ex Luerss. = Tectaria Cav. DRYOPTERIDACEAE (F)
Luerssenidendron Domin = Acradenia Kippist RUTACEAE (D)
Luetkea Bong. ROSACEAE (D)
Luetzelburgia Harms LEGUMINOSAE–PAPILIONOIDEAE (D)
Luffa Mill. CUCURBITACEAE (D)
Lugoa DC. COMPOSITAE (D)
Lugonia Wedd. ASCLEPIADACEAE (D)
Luina Benth. COMPOSITAE (D)
Luisia Gaudich. ORCHIDACEAE (M)
Lulia Zardini COMPOSITAE (D)
Luma A.Gray MYRTACEAE (D)
Lumanaja Blanco = Homonoia Lour. EUPHORBIACEAE (D)
Lumnitzera Willd. COMBRETACEAE (D)
Lunania Hook. FLACOURTIACEAE (D)
Lunaria L. CRUCIFERAE (D)
Lunaria Hill (SUH) = Botrychium Sw. OPHIOGLOSSACEAE (F)
Lunasia Blanco RUTACEAE (D)
Lunathyrium Koidz. WOODSIACEAE (F)
Lundellia Leonard = Holographis Nees ACANTHACEAE (D)
Lundellianthus H.Rob. COMPOSITAE (D)
Lundia DC. BIGNONIACEAE (D)
Lunellia Nieuwl. = Besseya Rydb. SCROPHULARIACEAE (D)
Luntia Neck. ex Raf. = Croton L. EUPHORBIACEAE (D)
Lunularia Batsch = Botrychium Sw. OPHIOGLOSSACEAE (F)
Lupinophyllum Hutch. = Tephrosia Pers. LEGUMINOSAE–PAPILIONOIDEAE (D)
Lupinus L. LEGUMINOSAE–PAPILIONOIDEAE (D)
Lupulina Noulet = Medicago L. LEGUMINOSAE–PAPILIONOIDEAE (D)
Luronium Raf. ALISMATACEAE (M)
Lustrinia Raf. = Justicia L. ACANTHACEAE (D)
Lutzia Gand. CRUCIFERAE (D)
Luvunga Buch.–Ham. ex Wight & Arn. RUTACEAE (D)
Luxemburgia A.St.–Hil. OCHNACEAE (D)
Luziola Juss. GRAMINEAE (M)
Luzonia Elmer LEGUMINOSAE–PAPILIONOIDEAE (D)
Luzula DC. JUNCACEAE (M)
Luzuriaga Ruíz & Pav. PHILESIACEAE (M)
Lyallia Hook.f. HECTORELLACEAE (D)
Lyauteya Maire = Cytisopsis Jaub. & Spach LEGUMINOSAE–PAPILIONOIDEAE (D)
Lycapsus Phil. COMPOSITAE (D)
Lycaste Lindl. ORCHIDACEAE (M)

Lychniothyrsus Lindau ACANTHACEAE (D)
Lychnis L. CARYOPHYLLACEAE (D)
Lychnodiscus Radlk. SAPINDACEAE (D)
Lychnophora Mart. COMPOSITAE (D)
Lychnophoriopsis Sch.Bip. = Lychnophora Mart. COMPOSITAE (D)
Lycianthes (Dunal) Hassl. SOLANACEAE (D)
Lyciodes Kuntze = Sideroxylon L. SAPOTACEAE (D)
Lyciopsis Spach = Fuchsia L. ONAGRACEAE (D)
Lyciopsis (Boiss.) Schweinf. (SUH) = Euphorbia L. EUPHORBIACEAE (D)
Lycioserissa Roem. & Schult. = Canthium Lam. RUBIACEAE (D)
Lycium L. SOLANACEAE (D)
Lycocarpus O.E.Schulz CRUCIFERAE (D)
Lycochloa Sam. GRAMINEAE (M)
Lycomela Fabr. = Lycopersicon Mill. SOLANACEAE (D)
Lycomormium Rchb.f. ORCHIDACEAE (M)
Lycopersicon Mill. SOLANACEAE (D)
Lycopodiastrum Holub ex R.D.Dixit = Lycopodium L. LYCOPODIACEAE (FA)
Lycopodiella Holub LYCOPODIACEAE (FA)
Lycopodiodes Dill. ex Kuntze = Selaginella P.Beauv. SELAGINELLACEAE (FA)
Lycopodioides Boehm. = Selaginella P.Beauv. SELAGINELLACEAE (FA)
Lycopodion Adans. = Selaginella P.Beauv. SELAGINELLACEAE (FA)
Lycopodium L. LYCOPODIACEAE (FA)
Lycopsis L. = Anchusa L. BORAGINACEAE (D)
Lycopus L. LABIATAE (D)
Lycoris Herb. AMARYLLIDACEAE (M)
Lycoseris Cass. COMPOSITAE (D)
Lycurus Kunth GRAMINEAE (M)
Lydenburgia N.Robson = Catha Forssk. ex Schreb. CELASTRACEAE (D)
Lygeum L. GRAMINEAE (M)
Lygia Fasano = Thymelaea Mill. THYMELAEACEAE (D)
Lyginia R.Br. RESTIONACEAE (M)
Lygisma Hook.f. ASCLEPIADACEAE (D)
Lygistrum P.Browne = Manettia L. RUBIACEAE (D)
Lygodesmia D.Don COMPOSITAE (D)
Lygodictyon J.Sm. ex Hook. = Lygodium Sw. SCHIZAEACEAE (F)
Lygodisodea Ruíz & Pav. = Paederia L. RUBIACEAE (D)
Lygodium Sw. SCHIZAEACEAE (F)
Lygos Adans. = Genista L. LEGUMINOSAE–PAPILIONOIDEAE (D)
Lymanbensonia Kimnach = Lepismium Pfeiff. CACTACEAE (D)
Lymania Read BROMELIACEAE (M)
Lyncea Cham. & Schltdl. = Melasma Bergius SCROPHULARIACEAE (D)
Lyonia Nutt. ERICACEAE (D)
Lyonothamnus A.Gray ROSACEAE (D)
Lyonsia R.Br. = Parsonsia R.Br. APOCYNACEAE (D)
Lyperanthus R.Br. ORCHIDACEAE (M)
Lyperia Benth. SCROPHULARIACEAE (D)
Lyraea Lindl. = Bulbophyllum Thouars ORCHIDACEAE (M)
Lyrocarpa Hook. & Harv. CRUCIFERAE (D)
Lyroglossa Schltr. = Stenorrhynchos Rich. ex Spreng. ORCHIDACEAE (M)
Lysiana Tiegh. LORANTHACEAE (D)
Lysias Salisb. ex Rydb. = Platanthera Rich. ORCHIDACEAE (M)
Lysicarpus F.Muell. MYRTACEAE (D)
Lysichiton Schott ARACEAE (M)
Lysiclesia A.C.Sm. = Orthaea Klotzsch ERICACEAE (D)
Lysidice Hance LEGUMINOSAE–CAESALPINIOIDEAE (D)

Lysiella Rydb. = Platanthera Rich. ORCHIDACEAE (M)
Lysiloma Benth. LEGUMINOSAE–MIMOSOIDEAE (D)
Lysimachia L. PRIMULACEAE (D)
Lysimnia Raf. = Brassavola R.Br. ORCHIDACEAE (M)
Lysinema R.Br. EPACRIDACEAE (D)
Lysionotus D.Don GESNERIACEAE (D)
Lysiosepalum F.Muell. STERCULIACEAE (D)
Lysiostyles Benth. CONVOLVULACEAE (D)
Lysiphyllum (Benth.) de Wit = Bauhinia L. LEGUMINOSAE–CAESALPINIOIDEAE (D)
Lysipomia Kunth CAMPANULACEAE (D)
Lytanthus Wettst. = Globularia L. GLOBULARIACEAE (D)
Lythrum L. LYTHRACEAE (D)
Lytocaryum Toledo PALMAE (M)

Maackia Rupr. LEGUMINOSAE–PAPILIONOIDEAE (D)
Maba J.R.Forst. & G.Forst. = Diospyros L. EBENACEAE (D)
Mabea Aubl. EUPHORBIACEAE (D)
Mabola Raf. = Diospyros L. EBENACEAE (D)
Maborea Aubl. = Phyllanthus L. EUPHORBIACEAE (D)
Mabrya Elisens SCROPHULARIACEAE (D)
Macadamia F.Muell. PROTEACEAE (D)
Macairea DC. MELASTOMATACEAE (D)
Macania Blanco = Platymitra Boerl. ANNONACEAE (D)
Macaranga Thouars EUPHORBIACEAE (D)
Macarenia P.Royen PODOSTEMACEAE (D)
Macarisia Thouars RHIZOPHORACEAE (D)
Macarthuria Hügel ex Endl. MOLLUGINACEAE (D)
Macbridea Elliott ex Nutt. LABIATAE (D)
Macbrideina Standl. RUBIACEAE (D)
Maccoya F.Muell. = Rochelia Rchb. BORAGINACEAE (D)
Macdonaldia Gunn. ex Lindl. = Thelymitra J.R.Forst. & G.Forst. ORCHIDACEAE (M)
Macdougalia A.Heller COMPOSITAE (D)
Macfadyena A.DC. BIGNONIACEAE (D)
Macgregoria F.Muell. STACKHOUSIACEAE (D)
Macgregorianthus Merr. = Enkleia Griff. THYMELAEACEAE (D)
Machadoa Welw. ex Hook.f. = Adenia Forssk. PASSIFLORACEAE (D)
Machaeranthera Nees COMPOSITAE (D)
Machaerina Vahl CYPERACEAE (M)
Machaerium Pers. LEGUMINOSAE–PAPILIONOIDEAE (D)
Machaerocarpus Small ALISMATACEAE (M)
Machaerocereus Britton & Rose = Stenocereus (A.Berger) Riccob. CACTACEAE (D)
Machaerophorus Schltdl. = Mathewsia Hook. & Arn. CRUCIFERAE (D)
Machairophyllum Schwantes AIZOACEAE (D)
Machaonia Bonpl. RUBIACEAE (D)
Machilus Nees = Persea Mill. LAURACEAE (D)
Mackaya Harv. ACANTHACEAE (D)
Mackeea H.E.Moore PALMAE (M)
Mackenziea Nees = Strobilanthes Blume ACANTHACEAE (D)
Mackinlaya F.Muell. ARALIACEAE (D)
Macklottia Korth. = Leptospermum J.R.Forst. & G.Forst. MYRTACEAE (D)
Macleania Hook. ERICACEAE (D)
Macleaya R.Br. PAPAVERACEAE (D)
Maclura Nutt. MORACEAE (D)
Maclurodendron T.G.Hartley RUTACEAE (D)
Maclurolyra Calderón & Soderstr. GRAMINEAE (M)

Macodes (Blume) Lindl. ORCHIDACEAE (M)
Macoubea Aubl. APOCYNACEAE (D)
Macowania Oliv. COMPOSITAE (D)
Macphersonia Blume SAPINDACEAE (D)
Macrachaenium Hook.f. COMPOSITAE (D)
Macradenia R.Br. ORCHIDACEAE (M)
Macraea Hook.f. COMPOSITAE (D)
Macraea Wight (SUH) = Phyllanthus L. EUPHORBIACEAE (D)
Macrandria Meisn. = Hedyotis L. RUBIACEAE (D)
Macranthera Nutt. ex Benth. SCROPHULARIACEAE (D)
Macranthisiphon Bureau ex K.Schum. BIGNONIACEAE (D)
Macroberlinia (Harms) Hauman = Berlinia Sol. ex Hook.f. LEGUMINOSAE–
    CAESALPINIOIDEAE (D)
Macrobia (Webb & Berthel.) G.Kunkel = Aichryson Webb & Berthel. CRASSULACEAE (D)
Macroblepharus Phil. = Eragrostis Wolf GRAMINEAE (M)
Macrobriza (Tzvelev) Tzvelev = Briza L. GRAMINEAE (M)
Macrocalyx Tiegh. = Aetanthus (Eichler) Engl. LORANTHACEAE (D)
Macrocalyx Costantin & J.Poiss. (SUH) = Megistostegium Hochr. MALVACEAE (D)
Macrocarpaea Gilg GENTIANACEAE (D)
Macrocarpium (Spach) Nakai = Cornus L. CORNACEAE (D)
Macrocatalpa (Griseb.) Britton = Catalpa Scop. BIGNONIACEAE (D)
Macrocentrum Hook.f. MELASTOMATACEAE (D)
Macrocentrum Phil. (SUH) = Habenaria Willd. ORCHIDACEAE (M)
Macrochaeta Steud. (SUS) = Pennisetum Rich. GRAMINEAE (M)
Macrochaetium Steud. = Tetraria P.Beauv. CYPERACEAE (M)
Macrochilus Knowles & Westc. = Miltonia Lindl. ORCHIDACEAE (M)
Macrochiton M.Roem. = Dysoxylum Blume MELIACEAE (D)
Macrochlaena Hand.–Mazz. = Nothosmyrnium Miq. UMBELLIFERAE (D)
Macrochlamys Decne. = Drymonia Mart. GESNERIACEAE (D)
Macrochloa Kunth = Stipa L. GRAMINEAE (M)
Macrochordion De Vriese = Aechmea Ruíz & Pav. BROMELIACEAE (M)
Macrocladus Griff. = Orania Zipp. PALMAE (M)
Macroclinidium Maxim. COMPOSITAE (D)
Macroclinium Barb.Rodr. = Notylia Lindl. ORCHIDACEAE (M)
Macrocnemum P.Browne RUBIACEAE (D)
Macrococculus Becc. MENISPERMACEAE (D)
Macrocroton Klotzsch = Croton L. EUPHORBIACEAE (D)
Macrodiscus Bureau = Distictis Mart. ex Meisn. BIGNONIACEAE (D)
Macroditassa Malme = Ditassa R.Br. ASCLEPIADACEAE (D)
Macroglena (C.Presl) Copel. = Trichomanes L. HYMENOPHYLLACEAE (F)
Macroglossum Copel. = Angiopteris Hoffm. MARATTIACEAE (F)
Macrohasseltia L.O.Williams TILIACEAE (D)
Macrolenes Naudin MELASTOMATACEAE (D)
Macrolepis A.Rich. = Bulbophyllum Thouars ORCHIDACEAE (M)
Macrolobium Schreb. LEGUMINOSAE–CAESALPINIOIDEAE (D)
Macromeria D.Don BORAGINACEAE (D)
Macromyrtus Miq. = Syzygium Gaertn. MYRTACEAE (D)
Macronax Raf. (SUS) = Arundinaria Michx. GRAMINEAE (M)
Macronema Nutt. = Ericameria Nutt. COMPOSITAE (D)
Macropanax Miq. ARALIACEAE (D)
Macropeplus Perkins MONIMIACEAE (D)
Macropetalum Burch. ex Decne. ASCLEPIADACEAE (D)
Macropharynx Rusby APOCYNACEAE (D)
Macrophloga Becc. = Chrysalidocarpus H.A.Wendl. PALMAE (M)
Macrophthalmia Gasp. = Ficus L. MORACEAE (D)

Macropidia J.L.Drumm. ex Harv. HAEMODORACEAE (M)
Macropiper Miq. PIPERACEAE (D)
Macroplacis Blume = Pternandra Jack MELASTOMATACEAE (D)
Macroplectrum Pfitzer = Angraecum Bory ORCHIDACEAE (M)
Macroplethus C.Presl = Belvisia Mirb. POLYPODIACEAE (F)
Macropodandra Gilg = Notobuxus Oliv. BUXACEAE (D)
Macropodanthus L.O.Williams ORCHIDACEAE (M)
Macropodiella Engl. PODOSTEMACEAE (D)
Macropodina R.M.King & H.Rob. COMPOSITAE (D)
Macropodium R.Br. CRUCIFERAE (D)
Macropsidium Blume = Myrtus L. MYRTACEAE (D)
Macropsychanthus Harms ex K.Schum. & Lauterb. LEGUMINOSAE–
    PAPILIONOIDEAE (D)
Macropteranthes F.Muell. COMBRETACEAE (D)
Macroptilium (Benth.) Urb. LEGUMINOSAE–PAPILIONOIDEAE (D)
Macrorhamnus Baill. = Colubrina Rich. ex Brongn. RHAMNACEAE (D)
Macrorungia C.B.Clarke = Anisotes Nees ACANTHACEAE (D)
Macrosamanea Britton & Rose LEGUMINOSAE–MIMOSOIDEAE (D)
Macroscepis Kunth ASCLEPIADACEAE (D)
Macrosciadium V.N.Tikhom. & Lavrova (UP) UMBELLIFERAE (D)
Macrosepalum Regel & Schmalh. = Sedum L. CRASSULACEAE (D)
Macrosiphon Miq. = Hindsia Benth. ex Lindl. RUBIACEAE (D)
Macrosiphonia Müll.Arg. APOCYNACEAE (D)
Macrosolen (Blume) Rchb. LORANTHACEAE (D)
Macrosphyra Hook.f. RUBIACEAE (D)
Macrostachya A.Rich. (SUI) = Enteropogon Nees GRAMINEAE (M)
Macrostegia Nees = Vitex L. VERBENACEAE (D)
Macrostelia Hochr. MALVACEAE (D)
Macrostemon Boriss. = Veronica L. SCROPHULARIACEAE (D)
Macrostigmatella Rauschert (SUS) = Eigia Soják CRUCIFERAE (D)
Macrostomium Blume = Dendrobium Sw. ORCHIDACEAE (M)
Macrostylis Breda (SUH) = Corymborkis Thouars ORCHIDACEAE (M)
Macrostylis Bartl. & H.L.Wendl. RUTACEAE (D)
Macrothelypteris (H.Ito) Ching THELYPTERIDACEAE (F)
Macrotomia DC. = Arnebia Forssk. BORAGINACEAE (D)
Macrotorus Perkins MONIMIACEAE (D)
Macrotropis DC. = Ormosia G.Jacks. LEGUMINOSAE–PAPILIONOIDEAE (D)
Macrotyloma (Wight & Arn.) Verdc. LEGUMINOSAE–PAPILIONOIDEAE (D)
Macroule Pierce = Ormosia G.Jacks. LEGUMINOSAE–PAPILIONOIDEAE (D)
Macrozamia Miq. ZAMIACEAE (G)
Macrozanonia (Cogn.) Cogn. = Alsomitra (Blume) M.Roem. CUCURBITACEAE (D)
Macuillamia Raf. = Bacopa Aubl. SCROPHULARIACEAE (D)
Macvaughiella R.M.King & H.Rob. COMPOSITAE (D)
Madarosperma Benth. ASCLEPIADACEAE (D)
Maddenia Hook.f. & Thomson ROSACEAE (D)
Madhuca Ham. ex J.F.Gmel. SAPOTACEAE (D)
Madia Molina COMPOSITAE (D)
Madronella Mill. = Monardella Benth. LABIATAE (D)
Maelenia Dumort. = Cattleya Lindl. ORCHIDACEAE (M)
Maerua Forssk. CAPPARACEAE (D)
Maesa Forssk. MYRSINACEAE (D)
Maesobotrya Benth. EUPHORBIACEAE (D)
Maesoluma Baill. = Pouteria Aubl. SAPOTACEAE (D)
Maesopsis Engl. RHAMNACEAE (D)
Maeviella Rossow SCROPHULARIACEAE (D)

Mafekingia Baill. = Raphionacme Harv. ASCLEPIADACEAE (D)
Mafureira Bertol. = Trichilia P.Browne MELIACEAE (D)
Maga Urb. = Montezuma DC. MALVACEAE (D)
Magadania Pimenov & Lavrova UMBELLIFERAE (D)
Magallana Cav. TROPAEOLACEAE (D)
Magdalenaea Brade SCROPHULARIACEAE (D)
Magnistipula Engl. CHRYSOBALANACEAE (D)
Magnolia L. MAGNOLIACEAE (D)
Magodendron Vink SAPOTACEAE (D)
Magonia A.St.-Hil. SAPINDACEAE (D)
Maguirea A.D.Hawkes = Dieffenbachia Schott ARACEAE (M)
Maguireanthus Wurdack MELASTOMATACEAE (D)
Maguireocharis Steyerm. RUBIACEAE (D)
Maguireothamnus Steyerm. RUBIACEAE (D)
Magydaris W.D.J.Koch ex DC. UMBELLIFERAE (D)
Mahafalia Jum. & H.Perrier ASCLEPIADACEAE (D)
Mahagoni Adans. = Swietenia Jacq. MELIACEAE (D)
Maharanga A.DC. BORAGINACEAE (D)
Mahawoa Schltr. (UP) ASCLEPIADACEAE (D)
Mahea Pierre = Manilkara Adans. SAPOTACEAE (D)
Mahernia L. = Hermannia L. STERCULIACEAE (D)
Mahonia Nutt. BERBERIDACEAE (D)
Mahurea Aubl. GUTTIFERAE (D)
Mahya Cordem. = Lepechinia Willd. LABIATAE (D)
Maianthemum G.H.Weber CONVALLARIACEAE (M)
Maidenia Domin (SUH) = Uldinia J.M.Black UMBELLIFERAE (D)
Maidenia Rendle HYDROCHARITACEAE (M)
Maierocactus E.C.Rost = Astrophytum Lem. CACTACEAE (D)
Maieta Aubl. MELASTOMATACEAE (D)
Maihuenia (F.A.C.Weber) K.Schum. CACTACEAE (D)
Maihueniopsis Speg. = Opuntia Mill. CACTACEAE (D)
Maillardia Frapp. & Duchartre = Trophis P.Browne MORACEAE (D)
Maillea Parl. = Phleum L. GRAMINEAE (M)
Maingaya Oliv. HAMAMELIDACEAE (D)
Maireana Moq. CHENOPODIACEAE (D)
Mairetis I.M.Johnst. BORAGINACEAE (D)
Mairia Nees COMPOSITAE (D)
Maizilla Schltdl. (SUS) = Paspalum L. GRAMINEAE (M)
Majepea Post & Kuntze (SUS) = Chionanthus L. OLEACEAE (D)
Majidea J.Kirk ex Oliv. SAPINDACEAE (D)
Majorana Mill. = Origanum L. LABIATAE (D)
Malabaila Hoffm. UMBELLIFERAE (D)
Malabathris Raf. = Otanthera Blume MELASTOMATACEAE (D)
Malacantha Pierre = Pouteria Aubl. SAPOTACEAE (D)
Malaccotristicha C.Cusset & G.Cusset PODOSTEMACEAE (D)
Malachadenia Lindl. = Bulbophyllum Thouars ORCHIDACEAE (M)
Malache Vogel = Pavonia Cav. MALVACEAE (D)
Malachra L. MALVACEAE (D)
Malacmaea Griseb. = Bunchosia Rich. ex Kunth MALPIGHIACEAE (D)
Malacocarpus Salm-Dyck (SUH) = Parodia Speg. CACTACEAE (D)
Malacocarpus Fisch. & C.A.Mey. ZYGOPHYLLACEAE (D)
Malacocera R.H.Anderson CHENOPODIACEAE (D)
Malacomeles (Decne.) Engl. ROSACEAE (D)
Malacothamnus Greene MALVACEAE (D)
Malacothrix DC. COMPOSITAE (D)

Malacurus Nevski = Leymus Hochst. GRAMINEAE (M)
Malagasia L.A.S.Johnson & B.G.Briggs PROTEACEAE (D)
Malaisia Blanco = Trophis P.Browne MORACEAE (D)
Malanea Aubl. RUBIACEAE (D)
Malania Chun & S.K.Lee OLACACEAE (D)
Malanthos Stapf = Catanthera F.Muell. MELASTOMATACEAE (D)
Malaxis Sol. ex Sw. ORCHIDACEAE (M)
Malcolmia R.Br. CRUCIFERAE (D)
Malea Lundell ERICACEAE (D)
Malephora N.E.Br. AIZOACEAE (D)
Malesherbia Ruíz & Pav. MALESHERBIACEAE (D)
Malidra Raf. = Syzygium Gaertn. MYRTACEAE (D)
Malinvaudia Fourn. ASCLEPIADACEAE (D)
Mallea A.Juss. = Cipadessa Blume MELIACEAE (D)
Malleastrum (Baill.) J.–F.Leroy MELIACEAE (D)
Malleola J.J.Sm. & Schltr. ORCHIDACEAE (M)
Malleostemon J.W.Green MYRTACEAE (D)
Mallophora Endl. VERBENACEAE (D)
Mallophyton Wurdack MELASTOMATACEAE (D)
Mallostoma Karst. = Arcytophyllum Willd. ex Schult. & Schult.f. RUBIACEAE (D)
Mallotonia (Griseb.) Britton BORAGINACEAE (D)
Mallotopus Franch. & Sav. = Arnica L. COMPOSITAE (D)
Mallotus Lour. EUPHORBIACEAE (D)
Malmea R.E.Fr. ANNONACEAE (D)
Malmeanthus R.M.King & H.Rob. COMPOSITAE (D)
Malope L. MALVACEAE (D)
Malortiea H.A.Wendl. = Reinhardtia Liebm. PALMAE (M)
Malosma (Nutt.) Raf. = Rhus L. ANACARDIACEAE (D)
Malouetia A.DC. APOCYNACEAE (D)
Malouetiella Pichon = Malouetia A.DC. APOCYNACEAE (D)
Maloutchia Warb. MYRISTICACEAE (D)
Malperia S.Watson COMPOSITAE (D)
Malpighia L. MALPIGHIACEAE (D)
Malpighiantha Rojas Acosta (UR) MALPIGHIACEAE (D)
Malpighiodes Nied. = Mascagnia (Bertero ex DC.) Colla MALPIGHIACEAE (D)
Maltebrunia Kunth GRAMINEAE (M)
Malus Mill. ROSACEAE (D)
Malva L. MALVACEAE (D)
Malvastrum A.Gray MALVACEAE (D)
Malvaviscus Fabr. MALVACEAE (D)
Malvella Jaub. & Spach MALVACEAE (D)
Malveopsis C.Presl = Anisodontea C.Presl MALVACEAE (D)
Malya Opiz (SUS) = Ventenata Koeler GRAMINEAE (M)
Mamboga Blanco = Mitragyna Korth. RUBIACEAE (D)
Mamillaria F.Rchb. (SUO) = Mammillaria Haw. CACTACEAE (D)
Mamillopsis E.Morren ex Britton & Rose = Mammillaria Haw. CACTACEAE (D)
Mammariella J.Shafter (SUI) = Mammillaria Haw. CACTACEAE (D)
Mammea L. GUTTIFERAE (D)
Mammilaria Torr. & A.Gray (SUO) = Mammillaria Haw. CACTACEAE (D)
Mammillaria Haw. CACTACEAE (D)
Mammilloydia Buxb. CACTACEAE (D)
Mamorea Sota = Thismia Griff. BURMANNIACEAE (M)
Mananthes Bremek. = Justicia L. ACANTHACEAE (D)
Manaosella J.C.Gomes BIGNONIACEAE (D)
Mancanilla Mill. = Hippomane L. EUPHORBIACEAE (D)

Mancinella Tussac = Hippomane L. EUPHORBIACEAE (D)
Mancoa Wedd. CRUCIFERAE (D)
Mandelorna Steud. = Vetiveria Bory GRAMINEAE (M)
Mandenovia Alava UMBELLIFERAE (D)
Mandevilla Lindl. APOCYNACEAE (D)
Mandioca Link = Manihot Mill. EUPHORBIACEAE (D)
Mandirola Decne. = Gloxinia L'Hér. GESNERIACEAE (D)
Mandonia Hassk. (SUH) = Tradescantia L. COMMELINACEAE (M)
Mandragora L. SOLANACEAE (D)
Manekia Trel. PIPERACEAE (D)
Manettia L. RUBIACEAE (D)
Manettia Boehm. (SUH) = Selago L. SCROPHULARIACEAE (D)
Manfreda Salisb. = Agave L. AGAVACEAE (M)
Manganaroa Speg. = Acacia Mill. LEGUMINOSAE–MIMOSOIDEAE (D)
Mangenotia Pichon ASCLEPIADACEAE (D)
Mangifera L. ANACARDIACEAE (D)
Mangium Rumph. ex Scop. = Rhizophora L. RHIZOPHORACEAE (D)
Manglesia Lindl. = Beaufortia R.Br. MYRTACEAE (D)
Manglietia Blume MAGNOLIACEAE (D)
Manglietiastrum Law = Magnolia L. MAGNOLIACEAE (D)
Mangonia Schott ARACEAE (M)
Manicaria Gaertn. PALMAE (M)
Manihot Mill. EUPHORBIACEAE (D)
Manihotoides D.J.Rogers & Appan EUPHORBIACEAE (D)
Manilkara Adans. SAPOTACEAE (D)
Manilkariopsis (Gilly) C.Lundell = Manilkara Adans. SAPOTACEAE (D)
Maniltoa Scheff. LEGUMINOSAE–CAESALPINIOIDEAE (D)
Manisuris L. GRAMINEAE (M)
Mannagettaea Harry Sm. SCROPHULARIACEAE (D)
Mannaphorus Raf. (SUS) = Fraxinus L. OLEACEAE (D)
Mannia Hook.f. = Pierreodendron Engl. SIMAROUBACEAE (D)
Manniella Rchb.f. ORCHIDACEAE (M)
Manniophyton Müll.Arg. EUPHORBIACEAE (D)
Manochlamys Aellen CHENOPODIACEAE (D)
Manoelia S.Bowdich = Withania Pauquy SOLANACEAE (D)
Manongarivea Choux = Lepisanthes Blume SAPINDACEAE (D)
Manostachya Bremek. RUBIACEAE (D)
Manotes Sol. ex Planch. CONNARACEAE (D)
Mansoa DC. BIGNONIACEAE (D)
Mansonia J.R.Drumm. ex Prain STERCULIACEAE (D)
Mantalania Capuron ex J.-F.Leroy RUBIACEAE (D)
Mantisalca Cass. COMPOSITAE (D)
Mantisia Sims ZINGIBERACEAE (M)
Manulea L. SCROPHULARIACEAE (D)
Manuleopsis Thell. ex Schinz SCROPHULARIACEAE (D)
Manungala Blanco = Quassia L. SIMAROUBACEAE (D)
Maoutia Wedd. URTICACEAE (D)
Mapania Aubl. CYPERACEAE (M)
Mapaniopsis C.B.Clarke CYPERACEAE (M)
Mapira Adans. (SUS) = Olyra L. GRAMINEAE (M)
Mapouria Aubl. = Psychotria L. RUBIACEAE (D)
Mappa Adr.Juss. = Macaranga Thouars EUPHORBIACEAE (D)
Mappia Jacq. ICACINACEAE (D)
Mappianthus Hand.-Mazz. ICACINACEAE (D)
Maprounea Aubl. EUPHORBIACEAE (D)

Maquira Aubl. MORACEAE (D)
Maracanthus Kuijt = Oryctina Tiegh. LORANTHACEAE (D)
Marah Kellogg CUCURBITACEAE (D)
Marahuacaea Maguire RAPATEACEAE (M)
Maralia Thouars = Polyscias J.R.Forst. & G.Forst. ARALIACEAE (D)
Maranta L. MARANTACEAE (M)
Maranthes Blume CHRYSOBALANACEAE (D)
Marantochloa Brongn. ex Gris MARANTACEAE (M)
Marara Karst. = Aiphanes Willd. PALMAE (M)
Marasmodes DC. COMPOSITAE (D)
Marathrum Bonpl. PODOSTEMACEAE (D)
Marattia Sw. MARATTIACEAE (F)
Marcania J.B.Imlay ACANTHACEAE (D)
Marcellia Baill. = Marcelliopsis Schinz AMARANTHACEAE (D)
Marcelliopsis Schinz AMARANTHACEAE (D)
Marcetella Svent. = Bencomia Webb & Berthel. ROSACEAE (D)
Marcetia DC. MELASTOMATACEAE (D)
Marcgravia L. MARCGRAVIACEAE (D)
Marcuccia Becc. = Enicosanthum Becc. ANNONACEAE (D)
Marenopuntia Backeb. = Opuntia Mill. CACTACEAE (D)
Marenteria Thouars = Uvaria L. ANNONACEAE (D)
Maresia Pomel CRUCIFERAE (D)
Mareya Baill. EUPHORBIACEAE (D)
Mareyopsis Pax & K.Hoffm. EUPHORBIACEAE (D)
Margaranthus Schltdl. SOLANACEAE (D)
Margaretta Oliv. ASCLEPIADACEAE (D)
Margaris Griseb. (SUI) = Margaritopsis Sauvalle RUBIACEAE (D)
Margaritaria L.f. EUPHORBIACEAE (D)
Margaritolobium Harms LEGUMINOSAE–PAPILIONOIDEAE (D)
Margaritopsis Sauvalle RUBIACEAE (D)
Margelliantha P.J.Cribb ORCHIDACEAE (M)
Marginaria Bory = Polypodium L. POLYPODIACEAE (F)
Marginariopsis C.Chr. POLYPODIACEAE (F)
Marginatocereus (Backeb.) Backeb. = Pachycereus (A.Berger) Britton & Rose
 CACTACEAE (D)
Margotia Boiss. UMBELLIFERAE (D)
Margyricarpus Ruíz & Pav. ROSACEAE (D)
Marianthus Hügel ex Endl. = Billardiera Sm. PITTOSPORACEAE (D)
Mariarisqueta Guinea = Cheirostylis Blume ORCHIDACEAE (M)
Marica Ker Gawl. (SUH) = Neomarica Sprague IRIDACEAE (M)
Marila Sw. GUTTIFERAE (D)
Marina Liebm. LEGUMINOSAE–PAPILIONOIDEAE (D)
Maripa Aubl. CONVOLVULACEAE (D)
Mariscopsis Cherm. = Queenslandiella Domin CYPERACEAE (M)
Marisculus Goetgh. = Alinula J.Raynal CYPERACEAE (M)
Mariscus Vahl = Cyperus L. CYPERACEAE (M)
Maritimocereus Akers = Cleistocactus Lem. CACTACEAE (D)
Markea Rich. SOLANACEAE (D)
Markhamia Seem. ex Baill. BIGNONIACEAE (D)
Marlierea Cambess. MYRTACEAE (D)
Marlieriopsis Kiaersk. = Blepharocalyx O.Berg MYRTACEAE (D)
Marlothia Engl. = Helinus E.Mey. ex Endl. RHAMNACEAE (D)
Marlothiella H.Wolff UMBELLIFERAE (D)
Marmaroxylon Killip LEGUMINOSAE–MIMOSOIDEAE (D)
Marmoritis Benth. LABIATAE (D)

Marniera Backeb. = Selenicereus (A.Berger) Britton & Rose CACTACEAE (D)
Marojejya Humbert PALMAE (M)
Marquesia Gilg DIPTEROCARPACEAE (D)
Marquisia A.Rich. = Coprosma J.R.Forst. & G.Forst. RUBIACEAE (D)
Marrubium L. LABIATAE (D)
Marsdenia R.Br. ASCLEPIADACEAE (D)
Marshallfieldia J.F.Macbr. = Adelobotrys DC. MELASTOMATACEAE (D)
Marshallia Schreb. COMPOSITAE (D)
Marshalljohnstonia Henrickson COMPOSITAE (D)
Marshallocereus Backeb. = Stenocereus (A.Berger) Riccob. CACTACEAE (D)
Marsilaea Neck. = Salvinia Séguier SALVINIACEAE (F)
Marsilea L. MARSILEACEAE (F)
Marsippospermum Desv. JUNCACEAE (M)
Marssonia Karst. = Napeanthus Gardner GESNERIACEAE (D)
Marsupiaria Hoehne = Maxillaria Ruíz & Pav. ORCHIDACEAE (M)
Marsypianthes Mart. ex Benth. LABIATAE (D)
Marsypopetalum Scheff. ANNONACEAE (D)
Martensia Giseke = Alpinia Roxb. ZINGIBERACEAE (M)
Martha F.Muell. = Posoqueria Aubl. RUBIACEAE (D)
Marthella Urb. BURMANNIACEAE (M)
Marticorenia Crisci COMPOSITAE (D)
Martiella Tiegh. = Psittacanthus Mart. LORANTHACEAE (D)
Martinella H.Lév. (SUH) = Neomartinella Pilg. CRUCIFERAE (D)
Martinella Baill. BIGNONIACEAE (D)
Martinezia Ruíz & Pav. = Prestoea Hook.f. PALMAE (M)
Martiodendron Gleason LEGUMINOSAE–CAESALPINIOIDEAE (D)
Martiusella Pierre = Chrysophyllum L. SAPOTACEAE (D)
Martiusia Benth. = Martiodendron Gleason LEGUMINOSAE–CAESALPINIOIDEAE (D)
Martretia Beille EUPHORBIACEAE (D)
Martynia L. PEDALIACEAE (D)
Marumia Blume = Macrolenes Naudin MELASTOMATACEAE (D)
Mascagnia (Bertero ex DC.) Colla MALPIGHIACEAE (D)
Mascarena L.H.Bailey = Hyophorbe Gaertn. PALMAE (M)
Mascarenhasia A.DC. APOCYNACEAE (D)
Maschalanthe Blume = Urophyllum Wall. RUBIACEAE (D)
Maschalanthus Nutt. = Savia Willd. EUPHORBIACEAE (D)
Maschalocephalus Gilg & K.Schum. RAPATEACEAE (M)
Maschalocorymbus Bremek. RUBIACEAE (D)
Maschalodesme K.Schum. & Lauterb. RUBIACEAE (D)
Maschalosorus Van den Bosch = Trichomanes L. HYMENOPHYLLACEAE (F)
Masdevallia Ruíz & Pav. ORCHIDACEAE (M)
Masmenia F.K.Mey. = Thlaspi L. CRUCIFERAE (D)
Masoala Jum. PALMAE (M)
Massangea E.Morren = Guzmania Ruíz & Pav. BROMELIACEAE (M)
Massartina Maire = Elizaldia Willk. BORAGINACEAE (D)
Massia Balansa = Eriachne R.Br. GRAMINEAE (M)
Mussvia Becc. = Cryptocarya R.Br. LAURACEAE (D)
Massonia Thunb. ex L.f. HYACINTHACEAE (M)
Massularia (K.Schum.) Hoyle RUBIACEAE (D)
Mastersia Benth. LEGUMINOSAE–PAPILIONOIDEAE (D)
Mastersiella Gilg–Ben. RESTIONACEAE (M)
Mastichodendron (Engl.) H.J.Lam = Sideroxylon L. SAPOTACEAE (D)
Mastigosciadium Rech.f. & Kuber UMBELLIFERAE (D)
Mastigostyla I.M.Johnst. IRIDACEAE (M)
Mastixia Blume CORNACEAE (D)

Mastixiodendron Melch. RUBIACEAE (D)
Mastosuke Raf. = Ficus L. MORACEAE (D)
Matayba Aubl. SAPINDACEAE (D)
Matelea Aubl. ASCLEPIADACEAE (D)
Mathaea Vell. = Schwenckia L. SOLANACEAE (D)
Mathewsia Hook. & Arn. CRUCIFERAE (D)
Mathiasella Constance & C.Hitchc. UMBELLIFERAE (D)
Mathieua Klotzsch AMARYLLIDACEAE (M)
Mathiola R.Br. = Matthiola R.Br. CRUCIFERAE (D)
Mathiolaria Chevall. = Matthiola R.Br. CRUCIFERAE (D)
Mathurina Balf.f. TURNERACEAE (D)
Matisia Bonpl. BOMBACACEAE (D)
Matonia R.Br. ex Wall. MATONIACEAE (F)
Matrella Pers. = Zoysia Willd. GRAMINEAE (M)
Matricaria L. COMPOSITAE (D)
Matsumurella Makino = Lamium L. LABIATAE (D)
Matsumuria Hemsl. = Titanotrichum Soler. GESNERIACEAE (D)
Matteuccia Tod. WOODSIACEAE (F)
Mattfeldanthus H.Rob. & R.M.King COMPOSITAE (D)
Mattfeldia Urb. COMPOSITAE (D)
Matthaea Blume MONIMIACEAE (D)
Matthiola R.Br. CRUCIFERAE (D)
Matthiola L. (SUH) = Guettarda L. RUBIACEAE (D)
Matthissonia Raddi = Schwenckia L. SOLANACEAE (D)
Mattia Schult. = Rindera Pall. BORAGINACEAE (D)
Mattiastrum (Boiss.) Brand = Paracaryum (A.DC.) Boiss. BORAGINACEAE (D)
Mattuschkaea Schreb. = Perama Aubl. RUBIACEAE (D)
Mattuschkea Batsch (SUO) = Perama Aubl. RUBIACEAE (D)
Matucana Britton & Rose = Oreocereus (A.Berger) Riccob. CACTACEAE (D)
Matudacalamus F.Maek. = Aulonemia Goudot GRAMINEAE (M)
Matudaea Lundell HAMAMELIDACEAE (D)
Matudanthus D.R.Hunt COMMELINACEAE (M)
Matudina R.M.King & H.Rob. COMPOSITAE (D)
Maturna Raf. = Gomesa R.Br. ORCHIDACEAE (M)
Maughania J.St.-Hil. = Flemingia Roxb. ex W.T.Aiton LEGUMINOSAE–
  PAPILIONOIDEAE (D)
Maughania N.E.Br. (SUH) = Diplosoma Schwantes AIZOACEAE (D)
Maughaniella L.Bolus = Diplosoma Schwantes AIZOACEAE (D)
Maundia F.Muell. JUNCAGINACEAE (M)
Maurandella (A.Gray) Rothm. SCROPHULARIACEAE (D)
Maurandya Ortega SCROPHULARIACEAE (D)
Mauranthe O.F.Cook = Chamaedorea Willd. PALMAE (M)
Mauria Kunth ANACARDIACEAE (D)
Mauritia L.f. PALMAE (M)
Mauritiella Burret PALMAE (M)
Maurocenia Mill. CELASTRACEAE (D)
Mausolea Poljakov COMPOSITAE (D)
Maxburretia Furtado PALMAE (M)
Maxia O.E.G.Nilsson = Montia L. PORTULACACEAE (D)
Maxillaria Ruíz & Pav. ORCHIDACEAE (M)
Maximiliana Mart. PALMAE (M)
Maximovsczia A.P.Khokhr. (SUH) = Scirpus L. CYPERACEAE (M)
Maximovicziella A.P.Khokhr. = Scirpus L. CYPERACEAE (M)
Maximowiczia Cogn. (SUS) = Ibervillea Greene CUCURBITACEAE (D)
Maxonia C.Chr. DRYOPTERIDACEAE (F)

Maxwellia Baill. STERCULIACEAE (D)
Mayaca Aubl. MAYACACEAE (M)
Mayanaea Lundell = Orthion Standl. & Steyerm. VIOLACEAE (D)
Mayepea Aubl. = Chionanthus L. OLEACEAE (D)
Mayna Aubl. FLACOURTIACEAE (D)
Mayodendron Kurz = Radermachera Zoll. & Moritzi BIGNONIACEAE (D)
Mays Mill. (SUS) = Zea L. GRAMINEAE (M)
Maytenus Molina CELASTRACEAE (D)
Mayzea Raf. (SUS) = Zea L. GRAMINEAE (M)
Mazaea Krug & Urb. RUBIACEAE (D)
Mazus Lour. SCROPHULARIACEAE (D)
Mcvaughia W.R.Anderson MALPIGHIACEAE (D)
Mearnsia Merr. = Metrosideros Banks ex Gaertn. MYRTACEAE (D)
Mecardonia Ruíz & Pav. SCROPHULARIACEAE (D)
Mechowia Schinz AMARANTHACEAE (D)
Meciclis Raf. = Coryanthes Hook. ORCHIDACEAE (M)
Meckelia (Mart. ex Adr.Juss.) Griseb. = Spachea A.Juss. MALPIGHIACEAE (D)
Mecodium C.Presl ex Copel. = Hymenophyllum Sm. HYMENOPHYLLACEAE (F)
Mecomischus Coss. ex Benth. COMPOSITAE (D)
Meconella Nutt. PAPAVERACEAE (D)
Meconopsis R.Vig. PAPAVERACEAE (D)
Mecopus Benn. LEGUMINOSAE–PAPILIONOIDEAE (D)
Mecosa Blume = Platanthera Rich. ORCHIDACEAE (M)
Mecosorus Klotzsch = Microgramma C.Presl POLYPODIACEAE (F)
Mecostylis Kurz ex Teijsm. & Binn. = Macaranga Thouars EUPHORBIACEAE (D)
Mecranium Hook.f. MELASTOMATACEAE (D)
Medea Klotzsch = Croton L. EUPHORBIACEAE (D)
Medemia Wurttemb. ex H.A.Wendl. PALMAE (M)
Medeola L. CONVALLARIACEAE (M)
Mediasia Pimenov UMBELLIFERAE (D)
Medica Cothen. = Tourrettia Foug. BIGNONIACEAE (D)
Medicago L. LEGUMINOSAE–PAPILIONOIDEAE (D)
Medicosma Hook.f. RUTACEAE (D)
Medinilla Gaudich. MELASTOMATACEAE (D)
Medinillopsis Cogn. = Plethiandra Hook.f. MELASTOMATACEAE (D)
Mediocactus Britton & Rose = Disocactus Lindl. CACTACEAE (D)
Mediocalcar J.J.Sm. ORCHIDACEAE (M)
Mediocereus Frič & Kreuz. (SUO) = Disocactus Lindl. CACTACEAE (D)
Mediolobivia Backeb. = Rebutia K.Schum. CACTACEAE (D)
Mediorebutia Frič (SUI) = Rebutia K.Schum. CACTACEAE (D)
Mediusella (Cavaco) Dorr SARCOLAENACEAE (D)
Medusagyne Baker MEDUSAGYNACEAE (D)
Medusandra Brenan MEDUSANDRACEAE (D)
Medusanthera Seem. ICACINACEAE (D)
Medusather Candargy (SUS) = Hordelymus (Jess.) Jess. ex Harz GRAMINEAE (M)
Medusea Haw. = Euphorbia L. EUPHORBIACEAE (D)
Meeboldia H.Wolff UMBELLIFERAE (D)
Meeboldia Pax & K.Hoffm. (SUI) = Boscia Lam. CAPPARACEAE (D)
Meeboldina Suess. RESTIONACEAE (M)
Meehania Britton LABIATAE (D)
Meehaniopsis Kudo = Glechoma L. LABIATAE (D)
Megabaria Pierre ex Hutch. = Spondianthus Engl. EUPHORBIACEAE (D)
Megacarpaea DC. CRUCIFERAE (D)
Megacarpha Hochst. = Oxyanthus DC. RUBIACEAE (D)
Megacaryon Boiss. – Echium L. BORAGINACEAE (D)

Megaclinium Lindl. = Bulbophyllum Thouars ORCHIDACEAE (M)
Megacodon (Hemsl.) Harry Sm. GENTIANACEAE (D)
Megadenia Maxim. CRUCIFERAE (D)
Megalachne Steud. GRAMINEAE (M)
Megalastrum Holttum DRYOPTERIDACEAE (F)
Megaleranthis Ohwi = Eranthis Salisb. RANUNCULACEAE (D)
Megalobivia Y.Ito (SUI) = Echinopsis Zucc. CACTACEAE (D)
Megalochlamys Lindau ACANTHACEAE (D)
Megalodonta Greene COMPOSITAE (D)
Megalonium (Berger) Kunkel = Aeonium Webb & Berthel. CRASSULACEAE (D)
Megalopanax Ekman ARALIACEAE (D)
Megaloprotachne C.E.Hubb. GRAMINEAE (M)
Megalopus K.Schum. = Psychotria L. RUBIACEAE (D)
Megalorchis H.Perrier ORCHIDACEAE (M)
Megalostoma Leonard ACANTHACEAE (D)
Megalostylis S.Moore EUPHORBIACEAE (D)
Megalotheca F.Muell. RESTIONACEAE (M)
Megalotus Garay ORCHIDACEAE (M)
Megaphrynium Milne–Redh. MARANTACEAE (M)
Megaphyllaea Hemsl. MELIACEAE (D)
Megaphyllum Spruce ex Baill. = Pentagonia Benth. RUBIACEAE (D)
Megapleilis Raf. = Sinningia Nees GESNERIACEAE (D)
Megapterium Spach = Oenothera L. ONAGRACEAE (D)
Megarrhena Schrad. ex Nees = Androtrichum (Brongn.) Brongn. CYPERACEAE (M)
Megarrhiza Torr. & A.Gray = Marah Kellogg CUCURBITACEAE (D)
Megasea Haw. = Bergenia Moench SAXIFRAGACEAE (D)
Megaskepasma Lindau ACANTHACEAE (D)
Megastachya P.Beauv. GRAMINEAE (M)
Megastigma Hook.f. RUTACEAE (D)
Megastoma Coss. & Durieu (SUH) = Ogastemma Brummitt BORAGINACEAE (D)
Megastylis Schltr. ORCHIDACEAE (M)
Megatritheca Cristóbal STERCULIACEAE (D)
Megistostegium Hochr. MALVACEAE (D)
Megistostigma Hook.f. EUPHORBIACEAE (D)
Megozipa Raf. = Utricularia L. LENTIBULARIACEAE (D)
Mehraea Á.Löve & D.Löve = Gentiana L. GENTIANACEAE (D)
Meialisa Raf. = Adriana Gaudich. EUPHORBIACEAE (D)
Meiandra Markgr. = Alloneuron Pilg. MELASTOMATACEAE (D)
Meiena Raf. = Dendrophthoe Mart. LORANTHACEAE (D)
Meineckia Baill. EUPHORBIACEAE (D)
Meiocarpidium Engl. & Diels ANNONACEAE (D)
Meiogyne Miq. ANNONACEAE (D)
Meioluma Baill. = Micropholis (Griseb.) Pierre SAPOTACEAE (D)
Meiomeria Standl. = Chenopodium L. CHENOPODIACEAE (D)
Meionandra Gauba = Valantia L. RUBIACEAE (D)
Meionectes R.Br. = Haloragis J.R.Forst. & G.Forst. HALORAGACEAE (D)
Meionula Raf. = Utricularia L. LENTIBULARIACEAE (D)
Meiostemon Exell & Stace COMBRETACEAE (D)
Meiota O.F.Cook (SUI) = Chamaedorea Willd. PALMAE (M)
Meiracyllium Rchb.f. ORCHIDACEAE (M)
Meisneria DC. = Siphanthera Pohl MELASTOMATACEAE (D)
Melachone Gilli = Amaracarpus Blume RUBIACEAE (D)
Meladendron St.–Lag. = Melaleuca L. MYRTACEAE (D)
Meladerma Kerr ASCLEPIADACEAE (D)
Melaleuca L. MYRTACEAE (D)

Melaleucon St.-Lag. = Melaleuca L. MYRTACEAE (D)
Melampodium L. COMPOSITAE (D)
Melampyrum L. SCROPHULARIACEAE (D)
Melananthus Walp. SOLANACEAE (D)
Melancium Naudin CUCURBITACEAE (D)
Melandrium Roehl. = Vaccaria Wolf CARYOPHYLLACEAE (D)
Melandryum Rchb. = Silene L. CARYOPHYLLACEAE (D)
Melanea Pers. (SUO) = Malanea Aubl. RUBIACEAE (D)
Melanidion Greene = Christolea Cambess. ex Jacquem. CRUCIFERAE (D)
Melanium P.Browne = Cuphea P.Browne LYTHRACEAE (D)
Melanocarpum Hook.f. = Pleuropetalum Hook.f. AMARANTHACEAE (D)
Melanocenchris Nees GRAMINEAE (M)
Melanochyla Hook.f. ANACARDIACEAE (D)
Melanococca Blume = Rhus L. ANACARDIACEAE (D)
Melanocommia Ridl. = Semecarpus L.f. ANACARDIACEAE (D)
Melanodendron DC. COMPOSITAE (D)
Melanodiscus Radlk. = Glenniea Hook.f. SAPINDACEAE (D)
Melanolepis Rchb.f. & Zoll. EUPHORBIACEAE (D)
Melanoleuca St.-Lag. = Melaleuca L. MYRTACEAE (D)
Melanophylla Baker MELANOPHYLLACEAE (D)
Melanopsidium Poit. ex DC. (SUH) = Alibertia A.Rich. ex DC. RUBIACEAE (D)
Melanopsidium Colla RUBIACEAE (D)
Melanorrhoea Wall. = Gluta L. ANACARDIACEAE (D)
Melanosciadium Boissieu UMBELLIFERAE (D)
Melanoselinum Hoffm. UMBELLIFERAE (D)
Melanospermum Hilliard SCROPHULARIACEAE (D)
Melanosticta DC. = Hoffmannseggia Cav. LEGUMINOSAE–CAESALPINIOIDEAE (D)
Melanoxylon Schott LEGUMINOSAE–CAESALPINIOIDEAE (D)
Melanthera Rohr COMPOSITAE (D)
Melanthes Blume = Breynia J.R.Forst. & G.Forst. EUPHORBIACEAE (D)
Melanthesa Blume (SUO) = Breynia J.R.Forst. & G.Forst. EUPHORBIACEAE (D)
Melanthesopsis Müll.Arg. = Breynia J.R.Forst. & G.Forst. EUPHORBIACEAE (D)
Melanthium L. MELANTHIACEAE (M)
Melasma Bergius SCROPHULARIACEAE (D)
Melasphaerula Ker Gawl. IRIDACEAE (M)
Melastoma L. MELASTOMATACFAE (D)
Melastomastrum Naudin MELASTOMATACEAE (D)
Melchiora Kobuski = Balthasaria Verdc. THEACEAE (D)
Melhania Forssk. STERCULIACEAE (D)
Melia L. MELIACEAE (D)
Meliandra Ducke = Votomita Aubl. MELASTOMATACEAE (D)
Melianthus L. MELIANTHACEAE (D)
Melica L. GRAMINEAE (M)
Melichrus R.Br. EPACRIDACEAE (D)
Melicoccus P.Browne SAPINDACEAE (D)
Melicope J.R.Forst. & G.Forst. RUTACEAE (D)
Melicytus J.R.Forst. & G.Forst. VIOLACEAE (D)
Melientha Pierre OPILIACEAE (D)
Melilotus Mill. LEGUMINOSAE–PAPILIONOIDEAE (D)
Melinia Decne. ASCLEPIADACEAE (D)
Melinis P.Beauv. GRAMINEAE (M)
Melinum Link (SUH) = Zizania L. GRAMINEAE (M)
Melio–Schinzia K.Schum. = Chisocheton Blume MELIACEAE (D)
Meliopsis Rchb. = Fraxinus L. OLEACEAE (D)
Meliosma Blume MELIOSMACEAE (D)

Melissa L. LABIATAE (D)
Melissitus Medik. = Trigonella L. LEGUMINOSAE–PAPILIONOIDEAE (D)
Melittacanthus S.Moore ACANTHACEAE (D)
Melittis L. LABIATAE (D)
Mellera S.Moore ACANTHACEAE (D)
Mellichampia A.Gray ex S.Watson ASCLEPIADACEAE (D)
Melliniella Harms LEGUMINOSAE–PAPILIONOIDEAE (D)
Melliodendron Hand.–Mazz. STYRACACEAE (D)
Mellissia Hook.f. SOLANACEAE (D)
Melloa Bureau BIGNONIACEAE (D)
Mellouia Gasp. = Cucurbita L. CUCURBITACEAE (D)
Melo Mill. = Cucumis L. CUCURBITACEAE (D)
Melocactus Link & Otto CACTACEAE (D)
Melocalamus Benth. GRAMINEAE (M)
Melocanna Trin. GRAMINEAE (M)
Melochia L. STERCULIACEAE (D)
Melodinus J.R.Forst. & G.Forst. APOCYNACEAE (D)
Melodorum (Dunal) Hook.f. & Thomson (SUH) = Fissistigma Griff. ANNONACEAE (D)
Melodorum Lour. ANNONACEAE (D)
Melolobium Eckl. & Zeyh. LEGUMINOSAE–PAPILIONOIDEAE (D)
Meloneura Raf. = Utricularia L. LENTIBULARIACEAE (D)
Melongena Mill. = Solanum L. SOLANACEAE (D)
Melopepo Mill. = Cucurbita L. CUCURBITACEAE (D)
Melosperma Benth. SCROPHULARIACEAE (D)
Melothria L. CUCURBITACEAE (D)
Melothrianthus Mart.Crov. CUCURBITACEAE (D)
Melvilla A.Anderson = Cuphea P.Browne LYTHRACEAE (D)
Memecylanthus Gilg & Schltr. = Wittsteinia F.Muell. ALSEUOSMIACEAE (D)
Memecylon L. MELASTOMATACEAE (D)
Memora Miers BIGNONIACEAE (D)
Memorialis Buch–Ham. ex Wedd. = Hyrtanandra Miq. URTICACEAE (D)
Menabea Baill. ASCLEPIADACEAE (D)
Menadena Raf. = Maxillaria Ruíz & Pav. ORCHIDACEAE (M)
Menadenium Raf. = Zygosepalum Rchb.f. ORCHIDACEAE (M)
Menarda Comm. ex Juss. = Phyllanthus L. EUPHORBIACEAE (D)
Mendoncella A.D.Hawkes ORCHIDACEAE (M)
Mendoncia Vell. ex Vand. ACANTHACEAE (D)
Mendoravia Capuron LEGUMINOSAE–CAESALPINIOIDEAE (D)
Menendezia Britton = Tetrazygia Rich. ex DC. MELASTOMATACEAE (D)
Menepetalum Loes. CELASTRACEAE (D)
Menephora Raf. = Paphiopedilum Pfitzer ORCHIDACEAE (M)
Menestoria DC. (UP) RUBIACEAE (D)
Meniocus Desv. = Alyssum L. CRUCIFERAE (D)
Meniscium Schreb. THELYPTERIDACEAE (F)
Meniscogyne Gagnep. URTICACEAE (D)
Menisorus Alston THELYPTERIDACEAE (F)
Menispermum L. MENISPERMACEAE (D)
Menkea Lehm. CRUCIFERAE (D)
Menodora Bonpl. OLEACEAE (D)
Menodoropsis (A.Gray) Small = Menodora Bonpl. OLEACEAE (D)
Menonvillea R.Br. ex DC. CRUCIFERAE (D)
Mentha L. LABIATAE (D)
Mentocalyx N.E.Br. = Gibbaeum Haw. AIZOACEAE (D)
Mentodendron Lundell = Pimenta Lindl. MYRTACEAE (D)
Mentzelia L. LOASACEAE (D)

Menyanthes L. MENYANTHACEAE (D)
Menziesia Sm. ERICACEAE (D)
Meoschium P.Beauv. = Ischaemum L. GRAMINEAE (M)
Mephitidia Reinw. ex Blume = Lasianthus Jack RUBIACEAE (D)
Meranthera Tiegh. = Psittacanthus Mart. LORANTHACEAE (D)
Merathrepta Raf. (SUS) = Danthonia DC. GRAMINEAE (M)
Mercadoa Naves = Doryxylon Zoll. EUPHORBIACEAE (D)
Merciera A.DC. CAMPANULACEAE (D)
Mercurialis L. EUPHORBIACEAE (D)
Mercuriastrum Fabr. = Acalypha L. EUPHORBIACEAE (D)
Merendera Ramond COLCHICACEAE (M)
Meresaldia Bullock ASCLEPIADACEAE (D)
Meriana Vent. (SUO) = Meriania Sw. MELASTOMATACEAE (D)
Meriandra Benth. LABIATAE (D)
Meriania Sw. MELASTOMATACEAE (D)
Merianthera Kuhlm. MELASTOMATACEAE (D)
Mericarpaea Boiss. RUBIACEAE (D)
Mericocalyx Bamps = Otiophora Zucc. RUBIACEAE (D)
Meringium C.Presl = Hymenophyllum Sm. HYMENOPHYLLACEAE (F)
Meringurus Murb. = Gaudinia P.Beauv. GRAMINEAE (M)
Merinogyne H.Wolff = Angoseseli Chiov. UMBELLIFERAE (D)
Merinthopodium Donn.Sm. = Markea Rich. SOLANACEAE (D)
Merinthosorus Copel. = Aglaomorpha Schott POLYPODIACEAE (F)
Meriolix Raf. ex Endl. = Calylophus Spach ONAGRACEAE (D)
Merisachne Steud. = Triplasis P.Beauv. GRAMINEAE (M)
Merismia Tiegh. (SUH) = Psittacanthus Mart. LORANTHACEAE (D)
Merismostigma S.Moore = Coelospermum Blume RUBIACEAE (D)
Meristotropis Fisch. & C.A.Mey. = Glycyrrhiza L. LEGUMINOSAE–PAPILIONOIDEAE (D)
Merleta Raf. = Croton L. EUPHORBIACEAE (D)
Merope M.Roem. RUTACEAE (D)
Merostachys Spreng. GRAMINEAE (M)
Merostela Pierre = Aglaia Lour. MELIACEAE (D)
Merremia Dennst. ex Endl. CONVOLVULACEAE (D)
Merrillanthus Chun & Tsiang ASCLEPIADACEAE (D)
Merrillia Swingle RUTACEAE (D)
Merrilliodendron Kaneh. ICACINACEAE (D)
Merrilliopanax H.L.Li ARALIACEAE (D)
Merrittia Merr. = Blumea DC. COMPOSITAE (D)
Mertensia Willd. (SUH) = Dicranopteris Bernh. GLEICHENIACEAE (F)
Mertensia Roth BORAGINACEAE (D)
Merumea Steyerm. RUBIACEAE (D)
Merwia Fedtsch. = Ferula L. UMBELLIFERAE (D)
Merwiopsis Safina (UP) UMBELLIFERAE (D)
Merxmuellera Conert = Rytidosperma Steud. GRAMINEAE (M)
Meryta J.R.Forst. & G.Forst. ARALIACEAE (D)
Mesadenella Pabst & Garay = Stenorrhynchos Rich. ex Spreng. ORCHIDACEAE (M)
Mesadenus Schltr. = Brachystele Schltr. ORCHIDACEAE (M)
Mesandrinia Raf. = Jatropha L. EUPHORBIACEAE (D)
Mesanthemum Koern. ERIOCAULACEAE (M)
Mesanthus Nees = Cannomois P.Beauv. ex Desv. RESTIONACEAE (M)
Mesaulosperma Slooten = Itoa Hemsl. FLACOURTIACEAE (D)
Mesechinopsis Y.Ito = Echinopsis Zucc. CACTACEAE (D)
Mesechites Müll.Arg. APOCYNACEAE (D)
Mesembryanthemum L. AIZOACEAE (D)
Mesochlaena R.Br. ex J.Sm. = Sphaerostephanos J.Sm. THELYPTERIDACEAE (F)

Mesoclastes Lindl. = Luisia Gaudich. ORCHIDACEAE (M)
Mesodactylis Wall. = Apostasia Blume ORCHIDACEAE (M)
Mesoglossum Halbinger ORCHIDACEAE (M)
Mesogyne Engl. MORACEAE (D)
Mesomelaena Nees CYPERACEAE (M)
Mesona Blume LABIATAE (D)
Mesoneuron Ching (SUH) = Mesophlebion Holttum THELYPTERIDACEAE (F)
Mesopanax R.Vig. = Dendropanax Decne. & Planch. ARALIACEAE (D)
Mesophlebion Holttum THELYPTERIDACEAE (F)
Mesoptera Raf. (SUH) = Liparis Rich. ORCHIDACEAE (M)
Mesoptera Hook.f. = Psydrax Gaertn. RUBIACEAE (D)
Mesopteris Ching = Amphineuron Holttum THELYPTERIDACEAE (F)
Mesoreanthus Greene = Pleiocardia Greene CRUCIFERAE (D)
Mesosetum Steud. GRAMINEAE (M)
Mesosorus Hassk. = Dicranopteris Bernh. GLEICHENIACEAE (F)
Mesosphaerum P.Browne = Hyptis Jacq. LABIATAE (D)
Mesospinidium Rchb.f. ORCHIDACEAE (M)
Mesostemma Vved. CARYOPHYLLACEAE (D)
Mesothema C.Presl = Blechnum L. BLECHNACEAE (F)
Mespilodaphne Nees = Ocotea Aubl. LAURACEAE (D)
Mespilus L. ROSACEAE (D)
Messerschmidia Hebenstr. (SUO) = Argusia Boehm. BORAGINACEAE (D)
Messerschmidia Roem. & Schult. (SUS) = Tournefortia L. BORAGINACEAE (D)
Messersmidia L. (SUS) = Argusia Boehm. BORAGINACEAE (D)
Mestoklema N.E.Br. ex Glen AIZOACEAE (D)
Mesua L. GUTTIFERAE (D)
Metabolos Blume = Hedyotis L. RUBIACEAE (D)
Metabriggsia W.T.Wang GESNERIACEAE (D)
Metachilum Lindl. = Appendicula Blume ORCHIDACEAE (M)
Metadina Bakh.f. RUBIACEAE (D)
Metaeritrichium W.T.Wang = Eritrichium Schrad. ex Gaudin BORAGINACEAE (D)
Metalasia R.Br. COMPOSITAE (D)
Metalepis Griseb. ASCLEPIADACEAE (D)
Metalonicera M.Wang & A.G.Gu = Lonicera L. CAPRIFOLIACEAE (D)
Metanarthecium Maxim. MELANTHIACEAE (M)
Metanemone W.T.Wang RANUNCULACEAE (D)
Metapetrocosmea W.T.Wang (UP) GESNERIACEAE (D)
Metaplexis R.Br. ASCLEPIADACEAE (D)
Metapolypodium Ching = Goniophlebium C.Presl POLYPODIACEAE (F)
Metaporana N.E.Br. CONVOLVULACEAE (D)
Metarungia Baden ACANTHACEAE (D)
Metasasa Lin = Acidosasa C.D.Chu C.S.Chao GRAMINEAE (M)
Metasequoia Miki ex Hu & Cheng TAXODIACEAE (G)
Metasocratea Dugand = Socratea Karst. PALMAE (M)
Metastachydium Airy Shaw ex C.Y.Wu & H.W.Li LABIATAE (D)
Metastachys Knorring (SUH) = Metastachydium Airy Shaw ex C.Y.Wu & H.W.Li LABIATAE (D)
Metastachys (Benth.) Tiegh. = Tristerix Mart. LORANTHACEAE (D)
Metastelma R.Br. ASCLEPIADACEAE (D)
Metastevia Grashoff COMPOSITAE (D)
Metathelypteris (H.Ito) Ching THELYPTERIDACEAE (F)
Metatrophis F.B.H.Br. (UP) URTICACEAE (D)
Metaxya C.Presl METAXYACEAE (F)
Metcalfia Conert GRAMINEAE (M)
Meteoromyrtus Gamble MYRTACEAE (D)

Meterana Raf. = Caperonia A.St.-Hil. EUPHORBIACEAE (D)
Meterostachys Nakai CRASSULACEAE (D)
Metharme Phil. ex Engl. ZYGOPHYLLACEAE (D)
Methysticodendron R.E.Schult. = Brugmansia Pers. SOLANACEAE (D)
Metopium P.Browne ANACARDIACEAE (D)
Metrodorea St.-Hil. RUTACEAE (D)
Metrosideros Banks ex Gaertn. MYRTACEAE (D)
Metroxylon Rottb. PALMAE (M)
Mettenia Griseb. EUPHORBIACEAE (D)
Metteniusa Karst. ICACINACEAE (D)
Metternichia J.C.Mikan SOLANACEAE (D)
Metula Tiegh. = Phragmanthera Tiegh. LORANTHACEAE (D)
Meum Mill. UMBELLIFERAE (D)
Mexacanthus T.F.Daniel ACANTHACEAE (D)
Mexerion Nesom COMPOSITAE (D)
Mexianthus B.L.Rob. COMPOSITAE (D)
Mexicoa Garay ORCHIDACEAE (M)
Meximalva Fryxell MALVACEAE (D)
Meyenia Schltdl. (SUH) = Cestrum L. SOLANACEAE (D)
Meyenia Nees ACANTHACEAE (D)
Meyenia Backeb. (SUH) = Weberbauerocereus Backeb. CACTACEAE (D)
Meyerophytum Schwantes AIZOACEAE (D)
Meyna Roxb. ex Link RUBIACEAE (D)
Meynia Schult. (SUO) = Meyna Roxb. ex Link RUBIACEAE (D)
Mezia Schwacke ex Nied. MALPIGHIACEAE (D)
Meziella Schindl. HALORAGACEAE (D)
Mezilaurus Kuntze ex Taub. LAURACEAE (D)
Meziothamnus Harms = Abromeitiella Mez BROMELIACEAE (M)
Mezleria C.Presl = Lobelia L. CAMPANULACEAE (D)
Mezobromelia L.B.Sm. BROMELIACEAE (M)
Mezochloa Butzin = Alloteropsis C.Presl GRAMINEAE (M)
Mezoneuron Desf. = Caesalpinia L. LEGUMINOSAE-CAESALPINIOIDEAE (D)
Mezzettia Becc. ANNONACEAE (D)
Mezzettiopsis Ridl. ANNONACEAE (D)
Mibora Adans. GRAMINEAE (M)
Micagrostis Juss. (SUI) = Mibora Adans. GRAMINEAE (M)
Michauxia L'Hér. CAMPANULACEAE (D)
Michelaria Dumort. = Bromus L. GRAMINEAE (M)
Michelia L. MAGNOLIACEAE (D)
Micheliella Briq. = Collinsonia L. LABIATAE (D)
Michelsonia Hauman LEGUMINOSAE-CAESALPINIOIDEAE (D)
Micholitzia N.E.Br. ASCLEPIADACEAE (D)
Miconia Ruíz & Pav. MELASTOMATACEAE (D)
Micractis DC. COMPOSITAE (D)
Micraeschynanthus Ridl. GESNERIACEAE (D)
Micraira F.Muell. GRAMINEAE (M)
Micrampelis Raf. = Echinocystis Torr. & A.Gray CUCURBITACEAE (D)
Micrandra Benn. & R.Br. (SUH) = Hevea Aubl. EUPHORBIACEAE (D)
Micrandra Benth. EUPHORBIACEAE (D)
Micrandropsis Rodrigues EUPHORBIACEAE (D)
Micrantha Dvořák == Hesperis L. CRUCIFERAE (D)
Micranthella Naudin = Tibouchina Aubl. MELASTOMATACEAE (D)
Micranthemum Michx. SCROPHULARIACEAE (D)
Micrantheum Desf. EUPHORBIACEAE (D)
Micranthocercus Backeb. CACTACEAE (D)

Micranthus (Pers.) Eckl. IRIDACEAE (M)
Micranthus J.C.Wendl. (SUH) = Phaulopsis Willd. ACANTHACEAE (D)
Micrargeria Benth. SCROPHULARIACEAE (D)
Micrargeriella R.E.Fr. SCROPHULARIACEAE (D)
Micrasepalum Urb. RUBIACEAE (D)
Micrechites Miq. APOCYNACEAE (D)
Microbahia Cockerell = Syntrichopappus A.Gray COMPOSITAE (D)
Microbambus K.Schum. = Guaduella Franch. GRAMINEAE (M)
Microberlinia A.Chev. LEGUMINOSAE–CAESALPINIOIDEAE (D)
Microbignonia Kraenzl. = Macfadyena A.DC. BIGNONIACEAE (D)
Microbiota Kom. CUPRESSACEAE (G)
Microblepharis (Wight & Arn.) M.Roem. = Adenia Forssk. PASSIFLORACEAE (D)
Microbriza Nicora & Rugolo GRAMINEAE (M)
Microbrochis C.Presl = Tectaria Cav. DRYOPTERIDACEAE (F)
Microcachrys Hook.f. PODOCARPACEAE (G)
Microcala Hoffmanns. & Link = Cicendia Adans. GENTIANACEAE (D)
Microcalamus Gamble (SUH) = Racemobambos Holttum GRAMINEAE (M)
Microcalamus Franch. GRAMINEAE (M)
Microcardamum O.E.Schulz CRUCIFERAE (D)
Microcarpaea R.Br. SCROPHULARIACEAE (D)
Microcaryum I.M.Johnst. BORAGINACEAE (D)
Microcasia Becc. = Bucephalandra Schott ARACEAE (M)
Microcephala Pobed. COMPOSITAE (D)
Microcharis Benth. = Indigofera L. LEGUMINOSAE–PAPILIONOIDEAE (D)
Microchilus C.Presl = Erythrodes Blume ORCHIDACEAE (M)
Microchlaena Ching (SUH) = Kuniwatsukia Pic.Serm. WOODSIACEAE (F)
Microchlaena Kuntze (SUO) = Ehrharta Thunb. GRAMINEAE (M)
Microchloa R.Br. GRAMINEAE (M)
Microchonea Pierre = Trachelospermum Lem. APOCYNACEAE (D)
Microcitrus Swingle RUTACEAE (D)
Microcnemum Ung.–Sternb. CHENOPODIACEAE (D)
Micrococca Benth. EUPHORBIACEAE (D)
Micrococos Phil. = Jubaea Kunth PALMAE (M)
Microcodon A.DC. CAMPANULACEAE (D)
Microcoelia Lindl. ORCHIDACEAE (M)
Microcoelum Burret & Potztal = Lytocaryum Toledo PALMAE (M)
Microconomorpha (Mez) Lundell = Cybianthus Mart. MYRSINACEAE (D)
Microcorys R.Br. LABIATAE (D)
Microcos L. TILIACEAE (D)
Microculcas Peter = Gonatopus Hook.f. ex Engl. ARACEAE (M)
Microcybe Turcz. RUTACEAE (D)
Microcycas (Miq.) A.DC. ZAMIACEAE (G)
Microdactylon Brandegee ASCLEPIADACEAE (D)
Microderis A.DC. = Leontodon L. COMPOSITAE (D)
Microdesmis Hook.f. ex Hook. PANDACEAE (D)
Microdon Choisy SCROPHULARIACEAE (D)
Microdracoides Hua CYPERACEAE (M)
Microelus Wight & Arn. = Bischofia Blume EUPHORBIACEAE (D)
Microepidendrum Brieger (SUI) = Epidendrum L. ORCHIDACEAE (M)
Microferus C.Presl = Crypsinus C.Presl POLYPODIACEAE (F)
Microglossa DC. COMPOSITAE (D)
Microgonium C.Presl = Trichomanes L. HYMENOPHYLLACEAE (F)
Microgramma C.Presl POLYPODIACEAE (F)
Microgynella Grau COMPOSITAE (D)
Microgynoecium Hook.f. CHENOPODIACEAE (D)

Microholmesia P.J.Cribb = Angraecopsis Kraenzl. ORCHIDACEAE (M)
Microjambosa Blume = Syzygium Gaertn. MYRTACEAE (D)
Microkentia H.A.Wendl. ex Hook.f. = Basselinia Vieill. PALMAE (M)
Microlaena R.Br. = Ehrharta Thunb. GRAMINEAE (M)
Microlecane Sch.Bip. ex Benth. = Bidens L. COMPOSITAE (D)
Microlepia C.Presl DENNSTAEDTIACEAE (F)
Microlepidium F.Muell. CRUCIFERAE (D)
Microlepis (DC.) Miq. MELASTOMATACEAE (D)
Microliabum Cabrera COMPOSITAE (D)
Microlicia D.Don MELASTOMATACEAE (D)
Microlobium Liebm. = Apoplanesia C.Presl LEGUMINOSAE–PAPILIONOIDEAE (D)
Microloma R.Br. ASCLEPIADACEAE (D)
Microlonchoides Candargy = Jurinea Cass. COMPOSITAE (D)
Microluma Baill. = Pouteria Aubl. SAPOTACEAE (D)
Micromeles Decne. = Sorbus L. ROSACEAE (D)
Micromelum Blume RUTACEAE (D)
Micromeria Benth. LABIATAE (D)
Micromonolepis Ulbr. = Monolepis Schrad. CHENOPODIACEAE (D)
Micromyrtus Benth. MYRTACEAE (D)
Micromystria O.E.Schulz = Arabidella (F.Muell.) O.E.Schulz CRUCIFERAE (D)
Micronoma H.Wendl. ex Hook.f. (UP) PALMAE (M)
Micronychia Oliv. ANACARDIACEAE (D)
Micropaegma Pichon = Mussatia Bureau ex Baill. BIGNONIACEAE (D)
Micropapyrus Suess. = Rhynchospora Vahl CYPERACEAE (M)
Microparacaryum (Popov ex Riedl) Hilger = Paracaryum (A.DC.) Boiss.
    BORAGINACEAE (D)
Micropeplis Bunge = Halogeton C.A.Mey. CHENOPODIACEAE (D)
Micropera Lindl. ORCHIDACEAE (M)
Micropetalum Poit. ex Baill. = Amanoa Aubl. EUPHORBIACEAE (D)
Microphacos Rydb. = Astragalus L. LEGUMINOSAE–PAPILIONOIDEAE (D)
Microphlebodium L.D.Gómez = Pleopeltis Humb. & Bonpl. ex Willd.
    POLYPODIACEAE (F)
Micropholis (Griseb.) Pierre SAPOTACEAE (D)
Microphyes Phil. CARYOPHYLLACEAE (D)
Microphysa Naudin (SUH) = Tococa Aubl. MELASTOMATACEAE (D)
Microphysa Schrenk RUBIACEAE (D)
Microphysca Naudin = Tococa Aubl. MELASTOMATACEAE (D)
Microphytanthe (Schltr.) Brieger = Dendrobium Sw. ORCHIDACEAE (M)
Micropleura Lag. UMBELLIFERAE (D)
Microplumeria Baill. APOCYNACEAE (D)
Micropodium Mett. = Diplora Baker ASPLENIACEAE (F)
Micropogon Pfeiff. (SUS) = Microchloa R.Br. GRAMINEAE (M)
Micropolypodium Hayata = Xiphopteris Kaulf. GRAMMITIDACEAE (F)
Micropora Hook.f. LAURACEAE (D)
Micropsis DC. COMPOSITAE (D)
Micropteris Desv. = Xiphopteris Kaulf. GRAMMITIDACEAE (F)
Micropteris J.Sm. (SUH) = Polypodium L. POLYPODIACEAE (F)
Micropterum Schwantes = Cleretum N.E.Br. AIZOACEAE (D)
Micropteryx Walp. = Erythrina L. LEGUMINOSAE–PAPILIONOIDEAE (D)
Micropuntia Daston = Opuntia Mill. CACTACEAE (D)
Micropus L. COMPOSITAE (D)
Micropyropsis Rom. Zarco & Cabezudo GRAMINEAE (M)
Micropyrum (Gaudin) Link GRAMINEAE (M)
Microrhamnus A.Gray = Condalia Cav. RHAMNACEAE (D)
Microrphium C.B.Clarke GENTIANACEAE (D)

Microsaccus Blume ORCHIDACEAE (M)
Microschizaea C.F.Reed = Schizaea Sm. SCHIZAEACEAE (F)
Microschoenus C.B.Clarke = Juncus L. JUNCACEAE (M)
Microschwenkia Benth. ex Hemsl. = Melananthus Walp. SOLANACEAE (D)
Microsciadium Hook.f. (SUH) = Azorella Lam. UMBELLIFERAE (D)
Microsciadium Boiss. UMBELLIFERAE (D)
Microsechium Naudin CUCURBITACEAE (D)
Microsemia Greene CRUCIFERAE (D)
Microsemma Labill. = Lethedon Spreng. THYMELAEACEAE (D)
Microsepala Miq. = Baccaurea Lour. EUPHORBIACEAE (D)
Microseris D.Don COMPOSITAE (D)
Microsideros Baum.–Bod. (SUI) = Metrosideros Banks ex Gaertn. MYRTACEAE (D)
Microsisymbrium O.E.Schulz CRUCIFERAE (D)
Microsorum Link POLYPODIACEAE (F)
Microsperma Hook. = Eucnide Zucc. LOASACEAE (D)
Microspermia Frič = Parodia Speg. CACTACEAE (D)
Microspermum Lag. COMPOSITAE (D)
Microsplenium Hook.f. (SUI) = Machaonia Bonpl. RUBIACEAE (D)
Microstachys A.Juss. = Sebastiania Spreng. EUPHORBIACEAE (D)
Microstaphyla C.Presl = Elaphoglossum Schott ex J.Sm. LOMARIOPSIDACEAE (F)
Microstegia C.Presl = Diplazium Sw. WOODSIACEAE (F)
Microstegia Pierre ex Harms = Gilletiodendron Vermoesen LEGUMINOSAE–
    CAESALPINIOIDEAE (D)
Microstegium Nees GRAMINEAE (M)
Microstegnus C.Presl = Cyathea Sm. CYATHEACEAE (F)
Microsteira Baker MALPIGHIACEAE (D)
Microstelma Baill. ASCLEPIADACEAE (D)
Microstemma R.Br. = Brachystelma R.Br. ASCLEPIADACEAE (D)
Microstemon Engl. = Pentaspadon Hook.f. ANACARDIACEAE (D)
Microstephanus N.E.Br. = Pleurostelma Baill. ASCLEPIADACEAE (D)
Microstephium Less. = Arctotheca J.C.Wendl. COMPOSITAE (D)
Microsteris Greene POLEMONIACEAE (D)
Microstigma Trautv. = Matthiola R.Br. CRUCIFERAE (D)
Microstrobilus Bremek. = Strobilanthes Blume ACANTHACEAE (D)
Microstrobos J.Garden & L.A.S.Johnson PODOCARPACEAE (G)
Microstylis (Nutt.) Eaton = Malaxis Sol. ex Sw. ORCHIDACEAE (M)
Microtatorchis Schltr. ORCHIDACEAE (M)
Microtea Sw. PHYTOLACCACEAE (D)
Microterangis Senghas ORCHIDACEAE (M)
Microtheca Schltr. = Cynorkis Thouars ORCHIDACEAE (M)
Microthelys Garay = Brachystele Schltr. ORCHIDACEAE (M)
Microthlaspi F.K.Mey. = Thlaspi L. CRUCIFERAE (D)
Microthuareia Thouars (SUS) = Thuarea Pers. GRAMINEAE (M)
Microtis R.Br. ORCHIDACEAE (M)
Microtoena Prain LABIATAE (D)
Microtrichia DC. = Grangea Adans. COMPOSITAE (D)
Microtrichomanes (Mett.) Copel. = Crepidomanes (C.Presl) C.Presl
    HYMENOPHYLLACEAE (F)
Microtropis Wall. ex Meisn. CELASTRACEAE (D)
Microula Benth. BORAGINACEAE (D)
Mida A.Cunn. ex Endl. SANTALACEAE (D)
Miegia Pers. (SUS) = Arundinaria Michx. GRAMINEAE (M)
Miersia Lindl. ALLIACEAE (M)
Miersiella Urb. BURMANNIACEAE (M)
Miersiophyton Engl. = Rhigiocarya Miers MENISPERMACEAE (D)

Migandra O.F.Cook = Chamaedorea Willd. PALMAE (M)
Mikania Willd. COMPOSITAE (D)
Mikaniopsis Milne-Redh. COMPOSITAE (D)
Mila Britton & Rose CACTACEAE (D)
Mildbraedia Pax EUPHORBIACEAE (D)
Mildbraediochloa Butzin = Melinis P.Beauv. GRAMINEAE (M)
Mildbraediodendron Harms LEGUMINOSAE–PAPILIONOIDEAE (D)
Mildella Trevis. ADIANTACEAE (F)
Miliarium Moench (SUS) = Milium L. GRAMINEAE (M)
Miliastrum Fabr. (SUI) = Setaria P.Beauv. GRAMINEAE (M)
Milicia T.R.Sim MORACEAE (D)
Milium Adans. (SUH) = Panicum L. GRAMINEAE (M)
Milium L. GRAMINEAE (M)
Miliusa Lesch. ex A.DC. ANNONACEAE (D)
Milla Cav. ALLIACEAE (M)
Milleria L. COMPOSITAE (D)
Millettia Wight & Arn. LEGUMINOSAE–PAPILIONOIDEAE (D)
Milligania Hook.f. ASTELIACEAE (M)
Millingtonia L.f. BIGNONIACEAE (D)
Millotia Cass. COMPOSITAE (D)
Millspaughia B.L.Rob. = Gymnopodium Rolfe POLYGONACEAE (D)
Milnea Roxb. = Aglaia Lour. MELIACEAE (D)
Miltianthus Bunge ZYGOPHYLLACEAE (D)
Miltitzia A.DC. HYDROPHYLLACEAE (D)
Miltonia Lindl. ORCHIDACEAE (M)
Miltonioides Brieger & Lückel ORCHIDACEAE (M)
Miltoniopsis God.–Leb. ORCHIDACEAE (M)
Milula Prain ALLIACEAE (M)
Mimela Phil. = Leucheria Lag. COMPOSITAE (D)
Mimetanthe Greene SCROPHULARIACEAE (D)
Mimetes Salisb. PROTEACEAE (D)
Mimetophytum L.Bolus = Mitrophyllum Schwantes AIZOACEAE (D)
Mimophytum Greenm. BORAGINACEAE (D)
Mimosa L. LEGUMINOSAE–MIMOSOIDEAE (D)
Mimosopsis Britton & Rose = Mimosa L. LEGUMINOSAE–MIMOSOIDEAE (D)
Mimozyganthus Burkart LEGUMINOSAE–MIMOSOIDEAE (D)
Mimulicalyx Tsoong SCROPHULARIACEAE (D)
Mimulopsis Schweinf. ACANTHACEAE (D)
Mimulus L. SCROPHULARIACEAE (D)
Mimusops L. SAPOTACEAE (D)
Mina Cerv. = Ipomoea L. CONVOLVULACEAE (D)
Minicolumna Brieger (SUI) = Epidendrum L. ORCHIDACEAE (M)
Minkelersia M.Martens & Gal. = Phaseolus L. LEGUMINOSAE–PAPILIONOIDEAE (D)
Minquartia Aubl. OLACACEAE (D)
Minthostachys (Benth.) Spach LABIATAE (D)
Minuartia L. CARYOPHYLLACEAE (D)
Minuopsis W.A.Weber = Minuartia L. CARYOPHYLLACEAE (D)
Minuria DC. COMPOSITAE (D)
Minutalia Fenzl = Antidesma L. EUPHORBIACEAE (D)
Minutia Vell. = Chionanthus L. OLEACEAE (D)
Mionandra Griseb. MALPIGHIACEAE (D)
Miphragtes Nieuwl. (SUI) = Phragmites Adans. GRAMINEAE (M)
Miquelia Meisn. ICACINACEAE (D)
Miquelia Arn. & Nees (SUH) = Garnotia Brongn. GRAMINEAE (M)
Miquelia Blume (SUH) = Stauranthera Benth. GESNERIACEAE (D)

Miqueliopuntia Frič ex F.Ritter = Opuntia Mill. CACTACEAE (D)
Mirabella F.Ritter = Cereus Mill. CACTACEAE (D)
Mirabellia Bert. ex Baill. = Dysopsis Baill. EUPHORBIACEAE (D)
Mirabilis L. NYCTAGINACEAE (D)
Miraglossum Kupicha ASCLEPIADACEAE (D)
Mirandaceltis A.J.Sharp = Aphananthe Planch. ULMACEAE (D)
Mirandea Rzed. ACANTHACEAE (D)
Mirbelia Sm. LEGUMINOSAE–PAPILIONOIDEAE (D)
Miricacalia Kitam. COMPOSITAE (D)
Mirmau Adans. = Selaginella P.Beauv. SELAGINELLACEAE (FA)
Mirmecodia Gaudich. (SUO) = Myrmecodia Jack RUBIACEAE (D)
Mirobalanus Rumph. = Phyllanthus L. EUPHORBIACEAE (D)
Misanteca Schltdl. & Cham. = Licaria Aubl. LAURACEAE (D)
Miscanthidium Stapf = Miscanthus Andersson GRAMINEAE (M)
Miscanthus Andersson GRAMINEAE (M)
Mischobulbon Schltr. (SUO) = Mischobulbum Schltr. ORCHIDACEAE (M)
Mischobulbum Schltr. ORCHIDACEAE (M)
Mischocarpus Blume SAPINDACEAE (D)
Mischocodon Radlk. = Mischocarpus Blume SAPINDACEAE (D)
Mischodon Thwaites EUPHORBIACEAE (D)
Mischogyne Exell ANNONACEAE (D)
Mischophloeus Scheff. = Areca L. PALMAE (M)
Mischopleura Wernham ex Ridl. ERICACEAE (D)
Misiessya Wedd. = Leucosyke Zoll. & Moritzi URTICACEAE (D)
Misodendron G.Don (SUO) = Misodendrum Banks ex DC. MISODENDRACEAE (D)
Misodendrum Banks ex DC. MISODENDRACEAE (D)
Misopates Raf. SCROPHULARIACEAE (D)
Mitchella L. RUBIACEAE (D)
Mitella L. SAXIFRAGACEAE (D)
Mitellastra Howell = Mitella L. SAXIFRAGACEAE (D)
Mitellopsis Meisn. = Mitella L. SAXIFRAGACEAE (D)
Mitolepis Balf.f. ASCLEPIADACEAE (D)
Mitopetalum Blume = Tainia Blume ORCHIDACEAE (M)
Mitophyllum Greene CRUCIFERAE (D)
Mitostemma Mast. PASSIFLORACEAE (D)
Mitostigma Blume (SUH) = Amitostigma Schltr. ORCHIDACEAE (M)
Mitostigma Decne. ASCLEPIADACEAE (D)
Mitozus Miers = Mandevilla Lindl. APOCYNACEAE (D)
Mitracarpus Zucc. RUBIACEAE (D)
Mitragyna Korth. RUBIACEAE (D)
Mitranthes O.Berg MYRTACEAE (D)
Mitrantia Peter G.Wilson & B.Hyland MYRTACEAE (D)
Mitraria Cav. GESNERIACEAE (D)
Mitrasacme Labill. LOGANIACEAE (D)
Mitrasacmopsis Jovet RUBIACEAE (D)
Mitrastemma Makino (SUI) = Mitrastemon Makino RAFFLESIACEAE (D)
Mitrastemon Makino RAFFLESIACEAE (D)
Mitrastigma Harv. = Psydrax Gaertn. RUBIACEAE (D)
Mitrastylus Alm & T.C.E.Fr. = Erica L. ERICACEAE (D)
Mitratheca K.Schum. = Oldenlandia L. RUBIACEAE (D)
Mitrella Miq. ANNONACEAE (D)
Mitreola L. LOGANIACEAE (D)
Mitreola Boehm. (SUH) = Ophiorrhiza L. RUBIACEAE (D)
Mitrephora (Blume) Hook.f. & Thomson ANNONACEAE (D)
Mitriostigma Hochst. RUBIACEAE (D)

Mitrocereus (Backeb.) Backeb. = Pachycereus (A.Berger) Britton & Rose CACTACEAE (D)
Mitrophyllum Schwantes AIZOACEAE (D)
Mitropsidium Burret = Psidium L. MYRTACEAE (D)
Mitrosicyos Maxim. = Actinostemma Griff. CUCURBITACEAE (D)
Mitwabochloa J.B.Phipps = Zonotriche (C.E.Hubb.) J.B.Phipps GRAMINEAE (M)
Mixandra Pierre = Diploknema Pierre SAPOTACEAE (D)
Miyakea Miyabe & Tatew. RANUNCULACEAE (D)
Miyamayomena Kitam. = Aster L. COMPOSITAE (D)
Mizonia A.Chev. = Pancratium L. AMARYLLIDACEAE (M)
Mkilua Verdc. ANNONACEAE (D)
Mnesithea Kunth GRAMINEAE (M)
Mnianthus Walp. = Dalzellia Wight PODOSTEMACEAE (D)
Mniarum J.R.Forst. & G.Forst. = Scleranthus L. ILLECEBRACEAE (D)
Mniochloa Chase GRAMINEAE (M)
Mniodes (A.Gray) Benth. COMPOSITAE (D)
Mniopsis Mart. & Zucc. PODOSTEMACEAE (D)
Mniothamnea (Oliv.) Nied. BRUNIACEAE (D)
Moacroton Croizat EUPHORBIACEAE (D)
Mobilabium Rupp ORCHIDACEAE (M)
Mocquerysia Hua FLACOURTIACEAE (D)
Modecca Lam. = Adenia Forssk. PASSIFLORACEAE (D)
Modestia Charadze & Tamamsch. = Anacantha (Iljin) Soják COMPOSITAE (D)
Modiola Moench MALVACEAE (D)
Modiolastrum K.Schum. MALVACEAE (D)
Moehringia L. CARYOPHYLLACEAE (D)
Moenchia Ehrh. CARYOPHYLLACEAE (D)
Moenchia Steud. (SUI) = Paspalum L. GRAMINEAE (M)
Moerenhoutia Blume ORCHIDACEAE (M)
Moeroris Raf. = Phyllanthus L. EUPHORBIACEAE (D)
Moghania J.St.–Hil. = Flemingia Roxb. ex W.T.Aiton LEGUMINOSAE–
    PAPILIONOIDEAE (D)
Mogiphanes Mart. = Alternanthera Forssk. AMARANTHACEAE (D)
Mogoltavia Korovin UMBELLIFERAE (D)
Mogorium Juss. = Jasminum L. OLEACEAE (D)
Mohavea A.Gray SCROPHULARIACEAE (D)
Mohlana Mart. = Hilleria Vell. PHYTOLACCACEAE (D)
Mohria Sw. SCHIZAEACEAE (F)
Moldavica Fabr. = Dracocephalum L. LABIATAE (D)
Moldenhauera Spreng. = Pyrenacantha Wight ICACINACEAE (D)
Moldenhawera Schrad. LEGUMINOSAE–CAESALPINIOIDEAE (D)
Moldenkea Traub = Hippeastrum Herb. AMARYLLIDACEAE (M)
Moldenkeanthus Morat = Paepalanthus Kunth ERIOCAULACEAE (M)
Molina Gay = Dysopsis Baill. EUPHORBIACEAE (D)
Molinadendron P.K.Endress HAMAMELIDACEAE (D)
Molinaea Bertero (SUH) = Jubaea Kunth PALMAE (M)
Molinaea Comm. ex Juss. SAPINDACEAE (D)
Molinerella Rouy = Periballia Trin. GRAMINEAE (M)
Molineria Parl. (SUH) = Periballia Trin. GRAMINEAE (M)
Molineria Colla HYPOXIDACEAE (M)
Molinia Schrank GRAMINEAE (M)
Moliniopsis Gand. (SUI) = Kengia Packer GRAMINEAE (M)
Moliniopsis Hayata = Molinia Schrank GRAMINEAE (M)
Mollera O.Hoffm. = Calostephane Benth. COMPOSITAE (D)
Mollia Mart. TILIACEAE (D)
Mollia J.F.Gmel. (SUH) = Baeckea L. MYRTACEAE (D)

Mollinedia Ruíz & Pav. MONIMIACEAE (D)
Mollugo L. MOLLUGINACEAE (D)
Molongum Pichon APOCYNACEAE (D)
Molopanthera Turcz. RUBIACEAE (D)
Molopospermum W.D.J.Koch UMBELLIFERAE (D)
Moltkia Lehm. BORAGINACEAE (D)
Moltkiopsis I.M.Johnst. BORAGINACEAE (D)
Moluccella L. LABIATAE (D)
Mommsenia Urb. & Ekman MELASTOMATACEAE (D)
Momordica L. CUCURBITACEAE (D)
Mona O.E.G.Nilsson PORTULACACEAE (D)
Monachanthus Lindl. = Catasetum Rich. ex Kunth ORCHIDACEAE (M)
Monachather Steud. GRAMINEAE (M)
Monachne P.Beauv. = Panicum L. GRAMINEAE (M)
Monachochlamys Baker = Mendoncia Vell. ex Vand. ACANTHACEAE (D)
Monachosorella Hayata = Ptilopteris Hance ADIANTACEAE (F)
Monachosorum Kunze ADIANTACEAE (F)
Monachyron Parl. = Melinis P.Beauv. GRAMINEAE (M)
Monactineirma Bory = Passiflora L. PASSIFLORACEAE (D)
Monactis Kunth COMPOSITAE (D)
Monadelphanthus Karst. = Capirona Spruce RUBIACEAE (D)
Monadenia Lindl. ORCHIDACEAE (M)
Monadenium Pax EUPHORBIACEAE (D)
Monandraira Desv. = Deschampsia P.Beauv. GRAMINEAE (M)
Monandriella Engl. = Ledermanniella Engl. PODOSTEMACEAE (D)
Monanthella A.Berger = Rosularia (DC.) Stapf CRASSULACEAE (D)
Monanthes Haw. CRASSULACEAE (D)
Monanthochloe Engelm. GRAMINEAE (M)
Monanthocitrus Tanaka RUTACEAE (D)
Monanthos (Schltr.) Brieger = Dendrobium Sw. ORCHIDACEAE (M)
Monanthotaxis Baill. ANNONACEAE (D)
Monarda L. LABIATAE (D)
Monardella Benth. LABIATAE (D)
Monarrhenus Cass. COMPOSITAE (D)
Monarthrocarpus Merr. = Desmodium Desv. LEGUMINOSAE–PAPILIONOIDEAE (D)
Monathera Raf. (SUS) = Ctenium Panz. GRAMINEAE (M)
Mondia Skeels ASCLEPIADACEAE (D)
Monebia L'Hér. = Azima Lam. SALVADORACEAE (D)
Monechma Hochst. ACANTHACEAE (D)
Monelytrum Hack. ex Schinz GRAMINEAE (M)
Monenteles Labill. COMPOSITAE (D)
Monerma P.Beauv. (SUS) = Lepturus R.Br. GRAMINEAE (M)
Moneses Salisb. ex Gray ERICACEAE (D)
Monguia Chapel. ex Baill. = Croton L. EUPHORBIACEAE (D)
Moniera Loefl. (SUO) = Monnieria Loefl. RUTACEAE (D)
Moniera P.Browne = Bacopa Aubl. SCROPHULARIACEAE (D)
Monilaria Schwantes AIZOACEAE (D)
Monilia Gray (SUO) = Molinia Schrank GRAMINEAE (M)
Monimia Thouars MONIMIACEAE (D)
Monimiastrum Guého & A.J.Scott MYRTACEAE (D)
Monimopetalum Rehder CELASTRACEAE (D)
Monium Stapf = Anadelphia Hack. GRAMINEAE (M)
Monixus Finet = Angraecum Bory ORCHIDACEAE (M)
Monizia Lowe UMBELLIFERAE (D)
Monnieria Loefl. RUTACEAE (D)

Monnina Ruíz & Pav. POLYGALACEAE (D)
Monocardia Pennell SCROPHULARIACEAE (D)
Monocarpia Miq. ANNONACEAE (D)
Monocelastrus F.T.Wang & T.Tang = Celastrus L. CELASTRACEAE (D)
Monocephalium S.Moore = Pyrenacantha Wight ICACINACEAE (D)
Monocera Elliott (SUS) = Ctenium Panz. GRAMINEAE (M)
Monocera Jack = Elaeocarpus L. ELAEOCARPACEAE (D)
Monochaete Döll = Gymnopogon P.Beauv. GRAMINEAE (M)
Monochaetum (DC.) Naudin MELASTOMATACEAE (D)
Monochasma Maxim. ex Franch. & Sav. SCROPHULARIACEAE (D)
Monochilus Wall. ex Lindl. (SUH) = Zeuxine Lindl. ORCHIDACEAE (M)
Monochilus Fisch. & C.A.Mey. VERBENACEAE (D)
Monochlaena Gaudich. = Didymochlaena Desv. DRYOPTERIDACEAE (F)
Monochoria C.Presl PONTEDERIACEAE (M)
Monocladus L.C.Chia & H.L.Fung = Bambusa Schreb. GRAMINEAE (M)
Monococcus F.Muell. PHYTOLACCACEAE (D)
Monocosmia Fenzl PORTULACACEAE (D)
Monocostus K.Schum. COSTACEAE (M)
Monocyclanthus Keay ANNONACEAE (D)
Monocymbium Stapf GRAMINEAE (M)
Monodia S.W.L.Jacobs GRAMINEAE (M)
Monodiella Maire (UP) GENTIANACEAE (D)
Monodora Dunal ANNONACEAE (D)
Monogereion G.M.Barroso & R.M.King COMPOSITAE (D)
Monogonia C.Presl = Pteris L. PTERIDACEAE (F)
Monogramma Comm. ex Schkuhr VITTARIACEAE (F)
Monolena Triana MELASTOMATACEAE (D)
Monolepis Schrad. CHENOPODIACEAE (D)
Monolopia DC. COMPOSITAE (D)
Monomelangium Hayata = Diplazium Sw. WOODSIACEAE (F)
Monomeria Lindl. ORCHIDACEAE (M)
Monomesia Raf. (SUS) = Tiquilia Pers. BORAGINACEAE (D)
Monoon Miq. = Polyalthia Blume ANNONACEAE (D)
Monopanax Regel = Oreopanax Decne. & Planch. ARALIACEAE (D)
Monopera Barringer SCROPHULARIACEAE (D)
Monopetalanthus Harms LEGUMINOSAE–CAESALPINIOIDEAE (D)
Monopholis S.F.Blake COMPOSITAE (D)
Monophrynium K.Schum. MARANTACEAE (M)
Monophyllaea R.Br. GESNERIACEAE (D)
Monophyllanthe K.Schum. MARANTACEAE (M)
Monophyllorchis Schltr. ORCHIDACEAE (M)
Monoplegma Piper = Oxyrhynchus Brandegee LEGUMINOSAE–PAPILIONOIDEAE (D)
Monopogon C.Presl = Tristachya Nees GRAMINEAE (M)
Monoporandra Thwaites = Stemonoporus Thwaites DIPTEROCARPACEAE (D)
Monoporus A.DC. MYRSINACEAE (D)
Monopsis Salisb. CAMPANULACEAE (D)
Monopteryx Spruce ex Benth. LEGUMINOSAE–PAPILIONOIDEAE (D)
Monoptilon Torr. & A.Gray COMPOSITAE (D)
Monopyle Moritz ex Benth. GESNERIACEAE (D)
Monopyrena Speg. VERBENACEAE (D)
Monorchis Agosti = Herminium L. ORCHIDACEAE (M)
Monosalpinx N.Hallé RUBIACEAE (D)
Monoschisma Brenan = Pseudopiptadenia Rauschert LEGUMINOSAE–
    MIMOSOIDEAE (D)
Monosemion Raf. = Amorpha L. LEGUMINOSAE–PAPILIONOIDEAE (D)

Monosepalum Schltr. ORCHIDACEAE (M)
Monosoma Griff. = Xylocarpus J.König MELIACEAE (D)
Monostachya Merr. = Rytidosperma Steud. GRAMINEAE (M)
Monostemon Henrard (SUI) = Microbriza Nicora & Rugolo GRAMINEAE (M)
Monostylis Tul. PODOSTEMACEAE (D)
Monotagma K.Schum. MARANTACEAE (M)
Monotaxis Brongn. EUPHORBIACEAE (D)
Monotes A.DC. DIPTEROCARPACEAE (D)
Monotheca A.DC. = Sideroxylon L. SAPOTACEAE (D)
Monothecium Hochst. ACANTHACEAE (D)
Monotoca R.Br. EPACRIDACEAE (D)
Monotrema Koern. RAPATEACEAE (M)
Monotris Lindl. = Holothrix Rich. ex Lindl. ORCHIDACEAE (M)
Monotropa L. ERICACEAE (D)
Monotropanthum Andres = Monotropastrum Andres ERICACEAE (D)
Monotropastrum Andres ERICACEAE (D)
Monotropsis Schwein. ex Elliott ERICACEAE (D)
Monoxora Wight = Rhodamnia Jack MYRTACEAE (D)
Monroa Torr. (SUO) = Munroa Torr. GRAMINEAE (M)
Monrosia Grondona = Polygala L. POLYGALACEAE (D)
Monsonia L. GERANIACEAE (D)
Monstera Adans. ARACEAE (M)
Montagueia Baker f. = Polyscias J.R.Forst. & G.Forst. ARALIACEAE (D)
Montamans Dwyer RUBIACEAE (D)
Montanoa Cerv. COMPOSITAE (D)
Montbretia DC. = Tritonia Ker–Gawl. IRIDACEAE (M)
Montbretiopsis L.Bolus = Tritonia Ker–Gawl. IRIDACEAE (M)
Monteiroa Krapov. MALVACEAE (D)
Montejacquia Roberty = Jacquemontia Choisy CONVOLVULACEAE (D)
Montelia A.Gray = Amaranthus L. AMARANTHACEAE (D)
Montezuma DC. MALVACEAE (D)
Montia L. PORTULACACEAE (D)
Montiastrum Rydb. = Montia L. PORTULACACEAE (D)
Montinia Thunb. MONTINIACEAE (D)
Montiopsis Kuntze PORTULACACEAE (D)
Montolivaea Rchb.f. = Habenaria Willd. ORCHIDACEAE (M)
Montrichardia Crueg. ARACEAE (M)
Montrouziera Planch. & Triana GUTTIFERAE (D)
Monttea Gay SCROPHULARIACEAE (D)
Monustes Raf. = Spiranthes Rich. ORCHIDACEAE (M)
Monvillea Britton & Rose = Acanthocereus (Engelm. ex A.Berger) Britton & Rose CACTACEAE (D)
Moonia Arn. COMPOSITAE (D)
Moorcroftia Choisy = Argyreia Lour. CONVOLVULACEAE (D)
Moorea Rolfe = Neomoorea Rolfe ORCHIDACEAE (M)
Moorea Lem. (SUH) = Cortaderia Stapf GRAMINEAE (M)
Mooria Montrouz. = Cloezia Brongn. & Gris MYRTACEAE (D)
Mopania Lundell = Manilkara Adans. SAPOTACEAE (D)
Moparia Britton & Rose = Caesalpinia L. LEGUMINOSAE–CAESALPINIOIDEAE (D)
Moquilea Aubl. = Licania Aubl. CHRYSOBALANACEAE (D)
Moquinia DC. COMPOSITAE (D)
Moquinia Spreng. (SUH) = Moquiniella Balle LORANTHACEAE (D)
Moquiniella Balle LORANTHACEAE (D)
Mora Benth. LEGUMINOSAE–CAESALPINIOIDEAE (D)
Moraea Mill. IRIDACEAE (M)

Morangaya G.D.Rowley = Echinocereus Engelm. CACTACEAE (D)
Moratia H.E.Moore PALMAE (M)
Morawetzia Backeb. = Oreocereus (A.Berger) Riccob. CACTACEAE (D)
Morelia A.Rich. ex DC. RUBIACEAE (D)
Morelotia Gaudich. CYPERACEAE (M)
Morenia Ruíz & Pav. = Chamaedorea Willd. PALMAE (M)
Morettia DC. CRUCIFERAE (D)
Morgania R.Br. = Stemodia L. SCROPHULARIACEAE (D)
Moricandia DC. CRUCIFERAE (D)
Moriera Boiss. CRUCIFERAE (D)
Morierina Vieill. RUBIACEAE (D)
Morina L. MORINACEAE (D)
Morinda L. RUBIACEAE (D)
Morindopsis Hook.f. RUBIACEAE (D)
Moringa Adans. MORINGACEAE (D)
Morisia J.Gay CRUCIFERAE (D)
Morisonia L. CAPPARACEAE (D)
Morithamnus R.M.King, H.Rob. & G.M.Barroso COMPOSITAE (D)
Moritzia DC. ex Meisn. BORAGINACEAE (D)
Morkillia Rose & Painter ZYGOPHYLLACEAE (D)
Morleya Woodson = Mortoniella Woodson APOCYNACEAE (D)
Mormodes Lindl. ORCHIDACEAE (M)
Mormolyca Fenzl ORCHIDACEAE (M)
Morocarpus Siebold & Zucc. (SUS) = Debregeasia Gaudich. URTICACEAE (D)
Morolobium Kosterm. = Archidendron F.Muell. LEGUMINOSAE–MIMOSOIDEAE (D)
Morongia Britton = Mimosa L. LEGUMINOSAE–MIMOSOIDEAE (D)
Moronobea Aubl. GUTTIFERAE (D)
Morrenia Lindl. ASCLEPIADACEAE (D)
Morrisiella Aellen = Atriplex L. CHENOPODIACEAE (D)
Morsacanthus Rizzini ACANTHACEAE (D)
Mortonia A.Gray CELASTRACEAE (D)
Mortoniella Woodson APOCYNACEAE (D)
Mortoniodendron Standl. & Steyerm. TILIACEAE (D)
Mortoniopteris Pic.Serm. = Crepidomanes (C.Presl) C.Presl HYMENOPHYLLACEAE (F)
Morus L. MORACEAE (D)
Moscharia Ruíz & Pav. COMPOSITAE (D)
Moschopsis Phil. CALYCERACEAE (D)
Moschosma Rchb. = Basilicum Moench LABIATAE (D)
Moschoxylum A.Juss. = Trichilia P.Browne MELIACEAE (D)
Mosdenia Stent GRAMINEAE (M)
Moseleya Hemsl. = Ellisiophyllum Maxim. SCROPHULARIACEAE (D)
Mosenodendron R.E.Fr. = Hornschuchia Nees ANNONACEAE (D)
Mosheovia Eig = Scrophularia L. SCROPHULARIACEAE (D)
Mosiera Small MYRTACEAE (D)
Mosla (Benth.) Buch.–Ham. ex Maxim. LABIATAE (D)
Mosquitoxylum Krug & Urb. ANACARDIACEAE (D)
Mossia N.E.Br. AIZOACEAE (D)
Mostacillastrum O.E.Schulz CRUCIFERAE (D)
Mostuea Didr. LOGANIACEAE (D)
Motandra A.DC. APOCYNACEAE (D)
Motherwellia F.Muell. ARALIACEAE (D)
Motleyia Johansson RUBIACEAE (D)
Moulinsia Raf. (SUH) = Aristida L. GRAMINEAE (M)
Moullava Adans. LEGUMINOSAE–CAESALPINIOIDEAE (D)
Moultonia Balf.f. & W.W.Sm. = Monophyllaea R.Br. GESNERIACEAE (D)

Moultonianthus Merr. EUPHORBIACEAE (D)
Mourera Aubl. PODOSTEMACEAE (D)
Mouretia Pit. RUBIACEAE (D)
Mouriri Aubl. MELASTOMATACEAE (D)
Mouriria Juss. (SUO) = Mouriri Aubl. MELASTOMATACEAE (D)
Mouroucoa Aubl. = Maripa Aubl. CONVOLVULACEAE (D)
Moussonia Regel GESNERIACEAE (D)
Moutabea Aubl. POLYGALACEAE (D)
Moya Griseb. = Maytenus Molina CELASTRACEAE (D)
Mozartia Urb. = Myrcia DC. ex Guill. MYRTACEAE (D)
Mozinna Ortega = Jatropha L. EUPHORBIACEAE (D)
Msuata O.Hoffm. COMPOSITAE (D)
Muantijamvella J.B.Phipps = Tristachya Nees GRAMINEAE (M)
Muantum Pichon = Beaumontia Wall. APOCYNACEAE (D)
Mucizonia (DC.) A.Berger CRASSULACEAE (D)
Mucoa Zarucchi APOCYNACEAE (D)
Mucronea Benth. POLYGONACEAE (D)
Mucuna Adans. LEGUMINOSAE–PAPILIONOIDEAE (D)
Muehlbergella Feer = Edraianthus (A.DC.) DC. CAMPANULACEAE (D)
Muehlenbeckia Meisn. POLYGONACEAE (D)
Muellera L.f. LEGUMINOSAE–PAPILIONOIDEAE (D)
Muelleranthus Hutch. LEGUMINOSAE–PAPILIONOIDEAE (D)
Muellerargia Cogn. CUCURBITACEAE (D)
Muellerina Tiegh. LORANTHACEAE (D)
Muellerolimon Lincz. PLUMBAGINACEAE (D)
Muenteria Seem. = Markhamia Seem. ex Baill. BIGNONIACEAE (D)
Muhlenbergia Schreb. GRAMINEAE (M)
Muilla S.Watson ex Benth. ALLIACEAE (M)
Muiria N.E.Br. AIZOACEAE (D)
Muiriantha C.A.Gardner RUTACEAE (D)
Mukdenia Koldz. SAXIFRAGACEAE (D)
Mukia Arn. CUCURBITACEAE (D)
Mulgedium Cass. COMPOSITAE (D)
Mulinum Pers. UMBELLIFERAE (D)
Multidentia Gilli RUBIACEAE (D)
Muluorchis J.J.Wood = Tropidia Lindl. ORCHIDACEAE (M)
Munbya Boiss. = Arnebia Forssk. BORAGINACEAE (D)
Mundia Kunth (SUS) = Nylandtia Dumort. POLYGALACEAE (D)
Mundulea (DC.) Benth. LEGUMINOSAE–PAPILIONOIDEAE (D)
Mungos Adans. = Ophiorrhiza L. RUBIACEAE (D)
Munnozia Ruíz & Pav. COMPOSITAE (D)
Munroa Torr. GRAMINEAE (M)
Munroidendron Sherff ARALIACEAE (D)
Munronia Wight MELIACEAE (D)
Muntafara Pichon = Tabernaemontana L. APOCYNACEAE (D)
Muntingia L. TILIACEAE (D)
Munzothamnus Raven COMPOSITAE (D)
Muraltia DC. POLYGALACEAE (D)
Muratina Maire = Salsola L. CHENOPODIACEAE (D)
Murbeckiella Rothm. CRUCIFERAE (D)
Murchisonia Brittan ANTHERICACEAE (M)
Murdannia Royle COMMELINACEAE (M)
Muretia Boiss. = Elaeosticta Fenzl UMBELLIFERAE (D)
Muricaria Desv. CRUCIFERAE (D)
Muricia Lour. = Momordica L. CUCURBITACEAE (D)

Muricococcum Chun & How = Cephalomappa Baill. EUPHORBIACEAE (D)
Muriea Hartog = Manilkara Adans. SAPOTACEAE (D)
Murieanthe (Baill.) Aubrév. = Manilkara Adans. SAPOTACEAE (D)
Murraya L. RUTACEAE (D)
Murrinea Raf. = Baeckea L. MYRTACEAE (D)
Murtekias Raf. = Euphorbia L. EUPHORBIACEAE (D)
Murtonia Craib = Desmodium Desv. LEGUMINOSAE–PAPILIONOIDEAE (D)
Murucuja Medik. = Passiflora L. PASSIFLORACEAE (D)
Musa L. MUSACEAE (M)
Musanga C.Sm. ex R.Br. CECROPIACEAE (D)
Muscari Mill. HYACINTHACEAE (M)
Muscarimia Kostel. HYACINTHACEAE (M)
Muschleria S.Moore COMPOSITAE (D)
Musella (Franch.) C.Y.Wu = Musa L. MUSACEAE (M)
Museniopsis (A.Gray) J.M.Coult. & Rose = Tauschia Schltdl. UMBELLIFERAE (D)
Musgravea F.Muell. PROTEACEAE (D)
Musilia Velen. = Rhanterium Desf. COMPOSITAE (D)
Musineon Raf. UMBELLIFERAE (D)
Mussaenda L. RUBIACEAE (D)
Mussaendopsis Baill. RUBIACEAE (D)
Mussatia Bureau ex Baill. BIGNONIACEAE (D)
Musschia Dumort. CAMPANULACEAE (D)
Mustelia Steud. (SUS) = Chusquea Kunth GRAMINEAE (M)
Mutisia L.f. COMPOSITAE (D)
Myagropsis Hort. ex O.E.Schulz = Sobolewskia M.Bieb. CRUCIFERAE (D)
Myagrum L. CRUCIFERAE (D)
Myanthus Lindl. = Catasetum Rich. ex Kunth ORCHIDACEAE (M)
Mycaranthes Blume (SUO)= Eria Lindl. ORCHIDACEAE (M)
Mycaridanthes Blume = Eria Lindl. ORCHIDACEAE (M)
Mycelis Cass. COMPOSITAE (D)
Mycerinus A.C.Sm. ERICACEAE (D)
Mycetia Reinw. RUBIACEAE (D)
Myconia Lapeyr. = Ramonda Rich. GESNERIACEAE (D)
Mygalurus Link (SUS) = Vulpia C.C.Gmelin GRAMINEAE (M)
Myginda Jacq. = Crossopetalum P.Browne CELASTRACEAE (D)
Myladenia Airy Shaw EUPHORBIACEAE (D)
Myllanthus R.S.Cowan = Raputia Aubl. RUTACEAE (D)
Myoda Lindl. = Ludisia A.Rich. ORCHIDACEAE (M)
Myodium Salisb. = Ophrys L. ORCHIDACEAE (M)
Myodocarpus Brongn. & Gris ARALIACEAE (D)
Myonima Comm. ex Juss. RUBIACEAE (D)
Myopordon Boiss. COMPOSITAE (D)
Myoporum Banks & Sol. ex G.Forst. MYOPORACEAE (D)
Myopsia C.Presl = Heterotoma Zucc. CAMPANULACEAE (D)
Myoschilos Ruíz & Pav. SANTALACEAE (D)
Myosotidium Hook. BORAGINACEAE (D)
Myosotis L. BORAGINACEAE (D)
Myosoton Moench CARYOPHYLLACEAE (D)
Myospyrum Lindl. (SUO) = Myxopyrum Blume OLEACEAE (D)
Myosurus L. RANUNCULACEAE (D)
Myoxanthus Poepp. & Endl. ORCHIDACEAE (M)
Myracrodruon Allemão = Astronium Jacq. ANACARDIACEAE (D)
Myrceugenella Kausel = Luma A.Gray MYRTACEAE (D)
Myrceugenia O.Berg MYRTACEAE (D)
Myrcia DC. ex Guill. MYRTACEAE (D)

Myrcialeucus Rojas Acosta = Eugenia L. MYRTACEAE (D)
Myrcianthes O.Berg MYRTACEAE (D)
Myrciaria O.Berg MYRTACEAE (D)
Myrciariopsis Kausel = Myrciaria O.Berg MYRTACEAE (D)
Myriachaeta Moritzi = Thysanolaena Nees GRAMINEAE (M)
Myriactis Less. COMPOSITAE (D)
Myrialepis Becc. PALMAE (M)
Myrianthemum Gilg = Medinilla Gaudich. MELASTOMATACEAE (D)
Myrianthus P.Beauv. CECROPIACEAE (D)
Myriaspora DC. MELASTOMATACEAE (D)
Myrica L. MYRICACEAE (D)
Myricanthe Airy Shaw EUPHORBIACEAE (D)
Myricaria Desv. TAMARICACEAE (D)
Myrinia Lilja = Fuchsia L. ONAGRACEAE (D)
Myriocarpa Benth. URTICACEAE (D)
Myriocephalus Benth. COMPOSITAE (D)
Myriocladus Swallen GRAMINEAE (M)
Myriodon (Copel.) Copel. = Hymenophyllum Sm. HYMENOPHYLLACEAE (F)
Myriogomphus Didr. = Croton L. EUPHORBIACEAE (D)
Myrioneuron R.Br. ex Hook. RUBIACEAE (D)
Myriophyllum L. HALORAGACEAE (D)
Myriopteris Fée = Cheilanthes Sw. ADIANTACEAE (F)
Myriopteron Griff. ASCLEPIADACEAE (D)
Myriostachya (Benth.) Hook.f. GRAMINEAE (M)
Myriotheca Comm. ex Juss. = Marattia Sw. MARATTIACEAE (F)
Myripnois Bunge COMPOSITAE (D)
Myristica Gronov. MYRISTICACEAE (D)
Myrmechis (Lindl.) Blume ORCHIDACEAE (M)
Myrmecodendron Britton & Rose = Acacia Mill. LEGUMINOSAE–MIMOSOIDEAE (D)
Myrmecodia Jack RUBIACEAE (D)
Myrmecoides Elmer (SUO) = Myrmecodia Jack RUBIACEAE (D)
Myrmeconauclea Merr. RUBIACEAE (D)
Myrmecophila Rolfe ORCHIDACEAE (M)
Myrmecophila Christ ex Nakai (SUH) = Lecanopteris Reinw. POLYPODIACEAE (F)
Myrmecopteris Pic.Serm. = Lecanopteris Reinw. POLYPODIACEAE (F)
Myrmecosicyos C.Jeffrey CUCURBITACEAE (D)
Myrmecostylum C.Presl = Hymenophyllum Sm. HYMENOPHYLLACEAE (F)
Myrmedoma Becc. = Myrmephytum Becc. RUBIACEAE (D)
Myrmephytum Becc. RUBIACEAE (D)
Myrmidone Mart. MELASTOMATACEAE (D)
Myrobalanifera Houtt. = Terminalia L. COMBRETACEAE (D)
Myrobalanus Gaertn. = Terminalia L. COMBRETACEAE (D)
Myrobroma Salisb. = Vanilla Mill. ORCHIDACEAE (M)
Myrocarpus Allemão LEGUMINOSAE–PAPILIONOIDEAE (D)
Myrosma L.f. MARANTACEAE (M)
Myrosmodes Rchb.f. ORCHIDACEAE (M)
Myrospermum Jacq. LEGUMINOSAE–PAPILIONOIDEAE (D)
Myrothamnus Welw. MYROTHAMNACEAE (D)
Myroxylon L.f. LEGUMINOSAE–PAPILIONOIDEAE (D)
Myrrhidendron J.M.Coult. & Rose UMBELLIFERAE (D)
Myrrhinium Schott MYRTACEAE (D)
Myrrhis Mill. UMBELLIFERAE (D)
Myrrhoides Heist. ex Fabr. UMBELLIFERAE (D)
Myrsine L. MYRSINACEAE (D)
Myrsiniluma Baill. = Pouteria Aubl. SAPOTACEAE (D)

Myrsiphyllum Willd. = Asparagus L. ASPARAGACEAE (M)
Myrstiphylla Raf. (SUO) = Psychotria L. RUBIACEAE (D)
Myrstiphyllum P.Browne = Psychotria L. RUBIACEAE (D)
Myrtama Ovcz. & Kinzik. TAMARICACEAE (D)
Myrtastrum Burret MYRTACEAE (D)
Myrtekmania Urb. = Pimenta Lindl. MYRTACEAE (D)
Myrtella F.Muell. MYRTACEAE (D)
Myrteola O.Berg MYRTACEAE (D)
Myrthoides Wolf = Syzygium Gaertn. MYRTACEAE (D)
Myrthus Scop. = Myrtus L. MYRTACEAE (D)
Myrtillocactus Console CACTACEAE (D)
Myrtillocereus Frič & Kreuz. (SUO) == Myrtillocactus Console CACTACEAE (D)
Myrtiluma Baill. = Pouteria Aubl. SAPOTACEAE (D)
Myrtobium Miq. = Lepidoceras Hook.f. EREMOLEPIDACEAE (D)
Myrtopsis Engl. RUTACEAE (D)
Myrtus L. MYRTACEAE (D)
Mystacidium Lindl. ORCHIDACEAE (M)
Mystirophora Nevski = Astragalus L. LEGUMINOSAE–PAPILIONOIDEAE (D)
Mystropetalon Harv. BALANOPHORACEAE (D)
Mystroxylon Eckl. & Zeyh. CELASTRACEAE (D)
Mytilaria Lecomte HAMAMELIDACEAE (D)
Myuropteris C.Chr. POLYPODIACEAE (F)
Myxopappus Källersjö COMPOSITAE (D)
Myxopyrum Blume OLEACEAE (D)
Myzodendron R.Br. (SUO) = Misodendrum Banks ex DC. MISODENDRACEAE (D)
Mzymtella Kolak. = Campanula L. CAMPANULACEAE (D)

Nabaluia Ames ORCHIDACEAE (M)
Nabalus Cass. COMPOSITAE (D)
Nabelekia Roshev. = Festuca L. GRAMINEAE (M)
Nablonium Cass. = Ammobium R.Br. ex Sims COMPOSITAE (D)
Nacibaea Poir. (SUO) = Manettia L. RUBIACEAE (D)
Nacibea Aubl. = Manettia L. RUBIACEAE (D)
Naegelia Regel = Smithiantha Kuntze GESNERIACEAE (D)
Nageia Gaertn. PODOCARPACEAE (G)
Nagelia Lindl. = Malacomeles (Decne.) Engl. ROSACEAE (D)
Nageliella L.O.Williams ORCHIDACEAE (M)
Nagelocarpus Bullock ERICACEAE (D)
Nahusia Schneev. = Fuchsia L. ONAGRACEAE (D)
Naiocrene (Torr. & A.Gray) Rydb. = Montia L. PORTULACACEAE (D)
Najas L. HYDROCHARITACEAE (M)
Naletonia Bremek. = Psychotria L. RUBIACEAE (D)
Nama L. HYDROPHYLLACEAE (D)
Namacodon Thulin CAMPANULACEAE (D)
Namaquanthus L.Bolus AIZOACEAE (D)
Namaquanula D.Müll.–Doblies & U.Müll.–Doblies AMARYLLIDACEAE (M)
Namation Brand SCROPHULARIACEAE (D)
Namibia (Schwantes) Dinter & Schwantes AIZOACEAE (D)
Nananthea DC. COMPOSITAE (D)
Nananthus N.E.Br. AIZOACEAE (D)
Nanarepenta Matuda DIOSCOREACEAE (M)
Nandina Thunb. BERBERIDACEAE (D)
Nandiroba Adans. (SUO) = Fevillea L. CUCURBITACEAE (D)
Nani Adans. = Xanthostemon F.Muell. MYRTACEAE (D)
Nannoglottis Maxim. COMPOSITAE (D)

Nannorrhops H.A.Wendl. PALMAE (M)
Nannothelypteris Holttum THELYPTERIDACEAE (F)
Nanochilus K.Schum. ZINGIBERACEAE (M)
Nanocnide Blume URTICACEAE (D)
Nanodea Banks ex C.F.Gaertn. SANTALACEAE (D)
Nanodes Lindl. ORCHIDACEAE (M)
Nanolirion Benth. = Caesia R.Br. ANTHERICACEAE (M)
Nanopetalum Hassk. = Cleistanthus Hook.f. ex Planch. EUPHORBIACEAE (D)
Nanophyton Less. CHENOPODIACEAE (D)
Nanorrhinum Betsche = Linaria Mill. SCROPHULARIACEAE (D)
Nanostelma Baill. = Tylophora R.Br. ASCLEPIADACEAE (D)
Nanothamnus Thomson COMPOSITAE (D)
Nanuza L.B.Sm. & Ayensu VELLOZIACEAE (M)
Napaea L. MALVACEAE (D)
Napeanthus Gardner GESNERIACEAE (D)
Napeodendron Ridl. = Walsura Roxb. MELIACEAE (D)
Napina Frič (SUI) = Neolloydia Britton & Rose CACTACEAE (D)
Napoleona P.Beauv. = Napoleonaea P.Beauv. LECYTHIDACEAE (D)
Napoleonaea P.Beauv. LECYTHIDACEAE (D)
Naravelia Adans. RANUNCULACEAE (D)
Narcissus L. AMARYLLIDACEAE (M)
Nardophyllum (Hook. & Arn.) Hook. & Arn. COMPOSITAE (D)
Nardostachys DC. VALERIANACEAE (D)
Narduretia Villar = Vulpia C.C.Gmelin GRAMINEAE (M)
Narduroides Rouy GRAMINEAE (M)
Nardurus (Bluff, Nees & Schauer) Rchb. = Vulpia C.C.Gmel. GRAMINEAE (M)
Nardus L. GRAMINEAE (M)
Narega Raf. = Catunaregam Wolf RUBIACEAE (D)
Naregamia Wight & Arn. MELIACEAE (D)
Narenga Bor = Saccharum L. GRAMINEAE (M)
Nargedia Bedd. RUBIACEAE (D)
Narica Raf. = Sarcoglottis C.Presl ORCHIDACEAE (M)
Naringi Adans. RUTACEAE (D)
Narthecium Huds. MELANTHIACEAE (M)
Narthex Falc. = Ferula L. UMBELLIFERAE (D)
Narum Adans. = Uvaria L. ANNONACEAE (D)
Naruma Raf. = Uvaria L. ANNONACEAE (D)
Narvalina Cass. COMPOSITAE (D)
Nashia Millsp. VERBENACEAE (D)
Nasonia Lindl. = Fernandezia Ruíz & Pav. ORCHIDACEAE (M)
Nassauvia Comm. ex Juss. COMPOSITAE (D)
Nassella (Trin.) E.Desv. GRAMINEAE (M)
Nastanthus Miers = Acarpha Griseb. CALYCERACEAE (D)
Nasturtiastrum (Gren. & Godr.) Gillet & Magne = Lepidium L. CRUCIFERAE (D)
Nasturtiicarpa Gilli = Calymmatium O.E.Schulz CRUCIFERAE (D)
Nasturtiopsis Boiss. CRUCIFERAE (D)
Nasturtium R.Br. = Rorippa Scop. CRUCIFERAE (D)
Nastus Lunell (SUS) = Cenchrus L. GRAMINEAE (M)
Nastus Juss. GRAMINEAE (M)
Natalanthe Sond. = Tricalysia A.Rich. ex DC. RUBIACEAE (D)
Nathaliella Fedtsch. SCROPHULARIACEAE (D)
Nathusia Hochst. = Schrebera Roxb. OLEACEAE (D)
Natschia Bubani (SUS) = Nardus L. GRAMINEAE (M)
Natsiatopsis Kurz ICACINACEAE (D)
Natsiatum Buch.–Ham. ex Arn. ICACINACEAE (D)

Nauclea L. RUBIACEAE (D)
Naucleopsis Miq. MORACEAE (D)
Naudinia Planch. & Linden RUTACEAE (D)
Naudiniella Krasser = Astronidium A.Gray MELASTOMATACEAE (D)
Nauenia Klotzsch = Lacaena Lindl. ORCHIDACEAE (M)
Naufraga Constance & Cannon UMBELLIFERAE (D)
Naumburgia Moench = Lysimachia L. PRIMULACEAE (D)
Nauplius (Cass.) Cass. COMPOSITAE (D)
Nautilocalyx Linden ex Hanst. GESNERIACEAE (D)
Nautochilus Bremek. == Orthosiphon Benth. LABIATAE (D)
Nautonia Decne. ASCLEPIADACEAE (D)
Nautophylla Guillaumin = Logania R.Br. LOGANIACEAE (D)
Navajoa Croizat = Pediocactus Britton & Rose CACTACEAE (D)
Navarretia Ruíz & Pav. POLEMONIACEAE (D)
Navia Schult.f. BROMELIACEAE (M)
Navicularia Raddi (SUH) = Ichnanthus P.Beauv. GRAMINEAE (M)
Naxiandra (Baill.) Krasser = Axinandra Thwaites CRYPTERONIACEAE (D)
Nayariophyton T.K.Paul (UP) MALVACEAE (D)
Nazia Adans. (SUS) = Tragus Haller GRAMINEAE (M)
Nealchornea Huber EUPHORBIACEAE (D)
Neamyza Tiegh. = Peraxilla Tiegh. LORANTHACEAE (D)
Neanotis W.H.Lewis RUBIACEAE (D)
Neanthe O.F.Cook (SUI) = Chamaedorea Willd. PALMAE (M)
Neatostema I.M.Johnst. BORAGINACEAE (D)
Nebelia Neck. ex Sweet BRUNIACEAE (D)
Neblinaea Maguire & Wurdack COMPOSITAE (D)
Neblinantha Maguire GENTIANACEAE (D)
Neblinanthera Wurdack MELASTOMATACEAE (D)
Neblinaria Maguire = Bonnetia Mart. GUTTIFERAE (D)
Neblinathamnus Steyerm. RUBIACEAE (D)
Necalistis Raf. = Ficus L. MORACEAE (D)
Necepsia Prain EUPHORBIACEAE (D)
Nechamandra Planch. HYDROCHARITACEAE (M)
Neckia Korth. OCHNACEAE (D)
Necramium Britton MELASTOMATACEAE (D)
Necranthus Gilli SCROPHULARIACEAE (D)
Nectandra Rol. ex Rottb. LAURACEAE (D)
Nectaropetalum Engl. ERYTHROXYLACEAE (D)
Nectaroscordum Lindl. ALLIACEAE (M)
Nectouxia Kunth SOLANACEAE (D)
Neea Ruíz & Pav. NYCTAGINACEAE (D)
Needhamia R.Br. = Needhamiella L.Watson EPACRIDACEAE (D)
Needhamiella L.Watson EPACRIDACEAE (D)
Neeopsis Lundell NYCTAGINACEAE (D)
Neeragrostis Bush = Eragrostis Wolf GRAMINEAE (M)
Neesenbeckia Levyns CYPERACEAE (M)
Neesia Blume BOMBACACEAE (D)
Neesiochloa Pilg. GRAMINEAE (M)
Negria Chiov. (SUH) = Lintonia Stapf GRAMINEAE (M)
Negria F.Muell. GESNERIACEAE (D)
Negripteris Pic.Serm. ADIANTACEAE (F)
Neillia D.Don ROSACEAE (D)
Neippergia C.Morren = Acineta Lindl. ORCHIDACEAE (M)
Neisosperma Raf. APOCYNACEAE (D)
Nelanaregam Adans. = Narcgamia Wight & Arn. MELIACEAE (D)

Neleixa Raf. = Faramea Aubl. RUBIACEAE (D)
Nelia Schwantes AIZOACEAE (D)
Nelipus Raf. = Utricularia L. LENTIBULARIACEAE (D)
Nelitrs Gaertn. (SUI) = Timonius DC. RUBIACEAE (D)
Nelitris Spreng. = Decaspermum J.R.Forst. & G.Forst. MYRTACEAE (D)
Nellica Raf. = Phyllanthus L. EUPHORBIACEAE (D)
Nelmesia Veken CYPERACEAE (M)
Nelsia Schinz AMARANTHACEAE (D)
Nelsonia R.Br. ACANTHACEAE (D)
Nelsonianthus H.Rob. & Brettell COMPOSITAE (D)
Nelumbo Adans. NELUMBONACEAE (D)
Nemacaulis Nutt. POLYGONACEAE (D)
Nemacladus Nutt. CAMPANULACEAE (D)
Nemaconia Knowles & Westc. = Ponera Lindl. ORCHIDACEAE (M)
Nemaluma Baill. = Pouteria Aubl. SAPOTACEAE (D)
Nemastachys Steud. = Microstegium Nees GRAMINEAE (M)
Nemastylis Nutt. IRIDACEAE (M)
Nematanthus Schrad. GESNERIACEAE (D)
Nematoceras Hook.f. = Corybas Salisb. ORCHIDACEAE (M)
Nematolepis Turcz. RUTACEAE (D)
Nematopoa C.E.Hubb. GRAMINEAE (M)
Nematopogon Bureau & K.Schum. = Digomphia Benth. BIGNONIACEAE (D)
Nematopteris Alderw. = Scleroglossum Alderw. GRAMMITIDACEAE (F)
Nematosciadium H.Wolff = Arracacia Bancr. UMBELLIFERAE (D)
Nematostemma Choux (UP) ASCLEPIADACEAE (D)
Nematostigma Planch. = Gironniera Gaudich. ULMACEAE (D)
Nematostylis Hook.f. RUBIACEAE (D)
Nematuris Turcz. = Ampelamus Raf. ASCLEPIADACEAE (D)
Nemcia Domin LEGUMINOSAE–PAPILIONOIDEAE (D)
Nemedra A.Juss. = Aglaia Lour. MELIACEAE (D)
Nemesia Vent. SCROPHULARIACEAE (D)
Nemodon Griff. = Lepistemon Blume CONVOLVULACEAE (D)
Nemopanthus Raf. AQUIFOLIACEAE (D)
Nemophila Nutt. HYDROPHYLLACEAE (D)
Nemosenecio (Kitam.) B.Nord. COMPOSITAE (D)
Nemoseris Greene = Rafinesquia Nutt. COMPOSITAE (D)
Nemostylis Steven = Phuopsis (Griseb.) Hook.f. RUBIACEAE (D)
Nemotopyxis Miq. = Ludwigia L. ONAGRACEAE (D)
Nemuaron Baill. MONIMIACEAE (D)
Nemum Desv. CYPERACEAE (M)
Nemuranthes Raf. = Habenaria Willd. ORCHIDACEAE (M)
Nenax Gaertn. RUBIACEAE (D)
Nenga H.A.Wendl. & Drude PALMAE (M)
Nengella Becc. = Gronophyllum Scheff. PALMAE (M)
Neo–Senea K.Schum. ex H.Pfeiff. = Lagenocarpus Nees CYPERACEAE (M)
Neo–Uvaria Airy Shaw ANNONACEAE (D)
Neo–urbania Fawc. & Rendle ORCHIDACEAE (M)
Neoabbottia Britton & Rose = Leptocereus (A.Berger) Britton & Rose CACTACEAE (D)
Neoacanthophora Bennet = Aralia L. ARALIACEAE (D)
Neoalsomitra Hutch. CUCURBITACEAE (D)
Neoancistrophyllum Rauschert (SUS) = Laccosperma (G.Mann & H.A.Wendl.) Drude PALMAE (M)
Neoapaloxylon Rauschert LEGUMINOSAE–CAESALPINIOIDEAE (D)
Neoastelia J.B.Williams ASTELIACEAE (M)
Neoathyrium Ching & Z.R.Wang = Cornopteris Nakai WOODSIACEAE (F)

Neoaulacolepis Rauschert = Calamagrostis Adans. GRAMINEAE (M)
Neobaclea Hochr. MALVACEAE (D)
Neobakeria Schltr. = Massonia Thunb. ex L.f. HYACINTHACEAE (M)
Neobalanocarpus P.S.Ashton DIPTEROCARPACEAE (D)
Neobambus Keng f. (SUI) == Sinobambusa Makino GRAMINEAE (M)
Neobaronia Baker = Phylloxylon Baill. LEGUMINOSAE–PAPILIONOIDEAE (D)
Neobartlettia Schltr. ORCHIDACEAE (M)
Neobassia A.J.Scott CHENOPODIACEAE (D)
Neobathiea Schltr. ORCHIDACEAE (M)
Neobaumannia Hutch. & Dalziel = Knoxia L. RUBIACEAE (D)
Neobeckia Greene = Rorippa Scop. CRUCIFERAE (D)
Neobeguea J.–F.Leroy MELIACEAE (D)
Neobenthamia Rolfe ORCHIDACEAE (M)
Neobertiera Wernham RUBIACEAE (D)
Neobesseya Britton & Rose = Escobaria Britton & Rose CACTACEAE (D)
Neobinghamia Backeb. = Haageocereus Backeb. CACTACEAE (D)
Neobiondia Pamp. = Saururus L. SAURURACEAE (D)
Neoblakea Standl. RUBIACEAE (D)
Neoboivinella Aubrév. & Pellegr. = Englerophytum K.Krause SAPOTACEAE (D)
Neobolusia Schltr. ORCHIDACEAE (M)
Neobotrydium Moldenke = Chenopodium L. CHENOPODIACEAE (D)
Neobouteloua Gould GRAMINEAE (M)
Neoboutonia Müll.Arg. EUPHORBIACEAE (D)
Neobracea Britton APOCYNACEAE (D)
Neobreonia Ridsdale RUBIACEAE (D)
Neobrittonia Hochr. MALVACEAE (D)
Neobuchia Urb. BOMBACACEAE (D)
Neobuxbaumia Backeb. CACTACEAE (D)
Neobyrnesia J.A.Armstr. RUTACEAE (D)
Neocabreria R.M.King & H.Rob. COMPOSITAE (D)
Neocallitropsis Florin CUPRESSACEAE (G)
Neocalyptrocalyx Hutch. = Capparis L. CAPPARACEAE (D)
Neocardenasia Backeb. = Neoraimondia Britton & Rose CACTACEAE (D)
Neocarya (DC.) Prance ex F.White CHRYSOBALANACEAE (D)
Neocastela Small = Castela Turpin SIMAROUBACEAE (D)
Neocentema Schinz AMARANTHACEAE (D)
Neochamaelea (Engl.) Erdtman = Cneorum L. CNEORACEAE (D)
Neocheiropteris Christ POLYPODIACEAE (F)
Neochevaliera A.Chev. & Beille = Chaetocarpus Thwaites EUPHORBIACEAE (D)
Neochevalierodendron J.Léonard LEGUMINOSAE–CAESALPINIOIDEAE (D)
Neochilenia Backeb. ex Doelz = Neoporteria Britton & Rose CACTACEAE (D)
Neocinnamomum H.Liou LAURACEAE (D)
Neoclemensia C.E.Carr ORCHIDACEAE (M)
Neocodon Kolak. & Serdyuk. = Campanula L. CAMPANULACEAE (D)
Neocogniauxia Schltr. ORCHIDACEAE (M)
Neocollettia Hemsl. LEGUMINOSAE–PAPILIONOIDEAE (D)
Neoconopodium (Kozo-Pol.) Pimenov & Kljuykov UMBELLIFERAE (D)
Neocouma Pierre APOCYNACEAE (D)
Neocracca Kuntze = Coursetia DC. LEGUMINOSAE–PAPILIONOIDEAE (D)
Neocryptodiscus Hedge & Lamond = Prangos Lindl. UMBELLIFERAE (D)
Neocuatrecasia R.M.King & H.Rob. COMPOSITAE (D)
Neocussonia Hutch. = Schefflera J.R.Forst. & G.Forst. ARALIACEAE (D)
Neodawsonia Backeb. = Cephalocereus Pfeiff. CACTACEAE (D)
Neodeutzia (Engl.) Small = Deutzia Thunb. HYDRANGEACEAE (D)
Neodielsia Harms LEGUMINOSAE–PAPILIONOIDEAE (D)

Neodissochaeta Bakh.f. = Dissochaeta Blume MELASTOMATACEAE (D)
Neodistemon Babu & A.N.Henry URTICACEAE (D)
Neodonnellia Rose = Tripogandra Raf. COMMELINACEAE (M)
Neodregea C.H.Wright COLCHICACEAE (M)
Neodriessenia M.P.Nayar MELASTOMATACEAE (D)
Neodryas Rchb.f. ORCHIDACEAE (M)
Neodunnia R.Vig. LEGUMINOSAE–PAPILIONOIDEAE (D)
Neodypsis Baill. PALMAE (M)
Neoeplingia Ramam., Hiriart & Medran LABIATAE (D)
Neoevansia W.T.Marshall = Peniocereus (A.Berger) Britton & Rose CACTACEAE (D)
Neofabricia Joy Thompson MYRTACEAE (D)
Neoferetia Baehni = Nothapodytes Blume ICACINACEAE (D)
Neofinetia Hu ORCHIDACEAE (M)
Neofranciella Guillaumin RUBIACEAE (D)
Neogaerrhinum Rothm. SCROPHULARIACEAE (D)
Neogaillonia Lincz. RUBIACEAE (D)
Neogardneria Schltr. ex Garay ORCHIDACEAE (M)
Neoglaziovia Mez BROMELIACEAE (M)
Neogleasonia Maguire = Bonnetia Mart. GUTTIFERAE (D)
Neogoetzia Pax = Bridelia Willd. EUPHORBIACEAE (D)
Neogoezia Hemsl. UMBELLIFERAE (D)
Neogomesia Castaneda = Ariocarpus Scheidw. CACTACEAE (D)
Neogomezia Buxb. (SUO) = Ariocarpus Scheidw. CACTACEAE (D)
Neogontscharovia Lincz. PLUMBAGINACEAE (D)
Neogoodenia C.A.Gardner & A.S.George = Goodenia Sm. GOODENIACEAE (D)
Neoguillauminia Croizat EUPHORBIACEAE (D)
Neogunnia Pax & K.Hoffm. = Gunniopsis Pax AIZOACEAE (D)
Neogymnantha Y.Ito (SUI) = Rebutia K.Schum. CACTACEAE (D)
Neogyna Rchb.f. ORCHIDACEAE (M)
Neogyne Rchb.f. (SUO) = Neogyna Rchb.f. ORCHIDACEAE (M)
Neohallia Hemsl. ACANTHACEAE (D)
Neoharmsia R.Vig. LEGUMINOSAE–PAPILIONOIDEAE (D)
Neohemsleya T.D.Penn. SAPOTACEAE (D)
Neohenricia L.Bolus AIZOACEAE (D)
Neohenrya Hemsl. = Tylophora R.Br. ASCLEPIADACEAE (D)
Neohickenia Frič = Parodia Speg. CACTACEAE (D)
Neohintonia R.M.King & H.Rob. COMPOSITAE (D)
Neoholstia Rauschert EUPHORBIACEAE (D)
Neohouzeaua A.Camus = Schizostachyum Nees GRAMINEAE (M)
Neohuberia Ledoux = Eschweilera Mart. ex DC. LECYTHIDACEAE (D)
Neohumbertiella Hochr. MALVACEAE (D)
Neohusnotia A.Camus = Acroceras Stapf GRAMINEAE (M)
Neohymenopogon Bennet RUBIACEAE (D)
Neohyptis J.K.Morton LABIATAE (D)
Neojatropha Pax = Mildbraedia Pax EUPHORBIACEAE (D)
Neojeffreya Cabrera COMPOSITAE (D)
Neojobertia Baill. BIGNONIACEAE (D)
Neojunghuhnia Koord. = Vaccinium L. ERICACEAE (D)
Neokeithia Steenis = Chilocarpus Blume APOCYNACEAE (D)
Neokoehleria Schltr. ORCHIDACEAE (M)
Neolabatia Aubrév. = Pouteria Aubl. SAPOTACEAE (D)
Neolacis Wedd. = Apinagia Tul. PODOSTEMACEAE (D)
Neolamarckia Bosser RUBIACEAE (D)
Neolauchea Kraenzl. = Isabelia Barb.Rodr. ORCHIDACEAE (M)
Neolaugeria Nicolson RUBIACEAE (D)

Neolehmannia Kraenzl. ORCHIDACEAE (M)
Neolemaireocereus Backeb. (SUS) = Stenocereus (A.Berger) Riccob. CACTACEAE (D)
Neolemonniera Heine SAPOTACEAE (D)
Neolepia W.A.Weber = Lepidium L. CRUCIFERAE (D)
Neolepisorus Ching = Neocheiropteris Christ POLYPODIACEAE (F)
Neoleroya Cavaco RUBIACEAE (D)
Neolindenia Baill. = Louteridium S.Watson ACANTHACEAE (D)
Neolindleya Kraenzl. = Platanthera Rich. ORCHIDACEAE (M)
Neolindleyella Fedde = Lindleya Kunth ROSACEAE (D)
Neolitsea (Benth.) Merr. LAURACEAE (D)
Neolloydia Britton & Rose CACTACEAE (D)
Neolobivia Y.Ito = Echinopsis Zucc. CACTACEAE (D)
Neolophocarpus E.G.Camus = Schoenus L. CYPERACEAE (M)
Neoluederitzia Schinz ZYGOPHYLLACEAE (D)
Neoluffa Chakrav. = Siraitia Merr. CUCURBITACEAE (D)
Neomacfadya Baill. BIGNONIACEAE (D)
Neomammillaria Britton & Rose (SUS) = Mammillaria Haw. CACTACEAE (D)
Neomandonia Hutch. = Tradescantia L. COMMELINACEAE (M)
Neomangenotia J.-F.Leroy = Commiphora Jacq. BURSERACEAE (D)
Neomanniophyton Pax & K.Hoffm. = Crotonogyne Muell.Arg. EUPHORBIACEAE (D)
Neomarica Sprague IRIDACEAE (M)
Neomartinella Pilg. CRUCIFERAE (D)
Neomazaea Urb. = Mazaea Krug & Urb. RUBIACEAE (D)
Neomezia Votsch = Deherainia Decne. THEOPHRASTACEAE (D)
Neomicrocalamus Keng f. = Racemobambos Holttum GRAMINEAE (M)
Neomillspaughia S.F.Blake POLYGONACEAE (D)
Neomimosa Britton & Rose = Mimosa L. LEGUMINOSAE–MIMOSOIDEAE (D)
Neomirandea R.M.King & H.Rob. COMPOSITAE (D)
Neomitranthes Legrand MYRTACEAE (D)
Neomolinia Honda = Diarrhena P.Beauv. GRAMINEAE (M)
Neomoorea Rolfe ORCHIDACEAE (M)
Neomortonia Wiehler GESNERIACEAE (D)
Neomphalea Pax & K.Hoffm. = Omphalea L. EUPHORBIACEAE (D)
Neomuellera Briq. = Plectranthus L'Hér. LABIATAE (D)
Neomyrtus Burret MYRTACEAE (D)
Neonauclea Merr. RUBIACEAE (D)
Neonelsonia J.M.Coult. & Rose UMBELLIFERAE (D)
Neonicholsonia Dammer PALMAE (M)
Neoniphopsis Nakai = Pyrrosia Mirb. POLYPODIACEAE (F)
Neonotonia Lackey LEGUMINOSAE–PAPILIONOIDEAE (D)
Neopalissya Pax = Necepsia Prain EUPHORBIACEAE (D)
Neopallasia Poljakov COMPOSITAE (D)
Neopanax Allan = Pseudopanax K.Koch ARALIACEAE (D)
Neoparrya Mathias UMBELLIFERAE (D)
Neopatersonia Schönl. HYACINTHACEAE (M)
Neopaulia Pimenov & Kljuykov (SUS) = Paulita Soják UMBELLIFERAE (D)
Neopaxia O.E.G.Nilsson PORTULACACEAE (D)
Neopeltandra Gamble (SUS) = Meineckia Baill. EUPHORBIACEAE (D)
Neopentanisia Verdc. RUBIACEAE (D)
Neopetalonema Brenan = Gravesia Naudin MELASTOMATACEAE (D)
Neophloga Baill. PALMAE (M)
Neophylum Tiegh. = Amyema Tiegh. LORANTHACEAE (D)
Neopicrorhiza D.Y.Hong SCROPHULARIACEAE (D)
Neopieris Britton = Lyonia Nutt. ERICACEAE (D)
Neopilea Leandri = Pilea Lindl. URTICACEAE (D)

Neoplatytaenia Geld. = Semenovia Regel & Herder UMBELLIFERAE (D)
Neopometia Aubrév. = Pradosia Liais SAPOTACEAE (D)
Neoporteria Backeb. (SUH) = Neoporteria Britton & Rose CACTACEAE (D)
Neoporteria Britton & Rose CACTACEAE (D)
Neopreissia Ulbr. = Atriplex L. CHENOPODIACEAE (D)
Neopringlea S.Watson FLACOURTIACEAE (D)
Neoptychocarpus Buchheim FLACOURTIACEAE (D)
Neopycnocoma Pax = Argomuellera Pax EUPHORBIACEAE (D)
Neoraimondia Britton & Rose CACTACEAE (D)
Neorapinia Moldenke VERBENACEAE (D)
Neoraputia Emmerich = Raputia Aubl. RUTACEAE (D)
Neorautanenia Schinz LEGUMINOSAE–PAPILIONOIDEAE (D)
Neoregelia L.B.Sm. BROMELIACEAE (M)
Neoregnellia Urb. STERCULIACEAE (D)
Neorhine Schwantes = Rhinephyllum N.E.Br. AIZOACEAE (D)
Neorites L.S.Sm. PROTEACEAE (D)
Neoroepera Müll.Arg. & F.Muell. EUPHORBIACEAE (D)
Neorosea N.Hallé = Tricalysia A.Rich. ex DC. RUBIACEAE (D)
Neorudolphia Britton LEGUMINOSAE–PAPILIONOIDEAE (D)
Neosabicea Wernham = Manettia L. RUBIACEAE (D)
Neosasamorpha Tatew. = Sasa Makino & Shib. GRAMINEAE (M)
Neoschimpera Hemsl. = Amaracarpus Blume RUBIACEAE (D)
Neoschischkinia Tzvelev = Agrostis L. GRAMINEAE (M)
Neoschroetera Briq. = Larrea Cav. ZYGOPHYLLACEAE (D)
Neoschumannia Schltr. ASCLEPIADACEAE (D)
Neosciadium Domin UMBELLIFERAE (D)
Neoscortechinia Pax EUPHORBIACEAE (D)
Neosepicaea Diels BIGNONIACEAE (D)
Neosieversia Bolle = Novosieversia F.Bolle ROSACEAE (D)
Neosinocalamus Keng f. = Dendrocalamus Nees GRAMINEAE (M)
Neosloetiopsis Engl. = Streblus Lour. MORACEAE (D)
Neosparton Griseb. VERBENACEAE (D)
Neosprucea Sleumer = Hasseltia Kunth TILIACEAE (D)
Neostachyanthus Exell & Mendonça = Stachyanthus Engl. ICACINACEAE (D)
Neostapfia Burtt Davy GRAMINEAE (M)
Neostapfiella A.Camus GRAMINEAE (M)
Neostenanthera Exell ANNONACEAE (D)
Neostrearia L.S.Sm. HAMAMELIDACEAE (D)
Neostricklandia Rauschert = Phaedranassa Herb. AMARYLLIDACEAE (M)
Neostyphonia Shafer = Rhus L. ANACARDIACEAE (D)
Neotainiopsis Bennet & Raizada = Eriodes Rolfe ORCHIDACEAE (M)
Neotanahashia Y.Ito = Neoporteria Britton & Rose CACTACEAE (D)
Neotatea Maguire GUTTIFERAE (D)
Neotchihatchewia Rauschert CRUCIFERAE (D)
Neotessmannia Burret TILIACEAE (D)
Neothorelia Gagnep. CAPPARACEAE (D)
Neothymopsis Britton & Millsp. = Thymopsis Benth. COMPOSITAE (D)
Neotina Capuron SAPINDACEAE (D)
Neotinea Rchb.f. ORCHIDACEAE (M)
Neotorularia Hedge & J.Léonard CRUCIFERAE (D)
Neotreleasea Rose (SUS) = Tradescantia L. COMMELINACEAE (M)
Neotrewia Pax & K.Hoffm. EUPHORBIACEAE (D)
Neotrigonostemon Pax & K.Hoffm. = Trigonostemon Blume EUPHORBIACEAE (D)
Neottia Guett. ORCHIDACEAE (M)
Neottianthe (Rchb.) Schltr. ORCHIDACEAE (M)

Neottidium Schltdl. = Neottia Guett. ORCHIDACEAE (M)
Neottopteris J.Sm. = Asplenium L. ASPLENIACEAE (F)
Neotuerckheimia Donn.Sm. BIGNONIACEAE (D)
Neoturczaninovia Kozo-Pol. (UP) UMBELLIFERAE (D)
Neotysonia Dalla Torre & Harms COMPOSITAE (D)
Neoveitchia Becc. PALMAE (M)
Neowashingtonia Sudw. = Washingtonia H.A.Wendl. PALMAE (M)
Neowawraea Rock = Flueggea Willd. EUPHORBIACEAE (D)
Neowerdermannia Frič CACTACEAE (D)
Neowilliamsia Garay ORCHIDACEAE (M)
Neowimmeria O.Deg. & I.Deg. = Lobelia L. CAMPANULACEAE (D)
Neowollastonia Wernham ex Ridl. = Melodinus J.R.Forst. & G.Forst. APOCYNACEAE (D)
Neowormia Hutch. & Summerh. = Dillenia L. DILLENIACEAE (D)
Neoxythece Aubrév. & Pellegr. = Pouteria Aubl. SAPOTACEAE (D)
Neozenkerina Mildbr. = Staurogyne Wall. ACANTHACEAE (D)
Nepa Webb = Ulex L. LEGUMINOSAE–PAPILIONOIDEAE (D)
Nepenthandra S.Moore = Trigonostemon Blume EUPHORBIACEAE (D)
Nepenthes L. NEPENTHACEAE (D)
Nepeta L. LABIATAE (D)
Nephelaphyllum Blume ORCHIDACEAE (M)
Nephelea R.M.Tryon = Cyathea Sm. CYATHEACEAE (F)
Nephelium L. SAPINDACEAE (D)
Nephelochloa Boiss. GRAMINEAE (M)
Nephopteris Lellinger ADIANTACEAE (F)
Nephradenia Decne. ASCLEPIADACEAE (D)
Nephrangis (Schltr.) Summerh. ORCHIDACEAE (M)
Nephranthera Hassk. = Renanthera Lour. ORCHIDACEAE (M)
Nephrocarpus Dammer = Basselinia Vieill. PALMAE (M)
Nephrocarya Candargy BORAGINACEAE (D)
Nephrodesmus Schindl. LEGUMINOSAE–PAPILIONOIDEAE (D)
Nephrodium Michx. = Dryopteris Adans. DRYOPTERIDACEAE (F)
Nephrolepis Schott OLEANDRACEAE (F)
Nephromeria (Benth.) Schindl. = Desmodium Desv. LEGUMINOSAE–
    PAPILIONOIDEAE (D)
Nephropetalum B.L.Rob. & Greenm. = Ayenia L. STERCULIACEAE (D)
Nephrophyllidium Gilg MENYANTHACEAE (D)
Nephrophyllum A.Rich. CONVOLVULACEAE (D)
Nephrosperma Balf.f. PALMAE (M)
Nephrostylus Gagnep. = Koilodepas Hassk. EUPHORBIACEAE (D)
Nephthytis Schott ARACEAE (M)
Nepsera Naudin MELASTOMATACEAE (D)
Neptunia Lour. LEGUMINOSAE–MIMOSOIDEAE (D)
Neraudia Gaudich. URTICACEAE (D)
Neriacanthus Benth. ACANTHACEAE (D)
Nerine Herb. AMARYLLIDACEAE (M)
Nerissa Raf. = Ponthieva R.Br. ORCHIDACEAE (M)
Nerisyrenia Greene CRUCIFERAE (D)
Nerium L. APOCYNACEAE (D)
Nernstia Urb. RUBIACEAE (D)
Nerophila Naudin MELASTOMATACEAE (D)
Nertera Banks & Sol. ex Gaertn. RUBIACEAE (D)
Nervilia Comm. ex Gaud. ORCHIDACEAE (M)
Nesaea Kunth LYTHRACEAE (D)
Nescidia A.Rich. = Coffea L. RUBIACEAE (D)
Nesiota Hook.f. RHAMNACEAE (D)

Neslia Desv. CRUCIFERAE (D)
Nesobium R.Phil. ex Fuentes = Parietaria L. URTICACEAE (D)
Nesocaryum I.M.Johnst. BORAGINACEAE (D)
Nesocodon Thulin CAMPANULACEAE (D)
Nesodaphne Hook.f. = Beilschmiedia Nees LAURACEAE (D)
Nesodoxa Calest. = Arthrophyllum Blume ARALIACEAE (D)
Nesodraba Greene = Draba L. CRUCIFERAE (D)
Nesogenes A.DC. NESOGENACEAE (D)
Nesogordonia Baill. STERCULIACEAE (D)
Nesohedyotis (Hook.f.) Bremek. RUBIACEAE (D)
Nesoluma Baill. SAPOTACEAE (D)
Nesopanax Seem. = Schefflera J.R.Forst. & G.Forst. ARALIACEAE (D)
Nesopteris Copel. = Cephalomanes C.Presl HYMENOPHYLLACEAE (F)
Nesoris Raf. = Pityrogramma Link ADIANTACEAE (F)
Nesothamnus Rydb. = Perityle Benth. COMPOSITAE (D)
Nesphostylis Verdc. LEGUMINOSAE–PAPILIONOIDEAE (D)
Nestegis Raf. OLEACEAE (D)
Nestlera Steud. (SUI) = Bouteloua Lag. GRAMINEAE (M)
Nestoria Urb. BIGNONIACEAE (D)
Nestronia Raf. SANTALACEAE (D)
Nettlera Raf. = Psychotria L. RUBIACEAE (D)
Nettoa Baill. = Grewia L. TILIACEAE (D)
Neuburgia Blume LOGANIACEAE (D)
Neudorfia Adans. = Nolana L.f. SOLANACEAE (D)
Neuhofia Stokes = Baeckea L. MYRTACEAE (D)
Neumannia A.Rich. = Aphloia (DC.) Bennett FLACOURTIACEAE (D)
Neuontobotrys O.E.Schulz CRUCIFERAE (D)
Neuracanthus Nees ACANTHACEAE (D)
Neurachne R.Br. GRAMINEAE (M)
Neuractis Cass. = Glossocardia Cass. COMPOSITAE (D)
Neurada L. NEURADACEAE (D)
Neuradopsis Bremek. & Oberm. NEURADACEAE (D)
Neurocalyx Hook. RUBIACEAE (D)
Neurocarpaea K.Schum. (SUO) = Nodocarpaea A.Gray RUBIACEAE (D)
Neurocarpaea R.Br. (SUI) = Pentas Benth. RUBIACEAE (D)
Neurodium Fée POLYPODIACEAE (F)
Neurogramma Link = Gymnopteris Bernh. ADIANTACEAE (F)
Neurolaena R.Br. COMPOSITAE (D)
Neurolakis Mattf. COMPOSITAE (D)
Neurolepis Meisn. GRAMINEAE (M)
Neurolobium Baill. = Diplorhynchus Welw. ex Ficalho & Hiern APOCYNACEAE (D)
Neuromanes Trevis. = Trichomanes L. HYMENOPHYLLACEAE (F)
Neuronia D.Don = Oleandra Cav. OLEANDRACEAE (F)
Neuropeltis Wall. CONVOLVULACEAE (D)
Neuropeltopsis Ooststr. CONVOLVULACEAE (D)
Neurophyllodes (A.Gray) O.Deg. = Geranium L. GERANIACEAE (D)
Neurophyllum C.Presl = Trichomanes L. HYMENOPHYLLACEAE (F)
Neuroplatyceros Fée = Platycerium Desv. POLYPODIACEAE (F)
Neuropoa Clayton GRAMINEAE (M)
Neuropteris Desv. = Saccoloma Kaulf. DENNSTAEDTIACEAE (F)
Neurosoria Mett. ex Kuhn = Cheilanthes Sw. ADIANTACEAE (F)
Neurosperma Raf. = Momordica L. CUCURBITACEAE (D)
Neurospermum Bartl. (SUO) = Momordica L. CUCURBITACEAE (D)
Neurotecoma K.Schum. = Spirotecoma Baill. ex Dalla Torre & Harms BIGNONIACEAE (D)
Neurotheca Salisb. ex Benth. GENTIANACEAE (D)

Neustanthus Benth. = Pueraria DC. LEGUMINOSAE–PAPILIONOIDEAE (D)
Neustruevia Juz. = Pseudomarrubium Popov LABIATAE (D)
Neuwiedia Blume ORCHIDACEAE (M)
Neves–Armondia K.Schum. = Pithecoctenium Mart. ex Meisn. BIGNONIACEAE (D)
Nevillea Esterh. & H.P.Linder RESTIONACEAE (M)
Neviusia A.Gray ROSACEAE (D)
Nevrilis Raf. = Millingtonia L.f. BIGNONIACEAE (D)
Nevrocallis Fée PTERIDACEAE (F)
Nevroctola Raf. (SUS) = Uniola L. GRAMINEAE (M)
Nevroloma Raf. = Glyceria R.Br. GRAMINEAE (M)
Nevrosperma Raf. (SUO) = Momordica L. CUCURBITACEAE (D)
Nevskiella Krecz. & Vved. = Bromus L. GRAMINEAE (M)
Newberrya Torr. = Hemitomes A.Gray ERICACEAE (D)
Newbouldia Seem. ex Bureau BIGNONIACEAE (D)
Newcastelia F.Muell. VERBENACEAE (D)
Newtonia Baill. LEGUMINOSAE–MIMOSOIDEAE (D)
Neyraudia Hook.f. GRAMINEAE (M)
Nhandiroba Adans. = Fevillea L. CUCURBITACEAE (D)
Nicandra Adans. SOLANACEAE (D)
Nicarago Britton & Rose = Caesalpinia L. LEGUMINOSAE–CAESALPINIOIDEAE (D)
Nichallea Bridson RUBIACEAE (D)
Nichelia Bullock (SUI) = Neoporteria Britton & Rose CACTACEAE (D)
Nicobariodendron Vasudeva Rao & Chakrab. (UP) CELASTRACEAE (D)
Nicodemia Ten. = Buddleja L. BUDDLEJACEAE (D)
Nicolaia Horan. = Etlingera Giseke ZINGIBERACEAE (M)
Nicolasia S.Moore COMPOSITAE (D)
Nicolletia A.Gray COMPOSITAE (D)
Nicoteba Lindau = Justicia L. ACANTHACEAE (D)
Nicotiana L. SOLANACEAE (D)
Nidema Britton & Millsp. ORCHIDACEAE (M)
Nidorella Cass. COMPOSITAE (D)
Nidularium Lem. BROMELIACEAE (M)
Nidus Riv. = Neottia Guett. ORCHIDACEAE (M)
Niebuhria DC. CAPPARACEAE (D)
Niedenzua Pax = Adenochlaena Boivin ex Baill. EUPHORBIACEAE (D)
Niederleinia Hieron. FRANKENIACEAE (D)
Niedzwedzkia Fedtsch. = Incarvillea Juss. BIGNONIACEAE (D)
Niemeyera F.Muell. (SUH) = Apostasia Blume ORCHIDACEAE (M)
Niemeyera F.Muell. SAPOTACEAE (D)
Nienokuea A.Chev. = Polystachya Hook. ORCHIDACEAE (M)
Nierembergia Ruíz & Pav. SOLANACEAE (D)
Nietneria Klotzsch ex Benth. MELANTHIACEAE (M)
Nietoa Seem. ex Schaffn. = Hanburia Seem. CUCURBITACEAE (D)
Nigella L. RANUNCULACEAE (D)
Nigrina L. = Melasma Bergius SCROPHULARIACEAE (D)
Nigritella Rich. ORCHIDACEAE (M)
Nigromnia Carolin = Scaevola L. GOODENIACEAE (D)
Nikitinia Iljin COMPOSITAE (D)
Nilgirianthus Bremek. = Strobilanthes Blume ACANTHACEAE (D)
Nimiria Prain ex Craib = Acacia Mill. LEGUMINOSAE–MIMOSOIDEAE (D)
Nimmoia Wight (SUH) = Aglaia Lour. MELIACEAE (D)
Niopa (Benth.) Britton & Rose = Anadenanthera Speg. LEGUMINOSAE–
    MIMOSOIDEAE (D)
Nipa Thunb. = Nypa Steck PALMAE (M)
Niphaea Lindl. GESNERIACEAE (D)

Niphidium J.Sm. POLYPODIACEAE (F)
Niphobolus Kaulf. = Pyrrosia Mirb. POLYPODIACEAE (F)
Niphogeton Schltdl. UMBELLIFERAE (D)
Niphopsis J.Sm. = Pyrrosia Mirb. POLYPODIACEAE (F)
Nipponanthemum (Kitam.) Kitam. COMPOSITAE (D)
Nipponobambusa Muroi = Sasa Makino & Shib. GRAMINEAE (M)
Nipponocalamus Nakai = Arundinaria Michx. GRAMINEAE (M)
Nipponorchis Masam. = Neofinetia Hu ORCHIDACEAE (M)
Nirarathamnos Balf.f. UMBELLIFERAE (D)
Niruri Adans. = Phyllanthus L. EUPHORBIACEAE (D)
Nisa Noronha ex Thouars = Homalium Jacq. FLACOURTIACEAE (D)
Nisomenes Raf. = Euphorbia L. EUPHORBIACEAE (D)
Nispero Aubrév. = Manilkara Adans. SAPOTACEAE (D)
Nissolia Jacq. LEGUMINOSAE–PAPILIONOIDEAE (D)
Nistarika B.K.Nayar, Madhus. & Molly = Leptochilus Kaulf. POLYPODIACEAE (F)
Nitraria L. ZYGOPHYLLACEAE (D)
Nitrophila S.Watson CHENOPODIACEAE (D)
Nivenia R.Br. (SUH) = Paranomus Salisb. PROTEACEAE (D)
Nivenia Vent. IRIDACEAE (M)
Niveophyllum Matuda = Hechtia Klotzsch BROMELIACEAE (M)
Nivieria Ser. = Triticum L. GRAMINEAE (M)
Noaea Moq. CHENOPODIACEAE (D)
Noahdendron Endress, B.Hyland & Tracey HAMAMELIDACEAE (D)
Nobeliodendron O.C.Schmidt = Licaria Aubl. LAURACEAE (D)
Nobula Adans. = Phyllis L. RUBIACEAE (D)
Noccaea Moench = Thlaspi L. CRUCIFERAE (D)
Noccidium F.K.Mey. = Thlaspi L. CRUCIFERAE (D)
Nodocarpaea A.Gray RUBIACEAE (D)
Nodonema B.L.Burtt GESNERIACEAE (D)
Nogalia Verdc. BORAGINACEAE (D)
Nogo Baehni = Lecomtedoxa (Pierre ex Engl.) Dubard SAPOTACEAE (D)
Nogra Merr. LEGUMINOSAE–PAPILIONOIDEAE (D)
Noisettia Kunth VIOLACEAE (D)
Nolana L.f. SOLANACEAE (D)
Noldeanthus Knobl. = Jasminum L. OLEACEAE (D)
Nolina Michx. DRACAENACEAE (M)
Nolletia Cass. COMPOSITAE (D)
Noltea Rchb. RHAMNACEAE (D)
Noltia Schumach. & Thonn. = Diospyros L. EBENACEAE (D)
Nomaphila Blume = Hygrophila R.Br. ACANTHACEAE (D)
Nomismia Wight & Arn. = Rhynchosia Lour. LEGUMINOSAE–PAPILIONOIDEAE (D)
Nomocharis Franch. LILIACEAE (M)
Nomosa I.M.Johnst. BORAGINACEAE (D)
Nonatelia Kuntze (SUH) = Lasianthus Jack RUBIACEAE (D)
Nonatelia Aubl. = Palicourea Aubl. RUBIACEAE (D)
Nonea Medik. BORAGINACEAE (D)
Nopalea Salm-Dyck = Opuntia Mill. CACTACEAE (D)
Nopalxochia Britton & Rose = Disocactus Lindl. CACTACEAE (D)
Norantea Aubl. MARCGRAVIACEAE (D)
Normanbokea Klad. & Buxb. = Neolloydia Britton & Rose CACTACEAE (D)
Normanboria Butzin = Acrachne Wight & Arn. ex Chiov. GRAMINEAE (M)
Normanbya F.Muell. ex Becc. PALMAE (M)
Normandia Hook.f. RUBIACEAE (D)
Normania Lowe = Solanum L. SOLANACEAE (D)
Norna Wahlenb. = Calypso Salisb. ORCHIDACEAE (M)

Noronhaea Post & Kuntze (SUO) = Noronhia Stadman ex Thouars OLEACEAE (D)
Noronhia Stadman ex Thouars OLEACEAE (D)
Norrisia Gardner LOGANIACEAE (D)
Northea Hook.f. (SUO) = Northia Hook.f. SAPOTACEAE (D)
Northia Hook.f. SAPOTACEAE (D)
Northiopsis Kaneh. = Manilkara Adans. SAPOTACEAE (D)
Nosema Prain LABIATAE (D)
Nostolachma T.Durand RUBIACEAE (D)
Notanthera (DC.) G.Don LORANTHACEAE (D)
Notechidnopsis Lavranos & Bleck ASCLEPIADACEAE (D)
Notelaea Vent. OLEACEAE (D)
Notelea Raf. (SUO) = Notelaea Vent. OLEACEAE (D)
Noterophila Mart. = Acisanthera P.Browne MELASTOMATACEAE (D)
Nothaphoebe Blume LAURACEAE (D)
Nothapodytes Blume ICACINACEAE (D)
Nothoalsomitra Telford CUCURBITACEAE (D)
Nothobaccharis R.M.King & H.Rob. COMPOSITAE (D)
Nothocalais Greene COMPOSITAE (D)
Nothocarpus Post & Kuntze = Nodocarpaea A.Gray RUBIACEAE (D)
Nothocestrum A.Gray SOLANACEAE (D)
Nothochelone (A.Gray) Straw SCROPHULARIACEAE (D)
Nothochilus Radlk. SCROPHULARIACEAE (D)
Nothocissus (Miq.) Latiff VITACEAE (D)
Nothocnestis Miq. = Bhesa Buch.–Ham. ex Arn. CELASTRACEAE (D)
Nothocnide Blume ex Chew URTICACEAE (D)
Nothodoritis Tsi ORCHIDACEAE (M)
Nothofagus Blume FAGACEAE (D)
Nothoholcus Nash (SUS) = Holcus L. GRAMINEAE (M)
Notholaena R.Br. = Cheilanthes Sw. ADIANTACEAE (F)
Notholcus Hitchc. (SUS) = Holcus L. GRAMINEAE (M)
Notholirion Wall. ex Boiss. LILIACEAE (M)
Nothomyrcia Kausel = Myrceugenia O.Berg MYRTACEAE (D)
Nothopanax Miq. = Polyscias J.R.Forst. & G.Forst. ARALIACEAE (D)
Nothopegia Blume ANACARDIACEAE (D)
Nothopegiopsis Lauterb. = Semecarpus L.f. ANACARDIACEAE (D)
Nothoperanema (Tagawa) Ching DRYOPTERIDACEAE (F)
Nothophlebia Standl. = Pentagonia Benth. RUBIACEAE (D)
Nothoprotium Miq. = Pentaspadon Hook.f. ANACARDIACEAE (D)
Nothoruellia Bremek. = Ruellia L. ACANTHACEAE (D)
Nothosaerva Wight AMARANTHACEAE (D)
Nothoscordum Kunth ALLIACEAE (M)
Nothosmyrnium Miq. UMBELLIFERAE (D)
Nothospondias Engl. SIMAROUBACEAE (D)
Nothostele Garay ORCHIDACEAE (M)
Nothotaxus Florin = Pseudotaxus W.C.Cheng TAXACEAE (G)
Nothotsuga H.H.Hu ex C.N.Page PINACEAE (G)
Noticastrum DC. COMPOSITAE (D)
Notiosciadium Speg. UMBELLIFERAE (D)
Notobasis (Cass.) Cass. COMPOSITAE (D)
Notobuxus Oliv. BUXACEAE (D)
Notocactus (K.Schum.) Frič = Parodia Speg. CACTACEAE (D)
Notocentrum Naudin = Meriania Sw. MELASTOMATACEAE (D)
Notoceras R.Br. CRUCIFERAE (D)
Notochaete Benth. LABIATAE (D)
Notochloe Domin GRAMINEAE (M)

Notodanthonia Zotov = Rytidosperma Steud. GRAMINEAE (M)
Notodon Urb. LEGUMINOSAE–PAPILIONOIDEAE (D)
Notodontia Pierre ex Pit. = Ophiorrhiza L. RUBIACEAE (D)
Notogramme C.Presl = Coniogramme Fée ADIANTACEAE (F)
Notolepeum Newman = Ceterach Willd. ASPLENIACEAE (F)
Notonema Raf. = Agrostis L. GRAMINEAE (M)
Notonia DC. = Kleinia Mill. COMPOSITAE (D)
Notoniopsis B.Nord. = Kleinia Mill. COMPOSITAE (D)
Notophaena Miers = Discaria Hook. RHAMNACEAE (D)
Notopleura (Hook.f.) Bremek. = Psychotria L. RUBIACEAE (D)
Notopora Hook.f. ERICACEAE (D)
Notoptera Urb. COMPOSITAE (D)
Notopterygium Boissieu UMBELLIFERAE (D)
Notosceptrum Benth. = Kniphofia Moench ASPHODELACEAE (M)
Notoseris C.Shih COMPOSITAE (D)
Notospartium Hook.f. LEGUMINOSAE–PAPILIONOIDEAE (D)
Notothixos Oliv. VISCACEAE (D)
Notothlaspi Hook.f. CRUCIFERAE (D)
Nototriche Turcz. MALVACEAE (D)
Nototrichium (A.Gray) W.F.Hillebr. AMARANTHACEAE (D)
Notoxylinon Lewton MALVACEAE (D)
Notylia Lindl. ORCHIDACEAE (M)
Nouelia Franch. COMPOSITAE (D)
Nouettea Pierre APOCYNACEAE (D)
Nouletia Endl. = Cuspidaria DC. BIGNONIACEAE (D)
Novatilia Wight (SUO) = Palicourea Aubl. RUBIACEAE (D)
Novenia Freire COMPOSITAE (D)
Novosieversia F.Bolle ROSACEAE (D)
Nowickea J.Martínez & J.A.McDonald PHYTOLACCACEAE (D)
Nowodworskya C.Presl = Polypogon Desf. GRAMINEAE (M)
Noyera Trécul = Perebea Aubl. MORACEAE (D)
Nucularia Batt. CHENOPODIACEAE (D)
Nudilus Raf. = Forestiera Poir. OLEACEAE (D)
Nuihonia Dop = Craibiodendron W.W.Sm. ERICACEAE (D)
Numaeacampa Gagnep. = Codonopsis Wall. CAMPANULACEAE (D)
Nunnezharia Ruíz & Pav. = Chamaedorea Willd. PALMAE (M)
Nunnezia Willd. = Chamaedorea Willd. PALMAE (M)
Nuphar Sm. NYMPHAEACEAE (D)
Nurmonia Harms = Turraea L. MELIACEAE (D)
Nuttallanthus D.A.Sutton SCROPHULARIACEAE (D)
Nuttallia Torr. & A.Gray = Oemleria Rchb. ROSACEAE (D)
Nuxia Comm. ex Lam. BUDDLEJACEAE (D)
Nuytsia R.Br. ex G.Don LORANTHACEAE (D)
Nyctaginia Choisy NYCTAGINACEAE (D)
Nyctanthes L. VERBENACEAE (D)
Nycteranthus Rothm. (SUS) = Phyllobolus N.E.Br. AIZOACEAE (D)
Nycterianthemum Haw. = Phyllobolus N.E.Br. AIZOACEAE (D)
Nycterinia D.Don = Zaluzianskya F.W.Schmidt SCROPHULARIACEAE (D)
Nycteristion Ruíz & Pav. = Chrysophyllum L. SAPOTACEAE (D)
Nycterium Vent. = Solanum L. SOLANACEAE (D)
Nycticalanthus Ducke RUTACEAE (D)
Nyctocalos Teijsm. & Binn. BIGNONIACEAE (D)
Nyctocereus (A.Berger) Britton & Rose = Peniocereus (A.Berger) Britton & Rose CACTACEAE (D)
Nyctosma Raf. = Epidendrum L. ORCHIDACEAE (M)

Nylandtia Dumort. POLYGALACEAE (D)
Nymania Lindb. MELIACEAE (D)
Nymania K.Schum. (SUH) = Phyllanthus L. EUPHORBIACEAE (D)
Nymphaea L. NYMPHAEACEAE (D)
Nymphanthus Lour. = Phyllanthus L. EUPHORBIACEAE (D)
Nymphoides Séguier MENYANTHACEAE (D)
Nypa Steck PALMAE (M)
Nyssa Gronov. ex L. CORNACEAE (D)
Nyssanthes R.Br. AMARANTHACEAE (D)
Nzidora A.Chev. (SUI) = Tridesmostemon Engl. SAPOTACEAE (D)

Oakes–Amesia C.Schweinf. & P.H.Allen ORCHIDACEAE (M)
Oaxacania B.L.Rob. & Greenm. COMPOSITAE (D)
Obbea Hook.f. = Bobea Gaudich. RUBIACEAE (D)
Oberonia Lindl. ORCHIDACEAE (M)
Obesia Haw. = Stapelia L. ASCLEPIADACEAE (D)
Obetia Gaudich. URTICACEAE (D)
Obione Gaertn. = Atriplex L. CHENOPODIACEAE (D)
Oblivia Strother COMPOSITAE (D)
Obolaria L. GENTIANACEAE (D)
Obolinga Barneby LEGUMINOSAE–MIMOSOIDEAE (D)
Obregonia Frič CACTACEAE (D)
Ocalia Klotzsch = Croton L. EUPHORBIACEAE (D)
Ocampoa A.Rich. & Galeotti = Cranichis Sw. ORCHIDACEAE (M)
Oceanopapaver Guillaumin CAPPARACEAE (D)
Ochagavia Phil. BROMELIACEAE (M)
Ochanostachys Mast. OLACACEAE (D)
Ochetocarpus Meyen = Scyphanthus Sweet LOASACEAE (D)
Ochetophila Poepp. ex Reissek. = Discaria Hook. RHAMNACEAE (D)
Ochlandra Thwaites GRAMINEAE (M)
Ochlogramma C.Presl = Diplazium Sw. WOODSIACEAE (F)
Ochna L. OCHNACEAE (D)
Ochocoa Pierre = Scyphocephalium Warb. MYRISTICACEAE (D)
Ochoterenaea F.A.Barkley ANACARDIACEAE (D)
Ochotia A.P.Khokhr. (UP) UMBELLIFERAE (D)
Ochotonophila Gilli CARYOPHYLLACEAE (D)
Ochradenus Delile RESEDACEAE (D)
Ochreinauclea Ridsdale & Bakh.f. RUBIACEAE (D)
Ochrocarpos Thouars = Garcinia L. GUTTIFERAE (D)
Ochrocephala Dittrich COMPOSITAE (D)
Ochroluma Baill. = Pouteria Aubl. SAPOTACEAE (D)
Ochroma Sw. BOMBACACEAE (D)
Ochronerium Baill. = Tabernaemontana L. APOCYNACEAE (D)
Ochropteris J.Sm. PTERIDACEAE (F)
Ochrosia Juss. APOCYNACEAE (D)
Ochrosperma Trudgen MYRTACEAE (D)
Ochrothallus Pierre ex Baill. = Niemeyera F.Muell. SAPOTACEAE (D)
Ochthephilus Wurdack MELASTOMATACEAE (D)
Ochthocharis Blume MELASTOMATACEAE (D)
Ochthochloa Edgew. GRAMINEAE (M)
Ochthocosmus Benth. IXONANTHACEAE (D)
Ochthodium DC. CRUCIFERAE (D)
Ocimastrum Rupr. = Circaea L. ONAGRACEAE (D)
Ocimum L. LABIATAE (D)
Oclemena Greene – Aster L. COMPOSITAE (D)

Ocotea Aubl. LAURACEAE (D)
Octadesmia Benth. = Dilomilis Raf. ORCHIDACEAE (M)
Octamyrtus Diels MYRTACEAE (D)
Octandrorchis Brieger (SUI) = Octomeria R.Br. ORCHIDACEAE (M)
Octarrhena Thwaites ORCHIDACEAE (M)
Octavia DC. = Lasianthus Jack RUBIACEAE (D)
Octoceras Bunge CRUCIFERAE (D)
Octodon Thonn. = Spermacoce L. RUBIACEAE (D)
Octoknema Pierre OLACACEAE (D)
Octolepis Oliv. THYMELAEACEAE (D)
Octolobus Welw. STERCULIACEAE (D)
Octomeles Miq. DATISCACEAE (D)
Octomeria R.Br. ORCHIDACEAE (M)
Octomeria Pfeiff. (SUO) = Otomeria Benth. RUBIACEAE (D)
Octomeron Robyns LABIATAE (D)
Octonum Raf. = Clidemia D.Don MELASTOMATACEAE (D)
Octopleura Griseb. = Ossaea DC. MELASTOMATACEAE (D)
Octopoma N.E.Br. AIZOACEAE (D)
Octosomatium Gagnep. = Trichodesma R.Br. BORAGINACEAE (D)
Octospermum Airy Shaw EUPHORBIACEAE (D)
Octotheca R.Vig. = Schefflera J.R.Forst. & G.Forst. ARALIACEAE (D)
Octotropis Bedd. RUBIACEAE (D)
Ocyroe Phil. = Nardophyllum (Hook. & Arn.) Hook. & Arn. COMPOSITAE (D)
Oddoniodendron De Wild. LEGUMINOSAE–CAESALPINIOIDEAE (D)
Odicardis Raf. SCROPHULARIACEAE (D)
Odina Roxb. = Lannea A.Rich. ANACARDIACEAE (D)
Odixia Orchard COMPOSITAE (D)
Odonectis Raf. = Isotria Raf. ORCHIDACEAE (M)
Odonellia K.R.Robertson CONVOLVULACEAE (D)
Odonostephana Alexander = Gonolobus Michx. ASCLEPIADACEAE (D)
Odontadenia Benth. APOCYNACEAE (D)
Odontandra Willd. ex Roem. & Schult. = Trichilia P.Browne MELIACEAE (D)
Odontanthera Wight ASCLEPIADACEAE (D)
Odonteilema Turcz. = Acalypha L. EUPHORBIACEAE (D)
Odontella Tiegh. (SUH) = Oncocalyx Tiegh. LORANTHACEAE (D)
Odontelytrum Hack. GRAMINEAE (M)
Odontites Ludw. SCROPHULARIACEAE (D)
Odontocarya Miers MENISPERMACEAE (D)
Odontochilus Blume = Anoectochilus Blume ORCHIDACEAE (M)
Odontocline B.Nord. COMPOSITAE (D)
Odontocyclus Turcz. = Draba L. CRUCIFERAE (D)
Odontoglossum Kunth ORCHIDACEAE (M)
Odontoloma J.Sm. = Lindsaea Dryand. ex Sm. DENNSTAEDTIACEAE (F)
Odontomanes C.Presl = Trichomanes L. HYMENOPHYLLACEAE (F)
Odontonema Nees ACANTHACEAE (D)
Odontonemella Lindau = Mackaya Harv. ACANTHACEAE (D)
Odontophorus N.E.Br. AIZOACEAE (D)
Odontophyllum Sreem. (SUH) = Aphelandra R.Br. ACANTHACEAE (D)
Odontopteris Bernh. = Lygodium Sw. SCHIZAEACEAE (F)
Odontorrhynchus M.N.Corrêa ORCHIDACEAE (M)
Odontosiphon M.Roem. = Trichilia P.Browne MELIACEAE (D)
Odontosoria Fée DENNSTAEDTIACEAE (F)
Odontostelma Rendle ASCLEPIADACEAE (D)
Odontostemum Baker = Odontostomum Torr. TECOPHILAEACEAE (M)
Odontostigma Zoll. & Moritzi = Gymnostachyum Nees ACANTHACEAE (D)

Odontostomum Torr. TECOPHILAEACEAE (M)
Odontostyles Breda = Bulbophyllum Thouars ORCHIDACEAE (M)
Odontotecoma Bureau & K.Schum. = Tabebuia Gomes ex DC. BIGNONIACEAE (D)
Odontotrichum Zucc. = Psacalium Cass. COMPOSITAE (D)
Odontychium K.Schum. = Alpinia Roxb. ZINGIBERACEAE (M)
Odosicyos Keraudren CUCURBITACEAE (D)
Odotalon Raf. = Argythamnia P.Browne EUPHORBIACEAE (D)
Odyendea Pierre ex Engl. SIMAROUBACEAE (D)
Odyssea Stapf GRAMINEAE (M)
Oeceoclades Lindl. ORCHIDACEAE (M)
Oecopetalum Greenm. & C.H.Thomps. ICACINACEAE (D)
Oedematopus Planch. & Triana GUTTIFERAE (D)
Oedera L. COMPOSITAE (D)
Oedibasis Kozo-Pol. UMBELLIFERAE (D)
Oedicephalus Nevski = Astragalus L. LEGUMINOSAE–PAPILIONOIDEAE (D)
Oedina Tiegh. LORANTHACEAE (D)
Oedipachne Link (SUI) = Eriochloa Kunth GRAMINEAE (M)
Oehmea Buxb. = Mammillaria Haw. CACTACEAE (D)
Oemleria Rchb. ROSACEAE (D)
Oenanthe L. UMBELLIFERAE (D)
Oenocarpus Mart. PALMAE (M)
Oenone Tul. = Apinagia Tul. PODOSTEMACEAE (D)
Oenosciadium Pomel = Oenanthe L. UMBELLIFERAE (D)
Oenostachys Bullock = Gladiolus L. IRIDACEAE (M)
Oenothera L. ONAGRACEAE (D)
Oenotheridium Reiche = Clarkia Pursh ONAGRACEAE (D)
Oenotrichia Copel. DENNSTAEDTIACEAE (F)
Oeonia Lindl. ORCHIDACEAE (M)
Oeoniella Schltr. ORCHIDACEAE (M)
Oeosporangium Vis. = Cheilanthes Sw. ADIANTACEAE (F)
Oerstedella Rchb.f. ORCHIDACEAE (M)
Oerstedianthus Lundell = Ardisia Sw. MYRSINACEAE (D)
Oerstedina Wiehler GESNERIACEAE (D)
Oetosis Neck. ex Greene (SUH) = Vittaria Sm. VITTARIACEAE (F)
Oetosis Neck. ex Kuntze = Pyrrosia Mirb. POLYPODIACEAE (F)
Ofaiston Raf. CHENOPODIACEAE (D)
Oftia Adans. SCROPHULARIACEAE (D)
Ogastemma Brummitt BORAGINACEAE (D)
Ogcodeia Bureau = Naucleopsis Miq. MORACEAE (D)
Ohigginsia Ruíz & Pav. = Hoffmannia Sw. RUBIACEAE (D)
Oianthus Benth. = Heterostemma Wight & Arn. ASCLEPIADACEAE (D)
Oiospermum Less. COMPOSITAE (D)
Oistanthera Markgr. = Tabernaemontana L. APOCYNACEAE (D)
Oistonema Schltr. ASCLEPIADACEAE (D)
Okenia Schltdl. & Cham. NYCTAGINACEAE (D)
Okoubaka Pellegr. & Normand SANTALACEAE (D)
Olax L. OLACACEAE (D)
Oldenburgia Less. COMPOSITAE (D)
Oldenlandia L. RUBIACEAE (D)
Oldenlandiopsis Terrell & W.H.Lewis RUBIACEAE (D)
Oldfieldia Benth. & Hook. EUPHORBIACEAE (D)
Olea L. OLEACEAE (D)
Oleandra Cav. OLEANDRACEAE (F)
Oleandropsis Copel. POLYPODIACEAE (F)
Olearia Moench COMPOSITAE (D)

Oleiocarpon Dwyer = Dipteryx Schreb. LEGUMINOSAE–PAPILIONOIDEAE (D)
Olfersia Raddi DRYOPTERIDACEAE (F)
Olgaea Iljin COMPOSITAE (D)
Olgasis Raf. = Oncidium Sw. ORCHIDACEAE (M)
Oligactis (Kunth) Cass. COMPOSITAE (D)
Oligandra Less. = Lucilia Cass. COMPOSITAE (D)
Oliganthes Cass. COMPOSITAE (D)
Oligarrhena R.Br. EPACRIDACEAE (D)
Oligobotrya Baker = Maianthemum G.H.Weber CONVALLARIACEAE (M)
Oligocarpus Less. = Osteospermum L. COMPOSITAE (D)
Oligoceras Gagnep. EUPHORBIACEAE (D)
Oligochaeta (DC.) K.Koch COMPOSITAE (D)
Oligocladus Chodat & Wilczek UMBELLIFERAE (D)
Oligocodon Keay RUBIACEAE (D)
Oligogynium Engl. = Nephthytis Schott ARACEAE (M)
Oligolobos Gagnep. HYDROCHARITACEAE (M)
Oligomeris Cambess. RESEDACEAE (D)
Oligophyton Linder ORCHIDACEAE (M)
Oligoscias Seem. = Polyscias J.R.Forst. & G.Forst. ARALIACEAE (D)
Oligospermum D.Y.Hong = Odicardis Raf. SCROPHULARIACEAE (D)
Oligostachyum Z.P.Wang & G.H.Ye = Arundinaria Michx. GRAMINEAE (M)
Oligostemon Benth. = Duparquetia Baill. LEGUMINOSAE–CAESALPINIOIDEAE (D)
Oligothrix DC. COMPOSITAE (D)
Olinia Thunb. OLINIACEAE (D)
Olisbaea Hook.f. (SUO) = Mouriri Aubl. MELASTOMATACEAE (D)
Olisbea DC. = Mouriri Aubl. MELASTOMATACEAE (D)
Olivaea Sch.Bip. ex Benth. COMPOSITAE (D)
Oliveranthus Rose = Echeveria DC. CRASSULACEAE (D)
Oliverella Tiegh. LORANTHACEAE (D)
Oliveria Vent. UMBELLIFERAE (D)
Oliveriana Rchb.f. ORCHIDACEAE (M)
Olmeca Soderstr. GRAMINEAE (M)
Olmedia Ruíz & Pav. = Trophis P.Browne MORACEAE (D)
Olmediella Baill. FLACOURTIACEAE (D)
Olmedioperebea Ducke = Maquira Aubl. MORACEAE (D)
Olmediophaena Karst. = Maquira Aubl. MORACEAE (D)
Olmediopsis Karst. = Pseudolmedia Trécul MORACEAE (D)
Olneya A.Gray LEGUMINOSAE–PAPILIONOIDEAE (D)
Olostyla DC. = Coelospermum Blume RUBIACEAE (D)
Olsynium Raf. IRIDACEAE (M)
Oluntos Raf. = Ficus L. MORACEAE (D)
Olympia Spach = Hypericum L. GUTTIFERAE (D)
Olymposciadium H.Wolff = Aegokeras Raf. UMBELLIFERAE (D)
Olyra L. GRAMINEAE (M)
Omalanthus A.Juss. (SUO) = Homalanthus A.Juss. EUPHORBIACEAE (D)
Omalocaldos Hook.f. (SUC) = Faramea Aubl. RUBIACEAE (D)
Omalocarpus Choux = Deinbollia Schumach. & Thonn. SAPINDACEAE (D)
Omalotheca Cass. COMPOSITAE (D)
Omania S.Moore SCROPHULARIACEAE (D)
Ombrocharis Hand.-Mazz. LABIATAE (D)
Ombrophytum Poepp. ex Endl. BALANOPHORACEAE (D)
Omeiocalamus Keng f. (SUI) = Arundinaria Michx. GRAMINEAE (M)
Omiltemia Standl. RUBIACEAE (D)
Ommatodium Lindl. = Pterygodium Sw. ORCHIDACEAE (M)
Omoea Blume ORCHIDACEAE (M)

Omphacomeria (Endl.) A.DC. SANTALACEAE (D)
Omphalandria P.Browne = Omphalea L. EUPHORBIACEAE (D)
Omphalea L. EUPHORBIACEAE (D)
Omphalobium Gaertn. = Connarus L. CONNARACEAE (D)
Omphalocarpum P.Beauv. SAPOTACEAE (D)
Omphalodes Mill. BORAGINACEAE (D)
Omphalogonus Baill. = Parquetina Baill. ASCLEPIADACEAE (D)
Omphalogramma (Franch.) Franch. PRIMULACEAE (D)
Omphalolappula Brand BORAGINACEAE (D)
Omphalopappus O.Hoffm. COMPOSITAE (D)
Omphalophthalmum Karst. = Matelea Aubl. ASCLEPIADACEAE (D)
Omphalopus Naudin MELASTOMATACEAE (D)
Omphalotrigonotis W.T.Wang = Trigonotis Steven BORAGINACEAE (D)
Omphalotrix Maxim. SCROPHULARIACEAE (D)
Ona Ravenna = Olsynium Raf. IRIDACEAE (M)
Onagra Mill. (SUS) = Oenothera L. ONAGRACEAE (D)
Oncella Tiegh. LORANTHACEAE (D)
Oncidium Sw. ORCHIDACEAE (M)
Oncinema Arn. ASCLEPIADACEAE (D)
Oncinocalyx F.Muell. VERBENACEAE (D)
Oncinotis Benth. APOCYNACEAE (D)
Oncoba Forssk. FLACOURTIACEAE (D)
Oncocalamus (G.Mann & H.A.Wendl.) Hook.f. PALMAE (M)
Oncocalyx Tiegh. LORANTHACEAE (D)
Oncocarpus A.Gray = Semecarpus L.f. ANACARDIACEAE (D)
Oncodia Lindl. = Brachtia Rchb.f. ORCHIDACEAE (M)
Oncodostigma Diels ANNONACEAE (D)
Oncosiphon Källersjö COMPOSITAE (D)
Oncosperma Blume PALMAE (M)
Oncostemma K.Schum. ASCLEPIADACEAE (D)
Oncostemum A.Juss. MYRSINACEAE (D)
Oncostylus (Schltdl.) Bolle = Geum L. ROSACEAE (D)
Oncotheca Baill. ONCOTHECACEAE (D)
Ondetia Benth. COMPOSITAE (D)
Ondinea Hartog NYMPHAEACEAE (D)
Ongokea Pierre OLACACEAE (D)
Onira Ravenna IRIDACEAE (M)
Onix Medik. = Astragalus L. LEGUMINOSAE–PAPILIONOIDEAE (D)
Onixotis Raf. COLCHICACEAE (M)
Onkeripus Raf. = Xylobium Lindl. ORCHIDACEAE (M)
Onobrychis Mill. LEGUMINOSAE–PAPILIONOIDEAE (D)
Onochiles Bubani = Alkanna Tausch BORAGINACEAE (D)
Onoclea L. WOODSIACEAE (F)
Onocleopsis Ballard WOODSIACEAE (F)
Onoea Franch. & Sav. (SUI) = Diarrhena P.Beauv. GRAMINEAE (M)
Onohualcoa Lundell = Mansoa DC. BIGNONIACEAE (D)
Ononis L. LEGUMINOSAE–PAPILIONOIDEAE (D)
Onopordum L. COMPOSITAE (D)
Onoseris Willd. COMPOSITAE (D)
Onosma L. BORAGINACEAE (D)
Onosmodium Michx. BORAGINACEAE (D)
Onosuris Raf. = Oenothera L. ONAGRACEAE (D)
Onosurus G.Don (SUO) = Oenothera L. ONAGRACEAE (D)
Onuris Phil. CRUCIFERAE (D)
Onus Gilli = Mellera S.Moore ACANTHACEAE (D)

Onychacanthus Nees = Bravaisia DC. ACANTHACEAE (D)
Onychium Blume (SUH) = Dendrobium Sw. ORCHIDACEAE (M)
Onychium Reinw. (SUH) = Lecanopteris Reinw. POLYPODIACEAE (F)
Onychium Kaulf. ADIANTACEAE (F)
Onychopetalum R.E.Fr. ANNONACEAE (D)
Onychosepalum Steud. RESTIONACEAE (M)
Oocarpon Micheli = Ludwigia L. ONAGRACEAE (D)
Oochlamys Fée = Amauropelta Kunze THELYPTERIDACEAE (F)
Oonopsis (Nutt.) Greene COMPOSITAE (D)
Oophytum N.E.Br. AIZOACEAE (D)
Oosterdickia Boehm. = Cunonia L. CUNONIACEAE (D)
Oothrinax O.F.Cook = Zombia L.H.Bailey PALMAE (M)
Opa Lour. = Syzygium Gaertn. MYRTACEAE (D)
Opanea Raf. = Rhodamnia Jack MYRTACEAE (D)
Oparanthus Sherff COMPOSITAE (D)
Opercularia Gaertn. RUBIACEAE (D)
Operculicarya H.Perrier ANACARDIACEAE (D)
Operculina Silva Manso CONVOLVULACEAE (D)
Ophellantha Standl. EUPHORBIACEAE (D)
Ophiala Desv. = Helminthostachys Kaulf. OPHIOGLOSSACEAE (F)
Ophidion Luer ORCHIDACEAE (M)
Ophiobotrys Gilg FLACOURTIACEAE (D)
Ophiocarpus (Bunge) Ikonn. = Astragalus L. LEGUMINOSAE–PAPILIONOIDEAE (D)
Ophiocaryon Endl. MELIOSMACEAE (D)
Ophiocaulon Hook.f. = Adenia Forssk. PASSIFLORACEAE (D)
Ophiocephalus Wiggins SCROPHULARIACEAE (D)
Ophiocolea H.Perrier BIGNONIACEAE (D)
Ophioderma (Blume) Endl. = Ophioglossum L. OPHIOGLOSSACEAE (F)
Ophioglossum L. OPHIOGLOSSACEAE (F)
Ophiomeris Miers = Thismia Griff. BURMANNIACEAE (M)
Ophione Schott = Dracontium L. ARACEAE (M)
Ophionella Bruyns ASCLEPIADACEAE (D)
Ophiopogon Ker Gawl. CONVALLARIACEAE (M)
Ophiopteris Reinw. = Oleandra Cav. OLEANDRACEAE (F)
Ophiorrhiza L. RUBIACEAE (D)
Ophiorrhiziphyllon Kurz ACANTHACEAE (D)
Ophiria Becc. = Pinanga Blume PALMAE (M)
Ophismenus Poir. (SUO) = Oplismenus P.Beauv. GRAMINEAE (M)
Ophiurinella Desv. = Stenotaphrum Trin. GRAMINEAE (M)
Ophiuros C.F.Gaertn. GRAMINEAE (M)
Ophrestia H.M.L.Forbes LEGUMINOSAE–PAPILIONOIDEAE (D)
Ophryococcus Oerst. RUBIACEAE (D)
Ophryosporus Meyen COMPOSITAE (D)
Ophrypetalum Diels ANNONACEAE (D)
Ophrys L. ORCHIDACEAE (M)
Ophthalmoblapton Allemão EUPHORBIACEAE (D)
Ophthalmophyllum Dinter & Schwantes = AIZOACEAE (D)
Opilia Roxb. OPILIACEAE (D)
Opisthiolepis L.S.Sm. PROTEACEAE (D)
Opisthocentra Hook.f. MELASTOMATACEAE (D)
Opisthopappus C.Shih COMPOSITAE (D)
Opithandra B.L.Burtt GESNERIACEAE (D)
Opizia C.Presl GRAMINEAE (M)
Oplexion Raf. = Lobostemon Lehm. BORAGINACEAE (D)
Oplismenopsis Parodi GRAMINEAE (M)

Oplismenus P.Beauv. GRAMINEAE (M)
Oplonia Raf. ACANTHACEAE (D)
Oplopanax (Torr. & A.Gray) Miq. ARALIACEAE (D)
Oplukion Raf. = Lycium L. SOLANACEAE (D)
Opocunonia Schltr. = Caldcluvia D.Don CUNONIACEAE (D)
Opoidea Lindl. = Peucedanum L. UMBELLIFERAE (D)
Opopanax W.D.J.Koch UMBELLIFERAE (D)
Opophytum N.E.Br. = Mesambryanthemum L. AIZOACEAE (D)
Opsago Raf. = Withania Pauquy SOLANACEAE (D)
Opsiandra O.F.Cook = Gaussia H.A.Wendl. PALMAE (M)
Opsianthes Lilja = Clarkia Pursh ONAGRACEAE (D)
Opuntia Mill. CACTACEAE (D)
Opuntiopsis Knebel (SUI) = Schlumbergera Lem. CACTACEAE (D)
Orania Zipp. PALMAE (M)
Oraniopsis J.Dransf., A.K.Irvine & N.Uhl PALMAE (M)
Oraoma Turcz. = Aglaia Lour. MELIACEAE (D)
Orbea Haw. ASCLEPIADACEAE (D)
Orbeanthus L.C. Leach ASCLEPIADACEAE (D)
Orbeopsis L.C. Leach ASCLEPIADACEAE (D)
Orbexilum Raf. LEGUMINOSAE–PAPILIONOIDEAE (D)
Orbicularia Baill. = Phyllanthus L. EUPHORBIACEAE (D)
Orbignya Mart. ex Endl. PALMAE (M)
Orchadocarpa Ridl. GESNERIACEAE (D)
Orchiastrum Séguier = Spiranthes Rich. ORCHIDACEAE (M)
Orchidantha N.E.Br. LOWIACEAE (M)
Orchidium Sw. = Calypso Salisb. ORCHIDACEAE (M)
Orchidocarpum Michx. = Asimina Adans. ANNONACEAE (D)
Orchidofunckia A.Rich. & Galeotti = Cryptarrhena R.Br. ORCHIDACEAE (M)
Orchidotypus Kraenzl. = Pachyphyllum Kunth ORCHIDACEAE (M)
Orchiodes Kuntze = Goodyera R.Br. ORCHIDACEAE (M)
Orchipeda Blume = Voacanga Thouars APOCYNACEAE (D)
Orchipedum Breda ORCHIDACEAE (M)
Orchis L. ORCHIDACEAE (M)
Orchyllium Barnhart = Utricularia L. LENTIBULARIACEAE (D)
Orcuttia Vasey GRAMINEAE (M)
Oreacanthus Benth. ACANTHACEAE (D)
Oreanthes Benth. ERICACEAE (D)
Oreas Cham. & Schltdl. = Aphragmus Andrz. ex DC. CRUCIFERAE (D)
Orectanthe Maguire XYRIDACEAE (M)
Oregandra Standl. RUBIACEAE (D)
Oreiostachys Gamble = Nastus Juss. GRAMINEAE (M)
Oreithales Schltdl. RANUNCULACEAE (D)
Oreobambos K.Schum. GRAMINEAE (M)
Oreoblastus Susl. = Christolea Cambess. ex Jacquem. CRUCIFERAE (D)
Oreobliton Durieu CHENOPODIACEAE (D)
Oreobolopsis Koyama & Guagl. CYPERACEAE (M)
Oreobolus R.Br. CYPERACEAE (M)
Oreobroma Howell = Lewisia Pursh PORTULACACEAE (D)
Oreocalamus Keng = Chimonobambusa Makino GRAMINEAE (M)
Oreocallis R.Br. PROTEACEAE (D)
Oreocallis Small (SUH) = Leucothoe D.Don ERICACEAE (D)
Oreocarya Greene = Cryptantha G.Don BORAGINACEAE (D)
Oreocaryon Kuntze ex K.Schum. = Cruckshanksia Hook. & Arn. RUBIACEAE (D)
Oreocereus (A.Berger) Riccob. CACTACEAE (D)
Oreocharis (Decne.) Lindl. (SUH) = Pseudomertensia Riedl BORAGINACEAE (D)

Oreocharis Benth. GESNERIACEAE (D)
Oreochloa Link GRAMINEAE (M)
Oreochrysum Rydb. COMPOSITAE (D)
Oreocnide Miq. URTICACEAE (D)
Oreocome Edgew. UMBELLIFERAE (D)
Oreocosmus Naudin = Tibouchina Aubl. MELASTOMATACEAE (D)
Oreodaphne Nees & Mart. = Ocotea Aubl. LAURACEAE (D)
Oreodendron C.T.White THYMELAEACEAE (D)
Oreodoxa Willd. = Prestoea Hook.f. PALMAE (M)
Oreogenia I.M.Johnst. (SUH) = Lasiocaryum I.M.Johnst. BORAGINACEAE (D)
Oreogeum (Ser.) Golubkova = Geum L. ROSACEAE (D)
Oreogrammitis Copel. = Grammitis Sw. GRAMMITIDACEAE (F)
Oreograstis K.Schum. = Carpha Banks & Sol. ex R.Br. CYPERACEAE (M)
Oreoherzogia W.Vent = Rhamnus L. RHAMNACEAE (D)
Oreoleysera Bremer COMPOSITAE (D)
Oreoloma Botsch. CRUCIFERAE (D)
Oreomitra Diels ANNONACEAE (D)
Oreomunnea Oerst. JUGLANDACEAE (D)
Oreomyrrhis Endl. UMBELLIFERAE (D)
Oreonana Jeps. UMBELLIFERAE (D)
Oreonesion J.Raynal GENTIANACEAE (D)
Oreopanax Decne. & Planch. ARALIACEAE (D)
Oreophylax Endl. = Gentianella Moench GENTIANACEAE (D)
Oreophysa (Bunge ex Boiss.) Bornm. LEGUMINOSAE–PAPILIONOIDEAE (D)
Oreophyton O.E.Schulz CRUCIFERAE (D)
Oreopoa Gand. (SUI) = Poa L. GRAMINEAE (M)
Oreopolus Schltdl. = Cruckshanksia Hook. & Arn. RUBIACEAE (D)
Oreoporanthera Hutch. EUPHORBIACEAE (D)
Oreopteris Holub THELYPTERIDACEAE (F)
Oreorchis Lindl. ORCHIDACEAE (M)
Oreorhamnus Ridl. = Rhamnus L. RHAMNACEAE (D)
Oreoschimperella Rauschert UMBELLIFERAE (D)
Oreosciadium Wedd. = Niphogeton Schltdl. UMBELLIFERAE (D)
Oreosedum Grulich = Sedum L. CRASSULACEAE (D)
Oreosolen Hook.f. SCROPHULARIACEAE (D)
Oreospacus Phil. = Satureja L. LABIATAE (D)
Oreosparte Schltr. ASCLEPIADACEAE (D)
Oreostemma Greene = Aster L. COMPOSITAE (D)
Oreostylidium Berggr. STYLIDIACEAE (D)
Oreosyce Hook.f. CUCURBITACEAE (D)
Oreothyrsus Lindau = Ptyssiglottis T.Anderson ACANTHACEAE (D)
Oreoxis Raf. UMBELLIFERAE (D)
Oresitrophe Bunge SAXIFRAGACEAE (D)
Orestias Ridl. ORCHIDACEAE (M)
Orfilea Baill. = Lautembergia Baill. EUPHORBIACEAE (D)
Orias Dode = Lagerstroemia L. LYTHRACEAE (D)
Oribasia Schreb. = Palicourea Aubl. RUBIACEAE (D)
Oricia Pierre RUTACEAE (D)
Oriciopsis Engl. RUTACEAE (D)
Origanum L. LABIATAE (D)
Orinocoa Raf. = Athenaea Sendtn. SOLANACEAE (D)
Orinus Hitchc. GRAMINEAE (M)
Orites R.Br. PROTEACEAE (D)
Orithalia Blume = Agalmyla Blume GESNERIACEAE (D)
Oritrephes Ridl. MELASTOMATACEAE (D)

Oritrophium (Kunth) Cuatrec. COMPOSITAE (D)
Orixa Thunb. RUTACEAE (D)
Orlaya Hoffm. UMBELLIFERAE (D)
Orleanesia Barb.Rodr. ORCHIDACEAE (M)
Ormenis Cass. = Chamaemelum Mill. COMPOSITAE (D)
Ormocarpopsis R.Vig. LEGUMINOSAE–PAPILIONOIDEAE (D)
Ormocarpum P.Beauv. LEGUMINOSAE–PAPILIONOIDEAE (D)
Ormoloma Maxon DENNSTAEDTIACEAE (F)
Ormopteris J.Sm. = Pellaea Link ADIANTACEAE (F)
Ormopterum Schischk. UMBELLIFERAE (D)
Ormosciadium Boiss. UMBELLIFERAE (D)
Ormosia G.Jacks. LEGUMINOSAE–PAPILIONOIDEAE (D)
Ormosiopsis Ducke = Ormosia G.Jacks. LEGUMINOSAE–PAPILIONOIDEAE (D)
Ormosolenia Tausch UMBELLIFERAE (D)
Ormostema Raf. = Dendrobium Sw. ORCHIDACEAE (M)
Ornanthes Raf. = Fraxinus L. OLEACEAE (D)
Ornichia Klack. GENTIANACEAE (D)
Ornitharium Lindl. & Paxton = Pteroceras Hasselt ex Hassk. ORCHIDACEAE (M)
Ornithidium R.Br. = Maxillaria Ruhz & Pav. ORCHIDACEAE (M)
Ornithoboea Parish ex C.B.Clarke GESNERIACEAE (D)
Ornithocarpa Rose CRUCIFERAE (D)
Ornithocephalochloa Kurz = Thuarea Pers. GRAMINEAE (M)
Ornithocephalus Hook. ORCHIDACEAE (M)
Ornithochilus (Lindl.) Benth. ORCHIDACEAE (M)
Ornithogalum L. HYACINTHACEAE (M)
Ornithoglossum Salisb. COLCHICACEAE (M)
Ornithophora Barb.Rodr. ORCHIDACEAE (M)
Ornithopteris Bernh. = Anemia Sw. SCHIZAEACEAE (F)
Ornithopteris (J.G.Agardh) J.Sm. (SUH) = Pteridium Gled. ex Scop.
    DENNSTAEDTIACEAE (F)
Ornithopus L. LEGUMINOSAE–PAPILIONOIDEAE (D)
Ornithospermum Dumoulin = Echinochloa P.Beauv. GRAMINEAE (M)
Ornithostaphylos Small ERICACEAE (D)
Ornithrophus Bojer ex Engl. (SUI) = Weinmannia L. CUNONIACEAE (D)
Ornus Boehm. = Fraxinus L. OLEACEAE (D)
Orobanche L. SCROPHULARIACEAE (D)
Orobus L. = Lathyrus L. LEGUMINOSAE–PAPILIONOIDEAE (D)
Orochaenactis Coville COMPOSITAE (D)
Orogenia S.Watson UMBELLIFERAE (D)
Orontium L. ARACEAE (M)
Oropetium Trin. GRAMINEAE (M)
Orophaca Nutt. = Astragalus L. LEGUMINOSAE–PAPILIONOIDEAE (D)
Orophea Blume ANNONACEAE (D)
Orophochilus Lindau ACANTHACEAE (D)
Orophoma Spruce = Mauritia L.f. PALMAE (M)
Orostachys Steud. (SUI) = Hordelymus (Jess.) Jess. ex Harz GRAMINEAE (M)
Orostachys Fisch. ex A.Berger CRASSULACEAE (D)
Orothamnus Pappe ex Hook. PROTEACEAE (D)
Oroxylum Vent. BIGNONIACEAE (D)
Oroya Britton & Rose = Oreocereus (A.Berger) Riccob. CACTACEAE (D)
Orphanidesia Boiss. & Balansa = Epigaea L. ERICACEAE (D)
Orphanodendron Barneby & J.W.Grimes LEGUMINOSAE–CAESALPINIOIDEAE (D)
Orphium E.Mey. GENTIANACEAE (D)
Orrhopygium A.Löve = Aegilops L. GRAMINEAE (M)
Orsidice Rchb.f. ▪ Thrixspermum Lour. ORCHIDACEAE (M)

Ortachne Nees ex Steud. GRAMINEAE (M)
Ortegia L. CARYOPHYLLACEAE (D)
Ortegocactus Alexander CACTACEAE (D)
Ortgiesia Regel = Aechmea Ruhz & Pav. BROMELIACEAE (M)
Orthaea Klotzsch ERICACEAE (D)
Orthandra Burret = Mortoniodendron Standl. & Steyerm. TILIACEAE (D)
Orthantha (Bernth.) Wettst. = Odontites Ludw. SCROPHULARIACEAE (D)
Orthanthe Lem. = Sinningia Nees GESNERIACEAE (D)
Orthanthella Rauschert = Odontites Ludw. SCROPHULARIACEAE (D)
Orthanthera Wight ASCLEPIADACEAE (D)
Orthechites Urb. = Secondatia A.DC. APOCYNACEAE (D)
Orthilia Raf. ERICACEAE (D)
Orthion Standl. & Steyerm. VIOLACEAE (D)
Orthiopteris Copel. DENNSTAEDTIACEAE (F)
Orthocarpus Nutt. SCROPHULARIACEAE (D)
Orthoceras R.Br. ORCHIDACEAE (M)
Orthochilus Hochst. ex A.Rich. = Eulophia R.Br. ex Lindl. ORCHIDACEAE (M)
Orthoclada P.Beauv. GRAMINEAE (M)
Orthodon Benth. & Oliv. = Mosla (Benth.) Buch.–Ham. ex Maxim. LABIATAE (D)
Orthogoneuron Gilg MELASTOMATACEAE (D)
Orthogramma C.Presl = Blechnum L. BLECHNACEAE (F)
Orthogynium Baill. MENISPERMACEAE (D)
Ortholobium Gagnep. = Archidendron F.Muell. LEGUMINOSAE–MIMOSOIDEAE (D)
Ortholoma Hanst. = Columnea L. GESNERIACEAE (D)
Orthomene Barneby & Krukoff MENISPERMACEAE (D)
Orthopenthea Rolfe = Disa Bergius ORCHIDACEAE (M)
Orthophytum Beer BROMELIACEAE (M)
Orthopichonia H.Huber APOCYNACEAE (D)
Orthopogon R.Br. = Oplismenus P.Beauv. GRAMINEAE (M)
Orthopterum L.Bolus AIZOACEAE (D)
Orthopterygium Hemsl. ANACARDIACEAE (D)
Orthoraphium Nees = Stipa L. GRAMINEAE (M)
Orthosia Decne. ASCLEPIADACEAE (D)
Orthosiphon Benth. LABIATAE (D)
Orthosphenia Standl. CELASTRACEAE (D)
Orthostemma Wall. ex Voigt (SUI) = Pentas Benth. RUBIACEAE (D)
Orthostemon O.Berg = Acca O.Berg MYRTACEAE (D)
Orthotactus Nees = Justicia L. ACANTHACEAE (D)
Orthotheca Pichon (SUH) = Cuspidaria DC. BIGNONIACEAE (D)
Orthothylax (Hook.f.) Skottsb. = Helmholtzia F.Muell. PHILYDRACEAE (M)
Orthrosanthus Sweet IRIDACEAE (M)
Orthurus Juz. ROSACEAE (D)
Ortiga Neck. (SUI) = Loasa Adans. LOASACEAE (D)
Ortmannia Opiz = Geodorum G.Jacks. ORCHIDACEAE (M)
Orumbella J.M.Coult. & Rose = Podistera S.Watson UMBELLIFERAE (D)
Orxera Raf. = Aerides Lour. ORCHIDACEAE (M)
Orychophragmus Bunge CRUCIFERAE (D)
Oryctanthus (Griseb.) Eichler LORANTHACEAE (D)
Oryctes S.Watson SOLANACEAE (D)
Oryctina Tiegh. LORANTHACEAE (D)
Orygia Forssk. = Corbichonia Scop. MOLLUGINACEAE (D)
Oryza L. GRAMINEAE (M)
Oryzidium C.E.Hubb. & Schweick. GRAMINEAE (M)
Oryzopsis Michx. GRAMINEAE (M)
Osa Aiello RUBIACEAE (D)

Osbeckia L. MELASTOMATACEAE (D)
Osbeckiastrum Naudin = Dissotis Benth. MELASTOMATACEAE (D)
Osbertia Greene COMPOSITAE (D)
Osbornia F.Muell. MYRTACEAE (D)
Oschatzia Walp. UMBELLIFERAE (D)
Oscularia Schwantes AIZOACEAE (D)
Oserya Tul. & Wedd. PODOSTEMACEAE (D)
Oshimella Masam. & Suzuki = Whytockia W.W.Sm. GESNERIACEAE (D)
Osmanthus Lour. OLEACEAE (D)
Osmaronia Greene = Oemleria Rchb. ROSACEAE (D)
Osmelia Thwaites FLACOURTIACEAE (D)
Osmhydrophora Barb.Rodr. = Tanaecium Sw. BIGNONIACEAE (D)
Osmiopsis R.M.King & H.Rob. COMPOSITAE (D)
Osmitopsis Cass. COMPOSITAE (D)
Osmoglossum (Schltr.) Schltr. ORCHIDACEAE (M)
Osmorhiza Raf. UMBELLIFERAE (D)
Osmoxylon Miq. ARALIACEAE (D)
Osmunda L. OSMUNDACEAE (F)
Osmundastrum C.Presl = Osmunda L. OSMUNDACEAE (F)
Osmundopteris (J.Milde) Small = Botrychium Sw. OPHIOGLOSSACEAE (F)
Ossaea DC. MELASTOMATACEAE (D)
Ossiculum P.J.Cribb & Laan ORCHIDACEAE (M)
Ossifraga Rumph. = Euphorbia L. EUPHORBIACEAE (D)
Ostenia Buchenau = Hydrocleys Rich. LIMNOCHARITACEAE (M)
Osteocarpum F.Muell. = Threlkeldia R.Br. CHENOPODIACEAE (D)
Osteocarpus Phil. = Nolana L.f. SOLANACEAE (D)
Osteomeles Lindl. ROSACEAE (D)
Osteophloeum Warb. MYRISTICACEAE (D)
Osteospermum L. COMPOSITAE (D)
Osterdamia Kuntze (SUS) = Zoysia Willd. GRAMINEAE (M)
Ostericum Hoffm. = Angelica L. UMBELLIFERAE (D)
Ostodes Blume EUPHORBIACEAE (D)
Ostrearia Baill. HAMAMELIDACEAE (D)
Ostrowskia Regel CAMPANULACEAE (D)
Ostrya Scop. CORYLACEAE (D)
Ostryocarpus Hook.f. LEGUMINOSAE-PAPILIONOIDEAE (D)
Ostryoderris Dunn = Aganope Miq. LEGUMINOSAE-PAPILIONOIDEAE (D)
Ostryopsis Decne. CORYLACEAE (D)
Osyricera Blume = Bulbophyllum Thouars ORCHIDACEAE (M)
Osyridicarpos A.DC. SANTALACEAE (D)
Osyris L. SANTALACEAE (D)
Otacanthus Lindl. SCROPHULARIACEAE (D)
Otachyrium Nees GRAMINEAE (M)
Otandra Salisb. = Geodorum G.Jacks. ORCHIDACEAE (M)
Otanthera Blume MELASTOMATACEAE (D)
Otanthus Hoffmanns. & Link COMPOSITAE (D)
Otatea (McClure & W.E.Sm.) Calderón & Soderstr. = Sinarundinaria Nakai
    GRAMINEAE (M)
Oteiza La Llave COMPOSITAE (D)
Otherodendron Makino = Microtropis Wall. ex Meisn. CELASTRACEAE (D)
Otholobium C.H.Stirt. LEGUMINOSAE-PAPILIONOIDEAE (D)
Othonna L. COMPOSITAE (D)
Otilix Raf. = Lycianthes (Dunal) Hassl. SOLANACEAE (D)
Otiophora Zucc. RUBIACEAE (D)
Otoba (DC.) Karst. MYRISTICACEAE (D)

Otocalyx Brandegee RUBIACEAE (D)
Otocarpus Durieu CRUCIFERAE (D)
Otocephalus Chiov. = Calanda K.Schum. RUBIACEAE (D)
Otochilus Lindl. ORCHIDACEAE (M)
Otoglossum (Schltr.) Garay & Dunst. ORCHIDACEAE (M)
Otomeria Benth. RUBIACEAE (D)
Otonephelium Radlk. SAPINDACEAE (D)
Otopappus Benth. COMPOSITAE (D)
Otopetalum Lehm. & Kraenzl. = Pleurothallis R.Br. ORCHIDACEAE (M)
Otophora Blume = Lepisanthes Blume SAPINDACEAE (D)
Otoptera DC. LEGUMINOSAE–PAPILIONOIDEAE (D)
Otospermum Willk. COMPOSITAE (D)
Otostegia Benth. LABIATAE (D)
Otostemma Blume = Hoya R.Br. ASCLEPIADACEAE (D)
Otostylis Schltr. ORCHIDACEAE (M)
Otoxalis Small = Oxalis L. OXALIDACEAE (D)
Ottelia Pers. HYDROCHARITACEAE (M)
Ottoa Kunth UMBELLIFERAE (D)
Ottochloa Dandy GRAMINEAE (M)
Ottonia Spreng. = Piper L. PIPERACEAE (D)
Ottoschmidtia Urb. RUBIACEAE (D)
Ottoschulzia Urb. ICACINACEAE (D)
Ottosonderia L.Bolus AIZOACEAE (D)
Oubanguia Baill. SCYTOPETALACEAE (D)
Oudneya R.Br. = Henophyton Coss. & Durieu CRUCIFERAE (D)
Ougeinia Benth. = Desmodium Desv. LEGUMINOSAE–PAPILIONOIDEAE (D)
Ouratea Aubl. OCHNACEAE (D)
Ourisia Comm. ex Juss. SCROPHULARIACEAE (D)
Ourisianthus Bonati = Artanema D.Don SCROPHULARIACEAE (D)
Ourouparia Aubl. = Uncaria Schreb. RUBIACEAE (D)
Outarda Dumort. (SUO) = Coutarea Aubl. RUBIACEAE (D)
Outreya Jaub. & Spach COMPOSITAE (D)
Ovaria Fabr. = Solanum L. SOLANACEAE (D)
Ovidia Meisn. THYMELAEACEAE (D)
Ovilla Adans. = Jasione L. CAMPANULACEAE (D)
Owataria Matsum. = Suregada Roxb. ex Rottler EUPHORBIACEAE (D)
Owenia F.Muell. MELIACEAE (D)
Oxalis L. OXALIDACEAE (D)
Oxalistylis Baill. = Phyllanthus L. EUPHORBIACEAE (D)
Oxandra A.Rich. ANNONACEAE (D)
Oxanthera Montrouz. RUTACEAE (D)
Oxera Labill. VERBENACEAE (D)
Oxyanthe Steud. = Phragmites Adans. GRAMINEAE (M)
Oxyanthera Brongn. = Thelasis Blume ORCHIDACEAE (M)
Oxyanthus DC. RUBIACEAE (D)
Oxybaphus L'Hér. ex Willd. NYCTAGINACEAE (D)
Oxycarpha S.F.Blake COMPOSITAE (D)
Oxycarpus Lour. = Garcinia L. GUTTIFERAE (D)
Oxycaryum Nees CYPERACEAE (M)
Oxyceros Lour. RUBIACEAE (D)
Oxychlamys Schltr. GESNERIACEAE (D)
Oxychloe Phil. JUNCACEAE (M)
Oxychloris Lazarides GRAMINEAE (M)
Oxycoccus Hill = Vaccinium L. ERICACEAE (D)
Oxydectes Kuntze = Croton L. EUPHORBIACEAE (D)

Oxydendrum DC. ERICACEAE (D)
Oxydenia Nutt. = Leptochloa P.Beauv. GRAMINEAE (M)
Oxyglottis (Bunge) Nevski = Astragalus L. LEGUMINOSAE–PAPILIONOIDEAE (D)
Oxygonium C.Presl = Diplazium Sw. WOODSIACEAE (F)
Oxygonum Burch. ex Campdera POLYGONACEAE (D)
Oxygraphis Bunge RANUNCULACEAE (D)
Oxygyne Schltr. BURMANNIACEAE (M)
Oxylaena Benth. ex Anderb. COMPOSITAE (D)
Oxylobium Andrews LEGUMINOSAE–PAPILIONOIDEAE (D)
Oxylobus (Moq. ex DC.) A.Gray COMPOSITAE (D)
Oxymeris DC. = Leandra Raddi MELASTOMATACEAE (D)
Oxymitra (Blume) Hook.f. & Thomson (SUH) = Friesodielsia Steenis ANNONACEAE (D)
Oxymyrrhine Schauer = Baeckea L. MYRTACEAE (D)
Oxyosmyles Speg. BORAGINACEAE (D)
Oxypappus Benth. COMPOSITAE (D)
Oxypetalum R.Br. ASCLEPIADACEAE (D)
Oxyphyllum Phil. COMPOSITAE (D)
Oxypolis Raf. UMBELLIFERAE (D)
Oxypteryx Greene = Asclepias L. ASCLEPIADACEAE (D)
Oxyrhachis Pilg. GRAMINEAE (M)
Oxyrhynchus Brandegee LEGUMINOSAE–PAPILIONOIDEAE (D)
Oxyria Hill POLYGONACEAE (D)
Oxys Mill. = Oxalis L. OXALIDACEAE (D)
Oxysepala Wight = Bulbophyllum Thouars ORCHIDACEAE (M)
Oxyspermum Eckl. & Zeyh. = Galopina Thunb. RUBIACEAE (D)
Oxyspora DC. MELASTOMATACEAE (D)
Oxystelma R.Br. ASCLEPIADACEAE (D)
Oxystemon Planch. & Triana = Clusia L. GUTTIFERAE (D)
Oxystigma Harms LEGUMINOSAE–CAESALPINIOIDEAE (D)
Oxystophyllum Blume = Dendrobium Sw. ORCHIDACEAE (M)
Oxystylis Torr. & Frém. CAPPARACEAE (D)
Oxytenanthera Munro GRAMINEAE (M)
Oxytenia Nutt. = Iva L. COMPOSITAE (D)
Oxytheca Nutt. POLYGONACEAE (D)
Oxythece Miq. = Pouteria Aubl. SAPOTACEAE (D)
Oxytropis DC. LEGUMINOSAE–PAPILIONOIDEAE (D)
Oyedaea DC. COMPOSITAE (D)
Ozodycus Raf. = Cucurbita L. CUCURBITACEAE (D)
Ozomelis Raf. = Mitella L. SAXIFRAGACEAE (D)
Ozoroa Delile ANACARDIACEAE (D)
Ozothamnus R.Br. COMPOSITAE (D)

Pabellonia Quezada & Martic. = Leucocoryne Lindl. ALLIACEAE (M)
Pabstia Garay ORCHIDACEAE (M)
Pabstiella Brieger & Senghas = Pleurothallis R.Br. ORCHIDACEAE (M)
Pachea Steud. (SUO) = Crypsis Aiton GRAMINEAE (M)
Pachecoa Standl. & Steyerm. LEGUMINOSAE–PAPILIONOIDEAE (D)
Pacheya Scop. = Coussarea Aubl. RUBIACEAE (D)
Pachira Aubl. BOMBACACEAE (D)
Pachites Lindl. ORCHIDACEAE (M)
Pachyanthus Rich. MELASTOMATACEAE (D)
Pachycarpus E.Mey. ASCLEPIADACEAE (D)
Pachycentria Blume MELASTOMATACEAE (D)
Pachycereus (A.Berger) Britton & Rose CACTACEAE (D)
Pachychilus Blume = Pachystoma Blume ORCHIDACEAE (M)

Pachychlamys Dyer ex Ridl. = Shorea Roxb. ex C.F.Gaertn. DIPTEROCARPACEAE (D)
Pachycladon Hook.f. CRUCIFERAE (D)
Pachycormus Coville ex Standl. ANACARDIACEAE (D)
Pachycornia Hook.f. CHENOPODIACEAE (D)
Pachyctenium Maire & Pamp. UMBELLIFERAE (D)
Pachycymbium L.C. Leach ASCLEPIADACEAE (D)
Pachyderma Blume = Olea L. OLEACEAE (D)
Pachydesmia Gleason = Miconia Ruíz & Pav. MELASTOMATACEAE (D)
Pachydiscus Gilg & Schltr. = Wittsteinia F.Muell. ALSEUOSMIACEAE (D)
Pachyelasma Harms LEGUMINOSAE–CAESALPINIOIDEAE (D)
Pachygone Miers MENISPERMACEAE (D)
Pachylaena D.Don ex Hook. & Arn. COMPOSITAE (D)
Pachylarnax Dandy MAGNOLIACEAE (D)
Pachylecythis Ledoux = Lecythis Loefl. LECYTHIDACEAE (D)
Pachylobus G.Don = Dacryodes Vahl BURSERACEAE (D)
Pachyloma Van den Bosch (SUH) = Hymenophyllum Sm. HYMENOPHYLLACEAE (F)
Pachyloma DC. MELASTOMATACEAE (D)
Pachylophus Spach = Oenothera L. ONAGRACEAE (D)
Pachymeria Benth. = Meriania Sw. MELASTOMATACEAE (D)
Pachymitus O.E.Schulz CRUCIFERAE (D)
Pachyne Salisb. = Phaius Lour. ORCHIDACEAE (M)
Pachynema R.Br. ex DC. DILLENIACEAE (D)
Pachyneurum Bunge CRUCIFERAE (D)
Pachynocarpus Hook.f. = Vatica L. DIPTEROCARPACEAE (D)
Pachypharynx Aellen = Atriplex L. CHENOPODIACEAE (D)
Pachyphragma (DC.) Rchb. CRUCIFERAE (D)
Pachyphyllum Kunth ORCHIDACEAE (M)
Pachyphytum Link, Klotzsch & Otto CRASSULACEAE (D)
Pachyplectron Schltr. ORCHIDACEAE (M)
Pachypleuria (C.Presl) C.Presl = Humata Cav. DAVALLIACEAE (F)
Pachypleurum Ledeb. UMBELLIFERAE (D)
Pachypodanthium Engl. & Diels ANNONACEAE (D)
Pachypodium Webb & Berthel. (SUH) = Sisymbrium L. CRUCIFERAE (D)
Pachypodium Nutt. (SUH) = Thelypodium Endl. CRUCIFERAE (D)
Pachypodium Lindl. APOCYNACEAE (D)
Pachyptera DC. = Mansoa DC. BIGNONIACEAE (D)
Pachypteris Kar. & Kir. = Pachypterygium Bunge CRUCIFERAE (D)
Pachypterygium Bunge CRUCIFERAE (D)
Pachyrhizanthe (Schltr.) Nakai = Cymbidium Sw. ORCHIDACEAE (M)
Pachyrhizus Rich. ex DC. LEGUMINOSAE–PAPILIONOIDEAE (D)
Pachyrhynchus DC. = Lucilia Cass. COMPOSITAE (D)
Pachysandra Michx. BUXACEAE (D)
Pachysanthus C.Presl = Rudgea Salisb. RUBIACEAE (D)
Pachysolen Phil. = Nolana L.f. SOLANACEAE (D)
Pachystachys Nees ACANTHACEAE (D)
Pachystegia Cheeseman = Olearia Moench COMPOSITAE (D)
Pachystela Pierre ex Engl. = Synsepalum (A.DC.) Daniell SAPOTACEAE (D)
Pachystele Schltr. ORCHIDACEAE (M)
Pachystelis Rauschert = Pachystele Schltr. ORCHIDACEAE (M)
Pachystelma Brandegee = Matelea Aubl. ASCLEPIADACEAE (D)
Pachystemon Blume = Macaranga Thouars EUPHORBIACEAE (D)
Pachystigma Hochst. RUBIACEAE (D)
Pachystima Raf. = Paxistima Raf. CELASTRACEAE (D)
Pachystoma Blume ORCHIDACEAE (M)
Pachystrobilus Bremek. = Strobilanthes Blume ACANTHACEAE (D)

Pachystroma Müll.Arg. EUPHORBIACEAE (D)
Pachystylidium Pax & K.Hoffm. EUPHORBIACEAE (D)
Pachystylis Blume = Pachystoma Blume ORCHIDACEAE (M)
Pachystylus K.Schum. RUBIACEAE (D)
Pachythamnus (R.M.King & H.Rob.) R.M.King & H.Rob. COMPOSITAE (D)
Pachytrophe Bureau = Streblus Lour. MORACEAE (D)
Packera Á.Löve & D.Löve COMPOSITAE (D)
Pacouria Aubl. APOCYNACEAE (D)
Pacourina Aubl. COMPOSITAE (D)
Padbruggea Miq. LEGUMINOSAE-PAPILIONOIDEAE (D)
Padellus Vassilcz. = Prunus L. ROSACEAE (D)
Padia Moritzi = Oryza L. GRAMINEAE (M)
Padus Mill. = Prunus L. ROSACEAE (D)
Paederia L. RUBIACEAE (D)
Paederota L. SCROPHULARIACEAE (D)
Paederotella (Wulf) Kem.–Nath. = Veronica L. SCROPHULARIACEAE (D)
Paedicalyx Pierre ex Pit. = Xanthophytum Reinw. ex Blume RUBIACEAE (D)
Paeonia L. PAEONIACEAE (D)
Paepalanthus Kunth ERIOCAULACEAE (M)
Paesia A.St.–Hil. DENNSTAEDTIACEAE (F)
Pagaea Griseb. = Irlbachia Mart. GENTIANACEAE (D)
Pagamea Aubl. RUBIACEAE (D)
Pagameopsis Steyerm. RUBIACEAE (D)
Pagella Schönl. = Crassula L. CRASSULACEAE (D)
Pagetia F.Muell. = Bosistoa F.Muell. RUTACEAE (D)
Pagiantha Markgr. = Tabernaemontana L. APOCYNACEAE (D)
Pahudia Miq. = Afzelia Sm. LEGUMINOSAE–CAESALPINIOIDEAE (D)
Painteria Britton & Rose = Havardia Small LEGUMINOSAE–MIMOSOIDEAE (D)
Paiva Vell. = Sabicea Aubl. RUBIACEAE (D)
Paivaea O.Berg = Campomanesia Ruíz & Pav. MYRTACEAE (D)
Paivaeusa Welw. = Oldfieldia Benth. & Hook.f. EUPHORBIACEAE (D)
Pajanelia DC. BIGNONIACEAE (D)
Pakaraimaea Maguire & P.S.Ashton DIPTEROCARPACEAE (D)
Paladelpha Pichon = Alstonia R.Br. APOCYNACEAE (D)
Palaeocyanus Dostál COMPOSITAE (D)
Palafoxia Lag. COMPOSITAE (D)
Palandra O.F.Cook = Phytelephas Ruíz & Pav. PALMAE (M)
Palanostigma Mart. ex Klotzsch = Croton L. EUPHORBIACEAE (D)
Palaoea Kaneh. = Tristiropsis Radlk. SAPINDACEAE (D)
Palaquium Blanco SAPOTACEAE (D)
Palaua Cav. MALVACEAE (D)
Palava Juss. (SUO) = Palaua Cav. MALVACEAE (D)
Paleaepappus Cabrera COMPOSITAE (D)
Palenga Thwaites = Drypetes Vahl EUPHORBIACEAE (D)
Paleodicraeia C.Cusset PODOSTEMACEAE (D)
Paletuviera Thouars ex DC. = Bruguiera Sav. RHIZOPHORACEAE (D)
Palhinhaea Franco & Vasc. = Lycopodiella Holub LYCOPODIACEAE (FA)
Paliavana Vell. ex Vand. GESNERIACEAE (D)
Palicourea Aubl. RUBIACEAE (D)
Palimbia Besser ex DC. UMBELLIFERAE (D)
Paliris Dumort. = Liparis Rich. ORCHIDACEAE (M)
Palisota Rchb. ex Endl. COMMELINACEAE (M)
Palissya Baill. (SUH) = Necepsia Prain EUPHORBIACEAE (D)
Paliurus Mill. RHAMNACEAE (D)
Pallasia Scop. (SUH) = Crypsis Alton GRAMINEAE (M)

Pallasia Klotzsch = Wittmackanthus Kuntze RUBIACEAE (D)
Pallavicinia De Not. = Cyphomandra Mart. ex Sendtn. SOLANACEAE (D)
Pallenis (Cass.) Cass. (SUS) = Asteriscus Mill. COMPOSITAE (D)
Palma Mill. = Phoenix L. PALMAE (M)
Palmerella A.Gray = Laurentia Adans. CAMPANULACEAE (D)
Palmeria F.Muell. MONIMIACEAE (D)
Palmerocassia Britton = Senna Mill. LEGUMINOSAE–CAESALPINIOIDEAE (D)
Palmervandenbrockia Gibbs = Polyscias J.R.Forst. & G.Forst. ARALIACEAE (D)
Palmjuncus Kuntze = Calamus L. PALMAE (M)
Palmoglossum Klotzsch ex Rchb.f. = Pleurothallis R.Br. ORCHIDACEAE (M)
Palmolmedia Ducke = Naucleopsis Miq. MORACEAE (D)
Palmorchis Barb.Rodr. ORCHIDACEAE (M)
Palmstruckia Sond. (SUH) = Thlaspeocarpa C.A.Sm. CRUCIFERAE (D)
Paloue Aubl. LEGUMINOSAE–CAESALPINIOIDEAE (D)
Palovea Juss. (SUO) = Paloue Aubl. LEGUMINOSAE–CAESALPINIOIDEAE (D)
Paloveopsis R.S.Cowan LEGUMINOSAE–CAESALPINIOIDEAE (D)
Paltonium C.Presl = Neurodium Fée POLYPODIACEAE (F)
Palumbina Rchb.f. ORCHIDACEAE (M)
Pamburus Swingle RUTACEAE (D)
Pamea Aubl. = Buchenavia Eichler COMBRETACEAE (D)
Pamianthe Stapf AMARYLLIDACEAE (M)
Pamphalea Lag. COMPOSITAE (D)
Pamphilia Mart. ex A.DC. STYRACACEAE (D)
Pamplethantha Bremek. RUBIACEAE (D)
Panamanthus Kuijt LORANTHACEAE (D)
Panax L. ARALIACEAE (D)
Pancheria Montrouz. (SUH) = Ixora L. RUBIACEAE (D)
Pancheria Brongn. & Gris CUNONIACEAE (D)
Panchezia B.D.Jacks. (SUO) = Ixora L. RUBIACEAE (D)
Pancicia Vis. & Schltdl. UMBELLIFERAE (D)
Pancovia Willd. SAPINDACEAE (D)
Pancratium L. AMARYLLIDACEAE (M)
Panda Pierre PANDACEAE (D)
Pandaca Thouars = Tabernaemontana L. APOCYNACEAE (D)
Pandacastrum Pichon = Tabernaemontana L. APOCYNACEAE (D)
Pandanus Parkinson PANDANACEAE (M)
Panderia Fisch. & C.A.Mey. CHENOPODIACEAE (D)
Pandiaka (Moq.) Hook.f. AMARANTHACEAE (D)
Pandorea (Endl.) Spach BIGNONIACEAE (D)
Paneion Lunell (SUS) = Poa L. GRAMINEAE (M)
Panetos Raf. = Houstonia L. RUBIACEAE (D)
Pangium Reinw. FLACOURTIACEAE (D)
Panicastrella Moench (SUS) = Echinaria Desf. GRAMINEAE (M)
Panicularia Fabr. (SUS) = Poa L. GRAMINEAE (M)
Panicularia Colla (SUH) = Thyrsopteris Kunze DICKSONIACEAE (F)
Paniculum Ard. (SUO) = Panicum L. GRAMINEAE (M)
Panicum L. GRAMINEAE (M)
Panisea (Lindl.) Lindl. ORCHIDACEAE (M)
Panke Willd. = Francoa Cav. SAXIFRAGACEAE (D)
Panopia Noronha ex Thouars = Macaranga Thouars EUPHORBIACEAE (D)
Panopsis Salisb. ex Knight PROTEACEAE (D)
Panphalea Lag. (SUO) = Pamphalea Lag. COMPOSITAE (D)
Panstrepis Raf. = Coryanthes Hook. ORCHIDACEAE (M)
Pantacantha Speg. SOLANACEAE (D)
Pantadenia Gagnep. EUPHORBIACEAE (D)

Pantathera Phil. = Megalachne Steud. GRAMINEAE (M)
Panterpa Miers = Arrabidaea DC. BIGNONIACEAE (D)
Pantorrhynchus Murb. = Trachystoma O.E.Schulz CRUCIFERAE (D)
Panurea Spruce ex Benth. LEGUMINOSAE–PAPILIONOIDEAE (D)
Panzeria J.F.Gmel. (SUH) = Lycium L. SOLANACEAE (D)
Panzeria Moench (SUH) = Panzerina Soják LABIATAE (D)
Panzerina Soják LABIATAE (D)
Paolia Chiov. = Coffea L. RUBIACEAE (D)
Papaver L. PAPAVERACEAE (D)
Paphia Seem. = Agapetes D.Don ex G.Don ERICACEAE (D)
Paphinia Lindl. ORCHIDACEAE (M)
Paphiopedilum Pfitzer ORCHIDACEAE (M)
Papilionanthe Schltr. ORCHIDACEAE (M)
Papiliopsis E.Morren = Oncidium Sw. ORCHIDACEAE (M)
Papillilabium Dockrill ORCHIDACEAE (M)
Pappagrostis Roshev. = Stephanachne Keng GRAMINEAE (M)
Pappea Eckl. & Zeyh. SAPINDACEAE (D)
Papperitzia Rchb.f. ORCHIDACEAE (M)
Pappobolus S.F.Blake COMPOSITAE (D)
Pappophorum Schreb. GRAMINEAE (M)
Pappostyles Pierre = Cremaspora Benth. RUBIACEAE (D)
Pappostylum Pierre (SUO) = Cremaspora Benth. RUBIACEAE (D)
Papuacedrus H.L.Li = Libocedrus Endl. CUPRESSACEAE (G)
Papuaea Schltr. ORCHIDACEAE (M)
Papualthia Diels ANNONACEAE (D)
Papuanthes Danser LORANTHACEAE (D)
Papuapteris C.Chr. DRYOPTERIDACEAE (F)
Papuastelma Bullock ASCLEPIADACEAE (D)
Papuechites Markgr. APOCYNACEAE (D)
Papuodendron C.T.White MALVACEAE (D)
Papuzilla Ridl. = Lepidium L. CRUCIFERAE (D)
Papyrius Lam. = Broussonetia L'Hér. ex Vent. MORACEAE (D)
Parabaena Miers MENISPERMACEAE (D)
Parabarium Pierre = Ecdysanthera Hook. & Arn. APOCYNACEAE (D)
Parabarleria Baill. = Barleria L. ACANTHACEAE (D)
Parabeaumontia Pichon = Vallaris Burm.f. APOCYNACEAE (D)
Parabenzoin Nakai = Lindera Thunb. LAURACEAE (D)
Paraberlinia Pellegr. = Julbernardia Pellegr. LEGUMINOSAE–CAESALPINIOIDEAE (D)
Parabesleria Oerst. = Besleria L. GESNERIACEAE (D)
Parabignonia Bureau ex K.Schum. BIGNONIACEAE (D)
Parablechnum C.Presl = Blechnum L. BLECHNACEAE (F)
Paraboea (C.B.Clarke) Ridl. GESNERIACEAE (D)
Parabotrys J.C.Muell. (SUO) = Xylopia L. ANNONACEAE (D)
Parabouchetia Baill. SOLANACEAE (D)
Paracalanthe Kudo = Calanthe Ker–Gawl. ORCHIDACEAE (M)
Paracaleana Blaxell ORCHIDACEAE (M)
Paracalia Cuatrec. COMPOSITAE (D)
Paracalyx Ali LEGUMINOSAE–PAPILIONOIDEAE (D)
Paracarpaea (K.Schum.) Pichon = Arrabidaea DC. BIGNONIACEAE (D)
Paracaryopsis R.R.Mill BORAGINACEAE (D)
Paracaryum (A.DC.) Boiss. BORAGINACEAE (D)
Paracasearia Boerl. = Drypetes Vahl EUPHORBIACEAE (D)
Paracautleya Ros.M.Sm. ZINGIBERACEAE (M)
Paracelastrus Miq. = Microtropis Wall. ex Meisn. CELASTRACEAE (D)
Paracelsea Zoll. (SUH) = Acalypha L. EUPHORBIACEAE (D)

Paracephaelis Baill. RUBIACEAE (D)
Paraceterach Copel. = Gymnopteris Bernh. ADIANTACEAE (F)
Parachampionella Bremek. = Strobilanthes Blume ACANTHACEAE (D)
Parachimarrhis Ducke RUBIACEAE (D)
Paraclarisia Ducke = Sorocea A.St.–Hil. MORACEAE (D)
Paracleisthus Gagnep. = Cleistanthus Hook.f. ex Planch. EUPHORBIACEAE (D)
Paracoffea J.–F.Leroy (SUI) = Psilanthus Hook.f. RUBIACEAE (D)
Paracolea Baill. = Phylloctenium Baill. BIGNONIACEAE (D)
Paracolpodium (Tsvelev) Tsvelev = Colpodium Trin. GRAMINEAE (M)
Paracorynanthe Capuron RUBIACEAE (D)
Paracroton Miq. = Fahrenheitia Rchb.f. & Zoll. EUPHORBIACEAE (D)
Paracryphia Baker f. PARACRYPHIACEAE (D)
Paractaenum P.Beauv. GRAMINEAE (M)
Paracynoglossum Popov = Cynoglossum L. BORAGINACEAE (D)
Paradaniellia Rolfe = Daniellia Benn. LEGUMINOSAE–CAESALPINIOIDEAE (D)
Paradavallodes Ching = Araiostegia Copel. DAVALLIACEAE (F)
Paradennstaedtia Tagawa = Dennstaedtia Bernh. DENNSTAEDTIACEAE (F)
Paradenocline Müll.Arg. = Adenocline Turcz. EUPHORBIACEAE (D)
Paraderris (Miq.) R.Geesink = Derris Lour. LEGUMINOSAE–PAPILIONOIDEAE (D)
Paradina Pierre ex Pit. = Mitragyna Korth. RUBIACEAE (D)
Paradisanthus Rchb.f. ORCHIDACEAE (M)
Paradisea Mazzuc. ASPHODELACEAE (M)
Paradolichandra Hassl. = Parabignonia Bureau ex K.Schum. BIGNONIACEAE (D)
Paradombeya Stapf STERCULIACEAE (D)
Paradrymonia Hanst. GESNERIACEAE (D)
Paradrypetes Kuhlm. EUPHORBIACEAE (D)
Paraeremostachys Adylov, Kamelin & Machm. = Eremostachys Bunge LABIATAE (D)
Parafestuca E.B.Alexeev GRAMINEAE (M)
Paragelonium Leandri = Aristogeitonia Prain EUPHORBIACEAE (D)
Paragenipa Baill. RUBIACEAE (D)
Parageum Nakai & H.Hara = Geum L. ROSACEAE (D)
Paraglycine F.J.Herm. = Ophrestia H.M.L.Forbes LEGUMINOSAE–
PAPILIONOIDEAE (D)
Paragnathis Spreng. = Diplomeris D.Don ORCHIDACEAE (M)
Paragoldfussia Bremek. = Strobilanthes Blume ACANTHACEAE (D)
Paragonia Bureau BIGNONIACEAE (D)
Paragophyton K.Schum. = Spermacoce L. RUBIACEAE (D)
Paragramma (Blume) T.Moore POLYPODIACEAE (F)
Paragrewia Gagnep. ex R.S.Rao = Leptonychia Turcz. STERCULIACEAE (D)
Paragulubia Burret = Gulubia Becc. PALMAE (M)
Paragutzlaffia H.P.Tsui = Strobilanthes Blume ACANTHACEAE (D)
Paragynoxys (Cuatrec.) Cuatrec. COMPOSITAE (D)
Parahancornia Ducke APOCYNACEAE (D)
Parahebe W.R.B.Oliv. SCROPHULARIACEAE (D)
Parahyparrhenia A.Camus GRAMINEAE (M)
Paraia Rohwer, H.G.Richt. & van der Werff LAURACEAE (D)
Paraixeris Nakai COMPOSITAE (D)
Parajaeschkea Burkill = Gentianella Moench GENTIANACEAE (D)
Parajubaea Burret PALMAE (M)
Parajusticia Benoist = Justicia L. ACANTHACEAE (D)
Parakaempferia A.S.Rao & D.M.Verma ZINGIBERACEAE (M)
Parakibara Philipson MONIMIACEAE (D)
Parakmeria Hu & Cheng = Magnolia L. MAGNOLIACEAE (D)
Paraknoxia Bremek. RUBIACEAE (D)
Parakohleria Wiehler GESNERIACEAE (D)

Paralabatia Pierre = Pouteria Aubl. SAPOTACEAE (D)
Paralamium Dunn LABIATAE (D)
Paralbizzia Kosterm. = Archidendron F.Muell. LEGUMINOSAE–MIMOSOIDEAE (D)
Paralepistemon Lejoly & Lisowski CONVOLVULACEAE (D)
Paraleptochilus Copel. = Leptochilus Kaulf. POLYPODIACEAE (F)
Paraligusticum V.N.Tikhom. UMBELLIFERAE (D)
Paralinospadix Burret = Calyptrocalyx Blume PALMAE (M)
Paralstonia Baill. = Alyxia Banks ex R.Br. APOCYNACEAE (D)
Paralychnophora MacLeish = Eremanthus Less. COMPOSITAE (D)
Paralyxia Baill. = Aspidosperma Mart. & Zucc. APOCYNACEAE (D)
Paramachaerium Ducke LEGUMINOSAE–PAPILIONOIDEAE (D)
Paramacrolobium J.Léonard LEGUMINOSAE–CAESALPINIOIDEAE (D)
Paramammea J.-F.Leroy = Mammea L. GUTTIFERAE (D)
Paramansoa Baill. = Arrabidaea DC. BIGNONIACEAE (D)
Paramapania Uittien CYPERACEAE (M)
Paramelhania Arènes STERCULIACEAE (D)
Parameria Benth. APOCYNACEAE (D)
Parameriopsis Pichon = Parameria Benth. APOCYNACEAE (D)
Paramichelia Hu = Michelia L. MAGNOLIACEAE (D)
Paramicropholis Aubrév. & Pellegr. = Micropholis (Griseb.) Pierre SAPOTACEAE (D)
Paramignya Wight RUTACEAE (D)
Paramitranthes Burret = Siphoneugena O.Berg MYRTACEAE (D)
Paramoltkia Greuter BORAGINACEAE (D)
Paramomum S.Q.Tong (UP) ZINGIBERACEAE (M)
Paramongaia Velarde AMARYLLIDACEAE (M)
Paramyrciaria Kausel MYRTACEAE (D)
Paranecepsia Radcl.-Sm. EUPHORBIACEAE (D)
Paranephelium Miq. SAPINDACEAE (D)
Paranephelius Poepp. COMPOSITAE (D)
Paraneurachne S.T.Blake GRAMINEAE (M)
Paranneslea Gagnep. THEACEAE (D)
Paranomus Salisb. PROTEACEAE (D)
Parantennaria Beauverd COMPOSITAE (D)
Paranthe O.F.Cook (SUI) = Chamaedorea Willd. PALMAE (M)
Parapachygone Forman MENISPERMACEAE (D)
Parapactis W.Zimmerm. = Epipactis Zinn ORCHIDACEAE (M)
Parapanax Miq. = Schefflera J.R.Forst. & G.Forst. ARALIACEAE (D)
Parapantadenia Capuron = Pantadenia Gagnep. EUPHORBIACEAE (D)
Parapentapanax Hutch. = Pentapanax Seem. ARALIACEAE (D)
Parapentas Bremek. RUBIACEAE (D)
Paraphalaenopsis A.D.Hawkes ORCHIDACEAE (M)
Paraphlomis Prain LABIATAE (D)
Parapholis C.E.Hubb. GRAMINEAE (M)
Paraphyadanthe Mildbr. = Caloncoba Gilg FLACOURTIACEAE (D)
Parapiptadenia Brenan LEGUMINOSAE–MIMOSOIDEAE (D)
Parapiqueria R.M.King & H.Rob. COMPOSITAE (D)
Parapodium E.Mey. ASCLEPIADACEAE (D)
Parapolystichum (Keyserl.) Ching = Lastreopsis Ching DRYOPTERIDACEAE (F)
Paraprenanthes Chang ex C.Shih COMPOSITAE (D)
Paraprotium Cuatrec. BURSERACEAE (D)
Parapteroceras Averyanov ORCHIDACEAE (M)
Parapteropyrum A.J.Li POLYGONACEAE (D)
Parapyrenaria H.T.Chang = Pyrenaria Blume THEACEAE (D)
Paraquilegia J.R.Drumm. & Hutch. RANUNCULACEAE (D)
Pararchidendron I.C.Nielsen LEGUMINOSAE–MIMOSOIDEAE (D)

Parardisia M.P.Nayar & Giri = Ardisia Sw. MYRSINACEAE (D)
Pararistolochia Hutch. & Dalziel ARISTOLOCHIACEAE (D)
Parartabotrys Miq. = Xylopia L. ANNONACEAE (D)
Parartocarpus Baill. MORACEAE (D)
Pararuellia Bremek. ACANTHACEAE (D)
Parasamanea Kosterm. = Albizia Durazz. LEGUMINOSAE–MIMOSOIDEAE (D)
Parasarcochilus Dockrill = Sarcochilus R.Br. ORCHIDACEAE (M)
Parasarcochilus Dockrill (SUH) = Pteroceras Hasselt ex Hassk. ORCHIDACEAE (M)
Parasassafras D.G.Long LAURACEAE (D)
Parascheelea Dugand = Orbignya Mart. ex Endl. PALMAE (M)
Parascopolia Baill. = Lycianthes (Dunal) Hassl. SOLANACEAE (D)
Paraselinum H.Wolff UMBELLIFERAE (D)
Parasenecio W.W.Sm. & Small COMPOSITAE (D)
Paraserianthes I.C.Nielsen LEGUMINOSAE–MIMOSOIDEAE (D)
Parashorea Kurz DIPTEROCARPACEAE (D)
Parasicyos Dieterle CUCURBITACEAE (D)
Parasilaus Leute UMBELLIFERAE (D)
Parasitaxus de Laub. PODOCARPACEAE (G)
Parasitipomoea Hayata = Ipomoea L. CONVOLVULACEAE (D)
Paraskevia W.Sauer & G.Sauer BORAGINACEAE (D)
Parasorus Alderw. DAVALLIACEAE (F)
Parasponia Miq. ULMACEAE (D)
Parastemon A.DC. CHRYSOBALANACEAE (D)
Parastranthus G.Don = Lobelia L. CAMPANULACEAE (D)
Parastrephia Nutt. COMPOSITAE (D)
Parastriga Mildbr. SCROPHULARIACEAE (D)
Parastrobilanthes Bremek. = Strobilanthes Blume ACANTHACEAE (D)
Parastyrax W.W.Sm. STYRACACEAE (D)
Parasympagis Bremek. = Strobilanthes Blume ACANTHACEAE (D)
Parasyringa W.W.Sm. = Ligustrum L. OLEACEAE (D)
Parasystasia Baill. = Asystasia Blume ACANTHACEAE (D)
Paratecoma Kuhlm. BIGNONIACEAE (D)
Paratephrosia Domin = Tephrosia Pers. LEGUMINOSAE–PAPILIONOIDEAE (D)
Parathelypteris (H.Ito) Ching THELYPTERIDACEAE (F)
Paratheria Griseb. GRAMINEAE (M)
Parathesis (A.DC.) Hook.f. MYRSINACEAE (D)
Parathyrium Holttum = Lunathyrium Koldz. WOODSIACEAE (F)
Paratriaina Bremek. RUBIACEAE (D)
Paratrophis Blume = Streblus Lour. MORACEAE (D)
Paratropia (Blume) DC. = Schefflera J.R.Forst. & G.Forst. ARALIACEAE (D)
Paravallaris Pierre = Kibatalia G.Don APOCYNACEAE (D)
Paravinia Hassk. (SUO) = Praravinia Korth. RUBIACEAE (D)
Paravitex H.R.Fletcher VERBENACEAE (D)
Pardanthopsis (Hance) Lenz IRIDACEAE (M)
Pardoglossum E.Barbier & Mathez BORAGINACEAE (D)
Parenterolobium Kosterm. = Albizia Durazz. LEGUMINOSAE–MIMOSOIDEAE (D)
Parentucellia Viv. SCROPHULARIACEAE (D)
Parestia C.Presl = Davallia Sm. DAVALLIACEAE (F)
Pareugenia Turrill = Syzygium Gaertn. MYRTACEAE (D)
Parhabenaria Gagnep. ORCHIDACEAE (M)
Pariana Aubl. GRAMINEAE (M)
Parietaria L. URTICACEAE (D)
Parinari Aubl. CHRYSOBALANACEAE (D)
Parinarium Juss. = Parinari Aubl. CHRYSOBALANACEAE (D)
Paris L. TRILLIACEAE (M)

Parishella A.Gray CAMPANULACEAE (D)
Parishia Hook.f. ANACARDIACEAE (D)
Pariti Adans. = Hibiscus L. MALVACEAE (D)
Parkeria Hook. = Ceratopteris Brongn. PARKERIACEAE (F)
Parkia R.Br. LEGUMINOSAE–MIMOSOIDEAE (D)
Parkinsonia L. LEGUMINOSAE–CAESALPINIOIDEAE (D)
Parlatorea Barb.Rodr. (SUH) = Sanderella Kuntze ORCHIDACEAE (M)
Parlatoria Boiss. CRUCIFERAE (D)
Parmentiera Raf. (SUH) = Solanum L. SOLANACEAE (D)
Parmentiera DC. BIGNONIACEAE (D)
Parnassia L. PARNASSIACEAE (D)
Parochetus Buch.–Ham. ex D.Don LEGUMINOSAE–PAPILIONOIDEAE (D)
Parodia Speg. CACTACEAE (D)
Parodianthus Tronc. VERBENACEAE (D)
Parodiella J.R.Reeder & C.G.Reeder (SUH) = Ortachne Nees ex Steud. GRAMINEAE (M)
Parodiochloa A.M.Molina (SUH) = Koeleria Pers. GRAMINEAE (M)
Parodiochloa C.E.Hubb. = Poa L. GRAMINEAE (M)
Parodiodendron Hunz. EUPHORBIACEAE (D)
Parodiodoxa O.E.Schulz CRUCIFERAE (D)
Parodiolyra Soderstr. & Zuloaga GRAMINEAE (M)
Parolinia Webb CRUCIFERAE (D)
Paronychia Mill. ILLECEBRACEAE (D)
Paropsia Noronha ex Thouars PASSIFLORACEAE (D)
Paropsiopsis Engl. PASSIFLORACEAE (D)
Paropyrum Ulbr. = Isopyrum L. RANUNCULACEAE (D)
Parosela Cav. = Dalea L. LEGUMINOSAE–PAPILIONOIDEAE (D)
Paroxygraphis W.W.Sm. RANUNCULACEAE (D)
Parquetina Baill. ASCLEPIADACEAE (D)
Parqui Adans. = Cestrum L. SOLANACEAE (D)
Parrasia Greene = Nerisyrenia Greene CRUCIFERAE (D)
Parrotia C.A.Mey. HAMAMELIDACEAE (D)
Parrotiopsis (Nied.) C.K.Schneid. HAMAMELIDACEAE (D)
Parrya R.Br. CRUCIFERAE (D)
Parryella Torr. & A.Gray LEGUMINOSAE–PAPILIONOIDEAE (D)
Parryodes Jafri CRUCIFERAE (D)
Parryopsis Botsch. CRUCIFERAE (D)
Parsana Parsa & Maleki = Laportea Gaudich. URTICACEAE (D)
Parsonsia R.Br. APOCYNACEAE (D)
Parsonsia P.Browne (SUH) = Cuphea P.Browne LYTHRACEAE (D)
Parthenice A.Gray COMPOSITAE (D)
Parthenium L. COMPOSITAE (D)
Parthenocissus Planch. VITACEAE (D)
Parvatia Decne. LARDIZABALACEAE (D)
Parviopuntia Soulaire & Marn.–Lap. (SUI) = Opuntia Mill. CACTACEAE (D)
Parvisedum R.T.Clausen = Sedella Britton & Rose CRASSULACEAE (D)
Parvotrisetum Chrtek = Trisetaria Forssk. GRAMINEAE (M)
Paryphantha Schauer – Thryptomene Endl. MYRTACEAE (D)
Paryphosphaera Karst. = Parkia R.Br. LEGUMINOSAE–MIMOSOIDEAE (D)
Pasaccardoa Kuntze COMPOSITAE (D)
Pasania Oerst. = Lithocarpus Blume FAGACEAE (D)
Pasaniopsis Kudo = Castanopsis (D.Don) Spach FAGACEAE (D)
Pascalia Ortega = Wedelia Jacq. COMPOSITAE (D)
Pascalium Cass. (SUO) = Psacalium Cass. COMPOSITAE (D)
Paschalococos J.Dransf. PALMAE (M)
Paschanthus Burch. – Adenia Forssk. PASSIFLORACEAE (D)

Pascopyrum Á.Löve = Elymus L. GRAMINEAE (M)
Pasithea D.Don ANTHERICACEAE (M)
Pasovia Karst. = Phthirusa Mart. LORANTHACEAE (D)
Paspalanthium Desv. = Paspalum L. GRAMINEAE (M)
Paspalidium Stapf GRAMINEAE (M)
Paspalum L. GRAMINEAE (M)
Passaea Adans. = Ononis L. LEGUMINOSAE–PAPILIONOIDEAE (D)
Passaea Baill. (SUH) = Bernardia Mill. EUPHORBIACEAE (D)
Passaveria Mart. & Eichler ex Miq. = Ecclinusa Mart. SAPOTACEAE (D)
Passerina L. THYMELAEACEAE (D)
Passiflora L. PASSIFLORACEAE (D)
Passowia Karst. (SUO) = Phthirusa Mart. LORANTHACEAE (D)
Pastinaca L. UMBELLIFERAE (D)
Pastinacopsis Golosk. UMBELLIFERAE (D)
Patabea Aubl. = Ixora L. RUBIACEAE (D)
Patagonula L. BORAGINACEAE (D)
Patania C.Presl = Dennstaedtia Bernh. DENNSTAEDTIACEAE (F)
Patascoya Urb. = Freziera Willd. THEACEAE (D)
Patellaria J.T.Williams, A.J.Scott & Ford–Lloyd (SUH) = Patellifolia A.J.Scott, Ford–Lloyd &
    J.T.Williams CHENOPODIACEAE (D)
Patellifolia A.J.Scott, Ford–Lloyd & J.T.Williams CHENOPODIACEAE (D)
Patersonia R.Br. IRIDACEAE (M)
Patima Aubl. = Sabicea Aubl. RUBIACEAE (D)
Patinoa Cuatrec. BOMBACACEAE (D)
Patis Ohwi = Stipa L. GRAMINEAE (M)
Patonia Wight = Xylopia L. ANNONACEAE (D)
Patosia Buchenau = Oxychloe Phil. JUNCACEAE (M)
Patrinia Juss. VALERIANACEAE (D)
Patrisia Rich. = Ryania Vahl FLACOURTIACEAE (D)
Patropyrum Á.Löve = Aegilops L. GRAMINEAE (M)
Patsjotti Adans. = Strumpfia Jacq. RUBIACEAE (D)
Pattalias S.Watson ASCLEPIADACEAE (D)
Pattonia Wight = Grammatophyllum Blume ORCHIDACEAE (M)
Paua Caball. = Andryala L. COMPOSITAE (D)
Pauella Ramam. & Sebastine = Theriophonum Blume ARACEAE (M)
Pauia Deb & Dutta SOLANACEAE (D)
Pauldopia Steenis BIGNONIACEAE (D)
Paulia Korovin (SUH) = Paulita Soják UMBELLIFERAE (D)
Paulita Soják UMBELLIFERAE (D)
Paullinia L. SAPINDACEAE (D)
Paulowilhelmia Hochst. (SUH) = Eremomastax Lindau ACANTHACEAE (D)
Paulownia Siebold & Zucc. SCROPHULARIACEAE (D)
Paulseniella Briq. = Elsholtzia Willd. LABIATAE (D)
Pauridia Harv. HYPOXIDACEAE (M)
Pauridiantha Hook.f. RUBIACEAE (D)
Paurolepis S.Moore = Gutenbergia Sch.Bip. COMPOSITAE (D)
Paurotis O.F.Cook = Acoelorrhaphe H.A.Wendl. PALMAE (M)
Pausandra Radlk. EUPHORBIACEAE (D)
Pausia Raf. (SUS) = Osmanthus Lour. OLEACEAE (D)
Pausinystalia Pierre ex Beille RUBIACEAE (D)
Pavate Adans. = Pavetta L. RUBIACEAE (D)
Pavetta L. RUBIACEAE (D)
Pavieasia Pierre SAPINDACEAE (D)
Pavinda Thunb. = Audouinia Brongn. BRUNIACEAE (D)
Pavonia Cav. MALVACEAE (D)

Paxia O.E.G.Nilsson (SUH) = Neopaxia O.E.G.Nilsson PORTULACACEAE (D)
Paxia Gilg = Rourea Aubl. CONNARACEAE (D)
Paxiodendron Engl. = Xymalos Baill. MONIMIACEAE (D)
Paxistima Raf. CELASTRACEAE (D)
Paxiuscula Herter = Ditaxis Vahl ex A.Juss. EUPHORBIACEAE (D)
Paxtonia Lindl. = Spathoglottis Blume ORCHIDACEAE (M)
Payena A.DC. SAPOTACEAE (D)
Payera Baill. RUBIACEAE (D)
Payeria Baill. = Turraea L. MELIACEAE (D)
Paypayrola Aubl. VIOLACEAE (D)
Pearcea Regel GESNERIACEAE (D)
Pearsonia Dümmer LEGUMINOSAE–PAPILIONOIDEAE (D)
Peccana Raf. = Euphorbia L. EUPHORBIACEAE (D)
Pechea Lapeyr. (SUI) = Crypsis Aiton GRAMINEAE (M)
Pecheya Scop. = Coussarea Aubl. RUBIACEAE (D)
Pechuel–Loeschea O.Hoffm. COMPOSITAE (D)
Peckeya Raf. (SUO) = Coussarea Aubl. RUBIACEAE (D)
Peckia Vell. = Cybianthus Mart. MYRSINACEAE (D)
Peckoltia Fourn. ASCLEPIADACEAE (D)
Pecluma M.G.Price POLYPODIACEAE (F)
Pecteilis Raf. ORCHIDACEAE (M)
Pectiantia Raf. = Mitella L. SAXIFRAGACEAE (D)
Pectinaria (Benth.) Hack. (SUH) = Eremochloa Büse GRAMINEAE (M)
Pectinaria Haw. ASCLEPIADACEAE (D)
Pectinaria Cordem. (SUH) = Angraecum Bory ORCHIDACEAE (M)
Pectinella J.M.Black = Amphibolis C.Agardh CYMODOCEACEAE (M)
Pectis L. COMPOSITAE (D)
Pectocarya DC. ex Meisn. BORAGINACEAE (D)
Pedaliodiscus Ihlenf. = Pedalium L. PEDALIACEAE (D)
Pedaliophyton Engl. = Pterodiscus Hook. PEDALIACEAE (D)
Pedalium L. PEDALIACEAE (D)
Peddiea Harv. ex Hook. THYMELAEACEAE (D)
Pederlea Raf. = Acnistus Schott SOLANACEAE (D)
Pedicellaria Schrank = Cleome L. CAPPARACEAE (D)
Pedicellarum M.Hotta ARACEAE (M)
Pediculariopsis Á.Löve & D.Löve = Pedicularis L. SCROPHULARIACEAE (D)
Pedicularis L. SCROPHULARIACEAE (D)
Pedilanthus Neck. ex Poit. EUPHORBIACEAE (D)
Pedilea Lindl. = Malaxis Sol. ex Sw. ORCHIDACEAE (M)
Pedilochilus Schltr. ORCHIDACEAE (M)
Pedilonum Blume = Dendrobium Sw. ORCHIDACEAE (M)
Pedinogyne Brand = Trigonotis Steven BORAGINACEAE (D)
Pedinopetalum Urb. & H.Wolff UMBELLIFERAE (D)
Pediocactus Britton & Rose CACTACEAE (D)
Pediomellum Rydb. = Orbexilum Raf. LEGUMINOSAE–PAPILIONOIDEAE (D)
Pediomelum Rydb. LEGUMINOSAE–PAPILIONOIDEAE (D)
Pedistylis Wiens LORANTHACEAE (D)
Peekelia Harms = Cajanus DC. LEGUMINOSAE–PAPILIONOIDEAE (D)
Peekeliodendron Sleumer = Merrilliodendron Kaneh. ICACINACEAE (D)
Peekeliopanax Harms = Gastonia Comm. ex Lam. ARALIACEAE (D)
Peersia L.Bolus = Rhinephyllum N.E.Br. AIZOACEAE (D)
Pegaeophyton Hayek & Hand.–Mazz. CRUCIFERAE (D)
Pegamea Vitman (SUO) = Pagamea Aubl. RUBIACEAE (D)
Peganum L. ZYGOPHYLLACEAE (D)
Pegia Colebr. ANACARDIACEAE (D)

Peglera Bolus = Nectaropetalum Engl. ERYTHROXYLACEAE (D)
Pegolettia Cass. COMPOSITAE (D)
Pehria Sprague LYTHRACEAE (D)
Peirescia Zucc. (SUO) = Pereskia Mill. CACTACEAE (D)
Peireskia Steud. (SUO) = Pereskia Mill. CACTACEAE (D)
Peireskiopsis Vaupel (SUO) = Pereskiopsis Britton & Rose CACTACEAE (D)
Peixotoa A.Juss. MALPIGHIACEAE (D)
Pelagatia O.E.Schulz = Weberbauera Gilg & Muschler CRUCIFERAE (D)
Pelagodendron Seem. RUBIACEAE (D)
Pelagodoxa Becc. PALMAE (M)
Pelaphia Banks & Sol. (SUI) = Coprosma J.R.Forst. & G.Forst. RUBIACEAE (D)
Pelargonium L'Hér. ex Aiton GERANIACEAE (D)
Pelatantheria Ridl. ORCHIDACEAE (M)
Pelea A.Gray = Melicope J.R.Forst. & G.Forst. RUTACEAE (D)
Pelecostemon Leonard ACANTHACEAE (D)
Pelecyphora Ehrenb. CACTACEAE (D)
Pelexia Poit. ex Lindl. ORCHIDACEAE (M)
Pelidnia Barnhart = Utricularia L. LENTIBULARIACEAE (D)
Peliosanthes Andrews CONVALLARIACEAE (M)
Peliostomum Benth. SCROPHULARIACEAE (D)
Peliotis E.Mey. = Lonchostoma Wikstr. BRUNIACEAE (D)
Peliotus E.Mey. = Lonchostoma Wikstr. BRUNIACEAE (D)
Pella Gaertn. = Ficus L. MORACEAE (D)
Pellacalyx Korth. RHIZOPHORACEAE (D)
Pellaea Link ADIANTACEAE (F)
Pellaeopsis J.Sm. = Pellaea Link ADIANTACEAE (F)
Pellegrinia Sleumer ERICACEAE (D)
Pellegriniodendron J.Léonard LEGUMINOSAE–CAESALPINIOIDEAE (D)
Pelletiera A.St.-Hil. PRIMULACEAE (D)
Pelliciera Planch. & Triana ex Benth. PELLICIERACEAE (D)
Pellionia Gaudich. = Elatostema J.R.Forst. & G.Forst. URTICACEAE (D)
Pelma Finet = Bulbophyllum Thouars ORCHIDACEAE (M)
Pelozia Rose = Lopezia Cav. ONAGRACEAE (D)
Peltaea (C.Presl) Standl. MALVACEAE (D)
Peltandra Raf. ARACEAE (M)
Peltandra Wight (SUH) = Meineckia Baill. EUPHORBIACEAE (D)
Peltanthera Benth. BUDDLEJACEAE (D)
Peltapteris Link LOMARIOPSIDACEAE (F)
Peltaria Jacq. CRUCIFERAE (D)
Peltariopsis (Boiss.) N.Busch CRUCIFERAE (D)
Peltastes Woodson APOCYNACEAE (D)
Peltiphyllum Engl. = Darmera Voss SAXIFRAGACEAE (D)
Peltoboykinia (Engl.) Hara SAXIFRAGACEAE (D)
Peltobractea Rusby = Peltaea (C.Presl) Standl. MALVACEAE (D)
Peltodon Pohl LABIATAE (D)
Peltogyne Vogel LEGUMINOSAE–CAESALPINIOIDEAE (D)
Peltophoropsis Chiov. = Parkinsonia L. LEGUMINOSAE–CAESALPINIOIDEAE (D)
Peltophorum (Vogel) Benth. LEGUMINOSAE–CAESALPINIOIDEAE (D)
Peltophorus Desv. (SUS) = Manisuris L. GRAMINEAE (M)
Peltophyllum Gardner TRIURIDACEAE (M)
Peltospermum Benth. = Sacosperma G.Taylor RUBIACEAE (D)
Peltostegia Turcz. = Peltaea (C.Presl) Standl. MALVACEAE (D)
Peltostigma Walp. RUTACEAE (D)
Pelucha S.Watson COMPOSITAE (D)
Pemphis J.R.Forst. & G.Forst. LYTHRACEAE (D)

Penaea L. PENAEACEAE (D)
Penar–Valli Adans. = Zanonia L. CUCURBITACEAE (D)
Penarvallia Post & Kuntze (SUO) = Zanonia L. CUCURBITACEAE (D)
Penelopeia Urb. CUCURBITACEAE (D)
Penianthus Miers MENISPERMACEAE (D)
Penicillaria Willd. (SUS) = Pennisetum Rich. GRAMINEAE (M)
Peniculifera Ridl. = Trigonopleura Hook.f. EUPHORBIACEAE (D)
Peniculus Swallen = Mesosetum Steud. GRAMINEAE (M)
Peniocereus (A.Berger) Britton & Rose CACTACEAE (D)
Peniophyllum Pennell = Oenothera L. ONAGRACEAE (D)
Pennantia J.R.Forst. & G.Forst. ICACINACEAE (D)
Pennellia Nieuwl. CRUCIFERAE (D)
Pennilabium J.J.Sm. ORCHIDACEAE (M)
Pennisetum Rich. GRAMINEAE (M)
Penstemon Schmidel SCROPHULARIACEAE (D)
Pentabothra Hook.f. ASCLEPIADACEAE (D)
Pentabrachion Müll.Arg. EUPHORBIACEAE (D)
Pentacaena Bartl. = Cardionema DC. ILLECEBRACEAE (D)
Pentacalia Cass. COMPOSITAE (D)
Pentacarpaea Hiern = Pentanisia Harv. RUBIACEAE (D)
Pentacarpus Post & Kuntze (SUO) = Pentanisia Harv. RUBIACEAE (D)
Pentace Hassk. TILIACEAE (D)
Pentaceras Hook.f. RUTACEAE (D)
Pentachaeta Nutt. COMPOSITAE (D)
Pentachlaena H.Perrier SARCOLAENACEAE (D)
Pentachondra R.Br. EPACRIDACEAE (D)
Pentaclathra Endl. = Polyclathra Bertol. CUCURBITACEAE (D)
Pentaclethra Benth. LEGUMINOSAE–MIMOSOIDEAE (D)
Pentacme A.DC. = Shorea Roxb. ex C.F.Gaertn. DIPTEROCARPACEAE (D)
Pentacraspedon Steud. = Amphipogon R.Br. GRAMINEAE (M)
Pentacrostigma K.Afzel. CONVOLVULACEAE (D)
Pentactina Nakai ROSACEAE (D)
Pentacyphus Schltr. ASCLEPIADACEAE (D)
Pentadenia Hanst. = Columnea L. GESNERIACEAE (D)
Pentadesma Sabine GUTTIFERAE (D)
Pentadiplandra Baill. CAPPARACEAE (D)
Pentadynamis R.Br. = Crotalaria L. LEGUMINOSAE–PAPILIONOIDEAE (D)
Pentaglottis Tausch BORAGINACEAE (D)
Pentagonanthus Bullock ASCLEPIADACEAE (D)
Pentagonaster Klotzsch = Kunzea Rchb. MYRTACEAE (D)
Pentagonia Benth. RUBIACEAE (D)
Pentagonia Fabr. (SUH) = Nicandra Adans. SOLANACEAE (D)
Pentagramma Yatsk., Windham & E.Wollenw. ADIANTACEAE (F)
Pentalinon Voigt APOCYNACEAE (D)
Pentaloncha Hook.f. RUBIACEAE (D)
Pentameria Klotzsch ex Baill. = Bridelia Willd. EUPHORBIACEAE (D)
Pentameris P.Beauv. GRAMINEAE (M)
Pentamerista Maguire TETRAMERISTACEAE (D)
Pentanema Cass. COMPOSITAE (D)
Pentanisia Harv. RUBIACEAE (D)
Pentanopsis Rendle RUBIACEAE (D)
Pentanura Blume ASCLEPIADACEAE (D)
Pentapanax Seem. ARALIACEAE (D)
Pentapeltis Bunge = Xanthosia Rudge UMBELLIFERAE (D)
Pentapera Klotzsch = Erica L. ERICACEAE (D)

Pentapetes L. STERCULIACEAE (D)
Pentaphalangium Warb. = Garcinia L. GUTTIFERAE (D)
Pentaphitrum Rchb. = Physalis L. SOLANACEAE (D)
Pentaphragma Wall. ex G.Don PENTAPHRAGMATACEAE (D)
Pentaphylax Gardner & Champ. PENTAPHYLACACEAE (D)
Pentaphylloides Duhamel = Potentilla L. ROSACEAE (D)
Pentaplaris L.O.Williams & Standl. TILIACEAE (D)
Pentapleura Hand.–Mazz. LABIATAE (D)
Pentapogon R.Br. GRAMINEAE (M)
Pentaptera Roxb. = Terminalia L. COMBRETACEAE (D)
Pentapterygium Klotzsch = Agapetes D.Don ex G.Don ERICACEAE (D)
Pentaptilon E.Pritz. GOODENIACEAE (D)
Pentarhaphia Lindl. = Gesneria L. GESNERIACEAE (D)
Pentarhizidium Hayata = Matteuccia Tod. WOODSIACEAE (F)
Pentarhopalopilia (Engl.) Hiepko OPILIACEAE (D)
Pentaria M.Roem. = Passiflora L. PASSIFLORACEAE (D)
Pentarrhaphis Kunth GRAMINEAE (M)
Pentarrhinum E.Mey. ASCLEPIADACEAE (D)
Pentas Benth. RUBIACEAE (D)
Pentasachme Wall. ex Wight (SUO) = Pentasacme Wall. ex Wight ASCLEPIADACEAE (D)
Pentasacme Wall. ex Wight ASCLEPIADACEAE (D)
Pentaschistis (Nees) Spach GRAMINEAE (M)
Pentascyphus Radlk. SAPINDACEAE (D)
Pentaspadon Hook.f. ANACARDIACEAE (D)
Pentaspatella Gleason = Sauvagesia L. OCHNACEAE (D)
Pentastachya Steud. (SUI) = Pennisetum Rich. GRAMINEAE (M)
Pentastelma Tsiang & P.T.Li ASCLEPIADACEAE (D)
Pentastemona Steenis STEMONACEAE (M)
Pentastemonodiscus Rech.f. CARYOPHYLLACEAE (D)
Pentasticha Turcz. = Fuirena Rottb. CYPERACEAE (M)
Pentatherum Nábělek = Agrostis L. GRAMINEAE (M)
Pentathymelaea Lecomte THYMELAEACEAE (D)
Pentatrichia Klatt COMPOSITAE (D)
Pentatropis R.Br. ex Wight & Arn. ASCLEPIADACEAE (D)
Penteca Raf. = Croton L. EUPHORBIACEAE (D)
Pentelesia Raf. = Arrabidaea DC. BIGNONIACEAE (D)
Penthea Lindl. = Disa Bergius ORCHIDACEAE (M)
Pentheriella O.Hoffm. & Muschl. = Heteromma Benth. COMPOSITAE (D)
Penthorum L. PENTHORACEAE (D)
Penthysa Raf. = Lobostemon Lehm. BORAGINACEAE (D)
Pentodon Hochst. RUBIACEAE (D)
Pentopetia Decne. ASCLEPIADACEAE (D)
Pentopetiopsis Costantin & Gallaud ASCLEPIADACEAE (D)
Pentossaea Judd MELASTOMATACEAE (D)
Pentstemon Aiton (SUO) = Penstemon Schmidel SCROPHULARIACEAE (D)
Pentstemonacanthus Nees (UP) ACANTHACEAE (D)
Pentulops Raf. = Maxillaria Ruíz & Pav. ORCHIDACEAE (M)
Pentzia Thunb. COMPOSITAE (D)
Peperomia Ruíz & Pav. PIPERACEAE (D)
Peplidium Delile SCROPHULARIACEAE (D)
Peplis L. = Lythrum L. LYTHRACEAE (D)
Peplonia Decne. ASCLEPIADACEAE (D)
Pepo Mill. = Cucurbita L. CUCURBITACEAE (D)
Peponia Naudin = Peponium Engl. CUCURBITACEAE (D)
Peponidium (Baill.) Arènes RUBIACEAE (D)

Peponiella Kuntze = Peponium Engl. CUCURBITACEAE (D)
Peponium Engl. CUCURBITACEAE (D)
Peponopsis Naudin CUCURBITACEAE (D)
Pera Mutis EUPHORBIACEAE (D)
Peracarpa Hook.f. & Thomson CAMPANULACEAE (D)
Perakanthus Robyns RUBIACEAE (D)
Perama Aubl. RUBIACEAE (D)
Peramium Salisb. ex Coult. = Goodyera R.Br. ORCHIDACEAE (M)
Peranema D.Don DRYOPTERIDACEAE (F)
Perantha Craib = Oreocharis Benth. GESNERIACEAE (D)
Perapentacoilanthus Rappa & Camarrone (SUS) = Mesembryanthemum L. AIZOACEAE (D)
Peraphyllum Nutt. ROSACEAE (D)
Peratanthe Urb. RUBIACEAE (D)
Peraxilla Tiegh. LORANTHACEAE (D)
Perdicesca Prov. = Mitchella L. RUBIACEAE (D)
Perdicium L. COMPOSITAE (D)
Perebea Aubl. MORACEAE (D)
Peregrina W.R.Anderson MALPIGHIACEAE (D)
Pereilema C.Presl GRAMINEAE (M)
Perella (Tiegh.) Tiegh. = Peraxilla Tiegh. LORANTHACEAE (D)
Perenideboles Ram.Goyena ACANTHACEAE (D)
Perescia Lem. (SUO) = Pereskia Mill. CACTACEAE (D)
Pereskia Mill. CACTACEAE (D)
Pereskiopsis Britton & Rose CACTACEAE (D)
Perezia Lag. COMPOSITAE (D)
Pergamena Finet = Dactylostalix Rchb.f. ORCHIDACEAE (M)
Pergularia L. ASCLEPIADACEAE (D)
Periandra Mart. ex Benth. LEGUMINOSAE–PAPILIONOIDEAE (D)
Perianthomega Bureau ex Baill. BIGNONIACEAE (D)
Perianthopodus Silva Manso = Cayaponia Silva Manso CUCURBITACEAE (D)
Perianthostelma Baill. = Cynanchum L. ASCLEPIADACEAE (D)
Periarrabidaea A.Samp. BIGNONIACEAE (D)
Periballia Trin. GRAMINEAE (M)
Periblema DC. = Boutonia DC. ACANTHACEAE (D)
Pericalia Cass. = Roldana La Llave & Lex. COMPOSITAE (D)
Pericallis D.Don COMPOSITAE (D)
Pericalymma (Endl.) Endl. MYRTACEAE (D)
Pericalymna Meisn. (SUO) = Pericalymma (Endl.) Endl. MYRTACEAE (D)
Pericalypta Benoist ACANTHACEAE (D)
Pericampylus Miers MENISPERMACEAE (D)
Perichasma Miers = Stephania Lour. MENISPERMACEAE (D)
Perichlaena Baill. BIGNONIACEAE (D)
Periclesia A.C.Sm. = Ceratostema Juss. ERICACEAE (D)
Pericome A.Gray COMPOSITAE (D)
Pericopsis Thwaites LEGUMINOSAE–PAPILIONOIDEAE (D)
Perictenia Miers = Odontadenia Benth. APOCYNACEAE (D)
Pericycla Blume = Licuala Thunb. PALMAE (M)
Perideridia Rchb. UMBELLIFERAE (D)
Peridiscus Benth. PERIDISCACEAE (D)
Peridium Schott = Pera Mutis EUPHORBIACEAE (D)
Periestes Baill. = Hypoestes Sol. ex R.Br. ACANTHACEAE (D)
Perieteris Raf. = Nicotiana L. SOLANACEAE (D)
Perigaria Spanoghe = Gustavia L. LECYTHIDACEAE (D)
Periglossum Decne. ASCLEPIADACEAE (D)

Perilepta Bremek. = Strobilanthes Blume ACANTHACEAE (D)
Perilimnastes Ridl. = Oritrephes Ridl. MELASTOMATACEAE (D)
Perilla L. LABIATAE (D)
Perillula Maxim. LABIATAE (D)
Periloba Raf. = Nolana L.f. SOLANACEAE (D)
Perilomia Kunth = Scutellaria L. LABIATAE (D)
Periomphale Baill. = Wittsteinia F.Muell. ALSEUOSMIACEAE (D)
Peripentadenia L.S.Sm. ELAEOCARPACEAE (D)
Peripeplus Pierre RUBIACEAE (D)
Periphragmos Ruíz & Pav. = Cantua Juss. ex Lam. POLEMONIACEAE (D)
Periplexis Wall. = Drypetes Vahl EUPHORBIACEAE (D)
Periploca L. ASCLEPIADACEAE (D)
Periptera DC. MALVACEAE (D)
Peripteris Raf. = Pteris L. PTERIDACEAE (F)
Peripterygia (Baill.) Loes. CELASTRACEAE (D)
Peripterygium Hassk. = Cardiopteris Wall. ex Royle CARDIOPTERIDACEAE (D)
Perispermum O.Deg. = Bonamia Thouars CONVOLVULACEAE (D)
Perissocarpa Steyerm. & Maquire OCHNACEAE (D)
Perissocoeleum Mathias & Constance UMBELLIFERAE (D)
Perissolobus N.E.Br. = Machairophyllum Schwantes AIZOACEAE (D)
Peristeranthus T.E.Hunt ORCHIDACEAE (M)
Peristeria Hook. ORCHIDACEAE (M)
Peristethium Tiegh. = Struthanthus Mart. LORANTHACEAE (D)
Peristrophe Nees ACANTHACEAE (D)
Peristylus Blume ORCHIDACEAE (M)
Peritassa Miers CELASTRACEAE (D)
Perithrix Pierre = Batesanthus N.E.Br. ASCLEPIADACEAE (D)
Perittostemma I.M.Johnst. BORAGINACEAE (D)
Perityle Benth. COMPOSITAE (D)
Perizoma (Miers) Lindl. = Salpichroa Miers SOLANACEAE (D)
Perlaria Fabr. (SUS) = Aegilops L. GRAMINEAE (M)
Permia Raf. = Entada Adans. LEGUMINOSAE–MIMOSOIDEAE (D)
Pernettya Gaudich. = Gaultheria Kalm ex L. ERICACEAE (D)
Pernettyopsis King & Gamble ERICACEAE (D)
Perobachne C.Presl = Themeda Forssk. GRAMINEAE (M)
Peronema Jack VERBENACEAE (D)
Perotis Aiton GRAMINEAE (M)
Perovskia Kar. LABIATAE (D)
Perplexia Iljin = Jurinea Cass. COMPOSITAE (D)
Perralderia Coss. COMPOSITAE (D)
Perralderiopsis Rauschert = Iphiona Cass. COMPOSITAE (D)
Perriera Courchet SIMAROUBACEAE (D)
Perrieranthus Hochr. = Perrierophytum Hochr. MALVACEAE (D)
Perrierastrum Guillaumin LABIATAE (D)
Perrierbambus A.Camus GRAMINEAE (M)
Perrieriella Schltr. ORCHIDACEAE (M)
Perrierodendron Cavaco SARCOLAENACEAE (D)
Perrierophytum Hochr. MALVACEAE (D)
Perrierosedum (A.Berger) H.Ohba CRASSULACEAE (D)
Perrottetia Kunth CELASTRACEAE (D)
Persea Mill. LAURACEAE (D)
Persica Mill. = Prunus L. ROSACEAE (D)
Persicaria (L.) Mill. POLYGONACEAE (D)
Personula Raf. = Utricularia L. LENTIBULARIACEAE (D)
Persoonia Sm. PROTEACEAE (D)

Persoonia Willd. (SUH) = Carapa Aubl. MELIACEAE (D)
Pertusadina Ridsdale RUBIACEAE (D)
Pertya Sch.Bip. COMPOSITAE (D)
Perula Schreb. = Pera Mutis EUPHORBIACEAE (D)
Perula Raf. (SUH) = Ficus L. MORACEAE (D)
Perularia Lindl. = Platanthera Rich. ORCHIDACEAE (M)
Perulifera A.Camus = Pseudechinolaena Stapf GRAMINEAE (M)
Peruvocereus Akers = Haageocereus Backeb. CACTACEAE (D)
Pervillaea Decne. = Toxocarpus Wight & Arn. ASCLEPIADACEAE (D)
Perxo Raf. = Basilicum Moench LABIATAE (D)
Perymeniopsis H.Rob. COMPOSITAE (D)
Perymenium Schrad. COMPOSITAE (D)
Pescatoria Rchb.f. ORCHIDACEAE (M)
Peschiera A.DC. = Tabernaemontana L. APOCYNACEAE (D)
Pesomeria Lindl. = Phaius Lour. ORCHIDACEAE (M)
Pessopteris Underw. & Maxon = Niphidium J.Sm. POLYPODIACEAE (F)
Pestalozzia Zoll. & Moritzi = Gynostemma Blume CUCURBITACEAE (D)
Petagnia Raf. (SUH) = Solanum L. SOLANACEAE (D)
Petagnia Guss. UMBELLIFERAE (D)
Petagomoa Bremek. = Psychotria L. RUBIACEAE (D)
Petalacte D.Don COMPOSITAE (D)
Petaladenium Ducke LEGUMINOSAE-PAPILIONOIDEAE (D)
Petalandra F.Muell. ex Boiss. = Euphorbia L. EUPHORBIACEAE (D)
Petalanthera Raf. (SUH) = Justicia L. ACANTHACEAE (D)
Petalanthera Nutt. (SUH) = Cevallia Lag. LOASACEAE (D)
Petalidium Nees ACANTHACEAE (D)
Petalocentrum Schltr. ORCHIDACEAE (M)
Petalochilus R.S.Rogers = Caladenia R.Br. ORCHIDACEAE (M)
Petalodactylis Arènes = Cassipourea Aubl. RHIZOPHORACEAE (D)
Petalodiscus Baill. = Savia Willd. EUPHORBIACEAE (D)
Petalogyne F.Muell. = Petalostylis R.Br. LEGUMINOSAE-CAESALPINIOIDEAE (D)
Petalolophus K.Schum. ANNONACEAE (D)
Petalonema Gilg (SUH) = Gravesia Naudin MELASTOMATACEAE (D)
Petalonema Schltr. = Quisumbingia Merr. ASCLEPIADACEAE (D)
Petalonyx A.Gray LOASACEAE (D)
Petalostelma E.Fourn. = Metastelma R.Br. ASCLEPIADACEAE (D)
Petalostemon Michx. = Dalea L. LEGUMINOSAE-PAPILIONOIDEAE (D)
Petalostigma F.Muell. EUPHORBIACEAE (D)
Petalostylis R.Br. LEGUMINOSAE-CAESALPINIOIDEAE (D)
Petamenes Salisb. ex J.W.Loudon = Gladiolus L. IRIDACEAE (M)
Petasites Mill. COMPOSITAE (D)
Petastoma Miers = Arrabidaea DC. BIGNONIACEAE (D)
Petchia Livera APOCYNACEAE (D)
Petelotiella Gagnep. URTICACEAE (D)
Petelotoma DC. = Carallia Roxb. RHIZOPHORACEAE (D)
Petenaea Lundell TILIACEAE (D)
Peteniodendron Lundell = Pouteria Aubl. SAPOTACEAE (D)
Peteravenia R.M.King & H.Rob. COMPOSITAE (D)
Peteria A.Gray LEGUMINOSAE-PAPILIONOIDEAE (D)
Peteria Raf. (SUO) = Rondeletia L. RUBIACEAE (D)
Petermannia F.Muell. PETERMANNIACEAE (M)
Peterodendron Sleumer FLACOURTIACEAE (D)
Petersia Welw. ex Benth. = Petersianthus Merr. LECYTHIDACEAE (D)
Petersianthus Merr. LECYTHIDACEAE (D)
Petesia P.Browne = Rondeletia L. RUBIACEAE (D)

Petilium Ludw. = Fritillaria L. LILIACEAE (M)
Petiniotia J.Léonard CRUCIFERAE (D)
Petitia Jacq. VERBENACEAE (D)
Petitiocodon Robbr. RUBIACEAE (D)
Petitmenginia Bonati SCROPHULARIACEAE (D)
Petiveria L. PHYTOLACCACEAE (D)
Petlomelia Nieuwl. = Fraxinus L. OLEACEAE (D)
Petopentia Bullock ASCLEPIADACEAE (D)
Petracanthus Nees = Gymnostachyum Nees ACANTHACEAE (D)
Petradoria Greene COMPOSITAE (D)
Petraeovitex Oliv. VERBENACEAE (D)
Petrea L. VERBENACEAE (D)
Petriella Zotov (SUH) = Ehrharta Thunb. GRAMINEAE (M)
Petrina J.B.Phipps = Danthoniopsis Stapf GRAMINEAE (M)
Petrobium R.Br. COMPOSITAE (D)
Petrocallis R.Br. CRUCIFERAE (D)
Petrocarya Schreb. = Parinari Aubl. CHRYSOBALANACEAE (D)
Petrocodon Hance GESNERIACEAE (D)
Petrocoptis A.Braun ex Endl. CARYOPHYLLACEAE (D)
Petrocosmea Oliv. GESNERIACEAE (D)
Petrodoxa Anthony = Beccarinda Kuntze GESNERIACEAE (D)
Petroedmondia Tamamsch. UMBELLIFERAE (D)
Petrogenia I.M.Johnst. CONVOLVULACEAE (D)
Petromarula Vent. ex Hedw.f. CAMPANULACEAE (D)
Petromecon Greene = Eschscholzia Cham. PAPAVERACEAE (D)
Petronymphe H.E.Moore ALLIACEAE (M)
Petrophile R.Br. ex Knight PROTEACEAE (D)
Petrophyes Webb & Berthel. = Monanthes Haw. CRASSULACEAE (D)
Petrophytum (Nutt.) Rydb. ROSACEAE (D)
Petrorhagia (Ser.) Link CARYOPHYLLACEAE (D)
Petrosavia Becc. MELANTHIACEAE (M)
Petrosciadium Edgew. = Eriocycla Lindl. UMBELLIFERAE (D)
Petrosedum Grulich = Sedum L. CRASSULACEAE (D)
Petroselinum Hill UMBELLIFERAE (D)
Petrosimonia Bunge CHENOPODIACEAE (D)
Petteria C.Presl LEGUMINOSAE–PAPILIONOIDEAE (D)
Petunga DC. = Hypobathrum Blume RUBIACEAE (D)
Petunia Juss. SOLANACEAE (D)
Peucedanum L. UMBELLIFERAE (D)
Peuceluma Baill. = Pouteria Aubl. SAPOTACEAE (D)
Peucephyllum A.Gray COMPOSITAE (D)
Peumus Molina MONIMIACEAE (D)
Peyritschia Fourn. GRAMINEAE (M)
Peyrousea DC. COMPOSITAE (D)
Pezisicarpus Vernet APOCYNACEAE (D)
Pfaffia Mart. AMARANTHACEAE (D)
Pfeiffera Salm-Dyck = Lepismium Pfeiff. CACTACEAE (D)
Pfeifferago Kuntze = Codia J.R.Forst. & G.Forst. CUNONIACEAE (D)
Phaca L. = Astragalus L. LEGUMINOSAE–PAPILIONOIDEAE (D)
Phacelia Juss. HYDROPHYLLACEAE (D)
Phacellanthus Siebold & Zucc. (SUH) = Gahnia J.R.Forst. & G.Forst. CYPERACEAE (M)
Phacellanthus Siebold & Zucc. SCROPHULARIACEAE (D)
Phacellaria Steud. (SUI) = Chloris Sw. GRAMINEAE (M)
Phacellaria Benth. SANTALACEAE (D)
Phacellothrix F.Muell. COMPOSITAE (D)

Phacelophrynium K.Schum. MARANTACEAE (M)
Phacelurus Griseb. GRAMINEAE (M)
Phacocapnos Bernh. = Cysticapnos Mill. PAPAVERACEAE (D)
Phacomene Rydb. = Astragalus L. LEGUMINOSAE–PAPILIONOIDEAE (D)
Phacopsis Rydb. = Astragalus L. LEGUMINOSAE–PAPILIONOIDEAE (D)
Phadrosanthus Neck. ex Raf. = Epidendrum L. ORCHIDACEAE (M)
Phaeanthus Hook.f. & Thomson ANNONACEAE (D)
Phaedra Klotzsch ex Endl. = Bernardia Mill. EUPHORBIACEAE (D)
Phaedranassa Herb. AMARYLLIDACEAE (M)
Phaedranthus Miers = Distictis Mart. ex Meisn. BIGNONIACEAE (D)
Phaenanthoecium C.E.Hubb. GRAMINEAE (M)
Phaenocoma D.Don COMPOSITAE (D)
Phaenohoffmannia Kuntze = Pearsonia Dümmer LEGUMINOSAE–
    PAPILIONOIDEAE (D)
Phaenosperma Munro ex Benth. GRAMINEAE (M)
Phaeocephalus S.Moore = Hymenolepis Cass. COMPOSITAE (D)
Phaeomeria Lindl. ex K.Schum. = Etlingera Giseke ZINGIBERACEAE (M)
Phaeoneuron Gilg = Ochthocharis Blume MELASTOMATACEAE (D)
Phaeonychium O.E.Schulz CRUCIFERAE (D)
Phaeoptilum Radlk. NYCTAGINACEAE (D)
Phaeosphaerion Hassk. = Commelina L. COMMELINACEAE (M)
Phaeostemma Fourn. ASCLEPIADACEAE (D)
Phaeostigma Muldashev COMPOSITAE (D)
Phaestoma Spach = Clarkia Pursh ONAGRACEAE (D)
Phagnalon Cass. COMPOSITAE (D)
Phainantha Gleason MELASTOMATACEAE (D)
Phaiophleps Raf. = Olsynium Raf. IRIDACEAE (M)
Phaius Lour. ORCHIDACEAE (M)
Phalacrachena Iljin COMPOSITAE (D)
Phalacraea DC. COMPOSITAE (D)
Phalacrocarpum (DC.) Willk. COMPOSITAE (D)
Phalacroderis DC. = Crepis L. COMPOSITAE (D)
Phalacroseris A.Gray COMPOSITAE (D)
Phalaenopsis Blume ORCHIDACEAE (M)
Phalarella Boiss. (SUI) = Phleum L. GRAMINEAE (M)
Phalaridantha St.–Lag. (SUS) = Phalaris L. GRAMINEAE (M)
Phalaridium Nees & Meyen = Dissanthelium Trin. GRAMINEAE (M)
Phalaris L. GRAMINEAE (M)
Phalaroides Wolf = Phalaris L. GRAMINEAE (M)
Phaleria Jack THYMELAEACEAE (D)
Phallaria Schumach. & Thonn. = Psydrax Gaertn. RUBIACEAE (D)
Phalocallis Herb. IRIDACEAE (M)
Phalona Dumort. (SUO) = Cynosurus L. GRAMINEAE (M)
Phanera Lour. = Bauhinia L. LEGUMINOSAE–CAESALPINIOIDEAE (D)
Phanerocalyx S.Moore = Heisteria Jacq. OLACACEAE (D)
Phanerodiscus Cavaco OLACACEAE (D)
Phaneroglossa B.Nord. COMPOSITAE (D)
Phanerogonocarpus Cavaco MONIMIACEAE (D)
Phanerophlebia C.Presl DRYOPTERIDACEAE (F)
Phanerophlebiopsis Ching DRYOPTERIDACEAE (F)
Phanerosorus Copel. MATONIACEAE (F)
Phanerostylis (A.Gray) R.M.King & H.Rob. COMPOSITAE (D)
Phania DC. COMPOSITAE (D)
Phaniasia Blume ex Miq. (SUI) = Habenaria Willd. ORCHIDACEAE (M)
Phanopyrum (Raf.) Nash = Panicum L. GRAMINEAE (M)

Phanrangia Tardieu = Mangifera L. ANACARDIACEAE (D)
Pharbitis Choisy = Ipomoea L. CONVOLVULACEAE (D)
Pharetrella Salisb. = Cyanella Royen ex L. TECOPHILAEACEAE (M)
Pharmaceum Kuntze = Astronia Blume MELASTOMATACEAE (D)
Pharmacosycea Miq. = Ficus L. MORACEAE (D)
Pharnaceum L. MOLLUGINACEAE (D)
Pharus P.Browne GRAMINEAE (M)
Phaseolus L. LEGUMINOSAE–PAPILIONOIDEAE (D)
Phaulanthus Ridl. = Anerincleistus Korth. MELASTOMATACEAE (D)
Phaulopsis Willd. ACANTHACEAE (D)
Phaulothamnus A.Gray ACHATOCARPACEAE (D)
Phaylopsis Willd. (SUO) = Phaulopsis Willd. ACANTHACEAE (D)
Phebalium Vent. RUTACEAE (D)
Phedimus Raf. = Sedum L. CRASSULACEAE (D)
Phegopteris Fée THELYPTERIDACEAE (F)
Pheidochloa S.T.Blake GRAMINEAE (M)
Pheidonocarpa L.E.Skog GESNERIACEAE (D)
Pheliandra Werderm. = Solanum L. SOLANACEAE (D)
Phelipaea Desf. (SUO) = Phelypaea L. SCROPHULARIACEAE (D)
Phelline Labill. PHELLINACEAE (D)
Phellocalyx Bridson RUBIACEAE (D)
Phellodendron Rupr. RUTACEAE (D)
Phellolophium Baker UMBELLIFERAE (D)
Phellopterus Benth. = Glehnia F.Schmidt ex Miq. UMBELLIFERAE (D)
Phellosperma Britton & Rose = Mammillaria Haw. CACTACEAE (D)
Phelpsiella Maguire RAPATEACEAE (M)
Phelypaea L. SCROPHULARIACEAE (D)
Phenakospermum Endl. STRELITZIACEAE (M)
Phenax Wedd. URTICACEAE (D)
Pherolobus N.E.Br. = Dorotheanthus Schwantes AIZOACEAE (D)
Pherosphaera Archer = Microcachrys Hook.f. PODOCARPACEAE (G)
Pherotrichis Decne. ASCLEPIADACEAE (D)
Phialacanthus Benth. ACANTHACEAE (D)
Phialanthus Griseb. RUBIACEAE (D)
Phialodiscus Radlk. = Blighia K.König SAPINDACEAE (D)
Phidiasia Urb. = Odontonema Nees ACANTHACEAE (D)
Philacra Dwyer OCHNACEAE (D)
Philactis Schrad. COMPOSITAE (D)
Philadelphus L. HYDRANGEACEAE (D)
Philastrea Pierre = Munronia Wight MELIACEAE (D)
Philbornea Hallier f. LINACEAE (D)
Philesia Comm. ex Juss. PHILESIACEAE (M)
Philgamia Baill. MALPIGHIACEAE (D)
Philibertella Vail = Funastrum Fourn. ASCLEPIADACEAE (D)
Philibertia Kunth ASCLEPIADACEAE (D)
Philippia Klotzsch = Erica L. ERICACEAE (D)
Philippiamra Kuntze (SUS) = Silvaea Phil. PORTULACACEAE (D)
Philippicereus Backeb. = Eulychnia Phil. CACTACEAE (D)
Philippiella Speg. ILLECEBRACEAE (D)
Philippinaea Schltr. & Ames ORCHIDACEAE (M)
Phillipsia Rolfe = Dyschoriste Nees ACANTHACEAE (D)
Phillyraea Moench (SUO) = Phillyrea L. OLEACEAE (D)
Phillyrea L. OLEACEAE (D)
Philodendron Schott ARACEAE (M)
Philodice Mart. ERIOCAULACEAE (M)

Philoglossa DC. COMPOSITAE (D)
Philonotion Schott = Schismatoglottis Zoll. & Moritzi ARACEAE (M)
Philotheca Rudge RUTACEAE (D)
Philoxerus R.Br. = Gomphrena L. AMARANTHACEAE (D)
Philydrella Caruel PHILYDRACEAE (M)
Philydrum Banks ex Gaertn. PHILYDRACEAE (M)
Philyra Klotzsch EUPHORBIACEAE (D)
Philyrea Blume (SUO) = Phillyrea L. OLEACEAE (D)
Philyrophyllum O.Hoffm. COMPOSITAE (D)
Phinaea Benth. GESNERIACEAE (D)
Phippsia (Trin.) R.Br. GRAMINEAE (M)
Phitopis Hook.f. RUBIACEAE (D)
Phlebiogonium Fée = Tectaria Cav. DRYOPTERIDACEAE (F)
Phlebiophragmus O.E.Schulz CRUCIFERAE (D)
Phlebiophyllum Van den Bosch = Polyphlebium Copel. HYMENOPHYLLACEAE (F)
Phlebocalymna Griff. ex Miers = Gonocaryum Miq. ICACINACEAE (D)
Phlebocarya R.Br. HAEMODORACEAE (M)
Phlebochiton Wall. = Pegia Colebr. ANACARDIACEAE (D)
Phlebodium (R.Br.) J.Sm. POLYPODIACEAE (F)
Phlebolithis Gaertn. (UP) SAPOTACEAE (D)
Phlebolobium O.E.Schulz CRUCIFERAE (D)
Phlebophyllum Nees = Strobilanthes Blume ACANTHACEAE (D)
Phlebotaenia Griseb. POLYGALACEAE (D)
Phlegmariurus (Herter) Holub = Huperzia Bernh. LYCOPODIACEAE (FA)
Phlegmatospermum O.E.Schulz CRUCIFERAE (D)
Phleum L. GRAMINEAE (M)
Phloeophila Hoehne & Schltr. = Pleurothallis R.Br. ORCHIDACEAE (M)
Phloga Noronha ex Hook.f. PALMAE (M)
Phlogacanthus Nees ACANTHACEAE (D)
Phlogella Baill. = Chrysalidocarpus H.A.Wendl. PALMAE (M)
Phlojodicarpus Turcz. ex Ledeb. UMBELLIFERAE (D)
Phlomidoschema (Benth.) Vved. LABIATAE (D)
Phlomis L. LABIATAE (D)
Phlomoides Moench = Phlomis L. LABIATAE (D)
Phlox L. POLEMONIACEAE (D)
Phlyarodoxa S.Moore = Ligustrum L. OLEACEAE (D)
Phlyctidocarpa Cannon & Theobald UMBELLIFERAE (D)
Phocea Seem. = Macaranga Thouars EUPHORBIACEAE (D)
Phoebanthus S.F.Blake COMPOSITAE (D)
Phoebe Nees LAURACEAE (D)
Phoenicanthemum (Blume) Blume = Helixanthera Lour. LORANTHACEAE (D)
Phoenicanthus Alston ANNONACEAE (D)
Phoenicaulis Nutt. CRUCIFERAE (D)
Phoenicocissus Mart. ex Meisn. = Lundia DC. BIGNONIACEAE (D)
Phoenicophorium H.A.Wendl. PALMAE (M)
Phoenicoseris (Skottsb.) Skottsb. = Dendroseris D.Don COMPOSITAE (D)
Phoenix L. PALMAE (M)
Pholacilia Griseb. = Trichilia P.Browne MELIACEAE (D)
Pholidia R.Br. = Eremophila R.Br. MYOPORACEAE (D)
Pholidocarpus Blume PALMAE (M)
Pholidostachys H.A.Wendl. ex Hook.f. PALMAE (M)
Pholidota Lindl. ex Hook. ORCHIDACEAE (M)
Pholisma Nutt. ex Hook. LENNOACEAE (D)
Pholistoma Lilja HYDROPHYLLACEAE (D)
Pholiurus Trin. GRAMINEAE (M)

Phoradendron Nutt. VISCACEAE (D)
Phormangis Schltr. = Ancistrorhynchus Finet ORCHIDACEAE (M)
Phormium J.R.Forst. & G.Forst. PHORMIACEAE (M)
Phornothamnus Baker = Gravesia Naudin MELASTOMATACEAE (D)
Phorolobus Desv. = Cryptogramma R.Br. ADIANTACEAE (F)
Phosanthus Raf. (SUI) = Isertia Schreb. RUBIACEAE (D)
Photinia Lindl. ROSACEAE (D)
Photinopteris J.Sm. = Aglaomorpha Schott POLYPODIACEAE (F)
Phragmanthera Tiegh. LORANTHACEAE (D)
Phragmipedium Rolfe ORCHIDACEAE (M)
Phragmites Adans. GRAMINEAE (M)
Phragmocarpidium Krapov. MALVACEAE (D)
Phragmocassia Britton & Rose = Senna Mill. LEGUMINOSAE–CAESALPINIOIDEAE (D)
Phragmorchis L.O.Williams ORCHIDACEAE (M)
Phragmotheca Cuatrec. BOMBACACEAE (D)
Phreatia Lindl. ORCHIDACEAE (M)
Phrissocarpus Miers = Tabernaemontana L. APOCYNACEAE (D)
Phrodus Miers SOLANACEAE (D)
Phryganocydia Mart. ex Bur. BIGNONIACEAE (D)
Phrygilanthus Eichler = Notanthera (DC.) G.Don LORANTHACEAE (D)
Phrygiobureaua Kuntze = Phryganocydia Mart. ex Bur. BIGNONIACEAE (D)
Phryma L. PHRYMACEAE (D)
Phryna (Boiss.) Pax & K.Hoffm. CARYOPHYLLACEAE (D)
Phryne Bubani CRUCIFERAE (D)
Phrynella Pax & K.Hoffm. (UP) CARYOPHYLLACEAE (D)
Phrynium Willd. MARANTACEAE (M)
Phtheirospermum Bunge ex Fisch. & C.A.Mey. SCROPHULARIACEAE (D)
Phthirusa Mart. LORANTHACEAE (D)
Phuodendron (Graebn.) Dalla Torre & Harms VALERIANACEAE (D)
Phuopsis (Griseb.) Hook.f. RUBIACEAE (D)
Phycella Lindl. = Hippeastrum Herb. AMARYLLIDACEAE (M)
Phycoschoenus (Asch.) Nakai = Cymodocea K.König CYMODOCEACEAE (M)
Phyganthus Poepp. & Endl. = Tecophilaea Bertero ex Colla TECOPHILAEACEAE (M)
Phygelius E.Mey. ex Benth. SCROPHULARIACEAE (D)
Phyla Lour. VERBENACEAE (D)
Phylacium Benn. LEGUMINOSAE–PAPILIONOIDEAE (D)
Phylanthera Noronha = Hypobathrum Blume RUBIACEAE (D)
Phylica L. RHAMNACEAE (D)
Phyllacantha Hook.f. (SUO) = Phyllacanthus Hook.f. RUBIACEAE (D)
Phyllacanthus Hook.f. RUBIACEAE (D)
Phyllachne J.R.Forst. & G.Forst. STYLIDIACEAE (D)
Phyllactis Pers. VALERIANACEAE (D)
Phyllagathis Blume MELASTOMATACEAE (D)
Phyllanoa Croizat EUPHORBIACEAE (D)
Phyllanthera Blume ASCLEPIADACEAE (D)
Phyllanthodendron Hemsl. = Phyllanthus L. EUPHORBIACEAE (D)
Phyllanthus L. EUPHORBIACEAE (D)
Phyllapophysis Mansf. = Catanthera F.Muell. MELASTOMATACEAE (D)
Phyllarthron DC. BIGNONIACEAE (D)
Phyllarthus Neck. ex M.Gómez (SUI) = Opuntia Mill. CACTACEAE (D)
Phyllaurea Lour. = Codiaeum A.Juss. EUPHORBIACEAE (D)
Phyllera Endl. (SUO) = Philyra Klotzsch EUPHORBIACEAE (D)
Phyllirea Duhamel (SUO) = Phillyrea L. OLEACEAE (D)
Phyllis L. RUBIACEAE (D)
Phyllitis Moench (SUH) = Asplenium L. ASPLENIACEAE (F)

Phyllitis Hill = Asplenium L. ASPLENIACEAE (F)
Phyllitis Raf. (SUH) = Pteris L. PTERIDACEAE (F)
Phyllitopsis Reichst. = Asplenium L. ASPLENIACEAE (F)
Phyllobaea Benth. (SUO) = Phylloboea Benth. GESNERIACEAE (D)
Phylloboea Benth. GESNERIACEAE (D)
Phyllobolus N.E.Br. AIZOACEAE (D)
Phyllobotryon Müll.Arg. FLACOURTIACEAE (D)
Phyllocactus Link (SUS) = Epiphyllum Haw. CACTACEAE (D)
Phyllocalyx O.Berg = Eugenia L. MYRTACEAE (D)
Phyllocara Guşul. = Anchusa L. BORAGINACEAE (D)
Phyllocarpus Riedel ex Tul. LEGUMINOSAE–CAESALPINIOIDEAE (D)
Phyllocephalum Blume COMPOSITAE (D)
Phyllocereus Miq. = Epiphyllum Haw. CACTACEAE (D)
Phyllocharis Diels (SUH) = Ruthiella Steenis CAMPANULACEAE (D)
Phyllochlamys Bureau = Streblus Lour. MORACEAE (D)
Phyllocladus Rich. ex Mirb. PHYLLOCLADACEAE (G)
Phylloclinium Baill. FLACOURTIACEAE (D)
Phyllocomos Mast. RESTIONACEAE (M)
Phyllocosmus Klotzsch IXONANTHACEAE (D)
Phyllocrater Wernham RUBIACEAE (D)
Phylloctenium Baill. BIGNONIACEAE (D)
Phyllodesmis Tiegh. = Taxillus Tiegh. LORANTHACEAE (D)
Phyllodium Desv. LEGUMINOSAE–PAPILIONOIDEAE (D)
Phyllodoce Salisb. ERICACEAE (D)
Phyllogeiton (Weberb.) Herzog = Berchemia Neck. ex DC. RHAMNACEAE (D)
Phylloglossum Kunze LYCOPODIACEAE (FA)
Phyllogonum Coville = Gilmania Coville POLYGONACEAE (D)
Phyllomelia Griseb. RUBIACEAE (D)
Phyllomphax Schltr. = Brachycorythis Lindl. ORCHIDACEAF (M)
Phyllonoma Willd. ex Schult. ESCALLONIACEAE (D)
Phyllophyton Kudo = Marmoritis Benth. LABIATAE (D)
Phyllopodium Benth. SCROPHULARIACEAE (D)
Phyllorachis Trimen GRAMINEAE (M)
Phyllorkis Thouars = Bulbophyllum Thouars ORCHIDACEAE (M)
Phylloscirpus C.B.Clarke CYPERACEAE (M)
Phyllosma Bolus RUTACEAE (D)
Phyllospadix Hook. ZOSTERACEAE (M)
Phyllostachys Siebold & Zucc. GRAMINEAE (M)
Phyllostegia Benth. LABIATAE (D)
Phyllostenonodaphne Kosterm. LAURACEAE (D)
Phyllostephanus Tiegh. = Aetanthus (Eichler) Engl. LORANTHACEAE (D)
Phyllostylon Capan. ex Benth. ULMACEAE (D)
Phyllota (DC.) Benth. LEGUMINOSAE–PAPILIONOIDEAE (D)
Phyllotrichum Thorel ex Lecomte SAPINDACEAE (D)
Phylloxylon Baill. LEGUMINOSAE–PAPILIONOIDEAE (D)
Phylohydrax Puff RUBIACEAE (D)
Phymaspermum Less. COMPOSITAE (D)
Phymatarum M.Hotta ARACEAE (M)
Phymatidium Lindl. ORCHIDACEAE (M)
Phymatocarpus F.Muell. MYRTACEAE (D)
Phymatodes C.Presl = Dipteris Reinw. DIPTERIDACEAE (F)
Phymatopsis J.Sm. = Phymatopteris Pic.Serm. POLYPODIACEAE (F)
Phymatopteris Pic.Serm. POLYPODIACEAE (F)
Phymatosorus Pic.Serm. POLYPODIACEAE (F)
Phymosia Desv. MALVACEAE (D)

Phyodina Raf. = Callisia Loefl. COMMELINACEAE (M)
Physacanthus Benth. ACANTHACEAE (D)
Physalastrum Monteiro = Sida L. MALVACEAE (D)
Physaliastrum Makino = Leucophysalis Rydb. SOLANACEAE (D)
Physalidium Fenzl = Graellsia Boiss. CRUCIFERAE (D)
Physalis L. SOLANACEAE (D)
Physalodes Boehm. = Nicandra Adans. SOLANACEAE (D)
Physaloides Moench = Withania Pauquy SOLANACEAE (D)
Physandra Botsch. = Salsola L. CHENOPODIACEAE (D)
Physanthillis Boiss. = Anthyllis L. LEGUMINOSAE–PAPILIONOIDEAE (D)
Physaria (Nutt.) A.Gray CRUCIFERAE (D)
Physedra Hook.f. = Coccinia Wight & Arn. CUCURBITACEAE (D)
Physematium Kaulf. = Woodsia R.Br. WOODSIACEAE (F)
Physena Noronha ex Thouars PHYSENACEAE (D)
Physetobasis Hassk. = Holarrhena R.Br. APOCYNACEAE (D)
Physinga Lindl. ORCHIDACEAE (M)
Physocalymma Pohl LYTHRACEAE (D)
Physocalymna DC. (SUO) = Physocalymma Pohl LYTHRACEAE (D)
Physocalyx Pohl SCROPHULARIACEAE (D)
Physocardamum Hedge CRUCIFERAE (D)
Physocarpus (Cambess.) Maxim. ROSACEAE (D)
Physocaulis (DC.) Tausch = Myrrhoides Heist. ex Fabr. UMBELLIFERAE (D)
Physoceras Schltr. ORCHIDACEAE (M)
Physochlaina G.Don SOLANACEAE (D)
Physodeira Hanst. = Episcia Mart. GESNERIACEAE (D)
Physodium C.Presl = Melochia L. STERCULIACEAE (D)
Physogyne Garay = Pseudogoodyera Schltr. ORCHIDACEAE (M)
Physokentia Becc. PALMAE (M)
Physoleucas Jaub. & Spach = Leucas R.Br. LABIATAE (D)
Physolophium Turcz. = Angelica L. UMBELLIFERAE (D)
Physoplexis (Endl.) Schur CAMPANULACEAE (D)
Physopodium Desv. = Combretum Loefl. COMBRETACEAE (D)
Physopsis Turcz. VERBENACEAE (D)
Physoptychis Boiss. CRUCIFERAE (D)
Physopyrum Popov POLYGONACEAE (D)
Physorhynchus Hook. CRUCIFERAE (D)
Physosiphon Lindl. = Pleurothallis R.Br. ORCHIDACEAE (M)
Physospermopsis H.Wolff UMBELLIFERAE (D)
Physospermum Cusson ex Juss. UMBELLIFERAE (D)
Physostegia Benth. LABIATAE (D)
Physostelma Wight ASCLEPIADACEAE (D)
Physostemon Mart. = Cleome L. CAPPARACEAE (D)
Physostigma Balf. LEGUMINOSAE–PAPILIONOIDEAE (D)
Physothallis Garay (SUH) = Pleurothallis R.Br. ORCHIDACEAE (M)
Physotrichia Hiern UMBELLIFERAE (D)
Physurus L. = Erythrodes Blume ORCHIDACEAE (M)
Phytelephas Ruíz & Pav. PALMAE (M)
Phyteuma L. CAMPANULACEAE (D)
Phyteumoides Smeathman ex DC. = Virectaria Bremek. RUBIACEAE (D)
Phytocrene Wall. ICACINACEAE (D)
Phytolacca L. PHYTOLACCACEAE (D)
Piaggiaea Chiov. = Wrightia R.Br. APOCYNACEAE (D)
Piaranthus R.Br. ASCLEPIADACEAE (D)
Picardaea Urb. RUBIACEAE (D)
Picconia DC. OLEACEAE (D)

Picea A.Dietr. PINACEAE (G)
Pichisermollia Mont.–Neto = Areca L. PALMAE (M)
Pichleria Stapf & Wettst. = Zosima Hoffm. UMBELLIFERAE (D)
Pichonia Pierre SAPOTACEAE (D)
Pickeringia Nutt. LEGUMINOSAE–PAPILIONOIDEAE (D)
Picnomon Adans. COMPOSITAE (D)
Picradeniopsis Rydb. = Bahia Lag. COMPOSITAE (D)
Picraena Lindl. = Picrasma Blume SIMAROUBACEAE (D)
Picralima Pierre APOCYNACEAE (D)
Picramnia Sw. SIMAROUBACEAE (D)
Picrasma Blume SIMAROUBACEAE (D)
Picria Lour. SCROPHULARIACEAE (D)
Picricarya Dennst. = Olea L. OLEACEAE (D)
Picris L. COMPOSITAE (D)
Picrocardia Radlk. = Soulamea Lam. SIMAROUBACEAE (D)
Picrodendron Griseb. EUPHORBIACEAE (D)
Picroderma Thorel ex Gagnep. = Trichilia P.Browne MELIACEAE (D)
Picrolemma Hook.f. SIMAROUBACEAE (D)
Picrorhiza Royle ex Benth. SCROPHULARIACEAE (D)
Picrosia D.Don COMPOSITAE (D)
Picrothamnus Nutt. COMPOSITAE (D)
Pictetia DC. LEGUMINOSAE–PAPILIONOIDEAE (D)
Piddingtonia A.DC. = Pratia Gaudich. CAMPANULACEAE (D)
Pierardia Raf. (SUH) = Dendrobium Sw. ORCHIDACEAE (M)
Pierardia Roxb. = Baccaurea Lour. EUPHORBIACEAE (D)
Pieris D.Don ERICACEAE (D)
Pierranthus Bonati SCROPHULARIACEAE (D)
Pierrea F.Heim = Hopea Roxb. DIPTEROCARPACEAE (D)
Pierrea Hance = Homalium Jacq. FLACOURTIACEAE (D)
Pierreodendron Engl. SIMAROUBACEAE (D)
Pierreodendron A.Chev. (SUH) = Letestua Lecomte SAPOTACEAE (D)
Pierrina Engl. SCYTOPETALACEAE (D)
Pigafetta (Blume) Becc. PALMAE (M)
Pigea DC. = Hybanthus Jacq. VIOLACEAE (D)
Pilea Lindl. URTICACEAE (D)
Pileanthus Labill. MYRTACEAE (D)
Pileocalyx Gasp. = Cucurbita L. CUCURBITACEAE (D)
Pileostegia Hook.f. & Thomson = Schizophragma Siebold & Zucc. HYDRANGEACEAE (D)
Pileus Ramírez = Jacaratia A.DC. CARICACEAE (D)
Pilgerochloa Eig = Ventenata Koeler GRAMINEAE (M)
Pilgerodendron Florin CUPRESSACEAE (G)
Pilidiostigma Burret MYRTACEAE (D)
Pilinophyton Klotzsch = Croton L. EUPHORBIACEAE (D)
Piliocalyx Brongn. & Gris MYRTACEAE (D)
Piliostigma Hochst. = Bauhinia L. LEGUMINOSAE–CAESALPINIOIDEAE (D)
Pillansia L.Bolus IRIDACEAE (M)
Piloblephis Raf. LABIATAE (D)
Pilocanthus B.W.Benson & Backeb. = Pediocactus Britton & Rose CACTACEAE (D)
Pilocarpus Vahl RUTACEAE (D)
Pilocereus K.Schum. (SUH) = Pilosocereus Byles & Rowley CACTACEAE (D)
Pilocereus Lem. (SUS) = Cephalocereus Pfeiff. CACTACEAE (D)
Pilocopiapoa F.Ritter = Copiapoa Britton & Rose CACTACEAE (D)
Pilocosta Almeda & Whiffin MELASTOMATACEAE (D)
Pilogyne Gagnep. (SUH) = Myrsine L. MYRSINACEAE (D)
Pilogyne Eckl. ex Schrad. = Zehneria Endl. CUCURBITACEAE (D)

Pilophora Jacq. = Manicaria Gaertn. PALMAE (M)
Pilophyllum Schltr. ORCHIDACEAE (M)
Pilopleura Schischk. UMBELLIFERAE (D)
Pilopsis Y.Ito (SUI) = Echinopsis Zucc. CACTACEAE (D)
Pilosella Hill COMPOSITAE (D)
Piloselloides (Less.) C.Jeffrey = Gerbera L. COMPOSITAE (D)
Pilosocereus Byles & Rowley CACTACEAE (D)
Pilosperma Planch. & Triana GUTTIFERAE (D)
Pilostemon Iljin COMPOSITAE (D)
Pilostigma Costantin (SUH) = Costantina Bullock ASCLEPIADACEAE (D)
Pilostigma Tiegh. = Amyema Tiegh. LORANTHACEAE (D)
Pilostyles Guill. RAFFLESIACEAE (D)
Pilothecium (Kiaersk.) Kausel = Myrtus L. MYRTACEAE (D)
Pilularia L. MARSILEACEAE (F)
Pilumna Lindl. = Trichopilia Lindl. ORCHIDACEAE (M)
Pimecaria Raf. = Ximenia L. OLACACEAE (D)
Pimelandra A.DC. = Ardisia Sw. MYRSINACEAE (D)
Pimelea Banks & Sol. THYMELAEACEAE (D)
Pimelodendron Hassk. EUPHORBIACEAE (D)
Pimenta Lindl. MYRTACEAE (D)
Pimentelia Wedd. RUBIACEAE (D)
Pimentus Raf. = Pimenta Lindl. MYRTACEAE (D)
Pimia Seem. STERCULIACEAE (D)
Pimpinella L. UMBELLIFERAE (D)
Pinacantha Gilli UMBELLIFERAE (D)
Pinacopodium Exell & Mendonça ERYTHROXYLACEAE (D)
Pinalia Lindl. = Eria Lindl. ORCHIDACEAE (M)
Pinanga Blume PALMAE (M)
Pinaropappus Less. COMPOSITAE (D)
Pinarophyllon Brandegee RUBIACEAE (D)
Pinckneya Michx. RUBIACEAE (D)
Pinda P.K.Mukh. & Constance UMBELLIFERAE (D)
Pindarea Barb.Rodr. = Attalea Kunth PALMAE (M)
Pineda Ruíz & Pav. FLACOURTIACEAE (D)
Pinelia Lindl. ORCHIDACEAE (M)
Pinelianthe Rauschert = Pinelia Lindl. ORCHIDACEAE (M)
Pinellia Ten. ARACEAE (M)
Pinguicula L. LENTIBULARIACEAE (D)
Pinillosia Ossa COMPOSITAE (D)
Pinknea Pers. (SUO) = Pinckneya Michx. RUBIACEAE (D)
Pinonia Gaudich. = Cibotium Kaulf. DICKSONIACEAE (F)
Pinosia Urb. CARYOPHYLLACEAE (D)
Pintoa Gay ZYGOPHYLLACEAE (D)
Pinus L. PINACEAE (G)
Pinzona Mart. & Zucc. DILLENIACEAE (D)
Pionandra Miers = Cyphomandra Mart. ex Sendtn. SOLANACEAE (D)
Pionocarpus S.F.Blake = Iostephane Benth. COMPOSITAE (D)
Piora J.Kost. COMPOSITAE (D)
Piper L. PIPERACEAE (D)
Piperanthera C.DC. = Peperomia Ruíz & Pav. PIPERACEAE (D)
Piperia Rydb. = Platanthera Rich. ORCHIDACEAE (M)
Pippenalia McVaugh COMPOSITAE (D)
Piptadenia Benth. LEGUMINOSAE–MIMOSOIDEAE (D)
Piptadeniastrum Brenan LEGUMINOSAE–MIMOSOIDEAE (D)
Piptadeniopsis Burkart LEGUMINOSAE–MIMOSOIDEAE (D)

Piptandra Turcz. = Scholtzia Schauer MYRTACEAE (D)
Piptanthocereus (A.Berger) Riccob. = Cereus Mill. CACTACEAE (D)
Piptanthus Sweet LEGUMINOSAE–PAPILIONOIDEAE (D)
Piptatherum P.Beauv. = Oryzopsis Michx. GRAMINEAE (M)
Piptocalyx Oliv. ex Benth. = Trimenia Seem. TRIMENIACEAE (D)
Piptocarpha R.Br. COMPOSITAE (D)
Piptochaetium J.Presl GRAMINEAE (M)
Piptochlamys C.A.Mey. = Thymelaea Mill. THYMELAEACEAE (D)
Piptocoma Cass. COMPOSITAE (D)
Piptolepis Sch.Bip. COMPOSITAE (D)
Piptophyllum C.E.Hubb. GRAMINEAE (M)
Piptoptera Bunge CHENOPODIACEAE (D)
Piptospatha N.E.Br. ARACEAE (M)
Piptostachya (C.E.Hubb.) J.B.Phipps = Zonotriche (C.E.Hubb.) J.B.Phipps
    GRAMINEAE (M)
Piptostigma Oliv. ANNONACEAE (D)
Piptothrix A.Gray COMPOSITAE (D)
Pipturus Wedd. URTICACEAE (D)
Piqueria Cav. COMPOSITAE (D)
Piqueriella R.M.King & H.Rob. COMPOSITAE (D)
Piqueriopsis R.M.King COMPOSITAE (D)
Piquetia (Pierre) Hallier f. = Camellia L. THEACEAE (D)
Piquetia N.E.Br. (SUH) = Erepsia N.E.Br. AIZOACEAE (D)
Piranhea Baill. EUPHORBIACEAE (D)
Piratinera Aubl. = Brosimum Sw. MORACEAE (D)
Pirazzia Chiov. = Matthiola R.Br. CRUCIFERAE (D)
Pirea T.Durand = Rorippa Scop. CRUCIFERAE (D)
Piresia Swallen GRAMINEAE (M)
Piresodendron Aubrév. = Pouteria Aubl. SAPOTACEAE (D)
Piriadacus Pichon = Cuspidaria DC. BIGNONIACEAE (D)
Pirigara Aubl. = Gustavia L. LECYTHIDACEAE (D)
Piringa Juss. = Gardenia Ellis RUBIACEAE (D)
Pirinia M.Král CARYOPHYLLACEAE (D)
Piriqueta Aubl. TURNERACEAE (D)
Pirottantha Speg. = Plathymenia Benth. LEGUMINOSAE–MIMOSOIDEAE (D)
Pisaura Bonato = Lopezia Cav. ONAGRACEAE (D)
Piscaria Piper = Eremocarpus Benth. EUPHORBIACEAE (D)
Piscidia L. LEGUMINOSAE–PAPILIONOIDEAE (D)
Piscipula Loefl. = Piscidia L. LEGUMINOSAE–PAPILIONOIDEAE (D)
Pisonia L. NYCTAGINACEAE (D)
Pisoniella (Heimerl) Standl. NYCTAGINACEAE (D)
Pisophaca Rydb. = Astragalus L. LEGUMINOSAE–PAPILIONOIDEAE (D)
Pisosperma Sond. = Kedrostis Medik. CUCURBITACEAE (D)
Pistacia L. ANACARDIACEAE (D)
Pistia L. ARACEAE (M)
Pistorinia DC. CRASSULACEAE (D)
Pisum L. LEGUMINOSAE–PAPILIONOIDEAE (D)
Pitardia Batt. ex Pit. LABIATAE (D)
Pitavia Molina RUTACEAE (D)
Pitcairnia L'Hér. BROMELIACEAE (M)
Pitcheria Nutt. = Rhynchosia Lour. LEGUMINOSAE–PAPILIONOIDEAE (D)
Pithecellobium Mart. LEGUMINOSAE–MIMOSOIDEAE (D)
Pithecoctenium Mart. ex Meisn. BIGNONIACEAE (D)
Pithecoseris Mart. ex DC. COMPOSITAE (D)
Pithecurus Kunth (SUI) = Schizachyrium Nees GRAMINEAE (M)

Pithocarpa Lindl. COMPOSITAE (D)
Pithodes O.F.Cook = Coccothrinax Sargent PALMAE (M)
Pitraea Turcz. VERBENACEAE (D)
Pittiera Cogn. = Polyclathra Bertol. CUCURBITACEAE (D)
Pittierella Schltr. = Cryptocentrum Benth. ORCHIDACEAE (M)
Pittierothamnus Steyerm. = Amphidasya Standl. RUBIACEAE (D)
Pittocaulon H.Rob. & Brettell COMPOSITAE (D)
Pittoniotis Griseb. RUBIACEAE (D)
Pittosporopsis Craib ICACINACEAE (D)
Pittosporum Banks ex Sol. PITTOSPORACEAE (D)
Pituranthos Viv. UMBELLIFERAE (D)
Pitygentias Gilg (SUS) = Gentianella Moench GENTIANACEAE (D)
Pityopsis Nutt. COMPOSITAE (D)
Pityopus Small ERICACEAE (D)
Pityothamnus Small = Asimina Adans. ANNONACEAE (D)
Pityphyllum Schltr. ORCHIDACEAE (M)
Pityranthe Thwaites TILIACEAE (D)
Pityrocarpa (Benth.) Britton & Rose = Piptadenia Benth. LEGUMINOSAE–
  MIMOSOIDEAE (D)
Pityrodia R.Br. VERBENACEAE (D)
Pityrogramma Link ADIANTACEAE (F)
Piuttia Mattei = Thalictrum L. RANUNCULACEAE (D)
Placea Miers AMARYLLIDACEAE (M)
Placocarpa Hook.f. RUBIACEAE (D)
Placodiscus Radlk. SAPINDACEAE (D)
Placodium Hook.f. = Plocama Aiton RUBIACEAE (D)
Placolobium Miq. = Ormosia G.Jacks. LEGUMINOSAE–PAPILIONOIDEAE (D)
Placoma J.F.Gmel. (SUO) = Plocama Aiton RUBIACEAE (D)
Placopoda Balf.f. RUBIACEAE (D)
Placospermum C.T.White & W.D.Francis PROTEACEAE (D)
Placostigma Blume = Podochilus Blume ORCHIDACEAE (M)
Placseptalia Espinosa = Ochagavia Phil. BROMELIACEAE (M)
Pladaroxylon (Endl.) Hook.f. COMPOSITAE (D)
Plaesiantha Hook.f. = Pellacalyx Korth. RHIZOPHORACEAE (D)
Plaesianthera (C.B.Clarke) Livera = Brillantaisia P.Beauv. ACANTHACEAE (D)
Plagiacanthus Nees = Justicia L. ACANTHACEAE (D)
Plagiantha Renvoize GRAMINEAE (M)
Plagianthera Rchb.f. & Zoll. = Mallotus Lour. EUPHORBIACEAE (D)
Plagianthus J.R.Forst. & G.Forst. MALVACEAE (D)
Plagiarthron P.A.Duvign. (SUI) = Loxodera Launert GRAMINEAE (M)
Plagiobasis Schrenk COMPOSITAE (D)
Plagiobothrys Fisch. & C.A.Mey. BORAGINACEAE (D)
Plagiocarpus Benth. LEGUMINOSAE–PAPILIONOIDEAE (D)
Plagioceltis Mildbr. ex Baehni = Ampelocera Klotzsch ULMACEAE (D)
Plagiocheilus Arn. ex DC. COMPOSITAE (D)
Plagiochloa Adamson & Sprague = Tribolium Desv. GRAMINEAE (M)
Plagiogyria (Kunze) Mett. PLAGIOGYRIACEAE (F)
Plagiolirion Baker AMARYLLIDACEAE (M)
Plagiolobium Sweet = Hovea R.Br. ex W.T.Aiton LEGUMINOSAE–
  PAPILIONOIDEAE (D)
Plagiolophus Greenm. COMPOSITAE (D)
Plagiolytrum Nees = Tripogon Roem. & Schult. GRAMINEAE (M)
Plagiopetalum Rehder MELASTOMATACEAE (D)
Plagiopteron Griff. PLAGIOPTERACEAE (D)
Plagiorhegma Maxim. BERBERIDACEAE (D)

Plagioscyphus Radlk. SAPINDACEAE (D)
Plagiosetum Benth. GRAMINEAE (M)
Plagiosiphon Harms LEGUMINOSAE–CAESALPINIOIDEAE (D)
Plagiospermum Oliv. (SUH) = Prinsepia Royle ROSACEAE (D)
Plagiostachys Ridl. ZINGIBERACEAE (M)
Plagiostyles Pierre EUPHORBIACEAE (D)
Plagiotaxis Wall. ex Kuntze = Chukrasia A.Juss. MELIACEAE (D)
Plagiotheca Chiov. = Isoglossa Oerst. ACANTHACEAE (D)
Plagius L'Hér. ex DC. COMPOSITAE (D)
Plakothira Florence LOASACEAE (D)
Planaltoa Taub. COMPOSITAE (D)
Plananthus P.Beauv. ex Mirb. = Huperzia Bernh. LYCOPODIACEAE (FA)
Planchonella Pierre = Pouteria Aubl. SAPOTACEAE (D)
Planchonia Blume LECYTHIDACEAE (D)
Planea Karis COMPOSITAE (D)
Planera J.F.Gmel. ULMACEAE (D)
Planichloa B.K.Simon = Ectrosia R.Br. GRAMINEAE (M)
Planodes Greene = Sibara Greene CRUCIFERAE (D)
Planotia Munro (SUS) = Neurolepis Meisn. GRAMINEAE (M)
Plantago L. PLANTAGINACEAE (D)
Plantinia Bubani (SUS) = Phleum L. GRAMINEAE (M)
Plarodrigoa Looser = Cristaria Cav. MALVACEAE (D)
Plastolaena Pierre ex Chev. = Schumanniophyton Harms RUBIACEAE (D)
Platanocarpum Korth. = Nauclea L. RUBIACEAE (D)
Platanocephalus Crantz (SUS) = Nauclea L. RUBIACEAE (D)
Platanthera Rich. ORCHIDACEAE (M)
Platanus L. PLATANACEAE (D)
Platea Blume ICACINACEAE (D)
Plateilema (A.Gray) Cockerell COMPOSITAE (D)
Platenia Karst. = Syagrus Mart. PALMAE (M)
Plathymenia Benth. LEGUMINOSAE–MIMOSOIDEAE (D)
Platolaria Raf. = Anemopaegma Mart. ex Meisn. BIGNONIACEAE (D)
Platonia Kunth (SUH) = Neurolepis Meisn. GRAMINEAE (M)
Platonia Mart. GUTTIFERAE (D)
Platostoma P.Beauv. LABIATAE (D)
Platyadenia B.L.Burtt GESNERIACEAE (D)
Platyaechmea (Baker) L.B.Sm. & Kress = Aechmea Ruíz & Pav. BROMELIACEAE (M)
Platycalyx N.E.Br. ERICACEAE (D)
Platycapnos (DC.) Bernh. PAPAVERACEAE (D)
Platycarpha Less. COMPOSITAE (D)
Platycarpum Bonpl. RUBIACEAE (D)
Platycarya Siebold & Zucc. JUGLANDACEAE (D)
Platycaulos Linder RESTIONACEAE (M)
Platycelyphium Harms LEGUMINOSAE–PAPILIONOIDEAE (D)
Platycentrum Naudin = Leandra Raddi MELASTOMATACEAE (D)
Platycerium Desv. POLYPODIACEAE (F)
Platychaete Boiss. = Pulicaria Gaertn. COMPOSITAE (D)
Platycladus Spach = Thuja L. CUPRESSACEAE (G)
Platyclinis Benth. = Dendrochilum Blume ORCHIDACEAE (M)
Platycodon A.DC. CAMPANULACEAE (D)
Platycoryne Rchb.f. ORCHIDACEAE (M)
Platycraspedum O.E.Schulz CRUCIFERAE (D)
Platycrater Siebold & Zucc. HYDRANGEACEAE (D)
Platycyamus Benth. LEGUMINOSAE–PAPILIONOIDEAE (D)
Platydesma H.Mann RUTACEAE (D)

Platyelasma Kitag. = Elsholtzia Willd. LABIATAE (D)
Platyglottis L.O.Williams ORCHIDACEAE (M)
Platygonia Naudin = Trichosanthes L. CUCURBITACEAE (D)
Platygyna Mercier EUPHORBIACEAE (D)
Platygyria Ching & S.K.Wu = Neocheiropteris Christ POLYPODIACEAE (F)
Platykeleba N.E.Br. ASCLEPIADACEAE (D)
Platylepis Kunth (SUH) = Ascolepis Nees ex Steud. CYPERACEAE (M)
Platylepis A.Rich. ORCHIDACEAE (M)
Platylobium Sm. LEGUMINOSAE–PAPILIONOIDEAE (D)
Platyloma J.Sm. = Pellaea Link ADIANTACEAE (F)
Platylophus D.Don CUNONIACEAE (D)
Platyluma Baill. = Micropholis (Griseb.) Pierre SAPOTACEAE (D)
Platymerium Bartl. ex DC. = Hypobathrum Blume RUBIACEAE (D)
Platymiscium Vogel LEGUMINOSAE–PAPILIONOIDEAE (D)
Platymitium Warb. = Dobera Juss. SALVADORACEAE (D)
Platymitra Boerl. ANNONACEAE (D)
Platynema Wight & Arn. = Tristellateia Thouars MALPIGHIACEAE (D)
Platyosprion (Maxim.) Maxim. = Cladrastis Raf. LEGUMINOSAE–PAPILIONOIDEAE (D)
Platypholis Maxim. SCROPHULARIACEAE (D)
Platypodanthera R.M.King & H.Rob. COMPOSITAE (D)
Platypodium Vogel LEGUMINOSAE–PAPILIONOIDEAE (D)
Platyptelea J.Drumm. ex Harv. = Aphanopetalum Endl. CUNONIACEAE (D)
Platypterocarpus Dunkley & Brenan CELASTRACEAE (D)
Platypus Small & Nash = Eulophia R.Br. ex Lindl. ORCHIDACEAE (M)
Platyraphe Miq. = Pimpinella L. UMBELLIFERAE (D)
Platyrhiza Barb.Rodr. ORCHIDACEAE (M)
Platysace Bunge UMBELLIFERAE (D)
Platyschkuhria (A.Gray) Rydb. COMPOSITAE (D)
Platysepalum Welw. ex Baker LEGUMINOSAE–PAPILIONOIDEAE (D)
Platysma Blume = Podochilus Blume ORCHIDACEAE (M)
Platyspermation Guillaumin ESCALLONIACEAE (D)
Platyspermum Hook. = Idahoa A.Nelson & J.F.Macbr. CRUCIFERAE (D)
Platystele Schltr. ORCHIDACEAE (M)
Platystemma Wall. GESNERIACEAE (D)
Platystemon Benth. PAPAVERACEAE (D)
Platystigma Benth. = Meconella Nutt. PAPAVERACEAE (D)
Platystylis Lindl. = Liparis Rich. ORCHIDACEAE (M)
Platytaenia Nevski & Vved. (SUH) = Semenovia Regel & Herder UMBELLIFERAE (D)
Platytaenia Kuhn = Taenitis Willd. ex Schkuhr ADIANTACEAE (F)
Platythea O.F.Cook (SUI) = Chamaedorea Willd. PALMAE (M)
Platytheca Steetz TREMANDRACEAE (D)
Platythelys Garay ORCHIDACEAE (M)
Platythyra N.E.Br. = Aptenia N.E.Br. AIZOACEAE (D)
Platytinospora (Engl.) Diels MENISPERMACEAE (D)
Platyzoma R.Br. PLATYZOMATACEAE (F)
Plazerium Kunth (SUI) = Eriochrysis P.Beauv. GRAMINEAE (M)
Plazia Ruíz & Pav. COMPOSITAE (D)
Plecosorus Fée DRYOPTERIDACEAE (F)
Plecospermum Trécul = Maclura Nutt. MORACEAE (D)
Plecostachys Hilliard & B.L.Burtt COMPOSITAE (D)
Plectaneia Thouars APOCYNACEAE (D)
Plectis O.F.Cook = Euterpe Mart. PALMAE (M)
Plectocephalus D.Don COMPOSITAE (D)
Plectocomia Mart. ex Blume PALMAE (M)
Plectocomiopsis Becc. PALMAE (M)

Plectoma Raf. = Utricularia L. LENTIBULARIACEAE (D)
Plectopoma Hanst. = Achimenes Pers. GESNERIACEAE (D)
Plectopteris Fée = Calymmodon C.Presl GRAMMITIDACEAE (F)
Plectorrhiza Dockrill ORCHIDACEAE (M)
Plectrachne Henrard GRAMINEAE (M)
Plectranthastrum T.C.E.Fr. = Alvesia Welw. LABIATAE (D)
Plectranthus L'Hér. LABIATAE (D)
Plectrelminthus Raf. ORCHIDACEAE (M)
Plectritis (Lindl.) DC. VALERIANACEAE (D)
Plectrocarpa Gillies ex Hook. & Arn. ZYGOPHYLLACEAE (D)
Plectronia L. = Olinia Thunb. OLINIACEAE (D)
Plectroniella Robyns RUBIACEAE (D)
Plectrophora H.Focke ORCHIDACEAE (M)
Plecturus Raf. = Tipularia Nutt. ORCHIDACEAE (M)
Pleea Michx. MELANTHIACEAE (M)
Plegmatolemma Bremek. = Justicia L. ACANTHACEAE (D)
Pleiadelphia Stapf = Elymandra Stapf GRAMINEAE (M)
Pleianthemum K.Schum. ex A.Chev. = Duboscia Bocquet TILIACEAE (D)
Pleimeris Raf. = Gardenia Ellis RUBIACEAE (D)
Pleioblastus Nakai = Arundinaria Michx. GRAMINEAE (M)
Pleiocardia Greene CRUCIFERAE (D)
Pleiocarpa Benth. APOCYNACEAE (D)
Pleiocarpidia K.Schum. RUBIACEAE (D)
Pleioceras Baill. APOCYNACEAE (D)
Pleiochasia (Kamieński) Barnhart = Utricularia L. LENTIBULARIACEAE (D)
Pleiochiton Naudin ex A.Gray MELASTOMATACEAE (D)
Pleiococca F.Muell. = Acronychia J.R.Forst. & G.Forst. RUTACEAE (D)
Pleiocoryne Rauschert RUBIACEAE (D)
Pleiocraterium Bremek. RUBIACEAE (D)
Pleiodon Rchb. (SUS) = Bouteloua Lag. GRAMINEAE (M)
Pleiogynium Engl. ANACARDIACEAE (D)
Pleiokirkia Capuron SIMAROUBACEAE (D)
Pleioluma Baill. = Pouteria Aubl. SAPOTACEAE (D)
Pleiomeris A.DC. MYRSINACEAE (D)
Pleione D.Don ORCHIDACEAE (M)
Pleioneura (C.E.Hubb.) J.B.Phipps (SUH) = Danthoniopsis Stapf GRAMINEAE (M)
Pleioneura Rech.f. CARYOPHYLLACEAE (D)
Pleiosepalum Hand.–Mazz. ROSACEAE (D)
Pleiospermium (Engl.) Swingle RUTACEAE (D)
Pleiospilos N.E.Br. AIZOACEAE (D)
Pleiospora Harv. = Pearsonia Dümmer LEGUMINOSAE–PAPILIONOIDEAE (D)
Pleiostachya K.Schum. MARANTACEAE (M)
Pleiostemon Sond. = Flueggea Willd. EUPHORBIACEAE (D)
Pleiotaenia J.M.Coult. & Rose = Polytaenia DC. UMBELLIFERAE (D)
Pleiotaxis Steetz COMPOSITAE (D)
Pleistachyopiper Trel. = Piper L. PIPERACEAE (D)
Plenasium C.Presl = Osmunda L. OSMUNDACEAE (F)
Plenckia Reissek CELASTRACEAE (D)
Pleocarphus D.Don = Jungia L.f. COMPOSITAE (D)
Pleocaulus Bremek. = Strobilanthes Blume ACANTHACEAE (D)
Pleocnemia C.Presl DRYOPTERIDACEAE (F)
Pleodendron Tiegh. CANELLACEAE (D)
Pleogyne Miers MENISPERMACEAE (D)
Pleomele Salisb. = Dracaena Vand. ex L. DRACAENACEAE (M)
Pleonotoma Miers BIGNONIACEAE (D)

Pleopadium Raf. = Croton L. EUPHORBIACEAE (D)
Pleopeltis Humb. & Bonpl. ex Willd. POLYPODIACEAE (F)
Pleopogon Nutt. = Lycurus Kunth GRAMINEAE (M)
Pleotheca Wall. (SUI) = Spiradiclis Blume RUBIACEAE (D)
Plerandra A.Gray = Schefflera J.R.Forst. & G.Forst. ARALIACEAE (D)
Plerandropsis R.Vig. = Trevesia Vis. ARALIACEAE (D)
Pleroma D.Don = Tibouchina Aubl. MELASTOMATACEAE (D)
Plesiatropha Pierre ex Hutch. = Mildbraedia Pax EUPHORBIACEAE (D)
Plesioneuron (Holttum) Holttum THELYPTERIDACEAE (F)
Plesisa Raf. = Utricularia L. LENTIBULARIACEAE (D)
Plesmonium Schott = Amorphophallus Blume ex Decne. ARACEAE (M)
Plethadenia Urb. RUTACEAE (D)
Plethiandra Hook.f. MELASTOMATACEAE (D)
Plethyrsis Raf. = Richardia L. RUBIACEAE (D)
Plettkea Mattf. ILLECEBRACEAE (D)
Pleuradena Raf. = Euphorbia L. EUPHORBIACEAE (D)
Pleurandropsis Baill. = Asterolasia F.Muell. RUTACEAE (D)
Pleuranthemum (Pichon) Pichon = Hunteria Roxb. APOCYNACEAE (D)
Pleuranthium Benth. = Epidendrum L. ORCHIDACEAE (M)
Pleuranthodendron L.O.Williams TILIACEAE (D)
Pleuranthodes Weberb. RHAMNACEAE (D)
Pleuranthodium (K.Schum.) Ros.M.Sm. ZINGIBERACEAE (M)
Pleuraphis Torr. = Hilaria Kunth GRAMINEAE (M)
Pleureia Raf. = Psychotria L. RUBIACEAE (D)
Pleuriarum Nakai = Arisaema Mart. ARACEAE (M)
Pleuricospora A.Gray ERICACEAE (D)
Pleuridium (C.Presl) Fée (SUH) = Niphidium J.Sm. POLYPODIACEAE (F)
Pleuripetalum T.Durand = Anaxagorea A.St.–Hil. ANNONACEAE (D)
Pleurisanthes Baill. ICACINACEAE (D)
Pleuroblepharis Baill. = Crossandra Salisb. ACANTHACEAE (D)
Pleurobotryum Barb.Rodr. = Pleurothallis R.Br. ORCHIDACEAE (M)
Pleurocalyptus Brongn. & Gris MYRTACEAE (D)
Pleurocarpaea Benth. COMPOSITAE (D)
Pleurocarpus Klotzsch = Cinchona L. RUBIACEAE (D)
Pleurocitrus Tanaka (SUI) = Citrus L. RUTACEAE (D)
Pleurocoffea Baill. = Coffea L. RUBIACEAE (D)
Pleurocoronis R.M.King & H.Rob. COMPOSITAE (D)
Pleuroderris Maxon DRYOPTERIDACEAE (F)
Pleurofossa Nakai ex H.Ito = Monogramma Comm. ex Schkuhr VITTARIACEAE (F)
Pleurogonium (C.Presl) Lindl. = Microgramma C.Presl POLYPODIACEAE (F)
Pleurogramme (Blume) C.Presl = Cochlidium Kaulf. GRAMMITIDACEAE (F)
Pleurogyna Eschsch. ex Cham. & Schltdl. = Lomatogonium A.Braun GENTIANACEAE (D)
Pleurogyne Eschsch. ex Griseb. = Swertia L. GENTIANACEAE (D)
Pleurogynella Ikonn. = Swertia L. GENTIANACEAE (D)
Pleuromanes (C.Presl) C.Presl = Crepidomanes (C.Presl) C.Presl
    HYMENOPHYLLACEAE (F)
Pleuropappus F.Muell. COMPOSITAE (D)
Pleuropetalum Hook.f. AMARANTHACEAE (D)
Pleurophora D.Don LYTHRACEAE (D)
Pleurophragma Rydb. CRUCIFERAE (D)
Pleurophyllum Hook.f. COMPOSITAE (D)
Pleuroplitis Trin. = Arthraxon P.Beauv. GRAMINEAE (M)
Pleuropogon R.Br. GRAMINEAE (M)
Pleuropterantha Franch. AMARANTHACEAE (D)
Pleuropteropyrum Gross = Aconogonon (Meisn.) Rchb. POLYGONACEAE (D)

Pleuroridgea Tiegh. = Brackenridgea A.Gray OCHNACEAE (D)
Pleurosorus Fée ASPLENIACEAE (F)
Pleurospa Raf. = Montrichardia Crueg. ARACEAE (M)
Pleurospermopsis C.Norman UMBELLIFERAE (D)
Pleurospermum Hoffm. UMBELLIFERAE (D)
Pleurostachys Brongn. CYPERACEAE (M)
Pleurostelma Schltr. (SUH) = Schlechterella K.Schum. ASCLEPIADACEAE (D)
Pleurostelma Baill. ASCLEPIADACEAE (D)
Pleurostima Raf. VELLOZIACEAE (M)
Pleurostylia Wight & Arn. CELASTRACEAE (D)
Pleurothallis R.Br. ORCHIDACEAE (M)
Pleurothallopsis Porto ORCHIDACEAE (M)
Pleurothyrium Nees LAURACEAE (D)
Plexaure Endl. = Phreatia Lindl. ORCHIDACEAE (M)
Plexipus Raf. = Chascanum E.Mey. VERBENACEAE (D)
Plicosepalus Tiegh. LORANTHACEAE (D)
Plicula Raf. = Acnistus Schott SOLANACEAE (D)
Plinia L. MYRTACEAE (D)
Plinthanthesis Steud. GRAMINEAE (M)
Plinthus Fenzl AIZOACEAE (D)
Pliocarpida Post & Kuntze (SUO) = Pleiocarpidia K.Schum. RUBIACEAE (D)
Ploca Lour. ex Gomes = Christia Moench LEGUMINOSAE–PAPILIONOIDEAE (D)
Plocama Aiton RUBIACEAE (D)
Plocaniophyllon Brandegee RUBIACEAE (D)
Plocoglottis Blume ORCHIDACEAE (M)
Plocosperma Benth. LOGANIACEAE (D)
Plocostemma Blume = Hoya R.Br. ASCLEPIADACEAE (D)
Ploiarium Korth. GUTTIFERAE (D)
Plotia Steud. (SUI) = Glyceria R.Br. GRAMINEAE (M)
Plowmania Hunz. & Subils SOLANACEAE (D)
Pluchea Cass. COMPOSITAE (D)
Plukenetia L. EUPHORBIACEAE (D)
Plumbagella Spach PLUMBAGINACEAE (D)
Plumbago L. PLUMBAGINACEAE (D)
Plumea Lunan = Guarea L. MELIACEAE (D)
Plumeria L. APOCYNACEAE (D)
Plumeriopsis Rusby & Woodson = Thevetia L. APOCYNACEAE (D)
Plummera A.Gray COMPOSITAE (D)
Plutarchia A.C.Sm. ERICACEAE (D)
Pneumatopteris Nakai THELYPTERIDACEAE (F)
Pneumonanthe Gled. = Gentiana L. GENTIANACEAE (D)
Poa L. GRAMINEAE (M)
Poacynum Baill. = Apocynum L. APOCYNACEAE (D)
Poaephyllum Ridl. ORCHIDACEAE (M)
Poagris Raf. (SUS) = Poa L. GRAMINEAE (M)
Poagrostis Stapf GRAMINEAE (M)
Poarion Rchb. (SUS) = Rostraria Trin. GRAMINEAE (M)
Poarium Desv. SCROPHULARIACEAE (D)
Pobeguinea (Stapf) Jacq.-Fél. = Anadelphia Hack. GRAMINEAE (M)
Pochota Ram.Goyena = Bombacopsis Pittier BOMBACACEAE (D)
Pocillaria Ridl. = Rhyticaryum Becc. ICACINACEAE (D)
Poculodiscus Danguy & Choux = Plagioscyphus Radlk. SAPINDACEAE (D)
Podachaenium Benth. ex Oerst. COMPOSITAE (D)
Podadenia Thwaites EUPHORBIACEAE (D)
Podaechmea (Mez) L.B.Sm. & Kress = Aechmea Ruíz & Pav. BROMELIACEAE (M)

Podagrostis (Griseb.) Scribn. & Merr. = Agrostis L. GRAMINEAE (M)
Podalyria Willd. LEGUMINOSAE–PAPILIONOIDEAE (D)
Podandra Baill. ASCLEPIADACEAE (D)
Podandrogyne Ducke CAPPARACEAE (D)
Podangis Schltr. ORCHIDACEAE (M)
Podanthera Wight = Epipogium Borkh. ORCHIDACEAE (M)
Podanthes Haw. = Orbea Haw. ASCLEPIADACEAE (D)
Podanthum Boiss. = Asyneuma Griseb. & Schenck CAMPANULACEAE (D)
Podanthus Lag. COMPOSITAE (D)
Podinapus Dulac (SUS) = Deschampsia P.Beauv. GRAMINEAE (M)
Podistera S.Watson UMBELLIFERAE (D)
Podocaelia (Benth.) A.Fern. & R.Fern. MELASTOMATACEAE (D)
Podocalyx Klotzsch EUPHORBIACEAE (D)
Podocarpium (Benth.) Y.C.Yang & P.H.Huang = Desmodium Desv. LEGUMINOSAE–
    PAPILIONOIDEAE (D)
Podocarpus L'Hér. ex Pers. PODOCARPACEAE (G)
Podochilus Blume ORCHIDACEAE (M)
Podochrosia Baill. = Rauvolfia L. APOCYNACEAE (D)
Podococcus G.Mann & H.A.Wendl. PALMAE (M)
Podocoma Cass. COMPOSITAE (D)
Podocytisus Boiss. & Heldr. LEGUMINOSAE–PAPILIONOIDEAE (D)
Podogynium Taub. = Zenkerella Taub. LEGUMINOSAE–CAESALPINIOIDEAE (D)
Podolasia N.E.Br. ARACEAE (M)
Podolepis Labill. COMPOSITAE (D)
Podolotus Royle = Lotus L. LEGUMINOSAE–PAPILIONOIDEAE (D)
Podoluma Baill. = Pouteria Aubl. SAPOTACEAE (D)
Podonephelium Baill. SAPINDACEAE (D)
Podonosma Boiss. = Onosma L. BORAGINACEAE (D)
Podoon Baill. = Dobinea Buch.–Ham. ex D.Don PODOACEAE (D)
Podopeltis Fée = Tectaria Cav. DRYOPTERIDACEAE (F)
Podopetalum F.Muell. = Ormosia G.Jacks. LEGUMINOSAE–PAPILIONOIDEAE (D)
Podophania Baill. = Hofmeisteria Walp. COMPOSITAE (D)
Podophorus Phil. GRAMINEAE (M)
Podophyllum L. BERBERIDACEAE (D)
Podopogon Raf. = Piptochaetium J.Presl GRAMINEAE (M)
Podopterus Bonpl. POLYGONACEAE (D)
Podorungia Baill. ACANTHACEAE (D)
Podosciadium A.Gray = Perideridia Rchb. UMBELLIFERAE (D)
Podosemum Desv. = Muhlenbergia Schreb. GRAMINEAE (M)
Podosorus Holttum POLYPODIACEAE (F)
Podospermum DC. = Scorzonera L. COMPOSITAE (D)
Podostachys Klotzsch = Croton L. EUPHORBIACEAE (D)
Podostelma K.Schum. ASCLEPIADACEAE (D)
Podostemma Greene = Asclepias L. ASCLEPIADACEAE (D)
Podostemum Michx. PODOSTEMACEAE (D)
Podostigma Elliott = Asclepias L. ASCLEPIADACEAE (D)
Podotheca Cass. COMPOSITAE (D)
Podranea Sprague BIGNONIACEAE (D)
Poecilandra Tul. OCHNACEAE (D)
Poecilanthe Benth. LEGUMINOSAE–PAPILIONOIDEAE (D)
Poecilocalyx Bremek. RUBIACEAE (D)
Poecilocarpus Nevski = Astragalus L. LEGUMINOSAE–PAPILIONOIDEAE (D)
Poecilochroma Miers = Saracha Ruíz & Pav. SOLANACEAE (D)
Poecilolepis Grau COMPOSITAE (D)
Poeciloneuron Bedd. GUTTIFERAE (D)

Poecilopteris C.Presl = Bolbitis Schott LOMARIOPSIDACEAE (F)
Poecilostachys Hack. GRAMINEAE (M)
Poederiopsis Rusby = Manettia L. RUBIACEAE (D)
Poellnitzia Uitew. ALOACEAE (M)
Poenosedum Holub = Rhodiola L. CRASSULACEAE (D)
Poeppigia Kunze ex Rchb. (SUH) = Tecophilaea Bertero ex Colla
    TECOPHILAEACEAE (M)
Poeppigia C.Presl LEGUMINOSAE–CAESALPINIOIDEAE (D)
Poga Pierre ANISOPHYLLEACEAE (D)
Pogenda Raf. = Olea L. OLEACEAE (D)
Poggea Gürke FLACOURTIACEAE (D)
Poggendorffia Karst. = Passiflora L. PASSIFLORACEAE (D)
Poggeophyton Pax = Erythrococca Benth. EUPHORBIACEAE (D)
Pogochilus Falc. = Galeola Lour. ORCHIDACEAE (M)
Pogochloa S.Moore = Gouinia E.Fourn. GRAMINEAE (M)
Pogogyne Benth. LABIATAE (D)
Pogonachne Bor GRAMINEAE (M)
Pogonanthera Blume MELASTOMATACEAE (D)
Pogonanthus Montrouz. = Morinda L. RUBIACEAE (D)
Pogonarthria Stapf GRAMINEAE (M)
Pogonatherum P.Beauv. GRAMINEAE (M)
Pogonia Juss. ORCHIDACEAE (M)
Pogoniopsis Rchb.f. ORCHIDACEAE (M)
Pogonochloa C.E.Hubb. GRAMINEAE (M)
Pogonolepis Steetz COMPOSITAE (D)
Pogonolobus F.Muell. (SUH) = Coelospermum Blume RUBIACEAE (D)
Pogononeura Napper GRAMINEAE (M)
Pogonophora Miers ex Benth. EUPHORBIACEAE (D)
Pogonophyllum Didr. = Micrandra Benth. EUPHORBIACEAE (D)
Pogonopsis C.Presl = Pogonatherum P.Beauv. GRAMINEAE (M)
Pogonopus Klotzsch RUBIACEAE (D)
Pogonorrhinum Betsche = Linaria Mill. SCROPHULARIACEAE (D)
Pogonospermum Hochst. = Monechma Hochst. ACANTHACEAE (D)
Pogonotium J.Dransf. PALMAE (M)
Pogonotrophe Miq. = Ficus L. MORACEAE (D)
Pogostemon Desf. LABIATAE (D)
Pohlidium Davidse, Soderstr. & R.P.Ellis GRAMINEAE (M)
Pohliella Engl. PODOSTEMACEAE (D)
Poicilla Griseb. = Jacaima Rendle ASCLEPIADACEAE (D)
Poicillopsis Schltr. ex Rendle ASCLEPIADACEAE (D)
Poidium Nees = Poa L. GRAMINEAE (M)
Poikilacanthus Lindau ACANTHACEAE (D)
Poikilogyne Baker f. MELASTOMATACEAE (D)
Poikilospermum Zipp. ex Miq. CECROPIACEAE (D)
Poilanedora Gagnep. CAPPARACEAE (D)
Poilaniella Gagnep. EUPHORBIACEAE (D)
Poilannammia C.Hansen MELASTOMATACEAE (D)
Poinciana L. = Caesalpinia L. LEGUMINOSAE–CAESALPINIOIDEAE (D)
Poincianella Britton & Rose = Caesalpinia L. LEGUMINOSAE–CAESALPINIOIDEAE (D)
Poinsettia Graham = Euphorbia L. EUPHORBIACEAE (D)
Poiretia J.F.Gmel. (SUH) = Houstonia L. RUBIACEAE (D)
Poiretia Vent. LEGUMINOSAE–PAPILIONOIDEAE (D)
Poissonella Pierre = Pouteria Aubl. SAPOTACEAE (D)
Poissonia Baill. = Coursetia DC. LEGUMINOSAE–PAPILIONOIDEAE (D)
Poitea Vent. LEGUMINOSAE–PAPILIONOIDEAE (D)

Poivrea Comm. ex DC. = Combretum Loefl. COMBRETACEAE (D)
Pojarkovia Askerova COMPOSITAE (D)
Pokornya Montrouz. = Lumnitzera Willd. COMBRETACEAE (D)
Polakia Stapf = Salvia L. LABIATAE (D)
Polakowskia Pittier = Sechium P.Browne CUCURBITACEAE (D)
Polanisia Raf. CAPPARACEAE (D)
Polaskia Backeb. CACTACEAE (D)
Polemannia Eckl. & Zeyh. UMBELLIFERAE (D)
Polemanniopsis B.L.Burtt UMBELLIFERAE (D)
Polemoniella A.Heller = Polemonium L. POLEMONIACEAE (D)
Polemonium L. POLEMONIACEAE (D)
Polevansia De Winter GRAMINEAE (M)
Polhillia C.H.Stirt. LEGUMINOSAE–PAPILIONOIDEAE (D)
Polianthes L. AGAVACEAE (M)
Poliomintha A.Gray LABIATAE (D)
Poliophyton O.E.Schulz CRUCIFERAE (D)
Poliothyrsis Oliv. FLACOURTIACEAE (D)
Pollalesta Kunth COMPOSITAE (D)
Pollia Thunb. COMMELINACEAE (M)
Pollichia Aiton ILLECEBRACEAE (D)
Pollinia Spreng. = Chrysopogon Trin. GRAMINEAE (M)
Pollinidium Haines = Eulaliopsis Honda GRAMINEAE (M)
Polliniopsis Hayata = Microstegium Nees GRAMINEAE (M)
Pollinirhiza Dulac = Listera R.Br. ORCHIDACEAE (M)
Polpoda C.Presl MOLLUGINACEAE (D)
Polyachyrus Lag. COMPOSITAE (D)
Polyadoa Stapf = Hunteria Roxb. APOCYNACEAE (D)
Polyalthia Blume ANNONACEAE (D)
Polyandra Leal EUPHORBIACEAE (D)
Polyandrococos Barb.Rodr. PALMAE (M)
Polyantherix Nees = Sitanion Raf. GRAMINEAE (M)
Polyanthina R.M.King & H.Rob. COMPOSITAE (D)
Polyanthus C.H.Hu & Y.C.Hu = Arundinaria Michx. GRAMINEAE (M)
Polyarrhena Cass. COMPOSITAE (D)
Polyaster Hook.f. RUTACEAE (D)
Polyaulax Backer ANNONACEAE (D)
Polybactrum Salisb. = Pseudorchis Séguier ORCHIDACEAE (M)
Polyboea Klotzsch ex Endl. = Bernardia Mill. EUPHORBIACEAE (D)
Polyboea Klotzsch (SUH) = Cavendishia Lindl. ERICACEAE (D)
Polybotrya Humb. & Bonpl. ex Willd. DRYOPTERIDACEAE (F)
Polycalymma F.Muell. & Sond. COMPOSITAE (D)
Polycampium C.Presl = Pyrrosia Mirb. POLYPODIACEAE (F)
Polycardia Juss. CELASTRACEAE (D)
Polycarena Benth. SCROPHULARIACEAE (D)
Polycarpaea Lam. CARYOPHYLLACEAE (D)
Polycarpon L. CARYOPHYLLACEAE (D)
Polycephalium Engl. ICACINACEAE (D)
Polyceratocarpus Engl. & Diels ANNONACEAE (D)
Polychilos Breda = Phalaenopsis Blume ORCHIDACEAE (M)
Polychrysum (Tzvelev) Kovalevsk. COMPOSITAE (D)
Polyclathra Bertol. CUCURBITACEAE (D)
Polyclita A.C.Sm. ERICACEAE (D)
Polycnemum L. CHENOPODIACEAE (D)
Polycocca Hill = Selaginella P.Beauv. SELAGINELLACEAE (FA)
Polycoryne Keay (SUH) = Pleiocoryne Rauschert RUBIACEAE (D)

Polyctenium Greene CRUCIFERAE (D)
Polycycliska Ridl. = Lerchea L. RUBIACEAE (D)
Polycycnis Rchb.f. ORCHIDACEAE (M)
Polydiclis (G.Don) Miers = Nicotiana L. SOLANACEAE (D)
Polydictyum C.Presl = Tectaria Cav. DRYOPTERIDACEAE (F)
Polydragma Hook.f. = Spathiostemon Blume EUPHORBIACEAE (D)
Polygala L. POLYGALACEAE (D)
Polygonanthus Ducke ANISOPHYLLEACEAE (D)
Polygonatum Mill. CONVALLARIACEAE (M)
Polygonella Michx. POLYGONACEAE (D)
Polygonum L. POLYGONACEAE (D)
Polylepis Ruíz & Pav. ROSACEAE (D)
Polylophium Boiss. UMBELLIFERAE (D)
Polylychnis Bremek. ACANTHACEAE (D)
Polymeria R.Br. CONVOLVULACEAE (D)
Polymita N.E.Br. AIZOACEAE (D)
Polymnia L. COMPOSITAE (D)
Polyneura Peter (SUH) = Panicum L. GRAMINEAE (M)
Polyodon Kunth = Bouteloua Lag. GRAMINEAE (M)
Polyosma Blume ESCALLONIACEAE (D)
Polyotidium Garay ORCHIDACEAE (M)
Polyozus Lour. = Psychotria L. RUBIACEAE (D)
Polyphlebium Copel. HYMENOPHYLLACEAE (F)
Polyphragmon Desf. = Timonius DC. RUBIACEAE (D)
Polypleurella Engl. PODOSTEMACEAE (D)
Polypleurum (Tul.) Warm. PODOSTEMACEAE (D)
Polypodiastrum Ching = Goniophlebium C.Presl POLYPODIACEAE (F)
Polypodioides Ching = Goniophlebium C.Presl POLYPODIACEAE (F)
Polypodiopsis Copel. = Polypodiopteris C.F.Reed POLYPODIACEAE (F)
Polypodiopteris C.F.Reed POLYPODIACEAE (F)
Polypodium L. POLYPODIACEAE (F)
Polypogon Desf. GRAMINEAE (M)
Polypompholyx Lehm. = Utricularia L. LENTIBULARIACEAE (D)
Polyporandra Becc. ICACINACEAE (D)
Polypremum L. BUDDLEJACEAE (D)
Polypsecadium O.E.Schulz CRUCIFERAE (D)
Polyradicion Garay ORCHIDACEAE (M)
Polyraphis (Trin.) Lindl. (SUS) = Pappophorum Schreb. GRAMINEAE (M)
Polyrhabda C.C.Towns. AMARANTHACEAE (D)
Polyrrhiza Pfitzer = Polyradicion Garay ORCHIDACEAE (M)
Polyschistis C.Presl = Pentarrhaphis Kunth GRAMINEAE (M)
Polyscias J.R.Forst. & G.Forst. ARALIACEAE (D)
Polysolen Rauschert (SUS) = Indopolysolenia Bennet RUBIACEAE (D)
Polysolenia Hook.f. = Indopolysolenia Bennet RUBIACEAE (D)
Polyspatha Benth. COMMELINACEAE (M)
Polysphaeria Hook.f. RUBIACEAE (D)
Polystachya Hook. ORCHIDACEAE (M)
Polystemma Decne. ASCLEPIADACEAE (D)
Polystemon D.Don (SUI) = Lamanonia Vell. CUNONIACEAE (D)
Polystemonanthus Harms LEGUMINOSAE–CAESALPINIOIDEAE (D)
Polystichopsis (J.Sm.) Holttum DRYOPTERIDACEAE (F)
Polystichum Roth DRYOPTERIDACEAE (F)
Polystylus Hasselt ex Hassk. = Phalaenopsis Blume ORCHIDACEAE (M)
Polytaenia DC. UMBELLIFERAE (D)
Polytaenium Desv. = Antrophyum Kaulf. VITTARIACEAE (F)

Polytaxis Bunge = Jurinea Cass. COMPOSITAE (D)
Polytepalum Suess. & Beyerle CARYOPHYLLACEAE (D)
Polythrix Nees = Crossandra Salisb. ACANTHACEAE (D)
Polythysania Hanst. = Drymonia Mart. GESNERIACEAE (D)
Polytoca R.Br. GRAMINEAE (M)
Polytoma Lour. ex Gomes (SUH) = Bletilla Rchb.f. ORCHIDACEAE (M)
Polytrema C.B.Clarke = Ptyssiglottis T.Anderson ACANTHACEAE (D)
Polytrias Hack. GRAMINEAE (M)
Polyura Hook.f. RUBIACEAE (D)
Polyxena Kunth HYACINTHACEAE (M)
Polyzone Endl. = Darwinia Rudge MYRTACEAE (D)
Polyzygus Dalzell UMBELLIFERAE (D)
Pomaderris Labill. RHAMNACEAE (D)
Pomangium Reinw. = Argostemma Wall. RUBIACEAE (D)
Pomasterion Miq. = Actinostemma Griff. CUCURBITACEAE (D)
Pomatium C.F.Gaertn. = Bertiera Aubl. RUBIACEAE (D)
Pomatocalpa Breda, Kuhl & Hasselt ORCHIDACEAE (M)
Pomatophytum M.E.Jones = Cheilanthes Sw. ADIANTACEAE (F)
Pomatosace Maxim. PRIMULACEAE (D)
Pomatostoma Stapf = Anerincleistus Korth. MELASTOMATACEAE (D)
Pomax Sol. ex DC. RUBIACEAE (D)
Pomazota Ridl. = Coptophyllum Korth. RUBIACEAE (D)
Pomelia Durando ex Pomel = Daucus L. UMBELLIFERAE (D)
Pometia Vell. (SUH) = Pradosia Liais SAPOTACEAE (D)
Pometia J.R.Forst. & G.Forst. SAPINDACEAE (D)
Pommereschea Wittm. ZINGIBERACEAE (M)
Pommereulla L.f. GRAMINEAE (M)
Pomphidea Miers = Ravenia Vell. RUTACEAE (D)
Ponapea Becc. = Ptychosperma Labill. PALMAE (M)
Ponceletia Thouars (SUH) = Spartina Schreb. GRAMINEAE (M)
Poncirus Raf. RUTACEAE (D)
Ponera Lindl. ORCHIDACEAE (M)
Ponerorchis Rchb.f. ORCHIDACEAE (M)
Pongamia Vent. LEGUMINOSAE–PAPILIONOIDEAE (D)
Pongamiopsis R.Vig. LEGUMINOSAE–PAPILIONOIDEAE (D)
Pongati Adans. = Sphenoclea Gaertn. CAMPANULACEAE (D)
Pongatium Juss. (SUO) = Sphenoclea Gaertn. CAMPANULACEAE (D)
Pongelia Raf. = Dolichandrone (Fenzl) Seem. BIGNONIACEAE (D)
Pongelion Adans. = Adenanthera L. LEGUMINOSAE–MIMOSOIDEAE (D)
Pontederia L. PONTEDERIACEAE (M)
Ponthieva R.Br. ORCHIDACEAE (M)
Pontopidana Scop. = Couroupita Aubl. LECYTHIDACEAE (D)
Pontya A.Chev. = Trilepisium Thouars MORACEAE (D)
Poortmannia Drake = Trianaea Planch. & Linden SOLANACEAE (D)
Poponax Raf. = Acacia Mill. LEGUMINOSAE–MIMOSOIDEAE (D)
Popoviocodonia Fedorov CAMPANULACEAE (D)
Popoviolimon Lincz. PLUMBAGINACEAE (D)
Popowia Endl. ANNONACEAE (D)
Poppya Neck. ex M.Roem. = Luffa Mill. CUCURBITACEAE (D)
Populina Baill. ACANTHACEAE (D)
Populus L. SALICACEAE (D)
Porana Burm.f. CONVOLVULACEAE (D)
Porandra D.Y.Hong = Amischotolype Hassk. COMMELINACEAE (M)
Poranopsis Roberty = Porana Burm.f. CONVOLVULACEAE (D)
Poranthera Rudge EUPHORBIACEAE (D)

Poranthera Raf. (SUH) = Sorghastrum Nash GRAMINEAE (M)
Poraqueiba Aubl. ICACINACEAE (D)
Poraresia Gleason = Pogonophora Miers ex Benth. EUPHORBIACEAE (D)
Porcelia Ruíz & Pav. ANNONACEAE (D)
Porfiria Boed. = Mammillaria Haw. CACTACEAE (D)
Porlieria Ruíz & Pav. ZYGOPHYLLACEAE (D)
Porocarpus Gaertn. = Timonius DC. RUBIACEAE (D)
Porocystis Radlk. SAPINDACEAE (D)
Porodittia G.Don (SUS) = Stemotria Wettst. & Harms ex Engl. SCROPHULARIACEAE (D)
Porolabium T.Tang & F.T.Wang ORCHIDACEAE (M)
Porophyllum Adans. COMPOSITAE (D)
Porospermum F.Muell. = Delarbrea Vieill. ARALIACEAE (D)
Porpax Lindl. ORCHIDACEAE (M)
Porphyranthus Engl. = Panda Pierre PANDACEAE (D)
Porphyrocodon Hook.f. = Cardamine L. CRUCIFERAE (D)
Porphyrocoma Scheidw. ex Hook. = Justicia L. ACANTHACEAE (D)
Porphyrodesme Schltr. ORCHIDACEAE (M)
Porphyroglottis Ridl. ORCHIDACEAE (M)
Porphyroscias Miq. = Angelica L. UMBELLIFERAE (D)
Porphyrospatha Engl. = Syngonium Schott ARACEAE (M)
Porphyrostachys Rchb.f. ORCHIDACEAE (M)
Porphyrostemma Benth. ex Oliv. COMPOSITAE (D)
Porroglossum Schltr. ORCHIDACEAE (M)
Porrorhachis Garay ORCHIDACEAE (M)
Porroteranthe Steud. = Glyceria R.Br. GRAMINEAE (M)
Porsildia Á.Löve & D.Löve = Minuartia L. CARYOPHYLLACEAE (D)
Portaea Ten. = Juanulloa Ruíz & Pav. SOLANACEAE (D)
Portea K.Koch BROMELIACEAE (M)
Portenschlagia Vis. (SUH) = Portenschlagiella Tutin UMBELLIFERAE (D)
Portenschlagiella Tutin UMBELLIFERAE (D)
Porterandia Ridl. RUBIACEAE (D)
Porteranthus Britton = Gillenia Moench ROSACEAE (D)
Porteresia Tateoka GRAMINEAE (M)
Portesia Cav. = Trichilia P.Browne MELIACEAE (D)
Portlandia P.Browne RUBIACEAE (D)
Portulaca L. PORTULACACEAE (D)
Portulacaria Jacq. PORTULACACEAE (D)
Posadaea Cogn. CUCURBITACEAE (D)
Posidonia K.König POSIDONIACEAE (M)
Poskea Vatke GLOBULARIACEAE (D)
Posoqueria Aubl. RUBIACEAE (D)
Posoria Raf. = Posoqueria Aubl. RUBIACEAE (D)
Possira Aubl. = Swartzia Schreb. LEGUMINOSAE–PAPILIONOIDEAE (D)
Postia Boiss. & Blanche (SUH) = Rhanteriopsis Rauschert COMPOSITAE (D)
Postiella Kljuykov (UP) UMBELLIFERAE (D)
Postuera Raf. = Notelaea Vent. OLEACEAE (D)
Potalia Aubl. LOGANIACEAE (D)
Potameia Thouars LAURACEAE (D)
Potamochloa Griff. (SUS) = Hygroryza Nees GRAMINEAE (M)
Potamoganos Sandwith BIGNONIACEAE (D)
Potamogeton L. POTAMOGETONACEAE (M)
Potamophila R.Br. GRAMINEAE (M)
Potamoxylon Raf. = Tabebuia Gomes ex DC. BIGNONIACEAE (D)
Potaninia Maxim. ROSACEAE (D)
Potarophytum Sandwith RAPATEACEAE (M)

Potentilla L. ROSACEAE (D)
Poteranthera Bong. MELASTOMATACEAE (D)
Poteridium Spach = Sanguisorba L. ROSACEAE (D)
Poterium L. = Sanguisorba L. ROSACEAE (D)
Pothoidium Schott ARACEAE (M)
Pothomorphe Miq. PIPERACEAE (D)
Pothos L. ARACEAE (M)
Pothuava Gaudich. = Aechmea Ruíz & Pav. BROMELIACEAE (M)
Potima R.A.Hedw. = Faramea Aubl. RUBIACEAE (D)
Potoxylon Kosterm. LAURACEAE (D)
Pottingeria Prain CELASTRACEAE (D)
Pottsia Hook. & Arn. APOCYNACEAE (D)
Pouchetia A.Rich. RUBIACEAE (D)
Poulsenia Eggers MORACEAE (D)
Pounguia Benoist = Whitfieldia Hook. ACANTHACEAE (D)
Poupartia Comm. ex Juss. ANACARDIACEAE (D)
Pourouma Aubl. CECROPIACEAE (D)
Pourthiaea Decne. = Photinia Lindl. ROSACEAE (D)
Pouteria Aubl. SAPOTACEAE (D)
Pouzolzia Gaudich. URTICACEAE (D)
Povedadaphne W.C.Burger = Ocotea Aubl. LAURACEAE (D)
Pozoa Lag. UMBELLIFERAE (D)
Pradosia Liais SAPOTACEAE (D)
Praecereus Buxb. = Cereus Mill. CACTACEAE (D)
Praecitrullus Pang. CUCURBITACEAE (D)
Prageluria N.E.Br. = Telosma Coville ASCLEPIADACEAE (D)
Pragmotessara Pierre = Euonymus L. CELASTRACEAE (D)
Pragmotropa Pierre = Euonymus L. CELASTRACEAE (D)
Prainea King ex Hook.f. MORACEAE (D)
Pranceacanthus Wassh. ACANTHACEAE (D)
Prangos Lindl. UMBELLIFERAE (D)
Praravinia Korth. RUBIACEAE (D)
Prasium L. LABIATAE (D)
Prasopepon Naudin = Cucurbitella Walp. CUCURBITACEAE (D)
Prasophyllum R.Br. ORCHIDACEAE (M)
Prasoxylon M.Roem. = Dysoxylum Blume MELIACEAE (D)
Pratia Gaudich. CAMPANULACEAE (D)
Pravinaria Bremek. RUBIACEAE (D)
Praxeliopsis G.M.Barroso COMPOSITAE (D)
Praxelis Cass. COMPOSITAE (D)
Preissia Opiz (SUI) = Avena L. GRAMINEAE (M)
Premna L. VERBENACEAE (D)
Prenanthella Rydb. COMPOSITAE (D)
Prenanthes L. COMPOSITAE (D)
Prenia N.E.Br. = Phyllobolus N.E.Br. AIZOACEAE (D)
Prepodesma N.E.Br. = Aloinopsis Schwantes N.E.Br. AIZOACEAE (D)
Preptanthe Rchb.f. = Calanthe Ker–Gawl. ORCHIDACEAE (M)
Prepusa Mart. GENTIANACEAE (D)
Prescottia Lindl. ORCHIDACEAE (M)
Preslia Opiz = Mentha L. LABIATAE (D)
Prestoea Hook.f. PALMAE (M)
Prestonia R.Br. APOCYNACEAE (D)
Pretrea J.Gay = Dicerocaryum Bojer PEDALIACEAE (D)
Pretreothamnus Engl. = Josephinia Vent. PEDALIACEAE (D)
Preussiella Gilg MELASTOMATACEAE (D)

Preussiodora Keay RUBIACEAE (D)
Prevostea Choisy = Calycobolus Willd. ex J.A.Schult. CONVOLVULACEAE (D)
Priamosia Urb. FLACOURTIACEAE (D)
Priestleya DC. LEGUMINOSAE–PAPILIONOIDEAE (D)
Prieurella Pierre = Chrysophyllum L. SAPOTACEAE (D)
Prieuria DC. = Ludwigia L. ONAGRACEAE (D)
Primula L. PRIMULACEAE (D)
Primularia Brenan = Cincinnobotrys Gilg MELASTOMATACEAE (D)
Primulina Hance GESNERIACEAE (D)
Princea Dubard & Dop = Triainolepis Hook.f. RUBIACEAE (D)
Principina Uittien CYPERACEAE (M)
Pringlea T.Anderson ex Hook.f. CRUCIFERAE (D)
Pringleochloa Scribn. GRAMINEAE (M)
Pringleophytum A.Gray = Holographis Nees ACANTHACEAE (D)
Prinsepia Royle ROSACEAE (D)
Printzia Cass. COMPOSITAE (D)
Prionachne Nees = Prionanthium Desv. GRAMINEAE (M)
Prionanthium Desv. GRAMINEAE (M)
Prionitis Oerst. = Barleria L. ACANTHACEAE (D)
Prionium E.Mey. JUNCACEAE (M)
Prionophyllum K.Koch = Dyckia Schult.f. BROMELIACEAE (M)
Prionoplectus Oerst. = Alloplectus Mart. GESNERIACEAE (D)
Prionopsis Nutt. COMPOSITAE (D)
Prionosciadium S.Watson UMBELLIFERAE (D)
Prionostemma Miers CELASTRACEAE (D)
Prionotes R.Br. EPACRIDACEAE (D)
Prionotrichon Botsch. & Vved. CRUCIFERAE (D)
Prioria Griseb. LEGUMINOSAE–CAESALPINIOIDEAE (D)
Priotropis Wight. & Arn. = Crotalaria L. LEGUMINOSAE–PAPILIONOIDEAE (D)
Prismatocarpus L'Hér. CAMPANULACEAE (D)
Prismatomeris Thwaites RUBIACEAE (D)
Pristidia Thwaites = Gaertnera Lam. RUBIACEAE (D)
Pristiglottis Cretz. & J.J.Sm. ORCHIDACEAE (M)
Pristimera Miers CELASTRACEAE (D)
Pritchardia Seem. & H.A.Wendl. PALMAE (M)
Pritchardiopsis Becc. PALMAE (M)
Pritzelago Kuntze CRUCIFERAE (D)
Pritzelia Schauer = Scholtzia Schauer MYRTACEAE (D)
Priva Adans. VERBENACEAE (D)
Proatriplex Stutz & G.L.Chu = Atriplex L. CHENOPODIACEAE (D)
Problastes Reinw. = Lumnitzera Willd. COMBRETACEAE (D)
Probletostemon K.Schum. = Tricalysia A.Rich. ex DC. RUBIACEAE (D)
Proboscidea Schmidel PEDALIACEAE (D)
Prochnyanthes S.Watson AGAVACEAE (M)
Prockia P.Browne ex L. TILIACEAE (D)
Prockiopsis Baill. FLACOURTIACEAE (D)
Proclesia Klotzsch = Cavendishia Lindl. ERICACEAE (D)
Procopiana Guşul. (SUO) = Symphytum L. BORAGINACEAE (D)
Procopiania Guşul. = Symphytum L. BORAGINACEAE (D)
Procrassula Griseb. = Sedum L. CRASSULACEAE (D)
Procris Comm. ex Juss. URTICACEAE (D)
Proferea C.Presl = Sphaerostephanos J.Sm. THELYPTERIDACEAE (F)
Proiphys Herb. AMARYLLIDACEAE (M)
Prolobus R.M.King & H.Rob. COMPOSITAE (D)
Prolongoa Boiss. COMPOSITAE (D)

Promenaea Lindl. ORCHIDACEAE (M)
Prometheum (A.Berger) H.Ohba CRASSULACEAE (D)
Pronaya Hügel PITTOSPORACEAE (D)
Pronephrium C.Presl THELYPTERIDACEAE (F)
Prosanerpis S.F.Blake = Clidemia D.Don MELASTOMATACEAE (D)
Prosaptia C.Presl GRAMMITIDACEAE (F)
Prosartema Gagnep. = Trigonostemon Blume EUPHORBIACEAE (D)
Proscephaleium Korth. RUBIACEAE (D)
Proserpinaca L. HALORAGACEAE (D)
Prosopanche Bary HYDNORACEAE (D)
Prosopidastrum Burkart LEGUMINOSAE–MIMOSOIDEAE (D)
Prosopis L. LEGUMINOSAE–MIMOSOIDEAE (D)
Prosopostelma Baill. ASCLEPIADACEAE (D)
Prosorus Dalzell = Margaritaria L.f. EUPHORBIACEAE (D)
Prosphysis Dulac (SUS) = Vulpia C.C.Gmelin GRAMINEAE (M)
Prosphytochloa Schweick. GRAMINEAE (M)
Prostanthera Labill. LABIATAE (D)
Prosthechea Knowles & Westc. = Epidendrum L. ORCHIDACEAE (M)
Prosthecidiscus Donn.Sm. ASCLEPIADACEAE (D)
Protamomum Ridl. = Orchidantha N.E.Br. LOWIACEAE (M)
Protangiopteris Hayata = Archangiopteris Christ & Giesenh. MARATTIACEAE (F)
Protarum Engl. ARACEAE (M)
Protasparagus Oberm. = Asparagus L. ASPARAGACEAE (M)
Protea L. PROTEACEAE (D)
Proteopsis Mart. & Zucc. ex Sch.Bip. COMPOSITAE (D)
Protium Burm.f. BURSERACEAE (D)
Protocyrtandra Hosok. = Cyrtandra J.R.Forst. & G.Forst. GESNERIACEAE (D)
Protogabunia Boiteau = Tabernaemontana L. APOCYNACEAE (D)
Protolindsaya Copel. = Tapeinidium (C.Presl) C.Chr. DENNSTAEDTIACEAE (F)
Protomarattia Hayata = Angiopteris Hoffm. MARATTIACEAE (F)
Protomegabaria Hutch. EUPHORBIACEAE (D)
Protorhus Engl. ANACARDIACEAE (D)
Protoschwenckia Soler. SOLANACEAE (D)
Protowoodsia Ching = Woodsia R.Br. WOODSIACEAE (F)
Proustia Lag. COMPOSITAE (D)
Provancheria B.Boivin (UP) CARYOPHYLLACEAE (D)
Prozetia Neck. (SUI) = Pouteria Aubl. SAPOTACEAE (D)
Prumnopitys Phil. PODOCARPACEAE (G)
Prunella L. LABIATAE (D)
Prunellopsis Kudo = Prunella L. LABIATAE (D)
Prunus L. ROSACEAE (D)
Pryona Miq. = Crudia Schreb. LEGUMINOSAE–CAESALPINIOIDEAE (D)
Przewalskia Maxim. SOLANACEAE (D)
Psacadocalymma Bremek. = Justicia L. ACANTHACEAE (D)
Psacadopaepale Bremek. = Strobilanthes Blume ACANTHACEAE (D)
Psacaliopsis H.Rob. & Brettell COMPOSITAE (D)
Psacalium Cass. COMPOSITAE (D)
Psamma P.Beauv. = Ammophila Host GRAMINEAE (M)
Psammagrostis C.A.Gardner & C.E.Hubb. GRAMINEAE (M)
Psammetes Hepper SCROPHULARIACEAE (D)
Psammiosorus C.Chr. OLEANDRACEAE (F)
Psammisia Klotzsch ERICACEAE (D)
Psammochloa Hitchc. GRAMINEAE (M)
Psammogeton Edgew. UMBELLIFERAE (D)
Psammomoya Diels & Loes. CELASTRACEAE (D)

Psammophila Schult. = Spartina Schreb. GRAMINEAE (M)
Psammophila Ikonn. (SUH) = Gypsophila L. CARYOPHYLLACEAE (D)
Psammophiliella Ikonn. = Gypsophila L. CARYOPHYLLACEAE (D)
Psammophora Dinter & Schwantes AIZOACEAE (D)
Psammopyrum Á.Löve = Elymus L. GRAMINEAE (M)
Psammotropha Eckl. & Zeyh. MOLLUGINACEAE (D)
Psathura Comm. ex Juss. RUBIACEAE (D)
Psathyranthus Ule LORANTHACEAE (D)
Psathyrostachys Nevski GRAMINEAE (M)
Psathyrotes A.Gray COMPOSITAE (D)
Psathyrotopsis Rydb. = Psathyrotes A.Gray COMPOSITAE (D)
Psednotrichia Hiern COMPOSITAE (D)
Pseudabutilon R.E.Fr. MALVACEAE (D)
Pseudacanthopale Benoist = Strobilanthopsis S.Moore ACANTHACEAE (D)
Pseudacoridium Ames ORCHIDACEAE (M)
Pseudactis S.Moore = Emilia (Cass.) Cass. COMPOSITAE (D)
Pseudaechmanthera Bremek. = Strobilanthes Blume ACANTHACEAE (D)
Pseudaechmea L.B.Sm. & Read BROMELIACEAE (M)
Pseudagrostistachys Pax & K.Hoffm. EUPHORBIACEAE (D)
Pseudaidia Tirveng. RUBIACEAE (D)
Pseudais Decne. THYMELAEACEAE (D)
Pseudalbizzia Britton & Rose = Albizia Durazz. LEGUMINOSAE–MIMOSOIDEAE (D)
Pseudaleia Thouars = Olax L. OLACACEAE (D)
Pseudalepyrum Dandy = Centrolepis Labill. CENTROLEPIDACEAE (M)
Pseudammi H.Wolff = Seseli L. UMBELLIFERAE (D)
Pseudanamomis Kausel MYRTACEAE (D)
Pseudananas Hassl. ex Harms BROMELIACEAE (M)
Pseudannona (Baill.) Saff. = Xylopia L. ANNONACEAE (D)
Pseudanthistiria (Hack.) Hook.f. GRAMINEAE (M)
Pseudanthus Sieber ex A.Spreng. EUPHORBIACEAE (D)
Pseudarabidella O.E.Schulz = Arabidella (F.Muell.) O.E.Schulz CRUCIFERAE (D)
Pseudarrhenatherum Rouy = Arrhenatherum P.Beauv. GRAMINEAE (M)
Pseudartabotrys Pellegr. ANNONACEAE (D)
Pseudarthria Wight & Arn. LEGUMINOSAE–PAPILIONOIDEAE (D)
Pseudathyrium Newman = Athyrium Roth WOODSIACEAE (F)
Pseudechinolaena Stapf GRAMINEAE (M)
Pseudelephantopus Rohr = Elephantopus L. COMPOSITAE (D)
Pseudelleanthus Brieger = Elleanthus C.Presl ORCHIDACEAE (M)
Pseudellipanthus Schellenb. = Ellipanthus Hook.f. CONNARACEAE (D)
Pseudeminia Verdc. LEGUMINOSAE–PAPILIONOIDEAE (D)
Pseudephedranthus Aristeg. ANNONACEAE (D)
Pseudepidendrum Rchb.f. = Epidendrum L. ORCHIDACEAE (M)
Pseuderanthemum Radlk. ACANTHACEAE (D)
Pseuderemostachys Popov LABIATAE (D)
Pseuderia Schltr. ORCHIDACEAE (M)
Pseuderiopsis Rchb.f. = Eriopsis Lindl. ORCHIDACEAE (M)
Pseuderucaria (Boiss.) O.E.Schulz CRUCIFERAE (D)
Pseudibatia Malme ASCLEPIADACEAE (D)
Pseudima Radlk. SAPINDACEAE (D)
Pseudiosma DC. (UP) RUTACEAE (D)
Pseudixora Miq. = Anomanthodia Hook.f. RUBIACEAE (D)
Pseudo–Barleria Oerst. = Barleria L. ACANTHACEAE (D)
Pseudoacanthocereus F.Ritter CACTACEAE (D)
Pseudoanastatica (Boiss.) Grossh. = Clypeola L. CRUCIFERAE (D)
Pseudobaccharis Cabrera COMPOSITAE (D)

Pseudobaeckea Nied. BRUNIACEAE (D)
Pseudobahia (A.Gray) Rydb. COMPOSITAE (D)
Pseudobarleria T.Anderson = Petalidium Nees ACANTHACEAE (D)
Pseudobartlettia Rydb. = Psathyrotes A.Gray COMPOSITAE (D)
Pseudobartsia D.Y.Hong SCROPHULARIACEAE (D)
Pseudobastardia Hassl. = Herissantia Medik. MALVACEAE (D)
Pseudobersama Verdc. MELIACEAE (D)
Pseudobesleria Oerst. = Besleria L. GESNERIACEAE (D)
Pseudobetckea (Hock) Lincz. VALERIANACEAE (D)
Pseudoblepharis Baill. (SUH) = Sclerochiton Harv. ACANTHACEAE (D)
Pseudoblepharispermum J.-P.Lebrun & Stork COMPOSITAE (D)
Pseudoboivinella Aubrév. & Pellegr. = Englerophytum K.Krause SAPOTACEAE (D)
Pseudobombax Dugand BOMBACACEAE (D)
Pseudobotrys Moeser ICACINACEAE (D)
Pseudobrachiaria Launert = Brachiaria (Trin.)Griseb. GRAMINEAE (M)
Pseudobrassaiopsis R.N.Banerjee = Brassaiopsis Decne. & Planch. ARALIACEAE (D)
Pseudobravoa Rose AGAVACEAE (M)
Pseudobraya Korsh. = Draba L. CRUCIFERAE (D)
Pseudobrazzeia Engl. = Brazzeia Baill. SCYTOPETALACEAE (D)
Pseudobrickellia R.M.King & H.Rob. COMPOSITAE (D)
Pseudobromus K.Schum. = Festuca L. GRAMINEAE (M)
Pseudobrownanthus Ihlenf. & Bittrich AIZOACEAE (D)
Pseudocadia Harms = Xanthocercis Baill. LEGUMINOSAE–PAPILIONOIDEAE (D)
Pseudocadiscus Lisowski COMPOSITAE (D)
Pseudocalymma A.Samp. & Kuhlm. = Mansoa DC. BIGNONIACEAE (D)
Pseudocalyx Radlk. ACANTHACEAE (D)
Pseudocamelina (Boiss.) N.Busch CRUCIFERAE (D)
Pseudocampanula Kołak. = Campanula L. CAMPANULACEAE (D)
Pseudocapsicum Medik. = Solanum L. SOLANACEAE (D)
Pseudocarapa Hemsl. MELIACEAE (D)
Pseudocarpidium Millsp. VERBENACEAE (D)
Pseudocarum C.Norman UMBELLIFERAE (D)
Pseudocaryophyllus O.Berg = Pimenta Lindl. MYRTACEAE (D)
Pseudocassia Britton & Rose = Senna Mill. LEGUMINOSAE–CAESALPINIOIDEAE (D)
Pseudocassine Bredell = Crocoxylon Eckl. & Zeyh. CELASTRACEAE (D)
Pseudocatalpa A.H.Gentry BIGNONIACEAE (D)
Pseudocedrela Harms MELIACEAE (D)
Pseudocentrum Lindl. ORCHIDACEAE (M)
Pseudochaetochloa Hitchc. GRAMINEAE (M)
Pseudochamaesphacos Parsa (UP) LABIATAE (D)
Pseudochimarrhis Ducke = Chimarrhis Jacq. RUBIACEAE (D)
Pseudochirita W.T.Wang GESNERIACEAE (D)
Pseudochrosia Blume = Ochrosia Juss. APOCYNACEAE (D)
Pseudocimum Bremek. = Endostemon N.E.Br. LABIATAE (D)
Pseudocinchona A.Chev. = Corynanthe Welw. RUBIACEAE (D)
Pseudocladia Pierre = Pouteria Aubl. SAPOTACEAE (D)
Pseudoclappia Rydb. COMPOSITAE (D)
Pseudoclausia Popov CRUCIFERAE (D)
Pseudocoix A.Camus GRAMINEAE (M)
Pseudocolysis L.D.Gómez = Polypodium L. POLYPODIACEAE (F)
Pseudoconnarus Radlk. CONNARACEAE (D)
Pseudoconyza Cuatrec. = Laggera Sch.Bip. ex Benth. COMPOSITAE (D)
Pseudocopaiva Britton & P.Wilson = Guibourtia Benn. LEGUMINOSAE–
    CAESALPINIOIDEAE (D)
Pseudocorchorus Capuron TILIACEAE (D)

Pseudocranichis Garay ORCHIDACEAE (M)
Pseudocroton Müll.Arg. EUPHORBIACEAE (D)
Pseudocryptocarya Teschner = Cryptocarya R.Br. LAURACEAE (D)
Pseudoctomeria Kraenzl. = Pleurothallis R.Br. ORCHIDACEAE (M)
Pseudocunila Brade = Hedeoma Pers. LABIATAE (D)
Pseudocyclanthera Mart.Crov. CUCURBITACEAE (D)
Pseudocyclosorus Ching THELYPTERIDACEAE (F)
Pseudocydonia (C.K.Schneid.) C.K.Schneid. ROSACEAE (D)
Pseudocymopterus J.M.Coult. & Rose UMBELLIFERAE (D)
Pseudocynometra Kuntze = Maniltoa Scheff. LEGUMINOSAE–CAESALPINIOIDEAE (D)
Pseudocystopteris Ching = Athyrium Roth WOODSIACEAE (F)
Pseudocytisus Kuntze = Vella DC. CRUCIFERAE (D)
Pseudodanthonia Bor & C.E.Hubb. GRAMINEAE (M)
Pseudodatura Zijp = Brugmansia Pers. SOLANACEAE (D)
Pseudodichanthium Bor GRAMINEAE (M)
Pseudodicliptera Benoist ACANTHACEAE (D)
Pseudodigera Chiov. = Digera Forssk. AMARANTHACEAE (D)
Pseudodiphasium Holub = Lycopodium L. LYCOPODIACEAE (FA)
Pseudodiphryllum Nevski = Platanthera Rich. ORCHIDACEAE (M)
Pseudodissochaeta M.P.Nayar MELASTOMATACEAE (D)
Pseudodracontium N.E.Br. ARACEAE (M)
Pseudodrynaria C.Chr. ex Ching = Aglaomorpha Schott POLYPODIACEAE (F)
Pseudoechinocereus Buining (SUI) = Cleistocactus Lem. CACTACEAE (D)
Pseudoechinopepon (Cogn.) Cockerell = Vaseyanthus Cogn. CUCURBITACEAE (D)
Pseudoentada Britton & Rose = Adenopodia C.Presl LEGUMINOSAE–
    MIMOSOIDEAE (D)
Pseudoeria (Schltr.) Schltr. = Pseuderia Schltr. ORCHIDACEAE (M)
Pseudoeriosema Hauman LEGUMINOSAE–PAPILIONOIDEAE (D)
Pseudoernestia (Cogn.) Krasser MELASTOMATACEAE (D)
Pseudoespostoa Backeb. = Espostoa Britton & Rose CACTACEAE (D)
Pseudoeugenia Scort. = Syzygium Gaertn. MYRTACEAE (D)
Pseudoeurya Yamam. = Eurya Thunb. THEACEAE (D)
Pseudoeurystyles Hoehne = Eurystyles Wawra ORCHIDACEAE (M)
Pseudoeverardia Gilly = Everardia Ridl. CYPERACEAE (M)
Pseudofortuynia Hedge CRUCIFERAE (D)
Pseudofumaria Medik. PAPAVERACEAE (D)
Pseudogaillonia Lincz. RUBIACEAE (D)
Pseudogaltonia (Kuntze) Engl. HYACINTHACEAE (M)
Pseudogardenia Keay RUBIACEAE (D)
Pseudogardnera Racib. = Gardneria Wall. LOGANIACEAE (D)
Pseudoglochidion Gamble = Glochidion J.R.Forst. & G.Forst. EUPHORBIACEAE (D)
Pseudoglycine F.J.Herm. = Ophrestia H.M.L.Forbes LEGUMINOSAE–
    PAPILIONOIDEAE (D)
Pseudognaphalium Kirp. COMPOSITAE (D)
Pseudognidia E.Phillips = Gnidia L. THYMELAEACEAE (D)
Pseudogomphrena R.E.Fr. AMARANTHACEAE (D)
Pseudogonocalyx Bisse & Berazaín ERICACEAE (D)
Pseudogoodyera Schltr. ORCHIDACEAE (M)
Pseudogynoxys (Greenm.) Cabrera COMPOSITAE (D)
Pseudohamelia Wernham RUBIACEAE (D)
Pseudohandelia Tzvelev COMPOSITAE (D)
Pseudohexadesmia Brieger (SUI) = Hexadesmia Brongn. ORCHIDACEAE (M)
Pseudohomalomena A.D.Hawkes = Zantedeschia Spreng. ARACEAE (M)
Pseudohydrosme Engl. ARACEAE (M)
Pseudojacobaea (Hook.f.) Mathur = Senecio L. COMPOSITAE (D)

Pseudokyrsteniopsis R.M.King & H.Rob. COMPOSITAE (D)
Pseudolabatia Aubrév. & Pellegr. = Pouteria Aubl. SAPOTACEAE (D)
Pseudolachnostylis Pax EUPHORBIACEAE (D)
Pseudolaelia Porto ORCHIDACEAE (M)
Pseudolarix Gordon PINACEAE (G)
Pseudolasiacis (A.Camus) A.Camus = Lasiacis (Griseb.) Hitchc. GRAMINEAE (M)
Pseudoligandra Dillon & Sagast. = Chionolaena DC. COMPOSITAE (D)
Pseudoliparis Finet = Malaxis Sol. ex Sw. ORCHIDACEAE (M)
Pseudolitchi Danguy & Choux = Stadmannia Lam. SAPINDACEAE (D)
Pseudolithos Bally ASCLEPIADACEAE (D)
Pseudolmedia Trécul MORACEAE (D)
Pseudolobivia (Backeb.) Backeb. = Echinopsis Zucc. CACTACEAE (D)
Pseudolopezia Rose = Lopezia Cav. ONAGRACEAE (D)
Pseudolophanthus Levin = Marmoritis Benth. LABIATAE (D)
Pseudolotus Rech.f. = Lotus L. LEGUMINOSAE–PAPILIONOIDEAE (D)
Pseudoloxocarya Linder RESTIONACEAE (M)
Pseudoludovia Harling = Sphaeradenia Harling CYCLANTHACEAE (M)
Pseudolycopodiella Holub = Lycopodiella Holub LYCOPODIACEAE (FA)
Pseudolycopodium Holub = Lycopodium L. LYCOPODIACEAE (FA)
Pseudolysimachion (W.D.J.Koch) Opiz SCROPHULARIACEAE (D)
Pseudo–Lysimachium (W.D.J.Koch) Opiz (SUO) = Pseudolysimachion (W.D.J.Koch) Opiz
    SCROPHULARIACEAE (D)
Pseudomachaerium Hassl. = Nissolia Jacq. LEGUMINOSAE–PAPILIONOIDEAE (D)
Pseudomacodes Rolfe = Macodes (Blume) Lindl. ORCHIDACEAE (M)
Pseudomacrolobium Hauman LEGUMINOSAE–CAESALPINIOIDEAE (D)
Pseudomammillaria Buxb. = Mammillaria Haw. CACTACEAE (D)
Pseudomantalania J.–F.Leroy RUBIACEAE (D)
Pseudomariscus Rauschert (SUS) = Courtoisina Soják CYPERACEAE (M)
Pseudomarrubium Popov LABIATAE (D)
Pseudomarsdenia Baill. ASCLEPIADACEAE (D)
Pseudomaxillaria Hoehne ORCHIDACEAE (M)
Pseudomelissitus Ovcz., Rassulova & Kinzik. = Medicago L. LEGUMINOSAE–
    PAPILIONOIDEAE (D)
Pseudomertensia Riedl BORAGINACEAE (D)
Pseudomitrocereus H.Bravo & Buxb. = Neobuxbaumia Backeb. CACTACEAE (D)
Pseudomorus Bureau = Streblus Lour. MORACEAE (D)
Pseudomuscari Garbari & Greuter HYACINTHACEAE (M)
Pseudomussaenda Wernham RUBIACEAE (D)
Pseudomyrcianthes Kausel = Myrcianthes O.Berg MYRTACEAE (D)
Pseudonemacladus McVaugh CAMPANULACEAE (D)
Pseudonephelium Radlk. = Dimocarpus Lour. SAPINDACEAE (D)
Pseudonesohedyotis Tennant RUBIACEAE (D)
Pseudonopalxochia Backeb. = Disocactus Lindl. CACTACEAE (D)
Pseudonoseris H.Rob. & Brettell COMPOSITAE (D)
Pseudoorleanesia Rauschert = Orleanesia Barb.Rodr. ORCHIDACEAE (M)
Pseudopachystela Aubrév. & Pellegr. = Synsepalum (A.DC.) Daniell SAPOTACEAE (D)
Pseudopaegma Urb. = Anemopaegma Mart. ex Meisn. BIGNONIACEAE (D)
Pseudopanax K.Koch ARALIACEAE (D)
Pseudopancovia Pellegr. SAPINDACEAE (D)
Pseudoparis H.Perrier COMMELINACEAE (M)
Pseudopavonia Hassl. = Pavonia Cav. MALVACEAE (D)
Pseudopectinaria Lavranos ASCLEPIADACEAE (D)
Pseudopentameris Conert GRAMINEAE (M)
Pseudopentatropis Costantin (UP) ASCLEPIADACEAE (D)
Pseudopeponidium Homolle ex Arènes = Pyrostria Comm. ex Juss. RUBIACEAE (D)

Pseudophacelurus A.Camus = Phacelurus Griseb. GRAMINEAE (M)
Pseudophegopteris Ching THELYPTERIDACEAE (F)
Pseudophleum Dogan = Phleum L. GRAMINEAE (M)
Pseudophoenix H.A.Wendl. ex Sarg. PALMAE (M)
Pseudopilocereus Buxb. = Pilosocereus Byles & Rowley CACTACEAE (D)
Pseudopinanga Burret = Pinanga Blume PALMAE (M)
Pseudopiptadenia Rauschert LEGUMINOSAE–MIMOSOIDEAE (D)
Pseudopipturus Skottsb. = Nothocnide Blume ex Chew URTICACEAE (D)
Pseudoplantago Suess. AMARANTHACEAE (D)
Pseudopogonatherum A.Camus = Eulalia Kunth GRAMINEAE (M)
Pseudoponera Brieger (SUI) = Ponera Lindl. ORCHIDACEAE (M)
Pseudoprimula (Pax) O.Schwarz (UP) PRIMULACEAE (D)
Pseudoprosopis Harms LEGUMINOSAE–MIMOSOIDEAE (D)
Pseudoprotorhus H.Perrier = Filicium Thwaites ex Benth. SAPINDACEAE (D)
Pseudopteris Baill. SAPINDACEAE (D)
Pseudopyxis Miq. RUBIACEAE (D)
Pseudorachicallis Post & Kuntze (SUO) = Arcytophyllum Willd. ex Schult. & Schult.f.
    RUBIACEAE (D)
Pseudoraphis Griff. GRAMINEAE (M)
Pseudorchis Séguier ORCHIDACEAE (M)
Pseudorhachicallis Hook.f. (SUH) = Arcytophyllum Willd. ex Schult. & Schult.f.
    RUBIACEAE (D)
Pseudorhipsalis Britton & Rose CACTACEAE (D)
Pseudorlaya (Murb.) Murb. UMBELLIFERAE (D)
Pseudorleanesia Rauschert = Orleanesia Barb.Rodr. ORCHIDACEAE (M)
Pseudorobanche Rouy SCROPHULARIACEAE (D)
Pseudoroegneria (Nevski) Á.Löve = Elymus L. GRAMINEAE (M)
Pseudorontium (A.Gray) Rothm. SCROPHULARIACEAE (D)
Pseudorosularia Gurgen. = Prometheum (A.Berger) H.Ohba CRASSULACEAE (D)
Pseudoruellia Benoist ACANTHACEAE (D)
Pseudoryza Griff. (SUS) = Leersia Sw. GRAMINEAE (M)
Pseudosabicea N.Hallé RUBIACEAE (D)
Pseudosagotia Secco EUPHORBIACEAE (D)
Pseudosalacia Codd CELASTRACEAE (D)
Pseudosamanea Harms = Albizia Durazz. LEGUMINOSAE–MIMOSOIDEAE (D)
Pseudosantalum Kuntze = Osmoxylon Miq. ARALIACEAE (D)
Pseudosaponaria (F.Williams) Ikonn. = Gypsophila L. CARYOPHYLLACEAE (D)
Pseudosarcolobus Costantin = Gymnema R.Br. ASCLEPIADACEAE (D)
Pseudosasa Makino ex Nakai GRAMINEAE (M)
Pseudosassafras Lecomte = Sassafras Nees LAURACEAE (D)
Pseudosbeckia A.Fern. & R.Fern. MELASTOMATACEAE (D)
Pseudoscabiosa Devesa = Scabiosa L. DIPSACACEAE (D)
Pseudoschoenus (C.B.Clarke) Oteng–Yeb. CYPERACEAE (M)
Pseudosciadium Baill. ARALIACEAE (D)
Pseudoscolopia Gilg FLACOURTIACEAE (D)
Pseudosecale (Godr.) Degen (SUS) = Dasypyrum (Coss. & Durieu) T.Durand
    GRAMINEAE (M)
Pseudosedum (Boiss.) A.Berger CRASSULACEAE (D)
Pseudoselinum C.Norman UMBELLIFERAE (D)
Pseudosempervivum (Boiss.) Grossh. = Cochlearia L. CRUCIFERAE (D)
Pseudosericocoma Cavaco AMARANTHACEAE (D)
Pseudosicydium Harms CUCURBITACEAE (D)
Pseudosindora Symington LEGUMINOSAE–CAESALPINIOIDEAE (D)
Pseudosmelia Sleumer FLACOURTIACEAE (D)
Pseudosmilax Hayata = Heterosmilax Kunth SMILACACEAE (M)

Pseudosmodingium Engl. ANACARDIACEAE (D)
Pseudosolisia Y.Ito (SUS) = Neolloydia Britton & Rose CACTACEAE (D)
Pseudosopubia Engl. SCROPHULARIACEAE (D)
Pseudosorghum A.Camus GRAMINEAE (M)
Pseudosorocea Baill. = Sorocea A.St.–Hil. MORACEAE (D)
Pseudospigelia Klett LOGANIACEAE (D)
Pseudospondias Engl. ANACARDIACEAE (D)
Pseudostachyum Munro = Schizostachyum Nees GRAMINEAE (M)
Pseudostelis Schltr. = Pleurothallis R.Br. ORCHIDACEAE (M)
Pseudostellaria Pax CARYOPHYLLACEAE (D)
Pseudostenomesson Velarde AMARYLLIDACEAE (M)
Pseudostenosiphonium Lindau = Strobilanthes Blume ACANTHACEAE (D)
Pseudostifftia H.Rob. COMPOSITAE (D)
Pseudostreblus Bureau = Streblus Lour. MORACEAE (D)
Pseudostreptogyne A.Camus = Streblochaete Pilg. GRAMINEAE (M)
Pseudostriga Bonati SCROPHULARIACEAE (D)
Pseudostrophis T.Durand & B.D.Jacks. = Streblus Lour. MORACEAE (D)
Pseudotaenidia Mack. UMBELLIFERAE (D)
Pseudotaxus W.C.Cheng TAXACEAE (G)
Pseudotectaria Tardieu DRYOPTERIDACEAE (F)
Pseudotephrocactus Frič = Opuntia Mill. CACTACEAE (D)
Pseudotragia Pax = Pterococcus Hassk. EUPHORBIACEAE (D)
Pseudotrewia Miq. = Wetria Baill. EUPHORBIACEAE (D)
Pseudotrimezia R.C.Foster IRIDACEAE (M)
Pseudotrophis Warb. = Streblus Lour. MORACEAE (D)
Pseudotsuga Carrière PINACEAE (G)
Pseudourceolina Vargas = Urceolina Rchb. AMARYLLIDACEAE (M)
Pseudovanilla Garay ORCHIDACEAE (M)
Pseudovesicaria (Boiss.) Rupr. CRUCIFERAE (D)
Pseudovigna (Harms) Verdc. LEGUMINOSAE–PAPILIONOIDEAE (D)
Pseudovossia A.Camus = Phacelurus Griseb. GRAMINEAE (M)
Pseudovouapa Britton & Rose = Macrolobium Schreb. LEGUMINOSAE–
    CAESALPINIOIDEAE (D)
Pseudoweinmannia Engl. CUNONIACEAE (D)
Pseudowillughbeia Markgr. = Melodinus J.R.Forst. & G.Forst. APOCYNACEAE (D)
Pseudowintera Dandy WINTERACEAE (D)
Pseudowolffia Hartog & Van der Plas = Wolffiella (Hegelm.) Hegelm. LEMNACEAE (M)
Pseudoxalis Rose = Oxalis L. OXALIDACEAE (D)
Pseudoxandra R.E.Fr. ANNONACEAE (D)
Pseudoxytenanthera Soderstr. & R.P.Ellis = Schizostachyum Nees GRAMINEAE (M)
Pseudoxythece Aubrév. = Pouteria Aubl. SAPOTACEAE (D)
Pseudozoysia Chiov. GRAMINEAE (M)
Pseudozygocactus Backeb. = Hatiora Britton & Rose CACTACEAE (D)
Pseuduvaria Miq. ANNONACEAE (D)
Pseusmagennetus Ruschenb. = Marsdenia R.Br. ASCLEPIADACEAE (D)
Psiadia Jacq. COMPOSITAE (D)
Psiadiella Humbert COMPOSITAE (D)
Psidiastrum Bello = Eugenia L. MYRTACEAE (D)
Psidiomyrtus Guillaumin = Rhodomyrtus (DC.) Rchb. MYRTACEAE (D)
Psidiopsis O.Berg = Psidium L. MYRTACEAE (D)
Psidium L. MYRTACEAE (D)
Psiguria Neck. ex Arn. CUCURBITACEAE (D)
Psila Phil. = Baccharis L. COMPOSITAE (D)
Psilactis A.Gray COMPOSITAE (D)
Psilantha (K.Koch) Tzvelev = Eragrostis Wolf GRAMINEAE (M)

Psilanthele Lindau ACANTHACEAE (D)
Psilanthopsis A.Chev. = Coffea L. RUBIACEAE (D)
Psilanthus Hook.f. RUBIACEAE (D)
Psilathera Link = Sesleria Scop. GRAMINEAE (M)
Psilobium Jack = Acranthera Arn. ex Meisn. RUBIACEAE (D)
Psilocarphus Nutt. COMPOSITAE (D)
Psilocarya Torr. = Rhynchospora Vahl CYPERACEAE (M)
Psilocaulon N.E.Br. AIZOACEAE (D)
Psilochilus Barb.Rodr. ORCHIDACEAE (M)
Psilochloa Launert = Panicum L. GRAMINEAE (M)
Psilodigera Suess. = Saltia R.Br. ex Moq. AMARANTHACEAE (D)
Psilodochea C.Presl = Angiopteris Hoffm. MARATTIACEAE (F)
Psiloesthes Benoist = Peristrophe Nees ACANTHACEAE (D)
Psilogramme Kuhn = Eriosorus Fée ADIANTACEAE (F)
Psilolaemus I.M.Johnst. BORAGINACEAE (D)
Psilolemma S.M.Phillips GRAMINEAE (M)
Psilonema C.A.Mey. = Alyssum L. CRUCIFERAE (D)
Psilopeganum Hemsl. RUTACEAE (D)
Psilopogon Hochst. = Microstegium Nees GRAMINEAE (M)
Psilostachys Hochst. (SUH) = Psilotrichum Blume AMARANTHACEAE (D)
Psilostachys Steud. (SUH) = Dimeria R.Br. GRAMINEAE (M)
Psilostachys Turcz. = Cleidion Blume EUPHORBIACEAE (D)
Psilostoma Klotzsch = Canthium Lam. RUBIACEAE (D)
Psilostrophe DC. COMPOSITAE (D)
Psilotrichopsis C.C.Towns. AMARANTHACEAE (D)
Psilotrichum Blume AMARANTHACEAE (D)
Psilotum Sw. PSILOTACEAE (FA)
Psiloxylon Thouars ex Tul. MYRTACEAE (D)
Psilurus Trin. GRAMINEAE (M)
Psittacanthus Mart. LORANTHACEAE (D)
Psittacoglossum La Llave & Lex. = Maxillaria Ruíz & Pav. ORCHIDACEAE (M)
Psomiocarpa C.Presl DRYOPTERIDACEAE (F)
Psophocarpus DC. LEGUMINOSAE–PAPILIONOIDEAE (D)
Psoralea L. LEGUMINOSAE–PAPILIONOIDEAE (D)
Psoralidium Rydb. LEGUMINOSAE–PAPILIONOIDEAE (D)
Psorobatus Rydb. = Errazurizia Phil. LEGUMINOSAE–PAPILIONOIDEAE (D)
Psorodendron Rydb. = Psorothamnus Rydb. LEGUMINOSAE–PAPILIONOIDEAE (D)
Psorospermum Spach GUTTIFERAE (D)
Psorothamnus Rydb. LEGUMINOSAE–PAPILIONOIDEAE (D)
Pstathura Raf. (SUO) = Psathura Comm. ex Juss. RUBIACEAE (D)
Psychanthus (K.Schum.) Ridl. (SUH) = Pleuranthodium (K.Schum.) Ros.M.Sm.
    ZINGIBERACEAE (M)
Psychechilos Breda = Zeuxine Lindl. ORCHIDACEAE (M)
Psychilis Raf. ORCHIDACEAE (M)
Psychine Desf. CRUCIFERAE (D)
Psychopsiella Lückel & Braem ORCHIDACEAE (M)
Psychopsis Raf. = Oncidium Sw. ORCHIDACEAE (M)
Psychotria L. RUBIACEAE (D)
Psychotrophum P.Browne = Psychotria L. RUBIACEAE (D)
Psychrogeton Boiss. COMPOSITAE (D)
Psychrophila (DC.) Bercht. & J.Presl RANUNCULACEAE (D)
Psychrophyton Beauverd COMPOSITAE (D)
Psydrax Gaertn. RUBIACEAE (D)
Psygmium C.Presl = Aglaomorpha Schott POLYPODIACEAE (F)
Psygmorchis Dodson & Dressler ORCHIDACEAE (M)

Psylliostachys (Jaub. & Spach) Nevski PLUMBAGINACEAE (D)
Psyllocarpus Mart. & Zucc. RUBIACEAE (D)
Ptaeroxylon Eckl. & Zeyh. PTAEROXYLACEAE (D)
Ptelea L. RUTACEAE (D)
Pteleocarpa Oliv. BORAGINACEAE (D)
Pteleopsis Engl. COMBRETACEAE (D)
Ptelidium Thouars CELASTRACEAE (D)
Pteracanthus (Nees) Bremek. = Strobilanthes Blume ACANTHACEAE (D)
Pterachenia (Benth.) Lipschitz COMPOSITAE (D)
Pteralyxia K.Schum. APOCYNACEAE (D)
Pterandra A.Juss. MALPIGHIACEAE (D)
Pteranthus Forssk. ILLECEBRACEAE (D)
Pteretis Hayata = Matteuccia Tod. WOODSIACEAE (F)
Pterichis Lindl. ORCHIDACEAE (M)
Pteridanetium Copel. = Anetium Splitg. VITTARIACEAE (F)
Pteridella Mett. ex Kuhn = Pellaea Link ADIANTACEAE (F)
Pteridium Raf. (SUH) = Pteris L. PTERIDACEAE (F)
Pteridium Gled. ex Scop. DENNSTAEDTIACEAE (F)
Pteridoblechnum Hennipman BLECHNACEAE (F)
Pteridocalyx Wernham RUBIACEAE (D)
Pteridophyllum Siebold & Zucc. PAPAVERACEAE (D)
Pteridrys C.Chr. & Ching DRYOPTERIDACEAE (F)
Pterigeron (DC.) Benth. = Streptoglossa Steetz ex F.Muell. COMPOSITAE (D)
Pteriglyphis Fée = Diplazium Sw. WOODSIACEAE (F)
Pterilema Reinw. = Engelhardtia Lesch. ex Blume JUGLANDACEAE (D)
Pterilis Raf. = Matteuccia Tod. WOODSIACEAE (F)
Pterilis Raf. (SUH) = Pteris L. PTERIDACEAE (F)
Pterinodes Siegesb. ex Kuntze = Matteuccia Tod. WOODSIACEAE (F)
Pteris L. PTERIDACEAE (F)
Pteris Gled. ex Scop. (SUH) = Dryopteris Adans. DRYOPTERIDACEAE (F)
Pterisanthes Blume VITACEAE (D)
Pterium Desv. = Lamarckia Moench GRAMINEAE (M)
Pternandra Jack MELASTOMATACEAE (D)
Pternopetalum Franch. UMBELLIFERAE (D)
Pterobesleria C.Morton = Besleria L. GESNERIACEAE (D)
Pterocactus K.Schum. CACTACEAE (D)
Pterocarpus Jacq. LEGUMINOSAE–PAPILIONOIDEAE (D)
Pterocarya Kunth JUGLANDACEAE (D)
Pterocassia Britton & Rose = Senna Mill. LEGUMINOSAE–CAESALPINIOIDEAE (D)
Pterocaulon Elliott COMPOSITAE (D)
Pterocelastrus Meisn. CELASTRACEAE (D)
Pteroceltis Maxim. ULMACEAE (D)
Pterocephalidium López Gonz. = Pterocephalus Adans. DIPSACACEAE (D)
Pterocephalus Adans. DIPSACACEAE (D)
Pteroceras Hasselt ex Hassk. ORCHIDACEAE (M)
Pterocereus MacDougall & Miranda CACTACEAE (D)
Pterochilus Hook. & Arn. = Maxillaria Ruíz & Pav. ORCHIDACEAE (M)
Pterochlaena Chiov. = Alloteropsis C.Presl GRAMINEAE (M)
Pterochlamys Roberty = Hildebrandtia Vatke CONVOLVULACEAE (D)
Pterochloris (A.Camus) A.Camus = Chloris Sw. GRAMINEAE (M)
Pterochrosia Baill. = Cerberiopsis Vieill. ex Pancher & Sebert APOCYNACEAE (D)
Pterocissus Urb. & Ekman VITACEAE (D)
Pterocladon Hook.f. = Miconia Ruíz & Pav. MELASTOMATACEAE (D)
Pterococcus Pall. (SUH) = Calligonum L. POLYGONACEAE (D)
Pterococcus Hassk. EUPHORBIACEAE (D)

Pterocyclus Klotzsch = Pleurospermum Hoffm. UMBELLIFERAE (D)
Pterocymbium R.Br. STERCULIACEAE (D)
Pterocypsela C.Shih COMPOSITAE (D)
Pterodiscus Hook. PEDALIACEAE (D)
Pterodon Vogel LEGUMINOSAE–PAPILIONOIDEAE (D)
Pterogaillonia Lincz. RUBIACEAE (D)
Pterogastra Naudin MELASTOMATACEAE (D)
Pteroglossa Schltr. = Stenorrhynchos Rich. ex Spreng. ORCHIDACEAE (M)
Pteroglossaspis Rchb.f. ORCHIDACEAE (M)
Pterogyne Tul. LEGUMINOSAE–CAESALPINIOIDEAE (D)
Pterolepis Endl. (SUH) = Osbeckia L. MELASTOMATACEAE (D)
Pterolepis (DC.) Miq. MELASTOMATACEAE (D)
Pterolobium R.Br. ex Wight & Arn. LEGUMINOSAE–CAESALPINIOIDEAE (D)
Pteroloma Desv. ex Benth. (SUH) = Tadehagi H.Ohashi LEGUMINOSAE–
    PAPILIONOIDEAE (D)
Pteroloma Hochst. & Steud. = Dipterygium Decne. CAPPARACEAE (D)
Pteromanes Pic.Serm. = Crepidomanes (C.Presl) C.Presl HYMENOPHYLLACEAE (F)
Pteromimosa Britton = Mimosa L. LEGUMINOSAE–MIMOSOIDEAE (D)
Pteromischus Pichon = Crescentia L. BIGNONIACEAE (D)
Pteroneura Fée = Humata Cav. DAVALLIACEAE (F)
Pteronia L. COMPOSITAE (D)
Pteropavonia Mattei = Pavonia Cav. MALVACEAE (D)
Pteropentacoilanthus Rappa & Camarrone = Mesembryanthemum L. AIZOACEAE (D)
Pteropepon (Cogn.) Cogn. CUCURBITACEAE (D)
Pteropetalum Pax = Euadenia Oliv. CAPPARACEAE (D)
Pterophacos Rydb. = Astragalus L. LEGUMINOSAE–PAPILIONOIDEAE (D)
Pterophora Harv. = Dregea E.Mey. ASCLEPIADACEAE (D)
Pterophylla D.Don = Weinmannia L. CUNONIACEAE (D)
Pteropodium DC. ex Meisn. = Jacaranda Juss. BIGNONIACEAE (D)
Pteropodium Steud. (SUI) = Calamagrostis Adans. GRAMINEAE (M)
Pteropsis Desv. = Pyrrosia Mirb. POLYPODIACEAE (F)
Pteroptychia Bremek. = Strobilanthes Blume ACANTHACEAE (D)
Pteropyrum Jaub. & Spach POLYGONACEAE (D)
Pterorhachis Harms MELIACEAE (D)
Pteroscleria Nees = Diplacrum R.Br. CYPERACEAE (M)
Pterosicyos Brandegee CUCURBITACEAE (D)
Pterosiphon Turcz. = Cedrela P.Browne MELIACEAE (D)
Pterospermum Schreb. STERCULIACEAE (D)
Pterospora Nutt. ERICACEAE (D)
Pterostegia Fisch. & C.A.Mey. POLYGONACEAE (D)
Pterostelma Wight = Hoya R.Br. ASCLEPIADACEAE (D)
Pterostemma Kraenzl. ORCHIDACEAE (M)
Pterostemon Schauer PTEROSTEMONACEAE (D)
Pterostephus C.Presl = Spermacoce L. RUBIACEAE (D)
Pterostylis R.Br. ORCHIDACEAE (M)
Pterostyrax Siebold & Zucc. STYRACACEAE (D)
Pterotaberna Stapf = Tabernaemontana L. APOCYNACEAE (D)
Pterothrix DC. COMPOSITAE (D)
Pterotropia W.F.Hillebr. = Tetraplasandra A.Gray ARALIACEAE (D)
Pteroxygonum Dammer & Diels = Fagopyrum Mill. POLYGONACEAE (D)
Pterozonium Fée ADIANTACEAE (F)
Pterygiella Oliv. SCROPHULARIACEAE (D)
Pterygiosperma O.E.Schulz CRUCIFERAE (D)
Pterygocalyx Maxim. GENTIANACEAE (D)
Pterygodium Sw. ORCHIDACEAE (M)

Pterygoloma Hanst. = Columnea L. GESNERIACEAE (D)
Pterygopappus Hook.f. COMPOSITAE (D)
Pterygopleurum Kitag. UMBELLIFERAE (D)
Pterygopodium Harms = Oxystigma Harms LEGUMINOSAE–CAESALPINIOIDEAE (D)
Pterygostachyum Steud. = Dimeria R.Br. GRAMINEAE (M)
Pterygostemon V.V.Botschantzeva CRUCIFERAE (D)
Pterygota Schott & Endl. STERCULIACEAE (D)
Pteryxia (Nutt.) J.M.Coult. & Rose UMBELLIFERAE (D)
Ptilagrostis Griseb. = Stipa L. GRAMINEAE (M)
Ptilanthelium Steud. CYPERACEAE (M)
Ptilanthus Gleason = Graffenrieda DC. MELASTOMATACEAE (D)
Ptilepida Raf. = Helenium L. COMPOSITAE (D)
Ptilimnium Raf. UMBELLIFERAE (D)
Ptilocalais A.Gray ex Greene = Microseris D.Don COMPOSITAE (D)
Ptilocalyx Torr. & A.Gray = Tiquilia Pers. BORAGINACEAE (D)
Ptilochaeta Turcz. MALPIGHIACEAE (D)
Ptilocnema D.Don = Pholidota Lindl. ex Hook. ORCHIDACEAE (M)
Ptiloneilema Steud. = Melanocenchris Nees GRAMINEAE (M)
Ptilonema Hook.f. (SUO) = Melanocenchris Nees GRAMINEAE (M)
Ptilophyllum Van den Bosch = Trichomanes L. HYMENOPHYLLACEAE (F)
Ptilopteris Hance ADIANTACEAE (F)
Ptilostemon Cass. COMPOSITAE (D)
Ptilotrichum C.A.Mey. = Alyssum L. CRUCIFERAE (D)
Ptilotus R.Br. AMARANTHACEAE (D)
Ptycanthera Decne. ASCLEPIADACEAE (D)
Ptychandra Scheff. = Heterospathe Scheff. PALMAE (M)
Ptychocarpus Kuhlm. = Neoptychocarpus Buchheim FLACOURTIACEAE (D)
Ptychococcus Becc. PALMAE (M)
Ptychodea Willd. ex Cham. & Schltdl. (SUI) = Sipanea Aubl. RUBIACEAE (D)
Ptychodon Klotzsch ex Rchb. = Lafoensia Vand. LYTHRACEAE (D)
Ptychogyne Pfitzer = Coelogyne Lindl. ORCHIDACEAE (M)
Ptycholobium Harms LEGUMINOSAE–PAPILIONOIDEAE (D)
Ptychomanes Hedw. = Hymenophyllum Sm. HYMENOPHYLLACEAE (F)
Ptychomeria Benth. = Gymnosiphon Blume BURMANNIACEAE (M)
Ptychopetalum Benth. OLACACEAE (D)
Ptychophyllum C.Presl = Hymenophyllum Sm. HYMENOPHYLLACEAE (F)
Ptychopyxis Miq. EUPHORBIACEAE (D)
Ptychoraphis Becc. = Rhopaloblaste Scheff. PALMAE (M)
Ptychosema Benth. LEGUMINOSAE–PAPILIONOIDEAE (D)
Ptychosperma Labill. PALMAE (M)
Ptychostigma Hochst. (SUI) = Galiniera Delile RUBIACEAE (D)
Ptychostylus Tiegh. = Struthanthus Mart. LORANTHACEAE (D)
Ptychotis W.D.J.Koch UMBELLIFERAE (D)
Ptyssiglottis T.Anderson ACANTHACEAE (D)
Pubeta L. = Duroia L.f. RUBIACEAE (D)
Pubistylus Thoth. RUBIACEAE (D)
Puccinellia Parl. GRAMINEAE (M)
Puccionia Chiov. CAPPARACEAE (D)
Puelia Franch. GRAMINEAE (M)
Pueraria DC. LEGUMINOSAE–PAPILIONOIDEAE (D)
Pugionium Gaertn. CRUCIFERAE (D)
Pukanthus Raf. = Grabowskia Schltdl. SOLANACEAE (D)
Pulchranthus V.M.Baum, Reveal & Nowicke ACANTHACEAE (D)
Pulicaria Gaertn. COMPOSITAE (D)
Puliculum Haines = Eulalia Kunth GRAMINEAE (M)

Pullea Schltr. CUNONIACEAE (D)
Pulmonaria L. BORAGINACEAE (D)
Pulsatilla Mill. RANUNCULACEAE (D)
Pultenaea Sm. LEGUMINOSAE–PAPILIONOIDEAE (D)
Pulvinaria Fourn. = Lhotzkyella Rauschert ASCLEPIADACEAE (D)
Puna Kiesling = Opuntia Mill. CACTACEAE (D)
Punctillaria N.E.Br. = Pleiospilos N.E.Br. AIZOACEAE (D)
Puneeria Stocks == Withania Pauquy SOLANACEAE (D)
Punica L. LYTHRACEAE (D)
Punicella Turcz. = Balaustion Hook. MYRTACEAE (D)
Punjuba Britton & Rose LEGUMINOSAE–MIMOSOIDEAE (D)
Puntia Hedge LABIATAE (D)
Pupalia Juss. AMARANTHACEAE (D)
Pupilla Rizzini = Justicia L. ACANTHACEAE (D)
Purdiaea Planch. CYRILLACEAE (D)
Purdieanthus Gilg = Lehmanniella Gilg GENTIANACEAE (D)
Purpurella Naudin = Tibouchina Aubl. MELASTOMATACEAE (D)
Purpureostemon Gugerli MYRTACEAE (D)
Purpusia Brandegee ROSACEAE (D)
Purshia DC. ex Poir. ROSACEAE (D)
Pusaetha Kuntze == Entada Adans. LEGUMINOSAE–MIMOSOIDEAE (D)
Puschkinia M.F.Adams HYACINTHACEAE (M)
Putoria Pers. RUBIACEAE (D)
Putranjiva Wall. EUPHORBIACEAE (D)
Putterlickia Endl. CELASTRACEAE (D)
Puya Molina BROMELIACEAE (M)
Pycnandra Benth. SAPOTACEAE (D)
Pycnanthemum Michx. LABIATAE (D)
Pycnanthus Warb. MYRISTICACEAE (D)
Pycnarrhena Miers ex Hook.f. & Thomson MENISPERMACEAE (D)
Pycnobolus Willd. ex O.E.Schulz = Eudema Humb. & Bonpl. CRUCIFERAE (D)
Pycnobotrya Benth. APOCYNACEAE (D)
Pycnobregma Baill. ASCLEPIADACEAE (D)
Pycnocephalum (Less.) DC. COMPOSITAE (D)
Pycnocoma Benth. EUPHORBIACEAE (D)
Pycnocomon Hoffmanns. & Link = Scabiosa L. DIPSACACEAE (D)
Pycnocycla Lindl. UMBELLIFERAE (D)
Pycnodoria C.Presl = Pteris L. PTERIDACEAE (F)
Pycnoloma C.Chr. POLYPODIACEAE (F)
Pycnoneurum Decne. ASCLEPIADACEAE (D)
Pycnophyllopsis Skottsb. CARYOPHYLLACEAE (D)
Pycnophyllum J.Remy CARYOPHYLLACEAE (D)
Pycnoplinthopsis Jafri CRUCIFERAE (D)
Pycnoplinthus O.E.Schulz CRUCIFERAE (D)
Pycnopteris T.Moore = Dryopteris Adans. DRYOPTERIDACEAE (F)
Pycnorhachis Benth. ASCLEPIADACEAE (D)
Pycnosandra Blume = Drypetes Vahl EUPHORBIACEAE (D)
Pycnospatha Thorel ex Gagnep. ARACEAE (M)
Pycnosphace Rydb. = Salvia L. LABIATAE (D)
Pycnosphaera Gilg GENTIANACEAE (D)
Pycnospora R.Br. ex Wight & Arn. LEGUMINOSAE–PAPILIONOIDEAE (D)
Pycnostachys Hook. LABIATAE (D)
Pycnostelma Bunge ex Decne. ASCLEPIADACEAE (D)
Pycnostylis Pierre = Triclisia Benth. MENISPERMACEAE (D)
Pycnothymus (Benth.) Small = Piloblephis Raf. LABIATAE (D)

Pycreus P.Beauv. CYPERACEAE (M)
Pygeum Gaertn. = Prunus L. ROSACEAE (D)
Pygmaeocereus J.H.Johnson & Backeb. = Echinopsis Zucc. CACTACEAE (D)
Pygmaeopremna Merr. = Premna L. VERBENACEAE (D)
Pygmaeorchis Brade ORCHIDACEAE (M)
Pygmaeothamnus Robyns RUBIACEAE (D)
Pygmea Hook.f. (SUH) = Chionohebe B.G.Briggs & Ehrend. SCROPHULARIACEAE (D)
Pynaertiodendron De Wild. = Cryptosepalum Benth. LEGUMINOSAE–
   CAESALPINIOIDEAE (D)
Pyracantha M.Roem. ROSACEAE (D)
Pyragma Noronha = Uvaria L. ANNONACEAE (D)
Pyragra Bremek. RUBIACEAE (D)
Pyramia Cham. = Cambessedesia DC. MELASTOMATACEAE (D)
Pyramidanthe Miq. ANNONACEAE (D)
Pyramidium Boiss. = Veselskya Opiz CRUCIFERAE (D)
Pyramidocarpus Oliv. = Dasylepis Oliv. FLACOURTIACEAE (D)
Pyramidoptera Boiss. UMBELLIFERAE (D)
Pyrenacantha Wight ICACINACEAE (D)
Pyrenaria Blume THEACEAE (D)
Pyrenocarpa H.T.Chang & R.H.Miau = Decaspermum J.R.Forst. & G.Forst.
   MYRTACEAE (D)
Pyrenoglyphis Karst. = Bactris Jacq. PALMAE (M)
Pyrethrum Zinn = Tanacetum L. COMPOSITAE (D)
Pyrgophyllum (Gagnep.) T.L.Wu & Z.Y.Chen ZINGIBERACEAE (M)
Pyriluma Baill. = Pouteria Aubl. SAPOTACEAE (D)
Pyrogennema Lunell = Epilobium L. ONAGRACEAE (D)
Pyrola L. ERICACEAE (D)
Pyrolirion Herb. AMARYLLIDACEAE (M)
Pyrospermum Miq. = Bhesa Buch.–Ham. ex Arn. CELASTRACEAE (D)
Pyrostegia C.Presl BIGNONIACEAE (D)
Pyrostria Comm. ex Juss. RUBIACEAE (D)
Pyrrhanthera Zotov GRAMINEAE (M)
Pyrrhanthus Jack = Lumnitzera Willd. COMBRETACEAE (D)
Pyrrheima Hassk. = Siderasis Raf. COMMELINACEAE (M)
Pyrrhocactus Backeb. & F.M.Knuth = Neoporteria Britton & Rose CACTACEAE (D)
Pyrrhopappus DC. COMPOSITAE (D)
Pyrrocoma Hook. COMPOSITAE (D)
Pyrrorhiza Maguire & Wurdack HAEMODORACEAE (M)
Pyrrosia Mirb. POLYPODIACEAE (F)
Pyrrothrix Bremek. = Strobilanthes Blume ACANTHACEAE (D)
Pyrularia Michx. SANTALACEAE (D)
Pyrus L. ROSACEAE (D)
Pythius B.D.Jacks. = Euphorbia L. EUPHORBIACEAE (D)
Pyxidanthera Michx. DIAPENSIACEAE (D)
Pyxidanthus Naudin = Blakea P.Browne MELASTOMATACEAE (D)
Pyxidaria Gled. = Trichomanes L. HYMENOPHYLLACEAE (F)

Qaisera Omer = Gentiana L. GENTIANACEAE (D)
Qiongzhuea C.J.Hsueh & T.P.Yi = Chimonobambusa Makino GRAMINEAE (M)
Quadrangula Baum.–Bod. (SUI) = Gymnostoma L.A.S.Johnson CASUARINACEAE (D)
Quadrella J.Presl = Capparis L. CAPPARACEAE (D)
Quadricasaea Woodson = Tabernaemontana L. APOCYNACEAE (D)
Quadricosta Dulac (SUS) = Ludwigia L. ONAGRACEAE (D)
Quadripterygium Tardieu (SUS) = Euonymus L. CELASTRACEAE (D)
Qualea Aubl. VOCHYSIACEAE (D)

Quamoclidion Choisy NYCTAGINACEAE (D)
Quamoclit Moench (SUH) = Ipomoea L. CONVOLVULACEAE (D)
Quamoclit Mill. = Ipomoea L. CONVOLVULACEAE (D)
Quapoya Aubl. GUTTIFERAE (D)
Quaqua N.E.Br. ASCLEPIADACEAE (D)
Quararibea Aubl. BOMBACACEAE (D)
Quartinia Endl. = Rotala L. LYTHRACEAE (D)
Quassia L. SIMAROUBACEAE (D)
Quaternella Pedersen AMARANTHACEAE (D)
Quebrachia Griseb. = Schinopsis Engl. ANACARDIACEAE (D)
Queenslandiella Domin CYPERACEAE (M)
Quekettia Lindl. ORCHIDACEAE (M)
Quelchia N.E.Br. COMPOSITAE (D)
Quelusia Vand. = Fuchsia L. ONAGRACEAE (D)
Quercifilix Copel. = Tectaria Cav. DRYOPTERIDACEAE (F)
Quercus L. FAGACEAE (D)
Queria L. = Minuartia L. CARYOPHYLLACEAE (D)
Quesnelia Gaudich. BROMELIACEAE (M)
Queteletia Blume = Orchipedum Breda ORCHIDACEAE (M)
Quetzalia Lundell CELASTRACEAE (D)
Quezelia H.Scholz (SUH) = Quezeliantha H.Scholz ex Rauschert CRUCIFERAE (D)
Quezeliantha H.Scholz ex Rauschert CRUCIFERAE (D)
Quiabentia Britton & Rose CACTACEAE (D)
Quidproquo Greuter & Burdet CRUCIFERAE (D)
Quiducia Gagnep. = Silvianthus Hook.f. CARLEMANNIACEAE (D)
Quiina Aubl. QUIINACEAE (D)
Quiliusa Hook.f. (SUO) = Fuchsia L. ONAGRACEAE (D)
Quillaja Molina ROSACEAE (D)
Quinchamalium Molina SANTALACEAE (D)
Quincula Raf. SOLANACEAE (D)
Quinetia Cass. COMPOSITAE (D)
Quinqueremulus Paul G.Wilson COMPOSITAE (D)
Quinquina Boehm. = Cinchona L. RUBIACEAE (D)
Quintilia Endl. = Stauranthera Benth. GESNERIACEAE (D)
Quintinia A.DC. ESCALLONIACEAE (D)
Quiotania Zarucchi APOCYNACEAE (D)
Quirosia Blanco = Crotalaria L. LEGUMINOSAE–PAPILIONOIDEAE (D)
Quisqualis L. COMBRETACEAE (D)
Quisqueya D.Don ORCHIDACEAE (M)
Quisumbingia Merr. ASCLEPIADACEAE (D)
Quivisia Cav. = Turraea L. MELIACEAE (D)
Quivisianthe Baill. MELIACEAE (D)

Rabdochloa P.Beauv. = Leptochloa P.Beauv. GRAMINEAE (M)
Rabdosia Hassk. = Isodon (Schrad. ex Benth.) Spach LABIATAE (D)
Rabdosiella Codd LABIATAE (D)
Rabelaisia Planch. = Lunasia Blanco RUTACEAE (D)
Rabenhorstia Rchb. = Berzelia Brongn. BRUNIACEAE (D)
Rabiea N.E.Br. AIZOACEAE (D)
Racapa M.Roem. = Carapa Aubl. MELIACEAE (D)
Racemobambos Holttum GRAMINEAE (M)
Rachicallis DC. (SUO) = Rhachicallis DC. RUBIACEAE (D)
Raclathris Raf. = Rochelia Rchb. BORAGINACEAE (D)
Racosperma Mart. = Acacia Mill. LEGUMINOSAE–MIMOSOIDEAE (D)
Radamaea Benth. SCROPHULARIACEAE (D)

Raddia Mazziari (SUH) = Crypsis Aiton GRAMINEAE (M)
Raddia Bertol. GRAMINEAE (M)
Raddiella Swallen GRAMINEAE (M)
Radermachera Zoll. & Moritzi BIGNONIACEAE (D)
Radia Noronha (SUI) = Mimusops L. SAPOTACEAE (D)
Radinocion Ridl. = Aerangis Rchb.f. ORCHIDACEAE (M)
Radinosiphon N.E.Br. IRIDACEAE (M)
Radiola Hill LINACEAE (D)
Radlkofera Gilg SAPINDACEAE (D)
Radlkoferella Pierre = Pouteria Aubl. SAPOTACEAE (D)
Radlkoferotoma Kuntze COMPOSITAE (D)
Radyera Bullock MALVACEAE (D)
Raffenaldia Godr. CRUCIFERAE (D)
Rafflesia R.Br. RAFFLESIACEAE (D)
Rafinesquia Nutt. COMPOSITAE (D)
Rafinesquia Raf. (SUH) = Jacaranda Juss. BIGNONIACEAE (D)
Rafnia Thunb. LEGUMINOSAE–PAPILIONOIDEAE (D)
Ragala Pierre = Chrysophyllum L. SAPOTACEAE (D)
Ragatelus C.Presl = Trichomanes L. HYMENOPHYLLACEAE (F)
Ragiopteris C.Presl = Onoclea L. WOODSIACEAE (F)
Rahowardiana D'Arcy SOLANACEAE (D)
Raillardella (A.Gray) Benth. COMPOSITAE (D)
Raillardia Gaudich. = Dubautia Gaudich. COMPOSITAE (D)
Raillardiopsis Rydb. COMPOSITAE (D)
Raimannia Rose = Oenothera L. ONAGRACEAE (D)
Raimondia Saff. ANNONACEAE (D)
Raimondianthus Harms = Chaetocalyx DC. LEGUMINOSAE–PAPILIONOIDEAE (D)
Raimundochloa A.M.Molina = Koeleria Pers. GRAMINEAE (M)
Rainiera Greene COMPOSITAE (D)
Rajania L. DIOSCOREACEAE (M)
Raleighia Gardner = Abatia Ruíz & Pav. FLACOURTIACEAE (D)
Ramatuela Kunth COMBRETACEAE (D)
Ramatuella Kunth (SUO) = Ramatuela Kunth COMBRETACEAE (D)
Ramelia Baill. = Bocquillonia Baill. EUPHORBIACEAE (D)
Rameya Baill. = Triclisia Benth. MENISPERMACEAE (D)
Ramirezella Rose LEGUMINOSAE–PAPILIONOIDEAE (D)
Ramischia Opiz ex Garcke = Orthilia Raf. ERICACEAE (D)
Ramisia Glaz. ex Baill. NYCTAGINACEAE (D)
Ramona Greene = Salvia L. LABIATAE (D)
Ramonda Rich. GESNERIACEAE (D)
Ramondia Mirb. = Lygodium Sw. SCHIZAEACEAE (F)
Ramonia Schltr. = Scaphyglottis Poepp. & Endl. ORCHIDACEAE (M)
Ramorinoa Speg. LEGUMINOSAE–PAPILIONOIDEAE (D)
Ramosia Merr. = Centotheca Desv. GRAMINEAE (M)
Ramosmania Tirveng. & Verdc. RUBIACEAE (D)
Ramphidia Miq. = Myrmechis (Lindl.) Blume ORCHIDACEAE (M)
Rampinia C.B.Clarke = Herpetospermum Wall. ex Hook.f. CUCURBITACEAE (D)
Ramsdenia Britton = Phyllanthus L. EUPHORBIACEAE (D)
Ramspekia Scop. = Posoqueria Aubl. RUBIACEAE (D)
Ramusia Nees = Peristrophe Nees ACANTHACEAE (D)
Ranalisma Stapf ALISMATACEAE (M)
Randia L. RUBIACEAE (D)
Randonia Coss. RESEDACEAE (D)
Ranevea L.H.Bailey = Ravenea C.D.Bouche PALMAE (M)
Rangaeris (Schltr.) Summerh. ORCHIDACEAE (M)

Rangia Griseb. (SUO) = Randia L. RUBIACEAE (D)
Rangium Juss. = Forsythia Vahl OLEACEAE (D)
Ranopisoa J.-F.Leroy SCROPHULARIACEAE (D)
Ranugia (Schltdl.) Post & Kuntze = Gurania (Schltdl.) Cogn. CUCURBITACEAE (D)
Ranunculus L. RANUNCULACEAE (D)
Ranzania T.Ito BERBERIDACEAE (D)
Raoulia Hook.f. ex Raoul COMPOSITAE (D)
Raouliopsis S.F.Blake COMPOSITAE (D)
Rapanea Aubl. MYRSINACEAE (D)
Raparia F.K.Mey. = Thlaspi L. CRUCIFERAE (D)
Rapatea Aubl. RAPATEACEAE (M)
Raphanistrocarpus (Baill.) Pax = Momordica L. CUCURBITACEAE (D)
Raphanocarpus Hook.f. = Momordica L. CUCURBITACEAE (D)
Raphanorhyncha Rollins CRUCIFERAE (D)
Raphanus L. CRUCIFERAE (D)
Raphia P.Beauv. PALMAE (M)
Raphiacme K.Schum. = Raphionacme Harv. ASCLEPIADACEAE (D)
Raphidiocystis Hook.f. CUCURBITACEAE (D)
Raphiocarpus Chun = Didissandra C.B.Clarke GESNERIACEAE (D)
Raphionacme Harv. ASCLEPIADACEAE (D)
Raphisanthe Lilja = Caiophora C.Presl LOASACEAE (D)
Raphistemma Wall. ASCLEPIADACEAE (D)
Rapicactus Buxb. & Oehme = Turbinicarpus (Backeb.) Buxb. & Backeb. CACTACEAE (D)
Rapinia Lour. = Sphenoclea Gaertn. CAMPANULACEAE (D)
Rapinia Montrouz. (SUH) = Neorapinia Moldenke VERBENACEAE (D)
Rapistrum Crantz CRUCIFERAE (D)
Rapona Baill. CONVOLVULACEAE (D)
Rapunculus Mill. = Phyteuma L. CAMPANULACEAE (D)
Rapuntia Chevall. = Campanula L. CAMPANULACEAE (D)
Rapuntium Post & Kuntze (SUO) = Campanula L. CAMPANULACEAE (D)
Rapuntium Mill. = Lobelia L. CAMPANULACEAE (D)
Raputia Aubl. RUTACEAE (D)
Raputiarana Emmerich = Raputia Aubl. RUTACEAE (D)
Raram Adans. (SUS) = Cenchrus L. GRAMINEAE (M)
Raritebe Wernham RUBIACEAE (D)
Raspailia C.Presl (SUI) = Polypogon Desf. GRAMINEAE (M)
Raspalia Brongn. BRUNIACEAE (D)
Rastrophyllum Wild & G.V.Pope COMPOSITAE (D)
Rathbunia Britton & Rose CACTACEAE (D)
Rathea Karst. = Synechanthus H.A.Wendl. PALMAE (M)
Ratibida Raf. COMPOSITAE (D)
Rattraya J.B.Phipps = Danthoniopsis Stapf GRAMINEAE (M)
Ratzeburgia Kunth GRAMINEAE (M)
Rauhia Traub AMARYLLIDACEAE (M)
Rauhiella Pabst & Braga ORCHIDACEAE (M)
Rauhocereus Backeb. = Weberbauerocereus Backeb. CACTACEAE (D)
Rauia Nees & Mart. RUTACEAE (D)
Raukana Seem. = Pseudopanax K.Koch ARALIACEAE (D)
Raulinoa R.S.Cowan RUTACEAE (D)
Raulinoreitzia R.M.King & H.Rob. COMPOSITAE (D)
Rautanenia Buchenau = Burnatia Micheli ALISMATACEAE (M)
Rauvolfia L. APOCYNACEAE (D)
Rauwenhoffia Scheff. = Melodorum Lour. ANNONACEAE (D)
Rauwolfia L. (SUO) = Rauvolfia L. APOCYNACEAE (D)
Ravenala Adans. STRELITZIACEAE (M)

Ravenea C.D.Bouché PALMAE (M)
Ravenia Vell. RUTACEAE (D)
Raveniopsis Gleason RUTACEAE (D)
Ravensara Sonn. LAURACEAE (D)
Ravnia Oerst. RUBIACEAE (D)
Rawsonia Harv. & Sond. FLACOURTIACEAE (D)
Raycadenco Dodson ORCHIDACEAE (M)
Rayera Gaudich. = Nolana L.f. SOLANACEAE (D)
Rayleya Cristóbal STERCULIACEAE (D)
Raynalia Soják = Alinula J.Raynal CYPERACEAE (M)
Razisea Oerst. ACANTHACEAE (D)
Razoumofskia Hoffm. = Arceuthobium M.Bieb. VISCACEAE (D)
Razumovia Spreng. = Centranthera R.Br. SCROPHULARIACEAE (D)
Rea Bertero ex Decne. = Dendroseris D.Don COMPOSITAE (D)
Readea Gillespie RUBIACEAE (D)
Reana Brign. = Zea L. GRAMINEAE (M)
Reaumuria L. TAMARICACEAE (D)
Rebentischia Opiz (SUS) = Trisetum Pers. GRAMINEAE (M)
Rebis Spach = Ribes L. GROSSULARIACEAE (D)
Reboudia Coss. & Durieu CRUCIFERAE (D)
Reboulea Kunth (SUH) = Sphenopholis Scribn. GRAMINEAE (M)
Rebulobivia Frič (SUI) = Echinopsis Zucc. CACTACEAE (D)
Rebutia K.Schum. CACTACEAE (D)
Recchia Moçino & Sessé ex DC. SIMAROUBACEAE (D)
Rechsteineria Regel = Sinningia Nees GESNERIACEAE (D)
Recordia Moldenke VERBENACEAE (D)
Recordoxylon Ducke LEGUMINOSAE–CAESALPINIOIDEAE (D)
Rectanthera O.Deg. = Callisia Loefl. COMMELINACEAE (M)
Rectomitra Blume = Pternandra Jack MELASTOMATACEAE (D)
Redfieldia Vasey GRAMINEAE (M)
Redia Casar. = Cleidion Blume EUPHORBIACEAE (D)
Redowskia Cham. & Schltdl. CRUCIFERAE (D)
Reederochloa Soderstr. & H.F.Decker GRAMINEAE (M)
Reedia F.Muell. CYPERACEAE (M)
Reediella Pic.Serm. = Crepidomanes (C.Presl) C.Presl HYMENOPHYLLACEAE (F)
Reedrollinsia J.W.Walker = Stenanona Standl. ANNONACEAE (D)
Reesia Ewart = Polycarpaea Lam. CARYOPHYLLACEAE (D)
Reevesia Lindl. STERCULIACEAE (D)
Regelia Hort. ex H.A.Wendl. (SUI) = Verschaffeltia H.A.Wendl. PALMAE (M)
Regelia Schauer MYRTACEAE (D)
Registaniella Rech.f. UMBELLIFERAE (D)
Regmus Dulac (SUS) = Circaea L. ONAGRACEAE (D)
Regnaldia Baill. = Chaetocarpus Thwaites EUPHORBIACEAE (D)
Regnellia Barb.Rodr. = Bletia Ruíz & Pav. ORCHIDACEAE (M)
Regnellidium Lindm. MARSILEACEAE (F)
Rehdera Moldenke VERBENACEAE (D)
Rehderodendron Hu STYRACACEAE (D)
Rehderophoenix Burret = Drymophloeʳˢ Zipp. PALMAE (M)
Rehia Fijten GRAMINEAE (M)
Rehmannia Libosch. ex Fisch. & C.A.Mey. SCROPHULARIACEAE (D)
Rehsonia Stritch = Wisteria Nutt. LEGUMINOSAE–PAPILIONOIDEAE (D)
Reichardia Roth COMPOSITAE (D)
Reichea Kausel = Myrcianthes O.Berg MYRTACEAE (D)
Reicheella Pax = Lyallia Hook.f. HECTORELLACEAE (D)
Reicheia Kausel (SUO) = Myrcianthes O.Berg MYRTACEAE (D)

Reichenbachanthus Barb.Rodr. ORCHIDACEAE (M)
Reichenbachia Spreng. NYCTAGINACEAE (D)
Reicheocactus Backeb. = Rebutia K.Schum. CACTACEAE (D)
Reidia Wight = Phyllanthus L. EUPHORBIACEAE (D)
Reifferscheidia C.Presl = Dillenia L. DILLENIACEAE (D)
Reimaria Fluegge = Paspalum L. GRAMINEAE (M)
Reimarochloa Hitchc. GRAMINEAE (M)
Reimbolea Debeaux (UR) GRAMINEAE (M)
Reineckea Karst. (SUH) = Synechanthus H.A.Wendl. PALMAE (M)
Reineckea Kunth CONVALLARIACEAE (M)
Reinhardtia Liebm. PALMAE (M)
Reinia Franch. = Itea L. ESCALLONIACEAE (D)
Reinwardtia Korth. (SUH) = Ternstroemia Mutis ex L.f. THEACEAE (D)
Reinwardtia Dumort. LINACEAE (D)
Reinwardtiodendron Koord. MELIACEAE (D)
Reissantia N.Hallé CELASTRACEAE (D)
Reissekia Endl. RHAMNACEAE (D)
Reissipa Steud. ex Klotzsch = Monotaxis Brongn. EUPHORBIACEAE (D)
Reitzia Swallen GRAMINEAE (M)
Rejoua Gaudich. = Tabernaemontana L. APOCYNACEAE (D)
Relbunium (Endl.) Hook.f. RUBIACEAE (D)
Relchela Steud. GRAMINEAE (M)
Reldia Wiehler GESNERIACEAE (D)
Relhania L'Hér. COMPOSITAE (D)
Remijia DC. RUBIACEAE (D)
Remirea Aubl. CYPERACEAE (M)
Remirema Kerr CONVOLVULACEAE (D)
Remusatia Schott ARACEAE (M)
Remya W.F.Hillebr. ex Benth. COMPOSITAE (D)
Renanthera Lour. ORCHIDACEAE (M)
Renantherella Ridl. ORCHIDACEAE (M)
Renarda Regel = Hymenolaena DC. UMBELLIFERAE (D)
Renata Ruschi ORCHIDACEAE (M)
Rendlia Chiov. = Microchloa R.Br. GRAMINEAE (M)
Renealmia L.f. ZINGIBERACEAE (M)
Renggeria Meisn. GUTTIFERAE (D)
Rengifa Poepp. & Endl. = Quapoya Aubl. GUTTIFERAE (D)
Rennellia Korth. RUBIACEAE (D)
Rennera Merxm. COMPOSITAE (D)
Renschia Vatke LABIATAE (D)
Rensonia S.F.Blake COMPOSITAE (D)
Rephesis Raf. = Ficus L. MORACEAE (D)
Reptonia A.DC. = Sideroxylon L. SAPOTACEAE (D)
Requienia DC. LEGUMINOSAE–PAPILIONOIDEAE (D)
Reseda L. RESEDACEAE (D)
Resia H.E.Moore GESNERIACEAE (D)
Resinaria Comm. ex Lam. = Terminalia L. COMBRETACEAE (D)
Resnova Van der Merwe = Drimiopsis Lindl. & Paxton HYACINTHACEAE (M)
Restella Pobed. THYMELAEACEAE (D)
Restiaria Lour. = Uncaria Schreb. RUBIACEAE (D)
Restio Rottb. RESTIONACEAE (M)
Restrepia Kunth ORCHIDACEAE (M)
Restrepiella Garay & Dunst. ORCHIDACEAE (M)
Restrepiopsis Luer ORCHIDACEAE (M)
Retama Raf. LEGUMINOSAE–PAPILIONOIDEAE (D)

Retanilla (DC.) Brongn. RHAMNACEAE (D)
Retiniphyllum Bonpl. RUBIACEAE (D)
Retinodendron Korth. = Vatica L. DIPTEROCARPACEAE (D)
Retispatha J.Dransf. PALMAE (M)
Retrophyllum C.N.Page PODOCARPACEAE (G)
Rettbergia Raddi = Chusquea Kunth GRAMINEAE (M)
Retzia Thunb. RETZIACEAE (D)
Reussia Dennst. (SUH) = Paederia L. RUBIACEAE (D)
Reussia Endl. PONTEDERIACEAE (M)
Reutealis Airy Shaw EUPHORBIACEAE (D)
Reutera Boiss. = Pimpinella L. UMBELLIFERAE (D)
Revealia R.M.King & H.Rob. COMPOSITAE (D)
Reverchonia A.Gray EUPHORBIACEAE (D)
Reya Kuntze = Burchardia R.Br. COLCHICACEAE (M)
Reyesia Gay SOLANACEAE (D)
Reymondia Karst. & Kuntze = Myoxanthus Poepp. & Endl. ORCHIDACEAE (M)
Reynaudia Kunth GRAMINEAE (M)
Reynoldsia A.Gray ARALIACEAE (D)
Reynosia Griseb. RHAMNACEAE (D)
Reynoutria Houtt. = Fallopia Adans. POLYGONACEAE (D)
Rhabdadenia Müll.Arg. APOCYNACEAE (D)
Rhabdia Mart. = Rotula Lour. BORAGINACEAE (D)
Rhabdocalyx Lindl. = Cordia L. BORAGINACEAE (D)
Rhabdocaulon (Benth.) Epling LABIATAE (D)
Rhabdodendron Gilg & Pilg. RHABDODENDRACEAE (D)
Rhabdophyllum Tiegh. = Gomphia Schreb. OCHNACEAE (D)
Rhabdosciadium Boiss. UMBELLIFERAE (D)
Rhabdostigma Hook.f. = Kraussia Harv. RUBIACEAE (D)
Rhabdothamnopsis Hemsl. GESNERIACEAE (D)
Rhabdothamnus A.Cunn. GESNERIACEAE (D)
Rhabdotosperma Hartl = Verbascum L. SCROPHULARIACEAE (D)
Rhachicallis DC. RUBIACEAE (D)
Rhachidosorus Ching = Diplazium Sw. WOODSIACEAE (F)
Rhachidospermum Vasey = Jouvea Fourn. GRAMINEAE (M)
Rhacodiscus Lindau = Justicia L. ACANTHACEAE (D)
Rhacoma P.Browne ex L. = Crossopetalum P.Browne CELASTRACEAE (D)
Rhadamanthus Salisb. HYACINTHACEAE (M)
Rhadinopus S.Moore RUBIACEAE (D)
Rhadinothamnus Paul G.Wilson RUTACEAE (D)
Rhaesteria Summerh. ORCHIDACEAE (M)
Rhagadiolus Scop. COMPOSITAE (D)
Rhagodia R.Br. CHENOPODIACEAE (D)
Rhammatophyllum O.E.Schulz CRUCIFERAE (D)
Rhamnella Miq. RHAMNACEAE (D)
Rhamnidium Reissek RHAMNACEAE (D)
Rhamnoluma Baill. = Pichonia Pierre SAPOTACEAE (D)
Rhamnoneuron Gilg THYMELAEACEAE (D)
Rhamnus L. RHAMNACEAE (D)
Rhamphicarpa Benth. SCROPHULARIACEAE (D)
Rhamphidia Lindl. = Hetaeria Blume ORCHIDACEAE (M)
Rhamphocarya Kuang = Carya Nutt. JUGLANDACEAE (D)
Rhamphogyne S.Moore COMPOSITAE (D)
Rhampholepis Stapf = Sacciolepis Nash GRAMINEAE (M)
Rhamphorhynchus Garay ORCHIDACEAE (M)
Rhanteriopsis Rauschert COMPOSITAE (D)

Rhanterium Desf. COMPOSITAE (D)
Rhaphidanthe Hiern ex Gürke = Diospyros L. EBENACEAE (D)
Rhaphidophora Hassk. ARACEAE (M)
Rhaphidophyton Iljin = Noaea Moq. CHENOPODIACEAE (D)
Rhaphidorhynchus Finet = Microcoelia Lindl. ORCHIDACEAE (M)
Rhaphidospora Nees = Justicia L. ACANTHACEAE (D)
Rhaphidura Bremek. RUBIACEAE (D)
Rhaphiodon Schauer LABIATAE (D)
Rhaphiolepis Lindl. ROSACEAE (D)
Rhaphiophallus Schott = Amorphophallus Blume ex Decne. ARACEAE (M)
Rhaphiostylis Planch. ex Benth. ICACINACEAE (D)
Rhaphis Lour. = Chrysopogon Trin. GRAMINEAE (M)
Rhaphispermum Benth. SCROPHULARIACEAE (D)
Rhaphithamnus Miers VERBENACEAE (D)
Rhapidophyllum H.A.Wendl. & Drude PALMAE (M)
Rhapis L.f. ex Aiton PALMAE (M)
Rhaponticum Ludw. (SUH) = Leuzea DC. COMPOSITAE (D)
Rhaptonema Miers MENISPERMACEAE (D)
Rhaptopetalum Oliv. SCYTOPETALACEAE (D)
Rhaptostylum Humb. & Bonpl. = Heisteria Jacq. OLACACEAE (D)
Rhazya Decne. APOCYNACEAE (D)
Rheedia L. = Garcinia L. GUTTIFERAE (D)
Rheithrophyllum Hassk. = Aeschynanthus Jack GESNERIACEAE (D)
Rhektophyllum N.E.Br. = Cercestis Schott ARACEAE (M)
Rheome Goldblatt IRIDACEAE (M)
Rheopteris Alston VITTARIACEAE (F)
Rhetinodendron Meisn. = Robinsonia DC. COMPOSITAE (D)
Rhetinosperma Radlk. = Chisocheton Blume MELIACEAE (D)
Rheum L. POLYGONACEAE (D)
Rhexia L. MELASTOMATACEAE (D)
Rhigiocarya Miers MENISPERMACEAE (D)
Rhigiophyllum Hochst. CAMPANULACEAE (D)
Rhigospira Miers APOCYNACEAE (D)
Rhigozum Burch. BIGNONIACEAE (D)
Rhinacanthus Nees ACANTHACEAE (D)
Rhinanthus L. SCROPHULARIACEAE (D)
Rhinephyllum N.E.Br. AIZOACEAE (D)
Rhinerrhiza Rupp ORCHIDACEAE (M)
Rhiniachne Steud. (SUI) = Thelepogon Roth ex Roem. & Schult. GRAMINEAE (M)
Rhinopetalum Fisch. ex D.Don = Fritillaria L. LILIACEAE (M)
Rhinopterys Nied. = Acridocarpus Guill. & Perr. MALPIGHIACEAE (D)
Rhiphidosperma G.Don (SUO) = Justicia L. ACANTHACEAE (D)
Rhipidantha Bremek. RUBIACEAE (D)
Rhipidia Markgr. = Condylocarpon Desf. APOCYNACEAE (D)
Rhipidocladum McClure GRAMINEAE (M)
Rhipidoglossum Schltr. = Diaphananthe Schltr. ORCHIDACEAE (M)
Rhipidopteris Fée = Peltapteris Link LOMARIOPSIDACEAE (F)
Rhipogonum J.R.Forst. & G.Forst. RHIPOGONACEAE (M)
Rhipsalidopsis Britton & Rose = Hatiora Britton & Rose CACTACEAE (D)
Rhipsalis Gaertn. CACTACEAE (D)
Rhizanota Lour. ex Gomes = Corchorus L. TILIACEAE (D)
Rhizanthella R.S.Rogers ORCHIDACEAE (M)
Rhizanthemum Tiegh. = Amyema Tiegh. LORANTHACEAE (D)
Rhizanthes Dumort. RAFFLESIACEAE (D)
Rhizobotrya Tausch CRUCIFERAE (D)

Rhizocephalum Wedd. CAMPANULACEAE (D)
Rhizocephalus Boiss. GRAMINEAE (M)
Rhizoglossum C.Presl = Ophioglossum L. OPHIOGLOSSACEAE (F)
Rhizomatopteris Khokhr. = Cystopteris Bernh. WOODSIACEAE (F)
Rhizomonanthes Danser LORANTHACEAE (D)
Rhizophora L. RHIZOPHORACEAE (D)
Rhizophyllum Newman = Phlebodium (R.Br.) J.Sm. POLYPODIACEAE (F)
Rhizosperma Meyen = Azolla Lam. AZOLLACEAE (F)
Rhodalsine J.Gay = Minuartia L. CARYOPHYLLACEAE (D)
Rhodamnia Jack MYRTACEAE (D)
Rhodanthe Lindl. COMPOSITAE (D)
Rhodiola L. CRASSULACEAE (D)
Rhodocactus (A.Berger) F.Knuth = Pereskia Mill. CACTACEAE (D)
Rhodocalyx Müll.Arg. APOCYNACEAE (D)
Rhodochiton Zucc. ex Otto & Dietr. SCROPHULARIACEAE (D)
Rhodocodon Baker HYACINTHACEAE (M)
Rhodocolea Baill. BIGNONIACEAE (D)
Rhodocoma Nees RESTIONACEAE (M)
Rhododendron L. ERICACEAE (D)
Rhododon Epling LABIATAE (D)
Rhodogeron Griseb. COMPOSITAE (D)
Rhodognaphalon (Ulbr.) Roberty BOMBACACEAE (D)
Rhodognaphalopsis A.Robyns = Bombax L. BOMBACACEAE (D)
Rhodohypoxis Nel HYPOXIDACEAE (M)
Rhodolaena Thouars SARCOLAENACEAE (D)
Rhodoleia Champ. ex Hook. HAMAMELIDACEAE (D)
Rhodomyrtus (DC.) Rchb. MYRTACEAE (D)
Rhodophiala C.Presl AMARYLLIDACEAE (M)
Rhodopis Urb. LEGUMINOSAE–PAPILIONOIDEAE (D)
Rhodosciadium S.Watson UMBELLIFERAE (D)
Rhodosepala Baker MELASTOMATACEAE (D)
Rhodospatha Poepp. ARACEAE (M)
Rhodosphaera Engl. ANACARDIACEAE (D)
Rhodostachys Phil. = Ochagavia Phil. BROMELIACEAE (M)
Rhodostemonodaphne Rohwer & Kubitzki LAURACEAE (D)
Rhodostoma Scheidw. = Palicourea Aubl. RUBIACEAE (D)
Rhodothamnus Rchb. ERICACEAE (D)
Rhodotypos Siebold & Zucc. ROSACEAE (D)
Rhoeo Hance = Tradescantia L. COMMELINACEAE (M)
Rhoiacarpos A.DC. SANTALACEAE (D)
Rhoicissus Planch. VITACEAE (D)
Rhoiptelea Diels & Hand.-Mazz. RHOIPTELEACEAE (D)
Rhombochlamys Lindau ACANTHACEAE (D)
Rhomboda Lindl. = Hetaeria Blume ORCHIDACEAE (M)
Rhombolytrum Link GRAMINEAE (M)
Rhombonema Schltr. = Parapodium E.Mey. ASCLEPIADACEAE (D)
Rhombophyllum (Schwantes) Schwantes AIZOACEAE (D)
Rhombospora Korth. = Greenea Wight & Arn. RUBIACEAE (D)
Rhoogeton Leeuwenb. GESNERIACEAE (D)
Rhopalephora Hassk. COMMELINACEAE (M)
Rhopaloblaste Scheff. PALMAE (M)
Rhopalobrachium Schltr. & K.Krause RUBIACEAE (D)
Rhopalocarpus Bojer SPHAEROSEPALACEAE (D)
Rhopalocarpus Teijsm. & Binn. ex Miq. (SUH) = Anaxagorea A.St.-Hil. ANNONACEAE (D)
Rhopalocnemis Jungh. BALANOPHORACEAE (D)

Rhopalopilia Pierre OPILIACEAE (D)
Rhopalopodium Ulbr. = Ranunculus L. RANUNCULACEAE (D)
Rhopalosciadium Rech.f. UMBELLIFERAE (D)
Rhopalostigma Phil. = Phrodus Miers SOLANACEAE (D)
Rhopalostylis H.A.Wendl. & Drude PALMAE (M)
Rhopalostylis Klotzsch ex Baill. (SUH) = Dalechampia L. EUPHORBIACEAE (D)
Rhopalota N.E.Br. = Crassula L. CRASSULACEAE (D)
Rhopium Schreb. = Phyllanthus L. EUPHORBIACEAE (D)
Rhus L. ANACARDIACEAE (D)
Rhyacophila Hochst. = Rotala L. LYTHRACEAE (D)
Rhynchadenia A.Rich. = Macradenia R.Br. ORCHIDACEAE (M)
Rhynchandra Rchb.f. = Corymborkis Thouars ORCHIDACEAE (M)
Rhynchanthera Blume (SUH) = Corymborkis Thouars ORCHIDACEAE (M)
Rhynchanthera DC. MELASTOMATACEAE (D)
Rhynchanthus Hook.f. ZINGIBERACEAE (M)
Rhyncharrhena F.Muell. ASCLEPIADACEAE (D)
Rhynchelythrum Nees (SUO) = Melinis P.Beauv. GRAMINEAE (M)
Rhynchelytrum Nees = Melinis P.Beauv. GRAMINEAE (M)
Rhynchocalyx Oliv. RHYNCHOCALYCACEAE (D)
Rhynchocarpa Backer ex Heyne (SUH) = Dialium L. LEGUMINOSAE–
    CAESALPINIOIDEAE (D)
Rhynchocarpa Schrad. ex Endl. = Kedrostis Medik. CUCURBITACEAE (D)
Rhynchocarpa Becc. (SUH) = Burretiokentia Pic.Serm. PALMAE (M)
Rhynchocladium T.Koyama CYPERACEAE (M)
Rhynchocorys Griseb. SCROPHULARIACEAE (D)
Rhynchodia Benth. APOCYNACEAE (D)
Rhynchodium C.Presl = Bituminaria Heist. ex Fabr. LEGUMINOSAE–PAPILIONOIDEAE (D)
Rhynchoglossum Blume GESNERIACEAE (D)
Rhynchogyna Seidenf. & Garay ORCHIDACEAE (M)
Rhyncholacis Tul. PODOSTEMACEAE (D)
Rhyncholaelia Schltr. ORCHIDACEAE (M)
Rhynchopera Klotzsch = Pleurothallis R.Br. ORCHIDACEAE (M)
Rhynchophora Arènes MALPIGHIACEAE (D)
Rhynchophreatia Schltr. ORCHIDACEAE (M)
Rhynchopyle Engl. = Piptospatha N.E.Br. ARACEAE (M)
Rhynchoryza Baill. GRAMINEAE (M)
Rhynchosia Lour. LEGUMINOSAE–PAPILIONOIDEAE (D)
Rhynchosida Fryxell MALVACEAE (D)
Rhynchosinapis Hayek = Coincya Rouy CRUCIFERAE (D)
Rhynchospermum Reinw. COMPOSITAE (D)
Rhynchospora Vahl CYPERACEAE (M)
Rhynchostele Rchb.f. = Leochilus Knowles & Westc. ORCHIDACEAE (M)
Rhynchostigma Benth. = Toxocarpus Wight & Arn. ASCLEPIADACEAE (D)
Rhynchostylis Blume ORCHIDACEAE (M)
Rhynchotechum Blume GESNERIACEAE (D)
Rhynchotheca Ruíz & Pav. GERANIACEAE (D)
Rhynchotropis Harms LEGUMINOSAE–PAPILIONOIDEAE (D)
Rhynea DC. (SUH) = Tenrhynea Hilliard & B.L.Burtt COMPOSITAE (D)
Rhysolepis S.F.Blake COMPOSITAE (D)
Rhysopterus J.M.Coult. & Rose UMBELLIFERAE (D)
Rhysospermum C.F.Gaertn. = Notelaea Vent. OLEACEAE (D)
Rhysotoechia Radlk. SAPINDACEAE (D)
Rhyssocarpus Endl. = Melanopsidium Colla RUBIACEAE (D)
Rhyssolobium E.Mey. ASCLEPIADACEAE (D)
Rhyssopterys Blume ex A.Juss. (SUO) = Ryssopterys Blume ex A.Juss. MALPIGHIACEAE (D)

Rhyssostelma Decne. ASCLEPIADACEAE (D)
Rhytachne Desv. ex Ham. GRAMINEAE (M)
Rhyticalymma Bremek. = Justicia L. ACANTHACEAE (D)
Rhyticarpus Sond. = Anginon Raf. UMBELLIFERAE (D)
Rhyticaryum Becc. ICACINACEAE (D)
Rhyticocos Becc. = Syagrus Mart. PALMAE (M)
Rhytidachne K.Schum. (SUO) = Rhytachne Desv. ex W.Ham. GRAMINEAE (M)
Rhytidanthera (Planch.) Tiegh. OCHNACEAE (D)
Rhytidocaulon Bally ASCLEPIADACEAE (D)
Rhytidomene Rydb. = Orbexilum Raf. LEGUMINOSAE–PAPILIONOIDEAE (D)
Rhytidophyllum Mart. GESNERIACEAE (D)
Rhytidostylis Rchb. (SUO) = Rytidostylis Hook. & Arn. CUCURBITACEAE (D)
Rhytidotus Hook.f. (SUO) = Bobea Gaudich. RUBIACEAE (D)
Rhytiglossa Nees ex Lindl. = Isoglossa Oerst. ACANTHACEAE (D)
Rhytis Lour. = Antidesma L. EUPHORBIACEAE (D)
Rhytispermum Link (SUS) = Alkanna Tausch BORAGINACEAE (D)
Ribes L. GROSSULARIACEAE (D)
Ribesium Medik. = Ribes L. GROSSULARIACEAE (D)
Ricardia Adans. (SUO) = Richardia L. RUBIACEAE (D)
Richaeia Thouars = Cassipourea Aubl. RHIZOPHORACEAE (D)
Richardella Pierre = Pouteria Aubl. SAPOTACEAE (D)
Richardia L. RUBIACEAE (D)
Richardia Kunth (SUH) = Zantedeschia Spreng. ARACEAE (M)
Richardsiella Elffers & Kenn.–O'Byrne GRAMINEAE (M)
Richardsonia Kunth = Richardia L. RUBIACEAE (D)
Richea Kuntze (SUO) = Cassipourea Aubl. RHIZOPHORACEAE (D)
Richea R.Br. EPACRIDACEAE (D)
Richella A.Gray ANNONACEAE (D)
Richeopsis Arènes = Scolopia Schreb. FLACOURTIACEAE (D)
Richeria Vahl EUPHORBIACEAE (D)
Richeriella Pax & K.Hoffm. EUPHORBIACEAE (D)
Richteria Kar. & Kir. COMPOSITAE (D)
Ricinella Müll.Arg. = Adelia L. EUPHORBIACEAE (D)
Ricinocarpos Desf. EUPHORBIACEAE (D)
Ricinocarpus Kuntze = Acalypha L. EUPHORBIACEAE (D)
Ricinodendron Müll.Arg. EUPHORBIACEAE (D)
Ricinoides Moench = Chrozophora Neck. ex Juss. EUPHORBIACEAE (D)
Ricinus L. EUPHORBIACEAE (D)
Ricotia L. CRUCIFERAE (D)
Ridleya (Hook.f.) Pfitzer = Thrixspermum Lour. ORCHIDACEAE (M)
Ridleyella Schltr. ORCHIDACEAE (M)
Ridolfia Moris UMBELLIFERAE (D)
Riedelia Kunth (SUI) = Arundinella Raddi GRAMINEAE (M)
Riedelia Oliv. ZINGIBERACEAE (M)
Riedeliella Harms LEGUMINOSAE–PAPILIONOIDEAE (D)
Riediea Mirb. = Matteuccia Tod. WOODSIACEAE (F)
Riencourtia Cass. COMPOSITAE (D)
Riesenbachia C.Presl = Lopezia Cav. ONAGRACEAE (D)
Rigidella Lindl. IRIDACEAE (M)
Rigiolepis Hook.f. = Vaccinium L. ERICACEAE (D)
Rigiopappus A.Gray COMPOSITAE (D)
Rigiostachys Planch. = Recchia Moçino & Sessé ex DC. SIMAROUBACEAE (D)
Rikliella J.Raynal = Lipocarpha R.Br. CYPERACEAE (M)
Rimacola Rupp ORCHIDACEAE (M)
Rimaria N.E.Br. (SUH) = Gibbaeum Haw. AIZOACEAE (D)

Rimaria L.Bolus = Vanheerdea L.Bolus ex H.E.K.Hartmann AIZOACEAE (D)
Rindera Pall. BORAGINACEAE (D)
Ringentiarum Nakai = Arisaema Mart. ARACEAE (M)
Rinorea Aubl. VIOLACEAE (D)
Rinoreocarpus Ducke VIOLACEAE (D)
Rinzia Schauer MYRTACEAE (D)
Riocreuxia Decne. ASCLEPIADACEAE (D)
Ripidium Bernh. = Schizaea Sm. SCHIZAEACEAE (F)
Ripidium Trin. (SUH) = Saccharum L. GRAMINEAE (M)
Ripogonum J.R.Forst. & G.Forst. (SUO) = Rhipogonum J.R.Forst. & G.Forst.
    RHIPOGONACEAE (M)
Riqueuria Ruíz & Pav. RUBIACEAE (D)
Riseleya Hemsl. = Drypetes Vahl EUPHORBIACEAE (D)
Risleya King & Pantl. ORCHIDACEAE (M)
Ristantia Peter G.Wilson & J.T.Waterh. MYRTACEAE (D)
Ritaia King & Pantl. = Ceratostylis Blume ORCHIDACEAE (M)
Ritchiea R.Br. ex G.Don CAPPARACEAE (D)
Ritchieophyton Pax = Givotia Griff. EUPHORBIACEAE (D)
Ritonia Benoist ACANTHACEAE (D)
Ritterocereus Backeb. = Stenocereus (A.Berger) Riccob. CACTACEAE (D)
Rivasgodaya Esteve = Genista L. LEGUMINOSAE–PAPILIONOIDEAE (D)
Rivea Choisy CONVOLVULACEAE (D)
Riveria Kunth = Swartzia Schreb. LEGUMINOSAE–PAPILIONOIDEAE (D)
Rivina L. PHYTOLACCACEAE (D)
Rivinoides Afzel. ex Prain = Erythrococca Benth. EUPHORBIACEAE (D)
Rizoa Cav. = Gardoquia Ruíz & Pav. LABIATAE (D)
Robbairea Boiss. CARYOPHYLLACEAE (D)
Robbia A.DC. = Malouetia A.DC. APOCYNACEAE (D)
Robertia Scop. = Sideroxylon L. SAPOTACEAE (D)
Robeschia Hochst. ex O.E.Schulz CRUCIFERAE (D)
Robinia L. LEGUMINOSAE–PAPILIONOIDEAE (D)
Robinsonella Rose & Baker f. MALVACEAE (D)
Robinsonia DC. COMPOSITAE (D)
Robinsoniodendron Merr. = Maoutia Wedd. URTICACEAE (D)
Robiquetia Gaudich. ORCHIDACEAE (M)
Roborowskia Batalin = Corydalis DC. PAPAVERACEAE (D)
Robynsia Hutch. RUBIACEAE (D)
Robynsiella Suess. = Centemopsis Schinz AMARANTHACEAE (D)
Robynsiochloa Jacq.-Fél. = Rottboellia L.f. GRAMINEAE (M)
Robynsiophyton R.Wilczek LEGUMINOSAE–PAPILIONOIDEAE (D)
Rochea DC. = Crassula L. CRASSULACEAE (D)
Rochefortia Sw. BORAGINACEAE (D)
Rochelia Rchb. BORAGINACEAE (D)
Rochetia Delile = Trichilia P.Browne MELIACEAE (D)
Rochonia DC. COMPOSITAE (D)
Rockia Heimerl = Pisonia L. NYCTAGINACEAE (D)
Rockinghamia Airy Shaw EUPHORBIACEAE (D)
Rodatia Raf. = Justicia L. ACANTHACEAE (D)
Rodentiophila F.Ritter ex Backeb. (SUI) = Eriosyce Phil. CACTACEAE (D)
Rodetia Moq. = Bosea L. AMARANTHACEAE (D)
Rodgersia A.Gray SAXIFRAGACEAE (D)
Rodrigoa Brass = Masdevallia Ruíz & Pav. ORCHIDACEAE (M)
Rodriguezia Ruíz & Pav. ORCHIDACEAE (M)
Rodrigueziella Kuntze ORCHIDACEAE (M)
Rodrigueziopsis Schltr. ORCHIDACEAE (M)

Roebelia Engel = Geonoma Willd. PALMAE (M)

Roegneria K.Koch = Elymus L. GRAMINEAE (M)

Roella L. CAMPANULACEAE (D)

Roemeria Roem. & Schult. (SUH) = Diarrhena P.Beauv. GRAMINEAE (M)

Roemeria Medik. PAPAVERACEAE (D)

Roemeria Thunb. (SUH) = Sideroxylon L. SAPOTACEAE (D)

Roentgenia Urb. BIGNONIACEAE (D)

Roepera A.Juss. = Zygophyllum L. ZYGOPHYLLACEAE (D)

Roeperia Spreng. = Ricinocarpos Desf. EUPHORBIACEAE (D)

Roeperocharis Rchb.f. ORCHIDACEAE (M)

Roettlera Vahl = Didymocarpus Wall. GESNERIACEAE (D)

Roezlia Regel = Monochaetum (DC.) Naudin MELASTOMATACEAE (D)

Roezliella Schltr. ORCHIDACEAE (M)

Rogeonella Chev. = Synsepalum (A.DC.) Daniell SAPOTACEAE (D)

Rogeria J.Gay ex Delile PEDALIACEAE (D)

Rogersonanthus Maguire & B.M.Boom GENTIANACEAE (D)

Roggeveldia Goldblatt IRIDACEAE (M)

Rogiera Planch. RUBIACEAE (D)

Rohdea Roth CONVALLARIACEAE (M)

Roia Scop. = Swietenia Jacq. MELIACEAE (D)

Roigella Borhidi & Fernández RUBIACEAE (D)

Roigia Britton = Phyllanthus L. EUPHORBIACEAE (D)

Rojasia Malme ASCLEPIADACEAE (D)

Rojasianthe Standl. & Steyerm. COMPOSITAE (D)

Rojasimalva Fryxell MALVACEAE (D)

Rojasiophytum Hassl. = Xylophragma Sprague BIGNONIACEAE (D)

Rojoc Adans. = Morinda L. RUBIACEAE (D)

Rolandra Rottb. COMPOSITAE (D)

Roldana La Llave COMPOSITAE (D)

Rolfea Zahlbr. = Palmorchis Barb.Rodr. ORCHIDACEAE (M)

Rolfeella Schltr. = Benthamia A.Rich. ORCHIDACEAE (M)

Rollandia Gaudich. CAMPANULACEAE (D)

Rollinia A.St.-Hil. ANNONACEAE (D)

Rolliniopsis Saff. = Rollinia A.St.-Hil. ANNONACEAE (D)

Rollinsia Al-Shehbaz CRUCIFERAE (D)

Rolsonia Rchb. = Ribes L. GROSSULARIACEAE (D)

Romanoa Trevis. EUPHORBIACEAE (D)

Romanowia Sander ex André = Ptychosperma Labill. PALMAE (M)

Romanschulzia O.E.Schulz CRUCIFERAE (D)

Romanzoffia Cham. HYDROPHYLLACEAE (D)

Romeroa Dugand BIGNONIACEAE (D)

Romnalda P.F.Stevens LOMANDRACEAE (M)

Romneya Harv. PAPAVERACEAE (D)

Romulea Maratti IRIDACEAE (M)

Ronabea Aubl. = Psychotria L. RUBIACEAE (D)

Ronabia St.-Lag. (SUO) = Psychotria L. RUBIACEAE (D)

Rondeletia L. RUBIACEAE (D)

Rondonanthus Herz. = Paepalanthus Kunth ERIOCAULACEAE (M)

Ronnbergia E.Morren & André BROMELIACEAE (M)

Ronnowia Buc'hoz = Omphalea L. EUPHORBIACEAE (D)

Roodia N.E.Br. = Argyroderma N.E.Br. AIZOACEAE (D)

Rooksbya Backeb. = Neobuxbaumia Backeb. CACTACEAE (D)

Rooseveltia O.F.Cook = Euterpe Mart. PALMAE (M)

Ropalopetalum Griff. = Artabotrys R.Br. ANNONACEAE (D)

Rophostemon Endl. (SUO) = Nervilia Comm. ex Gaud. ORCHIDACEAE (M)

Ropourea Aubl. = Diospyros L. EBENACEAE (D)
Roptrostemon Blume = Nervilia Comm. ex Gaud. ORCHIDACEAE (M)
Roraimanthus Gleason = Sauvagesia L. OCHNACEAE (D)
Rorida J.F.Gmel. = Cleome L. CAPPARACEAE (D)
Roridula Forssk. (SUH) = Cleome L. CAPPARACEAE (D)
Roridula Burm.f. ex L. RORIDULACEAE (D)
Roripella (Maire) Greuter & Burdet CRUCIFERAE (D)
Rorippa Scop. CRUCIFERAE (D)
Rosa L. ROSACEAE (D)
Rosanowia Regel = Sinningia Nees GESNERIACEAE (D)
Rosanthus Small = Gaudichaudia Kunth MALPIGHIACEAE (D)
Roscheria H.A.Wendl. ex Balf.f. PALMAE (M)
Roscoea Sm. ZINGIBERACEAE (M)
Rosea Klotzsch = Tricalysia A.Rich. ex DC. RUBIACEAE (D)
Roseanthus Cogn. = Polyclathra Bertol. CUCURBITACEAE (D)
Roseia Frič (SUI) = Coryphantha (Engelm.) Lem. CACTACEAE (D)
Rosenbergiodendron Fagerl. = Randia L. RUBIACEAE (D)
Rosenia Thunb. COMPOSITAE (D)
Rosenstockia Copel. HYMENOPHYLLACEAE (F)
Roseocactus A.Berger = Ariocarpus Scheidw. CACTACEAE (D)
Roseocereus Backeb. = Harrisia Britton CACTACEAE (D)
Roshevitzia Tzvelev = Eragrostis Wolf GRAMINEAE (M)
Rosifax C.C.Towns. AMARANTHACEAE (D)
Rosilla Less. = Dyssodia Cav. COMPOSITAE (D)
Rosmarinus L. LABIATAE (D)
Rossioglossum (Schltr.) Garay & G.C.Kenn. ORCHIDACEAE (M)
Rostellaria C.F.Gaertn. (UP) SAPOTACEAE (D)
Rostellaria Nees (SUH) = Justicia L. ACANTHACEAE (D)
Rostellularia Rchb. = Justicia L. ACANTHACEAE (D)
Rostkovia Desv. JUNCACEAE (M)
Rostraria Trin. GRAMINEAE (M)
Rostrinucula Kudo LABIATAE (D)
Rosularia (DC.) Stapf CRASSULACEAE (D)
Rotala L. LYTHRACEAE (D)
Rotang Adans. = Calamus L. PALMAE (M)
Rotanga Boehm. = Calamus L. PALMAE (M)
Rotantha Baker = Lawsonia L. LYTHRACEAE (D)
Rotantha Small (SUH) = Campanula L. CAMPANULACEAE (D)
Rotheria Meyen = Cruckshanksia Hook. & Arn. RUBIACEAE (D)
Rothia Borkh. (SUS) = Mibora Adans. GRAMINEAE (M)
Rothia Pers. LEGUMINOSAE–PAPILIONOIDEAE (D)
Rothmaleria Font Quer COMPOSITAE (D)
Rothmannia Thunb. RUBIACEAE (D)
Rothrockia A.Gray ASCLEPIADACEAE (D)
Rottboelia Scop. = Ximenia L. OLACACEAE (D)
Rottboellia L.f. GRAMINEAE (M)
Rottlera Willd. (1804) (SUH) = Mallotus Lour. EUPHORBIACEAE (D)
Rottlera Willd. (1797) = Trewia L. EUPHORBIACEAE (D)
Rotula Lour. BORAGINACEAE (D)
Roubieva Moq. CHENOPODIACEAE (D)
Roucheria Planch. LINACEAE (D)
Roulinia Decne. = Cynanchum L. ASCLEPIADACEAE (D)
Rouliniella Vail = Cynanchum L. ASCLEPIADACEAE (D)
Roupala Aubl. PROTEACEAE (D)
Roupellia Wall. & Hook. = Strophanthus DC. APOCYNACEAE (D)

Roupellina (Baill.) Pichon = Strophanthus DC. APOCYNACEAE (D)
Rourea Aubl. CONNARACEAE (D)
Roureopsis Planch. = Rourea Aubl. CONNARACEAE (D)
Roussea Sm. ESCALLONIACEAE (D)
Rousseauxia DC. MELASTOMATACEAE (D)
Rousselia Gaudich. URTICACEAE (D)
Rouxia Husn. = × Elyhordeum Tsitsin & Petrowa GRAMINEAE (M)
Rouya Coincy UMBELLIFERAE (D)
Roxburghia Roxb. = Stemona Lour. STEMONACEAE (M)
Roycea C.A.Gardner CHENOPODIACEAE (D)
Roydsia Roxb. = Stixis Lour. CAPPARACEAE (D)
Royena L. = Diospyros L. EBENACEAE (D)
Roylea Steud. (SUI) = Melanocenchris Nees GRAMINEAE (M)
Roylea Wall. ex Benth. LABIATAE (D)
Roystonea O.F.Cook PALMAE (M)
Ruagea Karst. MELIACEAE (D)
Rubachia O.Berg = Marlierea Cambess. MYRTACEAE (D)
Rubeola Hill (SUH) = Sherardia L. RUBIACEAE (D)
Rubeola Mill. = Crucianella L. RUBIACEAE (D)
Rubia L. RUBIACEAE (D)
Rubina Noronha = Antidesma L. EUPHORBIACEAE (D)
Rubioides Perkins = Opercularia Gaertn. RUBIACEAE (D)
Rubiteucris Kudo LABIATAE (D)
Rubrivena M.Král = Persicaria (L.) Mill. POLYGONACEAE (D)
Rubus L. ROSACEAE (D)
Rudbeckia Adans. (SUH) = Conocarpus L. COMBRETACEAE (D)
Rudbeckia L. COMPOSITAE (D)
Ruddia Yakovlev = Ormosia G.Jacks. LEGUMINOSAE–PAPILIONOIDEAE (D)
Rudgea Salisb. RUBIACEAE (D)
Rudolfiella Hoehne ORCHIDACEAE (M)
Rudua F.Maek. = Phaseolus L. LEGUMINOSAE–PAPILIONOIDEAE (D)
Ruehssia Karst. ex Schltdl. = Marsdenia R.Br. ASCLEPIADACEAE (D)
Ruellia L. ACANTHACEAE (D)
Ruelliola Baill. = Brillantaisia P.Beauv. ACANTHACEAE (D)
Ruelliopsis C.B.Clarke ACANTHACEAE (D)
Rufodorsia Wiehler GESNERIACEAE (D)
Rugelia Shuttlew. ex Chapm. COMPOSITAE (D)
Ruilopezia Cuatrec. COMPOSITAE (D)
Ruizia Cav. STERCULIACEAE (D)
Ruizodendron R.E.Fr. ANNONACEAE (D)
Ruizterania Marc.–Berti VOCHYSIACEAE (D)
Rulingia R.Br. STERCULIACEAE (D)
Rumex L. POLYGONACEAE (D)
Rumfordia DC. COMPOSITAE (D)
Rumia Hoffm. UMBELLIFERAE (D)
Rumicastrum Ulbr. PORTULACACEAE (D)
Rumicicarpus Chiov. = Triumfetta L. TILIACEAE (D)
Rumohra Raddi DAVALLIACEAE (F)
Runcina Allam. (UR) GRAMINEAE (M)
Rungia Nees ACANTHACEAE (D)
Runyonia Rose = Agave L. AGAVACEAE (M)
Rupestrina Prov. = Trisetum Pers. GRAMINEAE (M)
Rupicapnos Pomel PAPAVERACEAE (D)
Rupicola J.H.Maiden & E.Betche EPACRIDACEAE (D)
Rupiphila Pimenov & Lavrova UMBELLIFERAE (D)

Ruppia L. POTAMOGETONACEAE (M)
Ruprechtia C.A.Mey. POLYGONACEAE (D)
Rusbya Britton ERICACEAE (D)
Rusbyanthus Gilg = Macrocarpaea Gilg GENTIANACEAE (D)
Rusbyella Rolfe ex Rusby ORCHIDACEAE (M)
Ruschia Schwantes AIZOACEAE (D)
Ruschianthemum Friedrich AIZOACEAE (D)
Ruschianthus L.Bolus AIZOACEAE (D)
Ruscus L. RUSCACEAE (M)
Ruspolia Lindau ACANTHACEAE (D)
Russelia Jacq. SCROPHULARIACEAE (D)
Russowia C.Winkl. COMPOSITAE (D)
Rustia Klotzsch RUBIACEAE (D)
Ruta L. RUTACEAE (D)
Rutamuraria Ortega = Asplenium L. ASPLENIACEAE (F)
Rutaneblina Steyerm. & Luteyn RUTACEAE (D)
Rutea M.Roem. = Turraea L. MELIACEAE (D)
Ruthalicia C.Jeffrey CUCURBITACEAE (D)
Ruthea Bolle (SUH) = Rutheopsis A.Hansen & Kunkel UMBELLIFERAE (D)
Rutheopsis A.Hansen & Kunkel UMBELLIFERAE (D)
Ruthiella Steenis CAMPANULACEAE (D)
Rutidea DC. RUBIACEAE (D)
Rutidosis DC. COMPOSITAE (D)
Rutosma A.Gray = Thamnosma Torr. & Frém. RUTACEAE (D)
Ruttya Harv. ACANTHACEAE (D)
Ruyschia Jacq. MARCGRAVIACEAE (D)
Ryania Vahl FLACOURTIACEAE (D)
Rylstonea R.T.Baker = Homoranthus A.Cunn. ex Schauer MYRTACEAE (D)
Ryncholeucaena Britton & Rose = Leucaena Benth. LEGUMINOSAE–MIMOSOIDEAE (D)
Ryparosa Blume FLACOURTIACEAE (D)
Ryssopterys Blume ex A.Juss. MALPIGHIACEAE (D)
Rytidea Spreng. (SUO) = Rutidea DC. RUBIACEAE (D)
Rytidocarpus Coss. CRUCIFERAE (D)
Rytidosperma Steud. GRAMINEAE (M)
Rytidostylis Hook. & Arn. CUCURBITACEAE (D)
Rytidotus Hook.f. = Bobea Gaudich. RUBIACEAE (D)
Rytigynia Blume RUBIACEAE (D)
Rytilix C.Hitchc. (SUS) = Hackelochloa Kuntze GRAMINEAE (M)
Rzedowskia Medrano CELASTRACEAE (D)

Saba (Pichon) Pichon APOCYNACEAE (D)
Sabal Adans. PALMAE (M)
Sabatia Adans. GENTIANACEAE (D)
Sabaudia Buscal. & Muschler LABIATAE (D)
Sabaudiella Chiov. CONVOLVULACEAE (D)
Sabazia Cass. COMPOSITAE (D)
Sabia Colebr. SABIACEAE (D)
Sabicea Aubl. RUBIACEAE (D)
Sabina Mill. = Juniperus L. CUPRESSACEAE (G)
Sabinea DC. LEGUMINOSAE–PAPILIONOIDEAE (D)
Sabouraea Leandri = Talinella Baill. PORTULACACEAE (D)
Sabsab Adans. (SUS) = Paspalum L. GRAMINEAE (M)
Saccanthus Herz. = Basistemon Turcz. SCROPHULARIACEAE (D)
Saccardophytum Speg. = Benthamiella Speg. SOLANACEAE (D)
Saccellium Bonpl. BORAGINACEAE (D)

Saccharifera Stokes (SUS) = Saccharum L. GRAMINEAE (M)
Saccharum L. GRAMINEAE (M)
Saccia Naudin (UP) CONVOLVULACEAE (D)
Saccidium Lindl. = Holothrix Rich. ex Lindl. ORCHIDACEAE (M)
Saccifolium Maguire & Pires SACCIFOLIACEAE (D)
Sacciolepis Nash GRAMINEAE (M)
Saccocalyx Coss. & Durieu LABIATAE (D)
Saccochilus Blume = Thrixspermum Lour. ORCHIDACEAE (M)
Saccoglossum Schltr. ORCHIDACEAE (M)
Saccolabiopsis J.J.Sm. ORCHIDACEAE (M)
Saccolabium Blume ORCHIDACEAE (M)
Saccolaria Kuhlm. = Utricularia L. LENTIBULARIACEAE (D)
Saccolena Gleason = Salpinga Mart. ex DC. MELASTOMATACEAE (D)
Saccoloma Kaulf. DENNSTAEDTIACEAE (F)
Sacconia Endl. = Chione DC. RUBIACEAE (D)
Saccopetalum Benn. = Miliusa Lesch. ex A.DC. ANNONACEAE (D)
Saccoplectus Oerst. = Drymonia Mart. GESNERIACEAE (D)
Saccularia Kellogg = Gambelia Nutt. SCROPHULARIACEAE (D)
Sacculina Bosser = Utricularia L. LENTIBULARIACEAE (D)
Sachokiella Kolak. = Campanula L. CAMPANULACEAE (D)
Sachsia Griseb. COMPOSITAE (D)
Sacleuxia Baill. ASCLEPIADACEAE (D)
Sacodon Raf. = Cypripedium L. ORCHIDACEAE (M)
Sacoglottis Mart. HUMIRIACEAE (D)
Sacoila Raf. = Spiranthes Rich. ORCHIDACEAE (M)
Sacosperma G.Taylor RUBIACEAE (D)
Sadiria Mez MYRSINACEAE (D)
Sadleria Kaulf. BLECHNACEAE (F)
Saffordia Maxon ADIANTACEAE (F)
Saffordiella Merr. = Myrtella F.Muell. MYRTACEAE (D)
Sagenia C.Presl = Tectaria Cav. DRYOPTERIDACEAE (F)
Sagenopteris Trevis. = Pleocnemia C.Presl DRYOPTERIDACEAE (F)
Sageraea Dalzell ANNONACEAE (D)
Sageretia Brongn. RHAMNACEAE (D)
Sagina L. CARYOPHYLLACEAE (D)
Sagittaria L. ALISMATACEAE (M)
Sagittipetalum Merr. = Carallia Roxb. RHIZOPHORACEAE (D)
Saglorithys Rizzini = Justicia L. ACANTHACEAE (D)
Sagotanthus Tiegh. = Chaunochiton Benth. OLACACEAE (D)
Sagotia Baill. EUPHORBIACEAE (D)
Sagraea DC. = Clidemia D.Don MELASTOMATACEAE (D)
Saguaster Kuntze = Drymophloeus Zipp. PALMAE (M)
Saguerus Steck = Arenga Labill. PALMAE (M)
Sagus Gaertn. = Raphia P.Beauv. PALMAE (M)
Sagus Steck (SUH) = Metroxylon Rottb. PALMAE (M)
Sahagunia Liebm. = Clarisia Ruíz & Pav. MORACEAE (D)
Sahlbergia Neck. (SUI) = Gardenia Ellis RUBIACEAE (D)
Saintpaulia H.A.Wendl. GESNERIACEAE (D)
Saintpauliopsis Staner ACANTHACEAE (D)
Saipania Hosok. = Croton L. EUPHORBIACEAE (D)
Sairanthus G.Don = Nicotiana L. SOLANACEAE (D)
Sairocarpus D.A.Sutton SCROPHULARIACEAE (D)
Sajanella Soják UMBELLIFERAE (D)
Sajania Pimenov (SUH) = Sajanella Soják UMBELLIFERAE (D)
Sajorium Endl. = Pterococcus Hassk. EUPHORBIACEAE (D)

Sakakia Nakai (SUS) = Cleyera Thunb. THEACEAE (D)
Sakersia Hook.f. = Dichaetanthera Endl. MELASTOMATACEAE (D)
Sakoanala R.Vig. LEGUMINOSAE–PAPILIONOIDEAE (D)
Salacca Reinw. PALMAE (M)
Salacia L. CELASTRACEAE (D)
Salacicratea Loes. = Salacia L. CELASTRACEAE (D)
Salacighia Loes. CELASTRACEAE (D)
Salaciopsis Baker f. CELASTRACEAE (D)
Salacistis Rchb.f. = Goodyera R.Br. ORCHIDACEAE (M)
Salaxis Salisb. ERICACEAE (D)
Salazaria Torr. = Scutellaria L. LABIATAE (D)
Saldanha Vell. = Hillia Jacq. RUBIACEAE (D)
Saldanhaea Bureau = Cuspidaria DC. BIGNONIACEAE (D)
Saldania T.R.Sim = Ormocarpum P.Beauv. LEGUMINOSAE–PAPILIONOIDEAE (D)
Saldinia A.Rich. ex DC. RUBIACEAE (D)
Salicaria Adans. = Lythrum L. LYTHRACEAE (D)
Salicornia L. CHENOPODIACEAE (D)
Salisia Brongn. & Gris (SUH) = Xanthostemon F.Muell. MYRTACEAE (D)
Salisia Lindl. = Kunzea Rchb. MYRTACEAE (D)
Salisia Regel (SUH) = Gloxinia L'Hér. GESNERIACEAE (D)
Salix L. SALICACEAE (D)
Salmalia Schott & Endl. = Bombax L. BOMBACACEAE (D)
Salmasia Schreb. (SUS) = Hirtella L. CHRYSOBALANACEAE (D)
Salmasia Bubani (SUS) = Aira L. GRAMINEAE (M)
Salmea DC. COMPOSITAE (D)
Salmeopsis Benth. COMPOSITAE (D)
Salmiopuntia Frič (SUI) = Opuntia Mill. CACTACEAE (D)
Salmonia Scop. = Vochysia Aubl. VOCHYSIACEAE (D)
Salomonia Lour. POLYGALACEAE (D)
Salpianthus Bonpl. NYCTAGINACEAE (D)
Salpichlaena J.Sm. BLECHNACEAE (F)
Salpichroa Miers SOLANACEAE (D)
Salpiglossis Ruíz & Pav. SOLANACEAE (D)
Salpinctes Woodson APOCYNACEAE (D)
Salpinctium T.J.Edwards = Asystasia Blume ACANTHACEAE (D)
Salpinga Mart. ex DC. MELASTOMATACEAE (D)
Salpingantha Lem. (SUO) = Salpixantha Hook. ACANTHACEAE (D)
Salpingia (Torr. & A.Gray) Raim. = Calylophus Spach ONAGRACEAE (D)
Salpingolobivia Y.Ito = Echinopsis Zucc. CACTACEAE (D)
Salpingostylis Small = Calydorea Herb. IRIDACEAE (M)
Salpinxantha Urb. (SUO) = Salpixantha Hook. ACANTHACEAE (D)
Salpistele Dressler ORCHIDACEAE (M)
Salpixantha Hook. ACANTHACEAE (D)
Salsola L. CHENOPODIACEAE (D)
Saltera Bullock PENAEACEAE (D)
Saltia R.Br. ex Moq. AMARANTHACEAE (D)
Salvadora L. SALVADORACEAE (D)
Salvadoropsis H.Perrier CELASTRACEAE (D)
Salvertia A.St.-Hil. VOCHYSIACEAE (D)
Salvia L. LABIATAE (D)
Salviacanthus Lindau = Justicia L. ACANTHACEAE (D)
Salviastrum Scheele = Salvia L. LABIATAE (D)
Salvinia Séguier SALVINIACEAE (F)
Salweenia Baker f. LEGUMINOSAE–PAPILIONOIDEAE (D)
Salzmannia DC. RUBIACEAE (D)

Samadera Gaertn. = Quassia L. SIMAROUBACEAE (D)
Samaipaticereus Cárdenas CACTACEAE (D)
Samanea (DC.) Merr. = Albizia Durazz. LEGUMINOSAE–MIMOSOIDEAE (D)
Samara L. = Embelia Burm.f. MYRSINACEAE (D)
Samaropyxis Miq. = Hymenocardia Wall. ex Lindl. EUPHORBIACEAE (D)
Samarpsea Raf. = Fraxinus L. OLEACEAE (D)
Sambirania Tardieu = Lindsaea Dryand. ex Sm. DENNSTAEDTIACEAE (F)
Sambucus L. CAPRIFOLIACEAE (D)
Sameraria Desv. CRUCIFERAE (D)
Samolus L. PRIMULACEAE (D)
Sampaiella J.C.Gomes = Arrabidaea DC. BIGNONIACEAE (D)
Sampantaea Airy Shaw EUPHORBIACEAE (D)
Samuela Trel. AGAVACEAE (M)
Samuelssonia Urb. & Ekman ACANTHACEAE (D)
Samyda L. (SUH) = Guarea L. MELIACEAE (D)
Samyda Jacq. FLACOURTIACEAE (D)
Sanango Bunting & Duke BUDDLEJACEAE (D)
Sanblasia L.Andersson MARANTACEAE (M)
Sanchezia Ruíz & Pav. ACANTHACEAE (D)
Sanctambrosia Skottsb. ex Kuschel CARYOPHYLLACEAE (D)
Sandbergia Greene = Halimolobos Tausch CRUCIFERAE (D)
Sandemania Gleason MELASTOMATACEAE (D)
Sanderella Kuntze ORCHIDACEAE (M)
Sandersonia Hook. COLCHICACEAE (M)
Sandoricum Cav. MELIACEAE (D)
Sandwithia Lanj. EUPHORBIACEAE (D)
Sandwithiodoxa Aubrév. & Pellegr. = Pouteria Aubl. SAPOTACEAE (D)
Sanguinaria L. PAPAVERACEAE (D)
Sanguinaria Bubani (SUS) = Digitaria Haller GRAMINEAE (M)
Sanguinella Gleichen = Digitaria Haller GRAMINEAE (M)
Sanguisorba L. ROSACEAE (D)
Sanhilaria Baill. = Paragonia Bureau BIGNONIACEAE (D)
Sanicula L. UMBELLIFERAE (D)
Sanidophyllum Small = Hypericum L. GUTTIFERAE (D)
Saniella Hilliard & B.L.Burtt HYPOXIDACEAE (M)
Sansevieria Thunb. DRACAENACEAE (M)
Santalina Baill. = Tarenna Gaertn. RUBIACEAE (D)
Santalodes Kuntze (SUH) = Rourea Aubl. CONNARACEAE (D)
Santaloidella Schellenb. = Rourea Aubl. CONNARACEAE (D)
Santaloides Schellenb. = Rourea Aubl. CONNARACEAE (D)
Santalum L. SANTALACEAE (D)
Santapaua N.P.Balakr. & K.Subramanyam = Hygrophila R.Br. ACANTHACEAE (D)
Santia Wight & Arn. (SUH) = Lasianthus Jack RUBIACEAE (D)
Santia Savi = Polypogon Desf. GRAMINEAE (M)
Santiria Blume BURSERACEAE (D)
Santiriopsis Engl. = Santiria Blume BURSERACEAE (D)
Santisukia Brummitt BIGNONIACEAE (D)
Santolina L. COMPOSITAE (D)
Santomasia N.Robson GUTTIFERAE (D)
Santosia R.M.King & H.Rob. COMPOSITAE (D)
Sanvitalia Lam. COMPOSITAE (D)
Saphesia N.E.Br. AIZOACEAE (D)
Sapindopsis F.C.How & C.N.Ho (SUH) = Lepisanthes Blume SAPINDACEAE (D)
Sapindus L. SAPINDACEAE (D)
Sapiopsis Müll.Arg. = Sapium P.Browne EUPHORBIACEAE (D)

Sapium P.Browne EUPHORBIACEAE (D)
Saponaria L. CARYOPHYLLACEAE (D)
Saposhnikovia Schischk. UMBELLIFERAE (D)
Sapota Mill. = Manilkara Adans. SAPOTACEAE (D)
Sapphoa Urb. ACANTHACEAE (D)
Sapranthus Seem. ANNONACEAE (D)
Sapria Griff. RAFFLESIACEAE (D)
Saprosma Blume RUBIACEAE (D)
Sapucaya R.Knuth = Lecythis Loefl. LECYTHIDACEAE (D)
Saraca L. LEGUMINOSAE–CAESALPINIOIDEAE (D)
Saracha Ruíz & Pav. SOLANACEAE (D)
Saranthe (Regel & Koern.) Eichl. MARANTACEAE (M)
Sararanga Hemsl. PANDANACEAE (M)
Sarawakodendron Ding Hou CELASTRACEAE (D)
Sarcandra Gardner CHLORANTHACEAE (D)
Sarcanthidion Baill. = Citronella D.Don ICACINACEAE (D)
Sarcanthopsis Garay ORCHIDACEAE (M)
Sarcanthus Lindl. (SUH) = Cleisostoma Blume ORCHIDACEAE (M)
Sarcaulus Radlk. SAPOTACEAE (D)
Sarcobatus Nees CHENOPODIACEAE (D)
Sarcobodium Beer = Bulbophyllum Thouars ORCHIDACEAE (M)
Sarcobotrya R.Vig. = Kotschya Endl. LEGUMINOSAE–PAPILIONOIDEAE (D)
Sarcocapnos DC. PAPAVERACEAE (D)
Sarcocaulon (DC.) Sweet GERANIACEAE (D)
Sarcocephalus Afzel. ex Sabine RUBIACEAE (D)
Sarcochilus R.Br. ORCHIDACEAE (M)
Sarcochlamys Gaudich. URTICACEAE (D)
Sarcoclinium Wight = Agrostistachys Dalzell EUPHORBIACEAE (D)
Sarcococca Lindl. BUXACEAE (D)
Sarcocolla Kunth (SUH) = Saltera Bullock PENAEACEAE (D)
Sarcocornia A.J.Scott CHENOPODIACEAE (D)
Sarcodes Torr. ERICACEAE (D)
Sarcodraba Gilg & Muschl. CRUCIFERAE (D)
Sarcodrimys (Baill.) Baum.–Bod. (SUI) = Zygogynum Baill. WINTERACEAE (D)
Sarcodum Lour. LEGUMINOSAE–PAPILIONOIDEAE (D)
Sarcoglossum Beer = Cirrhaea Lindl. ORCHIDACEAE (M)
Sarcoglottis C.Presl ORCHIDACEAE (M)
Sarcoglyphis Garay ORCHIDACEAE (M)
Sarcolaena Thouars SARCOLAENACEAE (D)
Sarcolobus R.Br. ASCLEPIADACEAE (D)
Sarcolophium Troupin MENISPERMACEAE (D)
Sarcomelicope Engl. RUTACEAE (D)
Sarcomphalus P.Browne = Ziziphus Mill. RHAMNACEAE (D)
Sarcopetalum F.Muell. MENISPERMACEAE (D)
Sarcophagophilus Dinter = Quaqua N.E.Br. ASCLEPIADACEAE (D)
Sarcopharyngia (Stapf) Boiteau = Tabernaemontana L. APOCYNACEAE (D)
Sarcophrynium K.Schum. MARANTACEAE (M)
Sarcophysa Miers = Juanulloa Ruíz & Pav. SOLANACEAE (D)
Sarcophyte Sparrm. BALANOPHORACEAE (D)
Sarcophyton Garay ORCHIDACEAE (M)
Sarcopilea Urb. URTICACEAE (D)
Sarcopodium Lindl. = Epigeneium Gagnep. ORCHIDACEAE (M)
Sarcopoterium Spach ROSACEAE (D)
Sarcopteryx Radlk. SAPINDACEAE (D)
Sarcopygme Setch. & Christoph. RUBIACEAE (D)

Sarcopyramis Wall. MELASTOMATACEAE (D)
Sarcorhachis Trel. PIPERACEAE (D)
Sarcorhyna C.Presl (UP) SAPOTACEAE (D)
Sarcorhynchus Schltr. = Diaphananthe Schltr. ORCHIDACEAE (M)
Sarcorrhiza Bullock ASCLEPIADACEAE (D)
Sarcosperma Hook.f. SAPOTACEAE (D)
Sarcostemma R.Br. ASCLEPIADACEAE (D)
Sarcostigma Wight & Arn. ICACINACEAE (D)
Sarcostoma Blume ORCHIDACEAE (M)
Sarcostyles C.Presl ex DC. = Hydrangea L. HYDRANGEACEAE (D)
Sarcotheca Blume OXALIDACEAE (D)
Sarcotoechia Radlk. SAPINDACEAE (D)
Sarcoyucca (Engelm.) Lindinger = Yucca L. AGAVACEAE (M)
Sarcozona J.M.Black = Carpobrotus N.E.Br. AIZOACEAE (D)
Sarcozygium Bunge ZYGOPHYLLACEAE (D)
Sardinia Vell. = Guettarda L. RUBIACEAE (D)
Sarga Ewart = Sorghum Moench GRAMINEAE (M)
Sargentia S.Watson = Casimiroa La Llave RUTACEAE (D)
Sargentia H.A.Wendl. & Drude ex Salomon (SUH) = Pseudophoenix H.A.Wendl.
    ex Sarg. PALMAE (M)
Sargentodoxa Rehder & E.H.Wilson SARGENTODOXACEAE (D)
Saribus Blume = Livistona R.Br. PALMAE (M)
Sarinia O.F.Cook = Attalea Kunth PALMAE (M)
Sarissus Gaertn. = Hydrophylax L.f. RUBIACEAE (D)
Saritaea Dugand BIGNONIACEAE (D)
Sarmentaria Naudin = Adelobotrys DC. MELASTOMATACEAE (D)
Sarmienta Ruíz & Pav. GESNERIACEAE (D)
Sarojusticia Bremek. = Justicia L. ACANTHACEAE (D)
Sarothamnus Wimm. = Cytisus Desf. LEGUMINOSAE–PAPILIONOIDEAE (D)
Sarotheca Nees = Justicia L. ACANTHACEAE (D)
Sarothra L. = Hypericum L. GUTTIFERAE (D)
Sarothrochilus Schltr. = Staurochilus Ridl. ex Pfitzer ORCHIDACEAE (M)
Sarothrostachys Klotzsch = Sebastiania Spreng. EUPHORBIACEAE (D)
Sarracenella Luer = Pleurothallis R.Br. ORCHIDACEAE (M)
Sarracenia L. SARRACENIACEAE (D)
Sartidia De Winter GRAMINEAE (M)
Sartoria Boiss. & Heldr. LEGUMINOSAE–PAPILIONOIDEAE (D)
Sartorina R.M.King & H.Rob. COMPOSITAE (D)
Sartwellia A.Gray COMPOSITAE (D)
Saruma Oliv. ARISTOLOCHIACEAE (D)
Sarx St.John = Sicyos L. CUCURBITACEAE (D)
Sasa Makino & Shib. GRAMINEAE (M)
Sasaella Makino = Sasa Makino & Shib. GRAMINEAE (M)
Sasamorpha Nakai = Sasa Makino & Shib. GRAMINEAE (M)
Sassafras Nees LAURACEAE (D)
Sassafridium Meisn. = Cinnamomum Schaeffer LAURACEAE (D)
Sassia Molina = Oxalis L. OXALIDACEAE (D)
Satakentia H.E.Moore PALMAE (M)
Satanocrater Schweinf. ACANTHACEAE (D)
Satorkis Thouars = Coeloglossum Hartm. ORCHIDACEAE (M)
Sattadia Fourn. ASCLEPIADACEAE (D)
Satureja L. LABIATAE (D)
Satyria Klotzsch ERICACEAE (D)
Satyridium Lindl. ORCHIDACEAE (M)
Satyrium Sw. ORCHIDACEAE (M)

Saugetia Hitchc. & Chase = Enteropogon Nees GRAMINEAE (M)
Saundersia Rchb.f. ORCHIDACEAE (M)
Saurauia Willd. ACTINIDIACEAE (D)
Sauroglossum Lindl. ORCHIDACEAE (M)
Sauromatum Schott ARACEAE (M)
Sauropus Blume EUPHORBIACEAE (D)
Saururus L. SAURURACEAE (D)
Saussurea DC. COMPOSITAE (D)
Sautiera Decne. ACANTHACEAE (D)
Sauvagesia L. OCHNACEAE (D)
Sauvallea W.Wright COMMELINACEAE (M)
Sauvallella Rydb. LEGUMINOSAE–PAPILIONOIDEAE (D)
Savannosiphon Goldblatt & Marais IRIDACEAE (M)
Savastana Schrank (SUH) = Hierochloe R.Br. GRAMINEAE (M)
Savia Willd. EUPHORBIACEAE (D)
Savignya DC. CRUCIFERAE (D)
Saxegothaea Lindl. PODOCARPACEAE (G)
Saxicolella Engl. PODOSTEMACEAE (D)
Saxifraga L. SAXIFRAGACEAE (D)
Saxifragella Engl. SAXIFRAGACEAE (D)
Saxifragites Gagnep. = Distylium Siebold & Zucc. HAMAMELIDACEAE (D)
Saxifragodes D.M.Moore SAXIFRAGACEAE (D)
Saxifragopsis Small SAXIFRAGACEAE (D)
Saxiglossum Ching POLYPODIACEAE (F)
Saxofridericia R.H.Schomb. RAPATEACEAE (M)
Sayeria Kraenzl. = Dendrobium Sw. ORCHIDACEAE (M)
Scabiosa L. DIPSACACEAE (D)
Scabiosiopsis Rech.f. = Scabiosa L. DIPSACACEAE (D)
Scadoxus Raf. AMARYLLIDACEAE (M)
Scaevola L. GOODENIACEAE (D)
Scagea McPherson EUPHORBIACEAE (D)
Scalesia Arn. COMPOSITAE (D)
Scaligeria DC. UMBELLIFERAE (D)
Scambopus O.E.Schulz CRUCIFERAE (D)
Scandia J.W.Dawson UMBELLIFERAE (D)
Scandicium Thell. = Scandix L. UMBELLIFERAE (D)
Scandivepres Loes. CELASTRACEAE (D)
Scandix L. UMBELLIFERAE (D)
Scaphiophora Schltr. BURMANNIACEAE (M)
Scaphispatha Brongn. ex Schott ARACEAE (M)
Scaphium Schott & Endl. STERCULIACEAE (D)
Scaphocalyx Ridl. FLACOURTIACEAE (D)
Scaphochlamys Baker ZINGIBERACEAE (M)
Scaphopetalum Mast. STERCULIACEAE (D)
Scaphosepalum Pfitzer ORCHIDACEAE (M)
Scaphospermum Korovin (SUH) = Parasilaus Leute UMBELLIFERAE (D)
Scaphyglottis Poepp. & Endl. ORCHIDACEAE (M)
Scapicephalus Ovcz. & Chukav. BORAGINACEAE (D)
Scaredederis Thouars = Dendrobium Sw. ORCHIDACEAE (M)
Scariola F.W.Schmidt COMPOSITAE (D)
Scassellatia Chiov. = Lannea A.Rich. ANACARDIACEAE (D)
Sceletium N.E.Br. = Phyllobolus N.E.Br. AIZOACEAE (D)
Scelochiloides Dodson & M.W.Chase ORCHIDACEAE (M)
Scelochilus Klotzsch ORCHIDACEAE (M)
Scepa Lindl. = Aporusa Blume EUPHORBIACEAE (D)

Scepasma Blume = Phyllanthus L. EUPHORBIACEAE (D)
Scepseothamnus Cham. = Alibertia A.Rich. ex DC. RUBIACEAE (D)
Sceptridium Lyon = Botrychium Sw. OPHIOGLOSSACEAE (F)
Sceptrocnide Maxim. = Laportea Gaudich. URTICACEAE (D)
Schachtia Karst. RUBIACEAE (D)
Schaefferia Jacq. CELASTRACEAE (D)
Schaenomorphus Thorel ex Gagnep. = Tropidia Lindl. ORCHIDACEAE (M)
Schaetzellia Sch.Bip. (SUH) = Macvaughiella R.M.King & H.E.Rob. COMPOSITAE (D)
Schaffnera Benth. (SUH) = Schaffnerella Nash GRAMINEAE (M)
Schaffnerella Nash GRAMINEAE (M)
Schaffneria Fée ex T.Moore ASPLENIACEAE (F)
Schaueria Nees ACANTHACEAE (D)
Schaueria Hassk. (SUH) = Hyptis Jacq. LABIATAE (D)
Schedonnardus Steud. GRAMINEAE (M)
Schedonorus P.Beauv. = Festuca L. GRAMINEAE (M)
Scheelea Karst. PALMAE (M)
Scheeria Seem. = Achimenes Pers. GESNERIACEAE (D)
Schefferella Pierre = Burckella Pierre SAPOTACEAE (D)
Schefferomitra Diels ANNONACEAE (D)
Schefflera J.R.Forst. & G.Forst. ARALIACEAE (D)
Schefflerodendron Harms LEGUMINOSAE–PAPILIONOIDEAE (D)
Scheffleropsis Ridl. = Schefflera J.R.Forst. & G.Forst. ARALIACEAE (D)
Schelhammera R.Br. CONVALLARIACEAE (M)
Schellenbergia C.E.Parkinson = Vismianthus Mildbr. CONNARACEAE (D)
Schellingia Steud. = Aegopogon Humb. & Bonpl. ex Willd. GRAMINEAE (M)
Schellolepis J.Sm. = Goniophlebium C.Presl POLYPODIACEAE (F)
Schenckia K.Schum. = Deppea Cham. & Schltdl. RUBIACEAE (D)
Scherya R.M.King & H.Rob. COMPOSITAE (D)
Schetti Adans. = Ixora L. RUBIACEAE (D)
Scheuchzeria L. SCHEUCHZERIACEAE (M)
Schickendantzia Pax ALSTROEMERIACEAE (M)
Schickia Tischer (SUI) = Mesembryanthemum L. AIZOACEAE (D)
Schidiomyrtus Schauer = Baeckea L. MYRTACEAE (D)
Schiedea A.Rich. (SUH) = Machaonia Bonpl. RUBIACEAE (D)
Schiedea Bartl. (SUI) = Richardia L. RUBIACEAE (D)
Schiedea Cham. & Schltdl. CARYOPHYLLACEAE (D)
Schiedeella Schltr. ORCHIDACEAE (M)
Schiedophytum H.Wolff = Donnellsmithia J.M.Coult. & Rose UMBELLIFERAE (D)
Schiekia Meisn. HAEMODORACEAE (M)
Schima Reinw. ex Blume THEACEAE (D)
Schimpera Steud. & Hochst. ex Endl. CRUCIFERAE (D)
Schimperella H.Wolff = Oreoschimperella Rauschert UMBELLIFERAE (D)
Schimperina Tiegh. = Agelanthus Tiegh. LORANTHACEAE (D)
Schindleria H.Walter PHYTOLACCACEAE (D)
Schinopsis Engl. ANACARDIACEAE (D)
Schinus L. ANACARDIACEAE (D)
Schinzafra Kuntze = Thamnea Sol. ex Brongn. BRUNIACEAE (D)
Schinziella Gilg GENTIANACEAE (D)
Schinziophyton Hutch. ex Radcl.–Sm. EUPHORBIACEAE (D)
Schippia Burret PALMAE (M)
Schirostachyum De Vriese (SUO) = Schizostachyum Nees GRAMINEAE (M)
Schisandra Michx. SCHISANDRACEAE (D)
Schischkinia Iljin COMPOSITAE (D)
Schischkiniella Steenis = Silene L. CARYOPHYLLACEAE (D)
Schismatoclada Baker RUBIACEAE (D)

Schismatoglottis Zoll. & Moritzi ARACEAE (M)
Schismatopera Klotzsch = Pera Mutis EUPHORBIACEAE (D)
Schismocarpus S.F.Blake LOASACEAE (D)
Schismoceras C.Presl = Dendrobium Sw. ORCHIDACEAE (M)
Schismus P.Beauv. GRAMINEAE (M)
Schistachne Figari & De Not. = Stipagrostis Nees GRAMINEAE (M)
Schistocarpaea F.Muell. RHAMNACEAE (D)
Schistocarpha Less. COMPOSITAE (D)
Schistocaryum Franch. = Microula Benth. BORAGINACEAE (D)
Schistogyne Hook. & Arn. ASCLEPIADACEAE (D)
Schistonema Schltr. ASCLEPIADACEAE (D)
Schistophragma Benth. ex Endl. SCROPHULARIACEAE (D)
Schistophyllidium (Juz. ex Fedorov) Ikonn. = Potentilla L. ROSACEAE (D)
Schistostemon (Urb.) Cuatrec. HUMIRIACEAE (D)
Schistostephium Less. COMPOSITAE (D)
Schistostigma Lauterb. = Cleistanthus Hook.f. ex Planch. EUPHORBIACEAE (D)
Schistotylus Dockrill ORCHIDACEAE (M)
Schivereckia Andrz. ex DC. CRUCIFERAE (D)
Schizachne Hack. GRAMINEAE (M)
Schizachyrium Nees GRAMINEAE (M)
Schizaea Sm. SCHIZAEACEAE (F)
Schizangium Bartl. ex DC. = Mitracarpus Zucc. RUBIACEAE (D)
Schizanthus Ruíz & Pav. SOLANACEAE (D)
Schizeilema (Hook.f.) Domin UMBELLIFERAE (D)
Schizenterospermum Homolle ex Arènes RUBIACEAE (D)
Schizobasis Baker HYACINTHACEAE (M)
Schizoboea (Fritsch) B.L.Burtt GESNERIACEAE (D)
Schizocaena J.Sm. ex Hook. = Cyathea Sm. CYATHEACEAE (F)
Schizocalomyrtus Kausel = Calycorectes O.Berg MYRTACEAE (D)
Schizocalyx Wedd. RUBIACEAE (D)
Schizocalyx O.Berg (SUH) = Calycorectes O.Berg MYRTACEAE (D)
Schizocalyx Hochst. (SUH) = Dobera Juss. SALVADORACEAE (D)
Schizocapsa Hance = Tacca J.R.Forst. & G.Forst. TACCACEAE (M)
Schizocardia A.C.Sm. & Standl. = Purdiaea Planch. CYRILLACEAE (D)
Schizocarphus Van der Merwe = Scilla L. HYACINTHACEAE (M)
Schizocarpum Schrad. CUCURBITACEAE (D)
Schizocarya Spach = Gaura L. ONAGRACEAE (D)
Schizocasia Schott = Xenophya Schott ARACEAE (M)
Schizocentron Meisn. = Heterocentron Hook. & Arn. MELASTOMATACEAE (D)
Schizochilus Sond. ORCHIDACEAE (M)
Schizochiton Spreng. = Chisocheton Blume MELIACEAE (D)
Schizococcus Eastw. ERICACEAE (D)
Schizocodon Siebold & Zucc. DIAPENSIACEAE (D)
Schizocolea Bremek. RUBIACEAE (D)
Schizodium Lindl. ORCHIDACEAE (M)
Schizoglossum E.Mey. ASCLEPIADACEAE (D)
Schizogramma Link = Hemionitis L. ADIANTACEAE (F)
Schizogyne Ehrenb. ex Pax (SUH) = Acalypha L. EUPHORBIACEAE (D)
Schizogyne Cass. COMPOSITAE (D)
Schizolaena Thouars SARCOLAENACEAE (D)
Schizolegnia Alston = Lindsaea Dryand. ex Sm. DENNSTAEDTIACEAE (F)
Schizolepton Fée = Taenitis Willd. ex Schkuhr ADIANTACEAE (F)
Schizolobium Vogel LEGUMINOSAE–CAESALPINIOIDEAE (D)
Schizoloma Gaudich. = Lindsaea Dryand. ex Sm. DENNSTAEDTIACEAE (F)
Schizomeria D.Don CUNONIACEAE (D)

Schizomeryta R.Vig. = Meryta J.R.Forst. & G.Forst. ARALIACEAE (D)
Schizomussaenda H.L.Li = Mussaenda L. RUBIACEAE (D)
Schizonepeta (Benth.) Briq. LABIATAE (D)
Schizonotus A.Gray (SUH) = Asclepias L. ASCLEPIADACEAE (D)
Schizopedium Salisb. = Cypripedium L. ORCHIDACEAE (M)
Schizopepon Maxim. CUCURBITACEAE (D)
Schizopetalon Sims CRUCIFERAE (D)
Schizophragma Siebold & Zucc. HYDRANGEACEAE (D)
Schizopleura Endl. = Beaufortia R.Br. MYRTACEAE (D)
Schizopogon Rchb. ex Spreng. = Schizachyrium Nees GRAMINEAE (M)
Schizopremna Baill. = Faradaya F.Muell. VERBENACEAE (D)
Schizopsis Bureau = Tynnanthus Miers BIGNONIACEAE (D)
Schizoptera Turcz. COMPOSITAE (D)
Schizoscyphus K.Schum. ex Taub. = Maniltoa Scheff. LEGUMINOSAE–
  CAESALPINIOIDEAE (D)
Schizosepala G.M.Barroso SCROPHULARIACEAE (D)
Schizosiphon K.Schum. = Maniltoa Scheff. LEGUMINOSAE–CAESALPINIOIDEAE (D)
Schizospatha Furtado = Calamus L. PALMAE (M)
Schizospermum Boivin ex Baill. = Cremaspora Benth. RUBIACEAE (D)
Schizostachyum Nees GRAMINEAE (M)
Schizostege W.F.Hillebr. = Pteris L. PTERIDACEAE (F)
Schizostegeopsis Copel. = Pteris L. PTERIDACEAE (F)
Schizostephanus Hochst. ex K.Schum. = Cynanchum L. ASCLEPIADACEAE (D)
Schizostigma Arn. ex Meisn. RUBIACEAE (D)
Schizostigma Arn. (SUH) = Cucurbitella Walp. CUCURBITACEAE (D)
Schizostylis Backh. & Harv. IRIDACEAE (M)
Schizotorenia Yamaz. = Lindernia All. SCROPHULARIACEAE (D)
Schizotrichia Benth. COMPOSITAE (D)
Schizozygia Baill. APOCYNACEAE (D)
Schkuhria Roth COMPOSITAE (D)
Schlagintweitiella Ulbr. = Thalictrum L. RANUNCULACEAE (D)
Schlechtendalia Less. COMPOSITAE (D)
Schlechteranthus Schwantes AIZOACEAE (D)
Schlechterella K.Schum. ASCLEPIADACEAE (D)
Schlechterella Hoehne (SUH) = Rudolfiella Hoehne ORCHIDACEAE (M)
Schlechteria Bolus CRUCIFERAE (D)
Schlechteria Mast. (SUH) = Phyllocomos Mast. RESTIONACEAE (M)
Schlechterianthus Quisumb. = Quisumbingia Merr. ASCLEPIADACEAE (D)
Schlechterina Harms PASSIFLORACEAE (D)
Schlechterosciadium H.Wolff = Chamarea Eckl. & Zeyh. UMBELLIFERAE (D)
Schlegelia Miq. SCROPHULARIACEAE (D)
Schleichera Willd. SAPINDACEAE (D)
Schleinitzia Warb. LEGUMINOSAE–MIMOSOIDEAE (D)
Schleropelta Buckley = Hilaria Kunth GRAMINEAE (M)
Schliebenia Mildbr. = Isoglossa Oerst. ACANTHACEAE (D)
Schlimmia Planch. & Linden ORCHIDACEAE (M)
Schlumbergera E.Morren (SUH) = Guzmania Ruíz & Pav. BROMELIACEAE (M)
Schlumbergera Lem. CACTACEAE (D)
Schmalhausenia C.Winkl. COMPOSITAE (D)
Schmardaea Karst. MELIACEAE (D)
Schmidtia Tratt. (SUH) = Coleanthus J.Seidel GRAMINEAE (M)
Schmidtia Steud. ex J.A.Schmidt GRAMINEAE (M)
Schmidtottia Urb. RUBIACEAE (D)
Schmiedtia Raf. (SUO) = Coleanthus J.Seidel GRAMINEAE (M)
Schnabelia Hand.–Mazz. VERBENACEAE (D)

Schnella Raddi = Bauhinia L. LEGUMINOSAE–CAESALPINIOIDEAE (D)
Schnittspahnia Rchb. = Mitrella Miq. ANNONACEAE (D)
Schnizleinia Steud. (SUI) = Boissiera Hochst. ex Steud. GRAMINEAE (M)
Schoenanthus Adans. (SUS) = Ischaemum L. GRAMINEAE (M)
Schoenefeldia Kunth GRAMINEAE (M)
Schoenia Steetz COMPOSITAE (D)
Schoenlandia Cornu = Cyanastrum Oliv. CYANASTRACEAE (M)
Schoenleinia Klotzsch = Bathysa C.Presl RUBIACEAE (D)
Schoenleinia Klotzsch ex Lindl. (SUH) = Ponthieva R.Br. ORCHIDACEAE (M)
Schoenobiblus Mart. THYMELAEACEAE (D)
Schoenocaulon A.Gray MELANTHIACEAE (M)
Schoenocephalium Seub. RAPATEACEAE (M)
Schoenocrambe Greene CRUCIFERAE (D)
Schoenodendron Engl. = Microdracoides Hua CYPERACEAE (M)
Schoenodorus Roem. & Schult. (SUO) = Festuca L. GRAMINEAE (M)
Schoenoides Seberg = Oreobolus R.Br. CYPERACEAE (M)
Schoenolaena Bunge UMBELLIFERAE (D)
Schoenolirion Torr. ex E.M.Durand HYACINTHACEAE (M)
Schoenoplectus (Rchb.) Palla CYPERACEAE (M)
Schoenorchis Blume ORCHIDACEAE (M)
Schoenoxiphium Nees CYPERACEAE (M)
Schoenus L. CYPERACEAE (M)
Schoepfia Schreb. OLACACEAE (D)
Schoepfiopsis Miers = Schoepfia Schreb. OLACACEAE (D)
Scholleropsis H.Perrier PONTEDERIACEAE (M)
Scholtzia Schauer MYRTACEAE (D)
Schomburgkia Lindl. ORCHIDACEAE (M)
Schonlandia L.Bolus = Corpuscularia Schwantes AIZOACEAE (D)
Schorigeram Adans. = Tragia L. EUPHORBIACEAE (D)
Schotia Jacq. LEGUMINOSAE–CAESALPINIOIDEAE (D)
Schousboea Schumach. & Thonn. (SUH) = Alchornea Sw. EUPHORBIACEAE (D)
Schousboea Willd. = Combretum Loefl. COMBRETACEAE (D)
Schoutenia Korth. TILIACEAE (D)
Schouwia DC. CRUCIFERAE (D)
Schradera Willd. (SUH) = Croton L. EUPHORBIACEAE (D)
Schradera Vahl RUBIACEAE (D)
Schraderia Medik. = Salvia L. LABIATAE (D)
Schrameckia Danguy MONIMIACEAE (D)
Schrammia Britton & Rose = Hoffmannseggia Cav. LEGUMINOSAE–
    CAESALPINIOIDEAE (D)
Schranckiastrum Hassl. = Mimosa L. LEGUMINOSAE–MIMOSOIDEAE (D)
Schrankia Willd. LEGUMINOSAE–MIMOSOIDEAE (D)
Schrebera Roxb. OLEACEAE (D)
Schreibersia Pohl = Augusta Pohl RUBIACEAE (D)
Schreiteria Carolin PORTULACACEAE (D)
Schrenkia Fisch. & C.A.Mey. UMBELLIFERAE (D)
Schroeterella Briq. = Larrea Cav. ZYGOPHYLLACEAE (D)
Schtschurowskia Regel & Schmalh. UMBELLIFERAE (D)
Schubertia Mart. ASCLEPIADACEAE (D)
Schuechia Endl. = Qualea Aubl. VOCHYSIACEAE (D)
Schuermannia F.Muell. = Darwinia Rudge MYRTACEAE (D)
Schufia Spach = Fuchsia L. ONAGRACEAE (D)
Schultesia Mart. GENTIANACEAE (D)
Schultesia Spreng. (SUS) = Eustachys Desv. GRAMINEAE (M)
Schultesia Roth (SUH) = Wahlenbergia Schrad. ex Roth CAMPANULACEAE (D)

Schultesianthus Hunz. SOLANACEAE (D)
Schultesiophytum Harling CYCLANTHACEAE (M)
Schulzia Spreng. UMBELLIFERAE (D)
Schumacheria Vahl DILLENIACEAE (D)
Schumannia Kuntze = Ferula L. UMBELLIFERAE (D)
Schumannianthus Gagnep. MARANTACEAE (M)
Schumanniophyton Harms RUBIACEAE (D)
Schumeria Iljin COMPOSITAE (D)
Schunda–Pana Adans. = Caryota L. PALMAE (M)
Schuurmansia Blume OCHNACEAE (D)
Schuurmansiella Hallier f. OCHNACEAE (D)
Schwabea Endl. = Monechma Hochst. ACANTHACEAE (D)
Schwackaea Cogn. MELASTOMATACEAE (D)
Schwalbea L. SCROPHULARIACEAE (D)
Schwannia Endl. = Janusia A.Juss. MALPIGHIACEAE (D)
Schwantesia Dinter AIZOACEAE (D)
Schwantesia L.Bolus (SUH) = Mitrophyllum Schwantes AIZOACEAE (D)
Schwartzkopffia Kraenzl. ORCHIDACEAE (M)
Schweiggeria Spreng. VIOLACEAE (D)
Schweinfurthia A.Braun SCROPHULARIACEAE (D)
Schweinitzia Elliott ex Nutt. = Monotropsis Schwein. ex Elliott ERICACEAE (D)
Schwenckia L. SOLANACEAE (D)
Schwendenera K.Schum. RUBIACEAE (D)
Schwenkfelda Schreb. = Sabicea Aubl. RUBIACEAE (D)
Schwenkfeldia Willd. (SUO) = Sabicea Aubl. RUBIACEAE (D)
Schwenckiopsis Dammer SOLANACEAE (D)
Schwerinia Karst. = Meriania Sw. MELASTOMATACEAE (D)
Sciacassia Britton = Senna Mill. LEGUMINOSAE–CAESALPINIOIDEAE (D)
Sciadiodaphne Rchb. = Umbellularia (Nees) Nutt. LAURACEAE (D)
Sciadocalyx Regel = Kohleria Regel GESNERIACEAE (D)
Sciadocephala Mattf. COMPOSITAE (D)
Sciadodendron Griseb. ARALIACEAE (D)
Sciadonardus Steud. (SUI) = Gymnopogon P.Beauv. GRAMINEAE (M)
Sciadopanax Seem. = Polyscias J.R.Forst. & G.Forst. ARALIACEAE (D)
Sciadophyllum P.Browne = Schefflera J.R.Forst. & G.Forst. ARALIACEAE (D)
Sciadopitys Siebold & Zucc. TAXODIACEAE (G)
Sciadotenia Miers MENISPERMACEAE (D)
Sciaphila Blume TRIURIDACEAE (M)
Sciaphyllum Bremek. (UP) ACANTHACEAE (D)
Sciaplea Rauschert = Dialium L. LEGUMINOSAE–CAESALPINIOIDEAE (D)
Scilla L. HYACINTHACEAE (M)
Scindapsus Schott ARACEAE (M)
Scirpidiella Rauschert = Isolepis R.Br. CYPERACEAE (M)
Scirpobambus Kuntze (SUS) = Oxytenanthera Munro GRAMINEAE (M)
Scirpodendron Zipp. ex Kurz CYPERACEAE (M)
Scirpoides Séguier CYPERACEAE (M)
Scirpus L. CYPERACEAE (M)
Sclerachne Trin. (SUH) = Limnodea L.H.Dewey GRAMINEAE (M)
Sclerachne R.Br. GRAMINEAE (M)
Sclerandrium Stapf & C.E.Hubb. = Germainia Balansa & Poitr. GRAMINEAE (M)
Scleranthera Pichon = Wrightia R.Br. APOCYNACEAE (D)
Scleranthopsis Rech.f. CARYOPHYLLACEAE (D)
Scleranthus L. ILLECEBRACEAE (D)
Scleria Bergius CYPERACEAE (M)
Sclerobassia Ulbr. = Bassia All. CHENOPODIACEAE (D)

Scleroblitum Ulbr. = Chenopodium L. CHENOPODIACEAE (D)
Sclerocactus Britton & Rose CACTACEAE (D)
Sclerocalyx Nees = Ruellia L. ACANTHACEAE (D)
Sclerocarpus Jacq. COMPOSITAE (D)
Sclerocarya Hochst. ANACARDIACEAE (D)
Sclerocaryopsis Brand = Lappula Moench BORAGINACEAE (D)
Sclerocephalus Boiss. ILLECEBRACEAE (D)
Sclerochiton Harv. ACANTHACEAE (D)
Sclerochloa P.Beauv. GRAMINEAE (M)
Sclerochorton Boiss. UMBELLIFERAE (D)
Sclerocladus Raf. = Sideroxylon L. SAPOTACEAE (D)
Sclerococcus Bartl. (SUI) = Hedyotis L. RUBIACEAE (D)
Sclerocroton Hochst. = Sapium P.Browne EUPHORBIACEAE (D)
Sclerocyathium Prokh. = Euphorbia L. EUPHORBIACEAE (D)
Sclerodactylon Stapf GRAMINEAE (M)
Sclerodeyeuxia Pilg. = Calamagrostis Adans. GRAMINEAE (M)
Scleroglossum Alderw. GRAMMITIDACEAE (F)
Sclerolaena A.Camus (SUI) = Cyphochlaena Hack. GRAMINEAE (M)
Sclerolaena R.Br. CHENOPODIACEAE (D)
Sclerolepis Cass. COMPOSITAE (D)
Sclerolinon C.M.Rogers LINACEAE (D)
Sclerolobium Vogel LEGUMINOSAE–CAESALPINIOIDEAE (D)
Scleromelum K.Schum. & Lauterb. = Scleropyrum Arn. SANTALACEAE (D)
Scleromitrion Wight & Arn. = Hedyotis L. RUBIACEAE (D)
Scleronema Benth. BOMBACACEAE (D)
Sclerophylax Miers SOLANACEAE (D)
Sclerophyllum Griff. (SUH) = Porteresia Tateoka GRAMINEAE (M)
Scleropoa Griseb. = Catapodium Link GRAMINEAE (M)
Scleropogon Phil. GRAMINEAE (M)
Scleropterys Scheidw. = Cirrhaea Lindl. ORCHIDACEAE (M)
Scleropyrum Arn. SANTALACEAE (D)
Sclerorhachis (Rech.f.) Rech.f. COMPOSITAE (D)
Sclerosciadium W.D.J.Koch ex DC. = Capnophyllum Gaertn. UMBELLIFERAE (D)
Sclerosperma G.Mann & H.A.Wendl. PALMAE (M)
Sclerostachya (Hack.) A.Camus = Miscanthus Andersson GRAMINEAE (M)
Sclerostegia Paul G.Wilson CHENOPODIACEAE (D)
Sclerostephane Chiov. COMPOSITAE (D)
Sclerotheca A.DC. CAMPANULACEAE (D)
Sclerothrix C.Presl LOASACEAE (D)
Sclerotiaria Korovin UMBELLIFERAE (D)
Scleroxylon Bertol. (UP) SAPOTACEAE (D)
Sclerozus Raf. = Sideroxylon L. SAPOTACEAE (D)
Scobinaria Seibert BIGNONIACEAE (D)
Scoliaxon Payson CRUCIFERAE (D)
Scoliochilus Rchb.f. = Appendicula Blume ORCHIDACEAE (M)
Scoliopus Torr. TRILLIACEAE (M)
Scoliosorus T.Moore = Antrophyum Kaulf. VITTARIACEAE (F)
Scoliotheca Baill. = Monopyle Moritz ex Benth. GESNERIACEAE (D)
Scolochloa Mert. & W.D.J.Koch (SUS) = Arundo L. GRAMINEAE (M)
Scolochloa Link GRAMINEAE (M)
Scolopendrium Adans. = Asplenium L. ASPLENIACEAE (F)
Scolophyllum Yamaz. = Lindernia All. SCROPHULARIACEAE (D)
Scolopia Schreb. FLACOURTIACEAE (D)
Scolosanthus Vahl RUBIACEAE (D)
Scolymus L. COMPOSITAE (D)

Scoparebutia Frič & Kreuz. ex Buining = Echinopsis Zucc. CACTACEAE (D)
Scoparia L. SCROPHULARIACEAE (D)
Scopelogena L.Bolus AIZOACEAE (D)
Scopolia Jacq. SOLANACEAE (D)
Scopularia Lindl. = Holothrix Rich. ex Lindl. ORCHIDACEAE (M)
Scopulophila M.E.Jones ILLECEBRACEAE (D)
Scorodocarpus Becc. OLACACEAE (D)
Scorodophloeus Harms LEGUMINOSAE–CAESALPINIOIDEAE (D)
Scorodosma Bunge = Ferula L. UMBELLIFERAE (D)
Scorpiothyrsus H.L.Li MELASTOMATACEAE (D)
Scorpiurus L. LEGUMINOSAE–PAPILIONOIDEAE (D)
Scortechinia Hook.f. (SUH) = Neoscortechinia Pax EUPHORBIACEAE (D)
Scorzonella Nutt. COMPOSITAE (D)
Scorzonera L. COMPOSITAE (D)
Scotanthus Naudin = Gymnopetalum Arn. CUCURBITACEAE (D)
Scottellia Oliv. FLACOURTIACEAE (D)
Scottia Thunb. = Schotia Jacq. LEGUMINOSAE–CAESALPINIOIDEAE (D)
Scribneria Hack. GRAMINEAE (M)
Scrithacola Alava (UP) UMBELLIFERAE (D)
Scrobicaria Cass. COMPOSITAE (D)
Scrobicularia Mansf. = Poikilogyne Baker f. MELASTOMATACEAE (D)
Scrofella Maxim. SCROPHULARIACEAE (D)
Scrophularia L. SCROPHULARIACEAE (D)
Scrotochloa Judz. = Leptaspis R.Br. GRAMINEAE (M)
Scubulus Raf. = Solanum L. SOLANACEAE (D)
Scurrula L. LORANTHACEAE (D)
Scutachne Hitchc. & Chase GRAMINEAE (M)
Scutellaria L. LABIATAE (D)
Scutia (DC.) Brongn. RHAMNACEAE (D)
Scuticaria Lindl. ORCHIDACEAE (M)
Scutinanthe Thwaites BURSERACEAE (D)
Scutula Lour. = Memecylon L. MELASTOMATACEAE (D)
Scybalium Schott & Endl. BALANOPHORACEAE (D)
Scyopteris C.Presl = Pyrrosia Mirb. POLYPODIACEAE (F)
Scyphanthus Sweet LOASACEAE (D)
Scyphellandra Thwaites = Rinorea Aubl. VIOLACEAE (D)
Scyphiphora C.F.Gaertn. RUBIACEAE (D)
Scyphocephalium Warb. MYRISTICACEAE (D)
Scyphochlamys Balf.f. RUBIACEAE (D)
Scyphocoronis A.Gray COMPOSITAE (D)
Scyphofilix Thouars = Microlepia C.Presl DENNSTAEDTIACEAE (F)
Scyphogyne Decne. ERICACEAE (D)
Scypholepia J.Sm. = Microlepia C.Presl DENNSTAEDTIACEAE (F)
Scyphonychium Radlk. SAPINDACEAE (D)
Scyphopappus B.Nord. = Argyranthemum Webb COMPOSITAE (D)
Scyphopetalum Hiern = Paranephelium Miq. SAPINDACEAE (D)
Scyphopteris Raf. = Microlepia C.Presl DENNSTAEDTIACEAE (F)
Scyphostachys Thwaites RUBIACEAE (D)
Scyphostegia Stapf SCYPHOSTEGIACEAE (D)
Scyphostelma Baill. ASCLEPIADACEAE (D)
Scyphostigma M.Roem. = Turraea L. MELIACEAE (D)
Scyphostrychnos S.Moore = Strychnos L. LOGANIACEAE (D)
Scyphosyce Baill. MORACEAE (D)
Scyphularia Fée DAVALLIACEAE (F)
Scytanthus T.Anderson ex Benth. (SUH) = Thomandersia Baill. ACANTHACEAE (D)

Scytopetalum Pierre ex Engl. SCYTOPETALACEAE (D)
Seaforthia R.Br. = Ptychosperma Labill. PALMAE (M)
Searsia F.A.Barkley = Rhus L. ANACARDIACEAE (D)
Sebaea Sol. ex R.Br. GENTIANACEAE (D)
Sebastiania Spreng. EUPHORBIACEAE (D)
Sebastiano–Schaueria Nees ACANTHACEAE (D)
Sebertia Pierre = Niemeyera F.Muell. SAPOTACEAE (D)
Seborium Raf. = Sapium P.Browne EUPHORBIACEAE (D)
Secale L. GRAMINEAE (M)
Secalidium Schur (SUI) = Dasypyrum (Coss. & Durieu) T.Durand GRAMINEAE (M)
Secamone R.Br. ASCLEPIADACEAE (D)
Secamonopsis Jum. ASCLEPIADACEAE (D)
Sechiopsis Naudin CUCURBITACEAE (D)
Sechium P.Browne CUCURBITACEAE (D)
Secondatia A.DC. APOCYNACEAE (D)
Secula Small = Aeschynomene L. LEGUMINOSAE–PAPILIONOIDEAE (D)
Securidaca L. POLYGALACEAE (D)
Securigera DC. = Coronilla L. LEGUMINOSAE–PAPILIONOIDEAE (D)
Securinega Comm. ex Juss. EUPHORBIACEAE (D)
Sedastrum Rose = Sedum L. CRASSULACEAE (D)
Seddera Hochst. CONVOLVULACEAE (D)
Sedderopsis Roberty = Seddera Hochst. CONVOLVULACEAE (D)
Sedella Britton & Rose CRASSULACEAE (D)
Sedella Fourr. (SUI) = Sedum L. CRASSULACEAE (D)
Sedgwickia Griff. = Altingia Noronha HAMAMELIDACEAE (D)
Sedirea Garay & H.R.Sweet ORCHIDACEAE (M)
Sedopsis (Engl.) Exell & Mendonça = Portulaca L. PORTULACACEAE (D)
Sedum L. CRASSULACEAE (D)
Seemannantha Alef. = Tephrosia Pers. LEGUMINOSAE–PAPILIONOIDEAE (D)
Seemannaralia R.Vig. ARALIACEAE (D)
Seemannia Regel (SUH) = Gloxinia L'Hér. GESNERIACEAE (D)
Seemannia Hook. (SUH) = Pentagonia Benth. RUBIACEAE (D)
Seetzenia R.Br. ex Decne. ZYGOPHYLLACEAE (D)
Seguiera Rchb. ex Oliv. = Combretum Loefl. COMBRETACEAE (D)
Seguieria Loefl. PHYTOLACCACEAE (D)
Sehima Forssk. GRAMINEAE (M)
Seidelia Baill. EUPHORBIACEAE (D)
Seidenfadenia Garay ORCHIDACEAE (M)
Seidlitzia Bunge ex Boiss. CHENOPODIACEAE (D)
Selaginella P.Beauv. SELAGINELLACEAE (FA)
Selaginoides Séguier = Selaginella P.Beauv. SELAGINELLACEAE (FA)
Selago L. SCROPHULARIACEAE (D)
Selago P.Browne (SUH) = Selaginella P.Beauv. SELAGINELLACEAE (FA)
Selago Schur (SUH) = Huperzia Bernh. LYCOPODIACEAE (FA)
Selago Hill (SUH) = Huperzia Bernh. LYCOPODIACEAE (FA)
Selatium G.Don = Gentianella Moench GENTIANACEAE (D)
Selbya M.Roem. – Aglaia Lour. MELIACEAE (D)
Selenia Nutt. CRUCIFERAE (D)
Selenicereus (A.Berger) Britton & Rose CACTACEAE (D)
Selenipedium Rchb.f. ORCHIDACEAE (M)
Selenodesmium (Prantl) Copel. = Cephalomanes C.Presl HYMENOPHYLLACEAE (F)
Selenothamnus Melville = Lawrencia Hook. MALVACEAE (D)
Selera Ulbr. = Gossypium L. MALVACEAE (D)
Selinocarpus A.Gray NYCTAGINACEAE (D)
Selinopsis Coss. & Durieu ex Batt. & Trab. UMBELLIFERAE (D)

Selinum L. UMBELLIFERAE (D)
Selkirkia Hemsl. BORAGINACEAE (D)
Selleola Urb. = Minuartia L. CARYOPHYLLACEAE (D)
Selleophytum Urb. COMPOSITAE (D)
Selliera Cav. GOODENIACEAE (D)
Selliguea Bory POLYPODIACEAE (F)
Selloa Kunth COMPOSITAE (D)
Sellocharis Taub. LEGUMINOSAE–PAPILIONOIDEAE (D)
Selmation T.Durand = Metastelma R.Br. ASCLEPIADACEAE (D)
Selysia Cogn. CUCURBITACEAE (D)
Sematanthera Pierre ex Harms = Efulensia C.H.Wright PASSIFLORACEAE (D)
Semecarpus L.f. ANACARDIACEAE (D)
Semeiandra Hook. & Arn. = Lopezia Cav. ONAGRACEAE (D)
Semeiocardium Zoll. = Impatiens L. BALSAMINACEAE (D)
Semeiostachys Drobov = Elymus L. GRAMINEAE (M)
Semele Kunth RUSCACEAE (M)
Semenovia Regel & Herder UMBELLIFERAE (D)
Semetum Raf. = Lepidium L. CRUCIFERAE (D)
Semialarium N.Hallé CELASTRACEAE (D)
Semiaquilegia Makino RANUNCULACEAE (D)
Semiarundinaria Nakai GRAMINEAE (M)
Semibegoniella C.E.C.Fisch. = Begonia L. BEGONIACEAE (D)
Semicipium Pierre = Mimusops L. SAPOTACEAE (D)
Semiliquidambar H.T.Chang = Liquidambar L. HAMAMELIDACEAE (D)
Semilta Raf. = Croton L. EUPHORBIACEAE (D)
Semiphajus Gagnep. = Eulophia R.Br. ex Lindl. ORCHIDACEAE (M)
Semiramisia Klotzsch ERICACEAE (D)
Semnanthe N.E.Br. = Erepsia N.E.Br. AIZOACEAE (D)
Semnostachya Bremek. = Strobilanthes Blume ACANTHACEAE (D)
Semnothyrsus Bremek. = Strobilanthes Blume ACANTHACEAE (D)
Semonvillea J.Gay = Limeum L. MOLLUGINACEAE (D)
Sempervivella Stapf = Rosularia (DC.) Stapf CRASSULACEAE (D)
Sempervivum L. CRASSULACEAE (D)
Senaea Taub. GENTIANACEAE (D)
Senckenbergia Post & Kuntze = Cyphomeris Standl. NYCTAGINACEAE (D)
Senecio L. COMPOSITAE (D)
Senecioides Post & Kuntze (SUS) = Vernonia Schreb. COMPOSITAE (D)
Senefeldera Mart. EUPHORBIACEAE (D)
Senefelderopsis Steyerm. EUPHORBIACEAE (D)
Senegalia Raf. = Acacia Mill. LEGUMINOSAE–MIMOSOIDEAE (D)
Senisetum Honda = Agrostis L. GRAMINEAE (M)
Senites Adans. (SUS) = Zeugites P.Browne GRAMINEAE (M)
Senna Mill. LEGUMINOSAE–CAESALPINIOIDEAE (D)
Sennenia Sennen (SUI) = Trisetaria Forssk. GRAMINEAE (M)
Senniella Aellen CHENOPODIACEAE (D)
Senra Cav. MALVACEAE (D)
Sepalosaccus Schltr. ORCHIDACEAE (M)
Sepalosiphon Schltr. ORCHIDACEAE (M)
Separotheca Waterf. = Tradescantia L. COMMELINACEAE (M)
Sepikea Schltr. GESNERIACEAE (D)
Septimetula Tiegh. = Phragmanthera Tiegh. LORANTHACEAE (D)
Septogarcinia Kosterm. = Garcinia L. GUTTIFERAE (D)
Septotheca Ulbr. BOMBACACEAE (D)
Septulina Tiegh. LORANTHACEAE (D)
Sequoia Endl. TAXODIACEAE (G)

Sequoiadendron Buchholz TAXODIACEAE (G)
Seraphyta Fisch. & C.A.Mey. = Epidendrum L. ORCHIDACEAE (M)
Serapias L. ORCHIDACEAE (M)
Serapiastrum Kuntze = Serapias L. ORCHIDACEAE (M)
Serenoa Hook.f. PALMAE (M)
Seretoberlinia P.A.Duvign. = Julbernardia Pellegr. LEGUMINOSAE–
    CAESALPINIOIDEAE (D)
Sergia Fedorov CAMPANULACEAE (D)
Serialbizzia Kosterm. = Albizia Durazz. LEGUMINOSAE–MIMOSOIDEAE (D)
Serianthes Benth. LEGUMINOSAE–MIMOSOIDEAE (D)
Sericanthe Robbr. RUBIACEAE (D)
Sericeocassia Britton = Senna Mill. LEGUMINOSAE–CAESALPINIOIDEAE (D)
Sericocactus Y.Ito = Parodia Speg. CACTACEAE (D)
Sericocalyx Bremek. = Strobilanthes Blume ACANTHACEAE (D)
Sericocarpus Nees COMPOSITAE (D)
Sericocoma Fenzl AMARANTHACEAE (D)
Sericocomopsis Schinz AMARANTHACEAE (D)
Sericodes A.Gray ZYGOPHYLLACEAE (D)
Sericographis Nees = Justicia L. ACANTHACEAE (D)
Sericolea Schltr. ELAEOCARPACEAE (D)
Sericorema (Hook.f.) Lopr. AMARANTHACEAE (D)
Sericostachys Gilg & Lopr. AMARANTHACEAE (D)
Sericostoma Stocks BORAGINACEAE (D)
Sericotheca Raf. = Holodiscus (K.Koch) Maxim. ROSACEAE (D)
Sericrostis Raf. = Muhlenbergia Schreb. GRAMINEAE (M)
Sericura Hassk. = Pennisetum Rich. GRAMINEAE (M)
Serigrostis Steud. (SUO) = Muhlenbergia Schreb. GRAMINEAE (M)
Seringia J.Gay STERCULIACEAE (D)
Seriola L. = Hypochaeris L. COMPOSITAE (D)
Seriphidium (Bess.) Poljak. COMPOSITAE (D)
Serissa Comm. ex Juss. RUBIACEAE (D)
Serjania Mill. SAPINDACEAE (D)
Serophyton Benth. = Argythamnia P.Browne EUPHORBIACEAE (D)
Serpyllopsis Van den Bosch HYMENOPHYLLACEAE (F)
Serrafalcus Parl. = Bromus L. GRAMINEAE (M)
Serrastylis Rolfe = Macradenia R.Br. ORCHIDACEAE (M)
Serratula L. COMPOSITAE (D)
Serruria Salisb. PROTEACEAE (D)
Sersalisia R.Br. = Pouteria Aubl. SAPOTACEAE (D)
Sertifera Lindl. & Rchb.f. = Elleanthus C.Presl ORCHIDACEAE (M)
Sesamoides Ortega RESEDACEAE (D)
Sesamothamnus Welw. PEDALIACEAE (D)
Sesamum L. PEDALIACEAE (D)
Sesban Adans. = Sesbania Scop. LEGUMINOSAE–PAPILIONOIDEAE (D)
Sesbania Scop. LEGUMINOSAE–PAPILIONOIDEAE (D)
Seseli L. UMBELLIFERAE (D)
Seselopsis Schischk. UMBELLIFERAE (D)
Seshagiria Ansari & Hemadri ASCLEPIADACEAE (D)
Sesleria Scop. GRAMINEAE (M)
Sesleriella Deyl = Sesleria Scop. GRAMINEAE (M)
Sessea Ruíz & Pav. SOLANACEAE (D)
Sesseopsis Hassl. SOLANACEAE (D)
Sessilanthera Molseed & Cruden IRIDACEAE (M)
Sessilibulbum Brieger (SUI) = Scaphyglottis Poepp. & Endl. ORCHIDACEAE (M)
Sessilistigma Goldblatt IRIDACEAE (M)

Sestochilos Breda = Bulbophyllum Thouars ORCHIDACEAE (M)
Sesuvium L. AIZOACEAE (D)
Setaria P.Beauv. GRAMINEAE (M)
Setariopsis Scribn. ex Millsp. GRAMINEAE (M)
Setchellanthus Brandegee CAPPARACEAE (D)
Setcreasea K.Schum. & Sidow = Tradescantia L. COMMELINACEAE (M)
Setiacis S.L.Chen & Jin = Panicum L. GRAMINEAE (M)
Seticereus Backeb. = Cleistocactus Lem. CACTACEAE (D)
Seticleistocactus Backeb. = Cleistocactus Lem. CACTACEAE (D)
Setiechinopsis (Backeb.) de Haas = Echinopsis Zucc. CACTACEAE (D)
Setilobus Baill. BIGNONIACEAE (D)
Setirebutia Frič & Kreuz. (SUI) = Rebutia K.Schum. CACTACEAE (D)
Setiscapella Barnhart = Utricularia L. LENTIBULARIACEAE (D)
Setosa Ewart = Chamaeraphis R.Br. GRAMINEAE (M)
Sevada Moq. CHENOPODIACEAE (D)
Severinia Ten. RUTACEAE (D)
Sewerzowia Regel & Schmalh. = Astragalus L. LEGUMINOSAE–PAPILIONOIDEAE (D)
Seychellaria Hemsl. TRIURIDACEAE (M)
Seymeria Pursh SCROPHULARIACEAE (D)
Seymeriopsis Tzvelev = Seymeria Pursh SCROPHULARIACEAE (D)
Seyrigia Keraudren CUCURBITACEAE (D)
Shafera Greenm. COMPOSITAE (D)
Shaferocharis Urb. RUBIACEAE (D)
Shaferodendron Gilly = Manilkara Adans. SAPOTACEAE (D)
Shantzia Lewton = Azanza Alef. MALVACEAE (D)
Sheadendron G.Bertol = Combretum Loefl. COMBRETACEAE (D)
Sheareria S.Moore COMPOSITAE (D)
Sheilanthera I.Williams RUTACEAE (D)
Shepherdia Nutt. ELAEAGNACEAE (D)
Sherardia L. RUBIACEAE (D)
Sherbournia G.Don RUBIACEAE (D)
Shibataea Makino ex Nakai GRAMINEAE (M)
Shibateranthis Nakai = Eranthis Salisb. RANUNCULACEAE (D)
Shinnersia R.M.King & H.Rob. COMPOSITAE (D)
Shinnersoseris Tomb COMPOSITAE (D)
Shirakia Hurus. = Sapium P.Browne EUPHORBIACEAE (D)
Shiuyinghua Paclt SCROPHULARIACEAE (D)
Shorea Roxb. ex C.F.Gaertn. DIPTEROCARPACEAE (D)
Shortia Raf. (SUH) = Arabis L. CRUCIFERAE (D)
Shortia Torr. & A.Gray DIAPENSIACEAE (D)
Shoshonea Evert & Constance UMBELLIFERAE (D)
Shuteria Choisy (SUH) = Hewittia Wight & Arn. CONVOLVULACEAE (D)
Shuteria Wight & Arn. LEGUMINOSAE–PAPILIONOIDEAE (D)
Siagonanthus Poepp. & Endl. = Maxillaria Ruíz & Pav. ORCHIDACEAE (M)
Siamosia K.Larsen & Pedersen AMARANTHACEAE (D)
Sibangea Oliv. EUPHORBIACEAE (D)
Sibara Greene CRUCIFERAE (D)
Sibbaldia L. ROSACEAE (D)
Sibbaldianthe Juz. = Sibbaldia L. ROSACEAE (D)
Sibbaldiopsis Rydb. ROSACEAE (D)
Sibertia Steud. (SUO) = Bromus L. GRAMINEAE (M)
Sibiraea Maxim. ROSACEAE (D)
Sibthorpia L. SCROPHULARIACEAE (D)
Sicana Naudin CUCURBITACEAE (D)
Siccobaccatus P.J.Braun & E.Est.Pereira = Micranthocereus Backeb. CACTACEAE (D)

Sicelium P.Browne = Coccocypselum P.Browne RUBIACEAE (D)
Sickingia Willd. = Simira Aubl. RUBIACEAE (D)
Sicklera Sendtn. = Brachistus Miers SOLANACEAE (D)
Sickmannia Nees = Ficinia Schrad. CYPERACEAE (M)
Sicrea (Pierre) Hallier f. TILIACEAE (D)
Sicydium Schltdl. CUCURBITACEAE (D)
Sicyocarya (A.Gray) St.John = Sicyos L. CUCURBITACEAE (D)
Sicyocaulis Wiggins = Sicyos L. CUCURBITACEAE (D)
Sicyoides Mill. = Sicyos L. CUCURBITACEAE (D)
Sicyos L. CUCURBITACEAE (D)
Sicyosperma A.Gray CUCURBITACEAE (D)
Sicyus Clements (SUO) = Sicyos L. CUCURBITACEAE (D)
Sida L. MALVACEAE (D)
Sidalcea A.Gray MALVACEAE (D)
Sidastrum Baker f. = Sida L. MALVACEAE (D)
Siderasis Raf. COMMELINACEAE (M)
Sideria Ewart & A.H.K.Petrie = Melhania Forssk. STERCULIACEAE (D)
Sideritis L. LABIATAE (D)
Siderobombyx Bremek. RUBIACEAE (D)
Siderocarpos Small = Acacia Mill. LEGUMINOSAE–MIMOSOIDEAE (D)
Siderocarpus Pierre = Pouteria Aubl. SAPOTACEAE (D)
Siderodendrum Schreb. = Ixora L. RUBIACEAE (D)
Sideropogon Pichon = Arrabidaea DC. BIGNONIACEAE (D)
Sideroxyloides Jacq. = Ixora L. RUBIACEAE (D)
Sideroxylon L. SAPOTACEAE (D)
Sidopsis Rydb. = Malvastrum A.Gray MALVACEAE (D)
Siebera J.Gay COMPOSITAE (D)
Sieberia Spreng. = Platanthera Rich. ORCHIDACEAE (M)
Siegfriedia C.A.Gardner RHAMNACEAE (D)
Sieglingia Bernh. = Danthonia DC. GRAMINEAE (M)
Siemensia Urb. RUBIACEAE (D)
Sievekingia Rchb.f. ORCHIDACEAE (M)
Sieversia Willd. ROSACEAE (D)
Sigesbeckia L. COMPOSITAE (D)
Sigmatanthus Huber ex Emmerich = Raputia Aubl. RUTACEAE (D)
Sigmatochilus Rolfe = Chelonistele Pfitzer ORCHIDACEAE (M)
Sigmatogyne Pfitzer = Panisea (Lindl.) Lindl. ORCHIDACEAE (M)
Sigmatosiphon Engl. = Sesamothamnus Welw. PEDALIACEAE (D)
Sigmatostalix Rchb.f. ORCHIDACEAE (M)
Silaum Mill. UMBELLIFERAE (D)
Silaus Bernh. = Silaum Mill. UMBELLIFERAE (D)
Silene L. CARYOPHYLLACEAE (D)
Silentvalleya V.J.Nair, Sreek., Vajr. & Barghavan GRAMINEAE (M)
Siler Mill. = Laserpitium L. UMBELLIFERAE (D)
Silicularia Compton CRUCIFERAE (D)
Siliquamomum Baill. ZINGIBERACEAE (M)
Siloxerus Labill. COMPOSITAE (D)
Silphium L. COMPOSITAE (D)
Silvaea Phil. PORTULACACEAE (D)
Silvaea Hook. & Arn. (SUH) = Trigonostemon Blume EUPHORBIACEAE (D)
Silvia Benth. (SUH) = Silviella Pennell SCROPHULARIACEAE (D)
Silvia Allemão (SUH) = Mezilaurus Kuntze ex Taub. LAURACEAE (D)
Silvianthus Hook.f. CARLEMANNIACEAE (D)
Silviella Pennell SCROPHULARIACEAE (D)
Silvorchis J.J.Sm. ORCHIDACEAE (M)

Silybum Adans. COMPOSITAE (D)
Simaba Aubl. SIMAROUBACEAE (D)
Simarouba Aubl. SIMAROUBACEAE (D)
Simenia Szabo = Dipsacus L. DIPSACACEAE (D)
Simethis Kunth ASPHODELACEAE (M)
Simicratea N.Hallé CELASTRACEAE (D)
Similisinocarum Cauwet & Farille = Pimpinella L. UMBELLIFERAE (D)
Simira Aubl. RUBIACEAE (D)
Simirestis N.Hallé CELASTRACEAE (D)
Simmondsia Nutt. SIMMONDSIACEAE (D)
Simocheilus Klotzsch ERICACEAE (D)
Simonisia Nees = Justicia L. ACANTHACEAE (D)
Simplicia T.Kirk GRAMINEAE (M)
Simpsonia O.F.Cook = Thrinax Sw. PALMAE (M)
Simsia Pers. COMPOSITAE (D)
Sinacalia H.Rob. & Brettell COMPOSITAE (D)
Sinadoxa C.Y.Wu, Z.L.Wu & R.F.Huang ADOXACEAE (D)
Sinapidendron Lowe CRUCIFERAE (D)
Sinapis L. CRUCIFERAE (D)
Sinarundinaria Nakai GRAMINEAE (M)
Sinclairia Hook. & Arn. COMPOSITAE (D)
Sincoraea Ule = Orthophytum Beer BROMELIACEAE (M)
Sindechites Oliv. APOCYNACEAE (D)
Sindora Miq. LEGUMINOSAE–CAESALPINIOIDEAE (D)
Sindoropsis J.Léonard LEGUMINOSAE–CAESALPINIOIDEAE (D)
Sindroa Jum. = Orania Zipp. PALMAE (M)
Sineoperculum Jaarsveld = Dorotheanthus Schwantes AIZOACEAE (D)
Sinephropteris Mickel = Asplenium L. ASPLENIACEAE (F)
Sinia Diels OCHNACEAE (D)
Sinningia Nees GESNERIACEAE (D)
Sinoadina Ridsdale RUBIACEAE (D)
Sinoarundinaria Ohwi (SUS) = Phyllostachys Siebold & Zucc. GRAMINEAE (M)
Sinobacopa D.Y.Hong (UP) SCROPHULARIACEAE (D)
Sinobambusa Makino GRAMINEAE (M)
Sinoboea Chun = Ornithoboea Parish ex C.B.Clarke GESNERIACEAE (D)
Sinocalamus McClure = Dendrocalamus Nees GRAMINEAE (M)
Sinocalycanthus (C.C.Cheng & S.Y.Chang) C.C.Cheng & S.Y.Chang
    CALYCANTHACEAE (D)
Sinocarum H.Wolff UMBELLIFERAE (D)
Sinochasea Keng = Pseudodanthonia Bor & C.E.Hubb. GRAMINEAE (M)
Sinocrassula A.Berger CRASSULACEAE (D)
Sinodielsia H.Wolff UMBELLIFERAE (D)
Sinodolichos Verdc. LEGUMINOSAE–PAPILIONOIDEAE (D)
Sinofranchetia (Diels) Hemsl. LARDIZABALACEAE (D)
Sinoga S.T.Blake = Asteromyrtus Schauer MYRTACEAE (D)
Sinojackia Hu STYRACACEAE (D)
Sinojohnstonia Hu BORAGINACEAE (D)
Sinoleontopodium Y.L.Chen COMPOSITAE (D)
Sinolimprichtia H.Wolff UMBELLIFERAE (D)
Sinomenium Diels MENISPERMACEAE (D)
Sinomerrillia Hu = Neuropeltis Wall. CONVOLVULACEAE (D)
Sinopanax H.L.Li ARALIACEAE (D)
Sinopimelodendron Tsiang = Cleidiocarpon Airy Shaw EUPHORBIACEAE (D)
Sinoplagiospermum Rauschert = Prinsepia Royle ROSACEAE (D)
Sinopodophyllum Ying BERBERIDACEAE (D)

Sinopteris C.Chr. & Ching ADIANTACEAE (F)
Sinoradlkofera F.G.Mey. = Boniodendron Gagnep. SAPINDACEAE (D)
Sinorchis S.C.Chen ORCHIDACEAE (M)
Sinosassafras H.W.Li LAURACEAE (D)
Sinosenecio B.Nord. COMPOSITAE (D)
Sinosideroxylon (Engl.) Aubrév. = Sideroxylon L. SAPOTACEAE (D)
Sinowilsonia Hemsl. HAMAMELIDACEAE (D)
Sinthroblastes Bremek. = Strobilanthes Blume ACANTHACEAE (D)
Siolmatra Baill. CUCURBITACEAE (D)
Sipanea Aubl. RUBIACEAE (D)
Sipaneopsis Steyerm. RUBIACEAE (D)
Sipapoa Maguire = Diacidia Griseb. MALPIGHIACEAE (D)
Sipapoantha Maguire & B.M.Boom GENTIANACEAE (D)
Siparuna Aubl. MONIMIACEAE (D)
Siphanthera Pohl MELASTOMATACEAE (D)
Siphantheropsis Brade = Macairea DC. MELASTOMATACEAE (D)
Siphaulax Raf. = Nicotiana L. SOLANACEAE (D)
Siphoboea Baill. = Clerodendrum L. VERBENACEAE (D)
Siphocampylus Pohl CAMPANULACEAE (D)
Siphocodon Turcz. CAMPANULACEAE (D)
Siphocolea Baill. = Stereospermum Cham. BIGNONIACEAE (D)
Siphocranion Kudo = Hanceola Kudo LABIATAE (D)
Siphokentia Burret PALMAE (M)
Siphomeris Bojer = Paederia L. RUBIACEAE (D)
Siphonandra Turcz. (SUH) = Chiococca P.Browne RUBIACEAE (D)
Siphonandra Klotzsch ERICACEAE (D)
Siphonandrium K.Schum. RUBIACEAE (D)
Siphonanthus Schreb. ex Baill. = Hevea Aubl. EUPHORBIACEAE (D)
Siphonella (A.Gray) Heller (SUH) = Linanthus Benth. POLEMONIACEAE (D)
Siphonella Small = Fedia Gaertn. VALERIANACEAE (D)
Siphonema Raf. = Nierembergia Ruíz & Pav. SOLANACEAE (D)
Siphoneugena O.Berg MYRTACEAE (D)
Siphonia Rich. (SUS) = Hevea Aubl. EUPHORBIACEAE (D)
Siphonia Benth. (SUH) = Lindenia Benth. RUBIACEAE (D)
Siphonidium J.B.Armstr. = Euphrasia L. SCROPHULARIACEAE (D)
Siphonochilus J.M.Wood & Franks ZINGIBERACEAE (M)
Siphonodon Griff. CELASTRACEAE (D)
Siphonoglossa Oerst. ACANTHACEAE (D)
Siphonosmanthus Stapf = Osmanthus Lour. OLEACEAE (D)
Siphonostegia Benth. SCROPHULARIACEAE (D)
Siphonostelma Schltr. ASCLEPIADACEAE (D)
Siphonostema Griseb. = Ceratostema Juss. ERICACEAE (D)
Siphonostylis Wern.Schulze = Iris L. IRIDACEAE (M)
Siphonychia Torr. & A.Gray ILLECEBRACEAE (D)
Sipolisia Glaz. ex Oliv. COMPOSITAE (D)
Siraitia Merr. CUCURBITACEAE (D)
Sirhookera Kuntze ORCHIDACEAE (M)
Sison L. UMBELLIFERAE (D)
Sisymbrella Spach CRUCIFERAE (D)
Sisymbrianthus Chevall. = Rorippa Scop. CRUCIFERAE (D)
Sisymbriopsis Botsch. & Tzvelev CRUCIFERAE (D)
Sisymbrium L. CRUCIFERAE (D)
Sisyndite E.Mey. ex Sond. ZYGOPHYLLACEAE (D)
Sisyranthus E.Mey. ASCLEPIADACEAE (D)
Sisyrinchium L. IRIDACEAE (M)

Sisyrolepis Radlk. SAPINDACEAE (D)
Sitanion Raf. GRAMINEAE (M)
Sitella L.H.Bailey = Waltheria L. STERCULIACEAE (D)
Sitilias Raf. = Pyrrhopappus DC. COMPOSITAE (D)
Sitodium Parkinson = Artocarpus J.R.Forst. & G.Forst. MORACEAE (D)
Sitolobium Desv. = Dennstaedtia Bernh. DENNSTAEDTIACEAE (F)
Sitopsis (Jaub. & Spach) Á.Löve = Aegilops L. GRAMINEAE (M)
Sitospelos Adans. (SUS) = Elymus L. GRAMINEAE (M)
Sium L. UMBELLIFERAE (D)
Sixalix Raf. = Scabiosa L. DIPSACACEAE (D)
Skapanthus C.Y.Wu = Plectranthus L'Hér. LABIATAE (D)
Skeptrostachys Garay = Stenorrhynchos Rich. ex Spreng. ORCHIDACEAE (M)
Skiatophytum L.Bolus AIZOACEAE (D)
Skimmia Thunb. RUTACEAE (D)
Skinnera J.R.Forst. & G.Forst. = Fuchsia L. ONAGRACEAE (D)
Skinneria Choisy (SUH) = Merremia Dennst. ex Endl. CONVOLVULACEAE (D)
Skiophila Hanst. = Nautilocalyx Linden ex Hanst. GESNERIACEAE (D)
Skoliopterys Cuatrec. = Clonodia Griseb. MALPIGHIACEAE (D)
Skoliostigma Lauterb. = Spondias L. ANACARDIACEAE (D)
Skottsbergianthus Boelcke CRUCIFERAE (D)
Skottsbergiella Boelcke (SUH) = Skottsbergianthus Boelcke CRUCIFERAE (D)
Skottsbergiella Epling (SUH) = Cuminia Colla LABIATAE (D)
Skottsbergiliana St.John = Sicyos L. CUCURBITACEAE (D)
Skutchia Pax & K.Hoffm. ex C.Morton = Trophis P.Browne MORACEAE (D)
Skytanthus Meyen APOCYNACEAE (D)
Slackia Griff. (SUH) = Beccarinda Kuntze GESNERIACEAE (D)
Slackia Griff. = Iguanura Blume PALMAE (M)
Sladenia Kurz THEACEAE (D)
Sleumerodendron Virot PROTEACEAE (D)
Sloanea L. ELAEOCARPACEAE (D)
Sloetia Teijsm. & Binn. ex Kurz = Streblus Lour. MORACEAE (D)
Sloetiopsis Engl. = Streblus Lour. MORACEAE (D)
Smallanthus Mack. COMPOSITAE (D)
Smallia Nieuwl. (SUS) = Pteroglossaspis Rchb.f. ORCHIDACEAE (M)
Smeathmannia Sol. ex R.Br. PASSIFLORACEAE (D)
Smelophyllum Radlk. SAPINDACEAE (D)
Smelowskia C.A.Mey. CRUCIFERAE (D)
Smicrostigma N.E.Br. AIZOACEAE (D)
Smidetia Raf. (SUS) = Coleanthus J.Seidel GRAMINEAE (M)
Smilacina Desf. = Maianthemum G.H.Weber CONVALLARIACEAE (M)
Smilax L. SMILACACEAE (M)
Smirnowia Bunge LEGUMINOSAE–PAPILIONOIDEAE (D)
Smithia Aiton LEGUMINOSAE–PAPILIONOIDEAE (D)
Smithiantha Kuntze GESNERIACEAE (D)
Smithiella Dunn (SUH) = Aboriella Bennet URTICACEAE (D)
Smithiodendron Hu = Broussonetia L'Hér. ex Vent. MORACEAE (D)
Smithorchis T.Tang & F.T.Wang ORCHIDACEAE (M)
Smithsonia C.J.Saldanha ORCHIDACEAE (M)
Smitinandia Holttum ORCHIDACEAE (M)
Smodingium E.Mey. ex Sond. ANACARDIACEAE (D)
Smyrniopsis Boiss. UMBELLIFERAE (D)
Smyrnium L. UMBELLIFERAE (D)
Smythea Seem. RHAMNACEAE (D)
Snowdenia C.E.Hubb. GRAMINEAE (M)
Soala Blanco = Cyathocalyx Champ. ex Hook.f. & Thomson ANNONACEAE (D)

Soaresia Sch.Bip. COMPOSITAE (D)
Sobennikoffia Schltr. ORCHIDACEAE (M)
Sobolewskia M.Bieb. CRUCIFERAE (D)
Sobralia Ruíz & Pav. ORCHIDACEAE (M)
Socotora Balf.f. = Periploca L. ASCLEPIADACEAE (D)
Socotranthus Kuntze ASCLEPIADACEAE (D)
Socotria Levin LYTHRACEAE (D)
Socratea Karst. PALMAE (M)
Socratesia Klotzsch = Cavendishia Lindl. ERICACEAE (D)
Socratina Balle LORANTHACEAE (D)
Soderstromia C.Morton GRAMINEAE (M)
Sodiroa E.F.Andre = Guzmania Ruíz & Pav. BROMELIACEAE (M)
Sodiroella Schltr. = Stellilabium Schltr. ORCHIDACEAE (M)
Soehrensia Backeb. = Echinopsis Zucc. CACTACEAE (D)
Soemmeringia Mart. LEGUMINOSAE–PAPILIONOIDEAE (D)
Sogerianthe Danser LORANTHACEAE (D)
Sohnsia Airy Shaw GRAMINEAE (M)
Solanastrum Fabr. = Solanum L. SOLANACEAE (D)
Solandra Sw. SOLANACEAE (D)
Solanecio (Sch.Bip.) Walp. COMPOSITAE (D)
Solanoa Greene = Asclepias L. ASCLEPIADACEAE (D)
Solanocharis Bitter = Solanum L. SOLANACEAE (D)
Solanopsis Bitter = Solanum L. SOLANACEAE (D)
Solanopteris Copel. POLYPODIACEAE (F)
Solanum L. SOLANACEAE (D)
Solaria Phil. ALLIACEAE (M)
Soldanella L. PRIMULACEAE (D)
Soleirolia Gaudich. URTICACEAE (D)
Solena Lour. CUCURBITACEAE (D)
Solena Willd. (SUH) = Posoqueria Aubl. RUBIACEAE (D)
Solenachne Steud. = Spartina Schreb. GRAMINEAE (M)
Solenandra Hook.f. = Exostema (Pers.) Humb. & Bonpl. RUBIACEAE (D)
Solenangis Schltr. ORCHIDACEAE (M)
Solenanthus Ledeb. BORAGINACEAE (D)
Solenidiopsis Senghas ORCHIDACEAE (M)
Solenidium Lindl. ORCHIDACEAE (M)
Solenixora Baill. = Coffea L. RUBIACEAE (D)
Solenocalyx Tiegh. = Psittacanthus Mart. LORANTHACEAE (D)
Solenocarpus Wight & Arn. = Spondias L. ANACARDIACEAE (D)
Solenocentrum Schltr. ORCHIDACEAE (M)
Solenochasma Fenzl = Justicia L. ACANTHACEAE (D)
Solenogyne Cass. COMPOSITAE (D)
Solenomelus Miers IRIDACEAE (M)
Solenophora Benth. GESNERIACEAE (D)
Solenophyllum Baill. (SUI) = Monanthochloe Engelm. GRAMINEAE (M)
Solenopsis C.Presl = Laurentia Adans. CAMPANULACEAE (D)
Solenoruellia Baill. = Henrya Nees ACANTHACEAE (D)
Solenospermum Zoll. = Lophopetalum Wight ex Arn. CELASTRACEAE (D)
Solenostemma Hayne ASCLEPIADACEAE (D)
Solenostemon Thonn. LABIATAE (D)
Solfia Rech. = Drymophloeus Zipp. PALMAE (M)
Solidago L. COMPOSITAE (D)
Soliera Clos = Kurzamra Kuntze LABIATAE (D)
Solisia Britton & Rose = Mammillaria Haw. CACTACEAE (D)
Soliva Ruíz & Pav. COMPOSITAE (D)

Sollya Lindl. PITTOSPORACEAE (D)
Solms–Laubachia Muschl. CRUCIFERAE (D)
Solmsia Baill. THYMELAEACEAE (D)
Solonia Urb. MYRSINACEAE (D)
Sommera Schltdl. RUBIACEAE (D)
Sommerfeltia Less. COMPOSITAE (D)
Sommieria Becc. PALMAE (M)
Somphoxylon Eichler = Odontocarya Miers MENISPERMACEAE (D)
Sonchus L. COMPOSITAE (D)
Sonderina H.Wolff UMBELLIFERAE (D)
Sonderothamnus R.Dahlgren PENAEACEAE (D)
Sondottia P.S.Short COMPOSITAE (D)
Sonerila Roxb. MELASTOMATACEAE (D)
Sonneratia L.f. LYTHRACEAE (D)
Sooia Pócs = Epiclastopelma Lindau ACANTHACEAE (D)
Sooja Siebold = Chamaecrista Moench LEGUMINOSAE–CAESALPINIOIDEAE (D)
Sophiopsis O.E.Schulz CRUCIFERAE (D)
Sophoclesia Klotzsch = Sphyrospermum Poepp. & Endl. ERICACEAE (D)
Sophora L. LEGUMINOSAE–PAPILIONOIDEAE (D)
Sophronanthe Benth. = Gratiola L. SCROPHULARIACEAE (D)
Sophronia Lindl. = Sophronitis Lindl. ORCHIDACEAE (M)
Sophronitella Schltr. ORCHIDACEAE (M)
Sophronitis Lindl. ORCHIDACEAE (M)
Sopropis Britton & Rose = Prosopis L. LEGUMINOSAE–MIMOSOIDEAE (D)
Sopubia Buch.–Ham. ex D.Don SCROPHULARIACEAE (D)
Soranthus Ledeb. = Ferula L. UMBELLIFERAE (D)
Sorbaria (Ser. ex DC.) A.Braun ROSACEAE (D)
Sorbus L. ROSACEAE (D)
Sorema Lindl. = Nolana L.f. SOLANACEAE (D)
Sorghastrum Nash GRAMINEAE (M)
Sorghum Moench GRAMINEAE (M)
Sorgum Adans. (SUS) = Holcus L. GRAMINEAE (M)
Soridium Miers TRIURIDACEAE (M)
Sorindeia Thouars ANACARDIACEAE (D)
Sorindeiopsis Engl. = Sorindeia Thouars ANACARDIACEAE (D)
Sorocea A.St.–Hil. MORACEAE (D)
Sorocephalus R.Br. PROTEACEAE (D)
Sorolepidium Christ = Polystichum Roth DRYOPTERIDACEAE (F)
Soromanes Fée = Polybotrya Humb. & Bonpl. ex Willd. DRYOPTERIDACEAE (F)
Soroseris Stebb. COMPOSITAE (D)
Sorostachys Steud. = Cyperus L. CYPERACEAE (M)
Soterosanthus Lehm. ex Jenny ORCHIDACEAE (M)
Sotor Fenzl = Kigelia DC. BIGNONIACEAE (D)
Soulamea Lam. SIMAROUBACEAE (D)
Souliea Franch. RANUNCULACEAE (D)
Souroubea Aubl. MARCGRAVIACEAE (D)
Sowerbaea Sm. ANTHERICACEAE (M)
Soyauxia Oliv. MEDUSANDRACEAE (D)
Soymida A.Juss. MELIACEAE (D)
Spachea A.Juss. MALPIGHIACEAE (D)
Spachia Lilja = Fuchsia L. ONAGRACEAE (D)
Spallanzania DC. (SUH) = Mussaenda L. RUBIACEAE (D)
Spananthe Jacq. UMBELLIFERAE (D)
Spaniopappus B.L.Rob. COMPOSITAE (D)
Sparattanthelium Mart. HERNANDIACEAE (D)

Sparattosperma Mart. ex Meisn. BIGNONIACEAE (D)
Sparattosyce Bureau MORACEAE (D)
Sparaxis Ker Gawl. IRIDACEAE (M)
Sparganium L. TYPHACEAE (M)
Sparganophorus Boehm. = Struchium P.Browne COMPOSITAE (D)
Sparrea Hunz. & Dottori = Celtis L. ULMACEAE (D)
Sparrmannia L.f. TILIACEAE (D)
Sparteum P.Beauv. (SUI) = Stipa L. GRAMINEAE (M)
Spartidium Pomel LEGUMINOSAE–PAPILIONOIDEAE (D)
Spartina Schreb. GRAMINEAE (M)
Spartium L. LEGUMINOSAE–PAPILIONOIDEAE (D)
Spartochloa C.E.Hubb. GRAMINEAE (M)
Spartocytisus Webb & Berthel. = Cytisus Desf. LEGUMINOSAE–PAPILIONOIDEAE (D)
Spartothamnella Briq. VERBENACEAE (D)
Spartum P.Beauv. (SUI) = Lygeum L. GRAMINEAE (M)
Spatalla Salisb. PROTEACEAE (D)
Spatallopsis E.Phillips = Spatalla Salisb. PROTEACEAE (D)
Spathacanthus Baill. ACANTHACEAE (D)
Spathandra Guill. & Perr. MELASTOMATACEAE (D)
Spathantheum Schott ARACEAE (M)
Spathanthus Desv. RAPATEACEAE (M)
Spathelia L. RUTACEAE (D)
Spathia Ewart GRAMINEAE (M)
Spathicalyx J.C.Gomes BIGNONIACEAE (D)
Spathicarpa Hook. ARACEAE (M)
Spathichlamys R.Parker RUBIACEAE (D)
Spathidolepis Schltr. ASCLEPIADACEAE (D)
Spathiger Small = Epidendrum L. ORCHIDACEAE (M)
Spathionema Taub. LEGUMINOSAE–PAPILIONOIDEAE (D)
Spathiostemon Blume EUPHORBIACEAE (D)
Spathipappus Tzvelev = Tanacetum L. COMPOSITAE (D)
Spathiphyllum Schott ARACEAE (M)
Spathodea P.Beauv. BIGNONIACEAE (D)
Spathodeopsis Dop = Fernandoa Welw. ex Seem. BIGNONIACEAE (D)
Spathoglottis Blume ORCHIDACEAE (M)
Spatholirion Ridl. COMMELINACEAE (M)
Spatholobus Hassk. LEGUMINOSAE–PAPILIONOIDEAE (D)
Spathoscaphe Oerst. = Chamaedorea Willd. PALMAE (M)
Spathulata (Borisova) Á.Löve & D.Löve = Sedum L. CRASSULACEAE (D)
Spathulopetalum Chiov. = Caralluma R.Br. ASCLEPIADACEAE (D)
Specklinia Lindl. = Pleurothallis R.Br. ORCHIDACEAE (M)
Specularia A.DC. = Legousia Durande CAMPANULACEAE (D)
Speea Loes. ALLIACEAE (M)
Spegazzinia Backeb. (SUH) = Rebutia K.Schum. CACTACEAE (D)
Speirantha Baker CONVALLARIACEAE (M)
Speirema Hook.f. & Thomson = Pratia Gaudich. CAMPANULACEAE (D)
Spelta Wolf = Triticum L. GRAMINEAE (M)
Spenceria Trimen ROSACEAE (D)
Spennera Mart. ex DC. = Aciotis D.Don MELASTOMATACEAE (D)
Speranskia Baill. EUPHORBIACEAE (D)
Spergula L. CARYOPHYLLACEAE (D)
Spergularia (Pers.) J.Presl & C.Presl CARYOPHYLLACEAE (D)
Spermachiton Llanos = Sporobolus R.Br. GRAMINEAE (M)
Spermacoce L. RUBIACEAE (D)
Spermacoceodes Kuntze = Spermacoce L. RUBIACEAE (D)

Spermacon Raf. = Spermacoce L. RUBIACEAE (D)
Spermadictyon Roxb. RUBIACEAE (D)
Spermatochiton Pilg. (SUO) = Sporobolus R.Br. GRAMINEAE (M)
Spermaxyrum Labill. = Olax L. OLACACEAE (D)
Spermolepis Raf. UMBELLIFERAE (D)
Spermolepis Brongn. & Gris (SUH) = Arillastrum Panch. ex Baill. MYRTACEAE (D)
Sphacanthus Benoist ACANTHACEAE (D)
Sphacele Benth. = Lepechinia Willd. LABIATAE (D)
Sphacophyllum Benth. = Anisopappus Hook. & Arn. COMPOSITAE (D)
Sphaenolobium Pimenov UMBELLIFERAE (D)
Sphaeradenia Harling CYCLANTHACEAE (M)
Sphaeralcea A.St.–Hil. MALVACEAE (D)
Sphaeranthus L. COMPOSITAE (D)
Sphaerantia Peter G.Wilson & B.Hyland MYRTACEAE (D)
Sphaerella Bubani (SUS) = Airopsis Desv. GRAMINEAE (M)
Sphaereupatorium (O.Hoffm.) Kuntze ex B.L.Rob. COMPOSITAE (D)
Sphaerium Kuntze (SUS) = Coix L. GRAMINEAE (M)
Sphaerobambos S.Dransf. GRAMINEAE (M)
Sphaerocardamum Nees & Schauer CRUCIFERAE (D)
Sphaerocaryum Nees ex Hook.f. GRAMINEAE (M)
Sphaerocionium C.Presl HYMENOPHYLLACEAE (F)
Sphaeroclinium (DC.) Sch.Bip. COMPOSITAE (D)
Sphaerocodon Benth. ASCLEPIADACEAE (D)
Sphaerocoma T.Anderson ILLECEBRACEAE (D)
Sphaerocoryne (Boerl.) Scheff. ex Ridl. ANNONACEAE (D)
Sphaerocyperus Lye CYPERACEAE (M)
Sphaerodendron Seem. = Cussonia Thunb. ARALIACEAE (D)
Sphaerodiscus Nakai = Euonymus L. CELASTRACEAE (D)
Sphaerogyne Naudin = Tococa Aubl. MELASTOMATACEAE (D)
Sphaerolobium Sm. LEGUMINOSAE–PAPILIONOIDEAE (D)
Sphaeromariscus E.G.Camus = Cyperus L. CYPERACEAE (M)
Sphaeromeria Nutt. COMPOSITAE (D)
Sphaeromorphaea DC. COMPOSITAE (D)
Sphaerophora Blume = Morinda L. RUBIACEAE (D)
Sphaerophysa DC. LEGUMINOSAE–PAPILIONOIDEAE (D)
Sphaeropteris Bernh. = Cyathea Sm. CYATHEACEAE (F)
Sphaeropteris R.Br. ex Wall. (SUH) = Peranema D.Don DRYOPTERIDACEAE (F)
Sphaerosacme Wall. ex M.Roem. MELIACEAE (D)
Sphaerosciadium Pimenov & Kljuykov UMBELLIFERAE (D)
Sphaerosepalum Baker = Rhopalocarpus Bojer SPHAEROSEPALACEAE (D)
Sphaerosicyos Hook.f. = Lagenaria Ser. CUCURBITACEAE (D)
Sphaerostephanos J.Sm. THELYPTERIDACEAE (F)
Sphaerostichum C.Presl = Pyrrosia Mirb. POLYPODIACEAE (F)
Sphaerostigma (Ser.) Fisch. & C.A.Mey. = Camissonia Link ONAGRACEAE (D)
Sphaerostylis Baill. EUPHORBIACEAF (D)
Sphaerothalamus Hook.f. = Polyalthia Blume ANNONACEAE (D)
Sphaerothylax Bisch. ex Krauss PODOSTEMACEAE (D)
Sphaerotylos C.S.Chen = Sarcochlamys Gaudich. URTICACEAE (D)
Sphagneticola O.Hoffm. COMPOSITAE (D)
Sphalanthus Jack = Quisqualis L. COMBRETACEAE (D)
Sphallerocarpus Besser ex DC. UMBELLIFERAE (D)
Sphalmanthus N.E.Br. = Phyllobolus N.E.Br. AIZOACEAE (D)
Sphalmium B.G.Briggs, B.Hyland & L.A.S.Johnson PROTEACEAE (D)
Sphedamnocarpus Planch. ex Benth. MALPIGHIACEAE (D)
Sphenandra Benth. = Sutera Roth SCROPHULARIACEAE (D)

Sphenantha Schrad. = Cucurbita L. CUCURBITACEAE (D)
Spheneria Kuhlm. GRAMINEAE (M)
Sphenista Raf. = Hirtella L. CHRYSOBALANACEAE (D)
Sphenocarpus Korovin UMBELLIFERAE (D)
Sphenocentrum Pierre MENISPERMACEAE (D)
Sphenoclea Gaertn. CAMPANULACEAE (D)
Sphenodesme Jack VERBENACEAE (D)
Sphenogyne R.Br. = Ursinia Gaertn. COMPOSITAE (D)
Sphenomeris Maxon DENNSTAEDTIACEAE (F)
Sphenopholis Scribn. GRAMINEAE (M)
Sphenopus Trin. GRAMINEAE (M)
Sphenosciadium A.Gray UMBELLIFERAE (D)
Sphenostemon Baill. SPHENOSTEMONACEAE (D)
Sphenostigma Baker IRIDACEAE (M)
Sphenostylis E.Mey. LEGUMINOSAE–PAPILIONOIDEAE (D)
Sphenotoma R.Br. ex Sweet EPACRIDACEAE (D)
Spheroidea Dulac = Marsilea L. MARSILEACEAE (F)
Sphinctacanthus Benth. ACANTHACEAE (D)
Sphinctanthus Benth. RUBIACEAE (D)
Sphinctospermum Rose LEGUMINOSAE–PAPILIONOIDEAE (D)
Sphingiphila A.H.Gentry BIGNONIACEAE (D)
Sphragidia Thwaites = Drypetes Vahl EUPHORBIACEAE (D)
Sphyranthera Hook.f. EUPHORBIACEAE (D)
Sphyrarhynchus Mansf. ORCHIDACEAE (M)
Sphyrastylis Schltr. ORCHIDACEAE (M)
Sphyrospermum Poepp. & Endl. ERICACEAE (D)
Spicanta C.Presl = Blechnum L. BLECHNACEAE (F)
Spicantopsis Nakai = Blechnum L. BLECHNACEAE (F)
Spicillaria A.Rich. = Hypobathrum Blume RUBIACEAE (D)
Spiciviscum Engelm. = Phoradendron Nutt. VISCACEAE (D)
Spiculaea Lindl. ORCHIDACEAE (M)
Spielmannia Medik. (SUS) = Oftia Adans. SCROPHULARIACEAE (D)
Spigelia L. LOGANIACEAE (D)
Spiladocorys Ridl. = Pentasacme Wall. ex Wight ASCLEPIADACEAE (D)
Spilanthes Jacq. COMPOSITAE (D)
Spilocarpus Lem. = Tournefortia L. BORAGINACEAE (D)
Spiloxene Salisb. HYPOXIDACEAE (M)
Spinacia L. CHENOPODIACEAE (D)
Spinicalycium Frič (SUI) = Acanthocalycium Backeb. CACTACEAE (D)
Spinifex L. GRAMINEAE (M)
Spiniluma (Baill.) Aubrév. = Sideroxylon L. SAPOTACEAE (D)
Spiracantha Kunth COMPOSITAE (D)
Spiradiclis Blume RUBIACEAE (D)
Spiraea L. ROSACEAE (D)
Spiraeanthemum A.Gray CUNONIACEAE (D)
Spiraeanthus (Fisch. & C.A.Mey.) Maxim. ROSACEAE (D)
Spiraeopsis Miq. (SUI) = Caldcluvia D.Don CUNONIACEAE (D)
Spiranthera A.St.–Hil. RUTACEAE (D)
Spiranthera Bojer (SUH) = Merremia Dennst. ex Endl. CONVOLVULACEAE (D)
Spiranthes Rich. ORCHIDACEAE (M)
Spirella Costantin ASCLEPIADACEAE (D)
Spiroceratium H.Wolff = Pimpinella L. UMBELLIFERAE (D)
Spirochloe Lunell (SUS) = Schedonnardus Steud. GRAMINEAE (M)
Spirodela Schleid. LEMNACEAE (M)
Spirogardnera Stauffer SANTALACEAE (D)

Spirolobium Baill. APOCYNACEAE (D)
Spironema Lindl. (SUH) = Callisia Loefl. COMMELINACEAE (M)
Spiropetalum Gilg = Rourea Aubl. CONNARACEAE (D)
Spirorhynchus Kar. & Kir. CRUCIFERAE (D)
Spiroseris Rech.f. COMPOSITAE (D)
Spirospermum Thouars MENISPERMACEAE (D)
Spirostachys Ung.–Sternb. (SUH) = Heterostachys Ung.–Sternb. CHENOPODIACEAE (D)
Spirostachys S.Watson (SUH) = Allenrolfea Kuntze CHENOPODIACEAE (D)
Spirostachys Sond. EUPHORBIACEAE (D)
Spirostegia Ivanina SCROPHULARIACEAE (D)
Spirostigma Nees ACANTHACEAE (D)
Spirostylis C.Presl ex Schult. & Schult.f. = Struthanthus Mart. LORANTHACEAE (D)
Spirotecoma Baill. ex Dalla Torre & Harms BIGNONIACEAE (D)
Spirotheca Ulbr. = Ceiba Mill. BOMBACACEAE (D)
Spirotheros Raf. = Heteropogon Pers. GRAMINEAE (M)
Spirotropis Tul. LEGUMINOSAE–PAPILIONOIDEAE (D)
Spixia Leandro = Pera Mutis EUPHORBIACEAE (D)
Spodiopogon Trin. GRAMINEAE (M)
Spondianthus Engl. EUPHORBIACEAE (D)
Spondias L. ANACARDIACEAE (D)
Spondogona Raf. = Sideroxylon L. SAPOTACEAE (D)
Spongiocarpella Yakovlev & N.Ulziykh. LEGUMINOSAE–PAPILIONOIDEAE (D)
Spongiola J.J.Wood & A.L.Lamb ORCHIDACEAE (M)
Spongiosperma Zarucchi APOCYNACEAE (D)
Spongiosyndesmus Gilli = Ladyginia Lipsky UMBELLIFERAE (D)
Sponia Comm. ex Decne. = Trema Lour. ULMACEAE (D)
Sporadanthus F.Muell. RESTIONACEAE (M)
Sporichloe Piler (SUO) = Schedonnardus Steud. GRAMINEAE (M)
Sporobolus R.Br. GRAMINEAE (M)
Sporoxeia W.W.Sm. MELASTOMATACEAE (D)
Spraguea Torr. PORTULACACEAE (D)
Spragueanella Balle LORANTHACEAE (D)
Sprekelia Heist. AMARYLLIDACEAE (M)
Sprengelia Sm. EPACRIDACEAE (D)
Sprengeria Greene = Lepidium L. CRUCIFERAE (D)
Sprucea Benth. = Simira Aubl. RUBIACEAE (D)
Spruceanthus Sleumer = Hasseltia Kunth TILIACEAE (D)
Sprucella Pierre = Micropholis (Griseb.) Pierre SAPOTACEAE (D)
Sprucina Nied. = Jubelina A.Juss. MALPIGHIACEAE (D)
Spryginia Popov CRUCIFERAE (D)
Spuriodaucus C.Norman UMBELLIFERAE (D)
Spuriopimpinella Kitag. UMBELLIFERAE (D)
Spyridium Fenzl RHAMNACEAE (D)
Squamellaria Becc. RUBIACEAE (D)
Squamopappus Jansen, Harriman & Urbat COMPOSITAE (D)
Sredinskya (Stein) Fedorov = Primula L. PRIMULACEAE (D)
Sreemadhavana Rauschert = Aphelandra R.Br. ACANTHACEAE (D)
Staavia Dahl BRUNIACEAE (D)
Staberoha Kunth RESTIONACEAE (M)
Stachyandra J.–F.Leroy ex Radcl.–Sm. EUPHORBIACEAE (D)
Stachyanthus Engl. ICACINACEAE (D)
Stachyanthus Engl. (SUH) = Bulbophyllum Thouars ORCHIDACEAE (M)
Stachyarrhena Hook.f. RUBIACEAE (D)
Stachycarpus (Endl.) Tiegh. = Prumnopitys Phil. PODOCARPACEAE (G)
Stachycephalum Sch.Bip. ex Benth. COMPOSITAE (D)

Stachydeoma Small LABIATAE (D)
Stachygynandrum P.Beauv. = Selaginella P.Beauv. SELAGINELLACEAE (FA)
Stachyococcus Standl. RUBIACEAE (D)
Stachyophorbe (Liebm. ex Mart.) Liebm. = Chamaedorea Willd. PALMAE (M)
Stachyopsis Popov & Vved. LABIATAE (D)
Stachyothyrsus Harms LEGUMINOSAE–CAESALPINIOIDEAE (D)
Stachyphrynium K.Schum. MARANTACEAE (M)
Stachyphyllum Tiegh. = Antidaphne Poepp. & Endl. EREMOLEPIDACEAE (D)
Stachys L. LABIATAE (D)
Stachystemon Planch. EUPHORBIACEAE (D)
Stachytarpheta Vahl VERBENACEAE (D)
Stachyurus Siebold & Zucc. STACHYURACEAE (D)
Stackhousia Sm. STACKHOUSIACEAE (D)
Stadiochilus Ros.M.Sm. ZINGIBERACEAE (M)
Stadmannia Lam. SAPINDACEAE (D)
Staehelina L. COMPOSITAE (D)
Staelia Cham. & Schltdl. RUBIACEAE (D)
Stagmaria Jack = Gluta L. ANACARDIACEAE (D)
Stahelia Jonk. = Tapeinostemon Benth. GENTIANACEAE (D)
Stahlia Bello LEGUMINOSAE–CAESALPINIOIDEAE (D)
Stahlianthus Kuntze ZINGIBERACEAE (M)
Staintoniella Hara CRUCIFERAE (D)
Stalkya Garay ORCHIDACEAE (M)
Standleya Brade RUBIACEAE (D)
Standleyacanthus Leonard ACANTHACEAE (D)
Standleyanthus R.M.King & H.Rob. COMPOSITAE (D)
Stanfieldia Small COMPOSITAE (D)
Stanfieldiella Brenan COMMELINACEAE (M)
Stanfordia S.Watson = Caulanthus S.Watson CRUCIFERAE (D)
Stangea Graebn. VALERIANACEAE (D)
Stangeria T.Moore STANGERIACEAE (G)
Stanhopea Frost ex Hook. ORCHIDACEAE (M)
Stanhopeastrum Rchb.f. = Stanhopea Frost ex Hook. ORCHIDACEAE (M)
Stanleya Nutt. CRUCIFERAE (D)
Stanleyella Rydb. = Thelypodium Endl. CRUCIFERAE (D)
Stannia Karst. = Posoqueria Aubl. RUBIACEAE (D)
Stapelia L. ASCLEPIADACEAE (D)
Stapelianthus Choux ex A.C.White & B.Sloane ASCLEPIADACEAE (D)
Stapeliopsis Choux (SUH) = Stapelianthus Choux ex A.C.White & B.Sloane
    ASCLEPIADACEAE (D)
Stapeliopsis Pillans ASCLEPIADACEAE (D)
Stapfia Burtt Davy (SUH) = Neostapfia Burtt Davy GRAMINEAE (M)
Stapfiella Gilg TURNERACEAE (D)
Stapfiola Kuntze (SUS) = Desmostachya (Stapf) Stapf GRAMINEAE (M)
Stapfiophyton H.L.Li MELASTOMATACEAE (D)
Staphidiastrum Naudin = Clidemia D.Don MELASTOMATACEAE (D)
Staphidium Naudin = Clidemia D.Don MELASTOMATACEAE (D)
Staphylea L. STAPHYLEACEAE (D)
Staphylosyce Hook.f. = Coccinia Wight & Arn. CUCURBITACEAE (D)
Staphysora Pierre = Maesobotrya Benth. EUPHORBIACEAE (D)
Stathmostelma K.Schum. ASCLEPIADACEAE (D)
Statice L. = Limonium Mill. PLUMBAGINACEAE (D)
Staudtia Warb. MYRISTICACEAE (D)
Stauntonia DC. LARDIZABALACEAE (D)
Stauracanthus Link LEGUMINOSAE–PAPILIONOIDEAE (D)

Stauranthera Benth. GESNERIACEAE (D)
Stauranthus Liebm. RUTACEAE (D)
Staurites Rchb.f. = Phalaenopsis Blume ORCHIDACEAE (M)
Staurochilus Ridl. ex Pfitzer ORCHIDACEAE (M)
Staurochlamys Baker COMPOSITAE (D)
Stauroglottis Schauer = Phalaenopsis Blume ORCHIDACEAE (M)
Staurogyne Wall. ACANTHACEAE (D)
Staurogynopsis Mangenot & Ake Assi = Staurogyne Wall. ACANTHACEAE (D)
Staurophragma Fisch. & C.A.Mey. = Verbascum L. SCROPHULARIACEAE (D)
Stauropsis Rchb.f. = Trichoglottis Blume ORCHIDACEAE (M)
Staurospermum Thonn. = Mitracarpus Zucc. RUBIACEAE (D)
Staurostigma Scheidw. = Asterostigma Fisch. & C.A.Mey. ARACEAE (M)
Staurothylax Griff. = Phyllanthus L. EUPHORBIACEAE (D)
Stawellia F.Muell. ANTHERICACEAE (M)
Stayneria L.Bolus AIZOACEAE (D)
Stebbinsoseris K.L.Chambers COMPOSITAE (D)
Steenisia Bakh.f. RUBIACEAE (D)
Steenisioblechnum Hennipman BLECHNACEAE (F)
Stefanoffia H.Wolff UMBELLIFERAE (D)
Stegania R.Br. = Blechnum L. BLECHNACEAE (F)
Steganotaenia Hochst. UMBELLIFERAE (D)
Steganotropis Lehm. = Centrosema (DC.) Benth. LEGUMINOSAE–
    PAPILIONOIDEAE (D)
Steganthera Perkins MONIMIACEAE (D)
Steganthus Knobl. = Olea L. OLEACEAE (D)
Stegastrum Tiegh. (SUH) = Lepeostegeres Blume LORANTHACEAE (D)
Stegnocarpus Torr. & A.Gray = Tiquilia Pers. BORAGINACEAE (D)
Stegnogramma Blume THELYPTERIDACEAE (F)
Stegnosperma Benth. STEGNOSPERMATACEAE (D)
Stegolepis Klotzsch ex Koern. RAPATEACEAE (M)
Stegosia Lour. = Rottboellia L.f. GRAMINEAE (M)
Steigeria Müll.Arg. = Baloghia Endl. EUPHORBIACEAE (D)
Steinbachiella Harms = Diphysa Jacq. LEGUMINOSAE–PAPILIONOIDEAE (D)
Steinchisma Raf. GRAMINEAE (M)
Steinheilia Decne. = Odontanthera Wight ASCLEPIADACEAE (D)
Steinmannia Phil. (SUH) = Garaventia Looser ALLIACEAE (M)
Steirachne Ekman GRAMINEAE (M)
Steiractinia S.F.Blake COMPOSITAE (D)
Steirodiscus Less. COMPOSITAE (D)
Steironema Raf. = Lysimachia L. PRIMULACEAE (D)
Steiropteris (C.Chr.) Pic.Serm. THELYPTERIDACEAE (F)
Steirosanchezia Lindau = Sanchezia Ruíz & Pav. ACANTHACEAE (D)
Steirotis Raf. = Struthanthus Mart. LORANTHACEAE (D)
Stelechanteria Thouars ex Baill. = Drypetes Vahl EUPHORBIACEAE (D)
Stelechantha Bremek. RUBIACEAE (D)
Stelechocarpus Hook.f. & Thomson ANNONACEAE (D)
Steleocodon Gilli = Phalacraea DC. COMPOSITAE (D)
Steleostemma Schltr. ASCLEPIADACEAE (D)
Stelephuros Adans. (SUS) = Phleum L. GRAMINEAE (M)
Stelestylis Drude CYCLANTHACEAE (M)
Steliopsis Brieger = Stelis Sw. ORCHIDACEAE (M)
Stelis Sw. ORCHIDACEAE (M)
Stellaria L. CARYOPHYLLACEAE (D)
Stellariopsis (Baill.) Rydb. ROSACEAE (D)
Stellera L. THYMELAEACEAE (D)

Stelleropsis Pobed. THYMELAEACEAE (D)
Stelligera A.J.Scott CHENOPODIACEAE (D)
Stellilabium Schltr. ORCHIDACEAE (M)
Stellix Noronha = Psychotria L. RUBIACEAE (D)
Stellorchis Thouars (SUO) = Nervilia Comm. ex Gaud. ORCHIDACEAE (M)
Stellorkis Thouars = Nervilia Comm. ex Gaud. ORCHIDACEAE (M)
Stellularia Benth. = Buchnera L. SCROPHULARIACEAE (D)
Stelmacrypton Baill. = Pentanura Blume ASCLEPIADACEAE (D)
Stelmagonum Baill. (UP) ASCLEPIADACEAE (D)
Stelmanis Raf. (SUH) = Hedyotis L. RUBIACEAE (D)
Stelmation Fourn. = Metastelma R.Br. ASCLEPIADACEAE (D)
Stelmatocodon Schltr. ASCLEPIADACEAE (D)
Stelmatocrypton Baill. = Pentanura Blume ASCLEPIADACEAE (D)
Stelmotis Raf. = Hedyotis L. RUBIACEAE (D)
Stemmacantha Cass. = Leuzea DC. COMPOSITAE (D)
Stemmadenia Benth. APOCYNACEAE (D)
Stemmatella Wedd. ex Benth. = Galinsoga Ruíz & Pav. COMPOSITAE (D)
Stemmatium Phil. = Leucocoryne Lindl. ALLIACEAE (M)
Stemmatodaphne Gamble = Alseodaphne Nees LAURACEAE (D)
Stemmatophyllum Tiegh. = Amyema Tiegh. LORANTHACEAE (D)
Stemmatospermum P.Beauv. = Nastus Juss. GRAMINEAE (M)
Stemodia L. SCROPHULARIACEAE (D)
Stemodiopsis Engl. SCROPHULARIACEAE (D)
Stemona Lour. STEMONACEAE (M)
Stemonocoleus Harms LEGUMINOSAE–CAESALPINIOIDEAE (D)
Stemonoporus Thwaites DIPTEROCARPACEAE (D)
Stemonurus Blume ICACINACEAE (D)
Stemotria Wettst. & Harms ex Engl. SCROPHULARIACEAE (D)
Stenachaenium Benth. COMPOSITAE (D)
Stenadenium Pax = Monadenium Pax EUPHORBIACEAE (D)
Stenandriopsis S.Moore ACANTHACEAE (D)
Stenandrium Nees ACANTHACEAE (D)
Stenanona Standl. ANNONACEAE (D)
Stenanthella Rydb. = Stenanthium (A.Gray) Kunth MELANTHIACEAE (M)
Stenanthemum Reissek = Cryptandra Sm. RHAMNACEAE (D)
Stenanthera Engl. & Diels (SUH) = Neostenanthera Exell ANNONACEAE (D)
Stenanthera R.Br. = Astroloma R.Br. EPACRIDACEAE (D)
Stenanthium (A.Gray) Kunth MELANTHIACEAE (M)
Stenanthus Oerst. ex Hanst. = Columnea L. GESNERIACEAE (D)
Stenaria Raf. = Houstonia L. RUBIACEAE (D)
Stenia Lindl. ORCHIDACEAE (M)
Stenocactus (K.Schum.) A.W.Hill CACTACEAE (D)
Stenocalyx Turcz. (SUH) = Mezia Schwacke ex Nied. MALPIGHIACEAE (D)
Stenocalyx O.Berg = Eugenia L. MYRTACEAE (D)
Stenocarpha S.F.Blake COMPOSITAE (D)
Stenocarpus R.Br. PROTEACEAE (D)
Stenocephalum Sch.Bip. = Vernonia Schrcb. COMPOSITAE (D)
Stenocereus (A.Berger) Riccob. CACTACEAE (D)
Stenochasma Miq. = Broussonetia L'Hér. ex Vent. MORACEAE (D)
Stenochilus R.Br. = Eremophila R.Br. MYOPORACEAE (D)
Stenochlaena J.Sm. BLECHNACEAE (F)
Stenochloa Nutt. = Dissanthelium Trin. GRAMINEAE (M)
Stenocline DC. COMPOSITAE (D)
Stenocoelium Ledeb. UMBELLIFERAE (D)
Stenocoryne Lindl. = Bifrenaria Lindl. ORCHIDACEAE (M)

Stenodon Naudin MELASTOMATACEAE (D)
Stenodraba O.E.Schulz = Weberbauera Gilg & Muschler CRUCIFERAE (D)
Stenodrepanum Harms LEGUMINOSAE–CAESALPINIOIDEAE (D)
Stenofestuca (Honda) Nakai = Bromus L. GRAMINEAE (M)
Stenofilix Nakai = Cochlidium Kaulf. GRAMMITIDACEAE (F)
Stenogastra Hanst. = Sinningia Nees GESNERIACEAE (D)
Stenoglossum Kunth = Epidendrum L. ORCHIDACEAE (M)
Stenoglottis Lindl. ORCHIDACEAE (M)
Stenogonum Nutt. POLYGONACEAE (D)
Stenogyne Benth. LABIATAE (D)
Stenolepia Alderw. DRYOPTERIDACEAE (F)
Stenolobium D.Don = Cybistax Mart. ex Meisn. BIGNONIACEAE (D)
Stenolobus C.Presl = Davallia Sm. DAVALLIACEAE (F)
Stenoloma Fée = Sphenomeris Maxon DENNSTAEDTIACEAE (F)
Stenomeria Turcz. ASCLEPIADACEAE (D)
Stenomeris Planch. DIOSCOREACEAE (M)
Stenomesson Herb. AMARYLLIDACEAE (M)
Stenonema Hook. = Draba L. CRUCIFERAE (D)
Stenonia Didr. (SUH) = Ditaxis Vahl ex A.Juss. EUPHORBIACEAE (D)
Stenonia Baill. = Cleistanthus Hook.f. ex Planch. EUPHORBIACEAE (D)
Stenoniella Kuntze = Cleistanthus Hook.f. ex Planch. EUPHORBIACEAE (D)
Stenopadus S.F.Blake COMPOSITAE (D)
Stenopetalum R.Br. ex DC. CRUCIFERAE (D)
Stenophalium Anderb. COMPOSITAE (D)
Stenophragma Cĕlak. = Arabis L. CRUCIFERAE (D)
Stenophyllus Raf. = Bulbostylis Kunth CYPERACEAE (M)
Stenopolen Raf. = Stenia Lindl. ORCHIDACEAE (M)
Stenops B.Nord. COMPOSITAE (D)
Stenoptera C.Presl ORCHIDACEAE (M)
Stenorrhynchos Rich. ex Spreng. ORCHIDACEAE (M)
Stenoschista Bremek. = Ruellia L. ACANTHACEAE (D)
Stenosemia C.Presl = Tectaria Cav. DRYOPTERIDACEAE (F)
Stenosiphanthus A.Samp. = Arrabidaea DC. BIGNONIACEAE (D)
Stenosiphon Spach ONAGRACEAE (D)
Stenosiphonium Nees ACANTHACEAE (D)
Stenosolen (Müll.Arg.) Markgr. = Tabernaemontana L. APOCYNACEAE (D)
Stenosolenium Turcz. BORAGINACEAE (D)
Stenospermation Schott ARACEAE (M)
Stenospermum Sweet ex Heynh. = Kunzea Rchb. MYRTACEAE (D)
Stenostachys Turcz. = Hystrix Moench GRAMINEAE (M)
Stenostelma Schltr. ASCLEPIADACEAE (D)
Stenostephanus Nees ACANTHACEAE (D)
Stenostomum C.F.Gaertn. = Antirhea Comm. ex Juss. RUBIACEAE (D)
Stenotaphrum Trin. GRAMINEAE (M)
Stenothyrsus C.B.Clarke ACANTHACEAE (D)
Stenotopsis Rydb. = Ericameria Nutt. COMPOSITAE (D)
Stenotus Nutt. COMPOSITAE (D)
Stephanachne Keng GRAMINEAE (M)
Stephanandra Siebold & Zucc. ROSACEAE (D)
Stephania Lour. MENISPERMACEAE (D)
Stephaniscus Tiegh. = Englerina Tiegh. LORANTHACEAE (D)
Stephanium Schreb. = Palicourea Aubl. RUBIACEAE (D)
Stephanocaryum Popov BORAGINACEAE (D)
Stephanocereus A.Berger CACTACEAE (D)
Stephanochilus Coss. & Durieu ex Maire COMPOSITAE (D)

Stephanococcus Bremek. RUBIACEAE (D)
Stephanodaphne Baill. THYMELAEACEAE (D)
Stephanodoria Greene COMPOSITAE (D)
Stephanolepis S.Moore = Erlangea Sch.Bip. COMPOSITAE (D)
Stephanoluma Baill. = Micropholis (Griseb.) Pierre SAPOTACEAE (D)
Stephanomeria Nutt. COMPOSITAE (D)
Stephanopholis S.F.Blake = Chromolepis Benth. COMPOSITAE (D)
Stephanophysum Pohl = Ruellia L. ACANTHACEAE (D)
Stephanopodium Poepp. DICHAPETALACEAE (D)
Stephanorossia Chiov. = Oenanthe L. UMBELLIFERAE (D)
Stephanostachys (Klotzsch) Klotzsch ex O.E.Schulz = Chamaedorea Willd. PALMAE (M)
Stephanostegia Baill. APOCYNACEAE (D)
Stephanostema K.Schum. APOCYNACEAE (D)
Stephanotella E.Fourn. = Marsdenia R.Br. ASCLEPIADACEAE (D)
Stephanothelys Garay ORCHIDACEAE (M)
Stephanotis Thouars ASCLEPIADACEAE (D)
Stephanotrichum Naudin = Clidemia D.Don MELASTOMATACEAE (D)
Stephegyne Korth. = Mitragyna Korth. RUBIACEAE (D)
Steptorhamphus Bunge COMPOSITAE (D)
Sterculia L. STERCULIACEAE (D)
Stereocarpus (Pierre) Hallier f. = Camellia L. THEACEAE (D)
Stereocaryum Burret MYRTACEAE (D)
Stereochlaena Hack. GRAMINEAE (M)
Stereoderma Blume = Olea L. OLEACEAE (D)
Stereosandra Blume ORCHIDACEAE (M)
Stereospermum Cham. BIGNONIACEAE (D)
Stereoxylon Ruíz & Pav. = Escallonia Mutis ex L.f. ESCALLONIACEAE (D)
Sterigmanthe Klotzsch & Garcke = Euphorbia L. EUPHORBIACEAE (D)
Sterigmapetalum Kuhlm. RHIZOPHORACEAE (D)
Sterigmostemum M.Bieb. CRUCIFERAE (D)
Steriphoma Spreng. CAPPARACEAE (D)
Sternbergia Waldst. & Kit. AMARYLLIDACEAE (M)
Sterrhymenia Griseb. = Sclerophylax Miers SOLANACEAE (D)
Sterropetalum N.E.Br. = Nelia Schwantes AIZOACEAE (D)
Stethoma Raf. = Justicia L. ACANTHACEAE (D)
Stetsonia Britton & Rose CACTACEAE (D)
Steudelago Kuntze = Exostema (Pers.) Humb. & Bonpl. RUBIACEAE (D)
Steudelella Honda (SUS) = Sphaerocaryum Nees ex Hook.f. GRAMINEAE (M)
Steudnera K.Koch ARACEAE (M)
Stevenia Adams ex Fisch. CRUCIFERAE (D)
Steveniella Schltr. ORCHIDACEAE (M)
Stevenorchis Wankow & Kraenzl. = Steveniella Schltr. ORCHIDACEAE (M)
Stevensia Poit. RUBIACEAE (D)
Stevensonia Duncan ex Balf.f. = Phoenicophorium H.A.Wendl. PALMAE (M)
Stevia Cav. COMPOSITAE (D)
Steviopsis R.M.King & H.Rob. COMPOSITAE (D)
Stewartia L. (SUO) = Stuartia L. THEACEAE (D)
Stewartiella Nasir UMBELLIFERAE (D)
Steyerbromelia L.B.Sm. BROMELIACEAE (M)
Steyermarkia Standl. RUBIACEAE (D)
Steyermarkina R.M.King & H.Rob. COMPOSITAE (D)
Steyermarkochloa Davidse & R.P.Ellis GRAMINEAE (M)
Stibasia C.Presl = Marattia Sw. MARATTIACEAE (F)
Stiburus Stapf = Eragrostis Wolf GRAMINEAE (M)
Sticherus C.Presl GLEICHENIACEAE (F)

Stichianthus Valeton RUBIACEAE (D)
Stichoneuron Hook.f. STEMONACEAE (M)
Stichorkis Thouars = Liparis Rich. ORCHIDACEAE (M)
Stictocardia Hallier f. CONVOLVULACEAE (D)
Stictophyllum Dodson & M.W.Chase ORCHIDACEAE (M)
Stifftia J.C.Mikan COMPOSITAE (D)
Stigmanthus Lour. = Morinda L. RUBIACEAE (D)
Stigmaphyllon A.Juss. MALPIGHIACEAE (D)
Stigmatanthus Roem. & Schult. (SUO) = Morinda L. RUBIACEAE (D)
Stigmatella Eig (SUH) = Eigia Soják CRUCIFERAE (D)
Stigmatocarpum L.Bolus = Dorotheanthus Schwantes AIZOACEAE (D)
Stigmatodactylus Maxim. ex Makino ORCHIDACEAE (M)
Stigmatopteris C.Chr. DRYOPTERIDACEAE (F)
Stigmatorhynchus Schltr. ASCLEPIADACEAE (D)
Stigmatosema Garay ORCHIDACEAE (M)
Stilaginella Tul. = Hieronima Allemão EUPHORBIACEAE (D)
Stilago L. (SUS) = Antidesma L. EUPHORBIACEAE (D)
Stilbanthus Hook.f. AMARANTHACEAE (D)
Stilbe Bergius STILBACEAE (D)
Stilbocarpa (Hook.f.) Decne. & Planch. ARALIACEAE (D)
Stillingfleetia Bojer = Sapium P.Browne EUPHORBIACEAE (D)
Stillingia Garden ex L. EUPHORBIACEAE (D)
Stilpnogyne DC. COMPOSITAE (D)
Stilpnolepis I.M.Kraschen. COMPOSITAE (D)
Stilpnopappus Mart. ex DC. COMPOSITAE (D)
Stilpnophleum Nevski = Calamagrostis Adans. GRAMINEAE (M)
Stilpnophyllum Hook.f. RUBIACEAE (D)
Stilpnophyllum (Endl.) Drury (SUH) = Ficus L. MORACEAE (D)
Stilpnophyton Less. = Athanasia L. COMPOSITAE (D)
Stilpnophytum Less. (SUO) = Athanasia L. COMPOSITAE (D)
Stimegas Raf. = Cypripedium L. ORCHIDACEAE (M)
Stimenes Raf. = Nierembergia Ruíz & Pav. SOLANACEAE (D)
Stimomphis Raf. = Calibrachoa La Llave & Lex. SOLANACEAE (D)
Stimoryne Raf. = Petunia Juss. SOLANACEAE (D)
Stimpsonia C.Wright ex A.Gray PRIMULACEAE (D)
Stipa L. GRAMINEAE (M)
Stipagrostis Nees GRAMINEAE (M)
Stipavena Vierh. (SUS) = Helictotrichon Besser GRAMINEAE (M)
Stipecoma Müll.Arg. APOCYNACEAE (D)
Stipellaria Benth. (SUS) = Alchornea Sw. EUPHORBIACEAE (D)
Stiptanthus Briq. = Anisochilus Wall. ex Benth. LABIATAE (D)
Stipularia P.Beauv. RUBIACEAE (D)
Stipularia Delpino (SUH) = Thalictrum L. RANUNCULACEAE (D)
Stipulicida Michx. CARYOPHYLLACEAE (D)
Stirlingia Endl. PROTEACEAE (D)
Stironeuron Radlk. = Synsepalum (A.DC.) Daniell SAPOTACEAE (D)
Stisseria Scop. = Manilkara Adans. SAPOTACEAE (D)
Stixis Lour. CAPPARACEAE (D)
Stizolobium P.Browne = Mucuna Adans. LEGUMINOSAE–PAPILIONOIDEAE (D)
Stizophyllum Miers BIGNONIACEAE (D)
Stobaea Thunb. = Berkheya Ehrh. COMPOSITAE (D)
Stocksia Benth. SAPINDACEAE (D)
Stoebe L. COMPOSITAE (D)
Stoeberia Dinter & Schwantes AIZOACEAE (D)
Stoibrax Raf. UMBELLIFERAE (D)

Stokesia L'Hér. COMPOSITAE (D)
Stokoeanthus E.G.H.Oliv. ERICACEAE (D)
Stollaea Schltr. = Caldcluvia D.Don CUNONIACEAE (D)
Stolzia Schltr. ORCHIDACEAE (M)
Stomandra Standl. RUBIACEAE (D)
Stomatanthes R.M.King & H.Rob. COMPOSITAE (D)
Stomatium Schwantes AIZOACEAE (D)
Stomatocalyx Müll.Arg. = Pimelodendron Hassk. EUPHORBIACEAE (D)
Stomatochaeta (S.F.Blake) Maguire & Wurdack COMPOSITAE (D)
Stomatostemma N.E.Br. ASCLEPIADACEAE (D)
Stomoisia Raf. = Utricularia L. LENTIBULARIACEAE (D)
Stonesia G.Taylor PODOSTEMACEAE (D)
Storckiella Seem. LEGUMINOSAE–CAESALPINIOIDEAE (D)
Stormesia Kickx f. = Asplenium L. ASPLENIACEAE (F)
Stormia S.Moore = Cardiopetalum Schltdl. ANNONACEAE (D)
Storthocalyx Radlk. SAPINDACEAE (D)
Stracheya Benth. LEGUMINOSAE–PAPILIONOIDEAE (D)
Strailia T.Durand = Lecythis Loefl. LECYTHIDACEAE (D)
Strakaea C.Presl = Thottea Rottb. ARISTOLOCHIACEAE (D)
Stramentopappus H.Rob. & V.A.Funk COMPOSITAE (D)
Stramonium Mill. = Datura L. SOLANACEAE (D)
Strangea Meisn. PROTEACEAE (D)
Strangweja Bertol. = Bellevalia Lapeyr. HYACINTHACEAE (M)
Stranvaesia Lindl. = Photinia Lindl. ROSACEAE (D)
Strasburgeria Baill. STRASBURGERIACEAE (D)
Strateuma Raf. (SUH) = Zeuxine Lindl. ORCHIDACEAE (M)
Strateuma Salisb. = Orchis L. ORCHIDACEAE (M)
Stratiotes L. HYDROCHARITACEAE (M)
Straussia A.Gray = Psychotria L. RUBIACEAE (D)
Straussiella Hausskn. CRUCIFERAE (D)
Streblacanthus Kuntze ACANTHACEAE (D)
Streblochaete Pilg. GRAMINEAE (M)
Streblorrhiza Endl. LEGUMINOSAE–PAPILIONOIDEAE (D)
Streblosa Korth. RUBIACEAE (D)
Streblosiopsis Valeton RUBIACEAE (D)
Streblus Lour. MORACEAE (D)
Strelitzia Aiton STRELITZIACEAE (M)
Strempelia A.Rich. = Psychotria L. RUBIACEAE (D)
Strempeliopsis Benth. APOCYNACEAE (D)
Strephium Nees = Raddia Bertol. GRAMINEAE (M)
Strephonema Hook.f. COMBRETACEAE (D)
Strepsilobus Raf. = Entada Adans. LEGUMINOSAE–MIMOSOIDEAE (D)
Strepsimela Raf. = Helixanthera Lour. LORANTHACEAE (D)
Streptachne R.Br. = Aristida L. GRAMINEAE (M)
Streptanthella Rydb. CRUCIFERAE (D)
Streptanthera Sweet = Sparaxis Ker-Gawl. IRIDACEAE (M)
Streptanthus Nutt. CRUCIFERAE (D)
Streptia Döll (SUI) = Streptogyna P.Beauv. GRAMINEAE (M)
Streptocalyx Beer BROMELIACEAE (M)
Streptocarpus Lindl. GESNERIACEAE (D)
Streptocaulon Wight & Arn. ASCLEPIADACEAE (D)
Streptochaeta Schrad. ex Nees GRAMINEAE (M)
Streptoglossa Steetz ex F.Muell. COMPOSITAE (D)
Streptogyna P.Beauv. GRAMINEAE (M)
Streptolirion Edgew. COMMELINACEAE (M)

Streptoloma Bunge CRUCIFERAE (D)
Streptolophus D.K.Hughes GRAMINEAE (M)
Streptomanes K.Schum. ASCLEPIADACEAE (D)
Streptopetalum Hochst. TURNERACEAE (D)
Streptopus Michx. CONVALLARIACEAE (M)
Streptosiphon Mildbr. ACANTHACEAE (D)
Streptosolen Miers SOLANACEAE (D)
Streptostachys Desv. GRAMINEAE (M)
Streptostigma Regel = Exodeconus Raf. SOLANACEAE (D)
Streptothamnus F.Muell. FLACOURTIACEAE (D)
Streptotrachelus Greenm. = Laubertia A.DC. APOCYNACEAE (D)
Stricklandia Baker (SUH) = Phaedranassa Herb. AMARYLLIDACEAE (M)
Striga Lour. SCROPHULARIACEAE (D)
Strigina Engl. = Lindernia All. SCROPHULARIACEAE (D)
Strigosella Boiss. = Malcolmia R.Br. CRUCIFERAE (D)
Striolaria Ducke RUBIACEAE (D)
Strobidia Miq. = Alpinia Roxb. ZINGIBERACEAE (M)
Strobilacanthus Griseb. ACANTHACEAE (D)
Strobilanthes Blume ACANTHACEAE (D)
Strobilanthopsis S.Moore ACANTHACEAE (D)
Strobilocarpus Klotzsch = Grubbia Bergius GRUBBIACEAE (D)
Strobilopanax R.Vig. = Meryta J.R.Forst. & G.Forst. ARALIACEAE (D)
Strobilopsis Hilliard & B.L.Burtt SCROPHULARIACEAE (D)
Strobilorhachis Klotzsch = Aphelandra R.Br. ACANTHACEAE (D)
Strobopetalum N.E.Br. ASCLEPIADACEAE (D)
Stroganowia Kar. & Kir. CRUCIFERAE (D)
Stromanthe Sond. MARANTACEAE (M)
Stromatocactus Karw. ex Rümpler (SUI) = Ariocarpus Scheidw. CACTACEAE (D)
Stromatopteris Mett. STROMATOPTERIDACEAE (F)
Strombocactus Britton & Rose CACTACEAE (D)
Strombocarpa (Benth.) A.Gray = Prosopis L. LEGUMINOSAE–MIMOSOIDEAE (D)
Strombodurus Steud. (SUI) = Pentarrhaphis Kunth GRAMINEAE (M)
Strombosia Blume OLACACEAE (D)
Strombosiopsis Engl. OLACACEAE (D)
Strongylocalyx Blume = Syzygium Gaertn. MYRTACEAE (D)
Strongylocaryum Burret = Ptychosperma Labill. PALMAE (M)
Strongylodon Vogel LEGUMINOSAE–PAPILIONOIDEAE (D)
Strophacanthus Lindau = Isoglossa Oerst. ACANTHACEAE (D)
Strophanthus DC. APOCYNACEAE (D)
Strophioblachia Boerl. EUPHORBIACEAE (D)
Strophiodiscus Choux = Plagioscyphus Radlk. SAPINDACEAE (D)
Strophocactus Britton & Rose = Selenicereus (A.Berger) Britton & Rose CACTACEAE (D)
Strophocaulos Small = Convolvulus L. CONVOLVULACEAE (D)
Strophocereus Frič & Kreuz. (SUO) = Selenicereus (A.Berger) Britton & Rose
    CACTACEAE (D)
Stropholirion Torr. = Dichelostemma Kunth ALLIACEAE (M)
Strophostyles Elliott LEGUMINOSAE–PAPILIONOIDEAE (D)
Strotheria B.L.Turner COMPOSITAE (D)
Struchium P.Browne COMPOSITAE (D)
Strukeria Vell. = Vochysia Aubl. VOCHYSIACEAE (D)
Strumaria Jacq. ex Willd. AMARYLLIDACEAE (M)
Strumpfia Jacq. RUBIACEAE (D)
Struthanthus Mart. LORANTHACEAE (D)
Struthiola L. THYMELAEACEAE (D)
Struthiolopsis E.Phillips = Gnidia L. THYMELAEACEAE (D)

Struthiopteris Willd. (SUH) = Matteuccia Tod. WOODSIACEAE (F)
Struthiopteris Scop. = Blechnum L. BLECHNACEAE (F)
Struthopteris Bernh. = Osmunda L. OSMUNDACEAE (F)
Strychnopsis Baill. MENISPERMACEAE (D)
Strychnos L. LOGANIACEAE (D)
Stryphnodendron Mart. LEGUMINOSAE–MIMOSOIDEAE (D)
Stuartia L. THEACEAE (D)
Stuartina Sond. COMPOSITAE (D)
Stubendorffia Schrenk ex Fisch., C.A.Mey. & Avé–Lall. CRUCIFERAE (D)
Stuckertia Kuntze ASCLEPIADACEAE (D)
Stuckertiella Beauverd COMPOSITAE (D)
Stuebelia Pax = Belencita Karst. CAPPARACEAE (D)
Stuessya B.L.Turner & F.G.Davies COMPOSITAE (D)
Stuhlmannia Taub. LEGUMINOSAE–CAESALPINIOIDEAE (D)
Stultitia E.Phillips = Orbea Haw. ASCLEPIADACEAE (D)
Stupa Asch. (SUO) = Stipa L. GRAMINEAE (M)
Sturmia C.F.Gaertn. = Antirhea Comm. ex Juss. RUBIACEAE (D)
Sturmia Hoppe (SUS) = Mibora Adans. GRAMINEAE (M)
Sturmia Rchb.f. (SUH) = Liparis Rich. ORCHIDACEAE (M)
Stussenia C.Hansen MELASTOMATACEAE (D)
Stutzeria F.Muell. (SUI) = Pullea Schltr. CUNONIACEAE (D)
Styasasia S.Moore ACANTHACEAE (D)
Stygnanthe Hanst. = Columnea L. GESNERIACEAE (D)
Stylagrostis Mez = Calamagrostis Adans. GRAMINEAE (M)
Stylanthus Rchb.f. & Zoll. = Mallotus Lour. EUPHORBIACEAE (D)
Stylapterus A.Juss. PENAEACEAE (D)
Stylarthropus Baill. = Whitfieldia Hook. ACANTHACEAE (D)
Stylidium Sw. ex Willd. STYLIDIACEAE (D)
Stylisma Raf. CONVOLVULACEAE (D)
Stylites Amstutz ISOETACEAE (FA)
Stylobasium Desf. STYLOBASIACEAE (D)
Styloceras Kunth ex Juss. BUXACEAE (D)
Stylochaeton Lepr. ARACEAE (M)
Stylocline Nutt. COMPOSITAE (D)
Stylocoryna Cav. (SUH) = Aidia Lour. RUBIACEAE (D)
Stylocoryne Wight & Arn. (SUO) = Aidia Lour. RUBIACEAE (D)
Stylodiscus Benn. = Bischofia Blume EUPHORBIACEAE (D)
Stylodon Raf. VERBENACEAE (D)
Styloglossum Breda = Calanthe Ker–Gawl. ORCHIDACEAE (M)
Stylogyne A.DC. MYRSINACEAE (D)
Styloma O.F.Cook = Pritchardia Seem. & H.A.Wendl. PALMAE (M)
Stylomecon G.Taylor PAPAVERACEAE (D)
Stylophorum Nutt. PAPAVERACEAE (D)
Stylophyllum Britton & Rose = Dudleya Britton & Rose CRASSULACEAE (D)
Stylosanthes Sw. LEGUMINOSAE–PAPILIONOIDEAE (D)
Stylosiphonia Brandegee RUBIACEAE (D)
Stylotrichium Mattf. COMPOSITAE (D)
Stypandra R.Br. PHORMIACEAE (M)
Styphelia Sm. EPACRIDACEAE (D)
Styphnolobium Schott ex Endl. LEGUMINOSAE–PAPILIONOIDEAE (D)
Styppeiochloa De Winter GRAMINEAE (M)
Styrax L. STYRACACEAE (D)
Styrophyton S.Y.Hu = Allomorphia Blume MELASTOMATACEAE (D)
Styrosinia Raf. = Sinningia Nees GESNERIACEAE (D)
Suaeda Forssk. ex Scop. CHENOPODIACEAE (D)

Suardia Schrank = Melinis P.Beauv. GRAMINEAE (M)
Suarezia Dodson ORCHIDACEAE (M)
Suberanthus Borhidi & Fernández RUBIACEAE (D)
Sublimia Comm. ex Mart. = Hyophorbe Gaertn. PALMAE (M)
Submatucana Backeb. = Oreocereus (A.Berger) Riccob. CACTACEAE (D)
Subpilocereus Backeb. = Cereus Mill. CACTACEAE (D)
Subularia L. CRUCIFERAE (D)
Subulatopuntia Frič & Schelle = Opuntia Mill. CACTACEAE (D)
Succisa Haller DIPSACACEAE (D)
Succisella Beck DIPSACACEAE (D)
Succowia Medik. CRUCIFERAE (D)
Suchtelenia Kar. ex Meisn. BORAGINACEAE (D)
Suckleya A.Gray CHENOPODIACEAE (D)
Sucrea Soderstr. GRAMINEAE (M)
Suddia Renvoize GRAMINEAE (M)
Suessenguthia Merxm. ACANTHACEAE (D)
Suessenguthiella Friedrich MOLLUGINACEAE (D)
Suitenia Stokes (SUO) = Swietenia Jacq. MELIACEAE (D)
Suitramia Rchb. (SUO) = Svitramia Cham. MELASTOMATACEAE (D)
Suksdorfia A.Gray SAXIFRAGACEAE (D)
Sukunia A.C.Sm. RUBIACEAE (D)
Sulaimania Hedge & Rech.f. LABIATAE (D)
Sulcorebutia Backeb. = Rebutia K.Schum. CACTACEAE (D)
Sulipa Blanco = Gardenia Ellis RUBIACEAE (D)
Sulitia Merr. RUBIACEAE (D)
Sullivantia Torr. & A.Gray SAXIFRAGACEAE (D)
Sulpitia Raf. = Encyclia Hook. ORCHIDACEAE (M)
Sulzeria Roem. & Schult. (SUI) = Faramea Aubl. RUBIACEAE (D)
Sumatroscirpus Oteng-Yeb. CYPERACEAE (M)
Sumbavia Baill. = Doryxylon Zoll. EUPHORBIACEAE (D)
Sumbaviopsis J.J.Sm. EUPHORBIACEAE (D)
Summerhayesia P.J.Cribb ORCHIDACEAE (M)
Sunaptea Griff. = Vatica L. DIPTEROCARPACEAE (D)
Sundacarpus C.N.Page PODOCARPACEAE (G)
Sunipia Buch.–Ham. ex Lindl. ORCHIDACEAE (M)
Suregada Roxb. ex Rottler EUPHORBIACEAE (D)
Surenus Kuntze = Toona (Endl.) M.Roem. MELIACEAE (D)
Surfacea Moldenke = Premna L. VERBENACEAE (D)
Suriana L. SURIANACEAE (D)
Surwala M.Roem. = Walsura Roxb. MELIACEAE (D)
Susilkumara Bennet (SUS) = Alajja Ikonn. LABIATAE (D)
Susum Blume = Hanguana Blume HANGUANACEAE (M)
Sutera Roth SCROPHULARIACEAE (D)
Suteria DC. = Psychotria L. RUBIACEAE (D)
Sutherlandia R.Br. ex W.T.Aiton LEGUMINOSAE–PAPILIONOIDEAE (D)
Sutrina Lindl. ORCHIDACEAE (M)
Suttonia A.Rich. = Myrsine L. MYRSINACEAE (D)
Suzukia Kudo LABIATAE (D)
Svenkoeltzia Burns–Bal. = Stenorrhynchos Rich. ex Spreng. ORCHIDACEAE (M)
Svensonia Moldenke = Chascanum E.Mey. VERBENACEAE (D)
Sventenia Font Quer COMPOSITAE (D)
Svitramia Cham. MELASTOMATACEAE (D)
Swainsona Salisb. LEGUMINOSAE–PAPILIONOIDEAE (D)
Swallenia Soderstr. & H.F.Decker GRAMINEAE (M)
Swallenochloa McClure = Chusquea Kunth GRAMINEAE (M)

Swartsia J.F.Gmel. = Solandra Sw. SOLANACEAE (D)
Swartzia Schreb. LEGUMINOSAE–PAPILIONOIDEAE (D)
Sweetia Spreng. LEGUMINOSAE–PAPILIONOIDEAE (D)
Sweetiopsis Chodat & Hassk. = Riedeliella Harms LEGUMINOSAE–
    PAPILIONOIDEAE (D)
Swertia L. GENTIANACEAE (D)
Swertopsis Makino = Swertia L. GENTIANACEAE (D)
Swida Opiz = Cornus L. CORNACEAE (D)
Swietenia Jacq. MELIACEAE (D)
Swinglea Merr. RUTACEAE (D)
Swintonia Griff. ANACARDIACEAE (D)
Swynnertonia S.Moore ASCLEPIADACEAE (D)
Syagrus Mart. PALMAE (M)
Sychinium Desv. = Dorstenia L. MORACEAE (D)
Sycios Medik. (SUO) = Sicyos L. CUCURBITACEAE (D)
Sycocarpus Britton = Guarea L. MELIACEAE (D)
Sycodendron Rojas Acosta = Ficus L. MORACEAE (D)
Sycomorphe Miq. = Ficus L. MORACEAE (D)
Sycomorus Gasp. = Ficus L. MORACEAE (D)
Sycophila Welw. ex Tiegh. = Helixanthera Lour. LORANTHACEAE (D)
Sycopsis Oliv. HAMAMELIDACEAE (D)
Syena Schreb. = Mayaca Aubl. MAYACACEAE (M)
Sykesia Arn. = Gaertnera Lam. RUBIACEAE (D)
Sylitra E.Mey. (SUH) = Ptycholobium Harms LEGUMINOSAE–PAPILIONOIDEAE (D)
Syllepis Fourn. = Imperata Cyr. GRAMINEAE (M)
Syllisium Endl. (SUO) = Syzygium Gaertn. MYRTACEAE (D)
Syllysium Meyen & Schauer = Syzygium Gaertn. MYRTACEAE (D)
Sylvalismis Thouars = Calanthe Ker–Gawl. ORCHIDACEAE (M)
Symbasiandra Steud. (SUI) = Hilaria Kunth GRAMINEAE (M)
Symbegonia Warb. BEGONIACEAE (D)
Symbolanthus G.Don GENTIANACEAE (D)
Symingtonia Steenis (SUS) = Exbucklandia R.W.Br. HAMAMELIDACEAE (D)
Symmeria Benth. POLYGONACEAE (D)
Symonanthus Haegi SOLANACEAE (D)
Sympa Ravenna (UP) IRIDACEAE (M)
Sympagis (Nees) Bremek. = Strobilanthes Blume ACANTHACEAE (D)
Sympegma Bunge CHENOPODIACEAE (D)
Sympetalandra Stapf LEGUMINOSAE–CAESALPINIOIDEAE (D)
Sympetaleia A.Gray = Eucnide Zucc. LOASACEAE (D)
Symphiandra Steud. = Symphyandra A.DC. CAMPANULACEAE (D)
Symphionema R.Br. PROTEACEAE (D)
Symphonia L.f. GUTTIFERAE (D)
Symphorema Roxb. VERBENACEAE (D)
Symphoricarpos Duhamel CAPRIFOLIACEAE (D)
Symphostemon Hiern LABIATAE (D)
Symphyandra A.DC. CAMPANULACEAE (D)
Symphyglossum Schltr. ORCHIDACEAE (M)
Symphyllarion Gagnep. = Hedyotis L. RUBIACEAE (D)
Symphyllia Baill. = Epiprinus Griff. EUPHORBIACEAE (D)
Symphyllocarpus Maxim. COMPOSITAE (D)
Symphyllophyton Gilg GENTIANACEAE (D)
Symphyobasis K.Krause = Goodenia Sm. GOODENIACEAE (D)
Symphyochaeta (DC.) Skottsb. = Robinsonia DC. COMPOSITAE (D)
Symphyochlamys Gürke MALVACEAE (D)
Symphyogyne Burret = Maxburretia Furtado PALMAE (M)

Symphyoloma C.A.Mey. UMBELLIFERAE (D)
Symphyomyrtus Schauer = Eucalyptus L'Hér. MYRTACEAE (D)
Symphyopappus Turcz. COMPOSITAE (D)
Symphyosepalum Hand.–Mazz. ORCHIDACEAE (M)
Symphyostemon Miers ex Klatt (SUH) = Olsynium Raf. IRIDACEAE (M)
Symphysia C.Presl ERICACEAE (D)
Symphysicarpus Hassk. = Heterostemma Wight & Arn. ASCLEPIADACEAE (D)
Symphytonema Schltr. = Tanulepis Balf.f. ASCLEPIADACEAE (D)
Symphytosiphon Harms = Trichilia P.Browne MELIACEAE (D)
Symphytum L. BORAGINACEAE (D)
Sympieza Lichtenst. ex Roem. & Schult. ERICACEAE (D)
Symplectochilus Lindau = Anisotes Nees ACANTHACEAE (D)
Symplectrodia Lazarides GRAMINEAE (M)
Symplocarpus Salisb. ex Nutt. ARACEAE (M)
Symplococarpon Airy Shaw THEACEAE (D)
Symplocos Jacq. SYMPLOCACEAE (D)
Synadena Raf. = Phalaenopsis Blume ORCHIDACEAE (M)
Synadenium Boiss. EUPHORBIACEAE (D)
Synallodia Raf. = Swertia L. GENTIANACEAE (D)
Synammia J.Sm. POLYPODIACEAE (F)
Synandra Schrad. (SUH) = Aphelandra R.Br. ACANTHACEAE (D)
Synandra Nutt. LABIATAE (D)
Synandrina Standl. & L.O.Williams = Casearia Jacq. FLACOURTIACEAE (D)
Synandrodaphne Gilg THYMELAEACEAE (D)
Synandrogyne Buchet = Arophyton Jum. ARACEAE (M)
Synandropus A.C.Sm. MENISPERMACEAE (D)
Synandrospadix Engl. ARACEAE (M)
Synantherias Schott = Amorphophallus Blume ex Decne. ARACEAE (M)
Synanthes Burns–Bal., H.Rob. & M.S.Foster ORCHIDACEAE (M)
Synaphe Dulac (SUS) = Catapodium Link GRAMINEAE (M)
Synaphea R.Br. PROTEACEAE (D)
Synaphlebium J.Sm. = Lindsaea Dryand. ex Sm. DENNSTAEDTIACEAE (F)
Synapsis Griseb. BIGNONIACEAE (D)
Synaptantha Hook.f. RUBIACEAE (D)
Synaptea Kurz (SUO) = Vatica L. DIPTEROCARPACEAE (D)
Synaptolepis Oliv. THYMELAEACEAE (D)
Synaptophyllum N.E.Br. AIZOACEAE (D)
Synardisia (Mez) Lundell = Ardisia Sw. MYRSINACEAE (D)
Synarrhena Fisch. & C.A.Mey. = Manilkara Adans. SAPOTACEAE (D)
Synaspisma Endl. = Codiaeum A.Juss. EUPHORBIACEAE (D)
Synassa Lindl. = Sauroglossum Lindl. ORCHIDACEAE (M)
Synastemon F.Muell. = Sauropus Blume EUPHORBIACEAE (D)
Syncalathium Lipsch. COMPOSITAE (D)
Syncarpha DC. COMPOSITAE (D)
Syncarpia Ten. MYRTACEAE (D)
Syncephalum DC. COMPOSITAE (D)
Synchoriste Baill. = Lasiocladus Bojer ex Nees ACANTHACEAE (D)
Synclisia Benth. MENISPERMACEAE (D)
Syncolostemon E.Mey. LABIATAE (D)
Syncretocarpus S.F.Blake COMPOSITAE (D)
Syndesmanthus Klotzsch ERICACEAE (D)
Syndesmis Wall. = Gluta L. ANACARDIACEAE (D)
Syndesmon (Hoffmanns. ex Endl.) Britton = Anemonella Spach RANUNCULACEAE (D)
Syndiclis Hook.f. = Potameia Thouars LAURACEAE (D)
Syndyophyllum K.Schum. & Lauterb. EUPHORBIACEAE (D)

Synechanthus H.A.Wendl. PALMAE (M)
Synedrella Gaertn. COMPOSITAE (D)
Synedrellopsis Hieron. & Kuntze COMPOSITAE (D)
Syneilesis Maxim. COMPOSITAE (D)
Synelcosciadium Boiss. = Tordylium L. UMBELLIFERAE (D)
Synema Dulac = Mercurialis L. EUPHORBIACEAE (D)
Synepilaena Baill. = Kohleria Regel GESNERIACEAE (D)
Synexemia Raf. = Phyllanthus L. EUPHORBIACEAE (D)
Syngonanthus Ruhland ERIOCAULACEAE (M)
Syngonium Schott ARACEAE (M)
Syngramma J.Sm. ADIANTACEAE (F)
Syngrammatopsis Alston = Pterozonium Fée ADIANTACEAE (F)
Synima Radlk. SAPINDACEAE (D)
Synisoon Baill. = Retiniphyllum Bonpl. RUBIACEAE (D)
Synnema Benth. = Hygrophila R.Br. ACANTHACEAE (D)
Synnotia Sweet = Sparaxis Ker Gawl. IRIDACEAE (M)
Synochlamys Fée = Pellaea Link ADIANTACEAE (F)
Synoecia Miq. = Ficus L. MORACEAE (D)
Synoplectris Raf. = Sarcoglottis C.Presl ORCHIDACEAE (M)
Synosma Raf. ex Britton & A.Br. (SUS) = Hasteola Raf. COMPOSITAE (D)
Synostemon F.Muell. = Sauropus Blume EUPHORBIACEAE (D)
Synotis (C.B.Clarke) C.Jeffrey & Y.L.Chen COMPOSITAE (D)
Synotoma (G.Don) R.Schulz = Physoplexis (Endl.) Schur CAMPANULACEAE (D)
Synoum A.Juss. MELIACEAE (D)
Synptera Llanos = Trichoglottis Blume ORCHIDACEAE (M)
Synsepalum (A.DC.) Daniell SAPOTACEAE (D)
Synsiphon Regel = Colchicum L. COLCHICACEAE (M)
Synstemon Botsch. CRUCIFERAE (D)
Synstemonanthus Botsch. (SUS) = Synstemon Botsch. CRUCIFERAE (D)
Syntherisma Walter = Digitaria Haller GRAMINEAE (M)
Synthlipsis A.Gray CRUCIFERAE (D)
Synthyris Benth. SCROPHULARIACEAE (D)
Syntriandrium Engl. MENISPERMACEAE (D)
Syntrichopappus A.Gray COMPOSITAE (D)
Syntrinema H.Pfeiff. = Rhynchospora Vahl CYPERACEAE (M)
Synurus Iljin COMPOSITAE (D)
Sypharissa Salisb. HYACINTHACEAE (M)
Syreitschikovia Pavlov COMPOSITAE (D)
Syrenia Andrz. ex Besser CRUCIFERAE (D)
Syrenopsis Jaub. & Spach = Thlaspi L. CRUCIFERAE (D)
Syringa L. OLEACEAE (D)
Syringantha Standl. RUBIACEAE (D)
Syringidium Lindau (SUH) = Habracanthus Nees ACANTHACEAE (D)
Syringodea Hook.f. IRIDACEAE (M)
Syringodium Kütz. CYMODOCEACEAE (M)
Syrrheonema Miers MENISPERMACEAE (D)
Systeloglossum Schltr. ORCHIDACEAE (M)
Systenotheca Reveal & Hardham POLYGONACEAE (D)
Syzygiopsis Ducke = Pouteria Aubl. SAPOTACEAE (D)
Syzygium Gaertn. MYRTACEAE (D)
Szovitsia Fisch. & C.A.Mey. UMBELLIFERAE (D)

Tabacum Gilib. = Nicotiana L. SOLANACEAE (D)
Tabacus Moench = Nicotiana L. SOLANACEAE (D)
Tabascina Baill. = Justicia L. ACANTHACEAE (D)

Tabebuia Gomes ex DC. BIGNONIACEAE (D)
Taberna Miers = Tabernaemontana L. APOCYNACEAE (D)
Tabernaemontana L. APOCYNACEAE (D)
Tabernanthe Baill. APOCYNACEAE (D)
Tacarcuna Huft EUPHORBIACEAE (D)
Tacazzea Decne. ASCLEPIADACEAE (D)
Tacca J.R.Forst. & G.Forst. TACCACEAE (M)
Taccarum Brongn. ex Schott ARACEAE (M)
Tachia Aubl. GENTIANACEAE (D)
Tachiadenus Griseb. GENTIANACEAE (D)
Tachibota Aubl. = Hirtella L. CHRYSOBALANACEAE (D)
Tachigali Aubl. LEGUMINOSAE–CAESALPINIOIDEAE (D)
Tacinga Britton & Rose CACTACEAE (D)
Tacitus Moran = Graptopetalum Rose CRASSULACEAE (D)
Tacoanthus Baill. ACANTHACEAE (D)
Tacsonia Juss. = Passiflora L. PASSIFLORACEAE (D)
Tadehagi H.Ohashi LEGUMINOSAE–PAPILIONOIDEAE (D)
Taeckholmia Boulos = Sonchus L. COMPOSITAE (D)
Taeniandra Bremek. = Strobilanthes Blume ACANTHACEAE (D)
Taenianthera Burret = Geonoma Willd. PALMAE (M)
Taeniatherum Nevski GRAMINEAE (M)
Taenidia (Torr. & A.Gray) Drude UMBELLIFERAE (D)
Taeniochlaena Hook.f. = Rourea Aubl. CONNARACEAE (D)
Taeniophyllum Blume ORCHIDACEAE (M)
Taeniopleurum J.M.Coult. & Rose = Perideridia Rchb. UMBELLIFERAE (D)
Taeniopsis J.Sm. = Vittaria Sm. VITTARIACEAE (F)
Taeniopteris Hook. = Vittaria Sm. VITTARIACEAE (F)
Taeniorrhiza Summerh. ORCHIDACEAE (M)
Taeniosapium Müll.Arg. = Sapium P.Browne EUPHORBIACEAE (D)
Taenitis Willd. ex Schkuhr ADIANTACEAE (F)
Taetsia Medik. = Cordyline Comm. ex R.Br. AGAVACEAE (M)
Tagetes L. COMPOSITAE (D)
Taguaria Raf. = Gaiadendron G.Don LORANTHACEAE (D)
Tahitia Burret = Berrya Roxb. TILIACEAE (D)
Taihangia T.T.Yu & C.L.Li = Geum L. ROSACEAE (D)
Tainia Blume ORCHIDACEAE (M)
Tainionema Schltr. ASCLEPIADACEAE (D)
Tainiopsis Schltr. = Eriodes Rolfe ORCHIDACEAE (M)
Taitonia Yamam. = Gomphostemma Wall. ex Benth. LABIATAE (D)
Taiwania Hayata TAXODIACEAE (G)
Takasagoya Y.Kimura = Hypericum L. GUTTIFERAE (D)
Takeikadzuchia Kitag. & Kitam. = Olgaea Iljin COMPOSITAE (D)
Takhtajania Baranova & J.-F.Leroy WINTERACEAE (D)
Takhtajanianthus De = Rhanteriopsis Rauschert COMPOSITAE (D)
Talasium Spreng. = Panicum L. GRAMINEAE (M)
Talassia Korovin = Ferula L. UMBELLIFERAE (D)
Talauma Juss. = Magnolia L. MAGNOLIACEAE (D)
Talbotia S.Moore (SUH) = Afrofittonia Lindau ACANTHACEAE (D)
Talbotia Balf. VELLOZIACEAE (M)
Talbotiella Baker f. LEGUMINOSAE–CAESALPINIOIDEAE (D)
Talbotiopsis L.B.Sm. (SUS) = Talbotia Balf. VELLOZIACEAE (M)
Talguenea Miers ex Endl. RHAMNACEAE (D)
Taliera Mart. = Corypha L. PALMAE (M)
Talinaria Brandegee PORTULACACEAE (D)
Talinella Baill. PORTULACACEAE (D)

Talinopsis A.Gray PORTULACACEAE (D)
Talinum Adans. PORTULACACEAE (D)
Talisia Aubl. SAPINDACEAE (D)
Talisiopsis Radlk. = Zanha Hiern SAPINDACEAE (D)
Talpinaria Karst. = Pleurothallis R.Br. ORCHIDACEAE (M)
Tamamschjania Pimenov & Kljuykov UMBELLIFERAE (D)
Tamananthus Badillo COMPOSITAE (D)
Tamania Cuatrec. COMPOSITAE (D)
Tamaricaria Qaiser & Ali (SUS) = Myrtama Ovcz. & Kinzik. TAMARICACEAE (D)
Tamarindus L. LEGUMINOSAE–CAESALPINIOIDEAE (D)
Tamarix L. TAMARICACEAE (D)
Tamatavia Hook.f. = Chapelieria A.Rich. ex DC. RUBIACEAE (D)
Tamaulipa R.M.King & H.Rob. COMPOSITAE (D)
Tambourissa Sonn. MONIMIACEAE (D)
Tamilnadia Tirveng. & Sastre RUBIACEAE (D)
Tammsia Karst. RUBIACEAE (D)
Tamonea Aubl. VERBENACEAE (D)
Tamus L. DIOSCOREACEAE (M)
Tanacetum L. COMPOSITAE (D)
Tanaecium Sw. BIGNONIACEAE (D)
Tanakea Franch. & Sav. SAXIFRAGACEAE (D)
Tanaosolen N.E.Br. = Tritoniopsis L.Bolus IRIDACEAE (M)
Tanarius Kuntze = Macaranga Thouars EUPHORBIACEAE (D)
Tangaraca Adans. = Hamelia Jacq. RUBIACEAE (D)
Tangtsinia S.C.Chen = Cephalanthera Rich. ORCHIDACEAE (M)
Tanibouca Aubl. = Terminalia L. COMBRETACEAE (D)
Tankervillia Link = Phaius Lour. ORCHIDACEAE (M)
Tannodia Baill. EUPHORBIACEAE (D)
Tanquana H.E.K.Hartmann & Liede AIZOACEAE (D)
Tansaniochloa Rauschert = Setaria P.Beauv. GRAMINEAE (M)
Tanulepis Balf.f. ASCLEPIADACEAE (D)
Taonabo Aubl. = Ternstroemia Mutis ex L.f. THEACEAE (D)
Tapeinanthus Herb. (SUH) = Braxireon Raf. AMARYLLIDACEAE (M)
Tapeinanthus Boiss. ex Benth. (SUH) = Thuspeinanta T.Durand LABIATAE (D)
Tapeinia Comm. ex Juss. IRIDACEAE (M)
Tapeinidium (C.Presl) C.Chr. DENNSTAEDTIACEAE (F)
Tapeinochilos Miq. COSTACEAE (M)
Tapeinoglossum Schltr. ORCHIDACEAE (M)
Tapeinosperma Hook.f. MYRSINACEAE (D)
Tapeinostelma Schltr. = Brachystelma R.Br. ASCLEPIADACEAE (D)
Tapeinostemon Benth. GENTIANACEAE (D)
Tapeinotes DC. = Sinningia Nees GESNERIACEAE (D)
Tapesia C.F.Gaertn. = Hamelia Jacq. RUBIACEAE (D)
Taphrospermum C.A.Mey. CRUCIFERAE (D)
Tapina Mart. = Sinningia Nees GESNERIACEAE (D)
Tapinanthus (Blume) Blume LORANTHACEAE (D)
Tapinopentas Bremek. = Otomeria Benth. RUBIACEAE (D)
Tapinostemma Tiegh. = Plicosepalus Tiegh. LORANTHACEAE (D)
Tapiphyllum Robyns RUBIACEAE (D)
Tapirira Aubl. ANACARDIACEAE (D)
Tapirocarpus Sagot BURSERACEAE (D)
Tapiscia Oliv. STAPHYLEACEAE (D)
Taplinia Lander COMPOSITAE (D)
Tapogomea Aubl. = Psychotria L. RUBIACEAE (D)
Tapoides Airy Shaw EUPHORBIACEAE (D)

Tapomana Adans. (SUS) = Connarus L. CONNARACEAE (D)
Tapura Aubl. DICHAPETALACEAE (D)
Tara Molina = Caesalpinia L. LEGUMINOSAE–CAESALPINIOIDEAE (D)
Tarachia C.Presl = Asplenium L. ASPLENIACEAE (F)
Taraktogenos Hassk. = Hydnocarpus Gaertn. FLACOURTIACEAE (D)
Taralea Aubl. LEGUMINOSAE–PAPILIONOIDEAE (D)
Taramea Raf. (SUO) = Faramea Aubl. RUBIACEAE (D)
Tarasa Phil. MALVACEAE (D)
Taraxacum G.H.Weber ex Wigg. COMPOSITAE (D)
Taraxia (Nutt.) Raimann = Camissonia Link ONAGRACEAE (D)
Tarchonanthus L. COMPOSITAE (D)
Tardavel Adans. = Spermacoce L. RUBIACEAE (D)
Tarenna Gaertn. RUBIACEAE (D)
Tarennoidea Tirveng. & Sastre RUBIACEAE (D)
Tarigidia Stent GRAMINEAE (M)
Tarphochlamys Bremek. = Strobilanthes Blume ACANTHACEAE (D)
Tarrietia Blume = Heritiera Aiton STERCULIACEAE (D)
Tartonia Raf. = Thymelaea Mill. THYMELAEACEAE (D)
Taschneria C.Presl = Crepidomanes (C.Presl) C.Presl HYMENOPHYLLACEAE (F)
Tashiroea Matsum. = Bredia Blume MELASTOMATACEAE (D)
Tasmannia DC. = Drimys J.R.Forst. & G.Forst. WINTERACEAE (D)
Tassadia Decne. ASCLEPIADACEAE (D)
Tatea F.Muell. = Premna L. VERBENACEAE (D)
Tatea Seem. (SUH) = Bikkia Reinw. RUBIACEAE (D)
Tateanthus Gleason MELASTOMATACEAE (D)
Tatianyx Zuloaga & Soderstr. GRAMINEAE (M)
Tatina Raf. = Sideroxylon L. SAPOTACEAE (D)
Taubertia K.Schum. ex Taub. = Disciphania Eichler MENISPERMACEAE (D)
Tauroceras Britton & Rose = Acacia Mill. LEGUMINOSAE–MIMOSOIDEAE (D)
Taurostalix Rchb.f. = Bulbophyllum Thouars ORCHIDACEAE (M)
Tauscheria Fisch. ex DC. CRUCIFERAE (D)
Tauschia Schltdl. UMBELLIFERAE (D)
Tavaresia Welw. ex N.E.Br. = Decabelone Decne. ASCLEPIADACEAE (D)
Taverniera DC. LEGUMINOSAE–PAPILIONOIDEAE (D)
Taveunia Burret = Cyphosperma H.A.Wendl. ex Hook.f. PALMAE (M)
Taxillus Tiegh. LORANTHACEAE (D)
Taxodium Rich. TAXODIACEAE (G)
Taxotrophis Blume = Streblus Lour. MORACEAE (D)
Taxus L. TAXACEAE (G)
Tayloriophyton M.P.Nayar MELASTOMATACEAE (D)
Tchihatchewia Boiss. = Neotchihatchewia Rauschert CRUCIFERAE (D)
Teclea Delile RUTACEAE (D)
Tecleopsis Hoyle & Leakey = Vepris Comm. ex A.Juss. RUTACEAE (D)
Tecoma Juss. BIGNONIACEAE (D)
Tecomanthe Baill. BIGNONIACEAE (D)
Tecomaria (Endl.) Spach = Tecoma Juss. BIGNONIACEAE (D)
Tecomella Seem. BIGNONIACEAE (D)
Tecophilaea Bertero ex Colla TECOPHILAEACEAE (M)
Tectaria Cav. DRYOPTERIDACEAE (F)
Tectaridium Copel. DRYOPTERIDACEAE (F)
Tecticornia Hook.f. CHENOPODIACEAE (D)
Tectiphiala H.E.Moore PALMAE (M)
Tectona L.f. VERBENACEAE (D)
Tecunumania Standl. & Steyerm. CUCURBITACEAE (D)
Tedingea D.Müll.–Doblies & U.Müll.–Doblies AMARYLLIDACEAE (M)

Teedia Rudolphi SCROPHULARIACEAE (D)
Teesdalia R.Br. CRUCIFERAE (D)
Teesdaliopsis (Willk.) Rothm. CRUCIFERAE (D)
Teganium Schmidel = Nolana L.f. SOLANACEAE (D)
Tegicornia Paul G.Wilson CHENOPODIACEAE (D)
Tegularia Reinw. = Didymochlaena Desv. DRYOPTERIDACEAE (F)
Teichmeyeria Scop. = Gustavia L. LECYTHIDACEAE (D)
Teijsmanniodendron Koord. VERBENACEAE (D)
Teinosolen Hook.f. = Heterophyllaea Hook.f. RUBIACEAE (D)
Teinostachyum Munro = Schizostachyum Nees GRAMINEAE (M)
Teixeiranthus R.M.King & H.Rob. COMPOSITAE (D)
Telanthera R.Br. = Alternanthera Forssk. AMARANTHACEAE (D)
Telanthophora H.Rob. & Brettell COMPOSITAE (D)
Telectadium Baill. ASCLEPIADACEAE (D)
Telekia Baumg. COMPOSITAE (D)
Telemachia Urb. = Elaeodendron J.F.Jacq. CELASTRACEAE (D)
Teleozoma R.Br. = Ceratopteris Brongn. PARKERIACEAE (F)
Telephium L. MOLLUGINACEAE (D)
Telesilla Klotzsch ASCLEPIADACEAE (D)
Telesonix Raf. = Boykinia Nutt. SAXIFRAGACEAE (D)
Telfairia Hook. CUCURBITACEAE (D)
Teline Medik. = Genista L. LEGUMINOSAE–PAPILIONOIDEAE (D)
Teliostachya Nees ACANTHACEAE (D)
Telipogon Kunth ORCHIDACEAE (M)
Telitoxicum Moldenke MENISPERMACEAE (D)
Tellima R.Br. SAXIFRAGACEAE (D)
Telmatophila Mart. ex Baker COMPOSITAE (D)
Telminostelma Fourn. ASCLEPIADACEAE (D)
Telmissa Fenzl CRASSULACEAE (D)
Telogyne Baill. = Trigonostemon Blume EUPHORBIACEAE (D)
Telopea R.Br. PROTEACEAE (D)
Telosma Coville ASCLEPIADACEAE (D)
Teloxys Moq. CHENOPODIACEAE (D)
Tema Adans. = Echinochloa P.Beauv. GRAMINEAE (M)
Temmodaphne Kosterm. LAURACEAE (D)
Temnadenia Miers APOCYNACEAE (D)
Temnocalyx Robyns RUBIACEAE (D)
Temnopteryx Hook.f. RUBIACEAE (D)
Templetonia R.Br. ex W.T.Aiton LEGUMINOSAE–PAPILIONOIDEAE (D)
Temu O.Berg = Blepharocalyx O.Berg MYRTACEAE (D)
Tenagocharis Hochst. = Butomopsis Kunth LIMNOCHARITACEAE (M)
Tenaris E.Mey. ASCLEPIADACEAE (D)
Tengia Chun GESNERIACEAE (D)
Tennantia Verdc. RUBIACEAE (D)
Tenorea Gasp. (SUH) = Ficus L. MORACEAE (D)
Tenrhynea Hilliard & B.L.Burtt COMPOSITAE (D)
Teonongia Stapf = Streblus Lour. MORACEAE (D)
Tepesia C.F.Gaertn. = Hamelia Jacq. RUBIACEAE (D)
Tephrocactus Lem. = Opuntia Mill. CACTACEAE (D)
Tephroseris (Rchb.) Rchb. COMPOSITAE (D)
Tephrosia Pers. LEGUMINOSAE–PAPILIONOIDEAE (D)
Tepualia Griseb. MYRTACEAE (D)
Tepuia Camp ERICACEAE (D)
Tepuianthus Maguire & Steyerm. TEPUIANTHACEAE (D)
Teramnus P.Browne LEGUMINOSAE–PAPILIONOIDEAE (D)

Terana La Llave COMPOSITAE (D)
Teratophyllum Mett. ex Kuhn LOMARIOPSIDACEAE (F)
Terebraria Sessé ex Kunth = Neolaugeria Nicolson RUBIACEAE (D)
Teremis Raf. = Lycium L. SOLANACEAE (D)
Terminalia L. COMBRETACEAE (D)
Terminaliopsis Danguy COMBRETACEAE (D)
Terminthia Bernh. = Rhus L. ANACARDIACEAE (D)
Terminthodia Ridl. = Tetractomia Hook.f. RUTACEAE (D)
Terniola Tul. (SUS) = Dalzellia Wight PODOSTEMACEAE (D)
Terniopsis Chao = Dalzellia Wight PODOSTEMACEAE (D)
Ternstroemia Mutis ex L.f. THEACEAE (D)
Ternstroemiopsis Urb. THEACEAE (D)
Terrellia Lunell (SUS) = Elymus L. GRAMINEAE (M)
Tersonia Moq. GYROSTEMONACEAE (D)
Tertrea DC. = Machaonia Bonpl. RUBIACEAE (D)
Terua Standl. & F.J.Herm. = Lonchocarpus Kunth LEGUMINOSAE–
    PAPILIONOIDEAE (D)
Tessarandra Miers OLEACEAE (D)
Tessaria Ruíz & Pav. COMPOSITAE (D)
Tessiera DC. = Spermacoce L. RUBIACEAE (D)
Tessmannia Harms LEGUMINOSAE–CAESALPINIOIDEAE (D)
Tessmanniacanthus Mildbr. ACANTHACEAE (D)
Tessmannianthus Markgr. MELASTOMATACEAE (D)
Tessmanniodoxa Burret = Chelyocarpus Dammer PALMAE (M)
Tessmanniophoenix Burret = Chelyocarpus Dammer PALMAE (M)
Testudipes Markgr. = Tabernaemontana L. APOCYNACEAE (D)
Testulea Pellegr. OCHNACEAE (D)
Tetilla DC. SAXIFRAGACEAE (D)
Tetraberlinia (Harms) Hauman LEGUMINOSAE–CAESALPINIOIDEAE (D)
Tetracarpaea Hook. ESCALLONIACEAE (D)
Tetracarpaea Benth. (SUH) = Anisophyllea R.Br. ex Sabine ANISOPHYLLEACEAE (D)
Tetracarpidium Pax EUPHORBIACEAE (D)
Tetracarpus Post & Kuntze (SUO) = Tetracarpaea Hook. ESCALLONIACEAE (D)
Tetracentron Oliv. TETRACENTRACEAE (D)
Tetracera L. DILLENIACEAE (D)
Tetrachaete Chiov. GRAMINEAE (M)
Tetrachne Nees GRAMINEAE (M)
Tetrachondra Petrie ex Oliv. TETRACHONDRACEAE (D)
Tetrachyron Schltdl. COMPOSITAE (D)
Tetraclea A.Gray LABIATAE (D)
Tetraclinis Mast. CUPRESSACEAE (G)
Tetraclis Hiern EBENACEAE (D)
Tetracme Bunge CRUCIFERAE (D)
Tetracmidion Korsh. = Tetracme Bunge CRUCIFERAE (D)
Tetracoccus Engelm. ex Parry EUPHORBIACEAE (D)
Tetracoilanthus Rappa & Camarrone (SUS) = Aptenia N.E.Br. AIZOACEAE (D)
Tetractinostigma Hassk. = Aporusa Blume EUPHORBIACEAE (D)
Tetractomia Hook.f. RUTACEAE (D)
Tetracustelma Baill. = Matelea Aubl. ASCLEPIADACEAE (D)
Tetradapa Osbeck = Erythrina L. LEGUMINOSAE–PAPILIONOIDEAE (D)
Tetradema Schltr. = Agalmyla Blume GESNERIACEAE (D)
Tetradenia Benth. LABIATAE (D)
Tetradia R.Br. = Pterygota Schott & Endl. STERCULIACEAE (D)
Tetradiclis Steven ex M.Bieb. ZYGOPHYLLACEAE (D)
Tetradium Dulac (SUH) = Rhodiola L. CRASSULACEAE (D)

Tetradium Lour. RUTACEAE (D)
Tetradoa Pichon = Hunteria Roxb. APOCYNACEAE (D)
Tetradoxa C.Y.Wu ADOXACEAE (D)
Tetradyas Danser LORANTHACEAE (D)
Tetradymia DC. COMPOSITAE (D)
Tetraedrocarpus O.Schwarz = Echiochilon Desf. BORAGINACEAE (D)
Tetraena Maxim. ZYGOPHYLLACEAE (D)
Tetraeugenia Merr. = Syzygium Gaertn. MYRTACEAE (D)
Tetragamestus Rchb.f. ORCHIDACEAE (M)
Tetragastris Gaertn. BURSERACEAE (D)
Tetraglochidion K.Schum. = Glochidion J.R.Forst. & G.Forst. EUPHORBIACEAE (D)
Tetraglochidium Bremek. = Strobilanthes Blume ACANTHACEAE (D)
Tetraglochin Poepp. ROSACEAE (D)
Tetraglossa Bedd. = Cleidion Blume EUPHORBIACEAE (D)
Tetragoga Bremek. = Strobilanthes Blume ACANTHACEAE (D)
Tetragompha Bremek. = Strobilanthes Blume ACANTHACEAE (D)
Tetragonia L. AIZOACEAE (D)
Tetragonocalamus Nakai = Bambusa Schreb. GRAMINEAE (M)
Tetragonolobus Scop. = Lotus L. LEGUMINOSAE–PAPILIONOIDEAE (D)
Tetragonotheca L. COMPOSITAE (D)
Tetragyne Miq. = Microdesmis Hook.f. ex Hook. PANDACEAE (D)
Tetralix Griseb. TILIACEAE (D)
Tetralobus A.DC. = Utricularia L. LENTIBULARIACEAE (D)
Tetralocularia O'Don. CONVOLVULACEAE (D)
Tetralopha Hook.f. = Gynochthodes Blume RUBIACEAE (D)
Tetrameles R.Br. DATISCACEAE (D)
Tetrameranthus R.E.Fr. ANNONACEAE (D)
Tetramerista Miq. TETRAMERISTACEAE (D)
Tetramerium C.F.Gaertn. (SUH) = Faramea Aubl. RUBIACEAE (D)
Tetramerium Nees ACANTHACEAE (D)
Tetramicra Lindl. ORCHIDACEAE (M)
Tetramolopium Nees COMPOSITAE (D)
Tetranema Benth. SCROPHULARIACEAE (D)
Tetraneuris Greene COMPOSITAE (D)
Tetranthera Jacq. = Litsea Lam. LAURACEAE (D)
Tetranthus Sw. COMPOSITAE (D)
Tetrapanax (K.Koch) K.Koch ARALIACEAE (D)
Tetrapathaea Rchb. = Passiflora L. PASSIFLORACEAE (D)
Tetrapeltis Lindl. = Otochilus Lindl. ORCHIDACEAE (M)
Tetraperone Urb. COMPOSITAE (D)
Tetrapetalum Miq. ANNONACEAE (D)
Tetraphyllaster Gilg MELASTOMATACEAE (D)
Tetraphyllum Griff. ex C.B.Clarke GESNERIACEAE (D)
Tetraphysa Schltr. ASCLEPIADACEAE (D)
Tetrapilus Lour. OLEACEAE (D)
Tetraplandra Baill. EUPHORBIACEAE (D)
Tetraplasandra A.Gray ARALIACEAE (D)
Tetraplasia Rehder = Damnacanthus C.F.Gaertn. RUBIACEAE (D)
Tetrapleura Benth. LEGUMINOSAE–MIMOSOIDEAE (D)
Tetrapodenia Gleason = Burdachia Mart. ex A.Juss. MALPIGHIACEAE (D)
Tetrapogon Desf. GRAMINEAE (M)
Tetrapollinia Mauire & B.M.Boom GENTIANACEAE (D)
Tetrapoma Turcz. ex Fisch. & C.A.Mey. = Rorippa Scop. CRUCIFERAE (D)
Tetrapora Schauer = Baeckea L. MYRTACEAE (D)
Tetraptera Phil. = Caya Kunth MALVACEAE (D)

Tetrapterocarpon Humbert LEGUMINOSAE–CAESALPINIOIDEAE (D)
Tetrapterys Cav. MALPIGHIACEAE (D)
Tetrardisia Mez MYRSINACEAE (D)
Tetraria P.Beauv. CYPERACEAE (M)
Tetrariopsis C.B.Clarke = Tetraria P.Beauv. CYPERACEAE (M)
Tetrarrhena R.Br. = Ehrharta Thunb. GRAMINEAE (M)
Tetraselago Junell SCROPHULARIACEAE (D)
Tetrasida Ulbr. MALVACEAE (D)
Tetrasiphon Urb. CELASTRACEAE (D)
Tetraspidium Baker SCROPHULARIACEAE (D)
Tetraspora Miq. = Baeckea L. MYRTACEAE (D)
Tetrastemma Diels ex H.Winkl. = Uvariopsis Engl. ANNONACEAE (D)
Tetrastemon Hook. & Arn. = Myrrhinium Schott MYRTACEAE (D)
Tetrastichella Pichon = Arrabidaea DC. BIGNONIACEAE (D)
Tetrastigma (Miq.) Planch. VITACEAE (D)
Tetrastigma K.Schum. (SUH) = Schumanniophyton Harms RUBIACEAE (D)
Tetrastylidium Engl. OLACACEAE (D)
Tetrastylis Barb.Rodr. = Passiflora L. PASSIFLORACEAE (D)
Tetrasynandra Perkins MONIMIACEAE (D)
Tetrataenium (DC.) Manden. UMBELLIFERAE (D)
Tetrataxis Hook.f. LYTHRACEAE (D)
Tetrateleia Arwidsson = Cleome L. CAPPARACEAE (D)
Tetratelia Sond. = Cleome L. CAPPARACEAE (D)
Tetrathalamus Lauterb. = Zygogynum Baill. WINTERACEAE (D)
Tetratheca Sm. TREMANDRACEAE (D)
Tetrathylacium Poepp. FLACOURTIACEAE (D)
Tetrathyrium Benth. HAMAMELIDACEAE (D)
Tetraulacium Turcz. SCROPHULARIACEAE (D)
Tetrazygia Rich. ex DC. MELASTOMATACEAE (D)
Tetrazygiopsis Borhidi = Tetrazygia Rich. ex DC. MELASTOMATACEAE (D)
Tetroncium Willd. JUNCAGINACEAE (M)
Tetrorchidiopsis Rauschert = Tetrorchidium Poepp. EUPHORBIACEAE (D)
Tetrorchidium Poepp. EUPHORBIACEAE (D)
Tetrorum Rose = Sedum L. CRASSULACEAE (D)
Teucridium Hook.f. VERBENACEAE (D)
Teucrium L. LABIATAE (D)
Teuscheria Garay ORCHIDACEAE (M)
Texiera Jaub. & Spach (SUS) = Glastaria Boiss. CRUCIFERAE (D)
Textoria Miq. = Dendropanax Decne. & Planch. ARALIACEAE (D)
Teyleria Backer LEGUMINOSAE–PAPILIONOIDEAE (D)
Teysmannia Rchb. & Zoll. = Johannesteijsmannia H.E.Moore PALMAE (M)
Thaia Seidenf. ORCHIDACEAE (M)
Thailentadopsis Kosterm. = Havardia Small LEGUMINOSAE–MIMOSOIDEAE (D)
Thalassia Banks ex C.Koenig HYDROCHARITACEAE (M)
Thalassodendron Hartog CYMODOCEACEAE (M)
Thalestris Rizzini = Justicia L. ACANTHACEAE (D)
Thalia L. MARANTACEAE (M)
Thalictrum L. RANUNCULACEAE (D)
Thalysia Kuntze (SUS) = Zea L. GRAMINEAE (M)
Thaminophyllum Harv. COMPOSITAE (D)
Thamnea Sol. ex Brongn. BRUNIACEAE (D)
Thamnocalamus Munro GRAMINEAE (M)
Thamnocharis W.T.Wang (UP) GESNERIACEAE (D)
Thamnochortus Bergius RESTIONACEAE (M)
Thamnojusticia Mildbr. = Justicia L. ACANTHACEAE (D)

Thamnopteris (C.Presl) C.Presl = Asplenium L. ASPLENIACEAE (F)
Thamnosciadium Hartvig UMBELLIFERAE (D)
Thamnoseris F.Phil. COMPOSITAE (D)
Thamnosma Torr. & Frém. RUTACEAE (D)
Thamnus Klotzsch ERICACEAE (D)
Thapsia L. UMBELLIFERAE (D)
Tharpia Britton & Rose = Senna Mill. LEGUMINOSAE–CAESALPINIOIDEAE (D)
Thaspium Nutt. UMBELLIFERAE (D)
Thaumasianthes Danser LORANTHACEAE (D)
Thaumastochloa C.E.Hubb. GRAMINEAE (M)
Thaumatocaryon Baill. BORAGINACEAE (D)
Thaumatococcus Benth. MARANTACEAE (M)
Thaumatophyllum Schott = Philodendron Schott ARACEAE (M)
Thayeria Copel. = Aglaomorpha Schott POLYPODIACEAE (F)
Thea L. = Camellia L. THEACEAE (D)
Thecacoris A.Juss. EUPHORBIACEAE (D)
Thecagonum Babu = Oldenlandia L. RUBIACEAE (D)
Thecanthes Wikstr. THYMELAEACEAE (D)
Thecocarpus Boiss. UMBELLIFERAE (D)
Thecophyllum E.F.Andre = Guzmania Ruíz & Pav. BROMELIACEAE (M)
Thecopus Seidenf. ORCHIDACEAE (M)
Thecorchus Bremek. RUBIACEAE (D)
Thecostele Rchb.f. ORCHIDACEAE (M)
Theileamea Baill. = Chlamydacanthus Lindau ACANTHACEAE (D)
Theilera E.Phillips CAMPANULACEAE (D)
Thelasis Blume ORCHIDACEAE (M)
Thelecarpus Tiegh. = Phragmanthera Tiegh. LORANTHACEAE (D)
Thelechitonia Cuatrec. = Wedelia Jacq. COMPOSITAE (D)
Theleophyton (Hook.f.) Moq. CHENOPODIACEAE (D)
Thelepaepale Bremek. = Strobilanthes Blume ACANTHACEAE (D)
Thelepogon Roth ex Roem. & Schult. GRAMINEAE (M)
Thelesperma Less. COMPOSITAE (D)
Thelethylax C.Cusset PODOSTEMACEAE (D)
Theligonum L. THELIGONACEAE (D)
Thelionema R.J.F.Hend. PHORMIACEAE (M)
Thelipteris Adans. = Pteris L. PTERIDACEAE (F)
Thellungia Stapf ex Probst = Eragrostis Wolf GRAMINEAE (M)
Thellungiella O.E.Schulz CRUCIFERAE (D)
Thelocactus (K.Schum.) Britton & Rose CACTACEAE (D)
Thelocephala Y.Ito = Neoporteria Britton & Rose CACTACEAE (D)
Thelomastus Frič (SUI) = Thelocactus (K.Schum.) Britton & Rose CACTACEAE (D)
Thelychiton Endl. = Dendrobium Sw. ORCHIDACEAE (M)
Thelycrania (Dumort.) Fourr. = Cornus L. CORNACEAE (D)
Thelymitra J.R.Forst. & G.Forst. ORCHIDACEAE (M)
Thelypodiopsis Rydb. CRUCIFERAE (D)
Thelypodium Endl. CRUCIFERAE (D)
Thelypotzium Gagnep. = Andrachne L. EUPHORBIACEAE (D)
Thelypteris Schmidel THELYPTERIDACEAE (F)
Thelyra Thouars = Hirtella L. CHRYSOBALANACEAE (D)
Thelyschista Garay ORCHIDACEAE (M)
Themeda Forssk. GRAMINEAE (M)
Themistoclesia Klotzsch ERICACEAE (D)
Thenardia Kunth APOCYNACEAE (D)
Theobroma L. STERCULIACEAE (D)
Theodorea Barb.Rodr. = Rodrigueziella Kuntze ORCHIDACEAE (M)
Theophrasta L. THEOPHRASTACEAE (D)

Thereianthus G.J.Lewis IRIDACEAE (M)
Theresa Clos = Scutellaria L. LABIATAE (D)
Theresia K.Koch = Fritillaria L. LILIACEAE (M)
Theriophonum Blume ARACEAE (M)
Thermopsis R.Br. LEGUMINOSAE–PAPILIONOIDEAE (D)
Therocistus Holub = Tuberaria (Dunal) Spach CISTACEAE (D)
Therofon Raf. (SUS) = Boykinia Nutt. SAXIFRAGACEAE (D)
Therophon Rydb. (SUO) = Boykinia Nutt. SAXIFRAGACEAE (D)
Theropogon Maxim. CONVALLARIACEAE (M)
Therorhodion (Maxim.) Small ERICACEAE (D)
Thesidium Sond. SANTALACEAE (D)
Thesium L. SANTALACEAE (D)
Thespesia Sol. ex Corrêa MALVACEAE (D)
Thespesiopsis Exell & Hillc. MALVACEAE (D)
Thespidium F.Muell. ex Benth. COMPOSITAE (D)
Thespis DC. COMPOSITAE (D)
Thevenotia DC. COMPOSITAE (D)
Thevetia L. APOCYNACEAE (D)
Theyodis A.Rich. = Oldenlandia L. RUBIACEAE (D)
Thibaudia Ruíz & Pav. ERICACEAE (D)
Thicuania Raf. = Dendrobium Sw. ORCHIDACEAE (M)
Thiebautia Colla = Bletia Ruíz & Pav. ORCHIDACEAE (M)
Thieleodoxa Cham. = Alibertia A.Rich. ex DC. RUBIACEAE (D)
Thiersia Baill. = Faramea Aubl. RUBIACEAE (D)
Thilachium Lour. CAPPARACEAE (D)
Thilcum Molina = Fuchsia L. ONAGRACEAE (D)
Thiloa Eichler COMBRETACEAE (D)
Thinogeton Benth. = Exodeconus Raf. SOLANACEAE (D)
Thinopyrum Á.Löve = Elymus L. GRAMINEAE (M)
Thinouia Triana & Planch. SAPINDACEAE (D)
Thiollierea Montrouz. = Bikkia Reinw. RUBIACEAE (D)
Thiseltonia Hemsl. COMPOSITAE (D)
Thismia Griff. BURMANNIACEAE (M)
Thium Steud. = Astragalus L. LEGUMINOSAE–PAPILIONOIDEAE (D)
Thladiantha Bunge CUCURBITACEAE (D)
Thlaspeocarpa C.A.Sm. CRUCIFERAE (D)
Thlaspi L. CRUCIFERAE (D)
Thlaspiceras F.K.Mey. = Thlaspi L. CRUCIFERAE (D)
Thogsennia Aiello RUBIACEAE (D)
Thollonia Baill. = Icacina A.Juss. ICACINACEAE (D)
Thomandersia Baill. ACANTHACEAE (D)
Thomasia J.Gay STERCULIACEAE (D)
Thomassetia Hemsl. = Brexia Noronha ex Thouars ESCALLONIACEAE (D)
Thompsonella Britton & Rose CRASSULACEAE (D)
Thompsonia R.Br. = Deidamia Noronha ex Thouars PASSIFLORACEAE (D)
Thomsonia Wall. = Amorphophallus Blume ex Decne. ARACEAE (M)
Thonnera De Wild. = Uvariopsis Engl. ANNONACEAE (D)
Thonningia Vahl BALANOPHORACEAE (D)
Thoracocarpus Harling CYCLANTHACEAE (M)
Thoracosperma Klotzsch ERICACEAE (D)
Thoracostachyum Kurz = Mapania Aubl. CYPERACEAE (M)
Thorea Rouy (SUH) = Arrhenatherum P.Beauv. GRAMINEAE (M)
Thoreldora Pierre = Glycosmis Corrêa RUTACEAE (D)
Thorelia Gagnep. = Camchaya Gagnep. COMPOSITAE (D)
Thorella Briq. = Caropsis (Rouy & Camus) Rauschert UMBELLIFERAE (D)

Thoreochloa Holub (SUS) = Arrhenatherum P.Beauv. GRAMINEAE (M)
Thornbera Rydb. = Dalea L. LEGUMINOSAE–PAPILIONOIDEAE (D)
Thorncroftia N.E.Br. LABIATAE (D)
Thornea Breedlove & E.McClintock GUTTIFERAE (D)
Thorntonia Rchb. = Pavonia Cav. MALVACEAE (D)
Thorvaldsenia Liebm. = Chysis Lindl. ORCHIDACEAE (M)
Thottea Rottb. ARISTOLOCHIACEAE (D)
Thouarsia Kuntze (SUS) = Thuarea Pers. GRAMINEAE (M)
Thouarsiora Homolle ex Arènes = Ixora L. RUBIACEAE (D)
Thouinia Poit. SAPINDACEAE (D)
Thouinidium Radlk. SAPINDACEAE (D)
Thouvenotia Danguy = Beilschmiedia Nees LAURACEAE (D)
Thozetia F.Muell. ex Benth. ASCLEPIADACEAE (D)
Thrasya Kunth GRAMINEAE (M)
Thrasyopsis Parodi GRAMINEAE (M)
Thraulococcus Radlk. = Lepisanthes Blume SAPINDACEAE (D)
Threlkeldia R.Br. CHENOPODIACEAE (D)
Thrinax Sw. PALMAE (M)
Thrincoma O.F.Cook = Coccothrinax Sargent PALMAE (M)
Thringis O.F.Cook = Coccothrinax Sargent PALMAE (M)
Thrixanthocereus Backeb. = Espostoa Britton & Rose CACTACEAE (D)
Thrixgyne Keng = Duthiea Hack. GRAMINEAE (M)
Thrixspermum Lour. ORCHIDACEAE (M)
Thryallis L. (SUH) = Galphimia Cav. MALPIGHIACEAE (D)
Thryallis Mart. MALPIGHIACEAE (D)
Thryothamnus Phil. = Verbena L. VERBENACEAE (D)
Thryptomene Endl. MYRTACEAE (D)
Thuarea Pers. GRAMINEAE (M)
Thuessinkia Korth. ex Miq. = Caryota L. PALMAE (M)
Thuinia Raf. (SUO) = Chionanthus L. OLEACEAE (D)
Thuja L. CUPRESSACEAE (G)
Thujopsis Siebold & Zucc. ex Endl. CUPRESSACEAE (G)
Thulinia P.J.Cribb ORCHIDACEAE (M)
Thunbergia Retz. ACANTHACEAE (D)
Thunbergianthus Engl. SCROPHULARIACEAE (D)
Thunbergiella H.Wolff = Itasina Raf. UMBELLIFERAE (D)
Thunbgeria Montin = Gardenia Ellis RUBIACEAE (D)
Thunia Rchb.f. ORCHIDACEAE (M)
Thuranthos C.H.Wright HYACINTHACEAE (M)
Thurberia Benth. (SUH) = Limnodea L.H.Dewey GRAMINEAE (M)
Thurberia A.Gray = Gossypium L. MALVACEAE (D)
Thurnia Hook.f. THURNIACEAE (M)
Thurovia Rose = Gutierrezia Lag. COMPOSITAE (D)
Thurya Boiss. & Balansa CARYOPHYLLACEAE (D)
Thuspeinanta T.Durand LABIATAE (D)
Thyella Raf. = Jacquemontia Choisy CONVOLVULACEAE (D)
Thylacanthus Tul. LEGUMINOSAE–CAESALPINIOIDEAE (D)
Thylacis Gagnep. = Thrixspermum Lour. ORCHIDACEAE (M)
Thylacodraba (Nábělek) O.E.Schulz = Draba L. CRUCIFERAE (D)
Thylacophora Ridl. = Riedelia Oliv. ZINGIBERACEAE (M)
Thylacopteris Kunze ex J.Sm. POLYPODIACEAE (F)
Thylacospermum Fenzl CARYOPHYLLACEAE (D)
Thymbra L. LABIATAE (D)
Thymelaea Mill. THYMELAEACEAE (D)
Thymocarpus Nicolson, Stcycrm. & Sivad. MARANTACEAE (M)

Thymophylla Lag. COMPOSITAE (D)
Thymopsis Benth. COMPOSITAE (D)
Thymus L. LABIATAE (D)
Thyrasperma N.E.Br. = Hymenogyne Haw. AIZOACEAE (D)
Thyridachne C.E.Hubb. GRAMINEAE (M)
Thyridocalyx Bremek. RUBIACEAE (D)
Thyridolepis S.T.Blake GRAMINEAE (M)
Thyridostachyum Nees (SUS) = Mnesithea Kunth GRAMINEAE (M)
Thyrocarpus Hance BORAGINACEAE (D)
Thyrsacanthus Nees = Odontonema Nees ACANTHACEAE (D)
Thyrsanthella Pichon APOCYNACEAE (D)
Thyrsanthemum Pichon COMMELINACEAE (M)
Thyrsanthera Pierre ex Gagnep. EUPHORBIACEAE (D)
Thyrsia Stapf = Phacelurus Griseb. GRAMINEAE (M)
Thyrsodium Salzm. ex Benth. ANACARDIACEAE (D)
Thyrsopteris Kunze DICKSONIACEAE (F)
Thyrsosalacia Loes. CELASTRACEAE (D)
Thyrsostachys Gamble GRAMINEAE (M)
Thysanachne C.Presl = Arundinella Raddi GRAMINEAE (M)
Thysanella A.Gray = Polygonella Michx. POLYGONACEAE (D)
Thysanobotrya Alderw. = Cyathea Sm. CYATHEACEAE (F)
Thysanocarpus Hook. CRUCIFERAE (D)
Thysanochilus Falc. = Eulophia R.Br. ex Lindl. ORCHIDACEAE (M)
Thysanoglossa Porto ORCHIDACEAE (M)
Thysanolaena Nees GRAMINEAE (M)
Thysanosoria Gepp LOMARIOPSIDACEAE (F)
Thysanospermum Champ. ex Benth. = Coptosapelta Korth. RUBIACEAE (D)
Thysanostemon Maguire GUTTIFERAE (D)
Thysanostigma J.B.Imlay ACANTHACEAE (D)
Thysanotus R.Br. ANTHERICACEAE (M)
Thysanus Lour. = Cnestis Juss. CONNARACEAE (D)
Tianschaniella Fedtsch. ex Popov BORAGINACEAE (D)
Tiarella L. SAXIFRAGACEAE (D)
Tiarocarpus Rech.f. COMPOSITAE (D)
Tiarrhena (Maxim.) Nakai (SUO) = Miscanthus Andersson GRAMINEAE (M)
Tibestina Maire = Dicoma Cass. COMPOSITAE (D)
Tibetia (Ali) H.P.Tsui = Gueldenstaedtia Fisch. LEGUMINOSAE–PAPILIONOIDEAE (D)
Tibouchina Aubl. MELASTOMATACEAE (D)
Tibouchinopsis Markgr. MELASTOMATACEAE (D)
Ticodendron Gómez–Laur. & L.D.Gómez TICODENDRACEAE (D)
Ticoglossum Luc.Rodr. ex Halbinger ORCHIDACEAE (M)
Ticorea Aubl. RUTACEAE (D)
Tidestromia Standl. AMARANTHACEAE (D)
Tiedemannia DC. = Oxypolis Raf. UMBELLIFERAE (D)
Tieghemella Pierre SAPOTACEAE (D)
Tieghemia Balle = Oncocalyx Tiegh. LORANTHACEAE (D)
Tieghemopanax R.Vig. = Polyscias J.R.Forst. & G.Forst. ARALIACEAE (D)
Tienmuia Hu SCROPHULARIACEAE (D)
Tietkensia P.S.Short COMPOSITAE (D)
Tiglium Klotzsch = Croton L. EUPHORBIACEAE (D)
Tigridia Juss. IRIDACEAE (M)
Tigridiopalma C.Chen MELASTOMATACEAE (D)
Tikalia Lundell = Blomia Miranda SAPINDACEAE (D)
Tilco Adans. = Fuchsia L. ONAGRACEAE (D)
Tilia L. TILIACEAE (D)

Tiliacora Colebr. MENISPERMACEAE (D)
Tilingia Regel UMBELLIFERAE (D)
Tillaea L. = Crassula L. CRASSULACEAE (D)
Tillaeastrum Britton = Crassula L. CRASSULACEAE (D)
Tillandsia L. BROMELIACEAE (M)
Tillospermum Griff. = Kunzea Rchb. MYRTACEAE (D)
Tilmia O.F.Cook = Aiphanes Willd. PALMAE (M)
Timandra Klotzsch = Croton L. EUPHORBIACEAE (D)
Timeroya Benth. (SUO) = Pisonia L. NYCTAGINACEAE (D)
Timeroyea Montrouz. = Pisonia L. NYCTAGINACEAE (D)
Timonius DC. RUBIACEAE (D)
Timouria Roshev. = Stipa L. GRAMINEAE (M)
Tina Schult. SAPINDACEAE (D)
Tinaea Garzia (SUS) = Lamarckia Moench GRAMINEAE (M)
Tinantia Scheidw. COMMELINACEAE (M)
Tinea Biv. = Neotinea Rchb.f. ORCHIDACEAE (M)
Tinguarra Parl. UMBELLIFERAE (D)
Tinnea Kotschy & Peyr. LABIATAE (D)
Tinomiscium Miers ex Hook.f. & Thomson MENISPERMACEAE (D)
Tinopsis Radlk. SAPINDACEAE (D)
Tinospora Miers MENISPERMACEAE (D)
Tintinabulum Rydb. = Gilia Ruíz & Pav. POLEMONIACEAE (D)
Tintinnabularia Woodson APOCYNACEAE (D)
Tipuana (Benth.) Benth. LEGUMINOSAE–PAPILIONOIDEAE (D)
Tipularia Nutt. ORCHIDACEAE (M)
Tiquilia Pers. BORAGINACEAE (D)
Tiquiliopsis A.Heller = Tiquilia Pers. BORAGINACEAE (D)
Tirania Pierre CAPPARACEAE (D)
Tirpitzia Hallier f. LINACEAE (D)
Tirucalia Raf. (SUH) = Euphorbia L. EUPHORBIACEAE (D)
Tischleria Schwantes = Carruanthus (Schwantes) Schwantes AIZOACEAE (D)
Tisonia Baill. FLACOURTIACEAE (D)
Tisserantia Humbert = Sphaeranthus L. COMPOSITAE (D)
Tisserantiella Mimeur (SUH) = Thyridachne C.E.Hubb. GRAMINEAE (M)
Tisserantiodoxa Aubrév. & Pellegr. = Englerophytum K.Krause SAPOTACEAE (D)
Tisserantodendron Sillans = Fernandoa Welw. ex Seem. BIGNONIACEAE (D)
Titania Endl. = Oberonia Lindl. ORCHIDACEAE (M)
Titanopsis Schwantes AIZOACEAE (D)
Titanotrichum Soler. GESNERIACEAE (D)
Tithonia Desf. ex Juss. COMPOSITAE (D)
Tithymaloides Ortega = Pedilanthus Neck. ex Poit. EUPHORBIACEAE (D)
Tithymalopsis Klotzsch & Garcke = Euphorbia L. EUPHORBIACEAE (D)
Tithymalus Hill (SUH) = Euphorbia L. EUPHORBIACEAE (D)
Tithymalus Gaertn. = Pedilanthus Neck. ex Poit. EUPHORBIACEAE (D)
Tittmannia Brongn. BRUNIACEAE (D)
Tjongina Adans. = Baeckea L. MYRTACEAE (D)
Tmcsipteris Bernh. PSILOTACEAE (FA)
Tobagoa Urb. RUBIACEAE (D)
Tococa Aubl. MELASTOMATACEAE (D)
Tocoyena Aubl. RUBIACEAE (D)
Todaroa Parl. UMBELLIFERAE (D)
Todaroa A.Rich. & Galeotti (SUH) = Campylocentrum Benth. ORCHIDACEAE (M)
Toddalia Juss. RUTACEAE (D)
Toddaliopsis Engl. RUTACEAE (D)
Toddavaddiu Kuntze (SUS) = Biophytum DC. OXALIDACEAE (D)

Todea Willd. ex Bernh. OSMUNDACEAE (F)
Toechima Radlk. SAPINDACEAE (D)
Tofieldia Huds. MELANTHIACEAE (M)
Tolbonia Kuntze = Calotis R.Br. COMPOSITAE (D)
Tolmachevia Á.Löve & D.Löve = Rhodiola L. CRASSULACEAE (D)
Tolmiea Torr. & A.Gray SAXIFRAGACEAE (D)
Tolpis Adans. COMPOSITAE (D)
Tolumnia Raf. = Oncidium Sw. ORCHIDACEAE (M)
Tolypanthus (Blume) Blume LORANTHACEAE (D)
Tomex Forssk. = Dobera Juss. SALVADORACEAE (D)
Tommasinia Bertol. UMBELLIFERAE (D)
Tomotris Raf. = Corymborkis Thouars ORCHIDACEAE (M)
Tonalanthus Brandegee = Calea L. COMPOSITAE (D)
Tonduzia Pittier APOCYNACEAE (D)
Tonella Nutt. ex A.Gray SCROPHULARIACEAE (D)
Tonestus A.Nelson COMPOSITAE (D)
Tongoloa H.Wolff UMBELLIFERAE (D)
Tonina Aubl. ERIOCAULACEAE (M)
Tonningia Juss. (SUH) = Cyanotis D.Don COMMELINACEAE (M)
Tontanea Aubl. = Coccocypselum P.Browne RUBIACEAE (D)
Tontelea Aubl. CELASTRACEAE (D)
Toona (Endl.) M.Roem. MELIACEAE (D)
Topobea Aubl. MELASTOMATACEAE (D)
Toppingia O.Deg., I.Deg. & A.R.Sm. = Pseudophegopteris Ching
   THELYPTERIDACEAE (F)
Tordyliopsis DC. UMBELLIFERAE (D)
Tordylium L. UMBELLIFERAE (D)
Torenia L. SCROPHULARIACEAE (D)
Torfasadis Raf. = Euphorbia L. EUPHORBIACEAE (D)
Torgesia Bornm. = Crypsis Aiton GRAMINEAE (M)
Toricellia DC. (SUO) = Torricellia DC. TORRICELLIACEAE (D)
Torilis Adans. UMBELLIFERAE (D)
Tornabenea Parl. UMBELLIFERAE (D)
Torpesia (Endl.) M.Roem. = Trichilia P.Browne MELIACEAE (D)
Torralbasia Krug & Urb. CELASTRACEAE (D)
Torrenticola Domin PODOSTEMACEAE (D)
Torresea Allemão = Amburana Schwacke & Taub. LEGUMINOSAE–
   PAPILIONOIDEAE (D)
Torresia Ruíz & Pav. = Hierochloe R.Br. GRAMINEAE (M)
Torreya Arn. TAXACEAE (G)
Torreyochloa Church GRAMINEAE (M)
Torricellia DC. TORRICELLIACEAE (D)
Torrubia Vell. = Pisonia L. NYCTAGINACEAE (D)
Torrukia Vell. = Pisonia L. NYCTAGINACEAE (D)
Tortuella Urb. RUBIACEAE (D)
Torularia O.E.Schulz (SUH) = Neotorularia Hedge & J.Léonard CRUCIFERAE (D)
Torulinium Desv. = Cyperus L. CYPERACEAE (M)
Tosagris P.Beauv. (SUI) = Muhlenbergia Schreb. GRAMINEAE (M)
Toubaouate Aubrév. & Pellegr. = Didelotia Baill. LEGUMINOSAE–
   CAESALPINIOIDEAE (D)
Touchardia Gaudich. URTICACEAE (D)
Touchiroa Aubl. = Crudia Schreb. LEGUMINOSAE–CAESALPINIOIDEAE (D)
Toulichiba Adans. = Ormosia G.Jacks. LEGUMINOSAE–PAPILIONOIDEAE (D)
Toulicia Aubl. SAPINDACEAE (D)
Touloucouna M.Roem. = Carapa Aubl. MELIACEAE (D)

Toumeya Britton & Rose = Sclerocactus Britton & Rose CACTACEAE (D)
Tounatea Aubl. = Swartzia Schreb. LEGUMINOSAE–PAPILIONOIDEAE (D)
Tournefortia L. BORAGINACEAE (D)
Tournefortiopsis Rusby = Guettarda L. RUBIACEAE (D)
Tournesol Adans. = Chrozophora Neck. ex Juss. EUPHORBIACEAE (D)
Tournesolia Scop. = Chrozophora Neck. ex Juss. EUPHORBIACEAE (D)
Tourneuxia Coss. COMPOSITAE (D)
Tournonia Moq. BASELLACEAE (D)
Touroulia Aubl. QUIINACEAE (D)
Tourrettia Foug. BIGNONIACEAE (D)
Toussaintia Boutique ANNONACEAE (D)
Tovara Adans. = Persicaria (L.) Mill. POLYGONACEAE (D)
Tovaria Ruíz & Pav. TOVARIACEAE (D)
Tovarochloa Macfarlane & But GRAMINEAE (M)
Tovomita Aubl. GUTTIFERAE (D)
Tovomitidium Ducke GUTTIFERAE (D)
Tovomitopsis Planch. & Triana GUTTIFERAE (D)
Townsendia Hook. COMPOSITAE (D)
Townsonia Cheeseman ORCHIDACEAE (M)
Toxanthera Hook.f. = Kedrostis Medik. CUCURBITACEAE (D)
Toxanthes Turcz. COMPOSITAE (D)
Toxicodendron Mill. = Rhus L. ANACARDIACEAE (D)
Toxicodendrum Thunb. (SUH) = Hyaenanche Lamb. EUPHORBIACEAE (D)
Toxocarpus Wight & Arn. ASCLEPIADACEAE (D)
Toxophoenix Schott = Astrocaryum G.Mey. PALMAE (M)
Toxopteris Trevis. = Syngramma J.Sm. ADIANTACEAE (F)
Toxostigma A.Rich. = Arnebia Forssk. BORAGINACEAE (D)
Tozzettia Savi = Alopecurus L. GRAMINEAE (M)
Tozzia L. SCROPHULARIACEAE (D)
Trachelanthus Kunze BORAGINACEAE (D)
Tracheliopsis Buser = Campanula L. CAMPANULACEAE (D)
Trachelium L. CAMPANULACEAE (D)
Trachelosiphon Schltr. = Eurystyles Wawra ORCHIDACEAE (M)
Trachelospermum Lem. APOCYNACEAE (D)
Trachoma Garay ORCHIDACEAE (M)
Trachomitum Woodson = Apocynum L. APOCYNACEAE (D)
Trachyandra Kunth ASPHODELACEAE (M)
Trachycalymma (K.Schum.) Bullock ASCLEPIADACEAE (D)
Trachycarpus H.A.Wendl. PALMAE (M)
Trachycaryon Klotzsch = Adriana Gaudich. EUPHORBIACEAE (D)
Trachydium Lindl. UMBELLIFERAE (D)
Trachylobium Hayne = Hymenaea L. LEGUMINOSAE–CAESALPINIOIDEAE (D)
Trachymene Rudge UMBELLIFERAE (D)
Trachynia Link = Brachypodium P.Beauv. GRAMINEAE (M)
Trachynotia Michx. (SUS) = Spartina Schreb. GRAMINEAE (M)
Trachyozus Rchb. (SUS) = Trachys Pers. GRAMINEAE (M)
Trachyphrynium K.Schum. (SUH) = Hypselodelphys (K.Schum.) Milne–Redh.
    MARANTACEAE (M)
Trachyphrynium Benth. MARANTACEAE (M)
Trachyphytum Nutt. = Mentzelia L. LOASACEAE (D)
Trachypoa Bubani (SUS) = Dactylis L. GRAMINEAE (M)
Trachypogon Nees GRAMINEAE (M)
Trachypremnon Lindig = Cyathea Sm. CYATHEACEAE (F)
Trachypteris E.F.André ex H.Christ ADIANTACEAE (F)
Trachyrhizum (Schltr.) Brieger = Dendrobium Sw. ORCHIDACEAE (M)

Trachys Pers. GRAMINEAE (M)
Trachysciadium Eckl. & Zeyh. = Pimpinella L. UMBELLIFERAE (D)
Trachyspermum Link UMBELLIFERAE (D)
Trachystachys A.Dietr. (SUS) = Trachys Pers. GRAMINEAE (M)
Trachystemon D.Don BORAGINACEAE (D)
Trachystigma C.B.Clarke GESNERIACEAE (D)
Trachystoma O.E.Schulz CRUCIFERAE (D)
Trachystylis S.T.Blake CYPERACEAE (M)
Tractocopevodia Raizada & Naray. RUTACEAE (D)
Tracyina S.F.Blake COMPOSITAE (D)
Tradescantella Small = Callisia Loefl. COMMELINACEAE (M)
Tradescantia L. COMMELINACEAE (M)
Traganopsis Maire & Wilczek CHENOPODIACEAE (D)
Traganthus Klotzsch = Bernardia Mill. EUPHORBIACEAE (D)
Traganum Delile CHENOPODIACEAE (D)
Tragia L. EUPHORBIACEAE (D)
Tragiella Pax & K.Hoffm. EUPHORBIACEAE (D)
Tragiola Small & Pennell = Gratiola L. SCROPHULARIACEAE (D)
Tragiopsis Karst. = Sebastiania Spreng. EUPHORBIACEAE (D)
Tragiopsis Pomel (SUH) = Stoibrax Raf. UMBELLIFERAE (D)
Tragium Spreng. = Pimpinella L. UMBELLIFERAE (D)
Tragopogon L. COMPOSITAE (D)
Tragoselinum Hall. = Pimpinella L. UMBELLIFERAE (D)
Tragus Panz. (SUS) = Brachypodium P.Beauv. GRAMINEAE (M)
Tragus Haller GRAMINEAE (M)
Trailliaedoxa W.W.Sm. & Forrest RUBIACEAE (D)
Transcaucasia M.Hiroe (UP) UMBELLIFERAE (D)
Trapa L. TRAPACEAE (D)
Trapella Oliv. PEDALIACEAE (D)
Trattinnickia Willd. BURSERACEAE (D)
Traubia Moldenke AMARYLLIDACEAE (M)
Traunia K.Schum. = Toxocarpus Wight & Arn. ASCLEPIADACEAE (D)
Traunsteinera Rchb. ORCHIDACEAE (M)
Trautvetteria Fisch. & C.A.Mey. RANUNCULACEAE (D)
Traversia Hook.f. COMPOSITAE (D)
Traxara Raf. = Lobostemon Lehm. BORAGINACEAE (D)
Traxilum Raf. = Ehretia P.Browne BORAGINACEAE (D)
Trechonaetes Miers = Jaborosa Juss. SOLANACEAE (D)
Treculia Decne. ex Trécul MORACEAE (D)
Treichelia Vatke CAMPANULACEAE (D)
Treisia Haw. = Euphorbia L. EUPHORBIACEAE (D)
Treleasea Rose (SUH) = Tradescantia L. COMMELINACEAE (M)
Trema Lour. ULMACEAE (D)
Tremacanthus S.Moore ACANTHACEAE (D)
Tremacron Craib GESNERIACEAE (D)
Tremandra R.Br. ex DC. TREMANDRACEAE (D)
Tremastelma Raf. = Scabiosa L. DIPSACACEAE (D)
Trematocarpus Zahlbr. CAMPANULACEAE (D)
Trematolobelia Zahlbr. = Trematocarpus Zahlbr. CAMPANULACEAE (D)
Trematosperma Urb. = Pyrenacantha Wight ICACINACEAE (D)
Trembleya DC. MELASTOMATACEAE (D)
Tremotis Raf. = Ficus L. MORACEAE (D)
Tremularia Fabr. (SUS) = Briza L. GRAMINEAE (M)
Trepocarpus Nutt. ex DC. UMBELLIFERAE (D)
Tresanthera Karst. RUBIACEAE (D)

Tretorhiza Adans. = Gentiana L. GENTIANACEAE (D)
Treubania Tiegh. = Amylotheca Tiegh. LORANTHACEAE (D)
Treubella Tiegh. (SUH) = Amylotheca Tiegh. LORANTHACEAE (D)
Treubella Pierre = Palaquium Blanco SAPOTACEAE (D)
Treutlera Hook.f. ASCLEPIADACEAE (D)
Trevauxia Steud. (SUO) = Luffa Mill. CUCURBITACEAE (D)
Trevesia Vis. ARALIACEAE (D)
Trevia L. (SUO) = Trewia L. EUPHORBIACEAE (D)
Trevirana Willd. = Achimenes Pers. GESNERIACEAE (D)
Trevirania Heynh. = Psychotria L. RUBIACEAE (D)
Trevoa Miers ex Hook. RHAMNACEAE (D)
Trevoria F.Lehm. ORCHIDACEAE (M)
Trevouxia Scop. = Luffa Mill. CUCURBITACEAE (D)
Trewia L. EUPHORBIACEAE (D)
Triachyrum Hochst. = Sporobolus R.Br. GRAMINEAE (M)
Triactina Hook.f. & Thomson = Sedum L. CRASSULACEAE (D)
Triadenia Spach = Hypericum L. GUTTIFERAE (D)
Triadenum Raf. GUTTIFERAE (D)
Triadica Lour. = Sapium P.Browne EUPHORBIACEAE (D)
Triadodaphne Kosterm. LAURACEAE (D)
Triaena Kunth = Bouteloua Lag. GRAMINEAE (M)
Triaenacanthus Nees = Strobilanthes Blume ACANTHACEAE (D)
Triaenophora (Hook.f.) Soler. SCROPHULARIACEAE (D)
Triainolepis Hook.f. RUBIACEAE (D)
Trianaea Planch. & Linden SOLANACEAE (D)
Trianaeopiper Trel. PIPERACEAE (D)
Trianoptiles Fenzl ex Endl. CYPERACEAE (M)
Trianosperma (Torr. & A.Gray) Mart. = Cayaponia Silva Manso CUCURBITACEAE (D)
Trianthema L. AIZOACEAE (D)
Trianthera Wettst. (SUH) = Stemotria Wettst. & Harms ex Engl.
    SCROPHULARIACEAE (D)
Trianthium Desv. (SUI) = Chrysopogon Trin. GRAMINEAE (M)
Triaristella Brieger = Trisetella Luer ORCHIDACEAE (M)
Triaristellina Rauschert = Trisetella Luer ORCHIDACEAE (M)
Triarrhena (Maxim.) Nakai = Miscanthus Andersson GRAMINEAE (M)
Triarthron Baill. = Phthirusa Mart. LORANTHACEAE (D)
Trias Lindl. ORCHIDACEAE (M)
Triaspis Burch. MALPIGHIACEAE (D)
Triathera Desv. = Bouteloua Lag. GRAMINEAE (M)
Triatherus Raf. (SUS) = Ctenium Panz. GRAMINEAE (M)
Triavenopsis Candargy = Duthiea Hack. GRAMINEAE (M)
Tribeles Phil. ESCALLONIACEAE (D)
Triblemma (J.Sm.) Ching = Diplazium Sw. WOODSIACEAE (F)
Tribolium Desv. GRAMINEAE (M)
Tribonanthes Endl. HAEMODORACEAE (M)
Tribrachia Lindl. = Bulbophyllum Thouars ORCHIDACEAE (M)
Tribrachya Korth. = Rennellia Korth. RUBIACEAE (D)
Tribroma O.F.Cook = Theobroma L. STERCULIACEAE (D)
Tribulocarpus S.Moore AIZOACEAE (D)
Tribulopis R.Br. ZYGOPHYLLACEAE (D)
Tribulopsis R.Br. (SUO) = Tribulopis R.Br. ZYGOPHYLLACEAE (D)
Tribulus L. ZYGOPHYLLACEAE (D)
Tricalistra Ridl. CONVALLARIACEAE (M)
Tricalysia A.Rich. ex DC. RUBIACEAE (D)
Tricardia Torr. HYDROPHYLLACEAE (D)

Tricarium Lour. = Phyllanthus L. EUPHORBIACEAE (D)
Tricarpelema J.K.Morton COMMELINACEAE (M)
Tricera Schreb. = Buxus L. BUXACEAE (D)
Triceratella Brenan COMMELINACEAE (M)
Triceratia A.Rich. = Sicydium Schltdl. CUCURBITACEAE (D)
Triceratorhynchus Summerh. ORCHIDACEAE (M)
Tricerma Liebm. = Maytenus Molina CELASTRACEAE (D)
Triceros Griff. = Gomphogyne Griff. CUCURBITACEAE (D)
Trichacanthus Zoll. & Moritzi = Blepharis Juss. ACANTHACEAE (D)
Trichachne Nees = Digitaria Haller GRAMINEAE (M)
Trichadenia Thwaites FLACOURTIACEAE (D)
Trichaeta P.Beauv. = Trisetaria Forssk. GRAMINEAE (M)
Trichantha Karst. & Triana (SUH) = Bonamia Thouars CONVOLVULACEAE (D)
Trichantha Hook. = Columnea L. GESNERIACEAE (D)
Trichanthemis Regel & Schmalh. COMPOSITAE (D)
Trichanthera Kunth ACANTHACEAE (D)
Trichanthodium Sond. & F.Muell. COMPOSITAE (D)
Trichapium Gilli (UP) COMPOSITAE (D)
Tricherostigma Boiss. = Euphorbia L. EUPHORBIACEAE (D)
Trichilia P.Browne MELIACEAE (D)
Trichinium R.Br. = Ptilotus R.Br. AMARANTHACEAE (D)
Trichiocarpa (Hook.) J.Sm. = Tectaria Cav. DRYOPTERIDACEAE (F)
Trichiogramme Kuhn = Pterozonium Fée ADIANTACEAE (F)
Trichipteris C.Presl = Cyathea Sm. CYATHEACEAE (F)
Trichlora Baker ALLIACEAE (M)
Trichloris E.Fourn. ex Benth. GRAMINEAE (M)
Trichobasis Turcz. = Conothamnus Lindl. MYRTACEAE (D)
Trichocalyx Schauer (SUH) = Calytrix Labill. MYRTACEAE (D)
Trichocalyx Balf.f. ACANTHACEAE (D)
Trichocarya Miq. = Licania Aubl. CHRYSOBALANACEAE (D)
Trichocaulon N.E.Br. ASCLEPIADACEAE (D)
Trichocentrum Poepp. & Endl. ORCHIDACEAE (M)
Trichocereus (A.Berger) Riccob. = Echinopsis Zucc. CACTACEAE (D)
Trichoceros Kunth ORCHIDACEAE (M)
Trichochiton Kom. = Cryptospora Kar. & Kir. CRUCIFERAE (D)
Trichochlaena Kuntze (SUO) = Tricholaena Schrad. ex Schult. & Schult.f.
    GRAMINEAE (M)
Trichochloa DC. (SUS) = Muhlenbergia Schreb. GRAMINEAE (M)
Trichocladus Pers. HAMAMELIDACEAE (D)
Trichocline Cass. COMPOSITAE (D)
Trichocoronis A.Gray COMPOSITAE (D)
Trichocoryne S.F.Blake COMPOSITAE (D)
Trichocyamos Yakovlev = Ormosia G.Jacks. LEGUMINOSAE–PAPILIONOIDEAE (D)
Trichocyclus Dulac = Woodsia R.Br. WOODSIACEAE (F)
Trichodesma R.Br. BORAGINACEAE (D)
Trichodiadema Schwantes AIZOACEAE (D)
Trichodiclida Cerv. (UR) GRAMINEAE (M)
Trichodium Michx. = Agrostis L. GRAMINEAE (M)
Trichodon Benth. (SUO) = Phragmites Adans. GRAMINEAE (M)
Trichodrymonia Oerst. = Paradrymonia Hanst. GESNERIACEAE (D)
Trichodypsis Baill. = Dypsis Noronha ex Mart. PALMAE (M)
Trichogalium Fourr. = Galium L. RUBIACEAE (D)
Trichoglottis Blume ORCHIDACEAE (M)
Trichogonia (DC.) Gardner COMPOSITAE (D)
Trichogoniopsis R.M.King & H.Rob. COMPOSITAE (D)

Trichogyne Less. COMPOSITAE (D)
Tricholaena Schrad. ex Schult. & Schult.f. GRAMINEAE (M)
Tricholaser Gilli UMBELLIFERAE (D)
Tricholepidium Ching = Neocheiropteris Christ POLYPODIACEAE (F)
Tricholepis DC. COMPOSITAE (D)
Tricholobus Blume = Connarus L. CONNARACEAE (D)
Trichomanes L. HYMENOPHYLLACEAE (F)
Trichomanes Hill (SUH) = Asplenium L. ASPLENIACEAE (F)
Trichonema Ker Gawl. = Romulea Maratti IRIDACEAE (M)
Trichoneura Andersson GRAMINEAE (M)
Trichoneuron Ching WOODSIACEAE (F)
Trichoon Roth = Phragmites Adans. GRAMINEAE (M)
Trichopetalum Lindl. ANTHERICACEAE (M)
Trichophorum Pers. CYPERACEAE (M)
Trichopilia Lindl. ORCHIDACEAE (M)
Trichopodium C.Presl (SUH) = Marina Liebm. LEGUMINOSAE–PAPILIONOIDEAE (D)
Trichopodium Lindl. = Trichopus Gaertn. TRICHOPODACEAE (M)
Trichopteria Nees (SUO) = Trichopteryx Nees GRAMINEAE (M)
Trichopteryx Nees GRAMINEAE (M)
Trichoptilium A.Gray COMPOSITAE (D)
Trichopus Gaertn. TRICHOPODACEAE (M)
Trichopyrum Á.Löve = Elymus L. GRAMINEAE (M)
Trichosacme Zucc. ASCLEPIADACEAE (D)
Trichosalpinx Luer ORCHIDACEAE (M)
Trichosanchezia Mildbr. ACANTHACEAE (D)
Trichosandra Decne. ASCLEPIADACEAE (D)
Trichosantha Steud. (SUI) = Stipa L. GRAMINEAE (M)
Trichosanthes L. CUCURBITACEAE (D)
Trichoschoenus J.Raynal CYPERACEAE (M)
Trichoscypha Hook.f. ANACARDIACEAE (D)
Trichosia Blume = Eria Lindl. ORCHIDACEAE (M)
Trichosma Lindl. = Eria Lindl. ORCHIDACEAE (M)
Trichosorus Liebm. = Lophosoria C.Presl DICKSONIACEAE (F)
Trichospermum Blume TILIACEAE (D)
Trichospira Kunth COMPOSITAE (D)
Trichosporum D.Don = Aeschynanthus Jack GESNERIACEAE (D)
Trichostachys Hook.f. RUBIACEAE (D)
Trichostelma Baill. (UP) ASCLEPIADACEAE (D)
Trichostema L. LABIATAE (D)
Trichostephania Tardieu = Ellipanthus Hook.f. CONNARACEAE (D)
Trichostephanus Gilg FLACOURTIACEAE (D)
Trichosterigma Klotzsch & Garcke = Euphorbia L. EUPHORBIACEAE (D)
Trichostigma A.Rich. PHYTOLACCACEAE (D)
Trichostomantherium Domin = Melodinus J.R.Forst. & G.Forst. APOCYNACEAE (D)
Trichotolinum O.E.Schulz CRUCIFERAE (D)
Trichotosia Blume ORCHIDACEAE (M)
Trichovaselia Tiegh. = Elvasia DC. OCHNACEAE (D)
Trichuriella Bennet AMARANTHACEAE (D)
Trichurus C.C.Towns. (SUH) = Trichuriella Bennet AMARANTHACEAE (D)
Tricliceras Thonn. ex DC. TURNERACEAE (D)
Triclisia Benth. MENISPERMACEAE (D)
Tricochilus Ames = Dipodium R.Br. ORCHIDACEAE (M)
Tricomaria Gillies ex Hook. & Arn. MALPIGHIACEAE (D)
Tricomariopsis Dubard = Sphedamnocarpus Planch. ex Benth. MALPIGHIACEAE (D)
Tricoryne R.Br. ANTHERICACEAE (M)

Tricostularia Nees ex Lehm. CYPERACEAE (M)
Tricuspidaria Ruíz & Pav. = Crinodendron Molina ELAEOCARPACEAE (D)
Tricuspis P.Beauv. (SUH) = Tridens Roem. & Schult. GRAMINEAE (M)
Tricycla Cav. = Bougainvillea Comm. ex Juss. NYCTAGINACEAE (D)
Tricyclandra Keraudren CUCURBITACEAE (D)
Tricyrtis Wall. CONVALLARIACEAE (M)
Tridactyle Schltr. ORCHIDACEAE (M)
Tridactylina (DC.) Sch.Bip. COMPOSITAE (D)
Tridax L. COMPOSITAE (D)
Tridens Roem. & Schult. GRAMINEAE (M)
Tridentea Haw. ASCLEPIADACEAE (D)
Tridesmis Lour. = Croton L. EUPHORBIACEAE (D)
Tridesmostemon Engl. SAPOTACEAE (D)
Tridianisia Baill. = Cassinopsis Sond. ICACINACEAE (D)
Tridimeris Baill. ANNONACEAE (D)
Tridynamia Gagnep. = Porana Burm.f. CONVOLVULACEAE (D)
Trieenea Hilliard SCROPHULARIACEAE (D)
Trientalis L. PRIMULACEAE (D)
Trifidacanthus Merr. LEGUMINOSAE–PAPILIONOIDEAE (D)
Trifoliada Rojas Acosta (UR) OXALIDACEAE (D)
Trifolium L. LEGUMINOSAE–PAPILIONOIDEAE (D)
Trifurcia Herb. = Herbertia Sweet IRIDACEAE (M)
Trigella Salisb. = Cyanella Royen ex L. TECOPHILAEACEAE (M)
Triglochin L. JUNCAGINACEAE (M)
Triglossum Roem. & Schult. = Arundinaria Michx. GRAMINEAE (M)
Trigonachras Radlk. SAPINDACEAE (D)
Trigonanthe (Schltr.) Brieger (SUI) = Dryadella Luer ORCHIDACEAE (M)
Trigonella L. LEGUMINOSAE–PAPILIONOIDEAE (D)
Trigonia Aubl. TRIGONIACEAE (D)
Trigoniastrum Miq. TRIGONIACEAE (D)
Trigonidium Lindl. ORCHIDACEAE (M)
Trigoniodendron E.F.Guim. & Miguel TRIGONIACEAE (D)
Trigonobalanus Forman FAGACEAE (D)
Trigonocapnos Schltr. PAPAVERACEAE (D)
Trigonocaryum Trautv. BORAGINACEAE (D)
Trigonochlamys Hook.f. = Santiria Blume BURSERACEAE (D)
Trigonophyllum (Prantl) Pic.Serm. = Trichomanes L. HYMENOPHYLLACEAE (F)
Trigonopleura Hook.f. EUPHORBIACEAE (D)
Trigonopyren Bremek. = Psychotria L. RUBIACEAE (D)
Trigonosciadium Boiss. UMBELLIFERAE (D)
Trigonospermum Less. COMPOSITAE (D)
Trigonospora Holttum THELYPTERIDACEAE (F)
Trigonostemon Blume EUPHORBIACEAE (D)
Trigonotis Steven BORAGINACEAE (D)
Triguera Cav. SOLANACEAE (D)
Trigynaea Schltdl. ANNONACEAE (D)
Trigyneia Rchb. (SUO) = Trigynaea Schltdl. ANNONACEAE (D)
Trigynia Jacq.–Fél. = Leandra Raddi MELASTOMATACEAE (D)
Trikeraia Bor GRAMINEAE (M)
Trilepidea Tiegh. LORANTHACEAE (D)
Trilepis Nees CYPERACEAE (M)
Trilepisium Thouars MORACEAE (D)
Triliena Raf. = Acnistus Schott SOLANACEAE (D)
Trilisa (Cass.) Cass. COMPOSITAE (D)
Trillium L. TRILLIACEAE (M)

Trilobachne Schenck ex Henrard GRAMINEAE (M)
Trilobulina Raf. = Utricularia L. LENTIBULARIACEAE (D)
Trilopus Mitch. = Hamamelis L. HAMAMELIDACEAE (D)
Trimenia Seem. TRIMENIACEAE (D)
Trimeranthus Karst. = Chaetolepis (DC.) Miq. MELASTOMATACEAE (D)
Trimeria Harv. FLACOURTIACEAE (D)
Trimeris C.Presl CAMPANULACEAE (D)
Trimerisma C.Presl (SUS) = Platylophus D.Don CUNONIACEAE (D)
Trimeriza Lindl. = Thottea Rottb. ARISTOLOCHIACEAE (D)
Trimerocalyx (Murb.) Murb. = Linaria Mill. SCROPHULARIACEAE (D)
Trimezia Salisb. ex Herb. IRIDACEAE (M)
Trimorphandra Brongn. & Gris = Hibbertia Andrews DILLENIACEAE (D)
Trinia Hoffm. UMBELLIFERAE (D)
Triniella Calest. = Trinia Hoffm. UMBELLIFERAE (D)
Triniochloa Hitchc. GRAMINEAE (M)
Triniusa Steud. = Bromus L. GRAMINEAE (M)
Triodanis Raf. CAMPANULACEAE (D)
Triodia R.Br. GRAMINEAE (M)
Triodoglossum Bullock ASCLEPIADACEAE (D)
Triodon Baumg. (SUS) = Triodia R.Br. GRAMINEAE (M)
Triodon DC. = Diodia L. RUBIACEAE (D)
Triolena Naudin MELASTOMATACEAE (D)
Triomma Hook.f. BURSERACEAE (D)
Trioncinia (F.Muell.) Veldkamp COMPOSITAE (D)
Triopterys L. MALPIGHIACEAE (D)
Triorchis Agosti = Spiranthes Rich. ORCHIDACEAE (M)
Triorchos Small & Nash = Pteroglossaspis Rchb.f. ORCHIDACEAE (M)
Triosteum L. CAPRIFOLIACEAE (D)
Tripetalanthus A.Chev. = Plagiosiphon Harms LEGUMINOSAE–
    CAESALPINIOIDEAE (D)
Tripetaleia Siebold & Zucc. ERICACEAE (D)
Tripetalum K.Schum. = Garcinia L. GUTTIFERAE (D)
Triphasia Lour. RUTACEAE (D)
Triphelia R.Br. ex Endl. = Actinodium Schauer MYRTACEAE (D)
Triphlebia Stapf (SUH) = Eragrostis Wolf GRAMINEAE (M)
Triphlebium Baker = Diplora Baker ASPLENIACEAE (F)
Triphora Nutt. ORCHIDACEAE (M)
Triphylleion Suess. = Niphogeton Schltdl. UMBELLIFERAE (D)
Triphyophyllum Airy Shaw DIONCOPHYLLACEAE (D)
Triplachne Link GRAMINEAE (M)
Tripladenia D.Don CONVALLARIACEAE (M)
Triplandra Raf. = Croton L. EUPHORBIACEAE (D)
Triplarina Raf. = Baeckea L. MYRTACEAE (D)
Triplaris Loefl. ex. L. POLYGONACEAE (D)
Triplasandra Seem. = Tetraplasandra A.Gray ARALIACEAE (D)
Triplasis P.Beauv. GRAMINEAE (M)
Triplathera (Endl.) Lindl. = Bouteloua Lag. GRAMINEAE (M)
Triplectrum Wight & Arn. = Medinilla Gaudich. MELASTOMATACEAE (D)
Tripleura Lindl. = Zeuxine Lindl. ORCHIDACEAE (M)
Tripleurospermum Sch.Bip. COMPOSITAE (D)
Triplisomeris (Baill.) Aubrév. & Pellegr. = Anthonotha P.Beauv. LEGUMINOSAE–
    CAESALPINIOIDEAE (D)
Triplocephalum O.Hoffm. COMPOSITAE (D)
Triplochiton K.Schum. STERCULIACEAE (D)
Triplochlamys Ulbr. MALVACEAE (D)

Triplolepis Turcz. = Streptocaulon Wight & Arn. ASCLEPIADACEAE (D)
Triplopetalum Nyarady = Alyssum L. CRUCIFERAE (D)
Triplophyllum Holttum DRYOPTERIDACEAE (F)
Triplopogon Bor GRAMINEAE (M)
Triplorhiza Ehrh. = Pseudorchis Séguier ORCHIDACEAE (M)
Triplostegia Wall. ex DC. TRIPLOSTEGIACEAE (D)
Triplotaxis Hutch. = Vernonia Schreb. COMPOSITAE (D)
Tripodandra Baill. = Rhaptonema Miers MENISPERMACEAE (D)
Tripodanthera M.Roem. = Gymnopetalum Arn. CUCURBITACEAE (D)
Tripodanthus (Eichler) Tiegh. LORANTHACEAE (D)
Tripodion Medik. = Anthyllis L. LEGUMINOSAE–PAPILIONOIDEAE (D)
Tripogandra Raf. COMMELINACEAE (M)
Tripogon Roem. & Schult. GRAMINEAE (M)
Tripsacum L. GRAMINEAE (M)
Tripteris Less. = Osteospermum L. COMPOSITAE (D)
Tripterocalyx (Torr.) Hook. NYCTAGINACEAE (D)
Tripterococcus Endl. STACKHOUSIACEAE (D)
Tripterodendron Radlk. SAPINDACEAE (D)
Tripterospermum Blume GENTIANACEAE (D)
Tripterygium Hook.f. CELASTRACEAE (D)
Triptilion Ruíz & Pav. COMPOSITAE (D)
Triptilodiscus Turcz. COMPOSITAE (D)
Triraphis R.Br. GRAMINEAE (M)
Trirostellum Z.P.Wang & Xie = Gynostemma Blume CUCURBITACEAE (D)
Trisanthus Lour. = Centella L. UMBELLIFERAE (D)
Triscenia Griseb. GRAMINEAE (M)
Trisciadia Hook.f. = Coelospermum Blume RUBIACEAE (D)
Trisciadium Phil. = Huanaca Cav. UMBELLIFERAE (D)
Triscyphus Taub. ex Warm. = Thismia Griff. BURMANNIACEAE (M)
Trisema Hook.f. = Hibbertia Andrews DILLENIACEAE (D)
Trisepalum C.B.Clarke GESNERIACEAE (D)
Trisetaria Forssk. GRAMINEAE (M)
Trisetarium Poir. (SUS) = Trisetum Pers. GRAMINEAE (M)
Trisetella Luer ORCHIDACEAE (M)
Trisetobromus Nevski = Bromus L. GRAMINEAE (M)
Trisetum Pers. GRAMINEAE (M)
Trisiola Raf. (SUH) = Distichlis Raf. GRAMINEAE (M)
Trisiola Raf. (SUS) = Uniola L. GRAMINEAE (M)
Trismeria Fée = Pityrogramma Link ADIANTACEAE (F)
Trispermium Hill = Selaginella P.Beauv. SELAGINELLACEAE (FA)
Tristachya Nees GRAMINEAE (M)
Tristagma Poepp. ALLIACEAE (M)
Tristania Poir. (SUI) = Spartina Schreb. GRAMINEAE (M)
Tristania R.Br. MYRTACEAE (D)
Tristaniopsis Brongn. & Gris MYRTACEAE (D)
Tristeca P.Beauv. = Psilotum Sw. PSILOTACEAE (FA)
Tristegis Nees (SUI) = Melinis P.Beauv. GRAMINEAE (M)
Tristellateia Thouars MALPIGHIACEAE (D)
Tristemma Juss. MELASTOMATACEAE (D)
Tristemon Scheele = Cucurbita L. CUCURBITACEAE (D)
Tristemonanthus Loes. CELASTRACEAE (D)
Tristerix Mart. LORANTHACEAE (D)
Tristicha Thouars PODOSTEMACEAE (D)
Tristira Radlk. SAPINDACEAE (D)
Tristiropsis Radlk. SAPINDACEAE (D)

Tristylium Turcz. = Cleyera Thunb. THEACEAE (D)
Trisynsyne Baill. = Nothofagus Blume FAGACEAE (D)
Tritaxis Baill. = Trigonostemon Blume EUPHORBIACEAE (D)
Tritelandra Raf. = Epidendrum L. ORCHIDACEAE (M)
Triteleia Douglas ex Lindl. ALLIACEAE (M)
Triteleiopsis Hoover ALLIACEAE (M)
Trithecanthera Tiegh. LORANTHACEAE (D)
Trithrinax Mart. PALMAE (M)
Trithuria Hook.f. HYDATELLACEAE (M)
Triticum L. GRAMINEAE (M)
Tritomopterys (A.Juss. ex Endl.) Nied. = Gaudichaudia Kunth MALPIGHIACEAE (D)
Tritonia Ker Gawl. IRIDACEAE (M)
Tritoniopsis L.Bolus IRIDACEAE (M)
Triumfetta L. TILIACEAE (D)
Triumfettoides Rauschert = Triumfetta L. TILIACEAE (D)
Triunila Raf. (SUI) = Uniola L. GRAMINEAE (M)
Triuranthera Backer = Driessenia Korth. MELASTOMATACEAE (D)
Triuris Miers TRIURIDACEAE (M)
Triurocodon Schltr. = Thismia Griff. BURMANNIACEAE (M)
Trivalvaria (Miq.) Miq. ANNONACEAE (D)
Trixapias Raf. = Utricularia L. LENTIBULARIACEAE (D)
Trixis P.Browne COMPOSITAE (D)
Trixostis Raf. = Aristida L. GRAMINEAE (M)
Trizeuxis Lindl. ORCHIDACEAE (M)
Trochera Rich. = Ehrharta Thunb. GRAMINEAE (M)
Trochetia DC. STERCULIACEAE (D)
Trochetiopsis Marais STERCULIACEAE (D)
Trochilocactus Lindinger = Disocactus Lindl. CACTACEAE (D)
Trochisandra Bedd. = Bhesa Buch.-Ham. ex Arn. CELASTRACEAE (D)
Trochiscanthes W.D.J.Koch UMBELLIFERAE (D)
Trochiscus O.E.Schulz CRUCIFERAE (D)
Trochocarpa R.Br. EPACRIDACEAE (D)
Trochocodon Candargy (UP) CAMPANULACEAE (D)
Trochodendron Siebold & Zucc. TROCHODENDRACEAE (D)
Trochomeria Hook.f. CUCURBITACEAE (D)
Trochomeriopsis Cogn. CUCURBITACEAE (D)
Trochopteris Gardner = Anemia Sw. SCHIZAEACEAE (F)
Troglophyton Hilliard & B.L.Burtt COMPOSITAE (D)
Trogostolon Copel. DAVALLIACEAE (F)
Trollius L. RANUNCULACEAE (D)
Tromotriche Haw. ASCLEPIADACEAE (D)
Tromsdorffia Blume = Chirita Buch.-Ham. ex D.Don GESNERIACEAE (D)
Troostwykia Miq. = Agelaea Sol. ex Planch. CONNARACEAE (D)
Tropaeastrum Mabb. (SUO) = Trophaeastrum Sparre TROPAEOLACEAE (D)
Tropaeolum L. TROPAEOLACEAE (D)
Tropalanthe S.Moore = Pycnandra Benth. SAPOTACEAE (D)
Trophaeastrum Sparre TROPAEOLACEAE (D)
Trophianthus Scheidw. = Aspasia Lindl. ORCHIDACEAE (M)
Trophis P.Browne MORACEAE (D)
Trophisomia Rojas Acosta = Sorocea A.St.-Hil. MORACEAE (D)
Tropidia Lindl. ORCHIDACEAE (M)
Tropidocarpum Hook. CRUCIFERAE (D)
Tropidopetalum Turcz. = Bouea Meisn. ANACARDIACEAE (D)
Tropilis Raf. = Dendrobium Sw. ORCHIDACEAE (M)
Trouettia Pierre ex Baill. = Niemeyera F.Muell. SAPOTACEAE (D)

Trozelia Raf. = Acnistus Schott SOLANACEAE (D)
Trudelia Garay ORCHIDACEAE (M)
Trujanoa La Llave = Rhus L. ANACARDIACEAE (D)
Trukia Kaneh. RUBIACEAE (D)
Trungboa Rauschert SCROPHULARIACEAE (D)
Trybliocalyx Lindau = Chileranthemum Oerst. ACANTHACEAE (D)
Trychinolepis B.L.Rob. = Ophryosporus Meyen COMPOSITAE (D)
Trymalium Fenzl RHAMNACEAE (D)
Trymatococcus Poepp. & Endl. MORACEAE (D)
Tryonella Pic.Serm. = Doryopteris J.Sm. ADIANTACEAE (F)
Tryphane Rchb. = Minuartia L. CARYOPHYLLACEAE (D)
Tryphia Lindl. = Holothrix Rich. ex Lindl. ORCHIDACEAE (M)
Tryphostemma Harv. = Basananthe Peyr. PASSIFLORACEAE (D)
Tryptomene Walp. = Thryptomene Endl. MYRTACEAE (D)
Tryssophyton Wurdack MELASTOMATACEAE (D)
Tsaiorchis T.Tang & F.T.Wang ORCHIDACEAE (M)
Tschompskia Asch. & Graebn. (SUI) = Arundinaria Michx. GRAMINEAE (M)
Tsebona Capuron SAPOTACEAE (D)
Tsiangia But, H.H.Hsue & P.T.Li RUBIACEAE (D)
Tsimatimia Jum. & H.Perrier = Garcinia L. GUTTIFERAE (D)
Tsingya Capuron SAPINDACEAE (D)
Tsoongia Merr. VERBENACEAE (D)
Tsoongiodendron Chun = Michelia L. MAGNOLIACEAE (D)
Tsuga Carrière PINACEAE (G)
Tsusiophyllum Maxim. ERICACEAE (D)
Tuberaria (Dunal) Spach CISTACEAE (D)
Tuberolabium Yamam. ORCHIDACEAE (M)
Tuberostyles Benth. (SUO) = Tuberostylis Steetz COMPOSITAE (D)
Tuberostylis Steetz COMPOSITAE (D)
Tubiflora J.F.Gmel. = Elytraria Michx. ACANTHACEAE (D)
Tubilabium J.J.Sm. ORCHIDACEAE (M)
Tubocapsicum Makino SOLANACEAE (D)
Tuckermannia Nutt. = Coreopsis L. COMPOSITAE (D)
Tucma Ravenna = Ennealophus N.E.Br. IRIDACEAE (M)
Tuctoria J.R.Reeder GRAMINEAE (M)
Tuerckheimocharis Urb. SCROPHULARIACEAE (D)
Tugarinovia Iljin COMPOSITAE (D)
Tula Adans. = Nolana L.f. SOLANACEAE (D)
Tulasnea Naudin = Siphanthera Pohl MELASTOMATACEAE (D)
Tulasnea Wight (SUI) = Dalzellia Wight PODOSTEMACEAE (D)
Tulasneantha P.Royen PODOSTEMACEAE (D)
Tulbaghia L. ALLIACEAE (M)
Tulestea Aubrév. & Pellegr. = Synsepalum (A.DC.) Daniell SAPOTACEAE (D)
Tulexis Raf. = Brassavola R.Br. ORCHIDACEAE (M)
Tulipa L. LILIACEAE (M)
Tulisma Raf. = Sinningia Nees GESNERIACEAE (D)
Tulotis Raf. = Platanthera Rich. ORCHIDACEAE (M)
Tumalis Raf. = Euphorbia L. EUPHORBIACEAE (D)
Tumamoca Rose CUCURBITACEAE (D)
Tumidinodus H.W.Li = Anna Pellegr. GESNERIACEAE (D)
Tunaria Kuntze = Cantua Juss. ex Lam. POLEMONIACEAE (D)
Tunas Lunell = Opuntia Mill. CACTACEAE (D)
Tunica Mert. & W.D.J.Koch (SUH) = Petrorhagia (Ser.) Link CARYOPHYLLACEAE (D)
Tunica Ludw. = Dianthus L. CARYOPHYLLACEAE (D)
Tupa G.Don = Lobelia L. CAMPANULACEAE (D)

Tupeia Cham. & Schltdl. LORANTHACEAE (D)
Tupeianthus Takht. (SUO) = Tepuianthus Maguire & Steyerm. TEPUIANTHACEAE (D)
Tupidanthus Hook.f. & Thomson = Schefflera J.R.Forst. & G.Forst. ARALIACEAE (D)
Tupistra Ker Gawl. CONVALLARIACEAE (M)
Turaniphytum Poljakov COMPOSITAE (D)
Turbina Raf. CONVOLVULACEAE (D)
Turbinicarpus (Backeb.) Buxb. & Backeb. CACTACEAE (D)
Turczaninoviella Kozo-Pol. (UP) UMBELLIFERAE (D)
Turgenia Hoffm. UMBELLIFERAE (D)
Turgeniopsis Boiss. = Glochidotheca Fenzl UMBELLIFERAE (D)
Turia Forssk. ex J.F.Gmelin = Luffa Mill. CUCURBITACEAE (D)
Turnera L. TURNERACEAE (D)
Turpinia Vent. STAPHYLEACEAE (D)
Turraea L. MELIACEAE (D)
Turraeanthus Baill. MELIACEAE (D)
Turraya Wall. (SUI) = Leersia Sw. GRAMINEAE (M)
Turricula J.F.Macbr. HYDROPHYLLACEAE (D)
Turrigera Decne. = Tweedia Hook. & Arn. ASCLEPIADACEAE (D)
Turrillia A.C.Sm. (SUS) = Bleasdalea F.Muell. ex Domin PROTEACEAE (D)
Turrita Wallr. = Arabis L. CRUCIFERAE (D)
Turritis L. = Arabis L. CRUCIFERAE (D)
Turukhania Vassilcz. = Medicago L. LEGUMINOSAE-PAPILIONOIDEAE (D)
Tussaca Raf. = Goodyera R.Br. ORCHIDACEAE (M)
Tussacia Raf. ex Desv. (SUH) = Spiranthes Rich. ORCHIDACEAE (M)
Tussacia Benth. = Chrysothemis Decne. GESNERIACEAE (D)
Tussilago L. COMPOSITAE (D)
Tutcheria Dunn THEACEAE (D)
Tuxtla Villaseñor & Strother COMPOSITAE (D)
Tuyamaea Yamaz. = Lindernia All. SCROPHULARIACEAE (D)
Tweedia Hook. & Arn. ASCLEPIADACEAE (D)
Tydaea Decne. = Kohleria Regel GESNERIACEAE (D)
Tylanthera C.Hansen MELASTOMATACEAE (D)
Tylecarpus Engl. = Medusanthera Seem. ICACINACEAE (D)
Tylecodon Tölken CRASSULACEAE (D)
Tyleria Gleason OCHNACEAE (D)
Tyleropappus Greenm. COMPOSITAE (D)
Tylocarya Nelmes = Fimbristylis Vahl CYPERACEAE (M)
Tylochilus Nees = Cyrtopodium R.Br. ORCHIDACEAE (M)
Tylodontia Griseb. ASCLEPIADACEAE (D)
Tyloglossa Hochst. = Justicia L. ACANTHACEAE (D)
Tylopetalum Barneby & Krukoff = Sciadotenia Miers MENISPERMACEAE (D)
Tylophora R.Br. ASCLEPIADACEAE (D)
Tylophoropsis N.E.Br. ASCLEPIADACEAE (D)
Tylopsacas Leeuwenb. GESNERIACEAE (D)
Tylosema (Schweinf.) Torre & Hillcoat LEGUMINOSAE-CAESALPINIOIDEAE (D)
Tylosepalum Kurz ex Teijsm. & Binn. = Trigonostemon Blume EUPHORBIACEAE (D)
Tylospermu Leeuwenb. (SUH) = Tylopsacas Leeuwenb. GESNERIACEAE (D)
Tylosperma Botsch. = Potentilla L. ROSACEAE (D)
Tylostemon Engl. = Beilschmiedia Nees LAURACEAE (D)
Tylostigma Schltr. ORCHIDACEAE (M)
Tylostylis Blume = Eria Lindl. ORCHIDACEAE (M)
Tylothrasya Döll = Thrasya Kunth GRAMINEAE (M)
Tynanthus Miers (SUO) = Tynnanthus Miers BIGNONIACEAE (D)
Tynnanthus Miers BIGNONIACEAE (D)
Typha L. TYPHACEAE (M)

Typhoides Moench (SUS) = Phalaris L. GRAMINEAE (M)
Typhonium Schott ARACEAE (M)
Typhonodorum Schott ARACEAE (M)
Tyria Klotzsch ex Endl. = Bernardia Mill. EUPHORBIACEAE (D)
Tyrimnus (Cass.) Cass. COMPOSITAE (D)
Tysonia Bolus (SUH) = Afrotysonia Rauschert BORAGINACEAE (D)
Tytthostemma Nevski = Stellaria L. CARYOPHYLLACEAE (D)
Tzellemtinia Chiov. = Bridelia Willd. EUPHORBIACEAE (D)
Tzvelevia E.B.Alexeev = Festuca L. GRAMINEAE (M)

Uapaca Baill. EUPHORBIACEAE (D)
Ubochea Baill. = Stachytarpheta Vahl VERBENACEAE (D)
Ucriana Willd. (UP) RUBIACEAE (D)
Udani Adans. = Quisqualis L. COMBRETACEAE (D)
Udora Nutt. = Elodea Michx. HYDROCHARITACEAE (M)
Uebelinia Hochst. CARYOPHYLLACEAE (D)
Uebelmannia Buining CACTACEAE (D)
Uechtritzia Freyn COMPOSITAE (D)
Ugamia Pavlov COMPOSITAE (D)
Ugena Cav. = Lygodium Sw. SCHIZAEACEAE (F)
Ugni Turcz. MYRTACEAE (D)
Uittienia Steenis = Dialium L. LEGUMINOSAE–CAESALPINIOIDEAE (D)
Uladendron Marc.–Berti (UP) MALVACEAE (D)
Ulantha Hook. = Chloraea Lindl. ORCHIDACEAE (M)
Ulbrichia Urb. MALVACEAE (D)
Uldinia J.M.Black UMBELLIFERAE (D)
Uleanthus Harms LEGUMINOSAE–PAPILIONOIDEAE (D)
Ulearum Engl. ARACEAE (M)
Uleiorchis Hoehne ORCHIDACEAE (M)
Uleodendron Rauschert = Naucleopsis Miq. MORACEAE (D)
Uleophytum Hieron. COMPOSITAE (D)
Ulex L. LEGUMINOSAE–PAPILIONOIDEAE (D)
Ulleria Bremek. = Ruellia L. ACANTHACEAE (D)
Ulloa Pers. = Juanulloa Ruíz & Pav. SOLANACEAE (D)
Ullucus Caldas BASELLACEAE (D)
Ulmus L. ULMACEAE (D)
Ulostoma G.Don = Gentiana L. GENTIANACEAE (D)
Ulticona Raf. = Hebecladus Miers SOLANACEAE (D)
Ultragossypium Roberty = Gossypium L. MALVACEAE (D)
Ulugbekia Zak. = Arnebia Forssk. BORAGINACEAE (D)
Umbellularia (Nees) Nutt. LAURACEAE (D)
Umbilicus DC. CRASSULACEAE (D)
Umtiza T.R.Sim LEGUMINOSAE–CAESALPINIOIDEAE (D)
Uncaria Burch. (SUH) = Harpagophytum DC. ex Meisn. PEDALIACEAE (D)
Uncaria Schreb. RUBIACEAE (D)
Uncarina (Baill.) Stapf PEDALIACEAE (D)
Uncariopsis Karst. = Schradera Vahl RUBIACEAE (D)
Uncifera Lindl. ORCHIDACEAE (M)
Uncinaria Rchb. (SUO) = Uncaria Schreb. RUBIACEAE (D)
Uncinia Pers. CYPERACEAE (M)
Ungeria Schott & Endl. STERCULIACEAE (D)
Ungernia Bunge AMARYLLIDACEAE (M)
Ungnadia Endl. SAPINDACEAE (D)
Ungula Barlow = Amyema Tiegh. LORANTHACEAE (D)
Ungulipetalum Moldenke MENISPERMACEAE (D)

Unigenes E.Wimm. CAMPANULACEAE (D)
Uniola L. GRAMINEAE (M)
Unona L.f. = Xylopia L. ANNONACEAE (D)
Unonopsis R.E.Fr. ANNONACEAE (D)
Unxia L.f. COMPOSITAE (D)
Upuna Symington DIPTEROCARPACEAE (D)
Urachne Trin. (SUS) = Oryzopsis Michx. GRAMINEAE (M)
Uragoga Baill. = Psychotria L. RUBIACEAE (D)
Uralepis Nutt. = Triplasis P.Beauv. GRAMINEAE (M)
Uralepsis Nutt. (SUO) = Triplasis P.Beauv. GRAMINEAE (M)
Urandra Thwaites = Stemonurus Blume ICACINACEAE (D)
Uranodactylus Gilli = Winklera Regel CRUCIFERAE (D)
Uranthera Naudin = Acisanthera P.Browne MELASTOMATACEAE (D)
Uranthera Pax & K.Hoffm. (SUH) = Phyllanthus L. EUPHORBIACEAE (D)
Uranthoecium Stapf GRAMINEAE (M)
Uraria Desv. LEGUMINOSAE–PAPILIONOIDEAE (D)
Urariopsis Schindl. = Uraria Desv. LEGUMINOSAE–PAPILIONOIDEAE (D)
Urbananthus R.M.King & H.Rob. COMPOSITAE (D)
Urbanella Pierre = Pouteria Aubl. SAPOTACEAE (D)
Urbania Phil. VERBENACEAE (D)
Urbanodendron Mez LAURACEAE (D)
Urbanodoxa Muschl. CRUCIFERAE (D)
Urbanoguarea Harms = Guarea L. MELIACEAE (D)
Urbanolophium Melch. BIGNONIACEAE (D)
Urbanosciadium H.Wolff = Niphogeton Schltdl. UMBELLIFERAE (D)
Urbinella Greenm. COMPOSITAE (D)
Urbinia Rose = Echeveria DC. CRASSULACEAE (D)
Urceola Roxb. APOCYNACEAE (D)
Urceolaria Willd. ex Cothen. = Schradera Vahl RUBIACEAE (D)
Urceolina Rchb. AMARYLLIDACEAE (M)
Urechites Müll.Arg. = Pentalinon Voigt APOCYNACEAE (D)
Urelytrum Hack. GRAMINEAE (M)
Urena L. MALVACEAE (D)
Urera Gaudich. URTICACEAE (D)
Urginea Steinh. HYACINTHACEAE (M)
Urgineopsis Compton = Urginea Steinh. HYACINTHACEAE (M)
Uribea Dugand & Romero LEGUMINOSAE–PAPILIONOIDEAE (D)
Urinaria Medik. = Phyllanthus L. EUPHORBIACEAE (D)
Urmenetea Phil. COMPOSITAE (D)
Urnularia Stapf APOCYNACEAE (D)
Urobotrya Stapf OPILIACEAE (D)
Urocarpidium Ulbr. MALVACEAE (D)
Urocarpus J.Drumm. ex Harv. RUTACEAE (D)
Urochlaena Nees GRAMINEAE (M)
Urochloa P.Beauv. GRAMINEAE (M)
Urochondra C.E.Hubb. GRAMINEAE (M)
Urodesmium Naudin = Pachyloma DC. MELASTOMATACEAE (D)
Urodon Turcz. LEGUMINOSAE–PAPILIONOIDEAE (D)
Urogentias Gilg & Gilg–Ben. GENTIANACEAE (D)
Urolepis (DC.) R.M.King & H.Rob. COMPOSITAE (D)
Uromorus Bureau = Streblus Lour. MORACEAE (D)
Uromyrtus Burret MYRTACEAE (D)
Uropappus Nutt. COMPOSITAE (D)
Uropedium Lindl. = Phragmipedium Rolfe ORCHIDACEAE (M)
Urophyllum Wall. RUBIACEAE (D)

Urophysa Ulbr. RANUNCULACEAE (D)
Uroskinnera Lindl. SCROPHULARIACEAE (D)
Urospatha Schott ARACEAE (M)
Urospathella Bunting = Urospatha Schott ARACEAE (M)
Urospermum Scop. COMPOSITAE (D)
Urostachya (Lindl.) Brieger = Eria Lindl. ORCHIDACEAE (M)
Urostachys (E.Pritz.) Herter = Huperzia Bernh. LYCOPODIACEAE (FA)
Urostemon B.Nord. = Brachyglottis J.R.Forst. & G.Forst. COMPOSITAE (D)
Urostephanus B.L.Rob. & Greenm. ASCLEPIADACEAE (D)
Urostigma Gasp. = Ficus L. MORACEAE (D)
Urotheca Gilg = Gravesia Naudin MELASTOMATACEAE (D)
Ursia Vassilcz. = Trifolium L. LEGUMINOSAE–PAPILIONOIDEAE (D)
Ursinia Gaertn. COMPOSITAE (D)
Ursiniopsis E.Phillips = Ursinia Gaertn. COMPOSITAE (D)
Urtica L. URTICACEAE (D)
Uruparia Raf. = Uncaria Schreb. RUBIACEAE (D)
Urvillea Kunth SAPINDACEAE (D)
Usoricum Lunell (SUS) = Oenothera L. ONAGRACEAE (D)
Usteria Dennst. (SUH) = Acalypha L. EUPHORBIACEAE (D)
Usteria Willd. LOGANIACEAE (D)
Utahia Britton & Rose = Pediocactus Britton & Rose CACTACEAE (D)
Utleria Bedd. ex Benth. ASCLEPIADACEAE (D)
Utleya Wilbur & Luteyn ERICACEAE (D)
Utricularia L. LENTIBULARIACEAE (D)
Utsetela Pellegr. MORACEAE (D)
Uva Kuntze = Uvaria L. ANNONACEAE (D)
Uvaria L. ANNONACEAE (D)
Uvariastrum Engl. ANNONACEAE (D)
Uvariella Ridl. = Uvaria L. ANNONACEAE (D)
Uvariodendron (Engl. & Diels) R.E.Fr. ANNONACEAE (D)
Uvariopsis Engl. ANNONACEAE (D)
Uvularia L. CONVALLARIACEAE (M)

Vaccaria Wolf CARYOPHYLLACEAE (D)
Vacciniopsis Rusby = Disterigma (Klotzsch) Nied. ERICACEAE (D)
Vaccinium L. ERICACEAE (D)
Vachellia Wight & Arn. = Acacia Mill. LEGUMINOSAE–MIMOSOIDEAE (D)
Vagaria Herb. AMARYLLIDACEAE (M)
Vaginopteris Nakai = Vaginularia Fée VITTARIACEAE (F)
Vaginularia Fée VITTARIACEAE (F)
Vahadenia Stapf APOCYNACEAE (D)
Vahlia Thunb. VAHLIACEAE (D)
Vahlodea Fr. = Deschampsia P.Beauv. GRAMINEAE (M)
Vailia Rusby ASCLEPIADACEAE (D)
Vaillantia Hoffm. (SUO) = Valantia L. RUBIACEAE (D)
Valantia L. RUBIACEAE (D)
Valdivia C.Gay ex J.Remy ESCALLONIACEAE (D)
Valentina Speg. = Heliotropium L. BORAGINACEAE (D)
Valentiniella Speg. = Heliotropium L. BORAGINACEAE (D)
Valenzuelia Bertero ex Cambess. = Guindilia Gillies ex Hook. & Arn. SAPINDACEAE (D)
Valeriana L. VALERIANACEAE (D)
Valerianella Mill. VALERIANACEAE (D)
Valerioa Standl. & Steyerm. = Peltanthera Benth. BUDDLEJACEAE (D)
Valerioanthus Lundell MYRSINACEAE (D)
Valetonia T.Durand = Pleurisanthes Baill. ICACINACEAE (D)

Valikaha Adans. = Memecylon L. MELASTOMATACEAE (D)
Vallariopsis Woodson APOCYNACEAE (D)
Vallaris Burm.f. APOCYNACEAE (D)
Vallaris Raf. (SUH) = Euphorbia L. EUPHORBIACEAE (D)
Vallea Mutis ex L.f. ELAEOCARPACEAE (D)
Vallesia Ruíz & Pav. APOCYNACEAE (D)
Vallifilix Thouars = Lygodium Sw. SCHIZAEACEAE (F)
Vallisneria L. HYDROCHARITACEAE (M)
Vallota Salisb. ex Herb. = Cyrtanthus Aiton AMARYLLIDACEAE (M)
Valota Adans. (SUH) = Digitaria Haller GRAMINEAE (M)
Valteta Raf. = Iochroma Benth. SOLANACEAE (D)
Valvanthera C.T.White = Hernandia L. HERNANDIACEAE (D)
Van–Royena Aubrév. = Pouteria Aubl. SAPOTACEAE (D)
Vanasushava P.K.Mukh. & Constance UMBELLIFERAE (D)
Vanclevea Greene COMPOSITAE (D)
Vancouveria C.Morren & Decne. BERBERIDACEAE (D)
Vanda Jones ex R.Br. ORCHIDACEAE (M)
Vandasia Domin = Vandasina Rauschert LEGUMINOSAE–PAPILIONOIDEAE (D)
Vandasina Rauschert LEGUMINOSAE–PAPILIONOIDEAE (D)
Vandellia P.Browne ex L. = Lindernia All. SCROPHULARIACEAE (D)
Vandenboschia Copel. = Crepidomanes (C.Presl) C.Presl HYMENOPHYLLACEAE (F)
Vandera Raf. = Croton L. EUPHORBIACEAE (D)
Vanderystia De Wild. = Omphalocarpum P.Beauv. SAPOTACEAE (D)
Vandopsis Pfitzer ORCHIDACEAE (M)
Vanessa Raf. = Manettia L. RUBIACEAE (D)
Vangueria Comm. ex Juss. RUBIACEAE (D)
Vangueriella Verdc. RUBIACEAE (D)
Vangueriopsis Robyns RUBIACEAE (D)
Vanheerdia L.Bolus ex H.E.K.Hartmann AIZOACEAE (D)
Vanhouttea Lem. GESNERIACEAE (D)
Vania F.K.Mey. = Thlaspi L. CRUCIFERAE (D)
Vanieria Lour. = Maclura Nutt. MORACEAE (D)
Vanilla Mill. ORCHIDACEAE (M)
Vanillosmopsis Sch.Bip. COMPOSITAE (D)
Vaniotia H.Lév. = Petrocosmea Oliv. GESNERIACEAE (D)
Vanoverberghia Merr. ZINGIBERACEAE (M)
Vantanea Aubl. HUMIRIACEAE (D)
Vanwykia Wiens LORANTHACEAE (D)
Vanzijlia L.Bolus AIZOACEAE (D)
Varennea DC. = Eysenhardtia Kunth LEGUMINOSAE–PAPILIONOIDEAE (D)
Vargasiella C.Schweinf. ORCHIDACEAE (M)
Varilla A.Gray COMPOSITAE (D)
Varinga Raf. = Ficus L. MORACEAE (D)
Varnera L. = Gardenia Ellis RUBIACEAE (D)
Varthemia DC. COMPOSITAE (D)
Vasconcellia Mart. = Arrabidaea DC. BIGNONIACEAE (D)
Vaseya Thurb. = Muhlenbergia Schreb. GRAMINEAE (M)
Vaseyanthus Cogn. CUCURBITACEAE (D)
Vaseyochloa Hitchc. GRAMINEAE (M)
Vasivaea Baill. TILIACEAE (D)
Vasqueziella Dodson ORCHIDACEAE (M)
Vassilczenkoa Lincz. PLUMBAGINACEAE (D)
Vassobia Rusby SOLANACEAE (D)
Vatairea Aubl. LEGUMINOSAE–PAPILIONOIDEAE (D)
Vatairecopsis Ducke LEGUMINOSAE–PAPILIONOIDEAE (D)

Vateria L. DIPTEROCARPACEAE (D)
Vateriopsis F.Heim DIPTEROCARPACEAE (D)
Vatica L. DIPTEROCARPACEAE (D)
Vatovaea Chiov. LEGUMINOSAE–PAPILIONOIDEAE (D)
Vatricania Backeb. = Espostoa Britton & Rose CACTACEAE (D)
Vauanthes Haw. = Crassula L. CRASSULACEAE (D)
Vaughania S.Moore = Indigofera L. LEGUMINOSAE–PAPILIONOIDEAE (D)
Vaupelia Brand = Cystostemon Balf.f. BORAGINACEAE (D)
Vaupesia R.E.Schult. EUPHORBIACEAE (D)
Vauquelinia Corrêa ex Humb. & Bonpl. ROSACEAE (D)
Vausagesia Baill. = Sauvagesia L. OCHNACEAE (D)
Vavaea Benth. MELIACEAE (D)
Vavanga Rohr (SUI) = Vangueria Comm. ex Juss. RUBIACEAE (D)
Vavara Benoist (UP) ACANTHACEAE (D)
Vavilovia Fedorov LEGUMINOSAE–PAPILIONOIDEAE (D)
Veatchia A.Gray = Pachycormus Coville ex Standl. ANACARDIACEAE (D)
Veconcibea (Muell.Arg.) Pax & K.Hoffm. EUPHORBIACEAE (D)
Veeresia Monach. & Moldenke = Reevesia Lindl. STERCULIACEAE (D)
Vegaea Urb. MYRSINACEAE (D)
Veillonia H.E.Moore PALMAE (M)
Veitchia H.A.Wendl. PALMAE (M)
Velaea D.Dietr. = Arracacia Bancr. UMBELLIFERAE (D)
Velezia L. CARYOPHYLLACEAE (D)
Vella DC. CRUCIFERAE (D)
Velleia Sm. GOODENIACEAE (D)
Vellereophyton Hilliard & B.L.Burtt COMPOSITAE (D)
Vellosiella Baill. SCROPHULARIACEAE (D)
Vellozia Vand. VELLOZIACEAE (M)
Velpeaulia Gaudich. = Nolana L.f. SOLANACEAE (D)
Veltheimia Gled. HYACINTHACEAE (M)
Velvetia Tiegh. = Psittacanthus Mart. LORANTHACEAE (D)
Velvitsia Hiern = Melasma Bergius SCROPHULARIACEAE (D)
Venana Lam. = Brexia Noronha ex Thouars ESCALLONIACEAE (D)
Vendredia Baill. = Robinsonia DC. COMPOSITAE (D)
Venegasia DC. COMPOSITAE (D)
Venidium Less. = Arctotis L. COMPOSITAE (D)
Ventenata Koeler GRAMINEAE (M)
Ventenatia Tratt. = Pedilanthus Neck. ex Poit. EUPHORBIACEAE (D)
Ventilago Gaertn. RHAMNACEAE (D)
Ventricularia Garay ORCHIDACEAE (M)
Veprecella Naudin MELASTOMATACEAE (D)
Vepris Comm. ex A.Juss. RUTACEAE (D)
Veratrilla (Baill.) Franch. GENTIANACEAE (D)
Veratrum L. MELANTHIACEAE (M)
Verbascum L. SCROPHULARIACEAE (D)
Verbena L. VERBENACEAE (D)
Verbenoxylum Tronc. VERBENACEAE (D)
Verbesina L. COMPOSITAE (D)
Verdcourtia R.Wilczek = Dipogon Liebm. LEGUMINOSAE–PAPILIONOIDEAE (D)
Verhuellia Miq. PIPERACEAE (D)
Verinea Merino (SUH) = Melica L. GRAMINEAE (M)
Verlangia Neck. ex Raf. (UP) SAPOTACEAE (D)
Verlotia E.Fourn. = Marsdenia R.Br. ASCLEPIADACEAE (D)
Vermifrux J.B.Gillett = Lotus L. LEGUMINOSAE–PAPILIONOIDEAE (D)
Vernicia Lour. EUPHORBIACEAE (D)

Vernonia Schreb. COMPOSITAE (D)
Vernoniopsis Humbert COMPOSITAE (D)
Veronica L. SCROPHULARIACEAE (D)
Veronicastrum Heist. ex Fabr. SCROPHULARIACEAE (D)
Verreauxia Benth. GOODENIACEAE (D)
Verrucifera N.E.Br. = Titanopsis Schwantes AIZOACEAE (D)
Verrucularia A.Juss. MALPIGHIACEAE (D)
Verrucularina Rauschert = Verrucularia A.Juss. MALPIGHIACEAE (D)
Verschaffeltia H.A.Wendl. PALMAE (M)
Versteegia Valeton RUBIACEAE (D)
Verticordia DC. MYRTACEAE (D)
Verulamia DC. ex Poir. = Pavetta L. RUBIACEAE (D)
Veselskya Opiz CRUCIFERAE (D)
Veseyochloa J.B.Phipps = Tristachya Nees GRAMINEAE (M)
Vesicarex Steyerm. = Carex L. CYPERACEAE (M)
Vesicaria Adans. = Alyssoides Mill. CRUCIFERAE (D)
Vesiculina Raf. = Utricularia L. LENTIBULARIACEAE (D)
Vesselowskya Pamp. CUNONIACEAE (D)
Vestia Willd. SOLANACEAE (D)
Vetiveria Bory GRAMINEAE (M)
Vexatorella Rourke PROTEACEAE (D)
Vexibia Raf. = Sophora L. LEGUMINOSAE–PAPILIONOIDEAE (D)
Vexillabium F.Maek. ORCHIDACEAE (M)
Vexillifera Ducke = Dussia Krug & Urb. ex Taub. LEGUMINOSAE–
  PAPILIONOIDEAE (D)
Viburnum L. CAPRIFOLIACEAE (D)
Vicarya Wall. ex Voigt = Myriopteron Griff. ASCLEPIADACEAE (D)
Vicatia DC. UMBELLIFERAE (D)
Vicentia Allemão = Terminalia L. COMBRETACEAE (D)
Vicia L. LEGUMINOSAE–PAPILIONOIDEAE (D)
Vicilla Schur = Vicia L. LEGUMINOSAE–PAPILIONOIDEAE (D)
Vicioides Moench = Vicia L. LEGUMINOSAE–PAPILIONOIDEAE (D)
Vicoa Cass. = Pentanema Cass. COMPOSITAE (D)
Victoria Lindl. NYMPHAEACEAE (D)
Victorinia Léon = Cnidoscolus Pohl EUPHORBIACEAE (D)
Vidalia Fern.-Vill. = Mesua L. GUTTIFERAE (D)
Vidoricum Kuntze = Madhuca Ham. ex J.F.Gmel. SAPOTACEAE (D)
Vieillardorchis Kraenzl. = Goodyera R.Br. ORCHIDACEAE (M)
Vieraea Sch.Bip. COMPOSITAE (D)
Viereckia R.M.King & H.Rob. COMPOSITAE (D)
Vietnamosasa T.Q.Nguyen GRAMINEAE (M)
Vietsenia C.Hansen MELASTOMATACEAE (D)
Vigethia W.A.Weber COMPOSITAE (D)
Vigia Vell. = Fragariopsis A.St.-Hil. EUPHORBIACEAE (D)
Vigieria Vell. = Escallonia Mutis ex L.f. ESCALLONIACEAE (D)
Vigna Savi LEGUMINOSAE–PAPILIONOIDEAE (D)
Vignaldia A.Rich. = Pentas Benth. RUBIACEAE (D)
Vignaudia Schweinf. (SUO) = Pentas Benth. RUBIACEAE (D)
Vignopsis De Wild. = Psophocarpus DC. LEGUMINOSAE–PAPILIONOIDEAE (D)
Viguiera Kunth COMPOSITAE (D)
Viguierella A.Camus GRAMINEAE (M)
Vilfa Adans. = Agrostis L. GRAMINEAE (M)
Vilfagrostis Döll (SUI) = Eragrostis Wolf GRAMINEAE (M)
Villadia Rose CRASSULACEAE (D)
Villamilla Ruíz & Pav. = Trichostigma A.Rich. PHYTOLACCACEAE (D)

Villanova Pourr. ex Cutanda (SUI) = Flueggea Willd. EUPHORBIACEAE (D)
Villanova Lag. COMPOSITAE (D)
Villanova Ortega (SUH) = Parthenium L. COMPOSITAE (D)
Villaresia Ruíz & Pav. = Citronella D.Don ICACINACEAE (D)
Villaresiopsis Sleumer = Citronella D.Don ICACINACEAE (D)
Villaria Rolfe RUBIACEAE (D)
Villarsia Vent. MENYANTHACEAE (D)
Villebrunea Gaudich. = Oreocnide Miq. URTICACEAE (D)
Villocuspis (A.DC.) Aubrév. & Pellegr. = Chrysophyllum L. SAPOTACEAE (D)
Vilmorinia DC. = Poitea Vent. LEGUMINOSAE–PAPILIONOIDEAE (D)
Vilobia Strother COMPOSITAE (D)
Viminaria Sm. LEGUMINOSAE–PAPILIONOIDEAE (D)
Vinca L. APOCYNACEAE (D)
Vincentella Pierre = Synsepalum (A.DC.) Daniell SAPOTACEAE (D)
Vincentia Gaudich. = Machaerina Vahl CYPERACEAE (M)
Vincetoxicopsis Costantin ASCLEPIADACEAE (D)
Vincetoxicum Wolf ASCLEPIADACEAE (D)
Vindasia Benoist ACANTHACEAE (D)
Vinkia Meiden HALORAGACEAE (D)
Vinticena Steud. = Grewia L. TILIACEAE (D)
Viola L. VIOLACEAE (D)
Viorna Rchb. = Clematis L. RANUNCULACEAE (D)
Viposia Lundell = Plenckia Reissek CELASTRACEAE (D)
Virecta L.f. = Sipanea Aubl. RUBIACEAE (D)
Virecta Afzel. ex Sm. (SUH) = Virectaria Bremek. RUBIACEAE (D)
Virectaria Bremek. RUBIACEAE (D)
Vireya Raf. = Columnea L. GESNERIACEAE (D)
Virgilia Poir. LEGUMINOSAE–PAPILIONOIDEAE (D)
Virginia (DC.) Nicoli = Helichrysum Mill. COMPOSITAE (D)
Virgulaster Semple = Aster L. COMPOSITAE (D)
Viridivia J.H.Hemsl. & Verdc. PASSIFLORACEAE (D)
Virola Aubl. MYRISTICACEAE (D)
Viscainoa Greene ZYGOPHYLLACEAE (D)
Viscaria Roehl. CARYOPHYLLACEAE (D)
Viscoides Jacq. = Psychotria L. RUBIACEAE (D)
Viscum L. VISCACEAE (D)
Visiania Gasp. (SUH) = Ficus L. MORACEAE (D)
Visiania DC. = Ligustrum L. OLEACEAE (D)
Vismia Vand. GUTTIFERAE (D)
Vismianthus Mildbr. CONNARACEAE (D)
Visnea L.f. THEACEAE (D)
Vissadali Adans. = Knoxia L. RUBIACEAE (D)
Vitaliana Sesl. PRIMULACEAE (D)
Vitellaria C.F.Gaertn. SAPOTACEAE (D)
Vitellariopsis Baill. ex Dubard SAPOTACEAE (D)
Vitex L. VERBENACEAE (D)
Viticella Mitch. = Nemophila Nutt. HYDROPHYLLACEAE (D)
Viticipremna Lam. VERBENACEAE (D)
Vitiphoenix Becc. = Veitchia H.A.Wendl. PALMAE (M)
Vitis L. VITACEAE (D)
Vittadinia A.Rich. COMPOSITAE (D)
Vittaria Sm. VITTARIACEAE (F)
Vittetia R.M.King & H.Rob. COMPOSITAE (D)
Viviana Raf. (SUH) = Guettarda L. RUBIACEAE (D)
Viviana Colla = Melanopsidium Colla RUBIACEAE (D)

Viviania Cav. GERANIACEAE (D)
Vladimiria Iljin = Dolomiaea DC. COMPOSITAE (D)
Vleisia Toml. & Posl. ZANNICHELLIACEAE (M)
Voacanga Thouars APOCYNACEAE (D)
Voandzeia Thouars = Vigna Savi LEGUMINOSAE–PAPILIONOIDEAE (D)
Voanioala J.Dransf. PALMAE (M)
Voatamalo Capuron ex Bosser EUPHORBIACEAE (D)
Vochy Aubl. = Vochysia Aubl. VOCHYSIACEAE (D)
Vochysia Aubl. VOCHYSIACEAE (D)
Voelckeria Klotzsch & Karst. ex Endl. = Ternstroemia Mutis ex L.f. THEACEAE (D)
Vogelia Lam. (SUH) = Dyerophytum Kuntze PLUMBAGINACEAE (D)
Vogelia Medik. (SUH) = Neslia Desv. CRUCIFERAE (D)
Vogelocassia Britton = Senna Mill. LEGUMINOSAE–CAESALPINIOIDEAE (D)
Voharanga Costantin & Bois = Cynanchum L. ASCLEPIADACEAE (D)
Vohemaria Buchenau ASCLEPIADACEAE (D)
Voigtia Klotzsch = Bathysa C.Presl RUBIACEAE (D)
Voladeria Benoist = Oreobolus R.Br. CYPERACEAE (M)
Volkensia O.Hoffm. = Bothriocline Oliv. ex Benth. COMPOSITAE (D)
Volkensiella H.Wolff = Oenanthe L. UMBELLIFERAE (D)
Volkensinia Schinz AMARANTHACEAE (D)
Volkensiophyton Lindau = Lepidagathis Willd. ACANTHACEAE (D)
Volkiella Merxm. & Czech CYPERACEAE (M)
Volutaria Cass. COMPOSITAE (D)
Volvulopsis Roberty = Evolvulus L. CONVOLVULACEAE (D)
Volvulus Medik. = Calystegia R.Br. CONVOLVULACEAE (D)
Vonitra Becc. PALMAE (M)
Vonroemeria J.J.Sm. = Octarrhena Thwaites ORCHIDACEAE (M)
Vossia Wall. & Griff. GRAMINEAE (M)
Vossianthus Kuntze = Sparrmannia L.f. TILIACEAE (D)
Votomita Aubl. MELASTOMATACEAE (D)
Vouacapoua Aubl. LEGUMINOSAE–CAESALPINIOIDEAE (D)
Vouarana Aubl. SAPINDACEAE (D)
Vouay Aubl. = Geonoma Willd. PALMAE (M)
Voyria Aubl. GENTIANACEAE (D)
Voyriella Miq. GENTIANACEAE (D)
Vriesea Lindl. BROMELIACEAE (M)
Vrydagzynea Blume ORCHIDACEAE (M)
Vulpia C.C.Gmel. GRAMINEAE (M)
Vulpiella (Batt. & Trab.) Andr. GRAMINEAE (M)
Vvedenskya Korovin UMBELLIFERAE (D)
Vvedenskyella Botsch. CRUCIFERAE (D)

Wachendorfia Burm. HAEMODORACEAE (M)
Waddingtonia Phil. = Nicotiana L. SOLANACEAE (D)
Wadea Raf. = Cestrum L. SOLANACEAE (D)
Wagatea Dalzell = Moullava Adans. LEGUMINOSAE–CAESALPINIOIDEAE (D)
Wagenitzia Dostál COMPOSITAE (D)
Wahlenbergia Schrad. ex Roth CAMPANULACEAE (D)
Wahlenbergia Blume (SUH) = Tarenna Gaertn. RUBIACEAE (D)
Wailesia Lindl. = Dipodium R.Br. ORCHIDACEAE (M)
Waitzia J.C.Wendl. COMPOSITAE (D)
Wajira Thulin LEGUMINOSAE–PAPILIONOIDEAE (D)
Wakilia Gilli = Phaeonychium O.E.Schulz CRUCIFERAE (D)
Walafrida E.Mey. SCROPHULARIACEAE (D)
Walberia Mill. ex Ehret = Nolana L.f. SOLANACEAE (D)

Waldheimia Kar. & Kir. = Allardia Decne. COMPOSITAE (D)
Waldsteinia Willd. ROSACEAE (D)
Walidda (A.DC.) Pichon = Wrightia R.Br. APOCYNACEAE (D)
Walkeria A.Chev. = Lecomtedoxa (Pierre ex Engl.) Dubard SAPOTACEAE (D)
Wallacea Spruce ex Hook. OCHNACEAE (D)
Wallaceodendron Koord. LEGUMINOSAE–MIMOSOIDEAE (D)
Wallenia Sw. = Cybianthus Mart. MYRSINACEAE (D)
Walleniella P.Wilson MYRSINACEAE (D)
Walleria J.Kirk TECOPHILAEACEAE (M)
Wallichia Roxb. PALMAE (M)
Wallichia Reinw. ex Blume (SUH) = Urophyllum Wall. RUBIACEAE (D)
Wallrothia Spreng. = Bunium L. UMBELLIFERAE (D)
Walpersia Harv. = Phyllota (DC.) Benth. LEGUMINOSAE–PAPILIONOIDEAE (D)
Walsura Roxb. MELIACEAE (D)
Walteranthus Keighery GYROSTEMONACEAE (D)
Waltheria L. STERCULIACEAE (D)
Waluewa Regel = Leochilus Knowles & Westc. ORCHIDACEAE (M)
Wamalchitamia Strother COMPOSITAE (D)
Wangenheimia Moench GRAMINEAE (M)
Wangenheimia A.Dietr. (SUH) = Dendropanax Decne. & Planch. ARALIACEAE (D)
Wangerinia E.Franz = Microphyes Phil. CARYOPHYLLACEAE (D)
Warburgia Engl. CANELLACEAE (D)
Warburgina Eig RUBIACEAE (D)
Wardenia King = Brassaiopsis Decne. & Planch. ARALIACEAE (D)
Warea Nutt. CRUCIFERAE (D)
Warea C.B.Clarke (SUH) = Biswarea Cogn. CUCURBITACEAE (D)
Waria Aubl. (SUO) = Uvaria L. ANNONACEAE (D)
Warionia Benth. & Coss. COMPOSITAE (D)
Warmingia Rchb.f. ORCHIDACEAE (M)
Warneckea Gilg MELASTOMATACEAE (D)
Warneria Ellis (SUO) = Gardenia Ellis RUBIACEAE (D)
Warpuria Stapf = Podorungia Baill. ACANTHACEAE (D)
Warrea Lindl. ORCHIDACEAE (M)
Warreella Schltr. ORCHIDACEAE (M)
Warreopsis Garay ORCHIDACEAE (M)
Warscewiczella Rchb.f. = Chondrorhyncha Lindl. ORCHIDACEAE (M)
Warszewiczia Klotzsch RUBIACEAE (D)
Wartmannia Müll.Arg. = Homalanthus A.Juss. EUPHORBIACEAE (D)
Wasabia Matsum. CRUCIFERAE (D)
Wasatchia M.E.Jones (SUS) = Festuca L. GRAMINEAE (M)
Washingtonia Raf. (SUI) = Osmorhiza Raf. UMBELLIFERAE (D)
Washingtonia H.A.Wendl. PALMAE (M)
Waterhousea B.Hyland MYRTACEAE (D)
Watsonamra Kuntze = Pentagonia Benth. RUBIACEAE (D)
Watsonia Mill. IRIDACEAE (M)
Wattakaka Hassk. = Dregea E.Mey. ASCLEPIADACEAE (D)
Weatherbya Copel. = Lemmaphyllum C.Presl POLYPODIACEAE (F)
Webera J.F.Gmel. (SUH) = Bellucia Neck. ex Raf. MELASTOMATACEAE (D)
Webera Schreb. (SUS) = Tarenna Gaertn. RUBIACEAE (D)
Webera Cramer (UP) RUBIACEAE (D)
Weberaster Á.Löve & D.Löve = Aster L. COMPOSITAE (D)
Weberbauera Gilg & Muschler CRUCIFERAE (D)
Weberbauerella Ulbr. LEGUMINOSAE–PAPILIONOIDEAE (D)
Weberbauerocereus Backeb. CACTACEAE (D)
Weberiopuntia Frič = Opuntia Mill. CACTACEAE (D)

Weberocereus Britton & Rose CACTACEAE (D)
Websteria S.H.Wright CYPERACEAE (M)
Weddellina Tul. PODOSTEMACEAE (D)
Wedelia Loefl. (SUH) = Allionia L. NYCTAGINACEAE (D)
Wedelia Jacq. COMPOSITAE (D)
Wedeliella Cockerell = Allionia L. NYCTAGINACEAE (D)
Wehlia F.Muell. = Homalocalyx F.Muell. MYRTACEAE (D)
Weigela Thunb. CAPRIFOLIACEAE (D)
Weigeltia A.DC. = Cybianthus Mart. MYRSINACEAE (D)
Weihea Spreng. = Cassipourea Aubl. RHIZOPHORACEAE (D)
Weingaertneria Bernh. (SUS) = Corynephorus P.Beauv. GRAMINEAE (M)
Weingartia Werderm. = Rebutia K.Schum. CACTACEAE (D)
Weinmannia L. CUNONIACEAE (D)
Welchiodendron Peter G.Wilson & J.T.Waterh. MYRTACEAE (D)
Weldenia Schult.f. COMMELINACEAE (M)
Welfia H.A.Wendl. PALMAE (M)
Wellstedia Balf.f. BORAGINACEAE (D)
Welwitschia Rchb. (SUH) = Eriastrum Woot. & Standl. POLEMONIACEAE (D)
Welwitschia Hook.f. WELWITSCHIACEAE (G)
Welwitschiella O.Hoffm. COMPOSITAE (D)
Wenchengia C.Y.Wu & S.Chow LABIATAE (D)
Wendelboa Soest = Taraxacum G.H.Weber ex Wigg. COMPOSITAE (D)
Wenderothia Schltdl. = Canavalia DC. LEGUMINOSAE–PAPILIONOIDEAE (D)
Wendlandia Bartl. ex DC. RUBIACEAE (D)
Wendlandiella Dammer PALMAE (M)
Wendtia Meyen GERANIACEAE (D)
Wenzelia Merr. RUTACEAE (D)
Wercklea Pittier & Standl. MALVACEAE (D)
Werckleocereus Britton & Rose = Weberocereus Britton & Rose CACTACEAE (D)
Werdermannia O.E.Schulz CRUCIFERAE (D)
Werneria Kunth COMPOSITAE (D)
Wernhamia S.Moore RUBIACEAE (D)
Westia Vahl LEGUMINOSAE–CAESALPINIOIDEAE (D)
Westoniella Cuatrec. COMPOSITAE (D)
Westphalina A.Robyns & Bamps TILIACEAE (D)
Westringia Sm. LABIATAE (D)
Wetria Baill. EUPHORBIACEAE (D)
Wetriaria Pax = Argomuellera Pax EUPHORBIACEAE (D)
Wettinella O.F.Cook & Doyle = Wettinia Poepp. PALMAE (M)
Wettinia Poepp. PALMAE (M)
Wettiniicarpus Burret = Wettinia Poepp. PALMAE (M)
Wettsteinia Petrak = Carduus L. COMPOSITAE (D)
Wettsteiniola Suess. PODOSTEMACEAE (D)
Wheelerella G.B.Grant (SUS) = Cryptantha G.Don BORAGINACEAE (D)
Whipplea Torr. HYDRANGEACEAE (D)
White-Sloanea Chiov. ASCLEPIADACEAE (D)
Whiteheadia Harv. HYACINTHACEAE (M)
Whiteochloa C.E.Hubb. GRAMINEAE (M)
Whiteodendron Steenis MYRTACEAE (D)
Whitfieldia Hook. ACANTHACEAE (D)
Whitfordia Elmer = Whitfordiodendron Elmer LEGUMINOSAE–PAPILIONOIDEAE (D)
Whitfordiodendron Elmer LEGUMINOSAE–PAPILIONOIDEAE (D)
Whitia Blume = Cyrtandra J.R.Forst. & G.Forst. GESNERIACEAE (D)
Whitmorea Sleumer ICACINACEAE (D)
Whitneya A.Gray COMPOSITAE (D)

Whittonia Sandwith PERIDISCACEAE (D)
Whyanbeelia Airy Shaw & B.Hyland EUPHORBIACEAE (D)
Whytockia W.W.Sm. GESNERIACEAE (D)
Wiasemskya Klotzsch (UP) RUBIACEAE (D)
Wibelia Fée = Tapeinidium (C.Presl) C.Chr. DENNSTAEDTIACEAE (F)
Wibelia Bernh. (SUH) = Davallia Sm. DAVALLIACEAE (F)
Wiborgia Thunb. LEGUMINOSAE–PAPILIONOIDEAE (D)
Widdringtonia Endl. CUPRESSACEAE (G)
Widgrenia Malme ASCLEPIADACEAE (D)
Wiedemannia Fisch. & C.A.Mey. LABIATAE (D)
Wiegmannia Meyen = Hedyotis L. RUBIACEAE (D)
Wielandia Baill. EUPHORBIACEAE (D)
Wierzbickia Rchb. = Minuartia L. CARYOPHYLLACEAE (D)
Wiesneria Micheli ALISMATACEAE (M)
Wiestia Sch.Bip. = Lactuca L. COMPOSITAE (D)
Wiestia Boiss. (SUI) = Boissiera Hochst. ex Steud. GRAMINEAE (M)
Wigandia Kunth HYDROPHYLLACEAE (D)
Wigginsia D.M.Porter = Parodia Speg. CACTACEAE (D)
Wightia Wall. SCROPHULARIACEAE (D)
Wigmannia Walp. (SUO) = Hedyotis L. RUBIACEAE (D)
Wikstroemia Endl. THYMELAEACEAE (D)
Wikstroemia Schrad. (SUH) = Laplacea Kunth THEACEAE (D)
Wilbrandia Silva Manso CUCURBITACEAE (D)
Wilcoxia Britton & Rose = Echinocereus Engelm. CACTACEAE (D)
Wildemaniodoxa Aubrév. & Pellegr. = Englerophytum K.Krause SAPOTACEAE (D)
Wilhelminia Hochr. MALVACEAE (D)
Wilhelmsia W.Koch (SUH) = Rostraria Trin. GRAMINEAE (M)
Wilhelmsia Rchb. CARYOPHYLLACEAE (D)
Wilibald–schmidtia Conrad (SUS) = Danthonia DC. GRAMINEAE (M)
Wilibalda Roth (SUS) = Coleanthus J.Seidel GRAMINEAE (M)
Wilkesia A.Gray COMPOSITAE (D)
Wilkiea F.Muell. MONIMIACEAE (D)
Willardia Rose LEGUMINOSAE–PAPILIONOIDEAE (D)
Willbleibia Herter = Willkommia Hack. GRAMINEAE (M)
Willdenovia J.F.Gmel. = Rondeletia L. RUBIACEAE (D)
Willdenowia Thunb. RESTIONACEAE (M)
Williamia Baill. = Phyllanthus L. EUPHORBIACEAE (D)
Williamodendron Kubitzki & H.G.Richt. LAURACEAE (D)
Williamsia Merr. = Praravinia Korth. RUBIACEAE (D)
Willisia Warm. PODOSTEMACEAE (D)
Willkommia Hack. GRAMINEAE (M)
Willughbeia Roxb. APOCYNACEAE (D)
Willughbeiopsis Rauschert = Urnularia Stapf APOCYNACEAE (D)
Wilmattea Britton & Rose = Hylocereus (A.Berger) Britton & Rose CACTACEAE (D)
Wilsonia R.Br. CONVOLVULACEAE (D)
Wimmeria Schltdl. & Cham. CELASTRACEAE (D)
Winchia A.DC. = Alstonia R.Br. APOCYNACEAE (D)
Weindmannia P.Browne = Weinmannia L. CUNONIACEAE (D)
Windsoria Nutt. = Tridens Roem. & Schult. GRAMINEAE (M)
Windsorina Gleason RAPATEACEAE (M)
Winifredia L.A.S.Johnson & B.G.Briggs RESTIONACEAE (M)
Winklera Regel CRUCIFERAE (D)
Winklerella Engl. PODOSTEMACEAE (D)
Wintera G.Forst. (SUH) = Pseudowintera Dandy WINTERACEAE (D)
Wintera Murray (SUS) = Drimys J.R.Forst. & G.Forst. WINTERACEAE (D)

Winteria F.Ritter (SUH) = Cleistocactus Lem. CACTACEAE (D)
Winterocereus Backeb. (SUH) = Cleistocactus Lem. CACTACEAE (D)
Wirtgenia Döll (SUI) = Paspalum L. GRAMINEAE (M)
Wislizenia Engelm. CAPPARACEAE (D)
Wissadula Medik. MALVACEAE (D)
Wissmannia Burret = Livistona R.Br. PALMAE (M)
Wisteria Nutt. LEGUMINOSAE–PAPILIONOIDEAE (D)
Withania Pauquy SOLANACEAE (D)
Witheringia L'Hér. SOLANACEAE (D)
Witsenia Thunb. IRIDACEAE (M)
Wittea Kunth = Downingia Torr. CAMPANULACEAE (D)
Wittia K.Schum. (SUH) = Disocactus Lindl. CACTACEAE (D)
Wittiocactus Rauschert = Disocactus Lindl. CACTACEAE (D)
Wittmackanthus Kuntze RUBIACEAE (D)
Wittmackia Mez = Aechmea Ruíz & Pav. BROMELIACEAE (M)
Wittmannia Vahl = Vangueria Comm. ex Juss. RUBIACEAE (D)
Wittrockia Lindm. BROMELIACEAE (M)
Wittsteinia F.Muell. ALSEUOSMIACEAE (D)
Wodyetia Irvine PALMAE (M)
Woehleria Griseb. AMARANTHACEAE (D)
Woikoia Baehni = Pouteria Aubl. SAPOTACEAE (D)
Wolffia Horkel ex Schleid. LEMNACEAE (M)
Wolffiella (Hegelm.) Hegelm. LEMNACEAE (M)
Wolffiopsis Hartog & Van der Plas = Wolffiella (Hegelm.) Hegelm. LEMNACEAE (M)
Wolfia Post & Kuntze (SUH) = Orchidantha N.E.Br. LOWIACEAE (M)
Wollastonia DC. ex Decne. COMPOSITAE (D)
Woodburnia Prain ARALIACEAE (D)
Woodfordia Salisb. LYTHRACEAE (D)
Woodia Schltr. ASCLEPIADACEAE (D)
Woodiella Merr. = Woodiellantha Rauschert ANNONACEAE (D)
Woodiellantha Rauschert ANNONACEAE (D)
Woodrowia Stapf = Dimeria R.Br. GRAMINEAE (M)
Woodsia R.Br. WOODSIACEAE (F)
Woodsonia L.H.Bailey = Neonicholsonia Dammer PALMAE (M)
Woodwardia Sm. BLECHNACEAE (F)
Wooleya L.Bolus AIZOACEAE (D)
Woollsia F.Muell. EPACRIDACEAE (D)
Wootonella Standl. = Verbesina L. COMPOSITAE (D)
Wootonia Greene = Dicranocarpus A.Gray COMPOSITAE (D)
Wormskioldia Schumach. & Thonn. = Tricliceras Thonn. ex DC. TURNERACEAE (D)
Worsleya (Traub) Traub = Hippeastrum Herb. AMARYLLIDACEAE (M)
Woytkowskia Woodson APOCYNACEAE (D)
Wrightea Roxb. = Wallichia Roxb. PALMAE (M)
Wrightia R.Br. APOCYNACEAE (D)
Wrixonia F.Muell. LABIATAE (D)
Wulfenia Jacq. SCROPHULARIACEAE (D)
Wulfeniopsis D.Y.Hong SCROPHULARIACEAE (D)
Wulffia Neck. ex Cass. COMPOSITAE (D)
Wulfhorstia C.E.C.Fisch. = Entandrophragma C.E.C.Fisch. MELIACEAE (D)
Wullschlaegelia Rchb.f. ORCHIDACEAE (M)
Wunderlichia Riedel ex Benth. COMPOSITAE (D)
Wunschmannia Urb. BIGNONIACEAE (D)
Wurdackanthus Maguire GENTIANACEAE (D)
Wurdackia Moldenke = Paepalanthus Kunth ERIOCAULACEAE (M)
Wurmbea Thunb. COLCHICACEAE (M)

Wurtzia Baill. = Margaritaria L.f. EUPHORBIACEAE (D)
Wyethia Nutt. COMPOSITAE (D)

Xamesike Raf. = Euphorbia L. EUPHORBIACEAE (D)
Xananthes Raf. = Utricularia L. LENTIBULARIACEAE (D)
Xanthanthos St.–Lag. (SUO) = Anthoxanthum L. GRAMINEAE (M)
Xantheranthemum Lindau ACANTHACEAE (D)
Xanthisma DC. COMPOSITAE (D)
Xanthium L. COMPOSITAE (D)
Xanthobrychis Galushko = Onobrychis Mill. LEGUMINOSAE–PAPILIONOIDEAE (D)
Xanthocephalum Willd. COMPOSITAE (D)
Xanthoceras Bunge SAPINDACEAE (D)
Xanthocercis Baill. LEGUMINOSAE–PAPILIONOIDEAE (D)
Xanthochymus Roxb. = Garcinia L. GUTTIFERAE (D)
Xanthogalum Avé–Lall. = Angelica L. UMBELLIFERAE (D)
Xanthomyrtus Diels MYRTACEAE (D)
Xanthonanthus St.–Lag. (SUS) = Anthoxanthum L. GRAMINEAE (M)
Xanthopappus C.Winkl. COMPOSITAE (D)
Xanthophyllum Roxb. POLYGALACEAE (D)
Xanthophytopsis Pit. = Xanthophytum Reinw. ex Blume RUBIACEAE (D)
Xanthophytum Reinw. ex Blume RUBIACEAE (D)
Xanthorhiza Marshall RANUNCULACEAE (D)
Xanthorrhoea Sm. XANTHORRHOEACEAE (M)
Xanthosia Rudge UMBELLIFERAE (D)
Xanthosoma Schott ARACEAE (M)
Xanthostachya Bremek. = Strobilanthes Blume ACANTHACEAE (D)
Xanthostemon F.Muell. MYRTACEAE (D)
Xanthoxalis Small = Oxalis L. OXALIDACEAE (D)
Xantolis Raf. SAPOTACEAE (D)
Xantonnea Pierre ex Pit. RUBIACEAE (D)
Xantonneopsis Pit. RUBIACEAE (D)
Xaritonia Raf. = Oncidium Sw. ORCHIDACEAE (M)
Xatardia Meisn. & Zeyh. UMBELLIFERAE (D)
Xeilyathum Raf. = Oncidium Sw. ORCHIDACEAE (M)
Xenacanthus Bremek. = Strobilanthes Blume ACANTHACEAE (D)
Xenikophyton Garay ORCHIDACEAE (M)
Xenochloa Roem. & Schult. = Phragmites Adans. GRAMINEAE (M)
Xenodendron K.Schum. & Lauterb. = Acmena DC. MYRTACEAE (D)
Xenophya Schott = Alocasia (Schott) G.Don ARACEAE (M)
Xenostegia D.F.Austin & Staples CONVOLVULACEAE (D)
Xeranthemum L. COMPOSITAE (D)
Xeroaloysia Tronc. VERBENACEAE (D)
Xerocarpa Lam. = Teijsmanniodendron Koord. VERBENACEAE (D)
Xerocarpus Guill. & Perr. = Rothia Pers. LEGUMINOSAE–PAPILIONOIDEAE (D)
Xerocassia Britton & Rose = Senna Mill. LEGUMINOSAE–CAESALPINIOIDEAE (D)
Xerochlamys Baker = Leptolaena Thouars SARCOLAENACEAE (D)
Xerochloa R.Br. GRAMINEAE (M)
Xerocladia Harv. LEGUMINOSAE–MIMOSOIDEAE (D)
Xerococcus Oerst. = Hoffmannia Sw. RUBIACEAE (D)
Xerodanthia J.B.Phipps = Danthoniopsis Stapf GRAMINEAE (M)
Xeroderris Roberty LEGUMINOSAE–PAPILIONOIDEAE (D)
Xerodraba Skottsb. CRUCIFERAE (D)
Xerolekia Anderb. COMPOSITAE (D)
Xerolirion A.S.George LOMANDRACEAE (M)
Xeromphis Raf. = Catunaregam Wolf RUBIACEAE (D)

Xeronema Brongn. & Gris PHORMIACEAE (M)
Xerophyllum Michx. MELANTHIACEAE (M)
Xerophyta Juss. VELLOZIACEAE (M)
Xeroplana Briq. STILBACEAE (D)
Xerorchis Schltr. ORCHIDACEAE (M)
Xerosicyos Humbert CUCURBITACEAE (D)
Xerospermum Blume SAPINDACEAE (D)
Xerosphaera Soják = Trifolium L. LEGUMINOSAE–PAPILIONOIDEAE (D)
Xerospiraea Henrickson ROSACEAE (D)
Xerotes R.Br. = Lomandra Labill. LOMANDRACEAE (M)
Xerothamnella C.T.White ACANTHACEAE (D)
Xerotia Oliv. ILLECEBRACEAE (D)
Ximenia L. OLACACEAE (D)
Ximeniopsis Alain = Ximenia L. OLACACEAE (D)
Xiphagrostis Coville (SUI) = Miscanthus Andersson GRAMINEAE (M)
Xiphidium Aubl. HAEMODORACEAE (M)
Xiphium Mill. = Iris L. IRIDACEAE (M)
Xiphizusa Rchb.f. = Bulbophyllum Thouars ORCHIDACEAE (M)
Xiphophyllum Ehrh. = Cephalanthera Rich. ORCHIDACEAE (M)
Xiphopteris Kaulf. GRAMMITIDACEAE (F)
Xiphosium Griff. = Eria Lindl. ORCHIDACEAE (M)
Xizangia D.Y.Hong SCROPHULARIACEAE (D)
Xolisma Raf. = Lyonia Nutt. ERICACEAE (D)
Xolocotzia Miranda VERBENACEAE (D)
Xylanche Beck SCROPHULARIACEAE (D)
Xylanthemum Tzvelev COMPOSITAE (D)
Xylia Benth. LEGUMINOSAE–MIMOSOIDEAE (D)
Xylinabaria Pierre APOCYNACEAE (D)
Xylinabariopsis Pit. = Ecdysanthera Hook. & Arn. APOCYNACEAE (D)
Xylobium Lindl. ORCHIDACEAE (M)
Xylocalyx Balf.f. SCROPHULARIACEAE (D)
Xylocarpus J.König MELIACEAE (D)
Xylochlamys Domin = Amyema Tiegh. LORANTHACEAE (D)
Xylococcus Nutt. ERICACEAE (D)
Xylococcus R.Br. ex Britt. & S.Moore (SUH) = Petalostigma F.Muell.
  EUPHORBIACEAE (D)
Xylolobus Kuntze = Xylia Benth. LEGUMINOSAE–MIMOSOIDEAE (D)
Xylomelum Sm. PROTEACEAE (D)
Xylonagra Donn.Sm. & Rose ONAGRACEAE (D)
Xylonymus Kalkman ex Ding Hou CELASTRACEAE (D)
Xyloolaena Baill. SARCOLAENACEAE (D)
Xylophacos Rydb. = Astragalus L. LEGUMINOSAE–PAPILIONOIDEAE (D)
Xylophragma Sprague BIGNONIACEAE (D)
Xylophylla L. = Phyllanthus L. EUPHORBIACEAE (D)
Xylopia L. ANNONACEAE (D)
Xylopiastrum Roberty = Xylopia L. ANNONACEAE (D)
Xylopicron Adans. (SUO) = Xylopia L. ANNONACEAE (D)
Xylopicrum P.Browne = Xylopia L. ANNONACEAE (D)
Xylopleurum Spach = Oenothera L. ONAGRACEAE (D)
Xylosma G.Forst. FLACOURTIACEAE (D)
Xylosterculia Kosterm. = Sterculia L. STERCULIACEAE (D)
Xylothamia G.L.Nesom, Y.B.Suh, D.R.Morgan & B.B.Simpson COMPOSITAE (D)
Xylotheca Hochst. FLACOURTIACEAE (D)
Xylothermia Greene = Pickeringia Nutt. LEGUMINOSAE–PAPILIONOIDEAE (D)
Xymalos Baill. MONIMIACEAE (D)

Xyochlaena Stapf = Tricholaena Schrad. ex Schult. & Schult.f. GRAMINEAE (M)
Xyridopsis Welw. ex B.Nord. = Emilia (Cass.) Cass. COMPOSITAE (D)
Xyris L. XYRIDACEAE (M)
Xyropteris K.U.Kramer DENNSTAEDTIACEAE (F)
Xysmalobium R.Br. ASCLEPIADACEAE (D)
Xystidium Trin. = Perotis Aiton GRAMINEAE (M)
Xystrolobus Gagnep. = Ottelia Pers. HYDROCHARITACEAE (M)

Yabea Kozo–Pol. UMBELLIFERAE (D)
Yadakeya Makino (SUS) = Pseudosasa Makino ex Nakai GRAMINEAE (M)
Yakirra Lazarides & R.D.Webster GRAMINEAE (M)
Yangapa Raf. = Gardenia Ellis RUBIACEAE (D)
Yarina O.F.Cook = Phytelephas Ruíz & Pav. PALMAE (M)
Yeatesia Small ACANTHACEAE (D)
Yermo Dorn COMPOSITAE (D)
Yinquania Z.Y.Zhu = Cornus L. CORNACEAE (D)
Ynesa O.F.Cook = Attalea Kunth PALMAE (M)
Yoania Maxim. ORCHIDACEAE (M)
Yolanda Hoehne = Brachionidium Lindl. ORCHIDACEAE (M)
Youngia Cass. COMPOSITAE (D)
Ypsilandra Franch. MELANTHIACEAE (M)
Ypsilopus Summerh. ORCHIDACEAE (M)
Ystia Compère = Schizachyrium Nees GRAMINEAE (M)
Yua C.L.Li = Parthenocissus Planch. VITACEAE (D)
Yucaratonia Burkart LEGUMINOSAE–PAPILIONOIDEAE (D)
Yucca L. AGAVACEAE (M)
Yunckeria Lundell MYRSINACEAE (D)
Yungasocereus F.Ritter = Haageocereus Backeb. CACTACEAE (D)
Yunnanea Hu = Camellia L. THEACEAE (D)
Yunquea Skottsb. = Centaurodendron Johow COMPOSITAE (D)
Yushania Keng f. = Sinarundinaria Nakai GRAMINEAE (M)
Yutajea Steyerm. RUBIACEAE (D)
Yuyba (Barb.Rodr.) L.H.Bailey = Bactris Jacq. PALMAE (M)
Yvesia A.Camus GRAMINEAE (M)

Zaa Baill. = Phyllarthron DC. BIGNONIACEAE (D)
Zabelia (Rehder) Makino = Abelia R.Br. CAPRIFOLIACEAE (D)
Zacateza Bullock ASCLEPIADACEAE (D)
Zacintha Vell. (SUH) = Clavija Ruíz & Pav. THEOPHRASTACEAE (D)
Zacintha Mill. = Crepis L. COMPOSITAE (D)
Zaczatea Baill. = Raphionacme Harv. ASCLEPIADACEAE (D)
Zahlbrucknera Rchb. = Saxifraga L. SAXIFRAGACEAE (D)
Zalaccella Becc. = Calamus L. PALMAE (M)
Zaleja Burm.f. (SUO) = Zaleya Burm.f. AIZOACEAE (D)
Zaleya Burm.f. AIZOACEAE (D)
Zalitea Raf. = Euphorbia L. EUPHORBIACEAE (D)
Zalmaria B.D.Jacks. = Rondeletia L. RUBIACEAE (D)
Zaluzania Pers. COMPOSITAE (D)
Zaluzianskia Neck. = Marsilea L. MARSILEACEAE (F)
Zaluzianskya F.W.Schmidt SCROPHULARIACEAE (D)
Zamaria Raf. = Rondeletia L. RUBIACEAE (D)
Zamia L. ZAMIACEAE (G)
Zamioculcas Schott ARACEAE (M)
Zamzela Raf. = Hirtella L. CHRYSOBALANACEAE (D)
Zandera D.L.Schulz COMPOSITAE (D)

Zanha Hiern SAPINDACEAE (D)
Zannichellia L. ZANNICHELLIACEAE (M)
Zanonia L. CUCURBITACEAE (D)
Zantedeschia Spreng. ARACEAE (M)
Zanthoxylum L. RUTACEAE (D)
Zapoteca H.M.Hern. LEGUMINOSAE–MIMOSOIDEAE (D)
Zappania Scop. = Salvia L. LABIATAE (D)
Zarcoa Llanos = Glochidion J.R.Forst. & G.Forst. EUPHORBIACEAE (D)
Zataria Boiss. LABIATAE (D)
Zauschneria C.Presl = Epilobium L. ONAGRACEAE (D)
Zea L. GRAMINEAE (M)
Zebrina Schnizl. = Tradescantia L. COMMELINACEAE (M)
Zederachia Heist. ex Fabr. = Melia L. MELIACEAE (D)
Zederbauera H.P.Fuchs (UP) CRUCIFERAE (D)
Zehnderia C.Cusset PODOSTEMACEAE (D)
Zehneria Endl. CUCURBITACEAE (D)
Zehntnerella Britton & Rose = Facheiroa Britton & Rose CACTACEAE (D)
Zeia Lunell (SUS) = Triticum L. GRAMINEAE (M)
Zelkova Spach ULMACEAE (D)
Zelonops Raf. = Phoenix L. PALMAE (M)
Zenia Chun LEGUMINOSAE–CAESALPINIOIDEAE (D)
Zenkerella Taub. LEGUMINOSAE–CAESALPINIOIDEAE (D)
Zenkeria Trin. GRAMINEAE (M)
Zenkeria Rchb. (SUH) = Parmentiera DC. BIGNONIACEAE (D)
Zenkerina Engl. = Staurogyne Wall. ACANTHACEAE (D)
Zenkerodendron Gilg ex Jabl. = Cleistanthus Hook.f. ex Planch. EUPHORBIACEAE (D)
Zenkerophytum Engl. ex Diels = Syrrheonema Miers MENISPERMACEAE (D)
Zenobia D.Don ERICACEAE (D)
Zeocriton Wolf = Hordeum L. GRAMINEAE (M)
Zephyra D.Don TECOPHILAEACEAE (M)
Zephyranthella (Pax) Pax = Habranthus Herb. AMARYLLIDACEAE (M)
Zephyranthes Herb. AMARYLLIDACEAE (M)
Zeravschania Korovin UMBELLIFERAE (D)
Zerdana Boiss. CRUCIFERAE (D)
Zerna Panz. (SUS) = Vulpia C.C.Gmel. GRAMINEAE (M)
Zerumbet J.C.Wendl. = Alpinia Roxb. ZINGIBERACEAE (M)
Zetagyne Ridl. ORCHIDACEAE (M)
Zeugandra P.H.Davis CAMPANULACEAE (D)
Zeugites P.Browne GRAMINEAE (M)
Zeuktophyllum N.E.Br. AIZOACEAE (D)
Zeuxanthe Ridl. = Prismatomeris Thwaites RUBIACEAE (D)
Zeuxine Lindl. ORCHIDACEAE (M)
Zexmenia La Llave COMPOSITAE (D)
Zeydora Lour. ex Gomes = Pueraria DC. LEGUMINOSAE–PAPILIONOIDEAE (D)
Zeyhera Mart. (SUO) = Zeyheria Mart. BIGNONIACEAE (D)
Zeyherella (Pierre ex Engl.) Aubrev = Englerophytum K.Krause SAPOTACEAE (D)
Zeyheria Mart. BIGNONIACEAE (D)
Zeylanidium (Tul.) Engl. = Hydrobryum Endl. PODOSTEMACEAE (D)
Zhumeria Rech.f. & Wendelbo LABIATAE (D)
Zieria Sm. RUTACEAE (D)
Zieridium Baill. = Euodia J.R.Forst. & G.Forst. RUTACEAE (D)
Zigadenus Michx. MELANTHIACEAE (M)
Zilla Forssk. CRUCIFERAE (D)
Zimapania Engl. & Pax = Jatropha L. EUPHORBIACEAE (D)
Zimmermannia Pax EUPHORBIACEAE (D)

Zimmermanniopsis Radcl.–Sm. EUPHORBIACEAE (D)
Zingania A.Chev. = Didelotia Baill. LEGUMINOSAE–CAESALPINIOIDEAE (D)
Zingeria P.A.Smirn. GRAMINEAE (M)
Zingeriopsis Prob. = Zingeria P.A.Smirn. GRAMINEAE (M)
Zingiber Boehm. ZINGIBERACEAE (M)
Zinnia L. COMPOSITAE (D)
Zinowiewia Turcz. CELASTRACEAE (D)
Zippelia Blume PIPERACEAE (D)
Zizania L. GRAMINEAE (M)
Zizaniopsis Döll & Asch. GRAMINEAE (M)
Zizia W.D.J.Koch UMBELLIFERAE (D)
Ziziphora L. LABIATAE (D)
Ziziphus Mill. RHAMNACEAE (D)
Zoegea L. COMPOSITAE (D)
Zollernia Wied.–Neuw. & Nees LEGUMINOSAE–PAPILIONOIDEAE (D)
Zollingeria Kurz SAPINDACEAE (D)
Zombia L.H.Bailey PALMAE (M)
Zombitsia Keraudren CUCURBITACEAE (D)
Zomicarpa Schott ARACEAE (M)
Zomicarpella N.E.Br. ARACEAE (M)
Zonanthemis Greene = Hemizonia DC. COMPOSITAE (D)
Zonanthus Griseb. GENTIANACEAE (D)
Zonotriche (C.E.Hubb.) J.B.Phipps GRAMINEAE (M)
Zoophora Bernh. = Orchis L. ORCHIDACEAE (M)
Zootrophion Luer ORCHIDACEAE (M)
Zornia J.F.Gmel. LEGUMINOSAE–PAPILIONOIDEAE (D)
Zosima Hoffm. UMBELLIFERAE (D)
Zostera L. ZOSTERACEAE (M)
Zosterella Small PONTEDERIACEAE (M)
Zosterostylis Blume = Cryptostylis R.Br. ORCHIDACEAE (M)
Zoydia Pers. (SUO) = Zoysia Willd. GRAMINEAE (M)
Zoysia Willd. GRAMINEAE (M)
Zschokkea Müll.Arg. = Lacmellea Karst. APOCYNACEAE (D)
Zucca Comm. ex Juss. = Momordica L. CUCURBITACEAE (D)
Zuccagnia Cav. LEGUMINOSAE–CAESALPINIOIDEAE (D)
Zuccarinia Blume RUBIACEAE (D)
Zucchellia Decne. = Raphionacme Harv. ASCLEPIADACEAE (D)
Zuckertia Baill. = Tragia L. EUPHORBIACEAE (D)
Zuckia Standl. CHENOPODIACEAE (D)
Zuelania A.Rich. FLACOURTIACEAE (D)
Zunilia Lundell = Ardisia Sw. MYRSINACEAE (D)
Zurloa Ten. = Carapa Aubl. MELIACEAE (D)
Zwaardekronia Korth. = Psychotria L. RUBIACEAE (D)
Zwackhia Sendtn. ex Rchb. (SUH) = Halacsya Dörfl. BORAGINACEAE (D)
Zwingera Hofer = Nolana L.f. SOLANACEAE (D)
Zygadenia L.E.Bishop GRAMMITIDACEAE (F)
Zygalchemilla Rydb. = Alchemilla L. ROSACEAE (D)
Zygella S.Moore = Cypella Herb. IRIDACEAE (M)
Zygia P.Browne LEGUMINOSAE–MIMOSOIDEAE (D)
Zygocactus Schumann = Schlumbergera Lem. CACTACEAE (D)
Zygocereus Frič & Kreuz. (SUO) = Schlumbergera Lem. CACTACEAE (D)
Zygochloa S.T.Blake GRAMINEAE (M)
Zygodia Benth. = Baissea A.DC. APOCYNACEAE (D)
Zygoglossum Reinw. = Bulbophyllum Thouars ORCHIDACEAE (M)
Zygogynum Baill. WINTERACEAE (D)

Zygonerion Baill. = Strophanthus DC. APOCYNACEAE (D)
Zygoon Hiern = Tarenna Gaertn. RUBIACEAE (D)
Zygopetalum Hook. ORCHIDACEAE (M)
Zygophlebia L.E.Bishop GRAMMITIDACEAE (F)
Zygophyllidium (Boiss.) Small = Euphorbia L. EUPHORBIACEAE (D)
Zygophyllum L. ZYGOPHYLLACEAE (D)
Zygoruellia Baill. ACANTHACEAE (D)
Zygosepalum Rchb.f. ORCHIDACEAE (M)
Zygosicyos Humbert CUCURBITACEAE (D)
Zygospermum Thwaites ex Baill. = Margaritaria L.f. EUPHORBIACEAE (D)
Zygostates Lindl. ORCHIDACEAE (M)
Zygostelma Fourn. (SUH) = Lagoa T.Durand ASCLEPIADACEAE (D)
Zygostelma Benth. ASCLEPIADACEAE (D)
Zygostigma Griseb. GENTIANACEAE (D)
Zygotritonia Mildbr. IRIDACEAE (M)
Zymum Thouars = Tristellateia Thouars MALPIGHIACEAE (D)
Zyzyxia Strother COMPOSITAE (D)

PART 2

FAMILIES AND GENERA ALPHABETICALLY
IN MAJOR GROUPS

The vascular plants are here divided into five major groups, the Fern Allies, Ferns, Gymnosperms, Dicotyledons and Monocotyledons. These groups are adopted simply to give a convenient break-down, and no particular taxonomic rank is implied for them. Within each group the recognised families are listed alphabetically, and the recognised genera are listed similarly within each family.

An asterisk (*) against a generic name means that no material of that genus is known to be present in the Kew Herbarium.

Each family heading includes the name, its author and its year of publication. The authors and dates have been supplied by J.L. Reveal and R.D. Hoogland from their invaluable data-base on family names, and thanks are due to them for their kind and willing assistance. At the right hand margin, on the top line, is the family number used in the Kew Herbarium, purely for convenience within Kew.

On the second line the number of genera recognised in the family is stated, followed by a brief indication of the distribution of the family, and of its habit or, occasionally, habitat. Beneath that, except for the Gymnosperms, is an analysis of opinions of certain other significant authors concerning that family. These analyses are prepared in one format for Fern Allies and Ferns (data compiled by R.J. Johns), and in a different format for Dicotyledons and Monocotyledons (data compiled by R.K. Brummitt), as explained below.

FERN ALLIES AND FERNS

Numbers of genera recognised in the family by the following authors are quoted:

Cop. Copeland, E.B., *Genera Filicum,* 1947

Holtt. Holttum, R.E., various publications, including names of Pteridophyta in H.K. Airy Shaw, ed. 8 of J.C. Willis, *A Dictionary of the Flowering Plants and Ferns,* 1973, and unpublished notes.

Kub. Kubitzki,K., ed., *The Families and Genera of Vascular Plants,* vol. 1, 1990; Pteridophyta by K. U. Kramer and others.

PS Pichi Sermolli, R.E.G., Tentamen Pteridophytorum genera in taxonomicum ordineum redigendi. *Webbia* 31: 313–512 (1977).

Thus under Lycopodiaceae, for example, "Kub. 4; PS 7; Holtt. 2" implies that in Kubitzki 4 genera were recognised, in Pichi Sermolli 7 genera, and in Holttum's works 2 genera (Copeland did not include the Fern Allies).

Where family concepts different from those of Kew are involved, these are indicated using the words "separating", "including" or "within" as appropriate, while alternative names or spelling are preceded by "as". Families not recognised at Kew, or different names or spellings from those used at Kew, are always printed in *italic* type. Examples:

separating Under Psilotaceae, "PS 1, separating *Tmesipteridaceae* 1" means that Pichi Sermolli recognised Psilotaceae, and placed one genus in it, but also recognised Tmesipteridaceae with one genus, which is included in Psilotaceae by Kew.

<table>
<tr><td>including</td><td>Under Gleicheniaceae, "Kub. 5, including Stromatopteridaceae 1" means that Kubitzki recognised Gleicheniaceae with 5 genera, one of which is placed in the family Stromatopteridaceae by Kew but the latter family is not recognised by Kubitzki.</td></tr>
<tr><td>within</td><td>Under Stromatopteridaceae, "Kub. 1, within Gleicheniaceae" means that Kubitzki recognised one genus corresponding with Kew's concept of Stromatopteridaceae, but did not recognise the latter family and included it in Gleicheniaceae.</td></tr>
<tr><td>as</td><td>Under Loxsomataceae, "Cop. 2, as Loxsomaceae" means that Copeland recognised the family but spelled its name differently, and included two genera in it.</td></tr>
</table>

DICOTYLEDONS AND MONOCOTYLEDONS

For each family in these groups a chart is given showing where each of eight different systems of classification place the family. The eight systems are abbreviated as follows:

B&H	G. Bentham & J.D. Hooker
DT&H	C.G. de Dalla Torre & H. Harms [A. Engler's system]
Melc	H. Melchior [A. Engler's Syllabus ed. 12]
Thor	R.F. Thorne
Dahl	R.M.T. Dahlgren
Young	D.A. Young
Takh	A. Takhtajan
Cron	A. Cronquist

Full bibliographic details are given in Part 3, where the full system of each is presented in systematic order.

Higher groups, above the rank of order, are given to the left in upper case, for some authors (Bentham & Hooker, Young and Takhtajan) at two different levels in their hierarchy, separated by a comma. In a separate column to the right, in lower case, is given the order in which the family is placed, in some cases (Dalla Torre & Harms, Melchior, Thorne and Takhtajan) sometimes with suborder as well. Finally a number is given, which is the number of the family under the appropriate system in Part 3.

Differences in family concepts between each system and Kew are indicated using the words "separating", "including", "within" and "as" in a similar way to their use in Fern Allies and Ferns explained above. Here, however, the numbers are not numbers of genera but cross references to the family numbers in Part 3. Use of the different words may be explained by reference to Cunoniaceae as an example (pp.551–552).

<table>
<tr><td>within</td><td>For Bentham & Hooker, "(within Saxifragaceae, 59)" means they did not recognise Cunoniaceae but included it within their Saxifragaceae which is family 59 in their system.</td></tr>
<tr><td>including</td><td>For Dalla Torre & Harms, "(including Davidsoniaceae)" means that they did recognise Cunoniaceae and included within it a family, Davidsoniaceae, which is recognised as separate by Kew.</td></tr>
<tr><td>separating</td><td>For Thorne "(separating Baueraceae 216)" means that he did recognise Cunoniaceae and also recognised a separate family, Baueraceae, which is included in Cunoniaceae by Kew, the number 216 being again the number of Baueraceae in the Thorne system in Part 3. Where the separated family is placed in a different suprafamilial taxon, this is indicated, as with Young's "separating Baueraceae 300 to Saxifragales" (Cunoniaceae being referred to Cunoniales).</td></tr>
</table>

As in Fern Allies and Ferns above, "as" is used for alternative names or spellings, and family names not accepted at Kew are printed in *italics*.

It should be noted that a statement of inclusion of one family in another implies only that the type genus of the first family (but not necessarily the whole of that family) is placed in the second. Thus, for example, under Bixaceae, where it states that Bentham & Hooker included Flacourtiaceae here, it does not necessarily mean that the whole of the Flacourtiaceae in Kew's (or anyone else's) sense was included in Bixaceae, but it does mean that *Flacourtia* itself was included. (In fact some other genera of Flacourtiaceae as defined here were referred by Bentham & Hooker to Samydaceae). No attempt is made to account for transference of genera from one family to another unless such a genus is the type genus of a family name recognised either by the system concerned or by Kew. Thus, under Caprifoliaceae it is noted that Dahlgren separated the family Viburnaceae, but it does not indicate that *Viburnum* was placed in Adoxaceae by Thorne because neither Thorne nor Kew recognises Viburnaceae.

FERN ALLIES

EQUISETACEAE Michx. ex DC. 1804 — 253

1 genus. Widespread. Terrestrial

Holtt. 1; PS 1; Kub. 1.

Equisetum L.

ISOETACEAE Rchb. 1828 — 252

2 genera. Widespread. Wet places.

Holtt. 2; PS 2; Kub. 2.

Isoetes L.
Stylites Amstutz

LYCOPODIACEAE P.Beauv. ex Mirb. 1802 — 250

4 genera. Widespread. Terrestrial or epiphytic.

Holtt. 2; PS 7; Kub. 4.

Huperzia Bernh.
Lycopodiella Holub

Lycopodium L.
Phylloglossum Kunze

PSILOTACEAE Kanitz 1887 — 254

2 genera. Widespread tropics to warm-temperate. Terrestrial or epiphytic.

Copel. 2; Holtt. 1 (separating *Tmesipteridaceae*); PS 1 (separating *Tmesipteridaceae* 1); Kub. 2.

Psilotum Sw.
Tmesipteris Bernh.

SELAGINELLACEAE Mett. 1856 — 251

1 genus. Widespread. Terrestrial or rarely epiphytic.

Holtt. 1; PS 1; Kub. 1.

Selaginella P.Beauv.

FERNS

ACTINIOPTERIDACEAE Pic.Serm. 1962   312

1 genus. Africa, SW Asia to Australia. Terrestrial.

Copel. 1 (within Pteridaceae); Holtt. 1; PS 1; Kub. 1 (within Pteridaceae subfam. Taenitidoideae).

Actiniopteris Link

ADIANTACEAE (C.Presl) Ching 1940   313

33 genera. Widespread. Terrestrial.

Copel. 29 (within Pteridaceae); Holtt. 2 (separating *Gymnogrammaceae* 20 and *Negripteridaceae* 2, but including *Rheopteris* 1); PS 1 (separating *Hemionitidaceae* 21, *Taenitidaceae* 4, *Negripteridaceae* 2, *Sinopteridaceae* 15; Kub. 25 (within Pteridaceae subfam. Taenitidoideae 12, subfam. Cheilanthoideae 12 and subfam. Adiantoideae 1).

Adiantopsis Fée
Adiantum L.
Anogramma Link
Aspidotis (Nutt. ex Hook.) Copel.
Austrogramme Fourn.
Bommeria Fourn.
Cassebeera Kaulf.
Cerosora (Baker) Domin
Cheilanthes Sw.
Cheiloplecton Fée
Coniogramme Fée
Cryptogramma R.Br.
Doryopteris J.Sm.
Eriosorus Fée
Gymnopteris Bernh.
Hemionitis L.
Jamesonia Hook. & Grev.

Llavea Lag.
Mildella Trevis.
Monachosorum Kunze
Negripteris Pic.Serm.
Nephopteris Lellinger *
Onychium Kaulf.
Pellaea Link
Pentagramma Yatsk., Windham & E. Wollenw.
Pityrogramma Link
Pterozonium Fée
Ptilopteris Hance
Saffordia Maxon
Sinopteris C.Chr. & Ching
Syngramma J.Sm.
Taenitis Willd. ex Schkuhr
Trachypteris E.F.André ex H.Christ

ASPLENIACEAE Mett. ex A.B.Frank 1877   327

9 genera. Widespread. Terrestrial or epiphytic.

Copel. 9; Holtt. 12; PS 14; Kub. 1.

Antigramma C.Presl
Asplenium L.
Camptosorus Link
Ceterach Willd.
Diellia Brack.

Diplora Baker
Holodictyum Maxon
Pleurosorus Fée
Schaffneria Fée ex T.Moore

AZOLLACEAE Wettst. 1903 334

1 genus. Widespread. Free-floating aquatics.

Copel. 1 (within Salviniaceae); Holtt. 1; PS 1; Kub. 1.

Azolla Lam.

BLECHNACEAE (C.Presl) Copel. 1947 328

9 genera. Widespread. Terrestrial, sometimes climbing.

Copel. 8; Holtt. 8; PS 12; Kub. 9.

Blechnum L.
Brainea J.Sm.
Doodia R.Br.
Pteridoblechnum Hennipman
Sadleria Kaulf.

Salpichlaena J.Sm.
Steenisioblechnum Hennipman
Stenochlaena J.Sm.
Woodwardia Sm.

CHEIROPLEURIACEAE Nakai 1928 321.02

1 genus. Indochina to E. Malesia. Terrestrial or lithophytic.

Copel. 1 (within Polypodiaceae); Holtt. 1; PS 1; Kub. 1.

Cheiropleuria C.Presl

CYATHEACEAE Kaulf. 1827 318

1 genus. Tropical to warm-temperate. Tree ferns.

Copel. 7 (including Metaxyaceae 1, Lophosoriaceae 1); Holtt. 2; PS 6; Kub. 1.

Cyathea Sm.

DAVALLIACEAE Mett. ex A.B.Frank 1877 326

10 genera. Tropical and temperate. Terrestrial or lithophytic.

Copel. 12 (including Oleandraceae 4); Holtt. 10; PS 10; Kub. 5.

Araiostegia Copel.
Davallia Sm.
Davallodes (Copel.) Copel.
Gymnogrammitis Griff.
Humata Cav.

Leucostegia C.Presl
Parasorus Alderw.
Rumohra Raddi
Scyphularia Fée
Trogostolon Copel.

DENNSTAEDTIACEAE Lotsy 1901   322

18 genera. Widespread. Terrestrial.

Copel. 18 (within Pteridaceae); Holtt. 13 (separating *Lindsaeaceae* 10); PS 9 (separating *Hypolepidaceae* 6 and *Lindsaeaceae* 9); Kub. 16.

Blotiella R.M.Tryon
Dennstaedtia Bernh.
Histiopteris (J.G.Agardh) J.Sm.
Hypolepis Bernh.
Leptolepia Prantl
Lindsaea Dryand. ex Sm.
Lonchitis L.
Microlepia C.Presl
Odontosoria Fée

Oenotrichia Copel.
Ormoloma Maxon
Orthiopteris Copel.
Paesia A.St.-Hil.
Pteridium Gled. ex Scop.
Saccoloma Kaulf.
Sphenomeris Maxon
Tapeinidium (C.Presl) C.Chr.
Xyropteris K.U.Kramer

DICKSONIACEAE (C.Presl) Bower 1908   316

7 genera. Montane tropical and temperate. Epiphytes or lithophytes.

Copel. 5 (within Pteridaceae); Holtt. 5 (separating *Lophosoriaceae* 1); PS 3 (separating *Lophosoriaceae* 1, *Thyrsopteridaceae* 1, *Culcitaceae* 1); Kub. 6 (separating *Lophosoriaceae* 1).

Calochlaena (Maxon) R.A.White &
    M.D.Turner
Cibotium Kaulf.
Culcita C.Presl

Cystodium J.Sm.
Dicksonia L'Hér.
Lophosoria C.Presl
Thyrsopteris Kunze

DIPTERIDACEAE (Diels) Seward & E.Dale 1901   321.01

1 genus. India and China to Fiji. Terrestrial or rheophytic.

Copel. 1 (within Polypodiaceae); Holtt. 1; PS 1; Kub. 1.

Dipteris Reinw.

DRYOPTERIDACEAE Ching 1965   323.02

47 genera. Widespread. Terrestrial or rarely epiphytic.

Copel. 66, as *Aspidiaceae* (including Woodsiaceae 19); Holtt. 50; PS 52 (separating *Onocleaceae* 3); Kub. 45 (including Woodsiaceae 19).

Acropelta Nakai
Acrophorus C.Presl
Aenigmopteris Holttum
Amphiblestra C.Presl
Arachniodes Blume
Atalopteris Maxon & C.Chr.
Ataxipteris Holttum

Camptodium Fée
Chlamydogramme Holttum
Coveniella Tindale
Ctenitis (C.Chr.) C.Chr.
Cyclodium C.Presl
Cyclopeltis J.Sm.
Cyrtomium C.Presl

Diacalpe Blume
Didymochlaena Desv
Dryopolystichum Copel.
Dryopsis Holttum & P.J.Edwards
Dryopteris Adans.
Fadyenia Hook.
Heterogonium C.Presl
Hypoderris R.Br. ex Hook.
Lastreopsis Ching
Leptorumohra (H.Ito) H.Ito
Lithostegia Ching
Maxonia C.Chr.
Megalastrum Holttum
Nothoperanema (Tagawa) Ching
Olfersia Raddi
Papuapteris C.Chr.
Peranema D.Don

Phanerophlebia C.Presl
Phanerophlebiopsis Ching
Plecosorus Fée
Pleocnemia C.Presl
Pleuroderris Maxon
Polybotrya Humb. & Bonpl. ex Willd.
Polystichopsis (J.Sm.) Holttum
Polystichum Roth
Pseudotectaria Tardieu
Psomiocarpa C.Presl
Pteridrys C.Chr. & Ching
Stenolepia Alderw.
Stigmatopteris C.Chr.
Tectaria Cav.
Tectaridium Copel.
Triplophyllum Holttum

GLEICHENIACEAE (R.Br.) C.Presl 1825   304

4 genera. Widespread in tropics. Terrestrial, in dense stands.

Copel. 6 (including Platyzomataceae 1, Stromatopteridaceae 1); Holtt. 5 (including Stromatopteridaceae 1); PS 8; Kub. 5 (including Stromatopteridaceae 1).

Dicranopteris Bernh.
Diplopterygium (Diels) Nakai

Gleichenia Sm.
Sticherus C.Presl

GRAMMITIDACEAE (C.Presl) Ching 1940   330

14 genera. Widespread in tropics and S. temperate. Small epiphytes.

Copel. 12 (within Polypodiaceae); Holtt. 12; PS 12; Kub. 4.

Acrosorus Copel.
Adenophorus Gaudich.
Calymmodon C.Presl
Ceradenia L.E.Bishop
Cochlidium Kaulf.
Ctenopteris Blume ex Kunze
Glyphotaenium J.Sm.

Grammitis Sw.
Lellingeria A.R.Sm. & R.C. Moran
Prosaptia C.Presl
Scleroglossum Alderw.
Xiphopteris Kaulf.
Zygadenia L.E.Bishop
Zygophlebia L.E.Bishop

HYMENOPHYLLACEAE Link 1833   315

10 genera. Mostly tropical montane and S. temperate. Filmy ferns.

Copel. 34; Holtt. 34; PS 42; Kub. 8.

Cardiomanes C.Presl
Cephalomanes C.Presl

Crepidomanes (C.Presl) C.Presl
Hymenoglossum C.Presl

Hymenophyllum Sm.	**Serpyllopsis** Van den Bosch
Polyphlebium Copel.	**Sphaerocionium** C.Presl
Rosenstockia Copel.	**Trichomanes** L.

HYMENOPHYLLOPSIDACEAE C.Chr. ex Pic.Serm. 1970 317

1 genus. S. America. Lithophytic or terrestrial.

Copel. 1; Holtt. 1; PS 1; Kub. 1.

Hymenophyllopsis K.I.Goebel

LOMARIOPSIDACEAE Alston 1956 325

7 genera. Widespread tropical to warm-temperate. Epiphytes, climbers.

Copel. 10 (within *Aspidiaceae*); Holtt. 12; PS 8 (separating *Elaphoglossaceae* 3); Kub. 6.

Bolbitis Schott	**Peltapteris** Link
Elaphoglossum Schott ex J.Sm.	**Teratophyllum** Mett. ex Kuhn
Lomagramma J.Sm.	**Thysanosoria** Gepp
Lomariopsis Fée	

LOXSOMATACEAE C.Presl 1847 319

2 genera. C. and S. America, New Zealand. Terrestrial.

Copel. 2 (as Loxsomaceae); Holtt. 2 (as Loxsomaceae); PS 2 (as Loxsomaceae); Kub. 2 (as Loxomataceae).

Loxsoma R.Br. ex A.Cunn.
Loxsomopsis Christ

MARATTIACEAE Bercht. & J.Presl 301

4 genera. Tropics and subtropics. Terrestrial, often massive.

Copel. 6; Holtt. 1 (separating *Angiopteridaceae* 3, *Danaeaceae* 1, *Kaulfussiaceae* 1); PS 2 (separating *Angiopteridaceae* 3, *Danaeaceae* 1, *Christenseniaceae* 1); Kub. 4.

Angiopteris Hoffm.	**Danaea** Sm.
Christensenia Maxon	**Marattia** Sw.

MARSILEACEAE Mirb. 1802 332

3 genera. Mostly Africa, Australasia, New World, some temp. Wet places, some aquatic.

Copel. 3; Holtt. 3; PS 3; Kub. 3.

Marsilea L.
Pilularia L.
Regnellidium Lindm.

MATONIACEAE C.Presl 1847   306

2 genera. Malesia. Terrestrial or lithophytic on limestone.

Copel. 2; Holtt. 2; PS 2; Kub. 2.

Matonia R.Br. ex Wall.
Phanerosorus Copel.

METAXYACEAE Pic.Serm. 1970   320

1 genus. C. and S. America. Terrestrial.

Copel. 1 (within Cyatheaceae); Holtt. 1 (within *Lophosoriaceae* 1), PS 1; Kub. 1.

Metaxya C.Presl

OLEANDRACEAE (J.Sm.) Ching ex Pic.Serm. 1965   324

4 genera. Widespread tropical to warm-temperate. Terrestrial or epiphytic.

Copel. 4 (within Davalliaceae); Holtt. 4; PS 1 (separating *Nephrolepidaceae* 3); Kub. 3 (separating *Nephrolepidaceae* 1).

Arthropteris J.Sm.   **Oleandra** Cav.
Nephrolepis Schott   **Psammiosorus** C.Chr.

OPHIOGLOSSACEAE (R.Br.) C.Agardh 1822   300

3 genera. Widespread. Terrestrial or epiphytic.

Copel. 4; Holtt. 4; PS 4 (separating *Botrychiaceae* 2, *Helminthostachyaceae* 1); Kub. 3.

Botrychium Sw.
Helminthostachys Kaulf.
Ophioglossum L.

OSMUNDACEAE Gérardin & Desv. 1817   302

3 genera. Widespread. Terrestrial.

Copel. 3; Holtt. 3; PS 3; Kub. 3.

Leptopteris C.Presl
Osmunda L.
Todea Willd. ex Bernh.

PARKERIACEAE Hook. 1825 309

1 genus. Widespread in tropics. Floating aquatic.

Copel. 1; Holtt. 1; PS 1; Kub. 1 (within Pteridaceae subfam. Ceratopteridoideae).

Ceratopteris Brongn.

PLAGIOGYRIACEAE Bower 1926 303

1 genus. Tropics and subtropics excluding Africa. Terrestrial.

Copel. 1; Holtt. 1; PS 1; Kub. 1.

Plagiogyria (Kunze) Mett.

PLATYZOMATACEAE Nakai 1950 308

1 genus. N. and N.E. Australia. Terrestrial.

Copel. 1 (within Gleicheniaceae); Holtt. 1 (within *Gymnogrammaceae*); PS 1; Kub. 1 (within Pteridaceae subfam. Platyzomatoideae).

Platyzoma R.Br.

POLYPODIACEAE Bercht. & J.Presl 1820 331.02

47 genera. Tropical to temperate. Terrestrial or epiphytic.

Copel. 65 (including Grammitidaceae 12, Dipteridaceae 1, Cheiropleuriaceae 1); Holtt. 52; PS 63 (separating *Loxogrammaceae* 2); Kub. 29.

Aglaomorpha Schott
Anarthropteris Copel.
Arthromeris (T.Moore) J.Sm.
Belvisia Mirb.
Campyloneurum C.Presl
Christopteris Copel.
Colysis C.Presl
Crypsinus C.Presl
Diblemma J.Sm.
Dicranoglossum J.Sm.
Dictymia J.Sm.
Drymotaenium Makino
Drynaria (Bory) J.Sm.
Goniophlebium C.Presl
Grammatopteridium Alderw.
Holcosorus T.Moore
Hyalotrichopteris W.H.Wagner
Lecanopteris Reinw.
Lemmaphyllum C.Presl

Lepisorus (J.Sm.) Ching
Leptochilus Kaulf.
Loxogramme (Blume) C.Presl
Marginariopsis C.Chr.
Microgramma C.Presl
Microsorum Link
Myuropteris C.Chr.
Neocheiropteris Christ
Neurodium Fée
Niphidium J.Sm.
Oleandropsis Copel.
Paragramma (Blume) T.Moore
Pecluma M.G.Price
Phlebodium (R.Br.) J.Sm.
Phymatopteris Pic.Serm.
Phymatosorus Pic.Serm.
Platycerium Desv.
Pleopeltis Humb. & Bonpl. ex Willd.
Podosorus Holttum

Polypodiopteris C.F.Reed	**Selliguea** Bory
Polypodium L.	**Solanopteris** Copel.
Pycnoloma C.Chr.	**Synammia** J.Sm.
Pyrrosia Mirb.	**Thylacopteris** Kunze ex J.Sm.
Saxiglossum Ching	

PTERIDACEAE Spreng. ex Jameson 1821   311

6 genera. Widespread. Terrestrial.

Copel. 64 (including Actiniopteridaceae 1, Adiantaceae 29, Dennstaedtiaceae 18, Dicksoniaceae 5); Holtt. 9; PS 8; Kub. 44.

Acrostichum L.	**Nevrocallis** Fée
Afropteris Alston	**Ochropteris** J.Sm.
Anopteris Prantl ex Diels	**Pteris** L.

SALVINIACEAE T.Lestib. 1826   333

1 genus. Mostly tropical, subtropical. Free-floating aquatic.

Copel. 2 (including Azollaceae 1); Holtt. 1; PS 1; Kub. 1.

Salvinia Séguier

SCHIZAEACEAE Kaulf. 1827   307

4 genera. Tropics and S. warm-temp. Terrestrial or climbing.

Copel. 4; Holtt. 4; PS 2 (separating *Anemiaceae* 2, *Lygodiaceae* 1); Kub. 4.

Anemia Sw.	**Mohria** Sw.
Lygodium Sw.	**Schizaea** Sm.

STROMATOPTERIDACEAE (Nakai) Bierh. 1968   305

1 genus. New Caledonia. Terrestrial on serpentine.

Copel. 1 (within Gleicheniaceae); Holtt. 1 (within Gleicheniaceae); PS 1; Kub. 1 (within Gleicheniaceae).

Stromatopteris Mett.

THELYPTERIDACEAE Pic.Serm. 1970   329

30 genera. Tropical, subtropical, some temperate. Terrestrial.

Copel. 8 (within Aspidiaceae); Holtt. 31; PS 32; Kub. 5.

Amauropelta Kunze
Ampelopteris Kunze
Amphineuron Holttum
Chingia Holttum
Christella H.Lév.
Coryphopteris Holttum
Cyclogramma Tagawa
Cyclosorus Link
Glaphyropteridopsis Ching
Glaphyropteris (Fée) C.Presl ex Fée
Goniopteris C.Presl
Macrothelypteris (H.Ito) Ching
Meniscium Schreb.
Menisorus Alston
Mesophlebion Holttum

Metathelypteris (H.Ito) Ching
Nannothelypteris Holttum
Oreopteris Holub
Parathelypteris (H.Ito) Ching
Phegopteris Fée
Plesioneuron (Holttum) Holttum
Pneumatopteris Nakai
Pronephrium C.Presl
Pseudocyclosorus Ching
Pseudophegopteris Ching
Sphaerostephanos J.Sm.
Stegnogramma Blume
Steiropteris (C.Chr.) Pic.Serm.
Thelypteris Schmidel
Trigonospora Holttum

VITTARIACEAE (C.Presl) Ching 1940   314

8 genera. Tropical to warm-temperate. Epiphytes or lithophytes.

Copel. 9; Holtt. 9; PS 10; Kub. 6.

Ananthacorus Underw. & Maxon
Anetium Splitg.
Antrophyum Kaulf.
Hecistopteris J.Sm.

Monogramma Comm. ex Schkuhr
Rheopteris Alston
Vaginularia Fée
Vittaria Sm.

WOODSIACEAE (Diels) Herter 1949   323

20 genera. Temperate to montane tropics. Terrestrial or lithophytic.

Copel. 19 (within *Aspidiaceae*); Holtt. 10 (as *Athyriaceae*); PS 3 (separating *Athyriaceae* 23); Kub. 14 (within Dryopteridaceae subfam. Athyrioideae).

Acystopteris Nakai
Adenoderris J.Sm.
Anisocampium C.Presl
Athyrium Roth
Cheilanthopsis Hieron.
Cornopteris Nakai
Cystoathyrium Ching
Cystopteris Bernh.
Diplaziopsis C.Chr.
Diplazium Sw.

Gymnocarpium Newman
Hemidictyum C.Presl
Hypodematium Kunze
Kuniwatsukia Pic.Serm.
Lunathyrium Koidz.
Matteuccia Tod.
Onoclea L.
Onocleopsis Ballard
Trichoneuron Ching
Woodsia R.Br.

GYMNOSPERMS

ARAUCARIACEAE Henkel & W.Hochst. 1865   207

2 genera. S. America, Pacific, Australasia, S.E. Asia. Trees.

Agathis Salisb.
Araucaria Juss.

BOWENIACEAE D.W. Stev. 1981   219

1 genus. Northern Australia. Bipinnate cycads.

Bowenia Hook.

CEPHALOTAXACEAE Dumort. 1829   212

1 genus. Himalaya to Japan. Shrubs to trees.

Cephalotaxus Siebold & Zucc.

CUPRESSACEAE Rich. ex Bartl. 1830   210

18 genera. Widespread. Shrubs to trees.

Actinostrobus Miq.
Austrocedrus Florin & Boutelje
Callitris Vent.
Calocedrus Kurz
Chamaecyparis Spach
Cupressus L.
Diselma Hook.f.
Fitzroya Hook.f. ex Lindl.
Fokienia A.Henry & H.H.Thomas
Juniperus L.
Libocedrus Endl.
Microbiota Kom.
Neocallitropsis Florin
Pilgerodendron Florin
Tetraclinis Mast.
Thuja L.
Thujopsis Siebold & Zucc. ex Endl.
Widdringtonia Endl.

CYCADACEAE Pers. 1807   216

1 genus. Madagascar, tropical Asia to Australasia and Pacific. Cycads.

Cycas L.

EPHEDRACEAE Dumort. 1829   205

1 genus. Eurasia and New World. Shrubs.

Ephedra L.

GINKGOACEAE Engl. 1897   214

1 genus, monotypic. Eastern China. Tree.

Ginkgo L.

GNETACEAE Lindl. 1834   203

1 genus. Widespread in tropics. Mostly lianes, some shrubs or trees.

Gnetum L.

PHYLLOCLADACEAE (Pilg.) Keng 1973   211.02

1 genus. Malesia, Tasmania, New Zealand. Trees, shrubs.

Phyllocladus Rich. ex Mirb.

PINACEAE Lindl. 1836   208

12 genera. N. hemisphere to Malesia and C. America. Mostly trees.

Abies Mill.
Cathaya Chun & Kuang
Cedrus Trew
Hesperopeuce (Engelm.) Lemmon
Keteleeria Carrière
Larix Mill.

Nothotsuga H.H.Hu ex C.N.Page
Picea A.Dietr.
Pinus L.
Pseudolarix Gordon
Pseudotsuga Carrière
Tsuga Carrière

PODOCARPACEAE Endl. 1847   211.01

17 genera. S. hemisphere, north to Japan and C. America. Trees and shrubs.

Acmopyle Pilg.
Afrocarpus (Buchholz & E.Gray) C.N.Page
Dacrycarpus (Endl.) de Laub.
Dacrydium Lamb.
Falcatifolium de Laub.
Halocarpus Quinn
Lagarostrobos Quinn
Lepidothamnus Phil.
Microcachrys Hook.f.

Microstrobos J.Garden & L.A.S.Johnson
Nageia Gaertn.
Parasitaxus de Laub.
Podocarpus L'Hér. ex Pers.
Prumnopitys Phil.
Retrophyllum C.N.Page
Saxegothaea Lindl.
Sundacarpus C.N.Page

STANGERIACEAE (Pilg.) L.A.S. Johnson 1959   217

1 genus, monotypic. South Africa. Fern-like cycad.

Stangeria T.Moore

TAXACEAE Gray 1821   213

5 genera. N. hemisphere to Malesia and New Caledonia. Trees, shrubs.

Amentotaxus Pilg.
Austrotaxus Compton
Pseudotaxus W.C.Cheng

Taxus L.
Torreya Arn.

TAXODIACEAE Warm. 1884   209

10 genera. E. Asia, Tasmania, N. America. Trees.

Athrotaxis D.Don
Cryptomeria D.Don
Cunninghamia R.Br.
Glyptostrobus Endl.
Metasequoia Miki ex Hu & Cheng

Sciadopitys Siebold & Zucc.
Sequoia Endl.
Sequoiadendron Buchholz
Taiwania Hayata
Taxodium Rich.

WELWITSCHIACEAE Markgr. 1926   204

1 genus, monotypic. Angola, Namibia. Woody desert perennial.

Welwitschia Hook.f.

ZAMIACEAE Horan. 1834   218

8 genera. Africa, Australia, America. Cycads, often trees.

Ceratozamia Brongn.
Chigua D.W.Stev.
Dioon Lindl.
Encephalartos Lehm.

Lepidozamia Regel
Macrozamia Miq.
Microcycas (Miq.) A.DC.
Zamia L.

DICOTYLEDONS

ACANTHACEAE Juss. 1789

122

228 genera. Widespread trop., subtrop., rarely temp. Mostly herbs, rarely trees.

B&H GAMOPETALAE, BICARPELLATAE Personales, 122
DT&H METACHLAMYDEAE Tubiflorae, Acanthineae, 219
Melc SYMPETALAE Tubiflorae, Solanineae, 269
Thor GENTIANIFLORAE Bignoniales, 263
Dahl LAMIIFLORAE Scrophulariales, 362
Young ROSIDAE, GENTIANANAE Bignoniales, 361
Takh LAMIIDAE, LAMIANAE Scrophulariales, 409
  (including Cyclocheilaceae; separating *Mendonciaceae* 411, *Thunbergiaceae* 410)
Cron ASTERIDAE Scrophulariales, 299
  (separating *Mendonciaceae* 302)

Acanthopale C.B.Clarke
Acanthopsis Harv.
Acanthostelma Bidgood & Brummitt
Acanthura Lindau *
Acanthus L.
Achyrocalyx Benoist
Aechmanthera Nees
Afrofittonia Lindau
Ambongia Benoist *
Ancistranthus Lindau
Andrographis Wall. ex Nees
Angkalanthus Balf.f.
Anisacanthus Nees
Anisosepalum E.Hossain
Anisotes Nees
Anomacanthus R.D.Good
Anthacanthus Nees
Apassalus Kobuski
Aphanosperma T.F.Daniel
Aphelandra R.Br.
Aphelandrella Mildbr. *
Ascotheca Heine
Asystasia Blume
Asystasiella Lindau
Ballochia Balf.f.
Barleria L.
Barleriola Oerst.
Benoicanthus Heine & A.Raynal
Blechum P.Browne
Blepharis Juss.
Borneacanthus Bremek.
Boutonia DC.
Brachystephanus Nees
Bravaisia DC.
Brillantaisia P.Beauv.
Brunoniella Bremek.

Buceragenia Greenm.
Calacanthus T.Anderson ex Benth.
Calycacanthus K.Schum.
Camarotea Scott Elliot
Carlowrightia A.Gray
Celerina Benoist *
Centrilla Lindau
Cephalacanthus Lindau *
Chaetacanthus Nees
Chalarothyrsus Lindau
Chamaeranthemum Nees
Chileranthemum Oerst.
Chlamydacanthus Lindau
Chlamydocardia Lindau
Chlamydostachya Mildbr.
Chroesthes Benoist
Clinacanthus Nees
Clistax Mart.
Codonacanthus Nees
Cosmianthemum Bremek.
Crabbea Harv.
Crossandra Salisb.
Crossandrella C.B.Clarke
Cyclacanthus S.Moore
Cylindrosolenium Lindau
Cyphacanthus Leonard *
Cystacanthus T.Anderson
Danguya Benoist
Dasytropis Urb. *
Dichazothece Lindau
Dicladanthera F.Muell.
Dicliptera Juss.
Diotacanthus Benth.
Dischistocalyx T.Anderson ex Benth.
Duosperma Dayton
Dyschoriste Nees

Ecbolium Kurz
Echinacanthus Nees
Elytraria Michx.
Encephalosphaera Lindau
Epiclastopelma Lindau
Eranthemum L.
Eremomastax Lindau
Eusiphon Benoist
Filetia Miq.
Fittonia Coem.
Forcipella Baill. *
Geissomeria Lindl.
Glossochilus Nees
Golaea Chiov. *
Graphandra J.B.Imlay
Graptophyllum Nees
Gymnostachyum Nees
Gynocraterium Bremek.
Gypsacanthus Lott, Jaramillo & Rzed.
Habracanthus Nees
Haplanthodes Kuntze
Harpochilus Nees
Hemigraphis Nees
Henrya Nees
Herpetacanthus Nees
Heteradelphia Lindau
Holographis Nees
Hoverdenia Nees
Hulemacanthus S.Moore
Hygrophila R.Br.
Hypoestes Sol. ex R.Br.
Indoneesiella Sreem.
Ionacanthus Benoist
Isoglossa Oerst.
Isotheca Turrill
Jadunia Lindau
Juruasia Lindau
Justicia L.
Kalbreyeriella Lindau
Kolobochilus Lindau
Kosmosiphon Lindau
Kudoacanthus Hosok. *
Lankesteria Lindl.
Lasiocladus Bojer ex Nees
Leandriella Benoist *
Lepidagathis Willd.
Leptosiphonium F.Muell.
Leptostachya Nees
Linariantha B.L.Burtt & Ros.M.Sm.
Lindauea Rendle
Lophostachys Pohl
Louteridium S.Watson
Lychniothyrsus Lindau
Mackaya Harv.
Marcania J.B.Imlay

Megalochlamys Lindau
Megalostoma Leonard
Megaskepasma Lindau
Melittacanthus S.Moore *
Mellera S.Moore
Mendoncia Vell. ex Vand.
Metarungia Baden
Mexacanthus T.F.Daniel
Meyenia Nees
Mimulopsis Schweinf.
Mirandea Rzed.
Monechma Hochst.
Monothecium Hochst.
Morsacanthus Rizzini *
Nelsonia R.Br.
Neohallia Hemsl.
Neriacanthus Benth.
Neuracanthus Nees
Odontonema Nees
Ophiorrhiziphyllon Kurz
Oplonia Raf.
Oreacanthus Benth.
Orophochilus Lindau
Pachystachys Nees
Pararuellia Bremek.
Pelecostemon Leonard
Perenideboles Ram.Goyena
Pericalypta Benoist *
Peristrophe Nees
Petalidium Nees
Phaulopsis Willd.
Phialacanthus Benth.
Phlogacanthus Nees
Physacanthus Benth.
Podorungia Baill. *
Poikilacanthus Lindau
Polylychnis Bremek.
Populina Baill. *
Pranceacanthus Wassh.
Pseuderanthemum Radlk.
Pseudocalyx Radlk.
Pseudodicliptera Benoist
Pseudoruellia Benoist *
Psilanthele Lindau
Ptyssiglottis T.Anderson
Pulchranthus V.M.Baum, Reveal &
   Nowicke
Razisea Oerst.
Rhinacanthus Nees
Rhombochlamys Lindau
Ritonia Benoist *
Ruellia L.
Ruelliopsis C.B.Clarke
Rungia Nees
Ruspolia Lindau

Ruttya Harv.
Saintpauliopsis Staner
Salpixantha Hook.
Samuelssonia Urb. & Ekman
Sanchezia Ruíz & Pav.
Sapphoa Urb. *
Satanocrater Schweinf.
Sautiera Decne.
Schaueria Nees
Sclerochiton Harv.
Sebastiano–Schaueria Nees
Siphonoglossa Oerst.
Spathacanthus Baill.
Sphacanthus Benoist
Sphinctacanthus Benth.
Spirostigma Nees
Standleyacanthus Leonard *
Staurogyne Wall.
Stenandriopsis S.Moore
Stenandrium Nees
Stenosiphonium Nees
Stenostephanus Nees
Stenothyrsus C.B.Clarke
Streblacanthus Kuntze

Streptosiphon Mildbr.
Strobilacanthus Griseb.
Strobilanthes Blume
Strobilanthopsis S.Moore
Styasasia S.Moore
Suessenguthia Merxm.
Tacoanthus Baill.
Teliostachya Nees
Tessmanniacanthus Mildbr.
Tetramerium Nees
Thomandersia Baill.
Thunbergia Retz.
Thysanostigma J.B.Imlay
Tremacanthus S.Moore
Trichanthera Kunth
Trichocalyx Balf.f.
Trichosanchezia Mildbr. *
Vindasia Benoist *
Whitfieldia Hook.
Xantheranthemum Lindau
Xerothamnella C.T.White
Yeatesia Small
Zygoruellia Baill. *

ACERACEAE Juss. 1789　　　　　　　　　51.03

2 genera. N. temperate region and tropical mountains. to Malesia. Trees, shrubs.

B&H POLYPETALAE, DISCIFLORAE Sapindales (within Sapindaceae, 51)
DT&H ARCHICHLAMYDEAE Sapindales, Sapindineae, 113
Melc ARCHICHLAMYDEAE Sapindineae, Sapindineae, 145
Thor RUTIFLORAE Rutales, Sapindineae, 186
Dahl RUTIFLORAE Sapindales, 209
Young ROSIDAE, RUTANAE Sapindales, 259
Takh ROSIDAE, RUTANAE Sapindales, 246
Cron ROSIDAE Sapindales, 254

Acer L.
Dipteronia Oliv.

ACHARIACEAE Harms 1897　　　　　　　74.03

3 genera. South Africa. Herbs, subshrubs.

B&H POLYPETALAE, CALYCIFLORAE Passiflorales (within Passifloraceae, 74)
DT&H ARCHICHLAMYDEAE Parietales, Flacourtiineae, 154
Melc ARCHICHLAMYDEAE Violales, Flacourtiineae, 190
Thor VIOLIFLORAE Violales, Violineae, 132
Dahl VIOLIFLORAE Violales, 92
Young DILLENIIDAE, VIOLANAE Violales, 198

Takh DILLENIIDAE, VIOLANAE   Violales, 153
Cron DILLENIIDAE   Violales, 123

Acharia Thunb.
Ceratiosicyos Nees
Guthriea Bolus

ACHATOCARPACEAE Heimerl 1934   132.06

2 genera. Tropical America. Shrubs to small trees.

B&H MONOCHLAMYDEAE   Curvembryeae (within Amaranthaceae, 130)
DT&H ARCHICHLAMYDEAE   Centrospermae, Phytolaccineae (within
     Phytolaccaceae, 33)
Melc ARCHICHLAMYDEAE   Centrospermae, Phytolaccineae, 28
Thor CHENOPODIIFLORAE   Chenopodiales, Chenopodiineae, 83
Dahl CARYOPHYLLIFLORAE   Caryophyllales (within Phytolaccaceae, 39)
Young DILLENIIDAE, CARYOPHYLLANAE Chenopodiales, 161
Takh CARYOPHYLLIDAE, CARYOPHYLLANAE
     Caryophyllales, Caryophyllineae, 58
Cron CARYOPHYLLIDAE   Caryophyllales, 65

Achatocarpus Triana
Phaulothamnus A. Gray

ACTINIDIACEAE Hutch. 1926   28.05

3 genera. Warm-temp. Asia to Queensland, Fiji, trop. America. Trees, shrubs, lianes.

B&H POLYPETALAE, THALAMIFLORAE Guttiferales (within Theaceae, 28)
DT&H ARCHICHLAMYDEAE   Parietales, Theineae (within Dilleniaceae,
     130)
Melc ARCHICHLAMYDEAE   Guttiferales, Dilleniideae, 80
Thor THEIFLORAE   Theales, Theineae, 41
Dahl CORNIFLORAE   Ericales, 286
Young DILLENIIDAE, DILLENIANAE Theales, 104
Takh DILLENIIDAE, THEANAE   Actinidiales, 99
Cron DILLENIIDAE   Theales, 86

Actinidia Lindl.
Clematoclethra (Franch.) Maxim.
Saurauia Willd.

ADOXACEAE Trautv. 1853   83.04

3 genera. N. temperate region, China. Herbs.

B&H GAMOPETALAE, INFERAE   Rubiales (within Caprifoliaceae, 83)
DT&H METACHLAMYDEAE   Rubiales, 225
Melc SYMPETALAE   Dipsacales, 280
Thor CORNIFLORAE   Dipsacales, 292

Dahl   CORNIFLORAE   Cornales, 324
Young ROSIDAE, DIPSACANAE   Dipsacales, 382
Takh   ROSIDAE, CORNANAE   Dipsacales, 358
Cron   ASTERIDAE   Dipsacales, 314

Adoxa L.
Sinadoxa C.Y.Wu, Z.L.Wu & R..F.Huang *
Tetradoxa C.Y.Wu *

AEXTOXICACEAE Engl. & Gilg 1920   151.07

1 genus, monotypic. Chile. Tree.

B&H   MONOCHLAMYDEAE   Unisexuales (within Euphorbiaceae, 151)
DT&H   ARCHICHLAMYDEAE   Geraniales, Tricoccae (within Euphorbiaceae,
                                                                      97)

Melc   ARCHICHLAMYDEAE   Sapindales, Sapindineae, 151
Thor   MALVIFLORAE   Euphorbiales, 164
Dahl   MALVIFLORAE   Euphorbiales, 78
Young DILLENIIDAE, MALVANAE   Euphorbiales, 187
Takh   DILLENIIDAE, EUPHORBIANAE Euphorbiales, 193
Cron   ROSIDAE   Celastrales, 225

Aextoxicon Ruíz & Pav.

AGDESTIDACEAE (Heimerl) Nakai 1942   132.05

1 genus, monotypic. Mexico and W. Indies to Brazil. Climbing herb.

B&H   MONOCHLAMYDEAE   Curvembryeae (within Phytolaccaceae, 132)
DT&H   ARCHICHLAMYDEAE   Centrospermae, Phytolaccineae (within
                                                        Phytolaccaceae, 33)
Melc   ARCHICHLAMYDEAE   Centrospermae, Phytolaccineae (within
                                                        Phytolaccaceae, 26)
Thor   CHENOPODIIFLORAE   Chenopodiales, Chenopodiineae (within
                                                        Phytolaccaceae, 81)
Dahl   CARYOPHYLLIFLORAE   Caryophyllales (within Phytolaccaceae, 39)
Young DILLENIIDAE, CARYOPHYLLANAE Chenopodiales, 154
Takh   CARYOPHYLLIDAE, CARYOPHYLLANAE
                                Caryophyllales (within Phytolaccaceae, 57)
Cron   CARYOPHYLLIDAE   Caryophyllales (within Phytolaccaceae, 64)

Agdestis Moçino & Sessé ex DC.

AIZOACEAE F.Rudolphi 1830   79.01

128 genera. Trop. and subtrop., esp. S. Africa. Herbs, mostly succulent, rarely
subshrubs.

B&H   POLYPETALAE, CALYCIFLORAE   Ficoidales (as *Ficoideae*, 79)
       (including Gisekiaceae, Molluginaceae)

DT&H	ARCHICHLAMYDEAE (including Molluginaceae)	Centrospermae, Phytolaccineae, 34
Melc	ARCHICHLAMYDEAE	Centrospermae, Phytolaccineae, 31
Thor	CHENOPODIIFLORAE (including Molluginaceae)	Chenopodiales, Chenopodiineae, 85
Dahl	CARYOPHYLLIFLORAE	Caryophyllales, 44
Young	DILLENIIDAE, CARYOPHYLLANAE	Caryophyllales, 146
Takh	CARYOPHYLLIDAE, CARYOPHYLLANAE	Caryophyllales, Caryophyllineae, 61
Cron	CARYOPHYLLIDAE	Caryophyllales, 67

Acrodon N.E.Br.
Acrosanthes Eckl. & Zeyh.
Aethephyllum N.E.Br.
Aizoanthemum Dinter ex Friedrich
Aizoon L.
Aloinopsis Schwantes
Amphibolia L.Bolus
Antegibbaeum Schwantes ex C.Weber
Antimima N.E.Br.
Apatesia N.E.Br.
Aptenia N.E.Br.
Arenifera A.G.J.Herre
Argyroderma N.E.Br.
Aspazoma N.E.Br.
Astridia Dinter
Bergeranthus Schwantes
Berrisfordia L.Bolus
Bijlia N.E.Br.
Braunsia Schwantes
Brownanthus Schwantes
Carpanthea N.E.Br.
Carpobrotus N.E.Br.
Carruanthus (Schwantes) Schwantes
Caryotophora Leistner
Cephalophyllum (Haw.) N.E.Br.
Cerochlamys N.E.Br.
Chasmatophyllum (Schwantes) Dinter & Schwantes
Cheiridopsis N.E.Br.
Cleretum N.E.Br. *
Conicosia N.E.Br.
Conophytum N.E.Br.
Corpuscularia Schwantes
Cylindrophyllum Schwantes
Cypselea Turpin
Dactylopsis N.E.Br.
Delosperma N.E.Br.
Dicrocaulon N.E.Br.
Didymaotus N.E.Br.
Dinteranthus Schwantes
Diplosoma Schwantes
Disphyma N.E.Br.
Dorotheanthus Schwantes
Dracophilus (Schwantes) Dinter & Schwantes

Drosanthemopsis Rauschert
Drosanthemum Schwantes
Eberlanzia Schwantes
Ebracteola Dinter & Schwantes
Enarganthe N.E.Br.
Erepsia N.E.Br.
Esterhuysenia L.Bolus
Faucaria Schwantes
Fenestraria N.E.Br.
Frithia N.E.Br.
Galenia L.
Gibbaeum Haw.
Glottiphyllum Haw.
Gunniopsis Pax
Hallianthus H.E.K.Hartmann
Hereroa (Schwantes) Dinter & Schwantes
Herrea Schwantes
Herreanthus Schwantes
Hymenogyne Haw.
Imitaria N.E.Br.
Jacobsenia L.Bolus & Schwantes.
Jensenobotrya A.G.J.Herre
Jordaaniella H.E.K.Hartmann *
Juttadinteria Schwantes
Khadia N.E.Br.
Lampranthus N.E.Br.
Lapidaria (Dinter & Schwantes) N.E.Br.
Leipoldtia L.Bolus
Lithops N.E.Br.
Machairophyllum Schwantes
Malephora N.E.Br.
Mesembryanthemum L.
Mestoklema N.E.Br. ex Glen
Meyerophytum Schwantes
Mitrophyllum Schwantes
Monilaria Schwantes
Mossia N.E.Br.
Muiria N.E.Br.
Namaquanthus L.Bolus
Namibia (Schwantes) Dinter & Schwantes
Nananthus N.E.Br.
Nelia Schwantes
Neohenricia L.Bolus

Octopoma N.E.Br.
Odontophorus N.E.Br.
Oophytum N.E.Br.
Ophthalmophyllum Dinter & Schwantes
Orthopterum L.Bolus
Oscularia Schwantes
Ottosonderia L.Bolus
Phyllobolus N.E.Br.
Pleiospilos N.E.Br.
Plinthus Fenzl
Polymita N.E.Br.
Psammophora Dinter & Schwantes
Pseudobrownanthus Ihlenf. & Bittrich
Psilocaulon N.E.Br.
Rabiea N.E.Br.
Rhinephyllum N.E.Br.
Rhombophyllum (Schwantes)
   Schwantes
Ruschia Schwantes
Ruschianthemum Friedrich
Ruschianthus L.Bolus
Saphesia N.E.Br.

Schlechteranthus Schwantes
Schwantesia Dinter
Scopelogena L.Bolus
Sesuvium L.
Skiatophytum L.Bolus
Smicrostigma N.E.Br.
Stayneria L.Bolus
Stoeberia Dinter & Schwantes
Stomatium Schwantes
Synaptophyllum N.E.Br.
Tanquana H.E.K.Hartmann & Liede
Tetragonia L.
Titanopsis Schwantes
Trianthema L.
Tribulocarpus S.Moore
Trichodiadema Schwantes
Vanheerdea L.Bolus ex
   H.E.K.Hartmann
Vanzijlia L.Bolus
Wooleya L.Bolus
Zaleya Burm.f.
Zeuktophyllum N.E.Br.

AKANIACEAE Stapf 1912   51.06

1 genus, monotypic. E. Australia. Tree.

B&H   POLYPETALAE, DISCIFLORAE Sapindales (within Sapindaceae, 51)
DT&H ARCHICHLAMYDEAE   Sapindales, Sapindineae (within
                                                Sapindaceae, 115)
Melc ARCHICHLAMYDEAE   Rutales, Rutineae, 137
Thor RUTIFLORAE   Rutales, Sapindineae, 185
Dahl RUTIFLORAE   Sapindales, 210
Young ROSIDAE, RUTANAE   Sapindales, 257
Takh ROSIDAE, RUTANAE   Sapindales, 259
Cron ROSIDAE   Sapindales, 251

Akania Hook.f.

ALANGIACEAE DC. 1828   82.02

1 genus. Africa to China and Australia. Mostly trees.

B&H   POLYPETALAE, CALYCIFLORAE Umbellales (within Cornaceae, 82)
DT&H ARCHICHLAMYDEAE   Umbelliflorae (within Cornaceae, 181)
Melc ARCHICHLAMYDEAE   Umbelliflorae, 220
Thor CORNIFLORAE   Cornales, Cornineae, 285
Dahl CORNIFLORAE   Cornales, 299
Young ROSIDAE, CORNANAE   Cornales, 326
Takh ROSIDAE, CORNANAE   Cornales, 347
Cron ROSIDAE   Cornales, 201

Alangium Lam.

ALSEUOSMIACEAE Airy Shaw 1965   83.03

3 genera. Australia, New Zealand, New Caledonia. Shrubs.

B&H	GAMOPETALAE, INFERAE	Rubiales (within Caprifoliaceae, 83)
DT&H	METACHLAMYDEAE	Rubiales (within Caprifoliaceae, 224)
Melc	SYMPETALAE	Dipsacales (within Caprifoliaceae, 279)
Thor	ROSIFLORAE	Rosales, Saxifragineae (within Saxifragaceae, 208)
Dahl	CORNIFLORAE	Cornales, 318
Young	ROSIDAE, CORNANAE	Hydrangeales, 334
Takh	ROSIDAE, CORNANAE	Hydrangeales, Escalloniineae, 333
Cron	ROSIDAE	Rosales, 170

Alseuosmia A.Cunn.
Crispiloba Steenis
Wittsteinia F.Muell.

ALZATEACEAE S.Graham 1985   69.07

1 genus, monotypic. Peru, Bolivia. Tree.

B&H	POLYPETALAE, DISCIFLORAE	Celastrales (within Celastraceae, 47)
DT&H	ARCHICHLAMYDEAE	Sapindales, Celastrineae (within Celastraceae, 108)
Melc	Not mentioned	—
Thor	MYRTIFLORAE	Myrtales, Lythrineae, 241
Dahl	MYRTIFLORAE	Myrtales, 198
Young	Not mentioned	—
Takh	ROSIDAE, MYRTANAE	Myrtales, Myrtineae, 227
Cron	ROSIDAE	Myrtales (within Lythraceae 189 in 1981, as separate family in 1988)

Alzatea Ruíz & Pav.

AMARANTHACEAE Juss. 1789   130

70 genera. Widespread. Mostly herbs.

B&H	MONOCHLAMYDEAE (including Achatocarpaceae)	Curvembryeae, 130
DT&H	ARCHICHLAMYDEAE	Centrospermae, Chenopodiineae, 29
Melc	ARCHICHLAMYDEAE	Centrospermae, Chenopodiineae, 37
Thor	CHENOPODIIFLORAE	Chenopodiales, Chenopodiineae, 90
Dahl	CARYOPHYLLIFLORAE	Caryophyllales, 50
Young	DILLENIIDAE, CARYOPHYLLANAE	Chenopodiales, 160
Takh	CARYOPHYLLIDAE, CARYOPHYLLANAE	Caryophyllales, Chenopodiineae, 72
Cron	CARYOPHYLLIDAE	Caryophyllales, 71

Achyranthes L.
Achyropsis (Moq.) Hook.f.
Aerva Forssk.

Allmania R.Br. ex Wight
Allmaniopsis Suess.
Alternanthera Forssk.

Amaranthus L.
Arthraerua (Kuntze) Schinz
Blutaparon Raf.
Bosea L.
Calicorema Hook.f.
Celosia L.
Centema Hook.f.
Centemopsis Schinz
Centrostachys Wall.
Chamissoa Kunth
Charpentiera Gaudich.
Chionothrix Hook.f.
Cyathula Blume
Dasysphaera Volkens ex Gilg
Deeringia R.Br.
Digera Forssk.
Eriostylos C.C.Towns.
Froelichia Moench
Froelichiella R.E.Fr.
Gomphrena L.
Guilleminea Kunth
Henonia Moq.
Herbstia Sohmer
Hermbstaedtia Rchb.
Indobanalia A.N.Henry & B.Roy
Irenella Suess.
Iresine P.Browne
Kyphocarpa (Fenzl ex Endl.) Lopr.
Lagrezia Moq.
Leucosphaera Gilg
Lithophila Sw.
Lopriorea Schinz

Marcelliopsis Schinz
Mechowia Schinz
Nelsia Schinz
Neocentema Schinz
Nothosaerva Wight
Nototrichium (A.Gray) W.F.Hillebr.
Nyssanthes R.Br.
Pandiaka (Moq.) Hook.f.
Pfaffia Mart.
Pleuropetalum Hook.f.
Pleuropterantha Franch.
Polyrhabda C.C.Towns.
Pseudogomphrena R.E.Fr.
Pseudoplantago Suess. *
Pseudosericocoma Cavaco
Psilotrichopsis C.C.Towns.
Psilotrichum Blume
Ptilotus R.Br.
Pupalia Juss.
Quaternella Pedersen *
Rosifax C.C.Towns.
Saltia R.Br. ex Moq.
Sericocoma Fenzl
Sericocomopsis Schinz
Sericorema (Hook.f.) Lopr.
Sericostachys Gilg & Lopr.
Siamosia K.Larsen & Pedersen *
Stilbanthus Hook.f.
Tidestromia Standl.
Trichuriella Bennet
Volkensinia Schinz
Woehleria Griseb.

AMBORELLACEAE Pichon 1948   142.05

1 genus, monotypic. New Caledonia. Shrub.

B&H   MONOCHLAMYDEAE   Micrembryeae (within Monimiaceae, 142)
DT&H ARCHICHLAMYDEAE   Ranales, Magnoliineae (within Monimiaceae,
                                                                    51)

Melc   ARCHICHLAMYDEAE   Magnoliales, Laurineae, 51
Thor   ANNONIFLORAE   Annonales, Laurineae, 13
Dahl   MAGNOLIIFLORAE   Laurales, 17
Young MAGNOLIIDAE, RANUNCULANAE   Laurales, 17
Takh   MAGNOLIIDAE, MAGNOLIANAE   Laurales, 12
Cron   MAGNOLIIDAE   Laurales, 11

Amborella Baill.

ANACARDIACEAE Lindl. 1830   53.01

68 genera. Widespread, mostly tropical. Trees, shrubs, lianes.

B&H	POLYPETALAE, DISCIFLORAE	Sapindales, 53
	(including Corynocarpaceae)	
DT&H	ARCHICHLAMYDEAE	Sapindales, Anacardiineae, 103
	(including Podoaceae)	
Melc	ARCHICHLAMYDEAE	Sapindales, Anacardiineae, 144
	(including Podoaceae; separating *Julianiaceae* 153 to Julianiales)	
Thor	RUTIFLORAE	Rutales, Rutineae, 174
	(presumably including Podoaceae)	
Dahl	RUTIFLORAE	Sapindales, 204
Young	ROSIDAE, RUTANAE	Rutales, 251
	(including Podoaceae)	
Takh	ROSIDAE, RUTANAE	Rutales, 271
	(including Podoaceae)	
Cron	ROSIDAE	Sapindales, 256
	(including Podoaceae; separating *Julianiaceae* 257)	

Actinocheita F.A.Barkley
Amphipterygium Schiede ex Standl.
Anacardium L.
Androtium Stapf
Antrocaryon Pierre
Astronium Jacq.
Blepharocarya F.Muell.
Bonetiella Rzed.
Bouea Meisn.
Buchanania Spreng.
Campnosperma Thwaites
Cardenasiodendron F.A.Barkley *
Choerospondias B.L.Burtt & A.W.Hill
Comocladia P.Browne
Cotinus Mill.
Cyrtocarpa Kunth
Dracontomelon Blume
Drimycarpus Hook.f.
Euleria Urb. *
Euroschinus Hook.f.
Faguetia Marchand
Fegimanra Pierre
Gluta L.
Haematostaphis Hook.f.
Haplorhus Engl.
Harpephyllum Bernh. ex Krauss
Heeria Meisn.
Holigarna Buch.–Ham. ex Roxb.
Koordersiodendron Engl.
Lannea A.Rich.
Laurophyllus Thunb.
Lithraea Miers ex Hook. & Arn.
Loxopterygium Hook.f.
Loxostylis A.Spreng. ex Rchb.

Mangifera L.
Mauria Kunth
Melanochyla Hook.f.
Metopium P.Browne
Micronychia Oliv.
Mosquitoxylum Krug & Urb.
Nothopegia Blume
Ochoterenaea F.A.Barkley
Operculicarya H.Perrier
Orthopterygium Hemsl.
Ozoroa Delile
Pachycormus Coville ex Standl.
Parishia Hook.f.
Pegia Colebr.
Pentaspadon Hook.f.
Pistacia L.
Pleiogynium Engl.
Poupartia Comm. ex Juss.
Protorhus Engl.
Pseudosmodingium Engl.
Pseudospondias Engl.
Rhodosphaera Engl.
Rhus L.
Schinopsis Engl.
Schinus L.
Sclerocarya Hochst.
Semecarpus L.f.
Smodingium E.Mey. ex Sond.
Sorindeia Thouars
Spondias L.
Swintonia Griff.
Tapirira Aubl.
Thyrsodium Salzm. ex Benth.
Trichoscypha Hook.f.

ANCISTROCLADACEAE Planch. ex Walp. 29.02

1 genus. Tropical Africa, tropical Asia. Lianes.

B&H   POLYPET., THALAMIFLORAE   Guttiferales (within Dipterocarpaceae, 29)
DT&H ARCHICHLAMYDEAE   Parietales, Ancistrocladineae, 159
Melc ARCHICHLAMYDEAE   Guttiferales, Ancistrocladineae, 90
Thor THEIFLORAE   Theales, Scytopetalineae, 64
Dahl THEIFLORAE   Theales, 114
Young ROSIDAE, GERANIANAE   Linales, 271
Takh DILLENIIDAE, THEANAE   Ancistrocladales, 122
Cron DILLENIIDAE   Violales, 119

Ancistrocladus Wall.

ANISOPHYLLEACEAE Ridl. 1922 65.02

4 genera. Tropics. Trees, shrubs.

B&H   POLYPETALAE, CALYCIFLORAE   Myrtales (within Rhizophoraceae, 65)
DT&H ARCHICHLAMYDEAE   Myrtiflorae, Myrtineae (within
Rhizophoraceae, 171)
Melc ARCHICHLAMYDEAE   Myrtiflorae, Myrtineae (within
Rhizophoraceae, 212)
Thor CORNIFLORAE   Cornales, Rhizophorineae (within
Rhizophoraceae, 278)
Dahl CORNIFLORAE   Cornales, 319
Young (not mentioned, presumably within Rhizophoraceae, 322)
Takh ROSIDAE, MYRTANAE   Rhizophorales, 222
Cron ROSIDAE   Rosales, 169

Anisophyllea R.Br. ex Sabine   **Poga** Pierre
Combretocarpus Hook.f.   **Polygonanthus** Ducke

ANNONACEAE Juss. 1789 5.01

125 genera. Widespread trop. and subtrop., rarely warm-temp. Trees, shrubs, lianes.

B&H   POLYPETALAE, THALAMIFLORAE   Ranales, 5
        (including Eupomatiaceae)
DT&H ARCHICHLAMYDEAE   Ranales, Magnoliineae, 48
        (including Eupomatiaceae, Himantandraceae)
Melc ARCHICHLAMYDEAE   Magnoliales, Magnoliineae, 43
Thor ANNONIFLORAE   Annonales, Annonineae, 9
Dahl MAGNOLIIFLORAE   Annonales, 1
Young MAGNOLIIDAE, MAGNOLIANAE   Magnoliales, 5
Takh MAGNOLIIDAE, MAGNOLIANAE   Annonales, 5
Cron MAGNOLIIDAE   Magnoliales, 8

Afroguatteria Boutique   **Annona** L.
Alphonsea Hook.f. & Thomson   **Anomianthus** Zoll.
Ambavia Le Thomas   **Anonidium** Engl. & Diels
Anaxagorea A.St.–Hil.   **Artabotrys** R.Br.
Ancana F.Muell.   **Asimina** Adans.
Annickia Setten & Maas   **Asteranthe** Engl. & Diels

Balonga Le Thomas *
Bocagea A.St.–Hil.
Bocageopsis R.E.Fr.
Boutiquea Le Thomas
Cananga (DC.) Hook.f. & Thomson
Cardiopetalum Schltdl.
Chieniodendron Tsiang & P.T.Li
Cleistochlamys Oliv.
Cleistopholis Pierre ex Engl.
Cremastosperma R.E.Fr.
Cyathocalyx Champ. ex Hook.f. & Thomson
Cyathostemma Griff.
Cymbopetalum Benth.
Dasoclema J.Sinclair
Dasymaschalon (Hook.f. & Thomson) Dalla Torre & Harms
Deeringothamnus Small
Dendrokingstonia Rauschert
Dennettia Baker f.
Desmopsis Saff.
Desmos Lour.
Diclinanona Diels
Dielsiothamnus R.E.Fr.
Disepalum Hook.f.
Duckeanthus R.E.Fr.
Duguetia A.St.–Hil.
Ellipeia Hook.f. & Thomson
Ellipeiopsis R.E.Fr.
Enicosanthum Becc.
Ephedranthus S.Moore
Exellia Boutique
Fissistigma Griff.
Fitzalania F.Muell.
Friesodielsia Steenis
Froesiodendron R.E.Fr.
Fusaea (Baill.) Saff.
Gilbertiella Boutique
Goniothalamus (Blume) Hook.f. & Thomson
Greenwayodendron Verdc.
Guamia Merr.
Guatteria Ruíz & Pav.
Guatteriella R.E.Fr.
Guatteriopsis R.E.Fr.
Haplostichanthus F.Muell.
Heteropetalum Benth.
Hexalobus A.DC.
Hornschuchia Nees
Isolona Engl.
Letestudoxa Pellegr. *
Lettowianthus Diels
Malmea R.E.Fr.
Marsypopetalum Scheff.
Meiocarpidium Engl. & Diels

Meiogyne Miq.
Melodorum Lour.
Mezzettia Becc.
Mezzettiopsis Ridl.
Miliusa Lesch. ex A.DC.
Mischogyne Exell
Mitrella Miq.
Mitrephora (Blume) Hook.f. & Thomson
Mkilua Verdc.
Monanthotaxis Baill.
Monocarpia Miq.
Monocyclanthus Keay
Monodora Dunal
Neo–Uvaria Airy Shaw
Neostenanthera Exell
Oncodostigma Diels
Onychopetalum R.E.Fr.
Ophrypetalum Diels
Oreomitra Diels
Orophea Blume
Oxandra A.Rich.
Pachypodanthium Engl. & Diels
Papualthia Diels
Petalolophus K.Schum. *
Phaeanthus Hook.f. & Thomson
Phoenicanthus Alston
Piptostigma Oliv.
Platymitra Boerl.
Polyalthia Blume
Polyaulax Backer
Polyceratocarpus Engl. & Diels
Popowia Endl.
Porcelia Ruíz & Pav.
Pseudartabotrys Pellegr. *
Pseudephedranthus Aristeg.
Pseudoxandra R.E.Fr.
Pseuduvaria Miq.
Pyramidanthe Miq.
Raimondia Saff.
Richella A.Gray
Rollinia A.St.–Hil.
Ruizodendron R.E.Fr.
Sageraea Dalzell
Sapranthus Seem.
Schefferomitra Diels
Sphaerocoryne (Boerl.) Scheff. ex Ridl.
Stelechocarpus Hook.f. & Thomson
Stenanona Standl.
Tetrameranthus R.E.Fr.
Tetrapetalum Miq.
Toussaintia Boutique
Tridimeris Baill.
Trigynaea Schltdl.
Trivalvaria (Miq.) Miq.

Unonopsis R.E.Fr.
Uvaria L.
Uvariastrum Engl.
Uvariodendron (Engl. & Diels) R.E.Fr.

Uvariopsis Engl.
Woodiellantha Rauschert
Xylopia L.

APIACEAE — see UMBELLIFERAE

APOCYNACEAE Juss. 1789   106

168 genera. Widespread, mainly tropical. Mostly trees, shrubs, lianes.

B&H	GAMOPETALAE, BICARPELLATAE	Gentianales, 106
DT&H	METACHLAMYDEAE	Contortae, Gentianineae, 199
Melc	SYMPETALAE	Gentianales, 249
Thor	GENTIANIFLORAE	Gentianales, 253
	(including Asclepiadaceae)	
Dahl	GENTIANIFLORAE	Gentianales, 346
Young	ROSIDAE, GENTIANANAE	Gentianales, 350
Takh	LAMIIDAE, GENTIANANAE	Gentianales, 375
Cron	ASTERIDAE	Gentianales, 274

Acokanthera G.Don
Adenium Roem. & Schult.
Aganonerion Pierre ex Spire
Aganosma (Blume) G.Don
Alafia Thouars
Allamanda L.
Allomarkgrafia Woodson
Allowoodsonia Markgr.
Alstonia R.Br.
Alyxia Banks ex R.Br.
Amalocalyx Pierre
Ambelania Aubl.
Amsonia Walter
Ancylobothrys Pierre
Anechites Griseb.
Angadenia Miers
Anodendron A.DC.
Apocynum L.
Artia Guillaumin
Asketanthera Woodson
Aspidosperma Mart. & Zucc.
Baissea A.DC.
Beaumontia Wall.
Bousigonia Pierre
Cabucala Pichon
Callichilia Stapf
Calocrater K.Schum.
Cameraria L.
Carissa L.
Carruthersia Seem.
Carvalhoa K.Schum.
Catharanthus G.Don
Cerbera L.

Cerberiopsis Vieill. ex Pancher & Sebert
Chamaeclitandra (Stapf) Pichon
Chilocarpus Blume
Chonemorpha G.Don
Chunechites Tsiang
Cleghornia Wight
Clitandra Benth.
Condylocarpon Desf.
Couma Aubl.
Craspidospermum Bojer ex A.DC.
Crioceras Pierre
Cycladenia Benth.
Cyclocotyla Stapf
Cylindropsis Pierre
Delphyodon K.Schum.
Dewevrella De Wild.
Dictyophleba Pierre
Diplorhynchus Welw. ex Ficalho & Hiern
Dyera Hook.f.
Ecdysanthera Hook. & Arn.
Echites P.Browne
Elytropus Müll.Arg.
Epigynum Wight
Eucorymbia Stapf
Farquharia Stapf
Fernaldia Woodson
Forsteronia G.Mey.
Funtumia Stapf
Galactophora Woodson
Geissospermum Allemão
Gonioma E.Mey.

Grisseea Bakh.f.
Hancornia Gomes
Hanghomia Gagnep. & Thénint *
Haplophyton A.DC.
Himatanthus Willd. ex Schult.
Holarrhena R.Br.
Hunteria Roxb.
Ichnocarpus R.Br.
Isonema R.Br.
Ixodonerium Pit.
Kamettia Kostel.
Kibatalia G.Don
Kopsia Blume
Lacmellea Karst.
Landolphia P.Beauv.
Laubertia A.DC.
Laxoplumeria Markgr.
Lepinia Decne.
Lepiniopsis Valeton
Leuconotis Jack
Macoubea Aubl.
Macropharynx Rusby
Macrosiphonia Müll.Arg.
Malouetia A.DC.
Mandevilla Lindl.
Mascarenhasia A.DC.
Melodinus J.R.Forst. & G.Forst.
Mesechites Müll.Arg.
Micrechites Miq.
Microplumeria Baill.
Molongum Pichon
Mortoniella Woodson
Motandra A.DC.
Mucoa Zarucchi
Neisosperma Raf.
Neobracea Britton
Neocouma Pierre
Nerium L.
Nouettea Pierre
Ochrosia Juss.
Odontadenia Benth.
Oncinotis Benth.
Orthopichonia H.Huber
Pachypodium Lindl.
Pacouria Aubl.
Papuechites Markgr.
Parahancornia Ducke
Parameria Benth.
Parsonsia R.Br.
Peltastes Woodson
Pentalinon Voigt
Petchia Livera

Pezisicarpus Vernet *
Picralima Pierre
Plectaneia Thouars
Pleiocarpa Benth.
Pleioceras Baill.
Plumeria L.
Pottsia Hook. & Arn.
Prestonia R.Br.
Pteralyxia K.Schum.
Pycnobotrya Benth.
Quiotania Zarucchi
Rauvolfia L.
Rhabdadenia Müll.Arg.
Rhazya Decne.
Rhigospira Miers
Rhodocalyx Müll.Arg.
Rhynchodia Benth.
Saba (Pichon) Pichon
Salpinctes Woodson
Schizozygia Baill.
Secondatia A.DC.
Sindechites Oliv.
Skytanthus Meyen
Spirolobium Baill.
Spongiosperma Zarucchi
Stemmadenia Benth.
Stephanostegia Baill.
Stephanostema K.Schum.
Stipecoma Müll.Arg.
Strempeliopsis Benth.
Strophanthus DC.
Tabernaemontana L.
Tabernanthe Baill.
Temnadenia Miers
Thenardia Kunth
Thevetia L.
Thyrsanthella Pichon *
Tintinnabularia Woodson *
Tonduzia Pittier
Trachelospermum Lem.
Urceola Roxb.
Urnularia Stapf
Vahadenia Stapf
Vallariopsis Woodson
Vallaris Burm.f.
Vallesia Ruíz & Pav.
Vinca L.
Voacanga Thouars
Willughbeia Roxb.
Woytkowskia Woodson *
Wrightia R.Br.
Xylinabaria Pierre

AQUIFOLIACEAE Bartl. 1830   46.01

2 genera. Widespread. Trees, shrubs.

B&H	POLYPETALAE, DISCIFLORAE	Olacales (as *Ilicineae*), 46
DT&H	ARCHICHLAMYDEAE	Sapindales, Celastrineae, 107
	(including Phellinaceae, Sphenostemonaceae)	
Melc	ARCHICHLAMYDEAE .	Celastrales, Celastrineae, 156
	(including Phellinaceae, Sphenostemonaceae)	
Thor	THEIFLORAE	Theales, Theineae, 48
Dahl	CORNIFLORAE	Cornales, 310
Young	DILLENIIDAE, DILLENIANAE	Theales, 109
Takh	ROSIDAE, CELASTRANAE	Celastrales, Icacinineae, 298
Cron	ROSIDAE	Celastrales, 223
	(including Phellinaceae, Sphenostemonaceae)	

Ilex L.
Nemopanthus Raf.

ARALIACEAE Juss. 1789   81.01

47 genera. Mostly tropical. Trees, shrubs, climbers, rarely herbs.

B&H	POLYPETALAE, CALYCIFLORAE	Umbellales, 81
	(including Aralidiaceae, Helwingiaceae)	
DT&H	ARCHICHLAMYDEAE	Umbelliflorae, 179
	(including Aralidiaceae)	
Melc	ARCHICHLAMYDEAE	Umbelliflorae, 225
Thor	CORNIFLORAE	Araliales, 290
	(including Umbelliferae)	
Dahl	ARALIIFLORAE	Araliales, 267
Young	ROSIDAE, CORNANAE	Araliales, 342
Takh	ROSIDAE, CORNANAE	Apiales, 351
Cron	ROSIDAE	Apiales, 268

Anakasia Philipson
Apiopetalum Baill.
Aralia L.
Arthrophyllum Blume
Astrotricha DC.
Brassaiopsis Decne. & Planch.
Cephalaralia Harms
Cheirodendron Nutt. ex Seem.
Cuphocarpus Decne. & Planch.
Cussonia Thunb.
Delarbrea Vieill.
Dendropanax Decne. & Planch.
Eleutherococcus Maxim.
Fatsia Decne. & Planch.
Gamblea C.B.Clarke
Gastonia Comm. ex Lam.
Harmsiopanax Warb.
Hedera L.
Heteropanax Seem.

Hunaniopanax Qi & Cao *
Kalopanax Miq.
Mackinlaya F.Muell.
Macropanax Miq.
Megalopanax Ekman
Merrilliopanax H.L.Li
Meryta J.R.Forst. & G.Forst.
Motherwellia F.Muell.
Munroidendron Sherff
Myodocarpus Brongn. & Gris
Oplopanax (Torr. & A.Gray) Miq.
Oreopanax Decne. & Planch.
Osmoxylon Miq.
Panax L.
Pentapanax Seem.
Polyscias J.R.Forst. & G.Forst.
Pseudopanax K.Koch
Pseudosciadium Baill. *
Reynoldsia A.Gray

Schefflera J.R.Forst. & G.Forst.
Sciadodendron Griseb.
Seemannaralia R.Vig.
Sinopanax H.L.Li
Stilbocarpa (Hook.f.) Decne. & Planch.

Tetrapanax (K.Koch) K.Koch
Tetraplasandra A.Gray
Trevesia Vis.
Woodburnia Prain

ARALIDIACEAE Philipson & B.C.Stone 1980   81.03

1 genus, monotypic. Tropical Asia.

B&H	POLYPETALAE, CALYCIFLORAE	Umbellales (within Araliaceae, 81)
DT&H	ARCHICHLAMYDEAE	Umbelliflorae (within Araliaceae, 179)
Melc	Not mentioned	—
Thor	Not mentioned	—
Dahl	CORNIFLORAE	Cornales, 307
Young	Not mentioned	—
Takh	ROSIDAE, CORNANAE	Aralidiales, 348
Cron	ROSIDAE	Cornales (within Cornaceae, 203)

Aralidium Miq.

ARISTOLOCHIACEAE Juss. 1789   138

8 genera. Tropical to warm-temperate. Herbs, shrubs, lianes.

B&H	MONOCHLAMYDEAE	Multiovulatae Terrestres, 138
DT&H	ARCHICHLAMYDEAE	Aristolochiales, 24
Melc	ARCHICHLAMYDEAE	Aristolochiales, 72
Thor	ANNONIFLORAE	Annonales, Aristolochiineae,12
Dahl	MAGNOLIIFLORAE	Aristolochiales, 6
Young	MAGNOLIIDAE, RANUNCULANAE	Aristolochiales, 25
Takh	MAGNOLIIDAE, MAGNOLIANAE	Aristolochiales, 27
Cron	MAGNOLIIDAE	Aristolochiales, 22

Aristolochia L.
Asarum L.
Euglypha Chodat & Hassl.
Holostylis Duch.

Isotrema Raf.
Pararistolochia Hutch. & Dalziel
Saruma Oliv.
Thottea Rottb.

ASCLEPIADACEAE R.Br. 1810   107

315 genera. Widespread trop. and subtrop., occas. temp. Mostly herbs, climbers, lianes.

B&H	GAMOPETALAE, BICARPELLATAE	Gentianales 107
DT&H	METACHLAMYDEAE	Contortae, Gentianineae, 201
Melc	SYMPETALAE	Gentianales, 250
Thor	GENTIANIFLORAE	Gentianales (within Apocynaceae, 253)
Dahl	GENTIANIFLORAE	Gentianales, 347
Young	ROSIDAE, GENTIANANAE	Gentianales, 351
Takh	LAMIIDAE, GENTIANANAE	Gentianales, 376
Cron	ASTERIDAE	Gentianales, 275

Absolmsia Kuntze
Adelostemma Hook.f.
Aidomene Stopp *
Amblystigma Benth.
Ampelamus Raf.
Amphidetes Fourn.
Anatropanthus Schltr.
Anisopus N.E.Br.
Anisotoma Fenzl
Anomotassa K.Schum. *
Araujia Brot.
Asclepias L.
Aspidoglossum E.Mey.
Astephanus R.Br.
Asterostemma Decne.
Atherandra Decne.
Atherolepis Hook.f.
Atherostemon Blume
Baeolepis Decne. ex Moq.
Barjonia Decne.
Baroniella Costantin & Gallaud
Baseonema Schltr. & Rendle
Basistelma Bartlett *
Batesanthus N.E.Br.
Blepharodon Decne.
Blyttia Arn.
Brachystelma R.Br.
Calathostelma Fourn.
Calostigma Decne.
Calotropis R.Br.
Campestigma Pierre ex Costantin *
Camptocarpus Decne.
Caralluma R.Br.
Ceropegia L.
Chlorocyathus Oliv.
Cibirhiza Bruyns
Cionura Griseb.
Clemensiella Schltr.
Conomitra Fenzl
Cordylogyne E.Mey.
Corollonema Schltr. *
Cosmostigma Wight
Costantina Bullock *
Cryptolepis R.Br.
Cryptostegia R.Br.
Curinila Schult. *
Curroria Planch. ex Benth.
Cyathostelma Fourn.
Cynanchum L.
Cyprinia Browicz
Dalzielia Turrill
Decabelone Decne.
Decalepis Wight & Arn.
Decanema Decne.
Decastelma Schltr.

Dicarpophora Speg. *
Dictyanthus Decne.
Diplolepis R.Br.
Diplostigma K.Schum.
Dischidanthus Tsiang *
Dischidia R.Br.
Dischidiopsis Schltr.
Ditassa R.Br.
Dittoceras Hook.f.
Dolichostegia Schltr. *
Dorystephania Warb.
Dregea E.Mey.
Drepanostemma Jum. & H.Perrier
Duvalia Haw.
Duvaliandra M.G.Gilbert
Echidnopsis Hook.f.
Ecliptostelma Brandegee *
Ectadiopsis Benth.
Ectadium E.Mey.
Edithcolea N.E.Br.
Emicocarpus K.Schum. & Schltr.
Emplectanthus N.E.Br.
Epistemma D.V.Field & J.B.Hall
Eustegia R.Br.
Exolobus Fourn.
Fanninia Harv.
Fimbristemma Turcz.
Finlaysonia Wall.
Fischeria DC.
Fockea Endl.
Folotsia Costantin & Bois
Frerea Dalzell
Funastrum Fourn.
Genianthus Hook.f.
Glossonema Decne.
Glossostelma Schltr.
Gomphocarpus R.Br.
Gongronema (Endl.) Decne.
Gongylosperma King & Gamble
Gonioanthela Malme
Goniostemma Wight
Gonocrypta Baill.
Gonolobus Michx.
Graphistemma (Champ. ex Benth.)
    Champ. ex Benth.
Grisebachiella Lorentz
Gunnessia P.I.Forst.
Gymnanthera R.Br.
Gymnema R.Br.
Gymnemopsis Costantin
Harmandiella Costantin *
Harpanema Decne.
Heliostemma Woodson *
Hemidesmus R.Br.
Hemipogon Decne.

Heterostemma Wight & Arn.
Heynella Backer *
Himantostemma A.Gray
Holostemma R.Br.
Hoodia Sweet ex Decne.
Hoya R.Br.
Hoyella Ridl. *
Huernia R.Br.
Huerniopsis N.E.Br.
Hypolobus Fourn. *
Ibatia Decne.
Ischnolepis Jum. & H.Perrier
Ischnostemma King & Gamble
Jacaima Rendle
Janakia Joseph & Chandras.
Jobinia Fourn.
Kanahia R.Br.
Karimbolea Descoings *
Kinepetalum Schltr.
Labidostelma Schltr.
Lachnostoma Kunth
Lagoa T.Durand
Lavrania Plowes
Leichardtia R.Br.
Leptadenia R.Br.
Lhotzkyella Rauschert *
Loniceroides Bullock
Lorostelma Fourn.
Lugonia Wedd.
Lygisma Hook.f.
Macropetalum Burch. ex Decne.
Macroscepis Kunth
Madarosperma Benth.
Mahafalia Jum. & H.Perrier
Malinvaudia Fourn.
Mangenotia Pichon
Margaretta Oliv.
Marsdenia R.Br.
Matelea Aubl.
Meladerma Kerr
Melinia Decne.
Mellichampia A.Gray ex S.Watson
Menabea Baill.
Meresaldia Bullock
Merrillanthus Chun & Tsiang *
Metalepis Griseb.
Metaplexis R.Br.
Metastelma R.Br.
Micholitzia N.E.Br.
Microdactylon Brandegee
Microloma R.Br.
Microstelma Baill. *
Miraglossum Kupicha
Mitolepis Balf.f.
Mitostigma Decne.

Mondia Skeels
Morrenia Lindl.
Myriopteron Griff.
Nautonia Decne.
Neoschumannia Schltr.
Nephradenia Decne.
Notechidnopsis Lavranos & Bleck
Odontanthera Wight
Odontostelma Rendle
Oistonema Schltr.
Oncinema Arn.
Oncostemma K.Schum. *
Ophionella Bruyns
Orbea Haw.
Orbeanthus L.C.Leach
Orbeopsis L.C.Leach
Oreosparte Schltr. *
Orthanthera Wight
Orthosia Decne.
Oxypetalum R.Br.
Oxystelma R.Br.
Pachycarpus E.Mey.
Pachycymbium L.C.Leach
Papuastelma Bullock
Parapodium E.Mey.
Parquetina Baill.
Pattalias S.Watson
Peckoltia Fourn.
Pectinaria Haw.
Pentabothra Hook.f.
Pentacyphus Schltr. *
Pentagonanthus Bullock
Pentanura Blume
Pentarrhinum E.Mey.
Pentasacme Wall. ex Wight
Pentastelma Tsiang & P.T.Li *
Pentatropis R.Br. ex Wight & Arn.
Pentopetia Decne.
Pentopetiopsis Costantin & Gallaud *
Peplonia Decne.
Pergularia L.
Periglossum Decne.
Periploca L.
Petopentia Bullock
Phaeostemma Fourn.
Pherotrichis Decne.
Philibertia Kunth
Phyllanthera Blume
Physostelma Wight
Piaranthus R.Br.
Platykeleba N.E.Br.
Pleurostelma Baill.
Podandra Baill.
Podostelma K.Schum.
Poicillopsis Schltr. ex Rendle

Polystemma Decne.
Prosopostelma Baill.
Prosthecidiscus Donn.Sm.
Pseudibatia Malme
Pseudolithos Bally
Pseudomarsdenia Baill.
Pseudopectinaria Lavranos *
Ptycanthera Decne.
Pycnobregma Baill.
Pycnoneurum Decne.
Pycnorhachis Benth.
Pycnostelma Bunge ex Decne.
Quaqua N.E.Br.
Quisumbingia Merr. *
Raphionacme Harv.
Raphistemma Wall.
Rhyncharrhena F.Muell.
Rhyssolobium E.Mey.
Rhyssostelma Decne.
Rhytidocaulon Bally
Riocreuxia Decne.
Rojasia Malme
Rothrockia A.Gray
Sacleuxia Baill.
Sarcolobus R.Br.
Sarcorrhiza Bullock
Sarcostemma R.Br.
Sattadia Fourn.
Schistogyne Hook. & Arn.
Schistonema Schltr. *
Schizoglossum E.Mey.
Schlechterella K.Schum.
Schubertia Mart.
Scyphostelma Baill. *
Secamone R.Br.
Secamonopsis Jum.
Seshagiria Ansari & Hemadri
Siphonostelma Schltr.
Sisyranthus E.Mey.
Socotranthus Kuntze
Solenostemma Hayne
Spathidolepis Schltr.
Sphaerocodon Benth.
Spirella Costantin *
Stapelia L.
Stapelianthus Choux ex A.C.White & B.Sloane
Stapeliopsis Pillans
Stathmostelma K.Schum.

Steleostemma Schltr.
Stelmatocodon Schltr. *
Stenomeria Turcz.
Stenostelma Schltr.
Stephanotis Thouars
Stigmatorhynchus Schltr.
Stomatostemma N.E.Br.
Streptocaulon Wight & Arn.
Streptomanes K.Schum. *
Strobopetalum N.E.Br.
Stuckertia Kuntze *
Swynnertonia S.Moore
Tacazzea Decne.
Tainionema Schltr. *
Tanulepis Balf.f.
Tassadia Decne.
Telectadium Baill.
Telesilla Klotzsch *
Telminostelma Fourn.
Telosma Coville
Tenaris E.Mey.
Tetraphysa Schltr.
Thozetia F.Muell. ex Benth.
Toxocarpus Wight & Arn.
Trachycalymma (K.Schum.) Bullock
Treutlera Hook.f.
Trichocaulon N.E.Br.
Trichosacme Zucc.
Trichosandra Decne.
Tridentea Haw.
Triodoglossum Bullock
Tromotriche Haw.
Tweedia Hook. & Arn.
Tylodontia Griseb.
Tylophora R.Br.
Tylophoropsis N.E.Br.
Urostephanus B.L.Rob. & Greenm.
Utleria Bedd. ex Benth.
Vailia Rusby
Vincetoxicopsis Costantin *
Vincetoxicum Wolf
Vohemaria Buchenau
White–Sloanea Chiov.
Widgrenia Malme
Woodia Schltr.
Xysmalobium R.Br.
Zacateza Bullock
Zygostelma Benth.

ASTERACEAE — see COMPOSITAE

ASTEROPEIACEAE (Szyszyl.) Takht. ex Reveal & Hoogland 1990   17.08

1 genus. Madagascar. Trees, lianes.

B&H   POLYPETALAE, CALYCIFLORAE   Passiflorales (within *Samydaceae*, 71)
DT&H ARCHICHLAMYDEAE   Parietales, Theineae (within Theaceae, 136)
Melc ARCHICHLAMYDEAE   Guttiferales, Theineae (within Theaceae, 85)
Thor THEIFLORAE   Theales, Theineae (within Theaceae, 44)
Dahl (not mentioned, presumably within Theaceae, 124)
Young (not mentioned, presumably within Theaceae, 106)
Takh DILLENIIDAE, THEANAE   Theales, 110
Cron DILLENIIDAE   Theales (within Theaceae, 85)

Asteropeia Thouars

AUCUBACEAE J.Agardh 1858   82.07

1 genus. Himalaya to Japan. Shrubs, small trees.

B&H   POLYPETALAE, CALYCIFLORAE   Umbellales (within Cornaceae, 82)
DT&H ARCHICHLAMYDEAE   Umbelliflorae (within Cornaceae, 181)
Melc ARCHICHLAMYDEAE   Umbelliflorae (within Cornaceae, 223)
Thor CORNIFLORAE   Cornales, Cornineae, 287
Dahl CORNIFLORAE   Cornales, 306
Young (not mentioned, presumably within Cornaceae, 325)
Takh ROSIDAE, CORNANAE   Cornales, 345
Cron ROSIDAE   Cornales (within Cornaceae, 203)

Aucuba Thunb.

AUSTROBAILEYACEAE (Croizat) Croizat 1943   142.03

1 genus, monotypic. Queensland. Liane.

B&H   Not known   —
DT&H Not known   —
Melc ARCHICHLAMYDEAE   Magnoliales, Laurineae, 49
Thor ANNONIFLORAE   Annonales, Annonineae, 7
Dahl MAGNOLIIFLORAE   Annonales, 4
Young MAGNOLIIDAE, MAGNOLIANAE   Illiciales, 6
Takh MAGNOLIIDAE, MAGNOLIANAE   Austrobaileyales, 11
Cron MAGNOLIIDAE   Magnoliales, 5

Austrobaileya C.T.White

AVICENNIACEAE Endl. 1841   125.05

1 genus. Tropical to subtropical coasts. Mangroves.

B&H   GAMOPETALAE, BICARPELLATAE   Lamiales (within Verbenaceae, 125)
DT&H METACHLAMYDEAE   Tubiflorae, Verbenineae (within
                              Verbenaceae, 206)

| Melc | SYMPETALAE | Tubiflorae, Verbenineae (within Verbenaceae, 258) |

Thor GENTIANIFLORAE   Lamiales (within Verbenaceae, 265)
Dahl (not mentioned, presumably included in Verbenaceae, 364)
Young (not mentioned, presumably within Verbenaceae, 377)
Takh LAMIIDAE, LAMIANAE   Lamiales, Lamiineae (within Verbenaceae, 413)

Cron ASTERIDAE   Lamiales (within Verbenaceae, 286)

Avicennia L.

BALANITACEAE Endl. 1841   40.07

1 genus. Tropical Africa to Jordan, India and Burma. Trees, shrubs.

B&H POLYPETALAE, DISCIFLORAE Geraniales (within Simaroubaceae, 40)
DT&H ARCHICHLAMYDEAE   Geraniales, Geraniineae (within Zygophyllaceae, 85)

Melc ARCHICHLAMYDEAE   Geraniales, Geraniineae (within Zygophyllaceae, 126)

Thor RUTIFLORAE   Rutales, Rutineae (within Simaroubaceae, 169)

Dahl RUTIFLORAE   Geraniales, 233
Young Not mentioned   —
Takh ROSIDAE, RUTANAE   Rutales, 264
Cron ROSIDAE   Sapindales (within Zygophyllaceae, 262)

Balanites Delile

BALANOPACEAE Benth. 1880   152

1 genus. Queensland, New Caledonia, Fiji. Trees.

B&H MONOCHLAMYDEAE   Unisexuales, 152
DT&H ARCHICHLAMYDEAE   Balanopsidales (as *Balanopsidaceae*), 8
Melc ARCHICHLAMYDEAE   Balanopales, 4
Thor ROSIFLORAE   Pittosporales, Buxineae, 225
Dahl ROSIFLORAE   Balanopales, 145
Young ROSIDAE, ROSANAE   Balanopales, 283
Takh HAMAMELIDIDAE, HAMAMELIDANAE
    Balanopales, 86
Cron HAMAMELIDIDAE   Fagales, 60

Balanops Baill.

BALANOPHORACEAE Rich. 1822   150.01

17 genera. Tropics and subtropics. Root parasites.

B&H MONOCHLAMYDEAE   Achlamydosporeae, 150
    (including Cynomoriaceae)

DT&H	ARCHICHLAMYDEAE	Santalales, Balanophorineae, 23
Melc	ARCHICHLAMYDEAE	Balanophorales, 23
Thor	SANTALIFLORAE	Balanophorales, 122
Dahl	BALANOPHORIFLORAE	Balanophorales, 263
Young	ROSIDAE, SANTALANAE	Balanophorales 240
Takh	MAGNOLIIDAE, RAFFLESIANAE	Balanophorales, 37

    (separating *Dactylanthaceae* 33, *Latraeophilaceae* 35 (nom. inval.), *Lophophytaceae*
    36, *Sarcophytaceae* 34 and [Takh 1990] *Mystropetalaceae*)

| Cron | ROSIDAE | Santalales, 214 |

    (including Cynomoriaceae)

Balanophora J.R.Forst. & G.Forst.
Chlamydophytum Mildbr.
Corynaea Hook.f.
Dactylanthus Hook.f.
Ditepalanthus Fagerl.
Exorhopala Steenis
Hachettea Baill.
Helosis Rich.
Langsdorffia Mart.

Lathrophytum Eichler
Lophophytum Schott & Endl.
Mystropetalon Harv.
Ombrophytum Poepp. ex Endl.
Rhopalocnemis Jungh.
Sarcophyte Sparrm.
Scybalium Schott & Endl.
Thonningia Vahl

BALSAMINACEAE A.Rich. 1822   38.08

2 genera. Widespread except Australia and S. America. Herbs.

B&H	POLYPETALAE, DISCIFLORAE	Geraniales (within Geraniaceae, 38)
DT&H	ARCHICHLAMYDEAE	Sapindales, Balsaminineae, 118
Melc	ARCHICHLAMYDEAE	Sapindales, Balsamineae, 152
Thor	GERANIIFLORAE	Geraniales, Geraniineae, 104
Dahl	RUTIFLORAE	Balsaminales, 246
Young	ROSIDAE, GERANIANAE	Geraniales, 276
Takh	ROSIDAE, RUTANAE	Balsaminales, 289
Cron	ROSIDAE	Geraniales, 267

Hydrocera Blume
Impatiens L.

BARBEUIACEAE (H.Walter) Nakai 1942   132.02

1 genus, monotypic. Madagascar. Woody climber.

B&H	MONOCHLAMYDEAE	Curvembryeae (within Phytolaccaceae, 132)
DT&H	ARCHICHLAMYDEAE	Centrospermae, Phytolaccineae (within Phytolaccaceae, 33)
Melc	ARCHICHLAMYDEAE	Centrospermae, Phytolaccineae (within Phytolaccaceae, 26)
Thor	CHENOPODIIFLORAE	Chenopodiales, Chenopodiineae, 82
Dahl	Not mentioned	—
Young	DILLENIIDAE, CARYOPHYLLANAE	Chenopodiales, 151

Takh   CARYOPHYLLIDAE, CARYOPHYLLANAE
                                    Caryophyllales, Caryophyllineae, 59
Cron   CARYOPHYLLIDAE   Caryophyllales (within Phytolaccaceae, 64)

Barbeuia Thouars

BARBEYACEAE Rendle 1916   153.07

1 genus, monotypic. N.E. tropical Africa, Arabia. Small tree.

B&H   Not known   —
DT&H   ARCHICHLAMYDEAE   Urticales (within Ulmaceae, 13)
Melc   ARCHICHLAMYDEAE   Urticales (within Ulmaceae, 11)
Thor   INCERTAE SEDIS   349
Dahl   MALVIFLORAE   Urticales, 72
Young   ROSIDAE, GENTIANANAE   Oleales, 346
Takh   DILLENIIDAE, URTICANAE   Barbeyales, 189
Cron   HAMAMELIDIDAE   Urticales, 50

Barbeya Schweinf.

BASELLACEAE Moq. 1840   131.02

5 genera. Tropics and subtropics. Climbing herbs.

B&H   MONOCHLAMYDEAE   Curvembryeae (within Chenopodiaceae, 131)
DT&H   ARCHICHLAMYDEAE   Centrospermae, Portulacineae, 36
Melc   ARCHICHLAMYDEAE   Centrospermae, Portulacineae, 33
Thor   CHENOPODIIFLORAE   Chenopodiales, Portulacineae, 92
Dahl   CARYOPHYLLIFLORAE   Caryophyllales, 40
Young   DILLENIIDAE, CARYOPHYLLANAE   Chenopodiales, 162
Takh   CARYOPHYLLIDAE, CARYOPHYLLANAE
                                    Caryophyllales, Caryophyllineae, 67
Cron   CARYOPHYLLIDAE   Caryophyllales, 73

Anredera Juss.   **Tournonia** Moq.
Basella L.   **Ullucus** Caldas
Boussingaultia Kunth

BATACEAE Mart. ex Meisn. 1842   133

1 genus. New Guinea, Queensland, Pacific, coasts of N. and S. America. Shrubs.

B&H   MONOCHLAMYDEAE   Curvembryeae, 133
DT&H   ARCHICHLAMYDEAE   Centrospermae, Phytolaccineae, 31
Melc   ARCHICHLAMYDEAE   Batales, 100
Thor   RUTIFLORAE   Rutales, Sapindineae, 182
Dahl   VIOLIFLORAE   Capparales, 105
Young   DILLENIIDAE, VIOLANAE   Capparales, 210

| Takh | DILLENIIDAE, VIOLANAE | Batales, 169 |
| Cron | DILLENIIDAE | Batales, 138 |

Batis P.Browne

BEGONIACEAE C.Agardh 1825   76

3 genera. Tropics, subtropics. Herbs, sometimes subshrubs, climbers.

B&H	POLYPETALAE, CALYCIFLORAE	Passiflorales, 76
DT&H	ARCHICHLAMYDEAE	Parietales, Begoniineae, 158
Melc	ARCHICHLAMYDEAE	Violales, Begoniineae, 201
Thor	VIOLIFLORAE	Violales, Begoniineae, 138
Dahl	VIOLIFLORAE	Curcurbitales, 95
Young	DILLENIIDAE, VIOLANAE	Violales, 201
Takh	DILLENIIDAE, VIOLANAE	Begoniales, 162
Cron	DILLENIIDAE	Violales, 129

Begonia L.
Hillebrandia Oliv.
Symbegonia Warb.

BERBERIDACEAE Juss. 1789   7.01

18 genera. Widespread, esp. N. temp. Trees, shrubs, herbs.

B&H	POLYPETALAE, THALAMIFLORAE	Ranales, 7
	(including Lardizabalaceae)	
DT&H	ARCHICHLAMYDEAE	Ranales, Ranunculineae, 43
Melc	ARCHICHLAMYDEAE	Ranunculales, Ranunculineae, 62
Thor	ANNONIFLORAE	Berberidales, Berberidineae, 32
	(separating *Nandinaceae* 31)	
Dahl	RANUNCULIFLORAE	Ranunculales, 35
Young	MAGNOLIIDAE, RANUNCULANAE	Ranunculales, 31
	(separating *Nandinaceae* 30, *Podophyllaceae* 34)	
Takh	RANUNCULIDAE, RANUNCULANAE	Ranunculales, Berberidineae, 50
	(separating *Nandinaceae* 51)	
Cron	MAGNOLIIDAE	Ranunculales, 32

Aceranthus C.Morren & Decne.
Achlys DC.
Berberis L.
Bongardia C.A.Mey.
Caulophyllum Michx.
Diphylleia Michx.
Dysosma Woodson
Epimedium L.
Gymnospermium Spach

Jeffersonia Barton
Leontice L.
Mahonia Nutt.
Nandina Thunb.
Plagiorhegma Maxim.
Podophyllum L.
Ranzania T.Ito
Sinopodophyllum Ying
Vancouveria C.Morren & Decne.

BETULACEAE Gray 1821   159.01

2 genera. Temp. Eurasia to Indochina, Andes. Mostly trees.

B&H   MONOCHLAMYDEAE   Unisexuales (as *Cupuliferae*), 159
         (including Corylaceae, Fagaceae)
DT&H ARCHICHLAMYDEAE   Fagales, 11
         (including Corylaceae)
Melc ARCHICHLAMYDEAE   Fagales, 8
         (including Corylaceae)
Thor HAMAMELIDIFLORAE   Fagales, 202
         (including Corylaceae)
Dahl ROSIFLORAE   Fagales, 148
Young ROSIDAE, HAMAMELIDANAE   Fagales, 227
         (including Corylaceae)
Takh HAMAMELIDIDAE, HAMAMELIDANAE
         (including Corylaceae)   Betulales, 94
Cron HAMAMELIDIDAE   Fagales, 62
         (including Corylaceae)

Alnus Mill.
Betula L.

BIGNONIACEAE Juss. 1789   120.01

111 genera. Tropics, occas. warm-temp. Trees, lianes, rarely herbs.

B&H   GAMOPETALAE, BICARPELLATAE Personales, 120
DT&H METACHLAMYDEAE   Tubiflorae, Solanineae, 211
Melc SYMPETALAE   Tubiflorae, Solanineae, 267
Thor GENTIANIFLORAE   Bignoniales, 256
Dahl LAMIIFLORAE   Scrophulariales, 348
Young ROSIDAE, GENTIANANAE   Bignoniales, 354
Takh LAMIIDAE, LAMIANAE   Scrophulariales, 400
Cron ASTERIDAE   Scrophulariales, 301

Adenocalymma Mart. ex Meisn.
Amphilophium Kunth
Amphitecna Miers
Anemopaegma Mart. ex Meisn.
Argylia D.Don
Arrabidaea DC.
Astianthus D.Don
Bignonia L.
Callichlamys Miq.
Campsidium Seem.
Campsis Lour.
Catalpa Scop.
Catophractes D.Don
Ceratophytum Pittier
Chilopsis D.Don
Clytostoma Miers ex Bur.
Colea Bojer ex Meisn.
Crescentia L.

Cuspidaria DC.
Cybistax Mart. ex Meisn.
Cydista Miers
Delostoma D.Don
Deplanchea Vieill.
Digomphia Benth.
Dinklageodoxa Heine & Sandwith
Distictella Kuntze
Distictis Mart. ex Meisn.
Dolichandra Cham.
Dolichandrone (Fenzl) Seem.
Eccremocarpus Ruíz & Pav.
Ekmanianthe Urb.
Fernandoa Welw. ex Seem.
Fridericia Mart.
Gardnerodoxa Sandwith
Glaziova Bureau
Godmania Hemsl.

Haplolophium Cham.
Heterophragma DC.
Hieris Steenis
Incarvillea Juss.
Jacaranda Juss.
Kigelia DC.
Lamiodendron Steenis
Leucocalantha Barb.Rodr.
Lundia DC.
Macfadyena A.DC.
Macranthisiphon Bureau ex K.Schum.
Manaosella J.C.Gomes
Mansoa DC.
Markhamia Seem. ex Baill.
Martinella Baill.
Melloa Bureau
Memora Miers
Millingtonia L.f.
Mussatia Bureau ex Baill.
Neojobertia Baill.
Neomacfadya Baill.
Neosepicaea Diels
Neotuerckheimia Donn.Sm.
Nestoria Urb.
Newbouldia Seem. ex Bureau
Nyctocalos Teijsm. & Binn.
Ophiocolea H.Perrier
Oroxylum Vent.
Pajanelia DC.
Pandorea (Endl.) Spach
Parabignonia Bureau ex K.Schum.
Paragonia Bureau
Paratecoma Kuhlm.
Parmentiera DC.
Pauldopia Steenis
Perianthomega Bureau ex Baill.
Periarrabidaea A.Samp.
Perichlaena Baill.

Phryganocydia Mart. ex Bur.
Phyllarthron DC.
Phylloctenium Baill.
Pithecoctenium Mart. ex Meisn.
Pleonotoma Miers
Podranea Sprague
Potamoganos Sandwith
Pseudocatalpa A.H.Gentry
Pyrostegia C.Presl
Radermachera Zoll. & Moritzi
Rhigozum Burch.
Rhodocolea Baill.
Roentgenia Urb.
Romeroa Dugand
Santisukia Brummitt
Saritaea Dugand
Scobinaria Seibert
Setilobus Baill.
Sparattosperma Mart. ex Meisn.
Spathicalyx J.C.Gomes
Spathodea P.Beauv.
Sphingiphila A.H.Gentry *
Spirotecoma Baill. ex Dalla Torre & Harms
Stereospermum Cham.
Stizophyllum Miers
Synapsis Griseb.
Tabebuia Gomes ex DC.
Tanaecium Sw.
Tecoma Juss.
Tecomanthe Baill.
Tecomella Seem.
Tourrettia Foug.
Tynnanthus Miers
Urbanolophium Melch.
Wunschmannia Urb.
Xylophragma Sprague
Zeyheria Mart.

BIXACEAE Link 1831   17.01

1 genus, monotypic. Trop. America, introd. elsewhere. Shrubs to small trees.

B&H	POLYPETALAE, THALAMIFLORAE	Parietales, 17
	(including Cochlospermaceae, Flacourtiaceae, Peridiscaceae)	
DT&H	ARCHICHLAMYDEAE	Parietales, Cistineae, 144
	(including Sphaerosepalaceae)	
Melc	ARCHICHLAMYDEAE	Violales, Cistineae, 192
Thor	MALVIFLORAE	Malvales, 151
Dahl	MALVIFLORAE	Malvales, 59
Young	DILLENIIDAE, MALVANAE	Malvales, 172

Takh   DILLENIIDAE, MALVANAE   Bixales, 170
Cron   DILLENIIDAE   Violales, 109
         (including Cochlospermaceae)

Bixa L.

BOMBACACEAE Kunth 1822   31.02

30 genera. Tropics and subtropics. Trees, often pachycaul.

B&H   POLYPETALAE, THALAMIFLORAE Malvales (within Malvaceae, 31)
DT&H   ARCHICHLAMYDEAE   Malvales, Malvineae, 127
Melc   ARCHYCHLAMYDEAE   Malvales, Malvineae, 174
Thor   MALVIFLORAE   Malvales, 154
Dahl   MALVIFLORAE   Malvales, 68
Young DILLENIIDAE, MALVANAE   Malvales, 171
Takh   DILLENIIDAE, MALVANAE   Malvales, 181
Cron   DILLENIIDAE   Malvales, 101

Adansonia L.   **Huberodendron** Ducke
Aguiaria Ducke   **Kostermansia** Soegeng
Bernoullia Oliv.   **Lahia** Hassk.
Bombacopsis Pittier   **Matisia** Bonpl.
Bombax L.   **Neesia** Blume
Boschia Korth.   **Neobuchia** Urb.
Camptostemon Mast.   **Ochroma** Sw.
Catostemma Benth.   **Pachira** Aubl.
Cavanillesia Ruíz & Pav.   **Patinoa** Cuatrec.
Ceiba Mill.   **Phragmotheca** Cuatrec.
Chorisia Kunth   **Pseudobombax** Dugand
Coelostegia Benth.   **Quararibea** Aubl.
Durio Adans.   **Rhodognaphalon** (Ulbr.) Roberty
Eriotheca Schott & Endl.   **Scleronema** Benth.
Gyranthera Pittier *   **Septotheca** Ulbr.

BORAGINACEAE Juss. 1789   112.01

131 genera. Widespread. Trees, shrubs, herbs.

B&H   GAMOPETALAE, BICARPELLATAE Polemoniales, 112
DT&H   METACHLAMYDEAE   Tubiflorae, Boraginineae, 205
Melc   SYMPETALAE   Tubiflorae, Boraginineae, 256
Thor   SOLANIFLORAE   Solanales, Boraginineae, 269
Dahl   SOLANIFLORAE   Boraginales, 282
         (separating *Ehretiaceae* 281)
Young ROSIDAE, SOLANANAE   Boraginales, 368
Takh   LAMIIDAE, SOLANANAE   Boraginales, 393
         (separating *Cordiaceae* 392, *Ehretiaceae* 391, *Wellstediaceae* 394)
Cron   ASTERIDAE   Lamiales, 285

Actinocarya Benth.   **Alkanna** Tausch
Afrotysonia Rauschert   **Amblynotus** (A.DC.) I.M.Johnst.

Amphibologyne Brand *
Amsinckia Lehm.
Anchusa L.
Ancistrocarya Maxim.
Anoplocaryum Ledeb.
Antiotrema Hand.-Mazz.
Antiphytum DC. ex Meisn.
Argusia Boehm.
Arnebia Forssk.
Asperugo L.
Austrocynoglossum Popov ex R.R.Mill
Auxemma Miers
Borago L.
Bothriospermum Bunge
Bourreria P.Browne
Brachybotrys Maxim. ex Oliv.
Brandella R.R.Mill
Brunnera Steven
Buglossoides Moench
Caccinia Savi
Carmona Cav.
Cerinthe L.
Chamissoniophila Brand *
Chionocharis I.M.Johnst.
Choriantha Riedl
Coldenia L.
Cordia L.
Cortesia Cav.
Craniospermum Lehm.
Crucicaryum Brand *
Cryptantha G.Don
Cynoglossopsis Brand
Cynoglossum L.
Cynoglottis (Guşul.) Vural & Kit Tan
Cystostemon Balf.f.
Dasynotus I.M.Johnst. *
Decalepidanthus Riedl
Echiochilon Desf.
Echiochilopsis Caball. *
Echium L.
Ehretia P.Browne
Elizaldia Willk.
Embadium J.M.Black
Eritrichium Schrad. ex Gaudin
Gastrocotyle Bunge
Gyrocaryum B.Valdés
Hackelia Opiz
Halacsya Dörfl.
Halgania Gaudich.
Harpagonella A.Gray
Heliocarya Bunge
Heliotropium L.
Ivanjohnstonia Kazmi
Lacaitaea Brand
Lappula Moench

Lasiarrhenum I.M.Johnst.
Lasiocaryum I.M.Johnst.
Lepechiniella Popov
Lepidocordia Ducke
Lindelofia Lehm.
Lithodora Griseb.
Lithospermum L.
Lobostemon Lehm.
Macromeria D.Don
Maharanga A.DC.
Mairetis I.M.Johnst.
Mallotonia (Griseb.) Britton
Mertensia Roth
Microcaryum I.M.Johnst.
Microula Benth.
Mimophytum Greenm.
Moltkia Lehm.
Moltkiopsis I.M.Johnst.
Moritzia DC. ex Meisn.
Myosotidium Hook.
Myosotis L.
Neatostema I.M.Johnst.
Nephrocarya Candargy
Nesocaryum I.M.Johnst.
Nogalia Verdc.
Nomosa I.M.Johnst. *
Nonea Medik.
Ogastemma Brummitt
Omphalodes Mill.
Omphalolappula Brand
Onosma L.
Onosmodium Michx.
Oxyosmyles Speg.
Paracaryopsis R.R.Mill
Paracaryum (A.DC.) Boiss.
Paramoltkia Greuter
Paraskevia W.Sauer & G.Sauer *
Pardoglossum E.Barbier & Mathez
Patagonula L.
Pectocarya DC. ex Meisn.
Pentaglottis Tausch
Perittostemma I.M.Johnst. *
Plagiobothrys Fisch. & C.A.Mey.
Pseudomertensia Riedl
Psilolaemus I.M.Johnst. *
Pteleocarpa Oliv.
Pulmonaria L.
Rindera Pall.
Rochefortia Sw.
Rochelia Rchb.
Rotula Lour.
Saccellium Bonpl.
Scapicephalus Ovcz. & Chukav. *
Selkirkia Hemsl.
Sericostoma Stocks

Sinojohnstonia Hu
Solenanthus Ledeb.
Stenosolenium Turcz.
Stephanocaryum Popov *
Suchtelenia Kar. ex Meisn.
Symphytum L.
Thaumatocaryon Baill.
Thyrocarpus Hance
Tianschaniella Fedtsch. ex Popov

Tiquilia Pers.
Tournefortia L.
Trachelanthus Kunze
Trachystemon D.Don
Trichodesma R.Br.
Trigonocaryum Trautv.
Trigonotis Steven
Wellstedia Balf.f.

BRASSICACEAE — see CRUCIFERAE

BRETSCHNEIDERACEAE Engl. & Gilg 1924   51.08

1 genus, monotypic. China, Vietnam, Thailand. Tree.

B&H	Not known	—
DT&H	ARCHICHLAMYDEAE	Sapindales, Sapindineae (within Hippocastanaceae, 114)
Melc	ARCHICHLAMYDEAE	Sapindales, Sapindineae, 146
Thor	RUTIFLORAE	Rutales, Sapindineae, 188
Dahl	RUTIFLORAE	Sapindales, 211
Young	ROSIDAE, RUTANAE	Sapindales, 261
Takh	ROSIDAE, RUTANAE	Sapindales, 248
Cron	ROSIDAE	Sapindales, 250

Bretschneidera Hemsl.

BRUNELLIACEAE Engl. 1897   40.04

1 genus. Mexico to Peru and W. Indies. Trees.

B&H	POLYPETALAE, DISCIFLORAE	Geraniales (within Simaroubaceae, 40)
DT&H	ARCHICHLAMYDEAE	Rosales, Saxifragineae, 69
Melc	ARCHICHLAMYDEAE	Rosales, Saxifragineae, 107
Thor	ROSIFLORAE	Rosales, Cunoniineae, 218
Dahl	ROSIFLORAE	Cunoniales, 161
Young	ROSIDAE, ROSANAE	Cunoniales, 287
Takh	ROSIDAE, ROSANAE	Cunoniales, 199
Cron	ROSIDAE	Rosales, 156

Brunellia Ruíz & Pav.

BRUNIACEAE R.Br. ex DC. 1825   63

12 genera. South Africa. Mostly ericoid shrubs.

B&H	POLYPETALAE, CALYCIFLORAE	Rosales, 63
DT&H	ARCHICHLAMYDEAE	Rosales, Saxifragineae, 72
Melc	ARCHICHLAMYDEAE	Rosales, Saxifragineae, 113
Thor	ROSIFLORAE	Pittosporales, Bruniineae, 230
Dahl	ROSIFLORAE	Cunoniales, 161
Young	ROSIDAE, ROSANAE	Bruniales, 290
Takh	ROSIDAE, ROSANAE	Bruniales, 200
Cron	ROSIDAE	Rosales, 168

Audouinia Brongn.
Berzelia Brongn.
Brunia Lam.
Linconia L.
Lonchostoma Wikstr.
Mniothamnea (Oliv.) Nied.

Nebelia Neck. ex Sweet
Pseudobaeckea Nied.
Raspalia Brongn.
Staavia Dahl
Thamnea Sol. ex Brongn.
Tittmannia Brongn.

BUDDLEJACEAE K.Wilh. 1910 108.02

8 genera. Widespread, mostly tropical to warm-temp. Mostly trees, shrubs.

B&H	GAMOPETALAE, BICARPELLATAE	Gentianales (within Loganiaceae, 108)
DT&H	METACHLAMYDEAE	Contortae, Gentianineae (within Loganiaceae, 198)
Melc	SYMPETALAE	Tubiflorae, Solanineae, 264
Thor	GENTIANIFLORAE	Gentianales, 250
Dahl	LAMIIFLORAE	Scrophulariales, 351
Young	ROSIDAE, GENTIANANAE	Gentianales, 348
Takh	LAMIIDAE, LAMIANAE	Scrophulariales, 395
Cron	ASTERIDAE	Scrophulariales, 292

Androya H.Perrier
Buddleja L.
Emorya Torr.
Gomphostigma Turcz.

Nuxia Comm. ex Lam.
Peltanthera Benth.
Polypremum L.
Sanango Bunting & Duke

BURSERACEAE Kunth 1824 42

18 genera. Widespread in tropics. Trees, shrubs.

B&H	POLYPETALAE, DISCIFLORAE	Geraniales, 42
DT&H	ARCHICHLAMYDEAE	Geraniales, Geraniineae, 89
Melc	ARCHICHLAMYDEAE	Rutales, Rutineae, 135
Thor	RUTIFLORAE	Rutales, Rutineae, 173
Dahl	RUTIFLORAE	Rutales, 223
Young	ROSIDAE, RUTANAE	Rutales, 250
Takh	ROSIDAE, RUTANAE	Rutales, 270
Cron	ROSIDAE	Sapindales, 255

Aucoumea Pierre
Beiselia Forman
Boswellia Roxb. ex Colebr.
Bursera Jacq. ex L.

Canarium L.
Commiphora Jacq.
Crepidospermum Hook.f.
Dacryodes Vahl

Garuga Roxb.	**Scutinanthe** Thwaites
Haplolobus H.J.Lam	**Tapirocarpus** Sagot
Paraprotium Cuatrec.	**Tetragastris** Gaertn.
Protium Burm.f.	**Trattinnickia** Willd.
Santiria Blume	**Triomma** Hook.f.

BUXACEAE Dumort. 1822   151.02

5 genera. Widespread except Australasia and Pacific. Trees, shrubs, some herbs.

B&H	MONOCHLAMYDEAE	Unisexuales (within Euphorbiaceae, 151)
DT&H	ARCHICHLAMYDEAE	Sapindales, Buxineae, 99
	(including Simmondsiaceae)	
Melc	ARCHICHLAMYDEAE	Celastrales, Buxineae, 164
	(including Simmondsiaceae)	
Thor	ROSIFLORAE	Pittosporales, Buxineae, 222
Dahl	ROSIFLORAE	Buxales, 153
Young	ROSIDAE, ROSANAE	Buxales, 284
Takh	HAMAMELIDIDAE, HAMAMELIDANAE	
	(separating *Stylocerataceae* 90)	Buxales, 89
Cron	ROSIDAE	Euphorbiales, 229

Buxus L.	**Sarcococca** Lindl.
Notobuxus Oliv.	**Styloceras** Kunth ex Juss.
Pachysandra Michx.	

BYBLIDACEAE Domin 1922   61.03

1 genus. New Guinea, Australia. Insectivorous herbs.

B&H	POLYPETALAE, CALYCIFLORAE	Rosales (within Droseraceae, 61)
DT&H	ARCHICHLAMYDEAE	Sarraceniales (within Droseraceae, 62)
Melc	ARCHICHLAMYDEAE	Rosales, Saxifragineae, 111
Thor	ROSIFLORAE	Pittosporales, Pittosporineae, 227
Dahl	ARALIIFLORAE	Pittosporales, 266
Young	ROSIDAE, CORNANAE	Pittosporales, 339
Takh	ROSIDAE, CORNANAE	Byblidales, 354
Cron	ROSIDAE	Rosales, 163
	(including Roridulaceae)	

Byblis Salisb.

CABOMBACEAE A.Rich. 1828   8.02

2 genera. Tropical to warm-temperate. Aquatic herbs.

B&H	POLYPETALAE, THALAMIFLORAE	Ranales (within Nymphaeaceae, 8)
DT&H	ARCHICHLAMYDEAE	Ranales, Nymphaeineae (within Nymphaeaceae, 38)

Melc	ARCHICHLAMYDEAE	Ranunculales, Nymphaeineae (within
		Nymphaeaceae, 66)
Thor	NYMPHAEIFLORAE	Nymphaeales, 36
Dahl	NYMPHAEIFLORAE	Nymphaeales, 26
Young	MAGNOLIIDAE, NYMPHAEANAE	Nymphaeales, 11
Takh	MAGNOLIIDAE, NYMPHAEANAE	Nymphaeales, 39
Cron	MAGNOLIIDAE	Nymphaeales, 28

Brasenia Schreb.
Cabomba Aubl.

CACTACEAE Juss. 1789   78.01

100 genera. New World except *Rhipsalis*. Mostly spiny stem succulents.

B&H	POLYPETALAE, CALYCIFLORAE	Ficoidales, 78
DT&H	ARCHICHLAMYDEAE	Opuntiales, 160
Melc	ARCHICHLAMYDEAE	Cactales, 38
Thor	CHENOPODIIFLORAE	Chenopodiales, Portulacineae, 94
Dahl	CARYOPHYLLIFLORAE	Caryophyllales, 47
Young	DILLENIIDAE, CARYOPHYLLANAE	Caryophyllales, 147
Takh	CARYOPHYLLIDAE, CARYOPHYLLANAE	
		Caryophyllales, Caryophyllineae, 69
Cron	CARYOPHYLLIDAE	Caryophyllales, 69

Acanthocalycium Backeb.
Acanthocereus (Engelm. ex A.Berger)
  Britton & Rose
Ariocarpus Scheidw.
Armatocereus Backeb.
Arrojadoa Britton & Rose
Arthrocereus A.Berger & F.Knuth
Astrophytum Lem.
Austrocactus Britton & Rose
Aztekium Boed.
Bergerocactus Britton & Rose
Blossfeldia Werderm.
Brachycereus Britton & Rose
Brasilicereus Backeb.
Browningia Britton & Rose
Calymmanthium F.Ritter
Carnegiea Britton & Rose
Cephalocereus Pfeiff.
Cereus Mill.
Cipocereus F.Ritter
Cleistocactus Lem.
Coleocephalocereus Backeb.
Copiapoa Britton & Rose
Corryocactus Britton & Rose
Coryphantha (Engelm.) Lem.
Denmoza Britton & Rose
Discocactus Pfeiff.
Disocactus Lindl.

Echinocactus Link & Otto
Echinocereus Engelm.
Echinopsis Zucc.
Epiphyllum Haw.
Epithelantha F.A.C.Weber ex Britton &
  Rose
Eriosyce Phil.
Escobaria Britton & Rose
Escontria Rose
Espostoa Britton & Rose
Espostoopsis Buxb.
Eulychnia Phil.
Facheiroa Britton & Rose
Ferocactus Britton & Rose
Frailea Britton & Rose
Gymnocalycium Pfeiff.
Haageocereus Backeb.
Harrisia Britton
Hatiora Britton & Rose
Hylocereus (A.Berger) Britton & Rose
Jasminocereus Britton & Rose
Leocereus Britton & Rose
Lepismium Pfeiff.
Leptocereus (A.Berger) Britton & Rose
Leuchtenbergia Hook.
Lophophora J.M.Coult.
Maihuenia (F.A.C.Weber) K.Schum.
Mammillaria Haw.

Mammilloydia Buxb.
Melocactus Link & Otto
Micranthocereus Backeb.
Mila Britton & Rose
Myrtillocactus Console
Neobuxbaumia Backeb.
Neolloydia Britton & Rose
Neoporteria Britton & Rose
Neoraimondia Britton & Rose
Neowerdermannia Frič
Obregonia Frič
Opuntia Mill.
Oreocereus (A.Berger) Riccob.
Ortegocactus Alexander
Pachycereus (A.Berger) Britton & Rose
Parodia Speg.
Pediocactus Britton & Rose *
Pelecyphora Ehrenb.
Peniocereus (A.Berger) Britton & Rose
Pereskia Mill.
Pereskiopsis Britton & Rose
Pilosocereus Byles & Rowley
Polaskia Backeb.
Pseudoacanthocereus F.Ritter

Pseudorhipsalis Britton & Rose
Pterocactus K.Schum.
Pterocereus MacDougall & Miranda
Quiabentia Britton & Rose
Rathbunia Britton & Rose
Rebutia K.Schum.
Rhipsalis Gaertn.
Samaipaticereus Cárdenas
Schlumbergera Lem.
Sclerocactus Britton & Rose
Selenicereus (A.Berger) Britton & Rose
Stenocactus (K.Schum.) A.W.Hill
Stenocereus (A.Berger) Riccob.
Stephanocereus A.Berger
Stetsonia Britton & Rose
Strombocactus Britton & Rose
Tacinga Britton & Rose
Thelocactus (K.Schum.) Britton & Rose
Turbinicarpus (Backeb.) Buxb. &
    Backeb.
Uebelmannia Buining
Weberbauerocereus Backeb.
Weberocereus Britton & Rose

CALLITRICHACEAE Link 1821   64.04

1 genus. Widespread. Mostly aquatic herbs.

B&H   POLYPETALAE, CALYCIFLORAE   Rosales (within Haloragaceae, 64)
DT&H   ARCHICHLAMYDEAE   Geraniales, Incertae sedis, 98
Melc   SYMPETALAE   Tubiflorae, Verbenineae, 259
Thor   GENTIANIFLORAE   Lamiales, 267
Dahl   LAMIIFLORAE   Lamiales, 366
Young   ROSIDAE, LAMIANAE   Lamiales, 378
Takh   LAMIIDAE, LAMIANAE   Lamiales, Callitrichineae, 415
Cron   ASTERIDAE   Callitrichales, 289

Callitriche L.

CALYCANTHACEAE Lindl. 1819   3.01

3 genera. S. and E. USA, China. Shrubs, small trees.

B&H   POLYPETALAE, THALAMIFLORAE   Ranales, 3
DT&H   ARCHICHLAMYDEAE   Ranales, Magnoliineae, 46
Melc   ARCHICHLAMYDEAE   Magnoliales, Laurineae, 53
Thor   ANNONIFLORAE   Annonales, Laurineae, 19
        (including Idiospermaceae)
Dahl   MAGNOLIIFLORAE   Laurales, 21
        (including Idiospermaceae)

Young MAGNOLIIDAE, RANUNCULANAE Laurales, 15
Takh MAGNOLIIDAE, MAGNOLIANAE Laurales, 19
Cron MAGNOLIIDAE Magnoliales, 15

Calycanthus L.
Chimonanthus Lindl.

Sinocalycanthus (C.C.Cheng &
S.Y.Chang) C.C.Cheng & S.Y.Chang

CALYCERACEAE R.Br. ex Rich. 1820   87

6 genera. Tropical and temperate South America. Herbs.

B&H GAMOPETALAE, INFERAE Asterales, 87
DT&H METACHLAMYDEAE Campanulatae, Campanulineae, 232
Melc SYMPETALAE Campanulales, 289
Thor CORNIFLORAE Dipsacales, 295
Dahl CORNIFLORAE Dipsacales, 334
Young ROSIDAE, DIPSACANAE Dipsacales, 385
Takh ASTERIDAE, CAMPANULANAE Calycerales, 428
Cron ASTERIDAE Calycerales, 317

Acarpha Griseb.
Acicarpha Juss.
Boopis Juss.

Calycera Cav.
Gamocarpha DC.
Moschopsis Phil.

CAMPANULACEAE Juss. 1789   91.01

84 genera. Widespread. Mostly herbs.

B&H GAMOPETALAE, INFERAE Campanales, 91
    (including Pentaphragmataceae)
DT&H METACHLAMYDEAE Campanulatae, Campanulineae, 229
    (including Pentaphragmataceae)
Melc SYMPETALAE Campanulales, 283
    (separating *Sphenocleaceae* 284)
Thor SOLANIFLORAE Campanulales, 276
Dahl ASTERIFLORAE Campanulales, 270
    (separating *Lobeliaceae* 271)
Young ROSIDAE, SOLANANAE Campanulales, 372
Takh ASTERIDAE, CAMPANULANAE Campanulales, 419
    (separating *Cyphiaceae* 420, *Cyphocarpaceae* 423, *Lobeliaceae* 422, *Nemacladaceae* 421, *Sphenocleaceae* 418)
Cron ASTERIDAE Campanulales, 306
    (separating *Sphenocleaceae* 305)

Adenophora Fisch.
Apetahia Baill.
Asyneuma Griseb. & Schenck
Azorina Feer *
Berenice Tul.
Brighamia A.Gray
Burmeistera Karst. & Triana

Campanula L.
Canarina L.
Centropogon C.Presl
Clermontia Gaudich.
Codonopsis Wall.
Craterocapsa Hilliard & B.L.Burtt
Cryptocodon Fedorov *

Cyananthus Wall. ex Benth.
Cyanea Gaudich.
Cylindrocarpa Regel
Cyphia Bergius
Cyphocarpus Miers
Delissea Gaudich.
Dialypetalum Benth.
Diastatea Scheidw.
Dielsantha E.Wimm.
Diosphaera Buser
Dominella E.Wimm. *
Downingia Torr.
Echinocodon D.Y.Hong *
Edraianthus (A.DC.) DC.
Feeria Buser
Githopsis Nutt.
Grammatotheca C.Presl
Gunillaea Thulin
Hanabusaya Nakai
Heterochaenia A.DC.
Heterocodon Nutt.
Heterotoma Zucc.
Howellia A.Gray
Merciera A.DC.
Michauxia L'Hér.
Microcodon A.DC.
Monopsis Salisb.
Musschia Dumort.
Namacodon Thulin
Nemacladus Nutt.
Nesocodon Thulin

Ostrowskia Regel
Parishella A.Gray
Peracarpa Hook.f. & Thomson
Petromarula Vent. ex Hedw.f.
Physoplexis (Endl.) Schur
Phyteuma L.
Platycodon A.DC.
Popoviocodonia Fedorov
Pratia Gaudich.
Prismatocarpus L'Hér.
Pseudonemacladus McVaugh
Rhigiophyllum Hochst.
Rhizocephalum Wedd.
Roella L.
Rollandia Gaudich.
Ruthiella Steenis
Sclerotheca A.DC.
Sergia Fedorov *
Siphocampylus Pohl
Siphocodon Turcz.
Sphenoclea Gaertn.
Symphyandra A.DC.
Theilera E.Phillips
Trachelium L.
Treichelia Vatke
Trematocarpus Zahlbr.
Trimeris C.Presl
Triodanis Raf.
Unigenes E.Wimm.
Wahlenbergia Schrad. ex Roth
Zeugandra P.H.Davis

CANELLACEAE Mart. 1832

16

6 genera. Eastern Africa, Madagascar, and Florida to S. America. Trees.

B&H   POLYPETALAE, THALAMIFLORAE Parietales, 16
DT&H ARCHICHLAMYDEAE   Parietales, Flacourtiineae, 147
Melc   ARCHICHLAMYDEAE   Magnoliales, Magnoliineae, 46
Thor   ANNONIFLORAE   Annonales, Annonineae, 11
Dahl   MAGNOLIIFLORAE   Annonales, 5
Young MAGNOLIIDAE, MAGNOLIANAE Magnoliales, 4
Takh   MAGNOLIIDAE, MAGNOLIANAE Annonales, 6
Cron   MAGNOLIIDAE   Magnoliales,10

Canella P.Browne
Capsicodendron Hoehne
Cinnamodendron Endl.

Cinnamosma Baill.
Pleodendron Tiegh.
Warburgia Engl.

CANNABACEAE Endl. 1837

153.04

2 genera. Temperate Eurasia to S.E. Asia, N. America. Herbs, lianes.

B&H	MONOCHLAMYDEAE	Unisexuales (within Urticaceae, 153)
DT&H	ARCHICHLAMYDEAE	Urticales (within Moraceae, 14)
Melc	ARCHICHLAMYDEAE	Urticales (within Moraceae, 13)
Thor	MALVIFLORAE	Urticales, 158
Dahl	MALVIFLORAE	Urticales, 73
Young	DILLENIIDAE, MALVANAE	Urticales, 177
Takh	DILLENIIDAE, URTICANAE	Urticales, Urticineae, 186
Cron	HAMAMELIDIDAE	Urticales, 52

Cannabis L.
Humulus L.

CANOTIACEAE Airy Shaw 1965   47.04

2 genera. USA, Mexico. Shrubs, small trees.

B&H	POLYPETALAE, CALYCIFLORAE	Rosales (within Rosaceae, 58)
DT&H	ARCHICHLAMYDEAE	Sapindales, Celastrineae (within Celastraceae, 108)
Melc	(not mentioned, presumably within Celastraceae, 159)	
Thor	SANTALIFLORAE	Celastrales (within Celastraceae, 112)
Dahl	(not mentioned, presumably within Celastraceae, 249)	
Young	(not mentioned, presumably within Celastraceae, 229)	
Takh	ROSIDAE, CELASTRANAE	Celastrales, Celastrineae (within Celastraceae, 304)
Cron	ROSIDAE	Celastrales (within Celastraceae, 219)

Acanthothamnus Brandegee *
Canotia Torr.

CAPPARACEAE Juss. 1789   12.01

42 genera. Widespread trop., some warm-temp. Mostly shrubs, some herbs, trees.

B&H	POLYPETALAE, THALAMIFLORAE	Parietales, 12 (including Emblingiaceae, Tovariaceae)
DT&H	ARCHICHLAMYDEAE	Rhoeadales, Capparidineae, 57 (including Emblingiaceae; separating *Koeberliniaceae* 146 to Parietales, Cochlospermineae)
Melc	ARCHICHLAMYDEAE	Papaverales, Capparineae, 95 (including Emblingiaceae)
Thor	VIOLIFLORAE	Capparales, 142 (including Tovariaceae)
Dahl	VIOLIFLORAE	Capparales, 100
Young	DILLENIIDAE, VIOLANAE	Capparales, 207
Takh	DILLENIIDAE, VIOLANAE	Capparales, Capparineae, 163
Cron	DILLENIIDAE	Capparales, 133

Apophyllum F.Muell.
Atamisquea Miers ex Hook. & Arn.
Bachmannia Pax
Beautempsia Gaudich.

Belencita Karst.
Borthwickia W.W.Sm.
Boscia Lam.
Buchholzia Engl.

Buhsia Bunge
Cadaba Forssk.
Capparis L.
Cladostemon A.Braun & Vatke
Cleome L.
Cleomella DC.
Crateva L.
Cristatella Nutt.
Dactylaena Schrad. ex Schult.f.
Dhofaria A.G.Mill.
Dipterygium Decne.
Euadenia Oliv.
Forchhammeria Liebm.
Haptocarpum Ule
Isomeris Nutt.
Koeberlinia Zucc.
Maerua Forssk.

Morisonia L.
Neothorelia Gagnep.
Niebuhria DC.
Oceanopapaver Guillaumin
Oxystylis Torr. & Frem.
Pentadiplandra Baill.
Podandrogyne Ducke
Poilanedora Gagnep.
Polanisia Raf.
Puccionia Chiov.
Ritchiea R.Br. ex G.Don
Setchellanthus Brandegee *
Steriphoma Spreng.
Stixis Lour.
Thilachium Lour.
Tirania Pierre
Wislizenia Engelm.

CAPRIFOLIACEAE Juss. 1789   83.01

13 genera. North temperate, rarely to tropics. Mostly shrubs or lianes.

B&H   GAMOPETALAE, INFERAE   Rubiales, 83
      (including Adoxaceae, Alseuosmiaceae)
DT&H METACHLAMYDEAE   Rubiales, 224
      (including Alseuosmiaceae)
Melc   SYMPETALAE   Dipsacales, 279
      (including Alseuosmiaceae, Carlemanniaceae)
Thor   CORNIFLORAE   Dipsacales, 291
Dahl   CORNIFLORAE   Cornales, 322
      (separating *Sambucaceae* 321, *Viburnaceae* 323)
Young ROSIDAE, DIPSACANAE   Dipsacales, 380
      (separating *Sambucaceae* 381)
Takh   ROSIDAE, CORNANAE   Dipsacales, 355
      (separating *Sambucaceae* 357, *Viburnaceae* 356)
Cron   ASTERIDAE   Dipsacales, 313
      (including Carlemanniaceae)

Abelia R.Br.
Diervilla Mill.
Dipelta Maxim.
Heptacodium Rehder
Kolkwitzia Graebn.
Leycesteria Wall.
Linnaea L.

Lonicera L.
Sambucus L.
Symphoricarpos Duhamel
Triosteum L.
Viburnum L.
Weigela Thunb.

CARDIOPTERIDACEAE Blume 1849   45.04

1 genus. India to Solomons. Climbing herbs.

B&H   POLYPETALAE, DISCIFLORAE Olacales (within Olacaceae, 45)

DT&H	ARCHICHLAMYDEAE	Sapindales, Icacinineae (within Icacinaceae, 112)
Melc	ARCHICHLAMYDEAE	Celastrales, Icacinineae, 166
Thor	THEIFLORAE	Theales, Theineae, 52
Dahl	CORNIFLORAE	Cornales, 308
	(also doubtfully listed in SANTALIFLORAE Celastrales)	
Young	DILLENIIDAE, DILLENIANAE	Theales, 108
Takh	ROSIDAE, CELASTRANAE	Celastrales, Icacinineae, 302
Cron	ROSIDAE	Celastrales, 226

Cardiopteris Wall. ex Royle

CARICACEAE Dumort. 1829   74.04

4 genera. Trop. and subtrop. America, trop. Africa. Pachycaul trees, herbs.

B&H	POLYPETALAE, CALYCIFLORAE	Passiflorales (within Passifloraceae, 74)
DT&H	ARCHICHLAMYDEAE	Parietales, Papayineae, 155
Melc	ARCHICHLAMYDEAE	Violales, Caricineae 198
Thor	VIOLIFLORAE	Violales, Violineae, 133
Dahl	VIOLIFLORAE	Violales, 93
Young	DILLENIIDAE, VIOLANAE	Violales, 199
Takh	DILLENIIDAE, VIOLANAE	Violales, 154
Cron	DILLENIIDAE	Violales, 124

Carica L.　　　　　　　　　　**Jacaratia** A.DC.
Cylicomorpha Urb.　　　　　　**Jarilla** Rusby

CARLEMANNIACEAE Airy Shaw 1965   83.05

2 genera. Tropical Asia. Herbs to subshrubs.

B&H	GAMOPETALAE, INFERAE	Rubiales (within Rubiaceae, 84)
DT&H	METACHLAMYDEAE	Rubiales (within Rubiaceae, 223)
Melc.	SYMPETALAE	Dipsacales (within Caprifoliaceae, 279)
Thor	Not mentioned	—
Dahl	Not mentioned	—
Young	Not mentioned	—
Takh	LAMIIDAE, GENTIANANAE	Gentianales, 369
Cron	ASTERIDAE	Dipsacales (within Caprifoliaceae, 313)

Carlemannia Benth.
Silvianthus Hook.f.

CARYOCARACEAE Szyszyl. 1893   28.02

2 genera. Tropical America. Trees, some shrubs.

B&H	POLYPETALAE, THALAMIFLORAE	Guttiferales (within Theaceae, 28)
DT&H	ARCHICHLAMYDEAE	Parietales, Theineae, 133
Melc	ARCHICHLAMYDEAE	Guttiferales, Theineae, 86

Thor	THEIFLORAE	Theales, Theineae, 46
Dahl	THEIFLORAE	Theales, 118
Young	DILLENIIDAE, DILLENIANAE	Theales, 115
Takh	DILLENIIDAE, THEANAE	Theales, 109
Cron	DILLENIIDAE	Theales, 84

Anthodiscus G.Mey.
Caryocar L.

CARYOPHYLLACEAE Juss. 1789   22.01

66 genera. Widespread, mostly temperate. Mostly herbs.

B&H	POLYPETALAE, THALAMIFLORAE	Caryophyllineae, 22
DT&H	ARCHICHLAMYDEAE	Centrospermae, Caryophyllineae, 37
	(including Illecebraceae)	
Melc	ARCHICHLAMYDEAE	Centrospermae, Caryophyllineae, 34
	(including Hectorellaceae, Illecebraceae)	
Thor	CHENOPODIIFLORAE	Chenopodiales, Chenopodiineae, 86
	(including Illecebraceae)	
Dahl	CARYOPHYLLIFLORAE	Caryophyllales, 52
	(including Illecebraceae)	
Young	DILLENIIDAE, CARYOPHYLLANAE	Caryophyllales, 148
	(presumably including Illecebraceae)	
Takh	CARYOPHYLLIDAE, CARYOPHYLLANAE	
	(including Illecebraceae)	Caryophalles, Caryophyllineae, 71
Cron	CARYOPHYLLIDAE	Caryophyllales, 75
	(including Illecebraceae)	

Acanthophyllum C.A.Mey.
Agrostemma L.
Allochrusa Bunge ex Boiss.
Alsinidendron H.Mann
Ankyropetalum Fenzl
Arenaria L.
Bolanthus (Ser.) Rchb.
Brachystemma D.Don
Brewerina A.Gray
Bufonia L.
Cerastium L.
Cerdia Moçino & Sessé
Colobanthus Bartl.
Cucubalus L.
Cyathophylla Bocquet & Strid
Dianthus L.
Diaphanoptera Rech.f.
Drymaria Willd. ex Schult.
Drypis L.
Gypsophila L.
Habrosia Fenzl
Holosteum L.
Honckenya Ehrh.
Krauseola Pax & K.Hoffm.
Lepyrodiclis Fenzl
Loeflingia L.
Lychnis L.
Mesostemma Vved.
Microphyes Phil.
Minuartia L.
Moehringia L.
Moenchia Ehrh.
Myosoton Moench
Ochotonophila Gilli
Ortegia L.
Pentastemonodiscus Rech.f.
Petrocoptis A.Braun ex Endl.
Petrorhagia (Ser.) Link
Phryna (Boiss.) Pax & K.Hoffm.
Pinosia Urb.
Pirinia M.Král *
Pleioneura Rech.f.
Polycarpaea Lam.
Polycarpon L.
Polytepalum Suess. & Beyerle
Pseudostellaria Pax
Pycnophyllopsis Skottsb. *
Pycnophyllum J.Remy
Robbairea Boiss.
Sagina L.

Sanctambrosia Skottsb. ex Kuschel
Saponaria L.
Schiedea Cham. & Schltdl.
Scleranthopsis Rech.f.
Silene L.
Spergula L.
Spergularia (Pers.) J.Presl & C.Presl
Stellaria L.

Stipulicida Michx.
Thurya Boiss. & Balansa
Thylacospermum Fenzl
Uebelinia Hochst.
Vaccaria Wolf
Velezia L.
Viscaria Roehl.
Wilhelmsia Rchb.

CASUARINACEAE R.Br. 1814 158

4 genera. Tropical Asia to Australia and Pacific. Equisetoid trees, shrubs.

B&H	MONOCHLAMYDEAE	Unisexuales, 158
DT&H	ARCHICHLAMYDEAE	Verticillatae, 1
Melc	ARCHICHLAMYDEAE	Casuarinales, 1
Thor	HAMAMELIDIFLORAE	Casuarinales, 200
Dahl	ROSIFLORAE	Casuarinales, 152
Young	ROSIDAE, HAMAMELIDANAE	Casuarinales, 225
Takh	HAMAMELIDIDAE, HAMAMELIDANAE	
		Casuarinales, 92
Cron	HAMAMELIDIDAE	Casuarinales, 63

Allocasuarina L.A.S.Johnson
Casuarina L.

Ceuthostoma L.A.S.Johnson
Gymnostoma L.A.S.Johnson

CECROPIACEAE C.C.Berg 1978 153.06

6 genera. Widespread in tropics. Trees, shrubs, lianes.

B&H	MONOCHLAMYDEAE	Unisexuales (within Urticaceae, 153)
DT&H	ARCHICHLAMYDEAE	Urticales (within Moraceae, 14)
Mclc	ARCHICHLAMYDEAE	Urticales (within Moraceae, 13)
Thor	MALVIFLORAE	Urticales (within Urticaceae, 157)
Dahl	MALVIFLORAE	Urticales, 71
Young	DILLENIIDAE, MALVANAE	Urticales, 179
Takh	DILLENIIDAE, URTICANAE	Urticales, Urticineae, 187
Cron	HAMAMELIDIDAE	Urticales, 54

Cecropia Loefl.
Coussapoa Aubl.
Musanga C.Sm. ex R.Br.

Myrianthus P.Beauv.
Poikilospermum Zipp. ex Miq.
Pourouma Aubl.

CELASTRACEAE R.Br. 1814 47.01

85 genera. Widespread. Trees, shrubs, lianes.

B&H   POLYPETALAE, DISCIFLORAE Celastrales, 47
      (including Alzateaceae, Goupiaceae)

DT&H ARCHICHLAMYDEAE   Sapindales, Celastrineae, 108
     (including Alzateaceae, Canotiaceae, Goupiaceae; separating *Hippocrateaceae*)
Melc ARCHICHLAMYDEAE   Celastrales, Celastrineae, 159
     (including Goupiaceae, Lophopyxidaceae, ? Canotiaceae)
Thor SANTALIFLORAE   Celastrales, 112
     (including Canotiaceae, Goupiaceae)
Dahl SANTALIFLORAE   Celastrales, 249
     (including Goupiaceae, ? Canotiaceae)
Young ROSIDAE, SANTALANAE   Celastrales, 229
     (presumably including Canotiaceae, Goupiaceae)
Takh ROSIDAE, CELASTRANAE   Celastrales, Celastrineae, 304
     (including Canotiaceae; separating *Pottingeriaceae* 338 to CORNANAE
     Hydrangeales)
Cron ROSIDAE   Celastrales, 219
     (including Canotiaceae, Goupiaceae, Lophopyxidaceae; separating
     Hippocrateaceae 220)

Allocassine N.Robson
Anthodon Ruíz & Pav.
Apatophyllum McGill.
Apodostigma R.Wilczek
Arnicratea N.Hallé
Bequaertia R.Wilczek
Bhesa Buch.–Ham. ex Arn.
Brassiantha A.C.Sm.
Brexiella H.Perrier
Campylostemon Welw.
Cassine L.
Catha Forssk. ex Schreb.
Celastrus L.
Cheiloclinium Miers
Crocoxylon Eckl. & Zeyh.
Crossopetalum P.Browne
Cuervea Triana ex Miers
Denhamia Meisn.
Elachyptera A.C.Sm.
Elaeodendron J.F.Jacq.
Euonymus L.
Evonymopsis H.Perrier
Fraunhofera Mart.
Glyptopetalum Thwaites
Goniodiscus Kuhlm.
Gyminda Sarg.
Hartogiella Codd
Hartogiopsis H.Perrier
Hedraianthera F.Muell.
Helictonema Pierre
Hexaspora C.T.White
Hippocratea L.
Hylenaea Miers
Hypsophila F.Muell.
Katafa Costantin & J.Poiss.
Kokoona Thwaites
Loeseneriella A.C.Sm.
Lophopetalum Wight ex Arn.

Maurocenia Mill.
Maytenus Molina
Menepetalum Loes.
Microtropis Wall. ex Meisn.
Monimopetalum Rehder
Mortonia A.Gray
Mystroxylon Eckl. & Zeyh.
Orthosphenia Standl. *
Paxistima Raf.
Peripterygia (Baill.) Loes.
Peritassa Miers
Perrottetia Kunth
Platypterocarpus Dunkley & Brenan
Plenckia Reissek
Pleurostylia Wight & Arn.
Polycardia Juss.
Pottingeria Prain
Prionostemma Miers
Pristimera Miers
Psammomoya Diels & Loes.
Pseudosalacia Codd
Ptelidium Thouars *
Pterocelastrus Meisn.
Putterlickia Endl.
Quetzalia Lundell *
Reissantia N.Hallé
Rzedowskia Medrano
Salacia L.
Salacighia Loes.
Salaciopsis Baker f.
Salvadoropsis H.Perrier *
Sarawakodendron Ding Hou
Scandivepres Loes.
Schaefferia Jacq.
Semialarium N.Hallé
Simicratea N.Hallé
Simirestis N.Hallé
Siphonodon Griff.

Tetrasiphon Urb.	**Tristemonanthus** Loes.
Thyrsosalacia Loes.	**Wimmeria** Schltdl. & Cham.
Tontelea Aubl.	**Xylonymus** Kalkman ex Ding Hou
Torralbasia Krug & Urb.	**Zinowiewia** Turcz.
Tripterygium Hook.f.	

CEPHALOTACEAE Dumort. 1829 59.14

1 genus, monotypic. W. Australia. Insectivorous herb with pitchers.

B&H	POLYPETALAE, CALYCIFLORAE	Rosales (within Rosaceae, 58)
DT&H	ARCHICHLAMYDEAE	Rosales, Saxifragineae, 66
Melc	ARCHICHLAMYDEAE	Rosales, Saxifragineae, 105
Thor	ROSIFLORAE	Rosales, Saxifragineae, 207
Dahl	ROSIFLORAE	Saxifragales, 168
Young	ROSIDAE, ROSANAE	Saxifragales, 295
Takh	ROSIDAE, ROSANAE	Saxifragales, 204
Cron	ROSIDAE	Rosales, 172

Cephalotus Labill.

CERATOPHYLLACEAE Gray 1821 8.03

1 genus. Widespread. Aquatic herbs.

B&H	MONOCHLAMYDEAE	Ordines anomali, 163
DT&H	ARCHICHLAMYDEAE	Ranales, Nymphaeineae, 39
Melc	ARCHICHLAMYDEAE	Ranunculales, Nymphaeineae, 67
Thor	ANNONIFLORAE	Nelumbonales, 25
Dahl	NYMPHAEIFLORAE	Nymphaeales, 27
Young	MAGNOLIIDAE, RANUNCULANAE	Ranunculales, 28
Takh	MAGNOLIIDAE, NYMPHAEANAE	Ceratophyllales, 42
Cron	MAGNOLIIDAE	Nymphaeales, 29

Ceratophyllum L.

CERCIDIPHYLLACEAE Engl. 1909 4.09

1 genus, monotypic. China, Japan. Tree.

B&H	Not mentioned	—
DT&H	ARCHICHLAMYDEAE	Ranales, Trochodendrineae (within Trochodendraceae, 40)
Melc	ARCHICHLAMYDEAE	Magnoliales, Cercidiphyllineae, 60
Thor	HAMAMELIDIFLORAE	Hamamelidales, Trochodendrineae, 196
Dahl	ROSIFLORAE	Cercidiphyllales, 139
Young	ROSIDAE, HAMAMELIDANAE	Trochodendrales, 217
Takh	HAMAMELIDIDAE, TROCHODENDRANAE	Cercidiphyllales, 78
Cron	HAMAMELIDIDAE	Hamamelidales, 42

Cercidiphyllum Siebold & Zucc.

CHENOPODIACEAE Vent. 1799   131.01

113 genera. Widespread. Herbs, shrubs, rarely small trees.

B&H	MONOCHLAMYDEAE	Curvembryeae, 131
	(including Basellaceae)	
DT&H	ARCHICHLAMYDEAE	Centrospermae, Chenopodiineae, 28
	(including Halophytaceae)	
Melc	ARCHICHLAMYDEAE	Centrospermae, Chenopodiineae, 36
	(separating *Dysphaniaceae* 35)	
Thor	CHENOPODIIFLORAE	Chenopodiales, Chenopodiineae, 89
Dahl	CARYOPHYLLIFLORAE	Caryophyllales, 49
Young	DILLENIIDAE, CARYOPHYLLANAE	Chenopodiales, 159
Takh	CARYOPHYLLIDAE, CARYOPHYLLANAE	
		Caryophyllales, Chenopodiineae, 73
Cron	CARYOPHYLLIDAE	Caryophyllales, 70

Acroglochin Schrad. ex Schult.
Agathophora (Fenzl) Bunge
Agriophyllum M.Bieb.
Alexandra Bunge
Allenrolfea Kuntze
Anabasis L.
Anthochlamys Fenzl
Aphanisma Nutt. ex Moq.
Archiatriplex G.L.Chu *
Arthrocnemum Moq.
Arthrophytum Schrenk
Atriplex L.
Axyris L.
Babbagia F.Muell.
Bassia All.
Beta L.
Bienertia Bunge ex Boiss.
Borszczowia Bunge
Brachylepis C.A.Mey.
Camphorosma L.
Ceratocarpus L.
Chenopodium L.
Choriptera Botsch.
Climacoptera Botsch.
Corispermum L.
Cornulaca Delile
Cremnophyton Brullo & Pavone *
Cyathobasis Aellen
Cycloloma Moq.
Didymanthus Endl.
Dissocarpus F.Muell.
Dysphania R.Br.
Einadia Raf.
Enchylaena R.Br.
Eremophea Paul G.Wilson
Eriochiton (R.Anderson) A.J.Scott
Esfandiaria Charif & Aellen
Exomis Fenzl ex Moq.

Fadenia Aellen & C.C.Towns.
Fredolia (Coss. & Durieu ex Bunge) Ulbr.
Girgensohnia Bunge
Goerziella Urb. *
Grayia Hook. & Arn.
Gyroptera Botsch.
Hablitzia M.Bieb.
Halanthium K.Koch
Halarchon Bunge
Halimione Aellen
Halimocnemis C.A.Mey.
Halocharis Moq.
Halocnemum M.Bieb.
Halogeton C.A.Mey.
Halopeplis Bunge ex Ung.-Sternb.
Halosarcia Paul G.Wilson
Halostachys C.A.Mey. ex Schrenk
Halothamnus Jaub. & Spach
Haloxylon Bunge
Hammada Iljin
Helicilla Moq. *
Hemichroa R.Br.
Heterostachys Ung.-Sternb.
Holmbergia Hicken
Horaninovia Fisch. & C.A.Mey.
Kalidiopsis Aellen
Kalidium Moq.
Kirilowia Bunge
Kochia Roth
Krascheninnikovia Gueldenst.
Lagenantha Chiov.
Maireana Moq.
Malacocera R.H.Anderson
Manochlamys Aellen
Microcnemum Ung.-Sternb.
Microgynoecium Hook.f.
Monolepis Schrad.

Nanophyton Less.
Neobassia A.J.Scott
Nitrophila S.Watson
Noaea Moq.
Nucularia Batt.
Ofaiston Raf.
Oreobliton Durieu
Pachycornia Hook.f.
Panderia Fisch. & C.A.Mey.
Patellifolia A.J.Scott, Ford–Lloyd & J.T.Williams
Petrosimonia Bunge
Piptoptera Bunge
Polycnemum L.
Rhagodia R.Br.
Roubieva Moq.
Roycea C.A.Gardner
Salicornia L.
Salsola L.
Sarcobatus Nees

Sarcocornia A.J.Scott
Sclerolaena R.Br.
Sclerostegia Paul G.Wilson
Seidlitzia Bunge ex Boiss.
Senniella Aellen
Sevada Moq.
Spinacia L.
Stelligera A.J.Scott
Suaeda Forssk. ex Scop.
Suckleya A.Gray
Sympegma Bunge
Tecticornia Hook.f.
Tegicornia Paul G.Wilson
Teloxys Moq. *
Theleophyton (Hook.f.) Moq.
Threlkeldia R.Br.
Traganopsis Maire & Wilczek *
Traganum Delile
Zuckia Standl.

CHLORANTHACEAE Blume 1827   140

4 genera. Tropical and subtropical. Trees, shrubs, herbs.

B&H	MONOCHLAMYDEAE	Micrembryeae, 140
DT&H	ARCHICHLAMYDEAE	Piperales, 4
	(including Circaeasteraceae)	
Melc	ARCHICHLAMYDEAE	Piperales, 70
Thor	ANNONIFLORAE	Annonales, Laurineae, 15
Dahl	MAGNOLIIFLORAE	Chloranthales, 14
Young	MAGNOLIIDAE, RANUNCULANAE	Laurales, 18
Takh	MAGNOLIIDAE, MAGNOLIANAE	Chloranthales, 24
Cron	MAGNOLIIDAE	Piperales, 19

Ascarina J.R.Forst. & G.Forst.
Chloranthus Sw.

Hedyosmum Sw.
Sarcandra Gardner

CHRYSOBALANACEAE R.Br. 1818   58.02

17 genera. Tropics and subtropics. Trees, shrubs.

B&H	POLYPETALAE, CALYCIFLORAE	Rosales (within Rosaceae, 158)
DT&H	ARCHICHLAMYDEAE	Rosales, Rosineae (within Rosaceae, 76)
Melc	ARCHICHLAMYDEAE	Rosales, Rosineae, 116
	(including Stylobasiaceae)	
Thor	ROSIFLORAE	Rosales, Rosineae, 204
Dahl	MYRTIFLORAE	Chrysobalanales, 199
Young	ROSIDAE, ROSANAE	Rosales, 292
Takh	ROSIDAE, ROSANAE	Rosales, 218
Cron	ROSIDAE	Rosales, 177

Acioa Aubl.
Afrolicania Mildbr.
Atuna Raf.
Bafodeya Prance ex F.White
Chrysobalanus L.
Couepia Aubl.
Exellodendron Prance
Grangeria Comm. ex Juss.
Hirtella L.

Hunga Pancher ex Prance
Kostermanthus Prance
Licania Aubl.
Magnistipula Engl.
Maranthes Blume
Neocarya (DC.) Prance ex F.White
Parastemon A.DC.
Parinari Aubl.

CIRCAEASTERACEAE Hutch. 1926   1.02

1 genus, monotypic. Himalaya to N.W. China. Annual herb.

B&H	Not known	—
DT&H	ARCHICHLAMYDEAE	Piperales (within Chloranthaceae, 4)
Melc	ARCHICHLAMYDEAE	Ranunculales, Ranunculineae (within Ranunculaceae, 61)
Thor	ANNONIFLORAE	Berberidales, Berberidineae, 34
Dahl	RANUNCULIFLORAE	Ranunculales, 33
Young	MAGNOLIIDAE, RANUNCULANAE	Ranunculales, 38
Takh	RANUNCULIDAE, RANUNCULANAE	Ranunculales, Ranunculineae, 48
Cron	MAGNOLIIDAE	Ranunculales, 31

Circaeaster Maxim.

CISTACEAE Juss. 1789   14

7 genera. N. temp. region to N.E. trop. Afr. and trop. S. America. Shrubs, herbs.

B&H	POLYPETALAE, THALAMIFLORAE	Parietales, 14
DT&H	ARCHICHLAMYDEAE	Parietales, Cistineae, 143
Melc	ARCHICHLAMYDEAE	Violales, Cistineae, 191
Thor	MALVIFLORAE	Malvales, 153
Dahl	MALVIFLORAE	Malvales, 61
Young	DILLENIIDAE, MALVANAE	Malvales, 174
Takh	DILLENIIDAE, MALVANAE	Bixales, 172
Cron	DILLENIIDAE	Violales, 110

Cistus L.
Fumana (Dunal) Spach
Halimium (Dunal) Spach
Helianthemum Mill.

Hudsonia L.
Lechea L.
Tuberaria (Dunal) Spach

CLETHRACEAE Klotzsch 1851   93.03

1 genus. New World, Asia, Madeira. Shrubs, small trees.

B&H	GAMOPETALAE, HETEROMERAE	Ericales (within Ericaceae, 93)
DT&H	METACHLAMYDEAE	Ericales, 182

Melc	SYMPETALAE	Ericales, 228
Thor	THEIFLORAE	Theales, Clethrineae, 55
Dahl	CORNIFLORAE	Ericales, 287
Young	DILLENIIDAE, DILLENIANAE	Theales, 116
Takh	DILLENIIDAE, ERICANAE	Ericales, 126
Cron	DILLENIIDAE	Ericales, 140

Clethra L.

CLUSIACEAE — see GUTTIFERAE

CNEORACEAE Link 1831   40.02

1 genus. W. Mediterranean, Canary Is, Cuba. Shrubs.

B&H	POLYPETALAE, DISCIFLORAE	Geraniales (within Simaroubaceae, 40)
DT&H	ARCHICHLAMYDEAE	Geraniales, Geraniineae, 86
Melc	ARCHICHLAMYDEAE	Rutales, Rutineae, 132
Thor	RUTIFLORAE	Rutales, Rutineae, 168
Dahl	RUTIFLORAE	Rutales, 220
Young	ROSIDAE, RUTANAE	Rutales, 245
Takh	ROSIDAE, RUTANAE	Rutales, 258
Cron	ROSIDAE	Sapindales, 259

Cneorum L.

COBAEACEAE D.Don 1824   120.02

1 genus. Tropical America. Climbers.

B&H	GAMOPET., BICARPELLATAE	Polemoniales (within Polemoniaceae, 110)
DT&H	METACHLAMYDEAE	Tubiflorae, Convolvulineae (within Polemoniaceae, 203)
Melc	SYMPETALAE	Tubiflorae, Convolvulineae (within Polemoniaceae, 252)
Thor	SOLANIFLORAE	Solanales, Polemoniineae (within Polemoniaceae, 273)
Dahl	SOLANIFLORAE	Solanales, 278
Young	(not mentioned, presumably within Polemoniaceae, 365)	
Takh	LAMIIDAE, SOLANANAE	Polemoniales, 386
Cron	ASTERIDAE	Solanales (within Polemoniaceae, 282)

Cobaea Cav.

COCHLOSPERMACEAE Planch. 1847   17.02

2 genera. Trop. Afr. and Asia to Australia, trop. Amer. to SW USA. Trees to subshrubs.

| B&H | POLYPETALAE, THALAMIFLORAE | Parietales (within Bixaceae, 17) |

DT&H ARCHICHLAMYDEAE   Parietales, Cochlospermineae, 145
Melc ARCHICHLAMYDEAE   Violales, Cistineae, 194
Thor MALVIFLORAE   Malvales, 152
Dahl MALVIFLORAE   Malvales, 60
Young DILLENIIDAE, MALVANAE   Malvales, 173
Takh DILLENIIDAE, MALVANAE   Bixales, 170
Cron DILLENIIDAE   Violales (within Bixaceae, 109)

Amoreuxia Moçino & Sessé ex DC.
Cochlospermum Kunth

COLUMELLIACEAE D.Don 1828   59.15

1 genus. NW. South America. Shrubs.

B&H GAMOPETALAE, BICARPELLATAE Personales, 118
DT&H SYMPETALAE   Tubiflorae, Solanineae, 216
Melc SYMPETALAE   Tubiflorae, Solanineae, 273
Thor ROSIFLORAE   Rosales, Saxifragineae (within Saxifragaceae, 208)

Dahl CORNIFLORAE   Cornales, 316
Young ROSIDAE, CORNANAE   Hydrangeales, 333
Takh ROSIDAE, CORNANAE   Hydrangeales, Escalloniineae, 332
Cron ROSIDAE   Rosales, 165

Columellia Ruíz & Pav.

COMBRETACEAE R.Br. 1810   66

20 genera. Tropics and subtropics. Trees, shrubs, climbers, rarely herbs.

B&H POLYPETALAE, CALYCIFLORAE Myrtales, 66
DT&H ARCHICHLAMYDEAE   Myrtiflorae, Myrtineae, 172
Melc ARCHICHLAMYDEAE   Myrtiflorae, Myrtineae, 213
Thor MYRTIFLORAE   Myrtales, Lythrineae, 244
Dahl MYRTIFLORAE   Myrtales, 191
Young ROSIDAE, MYRTANAE   Myrtales, 314
Takh ROSIDAE, MYRTANAE   Myrtales, Myrtineae, 231
Cron ROSIDAE   Myrtales, 199

Anogeissus (DC.) Wall.
Buchenavia Eichler
Bucida L.
Calopyxis Tul.
Combretum Loefl.
Conocarpus L.
Dansiea Byrnes
Getonia Roxb.
Guiera Adans. ex Juss.
Laguncularia C.F.Gaertn.

Lumnitzera Willd.
Macropteranthes F.Muell.
Meiostemon Exell & Stace
Pteleopsis Engl.
Quisqualis L.
Ramatuela Kunth
Strephonema Hook.f.
Terminalia L.
Terminaliopsis Danguy
Thiloa Eichler

COMPOSITAE Giseke 1792

1509 genera. Widespread. Mostly herbs or shrubs, some trees.

B&H	GAMOPETALAE, INFERAE	Asterales, 88
DT&H	METACHLAMYDEAE	Campanulatae, Campanulineae, 233
Melc	SYMPETALAE	Campanulales, 290
Thor	ASTERIFLORAE	Asterales, 296
Dahl	ASTERIFLORAE	Asterales, 272
Young	ROSIDAE, ASTERANAE	Asterales, 386
Takh	ASTERIDAE, ASTERANAE	Asterales, 429
Cron	ASTERIDAE	Asterales, 318

Aaronsohnia Warb. & Eig
Abrotanella Cass.
Acamptopappus (A.Gray) A.Gray
Acanthocephalus Kar. & Kir.
Acanthocladium F.Muell.
Acanthodesmos C.D.Adams &
   duQuesnay
Acantholepis Less.
Acanthospermum Schrank
Acanthostyles R.M.King & H.Rob.
Achaetogeron A.Gray
Achillea L.
Achnophora F.Muell.
Achnopogon Maguire, Steyerm. &
   Wurdack
Achyrachaena Schauer
Achyrocline (Less.) DC.
Achyropappus Kunth
Achyrothalamus O.Hoffm.
Acmella Rich. ex Pers.
Acomis F.Muell.
Acourtia D.Don
Acrisione B.Nord.
Acritopappus R.M.King & H.Rob.
Acroclinium A.Gray
Acroptilon Cass.
Actinobole Endl.
Actinoseris (Endl.) Cabrera
Actinospermum Elliott
Adelostigma Steetz
Adenanthellum B.Nord.
Adenocaulon Hook.
Adenocritonia R.M.King & H.Rob.
Adenoglossa B.Nord.
Adenoon Dalzell
Adenopappus Benth.
Adenophyllum Pers.
Adenostemma J.R.Forst. & G.Forst.
Adenothamnus D.Keck
Aedesia O.Hoffm.
Aegopordon Boiss.
Acquatorium B.Nord.

Aetheorhiza Cass.
Ageratella A.Gray ex S.Watson
Ageratina Spach
Ageratinastrum Mattf.
Ageratum L.
Agiabampoa Rose ex O.Hoffm.
Agoseris Raf.
Agrianthus Mart. ex DC.
Ainsliaea DC.
Ajania Poljakov
Ajaniopsis C.Shih
Alatoseta Compton
Albertinia Spreng.
Alcantara Glaz. ex G.M.Barroso
Alciope DC. ex Lindl.
Aldama La Llave
Alepidocline S.F.Blake
Alfredia Cass.
Aliella Qaiser & Lack
Allagopappus Cass.
Allardia Decne.
Alloispermum Willd.
Allopterigeron Dunlop
Alomia Kunth
Alomiella R.M.King & H.Rob.
Alvordia Brandegee
Amauria Benth.
Amberboa (Pers.) Less.
Amblyocarpum Fisch. & C.A.Mey.
Amblyolepis DC.
Amblyopappus Hook. & Arn.
Amboroa Cabrera
Ambrosia L.
Ameghinoa Speg.
Amellus L.
Ammobium R.Br. ex Sims
Amolinia R.M.King & H.Rob.
Amphiachyris (DC.) Nutt.
Amphiglossa DC.
Amphipappus Torr. & A.Gray
Amphoricarpos Vis.
Anacantha (Iljin) Soják

Anacyclus L.
Anaphalioides (Benth.) Kirp.
Anaphalis DC.
Anastraphia D.Don
Anaxeton Gaertn.
Ancathia DC.
Ancistrophora A.Gray
Andryala L.
Angelphytum G.M.Barroso
Angianthus J.C.Wendl.
Anisochaeta DC.
Anisocoma Torr. & A.Gray
Anisopappus Hook. & Arn.
Anisothrix O.Hoffm. ex Kuntze
Anomostephium DC.
Antennaria Gaertn.
Anthemis L.
Antheropeas Rydb.
Antillia R.M.King & H.Rob.
Antiphiona Merxm.
Antithrixia DC.
Anura (Juz.) Tscherneva
Anvillea DC.
Apalochlamys (Cass.) Cass.
Apargidium Torr. & A.Gray
Aphanactis Wedd.
Aphanostephus DC.
Aphyllocladus Wedd.
Apodocephala Baker
Aposeris Neck. ex Cass.
Apostates Lander
Arbelaezaster Cuatrec.
Archibaccharis Heering
Arctanthemum (Tzvelev) Tzvelev
Arctium L.
Arctogeron DC.
Arctotheca J.C.Wendl.
Arctotis L.
Argyranthemum Webb
Argyroglottis Turcz.
Argyrophanes Schltdl.
Argyroxiphium DC.
Aristeguietia R.M.King & H.Rob.
Arnaldoa Cabrera *
Arnica L.
Arnicastrum Greenm.
Arnoglossum Raf.
Arnoseris Gaertn.
Arrhenechthites Mattf.
Arrojadocharis Mattf. *
Arrowsmithia DC.
Artemisia L.
Artemisiopsis S.Moore
Asanthus R.M.King & H.Rob.
Ascidiogyne Cuatrec. *

Aspilia Thouars
Asplundianthus R.M.King & H.Rob.
Aster L.
Asteridea Lindl.
Asteriscus Mill.
Asteromoea Blume
Astranthium Nutt.
Athanasia L.
Athrixia Kew Gawl.
Athroisma DC.
Atractylis L.
Atractylodes DC.
Atrichantha Hilliard & B.L.Burtt
Atrichoseris A.Gray
Austrobrickellia R.M.King & H.Rob.
Austrocritonia R.M.King & H.Rob.
Austroeupatorium R.M.King & H.Rob.
Austroliabum H.Rob. & Brettell
Austrosynotis C.Jeffrey
Avellara Blanca & Díaz Guard.
Axiniphyllum Benth.
Ayapana Spach
Ayapanopsis R.M.King & H.Rob.
Aylacophora Cabrera
Baccharis L.
Badilloa R.M.King & H.Rob.
Baeriopsis J.T.Howell
Bafutia C.D.Adams
Bahia Lag.
Bahianthus R.M.King & H.Rob.
Baileya Harv. & A.Gray
Balduina Nutt.
Balsamorhiza Hook. ex Nutt.
Baltimora L.
Barkleyanthus H.Rob. & Brettell
Barnadesia Mutis ex L.f.
Barroetea A.Gray
Barrosoa R.M.King & H.Rob.
Bartlettia A.Gray
Bartlettina R.M.King & H.Rob.
Basedowia E.Pritz.
Bebbia Greene
Bedfordia DC.
Bejaranoa R.M.King & H.Rob.
Bellida Ewart
Bellis L.
Bellium L.
Belloa J.Remy
Benitoa D.Keck
Berardia Vill.
Berkheya Ehrh.
Berlandiera DC.
Berroa Beauverd
Bidens L.
Bigelowia DC.

Bishopalea H.Rob.
Bishopanthus H.Rob.
Bishopiella R.M.King & H.Rob.
Bishovia R.M.King & H.Rob.
Blainvillea Cass.
Blakeanthus R.M.King & H.Rob.
Blakiella Cuatrec.
Blanchetia DC.
Blennosperma Less.
Blennospora A.Gray
Blepharipappus Hook.
Blepharispermum DC.
Blepharizonia (A.Gray) Greene
Blumea DC.
Blumeopsis Gagnep.
Boeberastrum (A.Gray) Rydb.
Boeberoides (DC.) Strother
Bolanosa A.Gray
Bolophyta Nutt.
Boltonia L'Hér.
Bombycilaena (DC.) Smoljan.
Borrichia Adans.
Bothriocline Oliv. ex Benth.
Brachanthemum DC.
Brachionostylum Mattf.
Brachyclados Gillies ex D.Don
Brachyglottis J.R.Forst. & G.Forst.
Brachylaena R.Br.
Brachyscome Cass.
Brachythrix Wild & G.V.Pope
Bracteantha Anderb.
Bradburia Torr. & A.Gray
Brickellia Elliott
Brickelliastrum R.M.King & H.Rob.
Bryomorphe Harv.
Buphthalmum L.
Burkartia Crisci
Cabreriella Cuatrec.
Cacalia L.
Cacaliopsis A.Gray
Cacosmia Kunth
Cadiscus E.Mey. ex DC.
Caesulia Roxb.
Calea L.
Calendula L.
Callicephalus C.A.Mey.
Callilepis DC.
Callistephus Cass.
Calocephalus R.Br.
Calomeria Vent.
Calopappus Meyen
Calostephane Benth.
Calotesta Karis
Calotis R.Br.
Calycadenia DC.

Calycocorsus F.W.Schmidt
Calycoseris A.Gray
Calyptocarpus Less.
Camchaya Gagnep.
Campovassouria R.M.King & H.Rob.
Camptacra N.T.Burb.
Campuloclinium DC.
Cancrinia Kar. & Kir.
Cancriniella Tzvelev *
Cardopatium Juss.
Carduncellus Adans.
Carduus L.
Carlina L.
Carminatia Moçino ex DC.
Carpesium L.
Carphephorus Cass.
Carphochaete A.Gray
Carramboa Cuatrec.
Carterothamnus R.M.King
Carthamus L.
Cassinia R.Br.
Castalis Cass.
Castenedia R.M.King & H.Rob.
Catamixis Thomson
Catananche L.
Catatia Humbert
Cavalcantia R.M.King & H.Rob.
Cavea W.W.Sm. & Small
Celmisia Cass.
Centaurea L.
Centaurodendron Johow
Centauropsis Bojer ex DC.
Centaurothamnus Wagenitz & Dittrich
Centipeda Lour.
Centratherum Cass.
Cephalipterum A.Gray
Cephalopappus Nees & Mart.
Cephalorrhynchus Boiss.
Cephalosorus A.Gray
Ceratogyne Turcz.
Ceruana Forssk.
Chacoa R.M.King & H.Rob.
Chaenactis DC.
Chaetadelpha A.Gray ex S.Watson
Chaetanthera Ruíz & Pav.
Chaetopappa DC.
Chaetospira S.F.Blake
Chaetymenia Hook. & Arn.
Chamaechaenactis Rydb.
Chamaegeron Schrenk
Chamaeleon Cass.
Chamaemelum Mill.
Chamaepus Wagenitz
Chaptalia Vent.
Chardinia Desf.

Cheirolophus Cass.
Chersodoma Phil.
Chevreulia Cass.
Chiliadenus Cass.
Chiliocephalum Benth.
Chiliophyllum Phil.
Chiliotrichiopsis Cabrera
Chiliotrichum Cass.
Chimantaea Maguire, Steyerm. &
    Wurdack
Chionolaena DC.
Chionopappus Benth.
Chlamydophora Ehrenb. ex Less.
Chloracantha G.L.Nesom, Y.B.Suh,
    D.R.Morgan, S.D.Sundb. &
    B.B.Simpson
Chondrilla L.
Chondropyxis D.A.Cooke
Chorisis DC.
Chresta Vell. ex DC.
Chromolaena DC.
Chromolepis Benth.
Chronopappus DC.
Chrysactinia A.Gray
Chrysactinium (Kunth) Wedd.
Chrysanthellum Rich.
Chrysanthemoides Fabr.
Chrysanthemum L.
Chrysocephalum Walp.
Chrysocoma L.
Chrysogonum L.
Chrysophthalmum Sch.Bip. ex Walp.
Chrysopsis (Nutt.) Elliott
Chrysothamnus Nutt.
Chthonocephalus Steetz
Chucoa Cabrera
Chuquiraga Juss.
Cicerbita Wallr.
Ciceronia Urb.
Cichorium L.
Cineraria L.
Cirsium Mill.
Cissampelopsis (DC.) Miq.
Cladanthus Cass.
Cladochaeta DC.
Clappia A.Gray
Clibadium L.
Cnicothamnus Griseb.
Cnicus L.
Coespeletia Cuatrec.
Coleocoma F.Muell.
Coleostephus Cass.
Colobanthera Humbert
Comaclinium Scheidw. & Planch.
Commidendrum DC.

Complaya Strother
Condylidium R.M.King & H.Rob.
Condylopodium R.M.King & H.Rob.
Conocliniopsis R.M.King & H.Rob.
Conoclinium DC.
Conyza Less.
Coreocarpus Benth.
Coreopsis L.
Corethamnium R.M.King & H.Rob.
Corethrogyne DC.
Correllia A.M.Powell *
Corymbium L.
Cosmos Cav.
Cotula L.
Coulterella Vasey & Rose
Cousinia Cass.
Cousiniopsis Nevski
Craspedia G.Forst.
Crassocephalum Moench
Cratystylis S.Moore
Cremanthodium Benth.
Crepidiastrum Nakai
Crepis L.
Critonia P.Browne
Critoniadelphus R.M.King & H.Rob.
Critoniella R.M.King & H.Rob.
Critoniopsis Sch.Bip.
Crocidium Hook.
Cronquistia R.M.King
Cronquistianthus R.M.King & H.Rob.
Croptilon Raf.
Crossostephium Less.
Crossothamnus R.M.King & H.Rob.
Crupina (Pers.) DC.
Cuatrecasasiella H.Rob.
Cuchumatanea Seid. & Beaman
Cullumia R.Br.
Cuspidia Gaertn.
Cyathocline Cass.
Cyathomone S.F.Blake
Cyclachaena Fres. ex Schltdl.
Cyclolepis Gillies ex D.Don
Cylindrocline Cass.
Cymbolaena Smoljan.
Cymbonotus Cass.
Cymbopappus B.Nord.
Cymophora B.L.Rob.
Cynara L.
Dacryotrichia Wild
Dahlia Cav.
Damnxanthodium Strother
Darwiniothamnus Harling
Dasycondylus R.M.King & H.Rob.
Dasyphyllum Kunth
Daveaua Willk. ex Mariz

Decachaeta DC.
Decastylocarpus Humbert
Decazesia F.Muell.
Delairea Lem.
Delamerea S.Moore
Delilia Spreng.
Dendranthema (DC.) Des Moulins
Dendrocacalia (Nakai) Tuyama
Dendrosenecio (Hauman ex Humbert) B.Nord.
Dendroseris D.Don
Denekia Thunb.
Desmanthodium Benth.
Dewildemania O.Hoffm.
Diacranthera R.M.King & H.Rob.
Dianthoseris Sch.Bip.
Diaperia Nutt.
Diaphractanthus Humbert
Dicercoclados C.Jeffrey & Y.L.Chen
Dichaetophora A.Gray *
Dichrocephala L'Hér. ex DC.
Dichromochlamys Dunlop
Dicoma Cass.
Dicoria Torr. & A.Gray
Dicranocarpus A.Gray
Didelta L'Hér.
Dielitzia P.S.Short
Digitacalia Pippen
Dimeresia A.Gray
Dimerostemma Cass.
Dimorphocoma F.Muell. & Tate
Dimorphotheca Moench
Dinoseris Griseb.
Diodontium F.Muell.
Diplazoptilon Y.Ling
Diplostephium Kunth
Dipterocome Fisch. & C.A.Mey.
Dipterocypsela S.F.Blake *
Disparago Gaertn.
Dissothrix A.Gray
Distephanus (Cass.) Cass.
Disynaphia Hook. & Arn. ex DC.
Dithyrostegia A.Gray
Dittrichia Greuter
Dolichlasium Lag.
Dolichoglottis B.Nord.
Dolichorrhiza (Pojark.) Galushko
Dolichothrix Hilliard & B.L.Burtt
Dolomiaea DC.
Doniophyton Wedd.
Doronicum L.
Dresslerothamnus H.Rob.
Dubautia Gaudich.
Dubyaea DC.
Dugaldia (Cass.) Cass.

Dugesia A.Gray
Duhaldea DC.
Duidaea S.F.Blake
Duseniella K.Schum.
Dymondia Compton
Dyscritogyne R.M.King & H.Rob.
Dyscritothamnus B.L.Rob.
Dysodiopsis (A.Gray) Rydb.
Dyssodia Cav.
Eastwoodia Brandegee
Eatonella A.Gray
Echinacea Moench
Echinops L.
Eclipta L.
Edmondia Cass.
Egletes Cass.
Eitenia R.M.King & H.Rob.
Ekmania Gleason
Elachanthus F.Muell.
Elaphandra Strother
Elephantopus L.
Eleutheranthera Poit. ex Bosc
Ellenbergia Cuatrec.
Elytropappus Cass.
Emilia (Cass.) Cass.
Emiliella S.Moore
Encelia Adans.
Enceliopsis (A.Gray) A.Nelson
Endocellion Turcz. ex Herder
Endopappus Sch.Bip.
Engelmannia A.Gray ex Nutt.
Engleria O.Hoffm.
Enydra Lour.
Epaltes Cass.
Epilasia (Bunge) Benth.
Episcothamnus H.Rob.
Epitriche Turcz.
Erato DC.
Erechtites Raf.
Eremanthus Less.
Eremosis (DC.) Gleason
Eremothamnus O.Hoffm.
Eriachaenium Sch.Bip.
Ericameria Nutt.
Ericentrodea S.F.Blake & Sherff
Erigeron L.
Eriocephalus L.
Eriochlamys Sond. & F.Muell.
Eriophyllum Lag.
Eriotrix Cass.
Erlangea Sch.Bip.
Erodiophyllum F.Muell.
Erymophyllum Paul G.Wilson
Eryngiophyllum Greenm.
Erythradenia (B.L.Rob.) R.M.King & H.Rob.

Erythrocephalum Benth.
Espejoa DC.
Espeletia Mutis ex Humb. & Bonpl.
Espeletiopsis Cuatrec.
Ethulia L.f.
Euchiton Cass.
Eumorphia DC.
Eupatoriastrum Greenm.
Eupatorina R.M.King & H.Rob.
Eupatoriopsis Hieron.
Eupatorium L.
Euphrosyne DC.
Eurydochus Maguire & Wurdack *
Euryops (Cass.) Cass.
Eutetras A.Gray
Euthamia (Nutt.) Elliott
Evacidium Pomel
Ewartia Beauverd
Ewartiothamnus Anderb.
Exomiocarpon Lawalrée
Faberia Hemsl.
Facelis Cass.
Farfugium Lindl.
Faujasia Cass.
Faxonia Brandegee
Feddea Urb.
Feldstonia P.S.Short
Felicia Cass.
Fenixia Merr.
Ferreyranthus H.Rob. & Brettell
Ferreyrella S.F.Blake
Filago L.
Filifolium Kitam.
Fitchia Hook.f.
Fitzwillia P.S.Short
Flaveria Juss.
Fleischmannia Sch.Bip.
Fleischmanniopsis R.M.King & H.Rob.
Florestina Cass.
Floscaldasia Cuatrec. *
Flosmutisia Cuatrec.
Flourensia DC.
Flyriella R.M.King & H.Rob.
Formania W.W.Sm. & Small
Foveolina Källersjö
Fulcaldea Poir.
Gaillardia Foug.
Galactites Moench
Galeana La Llave
Galeomma Rauschert
Galinsoga Ruíz & Pav.
Gamochaeta Wedd.
Gamochaetopsis Anderb. & Freire
Garberia A.Gray
Garcibarrigoa Cuatrec. *

Garcilassa Poepp.
Gardnerina R.M.King & H.Rob.
Garuleum Cass.
Gazania Gaertn.
Geigeria Griess.
Geissolepis B.L.Rob.
Geissopappus Benth.
Geraea Torr. & A.Gray
Gerbera L.
Geropogon L.
Gibbaria Cass.
Gilberta Turcz.
Gilruthia Ewart
Gladiopappus Humbert
Glossarion Maguire & Wurdack
Glossocardia Cass.
Glossopappus Kunze
Glyptopleura Eaton
Gnaphaliothamnus Kirp.
Gnaphalium L.
Gnephosis Cass.
Gochnatia Kunth
Goldmanella Greenm.
Gongrostylus R.M.King & H.Rob.
Gongylolepis R.H.Schomb.
Goniocaulon Cass.
Gonospermum Less.
Gorceixia Baker
Gorteria L.
Gossweilera S.Moore
Goyazianthus R.M.King & H.Rob.
Grangea Adans.
Grangeopsis Humbert
Graphistylis B.Nord.
Gratwickia F.Muell.
Grauanthus Fayed
Grazielia R.M.King & H.Rob.
Greenmaniella W.M.Sharp
Grindelia Willd.
Grisebachianthus R.M.King & H.Rob.
Grosvenoria R.M.King & H.Rob.
Guardiola Cerv. ex Humb. & Bonpl.
Guayania R.M.King & H.Rob.
Guevaria R.M.King & H.Rob.
Guizotia Cass.
Gundelia L.
Gundlachia A.Gray
Gutenbergia Sch.Bip.
Gutierrezia Lag.
Gymnarrhena Desf.
Gymnocondylus R.M.King & H.Rob.
Gymnocoronis DC.
Gymnodiscus Less.
Gymnolaena (DC.) Rydb.
Gymnopentzia Benth.

Gymnosperma Less.
Gymnostephium Less.
Gynoxys Cass.
Gynura Cass.
Gypothamnium Phil.
Gyptidium R.M.King & H.Rob.
Gyptis (Cass.) Cass.
Gyrodoma Wild
Haastia Hook.f.
Haeckeria F.Muell.
Haegiela P.S.Short
Handelia Heimerl
Haplocalymma S.F.Blake
Haplocarpha Less.
Haploesthes A.Gray
Haplopappus Cass.
Haplostephium Mart. ex DC.
Harleya S.F.Blake
Harnackia Urb. *
Hartwrightia A.Gray ex S.Watson
Hasteola Raf.
Hatschbachiella R.M.King & H.Rob.
Hazardia Greene
Hebeclinium DC.
Hecastocleis A.Gray
Hedypnois Mill.
Helenium L.
Helianthella Torr. & A.Gray
Helianthopsis H.Rob.
Helianthus L.
Helichrysopsis Kirp.
Helichrysum Mill.
Heliocauta Humphries
Heliomeris Nutt.
Heliopsis Pers.
Helogyne Nutt.
Hemisteptia Fisch. & C.A.Mey.
Hemizonia DC.
Henricksonia B.L.Turner
Heptanthus Griseb.
Herderia Cass.
Herodotia Urb. & Ekman
Hesperomannia A.Gray
Heteracia Fisch. & C.A.Mey.
Heteranthemis Schott
Heterocoma DC.
Heterocondylus R.M.King & H.Rob.
Heterocypsela H.Rob.
Heteroderis (Bunge) Boiss.
Heterolepis Cass.
Heteromera Pomel
Heteromma Benth.
Heteropappus Less.
Heteroplexis C.C.Chang *
Heterorhachis Sch.Bip. ex Walp.

Heterosperma Cav.
Heterothalamus Less.
Heterotheca Cass.
Hidalgoa La Llave
Hieracium L.
Hilliardia B.Nord.
Hinterhubera Sch.Bip.ex Wedd.
Hippia L.
Hippolytia Poljakov
Hirpicium Cass.
Hispidella Barnadez ex Lam.
Hoehnephytum Cabrera
Hoffmanniella Schltr. ex Lawalrée
Hofmeisteria Walp.
Holocarpha Greene
Holocheilus Cass.
Hololeion Kitam.
Holozonia Greene
Homogyne Cass.
Hoplophyllum DC.
Huarpea Cabrera
Hubertia Bory
Hughesia R.M.King & H.Rob.
Hulsea Torr. & A.Gray
Humeocline Anderb.
Hyalis D.Don ex Hook. & Arn.
Hyalochaete Dittrich & Rech.f.
Hyalochlamys A.Gray
Hyaloseris Griseb.
Hyalosperma Steetz
Hybridella Cass.
Hydrodyssodia B.L.Turner
Hydroidea Karis
Hydropectis Rydb.
Hymenocephalus Jaub. & Spach
Hymenoclea Torr. & A.Gray
Hymenolepis Cass.
Hymenonema Cass.
Hymenopappus L'Hér.
Hymenostemma Kunze ex Willk.
Hymenostephium Benth.
Hymenothrix A.Gray
Hymenoxys Cass.
Hyoseris L.
Hypacanthium Juz.
Hypelichrysum Kirp.
Hypericophyllum Steetz
Hypochaeris L.
Hysterionica Willd.
Hystrichophora Mattf.
Ichthyothere Mart.
Idiothamnus R.M.King & H.Rob.
Ifloga Cass.
Ighermia Wiklund *
Iltisia S.F.Blake

Imeria R.M.King & H.Rob.
Inezia E.Phillips
Inula L.
Inulanthera Källersjö
Iocenes B.Nord.
Iodocephalus Thorel ex Gagnep.
Iogeton Strother
Iostephane Benth.
Iphiona Cass.
Iphionopsis Anderb.
Iranecio B.Nord.
Irwinia Barroso
Ischnea F.Muell.
Isocarpha R.Br.
Isocoma Nutt.
Isoetopsis Turcz.
Isopappus Torr. & A.Gray
Isostigma Less.
Iva L.
Ixeridium (A.Gray) Tzvelev
Ixeris (Cass.) Cass.
Ixiochlamys F.Muell. & Sond.
Ixiolaena Benth.
Ixodia R.Br.
Jacmaia B.Nord.
Jaegeria Kunth
Jalcophila Dillon & Sagast.
Jaliscoa S.Watson
Jamesianthus S.F.Blake & Sherff
Jaramilloa R.M.King & H.Rob.
Jasonia (Cass.) Cass.
Jaumea Pers.
Jefea Strother
Jeffreya Wild
Jungia L.f.
Jurinea Cass.
Kalimeris (Cass.) Cass.
Karelinia Less.
Karvandarina Rech.f.
Kaschgaria Poljakov
Kaunia R.M.King & H.Rob.
Keysseria Lauterb.
Kinghamia C.Jeffrey
Kingianthus H.Rob.
Kippistia F.Muell.
Kirkianella Allan
Kleinia Mill.
Koanophyllon Arruda
Koehneola Urb.
Koelpinia Pall.
Krigia Schreb.
Kyrsteniopsis R.M.King & H.Rob.
Lachanodes DC.
Lachnophyllum Bunge
Lachnorhiza A.Rich.

Lachnospermum Willd.
Lactuca L.
Lactucosonchus (Sch.Bip.) Svent.
Laennecia Cass.
Laestadia Kunth ex Less.
Lagascea Cav.
Lagenophora Cass.
Laggera Sch.Bip. ex Benth.
Lagophylla Nutt.
Lamprachaenium Benth.
Lamprocephalus B.Nord.
Lamyropappus Knorring & Tamamsch.
Lamyropsis (Charadze) Dittrich
Langebergia Anderb.
Lantanopsis C.Wright
Lapsana L.
Lasianthaea DC.
Lasiocephalus Schltdl.
Lasiolaena R.M.King & H.Rob.
Lasiopogon Cass.
Lasiospermum Lag.
Lasthenia Cass.
Launaea Cass.
Lawrencella Lindl.
Layia Hook. & Arn. ex DC.
Lecocarpus Decne.
Leibnitzia Cass.
Leiboldia Schltdl. ex Gleason
Lembertia Greene
Lemooria P.S.Short
Leontodon L.
Leontopodium (Pers.) R.Br. ex Cass.
Lepidesmia Klatt
Lepidolopha C.Winkl.
Lepidolopsis Poljakov
Lepidonia S.F.Blake
Lepidophorum Neck. ex DC.
Lepidophyllum Cass.
Lepidospartum (A.Gray) A.Gray
Lepidostephium Oliv.
Leptinella Cass.
Leptocarpha DC.
Leptoclinium (Nutt.) Benth.
Leptorhynchos Less.
Leptotriche Turcz.
Lescaillea Griseb.
Lessingia Cham.
Leucactinia Rydb.
Leucanthemella Tzvelev
Leucanthemopsis (Giroux) Heywood
Leucanthemum Mill.
Leucheria Lag.
Leucoblepharis Arn.
Leucocyclus Boiss.
Leucogenes Beauverd

Leucopsis (DC.) Baker
Leucoptera B.Nord.
Leunisia Phil.
Leuzea DC.
Leysera L.
Liabum Adans.
Liatris Gaertn. ex Schreb.
Libanothamnus Ernst
Lidbeckia Bergius
Lifago Schweinf. & Muschler
Ligularia Cass.
Limbarda Adans.
Lindheimera A.Gray & Engelm.
Lipochaeta DC.
Lipskyella Juz.
Litothamnus R.M.King & H.Rob.
Litrisa Small
Llerasia Triana
Logfia Cass.
Lomatozona Baker
Lonas Adans.
Lopholaena DC.
Lophopappus Rusby
Lordhowea B.Nord.
Lorentzianthus R.M.King & H.Rob.
Loricaria Wedd.
Lourteigia R.M.King & H.Rob.
Loxothysanus B.L.Rob.
Lucilia Cass.
Luciliocline Anderb. & Freire
Lugoa DC.
Luina Benth.
Lulia Zardini
Lundellianthus H.Rob.
Lycapsus Phil.
Lychnophora Mart.
Lycoseris Cass.
Lygodesmia D.Don
Macdougalia A.Heller
Machaeranthera Nees
Macowania Oliv.
Macrachaenium Hook.f.
Macraea Hook.f.
Macropodina R.M.King & H.Rob.
Macvaughiella R.M.King & H.Rob.
Madia Molina
Mairia Nees
Malacothrix DC.
Malmeanthus R.M.King & H.Rob.
Malperia S.Watson
Mantisalca Cass.
Marasmodes DC.
Marshallia Schreb.
Marshalljohnstonia Henrickson
Marticorenia Crisci

Matricaria L.
Mattfeldanthus H.Rob. & R.M.King
Mattfeldia Urb. *
Matudina R.M.King & H.Rob.
Mausolea Poljakov
Mecomischus Coss. ex Benth.
Megalodonta Greene
Melampodium L.
Melanodendron DC.
Melanthera Rohr
Metalasia R.Br.
Metastevia Grashoff
Mexerion Nesom
Mexianthus B.L.Rob.
Micractis DC.
Microcephala Pobed.
Microglossa DC.
Microgynella Grau
Microliabum Cabrera
Micropsis DC.
Micropus L.
Microseris D.Don
Microspermum Lag.
Mikania Willd.
Mikaniopsis Milne-Redh.
Milleria L.
Millotia Cass.
Minuria DC.
Miricacalia Kitam.
Mniodes (A.Gray) Benth.
Monactis Kunth
Monarrhenus Cass.
Monenteles Labill.
Monogereion G.M.Barroso & R.M.King
Monolopia DC.
Monopholis S.F.Blake
Monoptilon Torr. & A.Gray
Montanoa Cerv.
Moonia Arn.
Moquinia DC.
Morithamnus R.M.King, H.Rob. &
  G.M.Barroso
Moscharia Ruíz & Pav.
Msuata O.Hoffm.
Mulgedium Cass.
Munnozia Ruíz & Pav.
Munzothamnus Raven *
Muschleria S.Moore
Mutisia L.f.
Mycelis Cass.
Myopordon Boiss.
Myriactis Less.
Myriocephalus Benth.
Myripnois Bunge
Myxopappus Källersjö

Nabalus Cass.
Nananthea DC.
Nannoglottis Maxim.
Nanothamnus Thomson
Nardophyllum (Hook. & Arn.) Hook. & Arn.
Narvalina Cass.
Nassauvia Comm. ex Juss.
Nauplius (Cass.) Cass.
Neblinaea Maguire & Wurdack
Nelsonianthus H.Rob. & Brettell
Nemosenecio (Kitam.) B.Nord.
Neocabreria R.M.King & H.Rob.
Neocuatrecasia R.M.King & H.Rob.
Neohintonia R.M.King & H.Rob.
Neojeffreya Cabrera
Neomirandea R.M.King & H.Rob.
Neopallasia Poljakov
Neotysonia Dalla Torre & Harms
Neurolaena R.Br.
Neurolakis Mattf.
Nicolasia S.Moore
Nicolletia A.Gray
Nidorella Cass.
Nikitinia Iljin
Nipponanthemum (Kitam.) Kitam.
Nolletia Cass.
Nothobaccharis R.M.King & H.Rob.
Nothocalais Greene
Noticastrum DC.
Notobasis (Cass.) Cass.
Notoptera Urb.
Notoseris C.Shih
Nouelia Franch.
Novenia Freire
Oaxacania B.L.Rob. & Greenm.
Oblivia Strother
Ochrocephala Dittrich
Odixia Orchard
Odontocline B.Nord.
Oedera L.
Oiospermum Less.
Oldenburgia Less.
Olearia Moench
Olgaea Iljin
Oligactis (Kunth) Cass.
Oliganthes Cass.
Oligochaeta (DC.) K.Koch
Oligothrix DC.
Olivaea Sch.Bip. ex Benth.
Omalotheca Cass.
Omphalopappus O.Hoffm.
Oncosiphon Källersjö
Ondetia Benth.
Onopordum L.

Onoseris Willd.
Oonopsis (Nutt.) Greene
Oparanthus Sherff
Ophryosporus Meyen
Opisthopappus C.Shih
Oreochrysum Rydb.
Oreoleysera Bremer
Oritrophium (Kunth) Cuatrec.
Orochaenactis Coville
Osbertia Greene
Osmiopsis R.M.King & H.Rob.
Osmitopsis Cass.
Osteospermum L.
Otanthus Hoffmanns. & Link
Oteiza La Llave
Othonna L.
Otopappus Benth.
Otospermum Willk.
Outreya Jaub. & Spach
Oxycarpha S.F.Blake
Oxylaena Benth. ex Anderb.
Oxylobus (Moq. ex DC.) A.Gray
Oxypappus Benth.
Oxyphyllum Phil.
Oyedaea DC.
Ozothamnus R.Br.
Pachylaena D.Don ex Hook. & Arn.
Pachythamnus (R.M.King & H.Rob.) R.M.King & H.Rob.
Packera Á.Löve & D.Löve
Pacourina Aubl.
Palaeocyanus Dostál
Palafoxia Lag.
Paleaepappus Cabrera
Pamphalea Lag.
Pappobolus S.F.Blake
Paracalia Cuatrec.
Paragynoxys (Cuatrec.) Cuatrec.
Paraixeris Nakai
Paranephelius Poepp.
Parantennaria Beauverd
Parapiqueria R.M.King & H.Rob.
Paraprenanthes Chang ex C.Shih
Parasenecio W.W.Sm. & Small
Parastrephia Nutt.
Parthenice A.Gray
Parthenium L.
Pasaccardoa Kuntze
Pechuel-Loeschea O.Hoffm.
Pectis L.
Pegolettia Cass.
Pelucha S.Watson
Pentacalia Cass.
Pentachaeta Nutt.
Pentanema Cass.

Pentatrichia Klatt
Pentzia Thunb.
Perdicium L.
Perezia Lag.
Pericallis D.Don
Pericome A.Gray
Perityle Benth.
Perralderia Coss.
Pertya Sch.Bip.
Perymeniopsis H.Rob.
Perymenium Schrad.
Petalacte D.Don
Petasites Mill.
Peteravenia R.M.King & H.Rob.
Petradoria Greene
Petrobium R.Br.
Peucephyllum A.Gray
Peyrousea DC.
Phacellothrix F.Muell.
Phaenocoma D.Don
Phaeostigma Muldashev
Phagnalon Cass.
Phalacrachena Iljin
Phalacraea DC.
Phalacrocarpum (DC.) Willk.
Phalacroseris A.Gray
Phaneroglossa B.Nord.
Phanerostylis (A.Gray) R.M.King &
  H.Rob.
Phania DC.
Philactis Schrad.
Philoglossa DC.
Philyrophyllum O.Hoffm.
Phoebanthus S.F.Blake
Phyllocephalum Blume
Phymaspermum Less.
Picnomon Adans.
Picris L.
Picrosia D.Don
Picrothamnus Nutt.
Pilosella Hill
Pilostemon Iljin *
Pinaropappus Less.
Pinillosia Ossa
Piora J.Kost.
Pippenalia McVaugh
Piptocarpha R.Br.
Piptocoma Cass.
Piptolepis Sch.Bip.
Piptothrix A.Gray
Piqueria Cav.
Piqueriella R.M.King & H.Rob.
Piqueriopsis R.M.King
Pithecoseris Mart. ex DC.
Pithocarpa Lindl.

Pittocaulon H.Rob. & Brettell
Pityopsis Nutt.
Pladaroxylon (Endl.) Hook.f.
Plagiobasis Schrenk
Plagiocheilus Arn. ex DC.
Plagiolophus Greenm.
Plagius L'Hér. ex DC.
Planaltoa Taub.
Planea Karis
Plateilema (A.Gray) Cockerell
Platycarpha Less.
Platypodanthera R.M.King & H.Rob.
Platyschkuhria (A.Gray) Rydb.
Plazia Ruíz & Pav.
Plecostachys Hilliard & B.L.Burtt
Plectocephalus D.Don
Pleiotaxis Steetz
Pleurocarpaea Benth.
Pleurocoronis R.M.King & H.Rob.
Pleuropappus F.Muell.
Pleurophyllum Hook.f.
Pluchea Cass.
Plummera A.Gray
Podachaenium Benth. ex Oerst.
Podanthus Lag.
Podocoma Cass.
Podolepis Labill.
Podotheca Cass.
Poecilolepis Grau
Pogonolepis Steetz
Pojarkovia Askerova
Pollalesta Kunth
Polyachyrus Lag.
Polyanthina R.M.King & H.Rob.
Polyarrhena Cass.
Polycalymma F.Muell. & Sond.
Polychrysum (Tzvelev) Kovalevsk.
Polymnia L.
Porophyllum Adans.
Porphyrostemma Benth. ex Oliv.
Praxeliopsis G.M.Barroso
Praxelis Cass.
Prenanthella Rydb.
Prenanthes L.
Printzia Cass.
Prionopsis Nutt.
Prolobus R.M.King & H.Rob.
Prolongoa Boiss.
Proteopsis Mart. & Zucc. ex Sch.Bip.
Proustia Lag.
Psacaliopsis H.Rob. & Brettell
Psacalium Cass.
Psathyrotes A.Gray
Psednotrichia Hiern
Pseudobaccharis Cabrera

Pseudobahia (A.Gray) Rydb.
Pseudoblepharispermum J.-P.Lebrun &
Stork
Pseudobrickellia R.M.King & H.Rob.
Pseudocadiscus Lisowski *
Pseudoclappia Rydb.
Pseudognaphalium Kirp.
Pseudogynoxys (Greenm.) Cabrera
Pseudohandelia Tzvelev
Pseudokyrsteniopsis R.M.King & H.Rob.
Pseudonoseris H.Rob. & Brettell
Pseudostifftia H.Rob.
Psiadia Jacq.
Psiadiella Humbert
Psilactis A.Gray
Psilocarphus Nutt.
Psilostrophe DC.
Psychrogeton Boiss.
Psychrophyton Beauverd
Pterachenia (Benth.) Lipschitz
Pterocaulon Elliott
Pterocypsela C.Shih
Pteronia L.
Pterothrix DC.
Pterygopappus Hook.f.
Ptilostemon Cass.
Pulicaria Gaertn.
Pycnocephalum (Less.) DC.
Pyrrhopappus DC.
Pyrrocoma Hook.
Quelchia N.E.Br.
Quinetia Cass.
Quinqueremulus Paul G.Wilson
Radlkoferotoma Kuntze *
Rafinesquia Nutt.
Raillardella (A.Gray) Benth.
Raillardiopsis Rydb.
Rainiera Greene
Raoulia Hook.f. ex Raoul
Raouliopsis S.F.Blake *
Rastrophyllum Wild & G.V.Pope
Ratibida Raf.
Raulinoreitzia R.M.King & H.Rob.
Reichardia Roth
Relhania L'Hér.
Remya W.F.Hillebr. ex Benth.
Rennera Merxm.
Rensonia S.F.Blake
Revealia R.M.King & H.Rob.
Rhagadiolus Scop.
Rhamphogyne S.Moore
Rhanteriopsis Rauschert
Rhanterium Desf.
Rhodanthe Lindl.
Rhodogeron Griseb.

Rhynchospermum Reinw.
Rhysolepis S.F.Blake
Richteria Kar. & Kir.
Riencourtia Cass.
Rigiopappus A.Gray
Robinsonia DC.
Rochonia DC.
Rojasianthe Standl. & Steyerm.
Rolandra Rottb.
Roldana La Llave
Rosenia Thunb.
Rothmaleria Font Quer
Rudbeckia L.
Rugelia Shuttlew. ex Chapm.
Ruilopezia Cuatrec.
Rumfordia DC.
Russowia C.Winkl.
Rutidosis DC.
Sabazia Cass.
Sachsia Griseb.
Salmea DC.
Salmeopsis Benth.
Santolina L.
Santosia R.M.King & H.Rob.
Sanvitalia Lam.
Sartorina R.M.King & H.Rob.
Sartwellia A.Gray
Saussurea DC.
Scalesia Arn.
Scariola F.W.Schmidt
Scherya R.M.King & H.Rob.
Schischkinia Iljin
Schistocarpha Less.
Schistostephium Less.
Schizogyne Cass.
Schizoptera Turcz.
Schizotrichia Benth.
Schkuhria Roth
Schlechtendalia Less.
Schmalhausenia C.Winkl.
Schoenia Steetz
Schumeria Iljin
Sciadocephala Mattf.
Sclerocarpus Jacq.
Sclerolepis Cass.
Sclerorhachis (Rech.f.) Rech.f.
Sclerostephane Chiov.
Scolymus L.
Scorzonella Nutt.
Scorzonera L.
Scrobicaria Cass.
Scyphocoronis A.Gray
Selleophytum Urb.
Selloa Kunth
Senecio L.

Sericocarpus Nees
Seriphidium (Bess.) Poljak.
Serratula L.
Shafera Greenm. *
Sheareria S.Moore
Shinnersia R.M.King & H.Rob.
Shinnersoseris Tomb
Siebera J.Gay
Sigesbeckia L.
Siloxerus Labill.
Silphium L.
Silybum Adans.
Simsia Pers.
Sinacalia H.Rob. & Brettell
Sinclairia Hook. & Arn.
Sinoleontopodium Y.L.Chen
Sinosenecio B.Nord.
Sipolisia Glaz. ex Oliv.
Smallanthus Mack.
Soaresia Sch.Bip.
Solanecio (Sch.Bip.) Walp.
Solenogyne Cass.
Solidago L.
Soliva Ruíz & Pav.
Sommerfeltia Less.
Sonchus L.
Sondottia P.S.Short
Soroseris Stebb.
Spaniopappus B.L.Rob.
Sphaeranthus L.
Sphaereupatorium (O.Hoffm.) Kuntze
ex B.L.Rob.
Sphaeroclinium (DC.) Sch.Bip.
Sphaeromeria Nutt.
Sphaeromorphaea DC.
Sphagneticola O.Hoffm.
Spilanthes Jacq.
Spiracantha Kunth
Spiroseris Rech.f.
Squamopappus Jansen, Harriman &
Urbat
Stachycephalum Sch.Bip. ex Benth.
Staehelina L.
Standleyanthus R.M.King & H.Rob.
Stanfieldia Small
Staurochlamys Baker
Stebbinsoseris K.L.Chambers
Steiractinia S.F.Blake
Steirodiscus Less.
Stenachaenium Benth.
Stenocarpha S.F.Blake
Stenocline DC.
Stenopadus S.F.Blake
Stenophalium Anderb.
Stenops B.Nord.

Stenotus Nutt.
Stephanochilus Coss. & Durieu ex Maire
Stephanodoria Greene
Stephanomeria Nutt.
Steptorhamphus Bunge
Stevia Cav.
Steviopsis R.M.King & H.Rob.
Steyermarkina R.M.King & H.Rob.
Stifftia J.C.Mikan
Stilpnogyne DC.
Stilpnolepis I.M.Kraschen.
Stilpnopappus Mart. ex DC.
Stoebe L.
Stokesia L'Hér.
Stomatanthes R.M.King & H.Rob.
Stomatochaeta (S.F.Blake) Maguire &
Wurdack
Stramentopappus H.Rob. & V.A.Funk
Streptoglossa Steetz ex F.Muell.
Strotheria B.L.Turner
Struchium P.Browne
Stuartina Sond.
Stuckertiella Beauverd
Stuessya B.L.Turner & F.G.Davies
Stylocline Nutt.
Stylotrichium Mattf.
Sventenia Font Quer
Symphyllocarpus Maxim.
Symphyopappus Turcz.
Syncalathium Lipsch.
Syncarpha DC.
Syncephalum DC.
Syncretocarpus S.F.Blake
Synedrella Gaertn.
Synedrellopsis Hieron. & Kuntze
Syneilesis Maxim.
Synotis (C.B.Clarke) C.Jeffrey &
Y.L.Chen
Syntrichopappus A.Gray
Synurus Iljin
Syreitschikovia Pavlov
Tagetes L.
Tamananthus Badillo *
Tamania Cuatrec.
Tamaulipa R.M.King & H.Rob.
Tanacetum L.
Taplinia Lander
Taraxacum G.H.Weber ex Wigg.
Tarchonanthus L.
Teixeiranthus R.M.King & H.Rob.
Telanthophora H.Rob. & Brettell
Telekia Baumg.
Telmatophila Mart. ex Baker
Tenrhynea Hilliard & B.L.Burtt
Tephroseris (Rchb.) Rchb.

Terana La Llave *
Tessaria Ruíz & Pav.
Tetrachyron Schltdl.
Tetradymia DC.
Tetragonotheca L.
Tetramolopium Nees
Tetraneuris Greene
Tetranthus Sw.
Tetraperone Urb.
Thaminophyllum Harv.
Thamnoseris F.Phil.
Thelesperma Less.
Thespidium F.Muell. ex Benth.
Thespis DC.
Thevenotia DC.
Thiseltonia Hemsl.
Thymophylla Lag.
Thymopsis Benth.
Tiarocarpus Rech.f.
Tietkensia P.S.Short
Tithonia Desf. ex Juss.
Tolpis Adans.
Tonestus A.Nelson
Tourneuxia Coss.
Townsendia Hook.
Toxanthes Turcz.
Tracyina S.F.Blake
Tragopogon L.
Traversia Hook.f.
Trichanthemis Regel & Schmalh.
Trichanthodium Sond. & F.Muell.
Trichocline Cass.
Trichocoronis A.Gray
Trichocoryne S.F.Blake *
Trichogonia (DC.) Gardner
Trichogoniopsis R.M.King & H.Rob.
Trichogyne Less.
Tricholepis DC.
Trichoptilium A.Gray
Trichospira Kunth
Tridactylina (DC.) Sch.Bip.
Tridax L.
Trigonospermum Less.
Trilisa (Cass.) Cass.
Trioncinia (F.Muell.) Veldkamp
Tripleurospermum Sch.Bip.
Triplocephalum O.Hoffm.
Triptilion Ruíz & Pav.
Triptilodiscus Turcz.
Trixis P.Browne
Troglophyton Hilliard & B.L.Burtt
Tuberostylis Steetz
Tugarinovia Iljin
Turaniphytum Poljakov
Tussilago L.

Tuxtla Villaseñor & Strother
Tyleropappus Greenm. *
Tyrimnus (Cass.) Cass.
Uechtritzia Freyn
Ugamia Pavlov *
Uleophytum Hieron.
Unxia L.f.
Urbananthus R.M.King & H.Rob.
Urbinella Greenm.
Urmenetea Phil.
Urolepis (DC.) R.M.King & H.Rob.
Uropappus Nutt.
Urospermum Scop.
Ursinia Gaertn.
Vanclevea Greene
Vanillosmopsis Sch.Bip.
Varilla A.Gray
Varthemia DC.
Vellereophyton Hilliard & B.L.Burtt
Venegasia DC.
Verbesina L.
Vernonia Schreb.
Vernoniopsis Humbert
Vieraea Sch.Bip.
Viereckia R.M.King & H.Rob.
Vigethia W.A.Weber
Viguiera Kunth
Villanova Lag.
Vilobia Strother
Vittadinia A.Rich.
Vittetia R.M.King & H.Rob.
Volutaria Cass.
Wagenitzia Dostál
Waitzia J.C.Wendl.
Wamalchitamia Strother
Warionia Benth. & Coss.
Wedelia Jacq.
Welwitschiella O.Hoffm.
Werneria Kunth
Westoniella Cuatrec.
Whitneya A.Gray
Wilkesia A.Gray
Wollastonia DC. ex Decne.
Wulffia Neck. ex Cass.
Wunderlichia Riedel ex Benth.
Wyethia Nutt.
Xanthisma DC.
Xanthium L.
Xanthocephalum Willd.
Xanthopappus C.Winkl.
Xeranthemum L.
Xerolekia Anderb.
Xylanthemum Tzvelev
Xylothamia G.L.Nesom, Y.B.Suh,
    D.R.Morgan & B.B.Simpson

Yermo Dorn *	**Zexmenia** La Llave
Youngia Cass.	**Zinnia** L.
Zaluzania Pers.	**Zoegea** L.
Zandera D.L.Schulz	**Zyzyxia** Strother

CONNARACEAE R.Br 1818   56

12 genera. Widespread in tropics. Trees, shrubs, lianes.

B&H	POLYPETALAE, CALYCIFLORAE	Rosales, 56
DT&H	ARCHICHLAMYDEAE	Rosales, Rosineae, 77
Melc	ARCHICHLAMYDEAE	Rosales, Leguminosineae, 117
Thor	RUTIFLORAE	Rutales, Fabineae, 190
Dahl	RUTIFLORAE	Sapindales, 217
Young	ROSIDAE, RUTANAE	Fabales, 265
Takh	ROSIDAE, RUTANAE	Connarales, 242
Cron	ROSIDAE	Rosales, 157

Agelaea Sol. ex Planch.	**Hemandradenia** Stapf
Burttia Baker f. & Exell	**Jollydora** Pierre ex Gilg
Cnestidium Planch.	**Manotes** Sol. ex Planch.
Cnestis Juss.	**Pseudoconnarus** Radlk.
Connarus L.	**Rourea** Aubl.
Ellipanthus Hook.f.	**Vismianthus** Mildbr.

CONVOLVULACEAE Juss. 1789   113.01

55 genera. Widespread. Mostly herbs, often climbing, rarely trees (*Humbertia*).

B&H	GAMOPETALAE, BICARPELLATAE	Polemoniales, 113
DT&H	METACHLAMYDEAE	Tubiflorae, Convolvulineae, 202
Melc	SYMPETALAE	Tubiflorae, Convolvulineae, 254
Thor	SOLANIFLORAE	Solanales, Solanineae, 272
Dahl	SOLANIFLORAE	Solanales, 276
	(separating *Cuscutaceae* 277)	
Young	ROSIDAE, SOLANANAE	Solanales, 364
Takh	LAMIIDAE, SOLANANAE	Convolvulales, 384
	(separating *Cuscutaceae* 385)	
Cron	ASTERIDAE	Solanales, 279
	(separating *Cuscutaceae* 280)	

Aniseia Choisy	**Cressa** L.
Argyreia Lour.	**Cuscuta** L.
Astripomoea A.D.J.Meeuse	**Decalobanthus** Ooststr. *
Blinkworthia Choisy	**Dichondra** J.R.Forst. & G.Forst.
Bonamia Thouars	**Dicranostyles** Benth.
Calycobolus Willd. ex J.A.Schult.	**Dipteropeltis** Hallier f.
Calystegia R.Br.	**Erycibe** Roxb.
Cardiochlamys Oliv.	**Evolvulus** L.
Cladostigma Radlk.	**Falkia** L.f.
Convolvulus L.	**Hewittia** Wight & Arn.
Cordisepalum Verdc.	**Hildebrandtia** Vatke

Humbertia Comm. ex Lam.
Hyalocystis Hallier f.
Ipomoea L.
Iseia O'Don.
Itzaea Standl. & Steyerm.
Jacquemontia Choisy
Lepistemon Blume
Lepistemonopsis Dammer
Lysiostyles Benth.
Maripa Aubl.
Merremia Dennst. ex Endl.
Metaporana N.E.Br.
Nephrophyllum A.Rich.
Neuropeltis Wall.
Neuropeltopsis Ooststr.
Odonellia K.R.Robertson
Operculina Silva Manso

Paralepistemon Lejoly & Lisowski
Pentacrostigma K.Afzel.
Petrogenia I.M.Johnst.
Polymeria R.Br.
Porana Burm.f.
Rapona Baill.
Remirema Kerr
Rivea Choisy
Sabaudiella Chiov.
Seddera Hochst.
Stictocardia Hallier f.
Stylisma Raf.
Tetralocularia O'Don.
Turbina Raf.
Wilsonia R.Br.
Xenostegia D.F.Austin & Staples

CORIARIACEAE DC. 1824   54

1 genus. Medit., trop. and E. Asia, New Zeal., Pacif., Mexico to Chile. Mostly shrubs.

B&H	POLYPETALAE, DISCIFLORAE	Ordines anomala, 54
DT&H	ARCHICHLAMYDEAE	Sapindales, Coriariineae, 100
Melc	ARCHICHLAMYDEAE	Sapindales, Coriariineae, 143
Thor	RUTIFLORAE	Rutales, Coriariineae, 179
Dahl	RUTIFLORAE	Sapindales, 203
Young	ROSIDAE, RUTANAE	Rutales, 246
Takh	ROSIDAE, RUTANAE	Coriariales, 273
Cron	MAGNOLIIDAE	Ranunculales, 36

Coriaria L.

CORNACEAE (Dumort.) Dumort. 1829   82.01

8 genera. N. temp. region to trop. Asia, Africa, rare S. America. Mostly trees, shrubs.

B&H   POLYPETALAE, CALYCIFLORAE
      (including Alangiaceae, Aucubaceae, Garryaceae, Griseliniaceae,
      Torricelliaceae)   Umbellales, 82
DT&H ARCHICHLAMYDEAE   Umbelliflorae, 181
      (including Alangiaceae, Aucubaceae, Garryaceae, Griseliniaceae,
      Helwingiaceae, Melanophyllaceae, Torricelliaceae)
Melc   ARCHICHLAMYDEAE   Umbelliflorae, 223
      (including Aucubaceae, Griseliniaceae, Helwingiaceae, Melanophyllaceae,
      Torricelliaceae; separating *Davidiaceae, Nyssaceae*)
Thor   CORNIFLORAE   Cornales, Cornineae, 284
      (separating *Nyssaceae* 283)
Dahl   CORNIFLORAE   Cornales, 301
      (separating *Davidiaceae* 302, *Nyssaceae* 300)
Young ROSIDAE, CORNANAE   Cornales, 325
      (probably including several families; separating *Davidiaceae* 324, *Nyssaceae* 323)

CORNACEAE (D)

Takh ROSIDAE, CORNANAE   Cornales, 342
      (separating *Curtisiaceae* 343, *Davidiaceae* 340, *Mastixiaceae* 344, *Nyssaceae* 341)
Cron ROSIDAE   Cornales, 203
      (including Aralidiaceae, Aucubaceae, Griseliniaceae, Helwingiaceae,
      Melanophyllaceae, Torricelliaceae; separating *Nyssaceae* 202 in 1981 but
      including it in 1988)

Afrocrania (Harms) Hutch.   **Davidia** Baill.
Camptotheca Decne.   **Diplopanax** Hand.–Mazz. *
Cornus L.   **Mastixia** Blume
Curtisia Aiton   **Nyssa** Gronov. ex L.

CORYLACEAE Mirb. 1815   159.02

4 genera. North temperate. Mostly trees.

B&H   MONOCHLAMYDEAE   Unisexuales (within Betulaceae, 159)
DT&H ARCHICHLAMYDEAE   Fagales (within Betulaceae, 11)
Melc ARCHICHLAMYDEAE   Fagales (within Betulaceae, 8)
Thor HAMAMELIDIFLORAE   Fagales (within Betulaceae, 202)
Dahl ROSIFLORAE   Fagales, 147
Young ROSIDAE, HAMAMELIDANAE Fagales (presumably within Betulaceae, 227)
Takh HAMAMELIDIDAE, HAMAMELIDANAE
                              Betulales (within Betulaceae, 94)
Cron HAMAMELIDIDAE   Fagales (within Betulaceae, 62)

Carpinus L.
Corylus L.
Ostrya Scop.
Ostryopsis Decne.

CORYNOCARPACEAE Engl. 1897   53.03

1 genus. New Guinea, Australia, New Zealand, S.W. Pacific. Trees, shrubs.

B&H   POLYPETALAE, DISCIFLORAE Sapindales (within Anacardiaceae, 53)
DT&H ARCHICHLAMYDEAE   Sapindales, Celastrineae, 106
Melc ARCHICHLAMYDEAE   Celastrales, Celastrineae, 157
Thor ROSIFLORAE   Rosales, Cunoniineae, 221
Dahl SANTALIFLORAE   Celastrales, 253
Young ROSIDAE, SANTALANAE   Celastrales, 232
Takh ROSIDAE, CELASTRANAE   Celastrales, Celastrineae, 309
Cron ROSIDAE   Cclastrales, 227

Corynocarpus J.R.Forst. & G.Forst.

CRASSULACEAE DC. 1805 60.01

37 genera. Widespread, except Australia. Succulent herbs, subshrubs.

B&H   POLYPETALAE, CALYCIFLORAE Rosales, 60
        (including Penthoraceae)
DT&H ARCHICHLAMYDEAE   Rosales, Saxifragineae, 65
        (including Penthoraceae)
Melc   ARCHICHLAMYDEAE   Rosales, Saxifragineae, 104
Thor   ROSIFLORAE   Rosales, Saxifragineae, 206
Dahl   ROSIFLORAE   Saxifragales, 169
Young ROSIDAE, ROSANAE   Saxifragales, 294
Takh   ROSIDAE, ROSANAE   Saxifragales, 203
Cron   ROSIDAE   Rosales, 171

Adromischus Lem.
Aeonium Webb & Berthel.
Aichryson Webb & Berthel.
Bryophyllum Salisb.
Chiastophyllum (Ledeb.) A.Berger
Cotyledon L.
Crassula L.
Cremnophila Rose
Diamorpha Nutt.
Dudleya Britton & Rose
Echeveria DC.
Graptopetalum Rose
Greenovia Webb & Berthel.
Hylotelephium H.Ohba
Hypagophytum A.Berger
Kalanchoe Adans.
Lenophyllum Rose
Meterostachys Nakai
Monanthes Haw.

Mucizonia (DC.) A.Berger
Orostachys Fisch. ex A.Berger
Pachyphytum Link, Klotzsch & Otto
Perrierosedum (A.Berger) H.Ohba
Pistorinia DC.
Prometheum (A.Berger) H.Ohba
Pseudosedum (Boiss.) A.Berger
Rhodiola L.
Rosularia (DC.) Stapf
Sedella Britton & Rose
Sedum L.
Sempervivum L.
Sinocrassula A.Berger
Telmissa Fenzl
Thompsonella Britton & Rose
Tylecodon Tölken
Umbilicus DC.
Villadia Rose

CROSSOSOMATACEAE Engl. 1897 2.02

3 genera. S.W. USA, Mexico. Shrubs.

B&H   POLYPETALAE, THALAMIFLORAE Ranales (within Dilleniaceae, 2)
DT&H ARCHICHLAMYDEAE   Rosales, Rosineae, 75
Melc   ARCHICHLAMYDEAE   Guttiferales, Dilleniineae, 77
Thor   ROSIFLORAE   Rosales, Rosineae, 205
Dahl   ROSIFLORAE   Rosales, 178
Young ROSIDAE, ROSANAE   Rosales, 293
Takh   ROSIDAE, ROSANAE   Crossosomatales, 220
Cron   ROSIDAE   Rosales, 176

Apacheria C.T.Mason *
Crossosoma Nutt.
Glossopetalon A.Gray

CRUCIFERAE Juss. 1789   11

381 genera. Widespread, mostly temp. and warm-temp. Herbs, some shrubs.

B&H	POLYPETALAE, THALAMIFLORAE	Parietales, 11
DT&H	ARCHICHLAMYDEAE	Rhoeadales, Capparidineae, 55
Melc	ARCHICHLAMYDEAE	Papaverales, Capparineae, 96
Thor	VIOLIFLORAE	Capparales, 143
Dahl	VIOLIFLORAE	Capparales, 101
Young	DILLENIIDAE, VIOLANAE	Capparales, 213
Takh	DILLENIIDAE, VIOLANAE	Capparales, Capparineae, 164
Cron	DILLENIIDAE	Capparales, 134

Acanthocardamum Thell. *
Achoriphragma Soják
Aethionema R.Br.
Agallis Phil. *
Agianthus Greene
Alliaria Heist. ex Fabr.
Alyssoides Mill.
Alyssopsis Boiss.
Alyssum L.
Ammosperma Hook.f.
Anastatica L.
Anchonium DC.
Andrzeiowskya Rchb.
Anelsonia J.F.Macbr. & Payson
Aphragmus Andrz. ex DC.
Aplanodes Marais
Arabidella (F.Muell.) O.E.Schulz
Arabis L.
Arcyosperma O.E.Schulz
Armoracia P.Gaertn., B.Mey. & Scherb.
Aschersoniodoxa Gilg & Muschler
Asperuginoides Rauschert
Asta Klotzsch ex O.E.Schulz
Atelanthera Hook.f. & Thomson
Athysanus Greene
Aubrieta Adans.
Aurinia Desv.
Ballantinia Hook.f. ex E.A.Shaw
Barbarea R.Br.
Berteroa DC.
Berteroella O.E.Schulz
Biscutella L.
Bivonaea DC.
Blennodia R.Br.
Boleum Desv.
Boreava Jaub. & Spach
Bornmuellera Hausskn.
Borodinia N.Busch *
Botschantzevia Nabiev
Brachycarpaea DC.
Brassica L.
Braya Sternb. & Hoppe

Brayopsis Gilg & Muschler
Brossardia Boiss.
Bunias L.
Cakile Mill.
Calepina Adans.
Calymmatium O.E.Schulz
Camelina Crantz
Camelinopsis A.G.Mill. *
Capsella Medik.
Cardamine L.
Carinavalva Ising *
Carrichtera DC.
Cartiera Greene
Catadysia O.E.Schulz *
Catenulina Soják *
Caulanthus S.Watson
Caulostramina Rollins
Ceratocnemum Coss. & Balansa
Ceriosperma (O.E.Schulz) Greuter & Burdet
Chalcanthus Boiss.
Chamira Thunb.
Chartoloma Bunge
Cheesemania O.E.Schulz
Chilocardamum O.E.Schulz
Chlorocrambe Rydb.
Chorispora R.Br. ex DC.
Christolea Cambess. ex Jacquem.
Chrysobraya Hara *
Chrysochamela (Fenzl) Boiss.
Cithareloma Bunge
Clastopus Bunge ex Boiss.
Clausia Trotzky ex Hayek
Clypeola L.
Cochlearia L.
Cochleariella Y.H.Zhang & R.Vogt *
Coelonema Maxim.
Coelophragmus O.E.Schulz
Coincya Rouy
Coluteocarpus Boiss.
Conringia Heist. ex Fabr.
Cordylocarpus Desf.

Coronopus Zinn
Crambe L.
Crambella Maire
Cremolobus DC.
Cryptospora Kar. & Kir.
Cuphonotus O.E.Schulz
Cusickiella Rollins
Cycloptychis E.Mey. ex Sond.
Cymatocarpus O.E.Schulz
Cyphocardamum Hedge *
Dactylocardamum Al–Shehbaz *
Decaptera Turcz. *
Degenia Hayek
Delpinophytum Speg.
Descurainia Webb & Berthel.
Diceratella Boiss.
Dichasianthus Ovcz. & Yanusov *
Dictyophragmus O.E.Schulz
Didesmus Desv.
Didymophysa Boiss.
Dielsiocharis O.E.Schulz
Dilophia Thomson
Dimorphocarpa Rollins
Dimorphostemon Kitag.
Diplotaxis DC.
Dipoma Franch.
Diptychocarpus Trautv.
Disaccanthus Greene
Discovium Raf.
Dithyrea Harv.
Dolichorhynchus Hedge & Kit Tan
Dontostemon Andrz. ex C.A.Mey.
Douepea Cambess.
Draba L.
Drabastrum (F.Muell.) O.E.Schulz
Drabopsis K.Koch
Dryopetalon A.Gray
Eigia Soják *
Elburzia Hedge
Enarthrocarpus Labill.
Englerocharis Muschler
Eremobium Boiss.
Eremoblastus Botsch. *
Eremodraba O.E.Schulz
Eremophyton Bég.
Ermaniopsis Hara *
Erophila DC.
Eruca Mill.
Erucaria Gaertn.
Erucastrum (DC.) C.Presl
Erysimum L.
Euclidium R.Br.
Eudema Humb. & Bonpl.
Euklisia Rydb. ex Small

Eurycarpus Botsch.
Eutrema R.Br.
Euzomodendron Coss.
Farsetia Turra
Fezia Pit.
Fibigia Medik.
Foleyola Maire *
Fortuynia Shuttlew. ex Boiss.
Galitzkya V.V.Botschantzeva
Geococcus J.L.Drumm. ex Harv.
Glaribraya Hara
Glastaria Boiss.
Glaucocarpum Rollins
Goldbachia DC.
Gorodkovia Botsch. & Karav. *
Graellsia Boiss.
Grammosperma O.E.Schulz
Guillenia Greene
Guiraoa Coss.
Gynophorea Gilli
Halimolobos Tausch
Harmsiodoxa O.E.Schulz
Hedinia Ostenf.
Hediniopsis Botsch. & Petrovsky *
Heldreichia Boiss.
Heliophila Burm.f. ex L.
Hemicrambe Webb
Hemilophia Franch.
Henophyton Coss. & Durieu
Hesperis L.
Heterodraba Greene
Hirschfeldia Moench
Hollermayera O.E.Schulz *
Hormathophylla Cullen & T.R.Dudley
Hornungia Rchb.
Horwoodia Turrill
Hugueninia Rchb.
Hutchinsiella O.E.Schulz
Hymenolobus Nutt.
Iberis L.
Icianthus Greene
Idahoa A.Nelson & J.F.Macbr.
Iodanthus (Torr. & A.Gray) Steud.
Ionopsidium Rchb.
Irenepharsus Hewson *
Isatis L.
Ischnocarpus O.E.Schulz
Iskandera N.Busch
Iti Garn.–Jones & P.N.John *
Ivania O.E.Schulz
Kardanoglyphos Schltdl.
Kernera Medik.
Kremeriella Maire
Lachnocapsa Balf.f.

Lachnoloma Bunge
Leavenworthia Torr.
Leiospora (C.A.Mey.) Vassilieva
Lepidium L.
Lepidostemon Hook.f. & Thomson
Lepidotrichum Velen. & Bornm.
Leptaleum DC.
Lesquerella S.Watson
Lignariella Baehni
Lobularia Desv.
Loxoptera O.E.Schulz *
Loxostemon Hook.f. & Thomson
Lunaria L.
Lutzia Gand.
Lycocarpus O.E.Schulz
Lyrocarpa Hook. & Harv.
Macropodium R.Br.
Malcolmia R.Br.
Mancoa Wedd.
Maresia Pomel
Mathewsia Hook. & Arn.
Matthiola R.Br.
Megacarpaea DC.
Megadenia Maxim.
Menkea Lehm.
Menonvillea R.Br. ex DC.
Microcardamum O.E.Schulz
Microlepidium F.Muell.
Microsemia Greene
Microsisymbrium O.E.Schulz
Mitophyllum Greene
Morettia DC.
Moricandia DC.
Moriera Boiss.
Morisia J.Gay
Mostacillastrum O.E.Schulz
Murbeckiella Rothm. *
Muricaria Desv.
Myagrum L.
Nasturtiopsis Boiss.
Neomartinella Pilg.
Neotchihatchewia Rauschert
Neotorularia Hedge & J.Léonard
Nerisyrenia Greene
Neslia Desv.
Neuontobotrys O.E.Schulz
Notoceras R.Br.
Notothlaspi Hook.f.
Ochthodium DC.
Octoceras Bunge
Onuris Phil.
Oreoloma Botsch. *
Oreophyton O.E.Schulz
Ornithocarpa Rose

Orychophragmus Bunge
Otocarpus Durieu
Pachycladon Hook.f.
Pachymitus O.E.Schulz
Pachyneurum Bunge
Pachyphragma (DC.) Rchb.
Pachypterygium Bunge
Parlatoria Boiss.
Parodiodoxa O.E.Schulz *
Parolinia Webb
Parrya R.Br.
Parryodes Jafri
Parryopsis Botsch.
Pegaeophyton Hayek & Hand.–Mazz.
Peltaria Jacq.
Peltariopsis (Boiss.) N.Busch
Pennellia Nieuwl.
Petiniotia J.Léonard
Petrocallis R.Br.
Phaeonychium O.E.Schulz
Phlebiophragmus O.E.Schulz
Phlebolobium O.E.Schulz
Phlegmatospermum O.E.Schulz
Phoenicaulis Nutt.
Phryne Bubani
Physaria (Nutt.) A.Gray
Physocardamum Hedge *
Physoptychis Boiss.
Physorhynchus Hook.
Platycraspedum O.E.Schulz
Pleiocardia Greene
Pleurophragma Rydb.
Poliophyton O.E.Schulz
Polyctenium Greene
Polypsecadium O.E.Schulz
Pringlea T.Anderson ex Hook.f.
Prionotrichon Botsch. & Vved.
Pritzelago Kuntze
Pseuderucaria (Boiss.) O.E.Schulz
Pseudocamelina (Boiss.) N.Busch
Pseudoclausia Popov
Pseudofortuynia Hedge
Pseudovesicaria (Boiss.) Rupr.
Psychine Desf.
Pterygiosperma O.E.Schulz
Pterygostemon V.V.Botschantzeva
Pugionium Gaertn.
Pycnoplinthopsis Jafri
Pycnoplinthus O.E.Schulz
Quezeliantha H.Scholz ex Rauschert
Quidproquo Greuter & Burdet
Raffenaldia Godr.
Raphanorhyncha Rollins *
Raphanus L.

Rapistrum Crantz
Reboudia Coss. & Durieu
Redowskia Cham. & Schltdl. *
Rhammatophyllum O.E.Schulz
Rhizobotrya Tausch
Ricotia L.
Robeschia Hochst. ex O.E.Schulz
Rollinsia Al-Shehbaz *
Romanschulzia O.E.Schulz
Roripella (Maire) Greuter & Burdet
Rorippa Scop.
Rytidocarpus Coss.
Sameraria Desv.
Sarcodraba Gilg & Muschler *
Savignya DC.
Scambopus O.E.Schulz
Schimpera Steud. & Hochst. ex Endl.
Schivereckia Andrz. ex DC.
Schizopetalon Sims
Schlechteria Bolus
Schoenocrambe Greene *
Schouwia DC.
Scoliaxon Payson
Selenia Nutt.
Sibara Greene
Silicularia Compton
Sinapidendron Lowe
Sinapis L.
Sisymbrella Spach
Sisymbriopsis Botsch. & Tzvelev *
Sisymbrium L.
Skottsbergianthus Boelcke
Smelowskia C.A.Mey.
Sobolewskia M.Bieb.
Solms-Laubachia Muschler
Sophiopsis O.E.Schulz
Sphaerocardamum Nees & Schauer
Spirorhynchus Kar. & Kir.
Spryginia Popov
Staintoniella Hara
Stanleya Nutt.
Stenopetalum R.Br. ex DC.

Sterigmostemum M.Bieb.
Stevenia Adams ex Fisch.
Straussiella Hausskn.
Streptanthella Rydb.
Streptanthus Nutt.
Streptoloma Bunge
Stroganowia Kar. & Kir.
Stubendorffia Schrenk ex Fisch.,
    C.A.Mey. & Avé-Lall.
Subularia L.
Succowia Medik.
Synstemon Botsch.
Synthlipsis A.Gray
Syrenia Andrz. ex Besser
Taphrospermum C.A.Mey.
Tauscheria Fisch. ex DC.
Teesdalia R.Br.
Teesdaliopsis (Willk.) Rothm.
Tetracme Bunge
Thellungiella O.E.Schulz
Thelypodiopsis Rydb. *
Thelypodium Endl.
Thlaspeocarpa C.A.Sm.
Thlaspi L.
Thysanocarpus Hook.
Trachystoma O.E.Schulz
Trichotolinum O.E.Schulz *
Trochiscus O.E.Schulz
Tropidocarpum Hook.
Urbanodoxa Muschler
Vella DC.
Veselskya Opiz
Vvedenskyella Botsch.
Warea Nutt.
Wasabia Matsum.
Weberbauera Gilg & Muschler
Werdermannia O.E.Schulz
Winklera Regel
Xerodraba Skottsb.
Zerdana Boiss.
Zilla Forssk.

CRYPTERONIACEAE A.DC. 1868   69.02

3 genera. Tropical Asia. Trees, shrubs.

B&H   POLYPETALAE, CALYCIFLORAE Myrtales (within Lythraceae, 69)
DT&H ARCHICHLAMYDEAE   Myrtiflorae, Myrtineae, 168
Melc   ARCHICHLAMYDEAE   Myrtiflorae, Myrtineae, 205
Thor   MYRTIFLORAE   Myrtales, Lythrineae, 242
Dahl   MYRTIFLORAE   Myrtales, 194
Young Not mentioned   —

CRYPTERONIACEAE (D)

Takh   ROSIDAE, MYRTANAE   Myrtales, Myrtineae, 232
Cron   ROSIDAE   Myrtales, 191

Axinandra Thwaites
Crypteronia Blume
Dactylocladus Oliv.

CTENOLOPHONACEAE (H.Winkl.) Exell & Mendonça 1951   34.05

1 genus. W. tropical Africa, Malesia. Trees.

B&H   Not known   —
DT&H   ARCHICHLAMYDEAE   Geraniales, Geraniineae (within Linaceae, 82)
Melc   ARCHICHLAMYDEAE   Geraniales, Geraniineae (within Linaceae, 127)
Thor   GERANIIFLORAE   Geraniales, Linineae, 96
Dahl   RUTIFLORAE   Geraniales, 237
Young (not mentioned, presumably within Linaceae, 270)
Takh   ROSIDAE, RUTANAE   Linales, 276
Cron   ROSIDAE   Linales (within *Hugoniaceae,* 239)

Ctenolophon Oliv.

CUCURBITACEAE Juss. 1789   75

120 genera. Widespread. Herbs, usually climbing, rarely woody.

B&H   POLYPETALAE, CALYCIFLORAE Passiflorales, 75
DT&H   METACHLAMYDEAE   Campanulatae, Cucurbitineae, 228
Melc   ARCHICHLAMYDEAE   Cucurbitales, 202
Thor   VIOLIFLORAE   Violales, Cucurbitineae, 137
Dahl   VIOLIFLORAE   Cucurbitales, 96
Young DILLENIIDAE, VIOLANAE   Violales, 200
Takh   DILLENIIDAE, VIOLANAE   Cucurbitales, 160
Cron   DILLENIIDAE   Violales, 127

Abobra Naudin
Acanthosicyos Welw. ex Hook.f.
Actinostemma Griff.
Alsomitra (Blume) M.Roem.
Ampelosicyos Thouars
Anacaona Alain
Apatzingania Dieterle
Apodanthera Arn.
Bambekea Cogn.
Benincasa Savi
Biswarea Cogn.
Bolbostemma Franquet
Brandegea Cogn.
Bryonia L.
Calycophysum Karst. & Triana
Cayaponia Silva Manso
Cephalopentandra Chiov.

Ceratosanthes Burm. ex Adans.
Chalema Dieterle
Cionosicyos Griseb.
Citrullus Schrad. ex Eckl. & Zeyh.
Coccinia Wight & Arn.
Cogniauxia Baill.
Corallocarpus Welw. ex Hook.f.
Cremastopus Paul G.Wilson
Ctenolepis Hook.f.
Cucumella Chiov.
Cucumeropsis Naudin
Cucumis L.
Cucurbita L.
Cucurbitella Walp.
Cyclanthera Schrad.
Cyclantheropsis Harms
Dactyliandra (Hook.f.) Hook.f.

550

Dendrosicyos Balf.f.
Dicoelospermum C.B.Clarke
Dieterlea Lott
Diplocyclos (Endl.) Post & Kuntze
Doyerea Grosourdy
Ecballium A.Rich.
Echinocystis Torr. & A.Gray
Echinopepon Naudin
Edgaria C.B.Clarke
Elateriopsis Ernst
Eureiandra Hook.f.
Fevillea L.
Gerrardanthus Harv. ex Hook.f.
Gomphogyne Griff.
Gurania (Schltdl.) Cogn.
Guraniopsis Cogn.
Gymnopetalum Arn.
Gynostemma Blume
Halosicyos Mart.Crov. *
Hanburia Seem.
Helmontia Cogn.
Hemsleya Cogn. ex F.B.Forbes &
   Hemsl.
Herpetospermum Wall. ex Hook.f.
Hodgsonia Hook.f. & Thomson
Ibervillea Greene
Indofevillea Chatterjee
Kedrostis Medik.
Lagenaria Ser.
Lemurosicyos Keraudren
Luffa Mill.
Marah Kellogg
Melancium Naudin
Melothria L.
Melothrianthus Mart.Crov.
Microsechium Naudin
Momordica L.
Muellerargia Cogn.
Mukia Arn.
Myrmecosicyos C.Jeffrey
Neoalsomitra Hutch.
Nothoalsomitra Telford
Odosicyos Keraudren *
Oreosyce Hook.f.

Parasicyos Dieterle
Penelopeia Urb.
Peponium Engl.
Peponopsis Naudin
Polyclathra Bertol.
Posadaea Cogn.
Praecitrullus Pang.
Pseudocyclanthera Mart.Crov.
Pseudosicydium Harms
Psiguria Neck. ex Arn.
Pteropepon (Cogn.) Cogn.
Pterosicyos Brandegee
Raphidiocystis Hook.f.
Ruthalicia C.Jeffrey
Rytidostylis Hook. & Arn.
Schizocarpum Schrad.
Schizopepon Maxim.
Sechiopsis Naudin
Sechium P.Browne
Selysia Cogn.
Seyrigia Keraudren
Sicana Naudin
Sicydium Schltdl.
Sicyos L.
Sicyosperma A.Gray
Siolmatra Baill.
Siraitia Merr.
Solena Lour.
Tecunumania Standl. & Steyerm.
Telfairia Hook.
Thladiantha Bunge
Trichosanthes L.
Tricyclandra Keraudren
Trochomeria Hook.f.
Trochomeriopsis Cogn.
Tumamoca Rose *
Vaseyanthus Cogn.
Wilbrandia Silva Manso
Xerosicyos Humbert
Zanonia L.
Zehneria Endl.
Zombitsia Keraudren *
Zygosicyos Humbert

CUNONIACEAE R.Br. 1814   59.09

23 genera. S. hem. (esp. Australasia, New Cal.) to Philippines, W. Indies. Woody.

B&H   POLYPETALAE, CALYCIFLORAE Rosales (within Saxifragaceae, 59)
DT&H ARCHICHLAMYDEAE   Rosales, Saxifragineae, 70
      (including Davidsoniaceae)
Melc   ARCHICHLAMYDEAE   Rosales, Saxifragineae, 108

Thor	ROSIFLORAE	Rosales, Cunoniineae, 215
	(separating *Baueraceae* 216)	
Dahl	ROSIFLORAE	Cunoniales, 156
	(separating *Baueraceae* 157)	
Young	ROSIDAE, ROSANAE	Cunoniales, 286
	(separating *Baueraceae* 300 to Saxifragales)	
Takh	ROSIDAE, ROSANAE	Cunoniales, 195
	(separating *Baueraceae* 196)	
Cron	ROSIDAE	Rosales, 159

Acrophyllum Benth.
Acsmithia Hoogland
Aistopetalum Schltr.
Anodopetalum A.Cunn. ex Endl.
Aphanopetalum Endl.
Bauera Banks ex Andr.
Caldcluvia D.Don
Callicoma Andrews
Ceratopetalum Sm.
Codia J.R.Forst. & G.Forst.
Cunonia L.
Geissois Labill.

Gillbeea F.Muell.
Gumillea Ruíz & Pav.
Lamanonia Vell.
Pancheria Brongn. & Gris
Platylophus D.Don
Pseudoweinmannia Engl.
Pullea Schltr.
Schizomeria D.Don
Spiraeanthemum A.Gray
Vesselowskya Pamp.
Weinmannia L.

CYCLOCHEILACEAE Marais 1981   125.07

2 genera. E. and NE. tropical Africa. Shrubs.

B&H	Not known	—
DT&H	METACHLAMYDEAE	Tubiflorae, Verbenineae (within Verbenaceae, 206)
Melc	Not mentioned	—
Thor	Not mentioned	—
Dahl	Not mentioned	—
Young	Not mentioned	—
Takh	LAMIIDAE, LAMIANAE	Scrophulariales (within Acanthaceae, 409)
Cron	Not mentioned	—

Asepalum Marais
Cyclocheilon Oliv.

CYNOMORIACEAE (C.Agardh) Lindl. 1825   150.02

1 genus. Mediterranean to N. Asia. Root parasites.

B&H	MONOCHLAMYDEAE	Achlamydosporeae (within Balanophoraceae, 150)
DT&H	ARCHICHLAMYDEAE	Myrtiflorae, Cynomoriineae, 178
Melc	ARCHICHLAMYDEAE	Myrtiflorae, Cynomoriineae, 219
Thor	SANTALIFLORAE	Balanophorales, 123
Dahl	BALANOPHORIFLORAE	Balanophorales, 262
Young	ROSIDAE, SANTALANAE	Balanophorales, 241

Takh   ROSIDAE, CORNANAE   Cynomoriales, 363
Cron   ROSIDAE   Santalales (within Balanophoraceae, 214)

Cynomorium L.

CYRILLACEAE Endl. 1841   46.02

3 genera. Trop to warm-temp. America. Trees, shrubs.

B&H   POLYPETALAE, DISCIFLORAE Olacales, 46a
DT&H   ARCHICHLAMYDEAE   Sapindales, Celastrineae, 104
Melc   ARCHICHLAMYDEAE   Celastrales, Celastrineae, 154
Thor   THEIFLORAE   Theales, Clethrineae, 56
Dahl   CORNIFLORAE   Ericales, 288
Young DILLENIIDAE, DILLENIANAE Theales, 117
Takh   DILLENIIDAE, ERICANAE   Ericales, 129
Cron   DILLENIIDAE   Ericales, 139

Cliftonia Banks ex Gaertn.f.
Cyrilla Garden ex L.
Purdiaea Planch.

DAPHNIPHYLLACEAE Muell. Arg. 1869   151.04

1 genus. India to Japan and New Guinea. Trees, shrubs.

B&H   MONOCHLAMYDEAE   Unisexuales (within Euphorbiaceae, 151)
DT&H   ARCHICHLAMYDEAE   Geraniales, Tricoccae (within Euphorbiaceae,
                                                                         97)
Melc   ARCHICHLAMYDEAE   Geraniales, Euphorbiineae, 130
Thor   ROSIFLORAE   Pittosporales, Buxineae, 224
Dahl   ROSIFLORAE   Buxales, 155
Young ROSIDAE, ROSANAE   Buxales, 285
Takh   HAMAMELIDIDAE, HAMAMELIDANAE
                                 Daphniphyllales, 85
Cron   HAMAMELIDIDAE   Daphniphyllales, 47

Daphniphyllum Blume

DATISCACEAE R.Br. 1830   77.01

3 genera. E. Medit. to Australia, S.W. USA and Mexico. Trees, herbs.

B&H   POLYPETALAE, CALYCIFLORAE Passiflorales, 77
DT&H   ARCHICHLAMYDEAE   Parietales, Datiscineae, 157
Melc   ARCHICHLAMYDEAE   Violales, Begoniineae, 200
Thor   VIOLIFLORAE   Violales, Begoniineae, 139
Dahl   VIOLIFLORAE   Cucurbitales, 94
Young DILLENIIDAE, VIOLANAE   Violales, 202

| Takh | DILLENIIDAE, VIOLANAE | Begoniales, 161 |
| Cron | DILLENIIDAE | Violales, 128 |

Datisca L.
Octomeles Miq.
Tetrameles R.Br.

DAVIDSONIACEAE Bange 1952   59.17

1 genus, monotypic. E. Australia. Small tree.

B&H	Not known	—
DT&H	ARCHICHLAMYDEAE	Rosales, Saxifragineae (within Cunoniaceae, 70)
Melc	ARCHICHLAMYDEAE	Rosales, Saxifragineae, 109
Thor	ROSIFLORAE	Rosales, Cunoniineae, 217
Dahl	ROSIFLORAE	Cunoniales, 159
Young	ROSIDAE, ROSANAE	Cunoniales, 289
Takh	ROSIDAE, ROSANAE	Cunoniales, 197
Cron	ROSIDAE	Rosales, 160

Davidsonia F.Muell.

DEGENERIACEAE I.W.Bailey & A.C.Sm. 1942   4.05

1 genus, 1 or 2 species. Fiji. Trees.

B&H	Not known	—
DT&H	Not known	—
Melc	ARCHICHLAMYDEAE	Magnoliales, Magnoliineae, 40
Thor	ANNONIFLORAE	Annonales, Annonineae, 5
Dahl	MAGNOLIIFLORAE	Magnoliales, 10
Young	MAGNOLIIDAE, RANUNCULANAE	Degeneriales, 12
Takh	MAGNOLIIDAE, MAGNOLIANAE	Magnoliales, 1
Cron	MAGNOLIIDAE	Magnoliales, 2

Degeneria I.W.Bailey & A.C.Sm.

DIALYPETALANTHACEAE Rizzini & Occhioni 1949   84.02

1 genus, monotypic. Brazil. Tree.

B&H	Not known	—
DT&H	Not known	—
Melc	ARCHICHLAMYDEAE	Myrtiflorae, Myrtineae, 207
Thor	GENTIANIFLORAE	Gentianales, 252
Dahl	GENTIANIFLORAE	Gentianales, 340
Young	Not mentioned	—
Takh	LAMIIDAE, GENTIANANAE	Gentianales, 370
Cron	ROSIDAE	Rosales, 161

Dialypetalanthus Kuhlm.

DIAPENSIACEAE (Link) Lindl. 1836   96

7 genera. N. temperate region to Himalaya. Herbs, subshrubs.

B&H   GAMOPETALAE, HETEROMERAE   Ericales, 96
DT&H   METACHLAMYDEAE   Ericales, 187
Melc   SYMPETALAE   Diapensiales, 227
Thor   ROSIFLORAE   Rosales, Saxifragineae, 214
Dahl   CORNIFLORAE   Ericales, 295
Young   DILLENIIDAE, DILLENIANAE   Diapensiales, 134
Takh   DILLENIIDAE, ERICANAE   Diapensiales, 132
Cron   DILLENIIDAE   Diapensiales, 147

Berneuxia Decne.   **Pyxidanthera** Michx.
Diapensia L.   **Schizocodon** Siebold & Zucc.
Diplarche Hook.f. & Thomson   **Shortia** Torr. & A.Gray
Galax Sims

DICHAPETALACEAE Baill. 1886   44

3 genera. Tropics to warm-temperate. Trees, shrubs, lianes.

B&H   POLYPETALAE, DISCIFLORAE   Geraniales (as *Chailletiaceae*) 44
DT&H   ARCHICHLAMYDEAE   Geraniales, Dichapetalineae, 96
Melc   ARCHICHLAMYDEAE   Thymelaeales, 179
Thor   MALVIFLORAE   Euphorbiales, 165
Dahl   MALVIFLORAE   Euphorbiales, 79
Young   DILLENIIDAE, MALVANAE   Euphorbiales, 189
Takh   DILLENIIDAE, EUPHORBIANAE   Euphorbiales, 192
Cron   ROSIDAE   Celastrales, 228

Dichapetalum Thouars
Stephanopodium Poepp.
Tapura Aubl.

DIDIEREACEAE Drake 1903   78.02

4 genera. Madagascar. Cactiform trees or shrubs.

B&H   Not known   —
DT&H   ARCHICHLAMYDEAE   Centrospermae, Chenopodiineae, post 28
Melc   ARCHICHLAMYDEAE   Centrospermae, post 37
Thor   CHENOPODIIFLORAE   Chenopodiales, Portulacineae, 93
Dahl   CARYOPHYLLIFLORAE   Caryophyllales, 46
Young   DILLENIIDAE, CARYOPHYLLANAE Chenopodiales, 152
Takh   CARYOPHYLLIDAE, CARYOPHYLLANAE
                                   Caryophyllales, Caryophyllineae, 70
Cron   CARYOPHYLLIDAE   Caryophyllales, 68

Alluaudia (Drake) Drake   **Decarya** Choux
Alluaudiopsis Humbert & Choux   **Didierea** Baill.

DIDYMELACEAE Léandri 1937   151.03

1 genus. Madagascar. Tree.

B&H	MONOCHLAMYDEAE	Unisexuales (? within Leitneriaceae, 155)
DT&H	INCERTAE SEDIS	—
Melc	ARCHICHLAMYDEAE	Leitneriales, 6
Thor	ROSIFLORAE	Pittosporales, Buxineae, 223
Dahl	ROSIFLORAE	Buxales, 154
Young	DILLENIIDAE, MALVANAE	Euphorbiales, 188
Takh	HAMAMELIDIDAE, HAMAMELIDANAE	
		Didymelales, 87
Cron	HAMAMELIDIDAE	Didymelales, 48

Didymeles Thouars

DIEGODENDRACEAE Capuron 1964   41.03

1 genus, monotypic. Madagascar. Shrub to small tree.

B&H	Not known	—
DT&H	Not known	—
Melc	Not known	—
Thor	THEIFLORAE	Theales, Scytopetalineae (within Sphaerosepalaceae, 61)
Dahl	THEIFLORAE	Theales (within Ochnaceae, 120)
Young	Not mentioned	—
Takh	DILLENIIDAE, THEANAE	Ochnales, 118
Cron	DILLENIIDAE	Theales (within Ochnaceae, 80)

Diegodendron Capuron

DILLENIACEAE Salisb. 1807   2.01

11 genera. Trop. and subtrop. to Tasmania. Trees, lianes, shrubs, occas. herbs.

B&H	POLYPETALAE, THALAMIFLORAE	Ranales, 2
	(including Crossosomataceae)	
DT&H	ARCHICHLAMYDEAE	Parietales, Theineae, 130
	(including Actinidiaceae)	
Melc	ARCHICHLAMYDEAE	Guttiferales, Dilleniineae, 75
Thor	THEIFLORAE	Theales, Dilleniineae, 40
Dahl	THEIFLORAE	Dilleniales, 108
Young	DILLENIIDAE, DILLENIANAE	Dilleniales, 103
Takh	DILLENIIDAE, DILLENIANAE	Dilleniales, 98
Cron	DILLENIIDAE	Dilleniales, 78

Acrotrema Jack   **Hibbertia** Andrews
Curatella Loefl.   **Pachynema** R.Br. ex DC.
Davilla Vand.   **Pinzona** Mart. & Zucc.
Didesmandra Stapf   **Schumacheria** Vahl
Dillenia L.   **Tetracera** L.
Doliocarpus Rol.

DIONCOPHYLLACEAE (Gilg) Airy Shaw 1952   61.04

3 genera. Guineo-Congolan Africa. Insectivorous lianes and shrubs.

B&H	Not known	—
DT&H	ARCHICHLAMYDEAE	Parietales, Flacourtiineae (within Flacourtiaceae, 149)
Melc	ARCHICHLAMYDEAE	Guttiferales, Ochnineae, 82
Thor	THEIFLORAE	Theales, Scytopetalineae, 65
Dahl	THEIFLORAE	Theales, 115
Young	DILLENIIDAE, DILLENIANAE	Scytopetalales, 126
Takh	DILLENIIDAE, VIOLANAE	Dioncophyllales, 159
Cron	DILLENIIDAE	Violales, 118

Dioncophyllum Baill.
Habropetalum Airy Shaw
Triphyophyllum Airy Shaw

DIPENTODONTACEAE Merr. 1941   45.06

1 genus, monotypic. W. China, Burma. Tree.

B&H	Not known	—
DT&H	Not known	—
Melc	ARCHICHLAMYDEAE	Santalales, Santalineae, 17
Thor	VIOLIFLORAE	Violales, Violineae, 125
Dahl	VIOLIFLORAE	Violales, 86
Young	DILLENIIDAE, VIOLANAE	Violales, 191
Takh	DILLENIIDAE, VIOLANAE	Violales, 148
Cron	ROSIDAE	Santalales, 206

Dipentodon Dunn

DIPSACACEAE Juss. 1789   86.01

7 genera. Eurasia and Africa. Herbs to subshrubs.

B&H	GAMOPETALAE, INFERAE (including Morinaceae, Triplostegiaceae)	Asterales, 86
DT&H	METACHLAMYDEAE (including Morinaceae)	Rubiales, 227
Melc	SYMPETALAE (including Morinaceae)	Dipsacales, 282
Thor	CORNIFLORAE (including Morinaceae, Triplostegiaceae)	Dipsacales, 294
Dahl	CORNIFLORAE	Dipsacales, 332
Young	ROSIDAE, DIPSACANAE (? including Morinaceae)	Dipsacales, 384
Takh	ROSIDAE, CORNANAE	Dipsacales, 361
Cron	ASTERIDAE (including Morinaceae)	Dipsacales, 316

Cephalaria Schrad. ex Roem. & Schult.
Dipsacus L.
Knautia L.
Pterocephalus Adans.

Scabiosa L.
Succisa Haller
Succisella Beck

DIPTEROCARPACEAE Blume 1825   29.01

16 genera. Old World trop. esp. Indomal., and (*Pakaraimaea*) Guyana Highl. Trees.

B&H POLYPETALAE, THALAMIFLORAE Guttiferales, 29
    (including Ancistrocladaceae)
DT&H ARCHICHLAMYDEAE   Parietales, Theineae, 138
Melc ARCHICHLAMYDEAE   Guttiferales, Ochnineae, 84
Thor MALVIFLORAE   Malvales, 149
Dahl MALVIFLORAE   Malvales, 66
Young DILLENIIDAE, MALVANAE   Malvales, 169
Takh DILLENIIDAE, MALVANAE   Malvales, 177
    (separating *Monotaceae* 176)
Cron DILLENIIDAE   Theales, 83

Anisoptera Korth.
Cotylelobium Pierre
Dipterocarpus C.F.Gaertn.
Dryobalanops C.F.Gaertn.
Hopea Roxb.
Marquesia Gilg
Monotes A.DC.
Neobalanocarpus P.S.Ashton

Pakaraimaea Maguire & P.S.Ashton
Parashorea Kurz
Shorea Roxb. ex C.F.Gaertn.
Stemonoporus Thwaites
Upuna Symington
Vateria L.
Vateriopsis F.Heim
Vatica L.

DROSERACEAE Salisb. 1808   61.01

4 genera. Widespread. Insectivorous herbs, rarely subshrubs.

B&H POLYPETALAE, CALYCIFLORAE Rosales, 61
    (including Byblidaceae, Roridulaceae)
DT&H ARCHICHLAMYDEAE   Sarraceniales, 62
    (including Byblidaceae, Roridulaceae)
Melc ARCHICHLAMYDEAE   Sarraceniales, 93
Thor ROSIDAE   Rosales, Saxifragineae, 211
Dahl ROSIFLORAE   Droserales, 170
Young ROSIDAE, ROSANAE   Saxifragales, 296
Takh ROSIDAE, ROSANAE   Droserales, 215
Cron DILLENIIDAE   Nepenthales, 106

Aldrovanda L.
Dionaea Ellis

Drosera L.
Drosophyllum Link

DUCKEODENDRACEAE Kuhlm. 1950   114.03

1 genus, monotypic. Brazil. Tree.

B&H	Not known	—
DT&H	Not known	—
Melc	SYMPETALAE	Tubiflorae, Solanineae, 263
Thor	SOLANIFLORAE	Solanales, Solanineae (within Solanaceae, 271)
Dahl	SOLANIFLORAE	Solanales (within Solanaceae, 273)
Young	Not mentioned	—
Takh	LAMIIDAE, SOLANANAE	Solanales, 381
Cron	ASTERIDAE	Solanales, 276

Duckeodendron Kuhlm.

EBENACEAE Gürke 1891   102

3 genera. Widespread tropics, occas. warm-temperate. Trees, shrubs.

B&H	GAMOPETALAE, HETEROMERAE	Ebenales, 102
DT&H	METACHLAMYDEAE	Ebenales, Diospyrineae, 193
Melc	SYMPETALAE	Ebenales, Ebenineae, 239
Thor	THEIFLORAE	Ebenales, Ebenineae, 73
Dahl	PRIMULIFLORAE	Ebenales, 136
Young	DILLENIIDAE, DILLENIANAE	Ebenales, 135
Takh	DILLENIIDAE, ERICANAE	Ebenales, 135
Cron	DILLENIIDAE	Ebenales, 149

Diospyros L.
Euclea L.
Tetraclis Hiern

ELAEAGNACEAE Juss. 1789   147

3 genera. Temperate N. hemisphere to trop. Asia and Australia. Shrubs, small trees.

B&H	MONOCHLAMYDEAE	Daphnales, 147
DT&H	ARCHICHLAMYDEAE	Myrtiflorae, Thymelaeineae, 165
Melc	ARCHICHLAMYDEAE	Thymelaeales, 181
Thor	MALVIFLORAE	Rhamnales, 160
Dahl	MALVIFLORAE	Elaeagnales, 83
Young	DILLENIIDAE, MALVANAE	Rhamnales, 182
Takh	ROSIDAE, RHAMNANAE	Elaeagnales, 320
Cron	ROSIDAE	Proteales, 183

Elaeagnus L.
Hippophae L.
Shepherdia Nutt.

ELAEOCARPACEAE Juss. ex DC. 1824   33.02

9 genera. Trop. and subtrop., excl. cont. Africa, occas. temp. Trees, shrubs.

B&H   POLYPETALAE, THALAMIFLORAE Malvales (within Tiliaceae, 33)
DT&H   ARCHICHLAMYDEAE   Malvales, Elaeocarpineae, 121
Melc   ARCHICHLAMYDEAE   Malvales, Elaeocarpineae, 170
Thor   MALVIFLORAE   Malvales, 146
Dahl   MALVIFLORAE   Malvales, 57
Young DILLENIIDAE, MALVANAE   Malvales, 166
Takh   DILLENIIDAE, MALVANAE   Malvales, 173
Cron   DILLENIIDAE   Malvales, 98

Aceratium DC.   **Peripentadenia** L.S.Sm.
Aristotelia L'Hér.   **Sericolea** Schltr.
Crinodendron Molina   **Sloanea** L.
Dubouzetia Pancher ex Brongn. & Gris **Vallea** Mutis ex L.f.
Elaeocarpus L.

ELATINACEAE Dumort. 1829   25

2 genera. Temperate and tropical. Herbs, subshrubs, often aquatic.

B&H   POLYPETALAE, THALAMIFLORAE Guttiferales, 25
DT&H   ARCHICHLAMYDEAE   Parietales, Tamaricineae, 139
Melc   ARCHICHLAMYDEAE   Violales, Tamaricineae, 197
Thor   THEIFLORAE   Theales, Hypericineae, 68
Dahl   THEIFLORAE   Theales, 127
Young DILLENIIDAE, DILLENIANAE   Hypericales, 129
Takh   DILLENIIDAE, THEANAE   Elatinales, 123
Cron   DILLENIIDAE   Theales, 94

Bergia L.
Elatine L.

EMBLINGIACEAE (Pax) Airy Shaw 1965   90.03

1 genus, monotypic. Western Australia. Prostrate subshrub.

B&H   POLYPETALAE, THALAMIFLORAE Parietales (within Capparaceae, 12)
DT&H   ARCHICHLAMYDEAE   Rhoeadales, Capparidineae (within
                                                    Capparaceae, 57)
Melc   ARCHICHLAMYDEAE   Papaverales, Capparineae (within
                                                    Capparaceae, 95)
Thor   RUTIFLORAE   Rutales, Sapindineae (within Sapindaceae,
                                                    180)
Dahl   RUTIFLORAE   Sapindales, 212
Young ROSIDAE, RUTANAE   Sapindales (within Sapindaceae, 253)
Takh   ROSIDAE, RUTANAE   Sapindales, 252
Cron   ROSIDAE   Polygalales (within Polygalaceae, 245)

Emblingia F.Muell.

EMPETRACEAE Gray 1821 162

3 genera. N. temperate, Andes, S. Atlantic. Shrubs, subshrubs.

B&H	MONOCHLAMYDEAE	Ordines anomali, 162
DT&H	ARCHICHLAMYDEAE	Sapindales, Empetrineae, 101
Melc	SYMPETALAE	Ericales, 231
Thor	THEIFLORAE	Ericales, 72
Dahl	CORNIFLORAE	Ericales, 290
Young	DILLENIIDAE, DILLENIANAE	Ericales, 133
Takh	DILLENIIDAE, ERICANAE	Ericales, 130
Cron	DILLENIIDAE	Ericales, 142

Ceratiola Michx.
Corema D.Don
Empetrum L.

EPACRIDACEAE R.Br. 1810 95

31 genera. Mostly Australasia, also trop. Asia, Pacific, S. America. Mostly shrubs.

B&H	GAMOPETALAE, HETEROMERAE	Ericales, 95
DT&H	METACHLAMYDEAE	Ericales, 186
Melc	SYMPETALAE	Ericales, 232
Thor	THEIFLORAE	Ericales, 71
Dahl	CORNIFLORAE	Ericales, 293
Young	DILLENIIDAE, DILLENIANAE	Ericales, 132
Takh	DILLENIIDAE, ERICANAE	Ericales, 128
Cron	DILLENIIDAE	Ericales, 143

Acrotriche R.Br.
Andersonia R.Br.
Archeria Hook.f.
Astroloma R.Br.
Brachyloma Sond.
Choristemon H.B.Williamson
Coleanthera Stschegl.
Conostephium Benth.
Cosmelia R.Br.
Cyathodes Labill.
Cyathopsis Brongn. & Gris
Decatoca F.Muell.
Dracophyllum Labill.
Epacris Cav.
Lebetanthus Endl.
Leucopogon R.Br.
Lissanthe R.Br.
Lysinema R.Br.
Melichrus R.Br.
Monotoca R.Br.
Needhamiella L.Watson
Oligarrhena R.Br.
Pentachondra R.Br.
Prionotes R.Br.
Richea R.Br.
Rupicola J.H.Maiden & E.Betche
Sphenotoma R.Br. ex Sweet
Sprengelia Sm.
Styphelia Sm.
Trochocarpa R.Br.
Woollsia F.Muell.

EREMOLEPIDACEAE Tiegh. ex Nakai 1952 148.02

3 genera. Tropical America. Hemiparasites on trees.

B&H	MONOCHLAMYDEAE	Achlamydosporeae (within Loranthaceae, 148)

DT&H	ARCHICHLAMYDEAE	Santalales, Loranthineae (within Loranthaceae, 17)
Melc	ARCHICHLAMYDEAE	Santalales, Loranthineae (within Loranthaceae, 22)
Thor	SANTALIFLORAE	Santalales, 118
Dahl	SANTALIFLORAE	Santalales, 259
Young	ROSIDAE, SANTALANAE	Santalales, 236
Takh	ROSIDAE, CELASTRANAE	Santalales, 318
Cron	ROSIDAE	Santalales, 213

Antidaphne Poepp. & Endl.
Eubrachion Hook.f.
Lepidoceras Hook.f.

EREMOSYNACEAE Dandy 1959   59.02

1 genus, monotypic. W. Australia. Annual herb.

B&H	POLYPETALAE, CALYCIFLORAE	Rosales (within Saxifragaceae, 59)
DT&H	ARCHICHLAMYDEAE	Rosales, Saxifragineae (within Saxifragaceae, 67)
Melc	ARCHICHLAMYDEAE	Rosales, Saxifragineae (within Saxifragaceae, 106)
Thor	ROSIFLORAE	Rosales, Saxifragineae (within Saxifragaceae, 208)
Dahl	CORNIFLORAE ?	Cornales ?, 327
Young	ROSIDAE, CORNANAE	Hydrangeales, 336
Takh	ROSIDAE	Saxifragales, 209
Cron	ROSIDAE	Rosales (within Saxifragaceae, 173)

Eremosyne Endl.

ERICACEAE Juss. 1789   93.01

116 genera. Widespread. Mostly shrubs.

B&H	GAMOPETALAE, HETEROMERAE	Ericales, 93
	(including Clethraceae; separating *Monotropaceae* 94, *Vacciniaceae* 92)	
DT&H	METACHLAMYDEAE	Ericales, 185
	(separating *Pyrolaceae* 183)	
Melc	SYMPETALAE	Ericales, 230
	(separating *Pyrolaceae* 229)	
Thor	THEIFLORAE	Ericales, 70
Dahl	CORNIFLORAE	Ericales, 289
	(separating *Monotropaceae* 291, *Pyrolaceae* 292)	
Young	DILLENIIDAE, DILLENIANAE	Ericales, 131
Takh	DILLENIIDAE, ERICANAE	Ericales, 127
Cron	DILLENIIDAE	Ericales, 144
	(separating *Monotropaceae* 146, *Pyrolaceae* 145)	

Acrostemon Klotzsch   **Allotropa** Torr. & A.Gray
Agapetes D.Don ex G.Don   **Andromeda** L.
Agarista D.Don ex G.Don   **Aniserica** N.E.Br.

Anomalanthus Klotzsch
Anthopteropsis A.C.Sm.
Anthopterus Hook.
Arachnocalyx Compton
Arbutus L.
Arctostaphylos Adans.
Bejaria Mutis ex L.
Botryostege Stapf
Bruckenthalia Rchb.
Bryanthus J.G.Gmel.
Calluna Salisb.
Calopteryx A.C.Sm.
Cassiope D.Don
Cavendishia Lindl.
Ceratostema Juss.
Chamaedaphne Moench
Cheilotheca Hook.f.
Chimaphila Pursh
Coccosperma Klotzsch
Codonostigma Klotzsch ex Benth.
Coilostigma Klotzsch
Comarostaphylis Zucc.
Costera J.J.Sm.
Craibiodendron W.W.Sm.
Daboecia D.Don
Demosthenesia A.C.Sm.
Didonica Luteyn & Wilbur *
Dimorphanthera (Drude) J.J.Sm.
Diogenesia Sleumer
Diplycosia Blume
Disterigma (Klotzsch) Nied.
Elliottia Mühl. ex Elliott
Enkianthus Lour.
Epigaea L.
Eremia D.Don
Eremiella Compton
Erica L.
Gaultheria Kalm ex L.
Gaylussacia Kunth
Gonocalyx Planch. & Linden
Grisebachia Klotzsch
Harrimanella Coville
Hemitomes A.Gray
Hexastemon Klotzsch
Kalmia L.
Kalmiopsis Rehder
Lateropora A.C.Sm.
Ledothamnus Meisn.
Leiophyllum (Pers.) R.G.Hedw.
Lepterica N.E.Br.
Leucothoe D.Don
Loiseleuria Desv.
Lyonia Nutt.
Macleania Hook.
Malea Lundell

Menziesia Sm.
Mischopleura Wernham ex Ridl.
Moneses Salisb. ex Gray
Monotropa L.
Monotropastrum Andres
Monotropsis Schwein. ex Elliott
Mycerinus A.C.Sm.
Nagelocarpus Bullock
Notopora Hook.f.
Oreanthes Benth.
Ornithostaphylos Small
Orthaea Klotzsch
Orthilia Raf.
Oxydendrum DC.
Pellegrinia Sleumer
Pernettyopsis King & Gamble
Phyllodoce Salisb.
Pieris D.Don
Pityopus Small
Platycalyx N.E.Br.
Pleuricospora A.Gray
Plutarchia A.C.Sm.
Polyclita A.C.Sm.
Psammisia Klotzsch
Pseudogonocalyx Bisse & Berazaín *
Pterospora Nutt.
Pyrola L.
Rhododendron L.
Rhodothamnus Rchb.
Rusbya Britton
Salaxis Salisb.
Sarcodes Torr.
Satyria Klotzsch
Schizococcus Eastw.
Scyphogyne Decne.
Semiramisia Klotzsch
Simocheilus Klotzsch
Siphonandra Klotzsch
Sphyrospermum Poepp. & Endl.
Stokoeanthus E.G.H.Oliv.
Symphysia C.Presl
Sympieza Lichtenst. ex Roem. & Schult.
Syndesmanthus Klotzsch
Tepuia Camp
Thamnus Klotzsch
Themistoclesia Klotzsch
Therorhodion (Maxim.) Small
Thibaudia Ruíz & Pav.
Thoracosperma Klotzsch
Tripetaleia Siebold & Zucc.
Tsusiophyllum Maxim.
Utleya Wilbur & Luteyn *
Vaccinium L.
Xylococcus Nutt.
Zenobia D.Don

ERYTHROXYLACEAE Kunth 1822   34.02

4 genera. Tropics and subtropics. Trees, shrubs.

B&H	POLYPETALAE, DISCIFLORAE	Geraniales (within Linaceae, 34)
DT&H	ARCHICHLAMYDEAE	Geraniales, Geraniineae, 84
Melc	ARCHICHLAMYDEAE	Geraniales, Geraniineae, 128
Thor	GERANIIFLORAE	Geraniales, Linineae, 98
Dahl	RUTIFLORAE	Geraniales, 234
Young	ROSIDAE, GERANIANAE	Linales, 272
Takh	ROSIDAE, RUTANAE	Linales, 279
Cron	ROSIDAE	Linales, 236

Aneulophus Benth.   **Nectaropetalum** Engl.
Erythroxylum P.Browne   **Pinacopodium** Exell & Mendonça

ESCALLONIACEAE R.Br. ex Dumort. 1829   59.07

20 genera. S. hemisphere to tropics, esp. S. America, Australasia. Trees, shrubs.

B&H	POLYPETALAE, CALYCIFLORAE	Rosales (within Saxifragaceae, 59)
DT&H	ARCHICHLAMYDEAE	Rosales, Saxifragineae (within Saxifragaceae, 67)
Melc	ARCHICHLAMYDEAE	Rosales, Saxifragineae (within Saxifragaceae, 106)
Thor	ROSIFLORAE	Rosales, Saxifragineae (within Saxifragaceae, 208)
Dahl	CORNIFLORAE	Cornales, 303

    (separating *Brexiaceae* 164 and *Iteaceae* 167 to ROSIFLORAE Saxifragales;
    and *Dulongiaceae* 325, *Tetracarpaeaceae* 329, *Tribelaceae* 326)

Young	ROSIDAE, CORNANAE	Hydrangeales, 330

    (separating *Brexiaceae* 305, *Iteaceae* 299 and *Tetracarpaeaceae* 303 to
    Saxifragales, and *Phyllonomaceae* [*Dulongiaceae*] 335, *Tribelaceae* 332)

Takh	ROSIDAE, CORNANAE	Hydrangeales, Escalloniineae, 325

    (separating *Brexiaceae* 308 to CELASTRANAE Celastrales, *Iteaceae* 207 to
    ROSANAE Saxifragales, *Tetracarpaeaceae* 324 to Tetracarpaeineae, and
    Argophyllaceae 326, *Carpodetaceae* 328, *Dulongiaceae* 334, *Polyosmataceae* 329
    and *Tribelaceae* 335)

Cron	ROSIDAE	Rosales (within Grossulariaceae, 166)

Abrophyllum Hook.f. ex Benth.   **Itea** L.
Anopterus Labill.   **Ixerba** A.Cunn.
Argophyllum J.R.Forst. & G.Forst.   **Phyllonoma** Willd. ex Schult.
Brexia Noronha ex Thouars   **Platyspermation** Guillaumin
Carpodetus J.R.Forst. & G.Forst.   **Polyosma** Blume
Choristylis Harv.   **Quintinia** A.DC.
Corokia A.Cunn.   **Roussea** Sm.
Cuttsia F.Muell.   **Tetracarpaea** Hook.
Escallonia Mutis ex L.f.   **Tribeles** Phil.
Forgesia Comm. ex Juss.   **Valdivia** C.Gay ex J.Remy

EUCOMMIACEAE Engl. 1909   153.03

1 genus, monotypic. China. Tree

B&H	Not known	—
DT&H	ARCHICHLAMYDEAE	Ranales, Trochodendrineae (within Trochodendraceae, 40)
Melc	ARCHICHLAMYDEAE	Urticales, 12
Thor	HAMAMELIDIFLORAE	Hamamelidales, Eucommiineae, 197
Dahl	CORNIFLORAE	Eucommiales, 296
Young	ROSIDAE, HAMAMELIDANAE	Eucommiales, 220
Takh	HAMAMELIDIDAE, EUCOMMIANAE	Eucommiales, 80
Cron	HAMAMELIDIDAE	Eucommiales, 49

Eucommia Oliv.

EUCRYPHIACEAE Endl. 1841   59.11

1 genus. Australia, Chile. Trees, shrubs.

B&H	POLYPETALAE, CALYCIFLORAE	Rosales (within Rosaceae, 58)
DT&H	ARCHICHLAMYDEAE	Parietales, Theineae, 131
Melc	ARCHICHLAMYDEAE	Guttiferales, Dilleniineae, 78
Thor	ROSIFLORAE	Rosales, Cunoniineae, 219
Dahl	ROSIFLORAE	Cunoniales, 160
Young	ROSIDAE, ROSANAE	Cunoniales, 288
Takh	ROSIDAE, ROSANAE	Cunoniales, 198
Cron	ROSIDAE	Rosales, 158

Eucryphia Cav.

EUPHORBIACEAE Juss. 1789   151.01

331 genera. Widespread. Trees, shrubs, herbs.

B&H	MONOCHLAMYDEAE	Unisexuales, 151
	(including Aextoxicaceae, Buxaceae, Daphniphyllaceae, Simmondsiaceae)	
DT&H	ARCHICHLAMYDEAE	Geraniales, Tricoccae, 97
	(including Aextoxicaceae, Daphniphyllaceae)	
Melc	ARCHICHLAMYDEAE	Geraniales, Euphorbiineae, 129
	(separating *Picrodendraceae* 134 to Rutales Rutineae)	
Thor	MALVIFLORAE	Euphorbiales, 161
Dahl	MALVIFLORAE	Euphorbiales, 75
Young	DILLENIIDAE, MALVANAE	Euphorbiales, 183
Takh	DILLENIIDAE, EUPHORBIANAE	Euphorbiales, 190
Cron	ROSIDAE	Euphorbiales, 232

Acalypha L.
Acidocroton Griseb.
Acidoton Sw.
Actephila Blume
Actinostemon Mart. ex Klotzsch
Adelia L.

Adenochlaena Boivin ex Baill.
Adenocline Turcz.
Adenopeltis Bertero ex A.Juss.
Adenophaedra (Müll.Arg.) Müll.Arg.
Adriana Gaudich.
Aerisilvaea Radcl.–Sm.

Afrotrewia Pax & K.Hoffm. *
Agrostistachys Dalzell
Alchornea Sw.
Alchorneopsis Müll.Arg.
Aleurites J.R.Forst. & G.Forst.
Algernonia Baill.
Alphandia Baill.
Amanoa Aubl.
Amperea A.Juss.
Amyrea Leandri
Andrachne L.
Androstachys Prain
Angostylis Benth.
Annesijoa Pax & K.Hoffm.
Anomalocalyx Ducke
Anthostema Juss.
Antidesma L.
Aparisthmium Endl.
Apodandra Pax & K.Hoffm.
Apodiscus Hutch.
Aporusa Blume
Argomuellera Pax
Argythamnia P.Browne
Aristogeitonia Prain
Ashtonia Airy Shaw
Astrocasia B.L.Rob. & Millsp.
Astrococcus Benth.
Austrobuxus Miq.
Avellanita Phil.
Baccaurea Lour.
Baliospermum Blume
Baloghia Endl.
Benoistia H.Perrier & Leandri *
Bernardia Mill.
Bertya Planch.
Beyeria Miq.
Bischofia Blume
Blachia Baill.
Blotia Leandri
Blumeodendron (Müll.Arg.) Kurz
Bocquillonia Baill.
Bonania A.Rich.
Borneodendron Airy Shaw
Botryophora Hook.f.
Breynia J.R.Forst. & G.Forst.
Bridelia Willd.
Calycopeplus Planch.
Caperonia A.St.-Hil.
Caryodendron Karst.
Cavacoa J.Léonard
Celaenodendron Standl.
Celianella Jabl.
Cephalocroton Hochst.
Cephalocrotonopsis Pax
Cephalomappa Baill.

Chaetocarpus Thwaites
Chascotheca Urb.
Cheilosa Blume
Chiropetalum A.Juss.
Chlamydojatropha Pax & K.Hoffm.
Chondrostylis Boerl.
Chonocentrum Pierre ex Pax & K.Hoffm.
Choriceras Baill.
Chorisandrachne Airy Shaw
Chrozophora Neck. ex Juss.
Cladogelonium Leandri
Cladogynos Zipp. ex Spanoghe
Claoxylon A.Juss.
Claoxylopsis Leandri *
Cleidiocarpon Airy Shaw
Cleidion Blume
Cleistanthus Hook.f. ex Planch.
Clutia L.
Cnesmone Blume
Cnidoscolus Pohl
Cocconerion Baill.
Codiaeum A.Juss.
Colliguaja Molina
Cometia Thouars ex Baill.
Conceveiba Aubl.
Cordemoya Baill.
Croizatia Steyerm. *
Croton L.
Crotonogyne Müll.Arg.
Crotonogynopsis Pax
Crotonopsis Michx.
Ctenomeria Harv.
Cubacroton Alain *
Cubanthus (Boiss.) Millsp.
Cyrtogonone Prain
Cyttaranthus J.Léonard
Dalechampia L.
Dalembertia Baill.
Danguyodrypetes Leandri
Dendrocousinsia Millsp.
Deuteromallotus Pax & K.Hoffm.
Deutzianthus Gagnep.
Dichostemma Pierre
Dicoelia Benth.
Didymocistus Kuhlm.
Dimorphocalyx Thwaites
Discocarpus Klotzsch
Discoclaoxylon (Müll.Arg.) Pax & K.Hoffm.
Discocleidion (Müll.Arg.) Pax & K.Hoffm.
Discoglypremna Prain
Dissiliaria F.Muell.
Ditaxis Vahl ex A.Juss.

Ditta Griseb.
Dodecastigma Ducke
Domohinea Leandri
Doryxylon Zoll.
Droceloncia J.Léonard
Drypetes Vahl
Duvigneaudia J.Léonard
Dysopsis Baill.
Elaeophora Ducke
Elaeophorbia Stapf
Elateriospermum Blume
Eleutherostigma Pax & K.Hoffm.
Endadenium L.C.Leach
Endospermum Benth.
Enriquebeltrania Rzed. *
Epiprinus Griff.
Eremocarpus Benth.
Erismanthus Wall. ex Müll.Arg.
Erythrococca Benth.
Euphorbia L.
Excoecaria L.
Fahrenheitia Rchb.f. & Zoll.
Flueggea Willd.
Fontainea Heckel
Fragariopsis A.St.–Hil.
Garcia Rohr
Gavarretia Baill.
Gitara Pax & K.Hoffm.
Givotia Griff.
Glochidion J.R.Forst. & G.Forst.
Glycydendron Ducke
Glyphostylus Gagnep.
Grimmeodendron Urb.
Grossera Pax
Gymnanthes Sw.
Haematostemon (Müll.Arg.) Pax & K.Hoffm.
Hamilcoa Prain
Hevea Aubl.
Hexaspermum Domin
Heywoodia T.R.Sim
Hieronima Allemão
Hippomane L.
Homalanthus A.Juss.
Homonoia Lour.
Hura L.
Hyaenanche Lamb.
Hylandia Airy Shaw
Hymenocardia Wall. ex Lindl.
Jablonskia G.L.Webster
Jatropha L.
Joannesia Vell.
Julocroton Mart.
Kairothamnus Airy Shaw
Keayodendron Leandri

Klaineanthus Pierre ex Prain
Kleinodendron L.B.Sm. & Downs
Koilodepas Hassk.
Lachnostylis Turcz.
Lasiococca Hook.f.
Lasiocroton Griseb.
Lautembergia Baill.
Leeuwenbergia Letouzey & N.Hallé
Leidesia Müll.Arg.
Leptonema A.Juss.
Leptopus Decne.
Leucocroton Griseb.
Lingelsheimia Pax
Lobanilia Radcl.–Sm.
Loerzingia Airy Shaw
Mabea Aubl.
Macaranga Thouars
Maesobotrya Benth.
Mallotus Lour.
Manihot Mill.
Manihotoides D.J.Rogers & Appan
Manniophyton Müll.Arg.
Maprounea Aubl.
Mareya Baill.
Mareyopsis Pax & K.Hoffm.
Margaritaria L.f.
Martretia Beille
Megalostylis S.Moore
Megistostigma Hook.f.
Meineckia Baill.
Melanolepis Rchb.f. & Zoll.
Mercurialis L.
Mettenia Griseb.
Micrandra Benth.
Micrandropsis Rodrigues *
Micrantheum Desf.
Micrococca Benth.
Mildbraedia Pax
Mischodon Thwaites
Moacroton Croizat
Monadenium Pax
Monotaxis Brongn.
Moultonianthus Merr.
Myladenia Airy Shaw
Myricanthe Airy Shaw
Nealchornea Huber
Necepsia Prain
Neoboutonia Müll.Arg.
Neoguillauminia Croizat
Neoholstia Rauschert
Neoroepera Müll.Arg. & F.Muell.
Neoscortechinia Pax
Neotrewia Pax & K.Hoffm.
Octospermum Airy Shaw
Oldfieldia Benth. & Hook.

Oligoceras Gagnep.
Omphalea L.
Ophellantha Standl. *
Ophthalmoblapton Allemão
Oreoporanthera Hutch.
Ostodes Blume
Pachystroma Müll.Arg.
Pachystylidium Pax & K.Hoffm.
Pantadenia Gagnep.
Paradrypetes Kuhlm.
Paranecepsia Radcl.–Sm.
Parodiodendron Hunz.
Pausandra Radlk.
Pedilanthus Neck. ex Poit.
Pentabrachion Müll.Arg.
Pera Mutis
Petalostigma F.Muell.
Philyra Klotzsch
Phyllanoa Croizat
Phyllanthus L.
Picrodendron Griseb.
Pimelodendron Hassk.
Piranhea Baill.
Plagiostyles Pierre
Platygyna Mercier
Plukenetia L.
Podadenia Thwaites
Podocalyx Klotzsch
Pogonophora Miers ex Benth.
Poilaniella Gagnep.
Polyandra Leal *
Poranthera Rudge
Protomegabaria Hutch.
Pseudagrostistachys Pax & K.Hoffm.
Pseudanthus Sieber ex A.Spreng.
Pseudocroton Müll.Arg.
Pseudolachnostylis Pax
Pseudosagotia Secco *
Pterococcus Hassk.
Ptychopyxis Miq.
Putranjiva Wall.
Pycnocoma Benth.
Reutealis Airy Shaw
Reverchonia A.Gray
Richeria Vahl
Richeriella Pax & K.Hoffm.
Ricinocarpos Desf.
Ricinodendron Müll.Arg.
Ricinus L.
Rockinghamia Airy Shaw
Romanoa Trevis.
Sagotia Baill.

Sampantaea Airy Shaw
Sandwithia Lanj.
Sapium P.Browne
Sauropus Blume
Savia Willd.
Scagea McPherson
Schinziophyton Hutch. ex Radcl.–Sm.
Sebastiania Spreng.
Securinega Comm. ex Juss.
Seidelia Baill.
Senefeldera Mart.
Senefelderopsis Steyerm.
Sibangea Oliv.
Spathiostemon Blume
Speranskia Baill.
Sphaerostylis Baill.
Sphyranthera Hook.f.
Spirostachys Sond.
Spondianthus Engl.
Stachyandra J.–F.Leroy ex Radcl.–Sm.
Stachystemon Planch.
Stillingia Garden ex L.
Strophioblachia Boerl.
Sumbaviopsis J.J.Sm.
Suregada Roxb. ex Rottler
Synadenium Boiss.
Syndyophyllum K.Schum. & Lauterb.
Tacarcuna Huft *
Tannodia Baill.
Tapoides Airy Shaw
Tetracarpidium Pax
Tetracoccus Engelm. ex Parry
Tetraplandra Baill.
Tetrorchidium Poepp.
Thecacoris A.Juss.
Thyrsanthera Pierre ex Gagnep.
Tragia L.
Tragiella Pax & K.Hoffm.
Trewia L.
Trigonopleura Hook.f.
Trigonostemon Blume
Uapaca Baill.
Vaupesia R.E.Schult.
Veconcibea (Müll.Arg.) Pax & K.Hoffm.
Vernicia Lour.
Voatamalo Capuron ex Bosser
Wetria Baill.
Whyanbeelia Airy Shaw & B.Hyland
Wielandia Baill.
Zimmermannia Pax
Zimmermanniopsis Radcl.–Sm.

EUPHRONIACEAE Marc.-Berti 1989 20.05

1 genus. Colombia, Venezuela, N. Brazil. Trees or shrubs.

B&H	POLYPET., THALAMIFLORAE	Polygalineae (within Vochysiaceae, 20a)
DT&H	ARCHICHLAMYDEAE	Rosales (within Rosaceae, 76)
	(but *Lightia* in Geraniales, Trigoniaceae, 192 !)	
Melc	ARCHICHLAMYDEAE	Rutales, Malpighiineae (within Trigoniaceae, 139)
Thor	Not mentioned	—
Dahl	Not mentioned	—
Young	Not mentioned	—
Takh	ROSIDAE	Polygalales (within Trigoniaceae, 292)
Cron	ROSIDAE	Polygalales (within Vochysiaceae, 242)

Euphronia Mart. & Zucc.

EUPOMATIACEAE Endl. 1841 5.02

1 genus. New Guinea, E. Australia. Shrubs to small trees.

B&H	POLYPETALAE, THALAMIFLORAE	Ranales (within Annonaceae, 5)
DT&H	ARCHICHLAMYDEAE	Ranales, Magnoliineae (within Annonaceae, 48)
Melc	ARCHICHLAMYDEAE	Magnoliales, Magnoliineae, 44
Thor	ANNONIFLORAE	Annonales, Annonineae, 8
Dahl	MAGNOLIIFLORAE	Annonales, 3
Young	MAGNOLIIDAE, MAGNOLIANAE	Magnoliales, 2
Takh	MAGNOLIIDAE, MAGNOLIANAE	Eupomatiales, 4
Cron	MAGNOLIIDAE	Magnoliales, 4

Eupomatia R.Br.

EUPTELEACEAE K.Wilh. 1910 4.10

1 genus. Assam to China, Japan.

B&H	POLYPETALAE, THALAMIFLORAE	Ranales (within Magnoliaceae, 4)
DT&H	ARCHICHLAMYDEAE	Ranales, Trochodendrineae (within Trochodendraceae, 40)
Melc	ARCHICHLAMYDEAE	Magnoliales, Eupteliineae, 59
Thor	HAMAMELIDIFLORAE	Hamamelidales, Trochodendrineae, 195
Dahl	ROSIFLORAE	Cercidiphyllales, 140
Young	ROSIDAE, HAMAMELIDANAE	Trochodendrales, 219
Takh	HAMAMELIDIDAE, TROCHODENDRANAE	Eupteleales, 79
Cron	HAMAMELIDIDAE	Hamamelidales, 43

Euptelea Siebold & Zucc.

FABACEAE — see LEGUMINOSAE

FAGACEAE Dumort. 1829 159.03

8 genera. Widespread except tropical Africa. Mostly trees.

B&H	MONOCHLAMYDEAE	Unisexuales (within Betulaceae, 159)
DT&H	ARCHICHLAMYDEAE	Fagales, 12
Melc	ARCHICHLAMYDEAE	Fagales, 9
Thor	HAMAMELIDIFLORAE	Fagales, 201
Dahl	ROSIFLORAE	Fagales, 146
Young	ROSIDAE, HAMAMELIDANAE	Fagales, 226.
Takh	HAMAMELIDIDAE, HAMAMELIDANAE	
		Fagales, 93
Cron	HAMAMELIDIDAE	Fagales, 61

Castanea Mill.
Castanopsis (D.Don) Spach
Chrysolepis Hjelmq.
Fagus L.

Lithocarpus Blume
Nothofagus Blume
Quercus L.
Trigonobalanus Forman

FLACOURTIACEAE Rich. ex DC. 1824 17.03

79 genera. Mostly tropics, occas. warm-temperate. Trees, shrubs.

B&H	POLYPETALAE, THALAMIFLORAE	Parietales (within Bixaceae, 17)
	(separating *Samydaceae* to CALYCIFLORAE Passiflorales)	
DT&H	ARCHICHLAMYDEAE	Parietales, Flacourtiineae, 149
	(including Dioncophyllaceae, Hoplestigmataceae, Peridiscaceae, Physenaceae, Plagiopteraceae)	
Melc	ARCHICHLAMYDEAE	Violales, Flacourtiineae, 182
	(including Lacistemataceae)	
Thor	VIOLIFLORAE	Violales, Violineae, 124
	(including Lacistemataceae)	
Dahl	VIOLIFLORAE	Violales, 84
	(including Lacistemataceae)	
Young	DILLENIIDAE, VIOLANAE	Violales 190
	(including Lacistemataceae)	
Takh	DILLENIIDAE, VIOLANAE	Violales, 143
	(separating *Aphloiaceae* 142, *Berberidopsidaceae* 141, *Kiggelariaceae* 144)	
Cron	DILLENIIDAE	Violales, 107
	(including Plagiopteraceae)	

Abatia Ruíz & Pav.
Ahernia Merr.
Aphaerema Miers
Aphloia (DC.) Bennett
Azara Ruíz & Pav.
Baileyoxylon C.T.White
Banara Aubl.
Bartholomaea Standl. & Steyerm.
Bembicia Oliv.
Bennettiodendron Merr.
Berberidopsis Hook.f.
Bivinia Jaub. ex Tul.
Buchnerodendron Gürke

Byrsanthus Guill.
Calantica Jaub. ex Tul.
Caloncoba Gilg
Camptostylus Gilg
Carpotroche Endl.
Carrierea Franch.
Casearia Jacq.
Chiangiodendron Wendt *
Chlorocarpa Alston
Dasylepis Oliv.
Dissomeria Hook.f. ex Benth.
Dovyalis E.Mey. ex Arn.
Eleutherandra Slooten

Erythrospermum Lam.
Euceraea Mart.
Flacourtia Comm. ex L'Hér.
Gerrardina Oliv.
Grandidiera Jaub.
Gynocardia R.Br.
Hecatostemon S.F.Blake
Hemiscolopia Slooten
Homalium Jacq.
Hydnocarpus Gaertn.
Idesia Maxim.
Itoa Hemsl.
Kiggelaria L.
Laetia L.
Lasiochlamys Pax & K.Hoffm.
Lindackeria C.Presl
Ludia Comm. ex Juss.
Lunania Hook.
Mayna Aubl.
Mocquerysia Hua
Neopringlea S.Watson
Neoptychocarpus Buchheim
Olmediella Baill.
Oncoba Forssk.
Ophiobotrys Gilg
Osmelia Thwaites
Pangium Reinw.

Peterodendron Sleumer
Phyllobotryon Müll.Arg.
Phylloclinium Baill.
Pineda Ruíz & Pav.
Poggea Gürke
Poliothyrsis Oliv.
Priamosia Urb.
Prockiopsis Baill.
Pseudoscolopia Gilg
Pseudosmelia Sleumer
Rawsonia Harv. & Sond.
Ryania Vahl
Ryparosa Blume
Samyda Jacq.
Scaphocalyx Ridl.
Scolopia Schreb.
Scottellia Oliv.
Streptothamnus F.Muell.
Tetrathylacium Poepp.
Tisonia Baill.
Trichadenia Thwaites
Trichostephanus Gilg
Trimeria Harv.
Xylosma G.Forst.
Xylotheca Hochst.
Zuelania A.Rich.

FOUQUIERIACEAE DC. 1828   24.02

1 genus. S.W. USA to Mexico. Shrubs, trees, cactus-like.

B&H	POLYPET,, THALAMIFLORAE	Caryophyllineae (within Tamaricaceae, 24)
DT&H	ARCHICHLAMYDEAE	Parietales, Fouquieriineae, 142
Melc	SYMPETALAE	Tubiflorae, Convolvulineae, 253
Thor	SOLANIFLORAE	Solanales, Fouquieriineae, 274
Dahl	CORNIFLORAE	Fouquieriales, 285
Young	ROSIDAE, SOLANANAE	Solanales, 366
Takh	DILLENIIDAE, VIOLANAE	Fouquieriales, 157
Cron	DILLENIIDAE	Violales, 125

Fouquieria Kunth

FRANKENIACEAE St.-Hil. ex Gray 1821   21

5 genera. Warm-temp. and subtrop. Herbs, shrubs.

B&H	POLYPETALAE, THALAMIFLORAE	Caryophyllineae, 21
DT&H	ARCHICHLAMYDEAE	Parietales, Tamaricineae, 140
Melc	ARCHICHLAMYDEAE	Vilales, Tamaricineae, 196
Thor	VIOLIFLORAE	Violales, Tamaricineae, 136
Dahl	VIOLIFLORAE	Tamaricales, 99

Young	DILLENIIDAE, VIOLANAE	Tamaricales, 206
Takh	DILLENIIDAE, VIOLANAE	Tamaricales, 156
Cron	DILLENIIDAE	Violales, 117

Anthobryum Phil.
Beatsonia Roxb.
Frankenia L.

Hypericopsis Boiss.
Niederleinia Hieron.

GARRYACEAE Lindl. 1834   82.03

1 genus. USA to Panama and W. Indies. Trees, shrubs.

B&H	POLYPETALAE, CALYCIFLORAE	Umbellales (within Cornaceae, 82)
DT&H	ARCHICHLAMYDEAE	Umbelliflorae (within Cornaceae, 181)
Melc	ARCHICHLAMYDEAE	Umbelliflorae, 224
Thor	CORNIFLORAE	Cornales, Cornineae, 286
Dahl	CORNIFLORAE	Cornales, 298
Young	ROSIDAE, CORNANAE	Cornales, 327
Takh	ROSIDAE, CORNANAE	Cornales, 346
Cron	ROSIDAE	Cornales, 204

Garrya Douglas ex Lindl.

GEISSOLOMATACEAE Endl. 1841   146.02

1 genus, monotypic. South Africa. Shrub.

B&H	MONOCHLAMYDEAE	Daphnales (within Penaeaceae, 146)
DT&H	ARCHICHLAMYDEAE	Myrtiflorae, Thymelaeineae, 161
Melc	ARCHICHLAMYDEAE	Thymelaeales, 177
Thor	ROSIFLORAE	Pittosporales, Bruniineae, 231
Dahl	ROSIFLORAE	Geissolomatales, 144
Young	ROSIDAE, HAMAMELIDANAE	Myrothamnales, 224
Takh	ROSIDAE, ROSANAE	Geissolomatales, 201
Cron	ROSIDAE	Celastrales, 218

Geissoloma Lindl. ex Kunth

GENTIANACEAE Juss. 1789   109.01

76 genera. Widespread. Herbs to rarely small trees.

B&H	GAMOPETALAE, BICARPELLATAE (including Menyanthaceae)	Gentianales, 109
DT&H	METACHLAMYDEAE (including Menyanthaceae)	Contortae, Gentianineae, 199
Melc	SYMPETALAE	Gentianales, 247
Thor	GENTIANIFLORAE (including Saccifoliaceae)	Gentianales, 254
Dahl	GENTIANIFLORAE	Gentianales, 344

Young ROSIDAE, GENTIANANAE   Gentianales, 352
Takh   LAMIIDAE, GENTIANANAE   Gentianales, 371
Cron   ASTERIDAE   Gentianales, 272

Bartonia Mühl. ex Willd.
Bisgoeppertia Kuntze
Blackstonia Huds.
Canscora Lam.
Celiantha Maguire
Centaurium Hill
Chironia L.
Chorisepalum Gleason & Wodehouse
Cicendia Adans.
Comastoma Toyok.
Congolanthus A.Raynal
Cotylanthera Blume
Coutoubea Aubl.
Cracosna Gagnep.
Crawfurdia Wall.
Curtia Cham. & Schltdl.
Deianira Cham. & Schltdl.
Djaloniella P.Taylor
Enicostema Blume
Eustoma Salisb.
Exaculum Caruel
Exacum L.
Faroa Welw.
Frasera Walter
Geniostemon Engelm. & A.Gray
Gentiana L.
Gentianella Moench
Gentianopsis Ma
Gentianothamnus Humbert
Halenia Borkh.
Hockinia Gardner
Hoppea Willd.
Irlbachia Mart.
Ixanthus Griseb.
Jaeschkea Kurz
Karina Boutique
Lapithea Griseb.
Latouchea Franch.

Lehmanniella Gilg
Lisianthius P.Browne
Lomatogoniopsis T.N.Ho & S.W.Liu *
Lomatogonium A.Braun
Macrocarpaea Gilg
Megacodon (Hemsl.) Harry Sm.
Microrphium C.B.Clarke
Neblinantha Maguire
Neurotheca Salisb. ex Benth.
Obolaria L.
Oreonesion J.Raynal
Ornichia Klack.
Orphium E.Mey.
Prepusa Mart.
Pterygocalyx Maxim.
Pycnosphaera Gilg
Rogersonanthus Maguire & B.M.Boom
Sabatia Adans.
Schinziella Gilg
Schultesia Mart.
Sebaea Sol. ex R.Br.
Senaea Taub.
Sipapoantha Maguire & B.M.Boom *
Swertia L.
Symbolanthus G.Don
Symphyllophyton Gilg
Tachia Aubl.
Tachiadenus Griseb.
Tapeinostemon Benth.
Tetrapollinia Mauire & B.M.Boom
Tripterospermum Blume
Urogentias Gilg & Gilg–Ben.
Veratrilla (Baill.) Franch.
Voyria Aubl.
Voyriella Miq.
Wurdackanthus Maguire *
Zonanthus Griseb.
Zygostigma Griseb.

GERANIACEAE Juss. 1789   38.01

11 genera. Widespread. Herbs, some shrubs.

B&H   POLYPETALAE, DISCIFLORAE Geraniales, 38
          (including Balsaminaceae, Limnanthaceae, Oxalidaceae, Tropaeolaceae)
DT&H ARCHICHLAMYDEAE   Geraniales, Geraniineae, 79
Melc   ARCHICHLAMYDEAE   Geraniales, Geraniineae, 124
Thor   GERANIIFLORAE   Geraniales, Geraniineae, 101
       (separating *Ledocarpaceae* 103, *Vivianiaceae* 102)

Dahl   RUTIFLORAE   Geraniales, 241
        (separating *Biebersteiniaceae* 244, *Dirachmaceae* 245, *Ledocarpaceae* 243,
        Vivianiaceae 242)
Young ROSIDAE, GERANIANAE   Geraniales, 275
Takh ROSIDAE, RUTANAE   Geraniales, 284
        (separating *Biebersteiniaceae* 283, *Dirachmaceae* 285, *Ledocarpaceae* 286,
        Rhynchothecaceae 287, *Vivianiaceae* 288)
Cron ROSIDAE   Geraniales, 264

Balbisia Cav.
Biebersteinia Stephan
Dirachma Schweinf. ex Balf.f.
Erodium L'Hér. ex Aiton
Geranium L.
Monsonia L.

Pelargonium L'Hér. ex Aiton
Rhynchotheca Ruíz & Pav.
Sarcocaulon (DC.) Sweet
Viviania Cav.
Wendtia Meyen

GESNERIACEAE Dumort. 1822   119

133 genera. Trop. and subtrop., occas. warm-temp. Herbs, shrubs, occas. trees.

B&H   GAMOPETALAE, BICARPELLATAE Personales, 119
DT&H METACHLAMYDEAE   Tubiflorae, Solanineae, 215
Melc SYMPETALAE   Tubiflorae, Solanineae, 272
Thor GENTIANIFLORAE   Bignoniales, 264
Dahl LAMIIFLORAE   Scrophulariales, 350
Young ROSIDAE, GENTIANANAE   Bignoniales, 362
Takh LAMIIDAE, LAMIANAE   Scrophulariales, 404
Cron ASTERIDAE   Scrophulariales, 298

Acanthonema Hook.f.
Achimenes Pers.
Aeschynanthus Jack
Agalmyla Blume
Allocheilos W.T.Wang
Alloplectus Mart.
Allostigma W.T.Wang *
Ancylostemon Craib
Anetanthus Hiern ex Benth.
Anna Pellegr.
Anodiscus Benth.
Asteranthera Hanst.
Beccarinda Kuntze
Bellonia L.
Besleria L.
Boea Comm. ex Lam.
Boeica T.Anderson ex C.B.Clarke
Bournea Oliv.
Briggsia Craib
Briggsiopsis K.Y.Pan
Capanea Decne.
Cathayanthe Chun *
Championia Gardner

Chirita Buch.–Ham. ex D.Don
Chiritopsis W.T.Wang *
Chrysothemis Decne.
Codonanthe (Mart.) Hanst.
Codonanthopsis Mansf.
Codonoboea Ridl.
Colpogyne B.L.Burtt
Columnea L.
Conandron Siebold & Zucc.
Coptocheile Hoffmanns. *
Corallodiscus Batalin
Coronanthera Vieill. ex C.B.Clarke
Corytoplectus Oerst.
Cremosperma Benth.
Cubitanthus Barringer *
Cyrtandra J.R.Forst. & G.Forst.
Dayaoshania W.T.Wang *
Depanthus S.Moore
Diastema Benth.
Didissandra C.B.Clarke
Didymocarpus Wall.
Dolicholoma D.Fang & W.T.Wang *
Drymonia Mart.

Episcia Mart.
Epithema Blume
Eucodonia Hanst.
Fieldia A.Cunn.
Gasteranthus Benth.
Gesneria L.
Gloxinia L'Hér.
Goyazia Taub.
Haberlea Friv.
Hemiboea C.B.Clarke
Heppiella Regel
Hexatheca C.B.Clarke
Hygea Hanst. *
Isometrum Craib
Jancaea Boiss.
Koellikeria Regel
Kohleria Regel
Lembocarpus Leeuwenb.
Lenbrassia G.W.Gillett
Leptoboea Benth.
Lietzia Regel
Linnaeopsis Engl.
Loxocarpus R.Br.
Loxonia Jack
Loxostigma C.B.Clarke
Lysionotus D.Don
Metabriggsia W.T.Wang *
Micraeschynanthus Ridl.
Mitraria Cav.
Monophyllaea R.Br.
Monopyle Moritz ex Benth.
Moussonia Regel
Napeanthus Gardner
Nautilocalyx Linden ex Hanst.
Negria F.Muell.
Nematanthus Schrad.
Neomortonia Wiehler
Niphaea Lindl.
Nodonema B.L.Burtt
Oerstedina Wiehler *
Opithandra B.L.Burtt
Orchadocarpa Ridl.
Oreocharis Benth.
Ornithoboea Parish ex C.B.Clarke

Oxychlamys Schltr.
Paliavana Vell. ex Vand.
Paraboea (C.B.Clarke) Ridl.
Paradrymonia Hanst.
Parakohleria Wiehler
Pearcea Regel
Petrocodon Hance
Petrocosmea Oliv.
Pheidonocarpa L.E.Skog *
Phinaea Benth.
Phylloboea Benth.
Platyadenia B.L.Burtt
Platystemma Wall.
Primulina Hance
Pseudochirita W.T.Wang *
Ramonda Rich.
Reldia Wiehler
Resia H.E.Moore
Rhabdothamnopsis Hemsl.
Rhabdothamnus A.Cunn.
Rhoogeton Leeuwenb.
Rhynchoglossum Blume
Rhynchotechum Blume
Rhytidophyllum Mart.
Rufodorsia Wiehler
Saintpaulia H.A.Wendl.
Sarmienta Ruíz & Pav.
Schizoboea (Fritsch) B.L.Burtt
Sepikea Schltr.
Sinningia Nees
Smithiantha Kuntze
Solenophora Benth.
Stauranthera Benth.
Streptocarpus Lindl.
Tengia Chun
Tetraphyllum Griff. ex C.B.Clarke
Titanotrichum Soler.
Trachystigma C.B.Clarke
Tremacron Craib
Trisepalum C.B.Clarke
Tylopsacas Leeuwenb.
Vanhouttea Lem.
Whytockia W.W.Sm.

GISEKIACEAE (Endl.) Nakai 1942   79.02

1 genus. Africa. Arabia, trop. Asia. Herbs.

B&H   POLYPETALAE, CALYCIFLORAE Ficoidales (within Aizoaceae, 79)
DT&H ARCHICHLAMYDEAE   Centrospermae, Phytolaccineae (within
                                                              Phytolaccaceae, 33)

Melc	ARCHICHLAMYDEAE	Centrospermae, Phytolaccineae (within Molluginaceae, 30)
Thor	CHENOPODIIFLORAE	Chenopodiales, Chenopodiineae (within Phytolaccaceae, 81)
Dahl	Not mentioned	—
Young	DILLENIIDAE, CARYOPHYLLANAE	Chenopodiales, 156
Takh	CARYOPHYLLIDAE, CARYOPHYLLANAE	Caryophyllales, Caryophyllineae (within Phytolaccaceae, 57)
Cron	CARYOPHYLLIDAE	Caryophyllales (within Phytolaccaceae, 64)

Gisekia L.

GLAUCIDIACEAE Tamura 1972   1.04

1 genus, monotypic. Japan. Rhizomatous herb.

B&H	POLYPETALAE, THALAMIFLORAE	Ranales (within Ranunculaceae, 1)
DT&H	ARCHICHLAMYDEAE	Ranales, Ranunculineae (within Ranunculaceae, 41)
Melc	ARCHICHLAMYDEAE	Ranunculales, Ranunculineae (within Ranunculaceae, 61)
Thor	ANNONIFLORAE	Paeoniales, 27
Dahl	RANUNCULIFLORAE	Ranunculales, 36
Young	MAGNOLIIDAE, RANUNCULANAE	Ranunculales, 36
Takh	RANUNCULIDAE, RANUNCULANAE	Glaucidiales, 52
Cron	MAGNOLIIDAE	Ranunculales (within Ranunculaceae, 30)

Glaucidium Siebold & Zucc.

GLOBULARIACEAE DC. 1805   116

2 genera. Macaronesia to Turkey, N.E. tropical Africa. Herbs, subshrubs.

B&H	GAMOPETALAE, BICARPELLATAE	Lamiales (within *Selaginaceae*, 124)
DT&H	METACHLAMYDEAE	Tubiflorae, Solanineae, 218
Melc	SYMPETALAE	Tubiflorae, Solanineae, 266
Thor	GENTIANIFLORAE	Bignoniales (within Scrophulariaceae, 260)
Dahl	LAMIIFLORAE	Scrophulariales, 353
Young	Not mentioned	—
Takh	LAMIIDAE, LAMIANAE	Scrophulariales, 397
Cron	ASTERIDAE	Scrophulariales, 295

Globularia L.
Poskea Vatke

GOETZEACEAE Miers ex Airy Shaw 1965   114.02

4 genera. West Indies. Shrubs, small trees.

| B&H | GAMOPETALAE, BICARPELLATAE | Polemoniales (within Solanaceae, 114) |

DT&H	METACHLAMYDEAE	Tubiflorae, Solanineae (within Solanaceae, 209)
Melc	SYMPETALAE	Tubiflorae, Solanineae (within Solanaceae, 262)
Thor	SOLANIFLORAE	Solanales, Solanineae (within Solanaceae, 271)
Dahl	SOLANIFLORAE	Solanales, 275
Young	Not mentioned	—
Takh	LAMIIDAE, SOLANANAE	Solanales, 383
Cron	ASTERIDAE	Solanales (within Solanaceae, 278)

Coeloneurum Radlk.
Espadaea A.Rich.

Goetzea Wydl.
Henoonia Griseb.

GOMORTEGACEAE Reiche 1896   142.04

1 genus, monotypic. Chile. Tree.

B&H	MONOCHLAMYDEAE	Daphnales, incertae sedis
DT&H	ARCHICHLAMYDEAE	Ranales, Magnoliineae, 50
Melc	ARCHICHLAMYDEAE	Magnoliales, Laurineae, 54
Thor	ANNONIFLORAE	Annonales, Laurineae, 18
Dahl	MAGNOLIIFLORAE	Laurales, 20
Young	MAGNOLIIDAE, RANUNCULANAE	Laurales, 20
Takh	MAGNOLIIDAE, MAGNOLIANAE	Laurales, 17
Cron	MAGNOLIIDAE	Laurales, 14

Gomortega Ruíz & Pav.

GOODENIACEAE R.Br. 1810   90.01

12 genera. Australasia, also trop. and E. Asia, S. Amer., Ind. Ocean shores. Mostly shrubs, herbs.

B&H	GAMOPETALAE, INFERAE	Campanales (as *Goodenovieae*) 90
DT&H	METACHLAMYDEAE	Campanulatae, Campanulineae, 230
Melc	SYMPETALAE (separating *Brunoniaceae* 287)	Campanulales, 286
Thor	SOLANIFLORAE	Campanulales, 277
Dahl	GENTIANIFLORAE	Goodeniales, 33
Young	ROSIDAE, SOLANANAE (separating *Brunoniaceae* 374)	Campanulales, 373
Takh	ASTERIDAE, CAMPANULANAE (separating *Brunoniaceae* 427)	Goodeniales, 426
Cron	ASTERIDAE (separating *Brunoniaceae* 309)	Campanulales, 310

Anthotium R.Br.
Brunonia Sm.
Coopernookia Carolin
Dampiera R.Br.
Diaspasis R.Br.
Goodenia Sm.

Leschenaultia R.Br.
Pentaptilon E.Pritz.
Scaevola L.
Selliera Cav.
Velleia Sm.
Verreauxia Benth.

GOUPIACEAE Miers 1862   47.03

1 genus. Northern South America. Trees, shrubs.

B&H	POLYPETALAE, DISCIFLORAE	Celastrales (within Celastraceae, 47)
DT&H	ARCHICHLAMYDEAE	Sapindales, Celastrineae (within Celastraceae, 108)
Melc	ARCHICHLAMYDEAE	Celastrales, Celastrineae (within Celastraceae, 159)
Thor	SANTALIFLORAE	Celastrales (within Celastraceae, 112)
Dahl	SANTALIFLORAE	Celastrales (within Celastraceae, 249)
Young	(not mentioned, presumably within Celastraceae, 229)	
Takh	ROSIDAE, CELASTRANAE	Celastrales, Celastrineae, 305
Cron	ROSIDAE,	Celastrales (within Celastraceae, 219)

Goupia Aubl.

GREYIACEAE Hutch. 1926   59.10

1 genus. South Africa. Shrubs, small trees.

B&H	POLYPETALAE, DISCIFLORAE	Sapindales (within Sapindaceae, 51)
DT&H	ARCHICHLAMYDEAE	Sapindales, Melianthineae (within Melianthaceae, 117)
Melc	ARCHICHLAMYDEAE	Sapindales, Sapindineae (within Melianthaceae, 150)
Thor	ROSIFLORAE	Rosales, Saxifragineae, 212
Dahl	ROSIFLORAE	Saxifragales, 166
Young	ROSIDAE, ROSANAE	Saxifragales, 297
Takh	ROSIDAE, ROSANAE	Saxifragales, 211
Cron	ROSIDAE	Rosales, 167

Greyia Hook. & Harv.

GRISELINIACEAE (Wanger.) Takht. 1987   82.08

1 genus. New Zealand, Chile, Argentina, S. Brazil. Trees, shrubs.

B&H	POLYPETALAE, CALYCIFLORAE	Umbellales (within Cornaceae, 82)
DT&H	ARCHICHLAMYDEAE	Umbelliflorae (within Cornaceae, 181)
Melc	ARCHICHLAMYDEAE	Umbelliflorae (within Cornaceae, 223)
Thor	ROSIFLORAE	Rosales, Saxifragineae (within Saxifragaceae, 208)
Dahl	Not mentioned	—
Young	(not mentioned, presumably within Cornaceae, 325)	
Takh	ROSIDAE, CORNANAE	Hydrangeales, Escalloniineae, 327
Cron	ROSIDAE	Cornales (within Cornaceae, 203)

Griselinia J.R.Forst. & G.Forst.

GROSSULARIACEAE DC.1805   59.12

1 genus. N. hemisphere, Andes. Shrubs.

B&H		Rosales (within Saxifragaceae, 59)
DT&H	ARCHICHLAMYDEAE	Rosales, Saxifragineae (within Saxifragaceae, 67)
Melc	ARCHICHLAMYDEAE	Rosales, Saxifragineae (within Saxifragaceae, 106)
Thor	ROSIFLORAE	Rosales, Saxifragineae (within Saxifragaceae, 208)
Dahl	ROSIFLORAE	Saxifragales, 165
Young	ROSIDAE, ROSANAE	Saxifragales (see index) 301
Takh	ROSIDAE, ROSANAE	Saxifragales, 206
Cron	ROSIDAE	Rosales, 166

(including Escalloniaceae, Montiniaceae, Pterostemonaceae)

Ribes L.

GRUBBIACEAE Endl. 1839   149.03

1 genus. South Africa. Ericoid shrubs.

B&H	MONOCHLAMYDEAE	Achlamydosporeae (within Santalaceae, 149)
DT&H	ARCHICHLAMYDEAE	Santalales, Santalineae, 20
Melc	ARCHICHLAMYDEAE	Santalales, Santalineae, 19
Thor	ROSIFLORAE	Pittosporales, Bruniineae, 232
Dahl	ROSIFLORAE	Cunoniales, 162
Young	ROSIDAE, CORNANAE	Pittosporales, 341
Takh	DILLENIIDAE, ERICANAE	Ericales, 131
Cron	DILLENIIDAE	Ericales, 141

Grubbia Bergius

GUNNERACEAE Meisn. 1841   64.03

1 genus. S. temperate, extending to tropics and Hawai'i. Herbs.

B&H	POLYPETALAE, CALYCIFLORAE	Rosales (within Haloragaceae, 64)
DT&H	ARCHICHLAMYDEAE	Myrtiflorae, Halorrhagidineae (within Haloragaceae, 177)
Melc	ARCHICHLAMYDEAE	Myrtiflorae, Myrtineae (within Haloragaceae, 216)
Thor	CORNIFLORAE	Cornales, Haloragineae, 281
Dahl	ROSIFLORAE	Gunnerales, 173
Young	ROSIDAE, MYRTANAE	Haloragales, 321
Takh	ROSIDAE, ROSANAE	Gunnerales, 216
Cron	ROSIDAE	Haloragales, 187

Gunnera L.

GUTTIFERAE Juss. 1789   27.01

47 genera. Widespread. Trees, shrubs, lianes, herbs.

B&H   POLYPETALAE, THALAMIFLORAE   Guttiferales, 27
         (including Quiinaceae; separating *Hypericaceae* 26)
DT&H ARCHICHLAMYDEAE   Parietales, Theineae, 137
         (including Medusagynaceae)
Melc   ARCHICHLAMYDEAE   Guttiferales, Theineae, 89
Thor   THEIFLORAE   Theales, Hypericineae, 67
Dahl   THEIFLORAE   Theales, 126
         (separating *Bonnetiaceae*, 125)
Young DILLENIIDAE, DILLENIANAE   Hypericales (as *Hypericaceae*) 128
Takh   DILLENIIDAE, THEANAE   Theales, 113
         (separating *Bonnetiaceae* 112)
Cron   DILLENIIDAE   Theales, 97

Allanblackia Oliv. ex Benth.	**Marila** Sw.
Archytaea Mart.	**Mesua** L.
Bonnetia Mart.	**Montrouziera** Planch. & Triana
Calophyllum L.	**Moronobea** Aubl.
Caraipa Aubl.	**Neotatea** Maguire
Chrysochlamys Poepp.	**Oedematopus** Planch. & Triana
Clusia L.	**Pentadesma** Sabine
Clusiella Planch. & Triana	**Pilosperma** Planch. & Triana
Cratoxylum Blume	**Platonia** Mart.
Dystovomita (Engl.) D'Arcy *	**Ploiarium** Korth.
Eliea Cambess.	**Poeciloneuron** Bedd.
Endodesmia Benth.	**Psorospermum** Spach
Garcinia L.	**Quapoya** Aubl.
Haploclathra Benth.	**Renggeria** Meisn.
Harungana Lam.	**Santomasia** N.Robson *
Havetia Kunth	**Symphonia** L.f.
Havetiopsis Planch. & Triana	**Thornea** Breedlove & E.McClintock
Hypericum L.	**Thysanostemon** Maguire
Kayea Wall.	**Tovomita** Aubl.
Kielmeyera Mart.	**Tovomitidium** Ducke
Lebrunia Staner	**Tovomitopsis** Planch. & Triana
Lorostemon Ducke	**Triadenum** Raf.
Mahurea Aubl.	**Vismia** Vand.
Mammea L.	

GYROSTEMONACEAE Endl. 1841   132.03

5 genera. Australia. Trees, shrubs.

B&H   MONOCHLAMYDEAE   Curvembryeae (within Phytolaccaceae, 132)
DT&H ARCHICHLAMYDEAE   Centrospermae, Phytolaccineae (within
                                          Phytolaccaceae, 33)
Melc   ARCHICHLAMYDEAE   Centrospermae, Phytolaccineae, 27
Thor   RUTIFLORAE   Rutales, Sapindineae, 181
Dahl   VIOLIFLORAE   Capparales, 104
Young DILLENIIDAE, VIOLANAE   Capparales, 211

Takh DILLENIIDAE, VIOLANAE Batales, 168
Cron DILLENIIDAE Batales, 137

Codonocarpus A.Cunn. ex Endl.
Cypselocarpus F.Muell.
Gyrostemon Desf.

Tersonia Moq.
Walteranthus Keighery *

HALOPHYTACEAE Soriano 1984 131.03

1 genus, monotypic. Argentina. Succulent annual herb.

B&H Not known —
DT&H ARCHICHLAMYDEAE Centrospermae, Chenopodiineae (within
  Chenopodiaceae, 28)

Melc Not mentioned —
Thor CHENOPODIIFLORAE Chenopodiales, Chenopodiineae, 87
Dahl CARYOPHYLLIFLORAE Caryophyllales, 45
Young DILLENIIDAE, CARYOPHYLLANAE Chenopodiales, 163
Takh CARYOPHYLLIDAE Caryophyllales, 68
Cron CARYOPHYLLIDAE Caryophyllales (within Chenopodiaceae, 70)

Halophytum Speg.

HALORAGACEAE R.Br. 1814 64.01

9 genera. Widespread, esp. Australasia. Herbs, often aquatic, rarely shrubs.

B&H POLYPETALAE, CALYCIFLORAE Rosales, 64
  (including Callitrichaceae, Gunneraceae, Hippuridaceae)
DT&H ARCHICHLAMYDEAE Myrtiflorae, Halorrhagidineae, 177
  (including Gunneraceae, Hippuridaceae)
Melc ARCHICHLAMYDEAE Myrtiflorae, Myrtineae, 216
  (including Gunneraceae)
Thor CORNIFLORAE Cornales, Haloragineae, 280
Dahl MYRTIFLORAE Haloragales, 183
Young ROSIDAE, MYRTANAE Haloragales, 320
Takh ROSIDAE, MYRTANAE Haloragales, 200
Cron ROSIDAE Haloragales, 186

Glischrocaryon Endl.
Gonocarpus Thunb.
Haloragis J.R.Forst. & G.Forst.
Haloragodendron Orchard
Laurembergia Bergius

Meziella Schindl.
Myriophyllum L.
Proserpinaca L.
Vinkia Meiden

HAMAMELIDACEAE R.Br. 1818 62.01

29 genera. Mostly subtropical, N. and S. Trees, shrubs.

B&H POLYPETALAE, CALYCIFLORAE Rosales, 62
  (including Myrothamnaceae)

DT&H	ARCHICHLAMYDEAE	Rosales, Saxifragineae, 73
Melc	ARCHICHLAMYDEAE	Rosales, Hamamelidineae, 102
Thor	HAMAMELIDIFLORAE	Hamamelidales, Hamamelidineae, 198
Dahl	ROSIFLORAE	Hamamelidales, 141
Young	ROSIDAE, HAMAMELIDANAE	Hamamelidales, 221
Takh	HAMAMELIDIDAE, HAMAMELIDANAE, Hamamelidales, 81	

(separating *Altingiaceae* 83, *Rhodoleiaceae* 82)

Cron	HAMAMELIDIDAE	Hamamelidales, 45

Altingia Noronha
Chunia H.T.Chang *
Corylopsis Siebold & Zucc.
Dicoryphe Thouars
Disanthus Maxim.
Distyliopsis P.K.Endress
Distylium Siebold & Zucc.
Embolanthera Merr.
Eustigma Gardner & Champ.
Exbucklandia R.W.Br.
Fortunearia Rehder & E.H.Wilson
Fothergilla L.
Hamamelis L.
Liquidambar L.
Loropetalum R.Br. ex Rchb.

Maingaya Oliv.
Matudaea Lundell
Molinadendron P.K.Endress
Mytilaria Lecomte
Neostrearia L.S.Sm.
Noahdendron Endress, B.Hyland &
    Tracey *
Ostrearia Baill.
Parrotia C.A.Mey.
Parrotiopsis (Nied.) C.K.Schneid.
Rhodoleia Champ. ex Hook.
Sinowilsonia Hemsl.
Sycopsis Oliv.
Tetrathyrium Benth.
Trichocladus Pers.

HECTORELLACEAE Philipson & Skipw. 1961   23.02

2 genera. New Zealand, Kerguelen. Caespitose herbs.

B&H	POLYPETALAE, THALAMIFLORAE	Caryophyllineae (within Portulacaceae, 23)
DT&H	ARCHICHLAMYDEAE	Centrospermae, Portulacineae (within Portulacaceae, 35)
Melc	ARCHICHLAMYDEAE	Centrospermae, Caryophyllineae (within Caryophyllaceae, 34)
Thor	CHENOPODIIFLORAE	Chenopodiales, Portulacineae (within Portulacaceae, 91)
Dahl	CARYOPHYLLIFLORAE	Caryophyllales, 48
Young	DILLENIIDAE, CARYOPHYLLANAE	Caryophyllales, 150
Takh	CARYOPHYLLIDAE, CARYOPHYLLANAE	
		Caryophyllales, Caryophyllineae, 66
Cron	CARYOPHYLLIDAE	Caryophyllales (within Portulacaceae, 72)

Hectorella Hook.f.
Lyallia Hook.f.

HELWINGIACEAE Decne. 1836   81.02

1 genus. Himalaya to Japan. Shrubs.

B&H	POLYPETALAE, CALYCIFLORAE	Umbellales (within Araliaceae, 81)
DT&H	ARCHICHLAMYDEAE	Umbelliflorae (within Cornaceae, 181)
Melc	ARCHICHLAMYDEAE	Umbelliflorae (within Cornaceae, 223)

Thor CORNIFLORAE
Dahl CORNIFLORAE
Young Not mentioned
Takh ROSIDAE, CORNANAE
Cron ROSIDAE

Araliales, 288
Cornales, 304
—
Apiales, 350
Cornales (within Cornaceae, 203)

Helwingia Willd.

HERNANDIACEAE Blume 1826   143.02

4 genera. Tropical. Trees, shrubs, lianes.

B&H   MONOCHLAMYDEAE
DT&H ARCHICHLAMYDEAE
Melc ARCHICHLAMYDEAE
Thor ANNONIFLORAE
Dahl MAGNOLIIFLORAE
Young MAGNOLIIDAE, RANUNCULANAE
Takh MAGNOLIIDAE, MAGNOLIANAE
      (separating *Gyrocarpaceae* 22)
Cron MAGNOLIIDAE

Daphnales (within Lauraceae, 143)
Ranales, Magnoliineae, 53
Magnoliales, Laurineae, 56
Annonales, Laurineae, 21
Laurales (within Lauraceae, 22)
Laurales, 22
Laurales, 18

Laurales, 18

Gyrocarpus Jacq.
Hernandia L.

Illigera Blume
Sparattanthelium Mart.

HIMANTANDRACEAE Diels 1917   4.04

1 genus. E. Malesia to N.E. Australia. Trees.

B&H   Not known
DT&H ARCHICHLAMYDEAE

Melc ARCHICHLAMYDEAE
Thor ANNONIFLORAE
Dahl MAGNOLIIFLORAE
Young MAGNOLIIDAE, RANUNCULANAE
Takh MAGNOLIIDAE, MAGNOLIANAE
Cron MAGNOLIIDAE

—
Ranales, Magnoliineae (within Annonaceae, 48)

Magnoliales, Magnoliineae, 41
Annonales, Annonineae, 6
Magnoliales, 11
Aristolochiales, 23
Magnoliales, 2
Magnoliales, 3

Galbulimima F.M.Bailey

HIPPOCASTANACEAE DC. 1824   51.02

2 genera. S. Europe, trop. and E. Asia, USA to S. America. Trees, shrubs.

B&H   POLYPETALAE, DISCIFLORAE
DT&H ARCHICHLAMYDEAE
      (including Bretschneideraceae)
Melc ARCHICHLAMYDEAE
Thor RUTIFLORAE

Sapindales (within Sapindaceae, 51)
Sapindales, Sapindineae, 114

Sapindales, Sapindineae, 148
Rutales, Sapindineae, 187

Dahl	RUTIFLORAE	Sapindales, 208
Young	ROSIDAE, RUTANAE	Sapindales, 260
Takh	ROSIDAE, RUTANAE	Sapindales, 247
Cron	ROSIDAE	Sapindales, 253

Aesculus L.
Billia Peyr.

HIPPURIDACEAE Link 1821   64.02

1 genus, ? monotypic. Widespread. Aquatic herbs.

B&H	POLYPETALAE, CALYCIFLORAE	Rosales (within Haloragaceae, 64)
DT&H	ARCHICHLAMYDEAE	Myrtiflorae, Halorrhagidineae (within Haloragaceae, 177)
Melc	ARCHICHLAMYDEAE	Myrtiflorae, Hippuridineae, 218
Thor	CORNIFLORAE	Cornales, Haloragineae, 282
Dahl	LAMIIFLORAE	Hippuridales, 363
Young	ROSIDAE, LAMIANAE	Hippuridales, 375
Takh	LAMIIDAE, LAMIANAE	Hippuridales, 412
Cron	ASTERIDAE	Callitrichales, 288

Hippuris L.

HOPLESTIGMATACEAE Gilg 1924   112.04

1 genus. Guineo-Congolan Africa. Trees.

B&H	Not known	—
DT&H	ARCHICHLAMYDEAE	Parietales, Flacourtiineae (within Flacourtiaceae, 149)
Melc	SYMPETALAE	Ebenales, Ebenineae, 243
Thor	INCERTAE SEDIS	350
Dahl	SOLANIFLORAE	Boraginales, 284
Young	ROSIDAE, SOLANANAE	Boraginales, 370
Takh	LAMIIDAE, SOLANANAE	Boraginales, 390
Cron	DILLENIIDAE	Violales, 126

Hoplestigma Pierre

HUACEAE A.Chev. 1947   34.06

2 genera. Guineo-Congolan Africa. Trees.

B&H	Not known	—
DT&H	ARCHICHLAMYDEAE	Malvales, Malvineae (within Sterculiaceae, 128)
Melc	UNCERTAIN, in note under Styracaceae	
Thor	MALVIFLORAE	Malvales, 145
Dahl	MALVIFLORAE	Malvales, 64

Young DILLENIIDAE, MALVANAE   Malvales, 165
Takh DILLENIIDAE, MALVANAE   Malvales, 183
Cron DILLENIIDAE   Violales, 111

Afrostyrax Perkins & Gilg
Hua Pierre ex De Wild.

HUMIRIACEAE A. Juss. 1829   35

8 genera. Trop. America, 1 sp. of *Sacoglottis* in Guineo-Congolan Africa. Trees, shrubs.

B&H   POLYPETALAE, DISCIFLORAE Geraniales, 35
DT&H ARCHICHLAMYDEAE   Geraniales, Geraniineae, 83
Melc ARCHICHLAMYDEAE   Geraniales, Geraniineae (within Linaceae,
                                                                     127)

Thor GERANIIFLORAE   Geraniales, Linineae, 95
Dahl RUTIFLORAE   Geraniales, 235
Young ROSIDAE, GERANIANAE   Linales, 269
Takh ROSIDAE, RUTANAE   Linales, 278
Cron ROSIDAE   Linales, 237

Duckesia Cuatrec.   **Hylocarpa** Cuatrec.
Endopleura Cuatrec.   **Sacoglottis** Mart.
Humiria Aubl.   **Schistostemon** (Urb.) Cuatrec.
Humiriastrum (Urb.) Cuatrec.   **Vantanea** Aubl.

HYDNORACEAE C. Agardh 1821   137.02

2 genera. Africa, South America. Root parasites.

B&H   MONOCHLAMYDEAE   Multiovulatae Terrestres (within
                                                        Rafflesiaceae, 137)

DT&H ARCHICHLAMYDEAE   Aristolochiales, 26
Melc ARCHICHLAMYDEAE   Aristolochiales, 74
Thor RAFFLESIIFLORAE   Rafflesiales, 39
Dahl MAGNOLIIFLORAE   Rafflesiales, 8
Young ROSIDAE, SANTALANAE   Rafflesiales, 243
Takh MAGNOLIIDAE, RAFFLESIANAE Hydnorales, 28
Cron ROSIDAE   Rafflesiales, 215

Hydnora Thunb.
Prosopanche Bary

HYDRANGEACEAE Dumort. 1829   59.06

16 genera. N. temperate to Malesia, Hawai'i. Shrubs, woody climbers, few herbs.

B&H   POLYPETALAE, CALYCIFLORAE Rosales (within Saxifragaceae, 59)
DT&H ARCHICHLAMYDEAE   Rosales, Saxifragineae (within
                                                        Saxifragaceae, 67)

Melc	ARCHICHLAMYDEAE	Rosales, Saxifragineae (within
		Saxifragaceae, 106)
Thor	ROSIFLORAE	Rosales, Saxifragineae (within
		Saxifragaceae, 208)
Dahl	CORNIFLORAE	Cornales, 320
Young	ROSIDAE, CORNANAE	Hydrangeales, 328
	(separating *Philadelphaceae* 329)	
Takh	ROSIDAE, CORNANAE	Hydrangeales, Hydrangeineae, 336
Cron	ROSIDAE	Rosales, 164

Broussaisia Gaudich.   **Fendlerella** A.Heller
Cardiandra Siebold & Zucc.   **Hydrangea** L.
Carpenteria Torr.   **Jamesia** Torr. & A.Gray
Decumaria L.   **Kirengeshoma** Yatabe
Deinanthe Maxim.   **Philadelphus** L.
Deutzia Thunb.   **Platycrater** Siebold & Zucc.
Dichroa Lour.   **Schizophragma** Siebold & Zucc.
Fendlera Engelm. & A.Gray   **Whipplea** Torr.

HYDROPHYLLACEAE R.Br. 1817   111

19 genera. Widespread, esp. N. and C. America. Mostly herbs.

B&H	GAMOPETALAE, BICARPELLATAE Polemoniales, 111	
DT&H	METACHLAMYDEAE	Tubiflorae, Boraginineae, 204
Melc	SYMPETALAE	Tubiflorae, Boraginineae, 255
Thor	SOLANIFLORAE	Solanales, Boraginineae, 268
Dahl	SOLANIFLORAE	Boraginales, 280
Young	ROSIDAE, SOLANANAE	Boraginales, 367
Takh	LAMIIDAE, SOLANANAE	Boraginales, 388
Cron	ASTERIDAE	Solanales, 283

Codon Royen ex L.   **Miltitzia** A.DC.
Draperia Torr.   **Nama** L.
Ellisia L.   **Nemophila** Nutt.
Emmenanthe Benth.   **Phacelia** Juss.
Eriodictyon Benth.   **Pholistoma** Lilja
Eucrypta Nutt.   **Romanzoffia** Cham.
Hesperochiron S.Watson   **Tricardia** Torr.
Hydrolea L.   **Turricula** J.F.Macbr.
Hydrophyllum L.   **Wigandia** Kunth
Lemmonia A.Gray

HYDROSTACHYACEAE Engl. 1898   135.02

1 genus. Trop. and South Africa, Madagascar. Lithophytes in rivers.

B&H	MONOCHLAMYDEAE	Multiovulatae Aquaticae (within
		Podostemaceae, 135)
DT&H	ARCHICHLAMYDEAE	Rosales, Podostemonineae, 64
Melc	ARCHICHLAMYDEAE	Hydrostachyales, 120
Thor	ROSIFLORAE	Pittosporales, Bruniineae, 234

Dahl LAMIIFLORAE Hydrostachyales, 367
Young ROSIDAE, LAMIANAE Hydrostachyales, 376
Takh LAMIIDAE, LAMIANAE Hydrostachyales, 416
Cron ASTERIDAE Callitrichales, 290

Hydrostachys Thouars

ICACINACEAE (Benth.) Miers 1851 45.03

53 genera. Tropics and subtropics. Trees, shrubs, lianes.

B&H POLYPETALAE, DISCIFLORAE Olacales (within Olacaceae, 45)
DT&H ARCHICHLAMYDEAE Sapindales, Icacinineae, 112
  (including Cardiopteridaceae, Lophopyxidaceae)
Melc ARCHICHLAMYDEAE Celastrales, Icacinineae, 165
Thor THEIFLORAE Theales, Theineae, 50
Dahl CORNIFLORAE Cornales, 314
Young DILLENIIDAE, DILLENIANAE Theales, 107
Takh ROSIDAE, CELASTRANAE Celastrales, Icacinineae, 300
Cron ROSIDAE Celastrales, 224

Alsodeiopsis Oliv.
Apodytes E.Mey. ex Arn.
Calatola Standl.
Cantleya Ridl.
Casimirella Hassl. *
Cassinopsis Sond.
Chlamydocarya Baill.
Citronella D.Don
Codiocarpus R.A.Howard
Dendrobangia Rusby
Desmostachys Miers
Discophora Miers
Emmotum Desv. ex Ham.
Gastrolepis Tiegh.
Gomphandra Wall. ex Lindl.
Gonocaryum Miq.
Grisollea Baill.
Hartleya Sleumer
Hosiea Hemsl. & E.H.Wilson
Humirianthera Huber
Icacina A.Juss.
Iodes Blume
Irvingbaileya R.A.Howard
Lasianthera P.Beauv.
Lavigeria Pierre
Leptaulus Benth.
Mappia Jacq.
Mappianthus Hand.–Mazz.
Medusanthera Seem.
Merrilliodendron Kaneh.
Metteniusa Karst.
Miquelia Meisn.
Natsiatopsis Kurz
Natsiatum Buch.–Ham. ex Arn.
Nothapodytes Blume
Oecopetalum Greenm. & C.H.Thomps.
Ottoschulzia Urb.
Pennantia J.R.Forst. & G.Forst.
Phytocrene Wall.
Pittosporopsis Craib
Platea Blume
Pleurisanthes Baill.
Polycephalium Engl.
Polyporandra Becc.
Poraqueiba Aubl.
Pseudobotrys Moeser
Pyrenacantha Wight
Rhaphiostylis Planch. ex Benth.
Rhyticaryum Becc.
Sarcostigma Wight & Arn.
Stachyanthus Engl.
Stemonurus Blume
Whitmorea Sleumer

IDIOSPERMACEAE S.T.Blake 1972 3.02

1 genus, monotypic. Queensland. Tree

B&H	Not known	—
DT&H	Not known	—
Melc	Not known	—
Thor	ANNONIFLORAE	Annonales, Laurineae (within Calycanthaceae, 19)
Dahl	MAGNOLIIFLORAE	Laurales (within Calycanthaceae, 21)
Young	MAGNOLIIDAE, RANUNCULANAE	Laurales, 14
Takh	MAGNOLIIDAE, MAGNOLIANAE	Laurales, 20
Cron	MAGNOLIIDAE	Laurales, 16

Idiospermum S.T.Blake

ILLECEBRACEAE R.Br. 1810   22.02

24 genera. Widespread. Mostly herbs, some dwarf shrubs.

B&H	MONOCHLAMYDEAE	Curvembryeae, 129
DT&H	ARCHICHLAMYDEAE	Centrospermae, Caryophyllineae (within Caryophyllaceae, 37)
Melc	ARCHICHLAMYDEAE	Centrospermae, Caryophyllineae (within Caryophyllaceae, 34)
Thor	CHENOPODIIFLORAE	Chenopodiales, Chenopodiineae (within Caryophyllaceae, 86)
Dahl	CARYOPHYLLIFLORAE	Caryophyllales (within Caryophyllaceae, 52)
Young	(not mentioned, presumably within Caryophyllaceae, 14)	
Takh	CARYOPHYLLIDAE, CARYOPHYLLANAE	Caryophyllales, Caryophyllineae (within Caryophyllaceae, 71)
Cron	CARYOPHYLLIDAE	Caryophyllales (within Caryophyllaceae, 75)

Achyronychia Torr. & A.Gray
Anychia Michx.
Cardionema DC.
Chaetonychia (DC.) Sweet
Cometes L.
Dicheranthus Webb
Geocarpon Mack. *
Gymnocarpos Forssk.
Haya Balf.f.
Herniaria L.
Illecebrum L.
Kabulia Bor & C.Fisch.

Lochia Balf.f.
Paronychia Mill.
Philippiella Speg.
Plettkea Mattf. *
Pollichia Aiton
Pteranthus Forssk.
Scleranthus L.
Sclerocephalus Boiss.
Scopulophila M.E.Jones
Siphonychia Torr. & A.Gray
Sphaerocoma T.Anderson
Xerotia Oliv.

ILLICIACEAE (DC.) A.C.Sm. 1947   4.03

1 genus. Trop. and subtrop. Asia, S.E. USA, Mexico, W. Indies. Trees, shrubs.

B&H	POLYPETALAE, THALAMIFLORAE	Ranales (within Magnoliaceae, 4)
DT&H	ARCHICHLAMYDEAE	Ranales, Magnoliineae (within Magnoliaceae, 45)
Melc	ARCHICHLAMYDEAE	Magnoliales, Illiciineae, 48
Thor	ANNONIFLORAE	Annonales, Illiciineae, 2
Dahl	MAGNOLIIFLORAE	Illiciales, 15

Young MAGNOLIIDAE, MAGNOLIANAE Illiciales, 7
Takh MAGNOLIIDAE, MAGNOLIANAE Illiciales, 9
Cron MAGNOLIIDAE   Illiciales, 23

Illicium L.

IRVINGIACEAE (Engl.) Exell & Mendonça 1951   34.04

3 genera. Tropical Africa, tropical S. America. Trees.

B&H POLYPETALAE, DISCIFLORAE Geraniales (within Simaroubaceae, 40)
DT&H ARCHICHLAMYDEAE   Geraniales, Geraniineae (within
                                            Simaroubaceae, 88)
Melc ARCHICHLAMYDEAE   Rutales, Rutineae (within
                                            Simaroubaceae, 133)
Thor RUTIFLORAE   Rutales, Rutineae (within
                                            Simaroubaceae, 169)
Dahl RUTIFLORAE   Rutales (within Simaroubaceae, 221)
Young (not mentioned, presumably within Simaroubaceae, 247)
Takh ROSIDAE, RUTANAE   Rutales, 260
Cron ROSIDAE   Sapindales (within Simaroubaceae, 258)

Desbordesia Pierre ex Tiegh.
Irvingia Hook.f.
Klainedoxa Pierre ex Engl.

IXONANTHACEAE (Benth.) Exell & Mendonça 1951   34.03

5 genera. Trop. and subtrop., not Madag., Australia or Pacific. Trees, shrubs

B&H POLYPETALAE, DISCIFLORAE Geraniales (within Linaceae, 34)
DT&H ARCHICHLAMYDEAE   Geraniales, Geraniineae (within
                                            Linaceae, 82)
Melc ARCHICHLAMYDEAE   Geraniales, Geraniineae (within
                                            Linaceae, 127)
Thor GERANIIFLORAE   Geraniales, Linineae (within Linaceae, 97)
Dahl RUTIFLORAE   Geraniales, 238
Young (not mentioned, presumably within Linaceae, 270)
Takh ROSIDAE, RUTANAE   Linales, 277
Cron ROSIDAE   Linales, 238

Allantospermum Forman   **Ochthocosmus** Benth.
Cyrillopsis Kuhlm.   **Phyllocosmus** Klotzsch
Ixonanthes Jack

JUGLANDACEAE A.Rich. ex Kunth 1824   156.01

8 genera. Mediterranean to trop. and E. Asia, N. and S. America. Mostly trees.

B&H MONOCHLAMYDEAE   Unisexuales, 156
DT&H ARCHICHLAMYDEAE   Juglandales, 10

JUGLANDACEAE (D)

Melc	ARCHICHLAMYDEAE	Juglandales, 3
Thor	RUTIFLORAE	Rutales, Juglandineae, 177
Dahl	ROSIFLORAE	Juglandales, 150
Young	ROSIDAE, RUTANAE	Juglandales, 267
Takh	HAMAMELIDIDAE, JUGLANDANAE	Juglandales, 97
Cron	HAMAMELIDIDAE	Juglandales, 58

Alfaroa Standl.
Carya Nutt.
Cyclocarya Iljinsk.
Engelhardtia Lesch. ex Blume

Juglans L.
Oreomunnea Oerst.
Platycarya Siebold & Zucc.
Pterocarya Kunth

KRAMERIACEAE Dumort. 1829   20.04

1 genus. Southern USA to Chile. Hemiparasitic shrubs, herbs, rarely trees.

B&H	POLYPETALAE, THALAMIFLORAE	Polygalineae (within Polygalaceae, 20)
DT&H	ARCHICHLAMYDEAE	Rosales, Rosineae (within Leguminosae-Caesalpinioideae, 78)
Melc	ARCHICHLAMYDEAE	Rosales, Leguminosineae, 119
Thor	GERANIIFLORAE	Geraniales, Polygalineae, 109
Dahl	RUTIFLORAE	Polygalales, 229
Young	ROSIDAE, GERANIANAE	Polygalales, 280
Takh	ROSIDAE, RUTANAE	Polygalales, 296
Cron	ROSIDAE	Polygalales, 247

Krameria L. ex Loefl.

LABIATAE Juss. 1789   126.01

212 genera. Widespread. Herbs, shrubs, rarely trees.

B&H	GAMOPETALAE, BICARPELLATAE	Lamiales, 126
DT&H	METACHLAMYDEAE	Tubiflorae, Verbenineae, 207
Melc	SYMPETALAE (including Tetrachondraceae)	Tubiflorae, Verbenineae, 260
Thor	GENTIANIFLORAE (including Tetrachondraceae)	Lamiales, 266
Dahl	LAMIIFLORAE (? including Tetrachondraceae)	Lamiales, 365
Young	ROSIDAE, LAMIANAE (presumably including Tetrachondraceae)	Lamiales, 379
Takh	LAMIIDAE, LAMIANAE (including Tetrachondraceae)	Lamiales, Lamiineae, 414
Cron	ASTERIDAE (including Tetrachondraceae)	Lamiales, 287

Acanthomintha (A.Gray) A.Gray
Achyrospermum Blume
Acinos Mill.
Acrocephalus Benth.
Acrotome Benth. ex Endl.

Acrymia Prain
Aeollanthus Mart. ex Spreng.
Agastache Gronov.
Ajuga L.
Ajugoides Makino

Alajja Ikonn.
Alvesia Welw.
Amethystea L.
Anisochilus Wall. ex Benth.
Anisomeles R.Br.
Asterohyptis Epling
Ballota L.
Basilicum Moench
Becium Lindl.
Benguellia G.Taylor
Blephilia Raf.
Bostrychanthera Benth.
Bovonia Chiov. *
Brazoria Engelm. & A.Gray
Bystropogon L'Hér.
Calamintha Mill.
Capitanopsis S.Moore
Capitanya Schweinf. ex Gürke
Catoferia (Benth.) Benth.
Cedronella Moench
Ceratanthus F.Muell. ex G.Taylor
Chaiturus Ehrh. ex Willd.
Chamaesphacos Schrenk ex Fisch. &
    C.A.Mey.
Chaunostoma Donn.Sm.
Chelonopsis Miq.
Cleonia L.
Clinopodium L.
Colebrookea Sm.
Collinsonia L.
Colquhounia Wall.
Comanthosphace S.Moore
Conradina A.Gray
Coridothymus Rchb.f.
Craniotome Rchb.
Cuminia Colla
Cunila Royen ex L.
Cyclotrichium (Boiss.) Manden. &
    Scheng.
Cymaria Benth.
Dauphinea Hedge
Dicerandra Benth.
Dorystaechas Boiss. & Heldr. ex Benth.
Dracocephalum L.
Drepanocaryum Pojark.
Elsholtzia Willd.
Endostemon N.E.Br.
Englerastrum Briq.
Eremostachys Bunge
Eriope Kunth ex Benth.
Eriophyton Benth.
Eriopidion Harley
Eriothymus (Benth.) Rchb.
Erythrochlamys Gürke
Euhesperida Brullo & Furnari

Eurysolen Prain
Fuerstia T.C.E.Fr.
Galeopsis L.
Gardoquia Ruíz & Pav.
Geniosporum Wall. ex Benth.
Glechoma L.
Glechon Spreng.
Gomphostemma Wall. ex Benth.
Gontscharovia Borisova
Hanceola Kudo
Haplostachys (A.Gray) W.F.Hillebr.
Haumaniastrum P.A.Duvign. & Plancke
Hedeoma Pers.
Hemiandra R.Br.
Hemigenia R.Br.
Hemizygia (Benth.) Briq.
Hesperozygis Epling
Heterolamium C.Y.Wu
Hoehnea Epling
Holocheila (Kudo) S.Chow
Holostylon Robyns & Lebrun
Horminum L.
Hoslundia Vahl
Hymenocrater Fisch. & C.A.Mey.
Hypenia (Mart. ex Benth.) Harley
Hypogomphia Bunge
Hyptidendron Harley
Hyptis Jacq.
Hyssopus L.
Isodictyophorus Briq.
Isodon (Schrad. ex Benth.) Spach
Isoleucas O.Schwartz *
Keiskea Miq.
Kudrjaschevia Pojark.
Kurzamra Kuntze
Lagochilus Bunge
Lallemantia Fisch. & C.A.Mey.
Lamium L.
Lavandula L.
Leocus A.Chev.
Leonotis (Pers.) R.Br.
Leonurus L.
Lepechinia Willd.
Leucas R.Br.
Leucosceptrum Sm.
Limniboza R.E.Fr.
Lophanthus Adans.
Loxocalyx Hemsl.
Lycopus L.
Macbridea Elliott ex Nutt.
Marmoritis Benth.
Marrubium L.
Marsypianthes Mart. ex Benth.
Meehania Britton
Melissa L.

Melittis L.
Mentha L.
Meriandra Benth.
Mesona Blume
Metastachydium Airy Shaw ex C.Y.Wu & H.W.Li
Microcorys R.Br.
Micromeria Benth.
Microtoena Prain
Minthostachys (Benth.) Spach
Moluccella L.
Monarda L.
Monardella Benth.
Mosla (Benth.) Buch.-Ham. ex Maxim.
Neoeplingia Ramam., Hiriart & Medran *
Neohyptis J.K.Morton
Nepeta L.
Nosema Prain
Notochaete Benth.
Ocimum L.
Octomeron Robyns
Ombrocharis Hand.-Mazz. *
Origanum L.
Orthosiphon Benth.
Otostegia Benth.
Panzerina Soják
Paralamium Dunn
Paraphlomis Prain
Peltodon Pohl
Pentapleura Hand.-Mazz. *
Perilla L.
Perillula Maxim.
Perovskia Kar.
Perrierastrum Guillaumin
Phlomidoschema (Benth.) Vved.
Phlomis L.
Phyllostegia Benth.
Physostegia Benth.
Piloblephis Raf.
Pitardia Batt. ex Pit.
Platostoma P.Beauv.
Plectranthus L'Hér.
Pogogyne Benth.
Pogostemon Desf.
Poliomintha A.Gray
Prasium L.
Prostanthera Labill.

Prunella L.
Pseuderemostachys Popov
Pseudomarrubium Popov *
Puntia Hedge *
Pycnanthemum Michx.
Pycnostachys Hook.
Rabdosiella Codd
Renschia Vatke
Rhabdocaulon (Benth.) Epling
Rhaphiodon Schauer
Rhododon Epling
Rosmarinus L.
Rostrinucula Kudo
Roylea Wall. ex Benth.
Rubiteucris Kudo
Sabaudia Buscal. & Muschler *
Saccocalyx Coss. & Durieu
Salvia L.
Satureja L.
Schizonepeta (Benth.) Briq.
Scutellaria L.
Sideritis L.
Solenostemon Thonn.
Stachydeoma Small
Stachyopsis Popov & Vved.
Stachys L.
Stenogyne Benth.
Sulaimania Hedge & Rech.f. *
Suzukia Kudo
Symphostemon Hiern
Synandra Nutt.
Syncolostemon E.Mey.
Tetraclea A.Gray
Tetradenia Benth.
Teucrium L.
Thorncroftia N.E.Br.
Thuspeinanta T.Durand
Thymbra L.
Thymus L.
Tinnea Kotschy & Peyr.
Trichostema L.
Wenchengia C.Y.Wu & S.Chow *
Westringia Sm.
Wiedemannia Fisch. & C.A.Mey.
Wrixonia F.Muell.
Zataria Boiss.
Zhumeria Rech.f. & Wendelbo
Ziziphora L.

LACISTEMATACEAE Mart. 1826   17.06

2 genera. Trop. America. Trees, shrubs.

B&H	MONOCHLAMYDEAE	Ordines anomali, 161
DT&H	ARCHICHLAMYDEAE	Piperales, 5
Melc	ARCHICHLAMYDEAE	Violales, Violineae (within Flacourtiaceae, 182)
Thor	VIOLIFLORAE	Violales, Violineae (within Flacourtiaceae, 124)
Dahl	VIOLIFLORAE	Violales (within Flacourtiaceae, 84)
Young	DILLENIIDAE, VIOLANAE	Violales (within Flacourtiaceae, 190)
Takh	DILLENIIDAE, VIOLANAE	Violales, 145
Cron	DILLENIIDAE	Violales, 112

Lacistema Sw.
Lozania S.Mutis ex Caldas

LACTORIDACEAE Engl. 1888   139.02

1 genus, monotypic. Juan Fernandez. Shrub.

B&H	MONOCHLAMYDEAE	Micrembryeae (within Piperaceae, 139)
DT&H	ARCHICHLAMYDEAE	Ranales, Magnoliineae, 47
Melc	ARCHICHLAMYDEAE	Piperales, 71
Thor	ANNONIFLORAE	Annonales, Laurineae, 16
Dahl	MAGNOLIIFLORAE	Lactoridales, 13
Young	MAGNOLIIDAE, RANUNCULANAE	Laurales, 13
Takh	MAGNOLIIDAE, MAGNOLIANAE	Lactoridales, 23
Cron	MAGNOLIIDAE	Magnoliales, 7

Lactoris Phil.

LAMIACEAE - See LABIATAE

LARDIZABALACEAE Decne. 1839   7.02

8 genera. Trop. to warm-temp. Asia, Chile. Lianes, rarely erect.

B&H	POLYPETALAE, THALAMIFLORAE	Ranales (within Berberidaceae, 7)
DT&H	ARCHICHLAMYDEAE	Ranales, Ranunculineae, 42
Melc	ARCHICHLAMYDEAE	Ranunculales, Ranunculineae, 64
Thor	ANNONIFLORAE	Berberidales, Berberidineae, 28
Dahl	RANUNCULIFLORAE	Ranunculales, 29
Young	MAGNOLIIDAE, RANUNCULANAE	Ranunculales, 29
Takh	RANUNCULIDAE, RANUNCULANAE	Ranunculales, Lardizabilineae, 44
Cron	MAGNOLIIDAE	Ranunculales, 34

Akebia Decne.
Boquila Decne.
Decaisnea Hook.f. & Thomson
Holboellia Wall.

Lardizabala Ruíz & Pav.
Parvatia Decne.
Sinofranchetia (Diels) Hemsl.
Stauntonia DC.

LAURACEAE Juss. 1789 143.01

49 genera. Tropical and warm-temperate. Trees, shrubs, rarely herbs.

B&H	MONOCHLAMYDEAE	Daphnales, 143
	(including Hernandiaceae)	
DT&H	ARCHICHLAMYDEAE	Ranales, Magnoliineae, 52
Melc	ARCHICHLAMYDEAE	Magnoliales, Laurineae, 55
Thor	ANNONIFLORAE	Annonales, Laurineae, 20
Dahl	MAGNOLIIFLORAE	Laurales, 22
	(including Hernandiaceae)	
Young	MAGNOLIIDAE, RANUNCULANAE	Laurales, 21
Takh	MAGNOLIIDAE, MAGNOLIANAE	Laurales, 21
Cron	MAGNOLIIDAE	Laurales, 17

Actinodaphne Nees
Aiouea Aubl.
Alseodaphne Nees
Aniba Aubl.
Apollonias Nees
Aspidostemon Rohwer & H.G.Richt. *
Beilschmiedia Nees
Caryodaphnopsis Airy Shaw
Cassytha L.
Chlorocardium Rohwer, H.G.Richt. &
    van der Werff
Cinnadenia Kosterm.
Cinnamomum Schaeffer
Cryptocarya R.Br.
Dehaasia Blume
Dicypellium Nees & Mart.
Dodecadenia Nees
Endiandra R.Br.
Endlicheria Nees
Eusideroxylon Teijsm. & Binn.
Hypodaphnis Stapf
Kubitzkia van der Werff
Laurus L.
Licaria Aubl.
Lindera Thunb.
Litsea Lam.
Mezilaurus Kuntze ex Taub.

Micropora Hook.f.
Nectandra Rol. ex Rottb.
Neocinnamomum H.Liou
Neolitsea (Benth.) Merr.
Nothaphoebe Blume
Ocotea Aubl.
Paraia Rohwer, H.G.Richt. & van der
    Werff *
Parasassafras D.G.Long
Persea Mill.
Phoebe Nees
Phyllostemonodaphne Kosterm.
Pleurothyrium Nees
Potameia Thouars
Potoxylon Kosterm.
Ravensara Sonn.
Rhodostemonodaphne Rohwer &
    Kubitzki
Sassafras Nees
Sinosassafras H.W.Li
Temmodaphne Kosterm.
Triadodaphne Kosterm.
Umbellularia (Nees) Nutt.
Urbanodendron Mez
Williamodendron Kubitzki &
    H.G.Richt.

LECYTHIDACEAE Poit. 1825 67.02

20 genera. Tropics, esp. New World. Trees, shrubs.

B&H	POLYPETALAE, CALYCIFLORAE	Myrtales (within Myrtaceae, 67)
DT&H	ARCHICHLAMYDEAE	Myrtiflorae, Myrtineae, 170
Melc	ARCHICHLAMYDEAE	Myrtiflorae, Myrtineae, 210
Thor	THEIFLORAE	Theales, Lecythidineae, 69
Dahl	THEIFLORAE	Theales, 123
Young	DILLENIIDAE, DILLENIANAE	Lecythidales, 130

Takh   DILLENIIDAE, LECYTHIDANAE Lecythidales, 124
Cron   DILLENIIDAE   Lecythidales, 103

Abdulmajidia Whitmore
Allantoma Miers
Asteranthos Desf.
Barringtonia J.R.Forst. & G.Forst.
Bertholletia Bonpl.
Careya Roxb.
Cariniana Casar.
Chydenanthus Miers
Corythophora R.Knuth
Couratari Aubl.

Couroupita Aubl.
Crateranthus Baker f.
Eschweilera Mart. ex DC.
Foetidia Comm. ex Lam.
Grias L.
Gustavia L.
Lecythis Loefl.
Napoleonaea P.Beauv.
Petersianthus Merr.
Planchonia Blume

LEEACEAE (DC.) Dumort. 1829   50.02

1 genus. Africa, Madagascar, trop. Asia. Trees, shrubs, herbs.

B&H   POLYPETALAE, DISCIFLORAE Celastrales (within Vitaceae, 50)
DT&H   ARCHICHLAMYDEAE   Rhamnales (within Vitaceae, 120)
Melc   ARCHICHLAMYDEAE   Rhamnales, 169
Thor   CORNIFLORAE   Cornales, Vitineae (within Vitaceae, 279)
Dahl   SANTALIFLORAE   Vitales (within Vitaceae, 254)
Young (not mentioned, presumably within Vitaceae, 233)
Takh   ROSIDAE, VITANAE   Vitales, 323
Cron   ROSIDAE   Rhamnales, 236

Leea L.

LEGUMINOSAE Juss. 1789   57

678 genera. Widespread. Trees, shrubs, climbers, herbs.

DT&H   ARCHICHLAMYDEAE   Rosales, Rosineae, 78
         (including Krameriaceae in Caesalpinioideae)
Melc   ARCHICHLAMYDEAE   Rosales, Leguminosineae, 118
Thor   RUTIFLORAE   Rutales, Fabineae, 191
Dahl   FABIFLORAE   Fabales, 200
         (separating *Caesalpiniaceae* 201, *Mimosaceae*, 202)
Young ROSIDAE, RUTANAE   Fabales, 262
         (separating *Caesalpiniaceae* 264, *Mimosaceae* 263)
Takh   ROSIDAE, FABANAE   Fabales, 241
Cron   ROSIDAE   Fabales, 182
         (separating *Caesalpiniaceae* 181, *Mimosaceae* 180)

LEGUMINOSAE-CAESALPINIOIDEAE   57.02

156 genera. Mostly in tropics. Trees, shrubs, climbers, some herbs.

Acrocarpus Wight ex Arn.
Adenolobus (Harv. ex Benth.) Torre & Hillc.
Afzelia Sm.
Amherstia Wall.
Androcalymma Dwyer
Anthonotha P.Beauv.
Aphanocalyx Oliv.
Apuleia Mart.
Arapatiella Rizzini & A.Mattos
Arcoa Urb.
Augouardia Pellegr.
Baikiaea Benth.
Balsamocarpon Clos
Batesia Spruce ex Benth.
Bathiaea Drake
Baudouinia Baill.
Bauhinia L.
Berlinia Sol. ex Hook.f.
Brachycylix (Harms) R.S.Cowan
Brachystegia Benth.
Brenierea Humbert
Brodriguesia R.S.Cowan
Brownea Jacq.
Browneopsis Huber *
Burkea Hook.
Bussea Harms
Caesalpinia L.
Campsiandra Benth.
Cassia L.
Cenostigma Tul.
Ceratonia L.
Cercis L.
Chamaecrista Moench
Chidlowia Hoyle
Colophospermum J.Léonard
Colvillea Bojer ex Hook.
Conzattia Rose
Copaifera L.
Cordeauxia Hemsl.
Crudia Schreb.
Cryptosepalum Benth.
Cynometra L.
Daniellia Benn.
Delonix Raf.
Detarium Juss.
Dialium L.
Dicorynia Benth.
Dicymbe Spruce ex Benth.
Didelotia Baill.
Dimorphandra Schott
Diptychandra Tul.
Distemonanthus Benth.
Duparquetia Baill.
Eligmocarpus Capuron

Elizabetha Schomb. ex Benth.
Endertia Steenis & de Wit
Englerodendron Harms
Eperua Aubl.
Erythrophleum Afzel. ex G.Don
Eurypetalum Harms
Gilbertiodendron J.Léonard
Gilletiodendron Vermoesen
Gleditsia L.
Goniorrhachis Taub.
Gossweilerodendron Harms
Griffonia Baill.
Guibourtia Benn.
Gymnocladus Lam.
Haematoxylum L.
Hardwickia Roxb.
Heterostemon Desf.
Hoffmannseggia Cav.
Humboldtia Vahl
Hylodendron Taub.,
Hymenaea L.
Hymenostegia (Benth.) Harms
Intsia Thouars
Isoberlinia Craib & Stapf ex Holland
Jacqueshuberia Ducke
Julbernardia Pellegr.
Kalappia Kosterm.
Kingiodendron Harms
Koompassia Maingay
Labichea Gaudich. ex DC.
Lebruniodendron J.Léonard
Lemuropisum H.Perrier
Leonardoxa Aubrév.
Leucostegane Prain
Librevillea Hoyle
Loesenera Harms
Lophocarpinia Burkart
Lysidice Hance
Macrolobium Schreb.
Maniltoa Scheff.
Martiodendron Gleason
Melanoxylon Schott
Mendoravia Capuron
Michelsonia Hauman
Microberlinia A.Chev.
Moldenhawera Schrad.
Monopetalanthus Harms
Mora Benth.
Moullava Adans.
Neoapaloxylon Rauschert
Neochevalierodendron J.Léonard
Oddoniodendron De Wild.
Orphanodendron Barneby & J.W.Grimes
Oxystigma Harms

Pachyelasma Harms
Paloue Aubl.
Paloveopsis R.S.Cowan
Paramacrolobium J.Léonard
Parkinsonia L.
Pellegriniodendron J.Léonard
Peltogyne Vogel
Peltophorum (Vogel) Benth.
Petalostylis R.Br.
Phyllocarpus Riedel ex Tul.
Plagiosiphon Harms
Poeppigia C.Presl
Polystemonanthus Harms
Prioria Griseb.
Pseudomacrolobium Hauman
Pseudosindora Symington
Pterogyne Tul.
Pterolobium R.Br. ex Wight & Arn.
Recordoxylon Ducke
Saraca L.
Schizolobium Vogel
Schotia Jacq.
Sclerolobium Vogel
Scorodophloeus Harms

Senna Mill.
Sindora Miq.
Sindoropsis J.Léonard
Stachyothyrsus Harms
Stahlia Bello
Stemonocoleus Harms
Stenodrepanum Harms
Storckiella Seem.
Stuhlmannia Taub.
Sympetalandra Stapf
Tachigali Aubl.
Talbotiella Baker f.
Tamarindus L.
Tessmannia Harms
Tetraberlinia (Harms) Hauman
Tetrapterocarpon Humbert
Thylacanthus Tul.
Tylosema (Schweinf.) Torre & Hillc.
Umtiza T.R.Sim
Vouacapoua Aubl.
Westia Vahl
Zenia Chun
Zenkerella Taub.
Zuccagnia Cav.

LEGUMINOSAE-MIMOSOIDEAE   57.03

66 genera. Mostly in tropics. Trees, shrubs, climbers, some herbs.

Abarema Pittier
Acacia Mill.
Adenanthera L.
Adenopodia C.Presl
Affonsea A.St.–Hil.
Albizia Durazz.
Amblygonocarpus Harms
Anadenanthera Speg.
Archidendron F.Muell.
Archidendropsis I.C.Nielsen
Aubrevillea Pellegr.
Calliandra Benth.
Calliandropsis H.M.Hern. & Guinet
Calpocalyx Harms
Cedrelinga Ducke
Chloroleucon (Benth.) Record
Cojoba Britton & Rose
Cylicodiscus Harms
Desmanthus Willd.
Dichrostachys (A.DC.) Wight & Arn.
Dinizia Ducke
Elephantorrhiza Benth.
Entada Adans.
Enterolobium Mart.

Faidherbia A.Chev.
Fillaeopsis Harms
Gagnebina Neck. ex DC.
Goldmania Rose ex Micheli
Havardia Small
Indopiptadenia Brenan
Inga Mill.
Lemurodendron Villiers & Guinet *
Leucaena Benth.
Lysiloma Benth.
Macrosamanea Britton & Rose
Marmaroxylon Killip
Mimosa L.
Mimozyganthus Burkart
Neptunia Lour.
Newtonia Baill.
Obolinga Barneby
Parapiptadenia Brenan
Pararchidendron I.C.Nielsen
Paraserianthes I.C.Nielsen
Parkia R.Br.
Pentaclethra Benth.
Piptadenia Benth.
Piptadeniastrum Brenan

Piptadeniopsis Burkart
Pithecellobium Mart.
Plathymenia Benth.
Prosopidastrum Burkart
Prosopis L.
Pseudopiptadenia Rauschert
Pseudoprosopis Harms
Punjuba Britton & Rose
Schleinitzia Warb.

Schrankia Willd.
Serianthes Benth.
Stryphnodendron Mart.
Tetrapleura Benth.
Wallaceodendron Koord.
Xerocladia Harv.
Xylia Benth.
Zapoteca H.M.Hern.
Zygia P.Browne

LEGUMINOSAE-PAPILIONOIDEAE 57.01

455 genera. Widespread. Trees, shrubs, climbers, herbs.

Abrus Adans.
Acosmium Schott
Adenocarpus DC.
Adenodolichos Harms
Adesmia DC.
Aenictophyton A.T.Lee
Aeschynomene L.
Afgekia Craib
Aganope Miq.
Airyantha Brummitt
Aldina Endl.
Alexa Moq.
Alhagi Gagnebin
Alistilus N.E.Br.
Almaleea Crisp & P.H.Weston
Alysicarpus Desv.
Amburana Schwacke & Taub.
Amicia Kunth
Ammodendron Fisch. ex DC.
Ammopiptanthus S.H.Cheng
Amorpha L.
Amphicarpaea Elliott ex Nutt.
Amphimas Pierre ex Harms
Amphithalea Eckl. & Zeyh.
Anagyris L.
Anarthrophyllum Benth.
Andira Juss.
Angylocalyx Taub.
Antheroporum Gagnep.
Anthyllis L.
Antopetitia A.Rich.
Aotus Sm.
Apios Fabr.
Apoplanesia C.Presl
Apurimacia Harms
Arachis L.
Argyrocytisus (Maire) Raynaud
Argyrolobium Eckl. & Zeyh.
Arthrocarpum Balf.f.

Arthroclianthus Baill.
Aspalathus L.
Astracantha Podlech
Astragalus L.
Ateleia (DC.) Benth.
Austrodolichos Verdc.
Austrosteenisia Geesink
Baphia Afzel. ex Lodd.
Baphiastrum Harms
Baphiopsis Benth. ex Baker
Baptisia Vent.
Baukea Vatke
Behaimia Griseb.
Belairia A.Rich.
Bembicidium Rydb.
Bergeronia Micheli
Biserrula L.
Bituminaria Heist. ex Fabr.
Bocoa Aubl.
Bolusafra Kuntze
Bolusanthus Harms
Bolusia Benth.
Bossiaea Vent.
Bowdichia Kunth
Bowringia Champ. ex Benth.
Brachysema R.Br.
Brongniartia Kunth
Brya P.Browne
Bryaspis P.A.Duvign.
Burgesia F.Muell.
Burkilliodendron Sastry
Burtonia R.Br.
Butea Roxb. ex Willd.
Cadia Forssk.
Cajanus DC.
Calicotome Link
Calispepla Vved. *
Callistachys Vent.
Calophaca Fisch. ex DC.

Calopogonium Desv.
Calpurnia E.Mey.
Camoensia Welw. ex Benth.
Camptosema Hook. & Arn.
Campylotropis Bunge
Canavalia DC.
Candolleodendron R.S.Cowan
Caragana Fabr.
Carmichaelia R.Br.
Carrissoa Baker f.
Cascaronia Griseb.
Castanospermum A.Cunn. ex Hook.
Centrolobium Mart. ex Benth.
Centrosema (DC.) Benth.
Chadsia Bojer
Chaetocalyx DC.
Chamaecytisus Link
Chapmannia Torr. & A.Gray
Chesneya Lindl. ex Endl.
Chordospartium Cheeseman
Chorizema Labill.
Christia Moench
Chrysoscias E.Mey.
Cicer L.
Cladrastis Raf.
Clathrotropis (Benth.) Harms
Cleobulia Mart. ex Benth.
Clianthus Sol. ex Lindl.
Clitoria L.
Clitoriopsis R.Wilczek
Cochlianthus Benth.
Codariocalyx Hassk.
Coelidium Vogel ex Walp.
Collaea DC.
Cologania Kunth
Colutea L.
Corallospartium J.B.Armstr.
Cordyla Lour.
Coronilla L.
Coroya Pierre
Corynella DC.
Coursetia DC.
Craibia Harms & Dunn
Cranocarpus Benth.
Craspedolobium Harms
Cratylia Mart. ex Benth.
Crotalaria L.
Cruddasia Prain
Cullen Medik.
Cupulanthus Hutch.
Cyamopsis DC.
Cyathostegia (Benth.) Schery
Cyclocarpa Afzel. ex Baker
Cyclolobium Benth.
Cyclopia Vent.

Cymbosema Benth.
Cytisophyllum O.Lang
Cytisopsis Jaub. & Spach
Cytisus Desf.
Dahlstedtia Malme
Dalbergia L.f.
Dalbergiella Baker f.
Dalea L.
Dalhousiea Wall. ex Benth.
Daviesia Sm.
Decorsea R.Vig.
Dendrolobium (Wight & Arn.) Benth.
Derris Lour.
Desmodiastrum (Prain) A.Pramanik & Thoth.
Desmodium Desv.
Dewevrea Micheli
Dicerma DC.
Dichilus DC.
Dicraeopetalum Harms
Dillwynia Sm.
Dioclea Kunth
Diphyllarium Gagnep.
Diphysa Jacq.
Diplotropis Benth.
Dipogon Liebm.
Dipteryx Schreb.
Discolobium Benth.
Disynstemon R.Vig.
Dolichopsis Hassl.
Dolichos L.
Droogmansia De Wild.
Dumasia DC.
Dunbaria Wight & Arn.
Dussia Krug & Urb. ex Taub.
Dysolobium (Benth.) Prain
Ebenus L.
Echinospartum (Spach) Fourr.
Eleiotis DC.
Eminia Taub.
Endosamara Geesink
Eremosparton Fisch. & C.A.Mey.
Erichsenia Hemsl.
Erinacea Adans.
Eriosema (DC.) Rchb.
Errazurizia Phil.
Erythrina L.
Etaballia Benth.
Euchilopsis F.Muell.
Euchresta Benn.
Eutaxia R.Br.
Eversmannia Bunge
Exostyles Schott
Eysenhardtia Kunth
Fiebrigiella Harms

Fissicalyx Benth.
Flemingia Roxb. ex W.T.Aiton
Fordia Hemsl.
Galactia P.Browne
Galega L.
Gastrolobium R.Br.
Geissaspis Wight & Arn.
Genista L.
Genistidium I.M.Johnst.
Geoffroea Jacq.
Gliricidia Kunth
Glottidium Desv.
Glycine Willd.
Glycyrrhiza L.
Gompholobium Sm.
Gonocytisus Spach
Goodia Salisb.
Grazielodendron H.C.Lima
Gueldenstaedtia Fisch.
Halimodendron Fisch. ex DC.
Hallia Thunb.
Hammatolobium Fenzl
Haplormosia Harms
Hardenbergia Benth.
Harleyodendron R.S.Cowan
Harpalyce Moçino & Sessé ex DC.
Hebestigma Urb.
Hedysarum L.
Herpyza Sauvalle
Hesperolaburnum Maire
Hesperothamnus Brandegee
Hippocrepis L.
Hoita Rydb.
Holocalyx Micheli
Hovea R.Br. ex W.T.Aiton
Humularia P.A.Duvign.
Hybosema Harms
Hymenocarpos Savi
Hymenolobium Benth.
Hypocalyptus Thunb.
Indigofera L.
Inocarpus J.R.Forst. & G.Forst.
Isotropis Benth.
Jacksonia R.Br. ex Sm.
Jansonia Kippist
Kennedia Vent.
Kotschya Endl.
Kummerowia Schindl.
Kunstleria Prain
Lablab Adans.
Laburnum Fabr.
Lamprolobium Benth.
Lathyrus L.
Latrobea Meisn.
Lebeckia Thunb.

Lecointea Ducke
Lennea Klotzsch
Lens Mill.
Leptoderris Dunn
Leptodesmia (Benth.) Benth.
Leptosema Benth.
Lespedeza Michx.
Lessertia DC.
Leucomphalos Benth. ex Planch.
Liparia L.
Lonchocarpus Kunth
Lotononis (DC.) Eckl. & Zeyh.
Lotus L.
Luetzelburgia Harms
Lupinus L.
Luzonia Elmer
Maackia Rupr.
Machaerium Pers.
Macropsychanthus Harms ex K.Schum.
   & Lauterb.
Macroptilium (Benth.) Urb.
Macrotyloma (Wight & Arn.) Verdc.
Margaritolobium Harms
Marina Liebm.
Mastersia Benth.
Mecopus Benn.
Medicago L.
Melilotus Mill.
Melliniella Harms
Melolobium Eckl. & Zeyh.
Mildbraediodendron Harms
Millettia Wight & Arn.
Mirbelia Sm.
Monopteryx Spruce ex Benth.
Mucuna Adans.
Muellera L.f.
Muelleranthus Hutch.
Mundulea (DC.) Benth.
Myrocarpus Allemão
Myrospermum Jacq.
Myroxylon L.f.
Nemcia Domin
Neocollettia Hemsl.
Neodielsia Harms
Neodunnia R.Vig.
Neoharmsia R.Vig.
Neonotonia Lackey
Neorautanenia Schinz
Neorudolphia Britton
Nephrodesmus Schindl.
Nesphostylis Verdc.
Nissolia Jacq.
Nogra Merr.
Notodon Urb.
Notospartium Hook.f.

Olneya A.Gray
Onobrychis Mill.
Ononis L.
Ophrestia H.M.L.Forbes
Orbexilum Raf.
Oreophysa (Bunge ex Boiss.) Bornm.
Ormocarpopsis R.Vig.
Ormocarpum P.Beauv.
Ormosia G.Jacks.
Ornithopus L.
Ostryocarpus Hook.f.
Otholobium C.H.Stirt.
Otoptera DC.
Oxylobium Andrews
Oxyrhynchus Brandegee
Oxytropis DC.
Pachecoa Standl. & Steyerm.
Pachyrhizus Rich. ex DC.
Padbruggea Miq.
Panurea Spruce ex Benth.
Paracalyx Ali
Paramachaerium Ducke
Parochetus Buch.–Ham. ex D.Don
Parryella Torr. & A.Gray
Pearsonia Dümmer
Pediomelum Rydb.
Periandra Mart. ex Benth.
Pericopsis Thwaites
Petaladenium Ducke
Peteria A.Gray
Petteria C.Presl
Phaseolus L.
Phylacium Benn.
Phyllodium Desv.
Phyllota (DC.) Benth.
Phylloxylon Baill.
Physostigma Balf.
Pickeringia Nutt.
Pictetia DC.
Piptanthus Sweet
Piscidia L.
Pisum L.
Plagiocarpus Benth.
Platycelyphium Harms
Platycyamus Benth.
Platylobium Sm.
Platymiscium Vogel
Platypodium Vogel
Platysepalum Welw. ex Baker
Podalyria Willd.
Podocytisus Boiss. & Heldr.
Poecilanthe Benth.
Poiretia Vent.
Poitea Vent.
Polhillia C.H.Stirt.

Pongamia Vent.
Pongamiopsis R.Vig.
Priestleya DC.
Pseudarthria Wight & Arn.
Pseudeminia Verdc.
Pseudoeriosema Hauman
Pseudovigna (Harms) Verdc.
Psophocarpus DC.
Psoralea L.
Psoralidium Rydb.
Psorothamnus Rydb.
Pterocarpus Jacq.
Pterodon Vogel
Ptycholobium Harms
Ptychosema Benth.
Pueraria DC.
Pultenaea Sm.
Pycnospora R.Br. ex Wight & Arn.
Rafnia Thunb.
Ramirezella Rose
Ramorinoa Speg.
Requienia DC.
Retama Raf.
Rhodopis Urb.
Rhynchosia Lour.
Rhynchotropis Harms
Riedeliella Harms
Robinia L.
Robynsiophyton R.Wilczek
Rothia Pers.
Sabinea DC.
Sakoanala R.Vig.
Salweenia Baker f.
Sarcodum Lour.
Sartoria Boiss. & Heldr.
Sauvallella Rydb.
Schefflerodendron Harms
Scorpiurus L.
Sellocharis Taub.
Sesbania Scop.
Shuteria Wight & Arn.
Sinodolichos Verdc.
Smirnowia Bunge
Smithia Aiton
Soemmeringia Mart.
Sophora L.
Spartidium Pomel
Spartium L.
Spathionema Taub.
Spatholobus Hassk.
Sphaerolobium Sm.
Sphaerophysa DC.
Sphenostylis E.Mey.
Sphinctospermum Rose
Spirotropis Tul.

Spongiocarpella Yakovlev & N.Ulziykh.
Stauracanthus Link
Stracheya Benth.
Streblorrhiza Endl.
Strongylodon Vogel
Strophostyles Elliott
Stylosanthes Sw.
Styphnolobium Schott ex Endl.
Sutherlandia R.Br. ex W.T.Aiton
Swainsona Salisb.
Swartzia Schreb.
Sweetia Spreng.
Tadehagi H.Ohashi
Taralea Aubl.
Taverniera DC.
Templetonia R.Br. ex W.T.Aiton
Tephrosia Pers.
Teramnus P.Browne
Teyleria Backer
Thermopsis R.Br.
Tipuana (Benth.) Benth.
Trifidacanthus Merr.
Trifolium L.
Trigonella L.
Uleanthus Harms

Ulex L.
Uraria Desv.
Uribea Dugand & Romero
Urodon Turcz.
Vandasina Rauschert
Vatairea Aubl.
Vataireopsis Ducke
Vatovaea Chiov.
Vavilovia Fedorov
Vicia L.
Vigna Savi
Viminaria Sm.
Virgilia Poir.
Wajira Thulin
Weberbauerella Ulbr.
Whitfordiodendron Elmer
Wiborgia Thunb.
Willardia Rose
Wisteria Nutt.
Xanthocercis Baill.
Xeroderris Roberty
Yucaratonia Burkart
Zollernia Wied.–Neuw. & Nees
Zornia J.F.Gmel.

LEITNERIACEAE Benth. 1880   155

1 genus, monotypic. S.E. USA. Shrub to small tree.

B&H	MONOCHLAMYDEAE	Unisexuales, 155
	(? including Didymelaceae)	
DT&H	ARCHICHLAMYDEAE	Leitneriales, 9
Melc	ARCHICHLAMYDEAE	Leitneriales, 5
Thor	RUTIFLORAE	Rutales, Rutineae, 175
Dahl	RUTIFLORAE	Sapindales, 205
Young	ROSIDAE, RUTANAE	Rutales, 252
Takh	ROSIDAE, RUTANAE	Leitneriales, 272
Cron	HAMAMELIDIDAE	Leitneriales, 56

Leitneria Chapm.

LENNOACEAE Solms-Laub. 1870   97

3 genera. S.W. USA to Venezuela. Root parasites.

B&H	GAMOPETALAE, HETEROMERAE	Ericales, 97
DT&H	METACHLAMYDEAE	Ericales, 184
Melc	SYMPETALAE	Tubiflorae, Boraginineae, 257
Thor	SOLANIFLORAE	Solanales, Boraginineae, 270
Dahl	SOLANIFLORAE	Boraginales, 283
Young	ROSIDAE, SOLANANAE	Boraginales, 369

Takh LAMIIDAE, SOLANANAE   Boraginales, 389
Cron ASTERIDAE   Lamiales, 284

Ammobroma Torr.
Lennoa Lex.
Pholisma Nutt. ex Hook.

LENTIBULARIACEAE Rich. 1808   117

3 genera. Widespread. Insectivorous herbs, some aquatic.

B&H   GAMOPETALAE, BICARPELLATAE Personales, 117
DT&H METACHLAMYDEAE   Tubiflorae, Solanineae, 217
Melc SYMPETALAE   Tubiflorae, Solanineae, 275
Thor GENTIANIFLORAE   Bignoniales, 262
Dahl LAMIIFLORAE   Scrophulariales, 358
Young ROSIDAE, GENTIANANAE   Bignoniales, 360
Takh LAMIIDAE, LAMIANAE   Scrophulariales, 406
Cron ASTERIDAE   Scrophulariales, 303

Genlisea A.St.–Hil.
Pinguicula L.
Utricularia L.

LEPIDOBOTRYACEAE J.Léonard 1950   38.05

1 genus, monotypic. Tropical Africa. Shrub.

B&H   Not known   —
DT&H ARCHICHLAMYDEAE   Geraniales, Geraniineae (within Linaceae,
                                                                          82)
Melc ARCHICHLAMYDEAE   Geraniales, Geraniineae (within Oxalidaceae,
                                                                         123)
Thor GERANIIFLORAE   Geraniales, Geraniineae (within Oxalidaceae,
                                                                         100)
Dahl RUTIFLORAE   Geraniales, 239
Young (not mentioned, presumably within Oxalidaceae, 274)
Takh ROSIDAE, RUTANAE   Geraniales, 281
Cron ROSIDAE   Geraniales (within Oxalidaceae, 263)

Lepidobotrys Engl.

LIMNANTHACEAE R.Br. 1833   38.03

2 genera. North America, esp. California. Annual herbs.

B&H   POLYPETALAE, DISCIFLORAE Geraniales (within Geraniaceae, 38)
DT&H ARCHICHLAMYDEAE   Sapindales, Limnanthineae, 102)
Melc ARCHICHLAMYDEAE   Geraniales, Limnanthineae, 122
Thor GERANIIFLORAE   Geraniales, Geraniineae, 106

Dahl	RUTIFLORAE	Tropaeolales, 247
Young	DILLENIIDAE, VIOLANAE	Tropaeolales, 215
Takh	ROSIDAE, RUTANAE	Limnanthales, 291
Cron	ROSIDAE	Geraniales, 265

Floerkea Willd.
Limnanthes R.Br.

LINACEAE DC. ex Gray 1821   34.01

14 genera. Widespread. Trees, shrubs, lianes, herbs.

B&H   POLYPETALAE, DISCIFLORAE Geraniales, 34
        (including Erythroxylaceae, Ixonanthaceae)
DT&H ARCHICHLAMYDEAE   Geraniales, Geraniineae, 82
        (including Ctenolophonaceae, Ixonanthaceae, Lepidobotryaceae)
Melc   ARCHICHLAMYDEAE   Geraniales, Geraniineae, 127
        (including Ctenolophonaceae, Humiriaceae,Ixonanthaceae)
Thor   GERANIIFLORAE   Geraniales, Linineae, 97
        (including Ixonanthaceae)
Dahl   RUTIFLORAE   Geraniales, 236
Young ROSIDAE, GERANIANAE   Linales, 270
        (presumably including Ctenolophonaceae, Ixonanthaceae)
Takh   ROSIDAE, RUTANAE   Linales, 275
        (separating *Hugoniaceae* 274)
Cron   ROSIDAE   Linales, 240
        (separating *Hugoniaceae* 239)

Anisadenia Wall. ex Meisn.
Cliococca Bab.
Durandea Planch.
Hebepetalum Benth.
Hesperolinon (A.Gray) Small
Hugonia L.
Indorouchera Hallier f.

Linum L.
Philbornea Hallier f.
Radiola Hill
Reinwardtia Dumort.
Roucheria Planch.
Sclerolinon C.M.Rogers
Tirpitzia Hallier f.

LISSOCARPACEAE Gilg 1924   103.03

1 genus. Tropical South America. Trees.

B&H   GAMOPETALAE, HETEROMERAE Ebenales (within Styracaceae, 103)
DT&H METACHLAMYDEAE   Ebenales, Diospyrineae (within
                                 Styracaceae, 194)

Melc	SYMPETALAE	Ebenales, Ebenineae, 241
Thor	THEIFLORAE	Ebenales, Ebenineae, 74
Dahl	PRIMULIFLORAE	Ebenales, 135
Young	DILLENIIDAE, DILLENIANAE	Ebenales, 138
Takh	DILLENIIDAE, ERICANAE	Ebenales, 134
Cron	DILLENIIDAE	Ebenales, 151

Lissocarpa Benth.

LOASACEAE Dumort. 1822   72

15 genera. Trop. and temp. America and (*Kissenia*) Arabia, Somalia, Namibia. Herbs to small trees.

B&H   POLYPETALAE, CALYCIFLORAE Passiflorales, 72
DT&H ARCHICHLAMYDEAE   Parietales, Loasineae, 156
Melc   ARCHICHLAMYDEAE   Violales, Loasineae, 199
Thor   LOASIFLORAE   Loasales, 235
Dahl   LOASIFLORAE   Loasales, 335
Young DILLENIIDAE, VIOLANAE   Loasales, 203
Takh   LAMIIDAE, LOASANAE   Loasales, 378
Cron   DILLENIIDAE   Violales, 130

Blumenbachia Schrad.   **Loasa** Adans.
Caiophora C.Presl   **Mentzelia** L.
Cevallia Lag.   **Petalonyx** A.Gray
Eucnide Zucc.   **Plakothira** Florence *
Fuertesia Urb.   **Schismocarpus** S.F.Blake
Gronovia L.   **Sclerothrix** C.Presl
Kissenia R.Br. ex Endl.   **Scyphanthus** Sweet
Klaprothia Kunth

LOGANIACEAE R.Br. ex Mart. 1827   108.01

20 genera. Widespread trop., rarely temp. Trees, shrubs, lianes, occas. herbs.

B&H   GAMOPETALAE, BICARPELLATAE Gentianales, 108
      (including Buddlejaceae)
DT&H METACHLAMYDEAE   Contortae, Gentianineae, 198
      (including Buddlejaceae, Retziaceae)
Melc   SYMPETALAE   Gentianales, 245
      (separating *Desfontainiaceae* 246)
Thor   GENTIANIFLORAE   Gentianales, 249
      (including Retziaceae)
Dahl   GENTIANIFLORAE   Gentianales, 339
      (separating *Desfontainiaceae* 338)
Young ROSIDAE, GENTIANANAE   Gentianales, 347
Takh   LAMIIDAE, GENTIANANAE   Gentianales, 365
      (separating *Desfontainiaceae* 364, *Plocospermataceae* 374, *Spigeliaceae* 366)
Cron   ASTERIDAE   Gentianales, 270

Anthocleista Afzel. ex R.Br.   **Mitreola** L.
Antonia Pohl   **Mostuea** Didr.
Bonyunia Schomb. ex Progel   **Neuburgia** Blume
Desfontainia Ruíz & Pav.   **Norrisia** Gardner
Fagraea Thunb.   **Plocosperma** Benth.
Gardneria Wall.   **Potalia** Aubl.
Gelsemium Juss.   **Pseudospigelia** Klett
Geniostoma J.R.Forst. & G.Forst.   **Spigelia** L.
Logania R.Br.   **Strychnos** L.
Mitrasacme Labill.   **Usteria** Willd.

LOPHOPYXIDACEAE (Engl.) H.Pfeiff. 1951   49.02

1 genus. Malaya to Pacific. Lianes or small trees.

B&H	Not known	—
DT&H	ARCHICHLAMYDEAE	Sapindales, Icacinineae (within Icacinaceae, 112)
Melc	ARCHICHLAMYDEAE	Celastrales, Celastrineae (within Celastraceae, 159)
Thor	SANTALIFLORAE	Celastrales, 113
Dahl	SANTALIFLORAE	Celastrales, 251
Young	ROSIDAE, SANTALANAE	Celastrales, 230
Takh	ROSIDAE, CELASTRANAE	Celastrales, Celastrineae, 306
Cron	ROSIDAE	Celastrales (within Celastraceae, 219)

Lophopyxis Hook.f.

LORANTHACEAE Juss. 1808   148.01

77 genera. Widespread. Hemiparasites usually on trees.

B&H	MONOCHLAMYDEAE (including Eremolepidaceae, Viscaceae)	Achlamydosporeae, 148
DT&H	ARCHICHLAMYDEAE (including Eremolepidaceae, Viscaceae)	Santalales, Loranthineae, 17
Melc	ARCHICHLAMYDEAE (including Eremolepidaceae, Viscaceae)	Santalales, Loranthineae, 22
Thor	SANTALIFLORAE	Santalales, 120
Dahl	SANTALIFLORAE	Santalales, 257
Young	ROSIDAE, SANTALANAE	Santalales, 238
Takh	ROSIDAE, CELASTRANAE	Santalales, 316
Cron	ROSIDAE	Santalales, 211

Actinanthella Balle
Aetanthus (Eichler) Engl.
Agelanthus Tiegh.
Alepis Tiegh.
Amyema Tiegh.
Amylotheca Tiegh.
Atkinsonia F.Muell.
Bakerella Tiegh.
Barathranthus (Korth.) Miq.
Benthamina Tiegh.
Berhautia Balle
Cecarria Barlow
Cladocolea Tiegh.
Cyne Danser
Dactyliophora Tiegh.
Decaisnina Tiegh.
Dendropemon (Blume) Rchb.
Dendrophthoe Mart.
Desmaria Tiegh.
Diplatia Tiegh.
Distrianthes Danser

Elytranthe (Blume) Blume
Emelianthe Danser
Englerina Tiegh.
Erianthemum Tiegh.
Gaiadendron G.Don
Globimetula Tiegh.
Helicanthes Danser
Helixanthera Lour.
Ileostylus Tiegh.
Ixocactus Rizzini
Kingella Tiegh.
Lampas Danser
Lepeostegeres Blume
Lepidaria Tiegh.
Ligaria Tiegh.
Loranthus Jacq.
Loxanthera (Blume) Blume
Lysiana Tiegh.
Macrosolen (Blume) Rchb.
Moquiniella Balle
Muellerina Tiegh.

Notanthera (DC.) G.Don
Nuytsia R.Br. ex G.Don
Oedina Tiegh.
Oliverella Tiegh.
Oncella Tiegh.
Oncocalyx Tiegh.
Oryctanthus (Griseb.) Eichler
Oryctina Tiegh.
Panamanthus Kuijt *
Papuanthes Danser
Pedistylis Wiens
Peraxilla Tiegh.
Phragmanthera Tiegh.
Phthirusa Mart.
Plicosepalus Tiegh.
Psathyranthus Ule
Psittacanthus Mart.
Rhizomonanthes Danser

Scurrula L.
Septulina Tiegh.
Socratina Balle
Sogerianthe Danser
Spragueanella Balle
Struthanthus Mart.
Tapinanthus (Blume) Blume
Taxillus Tiegh.
Tetradyas Danser
Thaumasianthes Danser
Tolypanthus (Blume) Blume
Trilepidea Tiegh.
Tripodanthus (Eichler) Tiegh.
Tristerix Mart.
Trithecanthera Tiegh.
Tupeia Cham. & Schltdl.
Vanwykia Wiens

LYTHRACEAE J.St.-Hil. 1805   69.01

30 genera. Widespread. Trees, shrubs, herbs.

B&H   POLYPETALAE, CALYCIFLORAE Myrtales, 69
         (including Crypteroniaceae, Oliniaceae)
DT&H ARCHICHLAMYDEAE   Myrtiflorae, Myrtineae, 166
         (including Rhynchocalycaceae; separating *Punicaceae, Sonneratiaceae*)
Melc ARCHICHLAMYDEAE   Myrtiflorae, Myrtineae, 203
         (separating *Punicaceae, Sonneratiaceae*)
Thor MYRTIFLORAE   Myrtales, Lythrineae, 236
Dahl MYRTIFLORAE   Myrtales, 190
Young ROSIDAE, MYRTANAE   Myrtales, 312
         (separating *Sonneratiaceae* 311)
Takh ROSIDAE, MYRTANAE   Myrtales, Myrtineae, 234
         (separating *Duabangaceae* 236, *Punicaceae* 235, *Sonneratiaceae* 237)
Cron ROSIDAE   Myrtales, 189
         (including Rhynchocalycaceae in 1981, not in 1988; separating *Punicaceae*
         195, *Sonneratiaceae* 188)

Adenaria Kunth
Ammannia L.
Capuronia Lourteig
Crenea Aubl.
Cuphea P.Browne
Decodon J.F.Gmel.
Diplusodon Pohl
Duabanga Buch.–Ham.
Galpinia N.E.Br.
Ginoria Jacq.
Haitia Urb.
Heimia Link
Hionanthera A.Fern. & Diniz
Koehneria S.A.Graham, Tobe & Baas
Lafoensia Vand.

Lagerstroemia L.
Lawsonia L.
Lourtella S.A.Graham, Baas & Tobe *
Lythrum L.
Nesaea Kunth
Pehria Sprague
Pemphis J.R.Forst. & G.Forst.
Physocalymma Pohl
Pleurophora D.Don
Punica L.
Rotala L.
Socotria Levin
Sonneratia L.f.
Tetrataxis Hook.f.
Woodfordia Salisb.

MAGNOLIACEAE Juss. 1789   4.01

7 genera. Widespread tropics to warm-temperate. Trees, shrubs.

B&H	POLYPETALAE, THALAMIFLORAE	Ranales, 4

        (including Eupteleaceae, Illiciaceae, Schisandraceae, Tetracentraceae, Trochodendraceae, Winteraceae)

DT&H ARCHICHLAMYDEAE   Ranales, Magnoliineae, 45

        (including Eupteleaceae, Illiciaceae, Schisandraceae, Tetracentraceae, Trochodendraceae, Winteraceae)

Melc	ARCHICHLAMYDEAE	Magnoliales, Magnoliineae, 39
Thor	ANNONIFLORAE	Annonales, Annonineae, 4
Dahl	MAGNOLIIFLORAE	Magnoliales, 12
Young	MAGNOLIIDAE, MAGNOLIANAE	Magnoliales, 1
Takh	MAGNOLIIDAE, MAGNOLIANAE	Magnoliales, 3
Cron	MAGNOLIIDAE	Magnoliales, 6

Elmerrillia Dandy   **Manglietia** Blume
Kmeria (Pierre) Dandy   **Michelia** L.
Liriodendron L.   **Pachylarnax** Dandy
Magnolia L.

MALESHERBIACEAE D.Don 1827   74.02

1 genus. Peru, Chile, Argentina. Herbs to subshrubs.

B&H	POLYPETALAE, CALYCIFLORAE	Passiflorales (within Passifloraceae, 74)
DT&H	ARCHICHLAMYDEAE	Parietales, Flacourtiineae, 152
Melc	ARCHICHLAMYDEAE	Violales, Flacourtiineae, 188
Thor	VIOLIFLORAE	Violales, Violineae, 131
Dahl	VIOLIFLORAE	Violales, 91
Young	DILLENIIDAE, VIOLANAE	Violales, 196
Takh	DILLENIIDAE, VIOLANAE	Violales, 152
Cron	DILLENIIDAE	Violales, 121

Malesherbia Ruíz & Pav.

MALPIGHIACEAE Juss. 1789   36

65 genera. Trop. and subtrop., esp. New World. Trees, shrubs, often climbing.

B&H	POLYPETALAE, DISCIFLORAE	Geraniales, 36
DT&H	ARCHICHLAMYDEAE	Geraniales, Malpighiineae, 91
Melc	ARCHICHLAMYDEAE	Rutales, Malpighiineae, 138
Thor	GERANIIFLORAE	Geraniales, Polygalineae, 107
Dahl	RUTIFLORAE	Polygalales, 225
Young	ROSIDAE, GERANIANAE	Polygalales, 277
Takh	ROSIDAE, RUTANAE	Polygalales, 292
Cron	ROSIDAE	Polygalales, 241

Acmanthera Griseb.   **Aspicarpa** Rich.
Acridocarpus Guill. & Perr.   **Aspidopterys** A.Juss.

Banisteriopsis C.B.Rob. ex Small
Barnebya W.R.Anderson & B.Gates *
Blepharandra Griseb.
Brachylophon Oliv.
Bunchosia Rich. ex Kunth
Burdachia Mart. ex A.Juss.
Byrsonima Rich. ex Kunth
Callaeum Small
Calyptostylis Arènes *
Camarea A.St.-Hil.
Caucanthus Forssk.
Clonodia Griseb.
Coleostachys A.Juss.
Cordobia Nied.
Diacidia Griseb.
Dicella Griseb.
Digoniopterys Arènes *
Dinemagonum A.Juss.
Dinemandra A.Juss.
Diplopterys A.Juss.
Echinopterys A.Juss.
Ectopopterys W.R.Anderson
Flabellaria Cav.
Flabellariopsis R.Wilczek
Gallardoa Hicken
Galphimia Cav.
Gaudichaudia Kunth
Glandonia Griseb.
Heladena A.Juss.
Henleophytum Karst.
Heteropterys Kunth

Hiptage Gaertn.
Hiraea Jacq.
Janusia A.Juss.
Jubelina A.Juss.
Lasiocarpus Liebm.
Lophanthera A.Juss.
Lophopterys A.Juss.
Malpighia L.
Mascagnia (Bertero ex DC.) Colla
Mcvaughia W.R.Anderson
Mezia Schwacke ex Nied.
Microsteira Baker
Mionandra Griseb.
Peixotoa A.Juss.
Peregrina W.R.Anderson
Philgamia Baill. *
Pterandra A.Juss.
Ptilochaeta Turcz.
Rhynchophora Arènes *
Ryssopterys Blume ex A.Juss.
Spachea A.Juss.
Sphedamnocarpus Planch. ex Benth.
Stigmaphyllon A.Juss.
Tetrapterys Cav.
Thryallis Mart.
Triaspis Burch.
Tricomaria Gillies ex Hook. & Arn.
Triopterys L.
Tristellateia Thouars
Verrucularia A.Juss.

MALVACEAE Juss. 1789   31.01

119 genera. Widespread. Herbs, shrubs, some trees.

B&H   POLYPETALAE, THALAMIFLORAE Malvales, 31
          (including Bombacaceae)
DT&H ARCHICHLAMYDEAE   Malvales, Malvineae, 125
Melc   ARCHICHLAMYDEAE   Malvales, Malvineae, 173
Thor   MALVIFLORAE   Malvales, 155
Dahl   MALVIFLORAE   Malvales, 67
Young DILLENIIDAE, MALVANAE   Malvales, 175
Takh   DILLENIIDAE, MALVANAE   Malvales, 182
Cron   DILLENIIDAE   Malvales, 102

Abelmoschus Medik.
Abutilon Mill.
Abutilothamnus Ulbr.
Acaulimalva Krapov.
Alcea L.
Allosidastrum (Hochr.) Krapov., Fryxell
    & D.M.Bates

Allowissadula D.M.Bates
Althaea L.
Alyogyne Alef.
Anisodontea C.Presl
Anoda Cav.
Anotea (DC.) Kunth
Asterotrichion Klotzsch

Azanza Alef.
Bakeridesia Hochr.
Bastardia Kunth
Bastardiastrum (Rose) D.M.Bates
Bastardiopsis (K.Schum.) Hassl.
Batesimalva Fryxell
Billieturnera Fryxell
Blanchetiastrum Hassl.
Bombycidendron Zoll. & Moritzi
Briquetia Hochr.
Callirhoe Nutt.
Calyculogygas Krapov.
Calyptraemalva Krapov. *
Cenocentrum Gagnep.
Cephalohibiscus Ulbr.
Cienfuegosia Cav.
Codonochlamys Ulbr.
Corynabutilon (K.Schum.) Kearney
Cristaria Cav.
Decaschistia Wight & Arn.
Dendrosida Fryxell
Dicellostyles Benth.
Dirhamphis Krapov.
Eremalche Greene
Erioxylum Rose & Standl.
Fioria Mattei
Fryxellia D.M.Bates *
Gaya Kunth
Goethea Nees
Gossypioides Skovst. ex J.B.Hutch.
Gossypium L.
Gynatrix Alef.
Hampea Schltdl.
Helicteropsis Hochr. *
Herissantia Medik.
Hibiscadelphus Rock
Hibiscus L.
Hochreutinera Krapov.
Hoheria A.Cunn.
Horsfordia A.Gray
Howittia F.Muell.
Humbertianthus Hochr. *
Humbertiella Hochr.
Julostylis Thwaites
Jumelleanthus Hochr. *
Kearnemalvastrum D.M.Bates
Kitaibela Willd.
Kokia Lewton
Kosteletzkya C.Presl
Krapovickasia Fryxell
Kydia Roxb.
Lagunaria (DC.) Rchb.
Lavatera L.

Lawrencia Hook.
Lebronnecia Fosberg
Lopimia Mart.
Macrostelia Hochr.
Malachra L.
Malacothamnus Greene
Malope L.
Malva L.
Malvastrum A.Gray
Malvaviscus Fabr.
Malvella Jaub. & Spach
Megistostegium Hochr.
Meximalva Fryxell
Modiola Moench
Modiolastrum K.Schum.
Monteiroa Krapov.
Montezuma DC.
Napaea L.
Neobaclea Hochr.
Neobrittonia Hochr.
Neohumbertiella Hochr.
Nototriche Turcz.
Notoxylinon Lewton
Palaua Cav.
Papuodendron C.T.White
Pavonia Cav.
Peltaea (C.Presl) Standl.
Periptera DC.
Perrierophytum Hochr.
Phragmocarpidium Krapov.
Phymosia Desv.
Plagianthus J.R.Forst. & G.Forst.
Pseudabutilon R.E.Fr.
Radyera Bullock
Rhynchosida Fryxell
Robinsonella Rose & Baker f.
Rojasimalva Fryxell
Senra Cav.
Sida L.
Sidalcea A.Gray
Sphaeralcea A.St.-Hil.
Symphyochlamys Gürke
Tarasa Phil.
Tetrasida Ulbr. *
Thespesia Sol. ex Corrêa
Thespesiopsis Exell & Hillc.
Triplochlamys Ulbr.
Ulbrichia Urb. *
Urena L.
Urocarpidium Ulbr.
Wercklea Pittier & Standl.
Wilhelminia Hochr. *
Wissadula Medik.

MARCGRAVIACEAE Choisy 1824   28.03

5 genera. Tropical America. Lianes, epiphytes, rarely erect.

B&H   POLYPETALAE, THALAMIFLORAE Guttiferales (within Theaceae, 28)
DT&H ARCHICHLAMYDEAE   Parietales, Theineae, 134
Melc ARCHICHLAMYDEAE   Guttiferales, Theineae, 87
Thor THEIFLORAE   Theales, Theineae, 53
Dahl THEIFLORAE   Theales, 112
Young DILLENIIDAE, DILLENIANAE Theales, 114
Takh DILLENIIDAE, THEANAE   Theales, 105
Cron DILLENIIDAE   Theales, 92

Caracasia Szyszyl.   **Ruyschia** Jacq.
Marcgravia L.   **Souroubea** Aubl.
Norantea Aubl.

MEDUSAGYNACEAE Engl. & Gilg 1924   28.11

1 genus, monotypic. Seychelles. Shrub.

B&H   Not known   —
DT&H ARCHICHLAMYDEAE   Parietales, Theineae (within Guttiferae, 137)
Melc ARCHICHLAMYDEAE   Guttiferales, Dilleniineae, 79
Thor THEIFLORAE   Theales, Scytopetalineae, 62
Dahl THEIFLORAE   Theales, 117
Young DILLENIIDAE, DILLENIANAE Scytopetalales, 124
Takh DILLENIIDAE, THEANAE   Medusagynales, 114
Cron DILLENIIDAE   Theales, 96

Medusagyne Baker

MEDUSANDRACEAE Brenan 1952   45.05

2 genera. Guineo-Congolan Africa. Trees, occas. shrubs.

B&H   Not known   —
DT&H Not known   —
Melc ARCHICHLAMYDEAE   Medusandrales, 24
Thor SANTALIFLORAE   Santalales, 116
      (with doubt 1980, not mentioned 1983)
Dahl CORNIFLORAE   Cornales
Young ROSIDAE, SANTALANAE   Celastrales, 228
Takh ROSIDAE, CELASTRANAE   Santalales, 313
Cron ROSIDAE   Santalales, 205

Medusandra Brenan
Soyauxia Oliv.

MELANOPHYLLACEAE Takht. ex Airy Shaw 1972   82.09

2 genera. Madagascar. Shrubs, small trees.

B&H	Not known	—
DT&H	ARCHICHLAMYDEAE	Umbelliflorae (within Cornaceae, 181)
Melc	ARCHICHLAMYDEAE	Umbelliflorae (within Cornaceae, 223)
Thor	ROSIFLORAE	Rosales, Saxifragineae (within Saxifragaceae, 208)
Dahl	CORNIFLORAE	Cornales (within Montiniaceae, 315)
Young	Not mentioned	—
Takh	ROSIDAE, CORNANAE	Hydrangeales, Escalloniineae, 331
Cron	ROSIDAE	Cornales (within Cornaceae, 203)

Kaliphora Hook.f.
Melanophylla Baker

MELASTOMATACEAE Juss. 1789   68

194 genera. Mostly trop. Shrubs, herbs, rarely trees or lianes.

B&H	POLYPETALAE, CALYCIFLORAE	Myrtales, 68
DT&H	ARCHICHLAMYDEAE	Myrtiflorae, Myrtineae, 174
Melc	ARCHICHLAMYDEAE	Myrtiflorae, Myrtineae, 211
Thor	MYRTIFLORAE	Myrtales, Lythrineae, 243
Dahl	MYRTIFLORAE	Myrtales, 192
	(separating *Memecylaceae* 193)	
Young	ROSIDAE, MYRTANAE	Myrtales, 318
Takh	ROSIDAE, MYRTANAE	Myrtales, Myrtineae, 233
Cron	ROSIDAE	Myrtales, 198

Acanthella Hook.f.
Aciotis D.Don
Acisanthera P.Browne
Adelobotrys DC.
Allomaieta Gleason
Allomorphia Blume
Alloneuron Pilg.
Amphiblemma Naudin
Amphorocalyx Baker
Anaectocalyx Triana ex Hook.f.
Anerincleistus Korth.
Antherotoma (Naudin) Hook.f.
Appendicularia DC.
Arthrostemma Pav. ex D.Don
Aschistanthera C.Hansen
Astrocalyx Merr.
Astronia Blume
Astronidium A.Gray
Axinaea Ruíz & Pav.
Barthea Hook.f.
Behuria Cham.
Bellucia Neck. ex Raf.
Benevidesia Saldanha & Cogn.

Bertolonia Raddi
Bisglaziovia Cogn.
Blakea P.Browne
Blastus Lour.
Boerlagea Cogn.
Boyania Wurdack
Brachyotum (DC.) Triana
Bredia Blume
Brittenia Cogn.
Bucquetia DC.
Cailliella Jacq.–Fél.
Calvoa Hook.f.
Calycogonium DC.
Cambessedesia DC.
Campimia Ridl.
Carionia Naudin
Castratella Naudin
Catanthera F.Muell.
Catocoryne Hook.f.
Centradenia G.Don
Centradeniastrum Cogn.
Centronia D.Don
Chaetolepis (DC.) Miq.

Chaetostoma DC.
Charianthus D.Don
Cincinnobotrys Gilg
Clidemia D.Don
Comolia DC.
Comoliopsis Wurdack
Conostegia D.Don
Creochiton Blume
Cyanandrium Stapf
Cyphostyla Gleason
Cyphotheca Diels
Dalenia Korth.
Desmoscelis Naudin
Dicellandra Hook.f.
Dichaetanthera Endl.
Dinophora Benth.
Dionycha Naudin
Dionychastrum A.Fern. & R.Fern.
Diplarpea Triana
Diplectria (Blume) Rchb.
Dissochaeta Blume
Dissotis Benth.
Dolichoura Brade
Driessenia Korth.
Ekmaniocharis Urb.
Enaulophyton Steenis
Eriocnema Naudin
Ernestia DC.
Feliciadamia Bullock
Fordiophyton Stapf
Fritzschia Cham.
Graffenrieda DC.
Gravesia Naudin
Guyonia Naudin
Henriettea DC.
Henriettella Naudin
Heterocentron Hook. & Arn.
Heterotrichum DC.
Huberia DC.
Huilaea Wurdack
Hypenanthe (Blume) Blume
Kendrickia Hook.f.
Kerriothyrsus C.Hansen
Killipia Gleason
Kirkbridea Wurdack
Lavoisiera DC.
Leandra Raddi
Lijndenia Zoll. & Moritzi
Lithobium Bong.
Llewelynia Pittier
Loreya DC.
Loricalepis Brade
Macairea DC.
Macrocentrum Hook.f.
Macrolenes Naudin

Maguireanthus Wurdack
Maieta Aubl.
Mallophyton Wurdack
Marcetia DC.
Mecranium Hook.f.
Medinilla Gaudich.
Melastoma L.
Melastomastrum Naudin
Memecylon L.
Meriania Sw.
Merianthera Kuhlm.
Miconia Ruíz & Pav.
Microlepis (DC.) Miq.
Microlicia D.Don
Mommsenia Urb. & Ekman
Monochaetum (DC.) Naudin
Monolena Triana
Mouriri Aubl.
Myriaspora DC.
Myrmidone Mart.
Neblinanthera Wurdack
Necramium Britton
Neodriessenia M.P.Nayar
Nepsera Naudin
Nerophila Naudin
Ochthephilus Wurdack
Ochthocharis Blume
Omphalopus Naudin
Opisthocentra Hook.f.
Oritrephes Ridl.
Orthogoneuron Gilg
Osbeckia L.
Ossaea DC.
Otanthera Blume
Oxyspora DC.
Pachyanthus Rich.
Pachycentria Blume
Pachyloma DC.
Pentossaea Judd
Phainantha Gleason
Phyllagathis Blume
Pilocosta Almeda & Whiffin
Plagiopetalum Rehder
Pleiochiton Naudin ex A.Gray
Plethiandra Hook.f.
Podocaelia (Benth.) A.Fern. & R.Fern.
Pogonanthera Blume
Poikilogyne Baker f.
Poilannammia C.Hansen *
Poteranthera Bong.
Preussiella Gilg
Pseudodissochaeta M.P.Nayar
Pseudoernestia (Cogn.) Krasser
Pseudosbeckia A.Fern. & R.Fern.
Pternandra Jack

Pterogastra Naudin
Pterolepis (DC.) Miq.
Rhexia L.
Rhodosepala Baker
Rhynchanthera DC.
Rousseauxia DC.
Salpinga Mart. ex DC.
Sandemania Gleason
Sarcopyramis Wall.
Schwackaea Cogn.
Scorpiothyrsus H.L.Li
Siphanthera Pohl
Sonerila Roxb.
Spathandra Guill. & Perr.
Sporoxeia W.W.Sm.
Stapfiophyton H.L.Li
Stenodon Naudin
Stussenia C.Hansen
Svitramia Cham.

Tateanthus Gleason
Tayloriophyton M.P.Nayar
Tessmannianthus Markgr.
Tetraphyllaster Gilg *
Tetrazygia Rich. ex DC.
Tibouchina Aubl.
Tibouchinopsis Markgr.
Tigridiopalma C.Chen *
Tococa Aubl.
Topobea Aubl.
Trembleya DC.
Triolena Naudin
Tristemma Juss.
Tryssophyton Wurdack
Tylanthera C.Hansen *
Veprecella Naudin
Vietsenia C.Hansen
Votomita Aubl.
Warneckea Gilg

MELIACEAE Juss. 1789   43.01

51 genera. Widespread trop. and subtrop., rarely temp. Trees, some shrubs.

B&H	POLYPETALAE, DISCIFLORAE	Geraniales, 43
DT&H	ARCHICHLAMYDEAE	Geraniales, Geraniineae, 90
	(including Ptaeroxylaceae)	
Melc	ARCHICHLAMYDEAE	Rutales, Rutineae, 136
	(including Ptaeroxylaceae)	
Thor	RUTIFLORAE	Rutales, Rutineae, 172
Dahl	RUTIFLORAE	Rutales, 224
Young	ROSIDAE, RUTANAE	Rutales, 248
Takh	ROSIDAE, RUTANAE	Rutales, 266
Cron	ROSIDAE	Sapindales, 260

Aglaia Lour.
Anthocarapa Pierre
Aphanamixis Blume
Astrotrichilia (Harms) T.D.Penn. &
   Styles
Azadirachta A.Juss.
Cabralea A.Juss.
Calodecaryia J.–F.Leroy
Capuronianthus J.–F.Leroy
Carapa Aubl.
Cedrela P.Browne
Chisocheton Blume
Chukrasia A.Juss.
Cipadessa Blume
Dysoxylum Blume
Ekebergia Sparrm.
Entandrophragma C.E.C.Fisch.
Guarea L.

Heckeldora Pierre
Humbertioturraea J.–F.Leroy
Khaya A.Juss.
Lansium Corrêa
Lepidotrichilia (Harms) J.–F.Leroy
Lovoa Harms
Malleastrum (Baill.) J.–F.Leroy
Megaphyllaea Hemsl.
Melia L.
Munronia Wight
Naregamia Wight & Arn.
Neobeguea J.–F.Leroy
Nymania Lindb.
Owenia F.Muell.
Pseudobersama Verdc.
Pseudocarapa Hemsl.
Pseudocedrela Harms
Pterorhachis Harms

Quivisianthe Baill.
Reinwardtiodendron Koord.
Ruagea Karst.
Sandoricum Cav.
Schmardaea Karst.
Soymida A.Juss.
Sphaerosacme Wall. ex M.Roem.
Swietenia Jacq.

Synoum A.Juss.
Toona (Endl.) M.Roem.
Trichilia P.Browne
Turraea L.
Turraeanthus Baill.
Vavaea Benth.
Walsura Roxb.
Xylocarpus J.König

MELIANTHACEAE Link 1831   51.04

2 genera. Tropical and South Africa. Trees, shrubs.

B&H	POLYPETALAE, DISCIFLORAE	Sapindales (within Sapindaceae, 51)
DT&H	ARCHICHLAMYDEAE	Sapindales, Melianthineae, 117
	(including Greyiaceae)	
Melc	ARCHICHLAMYDEAE	Sapindales, Sapindineae, 150
	(including Greyiaceae)	
Thor	RUTIFLORAE	Rutales, Sapindineae, 184
Dahl	RUTIFLORAE	Sapindales, 214
Young	ROSIDAE, RUTANAE	Sapindales, 256
Takh	ROSIDAE, RUTANAE	Sapindales, 249
Cron	ROSIDAE	Sapindales, 249

Bersama Fres.
Melianthus L.

MELIOSMACEAE Endl. 1841   52

2 genera. Trop. and E. Asia, trop. America. Trees, shrubs.

B&H	POLYPETALAE, DISCIFLORAE	Sapindales (within Sabiaceae, 52)
DT&H	ARCHICHLAMYDEAE	Sapindales, Sabiineae (within Sabiaceae, 116)
Melc	ARCHICHLAMYDEAE	Sapindales, Sapindineae (within Sabiaceae, 149)
Thor	RUTIFLORAE	Rutales, Sapindineae (within Sabiaceae, 183)
Dahl	RUTIFLORAE	Sapindales, 216
Young	ROSIDAE, RUTANAE	Sapindales (within Sabiaceae, 255)
Takh	ROSIDAE, RUTANAE	Sapindales (within Sabiaceae, 253)
Cron	MAGNOLIIDAE	Ranunculales (within Sabiaceae, 37)

Meliosma Blume
Ophiocaryon Endl.

MENISPERMACEAE Juss. 1789   6.01

73 genera. Trop. and subtrop., occas. temp. Mostly lianes, rarely trees, herbs.

B&H	POLYPETALAE, THALAMIFLORAE	Ranales, 6
DT&H	ARCHICHLAMYDEAE	Ranales, Ranunculineae, 44

Melc	ARCHICHLAMYDEAE	Ranunculales, Ranunculineae, 65
Thor	ANNONIFLORAE	Berberidales, Berberidineae, 30
Dahl	RANUNCULIFLORAE	Ranunculales, 31
Young	MAGNOLIIDAE, RANUNCULANAE	Ranunculales, 32
Takh	RANUNCULIDAE, RANUNCULANAE	Ranunculales, Menispermineae, 46
Cron	MAGNOLIIDAE	Ranunculales, 35

Abuta Aubl.
Albertisia Becc.
Anamirta Colebr.
Anisocycla Baill.
Anomospermum Miers
Antizoma Miers
Arcangelisia Becc.
Aspidocarya Hook.f. & Thomson
Beirnaertia Louis ex Troupin
Borismene Barneby
Burasaia Thouars
Calycocarpum Nutt. ex Spach
Carronia F.Muell.
Caryomene Barneby & Krukoff
Chasmanthera Hochst.
Chlaenandra Miq.
Chondrodendron Ruíz & Pav.
Cionomene Krukoff
Cissampelos L.
Cocculus DC.
Coscinium Colebr.
Curarea Barneby & Krukoff
Cyclea Arn. ex Wight
Dialytheca Exell & Mendonça
Dioscoreophyllum Engl.
Diploclisia Miers
Disciphania Eichler
Echinostephia (Diels) Domin
Elephantomene Barneby & Krukoff
Eleutharrhena Forman
Fibraurea Lour.
Haematocarpus Miers
Hyperbaena Miers ex Benth.
Hypserpa Miers
Jateorhiza Miers
Kolobopetalum Engl.
Legnephora Miers
Leptoterantha Louis ex Troupin

Limacia Lour.
Limaciopsis Engl.
Macrococculus Becc.
Menispermum L.
Odontocarya Miers
Orthogynium Baill.
Orthomene Barneby & Krukoff
Pachygone Miers
Parabaena Miers
Parapachygone Forman
Penianthus Miers
Pericampylus Miers
Platytinospora (Engl.) Diels
Pleogyne Miers
Pycnarrhena Miers ex Hook.f. & Thomson
Rhaptonema Miers
Rhigiocarya Miers
Sarcolophium Troupin
Sarcopetalum F.Muell.
Sciadotenia Miers
Sinomenium Diels
Sphenocentrum Pierre
Spirospermum Thouars
Stephania Lour.
Strychnopsis Baill.
Synandropus A.C.Sm. *
Synclisia Benth.
Syntriandrium Engl.
Syrrheonema Miers
Telitoxicum Moldenke
Tiliacora Colebr.
Tinomiscium Miers ex Hook.f. & Thomson
Tinospora Miers
Triclisia Benth.
Ungulipetalum Moldenke

MENYANTHACEAE (Dumort.) Dumort. 1829   109.02

5 genera. Widespread. Herbs, often aquatic.

B&H	GAMOPETALAE, BICARPELLATAE	Gentianales (within Gentianaceae, 109)
DT&H	METACHLAMYDEAE	Contortae, Gentianineae (within Gentianaceae, 199)

Melc   SYMPETALAE   Gentianales, 248
Thor   GENTIANIFLORAE   Gentianales, 255
Dahl   GENTIANIFLORAE   Gentianales, 343
Young   ROSIDAE, GENTIANANAE   Gentianales, 353
Takh   LAMIIDAE, GENTIANANAE   Gentianales, 373
Cron   ASTERIDAE   Solanales, 281

Liparophyllum Hook.f.   **Nymphoides** Séguier
Menyanthes L.   **Villarsia** Vent.
Nephrophyllidium Gilg

MISODENDRACEAE J.Agardh 1858   149.02

1 genus. South America. Hemiparasitic subshrubs on trees.

B&H   MONOCHLAMYDEAE   Achlamydosporeae (within Santalaceae, 149)
DT&H   ARCHICHLAMYDEAE   Santalales, Santalineae (as *Myzodendraceae*) 18
Melc   ARCHICHLAMYDEAE   Santalales, Santalineae, 21
Thor   SANTALIFLORAE   Santalales, 119
Dahl   SANTALIFLORAE   Santalales, 258
Young   ROSIDAE, SANTALANAE   Santalales, 237
Takh   ROSIDAE, CELASTRANAE   Santalales, 315
Cron   ROSIDAE   Santalales, 210

Misodendrum Banks ex DC.

MOLLUGINACEAE Hutch. 1926   79.03

15 genera. Mostly tropical, sometimes (*Corrigiola*) temperate. Mostly herbs.

B&H   POLYPETALAE, CALYCIFLORAE Ficoidales (within Aizoaceae, 79)
DT&H   ARCHICHLAMYDEAE   Centrospermae, Phytolaccineae (within
                                                        Aizoaceae, 34)
Melc   ARCHICHLAMYDEAE   Centrospermae, Phytolaccineae, 30
        (including Gisekiaceae)
Thor   CHENOPODIIFLORAE   Chenopodiales, Chenopodiineae (within
                                                        Aizoaceae, 85)
Dahl   CARYOPHYLLIFLORAE   Caryophyllales, 51
Young   DILLENIIDAE, CARYOPHYLLANAE Caryophyllales, 145
Takh   CARYOPHYLLIDAE, CARYOPHYLLANAE Caryophyllales, Caryophyllineae, 63
Cron   CARYOPHYLLIDAE   Caryophyllales, 74

Adenogramma Rchb.   **Macarthuria** Hügel ex Endl.
Coelanthum E.Mey. ex Fenzl   **Mollugo** L.
Corbichonia Scop.   **Pharnaceum** L.
Corrigiola L.   **Polpoda** C.Presl
Glinus L.   **Psammotropha** Eckl. & Zeyh.
Glischrothamnus Pilg.   **Suessenguthiella** Friedrich
Hypertelis E.Mey. ex Fenzl   **Telephium** L.
Limeum L.

MONIMIACEAE Juss. 1809 142.01

39 genera. Tropical to warm-temperate esp. southern. Trees, shrubs, lianes.

B&H MONOCHLAMYDEAE Micrembryeae, 142
(including Amborellaceae, Trimeniaceae)
DT&H ARCHICHLAMYDEAE Ranales, Magnoliineae, 51
(including Amborellaceae, Scyphostegiaceae, Trimeniaceae)
Melc ARCHICHLAMYDEAE Magnoliales, Laurineae, 52
Thor ANNONIFLORAE Annonales, Laurineae, 17
Dahl MAGNOLIIFLORAE Laurales, 19
Young MAGNOLIIDAE, RANUNCULANAE Laurales, 16
Takh MAGNOLIIDAE, MAGNOLIANAE Laurales, 14
(separating *Atherospermataceae* 15, *Siparunaceae* 16)
Cron MAGNOLIIDAE Laurales, 13

Anthobembix Perkins
Atherosperma Labill.
Austromatthaea L.S.Sm.
Bracteanthus Ducke
Carnegieodoxa Perkins
Daphnandra Benth.
Decarydendron Danguy
Doryphora Endl.
Dryadodaphne S.Moore
Ephippiandra Decne.
Faika Philipson *
Glossocalyx Benth.
Hedycarya J.R.Forst. & G.Forst.
Hedycaryopsis Danguy
Hennecartia Poiss.
Hortonia Wight
Kairoa Philipson
Kibara Endl.
Kibaropsis Vieill. ex Jérémie *
Laurelia Juss.

Lauterbachia Perkins
Levieria Becc.
Macropeplus Perkins
Macrotorus Perkins
Matthaea Blume
Mollinedia Ruíz & Pav.
Monimia Thouars
Nemuaron Baill.
Palmeria F.Muell.
Parakibara Philipson *
Peumus Molina
Phanerogonocarpus Cavaco
Schrameckia Danguy
Siparuna Aubl.
Steganthera Perkins
Tambourissa Sonn.
Tetrasynandra Perkins
Wilkiea F.Muell.
Xymalos Baill.

MONTINIACEAE Nakai 1966 70.03

2 genera. South Africa, East Africa, Madagascar. Shrubs, small trees.

B&H POLYPETALAE, CALYCIFLORAE Myrtales (within Onagraceae, 70)
DT&H ARCHICHLAMYDEAE Rosales, Saxifragineae (within Saxifragaceae, 67)
Melc ARCHICHLAMYDEAE Rosales, Saxigragineae (within Saxifragaceae, 106)
Thor ROSIFLORAE Rosales, Saxifragineae (within Saxifragaceae, 208)
Dahl CORNIFLORAE Cornales, 315
(including Melanophyllaceae)
Young ROSIDAE, CORNANAE Hydrangeales, 331

Takh ROSIDAE, CORNANAE   Hydrangeales, Escalloniineae, 330
Cron ROSIDAE   Rosales (within Grossulariaceae, 166)

Grevea Baill.
Montinia Thunb.

MORACEAE Link 1831   153.05

37 genera. Trop. and warm-temp., occas. temp. Trees, shrubs, lianes, herbs.

B&H   MONOCHLAMYDEAE   Unisexuales (within Urticaceae, 153)
DT&H ARCHICHLAMYDEAE   Urticales, 14
          (including Cannabaceae, Cecropiaceae)
Melc ARCHICHLAMYDEAE   Urticales, 13
          (including Cannabaceae, Cecropiaceae)
Thor MALVIFLORAE   Urticales (within Urticaceae, 157)
Dahl MALVIFLORAE   Urticales, 70
Young DILLENIIDAE, MALVANAE   Urticales, 180
Takh DILLENIIDAE, URTICANAE   Urticales, Urticineae, 185
Cron HAMAMELIDIDAE   Urticales, 53

Antiaris Lesch.
Antiaropsis K.Schum.
Artocarpus J.R.Forst. & G.Forst.
Bagassa Aubl.
Batocarpus Karst.
Bleekrodea Blume
Bosqueiopsis De Wild. & T.Durand
Brosimum Sw.
Broussonetia L'Hér. ex Vent.
Castilla Cerv.
Clarisia Ruíz & Pav.
Dorstenia L.
Fatoua Gaudich.
Ficus L.
Helianthostylis Baill.
Helicostylis Trécul
Hullettia King ex Hook.f.
Maclura Nutt.
Maquira Aubl.

Mesogyne Engl.
Milicia T.R.Sim
Morus L.
Naucleopsis Miq.
Parartocarpus Baill.
Perebea Aubl.
Poulsenia Eggers
Prainea King ex Hook.f.
Pseudolmedia Trécul
Scyphosyce Baill.
Sorocea A.St.–Hil.
Sparattosyce Bureau
Streblus Lour.
Treculia Decne. ex Trécul
Trilepisium Thouars
Trophis P.Browne
Trymatococcus Poepp. & Endl.
Utsetela Pellegr.

MORINACEAE Raf. 1820   86.02

1 genus. Eurasia, temperate and warm-temperate. Herbs.

B&H   GAMOPETALAE, INFERAE   Asterales (within Dipsacaceae, 86)
DT&H METACHLAMYDEAE   Rubiales (within Dipsacaceae, 227)
Melc SYMPETALAE   Dipsacales (within Dipsacaceae, 282)
Thor CORNIFLORAE   Dipsacales (within Dipsacaceae, 294)
Dahl CORNIFLORAE   Dipsacales, 333
Young (not mentioned, presumably within Dipsacaceae, 384)

Takh ROSIDAE, CORNANAE   Dipsacales, 362
Cron ASTERIDAE   Dipsacales (within Dipsacaceae, 316)

Morina L.

MORINGACEAE R.Br. ex Dumort. 1829   55

1 genus. Africa, Madagascar, India, naturalised elsewhere. Trees.

B&H   POLYPETALAE, DISCIFLORAE Ordines anomala, 55
DT&H   ARCHICHLAMYDEAE   Rhoeadales, Moringineae, 59
Melc   ARCHICHLAMYDEAE   Papaverales, Moringineae, 99
Thor   VIOLIFLORAE   Capparales, 140
Dahl   VIOLIFLORAE   Capparales, 106
Young DILLENIIDAE, VIOLANAE   Capparales, 209
Takh   DILLENIIDAE, VIOLANAE   Moringales, 167
Cron   DILLENIIDAE   Capparales, 135

Moringa Adans.

MYOPORACEAE R.Br. 1810   123

3 genera. Mauritius, E. Asia to Australasia, Pacific, trop. America. Trees, shrubs.

B&H   GAMOPETALAE, BICARPELLATAE Lamiales, 123
DT&H   METACHLAMYDEAE   Tubiflorae, Myoporineae, 220
Melc   SYMPETALAE   Tubiflorae, Myoporineae, 276
Thor   GENTIANIFLORAE   Bignoniales, 259
Dahl   LAMIIFLORAE   Scrophulariales, 349
Young ROSIDAE, GENTIANANAE   Bignoniales, 357
Takh   LAMIIDAE, LAMIANAE   Scrophulariales, 407
Cron   ASTERIDAE   Scrophulariales, 296

Bontia L.
Eremophila R.Br.
Myoporum Banks & Sol. ex G.Forst.

MYRICACEAE Blume 1829   157

3 genera. Widespread. Trees, shrubs.

B&H   MONOCHLAMYDEAE   Unisexuales, 157
DT&H   ARCHICHLAMYDEAE   Myricales, 7
Mclc   ARCHICHLAMYDEAE   Juglandales, 2
Thor   RUTIFLORAE   Rutales, Myricineae, 178
Dahl   ROSIFLORAE   Myricales, 151
Young ROSIDAE, RUTANAE   Myricales, 268
Takh   HAMAMELIDIDAE, JUGLANDANAE Myricales, 95
Cron   HAMAMELIDIDAE   Myricales, 59

Canacomyrica Guillaumin
Comptonia L'Hér. ex Aiton
Myrica L.

MYRISTICACEAE R.Br. 1810   141

18 genera. Tropical, mostly Asia and America. Trees.

B&H	MONOCHLAMYDEAE	Micrembryeae, 141
DT&H	ARCHICHLAMYDEAE	Ranales, Magnoliales, 49
Melc	ARCHICHLAMYDEAE	Magnoliales, Magnoliineae, 45
Thor	ANNONIFLORAE	Annonales, Annonineae, 10
Dahl	MAGNOLIIFLORAE	Annonales, 2
Young	MAGNOLIIDAE, RANUNCULANAE	Aristolochiales, 24
Takh	MAGNOLIIDAE, MAGNOLIANAE	Annonales, 7
Cron	MAGNOLIIDAE	Magnoliales, 9

Brochoneura Warb.
Cephalosphaera Warb.
Coelocaryon Warb.
Compsoneura (DC.) Warb.
Endocomia W.J.de Wilde
Gymnacranthera Warb.
Haematodendron Capuron
Horsfieldia Willd.
Iryanthera Warb.

Knema Lour.
Maloutchia Warb.
Myristica Gronov.
Osteophloeum Warb.
Otoba (DC.) Karst.
Pycnanthus Warb.
Scyphocephalium Warb.
Staudtia Warb.
Virola Aubl.

MYROTHAMNACEAE Nied. 1891   62.03

1 genus. Africa, Madagascar. Shrub, resurrecting.

B&H	POLYPETALAE, CALYCIFLORAE	Rosales (within Hamamelidaceae, 62)
DT&H	ARCHICHLAMYDEAE	Rosales, Saxifragineae, 71
Melc	ARCHICHLAMYDEAE	Rosales, Hamamelidineae, 103
Thor	ROSIFLORAE	Pittosporales, Bruniineae, 233
Dahl	ROSIFLORAE	Hamamelidales, 143
Young	ROSIDAE, HAMAMELIDANAE	Myrothamnales, 223
Takh	HAMAMELIDIDAE, HAMAMELIDANAE	
		Myrothamnales, 88
Cron	HAMAMELIDIDAE	Hamamelidales, 46

Myrothamnus Welw.

MYRSINACEAE R.Br. 1810   100.01

39 genera. Widespread trop., occas. temp. Mostly trees, shrubs, lianes.

B&H	GAMOPETALAE, HETEROMERAE	Primulales, 100
	(including Theophrastaceae)	
DT&H	METACHLAMYDEAE	Primulales, 189
Melc	SYMPETALAE	Primulales, 234
Thor	THEIFLORAE	Primulales, Primulineae, 77
	(including Theophrastaceae)	
Dahl	PRIMULIFLORAE	Primulales, 128
	(separating *Aegicerataceae* 129)	
Young	DILLENIIDAE, DILLENIANAE	Primulales, 140

MYRSINACEAE (D)

Takh DILLENIIDAE, ERICANAE   Primulales, 138
         (separating *Aegicerataceae* 139)
Cron DILLENIIDAE   Primulales, 154

Abromeitia Mez *
Aegiceras Gaertn.
Amblyanthopsis Mez
Amblyanthus A.DC.
Antistrophe A.DC.
Ardisia Sw.
Badula Juss.
Conandrium (K.Schum.) Mez
Ctenardisia Ducke
Cybianthus Mart.
Discocalyx (A.DC.) Mez
Elingamita G.T.S.Baylis
Embelia Burm.f.
Emblemantha B.C.Stone *
Fittingia Mez
Geissanthus Hook.f.
Gentlea Lundell
Graphardisia (Mez) Lundell *
Grenacheria Mez
Heberdenia Banks ex DC.

Hymenandra (DC.) Spach
Labisia Lindl.
Loheria Merr.
Maesa Forssk.
Monoporus A.DC.
Myrsine L.
Oncostemum A.Juss.
Parathesis (A.DC.) Hook.f.
Pleiomeris A.DC.
Rapanea Aubl.
Sadiria Mez
Solonia Urb. *
Stylogyne A.DC.
Tapeinosperma Hook.f.
Tetrardisia Mez
Valerioanthus Lundell *
Vegaea Urb. *
Walleniella P.Wilson *
Yunckeria Lundell

MYRTACEAE Juss. 1789   67.01

127 genera. Tropics, subtropics and S. temperate. Trees, shrubs.

B&H   POLYPETALAE, CALYCIFLORAE Myrtales, 67
         (including Lecythidaceae)
DT&H ARCHICHLAMYDEAE   Myrtiflorae, Myrtineae, 173
Melc ARCHICHLAMYDEAE   Myrtiflorae, Myrtineae, 206
Thor MYRTIFLORAE   Myrtales, Myrtineae, 246
Dahl MYRTIFLORAE   Myrtales, 187
         (separating *Psiloxylaceae* 185, *Heteropyxidaceae* 186)
Young ROSIDAE, MYRTANAE   Myrtales, 317
Takh ROSIDAE, MYRTANAE   Myrtales, Myrtineae, 226
         (separating *Psiloxylaceae* 224, *Heteropyxidaceae* 225)
Cron ROSIDAE   Myrtales, 194

Acca O.Berg
Accara Landrum
Acmena DC.
Acmenosperma Kausel
Actinodium Schauer
Agonis (DC.) Sweet
Allosyncarpia S.T.Blake
Amomyrtella Kausel
Amomyrtus (Burret) D.Legrand &
    Kausel
Angasomyrtus Trudgen & Keighery
Angophora Cav.

Archirhodomyrtus (Nied.) Burret
Arillastrum Panch. ex Baill.
Astartea DC.
Asteromyrtus Schauer
Austromyrtus (Nied.) Burret
Backhousia Hook. & Harv.
Baeckea L.
Balaustion Hook.
Barongia Peter G.Wilson & B.Hyland *
Basisperma C.T.White
Beaufortia R.Br.
Blepharocalyx O.Berg

Callistemon R.Br.
Calothamnus Labill.
Calycolpus O.Berg
Calycorectes O.Berg
Calyptranthes Sw.
Calyptrogenia Burret
Calythropsis C.A.Gardner *
Calytrix Labill.
Campomanesia Ruíz & Pav.
Carpolepis (J.W.Dawson) J.W.Dawson
Chamelaucium Desf.
Chamguava Landrum
Choricarpia Domin
Cleistocalyx Blume
Cloezia Brongn. & Gris
Conothamnus Lindl.
Corynanthera J.W.Green
Cupheanthus Seem.
Darwinia Rudge
Decaspermum J.R.Forst. & G.Forst.
Eremaea Lindl.
Eucalyptopsis C.T.White
Eucalyptus L'Hér.
Eugenia L.
Gomidesia O.Berg
Heteropyxis Harv.
Hexachlamys O.Berg
Homalocalyx F.Muell.
Homalospermum Schauer
Homoranthus A.Cunn. ex Schauer
Hottea Urb.
Hypocalymma (Endl.) Endl.
Kania Schltr.
Kjellbergiodendron Burret
Kunzea Rchb.
Lamarchea Gaudich.
Legrandia Kausel
Leptospermum J.R.Forst. & G.Forst.
Lindsayomyrtus B.Hyland & Steenis
Lophomyrtus Burret
Lophostemon Schott
Luma A.Gray
Lysicarpus F.Muell.
Malleostemon J.W.Green
Marlierea Cambess.
Melaleuca L.
Meteoromyrtus Gamble
Metrosideros Banks ex Gaertn.
Micromyrtus Benth.
Mitranthes O.Berg
Mitrantia Peter G.Wilson & B.Hyland *
Monimiastrum Guého & A.J.Scott
Mosiera Small

Myrceugenia O.Berg
Myrcia DC. ex Guill.
Myrcianthes O.Berg
Myrciaria O.Berg
Myrrhinium Schott
Myrtastrum Burret
Myrtella F.Muell.
Myrteola O.Berg
Myrtus L.
Neofabricia Joy Thompson
Neomitranthes Legrand
Neomyrtus Burret
Ochrosperma Trudgen
Octamyrtus Diels
Osbornia F.Muell.
Paramyrciaria Kausel
Pericalymma (Endl.) Endl.
Phymatocarpus F.Muell.
Pileanthus Labill.
Pilidiostigma Burret
Piliocalyx Brongn. & Gris
Pimenta Lindl.
Pleurocalyptus Brongn. & Gris
Plinia L.
Pseudanamomis Kausel
Psidium L.
Psiloxylon Thouars ex Tul.
Purpureostemon Gugerli
Regelia Schauer
Rhodamnia Jack
Rhodomyrtus (DC.) Rchb.
Rinzia Schauer
Ristantia Peter G.Wilson & J.T.Waterh.
Scholtzia Schauer
Siphoneugena O.Berg
Sphaerantia Peter G.Wilson & B.Hyland *
Stereocaryum Burret
Syncarpia Ten.
Syzygium Gaertn.
Tepualia Griseb.
Thryptomene Endl.
Tristania R.Br.
Tristaniopsis Brongn. & Gris
Ugni Turcz.
Uromyrtus Burret
Verticordia DC.
V'aterhousea B.Hyland
Welchiodendron Peter G.Wilson & J.T.Waterh.
Whiteodendron Steenis
Xanthomyrtus Diels
Xanthostemon F.Muell.

NELUMBONACEAE (DC.) Dumort. 1829   8.04

1 genus. Warm-temp. and trop. Asia to Australia, E. USA to Colombia. Aquatic.

B&H	POLYPETALAE, THALAMIFLORAE	Ranales (within Nymphaeaceae, 8)
DT&H	ARCHICHLAMYDEAE	Ranales, Nymphaeineae (within Nymphaeaceae, 38)
Melc	ARCHICHLAMYDEAE	Ranunculales, Nymphaeineae (within Nymphaeaceae, 66)
Thor	ANNONIFLORAE	Nelumbonales, 24
Dahl	MAGNOLIIFLORAE	Nelumbonales, 23
Young	MAGNOLIIDAE, MAGNOLIANAE	Illiciales, 9
Takh	MAGNOLIIDAE, NELUMBONANAE	Nelumbonales, 43
Cron	MAGNOLIIDAE	Nymphaeales, 25

Nelumbo Adans.

NEPENTHACEAE Dumort. 1829   136

1 genus. Madagascar and Seychelles to Australia and New Caledonia. Pitcher plants.

B&H	MONOCHLAMYDEAE	Multiovulatae Terrestres, 136
DT&H	ARCHICHLAMYDEAE	Sarraceniales, 61
Melc	ARCHICHLAMYDEAE	Sarraceniales, 92
Thor	THEIFLORAE	Theales, Nepenthineae, 66
Dahl	THEIFLORAE	Theales, 116
Young	DILLENIIDAE, DILLENIANAE	Nepenthales, 127
Takh	MAGNOLIIDAE, NEPENTHANAE	Nepenthales, 38
Cron	DILLENIIDAE	Nepenthales, 105

Nepenthes L.

NESOGENACEAE Marais 1981   125.06

1 genus. East Africa, Madagascar, Indian Ocean, S. Pacific. Herbs, subshrubs.

B&H	GAMOPETALAE, BICARPELLATAE	Lamiales (within Verbenaceae, 125)
DT&H	METACHLAMYDEAE	Tubiflorae, Verbenineae (within Verbenaceae, 206)
Melc	SYMPETALAE	Tubiflorae, Verbenineae (within Verbenaceae, 258)
Thor	(not mentioned, presumably included in Verbenaceae, 265)	
Dahl	(not mentioned, presumably included in Verbenaceae, 364)	
Young	(not mentioned, presumably within Verbenaceae, 377)	
Takh	LAMIIDAE, LAMIANAE	Lamiales, Lamiineae (within Verbenaceae, 413)
Cron	(not mentioned, presumably included in Verbenaceae, 286)	

Nesogenes A.DC.

NEURADACEAE Link 1831   58.04

3 genera. E. Medit. to India, southern Africa. Prostrate herbs.

B&H   POLYPETALAE, CALYCIFLORAE   Rosales (within Rosaceae, 58)
DT&H ARCHICHLAMYDEAE   Rosales, Rosineae (within Rosaceae, 76)
Melc ARCHICHLAMYDEAE   Rosales, Rosineae, 115
Thor ROSIFLORAE   Rosales, Rosineae (within Rosaceae, 203)
Dahl ROSIFLORAE   Rosales, 175
Young (not mentioned, presumably within Rosaceae, 291)
Takh ROSIDAE, ROSANAE   Rosales, 219
Cron ROSIDAE   Rosales, 175

Grielum L.
Neurada L.
Neuradopsis Bremek. & Oberm.

NYCTAGINACEAE Juss. 1789   128

38 genera. Tropics and subtropics, rarely temp. Trees, shrubs, herbs.

B&H   MONOCHLAMYDEAE   Curvembryeae, 128
DT&H ARCHICHLAMYDEAE   Centrospermae, Phytolaccineae, 30
Melc ARCHICHLAMYDEAE   Centrospermae, Phytolaccineae, 29
Thor CHENOPODIIFLORAE   Chenopodiales, Chenopodiineae, 88
Dahl CARYOPHYLLIFLORAE   Caryophyllales, 43
Young DILLENIIDAE, CARYOPHYLLANAE   Chenopodiales, 157
Takh CARYOPHYLLIDAE, CARYOPHYLLANAE
                                  Caryophyllales, Caryophyllineae, 60
Cron CARYOPHYLLIDAE   Caryophyllales, 66

Abronia Juss.
Acleisanthes A.Gray
Allionia L.
Allioniella Rydb.
Ammocodon Standl.
Andradea Allemão
Anulocaulis Standl.
Belemia Pires *
Boerhavia L.
Boldoa Cav. ex Lag.
Bougainvillea Comm. ex Juss.
Caribea Alain *
Cephalotomandra Karst. & Triana
Colignonia Endl.
Commicarpus Standl.
Cryptocarpus Kunth
Cuscatlania Standl.
Cyphomeris Standl.
Grajalesia Miranda *

Guapira Aubl.
Hesperonia Standl.
Izabalaea Lundell *
Leucaster Choisy
Mirabilis L.
Neea Ruíz & Pav.
Neeopsis Lundell
Nyctaginia Choisy
Okenia Schltdl. & Cham.
Oxybaphus L'Hér. ex Willd.
Phaeoptilum Radlk.
Pisonia L.
Pisoniella (Heimerl) Standl.
Quamoclidion Choisy *
Ramisia Glaz. ex Baill.
Reichenbachia Spreng.
Salpianthus Bonpl.
Selinocarpus A.Gray
Tripterocalyx (Torr.) Hook.

NYMPHAEACEAE Salisb. 1805   8.01

6 genera. Widespread. Aquatic herbs.

B&H   POLYPETALAE, THALAMIFLORAE   Ranales, 8
       (including Cabombaceae, Nelumbonaceae)
DT&H ARCHICHLAMYDEAE   Ranales, Nymphaeineae, 38
       (including Cabombaceae, Nelumbonaceae)
Melc   ARCHICHLAMYDEAE   Ranunculales, Nymphaeineae, 66
       (including Cabombaceae, Nelumbonaceae)
Thor   NYMPHAEIFLORAE   Nymphaeales, 37
Dahl   NYMPHAEIFLORAE   Nymphaeales, 28
Young MAGNOLIIDAE, NYMPHAEANAE   Nymphaeales, 10
Takh   MAGNOLIIDAE, NYMPHAEANAE   Nymphaeales, 40
       (separating *Barclayaceae* 41)
Cron   MAGNOLIIDAE   Nymphaeales, 26
       (separating *Barclayaceae* 27)

Barclaya Wall.   **Nymphaea** L.
Euryale Salisb.   **Ondinea** Hartog *
Nuphar Sm.   **Victoria** Lindl.

OCHNACEAE DC. 1811   41.01

26 genera. Widespread in tropics and subtropics. Trees, shrubs, few herbs.

B&H   POLYPETALAE, DISCIFLORAE Geraniales, 41
       (including Tetrameristaceae)
DT&H ARCHICHLAMYDEAE   Parietales, Theineae, 132
       (including Strasburgeriaceae)
Melc   ARCHICHLAMYDEAE   Guttiferales, Ochnineae, 81
Thor   THEIFLORAE   Theales, Scytopetalineae, 58
Dahl   THEIFLORAE   Theales, 120
       (including Diegodendraceae)
Young DILLENIIDAE, DILLENIANAE   Scytopetalales, 120
Takh   DILLENIIDAE, THEANAE   Ochnales, 115
       (separating *Lophiraceae* 116, *Sauvagesiaceae* 117)
Cron   DILLENIIDAE   Theales, 80
       (including Diegodendraceae, Strasburgeriaceae)

Adenarake Maguire & Wurdack   **Ochna** L.
Blastemanthus Planch.   **Ouratea** Aubl.
Brackenridgea A.Gray   **Perissocarpa** Steyerm. & Maquire *
Cespedesia Goudot   **Philacra** Dwyer
Elvasia DC.   **Poecilandra** Tul.
Euthemis Jack   **Rhytidanthera** (Planch.) Tiegh.
Fleurydora A.Chev.   **Sauvagesia** L.
Godoya Ruíz & Pav.   **Schuurmansia** Blume
Gomphia Schreb.   **Schuurmansiella** Hallier f.
Indosinia J.E.Vidal   **Sinia** Diels
Krukoviella A.C.Sm.   **Testulea** Pellegr.
Lophira Banks ex Gaertn.f.   **Tyleria** Gleason
Luxemburgia A.St.–Hil.   **Wallacea** Spruce ex Hook.

OLACACEAE Mirb. ex DC. 1824   45.01

27 genera. Trop. and subtrop. Mostly hemiparasitic trees, shrubs, lianes.

B&H	POLYPETALAE, DISCIFLORAE Olacales, 45	
	(including Cardiopteridaceae, Icacinaceae, Opiliaceae)	
DT&H	ARCHICHLAMYDEAE	Santalales, Santalineae, 22
Melc	ARCHICHLAMYDEAE	Santalales, Santalineae, 16
Thor	SANTALIFLORAE	Santalales, 115
	(including Opiliaceae)	
Dahl	SANTALIFLORAE	Santalales, 255
Young	ROSIDAE, SANTALANAE	Santalales, 234
	(presumably including Opiliaceae)	
Takh	ROSIDAE, CELASTRANAE,	Santalales, 310
	(separating *Octoknemataceae* 312)	
Cron	ROSIDAE	Santalales, 207

Anacolosa (Blume) Blume
Aptandra Miers
Brachynema Benth.
Cathedra Miers
Chaunochiton Benth.
Coula Baill.
Curupira G.A.Black
Diogoa Exell & Mendonça
Douradoa Sleumer *
Dulacia Vell.
Erythropalum Blume
Harmandia Pierre ex Baill.
Heisteria Jacq.
Malania Chun & S.K.Lee

Minquartia Aubl.
Ochanostachys Mast.
Octoknema Pierre
Olax L.
Ongokea Pierre
Phanerodiscus Cavaco
Ptychopetalum Benth.
Schoepfia Schreb.
Scorodocarpus Becc.
Strombosia Blume
Strombosiopsis Engl.
Tetrastylidium Engl.
Ximenia L.

OLEACEAE Hoffmanns. & Link 1813–1820   104

24 genera. Widespread tropics and warm-temperate. Trees, shrubs, climbers.

B&H	GAMOPETALAE, BICARPELLATAE	Gentianales, 104
DT&H	METACHLAMYDEAE	Contortae, Oleineae, 196
Melc	SYMPETALAE	Oleales, 244
Thor	GENTIANIFLORAE	Oleales, 248
Dahl	GENTIANIFLORAE	Oleales, 337
Young	ROSIDAE, GENTIANANAE	Oleales, 345
Takh	LAMIIDAE, GENTIANANAE	Oleales, 377
Cron	ASTERIDAE	Scrophulariales, 293

Abeliophyllum Nakai
Chionanthus L.
Comoranthus Knobl.
Fontanesia Labill.
Forestiera Poir.
Forsythia Vahl
Fraxinus L.
Haenianthus Griseb.
Hesperelaea A.Gray

Jasminum L.
Ligustrum L.
Menodora Bonpl.
Myxopyrum Blume
Nestegis Raf.
Noronhia Stadman ex Thouars
Notelaea Vent.
Olea L.
Osmanthus Lour.

Phillyrea L.
Picconia DC.
Schrebera Roxb.

Syringa L.
Tessarandra Miers
Tetrapilus Lour.

OLINIACEAE Harv. & Sond. 1862   69.05

1 genus. Africa, St. Helena. Trees, shrubs.

B&H	POLYPETALAE, CALYCIFLORAE	Myrtales (within Lythraceae, 69)
DT&H	ARCHICHLAMYDEAE	Myrtiflorae, Thymelaeineae, 163
Melc	ARCHICHLAMYDEAE	Myrtiflorae, Myrtineae, 215
Thor	MYRTIFLORAE	Myrtales, Lythrineae, 240
Dahl	MYRTIFLORAE	Myrtales, 195
Young	ROSIDAE, MYRTANAE	Myrtales, 315
Takh	ROSIDAE, MYRTANAE	Myrtales, Myrtineae, 230
Cron	ROSIDAE, MYRTANAE	Myrtales, 197

Olinia Thunb.

ONAGRACEAE Juss. 1789   70.01

17 genera. Widespread, esp. America. Herbs, rarely shrubs, trees.

B&H	POLYPETALAE, CALYCIFLORAE	Myrtales, 70
	(including Montiniaceae, Trapaceae)	
DT&H	ARCHICHLAMYDEAE	Myrtiflorae, Myrtineae, 175
Melc	ARCHICHLAMYDEAE	Myrtiflorae, Myrtineae, 214
Thor	MYRTIFLORAE	Myrtales, Onagrineae, 245
Dahl	MYRTIFLORAE	Myrtales, 188
Young	ROSIDAE, MYRTANAE	Myrtales, 319
Takh	ROSIDAE, MYRTANAE	Myrtales, Onagrineae, 238
Cron	ROSIDAE	Myrtales, 196

Boisduvalia Spach
Calylophus Spach
Camissonia Link
Circaea L.
Clarkia Pursh
Epilobium L.
Fuchsia L.
Gaura L.
Gayophytum A.Juss.

Gongylocarpus Schltdl. & Cham.
Hauya Moçino & Sessé ex DC.
Heterogaura Rothr.
Lopezia Cav.
Ludwigia L.
Oenothera L.
Stenosiphon Spach
Xylonagra Donn.Sm. & Rose

ONCOTHECACEAE Kobuski ex Airy Shaw 1965   46.03

1 genus. New Caledonia. Trees, shrubs.

B&H	Not known	—
DT&H	INCERTAE SEDIS, gen. 9750	—
Melc	Not mentioned	—

Thor	THEIFLORAE	Theales, Theineae, 47
Dahl	THEIFLORAE	Theales, 121
Young	DILLENIIDAE, DILLENIANAE	Theales, 111
Takh	DILLENIIDAE, THEANAE	Theales, 104
Cron	DILLENIIDAE	Theales, 91

Oncotheca Baill.

OPILIACEAE (Benth.) Valeton 1886   45.02

10 genera. Tropics and subtropics. Root parasite trees, shrubs, lianes.

B&H	POLYPETALAE, DISCIFLORAE	Olacales (within Olacaceae, 45)
DT&H	ARCHICHLAMYDEAE	Santalales, Santalineae, 21
Melc	ARCHICHLAMYDEAE	Santalales, Santalineae, 18
Thor	SANTALIFLORAE	Santalales (within Olacaceae, 115)
Dahl	SANTALIFLORAE	Santalales, 256
Young	(not mentioned, presumably within Olacaceae, 234)	
Takh	ROSIDAE, CELASTRANAE	Santalales, 311
Cron	ROSIDAE	Santalales, 208

Agonandra Miers ex Hook.f.
Cansjera Juss.
Champereia Griff.
Gjellerupia Lauterb.
Lepionurus Blume

Melientha Pierre
Opilia Roxb.
Pentarhopalopilia (Engl.) Hiepko
Rhopalopilia Pierre
Urobotrya Stapf

OXALIDACEAE R.Br 1818   38.06

6 genera. Widespread. Herbs, shrubs, some small trees.

B&H	POLYPETALAE, DISCIFLORAE	Geraniales (within Geraniaceae, 38)
DT&H	ARCHICHLAMYDEAE	Geraniales, Geraniineae, 80
Melc	ARCHICHLAMYDEAE	Geraniales, Geraniineae, 123
	(including Lepidobotryaceae)	
Thor	GERANIIFLORAE	Geraniales, Geraniineae, 100
	(including Lepidobotryaceae)	
Dahl	RUTIFLORAE	Geraniales, 240
Young	ROSIDAE, GERANIANAE	Geraniales, 274
	(presumably including Lepidobotryaceae)	
Takh	ROSIDAE, RUTANAE	Geraniales, 280
	(separating *Hypseocharitaceae* 282)	
Cron	ROSIDAE	Geraniales, 263
	(including Lepidobotryaceae)	

Averrhoa L.
Biophytum DC.
Dapania Korth.

Hypseocharis J.Remy
Oxalis L.
Sarcotheca Blume

PAEONIACEAE F.Rudolphi 1830   1.03

1 genus. North temperate to China. Herbs to subshrubs.

B&H   POLYPETALAE, THALAMIFLORAE Ranales (within Ranunculaceae, 1)
DT&H ARCHICHLAMYDEAE   Ranales, Ranunculineae (within
                                               Ranunculaceae, 41)
Melc ARCHICHLAMYDEAE   Guttiferales, Dilleniineae, 76
Thor ANNONIFLORAE   Paeoniales, 26
Dahl THEIFLORAE   Paeoniales, 109
Young MAGNOLIIDAE, RANUNCULANAE Ranunculales, 37
Takh RANUNCULIDAE, RANUNCULANAE Paeoniales, 53
Cron DILLENIIDAE   Dilleniales, 79

Paeonia L.

PANDACEAE Engl. & Gilg 1913   151.05

4 genera. Tropical Africa and Asia. Trees.

B&H   Not known   —
DT&H INCERTAE SEDIS   —
Melc ARCHICHLAMYDEAE   Celastrales, 158
Thor MALVIFLORAE   Euphorbiales, 162
Dahl MALVIFLORAE   Euphorbiales, 77
Young DILLENIIDAE, MALVANAE   Euphorbiales, 186
Takh DILLENIIDAE, EUPHORBIANAE Euphorbiales, 191
Cron ROSIDAE   Euphorbiales, 231

Centroplacus Pierre   **Microdesmis** Hook.f. ex Hook.
Galearia Zoll. & Moritzi   **Panda** Pierre

PAPAVERACEAE Juss. 1789   10

41 genera. Widespread, mostly N. temp. Herbs, shrubs.

B&H   POLYPETALAE, THALAMIFLORAE Parietales, 10
DT&H ARCHICHLAMYDEAE   Rhoeadales, Rhoeadineae, 54
Melc ARCHICHLAMYDEAE   Papaverales, Papaverineae, 94
Thor ANNONIFLORAE   Berberidales, Papaverineae, 35
Dahl RANUNCULIFLORAE   Papaverales, 37
       (separating *Fumariaceae* 38)
Young MAGNOLIIDAE, RANUNCULANAE Ranunculales, 40
       (separating *Fumariaceae* 41)
Takh RANUNCULIDAE, RANUNCULANAE Papaverales, 54
       (separating *Fumariaceae* 56, *Hypecoaceae* 55)
Cron MAGNOLIIDAE   Papaverales, 38
       (separating *Fumariaceae* 39)

Adlumia Raf. ex DC.   **Bocconia** L.
Arctomecon Torr. & Frém.   **Canbya** Parry ex A.Gray
Argemone L.   **Capnoides** Mill.

Ceratocapnos Durieu
Chelidonium L.
Corydalis DC.
Cryptocapnos Rech.f.
Cysticapnos Mill.
Dactylicapnos Wall.
Dendromecon Benth.
Dicentra Borkh. ex Bernh.
Dicranostigma Hook.f. & Thomson
Discocapnos Cham. & Schltdl.
Eomecon Hance
Eschscholzia Cham.
Fumaria L.
Fumariola Korsh.
Glaucium Mill.
Hesperomecon Greene
Hunnemannia Sweet
Hylomecon Maxim.

Hypecoum L.
Macleaya R.Br.
Meconella Nutt.
Meconopsis R.Vig.
Papaver L.
Platycapnos (DC.) Bernh.
Platystemon Benth.
Pseudofumaria Medik.
Pteridophyllum Siebold & Zucc.
Roemeria Medik.
Romneya Harv.
Rupicapnos Pomel
Sanguinaria L.
Sarcocapnos DC.
Stylomecon G.Taylor
Stylophorum Nutt.
Trigonocapnos Schltr.

PARACRYPHIACEAE Airy Shaw 1965   59.16

1 genus, monotypic? New Caledonia. Shrubs to small trees.

B&H	Not known	—
DT&H	Not known	—
Melc	Not known	—
Thor	THEIFLORAE	Theales, Theineae, 42
Dahl	CORNIFLORAE	Cornales, 311
Young	DILLENIIDAE, DILLENIANAE	Theales, 113
Takh	DILLENIIDAE, THEANAE	Paracryphiales, 100
Cron	DILLENIIDAE	Theales, 95

Paracryphia Baker f.

PARNASSIACEAE Gray 1821   59.04

2 genera. N. temperate region to Mexico, Chile, Uruguay. Herbs.

B&H	POLYPETALAE, CALYCIFLORAE	Rosales (within Saxifragaceae, 59)
DT&H	ARCHICHLAMYDEAE	Rosales, Saxifragineae (within Saxifragaceae, 67)
Melc	ARCHICHLAMYDEAE	Rosales, Saxifragineae (within Saxifragaceae, 106)
Thor	ROSIFLORAE	Rosales, Saxifragineae, 209
Dahl	ROSIFLORAE	Droserales 172
	(separating *Lepuropetalaceae* 171)	
Young	ROSIDAE, ROSANAE	Saxifragales, 307
Takh	ROSIDAE, ROSANAE	Saxifragales, 213
	(separating *Lepuropetalaceae* 214)	
Cron	ROSIDAE	Rosales (within Saxifragaceae, 173)

Lepuropetalon Elliott
Parnassia L.

PASSIFLORACEAE Juss. ex Kunth 1817   74.01

17 genera. Widespread trop. and warm-temp. Shrubs, herbs, rarely trees, often climbing.

B&H   POLYPETALAE, CALYCIFLORAE Passiflorales, 74
           (including Achariaceae, Caricaceae, Malesherbiaceae, Physenaceae)
DT&H ARCHICHLAMYDEAE   Parietales, Flacourtiineae, 153
Melc   ARCHICHLAMYDEAE   Violales, Flacourtiineae, 189
Thor   VIOLIFLORAE   Violales, Violineae, 129
Dahl   VIOLIFLORAE   Violales, 85
Young DILLENIIDAE, VIOLANAE   Violales, 197
Takh   DILLENIIDAE, VIOLANAE   Violales, 150
Cron   DILLENIIDAE   Violales, 122

Adenia Forssk.	**Hollrungia** K.Schum.
Ancistrothyrsus Harms	**Mitostemma** Mast.
Androsiphonia Stapf	**Paropsia** Noronha ex Thouars
Barteria Hook.f.	**Paropsiopsis** Engl.
Basananthe Peyr.	**Passiflora** L.
Crossostemma Planch. ex Benth.	**Schlechterina** Harms
Deidamia Noronha ex Thouars	**Smeathmannia** Sol. ex R.Br.
Dilkea Mast.	**Viridivia** J.H.Hemsl. & Verdc.
Efulensia C.H.Wright	

PEDALIACEAE R.Br. 1810   121

17 genera. Widespread, mainly tropical. Mostly herbs, shrubs.

B&H   GAMOPETALAE, BICARPELLATAE Personales, 121
DT&H METACHLAMYDEAE   Tubiflorae, Solanineae, 212
        (separating *Martyniaceae* 213)
Melc   SYMPETALAE   Tubiflorae, Solanineae, 270
        (separating *Martyniaceae* 271)
Thor   GENTIANIFLORAE   Bignoniales, 257
        (separating *Martyniaceae* 258)
Dahl   LAMIIFLORAE   Scrophulariales, 359
        (separating *Martyniaceae* 361, *Trapellaceae* 360)
Young ROSIDAE, GENTIANANAE   Bignoniales
        (separating *Martyniaceae* 356)
Takh   LAMIIDAE, LAMIANAE   Scrophulariales, 401
        (separating *Martyniaceae* 403, *Trapellaceae* 402)
Cron   ASTERIDAE   Scrophulariales, 300

Ceratotheca Endl.	**Pedalium** L.
Craniolaria L.	**Proboscidea** Schmidel
Dicerocaryum Bojer	**Pterodiscus** Hook.
Harpagophytum DC. ex Meisn.	**Rogeria** J.Gay ex Delile
Holubia Oliv.	**Sesamothamnus** Welw.
Ibicella Van Eselt.	**Sesamum** L.
Josephinia Vent.	**Trapella** Oliv.
Linariopsis Welw.	**Uncarina** (Baill.) Stapf
Martynia L.	

PELLICIERACEAE (Triana & Planch.) Beauvis. ex Bullock 1959 28.08

1 genus, monotypic. Costa Rica to Ecuador. Mangrove tree.

B&H POLYPETALAE, THALAMIFLORAE Guttiferales (within Theaceae, 28)
DT&H ARCHICHLAMYDEAE   Parietales, Theineae (within Theaceae, 136)
Melc ARCHICHLAMYDEAE   Guttiferales, Theineae (within Theaceae, 85)
Thor THEIFLORAE   Theales, Theineae (within Theaceae, 44)
Dahl THEIFLORAE   Theales (within Theaceae, 124)
Young (not mentioned, presumably within Theaceae, 106)
Takh DILLENIIDAE, THEANAE   Theales, 111
Cron DILLENIIDAE   Theales, 90

Pelliciera Planch. & Triana ex Benth.

PENAEACEAE Sweet ex Guill. 1828 146.01

7 genera. South Africa. Small shrubs, often ericoid.

B&H MONOCHLAMYDEAE   Daphnales, 146
       (including Geissolomataceae)
DT&H ARCHICHLAMYDEAE   Myrtiflorae, Thymelaeineae, 162
Melc ARCHICHLAMYDEAE   Thymelaeales, 178
Thor MYRTIFLORAE   Myrtales, Lythrineae, 238
Dahl MYRTIFLORAE   Myrtales, 196
Young ROSIDAE, MYRTANAE   Myrtales, 316
Takh ROSIDAE, MYRTANAE   Myrtales, Myrtineae, 229
Cron ROSIDAE   Myrtales, 190

Brachysiphon A.Juss.   **Saltera** Bullock
Endonema A.Juss.   **Sonderothamnus** R.Dahlgren
Glischrocolla (Endl.) A.DC.   **Stylapterus** A.Juss.
Penaea L.

PENTAPHRAGMATACEAE J.Agardh 1858 91.02

1 genus. Southern Asia. Succulent herbs.

B&H GAMOPETALAE, INFERAE   Campanales (within Campanulaceae, 91)
DT&H METACHLAMYDEAE   Campanulatae, Campanulineae (within
                                        Campanulaceae, 229)

Melc SYMPETALAE   Campanulales, 285
Thor SOLANIFLORAE   Campanulales, 275
Dahl ASTERIFLORAE   Campanulales, 269
Young ROSIDAE, SOLANANAE   Campanulales, 371
Takh ASTERIDAE, CAMPANULANAE Campanulales, 417
Cron ASTERIDAE   Campanulales, 304

Pentaphragma Wall. ex G.Don

PENTAPHYLACACEAE Engl. 1897 28.04

1 genus, monotypic? China to Sumatra. Tree.

B&H	POLYPETALAE, THALAMIFLORAE	Guttiferales (within Theaceae, 28)
DT&H	ARCHICHLAMYDEAE	Sapindales, Celastrineae, 105
Melc	ARCHICHLAMYDEAE	Celastrales, Celastrineae, 155
Thor	THEIFLORAE	Theales, Clethrineae, 54
Dahl	THEIFLORAE	Theales, 111
Young	DILLENIIDAE, DILLENIANAE	Theales, 118
Takh	DILLENIIDAE, THEANAE	Theales, 106
Cron	DILLENIIDAE	Theales, 88

Pentaphylax Gardner & Champ.

PENTHORACEAE Rydb. ex Britton 1901 60.02

1 genus. Tropical Asia, E. USA. Herbs.

B&H	POLYPETALAE, CALYCIFLORAE	Rosales (within Crassulaceae, 60)
DT&H	ARCHICHLAMYDEAE	Rosales, Saxifragineae (within Crassulaceae, 65)
Melc	ARCHICHLAMYDEAE	Rosales, Saxifragineae (within Saxifragaceae, 106)
Thor	ROSIFLORAE	Rosales, Saxifragineae (within Saxifragaceae, 208)
Dahl	ROSIFLORAE	Saxifragales (within Saxifragaceae, 163)
Young	ROSIDAE, ROSANAE	Saxifragales, 304
Takh	ROSIDAE, ROSANAE	Saxifragales, 202
Cron	ROSIDAE	Rosales (within Saxifragaceae, 173)

Penthorum L.

PERIDISCACEAE Kuhlm. 1950 17.04

2 genera. Tropical S. America. Trees.

B&H	POLYPETALAE, THALAMIFLORAE	Parietales (within Bixaceae, 17)
DT&H	ARCHICHLAMYDEAE	Parietales, Flacourtiineae (within Flacourtiaceae, 149)
Melc	ARCHICHLAMYDEAE	Violales, Flacourtiineae, 183
Thor	VIOLIFLORAE	Violales, Violineae, 126
Dahl	VIOLIFLORAE	Violales, 87
Young	DILLENIIDAE, VIOLANAE	Violales, 192
Takh	DILLENIIDAE, VIOLANAE	Violales, 146
Cron	DILLENIIDAE	Violales, 108

Peridiscus Benth.
Whittonia Sandwith

PHELLINACEAE (Loes.) Takht. 1967 46.04

1 genus. New Caledonia. Trees, shrubs.

B&H POLYPETALAE, DISCIFLORAE Geraniales (within Rutaceae, 39)
DT&H ARCHICHLAMYDEAE Sapindales, Celastrineae (within
Aquifoliaceae, 107)
Melc ARCHICHLAMYDEAE Celastraceae, Celastrineae (within
Aquifoliaceae, 156)
Thor THEIFLORAE Theales, Theineae, 49
Dahl CORNIFLORAE Cornales, 309
Young DILLENIIDAE, DILLENIANAE Theales, 110
Takh ROSIDAE, CELASTRANAE Celastrales, Icacinineae, 299
Cron ROSIDAE Celastrales (within Aquifoliaceae, 223)

Phelline Labill.

PHRYMACEAE Schauer 1847 125.02

1 genus. India to Japan, N. America. Herbs.

B&H GAMOPETALAE, BICARPELLATAE Lamiales (within Verbenaeae, 125)
DT&H METACHLAMYDEAE Tubiflorae, Phrymineae, 221
Melc SYMPETALAE Tubiflorae, Phrymineae, 277
Thor GENTIANIFLORAE Lamiales (within Verbenaceae, 265)
Dahl LAMIIFLORAE Lamiales (within Verbenaceae, 364)
Young (not mentioned, presumably within Verbenaceae, 377)
Takh LAMIIDAE, LAMIANAE Lamiales, Lamiineae (within
Verbenaceae, 413)
Cron ASTERIDAE Lamiales (within Verbenaceae, 286)

Phryma L.

PHYSENACEAE Takht. 1985 12.03

1 genus. Madagascar. Shrubs to trees.

B&H POLYPETALAE, CALYCIFLORAE Passiflorales (within Passifloraceae, 74)
DT&H ARCHICHLAMYDEAE Parietales, Flacourtiineae (within
Flacourtiaceae, 149)
Melc Not mentioned —
Thor INCERTAE SEDIS, at end
Dahl Not mentioned —
Young Not mentioned —
Takh ROSIDAE, RUTANAE Sapindales, 254
Cron Not mentioned in 1981, tentatively referred to HAMAMELIDIDAE in 1988

Physena Noronha ex Thouars

PHYTOLACCACEAE R.Br. 1818   132.01

15 genera. Tropics to warm-temperate. Mostly herbs, shrubs.

B&H   MONOCHLAMYDEAE   Curvembryeae, 132
       (including Agdestidaceae, Barbeuiaceae, Gyrostemonaceae,
       Stegnospermataceae)
DT&H ARCHICHLAMYDEAE   Centrospermae, Phytolaccineae, 33
       (including Achatocarpaceae, Agdestidaceae, Barbeuiaceae, Gisekiaceae,
       Gyrostemonaceae, Stegnospermataceae)
Melc   ARCHICHLAMYDEAE   Centrospermae, Phytolaccineae, 26
       (including Agdestidaceae, Barbeuiaceae, Stegnospermataceae)
Thor   CHENOPODIIFLORAE   Chenopodiales, Chenopodiineae, 81
       (including Agdestidaceae, Gisekiaceae)
Dahl   CARYOPHYLLIFLORAE   Caryophyllales, 39
       (including Achatocarpaceae, Agdestidaceae)
Young DILLENIIDAE, CARYOPHYLLANAE Chenopodiales, 158
       (separating *Petiveriaceae* 155)
Takh   CARYOPHYLLIDAE, CARYOPHYLLANAE Caryophyllales, Caryophyllineae, 57
       (including Agdestidaceae, Gisekiaceae)
Cron   CARYOPHYLLIDAE   Caryophyllales, 64
       (including Agdestidaceae, Barbeuiaceae, Gisekiaceae, Stegnospermataceae)

Anisomeria D.Don	**Nowickea** J.Mártinez & J.A.McDonald *
Ercilla A.Juss.	**Petiveria** L.
Gallesia Casar.	**Phytolacca** L.
Hilleria Vell.	**Rivina** L.
Ledenbergia Klotzsch ex Moq.	**Schindleria** H.Walter
Lophiocarpus Turcz.	**Seguieria** Loefl.
Microtea Sw.	**Trichostigma** A.Rich.
Monococcus F.Muell.	

PIPERACEAE C.Agardh 1824   139.01

10 genera. Mostly tropical. Herbs, shrubs, lianes, small trees.

B&H   MONOCHLAMYDEAE   Micrembryeae, 139
       (including Lactoridaceae, Saururaceae)
DT&H ARCHICHLAMYDEAE   Piperales, 3
Melc   ARCHICHLAMYDEAE   Piperales, 69
Thor   ANNONIFLORAE   Annonales, Piperineae, 23
Dahl   NYMPHAEIFLORAE   Piperales, 25
Young MAGNOLIIDAE, RANUNCULANAE Aristolochiales, 26
Takh   MAGNOLIIDAE, MAGNOLIANAE   Piperales, 26
Cron   MAGNOLIIDAE   Piperales, 21

Arctottonia Trel.	**Pothomorphe** Miq.
Macropiper Miq.	**Sarcorhachis** Trel.
Manekia Trel.	**Trianaeopiper** Trel.
Peperomia Ruíz & Pav.	**Verhuellia** Miq.
Piper L.	**Zippelia** Blume

PITTOSPORACEAE R.Br. 1814 18

9 genera. Trop. to warm-temp. Old World, esp. Australasia. Trees, shrubs.

B&H POLYPETALAE, THALAMIFLORAE Polygalineae, 18
DT&H ARCHICHLAMYDEAE   Rosales, Saxifragineae, 68
Melc ARCHICHLAMYDEAE   Rosales, Saxifragineae, 110
Thor ROSIFLORAE   Pittosporales, Pittosporineae, 226
Dahl ARALIIFLORAE   Pittosporales, 264
Young ROSIDAE, CORNANAE   Pittosporales, 338
Takh ROSIDAE, CORNANAE   Pittosporales, 353
Cron ROSIDAE   Rosales, 174

Bentleya E.M.Benn. *   **Hymenosporum** R.Br. ex F.Muell.
Billardiera Sm.   **Pittosporum** Banks ex Sol.
Bursaria Cav.   **Pronaya** Hügel
Cheiranthera A.Cunn. ex Brongn.   **Sollya** Lindl.
Citriobatus A.Cunn. ex Putterl.

PLAGIOPTERACEAE Airy Shaw 1965 33.04

1 genus, monotypic? Burma, Thailand, China. Lianes.

B&H POLYPETALAE, THALAMIFLORAE Malvales (within Tiliaceae, 33)
DT&H ARCHICHLAMYDEAE   Parietales, Flacourtiineae (within Flacourtiaceae, 149)

Melc Not mentioned   —
Thor MALVIFLORAE   Malvales, 147
Dahl MALVIFLORAE   Malvales, 58
Young DILLENIIDAE, MALVANAE   Malvales, 167
Takh DILLENIIDAE, MALVANAE   Malvales, 174
Cron DILLENIIDAE   Violales (within Flacourtiaceae, 107)

Plagiopteron Griff.

PLANTAGINACEAE Juss. 1789 127

3 genera. Widespread. Herbs, subshrubs.

B&H GAMOPETALAE, BICARPELLATAE Ordo anomalus, 127
DT&H METACHLAMYDEAE   Plantaginales, 222
Melc SYMPETALAE   Plantaginales, 278
Thor GENTIANIFLORAE   Bignoniales, 261
Dahl LAMIIFLORAE   Scrophulariales, 357
Young ROSIDAE, GENTIANANAE   Bignoniales, 359
Takh LAMIIDAE, LAMIANAE   Scrophulariales, 405
Cron ASTERIDAE   Plantaginales, 291

Bougueria Decne.
Littorella Bergius
Plantago L.

PLATANACEAE Lestib. ex Dumort. 1829   154

1 genus. E. Mediterranean, Indochina, S.E. USA, S.W. USA and Mexico. Trees.

B&H	MONOCHLAMYDEAE	Unisexuales, 154
DT&H	ARCHICHLAMYDEAE	Rosales, Rosineae, 74
Melc	ARCHICHLAMYDEAE	Rosales, Hamamelidineae, 101
Thor	HAMAMELIDIFLORAE	Hamamelidales, Hamamelidineae, 199
Dahl	ROSIFLORAE	Hamamelidales, 142
Young	ROSIDAE, HAMAMELIDANAE	Hamamelidales, 222
Takh	HAMAMELIDIDAE, HAMAMELIDANAE	
		Hamamelidales, 84
Cron	HAMAMELIDIDAE	Hamamelidales, 44

Platanus L.

PLUMBAGINACEAE Juss. 1789   98

25 genera. Widespread. Mostly herbs or subshrubs.

B&H	GAMOPETALAE, HETEROMERAE	Primulales, 98
DT&H	METACHLAMYDEAE	Primulales, 191
Melc	SYMPETALAE	Plumbaginales, 236
Thor	THEIFLORAE	Primulales, Plumbaginineae, 79
Dahl	PLUMBAGINIFLORAE	Plumbaginales 55
	(separating *Limoniaceae* 54)	
Young	DILLENIIDAE, DILLENIANAE	Plumbaginales, 143
Takh	CARYOPHYLLIDAE, PLUMBAGINANAE	
		Plumbaginales, 75
Cron	CARYOPHYLLIDAE	Plumbaginales, 77

Acantholimon Boiss.
Aegialitis R.Br.
Armeria Willd.
Bamiania Lincz. *
Bubania Girard
Bukiniczia Lincz.
Cephalorhizum Popov & Korovin
Ceratostigma Bunge
Chaetolimon (Bunge) Lincz.
Dictyolimon Rech.f.
Dyerophytum Kuntze
Ghaznianthus Lincz. *
Gladiolimon Mob.

Goniolimon Boiss.
Ikonnikovia Lincz.
Limoniastrum Fabr.
Limoniopsis Lincz.
Limonium Mill.
Muellerolimon Lincz.
Neogontscharovia Lincz. *
Plumbagella Spach
Plumbago L.
Popoviolimon Lincz. *
Psylliostachys (Jaub. & Spach) Nevski
Vassilczenkoa Lincz. *

PODOACEAE Baill. ex Franch. 1889   53.05

2 genera. Nepal to S. China and Thailand. Shrubs, herbs.

B&H	POLYPETALAE, DISCIFLORAE	Sapindales (within Sapindaceae, 53)
DT&H	ARCHICHLAMYDEAE	Sapindales, Anacardiineae (within
		Anacardiaceae, 103)

Melc   ARCHICHLAMYDEAE   Sapindales, Anacardiineae (within
                                                                Anacardiaceae, 144)

Thor   (not mentioned, presumably within Anacardiaceae, 174)
Dahl   RUTIFLORAE   Sapindales, 206
Young ROSIDAE, RUTANAE   Rutales (within Anacardiaceae, 251)
Takh   ROSIDAE, RUTANAE   Rutales (within Anacardiaceae, 271)
Cron   ROSIDAE   Sapindales (within Anacardiaceae, 256)

Campylopetalum Forman
Dobinea Buch.–Ham. ex D.Don

PODOSTEMACEAE Rich. ex C.Agardh 1822   135.01

50 genera. Tropical. Lithophytes in rivers.

B&H   MONOCHLAMYDEAE   Multiovulatae Aquaticae, 135
             (including Hydrostachyaceae)
DT&H   ARCHICHLAMYDEAE   Rosales, Podostemonineae, 63
Melc   ARCHICHLAMYDEAE   Podostemales, 121
Thor   ROSIFLORAE   Rosales, Saxifragineae, 213
Dahl   PODOSTEMIFLORAE   Podostemales, 181
Young ROSIDAE, ROSANAE   Podostemales, 309
Takh   ROSIDAE, ROSANAE   Podostemales, 221
Cron   ROSIDAE   Podostemales, 185

Angolaea Wedd.
Apinagia Tul.
Butumia G.Taylor *
Castelnavia Tul. & Wedd.
Ceratolacis (Tul.) Wedd.
Cladopus H.Moeller *
Dalzellia Wight
Devillea Tul. & Wedd. *
Dicraeanthus Engl.
Dicraeia Thouars
Diplobryum C.Cusset *
Djinga C.Cusset *
Endocaulos C.Cusset
Farmeria Willis ex Trimen
Griffithella (Tul.) Warm.
Heterotristicha Tobl. *
Hydrobryum Endl.
Indotristicha P.Royen
Jenmaniella Engl.
Ledermanniella Engl.
Leiothylax Warm.
Letestuella G.Taylor
Lonchostephus Tul.
Lophogyne Tul.
Macarenia P.Royen *

Macropodiella Engl.
Malaccotristicha C.Cusset & G.Cusset
Marathrum Bonpl.
Mniopsis Mart. & Zucc.
Monostylis Tul.
Mourera Aubl.
Oserya Tul. & Wedd.
Paleodicraeia C.Cusset *
Podostemum Michx.
Pohliella Engl.
Polypleurella Engl.
Polypleurum (Tul.) Warm.
Rhyncholacis Tul.
Saxicolella Engl.
Sphaerothylax Bisch. ex Krauss
Stonesia G.Taylor
Thelethylax C.Cusset
Torrenticola Domin
Tristicha Thouars
Tulasneantha P.Royen
Weddellina Tul.
Wettsteiniola Suess. *
Willisia Warm.
Winklerella Engl. *
Zehnderia C.Cusset *

POLEMONIACEAE Juss. 1789 110

19 genera. Temp. Eurasia, temp. and trop. America. Herbs, rarely small trees.

B&H GAMOPETALAE, BICARPELLATAE Polemoniales, 110
  (including Cobaeaceae)
DT&H METACHLAMYDEAE Tubiflorae, Convolvulineae, 203
  (including Cobaeaceae)
Melc SYMPETALAE Tubiflorae, Convolvulineae, 252
  (including Cobaeaceae)
Thor SOLANIFLORAE Solanales, Polemoniineae, 273
  (including Cobaeaceae)
Dahl SOLANIFLORAE Solanales, 279
Young ROSIDAE, SOLANANAE Solanales, 365
  (presumably including Cobaeaceae)
Takh LAMIIDAE, SOLANANAE Polemoniales, 387
Cron ASTERIDAE Solanales, 282
  (including Cobaeaceae)

Acanthogilia A.G.Day & Moran
Allophyllum (Nutt.) A.D.Grant &
  V.E.Grant
Bonplandia Cav.
Cantua Juss. ex Lam.
Collomia Nutt.
Eriastrum Woot. & Standl.
Gilia Ruíz & Pav.
Gymnosteris Greene
Huthia Brand

Ipomopsis Michx.
Langloisia Greene
Leptodactylon Hook. & Arn.
Linanthus Benth.
Loeselia L.
Loeseliastrum (Brand) Timbrook
Microsteris Greene
Navarretia Ruíz & Pav.
Phlox L.
Polemonium L.

POLYGALACEAE R.Br. 1814 20.01

19 genera. Widespread except N. Zealand and Polynesia. Trees, shrubs, lianes, herbs.

B&H POLYPETALAE, THALAMIFLORAE Polygalineae, 20
  (including Krameriaceae)
DT&H ARCHICHLAMYDEAE Geraniales, Polygalineae, 95
Melc ARCHICHLAMYDEAE Rutales, Polygalineae, 142
Thor GERANIIFLORAE Geraniales, Polygalineae, 108
Dahl RUTIFLORAE Polygalales, 228
Young ROSIDAE, GERANIANAE Polygalales, 278
Takh ROSIDAE, RUTANAE Polygalales, 295
Cron ROSIDAE Polygalales, 245
  (including Emblingiaceae; separating Xanthophyllaceae 246)

Atroxima Stapf
Badiera DC.
Balgoya Morat & Meijden
Barnhartia Gleason
Bredemeyera Willd.
Carpolobia G.Don
Comesperma Labill.
Diclidanthera Mart.
Epirhizanthes Blume
Eriandra P.Royen & Steenis

Monnina Ruíz & Pav.
Moutabea Aubl.
Muraltia DC.
Nylandtia Dumort.
Phlebotaenia Griseb.
Polygala L.
Salomonia Lour.
Securidaca L.
Xanthophyllum Roxb.

POLYGONACEAE Juss. 1789   134

49 genera. Widespread. Mostly herbs, some shrubs, lianes, trees.

B&H	MONOCHLAMYDEAE	Curvembryeae, 134
DT&H	ARCHICHLAMYDEAE	Polygonales, 27
Melc	ARCHICHLAMYDEAE	Polygonales, 25
Thor	THEIFLORAE	Polygonales, 80
Dahl	POLYGONIFLORAE	Polygonales, 53
Young	DILLENIIDAE, DILLENIANAE	Plumbaginales, 144
Takh	CARYOPHYLLIDAE, POLYGONANAE	Polygonales, 74
Cron	CARYOPHYLLIDAE	Polygonales, 76

Aconogonon (Meisn.) Rchb.
Afrobrunnichia Hutch. & Dalziel
Antigonon Endl.
Aristocapsa Reveal & Hardham
Atraphaxis L.
Bistorta (L.) Adans.
Brunnichia Banks ex Gaertn.
Calligonum L.
Centrostegia A.Gray
Chorizanthe R.Br. ex Benth.
Coccoloba P.Browne
Dedeckera Reveal & J.T.Howell
Dodecahema Reveal & Hardham
Emex Neck. ex Campdera
Enneatypus Herz. *
Eriogonum Michx.
Fagopyrum Mill.
Fallopia Adans.
Gilmania Coville
Goodmania Reveal & Ertter
Gymnopodium Rolfe
Harfordia Greene & Parry
Hollisteria S.Watson
Homalocladium (F.Muell.) L.H.Bailey
Koenigia L.

Lastarriaea J.Remy
Leptogonum Benth.
Mucronea Benth.
Muehlenbeckia Meisn.
Nemacaulis Nutt.
Neomillspaughia S.F.Blake
Oxygonum Burch. ex Campdera
Oxyria Hill
Oxytheca Nutt.
Parapteropyrum A.J.Li
Persicaria (L.) Mill.
Physopyrum Popov *
Podopterus Bonpl.
Polygonella Michx.
Polygonum L.
Pteropyrum Jaub. & Spach
Pterostegia Fisch. & C.A.Mey.
Rheum L.
Rumex L.
Ruprechtia C.A.Mey.
Stenogonum Nutt.
Symmeria Benth.
Systenotheca Reveal & Hardham
Triplaris Loefl. ex. L.

PORTULACACEAE Juss. 1789   23.01

25 genera. Widespread. Herbs, shrubs, often succulent.

B&H	POLYPETALAE, THALAMIFLORAE	Caryophyllineae, 23
	(including Hectorellaceae)	
DT&H	ARCHICHLAMYDEAE	Centrospermae, Portulacineae, 35
	(including Hectorellaceae)	
Melc	ARCHICHLAMYDEAE	Centrospermae, Portulacineae, 32
Thor	CHENOPODIIFLORAE	Chenopodiales, Portulacineae, 91
	(including Hectorellaceae)	
Dahl	CARYOPHYLLIFLORAE	Caryophyllales, 41
Young	DILLENIIDAE, CARYOPHYLLANAE	Caryophyllales, 149

PORTULACACEAE (D)

Takh   CARYOPHYLLIDAE, CARYOPHYLLANAE
                                   Caryophyllales, Caryophyllineae, 65
Cron   CARYOPHYLLIDAE   Caryophyllales, 72
            (including Hectorellaceae)

Amphipetalum Bacigalupo *
Anacampseros L.
Calandrinia Kunth
Calyptridium Nutt.
Calyptrotheca Gilg
Ceraria H.Pearson & Stephens
Claytonia L.
Grahamia Gillies ex Hook. & Arn.
Lenzia Phil.
Lewisia Pursh
Mona O.E.G.Nilsson
Monocosmia Fenzl
Montia L.

Montiopsis Kuntze *
Neopaxia O.E.G.Nilsson
Portulaca L.
Portulacaria Jacq.
Rumicastrum Ulbr.
Schreiteria Carolin
Silvaea Phil.
Spraguea Torr.
Talinaria Brandegee
Talinella Baill.
Talinopsis A.Gray
Talinum Adans.

PRIMULACEAE Vent. 1799   99

23 genera. Widespread, mostly N. temperate. Herbs.

B&H   GAMOPETALAE, HETEROMERAE Primulales, 99
DT&H   METACHLAMYDEAE   Primulales, 190
Melc   SYMPETALAE   Primulales, 235
Thor   THEIFLORAE   Primulales, Primulineae, 78
Dahl   PRIMULIFLORAE   Primulales, 131
            (? separating *Coridaceae* 132)
Young DILLENIIDAE, DILLENIANAE Primulales, 142
Takh   DILLENIIDAE, ERICANAE   Primulales, 140
Cron   DILLENIIDAE   Primulales, 155

Anagallis L.
Androsace L.
Ardisiandra Hook.f.
Asterolinon Hoffmanns. & Link
Bryocarpum Hook.f. & Thomson
Coris L.
Cortusa L.
Cyclamen L.
Dionysia Fenzl
Dodecatheon L.
Glaux L.
Hottonia L.

Kaufmannia Regel
Lysimachia L.
Omphalogramma (Franch.) Franch.
Pelletiera A.St.–Hil.
Pomatosace Maxim.
Primula L.
Samolus L.
Soldanella L.
Stimpsonia C.Wright ex A.Gray
Trientalis L.
Vitaliana Sesl.

PROTEACEAE Juss. 1789   144

69 genera. S. hemisphere to tropics, rarely N. hemisphere. Trees, shrubs.

B&H   MONOCHLAMYDEAE   Daphnales, 144
DT&H   ARCHICHLAMYDEAE   Proteales, 16

PROTEACEAE Juss. 1789   144

69 genera. S. hemisphere to tropics, rarely N. hemisphere. Trees, shrubs.

Melc	ARCHICHLAMYDEAE	Proteales, 15
Thor	PROTEIFLORAE	Proteales, 192
Dahl	PROTEIFLORAE	Proteales, 182
Young	ROSIDAE, ROSANAE	Proteales, 310
Takh	ROSIDAE, PROTEANAE	Proteales, 321
Cron	ROSIDAE	Proteales, 184

Adenanthos Labill.
Agastachys R.Br.
Alloxylon P.H.Weston & Crisp
Aulax Bergius
Austromuellera C.T.White
Banksia L.f.
Beauprea Brongn. & Gris
Beaupreopsis Virot
Bellendena R.Br.
Bleasdalea F.Muell. ex Domin
Brabejum L.
Buckinghamia F.Muell.
Cardwellia F.Muell.
Carnarvonia F.Muell.
Cenarrhenes Labill.
Conospermum Sm.
Darlingia F.Muell.
Diastella Salisb. ex Knight
Dilobeia Thouars
Dryandra R.Br.
Embothrium J.R.Forst. & G.Forst.
Euplassa Salisb. ex Knight
Faurea Harv.
Finschia Warb.
Franklandia R.Br.
Garnieria Brongn. & Gris
Gevuina Molina
Grevillea R.Br. ex Knight
Hakea Schrad.
Helicia Lour.
Heliciopsis Sleumer
Hicksbeachia F.Muell.
Hollandaea F.Muell.
Isopogon R.Br. ex Knight
Kermadecia Brongn. & Gris
Knightia R.Br.

Lambertia Sm.
Leucadendron R.Br.
Leucospermum R.Br.
Lomatia R.Br.
Macadamia F.Muell.
Malagasia L.A.S.Johnson & B.G.Briggs
Mimetes Salisb.
Musgravea F.Muell.
Neorites L.S.Sm.
Opisthiolepis L.S.Sm.
Oreocallis R.Br.
Orites R.Br.
Orothamnus Pappe ex Hook.
Panopsis Salisb. ex Knight
Paranomus Salisb.
Persoonia Sm.
Petrophile R.Br. ex Knight
Placospermum C.T.White &
    W.D.Francis
Protea L.
Roupala Aubl.
Serruria Salisb.
Sleumerodendron Virot
Sorocephalus R.Br.
Spatalla Salisb.
Sphalmium B.G.Briggs, B.Hyland &
    L.A.S.Johnson
Stenocarpus R.Br.
Stirlingia Endl.
Strangea Meisn.
Symphionema R.Br.
Synaphea R.Br.
Telopea R.Br.
Vexatorella Rourke
Xylomelum Sm.

PTAEROXYLACEAE J.-F.Leroy 1960   43.02

2 genera. Tropical and South Africa, Madagascar. Trees, shrubs.

B&H   POLYPETALAE, DISCIFLORAE Sapindales (within Sapindaceae, 51)
DT&H ARCHICHLAMYDEAE   Geraniales, Geraniineae (within Meliaceae, 90)

Melc	ARCHICHLAMYDEAE	Rutales, Rutineae (within Meliaceae, 136)
Thor	RUTIFLORAE	Rutales, Rutineae, 171
Dahl	RUTIFLORAE	Rutales, 219
Young	ROSIDAE, RUTANAE	Rutales, 249
Takh	ROSIDAE, RUTANAE	Rutales, 268
Cron	ROSIDAE	Sapindales (within Sapindaceae, 252)

Cedrelopsis Baill.
Ptaeroxylon Eckl. & Zeyh.

PTEROSTEMONACEAE Small 1905   59.18

1 genus. Mexico. Shrubs.

B&H	POLYPET., CALYCIFLORAE	Rosales (within Rosaceae, 58)
DT&H	ARCHICHLAMYDEAE	Rosales, Saxifragineae (within Saxifragaceae, 67)
Melc	ARCHICHLAMYDEAE	Rosales, Saxifragineae (within Saxifragaceae, 106)
Thor	ROSIFLORAE	Rosales, Saxifragineae (within Saxifragaceae, 208)
Dahl	CORNIFLORAE?	Cornales, 328
Young	ROSIDAE, ROSANAE	Saxifragales, 298
Takh	ROSIDAE, CORNANAE	Hydrangeales, Hydrangeineae, 339
Cron	ROSIDAE	Rosales (within Grossulariaceae, 166)

Pterostemon Schauer

QUIINACEAE Choisy ex Engl. 1888   27.02

4 genera. Tropical America. Trees, shrubs, lianes.

B&H	POLYPETALAE, THALAMIFLORAE	Guttiferales (within Guttiferae, 27)
DT&H	ARCHICHLAMYDEAE	Parietales, Theineae, 135
Melc	ARCHICHLAMYDEAE	Guttiferales, Theineae, 88
Thor	THEIFLORAE	Theales, Scytopetalineae, 59
Dahl	THEIFLORAE	Theales, 113
Young	DILLENIIDAE, DILLENIANAE	Scytopetalales, 121
Takh	DILLENIIDAE, THEANAE	Ochnales, 120
Cron	DILLENIIDAE	Theales, 93

Froesia Pires   **Quiina** Aubl.
Lacunaria Ducke   **Touroulia** Aubl.

RAFFLESIACEAE Dumort. 1829   137.01

9 genera. Tropics, occasionally warm-temperate. Root or stem parasites.

B&H	MONOCHLAMYDEAE (including Hydnoraceae)	Multiovulatae Terrestres (as *Cytinaceae*) 137
DT&H	ARCHICHLAMYDEAE	Aristolochiales, 25
Melc	ARCHICHLAMYDEAE	Aristolochiales, 73
Thor	RAFFLESIIFLORAE	Rafflesiales, 38

Dahl MAGNOLIIFLORAE   Rafflesiales, 7
Young ROSIDAE, SANTALANAE   Rafflesiales, 242
Takh MAGNOLIIDAE, RAFFLESIANAE Rafflesiales, 29
        (separating *Apodanthaceae* 30, *Cytinaceae* 32, *Mitrastemonaceae* 31)
Cron ROSIDAE   Rafflesiales, 217
        (separating *Mitrastemonaceae* 216)

Apodanthes Poit.   **Pilostyles** Guill.
Bdallophyton Eichler   **Rafflesia** R.Br.
Berlinianche (Harms) Vattimo   **Rhizanthes** Dumort.
Cytinus L.   **Sapria** Griff.
Mitrastemon Makino

RANUNCULACEAE Juss. 1789   1.01

52 genera. Widespread. Mostly herbs, some lianes (*Clematis*).

B&H   POLYPETALAE, THALAMIFLORAE Ranales, 1
        (including Glaucidiaceae, Paeoniaceae)
DT&H ARCHICHLAMYDEAE   Ranales, Ranunculineae, 41
        (including Glaucidiaceae, Paeoniaceae)
Melc ARCHICHLAMYDEAE   Ranunculales, Ranunculineae, 61
        (including Circaeasteraceae, Glaucidiaceae)
Thor ANNONIFLORAE   Berberidales, Berberidineae, 33
Dahl RANUNCULIFLORAE   Ranunculales, 34
        (separating *Kingdoniaceae* 32)
Young MAGNOLIIDAE, RANUNCULANAE   Ranunculales, 35
        (separating *Hydrastidaceae* 39)
Takh RANUNCULIDAE, RANUNCULANAE Ranunculales, Ranunculineae, 47
Cron MAGNOLIIDAE   Ranunculales, 30
        (including Glaucidiaceae)

Aconitum L.   **Enemion** Raf.
Actaea L.   **Eranthis** Salisb.
Adonis L.   **Halerpestes** Greene
Anemoclema (Franch.) W.T.Wang   **Hamadryas** Comm. ex Juss.
Anemone L.   **Helleborus** L.
Anemonella Spach   **Hepatica** Mill.
Anemonopsis Siebold & Zucc.   **Hydrastis** L.
Aquilegia L.   **Isopyrum** L.
Asteropyrum J.R.Drumm. & Hutch.   **Kingdonia** Balf.f. & W.W.Sm.
Barneoudia Gay   **Knowltonia** Salisb.
Beesia Balf.f. & W.W.Sm.   **Laccopetalum** Ulbr.
Calathodes Hook.f. & Thomson   **Leptopyrum** Rchb.
Callianthemum C.A.Mey.   **Metanemone** W.T.Wang *
Caltha L.   **Miyakea** Miyabe & Tatew. *
Ceratocephala Moench   **Myosurus** L.
Cimicifuga Wernisch.   **Naravelia** Adans.
Clematis L.   **Nigella** L.
Consolida Gray   **Oreithales** Schltdl.
Coptis Salisb.   **Oxygraphis** Bunge
Delphinium L.   **Paraquilegia** J.R.Drumm. & Hutch.
Dichocarpum W.T.Wang & Hsiao   **Paroxygraphis** W.W.Sm.

Psychrophila (DC.) Bercht. & J.Presl
Pulsatilla Mill.
Ranunculus L.
Semiaquilegia Makino
Souliea Franch.

Thalictrum L.
Trautvetteria Fisch. & C.A.Mey.
Trollius L.
Urophysa Ulbr.
Xanthorhiza Marshall

RESEDACEAE DC. ex Gray 1821   13

7 genera. Mostly N. temp., to India, S. Afr., Calif. Mostly herbs, some shrubs.

B&H   POLYPETALAE, THALAMIFLORAE   Parietales, 13
DT&H ARCHICHLAMYDEAE   Rhoeadales, Resedineae, 58
Melc   ARCHICHLAMYDEAE   Papaverales, Resedineae, 98
Thor   VIOLIFLORAE   Capparales, 141
Dahl   VIOLIFLORAE   Capparales, 103
Young DILLENIIDAE, VIOLANAE   Capparales, 212
Takh   DILLENIIDAE, VIOLANAE   Capparales, Resedineae, 166
Cron   DILLENIIDAE   Capparales, 136

Caylusea A.St.-Hil.
Homalodiscus Bunge ex Boiss.
Ochradenus Delile
Oligomeris Cambess.

Randonia Coss.
Reseda L.
Sesamoides Ortega

RETZIACEAE Bartl. 1830   108.03

1 genus, monotypic. South Africa. Shrub.

B&H   GAMOPETALAE, BICARPELLATAE Polemoniales (within Solanaceae, 114)
DT&H METACHLAMYDEAE   Contortae, Gentianineae (within
                                              Loganiaceae, 198)

Melc   Not mentioned   —
Thor   GENTIANIFLORAE   Gentianales (within Loganiaceae, 249)
Dahl   LAMIIFLORAE   Scrophulariales, 356
Young Not mentioned   —
Takh   LAMIIDAE, LAMIANAE   Scrophulariales, 398
Cron   ASTERIDAE   Gentianales, 271

Retzia Thunb.

RHABDODENDRACEAE (Huber) Prance 1968   39.02

1 genus. N. trop. South America. Shrubs to small trees.

B&H   Not known   —
DT&H ARCHICHLAMYDEAE   Geraniales, Geraniineae (within Rutaceae, 87)
Melc   ARCHICHLAMYDEAE   Rutales, Rutineae (within Rutaceae, 131)
Thor   RUTIFLORAE   Rutales, Rutineae (within Rutaceae, 167)
Dahl   ROSIFLORAE   Rosales, 180

Young ROSIDAE, RUTANAE   Rutales (within Rutaceae, 244)
Takh ROSIDAE, RUTANAE   Rutales, 256
Cron ROSIDAE   Rosales, 179

Rhabdodendron Gilg & Pilg.

RHAMNACEAE Juss. 1789   49.01

51 genera. Widespread. Trees, shrubs, lianes, few herbs.

B&H POLYPETALAE, DISCIFLORAE Celastrales, 49
DT&H ARCHICHLAMYDEAE   Rhamnales, 119
Melc ARCHICHLAMYDEAE   Rhamnales, 167
Thor MALVIFLORAE   Rhamnales, 159
Dahl MALVIFLORAE   Rhamnales, 82
Young DILLENIIDAE, MALVANAE Rhamnales, 181
Takh ROSIDAE, RHAMNANAE Rhamnales, 319
Cron ROSIDAE   Rhamnales, 233

Adolphia Meisn.
Alphitonia Reissek ex Endl.
Alvimiantha Grey–Wilson
Ampelozizyphus Ducke
Auerodendron Urb.
Bathiorhamnus Capuron
Berchemia Neck. ex DC.
Berchemiella Nakai
Blackallia C.A.Gardner
Ceanothus L.
Chaydaia Pit.
Colletia Comm. ex Juss.
Colubrina Rich. ex Brongn.
Condalia Cav.
Crumenaria Mart.
Cryptandra Sm.
Discaria Hook.
Doerpfeldia Urb.
Emmenosperma F.Muell.
Gouania Jacq.
Helinus E.Mey. ex Endl.
Hovenia Thunb.
Hybosperma Urb.
Karwinskia Zucc.
Kentrothamnus Suess. & Overk.
Krugiodendron Urb.

Lasiodiscus Hook.f.
Maesopsis Engl.
Nesiota Hook.f.
Noltea Rchb.
Paliurus Mill.
Phylica L.
Pleuranthodes Weberb.
Pomaderris Labill.
Reissekia Endl.
Retanilla (DC.) Brongn.
Reynosia Griseb.
Rhamnella Miq.
Rhamnidium Reissek
Rhamnus L.
Sageretia Brongn.
Schistocarpaea F.Muell.
Scutia (DC.) Brongn.
Siegfriedia C.A.Gardner
Smythea Seem.
Spyridium Fenzl
Talguenea Miers ex Endl.
Trevoa Miers ex Hook.
Trymalium Fenzl
Ventilago Gaertn.
Ziziphus Mill.

RHIZOPHORACEAE R.Br. 1814   65.01

14 genera. Tropics. Trees, shrubs, some mangroves.

B&H POLYPETALAE, CALYCIFLORAE Myrtales, 65
    (including Anisophylleaceae)

DT&H	ARCHICHLAMYDEAE	Myrtiflorae, Myrtineae, 171
	(including Anisophylleaceae)	
Melc	ARCHICHLAMYDEAE	Myrtiflorae, Myrtineae, 212
	(including Anisophylleaceae)	
Thor	CORNIFLORAE	Cornales, Rhizophorineae, 278
	(including Anisophylleaceae)	
Dahl	MYRTIFLORAE	Rhizophorales, 184
Young	ROSIDAE, MYRTANAE	Rhizophorales, 322
	(presumably including Anisophylleaceae)	
Takh	ROSIDAE, MYRTANAE	Rhizophorales, 223
Cron	ROSIDAE	Rhizophorales, 200

Anopyxis (Pierre) Engl.	**Crossostylis** J.R.Forst. & G.Forst.
Blepharistemma Wall. ex Benth.	**Gynotroches** Blume
Bruguiera Sav.	**Kandelia** (DC.) Wight & Arn.
Carallia Roxb.	**Macarisia** Thouars
Cassipourea Aubl.	**Pellacalyx** Korth.
Ceriops Arn.	**Rhizophora** L.
Comiphyton Floret *	**Sterigmapetalum** Kuhlm.

RHOIPTELEACEAE Hand.-Mazz. 1932   156.02

1 genus, monotypic. S.W. China, Vietnam. Tree.

B&H	Not known	—
DT&H	Not known	—
Melc	ARCHICHLAMYDEAE	Urticales, 10
Thor	RUTIFLORAE	Rutales, Juglandineae, 176
Dahl	ROSIFLORAE	Juglandales, 149
Young	ROSIDAE, RUTANAE	Juglandales, 266
Takh	HAMAMELIDIDAE, JUGLANDANAE	Rhoipteleales, 96
Cron	HAMAMELIDIDAE	Juglandales, 57

Rhoiptelea Diels & Hand.–Mazz.

RHYNCHOCALYCACEAE L.A.S. Johnson & B.G. Briggs 1985   69.08

1 genus, monotypic. South Africa (Natal, E. Cape Prov.). Tree.

B&H	Not known	—
DT&H	ARCHICHLAMYDEAE	Myrtiflorae, Myrtineae (within Lythraceae, 166)
Melc	Not mentioned	—
Thor	MYRTIFLORAE	Myrtales, Lythrineae, 237
Dahl	MYRTIFLORAE	Myrtales, 197
Young	Not mentioned	—
Takh	ROSIDAE, MYRTANAE	Myrtales, Myrtineae, 228
Cron	ROSIDAE	Myrtales
	(included in ? Lythraceae 189 in 1981, as separate family in 1988)	

Rhynchocalyx Oliv.

RORIDULACEAE Engl. & Gilg 1924   61.02

1 genus. South Africa. Insectivorous subshrubs.

B&H POLYPETALAE, CALYCIFLORAE Rosales (within Droseraceae, 61)
DT&H ARCHICHLAMYDEAE   Sarraceniales (within Droseraceae, 62)
Melc ARCHICHLAMYDEAE   Rosales, Saxifragineae, 112
Thor ROSIFLORAE   Pittosporales, Bruniineae, 229
Dahl CORNIFLORAE   Ericales, 294
Young ROSIDAE, CORNANAE   Pittosporales, 340
Takh ROSIDAE, CORNANAE   Hydrangeales, Hydrangeineae, 337
Cron ROSIDAE   Rosales (within Byblidaceae, 163)

Roridula Burm.f. ex L.

ROSACEAE Juss. 1789   58.01

97 genera. Widespread, esp. N. hemisphere. Trees, shrubs, herbs.

B&H POLYPETALAE, CALYCIFLORAE Rosales, 58
    (including Canotiaceae, Cephalotaceae, Chrysobalanaceae, Eucryphiaceae, Neuradaceae, Pterostemonaceae, Stylobasiaceae)
DT&H ARCHICHLAMYDEAE   Rosales, Rosineae, 76
    (including Chrysobalanaceae, Euphroniaceae, Neuradaceae, Stylobasiaceae)
Melc ARCHICHLAMYDEAE   Rosales, Rosineae, 114
Thor ROSIFLORAE   Rosales, Rosineae, 203
    (including Neuradaceae)
Dahl ROSIFLORAE   Rosales, 174
    (separating *Malaceae* 176, *Amygdalaceae* 177)
Young ROSIDAE, ROSANAE   Rosales, 291
    (presumably including Neuradaceae)
Takh ROSIDAE, ROSANAE   Rosales, 217
Cron ROSIDAE   Rosales, 174

Acaena Mutis ex L.
Adenostoma Hook. & Arn.
Agrimonia L.
Alchemilla L.
Amelanchier Medik.
Aphanes L.
Aremonia Neck. ex Nestl.
Aruncus L.
Bencomia Webb & Berthel.
Cercocarpus Kunth
Chaenomeles Lindl.
Chamaebatia Benth.
Chamaebatiaria (Porter) Maxim.
Chamaemeles Lindl.
Chamaerhodos Bunge
Cliffortia L.
Coleogyne Torr.
Coluria R.Br.
Cotoneaster Medik.
Cowania D.Don

Crataegus L.
Cydonia Mill.
Dichotomanthes Kurz
Docynia Decne.
Dryas L.
Drymocallis Fourr. ex Rydb.
Duchesnea Sm.
Eriobotrya Lindl.
Exochorda Lindl.
Fallugia Endl.
Filipendula Mill.
Fragaria L.
Geum L.
Gillenia Moench
Guamatela Donn.Sm.
Hagenia J.F.Gmel.
Hesperomeles Lindl.
Holodiscus (K.Koch) Maxim.
Horkelia Cham. & Schltdl.
Horkeliella (Rydb.) Rydb.

Ivesia Torr. & A.Gray
Kageneckia Ruíz & Pav.
Kelseya Rydb.
Kerria DC.
Leucosidea Eckl. & Zeyh.
Lindleya Kunth
Luetkea Bong.
Lyonothamnus A.Gray
Maddenia Hook.f. & Thomson
Malacomeles (Decne.) Engl.
Malus Mill.
Margyricarpus Ruíz & Pav.
Mespilus L.
Neillia D.Don
Neviusia A.Gray
Novosieversia F.Bolle
Oemleria Rchb.
Orthurus Juz.
Osteomeles Lindl.
Pentactina Nakai
Peraphyllum Nutt.
Petrophytum (Nutt.) Rydb.
Photinia Lindl.
Physocarpus (Cambess.) Maxim.
Pleiosepalum Hand.–Mazz. *
Polylepis Ruíz & Pav.
Potaninia Maxim.
Potentilla L.
Prinsepia Royle
Prunus L.

Pseudocydonia (C.K.Schneid.) C.K.Schneid.
Purpusia Brandegee
Purshia DC. ex Poir.
Pyracantha M.Roem.
Pyrus L.
Quillaja Molina
Rhaphiolepis Lindl.
Rhodotypos Siebold & Zucc.
Rosa L.
Rubus L.
Sanguisorba L.
Sarcopoterium Spach
Sibbaldia L.
Sibbaldiopsis Rydb.
Sibiraea Maxim.
Sieversia Willd.
Sorbaria (Ser. ex DC.) A.Braun
Sorbus L.
Spenceria Trimen
Spiraea L.
Spiraeanthus (Fisch. & C.A.Mey.) Maxim.
Stellariopsis (Baill.) Rydb.
Stephanandra Siebold & Zucc.
Tetraglochin Poepp.
Vauquelinia Corrêa ex Humb. & Bonpl.
Waldsteinia Willd.
Xerospiraea Henrickson *

RUBIACEAE Juss. 1789

84.01

606 genera. Widespread. Trees, shrubs, herbs.

B&H	GAMOPETALAE, INFERAE (including Carlemanniaceae)	Rubiales, 84
DT&H	METACHLAMYDEAE (including Carlemanniaceae)	Rubiales, 223
Melc	SYMPETALAE (separating *Henriqueziaceae* 268 to Tubiflorae Solanineae)	Gentianales, 251
Thor	GENTIANIFLORAE (including Theligonaceae)	Gentianales, 251
Dahl	GENTIANIFLORAE	Gentianales, 341
Young	ROSIDAE, GENTIANANAE (including Theligonaceae)	Gentianales, 349
Takh	LAMIIDAE, GENTIANANAE	Gentianales, 367
Cron	ASTERIDAE	Rubiales, 311

Acranthera Arn. ex Meisn.
Acrobotrys K.Schum. & K.Krause
Acunaeanthus Borhidi, Koml. & M.Moncada *

Adina Salisb.
Adinauclea Ridsdale
Agathisanthemum Klotzsch
Aidia Lour.

Aidiopsis Tirveng.
Airosperma K.Schum. & Lauterb.
Aitchisonia Hemsl. ex Aitch.
Alberta E.Mey.
Aleisanthia Ridl.
Alibertia A.Rich. ex DC.
Allaeophania Thwaites
Alleizettella Pit.
Allenanthus Standl.
Alseis Schott
Amaioua Aubl.
Amaracarpus Blume
Amphiasma Bremek.
Amphidasya Standl.
Ancylanthos Desf.
Anomanthodia Hook.f.
Antherostele Bremek.
Anthorrhiza C.Huxley & Jebb
Anthospermum L.
Antirhea Comm. ex Juss.
Aoranthe Somers
Aphaenandra Miq.
Aphanocarpus Steyerm.
Arachnothryx Planch.
Arcytophyllum Willd. ex Schult. &
    Schult.f.
Argocoffeopsis Lebrun
Argostemma Wall.
Ariadne Urb. *
Asemnantha Hook.f.
Asperula L.
Astiella Jovet *
Atractocarpus Schltr. & K.Krause
Atractogyne Pierre
Augusta Pohl
Aulacocalyx Hook.f.
Badusa A.Gray
Balmea Mártinez
Bathysa C.Presl
Batopedina Verdc.
Belonophora Hook.f.
Benkara Adans.
Benzonia Schumach. *
Berghesia Nees *
Bertiera Aubl.
Bikkia Reinw.
Blandibractea Wernham *
Blepharidium Standl.
Bobea Gaudich.
Boholia Merr.
Borojoa Cuatrec.
Bothriospora Hook.f.
Botryarrhena Ducke
Bouvardia Salisb.
Brachytome Hook.f.

Bradea Standl. ex Brade *
Brenania Keay
Breonadia Ridsdale
Breonia A.Rich. ex DC.
Burchellia R.Br.
Burttdavya Hoyle
Byrsophyllum Hook.f.
Calanda K.Schum.
Callipeltis Steven
Calochone Keay
Calycophyllum DC.
Calycosia A.Gray
Calycosiphonia Pierre ex Robbr.
Canephora Juss.
Canthium Lam.
Capirona Spruce
Captaincookia N.Hallé
Carpacoce Sond.
Carphalea Juss.
Carterella Terrell
Casasia A.Rich.
Catesbaea L.
Catunaregam Wolf
Cephalanthus L.
Cephalodendron Steyerm. *
Ceratopyxis Hook.f.
Ceriscoides (Hook.f.) Tirveng.
Ceuthocarpus Aiello
Chaetostachydium Airy Shaw *
Chalepophyllum Hook.f.
Chamaepentas Bremek.
Chapelieria A.Rich. ex DC.
Chassalia Comm. ex Poir.
Chazaliella E.Petit & Verdc.
Chimarrhis Jacq.
Chiococca P.Browne
Chione DC.
Chomelia Jacq.
Choulettia Pomel
Cigarrilla Aiello
Cinchona L.
Cladoceras Bremek.
Clarkella Hook.f.
Coccochondra Rauschert *
Coccocypselum P.Browne
Coddia Verdc.
Coelopyrena Valeton *
Coelospermum Blume
Coffea L.
Coleactina N.Hallé
Colletoecema E.Petit
Commitheca Bremek.
Condaminea DC.
Conostomium (Stapf) Cufod.
Coprosma J.R.Forst. & G.Forst.

Coptophyllum Korth.
Coptosapelta Korth.
Corynanthe Welw.
Coryphothamnus Steyerm.
Cosmibuena Ruíz & Pav.
Cosmocalyx Standl.
Coussarea Aubl.
Coutaportla Urb.
Coutarea Aubl.
Cowiea Wernham
Craterispermum Benth.
Cremaspora Benth.
Cremocarpon Baill.
Crobylanthe Bremek.
Crocyllis E.Mey. ex Hook.f.
Crossopteryx Fenzl
Crucianella L.
Cruciata Mill.
Cruckshanksia Hook. & Arn.
Crusea Cham. & Schltdl.
Cuatrecasasiodendron Standl. &
    Steyerm.
Cubanola Aiello
Cuviera DC.
Cyclophyllum Hook.f.
Damnacanthus C.F.Gaertn.
Danais Comm. ex Vent.
Deccania Tirveng.
Declieuxia Kunth
Dendrosipanea Ducke
Dentella J.R.Forst. & G.Forst.
Deppea Cham. & Schltdl.
Diacrodon Sprague
Dibrachionostylus Bremek.
Dichilanthe Thwaites
Dictyandra Welw. ex Hook.f.
Didymaea Hook.f.
Didymochlamys Hook.f.
Didymoecium Bremek. *
Didymopogon Bremek.
Didymosalpinx Keay
Diodia L.
Dioecrescis Tirveng.
Dioicodendron Steyerm.
Diplospora DC.
Diyaminauclea Ridsdale
Dolichodelphys K.Schum. & K.Krause
Dolicholobium A.Gray
Dolichometra K.Schum.
Doricera Verdc.
Duidania Standl.
Dunnia Tutcher
Duperrea Pierre ex Pit.
Duroia L.f.
Durringtonia R.J.F.Hend. & Guymer

Ecpoma K.Schum.
Eizia Standl.
Elaeagia Wedd.
Eleuthranthes F.Muell.
Emmenopterys Oliv.
Emmeorhiza Pohl ex Endl.
Eosanthe Urb. *
Eriosemopsis Robyns
Erithalis P.Browne
Ernodea Sw.
Euclinia Salisb.
Exostema (Pers.) Humb. & Bonpl.
Fadogia Schweinf.
Fadogiella Robyns
Fagerlindia Tirveng.
Faramea Aubl.
Ferdinandusa Pohl
Feretia Delile
Fergusonia Hook.f.
Fernelia Comm. ex Lam.
Flagenium Baill.
Flexanthera Rusby *
Gaertnera Lam.
Galiniera Delile
Galium L.
Gallienia Dubard & Dop *
Galopina Thunb.
Gardenia Ellis
Gardeniopsis Miq.
Genipa L.
Gentingia J.T. Johanss. & K.M.Wong
Geophila D.Don
Gillespiea A.C.Sm.
Gleasonia Standl.
Glionnetia Tirveng.
Glossostipula Lorence
Gomphocalyx Baker
Gonzalagunia Ruíz & Pav.
Gouldia A.Gray
Greenea Wight & Arn.
Greeniopsis Merr.
Guettarda L.
Gynochthodes Blume
Gynopachis Blume
Gyrostipula J.-F.Leroy
Habroneuron Standl.
Haldina Ridsdale
Hallea J.-F.Leroy
Hamelia Jacq.
Hayataella Masam. *
Hedstromia A.C.Sm.
Hedyotis L.
Hedythyrsus Bremek.
Heinsenia K.Schum.
Heinsia DC.

Hekistocarpa Hook.f.
Henriquezia Spruce ex Benth.
Heterophyllaea Hook.f.
Hillia Jacq.
Himalrandia Yamaz.
Hindsia Benth. ex Lindl.
Hintonia Bullock
Hippotis Ruíz & Pav.
Hodgkinsonia F.Muell.
Hoffmannia Sw.
Holstianthus Steyerm.
Homollea Arènes
Homolliella Arènes
Houstonia L.
Hutchinsonia Robyns
Hydnophytum Jack
Hydrophylax L.f.
Hymenocnemis Hook.f.
Hymenocoleus Robbr.
Hymenodictyon Wall.
Hyperacanthus E.Mey. ex Bridson
Hypobathrum Blume
Hyptianthera Wight & Arn.
Indopolysolenia Bennet
Isertia Schreb.
Isidorea A.Rich. ex DC.
Ixora L.
Jackiopsis Ridsdale
Janotia J.–F.Leroy
Jaubertia Guill.
Javorkaea Borhidi & Koml.
Joosia Karst.
Jovetia Guédès *
Kailarsenia Tirveng.
Kajewskiella Merr. & Perry
Keenania Hook.f.
Keetia E.Phillips
Kelloggia Torr. ex Benth.
Kerianthera J.H.Kirkbr.
Khasiaclunea Ridsdale
Klossia Ridl.
Knoxia L.
Kochummenia K.M.Wong
Kohautia Cham. & Schltdl.
Kraussia Harv.
Kutchubaea Fisch. ex DC.
Ladenbergia Klotzsch
Lagynias E.Mey. ex Robyns
Lamprothamnus Hiern
Lasianthus Jack
Lathraeocarpa Bremek. *
Lecananthus Jack
Lecariocalyx Bremek.
Lelya Bremek.
Lemyrea (A.Chev.) A.Chev. & Beille *

Lepidostoma Bremek. *
Leptactina Hook.f.
Leptodermis Wall.
Leptomischus Drake
Leptoscela Hook.f.
Leptostigma Arn.
Leptunis Steven
Lerchea L.
Leroya Cavaco
Leucocodon Gardner
Leucolophus Bremek.
Limnosipanea Hook.f.
Lindenia Benth.
Litosanthes Blume
Lucinaea DC.
Luculia Sweet
Lucya DC.
Ludekia Ridsdale
Macbrideina Standl.
Machaonia Bonpl.
Macrocnemum P.Browne
Macrosphyra Hook.f.
Maguireocharis Steyerm.
Maguireothamnus Steyerm.
Malanea Aubl.
Manettia L.
Manostachya Bremek.
Mantalania Capuron ex J.–F.Leroy
Margaritopsis Sauvalle
Maschalocorymbus Bremek.
Maschalodesme K.Schum. & Lauterb.
Massularia (K.Schum.) Hoyle
Mastixiodendron Melch.
Mazaea Krug & Urb.
Melanopsidium Colla
Mericarpaea Boiss.
Merumea Steyerm.
Metadina Bakh.f.
Meyna Roxb. ex Link
Micrasepalum Urb.
Microphysa Schrenk
Mitchella L.
Mitracarpus Zucc.
Mitragyna Korth.
Mitrasacmopsis Jovet
Mitriostigma Hochst.
Molopanthera Turcz.
Monosalpinx N.Hallé
Montamans Dwyer *
Morelia A.Rich. ex DC.
Morierina Vieill.
Morinda L.
Morindopsis Hook.f.
Motleyia Johansson
Mouretia Pit. *

653

Multidentia Gilli
Mussaenda L.
Mussaendopsis Baill.
Mycetia Reinw.
Myonima Comm. ex Juss.
Myrioneuron R.Br. ex Hook.
Myrmecodia Jack
Myrmeconauclea Merr.
Myrmephytum Becc. *
Nargedia Bedd.
Nauclea L.
Neanotis W.H.Lewis
Neblinathamnus Steyerm.
Nematostylis Hook.f.
Nenax Gaertn.
Neobertiera Wernham
Neoblakea Standl.
Neobreonia Ridsdale
Neofranciella Guillaumin *
Neogaillonia Lincz.
Neohymenopogon Bennet
Neolamarckia Bosser
Neolaugeria Nicolson
Neoleroya Cavaco
Neonauclea Merr.
Neopentanisia Verdc.
Nernstia Urb. *
Nertera Banks & Sol. ex Gaertn.
Nesohedyotis (Hook.f.) Bremek.
Neurocalyx Hook.
Nichallea Bridson
Nodocarpaea A.Gray
Normandia Hook.f.
Nostolachma T.Durand
Ochreinauclea Ridsdale & Bakh.f.
Octotropis Bedd.
Oldenlandia L.
Oldenlandiopsis Terrell & W.H.Lewis
Oligocodon Keay
Omiltemia Standl. *
Opercularia Gaertn.
Ophiorrhiza L.
Ophryococcus Oerst.
Oregandra Standl. *
Osa Aiello
Otiophora Zucc.
Otocalyx Brandegee *
Otomeria Benth.
Ottoschmidtia Urb.
Oxyanthus DC.
Oxyceros Lour.
Pachystigma Hochst.
Pachystylus K.Schum.
Paederia L.
Pagamea Aubl.

Pagameopsis Steyerm.
Palicourea Aubl.
Pamplethantha Bremek.
Paracephaelis Baill.
Parachimarrhis Ducke
Paracorynanthe Capuron
Paragenipa Baill. *
Paraknoxia Bremek.
Parapentas Bremek.
Paratriaina Bremek. *
Pauridiantha Hook.f.
Pausinystalia Pierre ex Beille
Pavetta L.
Payera Baill. *
Pelagodendron Seem.
Pentagonia Benth.
Pentaloncha Hook.f.
Pentanisia Harv.
Pentanopsis Rendle
Pentas Benth.
Pentodon Hochst.
Peponidium (Baill.) Arènes
Perakanthus Robyns
Perama Aubl.
Peratanthe Urb. *
Peripeplus Pierre
Pertusadina Ridsdale
Petitiocodon Robbr.
Phellocalyx Bridson
Phialanthus Griseb.
Phitopis Hook.f.
Phuopsis (Griseb.) Hook.f.
Phyllacanthus Hook.f.
Phyllis L.
Phyllocrater Wernham
Phyllomelia Griseb.
Phylohydrax Puff
Picardaea Urb.
Pimentelia Wedd.
Pinarophyllon Brandegee
Pinckneya Michx.
Pittoniotis Griseb.
Placocarpa Hook.f.
Placopoda Balf.f.
Platycarpum Bonpl.
Plectroniella Robyns
Pleiocarpidia K.Schum.
Pleiocoryne Rauschert
Pleiocraterium Bremek.
Plocama Aiton
Plocaniophyllon Brandegee
Poecilocalyx Bremek.
Pogonopus Klotzsch
Polysphaeria Hook.f.
Polyura Hook.f.

Pomax Sol. ex DC.
Porterandia Ridl.
Portlandia P.Browne
Posoqueria Aubl.
Pouchetia A.Rich.
Praravinia Korth.
Pravinaria Bremek. *
Preussiodora Keay
Prismatomeris Thwaites
Proscephaleium Korth.
Psathura Comm. ex Juss.
Pseudaidia Tirveng.
Pseudogaillonia Lincz.
Pseudogardenia Keay
Pseudohamelia Wernham *
Pseudomantalania J.–F.Leroy *
Pseudomussaenda Wernham
Pseudonesohedyotis Tennant
Pseudopyxis Miq.
Pseudosabicea N.Hallé
Psilanthus Hook.f.
Psychotria L.
Psydrax Gaertn.
Psyllocarpus Mart. & Zucc.
Pteridocalyx Wernham
Pterogaillonia Lincz.
Pubistylus Thoth. *
Putoria Pers.
Pygmaeothamnus Robyns
Pyragra Bremek.
Pyrostria Comm. ex Juss.
Ramosmania Tirveng. & Verdc.
Randia L.
Raritebe Wernham
Ravnia Oerst.
Readea Gillespie
Relbunium (Endl.) Hook.f.
Remijia DC.
Rennellia Korth.
Retiniphyllum Bonpl.
Rhachicallis DC.
Rhadinopus S.Moore
Rhaphidura Bremek.
Rhipidantha Bremek.
Rhopalobrachium Schltr. & K.Krause *
Richardia L.
Riqueuria Ruíz & Pav. *
Robynsia Hutch.
Rogiera Planch.
Roigella Borhidi & Fernández
Rondeletia L.
Rothmannia Thunb.
Rubia L.
Rudgea Salisb.
Rustia Klotzsch

Rutidea DC.
Rytigynia Blume
Sabicea Aubl.
Sacosperma G.Taylor
Saldinia A.Rich. ex DC.
Salzmannia DC.
Saprosma Blume
Sarcocephalus Afzel. ex Sabine
Sarcopygme Setch. & Christoph.
Schachtia Karst.
Schismatoclada Baker
Schizenterospermum Homolle ex Arènes
Schizocalyx Wedd.
Schizocolea Bremek.
Schizostigma Arn. ex Meisn.
Schmidtottia Urb.
Schradera Vahl
Schumanniophyton Harms
Schwendenera K.Schum.
Scolosanthus Vahl
Scyphiphora C.F.Gaertn.
Scyphochlamys Balf.f.
Scyphostachys Thwaites
Sericanthe Robbr.
Serissa Comm. ex Juss.
Shaferocharis Urb. *
Sherardia L.
Sherbournia G.Don
Siderobombyx Bremek. *
Siemensia Urb.
Simira Aubl.
Sinoadina Ridsdale
Sipanea Aubl.
Sipaneopsis Steyerm.
Siphonandrium K.Schum. *
Sommera Schltdl.
Spathichlamys R.Parker
Spermacoce L.
Spermadictyon Roxb.
Sphinctanthus Benth.
Spiradiclis Blume
Squamellaria Becc.
Stachyarrhena Hook.f.
Stachyococcus Standl.
Staelia Cham. & Schltdl.
Standleya Brade
Steenisia Bakh.f.
Stelechantha Bremek.
Stephanococcus Bremek.
Stevensia Poit.
Steyermarkia Standl.
Stichianthus Valeton
Stilpnophyllum Hook.f.
Stipularia P.Beauv.

Stomandra Standl.
Streblosa Korth.
Streblosiopsis Valeton *
Striolaria Ducke *
Strumpfia Jacq.
Stylosiphonia Brandegee
Suberanthus Borhidi & Fernández
Sukunia A.C.Sm.
Sulitia Merr. *
Synaptantha Hook.f.
Syringantha Standl. *
Tamilnadia Tirveng. & Sastre
Tammsia Karst.
Tapiphyllum Robyns
Tarenna Gaertn.
Tarennoidea Tirveng. & Sastre
Temnocalyx Robyns
Temnopteryx Hook.f.
Tennantia Verdc.
Thecorchus Bremek.
Thogsennia Aiello
Thyridocalyx Bremek.
Timonius DC.
Tobagoa Urb.
Tocoyena Aubl.
Tortuella Urb.

Trailliaedoxa W.W.Sm. & Forrest
Tresanthera Karst.
Triainolepis Hook.f.
Tricalysia A.Rich. ex DC.
Trichostachys Hook.f.
Trukia Kaneh.
Tsiangia But, H.H.Hsue & P.T.Li
Uncaria Schreb.
Urophyllum Wall.
Valantia L.
Vangueria Comm. ex Juss.
Vangueriella Verdc.
Vangueriopsis Robyns
Versteegia Valeton
Villaria Rolfe
Virectaria Bremek.
Warburgina Eig *
Warszewiczia Klotzsch
Wendlandia Bartl. ex DC.
Wernhamia S.Moore *
Wittmackanthus Kuntze
Xanthophytum Reinw. ex Blume
Xantonnea Pierre ex Pit.
Xantonneopsis Pit.
Yutajea Steyerm. *
Zuccarinia Blume

RUTACEAE Juss. 1789   39.01

158 genera. Widespread, mostly trop. and subtrop. Trees, shrubs, lianes, few herbs.

B&H	POLYPETALAE, DISCIFLORAE	Geraniales, 39
	(including Phellinaceae)	
DT&H	ARCHICHLAMYDEAE	Geraniales, Geraniineae, 87
	(including Rhabdodendraceae)	
Melc	ARCHICHLAMYDEAE	Rutales, Rutineae, 131
	(including Rhabdodendraceae)	
Thor	RUTIFLORAE	Rutales, Rutineae, 167
	(including Rhabdodendraceae)	
Dahl	RUTIFLORAE	Rutales, 218
Young	ROSIDAE, RUTANAE	Rutales, 244
	(including Rhabdodendraceae)	
Takh	ROSIDAE, RUTANAE	Rutales, 255
Cron	ROSIDAE	Sapindales, 261

Achuaria Gereau
Acmadenia Bartl. & H.L.Wendl.
Acradenia Kippist
Acronychia J.R.Forst. & G.Forst.
Adenandra Willd.
Adiscanthus Ducke
Aegle Corrêa

Aeglopsis Swingle
Afraegle (Swingle) Engl.
Agathosma Willd.
Almeidea A.St.-Hil.
Amyris P.Browne
Angostura Roem. & Schult.
Apocaulon R.S.Cowan

Araliopsis Engl.
Asterolasia F.Muell.
Atalantia Corrêa
Balfourodendron Corr.Mello ex Oliv.
Balsamocitrus Stapf
Boenninghausenia Rchb. ex Meisn.
Boninia Planch.
Boronella Baill.
Boronia Sm.
Bosistoa F.Muell.
Bouchardatia Baill. *
Brombya F.Muell.
Burkillanthus Swingle
Calodendrum Thunb.
Casimiroa La Llave
Chloroxylon DC.
Choisya Kunth
Chorilaena Endl.
Citropsis (Engl.) Swingle & M.Kellerm.
Citrus L.
Clausena Burm.f.
Clymenia Swingle
Cneoridium Hook.f.
Coleonema Bartl. & H.L.Wendl.
Comptonella Baker f.
Coombea P.Royen *
Correa Andrews
Crowea Sm.
Cyanothamnus Lindl.
Decagonocarpus Engl.
Decatropis Hook.f.
Decazyx Pittier & S.F.Blake
Dendrosma Pancher & Sebert
Dictamnus L.
Dictyoloma A.Juss.
Diosma L.
Diphasia Pierre
Diphasiopsis Mendonça
Diplolaena R.Br.
Drummondita Harv.
Dutaillyea Baill.
Echinocitrus Tanaka
Empleuridium Sond. & Harv.
Empleurum Aiton
Eremocitrus Swingle
Eriostemon Sm.
Erythrochiton Nees & Mart.
Esenbeckia Kunth
Euchaetis Bartl. & H.L.Wendl.
Euodia J.R.Forst. & G.Forst.
Euxylophora Huber
Evodiella Linden
Fagaropsis Mildbr. ex Siebenl.
Feroniella Swingle
Flindersia R.Br.

Fortunella Swingle
Galipea Aubl.
Geijera Schott
Geleznowia Turcz.
Glycosmis Corrêa
Halfordia F.Muell.
Haplophyllum A.Juss.
Helietta Tul.
Hortia Vand.
Ivodea Capuron
Kodalyodendron Borhidi & Acuña *
Leptothyrsa Hook.f.
Limnocitrus Swingle
Limonia L.
Lubaria Pittier
Lunasia Blanco
Luvunga Buch.–Ham. ex Wight & Arn.
Maclurodendron T.G.Hartley *
Macrostylis Bartl. & H.L.Wendl.
Medicosma Hook.f.
Megastigma Hook.f.
Melicope J.R.Forst. & G.Forst.
Merope M.Roem.
Merrillia Swingle
Metrodorea St.–Hil.
Microcitrus Swingle
Microcybe Turcz.
Micromelum Blume
Monanthocitrus Tanaka
Monnieria Loefl.
Muiriantha C.A.Gardner
Murraya L.
Myrtopsis Engl.
Naringi Adans.
Naudinia Planch. & Linden
Nematolepis Turcz.
Neobyrnesia J.A.Armstr.
Nycticalanthus Ducke
Oricia Pierre
Oriciopsis Engl.
Orixa Thunb.
Oxanthera Montrouz.
Pamburus Swingle
Paramignya Wight
Peltostigma Walp.
Pentaceras Hook.f.
Phebalium Vent.
Phellodendron Rupr.
Philotheca Rudge
Phyllosma Bolus
Pilocarpus Vahl
Pitavia Molina
Platydesma H.Mann
Pleiospermium (Engl.) Swingle
Plethadenia Urb.

Polyaster Hook.f.
Poncirus Raf.
Psilopeganum Hemsl.
Ptelea L.
Raputia Aubl.
Rauia Nees & Mart.
Raulinoa R.S.Cowan
Ravenia Vell.
Raveniopsis Gleason
Rhadinothamnus Paul G.Wilson
Ruta L.
Rutaneblina Steyerm. & Luteyn
Sarcomelicope Engl.
Severinia Ten.
Sheilanthera I.Williams
Skimmia Thunb.
Spathelia L.

Spiranthera A.St.–Hil.
Stauranthus Liebm.
Swinglea Merr.
Teclea Delile
Tetractomia Hook.f.
Tetradium Lour.
Thamnosma Torr. & Frém.
Ticorea Aubl.
Toddalia Juss.
Toddaliopsis Engl.
Tractocopevodia Raizada & Naray.
Triphasia Lour.
Urocarpus J.Drumm. ex Harv.
Vepris Comm. ex A.Juss.
Wenzelia Merr.
Zanthoxylum L.
Zieria Sm.

SABIACEAE Blume 1851   6.02

1 genus. Trop. and E. Asia to Solomons. Lianes, some shrubs.

B&H	POLYPETALAE, DISCIFLORAE (including Meliosmaceae)	Sapindales, 52
DT&H	ARCHICHLAMYDEAE (including Meliosmaceae)	Sapindales, Sabiineae, 116
Melc	ARCHICHLAMYDEAE (including Meliosmaceae)	Sapindales, Sapindineae, 149
Thor	RUTIFLORAE (including Meliosmaceae)	Rutales, Sapindineae, 183
Dahl	RUTIFLORAE	Sapindales, 215
Young	ROSIDAE, RUTANAE (including Meliosmaceae)	Sapindales, 255
Takh	ROSIDAE, RUTANAE (including Meliosmaceae)	Sapindales, 253
Cron	MAGNOLIIDAE (including Meliosmaceae)	Ranunculales, 37

Sabia Colebr.

SACCIFOLIACEAE Maguire & Pires 1978   109.03

1 genus, monotypic. Guyana Highlands of South America. Subshrub.

B&H	Not known	—
DT&H	Not known	—
Melc	Not known	—
Thor	GENTIANIFLORAE	Gentianales (within Gentianaceae, 254)
Dahl	GENTIANIFLORAE	Gentianales, 345
Young	Not mentioned	—

Takh   LAMIIDAE, GENTIANANAE   Gentianales, 372
Cron   ASTERIDAE   Gentianales, 273

Saccifolium Maguire & Pires

SALICACEAE Mirb. 1815   160

2 genera. Widespread except Australasia. Trees, shrubs.

B&H   MONOCHLAMYDEAE   Ordines anomali, 160
DT&H   ARCHICHLAMYDEAE   Salicales, 6
Melc   ARCHICHLAMYDEAE   Salicales, 7
Thor   VIOLIFLORAE   Violales, Salicineae, 134
Dahl   VIOLIFLORAE   Violales, 97
Young DILLENIIDAE, VIOLANAE   Salicales, 204
Takh   DILLENIIDAE, VIOLANAE   Salicales, 158
Cron   DILLENIIDAE   Salicales, 131

Populus L.
Salix L.

SALVADORACEAE Lindl. 1836   105

3 genera. Old World tropics and subtropics. Small trees, shrubs.

B&H   GAMOPETALAE, BICARPELLATAE Gentianales, 105
DT&H   METACHLAMYDEAE   Contortae, Oleineae, 197
Melc   ARCHICHLAMYDEAE   Celastrales, Celastrineae, 163
Thor   GENTIANIFLORAE   Oleales, 247
Dahl   VIOLIFLORAE   Salvadorales, 107
Young ROSIDAE, GENTIANANAE   Oleales, 344
Takh   ROSIDAE, CELASTRANAE   Celastrales, Celastrineae, 308
Cron   ROSIDAE   Celastrales, 222

Azima Lam.
Dobera Juss.
Salvadora L.

SANTALACEAE R.Br. 1810   149.01

37 genera. Widespread. Hemiparasitic trees, shrubs, herbs.

B&H   MONOCHLAMYDEAE   Achlamydosporeae, 149
       (including Grubbiaceae, Misodendraceae)
DT&H   ARCHICHLAMYDEAE   Santalales, Santalineae, 19
Melc   ARCHICHLAMYDEAE   Santalales, Santalineae, 20
Thor   SANTALIFLORAE   Santalales, 117
Dahl   SANTALIFLORAE   Santalales, 260
Young ROSIDAE, SANTALANAE   Santalales, 235
Takh   ROSIDAE, CELASTRANAE   Santalales, 314
Cron   ROSIDAE   Santalales, 209

Acanthosyris (Eichler) Griseb.
Amphorogyne Stauffer & Hürl.
Anthobolus R.Br.
Arjona Cav.
Buckleya Torr.
Cervantesia Ruíz & Pav.
Choretrum R.Br.
Cladomyza Danser
Colpoon Bergius
Comandra Nutt.
Daenikera Hürl. & Stauffer
Dendromyza Danser
Dendrotrophe Miq.
Dufrenoya Chatin
Elaphanthera N.Hallé
Exocarpos Labill.
Geocaulon Fernald *
Iodina Hook. & Arn. ex Meisn.
Kunkeliella Stearn *

Leptomeria R.Br.
Mida A.Cunn. ex Endl.
Myoschilos Ruíz & Pav.
Nanodea Banks ex C.F.Gaertn.
Nestronia Raf.
Okoubaka Pellegr. & Normand
Omphacomeria (Endl.) A.DC.
Osyridicarpos A.DC.
Osyris L.
Phacellaria Benth.
Pyrularia Michx.
Quinchamalium Molina
Rhoiacarpos A.DC.
Santalum L.
Scleropyrum Arn.
Spirogardnera Stauffer
Thesidium Sond.
Thesium L.

SAPINDACEAE Juss. 1789   51.01

134 genera. Trop. and subtrop., rarely temp. Trees, shrubs, climbers.

B&H   POLYPETALAE, DISCIFLORAE Sapindales, 51
       (including Aceraceae, Akaniaceae, Greyiaceae, Hippocastanaceae,
       Melianthaceae, Ptaeroxylaceae, Staphyleaceae)
DT&H ARCHICHLAMYDEAE   Sapindales, Sapindineae, 115
       (including Akaniaceae)
Melc   ARCHICHLAMYDEAE   Sapindales, Sapindineae, 147
Thor   RUTIFLORAE   Rutales, Sapindineae, 180
       (including Emblingiaceae, Stylobasiaceae)
Dahl   RUTIFLORAE   Sapindales, 207
Young ROSIDAE, RUTANAE   Sapindales, 253
       (including Emblingiaceae, Stylobasiaceae)
Takh   ROSIDAE, RUTANAE   Sapindales, 245
Cron   ROSIDAE   Sapindales, 252
       (including Ptaeroxylaceae)

Alectryon Gaertn.
Allophylus L.
Allosanthus Radlk.
Amesiodendron Hu
Aporrhiza Radlk.
Arfeuillea Pierre ex Radlk.
Arytera Blume
Atalaya Blume
Athyana (Griseb.) Radlk.
Averrhoidium Baill.
Beguea Capuron *
Bizonula Pellegr.
Blighia K.König
Blighiopsis Veken

Blomia Miranda
Boniodendron Gagnep.
Bottegoa Chiov.
Bridgesia Bertero ex Cambess.
Camptolepis Radlk.
Cardiospermum L.
Castanospora F.Muell.
Chonopetalum Radlk.
Chouxia Capuron
Chytranthus Hook.f.
Conchopetalum Radlk.
Cossinia Comm. ex Lam.
Cubilia Blume
Cupania L.

Cupaniopsis Radlk.
Deinbollia Schumach. & Thonn.
Delavaya Franch.
Diatenopteryx Radlk.
Dictyoneura Blume
Dilodendron Radlk.
Dimocarpus Lour.
Diploglottis Hook.f.
Diplokeleba N.E.Br.
Diplopeltis Endl.
Distichostemon F.Muell.
Dodonaea Mill.
Doratoxylon Thouars ex Hook.f.
Elattostachys Radlk.
Eriocoelum Hook.f.
Erythrophysa E.Mey. ex Arn.
Euchorium Ekman & Radlk.
Euphorianthus Radlk.
Eurycorymbus Hand.–Mazz.
Exothea Macfad.
Filicium Thwaites ex Benth.
Ganophyllum Blume
Glenniea Hook.f.
Gloeocarpus Radlk.
Gongrodiscus Radlk.
Gongrospermum Radlk.
Guindilia Gillies ex Hook. & Arn.
Guioa Cav.
Haplocoelum Radlk.
Harpullia Roxb.
Hippobromus Eckl. & Zeyh.
Hornea Baker
Houssayanthus Hunz.
Hypelate P.Browne
Hypseloderma Radlk.
Jagera Blume
Koelreuteria Laxm.
Laccodiscus Radlk.
Lecaniodiscus Planch. ex Benth.
Lepiderema Radlk.
Lepidopetalum Blume
Lepisanthes Blume
Litchi Sonn.
Llagunoa Ruíz & Pav.
Lophostigma Radlk.
Loxodiscus Hook.f.
Lychnodiscus Radlk.
Macphersonia Blume
Magonia A.St.–Hil.
Majidea J.Kirk ex Oliv.
Matayba Aubl.
Melicoccus P.Browne
Mischocarpus Blume

Molinaea Comm. ex Juss.
Neotina Capuron
Nephelium L.
Otonephelium Radlk.
Pancovia Willd.
Pappea Eckl. & Zeyh.
Paranephelium Miq.
Paullinia L.
Pavieasia Pierre
Pentascyphus Radlk.
Phyllotrichum Thorel ex Lecomte *
Placodiscus Radlk.
Plagioscyphus Radlk.
Podonephelium Baill.
Pometia J.R.Forst. & G.Forst.
Porocystis Radlk.
Pseudima Radlk.
Pseudopancovia Pellegr. *
Pseudopteris Baill.
Radlkofera Gilg
Rhysotoechia Radlk.
Sapindus L.
Sarcopteryx Radlk.
Sarcotoechia Radlk.
Schleichera Willd.
Scyphonychium Radlk.
Serjania Mill.
Sisyrolepis Radlk.
Smelophyllum Radlk.
Stadmannia Lam.
Stocksia Benth.
Storthocalyx Radlk.
Synima Radlk.
Talisia Aubl.
Thinouia Triana & Planch.
Thouinia Poit.
Thouinidium Radlk.
Tina Schult.
Tinopsis Radlk.
Toechima Radlk.
Toulicia Aubl.
Trigonachras Radlk.
Tripterodendron Radlk.
Tristira Radlk.
Tristiropsis Radlk.
Tsingya Capuron *
Ungnadia Endl.
Urvillea Kunth
Vouarana Aubl.
Xanthoceras Bunge
Xerospermum Blume
Zanha Hiern
Zollingeria Kurz

SAPOTACEAE Juss. 1789 101.01

53 genera. Widespread in tropics, rarely warm-temperate. Trees, shrubs.

B&H	GAMOPETALAE, HETEROMERAE	Ebenales, 101
DT&H	METACHLAMYDEAE	Ebenales, Sapotineae, 192
Melc	SYMPETALAE	Ebenales, Sapotineae, 237
	(separating *Sarcospermataceae*, 238)	
Thor	THEIFLORAE	Ebenales, Ebenineae, 75
Dahl	PRIMULIFLORAE	Ebenales, 133
Young	DILLENIIDAE, DILLENIANAE	Ebenales, 136
Takh	DILLENIIDAE, ERICANAE	Sapotales, 136
Cron	DILLENIIDAE	Ebenales, 148

Argania Roem. & Schult.
Aubregrinia Heine
Aulandra H.J.Lam
Autranella A.Chev.
Baillonella Pierre
Breviea Aubrév. & Pellegr.
Burckella Pierre
Capurodendron Aubrév.
Chromolucuma Ducke
Chrysophyllum L.
Delpydora Pierre
Diploknema Pierre
Diploon Cronquist
Eberhardtia Lecomte
Ecclinusa Mart.
Elaeoluma Baill.
Englerophytum K.Krause
Faucherea Lecomte
Gluema Aubrév. & Pellegr.
Inhambanella (Engl.) Dubard
Isonandra Wight
Labourdonnaisia Bojer
Labramia A.DC.
Lecomtedoxa (Pierre ex Engl.) Dubard
Leptostylis Benth.
Letestua Lecomte
Madhuca Ham. ex J.F.Gmel.

Magodendron Vink
Manilkara Adans.
Micropholis (Griseb.) Pierre
Mimusops L.
Neohemsleya T.D.Penn.
Neolemonniera Heine
Nesoluma Baill.
Niemeyera F.Muell.
Northia Hook.f.
Omphalocarpum P.Beauv.
Palaquium Blanco
Payena A.DC.
Pichonia Pierre
Pouteria Aubl.
Pradosia Liais
Pycnandra Benth.
Sarcaulus Radlk.
Sarcosperma Hook.f.
Sideroxylon L.
Synsepalum (A.DC.) Daniell
Tieghemella Pierre
Tridesmostemon Engl.
Tsebona Capuron
Vitellaria C.F.Gaertn.
Vitellariopsis Baill. ex Dubard
Xantolis Raf.

SARCOLAENACEAE Caruel 1881 30

9 genera. Madagascar. Trees, shrubs.

B&H	POLYPETALAE, THALAMIFLORAE	Guttiferales (as *Chlaenaceae*) 30
DT&H	ARCHICHLAMYDEAE	Malvales, Chlaenineae (as *Chlaenaceae*) 122
Melc	ARCHICHLAMYDEAE	Malvales, Sarcolaenineae, 171
Thor	MALVIFLORAE	Malvales, 150
Dahl	MALVIFLORAE	Malvales, 63
Young	DILLENIIDAE, MALVANAE	Malvales, 170
Takh	DILLENIIDAE, MALVANAE	Malvales, 178
Cron	DILLENIIDAE	Theales, 82

Eremolaena Baill.
Leptolaena Thouars
Mediusella (Cavaco) Dorr
Pentachlaena H.Perrier
Perrierodendron Cavaco

Rhodolaena Thouars
Sarcolaena Thouars
Schizolaena Thouars
Xyloolaena Baill.

SARGENTODOXACEAE Stapf ex Hutch. 1926   7.03

1 genus, monotypic. China, Indochina. Liane.

B&H	Not known	—
DT&H	Not known	—
Melc	ARCHICHLAMYDEAE	Ranunculales, Ranunculineae, 63
Thor	ANNONIFLORAE	Berberidales, Berberidineae, 29
Dahl	RANUNCULIFLORAE	Ranunculales, 30
Young	MAGNOLIIDAE, RANUNCULANAE Ranunculales, 33	
Takh	RANUNCULIDAE, RANUNCULANAE Ranunculales, Lardizabalineae, 45	
Cron	MAGNOLIIDAE	Ranunculales, 33

Sargentodoxa Rehder & E.H.Wilson

SARRACENIACEAE Dumort. 1829   9

3 genera. E. and Pacific N. America, Guyana Highlands. Pitcher plants.

B&H	POLYPETALAE, THALAMIFLORAE	Parietales, 9
DT&H	ARCHICHLAMYDEAE	Sarraceniales, 60
Melc	ARCHICHLAMYDEAE	Sarraceniales, 91
Thor	THEIFLORAE	Theales, Sarraceniineae, 57
Dahl	CORNIFLORAE	Sarraceniales, 297
Young	DILLENIIDAE, DILLENIANAE	Sarraceniales, 119
Takh	DILLENIIDAE, SARRACENIANAE	Sarraceniales, 125
Cron	DILLENIIDAE	Nepenthales, 104

Darlingtonia Torr.
Heliamphora Benth.
Sarracenia L.

SAURURACEAE Rich. ex E.Mey. 1827   139.03

5 genera. E. Asia, E. North America. Herbs.

B&H	MONOCHLAMYDEAE	Micrembryeae (within Piperaceae, 139)
DT&H	ARCHICHLAMYDEAE	Piperales, 2
Melc	ARCHICHLAMYDEAE	Piperales, 68
Thor	ANNONIFLORAE	Annonales, Piperineae, 22
Dahl	NYMPHAEIFLORAE	Piperales, 24
Young	MAGNOLIIDAE, RANUNCULANAE Aristolochiaceae, 27	
Takh	MAGNOLIIDAE, MAGNOLIANAE Piperales, 25	
Cron	MAGNOLIIDAE	Piperales, 20

Anemopsis Hook. & Arn.
Circaeocarpus C.Y.Wu *
Gymnotheca Decne.

Houttuynia Thunb.
Saururus L.

SAXIFRAGACEAE Juss. 1789   59.01

31 genera. N. temp., rare trop. mts. and S. hemisphere. Herbs.

B&H   POLYPETALAE, CALYCIFLORAE Rosales, 59
              (including Cephalotaceae, Cunoniaceae, Eremosynaceae, Escalloniaceae,
              Grossulariaceae, Hydrangeaceae, Parnassiaceae, Vahliaceae)
DT&H ARCHICHLAMYDEAE   Rosales, Saxifragineae, 67
              (including Eremosynaceae, Escalloniaceae, Grossulariaceae,
              Hydrangeaceae, Montiniaceae, Parnassiaceae, Pterostemonaceae,
              Vahliaceae)
Melc   ARCHICHLAMYDEAE   Rosales, Saxifragineae, 106
              (including Eremosynaceae, Escalloniaceae, Grossulariaceae,
              Hydrangeaceae, Montiniaceae, Parnassiaceae, Penthoraceae,
              Pterostemonaceae, Vahliaceae)
Thor   ROSIFLORAE   Rosales, Saxifragineae, 208
              (including Alseuosmiaceae, Columelliaceae, Eremosynaceae,
              Escalloniaceae, Griseliniaceae, Grossulariaceae, Hydrangeaceae, Melanophyllaceae,
              Montiniaceae, Penthoraceae, Pterostemonaceae, Vahliaceae)
Dahl   ROSIFLORAE   Saxifragales, 163
              (including Penthoraceae, Vahliaceae)
Young ROSIDAE, ROSANAE   Saxifragaceae, 302
              (separating *Francoaceae* 306)
Takh   ROSIDAE, ROSANAE   Saxifragales, 205
Cron   ROSIDAE   Rosales, 173
              (including Eremosynaceae, Parnassiaceae, Penthoraceae, Vahliaceae)

Astilbe Buch.–Ham. ex G.Don
Astilboides (Hemsl.) Engl.
Bensoniella C.Morton
Bergenia Moench
Bolandra A.Gray
Boykinia Nutt.
Chrysosplenium L.
Conimitella Rydb.
Darmera Voss
Elmera Rydb.
Francoa Cav.
Heuchera L.
Jepsonia Small
Leptarrhena R.Br.
Lithophragma (Nutt.) Torr. & A.Gray
Mitella L.

Mukdenia Koldz.
Oresitrophe Bunge
Peltoboykinia (Engl.) Hara
Rodgersia A.Gray
Saxifraga L.
Saxifragella Engl.
Saxifragodes D.M.Moore
Saxifragopsis Small
Suksdorfia A.Gray
Sullivantia Torr. & A.Gray
Tanakea Franch. & Sav.
Tellima R.Br.
Tetilla DC.
Tiarella L.
Tolmiea Torr. & A.Gray

SCHISANDRACEAE Blume 1830   4.06

2 genera. India to Moluccas and Japan, and S.E. USA. Lianes.

B&H   POLYPETALAE, THALAMIFLORAE Ranales (within Magnoliaceae, 4)

DT&H ARCHICHLAMYDEAE

Melc ARCHICHLAMYDEAE
Thor ANNONIFLORAE
Dahl MAGNOLIIFLORAE
Young MAGNOLIIDAE, MAGNOLIANAE
Takh MAGNOLIIDAE, MAGNOLIANAE
Cron MAGNOLIIDAE

Ranales, Magnoliineae (within
                    Magnoliaceae, 45)
Magnoliales, Illiciineae, 47
Annonales, Illiciineae, 3
Illiciales, 16
Illiciales, 8
Illiciales, 10
Illiciales, 24

Kadsura Juss.
Schisandra Michx.

SCROPHULARIACEAE Juss. 1789   115

292 genera. Widespread. Mostly herbs, some parasitic, or occas. trees (*Paulownia*).

B&H GAMOPETALAE, BICARPELLATAE Personales, 115
DT&H METACHLAMYDEAE
        (separating *Orobanchaceae* 214)
Melc SYMPETALAE
        (separating *Orobanchaceae*, 274)
Thor GENTIANIFLORAE
        (including Globulariaceae)
Dahl LAMIIFLORAE
        (separating *Selaginaceae* 354)
Young ROSIDAE, GENTIANANAE
Takh LAMIIDAE, LAMIANAE
        (separating *Spielmanniaceae* 408 (*Oftia*), nom. inval.)
Cron ASTERIDAE
        (separating *Orobanchaceae* 297)

Tubiflorae, Solanineae, 210

Tubiflorae, Solanineae, 265

Bignoniales, 260

Scrophulariales, 352

Bignoniales, 358
Scrophulariales, 396

Scrophulariales, 294

Acanthorrhinum Rothm.
Achetaria Cham. & Schltdl.
Adenosma R.Br.
Aeginetia L.
Agalinis Raf.
Agathelpis Choisy
Albraunia Speta
Alectra Thunb.
Alonsoa Ruíz & Pav.
Anarrhinum Desf.
Anastrabe E.Mey. ex Benth.
Ancistrostylis Yamaz. *
Angelonia Bonpl.
Anisantherina Pennell
Antherothamnus N.E.Br.
Anticharis Endl.
Antirrhinum L.
Aphyllon Mitch.
Aptosimum Burch. ex Benth.
Aragoa Kunth
Artanema D.Don
Asarina Mill.
Aureolaria Raf.

Bacopa Aubl.
Bampsia Lisowski & Mielcarek
Bartsia L.
Basistemon Turcz.
Baumia Engl. & Gilg
Benjaminia Mart. ex Benj.
Berendtiella Wettst. & Harms
Besseya Rydb.
Boschniakia C.A.Mey. ex Bong.
Bowkeria Harv.
Brandisia Hook.f. & Thomson
Braunblanquetia Eskuche *
Brookea Benth.
Bryodes Benth.
Buchnera L.
Bungea C.A.Mey.
Buttonia McKen ex Benth.
Bythophyton Hook.f.
Calceolaria L.
Campbellia Wight
Camptoloma Benth.
Campylanthus Roth
Capraria L.

Castilleja Mutis ex L.f.
Centranthera R.Br.
Chaenorhinum (DC.) Rchb.
Charadrophila Marloth
Chelone L.
Chenopodiopsis Hilliard & B.L.Burtt
Chionohebe B.G.Briggs & Ehrend.
Chionophila Benth.
Christisonia Gardner
Cistanche Hoffmanns. & Link
Clevelandia Greene
Cochlidiosperma (Rchb.) Rchb.
Collinsia Nutt.
Colpias E.Mey. ex Benth.
Conobea Aubl.
Conopholis Wallr.
Cordylanthus Nutt. ex Benth.
Craterostigma Hochst.
Cromidon Compton
Cycniopsis Engl.
Cycnium E.Mey. ex Benth.
Cymbalaria Hill
Cymbaria L.
Cyrtandromoea Zoll.
Dasistoma Raf.
Deinostema Yamaz.
Dermatobotrys Bolus
Detzneria Schltr. ex Diels
Diascia Link & Otto
Diclis Benth.
Digitalis L.
Dintera Stapf
Dischisma Choisy
Dizygostemon (Benth.) Radlk. ex Wettst.
Dodartia L.
Dopatrium Buch.–Ham.
Elacholoma F.Muell. & Tate
Elatinoides (Chavannes) Wettst.
Ellisiophyllum Maxim.
Encopella Pennell
Epifagus Nutt.
Epixiphium (A.Gray) Munz
Eremogeton Standl. & L.O.Williams
Erinus L.
Escobedia Ruíz & Pav.
Esterhazya J.C.Mikan
Euphrasia L.
Faxonanthus Greenm.
Freylinia Colla
Galvezia Dombey ex Juss.
Gambelia Nutt.
Gentrya Breedlove & Heckard
Geochorda Cham. & Schltdl.
Gerardiina Engl.
Ghikaea Volkens & Schweinf.

Gibsoniothamnus L.O.Williams
Gleadovia Gamble & Prain
Glekia Hilliard
Globulariopsis Compton *
Glossostigma Wight & Arn.
Glumicalyx Hiern
Gosela Choisy
Graderia Benth.
Gratiola L.
Halleria L.
Harveya Hook.
Hebe Comm. ex Juss.
Hebenstretia L.
Hedbergia Molau
Hemichaena Benth.
Hemimeris L.f.
Hemiphragma Wall.
Herpestis Gaertn.
Heteranthia Nees & Mart.
Hiernia S.Moore
Holmgrenanthe Elisens
Holzneria Speta
Howelliella Rothm.
Hydranthelium Kunth
Hydrotriche Zucc.
Hyobanche L.
Isoplexis (Lindl.) J.C.Loudon
Ixianthes Benth.
Jamesbrittenia Kuntze
Jerdonia Wight
Jovellana Ruíz & Pav.
Kashmiria D.Y.Hong
Keckiella Straw
Kickxia Dumort.
Kopsiopsis (Beck) Beck
Lafuentea Lag.
Lagotis Gaertn.
Lamourouxia Kunth
Lancea Hook.f. & Thomson
Lathraea L.
Leptorhabdos Schrenk
Lesquereuxia Boiss. & Reut.
Leucocarpus D.Don
Leucophyllum Bonpl.
Leucosalpa Scott Elliot
Limnophila R.Br.
Limosella L.
Linaria Mill.
Lindenbergia Lehm.
Lindernia All.
Lophospermum D.Don
Lyperia Benth.
Mabrya Elisens
Macranthera Nutt. ex Benth.
Maeviella Rossow *

Magdalenaea Brade *
Mannagettaea Harry Sm.
Manulea L.
Manuleopsis Thell. ex Schinz
Maurandella (A.Gray) Rothm.
Maurandya Ortega
Mazus Lour.
Mecardonia Ruíz & Pav.
Melampyrum L.
Melanospermum Hilliard
Melasma Bergius
Melosperma Benth.
Micranthemum Michx.
Micrargeria Benth.
Micrargeriella R.E.Fr.
Microcarpaea R.Br.
Microdon Choisy
Mimetanthe Greene
Mimulicalyx Tsoong *
Mimulus L.
Misopates Raf.
Mohavea A.Gray
Monocardia Pennell
Monochasma Maxim. ex Franch. & Sav.
Monopera Barringer
Monttea Gay
Namation Brand *
Nathaliella Fedtsch. *
Necranthus Gilli
Nemesia Vent.
Neogaerrhinum Rothm.
Neopicrorhiza D.Y.Hong
Nothochelone (A.Gray) Straw
Nothochilus Radlk.
Nuttallanthus D.A.Sutton
Odicardis Raf.
Odontites Ludw.
Oftia Adans.
Omania S.Moore
Omphalotrix Maxim.
Ophiocephalus Wiggins
Oreosolen Hook.f.
Orobanche L.
Orthocarpus Nutt.
Otacanthus Lindl.
Ourisia Comm. ex Juss.
Paederota L.
Parahebe W.R.B.Oliv.
Parastriga Mildbr.
Parentucellia Viv.
Paulownia Siebold & Zucc.
Pedicularis L.
Peliostomum Benth.
Penstemon Schmidel
Peplidium Delile

Petitmenginia Bonati
Phacellanthus Siebold & Zucc.
Phelypaea L.
Phtheirospermum Bunge ex Fisch. & C.A.Mey.
Phygelius E.Mey. ex Benth.
Phyllopodium Benth.
Physocalyx Pohl
Picria Lour.
Picrorhiza Royle ex Benth.
Pierranthus Bonati
Platypholis Maxim.
Poarium Desv.
Polycarena Benth.
Psammetes Hepper
Pseudobartsia D.Y.Hong *
Pseudolysimachion (W.D.J.Koch) Opiz
Pseudorobanche Rouy
Pseudorontium (A.Gray) Rothm.
Pseudosopubia Engl.
Pseudostriga Bonati
Pterygiella Oliv.
Radamaea Benth.
Ranopisoa J.-F.Leroy *
Rehmannia Libosch. ex Fisch. & C.A.Mey.
Rhamphicarpa Benth.
Rhaphispermum Benth.
Rhinanthus L.
Rhodochiton Zucc. ex Otto & Dietr.
Rhynchocorys Griseb.
Russelia Jacq.
Sairocarpus D.A.Sutton
Schistophragma Benth. ex Endl.
Schizosepala G.M.Barroso
Schlegelia Miq.
Schwalbea L.
Schweinfurthia A.Braun
Scoparia L.
Scrofella Maxim.
Scrophularia L.
Selago L.
Seymeria Pursh
Shiuyinghua Paclt
Sibthorpia L.
Silviella Pennell
Siphonostegia Benth.
Sopubia Buch.-Ham. ex D.Don
Spirostegia Ivanina
Stemodia L.
Stemodiopsis Engl.
Stemotria Wettst. & Harms ex Engl.
Striga Lour.
Strobilopsis Hilliard & B.L.Burtt
Sutera Roth

Synthyris Benth.
Teedia Rudolphi
Tetranema Benth.
Tetraselago Junell
Tetraspidium Baker
Tetraulacium Turcz.
Thunbergianthus Engl.
Tienmuia Hu
Tonella Nutt. ex A.Gray
Torenia L.
Tozzia L.
Triaenophora (Hook.f.) Soler.
Trieenea Hilliard
Trungboa Rauschert

Tuerckheimocharis Urb.
Uroskinnera Lindl.
Vellosiella Baill.
Verbascum L.
Veronica L.
Veronicastrum Heist. ex Fabr.
Walafrida E.Mey.
Wightia Wall.
Wulfenia Jacq.
Wulfeniopsis D.Y.Hong
Xizangia D.Y.Hong *
Xylanche Beck
Xylocalyx Balf.f.
Zaluzianskya F.W.Schmidt

SCYPHOSTEGIACEAE Hutch. 1926   17.05

1 genus, monotypic. Borneo. Tree.

B&H	Not known	—
DT&H	ARCHICHLAMYDEAE	Ranales, Magnoliineae (within Monimiaceae, 51)
Melc	ARCHICHLAMYDEAE	Violales, Flacourtiineae, 186
Thor	VIOLIFLORAE	Violales, Violineae, 127
Dahl	VIOLIFLORAE	Violales, 88
Young	DILLENIIDAE, VIOLANAE	Violales, 193
Takh	DILLENIIDAE, VIOLANAE	Violales, 149
Cron	DILLENIIDAE	Violales, 113

Scyphostegia Stapf

SCYTOPETALACEAE Engl. 1897   33.03

5 genera. Guineo-Congolan Africa. Trees, shrubs, lianes.

B&H	Not known	—
DT&H	ARCHICHLAMYDEAE	Malvales, Scytopetalineae, 129
Melc	ARCHICHLAMYDEAE	Malvales, Scytopetalineae, 176
Thor	THEIFLORAE	Theales, Scytopetalineae, 60
Dahl	THEIFLORAE	Theales, 122
Young	DILLENIIDAE, DILLENIANAE	Scytopetalales, 122
Takh	DILLENIIDAE, THEANAE	Ochnales, 121
Cron	DILLENIIDAE	Theales, 87

Brazzeia Baill.
Oubanguia Baill.
Pierrina Engl.

Rhaptopetalum Oliv.
Scytopetalum Pierre ex Engl.

SIMAROUBACEAE DC. 1811   40.01

24 genera. Tropics and subtropics, occas. temperate. Trees, shrubs.

B&H   POLYPETALAE, DISCIFLORAE Geraniales, 40
         (including Balanitaceae, Brunelliaceae, Cneoraceae, Irvingiaceae,
         Surianaceae)

DT&H	ARCHICHLAMYDEAE	Geraniales, Geraniineae, 88
	(including Irvingiaceae, Surianaceae)	
Melc	ARCHICHLAMYDEAE	Rutales, Rutineae, 133
	(including Irvingiaceae, Surianaceae)	
Thor	RUTIFLORAE	Rutales, Rutineae, 169
	(including Balanitaceae, Irvingiaceae)	
Dahl	RUTIFLORAE	Rutales, 221
	(including Irvingiaceae)	
Young	ROSIDAE, RUTANAE	Rutales, 247
	(presumably including Irvingiaceae)	
Takh	ROSIDAE, RUTANAE	Rutales, 259
	(separating *Kirkiaceae* 267)	
Cron	ROSIDAE	Sapindales, 258
	(including Irvingiaceae)	

Ailanthus Desf.
Alvaradoa Liebm.
Amaroria A.Gray
Brucea J.F.Mill.
Castela Turpin
Eurycoma Jack
Gymnostemon Aubrév. & Pellegr.
Hannoa Planch.
Harrisonia R.Br. ex Adr.Juss.
Kirkia Oliv.
Laumoniera Nooteboom *
Nothospondias Engl.

Odyendea Pierre ex Engl.
Perriera Courchet
Picramnia Sw.
Picrasma Blume
Picrolemma Hook.f.
Pierreodendron Engl.
Pleiokirkia Capuron *
Quassia L.
Recchia Moçino & Sessé ex DC.
Simaba Aubl.
Simarouba Aubl.
Soulamea Lam.

SIMMONDSIACEAE (Muell.Arg.) Tiegh. ex Reveal & Hoogland 1990 151.08

1 genus, monotypic. S.W. USA, N.W. Mexico. Shrub.

B&H	MONOCHLAMYDEAE	Unisexuales (within Euphorbiaceae, 151)
DT&H	ARCHICHLAMYDEAE	Sapindales, Buxineae (within Buxaceae, 99)
Melc	ARCHICHLAMYDEAE	Celastrales, Buxineae (within Buxaceae, 164)
Thor	MALVIFLORAE	Euphorbiales, 163
Dahl	MALVIFLORAE	Euphorbiales, 76
Young	DILLENIIDAE, MALVANAE	Euphorbiales, 185
Takh	HAMAMELIDIDAE, HAMAMELIDANAE	Simmondsiales, 91
Cron	ROSIDAE	Euphorbiales, 230

Simmondsia Nutt.

SOLANACEAE Juss. 1789 114.01

96 genera. Widespread. Mostly herbs, shrubs, occas. trees.

B&H	GAMOPETALAE, BICARPELLATAE	Polemoniales, 114
	(including Goetzeaceae, Retziaceae)	
DT&H	METACHLAMYDEAE	Tubiflorae, Solanineae, 209
	(including Goetzeaceae; separating *Nolanaceae* 208)	

Melc SYMPETALAE   Tubiflorae, Solanineae, 262
        (including Goetzeaceae; separating *Nolanaceae* 261)
Thor SOLANIFLORAE   Solanales, Solanineae, 271
        (including Duckeodendraceae, Goetzeaceae)
Dahl SOLANIFLORAE   Solanales, 273
        (including Duckeodendraceae; separating *Sclerophylacaceae* 274)
Young ROSIDAE, SOLANANAE   Solanales, 363
Takh LAMIIDAE, SOLANANAE   Solanales, 379
        (separating *Nolanaceae* 380, *Sclerophylacaceae* 382)
Cron ASTERDAE   Solanales, 278
        (separating *Nolanaceae* 277)

Acnistus Schott

Anisodus Link ex Spreng.

Anthocercis Labill.

Anthotroche Endl.

Archiphysalis Kuang

Athenaea Sendtn.

Atrichodendron Gagnep.

Atropa L.

Atropanthe Pascher

Aureliana Sendtn.

Benthamiella Speg.

Bouchetia Dunal

Brachistus Miers

Browallia L.

Brugmansia Pers.

Brunfelsia L.

Calibrachoa Cerv.

Capsicum L.

Cestrum L.

Chamaesaracha (A.Gray) Benth.

Combera Sandwith

Crenidium Haegi

Cuatresia Hunz.

Cyphanthera Miers

Cyphomandra Mart. ex Sendtn.

Datura L.

Deprea Raf.

Discopodium Hochst.

Dittostigma Phil.

Duboisia R.Br.

Dunalia Kunth

Exodeconus Raf.

Fabiana Ruíz & Pav.

Grabowskia Schltdl.

Grammosolen Haegi

Hawkesiophyton Hunz.

Hebecladus Miers

Hunzikeria D'Arcy

Hyoscyamus L.

Iochroma Benth.

Jaborosa Juss.

Jaltomata Schltdl. *

Juanulloa Ruíz & Pav.

Latua Phil.

Leptoglossis Benth.

Leucophysalis Rydb.

Lycianthes (Dunal) Hassl.

Lycium L.

Lycopersicon Mill.

Mandragora L.

Margaranthus Schltdl.

Markea Rich.

Melananthus Walp.

Mellissia Hook.f.

Metternichia J.C.Mikan

Nectouxia Kunth

Nicandra Adans.

Nicotiana L.

Nierembergia Ruíz & Pav.

Nolana L.f.

Nothocestrum A.Gray

Oryctes S.Watson

Pantacantha Speg.

Parabouchetia Baill.

Pauia Deb & Dutta *

Petunia Juss.

Phrodus Miers

Physalis L.

Physochlaina G.Don

Plowmania Hunz. & Subils

Protoschwenckia Soler.

Przewalskia Maxim.

Quincula Raf.

Rahowardiana D'Arcy *

Reyesia Gay

Salpichroa Miers

Salpiglossis Ruíz & Pav.

Saracha Ruíz & Pav.

Schizanthus Ruíz & Pav.

Schultesianthus Hunz.

Schwenckia L.

Schwenckiopsis Dammer

Sclerophylax Miers

Scopolia Jacq.

Sessea Ruíz & Pav.

Sesscopsis Hassl.

Solandra Sw.
Solanum L.
Streptosolen Miers
Symonanthus Haegi
Trianaea Planch. & Linden
Triguera Cav.

Tubocapsicum Makino
Vassobia Rusby
Vestia Willd.
Withania Pauquy
Witheringia L'Hér.

SPHAEROSEPALACEAE (Warb.) Tiegh. ex Bullock 1959   41.02

2 genera. Madagascar. Trees, shrubs.

B&H	Not placed	—
DT&H	ARCHICHLAMYDEAE	Parietales, Cochlospermineae (within Cochlospermaceae, 145)
Melc	ARCHICHLAMYDEAE	Violales, Cistineae, 193
Thor	THEIFLORAE (including Diegodendraceae)	Theales, Scytopetalineae, 61
Dahl	MALVIFLORAE	Malvales, 62
Young	DILLENIIDAE, DILLENIANAE	Scytopetalales, 123
Takh	DILLENIIDAE, MALVANAE	Malvales (as *Rhopalocarpaceae*) 179
Cron	DILLENIIDAE	Theales, 81

Dialyceras Capuron *
Rhopalocarpus Bojer

SPHENOSTEMONACEAE P.Royen & Airy Shaw 1972   45.07

1 genus. Malesia to Australia and New Caledonia. Trees, shrubs.

B&H	Not known	—
DT&H	ARCHICHLAMYDEAE	Sapindales, Celastrineae (within Aquifoliaceae, 107)
Melc	ARCHICHLAMYDEAE	Celastrales, Celastrineae (within Aquifoliaceae, 156)
Thor	THEIFLORAE	Theales, Theineae, 51
Dahl	CORNIFLORAE	Cornales, 312
Young	DILLENIIDAE, DILLENIANAE	Theales, 112
Takh	ROSIDAE, CELASTRANAE	Celastrales, Icacinineae, 301
Cron	ROSIDAE	Celastrales (within Aquifoliaceae, 223)

Sphenostemon Baill.

STACHYURACEAE J.Agardh 1858   28.07

1 genus. Himalaya to Japan. Shrubs, small trees.

B&H	POLYPETALAE, THALAMIFLORAE	Guttiferales (within Theaceae, 28)
DT&H	ARCHICHLAMYDEAE	Parietales, Flacourtiineae, 150
Melc	ARCHICHLAMYDEAE	Violales, Flacourtiineae, 185
Thor	THEIFLORAE	Theales, Theineae, 43

Dahl	THEIFLORAE	Theales, 110
Young	DILLENIIDAE, DILLENIANAE	Theales, 105
Takh	DILLENIIDAE, THEANAE	Theales, 101
Cron	DILLENIIDAE	Violales, 114

Stachyurus Siebold & Zucc.

STACKHOUSIACEAE R.Br. 1814   48

3 genera. Malesia to Australia, New Zealand, Pacific. Herbs.

B&H	POLYPETALAE, DISCIFLORAE	Celastrales, 48
DT&H	ARCHICHLAMYDEAE	Sapindales, Celastrineae, 110
Melc	ARCHICHLAMYDEAE	Celastrales, Celastrineae, 162
Thor	SANTALALES	Celastrales, 114
Dahl	SANTALIFLORAE	Celastrales, 250
Young	ROSIDAE, SANTALANAE	Celastrales, 231
Takh	ROSIDAE, CELASTRANAE	Celastrales, Celastrineae, 307
Cron	ROSIDAE	Celastrales, 221

Macgregoria F.Muell.
Stackhousia Sm.
Tripterococcus Endl.

STAPHYLEACEAE (DC.) Lindl. 1829   51.05

5 genera. N. temp. region, trop. Asia and America. Trees, shrubs.

B&H	POLYPETALAE, DISCIFLORAE	Sapindales (within Sapindaceae, 51)
DT&H	ARCHICHLAMYDEAE	Sapindales, Celastrineae, 111
Melc	ARCHICHLAMYDEAE	Celastrales, Celastrineae, 160
Thor	ROSIFLORAE	Rosales, Cunoniineae, 220
Dahl	RUTIFLORAE	Sapindales, 213
Young	ROSIDAE, RUTANAE	Sapindales, 258
Takh	ROSIDAE, RUTANAE	Sapindales, 243
	(separating *Tapisceaceae* 244)	
Cron	ROSIDAE	Sapindales, 248

Euscaphis Siebold & Zucc.   **Tapiscia** Oliv.
Huertea Ruíz & Pav.   **Turpinia** Vent.
Staphylea L.

STEGNOSPERMATACEAE (A.Rich.) Nakai 1942   132.04

1 genus. Mexico, C. America, W. Indies. Semi-scandent trees or lianes.

B&H	MONOCHLAMYDEAE	Curvembryeae (within Phytolaccaceae, 132)
DT&H	ARCHICHLAMYDEAE	Centrospermae, Phytolaccineae (within Phytolaccaceae, 33)
Melc	ARCHICHLAMYDEAE	Centrospermae, Phytolaccineae (within Phytolaccaceae, 26)

Thor CHENOPODIIFLORAE Chenopodiales, Chenopodiineae, 84
Dahl CARYOPHYLLIFLORAE Caryophyllales, 42
Young DILLENIIDAE, CARYOPHYLLANAE Chenopodiales, 153
Takh CARYOPHYLLIDAE Caryophyllales, 64
Cron CARYOPHYLLIDAE Caryophyllales (within Phytolaccaceae, 64)

Stegnosperma Benth.

STERCULIACEAE (DC.) Bartl. 1830 32

67 genera. Trop. and subtrop., occas. temp. Trees, shrubs, some herbs.

B&H POLYPETALAE, THALAMIFLORAE Malvales, 32
DT&H ARCHICHLAMYDEAE Malvales, Malvineae, 128
   (including Huaceae; separating *Triplochitonaceae* 126)
Melc ARCHICHLAMYDEAE Malvales, Malvineae, 175
Thor MALVIFLORAE Malvales, 144
Dahl MALVIFLORAE Malvales, 56
Young DILLENIIDAE, MALVANAE Malvales, 164
Takh DILLENIIDAE, MALVANAE Malvales, 180
Cron DILLENIIDAE Malvales, 100

Abroma Jacq.
Acropogon Schltr.
Aethiocarpa Vollesen
Astiria Lindl.
Ayenia L.
Brachychiton Schott & Endl.
Byttneria Loefl.
Cheirolaena Benth.
Chiranthodendron Larreat.
Cola Schott & Endl.
Commersonia J.R.Forst. & G.Forst.
Cotylonychia Stapf
Dicarpidium F.Muell.
Dombeya Cav.
Eriolaena DC.
Firmiana Marsili
Franciscodendron B.Hyland & Steenis
Fremontodendron Coville
Gilesia F.Muell.
Glossostemon Desf.
Guazuma Mill.
Guichenotia J.Gay
Hannafordia F.Muell.
Harmsia K.Schum.
Helicteres L.
Helmiopsiella Arènes
Helmiopsis H.Perrier
Heritiera Aiton
Hermannia L.
Herrania Goudot
Hildegardia Schott & Endl.

Keraudrenia J.Gay
Kleinhovia L.
Lasiopetalum Sm.
Leptonychia Turcz.
Leptonychiopsis Ridl. *
Lysiosepalum F.Muell.
Mansonia J.R.Drumm. ex Prain
Maxwellia Baill.
Megatritheca Cristóbal
Melhania Forssk.
Melochia L.
Neoregnellia Urb.
Nesogordonia Baill.
Octolobus Welw.
Paradombeya Stapf
Paramelhania Arènes
Pentapetes L.
Pimia Seem.
Pterocymbium R.Br.
Pterospermum Schreb.
Pterygota Schott & Endl.
Rayleya Cristóbal
Reevesia Lindl.
Ruizia Cav.
Rulingia R.Br.
Scaphium Schott & Endl.
Scaphopetalum Mast.
Seringia J.Gay
Sterculia L.
Theobroma L.
Thomasia J.Gay

Triplochiton K.Schum.
Trochetia DC.
Trochetiopsis Marais

Ungeria Schott & Endl.
Waltheria L.

STILBACEAE Kunth 1831   125.03

5 genera. South Africa. Shrubs.

B&H	GAMOPETALAE, BICARPELLATAE	Lamiales (within Verbenaceae, 125)
DT&H	METACHLAMYDEAE	Tubiflorae, Verbenineae (within Verbenaceae, 206)
Melc	SYMPETALAE	Tubiflorae, Verbenineae (within Verbenaceae, 258)
Thor	GENTIANIFLORAE	Lamiales (within Verbenaceae, 265)
Dahl	LAMIIFLORAE	Scrophulariales, 355
Young	(not mentioned, presumably within Verbenaceae, 377)	
Takh	LAMIIDAE, LAMIANAE	Scrophulariales, 399
Cron	ASTERIDAE	Lamiales (within Verbenaceae, 286)

Campylostachys Kunth
Eurylobium Hochst.
Euthystachys A.DC.

Stilbe Bergius
Xeroplana Briq.

STRASBURGERIACEAE Engl. & Gilg 1924   28.12

1 genus, monotypic. New Caledonia. Tree.

B&H	Not known	—
DT&H	ARCHICHLAMYDEAE	Parietales, Theineae (within Ochnaceae, 132)
Melc	ARCHICHLAMYDEAE	Guttiferales, Ochnineae, 83
Thor	THEIFLORAE	Theales, Scytopetalineae, 63
Dahl	THEIFLORAE	Theales, 119
Young	DILLENIIDAE, DILLENIANAE	Scytopetalales, 125
Takh	DILLENIIDAE, THEANAE	Ochnales, 119
Cron	DILLENIIDAE	Theales (within Ochnaceae, 80)

Strasburgeria Baill.

STYLIDIACEAE R.Br. 1810   89.01

6 genera. Mainly Australasia, also trop. Asia and temp. S. America. Mostly herbs.

B&H	GAMOPETALAE, INFERAE	Campanales, 89
DT&H	METACHLAMYDEAE	Campanulatea, Campanulineae, 231
Melc	SYMPETALAE	Campanulales, 288
Thor	ROSIFLORAE	Rosales, Saxifragineae, 210
Dahl	CORNIFLORAE	Cornales, 317
Young	ROSIDAE, CORNANAE	Hydrangeales, 337
Takh	ASTERIDAE, CAMPANULANAE	Stylidiales, 425
	(separating *Donatiaceae* 424)	

Cron ASTERIDAE   Campanulales, 307
        (separating *Donatiaceae* 308)

Donatia J.R.Forst. & G.Forst.   **Oreostylidium** Berggr.
Forstera L.f.   **Phyllachne** J.R.Forst. & G.Forst.
Levenhookia R.Br.   **Stylidium** Sw. ex Willd.

STYLOBASIACEAE J.Agardh 1858   58.03

1 genus. W. Australia. Small shrubs.

B&H POLYPET., CALYCIFLORAE Rosales (within Rosaceae, 58)
DT&H ARCHICHLAMYDEAE   Rosales, Rosineae (within Rosaceae, 76)
Melc ARCHICHLAMYDEAE   Rosales, Rosineae (within Chrysobalanaceae, 116)
Thor RUTIFLORAE   Rutales, Sapindineae (within Sapindaceae, 180)
Dahl ROSIFLORAE   Rosales (within Surianaceae, 179)
Young ROSIDAE, RUTANAE   Sapindales (within Sapindaceae, 253)
Takh ROSIDAE, RUTANAE   Sapindales, 251
Cron ROSIDAE   Rosales (within Surianaceae, 178)

Stylobasium Desf.

STYRACACEAE Dumort. 1829   103.01

11 genera. Tropical to warm-temperate. Trees, shrubs.

B&H GAMOPETALAE, HETEROMERAE Ebenales, 103
        (including Lissocarpaceae, Symplocaceae)
DT&H METACHLAMYDEAE   Ebenales, Diospyrineae, 194
        (including Lissocarpaceae)
Melc SYMPETALAE   Ebenales, Ebenineae, 240
Thor THEIFLORAE   Ebenales, Styracineae, 76
Dahl PRIMULIFLORAE   Ebenales, 134
Young DILLENIIDAE, DILLENIANAE Ebenales, 139
Takh DILLENIIDAE, ERICANAE Ebenales, 133
Cron DILLENIIDAE   Ebenales, 150

Alniphyllum Matsum.   **Parastyrax** W.W.Sm.
Bruinsmia Boerl. & Koord.   **Pterostyrax** Siebold & Zucc.
Halesia Ellis ex L.   **Rehderodendron** Hu
Huodendron Rehder   **Sinojackia** Hu
Melliodendron Hand.–Mazz.   **Styrax** L.
Pamphilia Mart. ex A.DC.

SURIANACEAE Arn. 1834   40.03

3 genera. Widespread in tropics. Trees, shrubs.

B&H POLYPETALAE, DISCIFLORAE Geraniales (within Simaroubaceae, 40)
DT&H ARCHICHLAMYDEAE   Geraniales, Geraniineae (within
                                        Simaroubaceae, 88)

675

Melc	ARCHICHLAMYDEAE	Rutales, Rutineae (within Simaroubaceae, 133)
Thor	RUTIFLORAE	Rutales, Fabineae, 189
Dahl	ROSIFLORAE	Rosales, 179
	(including Stylobasiaceae)	
Young	ROSIDAE, RUTANAE	Sapindales, 254
Takh	ROSIDAE, RUTANAE	Rutales, 261
Cron	ROSIDAE	Rosales, 178
	(including Stylobasiaceae)	

Cadellia F.Muell.
Guilfoylia F.Muell.
Suriana L.

SYMPLOCACEAE Desf. 1820 103.02

1 genus. Trop. and subtrop. Asia, Australia, Pacific, America. Mostly trees.

B&H	GAMOPETALAE, HETEROMERAE	Ebenales (within Styracaceae, 103)
DT&H	METACHLAMYDEAE	Ebenales, Diospyrineae, 195
Melc	SYMPETALAE	Ebenales, Ebenineae, 242
Thor	THEIFLORAE	Theales, Theineae, 45
Dahl	CORNIFLORAE	Cornales, 313
Young	DILLENIIDAE, DILLENIANAE	Ebenales, 137
Takh	DILLENIIDAE, THEANAE	Theales, 108
Cron	DILLENIIDAE	Ebenales, 152

Symplocos Jacq.

TAMARICACEAE Link 1821 24.01

5 genera. Eurasia and Africa, mostly warm-temp. Shrubs, small trees.

B&H	POLYPETALAE, THALAMIFLORAE	Caryophyllineae, 24
	(including Fouquieriaceae)	
DT&H	ARCHICHLAMYDEAE	Parietales, Tamaricineae, 141
Melc	ARCHICHLAMYDEAE	Violales, Tamaricineae, 195
Thor	VIOLIFLORAE	Violales, Tamaricineae, 135
Dahl	VIOLIFLORAE	Tamaricales, 98
Young	DILLENIIDAE, VIOLANAE	Tamaricales, 205
Takh	DILLENIIDAE, VIOLANAE	Tamaricales, 155
Cron	DILLENIIDAE	Violales, 116

Hololachna Ehrenb. **Reaumuria** L.
Myricaria Desv. **Tamarix** L.
Myrtama Ovcz. & Kinzik.

TEPUIANTHACEAE Maguire & Steyerm. 1981 51.07

1 genus. Colombia, Venezuela, N. Brazil. Trees, shrubs.

B&H Not known   —
DT&H Not known   —
Melc Not known   —
Thor RUTIFLORAE   Rutales, Rutineae, 170
Dahl RUTIFLORAE   Rutales, 222
Young Not mentioned   —
Takh ROSIDAE, RUTANAE   Rutales, 269
Cron ROSIDAE   Celastrales
      (tentative placement 1981, confirmed 1988)

Tepuianthus Maguire & Steyerm.

TETRACENTRACEAE A.C.Sm. 1945   4.08

1 genus, monotypic. Himalaya to Burma and China. Tree.

B&H POLYPET., THALAMIFLORAE Ranales (within Magnoliaceae, 4)
DT&H ARCHICHLAMYDEAE   Ranales, Magnoliineae (within Magnoliaceae, 45)
Melc ARCHICHLAMYDEAE   Magnoliales, Trochodendrineae, 57
Thor HAMAMELIDIFLORAE   Hamamelidales, Trochodendrineae, 194
Dahl ROSIFLORAE   Trochodendrales, 138
Young ROSIDAE, HAMAMELIDANAE Trochodendrales, 218
Takh HAMAMELIDIDAE, TROCHODENDRANAE
                                  Trochodendrales, 77
Cron HAMAMELIDIDAE   Trochodendrales, 40

Tetracentron Oliv.

TETRACHONDRACEAE Wettst. 1924   126.02

1 genus. New Zealand, Chile, Argentina. Creeping herbs.

B&H Not known   —
DT&H INCERTAE SEDIS   —
Melc SYMPETALAE   Tubiflorae, Verbenineae (within Labiatae, 260)
Thor GENTIANIFLORAE   Lamiales (within Labiatae, 266)
Dahl (not mentioned, presumably included in Lamiaceae, 365)
Young (not mentioned, presumably within Labiatae)
Takh LAMIIDAE, LAMIANAE   Lamiales, Lamiineae (within Labiatae, 414)
Cron ASTERIDAE   Lamiales (within Labiatae, 287)

Tetrachondra Petrie ex Oliv.

TETRAMERISTACEAE Hutch. 1959   28.10

2 genera. W. Malesia (*Tetramerista*), Guyana Highlands (*Pentamerista*). Trees, shrubs.

B&H POLYPETALAE, DISCIFLORAE Geraniales (within Ochnaceae, 41)
DT&H ARCHICHLAMYDEAE   Parietales, Theineae (within Theaceae, 136)
Melc ARCHICHLAMYDEAE   Guttiferales, Theineae (within Theaceae, 85)

Thor THEIFLORAE Theales, Theineae (within Theaceae, 44)
Dahl THEIFLORAE Theales (within Theaceae, 124)
Young (not mentioned, presumably within Theaceae, 106)
Takh DILLENIIDAE, THEANAE Theales, 107
Cron DILLENIIDAE Theales, 89

Pentamerista Maguire
Tetramerista Miq.

THEACEAE D.Don 1825 28.01

25 genera. Tropical to warm-temperate. Mostly trees, shrubs.

B&H POLYPETALAE, THALAMIFLORAE Guttiferales (as *Ternstroemiaceae*) 28
(including Actinidiaceae, Caryocaraceae, Marcgraviaceae, Pellicieraceae, Pentaphylacaceae, Stachyuraceae)
DT&H ARCHICHLAMYDEAE Parietales, Theineae, 136
(including Asteropeiaceae, Pellicieraceae, Tetrameristaceae)
Melc ARCHICHLAMYDEAE Guttiferales, Theineae, 85
(including Asteropeiaceae, Pellicieraceae, Tetrameristaceae)
Thor THEIFLORAE Theales, Theineae, 44
(including Asteropeiaceae, Pellicieraceae, Tetrameristaceae)
Dahl THEIFLORAE Theales, 124
(including Pellicieraceae, Tetrameristaceae)
Young DILLENIIDAE, DILLENIANAE Theales, 106
(presumably including Asteropeiaceae, Pellicieraceae, Tetrameristaceae)
Takh DILLENIIDAE, THEANAE Theales, 102
(separating *Sladeniaceae* 103)
Cron DILLENIIDAE Theales, 85
(including Asteropeiaceae)

Adinandra Jack
Anneslea Wall.
Apterosperma H.T.Chang *
Archboldiodendron Kobuski
Balthasaria Verdc.
Camellia L.
Cleyera Thunb.
Dankia Gagnep.
Eurya Thunb.
Ficalhoa Hiern
Franklinia W.Bartr. ex Marshall
Freziera Willd.
Gordonia Ellis

Killipiodendron Kobuski *
Laplacea Kunth
Paranneslea Gagnep.
Pyrenaria Blume
Schima Reinw. ex Blume
Sladenia Kurz
Stuartia L.
Symplococarpon Airy Shaw
Ternstroemia Mutis ex L.f.
Ternstroemiopsis Urb.
Tutcheria Dunn
Visnea L.f.

THELIGONACEAE Dumort. 1829 84.03

1 genus. Mediterranean to Japan. Herb.

B&H MONOCHLAMYDEAE Unisexuales (within Urticaceae, 153)
DT&H ARCHICHLAMYDEAE Centrospermae, Phytolaccineae (as
Cynocrambaceae) 39

Melc	ARCHICHLAMYDEAE	Myrtiflorae, Myrtineae, 217
Thor	GENTIANIFLORAE	Gentianales (within Rubiaceae, 251)
Dahl	GENTIANIFLORAE	Gentianales, 342
Young	ROSIDAE, GENTIANANAE	Gentianales (within Rubiaceae, 349)
Takh	LAMIIDAE, GENTIANANAE	Gentianales, 368
Cron	ASTERIDAE	Rubiales, 312

Theligonum L.

THEOPHRASTACEAE Link 1829   100.02

4 genera. Tropical America. Trees, shrubs.

B&H GAMOPETALAE, HETEROMERAE Primulales (within Myrsinaceae, 100)
DT&H METACHLAMYDEAE   Primulales, 188
Melc SYMPETALAE   Primulales, 233
Thor THEIFLORAE   Primulales, Primulineae (within Myrsinaceae, 77)
Dahl PRIMULIFLORAE   Primulales, 130
Young DILLENIIDAE, DILLENIANAE Primulales, 141
Takh DILLENIIDAE, ERICANAE   Primulales, 137
Cron DILLENIIDAE   Primulales, 153

Clavija Ruíz & Pav.   **Jacquinia** L.
Deherainia Decne.   **Theophrasta** L.

THYMELAEACEAE Juss. 1789   145.01

58 genera. Widespread. Trees, shrubs, lianes, herbs.

B&H MONOCHLAMYDEAE   Daphnales, 145
DT&H ARCHICHLAMYDEAE   Myrtiflorae, Thymelaeineae, 164
     (separating *Gonystylaceae* 123 to Malvales Malvineae)
Melc ARCHICHLAMYDEAE   Thymelaeales, 180
Thor MALVIFLORAE   Euphorbiales, 166
Dahl MALVIFLORAE   Thymelaeales, 80
     (? separating *Gonystylidaceae* 81)
Young DILLENIIDAE, MALVANAE   Euphorbiales, 184
Takh DILLENIIDAE, EUPHORBIANAE Thymelaeales, 194
Cron ROSIDAE   Myrtales, 192

Aetoxylon (Airy Shaw) Airy Shaw   **Dendrostellera** (C.A.Mey.) Tiegh.
Amyxa Tiegh.   **Diarthron** Turcz.
Aquilaria Lam.   **Dicranolepis** Planch.
Arnhemia Airy Shaw   **Dirca** L.
Basutica E.Phillips   **Drapetes** Banks ex Lam.
Craspedostoma Domke *   **Edgeworthia** Meisn.
Craterosiphon Engl. & Gilg   **Englerodaphne** Gilg
Cryptadenia Meisn.   **Enkleia** Griff.
Dais L.   **Eriosolena** Blume
Daphne L.   **Funifera** Leandro ex C.A.Mey.
Daphnimorpha Nakai *   **Gnidia** L.
Daphnopsis Mart.   **Gonystylus** Teijsm. & Binn.
Deltaria Steenis   **Goodallia** Benth.

Gyrinops Gaertn.
Jedda J.R.Clarkson
Kelleria Endl.
Lachnaea L.
Lagetta Juss.
Lasiadenia Benth.
Lethedon Spreng.
Linodendron Griseb.
Linostoma Wall. ex Endl.
Lophostoma (Meisn.) Meisn.
Octolepis Oliv.
Oreodendron C.T.White
Ovidia Meisn.
Passerina L.
Peddiea Harv. ex Hook.
Pentathymelaea Lecomte *

Phaleria Jack
Pimelea Banks & Sol.
Pseudais Decne.
Restella Pobed. *
Rhamnoneuron Gilg
Schoenobiblus Mart.
Solmsia Baill.
Stellera L.
Stelleropsis Pobed.
Stephanodaphne Baill.
Struthiola L.
Synandrodaphne Gilg
Synaptolepis Oliv.
Thecanthes Wikstr.
Thymelaea Mill.
Wikstroemia Endl.

TICODENDRACEAE Gómez–Laur. & L.D.Gómez 1991   153.08

1 genus, monotypic. Costa Rica, Nicaragua, Panama. Tree.

B&H	Not known	—
DT&H	Not known	—
Melc	Not known	—
Thor	Not known	—
Dahl	Not known	—
Young	Not known	—
Takh	Not known	—
Cron	Not known	—

Ticodendron Gómez–Laur. & L.D.Gómez

TILIACEAE Juss. 1789   33.01

53 genera. Widespread. Trees, shrubs, some herbs.

B&H	POLYPETALAE, THALAMIFLORAE	Malvales, 33
	(including Elaeocarpaceae, Plagiopteraceae)	
DT&H	ARCHICHLAMYDEAE	Malvales, Malvineae, 124
Melc	ARCHICHLAMYDEAE	Malvales, Malvineae, 172
Thor	MALVIFLORAE	Malvales, 148
Dahl	MALVIFLORAE	Malvales, 65
Young	DILLENIIDAE, MALVANAE	Malvales, 168
Takh	DILLENIIDAE, MALVANAE	Malvales, 175
Cron	DILLENIIDAE	Malvales, 99

Ancistrocarpus Oliv.
Apeiba Aubl.
Asterophorum Sprague
Berrya Roxb.
Brownlowia Roxb.
Burretiodendron Rehder

Carpodiptera Griseb.
Christiana DC.
Clappertonia Meisn.
Colona Cav.
Corchoropsis Siebold & Zucc.
Corchorus L.

Craigia W.W.Sm. & W.E.Evans
Desplatsia Bocquillon
Dicraspidia Standl.
Diplodiscus Turcz.
Duboscia Bocquet
Eleutherostylis Burret
Entelea R.Br.
Erinocarpus Nimmo ex J.Graham
Glyphaea Hook.f.
Goethalsia Pittier
Grewia L.
Hainania Merr.
Hasseltia Kunth
Hasseltiopsis Sleumer
Heliocarpus L.
Hydrogaster Kuhlm.
Jarandersonia Kosterm.
Luehea Willd.
Lueheopsis Burret
Macrohasseltia L.O.Williams
Microcos L.

Mollia Mart.
Mortoniodendron Standl. & Steyerm.
Muntingia L.
Neotessmannia Burret *
Pentace Hassk.
Pentaplaris L.O.Williams & Standl. *
Petenaea Lundell
Pityranthe Thwaites
Pleuranthodendron L.O.Williams
Prockia P.Browne ex L.
Pseudocorchorus Capuron
Schoutenia Korth.
Sicrea (Pierre) Hallier f.
Sparrmannia L.f.
Tetralix Griseb.
Tilia L.
Trichospermum Blume
Triumfetta L.
Vasivaea Baill.
Westphalina A.Robyns & Bamps

TORRICELLIACEAE (Wanger.) H.H.Hu 1934   82.06

1 genus. Himalaya, W. China. Small trees.

B&H	POLYPETALAE, CALYCIFLORAE	Umbellales (within Cornaceae, 82)
DT&H	ARCHICHLAMYDEAE	Umbelliflorae (within Cornaceae, 181)
Melc	ARCHICHLAMYDEAE	Umbelliflorae (within Cornaceae, 223)
Thor	CORNIFLORAE	Araliales, 289
Dahl	CORNIFLORAE	Cornales, 305
Young	Not mentioned, presumably within Cornaceae	
Takh	ROSIDAE, CORNANAE	Torricelliales, 349
Cron	ROSIDAE	Cornales (within Cornaceae, 203)

Torricellia DC.

TOVARIACEAE Pax 1891   12.02

1 genus. New World tropics and warm-temperate. Shrubs, herbs.

B&H	POLYPETALAE, THALAMIFLORAE	Parietales (within Capparaceae, 12)
DT&H	ARCHICHLAMYDEAE	Rhoeadales, Capparidineae, 56
Melc	ARCHICHLAMYDEAE	Papaverales, Capparineae, 97
Thor	VIOLIFLORAE	Capparales (within Capparaceae, 142)
Dahl	VIOLIFLORAE	Capparales, 102
Young	DILLENIIDAE, VIOLANAE	Capparales, 208
Takh	DILLENIIDAE, VIOLANAE	Capparales, Capparineae, 165
Cron	DILLENIIDAE	Capparales, 132

Tovaria Ruíz & Pav.

TRAPACEAE Dumort. 1829   70.02

1 genus. Warm-temperate to tropical Eurasia, Africa. Aquatic herbs.

B&H	POLYPETALAE, CALYCIFLORAE	Myrtales (within Onagraceae, 70)
DT&H	ARCHICHLAMYDEAE	Myrtiflorae, Myrtineae (as *Hydrocaryaceae)* 176
Melc	ARCHICHLAMYDEAE	Myrtiflorae, Myrtineae, 204
Thor	MYRTIFLORAE	Myrtales, Lythrineae, 239
Dahl	MYRTIFLORAE	Myrtales, 189
Young	ROSIDAE, MYRTANAE	Myrtales, 313
Takh	ROSIDAE, MYRTANAE	Myrtales, Trapineae, 239
Cron	ROSIDAE	Myrtales, 193

Trapa L.

TREMANDRACEAE R.Br. ex DC. 1824   19

3 genera. Australia. Small shrubs, often ericoid.

B&H	POLYPETALAE, THALAMIFLORAE	Polygalineae, 19
DT&H	ARCHICHLAMYDEAE	Geraniales, Polygalineae, 94
Melc	ARCHICHLAMYDEAE	Rutales, Polygalineae, 141
Thor	ROSIFLORAE	Pittosporales, Pittosporineae, 228
Dahl	ARALIIFLORAE	Pittosporales, 265
Young	ROSIDAE, GERANIANAE	Polygalales, 279
Takh	ROSIDAE, RUTANAE	Polygalales, 297
Cron	ROSIDAE	Polygalales, 244

Platytheca Steetz
Tetratheca Sm.
Tremandra R.Br. ex DC.

TRIGONIACEAE Endl. 1841   20.03

4 genera. Madagascar, W. Malesia, trop. America. Trees, shrubs, lianes.

B&H	POLYPETALAE, THALAMIFLORAE	Polygalineae (within Vochysiaceae, 20a)
DT&H	ARCHICHLAMYDEAE (? including Euphroniaceae)	Geraniales, Malpighiineae, 92
Melc	ARCHICHLAMYDEAE (including Euphroniaceae)	Rutales, Polygalineae, 139
Thor	GERANIIFLORAE	Geraniales, Polygalineae, 110
Dahl	RUTIFLORAE	Polygalales, 226
Young	ROSIDAE, GERANIANAE	Polygalales, 281
Takh	ROSIDAE, RUTANAE (including Euphroniaceae)	Polygalales, 293
Cron	ROSIDAE	Polygalales, 243

Humbertiodendron Leandri   **Trigoniastrum** Miq.
Trigonia Aubl.   **Trigoniodendron** E.F.Guim. & Miguel

TRIMENIACEAE (Perkins & Gilg) Gibbs 1917   142.02

1 genus. Malesia to Australia and Pacific. Trees, shrubs.

B&H	MONOCHLAMYDEAE	Micrembryeae (within Monimiaceae, 142)
DT&H	ARCHICHLAMYDEAE	Ranales, Magnoliineae (within Monimiaceae, 51)
Melc	ARCHICHLAMYDEAE	Magnoliales, Laurineae, 50
Thor	ANNONIFLORAE	Annonales, Laurineae, 14
Dahl	MAGNOLIIFLORAE	Laurales, 18
Young	MAGNOLIIDAE, RANUNCULANAE	Laurales, 19
Takh	MAGNOLIIDAE, MAGNOLIANAE	Laurales, 13
Cron	MAGNOLIIDAE	Laurales, 12

Trimenia Seem.

TRIPLOSTEGIACEAE (Höck) Bobrov ex Airy Shaw 1965   86.03

1 genus. Himalaya to China and New Guinea. Herbs.

B&H	GAMOPETALAE, INFERAE	Asterales (within Dipsacaceae, 86)
DT&H	METACHLAMYDEAE	Rubiales (within Valerianaceae, 226)
Melc	SYMPETALAE	Dipsacales (within Valerianaceae 281)
Thor	CORNIFLORAE	Dipsacales (within Dipsacaceae, 294)
Dahl	CORNIFLORAE	Dipsacales, 331
Young	Not mentioned	—
Takh	ROSIDAE, CORNANAE	Dipsacales, 360
Cron	ASTERIDAE	Dipsacales (within Valerianaceae, 315)

Triplostegia Wall. ex DC.

TROCHODENDRACEAE Prantl 1888   4.07

1 genus, monotypic. Korea, Japan, Taiwan. Tree.

B&H	POLYPET., THALAMIFLORAE	Ranales (within Magnoliaceae, 4)
DT&H	ARCHICHLAMYDEAE	Ranales, Trochodendrineae, 40
	(including Cercidiphyllaceae, Eucommiaceae, Eupteleaceae)	
Melc	ARCHICHLAMYDEAE	Magnoliales, Trochodendrineae, 58
Thor	HAMAMELIDIFLORAE	Hamamelidales, Trochodendrineae, 193
Dahl	ROSIFLORAE	Trochodendrales, 137
Young	ROSIDAE, HAMAMELIDANAE	Trochodendrales, 216
Takh	HAMAMELIDIDAE, TROCHODENDRANAE	
		Trochodendrales, 76
Cron	HAMAMELIDIDAE	Trochodendrales, 41

Trochodendron Siebold & Zucc.

TROPAEOLACEAE Juss. ex DC. 1824   38.02

3 genera. Mexico to temp. S. America. Herbs, mostly climbing.

B&H	POLYPETALAE, DISCIFLORAE	Geraniales (within Geraniaceae, 38)
DT&H	ARCHICHLAMYDEAE	Geraniales, Geraniineae, 81

Melc	ARCHICHLAMYDEAE	Geraniales, Geraniineae, 125
Thor	GERANIIFLORAE	Geraniales, Geraniineae, 105
Dahl	RUTIFLORAE	Tropaeolales, 248
Young	DILLENIIDAE, VIOLANAE	Tropaeolales, 214
Takh	ROSIDAE, RUTANAE	Tropaeolales, 290
Cron	ROSIDAE	Geraniales, 266

Magallana Cav.
Tropaeolum L.
Trophaeastrum Sparre *

TURNERACEAE Kunth ex DC. 1828   73

10 genera. Trop. and warm-temp. America, Africa, Indian Ocean. Trees, shrubs, herbs.

B&H	POLYPET., CALYCIFLORAE	Passiflorales, 73
DT&H	ARCHICHLAMYDEAE	Parietales, Flacourtiineae, 151
Melc	ARCHICHLAMYDEAE	Violales, Flacourtiineae, 187
Thor	VIOLIFLORAE	Violales, Violineae, 130
Dahl	VIOLIFLORAE	Violales, 90
Young	DILLENIIDAE, VIOLANAE	Violales, 195
Takh	DILLENIIDAE, VIOLANAE	Violales, 151
Cron	DILLENIIDAE	Violales, 120

Adenoa Arbo *
Erblichia Seem.
Hyalocalyx Rolfe
Loewia Urb.
Mathurina Balf.f.

Piriqueta Aubl.
Stapfiella Gilg
Streptopetalum Hochst.
Tricliceras Thonn. ex DC.
Turnera L.

ULMACEAE Mirb. 1815   153.02

15 genera. Widespread. Trees, shrubs.

B&H	MONOCHLAMYDEAE	Unisexuales (within Urticaceae, 153)
DT&H	ARCHICHLAMYDEAE (including Barbeyaceae)	Urticales, 13
Melc	ARCHICHLAMYDEAE (including Barbeyaceae)	Urticales, 11
Thor	MALVIFLORAE	Urticales, 156
Dahl	MALVIFLORAE	Urticales, 69
Young	DILLENIIDAE, MALVANAE	Urticales, 176
Takh	DILLENIIDAE, URTICANAE	Urticales, Ulmineae, 184
Cron	HAMAMELIDIDAE	Urticales, 51

Ampelocera Klotzsch
Aphananthe Planch.
Celtis L.
Chaetachme Planch.
Gironniera Gaudich.
Hemiptelea Planch.
Holoptelea Planch.
Lozanella Greenm.

Parasponia Miq.
Phyllostylon Capan. ex Benth.
Planera J.F.Gmel.
Pteroceltis Maxim.
Trema Lour.
Ulmus L.
Zelkova Spach

UMBELLIFERAE Juss. 1789   80

428 genera. Widespread. Herbs to rarely (e.g. *Steganotaenia)* trees.

B&H	POLYPETALAE, CALYCIFLORAE	Umbellales, 80
DT&H	ARCHICHLAMYDEAE	Umbelliflorae, 180
Melc	ARCHICHLAMYDEAE	Umbelliflorae, 226
Thor	CORNIFLORAE	Araliales (within Araliaceae, 290)
Dahl	ARALIIFLORAE	Araliales, 268
Young	ROSIDAE, CORNANAE	Araliales, 343
Takh	ROSIDAE, CORNANAE	Apiales, 352
Cron	ROSIDAE	Apiales, 269

Aciphylla J.R.Forst. & G.Forst.
Acronema Falc. ex Edgew.
Actinanthus Ehrenb.
Actinolema Fenzl
Actinotus Labill.
Adenosciadium H.Wolff
Aegokeras Raf.
Aegopodium L.
Aethusa L.
Aframmi C.Norman
Afrocarum Rauschert
Afroligusticum C.Norman
Afrosison H.Wolff
Agasyllis Spreng.
Agrocharis Hochst.
Ainsworthia Boiss.
Alepidea F.Delaroche
Aletes J.M.Coult. & Rose
Alococarpum Riedl & Kuber
Alposelinum Pimenov *
Ammi L.
Ammiopsis Boiss.
Ammodaucus Coss. & Durieu
Ammoides Adans.
Ammoselinum Torr. & A.Gray
Anethum L.
Angelica L.
Angelocarpa Rupr.
Anginon Raf.
Angoseseli Chiov.
Anisopoda Baker
Anisosciadium DC.
Anisotome Hook.f.
Annesorhiza Cham. & Schltdl.
Anthriscus Pers.
Aphanopleura Boiss.
Apiastrum Nutt.
Apium L.
Apodicarpum Makino
Arafoe Pimenov & Lavrova
Arctopus L.
Arracacia Bancr.

Artedia L.
Asciadium Griseb.
Asteriscium Cham. & Schltdl.
Astomaea Rchb.
Astrantia L.
Astrodaucus Drude
Astydamia DC.
Athamanta L.
Aulacospermum Ledeb.
Austropeucedanum Mathias &
    Constance
Autumnalia Pimenov *
Azilia Hedge & Lamond
Azorella Lam.
Berula Besser & W.D.J.Koch
Bifora Hoffm.
Bilacunaria Pimenov & V.N.Tikhom.
Bolax Comm. ex Juss.
Bonannia Guss.
Bowlesia Ruíz & Pav.
Bunium L.
Bupleurum L.
Cachrys L.
Calyptrosciadium Rech.f. & Kuber *
Capnophyllum Gaertn.
Carlesia Dunn
Caropsis (Rouy & Camus) Rauschert
Carum L.
Caucalis L.
Cenolophium W.D.J.Koch ex DC.
Centella L.
Cephalopodium Korovin *
Chaerophyllopsis Boissieu
Chaerophyllum L.
Chaetosciadium Boiss.
Chamaele Miq.
Chamaesciadium C.A.Mey.
Chamaesium H.Wolff
Chamarea Eckl. & Zeyh.
Changium H.Wolff *
Chlaenosciadium C.Norman
Choritaenia Benth.

Chuanminshen M.L.Sheh & R.H.Shan *
Chymsydia Albov
Cicuta L.
Cnidiocarpa Pimenov *
Cnidium Cusson ex Juss.
Coaxana J.M.Coult. & Rose
Conioselinum Fisch. ex Hoffm.
Conium L.
Conopodium W.D.J.Koch
Coriandrum L.
Cortia DC.
Cortiella C.Norman
Cotopaxia Mathias & Constance
Coulterophytum B.L.Rob.
Coxella Cheeseman & Hemsl.
Crithmum L.
Cryptotaenia DC.
Cuminum L.
Cyathoselinum Benth.
Cyclorhiza M.L.Sheh & R.H.Shan *
Cyclospermum Lag.
Cymbocarpum DC. ex C.A.Mey.
Cymopterus Raf.
Cynosciadium DC.
Dactylaea Fedde ex H.Wolff *
Dasispermum Raf.
Daucosma Engelm. & A.Gray
Daucus L.
Demavendia Pimenov
Dethawia Endl.
Dichosciadium Domin
Dickinsia Franch.
Dicyclophora Boiss.
Dimorphosciadium Pimenov
Diplaspis Hook.f.
Diplolophium Turcz.
Diplotaenia Boiss.
Diposis DC.
Distichoselinum Garc.Mart. & Silvestre
Domeykoa Phil.
Donnellsmithia J.M.Coult. & Rose
Dorema D.Don
Dracosciadium Hilliard & B.L.Burtt
Drusa DC.
Ducrosia Boiss.
Dystaenia Kitag. *
Echinophora L.
Elaeoselinum W.D.J.Koch ex DC.
Elaeosticta Fenzl
Eleutherospermum K.Koch
Enantiophylla J.M.Coult. & Rose
Endressia J.Gay
Eremocharis Phil.
Eremodaucus Bunge

Ergocarpon C.C.Towns.
Erigenia Nutt.
Eriocycla Lindl.
Eriosynaphe DC.
Eryngium L.
Erythroselinum Chiov.
Eurytaenia Torr. & A.Gray
Exoacantha Labill.
Falcaria Fabr.
Fergania Pimenov *
Ferula L.
Ferulago W.D.J.Koch
Ferulopsis Kitag. *
Foeniculum Mill.
Frommia H.Wolff
Froriepia K.Koch
Fuernrohria K.Koch
Galagania Lipsky
Geocaryum Coss.
Gingidia J.W.Dawson
Glaucosciadium B.L.Burtt & P.H.Davis
Glehnia F.Schmidt ex Miq.
Glia Sond.
Glochidotheca Fenzl
Gongylosciadium Rech.f.
Grafia Rchb.
Grammosciadium DC.
Guillonea Coss.
Gymnophyton Clos
Hacquetia Neck. ex DC.
Halosciastrum Koidz. *
Haplosciadium Hochst.
Haplosphaera Hand.–Mazz.
Harbouria J.M.Coult. & Rose
Harrysmithia H.Wolff
Haussknechtia Boiss.
Hellenocarum H.Wolff
Heptaptera Margot & Reut.
Heracleum L.
Hermas L.
Heteromorpha Cham. & Schltdl.
Hladnikia Rchb.
Hohenackeria Fisch. & C.A.Mey.
Homalocarpus Hook. & Arn.
Homalosciadium Domin
Horstrissea Greuter, Gerstb. & Egli *
Huanaca Cav.
Hyalolaena Bunge *
Hydrocotyle L.
Hymenolaena DC.
Imperatoria L.
Itasina Raf.
Johrenia DC.
Johreniopsis Pimenov

Kadenia Lavrova & V.N.Tikhom. *
Kafirnigania Kamelin & Kinzik. *
Kalakia Alava
Kandaharia Alava *
Karatavia Pimenov & Lavrova *
Karnataka P.K.Mukh. & Constance
Kedarnatha P.K.Mukh. & Constance
Keraymonia Farille
Kitagawia Pimenov
Klotzschia Cham.
Komarovia Korovin *
Korshinskia Lipsky
Kosopoljanskia Korovin *
Kozlovia Lipsky
Krasnovia Popov ex Schischk.
Krubera Hoffm.
Kundmannia Scop.
Ladyginia Lipsky
Lagoecia L.
Lalldhwojia Farille *
Laretia Gillies & Hook.
Laser P.Gaertn., B.Mey. & Scherb.
Laserpitium L.
Lecokia DC.
Ledebouriella H.Wolff
Lefebvrea A.Rich.
Leibergia J.M.Coult. & Rose
Leutea Pimenov
Levisticum Hill
Lichtensteinia Cham. & Schltdl.
Lignocarpa J.W.Dawson
Ligusticella J.M.Coult. & Rose
Ligusticopsis Leute
Ligusticum L.
Lilaeopsis Greene
Limnosciadium Mathias & Constance
Lipskya (Kozo–Pol.) Nevski *
Lisaea Boiss.
Lomatium Raf.
Lomatocarpa Pimenov *
Magadania Pimenov & Lavrova *
Magydaris W.D.J.Koch ex DC.
Malabaila Hoffm.
Mandenovia Alava *
Margotia Boiss.
Marlothiella H.Wolff
Mastigosciadium Rech.f. & Kuber *
Mathiasella Constance & C.Hitchc.
Mediasia Pimenov
Meeboldia H.Wolff *
Melanosciadium Boissieu
Melanoselinum Hoffm.
Meum Mill.
Micropleura Lag.
Microsciadium Boiss.

Mogoltavia Korovin
Molopospermum W.D.J.Koch
Monizia Lowe
Mulinum Pers.
Musineon Raf.
Myrrhidendron J.M.Coult. & Rose
Myrrhis Mill.
Myrrhoides Heist. ex Fabr.
Naufraga Constance & Cannon
Neoconopodium (Kozo–Pol.) Pimenov
  & Kljuykov
Neogoezia Hemsl.
Neonelsonia J.M.Coult. & Rose
Neoparrya Mathias
Neosciadium Domin
Niphogeton Schltdl.
Nirarathamnos Balf.f.
Nothosmyrnium Miq.
Notiosciadium Speg.
Notopterygium Boissieu
Oedibasis Kozo–Pol.
Oenanthe L.
Oligocladus Chodat & Wilczek
Oliveria Vent.
Opopanax W.D.J.Koch
Oreocome Edgew.
Oreomyrrhis Endl.
Oreonana Jeps.
Oreoschimperella Rauschert
Oreoxis Raf.
Orlaya Hoffm.
Ormopterum Schischk. *
Ormosciadium Boiss.
Ormosolenia Tausch
Orogenia S.Watson
Oschatzia Walp.
Osmorhiza Raf.
Ottoa Kunth
Oxypolis Raf.
Pachyctenium Maire & Pamp.
Pachypleurum Ledeb.
Palimbia Besser ex DC.
Pancicia Vis. & Schltdl.
Paraligusticum V.N.Tikhom.
Paraselinum H.Wolff *
Parasilaus Leute *
Pastinaca L.
Pastinacopsis Golosk. *
Paulita Soják
Pedinopetalum Urb. & H.Wolff *
Perideridia Rchb.
Perissocoeleum Mathias & Constance
Petagnia Guss.
Petroedmondia Tamamsch.
Petroselinum Hill

687

Peucedanum L.
Phellolophium Baker
Phlojodicarpus Turcz. ex Ledeb.
Phlyctidocarpa Cannon & Theobald
Physospermopsis H.Wolff
Physospermum Cusson ex Juss.
Physotrichia Hiern
Pilopleura Schischk.
Pimpinella L.
Pinacantha Gilli *
Pinda P.K.Mukh. & Constance
Pituranthos Viv.
Platysace Bunge
Pleurospermopsis C.Norman
Pleurospermum Hoffm.
Podistera S.Watson
Polemannia Eckl. & Zeyh.
Polemanniopsis B.L.Burtt
Polylophium Boiss.
Polytaenia DC.
Polyzygus Dalzell
Portenschlagiella Tutin
Pozoa Lag.
Prangos Lindl.
Prionosciadium S.Watson
Psammogeton Edgew.
Pseudocarum C.Norman
Pseudocymopterus J.M.Coult. & Rose
Pseudorlaya (Murb.) Murb.
Pseudoselinum C.Norman
Pseudotaenidia Mack. *
Pternopetalum Franch.
Pterygopleurum Kitag. *
Pteryxia (Nutt.) J.M.Coult. & Rose
Ptilimnium Raf.
Ptychotis W.D.J.Koch
Pycnocycla Lindl.
Pyramidoptera Boiss.
Registaniella Rech.f.
Rhabdosciadium Boiss.
Rhodosciadium S.Watson
Rhopalosciadium Rech.f. *
Rhysopterus J.M.Coult. & Rose
Ridolfia Moris
Rouya Coincy
Rumia Hoffm.
Rupiphila Pimenov & Lavrova *
Rutheopsis A.Hansen & Kunkel
Sajanella Soják *
Sanicula L.
Saposhnikovia Schischk.
Scaligeria DC.
Scandia J.W.Dawson
Scandix L.

Schizeilema (Hook.f.) Domin
Schoenolaena Bunge
Schrenkia Fisch. & C.A.Mey.
Schtschurowskia Regel & Schmalh. *
Schulzia Spreng.
Sclerochorton Boiss.
Sclerotiaria Korovin *
Selinopsis Coss. & Durieu ex Batt. &
   Trab.
Selinum L.
Semenovia Regel & Herder
Seseli L.
Seselopsis Schischk. *
Shoshonea Evert & Constance
Silaum Mill.
Sinocarum H.Wolff
Sinodielsia H.Wolff *
Sinolimprichtia H.Wolff
Sison L.
Sium L.
Smyrniopsis Boiss.
Smyrnium L.
Sonderina H.Wolff
Spananthe Jacq.
Spermolepis Raf.
Sphaenolobium Pimenov
Sphaerosciadium Pimenov & Kljuykov *
Sphallerocarpus Besser ex DC.
Sphenocarpus Korovin *
Sphenosciadium A.Gray
Spuriodaucus C.Norman
Spuriopimpinella Kitag. *
Stefanoffia H.Wolff
Steganotaenia Hochst.
Stenocoelium Ledeb.
Stewartiella Nasir *
Stoibrax Raf.
Symphyoloma C.A.Mey.
Szovitsia Fisch. & C.A.Mey.
Taenidia (Torr. & A.Gray) Drude
Tamamschjania Pimenov & Kljuykov *
Tauschia Schltdl.
Tetrataenium (DC.) Manden.
Thamnosciadium Hartvig *
Thapsia L.
Thaspium Nutt.
Thecocarpus Boiss.
Tilingia Regel
Tinguarra Parl.
Todaroa Parl.
Tommasinia Bertol.
Tongoloa H.Wolff
Tordyliopsis DC.
Tordylium L.

Torilis Adans.
Tornabenea Parl.
Trachydium Lindl.
Trachymene Rudge
Trachyspermum Link
Trepocarpus Nutt. ex DC.
Tricholaser Gilli
Trigonosciadium Boiss.
Trinia Hoffm.
Trochiscanthes W.D.J.Koch
Turgenia Hoffm.

Uldinia J.M.Black
Vanasushava P.K.Mukh. & Constance
Vicatia DC.
Vvedenskya Korovin *
Xanthosia Rudge
Xatardia Meisn. & Zeyh.
Yabea Kozo–Pol.
Zeravschania Korovin
Zizia W.D.J.Koch
Zosima Hoffm.

URTICACEAE Juss. 1789   153.01

48 genera. Widespread. Herbs, shrubs, occas. lianes, trees.

B&H	MONOCHLAMYDEAE	Unisexuales, 153

(including Cannabaceae, Cecropiaceae, Moraceae, Theligonaceae, Ulmaceae)

DT&H	ARCHICHLAMYDEAE	Urticales, 15
Melc	ARCHICHLAMYDEAE	Urticales, 14
Thor	MALVIFLORAE	Urticales, 157

(including Cecropiaceae, Moraceae)

Dahl	MALVIFLORAE	Urticales, 74
Young	DILLENIIDAE, MALVANAE	Urticales, 178
Takh	DILLENIIDAE, URTICANAE	Urticales, Urticineae, 188
Cron	HAMAMELIDIDAE	Urticales, 55

Aboriella Bennet
Achudemia Blume
Archiboehmeria C.J.Chen
Astrothalamus C.B.Rob.
Australina Gaudich.
Boehmeria Jacq.
Chamabainia Wight
Cypholophus Wedd.
Debregeasia Gaudich.
Dendrocnide Miq.
Didymodoxa E.Mey. ex Wedd.
Discocnide Chew
Droguetia Gaudich.
Elatostema J.R.Forst. & G.Forst.
Forsskaolea L.
Gesnouinia Gaudich.
Gibbsia Rendle
Girardinia Gaudich.
Gyrotaenia Griseb.
Hemistylus Benth.
Hesperocnide Torr.
Hyrtanandra Miq.
Laportea Gaudich.
Lecanthus Wedd.

Leucosyke Zoll. & Moritzi
Maoutia Wedd.
Meniscogyne Gagnep.
Myriocarpa Benth.
Nanocnide Blume
Neodistemon Babu & A.N.Henry
Neraudia Gaudich.
Nothocnide Blume ex Chew
Obetia Gaudich.
Oreocnide Miq.
Parietaria L.
Petelotiella Gagnep. *
Phenax Wedd.
Pilea Lindl.
Pipturus Wedd.
Pouzolzia Gaudich.
Procris Comm. ex Juss.
Rousselia Gaudich.
Sarcochlamys Gaudich.
Sarcopilea Urb.
Soleirolia Gaudich.
Touchardia Gaudich.
Urera Gaudich.
Urtica L.

VAHLIACEAE Dandy 1959   59.03

1 genus. Africa to N.W. India, Indochina. Herbs.

B&H	POLYPETALAE, CALYCIFLORAE	Rosales (within Saxifragaceae, 59)
DT&H	ARCHICHLAMYDEAE	Rosales, Saxifragineae (within Saxifragaceae, 67)
Melc	ARCHICHLAMYDEAE	Rosales, Saxifragineae (within Saxifragaceae, 106)
Thor	ROSIFLORAE	Rosales, Saxifragineae (within Saxifragaceae, 208)
Dahl	ROSIFLORAE	Saxifragales (within Saxifragaceae, 163)
Young	ROSIDAE, ROSANAE	Saxifragales, 308
Takh	ROSIDAE, ROSANAE	Saxifragales, 208
Cron	ROSIDAE	Rosales (within Saxifragaceae, 173)

Vahlia Thunb.

VALERIANACEAE Batsch 1802   85

15 genera. Widespread. Mostly herbs.

B&H	GAMOPETALAE, INFERAE	Asterales, 85
DT&H	METACHLAMYDEAE (including Triplostegiaceae)	Rubiales, 226
Melc	SYMPETALAE (including Triplostegiaceae)	Dipsacales, 281
Thor	CORNIFLORAE	Dipsacales, 293
Dahl	CORNIFLORAE	Dipsacales, 330
Young	ROSIDAE, DIPSACANAE	Dipsacales, 383
Takh	ROSIDAE, CORNANAE	Dipsacales, 359
Cron	ASTERIDAE (including Triplostegiaceae)	Dipsacales, 315

Aligera Suksd.
Aretiastrum (DC.) Spach
Astrephia Dufr.
Belonanthus Graebn.
Centranthus Lam. & DC.
Fedia Gaertn.
Nardostachys DC.
Patrinia Juss.

Phuodendron (Graebn.) Dalla Torre & Harms
Phyllactis Pers.
Plectritis (Lindl.) DC.
Pseudobetckea (Hock) Lincz.
Stangea Graebn.
Valeriana L.
Valerianella Mill.

VERBENACEAE J.St.-Hil. 1805   125.01

86 genera. Widespread, mostly tropical. Trees, shrubs, herbs.

B&H	GAMOPETALAE, BICARPELLATAE (including Avicenniaceae, Nesogenaceae, Phrymaceae, Stilbaceae)	Lamiales, 125
DT&H	METACHLAMYDEAE (including Avicenniaceae, Cyclocheilaceae, Nesogenaceae, Stilbaceae)	Tubiflorae Verbenineae, 206
Melc	SYMPETALAE (including Avicenniaceae, Nesogenaceae, Stilbaceae)	Tubiflorae, Verbenineae, 258

Thor GENTIANIFLORAE   Lamiales, 265
       (including Avicenniaceae, Phrymaceae, Stilbaceae)
Dahl LAMIIFLORAE   Lamiales, 364
       (including Phrymaceae and ?Avicenniaceae, ?Cyclocheilaceae,
       ?Nesogenaceae)
Young ROSIDAE, LAMIANAE   Lamiales, 377
       (presumably including Phrymaceae, Stilbaceae, Avicenniaceae,
       Nesogenaceae)
Takh LAMIIDAE, LAMIANAE   Lamiales, Lamiineae, 413
       (including Avicenniaceae, Nesogenaceae, Phrymaceae)
Cron ASTERIDAE   Lamiales, 286
       (including Avicenniaceae, Phrymaceae, Stilbaceae)

Acantholippia Griseb. *
Adelosa Blume
Aegiphila Jacq.
Aloysia Juss.
Amasonia L.f.
Archboldia E.Beer & H.J.Lam *
Baillonia Bocquillon
Bouchea Cham.
Callicarpa L.
Caryopteris Bunge
Casselia Nees & Mart.
Castelia Cav.
Chascanum E.Mey.
Chloanthes R.Br.
Citharexylum Mill.
Clerodendrum L.
Coelocarpum Balf.f.
Congea Roxb.
Cornutia L.
Cyanostegia Turcz.
Dicrastylis J.L.Drumm. ex Harv.
Dimetra Kerr
Diostea Miers
Dipyrena Hook.
Duranta L.
Faradaya F.Muell.
Garrettia H.R.Fletcher
Glandularia J.F.Gmel.
Glossocarya Wall. ex Griff.
Gmelina L.
Hemiphora (F.Muell.) F.Muell.
Hierobotana Briq.
Holmskioldia Retz.
Hosea Ridl.
Huxleya Ewart
Hymenopyramis Wall. ex Griff.
Junellia Moldenke
Karomia Dop
Lachnostachys Hook.
Lampayo Phil.
Lantana L.
Lippia L.
Mallophora Endl.

Monochilus Fisch. & C.A.Mey.
Monopyrena Speg.
Nashia Millsp. *
Neorapinia Moldenke
Neosparton Griseb.
Newcastelia F.Muell.
Nyctanthes L.
Oncinocalyx F.Muell.
Oxera Labill.
Paravitex H.R.Fletcher
Parodianthus Tronc.
Peronema Jack
Petitia Jacq.
Petraeovitex Oliv.
Petrea L.
Phyla Lour.
Physopsis Turcz.
Pitraea Turcz. *
Pityrodia R.Br.
Premna L.
Priva Adans.
Pseudocarpidium Millsp.
Recordia Moldenke
Rehdera Moldenke
Rhaphithamnus Miers
Schnabelia Hand.–Mazz.
Spartothamnella Briq.
Sphenodesme Jack
Stachytarpheta Vahl
Stylodon Raf. *
Symphorema Roxb.
Tamonea Aubl.
Tectona L.f.
Teijsmanniodendron Koord.
Teucridium Hook.f.
Tsoongia Merr.
Urbania Phil. *
Verbena L.
Verbenoxylum Tronc. *
Vitex L.
Viticipremna Lam.
Xeroaloysia Tronc.
Xolocotzia Miranda *

VIOLACEAE Batsch 1802   15

20 genera. Widespread. Trees, shrubs, lianes, herbs.

B&H   POLYPETALAE, THALAMIFLORAE   Parietales, 15
DT&H ARCHICHLAMYDEAE   Parietales, Flacourtiineae, 148
Melc   ARCHICHLAMYDEAE   Violales, Flacourtiineae, 184
Thor   VIOLIFLORAE   Violales, Violineae, 128
Dahl   VIOLIFLORAE   Violales, 89
Young DILLENIIDAE, VIOLANAE   Violales, 194
Takh   DILLENIIDAE, VIOLANAE   Violales, 147
Cron   DILLENIIDAE   Violales, 115

Acentra Phil.
Agatea A.Gray
Allexis Pierre
Amphirrhox Spreng.
Anchietea A.St.–Hil.
Corynostylis Mart.
Decorsella A.Chev.
Fusispermum Cuatrec. *
Gloeospermum Triana & Planch.
Hybanthus Jacq.

Isodendrion A.Gray
Leonia Ruíz & Pav.
Melicytus J.R.Forst. & G.Forst.
Noisettia Kunth
Orthion Standl. & Steyerm. *
Paypayrola Aubl.
Rinorea Aubl.
Rinoreocarpus Ducke
Schweiggeria Spreng.
Viola L.

VISCACEAE Batsch 1802   148.03

7 genera. Widespread. Hemiparasites on trees.

B&H   MONOCHLAMYDEAE   Achlamydosporeae (within
                                       Loranthaceae, 148)
DT&H ARCHICHLAMYDEAE   Santalales, Loranthineae (within
                                       Loranthaceae, 17)
Melc   ARCHICHLAMYDEAE   Santalales, Loranthineae (within
                                       Loranthaceae, 22)
Thor   SANTALIFLORAE   Santalales, 121
Dahl   SANTALIFLORAE   Santalales, 261
Young ROSIDAE, SANTALANAE   Santalales, 239
Takh   ROSIDAE, CELASTRANAE   Santalales, 317
Cron   ROSIDAE   Santalales, 212

Arceuthobium M.Bieb.
Dendrophthora Eichler
Ginalloa Korth.
Korthalsella Tiegh.

Notothixos Oliv.
Phoradendron Nutt.
Viscum L.

VITACEAE Juss. 1789   50.01

14 genera. Tropics to warm-temperate. Mostly climbing shrubs, some herbs.

B&H   POLYPETALAE, DISCIFLORAE Celastrales (as *Ampelideae*) 50
          (including Leeaceae)
DT&H ARCHICHLAMYDEAE   Rhamnales, 120
          (including Leeaceae)

Melc	ARCHICHLAMYDEAE	Rhamnales, 168
Thor	CORNIFLORAE	Cornales, Vitineae, 279
	(including Leeaceae)	
Dahl	SANTALIFLORAE	Vitales, 254
	(including Leeaceae)	
Young	ROSIDAE, SANTALANAE	Vitales, 233
Takh	ROSIDAE, VITANAE	Vitales, 322
Cron	ROSIDAE	Rhamnales, 235

Acareosperma Gagnep. *
Ampelocissus Planch.
Ampelopsis Michx.
Cayratia Juss.
Cissus L.
Clematicissus Planch.
Cyphostemma (Planch.) Alston

Nothocissus (Miq.) Latiff
Parthenocissus Planch.
Pterisanthes Blume
Pterocissus Urb. & Ekman
Rhoicissus Planch.
Tetrastigma (Miq.) Planch.
Vitis L.

VOCHYSIACEAE St.-Hil. 1820   20.02

7 genera. W. Africa (*Erismadelphus*), trop. America. Trees, shrubs, lianes, herbs.

B&H	POLYPETALAE, THALAMIFLORAE	Polygalineae, 20a
	(including Euphroniaceae, Trigoniaceae)	
DT&H	ARCHICHLAMYDEAE	Geraniales, Malpighiineae, 93
Melc	ARCHICHLAMYDEAE	Rutales, Melpighiineae, 140
Thor	GERANIIFLORAE	Geraniales, Polygalineae, 111
Dahl	RUTIFLORAE	Polygalales, 227
Young	ROSIDAE, GERANIANAE	Polygalales, 282
Takh	ROSIDAE, RUTANAE	Polygalales, 294
Cron	ROSIDAE	Polygalales, 242
	(including Euphroniaceae)	

Callisthene Mart.
Erisma Rudge
Erismadelphus Mildbr.
Qualea Aubl.

Ruizterania Marc.–Berti
Salvertia A.St.–Hil.
Vochysia Aubl.

WINTERACEAE R.Br. ex Lindl. 1830   4.02

4 genera. Madag., Malesia to Austral., Pacific, C. and S. America. Trees shrubs.

B&H	POLYPETALAE, THALAMIFLORAE	Ranales (within Magnoliaceae, 4)
DT&H	ARCHICHLAMYDEAE	Ranales, Magnoliineae (within Magnoliaceae, 45)
Melc	ARCHICHLAMYDEAE	Magnoliales, Magnoliineae, 42
Thor	ANNONIFLORAE	Annonales, Winterineae, 1
Dahl	MAGNOLIIFLORAE	Magnoliales, 9
Young	MAGNOLIIDAE, MAGNOLIANAE	Magnoliales, 3
Takh	MAGNOLIIDAE, MAGNOLIANAE	Winterales, 8
Cron	MAGNOLIIDAE	Magnoliales, 1

Drimys J.R.Forst. & G.Forst.
Pseudowintera Dandy

Takhtajania Baranova & J.–F.Leroy *
Zygogynum Baill.

ZYGOPHYLLACEAE R.Br. 1814   37

27 genera. Widespread trop. to warm-temp. Mostly shrubs, some trees, herbs.

B&H	POLYPETALAE, DISCIFLORAE	Geraniales, 37
DT&H	ARCHICHLAMYDEAE	Geraniales, Geraniineae, 85
	(including Balanitaceae)	
Melc	ARCHICHLAMYDEAE	Geraniales, Geraniineae, 126
	(including Balanitaceae)	
Thor	GERANIIFLORAE	Geraniales, Linineae, 99
Dahl	RUTIFLORAE	Geraniales, 230
	(separating *Nitrariaceae* 231, *Peganaceae* 232)	
Young	ROSIDAE, GERANIANAE	Linales, 273
Takh	ROSIDAE, RUTANAE	Rutales, 262
	(separating *Tetradiclidaceae* 257, *Nitrariaceae* 263, *Peganaceae* 265)	
Cron	ROSIDAE	Sapindales, 262
	(including Balanitaceae)	

Augea Thunb.
Bulnesia Gay
Fagonia L.
Guaiacum L.
Kallstroemia Scop.
Kelleronia Schinz
Larrea Cav.
Malacocarpus Fisch. & C.A.Mey.
Metharme Phil. ex Engl.
Miltianthus Bunge
Morkillia Rose & Painter
Neoluederitzia Schinz
Nitraria L.
Peganum L.

Pintoa Gay
Plectrocarpa Gillies ex Hook. & Arn.
Porlieria Ruíz & Pav.
Sarcozygium Bunge
Seetzenia R.Br. ex Decne.
Sericodes A.Gray
Sisyndite E.Mey. ex Sond.
Tetradiclis Steven ex M.Bieb.
Tetraena Maxim.
Tribulopis R.Br.
Tribulus L.
Viscainoa Greene
Zygophyllum L.

MONOCOTYLEDONS

ACORACEAE Martinov 1820   191.02

1 genus. North temperate to tropics. Rhizomatous herbs.

B&H	(MONOCOTS)	Nudiflorae (within Araceae, 191)
DT&H	(MONOCOTS)	Spathiflorae (within Araceae, 249)
Melc	(MONOCOTS)	Spathiflorae (within Araceae, 332)
Thor	ARIFLORAE	Arales (within Araceae, 319)
Dahl	ARIFLORAE	Arales (within Araceae, 420)
Young	(not mentioned, presumably within Araceae, 55)	
Takh	ARECIDAE, ARANAE	Arales (within Araceae, 530)
Cron	ARECIDAE	Arales (within Araceae 338 in 1981, as separate family in 1988)

Acorus L.

AGAVACEAE Endl. 1841   175.11

12 genera. New World and (*Cordyline*) India to Australia. Rosette herbs to trees.

B&H	(MONOCOTS)	Epigynae (within Amaryllidaceae, 174)
DT&H	(MONOCOTS)	Liliiflorae, Liliineae (within Amaryllidaceae, 266)
Melc	(MONOCOTS) (including Doryanthaceae)	Liliiflorae, Liliineae, 304
Thor	LILIIFLORAE	Liliales, Liliineae (within Liliaceae, 297)
Dahl	LILIIFLORAE	Asparagales, 389
Young	LILIIDAE, LILIANAE	Liliales, 71
Takh	LILIIDAE, LILIANAE	Amaryllidales, 468
Cron	LILIIDAE (including Doryanthaceae, Dracaenaceae, Phormiaceae)	Liliales, 373

Agave L.
Beschorneria Kunth
Bravoa Lex.
Clistoyucca (Engelm.) Trel.
Cordyline Comm. ex R.Br.
Furcraea Vent.

Hesperaloe Engelm.
Polianthes L.
Prochnyanthes S.Watson
Pseudobravoa Rose
Samuela Trel.
Yucca L.

ALISMATACEAE Vent. 1799   194.01

12 genera. Widespread. Herbs, often aquatic.

B&H	(MONOCOTS) (including Butomaceae, Limnocharitaceae)	Apocarpae, 194
DT&H	(MONOCOTS)	Helobiae, Alismineae, 241
Melc	(MONOCOTS)	Helobiae, Alismatineae, 291
Thor	ALISMATIFLORAE (including Limnocharitaceae)	Alismatales, 309

ALISMATACEAE (M)

Dahl	ALISMATIFLORAE	Alismatales,	426
Young	LILIIDAE, ALISMATANAE	Alismatales,	42
	(including Limnocharitaceae)		
Takh	ALISMATIDAE, BUTOMANAE	Alismatales,	435
Cron	ALISMATIDAE	Alismatales,	321

Alisma L.
Baldellia Parl.
Burnatia Micheli
Caldesia Parl.
Damasonium Mill.
Echinodorus Rich. ex Engelm.

Limnophyton Miq.
Luronium Raf.
Machaerocarpus Small
Ranalisma Stapf
Sagittaria L.
Wiesneria Micheli

ALLIACEAE J. Agardh 1858   175.27

31 genera. Widespread. Bulbous, cormous or rhizomatous herbs.

B&H	(MONOCOTS)	Coronarieae (within Liliaceae, 178)
DT&H	(MONOCOTS)	Liliiflorae, Liliineae (within Liliaceae, 264)
Melc	(MONOCOTS)	Liliiflorae, Liliineae (within Liliaceae, 301)
Thor	LILIIFLORAE	Liliales, Liliineae (within Liliaceae, 297)
Dahl	LILIIFLORAE	Asparagales, 403
Young	(not mentioned, presumably within Liliaceae, 63)	
Takh	LILIIDEAE, LILIANAE	Amaryllidales, 465
Cron	LILIIDAE	Liliales (within Liliaceae, 369)

Agapanthus L'Hér.
Allium L.
Ancrumia Harv. ex Baker
Androstephium Torr.
Bessera Schult.
Bloomeria Kellogg
Brodiaea Sm.
Caloscordum Herb.
Dandya H.E.Moore *
Dichelostemma Kunth
Erinna Phil. *
Garaventia Looser
Gethyum Phil. *
Gilliesia Lindl.
Ipheion Raf.
Latace Phil.

Leucocoryne Lindl.
Miersia Lindl.
Milla Cav.
Milula Prain
Muilla S.Watson ex Benth.
Nectaroscordum Lindl.
Nothoscordum Kunth
Petronymphe H.E.Moore
Solaria Phil.
Speea Loes. *
Trichlora Baker
Tristagma Poepp.
Triteleia Douglas ex Lindl.
Triteleiopsis Hoover
Tulbaghia L.

ALOACEAE Batsch 1802   175.22

5 genera. Arabia, Africa, Madagascar and Mascarenes. Rosette herbs to arborescent.

B&H	(MONOCOTS)	Coronarieae (within Liliaceae, 178)
DT&H	(MONOCOTS)	Liliiflorae, Liliineae (within Liliaceae, 264)
Melc	(MONOCOTS)	Liliiflorae, Liliineae (within Liliaceae, 301)
Thor	LILIIFLORAE	Liliales, Liliineae (within Liliaceae, 297)
Dahl	LILIIFLORAE	Asparagales (within Asphodelaceae, 398)

Young	LILIIDAE, LILIANAE	Liliales, 72
Takh	LILIIDAE, LILIANAE	Amaryllidales (within Asphodelaceae, 460)
Cron	LILIIDAE	Liliales, 372

Aloe L.
Gasteria Duval
Haworthia Duval

Lomatophyllum Willd.
Poellnitzia Uitew.

ALSTROEMERIACEAE Dumort. 1829   178.01

4 genera. Mexico to S. America. Rhizomatous herbs, some climbing.

B&H	(MONOCOTS)	Epigynae (within Amaryllidaceae, 174)
DT&H	(MONOCOTS)	Liliiflorae, Liliineae (within Amaryllidaceae, 266)
Melc	(MONOCOTS)	Liliiflorae, Liliineae (within Liliaceae, 301)
Thor	LILIIFLORAE	Liliales, Dioscoreineae (within Philesiaceae, 298)
Dahl	LILIIFLORAE	Liliales, 410
Young	(not mentioned, presumably within Liliaceae, 63)	
Takh	LILIIDAE, LILIANAE	Alstroemeriales, 494
Cron	LILIIDAE	Liliales (within Liliaceae, 389)

Alstroemeria L.
Bomarea Mirb.

Leontochir Phil.
Schickendantzia Pax

AMARYLLIDACEAE Juss. 1789   175.28

65 genera. Widespread. Mostly bulbous, or rarely rhizomatous, herbs.

B&H	(MONOCOTS)	Epigynae, 174
	(including Agavaceae, Alstroemeriaceae, Doryanthaceae, Hypoxidaceae, Ixioliriaceae, Velloziaceae)	
DT&H	(MONOCOTS)	Liliiflorae, Liliineae, 266
	(including Agavaceae, Alstroemeriaceae, Doryanthaceae, Hypoxidaceae, Ixioliriaceae, Lanariaceae, Tecophilaeaceae)	
Melc	(MONOCOTS)	Liliiflorae, Liliineae, 307
	(including Ixioliriaceae)	
Thor	LILIIFLORAE	Liliales, Liliineae (within Liliaceae, 297)
Dahl	LILIIFLORAE	Asparagales, 404
Young	(not mentioned, presumably within Liliaceae, 63)	
Takh	LILIIDAE, LILIANAE	Amaryllidales, 473
Cron	LILIIDAE	Liliales (within Liliaceae, 369)

Amaryllis L.
Ammocharis Herb.
Apodolirion Baker
Bokkeveldia D.Müll.–Doblies &
   U.Müll.–Doblies
Boophone Herb.
Braxireon Raf.
Brunsvigia Heist.

Caliphruria Herb.
Calostemma R.Br.
Carpolyza Salisb.
Castellanoa Traub *
Chapmanolirion Dinter
Chlidanthus Herb.
Clivia Lindl.
Crinum L.

Cryptostephanus Welw. ex Baker
Cybistetes Milne–Redh. & Schweick.
Cyrtanthus Aiton
Elisena Herb.
Eucharis Planch. & Linden
Eucrosia Ker Gawl.
Eustephia Cav.
Famatina Ravenna
Galanthus L.
Gemmaria Salisb.
Gethyllis L.
Griffinia Ker Gawl.
Habranthus Herb.
Haemanthus L.
Hannonia Braun–Blanq. & Maire
Haylockia Herb.
Hessea Herb.
Hieronymiella Pax
Hippeastrum Herb.
Hyline Herb.
Hymenocallis Salisb.
Lapiedra Lag.
Leucojum L.
Lycoris Herb.
Mathieua Klotzsch
Namaquanula D.Müll.–Doblies &
    U.Müll.–Doblies

Narcissus L.
Nerine Herb.
Pamianthe Stapf
Pancratium L.
Paramongaia Velarde
Phaedranassa Herb.
Placea Miers
Plagiolirion Baker
Proiphys Herb.
Pseudostenomesson Velarde
Pyrolirion Herb.
Rauhia Traub
Rhodophiala C.Presl
Scadoxus Raf.
Sprekelia Heist.
Stenomesson Herb.
Sternbergia Waldst. & Kit.
Strumaria Jacq. ex Willd.
Tedingea D.Müll.–Doblies &
    U.Müll.–Doblies
Traubia Moldenke
Ungernia Bunge
Urceolina Rchb.
Vagaria Herb.
Zephyranthes Herb.

ANARTHRIACEAE D.F. Cutler & Airy Shaw 1965   198.03

1 genus. Western Australia. Rhizomatous herbs.

B&H	(MONOCOTS)	Glumaceae (within Restionaceae, 198)
DT&H	(MONOCOTS)	Farinosae, Enantioblastae (within Restionaceae, 252)
Melc	(MONOCOTS)	Commelinales, Restionineae (within Restionaceae, 326)
Thor	COMMELINIFLORAE	Commelinales, Flagellariineae (within Restionaceae, 338)
Dahl	COMMELINIFLORAE	Poales, 464
Young	(not mentioned, presumably within Restionaceae, 87)	
Takh	LILIIDAE, COMMELINANAE	Restionales, Restionineae, 523
Cron	COMMELINIDAE	Restionales (within Restionaceae, 347)

Anarthria R.Br.

ANTHERICACEAE J. Agardh 1858   175.23

29 genera. Old World, esp. Australasia, and 5 genera in New World. Rhizom. herbs.

B&H	(MONOCOTS)	Coronarieae (within Liliaceae, 178)
DT&H	(MONOCOTS)	Liliiflorae, Liliineae (within Liliaceae, 264)

Melc	(MONOCOTS)	Commelinales, Restionineae (within Restionaceae, 326)
Thor	COMMELINIFLORAE	Commelinales, Flagellariineae (within Restionaceae, 338)
Dahl	COMMELINIFLORAE	Poales, 464
Young	(not mentioned, presumably within Restionaceae, 87)	
Takh	LILIIDAE, COMMELINANAE	Restionales, Restionineae, 523
Cron	COMMELINIDAE	Restionales (within Restionaceae, 347)

Anarthria R.Br.

ANTHERICACEAE J. Agardh 1858   175.23

29 genera. Old World, esp. Australasia, and 5 genera in New World. Rhizom. herbs.

B&H	(MONOCOTS)	Coronarieae (within Liliaceae, 178)
DT&H	(MONOCOTS)	Liliiflorae, Liliineae (within Liliaceae, 264)
Melc	(MONOCOTS)	Liliiflorae, Liliineae (within Liliaceae, 301)
Thor	LILIIFLORAE	Liliales, Liliineae (within Liliaceae, 297)
Dahl	LILIIFLORAE	Asparagales, 399
Young	(not mentioned, presumably within Liliaceae, 63)	
Takh	LILIIDAE, LILIANAE	Amaryllidales (within Asphodelaceae, 460)
Cron	LILIIDAE	Liliales (within Liliaceae, 369)

Alania Endl.
Anemarrhena Bunge
Anthericum L.
Arnocrinum Endl. & Lehm.
Arthropodium R.Br.
Borya Labill.
Caesia R.Br.
Chamaescilla F.Muell. ex Benth.
Chlorophytum Ker Gawl.
Comospermum Rauschert
Corynotheca F.Muell. ex Benth.
Dichopogon Kunth
Diora Ravenna
Echeandia Ortega
Eremocrinum M.E.Jones

Hagenbachia Nees & Mart.
Hensmania W.Fitzg.
Herpolirion Hook.f.
Hodgsoniola F.Muell.
Johnsonia R.Br.
Laxmannia R.Br.
Leucocrinum Nutt. ex A.Gray
Murchisonia Brittan
Pasithea D.Don
Sowerbaea Sm.
Stawellia F.Muell.
Thysanotus R.Br.
Trichopetalum Lindl.
Tricoryne R.Br.

APHYLLANTHACEAE Burnett 1835   175.24

1 genus, monotypic. S. Europe, N. Africa. Rhizomatous herb.

B&H	(MONOCOTS)	Coronarieae (within Liliaceae, 178)
DT&H	(MONOCOTS)	Liliiflorae, Liliineae (within Liliaceae, 264)
Melc	(MONOCOTS)	Liliiflorae, Liliineae (within Liliaceae, 301)
Thor	LILIIFLORAE	Liliales, Liliineae (within Liliaceae, 297)
Dahl	LILIIFLORAE	Asparagales, 400
Young	(not mentioned, presumably within Liliaceae, 63)	
Takh	LILIIDAE, LILIANAE	Amaryllidales, 463
Cron	LILIIDAE	Liliales (within Liliaceae, 369)

Aphyllanthes L.

APONOGETONACEAE J. Agardh 1858 195.05

1 genus. Old World tropics and S. Africa. Aquatic herbs.

B&H	(MONOCOTS)	Apocarpae (within *Najadaceae*, 195)
DT&H	(MONOCOTS)	Helobiae, Potamogetonineae, 239
Melc	(MONOCOTS)	Helobiae, Potamogetonineae, 295
Thor	ALISMATIFLORAE	Zosterales, Aponogetonineae, 311
Dahl	ALISMATIFLORAE	Alismatales, 423
Young	LILIIDAE, ALISMATANAE	Zosterales, 45
Takh	ALISMATIDAE, NAJADANAE	Aponogetonales, 436
Cron	ALISMATIDAE	Najadales, 323

Aponogeton L.f.

ARACEAE Juss. 1789 191.01

105 genera. Widespread, esp. trop. Mostly rhizomatous herbs, some climbers.

B&H	(MONOCOTS) (including Acoraceae)	Nudiflorae, 191
DT&H	(MONOCOTS) (including Acoraceae)	Spathiflorae, 249
Melc	(MONOCOTS) (including Acoraceae)	Spathiflorae, 332
Thor	ARIFLORAE (including Acoraceae)	Arales, 319
Dahl	ARIFLORAE (including Acoraceae)	Arales, 420
Young	LILIIDAE, ARANAE (including Acoraceae)	Arales, 55
Takh	ARECIDAE, ARANAE (including Acoraceae)	Arales, 530
Cron	ARECIDAE	Arales, 338
	(including Acoraceae in 1981, not in 1988)	

Aglaodorum Schott
Aglaonema Schott
Alloschemone Schott
Alocasia (Schott) G.Don
Ambrosina Bassi
Amorphophallus Blume ex Decne.
Amydrium Schott
Anadendrum Schott
Anaphyllopsis A.Hay
Anaphyllum Schott
Anchomanes Schott
Anthurium Schott
Anubias Schott
Aridarum Ridl.
Ariopsis Nimmo
Arisaema Mart.
Arisarum Mill.
Arophyton Jum.

Arum L.
Asterostigma Fisch. & C.A.Mey.
Biarum Schott
Bognera Mayo & Nicolson
Bucephalandra Schott
Caladium Vent.
Calla L.
Callopsis Engl.
Carlephyton Jum.
Cercestis Schott
Chlorospatha Engl.
Colletogyne Buchet
Colocasia Schott
Cryptocoryne Fisch. ex Wydl.
Culcasia P.Beauv.
Cyrtosperma Griff.
Dieffenbachia Schott
Dracontioides Engl.

Dracontium L.
Dracunculus Mill.
Eminium (Blume) Schott
Epipremnum Schott
Filarum Nicolson
Furtadoa M.Hotta
Gearum N.E.Br.
Gonatanthus Klotzsch
Gonatopus Hook.f. ex Engl.
Gorgonidium Schott
Gymnostachys R.Br.
Hapaline Schott
Helicodiceros Schott ex K.Koch
Heteroaridarum M.Hotta
Heteropsis Kunth
Holochlamys Engl.
Homalomena Schott
Hottarum Bogner & Nicolson
Jasarum Bunting
Lagenandra Dalzell
Lasia Lour.
Lasimorpha Schott
Lysichiton Schott
Mangonia Schott
Monstera Adans.
Montrichardia Crueg.
Nephthytis Schott
Orontium L.
Pedicellarum M.Hotta
Peltandra Raf.
Philodendron Schott
Phymatarum M.Hotta
Pinellia Ten.
Piptospatha N.E.Br.
Pistia L.

Podolasia N.E.Br.
Pothoidium Schott
Pothos L.
Protarum Engl.
Pseudodracontium N.E.Br.
Pseudohydrosme Engl.
Pycnospatha Thorel ex Gagnep.
Remusatia Schott
Rhaphidophora Hassk.
Rhodospatha Poepp.
Sauromatum Schott
Scaphispatha Brongn. ex Schott
Schismatoglottis Zoll. & Moritzi
Scindapsus Schott
Spathantheum Schott
Spathicarpa Hook.
Spathiphyllum Schott
Stenospermation Schott
Steudnera K.Koch
Stylochaeton Lepr.
Symplocarpus Salisb. ex Nutt.
Synandrospadix Engl.
Syngonium Schott
Taccarum Brongn. ex Schott
Theriophonum Blume
Typhonium Schott
Typhonodorum Schott
Ulearum Engl.
Urospatha Schott
Xanthosoma Schott
Zamioculcas Schott
Zantedeschia Spreng.
Zomicarpa Schott
Zomicarpella N.E.Br.

ARECACEAE – see PALMAE

ASPARAGACEAE Juss. 1789   175.03

1 genus. Widespread in Old World. Herbs, shrubs, vines.

B&H (MONOCOTS)   Coronarieae (within Liliaceae, 178)
DT&H (MONOCOTS   Liliiflorae, Liliineae (within Liliaceae, 264)
Melc (MONOCOTS)   Liliiflorae, Liliineae (within Liliaceae, 301)
Thor LILIIFLORAE   Liliales, Liliineae (within Liliaceae, 297)
Dahl LILIIFLORAE   Asparagales, 378
Young (not mentioned, presumably within Liliaceae, 63)
Takh LILIIDAE, LILIANAE   Asparagales, 477
Cron LILIIDAE   Liliales (within Liliaceae, 369)

Asparagus L.

ASPHODELACEAE Juss. 1789   175.21

11 genera. Europe, Africa, Asia (mostly temp.), New Zealand (*Bulbinella*) and Mexico (*Hemiphylacus*). Rhizomatous herbs.

B&H	(MONOCOTS)	Coronarieae (within Liliaceae, 178)
DT&H	(MONOCOTS)	Liliiflorae, Liliineae (within Liliaceae, 264)
Melc	(MONOCOTS)	Liliiflorae, Liliineae (within Liliaceae, 301)
Thor	LILIIFLORAE	Liliales, Liliineae (within Liliaceae, 297)
Dahl	LILIIFLORAE	Asparagales, 398
	(including Aloaceae)	
Young	(not mentioned, presumably within Liliaceae, 63)	
Takh	LILIIDAE, LILIANAE	Amaryllidales, 460
	(including Aloaceae, Anthericaceae)	
Cron	LILIIDAE	Liliales (within Liliaceae, 369)

Asphodeline Rchb.   **Jodrellia** Baijnath
Asphodelus L.   **Kniphofia** Moench
Bulbine Wolf   **Paradisea** Mazzuc.
Bulbinella Kunth   **Simethis** Kunth
Eremurus M.Bieb.   **Trachyandra** Kunth
Hemiphylacus S.Watson

ASTELIACEAE Dumort. 1829   175.07

4 genera. Mascarenes, New Guinea, Australasia, Pacific, Chile. Herbs, shrubs, trees.

B&H	(MONOCOTS)	Coronarieae (within Liliaceae, 178)
DT&H	(MONOCOTS)	Liliiflorae, Liliineae (within Liliaceae, 264)
Melc	(MONOCOTS)	Liliiflorae, Liliineae (within Liliaceae, 301)
Thor	LILIIFLORAE	Liliales, Liliineae (within Liliaceae, 297)
Dahl	LILIIFLORAE	Asparagales, 383
Young	(not mentioned, presumably within Liliaceae, 63)	
Takh	LILIIDAE, LILIANAE	Asparagales, 481
Cron	LILIIDAE	Liliales (within Liliaceae, 369)

Astelia Banks & Sol. ex R.Br.   **Milligania** Hook.f.
Collospermum Skottsb.   **Neoastelia** J.B.Williams

BLANDFORDIACEAE R. Dahlgren & Clifford 1985   175.09

1 genus. Eastern Australia. Rhizomatous herbs.

B&H	(MONOCOTS)	Coronarieae (within Liliaceae, 185)
DT&H	(MONOCOTS)	Liliales, Liliineae (within Liliaceae, 264)
Melc	(not mentioned, presumably included in Liliaceae, 301)	
Thor	LILIIFLORAE	Liliales, Liliineae (within Liliaceae, 297)
Dahl	LILIIFLORAE	Asparagales, 387
Young	(not mentioned, presumably within Liliaceae, 63)	
Takh	LILIIDAE, LILIANAE	Amaryllidales, 471
Cron	LILIIDAE	Liliales (within Liliaceae, 369)

Blandfordia Sm.

BROMELIACEAE Juss. 1789 171

51 genera. New World trop., 1 sp. in W. trop. Afr. Mostly rosette herbs, rarely pachycaul.

B&H	(MONOCOTS)	Epignae, 171
DT&H	(MONOCOTS)	Farinosae, Bromeliineae, 258
Melc	(MONOCOTS)	Bromeliales, 320
Thor	COMMELINIFLORAE	Commelinales, Bromeliineae, 325
Dahl	BROMELIIFLORAE	Bromeliales, 437
Young	LILIIDAE, ZINGIBERANAE	Bromeliales, 94
Takh	LILIIDAE, BROMELIANAE	Bromeliales, 501
Cron	ZINGIBERIDAE	Bromeliales, 356

Abromeitiella Mez
Acanthostachys Klotzsch
Aechmea Ruíz & Pav.
Ananas Mill.
Andrea Mez *
Androlepis Brongn. ex Houllet
Araeococcus Brongn.
Ayensua L.B.Sm.
Billbergia Thunb.
Brewcaria L.B.Sm., Steyerm. & H.E.
    Rob.*
Brocchinia Schult.f.
Bromelia L.
Canistrum E.Morren
Catopsis Griseb.
Connellia N.E.Br.
Cottendorfia Schult.f.
Cryptanthus Otto & A.Dietr.
Deuterocohnia Mez
Disteganthus Lem. *
Dyckia Schult.f.
Encholirium Mart. ex Schult.f.
Fascicularia Mez
Fernseea Baker
Fosterella L.B.Sm.
Glomeropitcairnia Mez

Greigia Regel
Guzmania Ruíz & Pav.
Hechtia Klotzsch
Hohenbergia Schult.f.
Hohenbergiopsis L.B.Sm. & Read
Lindmania Mez
Lymania Read *
Mezobromelia L.B.Sm. *
Navia Schult.f.
Neoglaziovia Mez
Neoregelia L.B.Sm.
Nidularium Lem.
Ochagavia Phil.
Orthophytum Beer
Pitcairnia L'Hér.
Portea K.Koch
Pseudaechmea L.B.Sm. & Read *
Pseudananas Hassl. ex Harms
Puya Molina
Quesnelia Gaudich.
Ronnbergia E.Morren & André
Steyerbromelia L.B.Sm. *
Streptocalyx Beer
Tillandsia L.
Vriesea Lindl.
Wittrockia Lindm.

BURMANNIACEAE Blume 1827 177.01

15 genera. Widespread except Europe, mostly tropical. Herbs, some saprophytic.

B&H	(MONOCOTS)	Microspermae, 168
	(including Corsiaceae)	
DT&H	(MONOCOTS)	Microspermae, Burmanniineae, 275
	(including Corsiaceae)	
Melc	(MONOCOTS)	Liliiflorae, Burmanniineae, 315
Thor	LILIIFLORAE	Liliales, Iridineae, 305
	(including Corsiaceae)	
Dahl	LILIIFLORAE	Burmanniales, 407
	(separating *Thismiaceae* 408)	
Young	LILIIDAE, LILIANAE	Burmanniales, 77

BURMANNIACEAE (M)

| Takh | LILIIDAE, LILIANAE | Burmanniales, 458 |
| Cron | LILIIDAE | Orchidales, 381 |

Afrothismia (Engl.) Schltr.
Apteria Nutt.
Burmannia L.
Campylosiphon Benth.
Cymbocarpa Miers
Dictyostega Miers
Geomitra Becc.
Gymnosiphon Blume

Haplothismia Airy Shaw
Hexapterella Urb.
Marthella Urb.
Miersiella Urb.
Oxygyne Schltr. *
Scaphiophora Schltr. *
Thismia Griff.

BUTOMACEAE Rich. 1815–16   194.02

1 genus, monotypic. Temperate Europe and Asia. Herb.

B&H	(MONOCOTS)	Apocarpae (within Alismataceae, 194)
DT&H	(MONOCOTS)	Helobiae, Butomineae, 242
	(including Limnocharitaceae)	
Melc	(MONOCOTS)	Helobiae, Alismatineae, 292
	(including Limnocharitaceae)	
Thor	ALISMATIFLORAE	Alismatales, 308
Dahl	ALISMATIFLORAE	Alismatales, 424
Young	LILIIDAE, ALISMATANAE	Alismatales, 43
Takh	ALISMATIDAE, BUTOMANAE	Butomales, 430
Cron	ALISMATIDAE	Alismatales, 319

Butomus L.

CALECTASIACEAE Endl. 1837   175.31

1 genus. Australia. Rhizomatous subshrubs.

B&H	(MONOCOTS)	Calycineae (within Juncaceae, 186)
DT&H	(MONOCOTS)	Liliiflorae, Liliineae (within Liliaceae, 264)
Melc	(MONOCOTS)	Liliiflorae, Liliineae (within Xanthorrhoeaceae, 302)
Thor	LILIIFLORAE	Liliales, Liliineae (within Liliaceae, 297)
Dahl	LILIIFLORAE	Asparagales, 386
Young	(not mentioned, presumably within Xanthorrhoeaceae, 73)	
Takh	LILIIDAE, LILIANAE	Amaryllidales (within Dasypogonaceae, 462)
Cron	LILIIDAE	Liliales (within Xanthorrhoeaceae, 374)

Calectasia R.Br.

CANNACEAE Juss. 1789   170.06

1 genus. New World tropics. Rhizomatous herbs.

B&H	(MONOCOTS)	Epigynae (within Scitamineae, 170)
DT&H	(MONOCOTS)	Scitamineae, 273
Melc	(MONOCOTS)	Scitamineae, 340
Thor	COMMELINIFLORAE	Zingiberales, Zingiberineae, 347
Dahl	ZINGIBERIFLORAE	Zingiberales, 449
Young	LILIIDAE, ZINGIBERANAE	Zingiberales, 101
Takh	LILIIDAE, ZINGIBERANAE	Zingiberales, 509
Cron	ZINGIBERIDAE	Zingiberales, 363

Canna L.

CENTROLEPIDACEAE Endl. 1836   197.01

4 genera. Trop. Asia to Pacific and temp. S. America. Small annual or rhizomatous herbs.

B&H	(MONOCOTS)	Glumaceae, 197
DT&H	(MONOCOTS)	Farinosae, Enantioblastae, 253
	(including Hydatellaceae)	
Melc	(MONOCOTS)	Commelinales, Restionineae, 327
	(including Hydatellaceae)	
Thor	COMMELINIFLORAE	Commelinales, Flagellariineae, 339
Dahl	COMMELINIFLORAE	Poales, 466
Young	LILIIDAE, COMMELINANAE	Restionales, 88
Takh	LILIIDAE, COMMELINANAE	Restionales, Centrolepidineae, 525
Cron	COMMELINIDAE	Restionales, 348

Aphelia R.Br.
Brizula Hieron.

Centrolepis Labill.
Gaimardia Gaudich.

COLCHICACEAE DC. 1805   178.02

16 genera. Widespread in Old World. Cormous herbs or climbers.

B&H	(MONOCOTS)	Coronarieae (within Liliaceae, 178)
DT&H	(MONOCOTS)	Liliiflorae, Liliineae (within Liliaceae, 264)
Melc	(MONOCOTS)	Liliiflorae, Liliineae (within Liliaceae, 301)
Thor	LILIIFLORAE	Liliales, Liliineae (within Liliaceae, 297)
Dahl	LILIIFLORAE	Liliales, 411
Young	(not mentioned, presumably within Liliaceae, 63)	
Takh	LILIIDAE, LILIANAE	Liliales (within Melanthiaceae, 449)
Cron	LILIIDAE	Liliales (within Liliaceae, 369)

Androcymbium Willd.
Baeometra Salisb. ex Endl.
Bulbocodium L.
Burchardia R.Br.
Camptorrhiza Hutch.
Colchicum L.
Gloriosa L.
Hexacyrtis Dinter

Iphigenia Kunth
Littonia Hook.
Merendera Ramond
Neodregea C.H.Wright
Onixotis Raf.
Ornithoglossum Salisb.
Sandersonia Hook.
Wurmbea Thunb.

COMMELINACEAE R.Br. 1810   183.01

38 genera. Widespread in tropics, rarely temperate. Herbs, sometimes climbing.

B&H	(MONOCOTS)	Coronarieae, 183
DT&H	(MONOCOTS)	Farinosae, Commelinineae, 259
Melc	(MONOCOTS)	Commelinales, Commelinineae, 321
Thor	COMMELINIFLORAE	Commelinales, Commelinineae, 334
Dahl	COMMELINIFLORAE	Commelinales, 451
Young	LILIIDAE, COMMELINANAE	Commelinales, 83
Takh	LILIIDAE, COMMELINANAE	Commelinales, Commelinineae, 515
Cron	COMMELINIDAE	Commelinales, 343

Aetheolirion Forman
Amischotolype Hassk.
Aneilema R.Br.
Anthericopsis Engl.
Belosynapsis Hassk.
Buforrestia C.B.Clarke
Callisia Loefl.
Cartonema R.Br.
Cochliostema Lem.
Coleotrype C.B.Clarke
Commelina L.
Cyanotis D.Don
Dichorisandra J.C.Mikan
Dictyospermum Wight
Elasis D.R.Hunt
Floscopa Lour.
Geogenanthus Ule
Gibasis Raf.
Gibasoides D.R.Hunt

Matudanthus D.R.Hunt
Murdannia Royle
Palisota Rchb. ex Endl.
Pollia Thunb.
Polyspatha Benth.
Pseudoparis H.Perrier
Rhopalephora Hassk.
Sauvallea W.Wright
Siderasis Raf.
Spatholirion Ridl.
Stanfieldiella Brenan
Streptolirion Edgew.
Thyrsanthemum Pichon
Tinantia Scheidw.
Tradescantia L.
Tricarpelema J.K.Morton
Triceratella Brenan
Tripogandra Raf.
Weldenia Schult.f.

CONVALLARIACEAE Horan. 1834   175.02

25 genera. Predominantly North temperate. Rhizomatous herbs.

B&H	(MONOCOTS)	Coronarieae (within Liliaceae, 178)
DT&H	(MONOCOTS)	Liliiflorae, Liliineae (within Liliaceae, 264)
Melc	(MONOCOTS)	Liliiflorae, Liliineae (within Liliaceae, 301)
Thor	LILIIFLORAE	Liliales, Liliineae (within Liliaceae, 297)
Dahl	LILIIFLORAE	Asparagales, 377
	(separating *Uvulariaceae* 412 to Liliales)	
Young	(not mentioned, presumably within Liliaceae, 63)	
Takh	LILIIDAE, LILIANAE	Asparagales, 475
	(separating *Medeolaceae* 456 to Liliales)	
Cron	LILIIDAE	Liliales (within Liliaceae, 369)

Aspidistra Ker Gawl.
Clintonia Raf.
Convallaria L.
Disporopsis Hance
Disporum Salisb. ex D.Don
Drymophila R.Br.

Gonioscypha Baker
Kuntheria Conran & Clifford
Liriope Lour.
Maianthemum G.H.Weber
Medeola L.
Ophiopogon Ker Gawl.

Peliosanthes Andrews
Polygonatum Mill.
Reineckea Kunth
Rohdea Roth
Schelhammera R.Br.
Speirantha Baker
Streptopus Michx.

Theropogon Maxim.
Tricalistra Ridl.
Tricyrtis Wall.
Tripladenia D.Don
Tupistra Ker Gawl.
Uvularia L.

CORSIACEAE Becc. 1878   177.02

2 genera. New Guinea, Australasia, Pacific, Chile. Rhizomatous saprophytes.

B&H	(MONOCOTS)	Microspermae (within Burmanniaceae, 168)
DT&H	(MONOCOTS)	Liliiflorae, Burmanniineae (within Burmanniaceae, 275)
Melc	(MONOCOTS)	Liliiflorae, Burmanniineae, 316
Thor	LILIIFLORAE	Liliales, Iridineae (within Burmanniaceae, 305)
Dahl	LILIIFLORAE	Burmanniales, 409
Young	LILIIDAE, LILIANAE	Burmanniales, 78
Takh	LILIIDAE, LILIANAE	Burmanniales, 458
Cron	LILIIDAE	Orchidales, 382

Arachnitis Phil.
Corsia Becc.

COSTACEAE (Meisn.) Nakai 1941   170.02

4 genera. Widespread in tropics. Rhizomatous herbs.

B&H	(MONOCOTS)	Epigynae (within Scitamineae, 170)
DT&H	(MONOCOTS)	Scitamineae (within Zingiberaceae, 272)
Melc	(MONOCOTS)	Scitamineae (within Zingiberaceae, 339)
Thor	COMMELINIFLORAE	Zingiberales, Zingiberineae, 346
Dahl	ZINGIBERIFLORAE	Zingiberales, 448
Young	LILIIDAE, ZINGIBERANAE	Zingiberales, 100
Takh	LILIIDAE, ZINGIBERANAE	Zingiberales, 508
Cron	ZINGIBERIDAE	Zingiberales, 362

Costus L.
Dimerocostus Kuntze

Monocostus K.Schum.
Tapeinochilos Miq.

CYANASTRACEAE Engl. 1900   175.15

1 genus. Tropical Africa. Cormous herbs.

B&H	Not known	—
DT&H	(MONOCOTS)	Farinosae, Pontederiineae (within Pontederiaceae, 260)
Melc	(MONOCOTS)	Liliiflorae, Liliineae, 306

Thor	LILIIFLORAE	Liliales, Liliineae (within Liliaceae, 297)
Dahl	LILIIFLORAE	Asparagales, 392
Young	LILIIDAE, LILIANAE	Liliales, 67
Takh	LILIIDAE, LILIANAE	Liliales, 454
Cron	LILIIDAE	Liliales, 368

Cyanastrum Oliv.

CYCLANTHACEAE Poit. ex A.Rich. 1824   189

12 genera. Tropical New World. Rhizomatous herbs, shrubs, lianes.

B&H	(MONOCOTS)	Nudiflorae, 189
DT&H	(MONOCOTS)	Synanthae, 248
Melc	(MONOCOTS)	Synanthae, 331
Thor	CYCLANTHIFLORAE	Cyclanthales, 321
Dahl	CYCLANTHIFLORAE	Cyclanthales, 46
Young	LILIIDAE, ARECANAE	Cyclanthales, 58
Takh	ARECIDAE, ARECANAE	Cyclanthales, 528
Cron	ARECIDAE	Cyclanthales, 336

Asplundia Harling
Carludovica Ruíz & Pav.
Chorigyne R.Erikss.
Cyclanthus Poit.
Dianthoveus Hammel & Wilder *
Dicranopygium Harling

Evodianthus Oerst.
Ludovia Brongn.
Schultesiophytum Harling *
Sphaeradenia Harling
Stelestylis Drude
Thoracocarpus Harling

CYMODOCEACEAE N. Taylor 1909   195.11

5 genera. Tropical and subtropical coasts. Marine aquatics.

B&H	(MONOCOTS)	Apocarpae (within *Najadaceae*, 195)
DT&H	(MONOCOTS)	Helobiae, Potamogetonineae (within Potamogetonaceae, 237)
Melc	(MONOCOTS)	Helobiae, Potamogetonineae (within Zannichelliaceae, 298)
Thor	ALISMATIFLORAE	Zosterales, Potamogetonineae, 316
Dahl	ALISMATIFLORAE	Najadales, 434
Young	LILIIDAE, ALISMATANAE	Zosterales, 51
Takh	ALISMATIDAE, NAJADANAE	Cymodoceales, 446
Cron	ALISMATIDAE	Najadales, 331

Amphibolis C.Agardh
Cymodocea K.König
Halodule Endl.

Syringodium Kütz.
Thalassodendron Hartog

CYPERACEAE Juss. 1789   199

102 genera. Widespread. Herbs.

B&H (MONOCOTS)
DT&H (MONOCOTS)
Melc (MONOCOTS)
Thor COMMELINIFLORAE
Dahl COMMELINIFLORAE
Young LILIIDAE, COMMELINANAE
Takh LILIIDAE, JUNCANAE
Cron COMMELINIDAE

Glumaceae, 199
Glumiflorae, 246
Cyperales, 337
Commelinales, Juncineae, 333
Cyperales, 459
Cyperales, 92
Cyperales, 513
Cyperales, 351

Actinoschoenus Benth.
Actinoscirpus (Ohwi) R.W.Haines & Lye
Afrotrilepis (Gilly) J.Raynal
Alinula J.Raynal
Amphiscirpus Oteng–Yeb.
Androtrichum (Brongn.) Brongn.
Arthrostylis R.Br.
Ascolepis Nees ex Steud.
Becquerelia Brongn.
Bisboeckelera Kuntze
Blysmus Panz. ex Schult.
Bolboschoenus (Asch.) Palla
Bulbostylis Kunth
Calyptrocarya Nees
Capitularina Kern
Carex L.
Carpha Banks & Sol. ex R.Br.
Caustis R.Br.
Cephalocarpus Nees
Chillania Roiv. *
Chorizandra R.Br.
Chrysitrix L.
Cladium P.Browne
Coleochloa Gilly
Costularia C.B.Clarke
Courtoisina Soják
Crosslandia W.Fitzg.
Cyathochaeta Nees
Cymophyllus Mack. ex Britton & A.Br.
Cyperus L.
Desmoschoenus Hook.f.
Didymiandrum Gilly
Diplacrum R.Br.
Diplasia Pers.
Dulichium Pers.
Egleria G.Eiten
Eleocharis R.Br.
Epischoenus C.B.Clarke
Eriophorum L.
Evandra R.Br.
Everardia Ridl.
Exocarya Benth.
Exochogyne C.B.Clarke
Ficinia Schrad.
Fimbristylis Vahl
Fuirena Rottb.

Gahnia J.R.Forst. & G.Forst.
Gymnoschoenus Nees
Hellmuthia Steud.
Hypolytrum Rich. ex Pers.
Isolepis R.Br.
Kobresia Willd.
Koyamaea W.W.Thomas & Davidse
Kyllinga Rottb.
Kyllingiella R.W.Haines & Lye
Lagenocarpus Nees
Lepidosperma Labill.
Lepironia Rich.
Lipocarpha R.Br.
Machaerina Vahl
Mapania Aubl.
Mapaniopsis C.B.Clarke
Mesomelaena Nees
Microdracoides Hua
Morelotia Gaudich.
Neesenbeckia Levyns
Nelmesia Veken
Nemum Desv.
Oreobolopsis Koyama & Guagl. *
Oreobolus R.Br.
Oxycaryum Nees
Paramapania Uittien
Phylloscirpus C.B.Clarke
Pleurostachys Brongn.
Principina Uittien
Pseudoschoenus (C.B.Clarke)
    Oteng–Yeb.
Ptilanthelium Steud.
Pycreus P.Beauv.
Queenslandiella Domin
Reedia F.Muell.
Remirea Aubl.
Rhynchocladium T.Koyama
Rhynchospora Vahl
Schoenoplectus (Rchb.) Palla
Schoenoxiphium Nees
Schoenus L.
Scirpodendron Zipp. ex Kurz
Scirpoides Séguier
Scirpus L.
Scleria Bergius
Sphaerocyperus Lye

Sumatroscirpus Oteng-Yeb.
Tetraria P.Beauv.
Trachystylis S.T.Blake
Trianoptiles Fenzl ex Endl.
Trichophorum Pers.
Trichoschoenus J.Raynal

Tricostularia Nees ex Lehm.
Trilepis Nees
Uncinia Pers.
Volkiella Merxm. & Czech
Websteria S.H.Wright

DASYPOGONACEAE Dumort. 1829   175.29

2 genera. Australia. Herbs, sometimes arborescent.

B&H   (MONOCOTS)   Calycineae (within Juncaceae, 186)
DT&H   (MONOCOTS)   Liliiflorae, Liliineae (within Liliaceae, 264)
Melc   (MONOCOTS)   Liliiflorae, Liliineae (within
                                              Xanthorrhoeaceae, 302)
Thor   LILIIFLORAE   Liliales, Liliineae (within Liliaceae, 297)
Dahl   LILIIFLORAE   Asparagales, 385
          (including Lomandraceae)
Young (not mentioned, presumably within Xanthorrhoeaceae, 73)
Takh   LILIIDAE, LILIANAE   Amaryllidales, 462
          (including Calectasiaceae, Lomandraceae)
Cron   LILIIDAE   Liliales (within Xanthorrhoeaceae, 374)

Dasypogon R.Br.
Kingia R.Br.

DIOSCOREACEAE R. Br. 1810   174.02

8 genera. Widespread in tropics and subtropics, rarely temperate. Herbs, shrubs, vines.

B&H   (MONOCOTS)   Epigynae, 176
          (including Petermanniaceae, Trichopodaceae)
DT&H   (MONOCOTS)   Liliiflorae, Liliineae, 269
          (including Petermanniaceae, Trichopodaceae)
Melc   (MONOCOTS)   Liliiflorae, Liliineae, 311
          (including Trichopodaceae)
Thor   LILIIFLORAE   Liliales, Dioscoreineae, 301
Dahl   LILIIFLORAE   Dioscoreales, 369
Young LILIIDAE, LILIANAE   Dioscoreales, 60
Takh   LILIIDAE, LILIANAE   Dioscoreales, 490
          (separating *Stenomeridaceae* 489)
Cron   LILIIDAE   Liliales, 379
          (including Trichopodaceae)

Avetra H.Perrier
Borderea Miègev.
Dioscorea L.
Epipetrum Phil.

Nanarepenta Matuda *
Rajania L.
Stenomeris Planch.
Tamus L.

DORYANTHACEAE R. Dahlgren & Clifford 1985   175.19

1 genus. Australia. Giant rosettes.

B&H (MONOCOTS)	Epigynae (within Amaryllidaceae, 174)
DT&H (MONOCOTS)	Liliiflorae, Liliineae (within Amaryllidaceae, 266)
Melc (MONOCOTS)	Liliiflorae, Liliineae (within Agavaceae, 304)
Thor LILIIFLORAE	Liliales, Liliineae (within Liliaceae, 297)
Dahl LILIIFLORAE	Asparagales, 396
Young (not mentioned, presumably within Agavaceae, 71)	
Takh LILIIDAE, LILIANAE	Amaryllidales, 472
Cron LILIIDAE	Liliales (within Agavaceae, 373)

Doryanthes Corrêa

DRACAENACEAE Salisb. 1866   175.06

6 genera. Tropics and subtropics, to S.W. USA. Rhizomatous herbs to trees.

B&H (MONOCOTS)	Coronarieae (within Liliaceae, 178)
DT&H (MONOCOTS)	Liliiflorae, Liliineae (within Liliaceae, 264)
Melc (MONOCOTS)	Liliiflorae, Liliineae (within Liliaceae, 301)
Thor LILIIFLORAE	Liliales, Liliineae (within Liliaceae, 297)
Dahl LILIIFLORAE	Asparagales, 381
(separating *Nolinaceae*)	
Young (not mentioned, presumably within Agavaceae, 71)	
Takh LILIIDAE, LILIANAE	Asparagales, 478
(separating *Nolinaceae*)	
Cron LILIIDAE	Liliales (within Agavaceae, 373)

Beaucarnea Lem.	**Dracaena** Vand. ex L.
Calibanus Rose	**Nolina** Michx.
Dasylirion Zucc.	**Sansevieria** Thunb.

ECDEIOCOLEACEAE D.F. Cutler & Airy Shaw 1965   198.02

1 genus, monotypic. Western Australia. Rhizom. herbs.

B&H (MONOCOTS)	Glumaceae (within Restionaceae, 198)
DT&H (MONOCOTS)	Farinosae, Enantioblastae (within Restionaceae, 252)
Melc (not mentioned, presumably within Restionaceae, 326)	
Thor COMMELINIFLORAE	Commelinales, Flagellariineae (within Restionaceae, 338)
Dahl COMMELINIFLORAE	Poales, 463
Young LILIIDAE, COMMELINANAE	Restionales, 89
Takh LILIIDAE, COMMELINANAE	Restionales, Restionineae, 524
Cron COMMELINIDAE	Restionales (within Restionaceae, 347)

Ecdeiocolea F.Muell.

ERIOCAULACEAE P. Beauv. ex Desv. 1828 196

9 genera. Widespread. Mostly rosette herbs.

B&H	(MONOCOTS)	Glumaceae, 196
DT&H	(MONOCOTS)	Farinosae, Enantioblastae, 256
Melc	(MONOCOTS)	Commelinales, Eriocaulineae, 325
Thor	COMMELINIFLORAE	Commelinales, Eriocaulineae, 336
Dahl	COMMELINIFLORAE	Commelinales, 455
Young	LILIIDAE, COMMELINANAE	Eriocaulales, 84
Takh	LILIIDAE, COMMELINANAE	Commelinales, Eriocaulineae, 519
Cron	COMMELINIDAE	Eriocaulales, 344

Blastocaulon Ruhland
Eriocaulon L.
Lachnocaulon Kunth
Leiothrix Ruhland
Mesanthemum Koern.

Paepalanthus Kunth
Philodice Mart.
Syngonanthus Ruhland
Tonina Aubl.

ERIOSPERMACEAE Endl. 1841 175.16

1 genus. Tropical and southern Africa. Tuberous herbs.

B&H	(MONOCOTS)	Coronarieae (within Liliaceae, 178)
DT&H	(MONOCOTS)	Liliiflorae, Liliineae (within Liliaceae, 264)
Melc	(MONOCOTS)	Liliiflorae, Liliineae (within Liliaceae, 301)
Thor	LILIIFLORAE	Liliales, Liliineae (within Liliaceae, 297)
Dahl	LILIIFLORAE	Asparagales, 393
Young	(not mentioned, presumably within Liliaceae, 63)	
Takh	LILIIDAE, LILIANAE	Liliales, 455
Cron	LILIIDAE	Liliales (within Liliaceae, 369)

Eriospermum Jacq. ex Willd.

FLAGELLARIACEAE Dumort. 1829 185.01

1 genus. Old World tropics. Lianes.

B&H	(MONOCOTS)	Calycineae, 185
	(including Hanguanaceae, Joinvilleaceae)	
DT&H	(MONOCOTS)	Farinosae, Flagellariineae, 251
	(including Hanguanaceae, Joinvilleaceae)	
Melc	(MONOCOTS)	Commelinales, Flagellariineae, 328
	(including Joinvilleaceae)	
Thor	COMMELINIFLORAE	Commelinales, Flagellariineae, 337
	(including Joinvilleaceae)	
Dahl	COMMELINIFLORAE	Poales, 460
Young	LILIIDAE, COMMELINANAE	Restionales, 85
Takh	LILIIDAE, COMMELINANAE	Restionales, Flagellariineae, 520
Cron	COMMELINIDAE	Restionales, 345

Flagellaria L.

GRAMINEAE Juss. 1789

657 genera. Widespread. Herbs or bamboos.

B&H (MONOCOTS)	Glumaceae, 200
DT&H (MONOCOTS)	Glumiflorae, 245
Melc (MONOCOTS)	Graminales, 329
Thor COMMELINIFLORAE	Commelinales, Poineae, 340
Dahl COMMELINIFLORAE	Poales, 462
Young LILIIDAE, COMMELINANAE	Poales, 93
Takh LILIIDAE, COMMELINANAE	Poales, 526
Cron COMMELINIDAE	Cyperales, 352

Aciachne Benth.
Acidosasa C.D.Chu C.S.Chao
Acostia Swallen
Acrachne Wight & Arn. ex Chiov.
Acritochaete Pilg.
Acroceras Stapf
Actinocladum McClure ex Soderstr.
Aegilops L.
Aegopogon Humb. & Bonpl. ex Willd.
Aeluropus Trin.
Afrotrichloris Chiov.
Agenium Nees
Agropyron Gaertn.
Agropyropsis (Batt. & Trab.) A.Camus
Agrostis L.
Aira L.
Airopsis Desv.
Alloeochaete C.E.Hubb.
Allolepis Soderstr. & H.F.Decker
Alloteropsis C.Presl
Alopecurus L.
Alvimia Calderón ex Soderstr. & Londoño
Ammochloa Boiss.
Ammophila Host
Ampelodesmos Link
Amphicarpum Kunth
Amphipogon R.Br.
Anadelphia Hack.
Ancistrachne S.T.Blake
Ancistragrostis S.T.Blake
Andropogon L.
Andropterum Stapf
Anisopogon R.Br.
Anomochloa Brongn.
Anthaenantiopsis Mex ex Pilger
Anthenantia P.Beauv.
Anthephora Schreb.
Anthochloa Nees & Meyen
Anthoxanthum L.
Antinoria Parl.
Apera Adans.

Aphanelytrum Hack.
Apluda L.
Apochiton C.E.Hubb.
Apoclada McClure
Apocopis Nees
Arberella Soderstr. & Calderón
Arctagrostis Griseb.
Arctophila (Rupr.) N.J.Andersson
Aristida L.
Arrhenatherum P.Beauv.
Arthragrostis Lazarides
Arthraxon P.Beauv.
Arthropogon Nees
Arthrostylidium Rupr.
Arundinaria Michx.
Arundinella Raddi
Arundo L.
Arundoclaytonia Davidse & R.P.Ellis
Asthenochloa Büse
Astrebla F.Muell.
Athroostachys Benth.
Atractantha McClure
Aulonemia Goudot
Austrochloris Lazarides
Austrofestuca (Tzvelev) Alexeev
Avena L.
Axonopus P.Beauv.
Bambusa Schreb.
Baptorhachis Clayton & Renvoize
Beckmannia Host
Bewsia Goossens
Bhidea Stapf ex Bor
Blepharidachne Hack.
Blepharoneuron Nash
Boissiera Hochst. ex Steud.
Bothriochloa Kuntze
Bouteloua Lag.
Brachiaria (Trin.)Griseb.
Brachyachne (Benth.) Stapf
Brachychloa S.M.Phillips
Brachyelytrum P.Beauv.
Brachypodium P.Beauv.

Briza L.
Bromuniola Stapf & C.E.Hubb.
Bromus L.
Brylkinia F.Schmidt
Buchloe Engelm.
Buchlomimus J.R.Reeder, C.G.Reeder &
    Rzed.
Buergersiochloa Pilg.
Calamagrostis Adans.
Calamovilfa (A.Gray) Hack.
Calderonella Soderstr. & H.F.Decker
Calyptochloa C.E.Hubb.
Capillipedium Stapf
Castellia Tineo
Catabrosa P.Beauv.
Catalepis Stapf & Stent
Catapodium Link
Cathestecum C.Presl
Cenchrus L.
Centotheca Desv.
Centrochloa Swallen
Centropodia Rchb.
Chaetium Nees
Chaetobromus Nees
Chaetopoa C.E.Hubb.
Chaetopogon Janchen
Chamaeraphis R.Br.
Chandrasekharania V.J.Nair,
    V.S.Ramach. & Sreek.
Chasmanthium Link
Chasmopodium Stapf
Chevalierella A.Camus
Chikusichloa Koidz.
Chimonobambusa Makino
Chionachne R.Br.
Chionochloa Zotov
Chloris Sw.
Chlorocalymma Clayton
Chondrosum Desv.
Chrysochloa Swallen
Chrysopogon Trin.
Chusquea Kunth
Cinna L.
Cladoraphis Franch.
Clausopicula Lazarides
Cleistachne Benth.
Cleistochloa C.E.Hubb.
Coelachne R.Br.
Coelachyrum Hochst. & Nees
Coelorachis Brongn.
Coix L.
Colanthelia McClure & E.W.Sm.
Coleanthus J.Seidel
Colpodium Trin.
Cornucopiae L.

Cortaderia Stapf
Corynephorus P.Beauv.
Cottea Kunth
Craspedorhachis Benth.
Criciuma Soderstr. & Londoño
Crinipes Hochst.
Crithopsis Jaub. & Spach
Crypsis Aiton
Cryptochloa Swallen
Ctenium Panz.
Cutandia Willk.
Cyathopus Stapf
Cyclostachya J.R.Reeder & C.G.Reeder
Cymbopogon Spreng.
Cynodon Rich.
Cynosurus L.
Cyperochloa Lazarides & L.Watson
Cyphochlaena Hack.
Cyrtococcum Stapf
Dactylis L.
Dactyloctenium Willd.
Daknopholis Clayton
Danthonia DC.
Danthonidium C.E.Hubb.
Danthoniopsis Stapf
Dasypyrum (Coss. & Durieu) T.Durand
Decaryella A.Camus *
Decaryochloa A.Camus
Dendrocalamus Nees
Deschampsia P.Beauv.
Desmazeria Dumort.
Desmostachya (Stapf) Stapf
Diandrolyra Stapf
Diarrhena P.Beauv.
Dichaetaria Nees ex Steud.
Dichanthium Willemet
Dichelachne Endl.
Dielsiochloa Pilg.
Digitaria Haller
Dignathia Stapf
Diheteropogon (Hack.) Stapf
Dilophotriche (C.E.Hubb.) Jacq.–Fél.
Dimeria R.Br.
Dinebra Jacq.
Dinochloa Büse
Diplopogon R.Br.
Dissanthelium Trin.
Dissochondrus (W.F.Hillebr.) Kuntze
Distichlis Raf.
Drake–Brockmania Stapf
Dregeochloa Conert
Dryopoa Vickery
Dupontia R.Br.
Duthiea Hack.
Eccoptocarpha Launert

Echinaria Desf.
Echinochloa P.Beauv.
Echinolaena Desv.
Echinopogon P.Beauv.
Ectrosia R.Br.
Ectrosiopsis (Ohwi) Jansen
Ehrharta Thunb.
Ekmanochloa Hitchc.
Eleusine Gaertn.
Elionurus Humb. & Bonpl. ex Willd.
Elymandra Stapf
Elymus L.
Elytrophorus P.Beauv.
Elytrostachys McClure
Enneapogon Desv. ex P.Beauv.
Enteropogon Nees
Entolasia Stapf
Entoplocamia Stapf
Eragrostiella Bor
Eragrostis Wolf
Eremitis Döll
Eremocaulon Soderstr. & Londoño
Eremochloa Büse
Eremopoa Roshev.
Eremopyrum (Ledeb.) Jaub. & Spach
Eriachne R.Br.
Erianthecium Parodi
Eriochloa Kunth
Eriochrysis P.Beauv.
Erioneuron Nash
Euclasta Franch.
Eulalia Kunth
Eulaliopsis Honda
Eustachys Desv.
Euthryptochloa Cope
Exotheca Andersson
Farrago Clayton
Festuca L.
Fingerhuthia Nees
Froesiochloa G.A.Black
Garnotia Brongn.
Gastridium P.Beauv.
Gaudinia P.Beauv.
Germainia Balansa & Poitr.
Gigantochloa Kurz ex Munro
Gilgiochloa Pilg.
Glaziophyton Franch.
Glyceria R.Br.
Glyphochloa Clayton
Gouinia E.Fourn.
Graphephorum Desv.
Greslania Balansa
Griffithsochloa G.J.Pierce
Guaduella Franch.
Gymnopogon P.Beauv.

Gynerium P.Beauv.
Habrochloa C.E.Hubb.
Hackelochloa Kuntze
Hainardia Greuter
Hakonechloa Makino ex Honda
Halopyrum Stapf
Harpachne Hochst. ex A.Rich.
Harpochloa Kunth
Helictotrichon Besser
Hemarthria R.Br.
Hemisorghum C.E.Hubb. ex Bor
Henrardia C.E.Hubb.
Heterachne Benth.
Heteranthelium Jaub. & Spach
Heteranthoecia Stapf
Heteropholis C.E.Hubb.
Heteropogon Pers.
Hickelia A.Camus
Hierochloe R.Br.
Hilaria Kunth
Hitchcockella A.Camus
Holcolemma Stapf & C.E.Hubb.
Holcus L.
Homolepis Chase
Homopholis C.E.Hubb.
Homozeugos Stapf
Hordelymus (Jess.) Jess. ex Harz
Hordeum L.
Hubbardia Bor
Hubbardochloa Auquier
Humbertochloa A.Camus & Stapf
Hydrothauma C.E.Hubb.
Hygrochloa Lazarides
Hygroryza Nees
Hylebates Chippindall
Hymenachne P.Beauv.
Hyparrhenia Fourn.
Hyperthelia Clayton
Hypseochloa C.E.Hubb.
Hystrix Moench
Ichnanthus P.Beauv.
Imperata Cyr.
Indocalamus Nakai
Indopoa Bor
Indosasa McClure
Isachne R.Br.
Ischaemum L.
Iseilema Andersson
Ixophorus Schltdl.
Jansenella Bor
Jouvea Fourn.
Kampochloa Clayton
Kaokochloa De Winter
Kengia Packer
Kerriochloa C.E.Hubb.

Koeleria Pers.
Lagurus L.
Lamarckia Moench
Lamprothyrsus Pilg.
Lasiacis (Griseb.) Hitchc.
Lasiurus Boiss.
Lecomtella A.Camus
Leersia Sw.
Leptagrostis C.E.Hubb.
Leptaspis R.Br.
Leptocarydion Stapf
Leptochloa P.Beauv.
Leptocoryphium Nees
Leptothrium Kunth
Lepturidium Hitchc. & Ekman
Lepturopetium Morat
Lepturus R.Br.
Leymus Hochst.
Libyella Pamp.
Limnas Trin.
Limnodea L.H.Dewey
Limnopoa C.E.Hubb.
Lindbergella Bor
Lintonia Stapf
Lithachne P.Beauv.
Littledalea Hemsl.
Loliolum Krecz. & Bobrov
Lolium L.
Lophacme Stapf
Lophatherum Brongn.
Lopholepis Decne.
Lophopogon Hack.
Loudetia Hochst. ex Steud.
Loudetiopsis Conert
Louisiella C.E.Hubb. & J.Léonard
Loxodera Launert
Luziola Juss.
Lycochloa Sam.
Lycurus Kunth
Lygeum L.
Maclurolyra Calderón & Soderstr.
Maltebrunia Kunth
Manisuris L.
Megalachne Steud.
Megaloprotachne C.E.Hubb.
Megastachya P.Beauv.
Melanocenchris Nees
Melica L.
Melinis P.Beauv.
Melocalamus Benth.
Melocanna Trin.
Merostachys Spreng.
Mesosetum Steud.
Metcalfia Conert
Mibora Adans.

Micraira F.Muell.
Microbriza Nicora & Rugolo
Microcalamus Franch.
Microchloa R.Br.
Micropyropsis Rom.Zarco & Cabezudo
Micropyrum (Gaudin) Link
Microstegium Nees
Milium L.
Miscanthus Andersson
Mnesithea Kunth
Mniochloa Chase
Molinia Schrank
Monachather Steud.
Monanthochloe Engelm.
Monelytrum Hack. ex Schinz
Monocymbium Stapf
Monodia S.W.L.Jacobs
Mosdenia Stent
Muhlenbergia Schreb.
Munroa Torr.
Myriocladus Swallen
Myriostachya (Benth.) Hook.f.
Narduroides Rouy
Nardus L.
Nassella (Trin.) E.Desv.
Nastus Juss.
Neesiochloa Pilg.
Nematopoa C.E.Hubb.
Neobouteloua Gould
Neostapfia Burtt Davy
Neostapfiella A.Camus
Nephelochloa Boiss.
Neurachne R.Br.
Neurolepis Meisn.
Neuropoa Clayton
Neyraudia Hook.f.
Notochloe Domin
Ochlandra Thwaites
Ochthochloa Edgew.
Odontelytrum Hack.
Odyssea Stapf
Olmeca Soderstr.
Olyra L.
Ophiuros C.F.Gaertn.
Opizia C.Presl
Oplismenopsis Parodi
Oplismenus P.Beauv.
Orcuttia Vasey
Oreobambos K.Schum.
Oreochloa Link
Orinus Hitchc.
Oropetium Trin.
Ortachne Nees ex Steud.
Orthoclada P.Beauv.
Oryza L.

Oryzidium C.E.Hubb. & Schweick.
Oryzopsis Michx.
Otachyrium Nees
Ottochloa Dandy
Oxychloris Lazarides
Oxyrhachis Pilg.
Oxytenanthera Munro
Panicum L.
Pappophorum Schreb.
Paractaenum P.Beauv.
Parafestuca E.B.Alexeev
Parahyparrhenia A.Camus
Paraneurachne S.T.Blake
Parapholis C.E.Hubb.
Paratheria Griseb.
Pariana Aubl.
Parodiolyra Soderstr. & Zuloaga
Paspalidium Stapf
Paspalum L.
Pennisetum Rich.
Pentameris P.Beauv.
Pentapogon R.Br.
Pentarrhaphis Kunth
Pentaschistis (Nees) Spach
Pereilema C.Presl
Periballia Trin.
Perotis Aiton
Perrierbambus A.Camus
Peyritschia Fourn.
Phacelurus Griseb.
Phaenanthoecium C.E.Hubb.
Phaenosperma Munro ex Benth.
Phalaris L.
Pharus P.Browne
Pheidochloa S.T.Blake
Phippsia (Trin.) R.Br.
Phleum L.
Pholiurus Trin.
Phragmites Adans.
Phyllorachis Trimen
Phyllostachys Siebold & Zucc.
Piptochaetium J.Presl
Piptophyllum C.E.Hubb.
Piresia Swallen
Plagiantha Renvoize
Plagiosetum Benth.
Plectrachne Henrard
Pleuropogon R.Br.
Plinthanthesis Steud.
Poa L.
Poagrostis Stapf
Podophorus Phil.
Poecilostachys Hack.
Pogonachne Bor
Pogonarthria Stapf

Pogonatherum P.Beauv.
Pogonochloa C.E.Hubb.
Pogononeura Napper
Pohlidium Davidse, Soderstr. & R.P.Ellis
Polevansia De Winter
Polypogon Desf.
Polytoca R.Br.
Polytrias Hack.
Pommereulla L.f.
Porteresia Tateoka
Potamophila R.Br.
Pringleochloa Scribn.
Prionanthium Desv.
Prosphytochloa Schweick.
Psammagrostis C.A.Gardner &
    C.E.Hubb.
Psammochloa Hitchc.
Psathyrostachys Nevski
Pseudanthistiria (Hack.) Hook.f.
Pseudechinolaena Stapf
Pseudochaetochloa Hitchc.
Pseudocoix A.Camus
Pseudodanthonia Bor & C.E.Hubb.
Pseudodichanthium Bor
Pseudopentameris Conert
Pseudoraphis Griff.
Pseudosasa Makino ex Nakai
Pseudosorghum A.Camus
Pscudozoysia Chiov. *
Psilolemma S.M.Phillips
Psilurus Trin.
Puccinellia Parl.
Puelia Franch.
Pyrrhanthera Zotov
Racemobambos Holttum
Raddia Bertol.
Raddiella Swallen
Ratzeburgia Kunth
Redfieldia Vasey
Reederochloa Soderstr. & H.F.Decker
Rehia Fijten
Reimarochloa Hitchc.
Reitzia Swallen
Relchela Steud.
Reynaudia Kunth
Rhipidocladum McClure
Rhizocephalus Boiss.
Rhombolytrum Link
Rhynchoryza Baill.
Rhytachne Desv. ex Ham.
Richardsiella Elffers & Kenn.–O'Byrne
Rostraria Trin.
Rottboellia L.f.
Rytidosperma Steud.
Saccharum L.

717

Sacciolepis Nash
Sartidia De Winter
Sasa Makino & Shib.
Schaffnerella Nash
Schedonnardus Steud.
Schismus P.Beauv.
Schizachne Hack.
Schizachyrium Nees
Schizostachyum Nees
Schmidtia Steud. ex J.A.Schmidt
Schoenefeldia Kunth
Sclerachne R.Br.
Sclerochloa P.Beauv.
Sclerodactylon Stapf
Scleropogon Phil.
Scolochloa Link
Scribneria Hack.
Scutachne Hitchc. & Chase
Secale L.
Sehima Forssk.
Semiarundinaria Nakai
Sesleria Scop.
Setaria P.Beauv.
Setariopsis Scribn. ex Millsp.
Shibataea Makino ex Nakai
Silentvalleya V.J.Nair, Sreek., Vajr. &
 Barghavan
Simplicia T.Kirk
Sinarundinaria Nakai
Sinobambusa Makino
Sitanion Raf.
Snowdenia C.E.Hubb.
Soderstr.ia C.Morton
Sohnsia Airy Shaw
Sorghastrum Nash
Sorghum Moench
Spartina Schreb.
Spartochloa C.E.Hubb.
Spathia Ewart
Sphaerobambos S.Dransf.
Sphaerocaryum Nees ex Hook.f.
Spheneria Kuhlm.
Sphenopholis Scribn.
Sphenopus Trin.
Spinifex L.
Spodiopogon Trin.
Sporobolus R.Br.
Steinchisma Raf.
Steirachne Ekman
Stenotaphrum Trin.
Stephanachne Keng
Stereochlaena Hack.
Steyermarkochloa Davidse & R.P.Ellis
Stipa L.
Stipagrostis Nees

Streblochaete Pilg.
Streptochaeta Schrad. ex Nees
Streptogyna P.Beauv.
Streptolophus D.K.Hughes
Streptostachys Desv.
Styppeiochloa De Winter
Sucrea Soderstr.
Suddia Renvoize
Swallenia Soderstr. & H.F.Decker
Symplectrodia Lazarides
Taeniatherum Nevski
Tarigidia Stent
Tatianyx Zuloaga & Soderstr.
Tetrachaete Chiov.
Tetrachne Nees
Tetrapogon Desf.
Thamnocalamus Munro
Thaumastochloa C.E.Hubb.
Thelepogon Roth ex Roem. & Schult.
Themeda Forssk.
Thrasya Kunth
Thrasyopsis Parodi
Thuarea Pers.
Thyridachne C.E.Hubb.
Thyridolepis S.T.Blake
Thyrsostachys Gamble
Thysanolaena Nees
Torreyochloa Church
Tovarochloa Macfarlane & But *
Trachypogon Nees
Trachys Pers.
Tragus Haller
Tribolium Desv.
Trichloris E.Fourn. ex Benth.
Tricholaena Schrad. ex Schult. &
 Schult.f.
Trichoneura Andersson
Trichopteryx Nees
Tridens Roem. & Schult.
Trikeraia Bor
Trilobachne Schenck ex Henrard
Triniochloa Hitchc.
Triodia R.Br.
Triplachne Link
Triplasis P.Beauv.
Triplopogon Bor
Tripogon Roem. & Schult.
Tripsacum L.
Triraphis R.Br.
Triscenia Griseb.
Trisetaria Forssk.
Trisetum Pers.
Tristachya Nees
Triticum L.
Tuctoria J.R.Reeder

Uniola L.
Uranthoecium Stapf
Urelytrum Hack.
Urochlaena Nees
Urochloa P.Beauv.
Urochondra C.E.Hubb.
Vaseyochloa Hitchc.
Ventenata Koeler
Vetiveria Bory
Vietnamosasa T.Q.Nguyen *
Viguierella A.Camus
Vossia Wall. & Griff.
Vulpia C.C.Gmel.
Vulpiella (Batt. & Trab.) Andr.
Wangenheimia Moench

Whiteochloa C.E.Hubb.
Willkommia Hack.
Xerochloa R.Br.
Yakirra Lazarides & R.D.Webster
Yvesia A.Camus
Zea L.
Zenkeria Trin.
Zeugites P.Browne
Zingeria P.A.Smirn.
Zizania L.
Zizaniopsis Döll & Asch.
Zonotriche (C.E.Hubb.) J.B.Phipps
Zoysia Willd.
Zygochloa S.T.Blake

HAEMODORACEAE R.Br. 1810 172.01

14 genera. S. Africa, New Guinea, Australia, New World. Rhizom. or tuberous herbs.

B&H	(MONOCOTS)	Epigynae, 172
	(including Lanariaceae, Tecophilaeaceae)	
DT&H	(MONOCOTS)	Liliiflorae, Liliineae, 265
Melc	(MONOCOTS)	Liliiflorae, Liliineae, 305
	(including Tecophilaeaceae)	
Thor	COMMELINIFLORAE	Commelinales, Pontederiineae, 331
Dahl	BROMELIIFLORAE	Haemodorales, 439
Young	LILIIDAE, LILIANAE	Liliales, 66
Takh	LILIIDAE, LILIANAE	Haemodorales, 495
	(separating *Conostylidaceae* 496)	
Cron	LILIIDAE	Liliales, 367

Anigozanthos Labill.
Barberetta Harv.
Blancoa Lindl.
Conostylis R.Br.
Dilatris Bergius
Haemodorum Sm.
Lachnanthes Elliott

Macropidia J.L.Drumm. ex Harv.
Phlebocarya R.Br.
Pyrrorhiza Maguire & Wurdack
Schiekia Meisn.
Tribonanthes Endl.
Wachendorfia Burm.
Xiphidium Aubl.

HANGUANACEAE Airy Shaw 1965 175.08

1 genus. Tropical Asia. Rhizomatous herbs.

B&H	(MONOCOTS)	Calycineae (within Flagellariaceae, 185)
DT&H	(MONOCOTS)	Farinosae, Flagellariineae (within Flagellariaceae, 264)
Melc	(MONOCOTS)	Liliiflorae, Liliineae (within ?Xanthorrhoeaceae, 302)
Thor	LILIIFLORAE	Liliales, Liliineae (within Liliaceae, 297)
Dahl	LILIIFLORAE	Asparagales, 384
Young	LILIIDAE, LILIANAE	Liliales, 74

Takh	LILIIDAE, LILIANAE	Asparagales, 482
Cron	LILIIDAE	Liliales, 375

Hanguana Blume

HELICONIACEAE (A. Rich.) Nakai 1941   170.08

1 genus. Tropical America to Pacific. Rhizomatous herbs.

B&H	(MONOCOTS)	Epignyae (within Scitamineae, 170)
DT&H	(MONOCOTS)	Scitamineae (within Musaceae, 271)
Melc	(MONOCOTS)	Scitamineae (within Musaceae, 338)
Thor	COMMELINIFLORAE	Zingiberales, Musineae, 343
Dahl	ZINGIBERIFLORAE	Zingiberales, 445
Young	LILIIDAE, ZINGIBERANAE	Zingiberales, 97
Takh	LILIIDAE, ZINGIBERANAE	Zingiberales, 505
Cron	ZINGIBERIDAE	Zingiberales, 358

Heliconia L.

HEMEROCALLIDACEAE R.Br. 1810   175.20

1 genus. Central Europe to Japan. Herbs.

B&H	(MONOCOTS)	Coronarieae (within Liliaceae, 178)
DT&H	(MONOCOTS)	Liliiflorae, Liliineae (within Liliaceae, 264)
Melc	(MONOCOTS)	Liliiflorae, Liliineae (within Liliaceae, 301)
Thor	LILIIFLORAE	Liliales, Liliineae (within Liliaceae, 297)
Dahl	LILIIFLORAE	Asparagales, 397
Young	(not mentioned, presumably within Liliaceae, 63)	
Takh	LILIIDAE, LILIANAE	Amaryllidales, 469
Cron	LILIIDAE	Liliales (within Liliaceae, 369)

Hemerocallis L.

HERRERIACEAE Endl. 1841   175.05

2 genera. Madagascar, S. America. Climbing subshrubs.

B&H	(MONOCOTS)	Coronarieae (within Liliaceae, 178)
DT&H	(MONOCOTS)	Liliiflorae, Liliineae (within Liliaceae, 264)
Melc	(MONOCOTS)	Liliiflorae, Liliineae (within Liliaceae, 301)
Thor	LILIIFLORAE	Liliales, Liliineae (within Liliaceae, 297)
Dahl	LILIIFLORAE	Asparagales, 380
Young	(not mentioned, presumably within Liliaceae, 63)	
Takh	LILIIDAE, LILIANAE	Asparagales, 480
Cron	LILIIDAE	Liliales (within Liliaceae, 369)

Herreria Ruíz & Pav.
Herreriopsis H Perrier

HOSTACEAE Mathew 1988 175.25

1 genus. China to Japan. Rhizomatous herbs.

B&H	(MONOCOTS)	Coronarieae (within Liliaceae, 178)
DT&H	(MONOCOTS)	Liliiflorae, Liliineae (within Liliaceae, 264)
Melc	(MONOCOTS)	Liliiflorae, Liliineae (within Liliaceae, 301)
Thor	LILIIFLORAE	Liliales, Liliineae (within Liliaceae, 297)
Dahl	LILIIFLORAE	Asparagales (as *Funkiaceae*) 401
Young	(not mentioned, presumably within Liliaceae, 63)	
Takh	LILIIDAE, LILIANAE	Amaryllidales (as *Funkiaceae*) 467
Cron	LILIIDAE	Liliales (within Liliaceae, 369)

Hosta Tratt.

HYACINTHACEAE Batsch ex Borckh. 1797 175.26

45 genera. Eurasia, Africa, N. America. Mostly bulbous, some rhizomatous.

B&H	(MONOCOTS)	Coronarieae (within Liliaceae, 178)
DT&H	(MONOCOTS)	Liliiflorae, Liliineae (within Liliaceae, 264)
Melc	(MONOCOTS)	Liliiflorae, Liliineae (within Liliaceae, 301)
Thor	LILIIFLORAE	Liliales, Liliineae (within Liliaceae, 297)
Dahl	LILIIFLORAE	Asparagales, 402
Young	(not mentioned, presumably within Liliaceae, 63)	
Takh	LILIIDAE, LILIANAE	Amaryllidales, 464
	(separating *Hesperocallidaceae* 466)	
Cron	LILIIDAE	Liliales (within Liliaceae, 369)

Albuca L.
Alrawia (Wendelbo) K.Persson & Wendelbo
Amphisiphon W.F.Barker
Androsiphon Schltr.
Bellevalia Lapeyr.
Bowiea Harv. ex Hook.f.
Brimeura Salisb.
Camassia Lindl.
Chionodoxa Boiss.
Chlorogalum (Lindl.) Kunth
Daubenya Lindl.
Dipcadi Medik.
Drimia Jacq. ex Willd.
Drimiopsis Lindl. & Paxton
Eucomis L'Hér.
Fortunatia J.F.Macbr.
Galtonia Decne.
Hastingsia S.Watson
Hesperocallis A.Gray
Hyacinthella Schur
Hyacinthoides Heist. ex Fabr.
Hyacinthus L.

Lachenalia J.F.Jacq. ex Murray
Ledebouria Roth
Leopoldia Parl.
Litanthus Harv.
Massonia Thunb. ex L.f.
Muscari Mill.
Muscarimia Kostel.
Neopatersonia Schönl.
Ornithogalum L.
Polyxena Kunth
Pseudogaltonia (Kuntze) Engl.
Pseudomuscari Garbari & Greuter
Puschkinia M.F.Adams
Rhadamanthus Salisb.
Rhodocodon Baker
Schizobasis Baker
Schoenolirion Torr. ex E.M.Durand
Scilla L.
Sypharissa Salisb.
Thuranthos C.H.Wright
Urginea Steinh.
Veltheimia Gled.
Whiteheadia Harv.

HYDATELLACEAE U.Hamann 1976   197.02

2 genera. Australia, New Zealand. Small annual or rhizomatous subaquatic herbs.

B&H	Not known	—
DT&H	(MONOCOTS)	Farinosae, Enantioblastae (within Centrolepidaceae, 253)
Melc	(MONOCOTS)	Commelinales, Restionineae (within Centrolepidaceae, 327)
Thor	INCERTAE SEDIS, 351	
Dahl	COMMELINIFLORAE	Hydatellales, 456
Young	LILIIDAE, COMMELINANE	Restionales, 90
Takh	LILIIDAE, HYDATELLANAE	Hydatellales, 514
Cron	COMMELINIDAE	Hydatellales, 353

Hydatella Diels
Trithuria Hook.f.

HYDROCHARITACEAE Juss. 1789   194.04

20 genera. Widespread. Aquatic, sometimes marine, herbs.

B&H	(MONOCOTS) (separating *Najadaceae*, 195)	Microspermae, 167
DT&H	(MONOCOTS) (separating *Najadaceae*, 238)	Helobiae, Butomineae, 243
Melc	(MONOCOTS) (separating *Najadaceae* 299)	Helobiae, Hydrocharitineae, 293
Thor	ALISMATIFLORAE (separating *Najadaceae*, 318)	Alismatales, 310
Dahl	ALISMATIFLORAE (separating *Najadaceae* 435)	Alismatales, 427
Young	LILIIDAE, ALISMATANAE (separating *Najadaceae* 53)	Alismatales, 44
Takh	ALISMATIDAE, BUTOMANAE (separating *Thalassiaceae* 432, *Halophilaceae* 433, *Najadaceae* 447)	Hydrocharitales, 431
Cron	ALISMATIDAE (separating *Najadaceae* 328)	Hydrocharitales, 322

Apalanthe Planch.
Appertiella C.D.K.Cook & Triest
Blyxa Noronha ex Thouars
Egeria Planch.
Elodea Michx.
Enhalus Rich.
Enhydrias Ridl.
Halophila Thouars
Hydrilla Rich.
Hydrocharis L.

Lagarosiphon Harv.
Limnobium Rich.
Maidenia Rendle
Najas L.
Nechamandra Planch.
Oligolobos Gagnep.
Ottelia Pers.
Stratiotes L.
Thalassia Banks ex C.Koenig
Vallisneria L.

HYPOXIDACEAE R.Br. 1814   175.12

9 genera. New World, Old World tropics and S. Africa. Rhizomatous or cormous herbs.

B&H　(MONOCOTS)
DT&H　(MONOCOTS)

Melc　(MONOCOTS)
Thor　LILIIFLORAE
Dahl　LILIIFLORAE
Young　(not mentioned, presumably within Liliaceae, 63)
Takh　LILIIDAE, LILIANAE
Cron　LILIIDAE

Curculigo Gaertn.
Empodium Salisb.
Hypoxidia Friedmann
Hypoxis L.
Molineria Colla

Epigynae (within Amaryllidaceae, 174)
Liliiflorae, Liliineae (within
　　　　　　　　　Amaryllidaceae, 266)
Liliiflorae, Liliineae, 308
Liliales, Liliineae (within Liliaceae, 297)
Asparagales, 390

Haemodorales, 497
Liliales (within Liliaceae, 369)

Pauridia Harv.
Rhodohypoxis Nel
Saniella Hilliard & B.L.Burtt
Spiloxene Salisb.

IRIDACEAE Juss. 1789

178.05

88 genera. Widespread. Rhizomatous or cormous herbs.

B&H　(MONOCOTS)
DT&H　(MONOCOTS)
Melc　(MONOCOTS)
　　　(separating *Geosiridaceae*)
Thor　LILIIFLORAE
Dahl　LILIIFLORAE
　　　(separating *Geosiridaceae*)
Young LILIIDAE, LILIANAE
　　　(separating *Geosiridaceae*)
Takh　LILIIDAE, LILIANAE
　　　(separating *Geosiridaceae*)
Cron　LILIIDAE
　　　(separating *Geosiridaceae*)

Epigynae, 173
Liliiflorae, Iridineae, 270
Liliiflorae, Iridineae, 313

Liliales, Iridineae, 304
Liliales, 416

Liliales, 68

Liliales, 451

Liliales, 370

Ainea Ravenna *
Alophia Herb.
Anapalina N.E.Br.
Anomatheca Ker Gawl.
Antholyza L.
Aristea Sol. ex Aiton
Babiana Ker Gawl. ex Sims
Barnardiella Goldblatt
Belamcanda Adans.
Bobartia L.
Calydorea Herb.
Cardenanthus R.C.Foster
Chasmanthe N.E.Br.
Cipura Aubl.
Cobana Ravenna
Crocosmia Planch.
Crocus L.
Cypella Herb.
Devia Goldblatt & J.C.Manning

Dierama K.Koch
Dietes Salisb. ex Klatt
Diplarrhena Labill.
Duthiastrum M.P.de Vos
Eleutherine Herb.
Ennealophus N.E.Br.
Eurynotia R.C.Foster *
Ferraria Burm. ex Mill.
Fosteria Molseed
Freesia Eckl. ex Klatt
Galaxia Thunb.
Geissorhiza Ker Gawl.
Gelasine Herb.
Geosiris Baill.
Gladiolus L.
Gynandriris Parl.
Herbertia Sweet
Hermodactylus Mill.
Hesperantha Ker Gawl.

Hesperoxiphion Baker
Hexaglottis Vent.
Homeria Vent.
Iris L.
Isophysis T.Moore
Ixia L.
Kelissa Ravenna *
Klattia Baker
Lapeirousia Pourr.
Larentia Klatt
Lethia Ravenna
Libertia Spreng.
Mastigostyla I.M.Johnst.
Melasphaerula Ker Gawl.
Micranthus (Pers.) Eckl.
Moraea Mill.
Nemastylis Nutt.
Neomarica Sprague
Nivenia Vent.
Olsynium Raf.
Onira Ravenna
Orthrosanthus Sweet
Pardanthopsis (Hance) Lenz
Patersonia R.Br.
Phalocallis Herb.

Pillansia L.Bolus
Pseudotrimezia R.C.Foster
Radinosiphon N.E.Br.
Rheome Goldblatt
Rigidella Lindl.
Roggeveldia Goldblatt
Romulea Maratti
Savannosiphon Goldblatt & Marais
Schizostylis Backh. & Harv.
Sessilanthera Molseed & Cruden
Sessilistigma Goldblatt
Sisyrinchium L.
Solenomelus Miers
Sparaxis Ker Gawl.
Sphenostigma Baker
Syringodea Hook.f.
Tapeinia Comm. ex Juss.
Thereianthus G.J.Lewis
Tigridia Juss.
Trimezia Salisb. ex Herb.
Tritonia Ker Gawl.
Tritoniopsis L.Bolus
Watsonia Mill.
Witsenia Thunb.
Zygotritonia Mildbr.

IXIOLIRIACEAE (Pax) Nakai 1943   175.17

1 genus. Turkey to Siberia, Cormous herbs.

B&H	(MONOCOTS)	Epigynae (within Amaryllidaceae, 178)
DT&H	(MONOCOTS)	Liliiflorae, Liliineae (within Amaryllidaceae, 266)
Melc	(MONOCOTS)	Liliiflorae, Liliineae (within Amaryllidaceae, 307)
Thor	LILIIFLORAE	Liliales, Liliineae (within Liliaceae, 297)
Dahl	LILIIFLORAE	Asparagales, 394
Young	(not mentioned, presumably within Liliaceae, 63)	
Takh	LILIIDAE, LILIANAE	Amaryllidales, 474
Cron	LILIIDAE	Liliales (within Liliaceae, 369)

Ixiolirion Herb.

JOINVILLEACEAE Toml. & A.C. Sm. 1970   185.02

1 genus. Malesia to Pacific. Large erect herbs.

B&H	(MONOCOTS)	Calcycineae (within Flagellariaceae, 185)
DT&H	(MONOCOTS)	Farinosae, Flagellariineae (within Flagellariaceae, 251)
Melc	(MONOCOTS)	Commelinales, Flagellariineae (within Flagellariaceae, 328)

Thor COMMELINIFLORAE

Dahl COMMELINIFLORAE
Young LILIIDAE, COMMELINANAE
Takh LILIIDAE, COMMELINANAE
Cron COMMELINIDAE

Joinvillea Gaudich. ex Brongn. & Gris

Commelinales, Flagellariineae (within
Flagellariaceae, 337)

Poales, 461
Restionales, 86
Restionales, Flagellariineae, 521
Restionales, 346

JUNCACEAE Juss. 1789

186.01

7 genera. Widespread. Annual to rhizomatous herbs, rarely woody.

B&H (MONOCOTS)
      (including Calectasiaceae, Dasypogonaceae, Lomandraceae, Thurniaceae,
      Xanthorrhoeaceae)
DT&H (MONOCOTS)
      (including Thurniaceae)
Melc (MONOCOTS)
Thor COMMELINIFLORAE
      (including Thurniaceae)
Dahl COMMELINIFLORAE
Young LILIIDAE, COMMELINANAE
      (including Thurniaceae)
Takh LILIIDAE, JUNCANAE
Cron COMMELINIDAE

Calycineae, 186

Liliiflorae, Juncineae, 262

Juncales, 318
Commelinales, Juncineae, 332

Cyperales, 457
Juncales, 91

Juncales, 511
Juncales, 349

Distichia Nees & Meyen
Juncus L.
Luzula DC.
Marsippospermum Desv.

Oxychloe Phil.
Prionium E.Mey.
Rostkovia Desv.

JUNCAGINACEAE Rich. 1808

195.02

3 genera. N and S temperate or cold. Rhizomatous herbs.

B&H (MONOCOTS)
DT&H (MONOCOTS)
      (including Lilaeaceae, Scheuchzeriaceae)
Melc (MONOCOTS)
      (including Lilaeaceae)
Thor ALISMATIFLORAE
      (including Lilaeaceae, Scheuchzeriaceae)
Dahl ALISMATIFLORAE
      (including Lilaeaceae)
Young LILIIDAE, ALISMATANAE
      (including Lilaeaceae)
Takh ALISMATIDAE, NAJADANAE
      (separating *Maundiaceae*)
Cron ALISMATIDAE
      (including Lilaeaceae)

Apocarpae (within *Najadaceae*, 195)
Helobiae, Potamogetonineae, 240

Helobiae, Potamogetonineae, 296

Zosterales, Potamogetonineae, 312

Najadales, 429

Zosterales, 47

Juncaginales, 438

Najadales, 325

Maundia F.Muell.
Tetroncium Willd.
Triglochin L.

LANARIACEAE H. Huber ex R. Dahlgren 1988   175.14

1 genus, monotypic. South Africa. Herb.

B&H	(MONOCOTS)	Epigynae (within Haemodoraceae, 172)
DT&H	(MONOCOTS	Liliiflorae, Liliineae (within Amaryllidaceae, 266)
Melc	Not mentioned	—
Thor	LILIIFLORAE	Liliales, Liliineae (within Liliaceae, 297)
Dahl	LILIIFLORAE	Asparagales (within Tecophilaeaceae, 391)
Young	Not mentioned	—
Takh	LILIIDAE, LILIANAE	Liliales (within Tecophilaeaceae, 453)
Cron	Not mentioned	—

Lanaria Aiton

LEMNACEAE Gray 1821   192

4 genera. Widespread. Minute free-floating aquatics.

B&H	(MONOCOTS)	Nudiflorae, 192
DT&H	(MONOCOTS)	Spathiflorae, 250
Melc	(MONOCOTS)	Spathiflorae, 333
Thor	ARIFLORAE	Arales, 320
Dahl	ARIFLORAE	Arales, 421
Young	LILIIDAE, ARANAE	Arales, 56
Takh	ARECIDAE, ARANAE	Arales, 531
Cron	ARECIDAE	Arales, 339

Lemna L.   **Wolffia** Horkel ex Schleid.
Spirodela Schleid.   **Wolffiella** (Hegelm.) Hegelm.

LILAEACEAE Dumort. 1829   195.04

1 genus, monotypic. British Columbia and Idaho to Mexico and high Andes. Herb.

B&H	(MONOCOTS)	Apocarpae (within *Najadaceae*, 195)
DT&H	(MONOCOTS)	Helobiae, Potamogetonineae (within Juncaginaceae, 240)
Melc	(MONOCOTS)	Helobiae, Potamogetonineae (within Juncaginaceae, 296)
Thor	ALISMATIFLORAE	Zosterales, Potamogetonineae (within Juncaginaceae, 312)
Dahl	ALISMATIFLORAE	Najadales (within Juncaginaceae, 429)
Young	LILIIDAE, ALISMATANAE	Zosterales (within Juncaginaceae, 47)
Takh	ALISMATIDAE, NAJADANAE	Juncaginales, 439
Cron	ALISMATIDAE	Najadales (within Juncaginaceae, 325)

Lilaea Bonpl.

LILIACEAE Juss. 1789   178.03

11 genera. Widespread but mostly N. temp. Bulbous herbs.

B&H   (MONOCOTS)   Coronarieae 178
      (including Alliaceae, Aloaceae, Anthericaceae, Aphyllanthaceae, Aspara-
      gaceae, Asphodelaceae, Asteliaceae, Blandfordiaceae, Colchicaceae, Conval-
      lariaceae, Dracaenaceae, Eriospermaceae, Hemerocallidaceae, Herreriaceae,
      Hostaceae, Hyacinthaceae, Melanthiaceae, Philesiaceae, Phormiaceae,
      Rhipogonaceae, Ruscaceae, Smilacaceae, Trilliaceae)
DT&H (MONOCOTS)   Liliiflorae, Liliineae 264)
      (including all those included by B&H above, plus Calectasiaceae, Dasypo-
      gonaceae, Lomandraceae and Xanthorrhoeaceae)
Melc   (MONOCOTS)   Liliiflorae, Liliineae, 301
      (including all those included by B&H above, plus Alstroemeriaceae and
      Petermanniaceae)
Thor   LILIIFLORAE   Liliales, Liliinae, 297)
      (including all those included by B&H above except Philesiaceae, Rhipo-
      gonaceae and Smilacaceae, plus Agavaceae, Amaryllidaceae, Calectasiaceae,
      Cyanastraceae, Dasypogonaceae, Doryanthaceae, Hanguanaceae, Hypoxi-
      daceae, Ixioliriaceae, Lanariaceae, Lomandraceae, Tecophilaeaceae and
      Xanthorrhoeaceae)
Dahl   LILIIFLORAE   Liliales, 414
      (separating *Calochortaceae*)
Young LILIIDAE, LILIANAE   Liliales, 63
      (with same circumscription as Cronquist below)
Takh   LILIIDAE, LILIANAE   Liliales, 457
      (separating *Calochortaceae*)
Cron   LILIIDAE   Liliales, 369
      (including all those included by B&H above except Aloaceae, Dracaenaceae,
      Philesiaceae, Phormiaceae, Rhipogonaceae and Smilacaceae, plus Alstroem-
      eriaceae, Amaryllidaceae, Hypoxidaceae, Ixioliriaceae and Tecophilaeaceae)

Calochortus Pursh   **Lilium** L.
Cardiocrinum (Endl.) Lindl.   **Lloydia** Salisb. ex Rchb.
Erythronium L.   **Nomocharis** Franch.
Fritillaria L.   **Notholirion** Wall. ex Boiss.
Gagea Salisb.   **Tulipa** L.
Korolkowia Regel

LIMNOCHARITACEAE Takht. ex Cronquist 1981   194.03

3 genera. Tropics and subtropics. Aquatic or semi-aquatic herbs.

B&H   (MONOCOTS)   Apocarpae (within Alismataceae, 194)
DT&H (MONOCOTS)   Helobiae, Butomineae (within Butomaceae,
                                                                    242)
Melc   (MONOCOTS)   Helobiae, Alismatineae (within Butomaceae,
                                                                    292)
Thor   ALISMATIFLORAE   Alismatales (within Alismataceae, 309)
Dahl   ALISMATAFLORAE   Alismatales, 425
Young LILIIDAE, ALISMATANAE   Alismatales (within Alismataceae, 42)

Takh   ALISMATIDAE, BUTOMANAE   Alismatales, 434
Cron   ALISMATIDAE   Alismatales, 320

Butomopsis Kunth
Hydrocleys Rich.
Limnocharis Bonpl.

LOMANDRACEAE Lotsy 1911   175.30

6 genera. New Guinea, Australia, New Caledonia. Tufted perennial herbs.

B&H   (MONOCOTS)   Calycineae (within Juncaceae, 186)
DT&H   (MONOCOTS)   Liliiflorae, Liliineae (within Liliaceae, 264)
Melc   (MONOCOTS)   Liliiflorae, Liliineae
                                    (within Xanthorrhoeaceae, 302)
Thor   LILIIFLORAE   Liliales, Liliineae (within Liliaceae, 297)
Dahl   LILIIFLORAE   Asparagales (within Dasypogonaceae, 385)
Young (not mentioned, presumably within Xanthorrhoeaceae, 73)
Takh   LILIIDAE, LILIANAE   Amaryllidales (within Dasypogonaceae, 462)
Cron   LILIIDAE   Liliales (within Xanthorrhoeaceae, 374)

Acanthocarpus Lehm.   **Lomandra** Labill.
Baxteria R.Br. ex Hook.   **Romnalda** P.F.Stevens
Chamaexeros Benth.   **Xerolirion** A.S.George

LOWIACEAE Ridl. 1924   170.05

1 genus. South China and Indochina to Borneo. Rhizomatous herbs.

B&H   (MONOCOTS)   Epigynae (within Scitamineae, 170)
DT&H   (MONOCOTS)   Scitamineae (within Musaceae, 271)
Melc   (MONOCOTS)   Scitamineae, 342
Thor   COMMELINIFLORAE   Zingiberales, Musineae, 344
Dahl   ZINGIBERIFLORAE   Zingiberales, 443
Young LILIIDAE, ZINGIBERANAE   Zingiberales, 98
Takh   LILIIDAE, ZINGIBERANAE   Zingiberales, 506
Cron   ZINGIBERIDAE   Zingiberales, 360

Orchidantha N.E.Br.

MARANTACEAE O. Petersen 1888   170.03

32 genera. Tropical, especially New World. Rhizomatous herbs to lianes.

B&H   (MONOCOTS)   Epigynae (within *Scitamineae*, 170)
DT&H   (MONOCOTS)   Scitamineae, 274
Melc   (MONOCOTS)   Scitamineae, 341
Thor   COMMELINIFLORAE   Zingiberales, Zingiberineae, 348
Dahl   ZINGIBERIFLORAE   Zingiberales, 450
Young LILIIDAE, ZINGIBERANAE   Zingiberales, 102

| Takh | LILIIDAE, ZINGIBERANAE | Zingiberales, 510 |
| Cron | ZINGIBERIDAE | Zingiberales, 364 |

Afrocalathea K.Schum.
Ataenidia Gagnep.
Calathea G.Mey.
Cominsia Hemsl.
Ctenanthe Eichler
Donax Lour.
Halopegia K.Schum.
Haumania J.Léonard
Hylaeanthe Jonk.–Verh. & Jonk.
Hypselodelphys (K.Schum.)
   Milne–Redh.
Ischnosiphon Koern.
Koernickanthe L.Andersson
Maranta L.
Marantochloa Brongn. ex Gris
Megaphrynium Milne–Redh.
Monophrynium K.Schum.

Monophyllanthe K.Schum.
Monotagma K.Schum.
Myrosma L.f.
Phacelophrynium K.Schum.
Phrynium Willd.
Pleiostachya K.Schum.
Sanblasia L.Andersson *
Saranthe (Regel & Koern.) Eichl.
Sarcophrynium K.Schum.
Schumannianthus Gagnep.
Stachyphrynium K.Schum.
Stromanthe Sond.
Thalia L.
Thaumatococcus Benth.
Thymocarpus Nicolson, Steyerm. &
   Sivad.
Trachyphrynium Benth.

MAYACACEAE Kunth 1842   182

1 genus. W.C. trop. Africa, N. and S. America. Aquatic herbs.

B&H	(MONOCOTS)	Coronarieae, 182
DT&H	(MONOCOTS)	Farinosae, Enantioblastae, 254
Melc	(MONOCOTS)	Commelinales, Commelinineae, 322
Thor	COMMELINIFLORAE	Commelinales, Commelinineae, 335
Dahl	COMMELINIFLORAE	Commelinales, 452
Young	LILIIDAE, COMMELINANAE	Commelinales, 82
Takh	LILIIDAE, COMMELINANAE	Commelinales, Commelinineae, 516
Cron	COMMELINIDAE	Commelinales, 342

Mayaca Aubl.

MELANTHIACEAE Batsch 1802   176

24 genera. Widespread except Africa, mostly north temperate. Rhizomatous herbs.

B&H	(MONOCOTS)	Coronarieae (within Liliaceae, 178)
DT&H	(MONOCOTS)	Liliiflorae, Liliineae (within Liliaceae, 264)
Melc	(MONOCOTS)	LIliiflorae, Liliineae (within Liliaceae, 301)
Thor	LILIIFLORAE	Liliales, Liliineae (within Liliaceae, 297)
Dahl	LILIIFLORAE	Melanthiales, 405

    (separating *Campynemaceae* 406)
Young (not mentioned, presumably within Liliaceae 63)

| Takh | LILIIDAE, LILIANAE | Liliales, 449 |

    (including Colchicaceae)

| Cron | LILIIDAE | Liliales (within Liliaceae 369) |

    (separating *Petrosaviaceae* 333 to ALISMATIDAE Triuridales)

Aletris L.
Amianthium A.Gray

Campynema Labill.
Campynemanthe Baill.

Chamaelirium Willd.	**Nietneria** Klotzsch ex Benth.
Chionographis Maxim.	**Petrosavia** Becc.
Harperocallis McDaniel *	**Pleea** Michx.
Helonias L.	**Schoenocaulon** A.Gray
Heloniopsis A.Gray	**Stenanthium** (A.Gray) Kunth
Japonolirion Nakai	**Tofieldia** Huds.
Lophiola Ker Gawl.	**Veratrum** L.
Melanthium L.	**Xerophyllum** Michx.
Metanarthecium Maxim.	**Ypsilandra** Franch.
Narthecium Huds.	**Zigadenus** Michx.

MUSACEAE Juss. 1789 170.04

2 genera. Old World tropics. Giant herbs with pseudostems.

B&H	(MONOCOTS)	Epigynae (within *Scitamineae*, 170)
DT&H	(MONOCOTS)	Scitamineae, 271
	(including Heliconiaceae, Lowiaceae, Strelitziaceae)	
Melc	(MONOCOTS)	Scitamineae, 338
	(including Heliconiaceae, Lowiaceae, Strelitziaceae)	
Thor	COMMELINIFLORAE	Zingiberales, Musineae, 341
Dahl	ZINGIBERIFLORAE	Zingiberales, 444
Young	LILIIDAE, ZINGIBERANAE	Zingiberales, 95
Takh	LILIIDAE, ZINGIBERANAE	Zingiberales, 504
Cron	ZINGIBERIDAE	Zingiberales, 359

Ensete Horan.
Musa L.

ORCHIDACEAE Juss. 1789 169

835 genera. Widespread. Perennial herbs, often epiphytic, some saprophytes.

B&H	(MONOCOTS)	Microspermae, 169
DT&H	(MONOCOTS)	Microspermae, Gynandrae, 276
Melc	(MONOCOTS)	Microspermae, 343
Thor	LILIIFLORAE	Liliales, Orchidineae, 306
Dahl	LILIIFLORAE	Liliales, 419
	(separating *Apostasiaceae* 417, *Cypripediaceae* 418)	
Young	LILIIDAE, LILIANAE	Orchidales, 79
Takh	LILIIDAE, LILIANAE	Orchidales, 498
Cron	LILIIDAE	Orchidales, 383

Aa Rchb.f.	**Acostaea** Schltr.
Abdominea J.J.Sm.	**Acriopsis** Blume
Acampe Lindl.	**Acrolophia** Pfitzer
Acanthephippium Blume	**Acrorchis** Dressler *
Aceras R.Br.	**Ada** Lindl.
Aceratorchis Schltr.	**Adenochilus** Hook.f.
Acianthus R.Br.	**Adenoncos** Blume
Acineta Lindl.	**Adrorhizon** Hook.f.
Acoridium Nees & Meyen	**Aerangis** Rchb.f.

Aeranthes Lindl.
Aerides Lour.
Aganisia Lindl.
Aglossorhyncha Schltr.
Agrostophyllum Blume
Alamania Lex.
Altensteinia Kunth
Amblostoma Scheidw.
Ambrella H.Perrier
Amerorchis Hultén
Amesiella Schltr. ex Garay
Amitostigma Schltr.
Amparoa Schltr.
Amphigena Rolfe
Anacamptis Rich.
Ancistrochilus Rolfe
Ancistrorhynchus Finet
Androchilus Liebm. ex Hartman
Androcorys Schltr.
Angraecopsis Kraenzl.
Angraecum Bory
Anguloa Ruíz & Pav.
Ania Lindl.
Anneliesia Brieger & Lückel
Anochilus Rolfe
Anoectochilus Blume
Ansellia Lindl.
Anthogonium Wall. ex Lindl.
Anthosiphon Schltr.
Antillanorchis Garay
Aphyllorchis Blume
Aplectrum (Nutt.) Torr.
Aporostylis Rupp & Hatch
Apostasia Blume
Appendicula Blume
Aracamunia Carnevali & I.Ramírez *
Arachnis Blume
Archineottia S.C.Chen
Arethusa L.
Armodorum Breda
Arnottia A.Rich.
Arpophyllum Lex.
Arthrochilus F.Muell.
Artorima Dressler & G.E.Pollard
Arundina Blume
Ascidieria Seidenf.
Ascocentrum Schltr. ex J.J.Sm.
Ascochilopsis C.E.Carr
Ascochilus Ridl.
Ascoglossum Schltr.
Aspasia Lindl.
Aspidogyne Garay
Aulosepalum Garay
Auxopus Schltr.
Baptistonia Barb.Rodr.

Barbosella Schltr.
Barbrodria Luer
Barkeria Knowles & Westc.
Barlia Parl.
Barombia Schltr.
Bartholina R.Br.
Basiphyllaea Schltr.
Baskervillea Lindl.
Batemannia Lindl.
Beclardia A.Rich.
Benthamia A.Rich.
Benzingia Dodson *
Biermannia King & Pantl.
Bifrenaria Lindl.
Binotia Rolfe
Bipinnula Comm. ex Juss.
Bletia Ruíz & Pav.
Bletilla Rchb.f.
Bogoria J.J.Sm.
Bollea Rchb.f.
Bolusiella Schltr.
Bonatea Willd.
Bonniera Cordem.
Bothriochilus Lem.
Brachionidium Lindl.
Brachtia Rchb.f.
Brachycorythis Lindl.
Brachypeza Garay
Brachystele Schltr.
Bracisepalum J.J.Sm.
Brassavola R.Br.
Brassia R.Br.
Briegeria Senghas *
Bromheadia Lindl.
Broughtonia R.Br.
Brownleea Harv. ex Lindl.
Buchtienia Schltr.
Bulbophyllum Thouars
Bulleyia Schltr.
Burnettia Lindl.
Cadetia Gaudich.
Caladenia R.Br.
Calanthe Ker Gawl.
Caleana R.Br.
Calochilus R.Br.
Calopogon R.Br.
Caluera Dodson & Determann *
Calymmanthera Schltr.
Calypso Salisb.
Calyptrochilum Kraenzl.
Campylocentrum Benth.
Capanemia Barb.Rodr.
Cardiochilos P.J.Cribb
Catasetum Rich. ex Kunth
Cattleya Lindl.

Cattleyopsis Lem.
Caucaea Schltr.
Caularthron Raf.
Centroglossa Barb.Rodr.
Centrostigma Schltr.
Cephalanthera Rich.
Cephalantheropsis Guillaumin
Ceratandra Eckl. ex F.A.Bauer
Ceratocentron Senghas
Ceratochilus Blume
Ceratostylis Blume
Chaenanthe Lindl.
Chamaeangis Schltr.
Chamaeanthus Schltr. ex J.J.Sm.
Chamaegastrodia Makino & F.Maek.
Chamelophyton Garay
Chamorchis Rich.
Changnienia S.S.Chien
Chaseella Summerh.
Chaubardia Rchb.f.
Chaubardiella Garay
Chauliodon Summerh.
Cheiradenia Lindl.
Cheirostylis Blume
Chelonistele Pfitzer
Chiloglottis R.Br.
Chiloschista Lindl.
Chitonanthera Schltr.
Chitonochilus Schltr.
Chloraea Lindl.
Chlorosa Blume
Chondrorhyncha Lindl.
Chroniochilus J.J.Sm.
Chrysocycnis Linden & Rchb.f.
Chrysoglossum Blume
Chusua Nevski
Chysis Lindl.
Chytroglossa Rchb.f.
Cirrhaea Lindl.
Cischweinfia Dressler & N.H.Williams
Claderia Hook.f.
Cleisocentron Brühl
Cleisomeria Lindl. ex G.Don
Cleisostoma Blume
Cleistes Rich. ex Lindl.
Clematepistephium N.Hallé
Clowesia Lindl.
Cochleanthes Raf.
Cochlioda Lindl.
Codonorchis Lindl.
Codonosiphon Schltr.
Coelia Lindl.
Coeliopsis Rchb.f.
Coeloglossum Hartm.
Coelogyne Lindl.

Cohniella Pfitzer
Coilochilus Schltr.
Collabium Blume
Comparettia Poepp. & Endl.
Comperia K.Koch
Condylago Luer
Constantia Barb.Rodr.
Corallorrhiza Gagnebin
Cordiglottis J.J.Sm.
Coryanthes Hook.
Corybas Salisb.
Corycium Sw.
Corymborkis Thouars
Cottonia Wight
Cranichis Sw.
Cremastra Lindl.
Cribbia Senghas
Crybe Lindl.
Cryptarrhena R.Br.
Cryptocentrum Benth.
Cryptochilus Wall.
Cryptophoranthus Barb.Rodr.
Cryptopus Lindl.
Cryptopylos Garay
Cryptostylis R.Br.
Cuitlauzina La Llave & Lex.
Cyanaeorchis Barb.Rodr.
Cybebus Garay *
Cyclopogon C.Presl
Cycnoches Lindl.
Cymbidiella Rolfe
Cymbidium Sw.
Cynorkis Thouars
Cyphochilus Schltr.
Cypholoron Dodson & Dressler
Cypripedium L.
Cyrtidiorchis Rauschert
Cyrtopodium R.Br.
Cyrtorchis Schltr.
Cyrtosia Blume
Cyrtostylis R.Br.
Cystorchis Blume
Dactylorhiza Neck. ex Nevski
Dactylorhynchus Schltr.
Dactylostalix Rchb.f.
Darwiniera Braas & Lückel *
Degranvillea Determann
Deiregyne Schltr.
Dendrobium Sw.
Dendrochilum Blume
Dendrophylax Rchb.f.
Diadenium Poepp. & Endl.
Diaphananthe Schltr.
Diceratostele Summerh.
Dicerostylis Blume

Dichaea Lindl.
Dichromanthus Garay *
Dickasonia L.O.Williams
Dictyophyllaria Garay
Didiciea King & Prain
Didymoplexiella Garay
Didymoplexis Griff.
Diglyphosa Blume
Dignathe Lindl.
Dilochia Lindl.
Dilomilis Raf.
Dimerandra Schltr.
Dimorphorchis Rolfe
Dinema Lindl.
Dinklageella Mansf.
Diothonea Lindl.
Diphylax Hook.f.
Diplandrorchis S.C.Chen
Diplocaulobium (Rchb.f.) Kraenzl.
Diplocentrum Lindl.
Diplolabellum F.Maek.
Diplomeris D.Don
Diploprora Hook.f.
Dipodium R.Br.
Dipteranthus Barb.Rodr.
Dipterostele Schltr.
Disa Bergius
Discyphus Schltr.
Disperis Sw.
Distylodon Summerh.
Dithrix Schltr.
Diuris Sm.
Dodsonia Ackerman
Domingoa Schltr.
Doritis Lindl.
Dossinia C.Morren
Dracula Luer
Drakaea Lindl.
Dresslerella Luer
Dressleria Dodson
Dryadella Luer
Dryadorchis Schltr.
Drymoanthus Nicholls
Drymoda Lindl.
Duckeella Porto
Dunstervillea Garay
Dyakia Christenson
Earina Lindl.
Eggelingia Summerh.
Eleorchis F.Maek.
Elleanthus C.Presl
Elythranthera (Endl.) A.S.George
Embreea Dodson
Encyclia Hook.
Endresiella Schltr.

Entomophobia de Vogel
Eparmatostigma Garay
Ephippianthus Rchb.f.
Epiblastus Schltr.
Epiblema R.Br.
Epidanthus L.O.Williams
Epidendropsis Garay & Dunst.
Epidendrum L.
Epigeneium Gagnep.
Epipactis Zinn
Epipogium Borkh.
Epistephium Kunth
Eria Lindl.
Eriaxis Rchb.f.
Eriochilus R.Br.
Eriodes Rolfe
Eriopsis Lindl.
Erycina Lindl.
Erythrodes Blume
Erythrorchis Blume
Esmeralda Rchb.f.
Euanthe Schltr.
Eucosia Blume
Eulophia R.Br. ex Lindl.
Eulophiella Rolfe
Eurycentrum Schltr.
Eurychone Schltr.
Eurystyles Wawra
Evrardianthe Rauschert
Fernandezia Ruíz & Pav.
Flickingeria A.D.Hawkes
Forficaria Lindl.
Fregea Rchb.f.
Frondaria Luer
Fuertesiella Schltr.
Galeandra Lindl.
Galearis Raf.
Galeola Lour.
Gastrochilus D.Don
Gastrodia R.Br.
Gavilea Poepp.
Geesinkorchis de Vogel
Gennaria Parl.
Genoplesium R.Br.
Genyorchis Schltr.
Geoblasta Barb.Rodr.
Geodorum G.Jacks.
Glomera Blume
Glossodia R.Br.
Glossorhyncha Ridl.
Gomesa R.Br.
Gomphichis Lindl.
Gonatostylis Schltr.
Gongora Ruíz & Pav.
Goniochilus M.W.Chase

Goodyera R.Br.
Govenia Lindl.
Grammangis Rchb.f.
Grammatophyllum Blume
Graphorkis Thouars
Greenwoodia Burns–Bal.
Grobya Lindl.
Grosourdya Rchb.f.
Gunnarella Senghas
Gymnadenia R.Br.
Gymnochilus Blume
Gynoglottis J.J.Sm.
Habenaria Willd.
Hagsatera Tamayo *
Hammarbya Kuntze
Hancockia Rolfe
Hapalorchis Schltr.
Haraella Kudo
Harrisella Fawc. & Rendle
Hederorkis Thouars
Helcia Lindl.
Helleriella A.D.Hawkes
Helonoma Garay *
Hemipilia Lindl.
Hemiscleria Lindl.
Herminium L.
Herpysma Lindl.
Herschelianthe Rauschert
Hetaeria Blume
Hexadesmia Brongn.
Hexalectris Raf.
Hexisea Lindl.
Himantoglossum K.Koch
Hintonella Ames
Hippeophyllum Schltr.
Hirtzia Dodson *
Hispaniella Braem
Hoehneella Ruschi
Hofmeisterella Rchb.f.
Holcoglossum Schltr.
Holopogon Kom. & Nevski
Holothrix Rich. ex Lindl.
Homalopetalum Rolfe
Horichia Jenny
Hormidium (Lindl.) Heynh.
Horvatia Garay
Houlletia Brongn.
Huntleya Bateman ex Lindl.
Huttonaea Harv.
Hybochilus Schltr.
Hygrochilus Pfitzer
Hylophila Lindl.
Hymenorchis Schltr.
Imerinaea Schltr.
Ionopsis Kunth

Ipsea Lindl.
Isabelia Barb.Rodr.
Ischnocentrum Schltr.
Ischnogyne Schltr.
Isochilus R.Br.
Isotria Raf.
Jacquiniella Schltr.
Jejosephia A.N.Rao & Mani
Jumellea Schltr.
Kalopternix Garay & Dunst.
Kefersteinia Rchb.f.
Kegeliella Mansf.
Kingidium P.F.Hunt
Koellensteinia Rchb.f.
Konantzia Dodson & N.H.Williams
Kreodanthus Garay
Kuhlhasseltia J.J.Sm.
Lacaena Lindl.
Laelia Lindl.
Laeliopsis Lindl. & Paxton
Lanium (Lindl.) Benth.
Lecanorchis Blume
Lemboglossum Halbinger
Lemurella Schltr.
Lemurorchis Kraenzl.
Leochilus Knowles & Westc.
Lepanthes Sw.
Lepanthopsis Ames
Lepidogyne Blume
Leporella A.S.George
Leptotes Lindl.
Lesliea Seidenf.
Leucohyle Klotzsch
Ligeophila Garay
Limodorum Boehm.
Lindsayella Ames & C.Schweinf.
Liparis Rich.
Listera R.Br.
Listrostachys Rchb.
Lockhartia Hook.
Loefgrenianthus Hoehne
Ludisia A.Rich.
Lueddemannia Rchb.f.
Luisia Gaudich.
Lycaste Lindl.
Lycomormium Rchb.f.
Lyperanthus R.Br.
Macodes (Blume) Lindl.
Macradenia R.Br.
Macropodanthus L.O.Williams
Malaxis Sol. ex Sw.
Malleola J.J.Sm. & Schltr.
Manniella Rchb.f.
Margelliantha P.J.Cribb
Masdevallia Ruíz & Pav.

Maxillaria Ruíz & Pav.
Mediocalcar J.J.Sm.
Megalorchis H.Perrier
Megalotus Garay
Megastylis Schltr.
Meiracyllium Rchb.f.
Mendoncella A.D.Hawkes
Mesoglossum Halbinger
Mesospinidium Rchb.f.
Mexicoa Garay
Microcoelia Lindl.
Micropera Lindl.
Microsaccus Blume
Microtatorchis Schltr.
Microterangis Senghas
Microtis R.Br.
Miltonia Lindl.
Miltonioides Brieger & Lückel
Miltoniopsis God.–Leb.
Mischobulbum Schltr.
Mobilabium Rupp
Moerenhoutia Blume
Monadenia Lindl.
Monomeria Lindl.
Monophyllorchis Schltr.
Monosepalum Schltr.
Mormodes Lindl.
Mormolyca Fenzl
Myoxanthus Poepp. & Endl.
Myrmechis (Lindl.) Blume
Myrmecophila Rolfe
Myrosmodes Rchb.f.
Mystacidium Lindl.
Nabaluia Ames
Nageliella L.O.Williams
Nanodes Lindl.
Neo–urbania Fawc. & Rendle
Neobartlettia Schltr.
Neobathiea Schltr.
Neobenthamia Rolfe
Neobolusia Schltr.
Neoclemensia C.E.Carr
Neocogniauxia Schltr.
Neodryas Rchb.f.
Neofinetia Hu
Neogardneria Schltr. ex Garay
Neogyna Rchb.f.
Neokoehleria Schltr.
Neolehmannia Kraenzl.
Neomoorea Rolfe
Neotinea Rchb.f.
Neottia Guett.
Neottianthe (Rchb.) Schltr.
Neowilliamsia Garay
Nephelaphyllum Blume

Nephrangis (Schltr.) Summerh.
Nervilia Comm. ex Gaud.
Neuwiedia Blume
Nidema Britton & Millsp.
Nigritella Rich.
Nothodoritis Tsi
Nothostele Garay *
Notylia Lindl.
Oakes–Amesia C.Schweinf. & P.H.Allen
Oberonia Lindl.
Octarrhena Thwaites
Octomeria R.Br.
Odontoglossum Kunth
Odontorrhynchus M.N.Corrêa
Oeceoclades Lindl.
Oeonia Lindl.
Oeoniella Schltr.
Oerstedella Rchb.f.
Oligophyton Linder
Oliveriana Rchb.f.
Omoea Blume
Oncidium Sw.
Ophidion Luer *
Ophrys L.
Orchipedum Breda
Orchis L.
Oreorchis Lindl.
Orestias Ridl.
Orleanesia Barb.Rodr.
Ornithocephalus Hook.
Ornithochilus (Lindl.) Benth.
Ornithophora Barb.Rodr.
Orthoceras R.Br.
Osmoglossum (Schltr.) Schltr.
Ossiculum P.J.Cribb & Laan
Otochilus Lindl.
Otoglossum (Schltr.) Garay & Dunst.
Otostylis Schltr.
Pabstia Garay
Pachites Lindl.
Pachyphyllum Kunth
Pachyplectron Schltr.
Pachystele Schltr.
Pachystoma Blume
Palmorchis Barb.Rodr.
Palumbina Rchb.f.
Panisea (Lindl.) Lindl.
Paphinia Lindl.
Paphiopedilum Pfitzer
Papilionanthe Schltr.
Papillilabium Dockrill
Papperitzia Rchb.f.
Papuaea Schltr.
Paracaleana Blaxell
Paradisanthus Rchb.f.

Paraphalaenopsis A.D.Hawkes
Parapteroceras Averyanov
Parhabenaria Gagnep.
Pecteilis Raf.
Pedilochilus Schltr.
Pelatantheria Ridl.
Pelexia Poit. ex Lindl.
Pennilabium J.J.Sm.
Peristeranthus T.E.Hunt
Peristeria Hook.
Peristylus Blume
Perrieriella Schltr.
Pescatoria Rchb.f.
Petalocentrum Schltr.
Phaius Lour.
Phalaenopsis Blume
Philippinaea Schltr. & Ames
Pholidota Lindl. ex Hook.
Phragmipedium Rolfe
Phragmorchis L.O.Williams
Phreatia Lindl.
Phymatidium Lindl.
Physinga Lindl.
Physoceras Schltr.
Pilophyllum Schltr.
Pinelia Lindl.
Pityphyllum Schltr.
Platanthera Rich.
Platycoryne Rchb.f.
Platyglottis L.O.Williams
Platylepis A.Rich.
Platyrhiza Barb.Rodr.
Platystele Schltr.
Platythelys Garay
Plectorrhiza Dockrill
Plectrelminthus Raf.
Plectrophora H.Focke
Pleione D.Don
Pleurothallis R.Br.
Pleurothallopsis Porto
Plocoglottis Blume
Poaephyllum Ridl.
Podangis Schltr.
Podochilus Blume
Pogonia Juss.
Pogoniopsis Rchb.f.
Polycycnis Rchb.f.
Polyotidium Garay
Polyradicion Garay
Polystachya Hook.
Pomatocalpa Breda, Kuhl & Hasselt
Ponera Lindl.
Ponerorchis Rchb.f.
Ponthieva R.Br.
Porolabium T.Tang & F.T.Wang

Porpax Lindl.
Porphyrodesme Schltr.
Porphyroglottis Ridl.
Porphyrostachys Rchb.f.
Porroglossum Schltr.
Porrorhachis Garay
Prasophyllum R.Br.
Prescottia Lindl.
Pristiglottis Cretz. & J.J.Sm.
Promenaea Lindl.
Pseudacoridium Ames
Pseuderia Schltr.
Pseudocentrum Lindl.
Pseudocranichis Garay *
Pseudogoodyera Schltr.
Pseudolaelia Porto
Pseudomaxillaria Hoehne
Pseudorchis Séguier
Pseudovanilla Garay
Psilochilus Barb.Rodr.
Psychilis Raf.
Psychopsiella Lückel & Braem
Psygmorchis Dodson & Dressler
Pterichis Lindl.
Pteroceras Hasselt ex Hassk.
Pteroglossaspis Rchb.f.
Pterostemma Kraenzl.
Pterostylis R.Br.
Pterygodium Sw.
Pygmaeorchis Brade
Quekettia Lindl.
Quisqueya D.Don
Rangaeris (Schltr.) Summerh.
Rauhiella Pabst & Braga
Raycadenco Dodson *
Reichenbachanthus Barb.Rodr.
Renanthera Lour.
Renantherella Ridl.
Renata Ruschi
Restrepia Kunth
Restrepiella Garay & Dunst.
Restrepiopsis Luer
Rhaesteria Summerh.
Rhamphorhynchus Garay
Rhinerrhiza Rupp
Rhizanthella R.S.Rogers
Rhynchogyna Seidenf. & Garay
Rhyncholaelia Schltr.
Rhynchophreatia Schltr.
Rhynchostylis Blume
Ridleyella Schltr.
Rimacola Rupp
Risleya King & Pantl.
Robiquetia Gaudich.
Rodriguezia Ruíz & Pav.

736

Rodrigueziella Kuntze
Rodrigueziopsis Schltr.
Roeperocharis Rchb.f.
Roezliella Schltr.
Rossioglossum (Schltr.) Garay &
  G.C.Kenn.
Rudolfiella Hoehne
Rusbyella Rolfe ex Rusby
Saccoglossum Schltr.
Saccolabiopsis J.J.Sm.
Saccolabium Blume
Salpistele Dressler
Sanderella Kuntze
Sarcanthopsis Garay
Sarcochilus R.Br.
Sarcoglottis C.Presl
Sarcoglyphis Garay
Sarcophyton Garay
Sarcostoma Blume
Satyridium Lindl.
Satyrium Sw.
Saundersia Rchb.f.
Sauroglossum Lindl.
Scaphosepalum Pfitzer
Scaphyglottis Poepp. & Endl.
Scelochiloides Dodson & M.W.Chase *
Scelochilus Klotzsch
Schiedeella Schltr.
Schistotylus Dockrill
Schizochilus Sond.
Schizodium Lindl.
Schlimmia Planch. & Linden
Schoenorchis Blume
Schomburgkia Lindl.
Schwartzkopffia Kraenzl.
Scuticaria Lindl.
Sedirea Garay & H.R.Sweet
Seidenfadenia Garay
Selenipedium Rchb.f.
Sepalosaccus Schltr.
Sepalosiphon Schltr.
Serapias L.
Sievekingia Rchb.f.
Sigmatostalix Rchb.f.
Silvorchis J.J.Sm.
Sinorchis S.C.Chen
Sirhookera Kuntze
Smithorchis T.Tang & F.T.Wang
Smithsonia C.J.Saldanha
Smitinandia Holttum
Sobennikoffia Schltr.
Sobralia Ruíz & Pav.
Solenangis Schltr.
Solenidiopsis Senghas *
Solenidium Lindl.

Solenocentrum Schltr.
Sophronitella Schltr.
Sophronitis Lindl.
Soterosanthus Lehm. ex Jenny
Spathoglottis Blume
Sphyrarhynchus Mansf.
Sphyrastylis Schltr.
Spiculaea Lindl.
Spiranthes Rich.
Spongiola J.J.Wood & A.L.Lamb
Stalkya Garay *
Stanhopea Frost ex Hook.
Staurochilus Ridl. ex Pfitzer
Stelis Sw.
Stellilabium Schltr.
Stenia Lindl.
Stenoglottis Lindl.
Stenoptera C.Presl
Stenorrhynchos Rich. ex Spreng.
Stephanothelys Garay
Stereosandra Blume
Steveniella Schltr.
Stictophyllum Dodson & M.W.Chase *
Stigmatodactylus Maxim. ex Makino
Stigmatosema Garay *
Stolzia Schltr.
Suarezia Dodson *
Summerhayesia P.J.Cribb
Sunipia Buch.-Ham. ex Lindl.
Sutrina Lindl.
Symphyglossum Schltr.
Symphyosepalum Hand.-Mazz.
Synanthes Burns-Bal., H.Rob. &
  M.S.Foster
Systeloglossum Schltr.
Taeniophyllum Blume
Taeniorrhiza Summerh.
Tainia Blume
Tapeinoglossum Schltr.
Telipogon Kunth
Tetragamestus Rchb.f.
Tetramicra Lindl.
Teuscheria Garay
Thaia Seidenf.
Thecopus Seidenf.
Thecostele Rchb.f.
Thelasis Blume
Thelymitra J.R.Forst. & G.Forst.
Thelyschista Garay *
Thrixspermum Lour.
Thulinia P.J.Cribb
Thunia Rchb.f.
Thysanoglossa Porto
Ticoglossum Luc.Rodr. ex Halbinger
Tipularia Nutt.

Townsonia Cheeseman
Trachoma Garay
Traunsteinera Rchb.
Trevoria F.Lehm.
Trias Lindl.
Triceratorhynchus Summerh.
Trichocentrum Poepp. & Endl.
Trichoceros Kunth
Trichoglottis Blume
Trichopilia Lindl.
Trichosalpinx Luer
Trichotosia Blume
Tridactyle Schltr.
Trigonidium Lindl.
Triphora Nutt.
Trisetella Luer
Trizeuxis Lindl.
Tropidia Lindl.
Trudelia Garay
Tsaiorchis T.Tang & F.T.Wang
Tuberolabium Yamam.
Tubilabium J.J.Sm.
Tylostigma Schltr.
Uleiorchis Hoehne
Uncifera Lindl.

Vanda Jones ex R.Br.
Vandopsis Pfitzer
Vanilla Mill.
Vargasiella C.Schweinf.
Vasqueziella Dodson *
Ventricularia Garay
Vexillabium F.Maek.
Vrydagzynea Blume
Warmingia Rchb.f.
Warrea Lindl.
Warreella Schltr.
Warreopsis Garay
Wullschlaegelia Rchb.f.
Xenikophyton Garay
Xerorchis Schltr.
Xylobium Lindl.
Yoania Maxim.
Ypsilopus Summerh.
Zetagyne Ridl.
Zeuxine Lindl.
Zootrophion Luer *
Zygopetalum Hook.
Zygosepalum Rchb.f.
Zygostates Lindl.

PALMAE Juss. 1789   187

202 genera. Widespread trop. and subtrop., rarely temp. Trees, subshrubs, lianes.

B&H	(MONOCOTS)	Calycineae, 187
DT&H	(MONOCOTS)	Principes, 247
Melc	(MONOCOTS)	Principes, 330
Thor	ARECIFLORAE	Arecales, 323
Dahl	ARECIFLORAE	Arecales, 468
Young	LILIIDAE, ARECANAE	Arecales, 57
Takh	ARECIDAE, ARECANAE	Arecales, 527
Cron	ARECIDAE	Arecales, 335

Acanthophoenix H.A.Wendl.
Acoelorrhaphe H.A.Wendl.
Acrocomia Mart.
Actinokentia Dammer
Actinorhytis H.A.Wendl. & Drude
Aiphanes Willd.
Allagoptera Nees
Alloschmidia H.E.Moore
Alsmithia H.E.Moore
Ammandra O.F.Cook
Aphandra Barfod
Archontophoenix H.A.Wendl.
    & Drude
Areca L.

Arenga Labill.
Asterogyne H.A.Wendl.
Astrocaryum G.Mey.
Attalea Kunth
Bactris Jacq.
Balaka Becc.
Barcella (Trail) Trail ex Drude
Basselinia Vieill.
Beccariophoenix Jum. & H.Perrier
Bentinckia Berry ex Roxb.
Bismarckia Hildebrandt & H.A.Wendl.
Borassodendron Becc.
Borassus L.
Brahea Mart. ex Endl.

Brassiophoenix Burret
Brongniartikentia Becc.
Burretiokentia Pic.Serm.
Butia (Becc.) Becc.
Calamus L.
Calospatha Becc.
Calyptrocalyx Blume
Calyptrogyne H.A.Wendl.
Calyptronoma Griseb.
Campecarpus H.A.Wendl. ex Becc.
Carpentaria Becc.
Carpoxylon H.A.Wendl. & Drude
Caryota L.
Catoblastus H.A.Wendl.
Ceratolobus Blume
Ceroxylon Bonpl. ex DC.
Chamaedorea Willd.
Chamaerops L.
Chambeyronia Vieill.
Chelyocarpus Dammer
Chrysalidocarpus H.A.Wendl.
Chuniophoenix Burret
Clinosperma Becc.
Clinostigma H.A.Wendl.
Coccothrinax Sarg.
Cocos L.
Colpothrinax Griseb. & H.A.Wendl.
Copernicia Mart. ex Endl.
Corypha L.
Cryosophila Blume
Cyphokentia Brongn.
Cyphophoenix H.A.Wendl. ex Hook.f.
Cyphosperma H.A.Wendl. ex Hook.f.
Cyrtostachys Blume
Daemonorops Blume
Deckenia H.A.Wendl. ex Seem.
Desmoncus Mart.
Dictyocaryum H.A.Wendl.
Dictyosperma H.A.Wendl. & Drude
Drymophloeus Zipp.
Dypsis Noronha ex Mart.
Elaeis Jacq.
Eleiodoxa (Becc.) Burret
Eremospatha (G.Mann & H.A.Wendl.)
    H.A.Wendl.
Eugeissona Griff.
Euterpe Mart.
Gastrococos Morales
Gaussia H.A.Wendl.
Geonoma Willd.
Goniocladus Burret
Gronophyllum Scheff.
Guihaia J.Dransf., S.K.Lee & F.N.Wei
Gulubia Becc.
Halmoorea J.Dransf. & N.W.Uhl

Hedyscepe H.A.Wendl. & Drude
Heterospathe Scheff.
Howea Becc.
Hydriastele H.A.Wendl. & Drude
Hyophorbe Gaertn.
Hyospathe Mart.
Hyphaene Gaertn.
Iguanura Blume
Iriartea Ruíz & Pav.
Iriartella H.A.Wendl.
Itaya H.E.Moore
Johannesteijsmannia H.E.Moore
Juania Drude
Jubaea Kunth
Jubaeopsis Becc.
Kentiopsis Brongn.
Kerriodoxa J.Dransf.
Korthalsia Blume
Laccospadix Drude & H.A.Wendl.
Laccosperma (G.Mann & H.A.Wendl.)
    Drude
Latania Comm. ex Juss.
Lavoixia H.E.Moore
Lemurophoenix J.Dransf.
Leopoldinia Mart.
Lepidocaryum Mart.
Lepidorrhachis (H.A.Wendl. & Drude)
    O.F.Cook
Licuala Thunb.
Linospadix H.A.Wendl.
Livistona R.Br.
Lodoicea Comm. ex DC.
Louvelia Jum. & H.Perrier
Loxococcus H.A.Wendl. & Drude
Lytocaryum Toledo
Mackeea H.E.Moore
Manicaria Gaertn.
Marojejya Humbert
Masoala Jum.
Mauritia L.f.
Mauritiella Burret
Maxburretia Furtado
Maximiliana Mart.
Medemia Wurttemb. ex H.A.Wendl.
Metroxylon Rottb.
Moratia H.E.Moore
Myrialepis Becc.
Nannorrhops H.A.Wendl.
Nenga H.A.Wendl. & Drude
Neodypsis Baill.
Neonicholsonia Dammer
Neophloga Baill.
Neoveitchia Becc.
Nephrosperma Balf.f.
Normanbya F.Muell. ex Becc.

Nypa Steck
Oenocarpus Mart.
Oncocalamus (G.Mann & H.A.Wendl.) Hook.f.
Oncosperma Blume
Orania Zipp.
Oraniopsis J.Dransf., A.K.Irvine & N.Uhl
Orbignya Mart. ex Endl.
Parajubaea Burret
Paschalococos J.Dransf.
Pelagodoxa Becc.
Phloga Noronha ex Hook.f.
Phoenicophorium H.A.Wendl.
Phoenix L.
Pholidocarpus Blume
Pholidostachys H.A.Wendl. ex Hook.f.
Physokentia Becc.
Phytelephas Ruíz & Pav.
Pigafetta (Blume) Becc.
Pinanga Blume
Plectocomia Mart. ex Blume
Plectocomiopsis Becc.
Podococcus G.Mann & H.A.Wendl.
Pogonotium J.Dransf.
Polyandrococos Barb.Rodr.
Prestoea Hook.f.
Pritchardia Seem. & H.A.Wendl.
Pritchardiopsis Becc.
Pseudophoenix H.A.Wendl. ex Sarg.
Ptychococcus Becc.
Ptychosperma Labill.
Raphia P.Beauv.
Ravenea C.D.Bouché
Reinhardtia Liebm.

Retispatha J.Dransf.
Rhapidophyllum H.A.Wendl. & Drude
Rhapis L.f. ex Aiton
Rhopaloblaste Scheff.
Rhopalostylis H.A.Wendl. & Drude
Roscheria H.A.Wendl. ex Balf.f.
Roystonea O.F.Cook
Sabal Adans.
Salacca Reinw.
Satakentia H.E.Moore
Scheelea Karst.
Schippia Burret
Sclerosperma G.Mann & H.A.Wendl.
Serenoa Hook.f.
Siphokentia Burret
Socratea Karst.
Sommieria Becc.
Syagrus Mart.
Synechanthus H.A.Wendl.
Tectiphiala H.E.Moore
Thrinax Sw.
Trachycarpus H.A.Wendl.
Trithrinax Mart.
Veillonia H.E.Moore
Veitchia H.A.Wendl.
Verschaffeltia H.A.Wendl.
Voanioala J.Dransf.
Vonitra Becc.
Wallichia Roxb.
Washingtonia H.A.Wendl.
Welfia H.A.Wendl.
Wendlandiella Dammer
Wettinia Poepp.
Wodyetia Irvine
Zombia L.H.Bailey

PANDANACEAE R.Br. 1810   188

3 genera. Old World trop. to New Zealand. Trees, shrubs, lianes.

B&H	(MONOCOTS)	Nudiflorae, 188
DT&H	(MONOCOTS)	Pandanales, 235
Melc	(MONOCOTS)	Pandanales, 334
Thor	PANDANIFLORAE	Pandanales, 322
Dahl	PANDANIFLORAE	Pandanales, 469
Young	LILIIDAE, ARECANAE	Pandanales, 59
Takh	ARECIDAE, PANDANANAE	Pandanales, 529
Cron	ARECIDAE	Pandanales, 337

Freycinetia Gaudich.
Pandanus Parkinson
Sararanga Hemsl.

PETERMANNIACEAE Hutch. 1934   174.08

1 genus. Eastern Australia. Lianes.

B&H (MONOCOTS)	Epigynae (within Dioscoreaceae, 176)
DT&H (MONOCOTS)	Liliiflorae, Liliineae (within
	Dioscoreaceae, 264)
Melc (MONOCOTS)	Liliiflorae, Liliineae (within Liliaceae, 301)
Thor LILIIFLORAE	Liliales, Dioscoreineae (within
	Smilacaceae, 299)
Dahl LILIIFLORAE	Dioscoreales, 374

Young (not mentioned, presumably within Smilacaceae, 75)

Takh LILIIDAE, LILIANAE	Smilacales, 485
Cron LILIIDAE	Liliales (within Smilacaceae, 378)

Petermannia F.Muell.

PHILESIACEAE Dumort. 1829   175.01

6 genera. Temperate S. hemisphere to Malesia and Pacific. Subshrubs, vines.

B&H (MONOCOTS)	Coronarieae (within Liliaceae, 178)
DT&H (MONOCOTS)	Liliiflorae, Liliineae (within Liliaceae, 264)
Melc (MONOCOTS)	Liliiflorae, Liliineae (within Liliaceae, 301)
Thor LILIIFLORAE	Liliales, Dioscoreineae, 298
(including Alstroemeriaceae)	
Dahl LILIIFLORAE	Asparagales, 375
(separating *Luzuriagaceae*, 376)	

Young (not mentioned, presumably within Smilacaceae, 75)

Takh LILIIDAE, LILIANAE	Smilacales, 484
(separating *Luzuriagaceae* 483)	
Cron LILIIDAE	Liliales (within Smilacaceae, 378)

Behnia Didr.	**Lapageria** Ruíz & Pav.
Eustrephus R.Br.	**Luzuriaga** Ruíz & Pav.
Geitonoplesium A.Cunn. ex R.Br.	**Philesia** Comm. ex Juss.

PHILYDRACEAE Link 1821   180

3 genera. Trop. and east Asia to Australia. Rhizomatous or cormous herbs.

B&H (MONOCOTS)	Coronarieae, 180
DT&H (MONOCOTS)	Farinosae, Philydrineae, 261
Melc (MONOCOTS)	Liliiflorae, Philydrineae, 317
Thor COMMELINIFLORAE	Commelinales, Pontederiineae, 330
Dahl BROMELIIFLORAE	Philydrales, 438
Young LILIIDAE, LILIANAE	Liliales, 64
Takh LILIIDAE, PONTEDERIANAE	Philydrales, 500
Cron LILIIDAE	Liliales, 365

Helmholtzia F.Muell.
Philydrella Caruel
Philydrum Banks ex Gaertn.

PHORMIACEAE J. Agardh 1858 175.18

7 genera. Old World trop., esp. Australasia and S. America. Rhizom. herbs to subshrubs.

B&H   (MONOCOTS)   Coronarieae (within Liliaceae, 178)
DT&H (MONOCOTS)   Liliiflorae, Liliineae (within Liliaceae, 264)
Melc   (MONOCOTS)   Liliiflorae, Liliineae (within Liliaceae, 301)
Thor   LILIIFLORAE   Liliales, Liliineae (within Liliaceae, 297)
Dahl   LILIIFLORAE   Asparagales, 395
Young (not mentioned, presumably within Agavaceae, 71)
Takh   LILIIDAE, LILIANAE   Amaryllidales, 470
Cron   LILIIDAE   Liliales (within Agavaceae, 373)

Agrostocrinum F.Muell.   **Stypandra** R.Br.
Dianella Lam.   **Thelionema** R.J.F.Hend.
Eccremis Baker   **Xeronema** Brongn. & Gris
Phormium J.R.Forst. & G.Forst.

POACEAE — see GRAMINEAE

PONTEDERIACEAE Kunth 1816 179

9 genera. Widespread in tropics, esp. New World, rarely temp. Mostly rhizom. aquatics.

B&H   (MONOCOTS)   Coronarieae, 179
DT&H (MONOCOTS)   Farinosae, Pontederiineae, 260
         (including Cyanastraceae)
Melc   (MONOCOTS)   Liliiflorae, Pontederiineae, 312
Thor   COMMELINIFLORAE   Commelinales, Pontederiineae, 329
Dahl   BROMELIIFLORAE   Pontederiales, 440
Young LILIIDAE, LILIANAE   Liliales, 65
Takh   LILIIDAE, PONTEDERIANAE Pontederiales, 499
Cron   LILIIDAE   Liliales, 366

Eichhornia Kunth   **Pontederia** L.
Eurystemon Alexander   **Reussia** Endl.
Heteranthera Ruíz & Pav.   **Scholleropsis** H.Perrier
Hydrothrix Hook.f.   **Zosterella** Small
Monochoria C.Presl

POSIDONIACEAE Hutch. 1934 195.08

1 genus. Mediterranean, Australia. Marine aquatics.

B&H   (MONOCOTS)   Apocarpae (within *Najadaceae*, 195)
DT&H (MONOCOTS)   Helobiae, Potamogetonineae (within
                                                         Potamogetonaceae, 237)
Melc   (MONOCOTS)   Helobiae, Potamogetonineae (within
                                                         Potamogetonaceae, 297)
Thor   ALISMATIFLORAE   Zosterales, Potamogetonineae, 314

Dahl   ALISMATIFLORAE   Najadales, 431
Young LILIIDAE, ALISMATANAE   Zosterales, 48
Takh   ALISMATIDAE, NAJADANAE Posidoniales, 443
Cron   ALISMATIDAE   Najadales, 330

Posidonia K.König

POTAMOGETONACEAE Dumort. 1829   195.06

3 genera. Widespread. Aquatic herbs.

B&H   (MONOCOTS)   Apocarpae (within *Najadaceae*, 195)
DT&H (MONOCOTS)   Helobiae, Potamogetonineae, 237
         (including Cymodoceaceae, Posidoniaceae, Zannichelliaceae, Zosteraceae)
Melc   (MONOCOTS)   Helobiae, Potamogetonineae, 297
         (including Posidoniaceae, Zosteraceae)
Thor   ALISMATIFLORAE   Zosterales, Potamogetonineae, 313
Dahl   ALISMATIFLORAE   Najadales, 430
Young LILIIDAE, ALISMATANAE   Zosterales, 49
Takh   ALISMATIDAE, NAJADANAE Potamogetonales, 441
         (separating *Ruppiaceae* 442)
Cron   ALISMATIDAE   Najadales, 326
         (separating *Ruppiaceae* 327)

Groenlandia J.Gay
Potamogeton L.
Ruppia L.

RAPATEACEAE Dumort. 1829   184

17 genera. Trop. S. Amer. and (*Maschalocephalus*) W. trop. Afr. Mostly large rhizom. herbs.

B&H   (MONOCOTS)   Coronarieae, 184
DT&H (MONOCOTS)   Farinosae, Bromeliineae, 257
Melc   (MONOCOTS)   Commelinales, Commelinineae, 324
Thor   COMMELINIFLORAE   Commelinales, Bromeliineae, 326
Dahl   COMMELINIFLORAE   Commelinales, 454
Young LILIIDAE, COMMELINANAE Commelinales, 80
Takh   LILIIDAE, COMMELINANAE Commelinales, Xyridineae, 518
Cron   COMMELINIDAE   Commelinales, 340

Amphiphyllum Gleason   **Phelpsiella** Maguire
Cephalostemon R.H.Schomb.   **Potarophytum** Sandwith
Duckea Maguire   **Rapatea** Aubl.
Epidryos Maguire   **Saxofridericia** R.H.Schomb.
Guacamaya Maguire   **Schoenocephalium** Seub.
Kunhardtia Maguire   **Spathanthus** Desv.
Marahuacaea Maguire   **Stegolepis** Klotzsch ex Koern.
Maschalocephalus Gilg & K.Schum. **Windsorina** Gleason
Monotrema Koern.

RESTIONACEAE R.Br. 1810   198.01

40 genera. S. hemisphere, rarely tropics. Rush-like herbs, some climbers.

B&H (MONOCOTS)   Glumaceae, 198
        (including Anarthriaceae, Ecdeiocoleaceae)
DT&H (MONOCOTS)   Farinosae, Enantioblastae, 252
        (including Anarthriaceae and Ecdeiocoleaceae)
Melc (MONOCOTS)   Commelinales, Restionineae, 326
        (including Anarthriaceae, ?Ecdeiocoleaceae)
Thor COMMELINIFLORAE   Commelinales, Flagellariineae, 338
        (including Anarthriaceae, Ecdeiocoleaceae)
Dahl COMMELINIFLORAE   Poales, 465
Young LILIIDAE, COMMELINANAE   Restionales, 87
Takh LILIIDAE, COMMELINANAE   Restionales, Restioninaeae, 522
Cron COMMELINIDAE   Restionales, 347
        (including Anarthriaceae, Ecdeiocoleaceae)

Alexgeorgea Carlquist
Anthochortus Nees
Askidiosperma Steud. *
Calopsis P.Beauv. ex Desv.
Calorophus Labill.
Cannomois P.Beauv. ex Desv.
Ceratocaryum Nees
Chaetanthus R.Br.
Chondropetalum Rottb.
Coleocarya S.T.Blake
Dielsia Gilg
Dovea Kunth
Elegia L.
Empodisma L.A.S.Johnson & D.F.Cutler
Harperia W.Fitzg.
Hopkinsia W.Fitzg.
Hydrophilus Linder
Hypodiscus Nees
Hypolaena R.Br.
Ischyrolepis Steud.

Lepidobolus Nees
Leptocarpus R.Br.
Lepyrodia R.Br.
Loxocarya R.Br.
Lyginia R.Br.
Mastersiella Gilg–Ben.
Meeboldina Suess.
Megalotheca F.Muell.
Nevillea Esterh. & H.P.Linder
Onychosepalum Steud.
Phyllocomos Mast.
Platycaulos Linder
Pseudoloxocarya Linder
Restio Rottb.
Rhodocoma Nees
Sporadanthus F.Muell.
Staberoha Kunth
Thamnochortus Bergius
Willdenowia Thunb.
Winifredia L.A.S.Johnson & B.G.Briggs *

RHIPOGONACEAE Conran & Clifford 1985   174.05

1 genus. New Guinea, Australia, New Zealand. Erect or climbing shrubs.

B&H (MONOCOTS)   Coronarieae (within Liliaceae, 178)
DT&H (MONOCOTS)   Liliiflorae, Liliineae (within Liliaceae, 264)
Melc (MONOCOTS)   Liliiflorae, Liliineae (within Liliaceae, 301)
Thor LILIIFLORAE   Liliales, Dioscoreincae (within
                                        Smilacaceae, 299)
Dahl LILIIFLORAE   Dioscoreales (within Smilacaceae, 371)
Young (not mentioned, presumably within Smilacaceae, 75)
Takh LILIIDAE, LILIANAE   Smilacales, 486
Cron LILIIDAE   Liliales (within Smilacaceae, 378)

Rhipogonum J.R.Forst. & G.Forst.

RUSCACEAE Spreng. ex Hutch. 1934 175.04

3 genera. Macaronesia, Europe, Mediterranean, W. Asia. Subshrubs, vines.

B&H (MONOCOTS)	Coronarieae (within Liliaceae, 264)
DT&H (MONOCOTS)	Liliiflorae, Liliineae (within Liliaceae, 264)
Melc (MONOCOTS)	Liliiflorae, Liliineae (within Liliaceae, 301)
Thor LILIIFLORAE	Liliales, Liliineae (within Liliaceae, 297)
Dahl LILIIFLORAE	Asparagales, 379
Young (not mentioned, presumably within Liliaceae, 63)	
Takh LILIIDAE, LILIANAE	Asparagales, 476
Cron LILIIDAE	Liliales (within Liliaceae, 369)

Danae Medik.
Ruscus L.
Semele Kunth

SCHEUCHZERIACEAE F. Rudolphi 1830 195.03

1 genus, monotypic. North temperate to arctic. Herb.

B&H (MONOCOTS)	Apocarpae (within *Najadaceae*, 195)
DT&H (MONOCOTS)	Helobiae, Potamogetonineae (within Juncaginaceae, 240)
Melc (MONOCOTS)	Helobiae, Scheuchzeriineae, 294
Thor ALISMATIFLORAE	Zosterales, Potamogetonineae (within Juncaginaceae, 312)
Dahl ALISMATIFLORAE	Najadales, 428
Young LILIIDAE, ALISMATANAE	Zosterales, 46
Takh ALISMATIDAE, NAJADANAE	Scheuchzeriales, 437
Cron ALISMATIDAE	Najadales, 324

Scheuchzeria L.

SMILACACEAE Vent. 1799 174.07

2 genera. Widespread, mostly tropical. Herbs, shrubs, woody vines.

B&H (MONOCOTS)	Coronarieae (within Liliaceae, 178)
DT&H (MONOCOTS)	Liliiflorae, Liliineae (within Liliaceae, 264)
Melc (MONOCOTS)	Liliiflorae, Liliineae (within Liliaceae, 301)
Thor LILIIFLORAE	Liliales, Dioscoreineae, 299
(including Petermanniaceae, Rhipogonaceae)	
Dahl LILIIFLORAE	Dioscoreales, 373
(including Rhipogonaceae)	
Young LILIIDAE, LILIANAE	Liliales, 75
Takh LILIIDAE, LILIANAE	Smilacales, 487
Cron LILIIDAE	Liliales, 378
(including Petermanniaceae, Philesiaceae, Rhipogonaceae)	

Heterosmilax Kunth
Smilax L.

STEMONACEAE Engl. 1887 174.04

4 genera. Trop. Asia to Japan and Australia, eastern N. Amer. Herbs, vines, subshrubs.

B&H	(MONOCOTS)	Coronarieae (as *Roxburghiaceae*) 177
DT&H	(MONOCOTS)	Liliiflorae, Liliineae, 263
Melc	(MONOCOTS)	Liliiflorae, Liliineae, 303
Thor	LILIIFLORAE	Liliales, Dioscoreineae, 300
Dahl	LILIIFLORAE	Dioscoreales, 371
Young	LILIIDAE, LILIANAE	Dioscoreales, 61
Takh	LILIIDAE, LILIANAE	Dioscoreales, 491
Cron	LILIIDAE	Liliales, 177

Croomia Torr.　　　　　　　　　　**Stemona** Lour.
Pentastemona Steenis　　　　　　　**Stichoneuron** Hook.f.

STRELITZIACEAE (K. Schum.) Hutch. 1934 170.07

3 genera. S. Africa, Madagascar, S. America. Giant herbs to trees.

B&H	(MONOCOTS)	Epigynae (within Scitamineae, 170)
DT&H	(MONOCOTS)	Scitamineae (within Musaceae, 271)
Melc	(MONOCOTS)	Scitamineae (within Musaceae, 338)
Thor	COMMELINIFLORAE	Zingiberales, Musineae, 342
Dahl	ZINGIBERIFLORAE	Zingiberales, 446
Young	LILIIDAE, ZINGIBERANAE	Zingiberales, 96
Takh	LILIIDAE, ZINGIBERANAE	Zingiberales, 503
Cron	ZINGIBERIDAE	Zingiberales, 357

Phenakospermum Endl.
Ravenala Adans.
Strelitzia Aiton

TACCACEAE Dumort. 1829 174.03

1 genus. Widespread in tropics, especially S.E. Asia. Rhizomatous or tuberous herbs.

B&H	(MONOCOTS)	Epigynae, 175
DT&H	(MONOCOTS	Liliiflorae, Liliineae, 268
Melc	(MONOCOTS)	Liliiflorae, Liliineae, 310
Thor	LILIIFLORAE	Liliales, Droscoreineae, 303
Dahl	LILIIFLORAE	Dioscoreales, 370
Young	LILIIDAE, LILIANAE	Dioscoreales, 62
Takh	LILIIDAE, LILIANAE,	Taccales, 493
Cron	LILIIDAE	Liliales, 376

Tacca J.R.Forst. & G.Forst.

TECOPHILAEACEAE Leyb. 1862   175.13

6 genera. Africa, S. America to California. Rhizomatous or cormous herbs.

B&H	(MONOCOTS)	Epigynae (within Haemodoraceae, 172)
DT&H	(MONOCOTS	Liliiflorae, Liliineae (within Amaryllidaceae, 266)
Melc	(MONOCOTS)	Liliiflorae, Liliineae (within Haemodoraceae, 305)
Thor	LILIIFLORAE	Liliales, Liliineae (within Liliaceae, 297)
Dahl	LILIIFLORAE (including Lanariaceae)	Asparagales, 391

Young (not mentioned, presumably within Liliaceae, 63)

Takh	LILIIDAE, LILIANAE	Liliales, 453
Cron	LILIIDAE	Liliales (within Liliaceae, 369)

Conanthera Ruíz & Pav.
Cyanella Royen ex L.
Odontostomum Torr.

Tecophilaea Bertero ex Colla
Walleria J.Kirk
Zephyra D.Don

THURNIACEAE Engl. 1907   186.02

1 genus. Northern S. America. Rhizomatous herbs.

B&H	(MONOCOTS)	Calycineae (within Juncaceae, 186)
DT&H	(MONOCOTS)	Liliiflorae, Juncineae (within Juncaceae, 262)
Melc	(MONOCOTS)	Juncales, 319)
Thor	COMMELINIFLORAE	Commelinales, Juncineae (within Juncaceae, 332)
Dahl	COMMELINIFLORAE	Cyperales, 458
Young	LILIIDAE, COMMELINANAE	Juncales (within Juncaceae, 91)
Takh	LILIIDAE, JUNCANAE	Juncales, 512
Cron	COMMELINIDAE	Juncales, 350

Thurnia Hook.f.

TRICHOPODACEAE Hutch. 1934   174.01

1 genus, monotypic. India, Sri Lanka, Malay Peninsula. Rhizomatous herbs.

B&H	(MONOCOTS)	Epigynae (within Dioscoreaceae, 176)
DT&H	(MONOCOTS)	Liliifloraee, Liliineae (within Dioscoreaceae, 269)
Melc	(MONOCOTS)	Liliiflorae, Liliineae (within Dioscoreaceae, 311)
Thor	LILIIFLORAE	Liliales, Dioscoreineae, 302
Dahl	LILIIFLORAE	Dioscoreales, 368

Young (not mentioned, presumably within Dioscoreaceae, 60)

Takh	LILIIDAE, LILIANAE	Dioscoreales, 488
Cron	LILIIDAE	Liliales (within Dioscoreaceae, 379)

Trichopus Gaertn.

TRILLIACEAE Lindl. 1846   174.06

5 genera. North temperate. Rhizomatous herbs.

B&H	(MONOCOTS)	Coronarieae (within Liliaceae, 178)
DT&H	(MONOCOTS)	Liliiflorae, Liliineae (within Liliaceae, 264)
Melc	(MONOCOTS)	Liliiflorae, Liliineae (within Liliaceae, 301)
Thor	LILIIFLORAE	Liliales (within Liliaceae, 297)
Dahl	LILIIFLORAE	Dioscoreales, 372
Young	(not mentioned, presumably within Liliaceae, 63)	
Takh	LILIIDAE, LILIANAE	Dioscoreales, 492
Cron	LILIIDAE	Liliales (within Liliaceae, 369)

Daiswa Raf.   **Scoliopus** Torr.
Kinugasa Tatew. & Suto   **Trillium** L.
Paris L.

TRIURIDACEAE Gardner 1843   193

6 genera. Widespread in tropics, extending to Japan. Saprophytic herbs.

B&H	(MONOCOTS)	Apocarpae, 193
DT&H	(MONOCOTS)	Triuridales, 244
Melc	(MONOCOTS)	Triuridales, 300
Thor	TRIURIDIFLORAE	Triuridales, 307
Dahl	TRIURIDIFLORAE	Triuridales, 422
Young	LILIIDAE, TRIURIDANAE	Triuridales, 54
Takh	TRIURIDIDAE, TRIURIDANAE	Triuridales, 448
Cron	ALISMATIDAE	Triuridales, 334

Andruris Schltr.   **Seychellaria** Hemsl.
Peltophyllum Gardner   **Soridium** Miers
Sciaphila Blume   **Triuris** Miers

TYPHACEAE Juss. 1789   190.01

2 genera. Widespread. Rhizomatous herbs, some aquatic.

B&H	(MONOCOTS)	Nudiflorae, 190
DT&H	(MONOCOTS)	Pandanales, 234
	(separating *Sparganiaceae*, 236)	
Melc	(MONOCOTS)	Pandanales, 366
	(separating *Sparganiaceae*, 335)	
Thor	TYPHIFLORAE	Typhales, 324
Dahl	BROMELIIFLORAE	Tyhales, 442
	(separating *Sparganiaceae*, 441)	
Young	LILIIDAE, LILIANAE	Typhales, 76
Takh	ARECIDAE, TYPHANAE	Typhales, 533
	(separating *Sparganiaceae* 532)	
Cron	COMMELINIDAE	Typhales, 355
	(separating *Sparganiaceae* 354)	

Sparganium L.
Typha L.

VELLOZIACEAE Endl. 1841 173

10 genera. Arabia, Africa, Madagascar, S. America and (*Acanthochlamys*) China (Sichuan).
Herbs to subarborescent.

B&H (MONOCOTS) Epigynae (within Amaryllidaceae, 174)
DT&H (MONOCOTS) Liliiflorae, Liliineae, 267
Melc (MONOCOTS) Liliiflorae, Liliineae, 309
Thor COMMELINIFLORAE Commelinales, Velloziineae, 328
Dahl BROMELIIFLORAE Velloziales, 436
Young LILIIDAE, LILIANAE Liliales, 70
Takh LILIIDAE, BROMELIANAE Velloziales, 502
Cron LILIIDAE Liliales, 371

Acanthochlamys P.C.Kao * **Nanuza** L.B.Sm. & Ayensu
Aylthonia N.L.Menezes **Pleurostima** Raf.
Barbacenia Vand. **Talbotia** Balf.
Barbaceniopsis L.B.Sm. **Vellozia** Vand.
Burlemarxia N.L.Menezes & Semir **Xerophyta** Juss.

XANTHORRHOEACEAE Dumort. 1829 175.10

1 genus. Australia. Robust rhizomatous herbs or subarborescent.

B&H (MONOCOTS) Calycineae (within Juncaceae, 186)
DT&H (MONOCOTS) Liliiflorae, Liliineae (within Liliaceae, 264)
Melc (MONOCOTS) Liliiflorae, Liliineae, 302
    (including Calectasiaceae, Dasypogonaceae, Lomandraceae, and possibly
    including Hanguanaceae)
Thor LILIIFLORAE Liliales, Liliineae (within Liliaceae, 297)
Dahl LILIIFLORAE Asparagales, 388
Young LILIIDAE, LILIANAE Liliales, 73
    (presumably including Calectasiaceae, Dasypogonaceae, Lomandraceae)
Takh LILIIDAE, LILIANAE Amaryllidales, 461
Cron LILIIDAE Liliales, 374
    (including Calectasiacee, Dasypogonoaceae, Lomandraceae)

Xanthorrhoea Sm.

XYRIDACEAE C. Agardh 1823 181

5 genera. Widespread in tropics and subtropics. Rhizomatous or cormous herbs.

B&H (MONOCOTS) Coronarieae, 181
DT&H (MONOCOTS) Farinosae, Enantioblastae, 255
Melc (MONOCOTS) Commelinales, Commelinineae, 323
Thor COMMELINIFLORAE Commelinales, Bromeliineae, 327
Dahl COMMELINIFLORAE Commelinales, 453
Young LILIIDAE, COMMELINANAE Commelinales, 81
Takh LILIIDAE, COMMELINANAE Commelinales, Xyridineae, 517
Cron COMMELINIDAE Commelinales, 341

Abolboda Bonpl. **Orectanthe** Maguire
Achlyphila Maguire & Wurdack **Xyris** L.
Aratitiyopea Steyerm. *

ZANNICHELLIACEAE Dumort. 1829   195.09

4 genera. Widespread. Aquatic herbs.

B&H	(MONOCOTS)	Apocarpae (within *Najadaceae*, 195)
DT&H	(MONOCOTS)	Helobiae, Potamogetonineae (within
		Potamogetonaceae, 237)
Melc	(MONOCOTS)	Helobiae, Potamogetonineae, 298
	(including Cymodoceaceae)	
Thor	ALISMATIFLORAE	Zosterales, Potamogetonineae, 315
Dahl	ALISMATIFLORAE	Najadales, 433
Young	LILIIDAE, ALISMATANAE	Zosterales, 50
Takh	ALISMATIDAE, NAJADANAE	Cymodoceales, 445
Cron	ALISMATIDAE	Najadales, 329

Althenia F.Petit   **Vleisia** Toml. & Posl.
Lepilaena J.L.Drumm. ex Harv.   **Zannichellia** L.

ZINGIBERACEAE Lindl. 1835   170.01

46 genera. Widespread in tropics. Rhizomatous herbs.

B&H	(MONOCOTS)	Epigynae (as *Scitamineae*) 170
	(including Cannaceae, Costaceae, Heliconiaceae, Lowiaceae, Marantaceae, Musaceae, Strelitziaceae)	
DT&H	(MONOCOTS)	Scitamineae, 272
	(including Costaceae)	
Melc	(MONOCOTS)	Scitamineae, 339
	(including Costaceae)	
Thor	COMMELINIFLORAE	Zingiberales, Zingiberineae, 345
Dahl	ZINGIBERIFLORAE	Zingiberales, 447
Young	LILIIDAE, ZINGIBERANAE	Zingiberales, 99
Takh	LILIIDAE, ZINGIBERANAE	Zingiberales, 507
Cron	ZINGIBERIDAE	Zingiberales, 361

Aframomum K.Schum.   **Haniffia** Holttum
Alpinia Roxb.   **Haplochorema** K.Schum.
Amomum Roxb.   **Hedychium** J.König
Aulotandra Gagnep.   **Hemiorchis** Kurz
Boesenbergia Kuntze   **Hitchenia** Wall.
Burbidgea Hook.f.   **Hornstedtia** Retz.
Camptandra Ridl.   **Kaempferia** L.
Caulokaempferia K.Larsen   **Leptosolena** C.Presl
Cautleya (Benth.) Hook.f.   **Mantisia** Sims
Curcuma L.   **Nanochilus** K.Schum.
Curcumorpha A.S.Rao & D.M.Verma   **Paracautleya** Ros.M.Sm. *
Cyphostigma Benth.   **Parakaempferia** A.S.Rao & D.M.Verma
Elettaria Maton   **Plagiostachys** Ridl.
Elettariopsis Baker   **Pleuranthodium** (K.Schum.) Ros.M.Sm.
Etlingera Giseke   **Pommereschea** Wittm.
Gagnepainia K.Schum.   **Pyrgophyllum** (Gagnep.) T.L.Wu &
Geocharis (K.Schum.) Ridl.   Z.Y.Chen
Geostachys (Baker) Ridl.   **Renealmia** L.f.
Globba L.   **Rhynchanthus** Hook.f.

Riedelia Oliv.
Roscoea Sm.
Scaphochlamys Baker
Siliquamomum Baill.
Siphonochilus J.M.Wood & Franks

Stadiochilus Ros.M.Sm. *
Stahlianthus Kuntze
Vanoverberghia Merr.
Zingiber Boehm.

ZOSTERACEAE Dumort. 1829   195.10

3 genera. N. and S. temperate coasts, rarely tropical. Marine aquatics.

B&H   (MONOCOTS)
DT&H (MONOCOTS)

Melc   (MONOCOTS)

Thor ALISMATIFLORAE
Dahl ALISMATIFLORAE
Young LILIIDAE, ALISMATANAE
Takh ASLIMATIDAE, NAJADANAE
Cron ALISMATIDAE

Apocarpae (within *Najadaceae*, 195)
Helobiae, Potamogetonineae (within
                 Potamogetonaceae, 237)
Helobiae, Potamogetonineae (within
                 Potamogetonaceae, 297)
Zosterales, Zosterineae, 317
Najadales, 432
Zosterales, 52
Zosterales, 444
Najadales, 332

Heterozostera (Setch.) Hartog
Phyllospadix Hook.
Zostera L.

PART 3

EIGHT SYSTEMS OF CLASSIFICATION OF THE FLOWERING PLANTS

In this part, the eight systems included in the tables at the head of each flowering plant family in Part 2 are presented in their own systematic order. The only change of sequence made is in the Dalla Torre & Harms system where the Monocotyledons are here placed after the Dicotyledons to make the sequence more easily comparable with other systems.

As in Part 2, names of taxa above the rank of order are given in bold upper case, while names of orders and suborders are given in bold lower case.

In the original publications, only one of the eight systems — that of Bentham & Hooker — had a single sequential numbering of the familes from beginning to end of the system. The original numbers of Bentham & Hooker are retained here, with extra numbers interpolated for Vochysiaceae and Cyrillaceae which originally escaped numbering. For the other seven systems, sequential numbering from beginning to end has been added here for ease of cross reference with Part 2. The idea of this convenient numerical reference is taken up from H.G. Bedell & J.L. Reveal in Phytologia 51: 65–156 (1982).

Families recognised in the system presented but not recognised at Kew are numbered and are printed in *italic* type, as in Part 2. Families recognised at Kew but not in the system presented are not numbered and are placed in brackets after the appropriate family in which the system placed them. Thus, for example, in the Dahlgren system, "21. Calycanthaceae (incl. Idiospermaceae)" means that the family Idiospermaceae is recognised by Kew but was not recognised by Dahlgren who included it in his family 21. Calycanthaceae. It should again be noted, as in Part 2, that in this context inclusion of one family in another is determined on the basis only of inclusion of the type genus and not necessarily of other genera.

THE BENTHAM & HOOKER SYSTEM

Taken from G. Bentham & J.D. Hooker, *Genera Plantarum* (1862–1883). The terminations of the family names have here been brought into line with the International Code (e.g. Berberidaceae rather than Berberideae). The original numbering of the families has been retained.

DICOTYLEDONES

POLYPETALAE

THALAMIFLORAE
Ranales
> 1 Ranunculaceae (incl. Glaucidiaceae, Paeoniaceae), 2 Dilleniaceae (incl. Crossosomataceae); 3 Calycanthaceae, 4 Magnoliaceae (incl. Eupteleaceae, Illiciaceae, Schisandraceae, Tetracentraceae, Trochodendraceae, Winteraceae), 5 Annonaceae (incl. Eupomatiaceae), 6 Menispermaceae, 7 Berberidaceae (incl. Lardizabalaceae); 8 Nymphaeaceae (incl. Cabombaceae, Nelumbonaceae)

Parietales
> 9 Sarraceniaceae, 10 Papaveraceae; 11 Cruciferae, 12 Capparaceae (incl. Emblingiaceae, Tovariaceae), 13 Resedaceae; 14 Cistaceae, 15 Violaceae, 16 Canellaceae, 17 Bixaceae (incl. Cochlospermaceae, Flacourtiaceae, Peridiscaceae)

Polygalineae
> 18 Pittosporaceae, 19 Tremandraceae; 20 Polygalaceae (incl. Krameriaceae), 20a Vochysiaceae (incl. Euphroniaceae, Trigoniaceae)

Caryophyllineae
> 21 Frankeniaceae, 22 Caryophyllaceae, 23 Portulacaceae, (incl. Hectorellaceae), 24 Tamaricaceae (incl. Fouquieriaceae)

Guttiferales
> 25 Elatinaceae, 26 *Hypericaceae*, 27 Guttiferae (incl. Quiinaceae), 28 Ternstroemiaceae [Theaceae] (incl. Actinidiaceae, Caryocaraceae, Marcgraviaceae, Pellicieraceae, Pentaphylacaceae, Stachyuraceae), 29 Dipterocarpaceae (incl. Ancistrocladaceae), 30 Chlaenaceae [Sarcolaenaceae]

Malvales
> 31 Malvaceae (incl. Bombacaceae), 32 Sterculiaceae, 33 Tiliaceae (incl. Elaeocarpaceae, Plagiopteraceae)

DISCIFLORAE
Geraniales
> 34 Linaceae (incl. Erythroxylaceae, Ixonanthaceae), 35 Humiriaceae, 36 Malpighiaceae, 37 Zygophyllaceae, 38 Geraniaceae (incl. Balsaminaceae, Limnanthaceae, Oxalidaceae, Tropaeolaceae), 39 Rutaceae (incl. Phellinaceae), 40 Simarubaceae (incl. Balanitaceae, Brunelliaceae, Cneoraceae, Irvingiaceae, Surianaceae), 41 Ochnaceae (incl. Tetrameristaceae), 42 Burseraceae, 43 Meliaceae, 44 Chailletiaceae [Dichapetalaceae]

Olacales
> 45 Olacaceae (incl. Cardiopteridaceae, Icacinaceae, Opiliaceae), 46 Ilicineae [Aquifoliaceae], 46a Cyrillaceae

Celastrales
> 47 Celastraceae, (incl. Alzateaceae, Goupiaceae), 48 Stackhousiaceae, 49 Rhamnaceae, 50 Ampelidaceae [Vitaceae] (incl. Leeaceae)

Sapindales

51 Sapindaceae (incl. Aceraceae, Akaniaceae, Greyiaceae, Hippocastanaceae, Melianthaceae, Ptaeroxylaceae, Staphyleaceae), 52 Sabiaceae (incl. Meliosmaceae), 53 Anacardiaceae (incl. Corynocarpaceae)

Ordines anomala

54 Coriariaceae, 55 Moringaceae

CALYCIFLORAE

Rosales

56 Connaraceae, 57 Leguminosae, 58 Rosaceae (incl. Canotiaceae, Cephalotaceae, Chrysobalanaceae, Eucryphiaceae, Neuradaceae, Pterostemonaceae, Stylobasiaceae), 59 Saxifragaceae (incl. Cephalotaceae, Cunoniaceae, Eremosynaceae, Escalloniaceae, Grossulariaceae, Hydrangeaceae, Parnassiaceae, Vahliaceae), 60 Crassulaceae (incl. Penthoraceae); 61 Droseraceae (incl. Byblidaceae, Roridulaceae); 62 Hamamelidaceae (incl. Myrothamnaceae), 63 Bruniaceae, 64 Haloragaceae (incl. Callitrichaceae, Gunneraceae, Hippuridaceae)

Myrtales

65 Rhizophoraceae (incl. Anisophylleaceae), 66 Combretaceae; 67 Myrtaceae (incl. Lecythidaceae), 68 Melastomaceae, 69 Lythraceae (incl. Crypteroniaceae, Oliniaceae), 70 Onagraceae (incl. Montiniaceae, Trapaceae)

Passiflorales

71 *Samydaceae* (incl. Asteropeiaceae), 72 Loasaceae, 73 Turneraceae, 74 Passifloraceae (incl. Achariaceae, Caricaceae, Malesherbiaceae, Physenaceae), 75 Cucurbitaceae, 76 Begoniaceae, 77 Datiscaceae

Ficoidales

78 Cactaceae, 79 Ficoidaceae [Aizoaceae] (incl. Gisekiaceae, Molluginaceae)

Umbellales

80 Umbelliferae, 81 Araliaceae (incl. Aralidiaceae, Helwingiaceae), 82 Cornaceae (incl. Alangiaceae, Aucubaceae, Garryaceae, Griseliniaceae, Torricelliaceae)

GAMOPETALAE

INFERAE

Rubiales

83 Caprifoliaceae (incl. Adoxaceae, Alseuosmiaceae), 84 Rubiaceae (incl. Carlemanniaceae)

Asterales

85 Valerianaceae, 86 Dipsacaceae (incl. Morinaceae, Triplostegiaceae); 87 Calyceraceae, 88 Compositae

Campanales

89 Stylidiaceae, 90 Goodenovieae [Goodeniaceae], 91 Campanulaceae (incl. Pentaphragmataceae)

HETEROMERAE

Ericales

91 *Vacciniaceae*, 93 Ericaceae (incl. Clethraceae), 94 *Monotropaceae*, 95 Epacridaceae, 96 Diapensiaceae, 97 Lennoaceae

Primulales

98 Plumbaginaceae, 99 Primulaceae, 100 Myrsinaceae (incl. Theophrastaceae)

Ebenales

101 Sapotaceae, 102 Ebenaceae, 103 Styracaceae (incl. Lissocarpaceae, Symplocaceae)

BICARPELLATAE

Gentianales

104 Oleaceae, 105 Salvadoraceae, 106 Apocynaceae, 107 Asclepiadaceae, 108 Loganiaceae (incl. Buddlejaceae), 109 Gentianaceae (incl. Menyanthaceae)

Polemoniales

110 Polemoniaceae (incl. Cobaeaceae), 111 Hydrophyllaceae, 112 Boraginaceae, 113 Convolvulaceae, 114 Solanaceae (incl. Goetzeaceae, Retziaceae)

Personales

115 Scrophulariaceae, 116 *Orobanchaceae*, 117 Lentibulariaceae, 118 Columelliaceae, 119 Gesneriaceae, 120 Bignoniaceae, 121 Pedaliaceae, 122 Acanthaceae

Lamiales

123 Myoporaceae, 124 *Selaginaceae*; 125 Verbenaceae (incl. Avicenniaceae, Nesogenaceae, Phrymaceae, Stilbaceae), 126 Labiatae

Ordo anomalus

127 Plantaginaceae

MONOCHLAMYDEAE

Curvembryeae

128 Nyctaginaceae, 129 Illecebraceae, 130 Amarantaceae (incl. Achatocarpaceae), 131 Chenopodiaceae (incl. Basellaceae), 132 Phytolaccaceae (incl. Agdestidaceae, Barbeuiaceae, Gyrostemonaceae, Stegnospermataceae), 133 Batidaceae, 134 Polygonaceae

Multiovulatae Aquaticae

135 Podostemaceae (incl. Hydrostachyaceae), 136 Nepenthaceae, 137 Cytinaceae [Rafflesiaceae] (incl. Hydnoraceae), 138 Aristolochiaceae

Micrembryeae

139 Piperaceae (incl. Lactoridaceae, Saururaceae); 140 Chloranthaceae, 141 Myristicaceae, 142 Monimiaceae (incl. Amborellaceae, Trimeniaceae)

Daphnales

143 Lauraceae (incl. Hernandiaceae), 144 Proteaceae, 145 Thymelaeaceae, 146 Penaeaceae (incl. Geissolomataceae), 147 Elaeagnaceae

Achlamydosporeae

148 Loranthaceae (incl. Eremolepidaceae, Viscaceae), 149 Santalaceae (incl. Grubbiaceae, Misodendraceae), 150 Balanophoraceae (incl. Cynomoriaceae)

Unisexuales

151 Euphorbiaceae (incl. Aextoxicaceae, Buxaceae, Daphniphyllaceae, Simmondsiaceae), 152 Balanopseae [Balanopaceae], 153 Urticaceae (incl. Cannabaceae, Cecropiaceae, Moraceae, Theligonaceae, Ulmaceae), 154 Platanaceae, 155 Leitneriaceae (? incl. Didymelaceae), 156 Juglandaceae, 157 Myricaceae, 158 Casuarinaceae, 159 Cupuliferae [Betulaceae] (incl. Corylaceae, Fagaceae)

Ordines anomali

160 Salicaceae, 161 Lacistemaceae, 162 Empetraceae, 163 Ceratophyllaceae

[FAMILIES 164–166 — GYMNOSPERMAE]

MONOCOTYLEDONES

Microspermae
167 Hydrocharitaceae, 168 Burmanniaceae (incl. Corsiaceae), 169 Orchidaceae

Epigynae
170 Scitamineae [Zingiberaceae] (incl. Cannaceae, Costaceae, Heliconiaceae, Lowiaceae, Marantaceae, Musaceae, Strelitziaceae), 171 Bromeliaceae, 172 Haemodoraceae (incl. Lanariaceae, Tecophilaeaceae), 173 Iridaceae, 174 Amaryllidaceae (incl. Agavaceae, Alstroemeriaceae, Doryanthaceae, Hypoxidaceae, Ixioliriaceae, Velloziaceae), 175 Taccaceae, 176 Dioscoreaceae (incl. Petermanniaceae, Trichopodaceae)

Coronarieae
177 Roxburghiaceae [Stemonaceae], 178 Liliaceae (incl. Alliaceae, Aloaceae, Anthericaceae, Aphyllanthaceae, Asparagaceae, Asphodelaceae, Asteliaceae, Blandfordiaceae, Colchicaceae, Convallariaceae, Dracaenaceae, Eriospermaceae, Hemerocallidaceae, Herreriaceae, Hostaceae, Hyacinthaceae, Melianthaceae, Philesiaceae, Phormiaceae, Rhipogonaceae, Ruscaceae, Smilacaceae, Trilliaceae), 179 Pontederiaceae, 180 Phylidraceae, 181 Xyridaceae, 182 Mayacaceae, 183 Commelinaceae, 184 Rapateaceae

Calycineae
185 Flagellariaceae (incl. Hanguanaceae, Joinvilleaceae), 186 Juncaceae (incl. Calectasiaceae, Dasypogonaceae, Lomandraceae, Thurniaceae, Xanthorrhoeaceae), 187 Palmae

Nudiflorae
188 Pandanaceae, 189 Cyclanthaceae, 190 Typhaceae, 191 Araceae (incl. Acoraceae) 192 Lemnaceae

Apocarpae
193 Triuridaceae, 194 Alismataceae (incl. Butomaceae, Limnocharitaceae), 195 Naiadaceae

Glumaceae
196 Eriocaulaceae, 197 Centrolepidaceae, 198 Restionaceae (incl. Anarthriaceae, Ecdeiocoleaceae), 199 Cyperaceae, 200 Gramineae

THE ENGLER SYSTEM OF DALLA TORRE & HARMS

Taken from C.G. de Dalla Torre & H. Harms, *Genera Siphonogamarum ad Systema Englerianum Conscripta* (1900–1907). This is a numbered synopsis, with extensive bibliographical references and synonymy, of the system elaborated by A. Engler in his various works, particularly as summarised in his *Syllabus* (1892 and successive editions). The Monocots, placed by Dalla Torre & Harms before the Dicots, have here been placed after them to make the sequence more comparable with other systems.

DICOTYLEDONEAE

ARCHICHLAMYDEAE

Verticillatae
> 1 Casuarinaceae

Piperales
> 2 Saururaceae, 3 Piperaceae, 4 Chloranthaceae (incl. Circaeasteraceae), 5 Lacistemaceae

Salicales
> 6 Salicaceae

Myricales
> 7 Myricaceae

Balanopsidales
> 8 Balanopsidaceae

Leitneriales
> 9 Leitneriaceae

Juglandales
> 10 Juglandaceae

Fagales
> 11 Betulaceae (incl. Corylaceae), 12 Fagaceae

Urticales
> 13 Ulmaceae (incl. Barbeyaceae), 14 Moraceae (incl. Cannabaceae, Cecropiaceae), 15 Urticaceae

Proteales
> 16 Proteaceae

Santalales
> **Loranthineae**
> 17 Loranthaceae (incl. Eremolepidaceae, Viscaceae)
> **Santalineae**
> 18 Myzodendraceae, 19 Santalaceae, 20 Grubbiaceae, 21 Opiliaceae, 22 Olacaceae
> **Balanophorineae**
> 23 Balanophoraceae

Aristolochiales
> 24 Aristolochiaceae, 25 Rafflesiaceae, 26 Hydnoraceae

Polygonales
> 27 Polygonaceae

Centrospermae
> **Chenopodiineae**
> 28 Chenopodiaceae (incl. Halophytaceae), [Suppl.] Didiereaceae, 29 Amarantaceae

Phytolaccineae
30 Nyctaginaceae, 31 Batidaceae, 32 Cynocrambaceae [Theligonaceae], 33 Phytolaccaceae (incl. Achatocarpaceae, Agdestidaceae, Barbeuiaceae, Gisekiaceae, Gyrostemonaceae, Stegnospermataceae), 34 Aizoaceae (incl. Molluginaceae)

Portulacineae
35 Portulacaceae (incl. Hectorellaceae), 36 Basellaceae

Caryophyllineae
37 Caryophyllaceae (incl. Illecebraceae)

Ranales

Nymphaeineae
38 Nymphaeaceae (incl. Cabombaceae, Nelumbonaceae), 39 Ceratophyllaceae

Trochodendrineae
40 Trochodendraceae (incl. Cercidiphyllaceae, Eucommiaceae, Eupteleaceae)

Ranunculineae
41 Ranunculaceae (incl. Glaucidiaceae, Paeoniaceae), 42 Lardizabalaceae, 43 Berberidaceae, 44 Menispermaceae

Magnoliineae
45 Magnoliaceae (incl. Eupteleaceae, Illiciaceae, Schisandraceae, Tetracentraceae, Trochodendraceae, Winteraceae), 46 Calycanthaceae, 47 Lactoridaceae, 48 Annonaceae (incl. Himantandraceae, Eupomatiaceae), 49 Myristicaceae, 50 Gomortegaceae, 51 Monimiaceae (incl. Amborellaceae, Scyphostegiaceae, Trimeniaceae), 52 Lauraceae, 53 Hernandiaceae

Rhoeadales

Rhoeadineae
54 Papaveraceae

Capparidineae
55 Cruciferae, 56 Tovariaceae, 57 Capparidaceae (incl. Emblingiaceae)

Resedineae
58 Resedaceae

Moringineae
59 Moringaceae

Sarraceniales
60 Sarraceniaceae, 61 Nepenthaceae, 62 Droseraceae (incl. Byblidaceae, Roridulaceae)

Rosales

Podostemonineae
63 Podostemonaceae, 64 Hydrostachyaceae

Saxifragineae
65 Crassulaceae (incl. Penthoraceae), 66 Cephalotaceae, 67 Saxifragaceae (incl. Eremosynaceae, Escalloniaceae, Grossulariaceae, Hydrangeaceae, Montiniaceae, Parnassiaceae, Pterostemonaceae, Vahliaceae), 68 Pittosporaceae, 69 Brunelliaceae, 70 Cunoniaceae (incl. Davidsoniaceae), 71 Myrothamnaceae, 72 Bruniaceae, 73 Hamamelidaceae

Rosineae
74 Platanaceae, 75 Crossosomataceae, 76 Rosaceae (incl. Chrysobalanaceae, Euphroniaceae, Neuradaceae, Stylobasiaceae), 77 Connaraceae, 78 Leguminosae (incl. Krameriaceae)

Geraniales

Geraniineae
79 Geraniaceae, 80 Oxalidaceae, 81 Tropaeolaceae, 82 Linaceae (incl. Ctenolophonaceae, Ixonanthaceae, Lepidobotryaceae), 83 Humiriaceae, 84 Erythroxylaceae, 85 Zygophyllaceae (incl. Balanitaceae), 86 Cneoraceae, 87 Rutaceae (incl. Rhabdodendraceae) 88 Simarubaceae (incl. Irvingiaceae, Surianaceae), 89 Burseraceae, 90 Meliaceae (incl. Ptaeroxylaceae)

Malpighiineae
> 91 Malpighiaceae, 92 Trigoniaceae (? incl. Euphroniaceae), 93 Vochysiaceae

Polygalineae
> 94 Tremandraceae, 95 Polygalaceae

Dichapetalineae
> 96 Dichapetalaceae

Tricoccae
> 97 Euphorbiaceae (incl. Aextoxicaceae, Daphniphyllaceae)

Incertae sedis
> 98 Callitrichaceae

Sapindales

Buxineae
> 99 Buxaceae (incl. Simmondsiaceae)

Coriariineae
> 100 Coriariaceae

Empetrineae
> 101 Empetraceae

Limnanthineae
> 102 Limnanthaceae

Anacardiineae
> 103 Anacardiaceae (incl. Podoaceae)

Celastrineae
> 104 Cyrillaceae, 105 Pentaphylacaceae, 106 Corynocarpaceae, 107 Aquifoliaceae (incl. Phellinaceae, Sphenostemonaceae), 108 Celastraceae (incl. Canotiaceae, Goupiaceae), 109 Hippocrateaceae, 110 Stackhousiaceae, 111 Staphyleaceae

Icacinineae
> 112 Icacinaceae (incl. Cardiopteridaceae, Lophopyxidaceae)

Sapindineae
> 113 Aceraceae, 114 Hippocastanaceae (incl. Bretschneideraceae), 115 Sapindaceae (incl. Akaniaceae)

Sabiineae
> 116 Sabiaceae (incl. Meliosmaceae)

Melianthineae
> 117 Melianthaceae (incl. Greyiaceae)

Balsaminineae
> 118 Balsaminaceae

Rhamnales
> 119 Rhamnaceae, 120 Vitaceae (incl. Leeaceae)

Malvales

Elaeocarpineae
> 121 Elaeocarpaceae

Chlaenineae
> 122 Chlaenaceae [Sarcolaenaceae]

Malvineae
> 123 *Gonystylaceae*, 124 Tiliaceae, 125 Malvaceae, 126 *Triplochitonaceae*, 127 Bombacaceae, 128 Sterculiaceae (incl. Huaceae)

Scytopetalineae
> 129 Scytopetalaceae

Parietales

Theineae
> 130 Dilleniaceae (incl. Actinidiaceae), 131 Eucryphiaceae, 132 Ochnaceae (incl. Strasburgeriaceae), 133 Caryocaraceae, 134 Marcgraviaceae, 135 Quiinaceae, 136 Theaceae (incl. Asteropeiaceae, Pellicieraceae, Tetrameristaceae), 137 Guttiferae (incl. Medusagynaceae), 138 Dipterocarpaceae

Tamaricineae
139 Elatinaceae, 140 Frankeniaceae, 141 Tamaricaceae
Fouquieriineae
142 Fouquieriaceae
Cistineae
143 Cistaceae, 144 Bixaceae (incl. Sphaerosepalaceae)
Cochlospermineae
145 Cochlospermaceae, 146 *Koeberliniaceae*
Flacourtiineae
147 Canellaceae, 148 Violaceae, 149 Flacourtiaceae (incl. Dioncophyllaceae, Hoplestigmataceae, Peridiscaceae, Physenaceae, Plagiopteraceae), 150 Stachyuraceae, 151 Turneraceae, 152 Malesherbiaceae, 153 Passifloraceae, 154 Achariaceae
Papayineae
155 Caricaceae
Loasineae
156 Loasaceae
Datiscineae
157 Datiscaceae
Begoniineae
158 Begoniaceae
Ancistrocladineae
159 Ancistrocladaceae
Opuntiales
160 Cactaceae
Myrtiflorae
  Thymelaeineae
161 Geissolomaceae, 162 Penaeaceae, 163 Oliniaceae, 164 Thymelaeaceae, 165 Elaeagnaceae
  Myrtineae
166 Lythraceae (incl. Rhynchocalycaceae), 167 *Sonneratiaceae*, 168 Crypteroniaceae, 169 *Punicaceae*, 170 Lecythidaceae, 171 Rhizophoraceae (incl. Anisophylleaceae), 172 Combretaceae, 173 Myrtaceae, 174 Melastomataceae, 175 Onagraceae, 176 Hydrocaryaceae [Trapaceae]
  Halorrhagidineae
177 Halorrhagidaceae [Haloragaceae] (incl. Gunneraceae, Hippuridaceae)
  Cynomoriineae
178 Cynomoriaceae
Umbelliflorae
179 Araliaceae (incl. Aralidiaceae), 180 Umbelliferae, 181 Cornaceae (incl. Alangiaceae, Aucubaceae, Garryaceae, Griseliniaceae, Helwingiaceae, Melanophyllaceae, Torricelliaceae)

METACHLAMYDEAE

Ericales
182 Clethraceae, 183 *Pirolaceae*, 184 Lennoaceae, 185 Ericaceae, 186 Epacridaceae, 187 Diapensiaceae
Primulales
188 Theophrastaceae, 189 Myrsinaceae, 190 Primulaceae, 191 Plumbaginaceae
Ebenales
  Sapotineae
192 Sapotaceae

Diospyrineae

193 Ebenaceae, 194 Styracaceae (incl. Lissocarpaceae), 195 Symplocaceae

Contortae

Oleineae

196 Oleaceae, 197 Salvadoraceae

Gentianinae

198 Loganiaceae (incl. Buddlejaceae, Retziaceae), 199 Gentianaceae (incl. Menyanthaceae), 200 Apocynaceae, 201 Asclepiadaceae

Tubiflorae

Convolvulineae

202 Convolvulaceae, 203 Polemoniaceae

Borraginineae

204 Hydrophyllaceae, 205 Borraginaceae

Verbenineae

206 Verbenaceae (incl. Avicenniaceae, Cyclocheilaceae, Nesogenaceae, Stilbaceae), 207 Labiatae

Solanineae

208 *Nolanaceae,* 209 Solanaceae (incl. Goetzeaceae), 210 Scrophulariaceae, 211 Bignoniaceae, 212 Pedaliaceae, 213 *Martyniaceae,* 214 *Orobanchaceae,* 215 Gesneriaceae, 216 Columelliaceae, 217 Lentibulariaceae, 218 Globulariaceae

Acanthineae

219 Acanthaceae

Myoporineae

220 Myoporaceae

Phrymineae

221 Phrymaceae

Plantaginales

222 Plantaginaceae

Rubiales

223 Rubiaceae (incl. Carlemanniaceae), 224 Caprifoliaceae (incl. Alseuosmiaceae), 225 Adoxaceae, 226 Valerianaceae (incl. Triplostegiaceae), 227 Dipsacaceae (incl. Morinaceae)

Campanulatae

Cucurbitineae

228 Cucurbitaceae

Campanulineae

229 Campanulaceae (incl. Pentaphragmataceae), 230 Goodeniaceae, 231 Stylidiaceae, 232 Calyceraceae, 233 Compositae

MONOCOTYLEDONEAE

Pandanales

234 Typhaceae, 235 Pandanaceae, 236 *Sparganiaceae*

Helobiae

Potamogetoninae

237 Potamogetonaceae (incl. Cymodoceaceae, Posidoniaceae, Zannichelliaceae, Zosteraceae), 238 *Najadaceae,* 239 Aponogetonaceae, 240 Juncaginaceae (incl. Lilaeaceae, Scheuchzeriaceae)

Alismineae

241 Alismaceae

Butomineae

242 Butomaceae (incl. Limnocharitaceae), 243 Hydrocharitaceae

Triuridales
244 Triuridaceae
Glumiflorae
245 Gramineae, 246 Cyperaceae
Principes
247 Palmae
Synanthae
248 Cyclanthaceae
Spathiflorae
249 Araceae (incl. Acoraceae), 250 Lemnaceae
Farinosae
  Flagellariineae
    251 Flagellariaceae (incl. Hanguanaceae, Joinvilleaceae)
  Enantioblastae
    252 Restionaceae (incl. Anarthriaceae, Ecdeiocoleaceae), 253 Centrolepidaceae
    (incl. Hydatellaceae), 254 Mayacaceae, 255 Xyridaceae, 256 Eriocaulaceae
  Bromeliineae
    257 Rapateaceae, 258 Bromeliaceae
  Commelinineae
    259 Commelinaceae
  Pontederiineae
    260 Pontederiaceae (incl. Cyanastraceae)
  Philydrineae
    261 Philydraceae
Liliflorae
  Juncineae
    262 Juncaceae (incl. Thurniaceae)
  Liliineae
    263 Stemonaceae, 264 Liliaceae (incl. Alliaceae, Aloaceae, Anthericaceae,
    Aphyllanthaceae, Asparagaceae, Asphodelaceae, Asteliaceae, Blandfordiaceae,
    Calectasiaceae, Colchicaceae, Convallariaceae, Dasypogonaceae, Dracaenaceae,
    Eriospermaceae, Hemerocallidaceae, Herreriaceae, Hostaceae, Hyacinthaceae,
    Lomandraceae, Melanthiaceae, Philesiaceae, Phormiaceae, Rhipogonaceae,
    Ruscaceae, Smilacaceae, Trilliaceae, Xanthorrhoeaceae), 265 Haemodoraceae,
    266 Amaryllidaceae (incl. Agavaceae, Alstroemeriaceae, Doryanthaceae,
    Hypoxidaceae, Ixioliriaceae, Lanariaceae, Tecophilaeaceae), 267 Velloziaceae,
    268 Taccaceae, 269 Dioscoreaceae (incl. Petermanniaceae, Trichopodaceae)
  Iridineae
    270 Iridaceae
Scitamineae
    271 Musaceae (incl. Heliconiaceae, Lowiaceae, Strelitziaceae), 272 Zingiberaceae
    (incl. Costaceae), 273 Cannaceae, 274 Marantaceae
Microspermae
  Burmanniineae
    275 Burmanniaceae (incl. Corsiaceae)
  Gynandrae
    276 Orchidaceae

THE ENGLER SYSTEM OF MELCHIOR

Taken from A. Engler's *Syllabus der Pflanzenfamilien* ed. 12, vol. 2, Angiospermen (1964), edited by H. Melchior, with contributions by G. Buchheim, W. Schultze-Motel, Th. Eckardt, H. Scholz, G. Wagenitz, U. Hamann and E. Potztal.

DICOTYLEDONEAE

ARCHICHLAMYDEAE

Casuarinales
1 Casuarinaceae
Juglandales
2 Myricaceae, 3 Juglandaceae
Balanopales
4 Balanopaceae
Leitneriales
5 Leitneriaceae, 6 Didymelaceae
Salicales
7 Salicaceae
Fagales
8 Betulaceae (incl. Corylaceae), 9 Fagaceae
Urticales
10 Rhoipteleaceae, 11 Ulmaceae (incl. Barbeyaceae), 12 Eucommiaceae, 13 Moraceae (incl. Cannabaceae, Cecropiaceae), 14 Urticaceae
Proteales
15 Proteaceae
Santalales
  Santalineae
16 Olacaceae, 17 Dipentodontaceae, 18 Opiliaceae, 19 Grubbiaceae, 20 Santalaceae, 21 Misodendraceae
  Loranthineae
22 Loranthaceae (incl. Eremolepidaceae, Viscaceae)
Balanophorales
23 Balanophoraceae
Medusandrales
24 Medusandraceae
Polygonales
25 Polygonaceae
Centrospermae
  Phytolaccineae
26 Phytolaccaceae (incl. Agdestidaceae, Barbeuiaceae, Stegnospermataceae), 27 Gyrostemonaceae, 28 Achatocarpaceae, 29 Nyctaginaceae, 30 Molluginaceae (incl. Gisekiaceae), 31 Aizoaceae
  Portulacineae
32 Portulacaceae, 33 Basellaceae
  Caryophyllineae
34 Caryophyllaceae (incl. Hectorellaceae, Illecebraceae)
  Chenopodiineae
35 *Dysphaniaceae*, 36 Chenopodiaceae, 37 Amaranthaceae
Anhang: Didiereaceae

Cactales
  38 Cactaceae
Magnoliales
  Magnoliineae
    39 Magnoliaceae, 40 Degeneriaceae, 41 Himantandraceae, 42 Winteraceae, 43 Annonaceae, 44 Eupomatiaceae, 45 Myristicaceae, 46 Canellaceae
  Illiciineae
    47 Schisandraceae, 48 Illiciaceae
  Laurineae
    49 Austrobaileyaceae, 50 Trimeniaceae, 51 Amborellaceae, 52 Monomiaceae, 53 Calycanthaceae, 54 Gomortegaceae, 55 Lauraceae, 56 Hernandiaceae
  Trochodendrineae
    57 Tetracentraceae, 58 Trochodendraceae
  Eupteleineae
    59 Eupteleaceae
  Cercidyphyllineae
    60 Cercidiphyllaceae
Ranunculales
  Ranunculineae
    61 Ranunculaceae (incl. Circaeasteraceae, Glaucidiaceae), 62 Berberidaceae, 63 Sargentodoxaceae, 64 Lardizabalaceae, 65 Menispermaceae
  Nymphaeineae
    66 Nymphaeaceae (incl. Cabombaceae, Nelumbonaceae), 67 Ceratophyllaceae
Piperales
    68 Saururaceae, 69 Piperaceae, 70 Chloranthaceae, 71 Lactoridaceae
Aristolochiales
    72 Aristolochiaceae, 73 Rafflesiaceae, 74 Hydnoraceae
Guttiferales
  Dilleniineae
    75 Dilleniaceae, 76 Paeoniaceae, 77 Crossosomataceae, 78 Eucryphiaceae, 79 Medusagynaceae, 80 Actinidiaceae
  Ochnineae
    81 Ochnaceae, 82 Dioncophyllaceae, 83 Strasburgeriaceae, 84 Dipterocarpaceae
  Theineae
    85 Theaceae (incl. Asteropeiaceae, Pellicieraceae, Tetrameristaceae), 86 Caryocaraceae, 87 Marcgraviaceae, 88 Quiinaceae, 89 Guttiferae
  Ancistrocladineae
    90 Ancistrocladaceae
Sarraceniales
    91 Sarraceniaceae, 92 Nepenthaceae, 93 Droseraceae
Papaverales
  Papaverineae
    94 Papaveraceae
  Capparineae
    95 Capparaceae (incl. Emblingiaceae), 96 Cruciferae, 97 Tovariaceae
  Resedineae
    98 Resedaceae
  Moringineae
    99 Moringaceae
Batales
    100 Bataceae
Rosales
  Hamamelidineae
    101 Platanaceae. 102 Hamamelidaceae, 103 Myrothamnaceae

Saxifragineae

104 Crassulaceae, 105 Cephalotaceae, 106 Saxifragaceae (incl. Eremosynaceae, Escalloniaceae, Grossulariaceae, Hydrangeaceae, Montiniaceae, Parnassiaceae, Penthoraceae, Pterostemonaceae, Vahliaceae), 107 Brunelliaceae, 108 Cunoniaceae, 109 Davidsoniaceae, 110 Pittosporaceae, 111 Byblidaceae, 112 Roridulaceae, 113 Bruniaceae

Rosineae

114 Rosaceae, 115 Neuradaceae, 116 Chrysobalanaceae (incl. Stylobasiaceae)

Leguminosineae

117 Connaraceae, 118 Leguminosae, 119 Krameriaceae

Hydrostachyales

120 Hydrostachyaceae

Podostemales

121 Podostemaceae

Geraniales

Limnanthineae

122 Limnanthaceae

Geraniineae

123 Oxalidaceae (incl. Lepidobotryaceae), 124 Geraniaceae, 125 Tropaeolaceae, 126 Zygophyllaceae (incl. Balanitaceae), 127 Linaceae (incl. Ctenolophonaceae, Humiriaceae, Ixonanthaceae), 128 Erythroxylaceae

Euphorbiineae

129 Euphorbiaceae, 130 Daphniphyllaceae

Rutales

Rutineae

131 Rutaceae (incl. Rhabdodendraceae), 132 Cneoraceae, 133 Simaroubaceae (incl. Irvingiaceae, Surianaceae), 134 *Picrodendraceae*, 135 Burseraceae, 136 Meliaceae (incl. Ptaeroxylaceae), 137 Akaniaceae

Malpighiineae

138 Malpighiaceae, 139 Trigoniaceae (incl. Euphroniaceae), 140 Vochysiaceae

Polygalineae

141 Tremandraceae, 142 Polygalaceae

Sapindales

Coriariineae

143 Coriariaceae

Anacardiineae

144 Anacardiaceae (incl. Podoaceae)

Sapindineae

145 Aceraceae, 146 Bretschneideraceae, 147 Sapindaceae, 148 Hippocastanaceae, 149 Sabiaceae (incl. Meliosmaceae), 150 Melianthaceae (incl. Greyiaceae), 151 Aextoxicaceae

Balsamineae

152 Balsaminaceae

Julianiales

153 *Julianiaceae*

Celastrales

Celastrineae

154 Cyrillaceae, 155 Pentaphylacaceae, 156 Aquifoliaceae (incl. Phellinaceae, Sphenostemonaceae), 157 Corynocarpaceae, 158 Pandaceae, 159 Celastraceae (incl. Goupiaceae, Lophopyxidaceae, ? Canotiaceae), 160 Staphyleaceae, 161 *Hippocrateaceae*, 162 Stackhousiaceae, 163 Salvadoraceae

Buxineae

164 Buxaceae (incl. Simmondsiaceae)

Icacinineae

165 Icacinaceae, 166 Cardiopteridaceae

Rhamnales
167 Rhamnaceae, 168 Vitaceae, 169 Leeaceae

Malvales

Elaeocarpineae
170 Elaeocarpaceae

Sarcolaenineae
171 Sarcolaenaceae

Malvineae
172 Tiliaceae, 173 Malvaceae, 174 Bombacaceae, 175 Sterculiaceae

Scytopetalineae
176 Scytopetalaceae

Thymelaeales
177 Geissolomataceae, 178 Penaeaceae, 179 Dichapetalaceae, 180 Thymelaeaceae, 181 Elaeagnaceae

Violales

Flacourtiineae
182 Flacourtiaceae (incl. Lacistemataceae), 183 Peridiscaceae, 184 Violaceae,
185 Stachyuraceae, 186 Scyphostegiaceae, 187 Turneraceae, 188 Malesherbiaceae, 189 Passifloraceae, 190 Achariaceae

Cistineae
191 Cistaceae, 192 Bixaceae, 193 Sphaerosepalaceae, 194 Cochlospermaceae

Tamaricineae
195 Tamaricaceae, 196 Frankeniaceae, 197 Elatinaceae

Caricineae
198 Caricaceae

Loasineae
199 Loasaceae

Begoniineae
200 Datiscaceae, 201 Begoniaceae

Cucurbitales
202 Cucurbitaceae

Myrtiflorae

Myrtineae
203 Lythraceae, 204 Trapaceae, 205 Crypteroniaceae, 206 Myrtaceae, 207
Dialypetalanthaceae, 208 *Sonneratiaceae*, 209 *Punicaceae*, 210 Lecythidaceae, 211
Melastomataceae, 212 Rhizophoraceae (incl. Anisophylleaceae), 213 Combretaceae, 214 Onagraceae, 215 Oliniaceae, 216 Haloragaceae (incl. Gunneraceae),
217 Theligonaceae

Hippuridineae
218 Hippuridaceae

Cynomoriineae
219 Cynomoriaceae

Umbelliflorae
220 Alangiaceae, 221 *Nyssaceae*, 222 *Davidiaceae*, 223 Cornaceae (incl.
Aucubaceae, Griseliniaceae, Helwingiaceae, Melanophyllaceae, Torricelliaceae), 224 Garryaceae, 225 Araliaceae, 226 Umbelliferae

SYMPETALAE

Diapensiales
227 Diapensiaceae

Ericales
228 Clethraceae, 229 *Pyrolaceae*, 230 Ericaceae, 231 Empetraceae, 232
Epacridaceae

Primulales
233 Theophrastaceae, 234 Myrsinaceae, 235 Primulaceae

Plumbaginales
236 Plumbaginaceae

Ebenales
Sapotineae
237 Sapotaceae, 238 *Sarcospermataceae*

Ebenineae
239 Ebenaceae, 240 Styracaceae, 241 Lissocarpaceae, 242 Symplocaceae, 243 Hoplestigmataceae

Oleales
244 Oleaceae

Gentianales
245 Loganiaceae, 246 *Desfontainiaceae*, 247 Gentianaceae, 248 Menyanthaceae, 249 Apocynaceae, 250 Asclepiadaceae, 251 Rubiaceae

Tubiflorae
Convolvulineae
252 Polemoniaceae (incl. Cobaeaceae), 253 Fouquieriaceae, 254 Convolvulaceae

Boraginineae
255 Hydrophyllaceae, 256 Boraginaceae, 257 Lennoaceae

Verbenineae
258 Verbenaceae (incl. Avicenniaceae, Nesogenaceae, Stilbaceae), 259 Callitrichaceae, 260 Labiatae (incl. Tetrachondraceae)

Solanineae
261 *Nolanaceae*, 262 Solanaceae (incl. Goetzeaceae), 263 Duckeodendraceae, 264 Buddlejaceae, 265 Scrophulariaceae, 266 Globulariaceae, 267 Bignoniaceae, 268 *Henriqueziaceae*, 269 Acanthaceae, 270 Pedaliaceae, 271 *Martyniaceae*, 272 Gesneriaceae, 273 Columelliaceae, 274 *Orobanchaceae*, 275 Lentibulariaceae

Myoporineae
276 Myoporaceae

Phrymineae
277 Phrymaceae

Plantaginales
278 Plantaginaceae

Dipsacales
279 Caprifoliaceae (incl. Alseuosmiaceae, Carlemanniaceae), 280 Adoxaceae, 281 Valerianaceae (incl. Triplostegiaceae), 282 Dipsacaceae (incl. Morinaceae)

Campanulales
283 Campanulaceae, 284 *Sphenocleaceae*, 285 Pentaphragmataceae, 286 Goodeniaceae, 287 *Brunoniaceae*, 288 Stylidiaceae, 289 Calyceraceae, 290 Compositae

MONOCOTYLEDONEAE

Helobiae
Alismatineae
291 Alismataceae, 292 Butomaceae (incl. Limnocharitaceae)

Hydrocharitineae
293 Hydrocharitaceae

Scheuchzeriineae
294 Scheuchzeriaceae

Potamogetonineae
295 Aponogetonaceae, 296 Juncaginaceae (incl. Lilaeaceae), 297 Potamogetonaceae (incl. Posidoniaceae, Zosteraceae), 298 Zannichelliaceae (incl. Cymodoceaceae), 299 *Najadaceae*

Triuridales
300 Triuridaceae
Lilliflorae
Liliineae
301 Liliaceae (incl. Alliaceae, Aloaceae, Alstroemeriaceae, Anthericaceae, Aphyllanthaceae, Asparagaceae, Asphodelaceae, Asteliaceae, Blandfordiaceae, Colchicaceae, Convallariaceae, Dracaenaceae, Eriospermaceae, Hemero-callidaceae, Herreriaceae, Hostaceae, Hyacinthaceae, Melanthiaceae, Peterman-niaceae, Philesiaceae, Phormiaceae, Rhipogonaceae, Ruscaceae, Smilacaceae, Trilliaceae), 302 Xanthorrhoeaceae (incl. Calectasiaceae, Dasypogonaceae, Lomandraceae, ? Hanguanaceae), 303 Stemonaceae, 304 Agavaceae (incl. Doryanthaceae), 305 Haemodoraceae (incl. Tecophilaeaceae), 306 Cyan-astraceae, 307 Amaryllidaceae (incl. Ixioliriaceae), 308 Hypoxidaceae, 309 Velloziaceae, 310 Taccaceae, 311 Dioscoreaceae (incl. Trichopodaceae)
Pontederiineae
312 Pontederiaceae
Iridineae
313 Iridaceae, 314 *Geosiridaceae*
Burmanniineae
315 Burmanniaceae, 316 Corsiaceae
Philydrineae
317 Philydraceae
Juncales
318 Juncaceae, 319 Thurniaceae
Bromeliales
320 Bromeliaceae
Commelinales
Commelinineae
321 Commelinaceae, 322 Mayacaceae, 323 Xyridaceae, 324 Rapateaceae
Eriocaulineae
325 Eriocaulaceae
Restionineae
326 Restionaceae (incl. Anarthriaceae, ? Ecdeiocoleaceae), 327 Centrolepid-aceae (incl. Hydatellaceae)
Flagellariineae
328 Flagellariaceae (incl. Joinvilleaceae)
Graminales
329 Gramineae
Principes
330 Palmae
Synanthae
331 Cyclanthaceae
Spathiflorae
332 Araceae (incl. Acoraceae), 333 Lemnaceae
Pandanales
334 Pandanaceae, 335 *Sparganiaceae*, 336 Typhaceae
Cyperales
337 Cyperaceae
Scitamineae
338 Musaceae (incl. Heliconiaceae, Lowiaceae, Strelitziaceae), 339 Zingiber-aceae (incl. Costaceae), 340 Cannaceae, 341 Marantaceae, 342 Lowiaceae
Microspermae
343 Orchidaceae

THE THORNE SYSTEM

Taken from pp. 102–111 of R. F. Thorne, Proposed new realignments in the angiosperms, in *Nordic J. Bot.* 3:85–117 (1983). Earlier versions appear in (1) Synopsis of a putatively phylogenetic classification of the flowering plants, in *Aliso* 6(4):57–66 (1968), (2) A phytogenetic classification of the angiospermae, in *Evolutionary Biology* 9:35–106 (1976), and (3) Phytochemistry and angiosperm phylogeny: a summary statement, in D. A. Young & D. S. Seigler, eds., Phytochemistry and Angiosperm Phylogeny: 233–295 (1981).

DICOTYLEDONEAE

ANNONIFLORAE

Annonales
  Winterineae
    1 Winteraceae
  Illiciineae
    2 Illiciaceae, 3 Schisandraceae
  Annonineae
    4 Magnoliaceae, 5 Degeneriaceae, 6 Himantandraceae, 7 Austrobaileyaceae, 8 Eupomatiaceae, 9 Annonaceae, 10 Myristicaceae, 11 Canellaceae
  Aristolochiineae
    12 Aristolochiaceae
  Laurineae
    13 Amborellaceae, 14 Trimeniaceae, 15 Chloranthaceae, 16 Lactoridaceae, 17 Monimiaceae, 18 Gomortegaceae, 19 Calycanthaceae (incl. Idiospermaceae), 20 Lauraceae, 21 Hernandiaceae
  Piperineae
    22 Saururaceae, 23 Piperaceae
Nelumbonales
    24 Nelumbonaceae, 25 Ceratophyllaceae
Paeoniales
    26 Paeoniaceae, 27 Glaucidiaceae
Berberidales
  Berberidineae
    28 Lardizabalaceae, 29 Sargentodoxaceae, 30 Menispermaceae, 31 *Nandinaceae*, 32 Berberidaceae, 33 Ranunculaceae, 34 Circaeasteraceae
  Papaverineae
    35 Papaveraceae

NYMPHAEIFLORAE

Nymphaeales
    36 Cabombaceae, 37 Nymphaeaceae

RAFFLESIIFLORAE

Rafflesiales
    38 Rafflesiaceae, 39 Hydnoraceae

THEIFLORAE

Theales
  Dilleniineae
    40 Dilleniaceae

Theineae
41 Actinidiaceae, 42 Paracryphiaceae, 43 Stachyuraceae, 44 Theaceae (incl. Asteropeiaceae, Pellicieraceae, Tetrameristaceae), 45 Symplocaceae, 46 Caryocaraceae, 47 Oncothecaceae, 48 Aquifoliaceae, 49 Phellinaceae, 50 Icacinaceae, 51 Sphenostemonaceae, 52 Cardiopteridaceae, 53 Marcgraviaceae

Clethrineae
54 Pentaphylacaceae, 55 Clethraceae, 56 Cyrillaceae

Sarraceniineae
57 Sarraceniaceae

Scytopetalineae
58 Ochnaceae, 59 Quiinaceae, 60 Scytopetalaceae, 61 Sphaerosepalaceae (incl. Diegodendraceae), 62 Medusagynaceae, 63 Strasburgeriaceae, 64 Ancistrocladaceae, 65 Dioncophyllaceae

Nepenthineae
66 Nepenthaceae

Hypericineae
67 Clusiaceae, 68 Elatinaceae

Lecythidineae
69 Lecythidaceae

Ericales
70 Ericaceae, 71 Epacridaceae, 72 Empetraceae

Ebenales
Ebenineae
73 Ebenaceae, 74 Lissocarpaceae, 75 Sapotaceae

Styracineae
76 Styracaceae

Primulales
Primulineae
77 Myrsinaceae (incl. Theophrastaceae), 78 Primulaceae

Plumbaginineae
79 Plumbaginaceae

Polygonales
80 Polygonaceae

CHENOPODIIFLORAE

Chenopodiales
Chenopodiineae
81 Phytolaccaceae (incl. Agdestidaceae, Gisekiaceae), 82 Barbeuiaceae, 83 Achatocarpaceae, 84 Stegnospermataceae, 85 Aizoaceae (incl. Molluginaceae), 86 Caryophyllaceae (incl. Illecebraceae), 87 Halophytaceae, 88 Nyctaginaceae, 87 Chenopodiaceae, 90 Amaranthaceae

Portulacineae
91 Portulacaceae (incl. Hectorellaceae), 92 Basellaceae, 93 Didiereaceae, 94 Cactaceae

GERANIIFLORAE

Geraniales
Linineae
95 Humiriaceae, 96 Ctenolophonaceae, 97 Linaceae (incl. Ixonanthaceae), 98 Erythroxylaceae, 99 Zygophyllaceae

Geraniineae
100 Oxalidaceae (incl. Lepidobotryaceae), 101 Geraniaceae, 102 *Vivianiaceae*, 103 *Ledocarpaceae*, 104 Balsaminaceae, 105 Tropaeolaceae, 106 Limnanthaceae

Polygalineae
107 Malpighiaceae, 108 Polygalaceae, 109 Krameriaceae, 110 Trigoniaceae, 111 Vochysiaceae

SANTALIFLORAE

Celastrales
112 Celastraceae (incl. Canotiaceae, Goupiaceae), 113 Lophopyxidaceae, 114 Stackhousiaceae
Santalales
115 Olacaceae (incl. Opiliaceae), 116 Medusandraceae, 117 Santalaceae, 118 Eremolepidaceae, 119 Misodendraceae, 120 Loranthaceae, 121 Viscaceae
Balanophorales
122 Balanophoraceae, 123 Cynomoriaceae

VIOLIFLORAE

Violales
Violineae
124 Flacourtiaceae (incl. Lacistemataceae), 125 Dipentodontaceae, 126 Peridiscaceae, 127 Scyphostegiaceae, 128 Violaceae, 129 Passifloraceae, 130 Turneraceae, 131 Malesherbiaceae, 132 Achariaceae, 133 Caricaceae
Salicineae
134 Salicaceae
Tamaricineae
135 Tamaricaceae, 136 Frankeniaceae
Cucurbitineae
137 Cucurbitaceae
Begoniineae
138 Begoniaceae, 139 Datiscaceae
Capparales
140 Moringaceae, 141 Resedaceae, 142 Capparaceae (incl. Tovariaceae), 143 Brassicaceae

MALVIFLORAE

Malvales
144 Sterculiaceae, 145 Huaceae, 146 Elaeocarpaceae, 147 Plagiopteraceae, 148 Tiliaceae, 149 Dipterocarpaceae, 150 Sarcolaenaceae, 151 Bixaceae, 152 Cochlospermaceae, 153 Cistaceae, 154 Bombacaceae, 155 Malvaceae
Urticales
156 Ulmaceae, 157 Urticaceae (incl. Cecropiaceae, Moraceae), 158 Cannabaceae
Rhamnales
159 Rhamnaceae, 160 Elaeagnaceae
Euphorbiales
161 Euphorbiaceae, 162 Pandaceae, 163 Simmondsiaceae, 164 Aextoxicaceae, 165 Dichapetalaceae, 166 Thymelaeaceae

RUTIFLORAE

Rutales
Rutineae
167 Rutaceae (incl. Rhabdodendraceae), 168 Cneoraceae, 169 Simaroubaceae (incl. Balanitaceae, Irvingiaceae), 170 Tepuianthaceae, 171 Ptaeroxylaceae, 172 Meliaceae, 173 Burseraceae, 174 Anacardiaceae (? incl. Podoaceae), 175 Leitneriaceae
Juglandineae
176 Rholpteleaceae, 177 Juglandaceae

Myricineae
178 Myricaceae
Coriariineae
179 Coriariaceae
Sapindineae
180 Sapindaceae (incl. Emblingiaceae, Stylobasiaceae), 181 Gyrostemonaceae, 182 Bataceae, 183 Sabiaceae (incl. Meliosmaceae), 184 Melianthaceae, 185 Akaniaceae, 186 Aceraceae, 187 Hippocastanaceae, 188 Bretschneideraceae
Fabineae
189 Surianaceae, 190 Connaraceae, 191 Fabaceae

PROTEIFLORAE

Proteales
192 Proteaceae

HAMAMELIDIFLORAE

Hamamelidales
Trochodendrineae
193 Trochodendraceae, 194 Tetracentraceae, 195 Eupteleaceae, 196 Cercidiphyllaceae
Eucommiineae
197 Eucommiaceae
Hamamelidineae
198 Hamamelidaceae, 199 Platanaceae
Casuarinales
200 Casuarinaceae
Fagales
201 Fagaceae, 202 Betulaceae (incl. Corylaccac)

ROSIFLORAE

Rosales
Rosineae
203 Rosaceae (incl. Neuradaceae), 204 Chrysobalanaceae, 205 Crossosomataceae
Saxifragineae
206 Crassulaceae, 207 Cephalotaceae, 208 Saxifragaceae (incl. Alseuosmiaceae, Columelliaceae, Eremosynaceae, Escalloniaceae, Griseliniaceae, Grossulariaceae, Hydrangeaceae, Melanophyllaceae, Montiniaceae, Penthoraceae, Pterostemonaceae, Vahliaceae), 209 Parnassiaceae, 210 Stylidiaceae, 211 Droseraceae, 212 Greyiaceae, 213 Podostemaceae, 214 Diapensiaceae
Cunoniineae
215 Cunoniaceae, 216 *Baueraceae*, 217 Davidsoniaceae, 218 Brunelliaceae, 219 Eucryphiaceae, 220 Staphyleaceae, 221 Corynocarpaceae
Pittosporales
Buxineae
222 Buxaceae, 223 Didymelaceae, 224 Daphniphyllaceae, 225 Balanopaceae
Pittosporineae
226 Pittosporaceae, 227 Byblidaceae, 228 Tremandraceae
Bruniineae
229 Roridulaceae, 230 Bruniaceae, 231 Geissolomataceae, 232 Grubbiaceae, 233 Myrothamnaceae, 234 Hydrostachyaceae

LOASIFLORAE

Loasales
235 Loasaceae

MYRTIFLORAE

Myrtales
Lythrineae
236 Lythraceae, 237 Rhynchocalycaceae, 238 Penaeaceae, 239 Trapaceae, 240 Oliniaceae, 241 Alzateaceae, 242 Crypteroniaceae, 243 Melastomataceae, 244 Combretaceae
Onagrineae
245 Onagraceae
Myrtineae
246 Myrtaceae

GENTIANIFLORAE

Oleales
247 Salvadoraceae, 248 Oleaceae
Gentianales
249 Loganiaceae (incl. Retziaceae), 250 Buddlejaceae, 251 Rubiaceae (incl. Theligonaceae), 252 Dialypetalanthaceae, 253 Apocynaceae (incl. Asclepiadaceae), 254 Gentianaceae (incl. Saccifoliaceae), 255 Menyanthaceae
Bignoniales
256 Bignoniaceae, 257 Pedaliaceae, 258 *Martyniaceae*, 259 Myoporaceae, 260 Scrophulariaceae (incl. Globulariaceae), 261 Plantaginaceae, 262 Lentibulariaceae, 263 Acanthaceae, 264 Gesneriaceae
Lamiales
265 Verbenaceae (incl. Avicenniaceae, Phrymaceae, Stilbaceae), 266 Lamiaceae (incl. Tetrachondraceae), 267 Callitrichaceae

SOLANIFLORAE

Solanales
Boraginineae
268 Hydrophyllaceae, 269 Boraginaceae, 270 Lennoaceae
Solanineae
271 Solanaceae (incl. Duckeodendraceae, Goetzeaceae), 272 Convolvulaceae
Polemoniineae
273 Polemoniaceae (incl. Cobaeaceae)
Fouquieriineae
274 Fouquieriaceae
Campanulales
275 Pentaphragmataceae, 276 Campanulaceae, 277 Goodeniaceae

CORNIFLORAE

Cornales
Rhizophorineae
278 Rhizophoraceae (incl. Anisophylleaceae)
Vitineae
279 Vitaceae (incl. Leeaceae)
Haloragineae
280 Haloragaceae, 281 Gunneraceae, 282 Hippuridaceae
Cornineae
283 *Nyssaceae*, 284 Cornaceae, 285 Alangiaceae, 286 Garryaceae, 287 Aucubaceae
Araliales
288 Helwingiaceae, 289 Torricelliaceae, 290 Araliaceae (incl. Umbelliferae)
Dipsacales
291 Caprifoliaceae, 292 Adoxaceae, 293 Valerianaceae, 294 Dipsacaceae (incl. Morinaceae, Triplostegiaceae), 295 Calyceraceae

ASTERIFLORAE

Asterales
296 Asteraceae

MONOCOTYLEDONEAE

LILIIFLORAE

Liliales
Liliineae
297 Liliaceae (incl. Alliaceae, Aloaceae, Amaryllidaceae, Anthericaceae, Aphyllanthaceae, Asparagaceae, Asphodelaceae, Asteliaceae, Blandfordiaceae, Calectasiaceae, Colchicaceae, Convallariaceae, Cyanastraceae, Dasypogonaceae, Doryanthaceae, Dracaenaceae, Eriospermaceae, Hanguanaceae, Hemerocallidaceae, Herreriaceae, Hostaceae, Hyacinthaceae, Hypoxidaceae, Ixioliriaceae, Lanariaceae, Lomandraceae, Melanthiaceae, Phormiaceae, Ruscaceae, Tecophilaeaceae, Trilliaceae, Xanthorrhoeaceae)
Dioscoreineae
298 Philesiaceae (incl. Alstroemeriaceae), 299 Smilacaceae (incl. Petermanniaceae, Rhipogonaceae), 300 Stemonaceae, 301 Dioscoreaceae, 302 Trichopodaceae, 303 Taccaceae
Iridineae
304 Iridaceae, 305 Burmanniaceae (incl. Corsiaceae)
Orchidineae
306 Orchidaceae

TRIURIDIFLORAE

Triuridales
307 Triuridaceae

ALISMATIFLORAE

Alismatales
308 Butomaceae, 309 Alismataceae (incl. Limnocharitaceae), 310 Hydrocharitaceae
Zosterales
Aponogetonineae
311 Aponogetonaceae
Potamogetonineae
312 Juncaginaceae (incl. Lilaeaceae, Scheuchzeriaceae), 313 Potamogetonaceae, 314 Posidoniaceae, 315 Zannichelliaceae, 316 Cymodoceaceae
Zosterineae
317 Zosteraceae
Najadales
318 *Najadaceae*

ARIFLORAE

Arales
319 Araceae, 320 Lemnaceae

CYCLANTHIFLORAE

Cyclanthales
321 Cyclanthaceae

PANDANIFLORAE

Pandanales
322 Pandanaceae

ARECIFLORAE

Arecales
323 Arecaceae

TYPHIFLORAE

Typhales
324 Typhaceae

COMMELINIFLORAE

Commelinales
Bromeliineae
325 Bromeliaceae, 326 Rapateaceae, 327 Xyridaceae
Velloziineae
328 Velloziaceae
Pontederiineae
329 Pontederiaceae, 330 Philydraceae, 331 Haemodoraceae
Juncineae
332 Juncaceae (incl. Thurniaceae), 333 Cyperaceae
Commelinineae
334 Commelinaceae, 335 Mayacaceae
Eriocaulineae
336 Eriocaulaceae
Flagellariineae
337 Flagellariaceae (incl. Joinvilleaceae), 338 Restionaceae (incl. Anarthriaceae, Ecdeiocoleaceae), 339 Centrolepidaceae
Poineae
340 Poaceae
Zingiberales
Musineae
341 Musaceae, 342 Strelitziaceae, 343 Heliconiaceae, 344 Lowiaceae
Zingiberineae
345 Zingiberaceae, 346 Costaceae, 347 Cannaceae, 348 Marantaceae

INCERTAE SEDIS

349 Barbeyaceae, 350 Hoplestigmataceae, 351 Hydatellaceae, (also incl. Physenaceae)

THE DAHLGREN SYSTEM

Dicotyledons taken from pp. 143–146 in R.M.T. Dahlgren, General aspects of angiosperm evolution and macrosystematics, in *Nordic J. Bot.* 3: 119–149 (1983), and Monocotyledons from R.M.T. Dahlgren, H.T. Clifford & P.F. Yeo, *The Families of the Monocotyledons* (1985). Earlier versions by R.M.T. Dahlgren appear in (1) A system of classification of the angiosperms to be used to demonstrate the distribution of characters, in *Bot. Notiser* 128: 119–147 (1975), (2) A revised system of classification of the angiosperms, in *Bot. J. Linn. Soc.* 80: 91–124 (1980), and (3) pp. 200–204 in A revised classification of the angiosperms with comments on correlation between chemical and other characters, in D.A. Young & D.S. Seigler, eds., *Phytochemistry and Angiosperm Phylogeny:* 149–204 (1981).

DICOTYLEDONEAE

MAGNOLIIFLORAE

Annonales
1 Annonaceae, 2 Myristicaceae, 3 Eupomatiaceae, 4 Austrobaileyaceae, 5 Canellaceae
Aristolochiales
6 Aristolochiaceae
Rafflesiales
7 Rafflesiaceae, 8 Hydnoraceae
Magnoliales
9 Winteraceae, 10 Degeneriaceae, 11 Himantandraceae, 12 Magnoliaceae
Lactoridales
13 Lactoridaceae
Chloranthales
14 Chloranthaceae
Illiciales
15 Illiciaceae, 16 Schisandraceae
Laurales
17 Amborellaceae, 18 Trimeniaceae, 19 Monimiaceae, 20 Gomortegaceae, 21 Calycanthaceae (incl. Idiospermaceae), 22 Lauraceae (incl. Hernandiaceae)
Nelumbonales
23 Nelumbonaceae

NYMPHAEIFLORAE

Piperales
24 Saururaceae, 25 Piperaceae
Nymphaeales
26 Cabombaceae, 27 Ceratophyllaceae, 28 Nymphaeaceae

RANUNCULIFLORAE

Ranunculales
29 Lardizabalaceae, 30 Sargentodoxaceae, 31 Menispermaceae, 32 *Kingdoniaceae*, 33 Circaeasteraceae, 34 Ranunculaceae, 35 Berberidaceae, 36 Glaucidiaceae
Papaverales
37 Papaveraceae, 38 *Fumariaceae*

CARYOPHYLLIFLORAE

Caryophyllales
39 Phytolaccaceae (incl. Achatocarpaceae, Agdestidaceae), 40 Basellaceae, 41 Portulacaceae, 42 Stegnospermataceae, 43 Nyctaginaceae, 44 Aizoaceae, 45 Halophytaceae, 46 Didiereaceae, 47 Cactaceae, 48 Hectorellaceae, 49 Chenopodiaceae, 50 Amaranthaceae, 51 Molluginaceae, 52 Caryophyllaceae (incl. Illecebraceae)

POLYGONIFLORAE

Polygonales
53 Polygonaceae

PLUMBAGINIFLORAE

Plumbaginales
54 *Limoniaceae*, 55 Plumbaginaceae

MALVIFLORAE

Malvales
56 Sterculiaceae, 57 Elaeocarpaceae, 58 Plagiopteraceae, 59 Bixaceae, 60 Cochlospermaceae, 61 Cistaceae, 62 Sphaerosepalaceae, 63 Sarcolaenaceae, 64 Huaceae, 65 Tiliaceae, 66 Dipterocarpaceae, 67 Malvaceae, 68 Bombacaceae
Urticales
69 Ulmaceae, 70 Moraceae, 71 Cecropiaceae, 72 Barbeyaceae, 73 Cannabaceae, 74 Urticaceae
Euphorbiales
75 Euphorbiaceae, 76 Simmondsiaceae, 77 Pandaceae, 78 Aextoxicaceae, 79 Dichapetalaceae
Thymelaeales
80 Thymelaeaceae, 81 *Gonystylidaceae*
Rhamnales
82 Rhamnaceae
Elaeagnales
83 Elaeagnaceae

VIOLIFLORAE

Violales
84 Flacourtiaceae (incl. Lacistemataceae), 85 Passifloraceae, 86 Dipentodontaceae, 87 Peridiscaceae, 88 Scyphostegiaceae, 89 Violaceae, 90 Turneraceae, 91 Malesherbiaceae, 92 Achariaceae, 93 Caricaceae
Cucurbitales
94 Datiscaceae, 95 Begoniaceae, 96 Cucurbitaceae
Salicales
97 Salicaceae
Tamaricales
98 Tamaricaceae, 99 Frankeniaceae
Capparales
100 Capparaceae, 101 Brassicaceae, 102 Tovariaceae, 103 Resedaceae, 104 Gyrostemonaceae, 105 Batidaceae, 106 Moringaceae
Salvadorales
107 Salvadoraceae

THEIFLORAE

Dilleniales
108 Dilleniaceae
Paeoniales
109 Paeoniaceae
Theales
110 Stachyuraceae, 111 Pentaphylacaceae, 112 Marcgraviaceae, 113 Quiinaceae, 114, Ancistrocladaceae, 115 Dioncophyllaceae, 116 Nepenthaceae, 117 Medusagynaceae, 118 Caryocaraceae, 119 Strasburgeriaceae, 120 Ochnaceae (incl. Diegodendraceae), 121 Oncothecaceae, 122 Scytopetalaceae, 123 Lecythidaceae, 124 Theaceae (incl. Pellicieraceae, Tetrameristaceae), 125 *Bonnetiaceae*, 126 Clusiaceae, 127 Elatinaceae

PRIMULIFLORAE

Primulales
128 Myrsinaceae, 129 *Aegicerataceae*, 130 Theophrastaceae, 131 Primulaceae, 132 *Coridaceae*
Ebenales
133 Sapotaceae, 134 Styracaceae, 135 Lissocarpaceae, 136 Ebenaceae

ROSIFLORAE

Trochodendrales
137 Trochodendraceae, 138 Tetracentraceae
Cercidiphyllales
139 Cercidiphyllaceae, 140 Eupteleaceae
Hamamelidales
141 Hamamelidaceae, 142 Platanaceae, 143 Myrothamnaceae
Geissolomatales
144 Geissolomataceae
Balanopales
145 Balanopaceae
Fagales
146 Fagaceae, 147 Corylaceae, 148 Betulaceae
Juglandales
149 Rhoipteleaceae, 150 Juglandaceae
Myricales
151 Myricaceae
Casuarinales
152 Casuarinaceae
Buxales
153 Buxaceae, 154 Didymelaceae, 155 Daphniphyllaceae
Cunoniales
156 Cunoniaceae, 157 *Baueraceae*, 158 Brunelliaceae, 159 Davidsoniaceae, 160 Eucryphiaceae, 161 Bruniaceae, 162 Grubbiaceae
Saxifragales
163 Saxifragaceae (incl. Penthoraceae, Vahliaceae) 164 *Brexiaceae*, 165 Grossulariaceae, 166 Greyiaceae, 167 *Iteaceae*, 168 Cephalotaceae, 169 Crassulaceae
Droserales
170 Droseraceae, 171 *Lepuropetalaceae*, 172 Parnassiaceae
Gunnerales
173 Gunneraceae
Rosales
174 Rosaceae, 175 Neuradaceae, 176 *Malaceae*, 177 *Amygdalaceae*, 178 Crossosomataceae, 179 Surianaceae (incl. Stylobasiaceae), 180 Rhabdodendraceae

PODOSTEMIFLORAE

Podostemales
    181 Podostemaceae

PROTEIFLORAE

Proteales
    182 Proteaceae

MYRTIFLORAE

Haloragales
    183 Haloragaceae
Rhizophorales
    184 Rhizophoraceae
Myrtales
    185 *Psiloxylaceae*, 186 *Heteropyxidaceae*, 187 Myrtaceae, 188 Onagraceae, 189 Trapaceae, 190 Lythraceae, 191 Combretaceae, 192 Melastomataceae, 193 *Memecylaceae*, 194 Crypteroniaceae, 195 Oliniaceae, 196 Penaeaceae, 197 Rhynchocalycaceae, 198 Alzateaceae
Chrysobalanales
    199 Chrysobalanaceae

FABIFLORAE

Fabales
    200 *Mimosaceae*, 201 *Caesalpiniaceae*, 202 Fabaceae

RUTIFLORAE

Sapindales
    203 Coriariaceae, 204 Anacardiaceae, 25 Leitneriaceae, 206 Podoaceae, 207 Sapindaceae, 208 Hippocastanaceae, 209 Aceraceae, 210 Akaniaceae, 211 Bretschneideraceae, 212 Emblingiaceae, 213 Staphyleaceae, 214 Melianthaceae, 215 Sabiaceae, 216 Meliosmaceae, 217 Connaraceae
Rutales
    218 Rutaceae, 219 Ptaeroxylaceae, 220 Cneoraceae, 221 Simaroubaceae (incl. Irvingiaceae), 222 Tepuianthaceae, 223 Burseraceae, 224 Meliaceae
Polygalales
    225 Malpighiaceae, 226 Trigoniaceae, 227 Vochysiaceae, 228 Polygalaceae, 229 Krameriaceae
Geraniales
    230 Zygophyllaceae, 231 *Nitrariaceae*, 232 *Peganaceae*, 233 Balanitaceae, 234 Erythroxylaceae, 235 Humiriaceae, 236 Linaceae, 237 Ctenolophonaceae, 238 Ixonanthaceae, 239 Lepidobotryaceae, 240 Oxalidaceae, 241 Geraniaceae, 242 *Vivianiaceae*, 243 *Ledocarpaceae*, 244 *Biebersteiniaceae*, 245 *Dirachmaceae*
Balsaminales
    246 Balsaminaceae
Tropaeolales
    247 Limnanthaceae, 248 Tropaeolaceae

SANTALIFLORAE

Celastrales
    249 Celastraceae (incl. Goupiaceae, ? Canotiaceae), 250 Stackhousiaceae, 251 Lophopyxidaceae, 252 Cardiopteridaceae, 253 Corynocarpaceae
Vitales
    254 Vitaceae (incl. Leeaceae)

Santalales
255 Olacaceae, 256 Opiliaceae, 257 Loranthaceae, 258 Misodendraceae, 259 Eremolepidaceae, 260 Santalaceae, 261 Viscaceae

BALANOPHORIFLORAE

Balanophorales
262 Cynomoriaceae, 263 Balanophoraceae

ARALIIFLORAE

Pittosporales
264 Pittosporaceae, 265 Tremandraceae, 266 Byblidaceae
Araliales
267 Araliacaeae, 268 Apiaceae

ASTERIFLORAE

Campanulales
269 Pentaphragmataceae, 270 Campanulaceae, 271 *Lobeliaceae*
Asterales
272 Asteraceae

SOLANIFLORAE

Solanales
273 Solanaceae (incl. Duckeodendraceae), 274 *Sclerophylacaceae*, 275 Goetzeaceae, 276 Convolvulaceae, 277 *Cuscutaceae*, 278 Cobaeaceae, 279 Polemoniaceae
Boraginales
280 Hydrophyllaceae, 281 *Ehretiaceae*, 282 Boraginaceae, 283 Lennoaceae, 284 Hoplestigmataceae

CORNIFLORAE

Fouquieriales
285 Fouquieriaceae
Ericales
286 Actinidiaceae, 287 Clethraceae, 288 Cyrillaceae, 289 Ericaceae, 290 Empetraceae, 291 *Monotropaceae*, 292 *Pyrolaceae*, 293 Epacridaceae, 294 Roridulaceae, 295 Diapensiaceae
Eucommiales
296 Eucommiaceae
Sarraceniales
297 Sarraceniaceae
Cornales
298 Garryaceae, 299 Alangiaceae, 300 *Nyssaceae*, 301 Cornaceae, 302 *Davidiaceae*, 303 Escalloniaceae, 304 Helwingiaceae, 305 Torricelliaceae, 306 Aucubaceae, 307 Aralidiaceae, 308 Cardiopteridaceae, 309 Phellinaceae, 310 Aquifoliaceae, 311 Paracryphiaceae, 312 Sphenostemonaceae, 313 Symplocaceae, 314 Icacinaceae, 315 Montiniaceae (incl. Melanophyllaceae), 316 Columelliaceae, 317 Stylidiaceae, 318 Alseuosmiaceae, 319 Anisophylleaceae, 320 Hydrangeaceae, 321 *Sambucaceae*, 322 Caprifoliaceae, 323 *Viburnaceae*, 324 Adoxaceae, 325 *Dulongiaceae*, 326 *Tribelaceae*, 327 Eremosynaceae, 328 Pterostemonaceae, 329 *Tetracarpaeaceae*
Dipsacales
330 Valerianaceae, 331 Triplostegiaceae, 332 Dipsacaceae, 333 Morinaceae, 334 Calyceraceae

LOASIFLORAE

Loasales
335 Loasaceae

GENTIANIFLORAE

Goodeniales
336 Goodeniaceae
Oleales
337 Oleaceae
Gentianales
338 *Desfontainiaceae*, 339 Loganiaceae, 340 Dialypetalanthaceae, 341 Rubiaceae, 342 Theligonaceae, 343 Menyanthaceae, 344 Gentianaceae, 345 Saccifoliaceae, 346 Apocynaceae, 347 Asclepiadaceae

LAMIIFLORAE

Scrophulariales
348 Bignoniaceae, 349 Myoporaceae, 350 Gesneriaceae, 351 Buddlejaceae, 352 Scrophulariaceae, 353 Globulariaceae, 354 *Selaginaceae*, 355 Stilbaceae, 356 Retziaceae, 357 Plantaginaceae, 358 Lentibulariaceae, 359 Pedaliaceae, 360 *Trapellaceae*, 361 *Martyniaceae*, 362 Acanthaceae
Hippuridales
363 Hippuridaceae
Lamiales
364 Verbenaceae (incl. Phrymaceae, and ? Avicenniaceae, ? Cyclocheilaceae, ? Nesogenaceae), 365 Lamiaceae (? incl. Tetrachondraceae), 366 Callitrichaceae
Hydrostachyales
367 Hydrostachyaceae

MONOCOTYLEDONEAE

LILIIFLORAE

Dioscoreales
368 Trichopodaceae, 369 Dioscoreaceae, 370 Taccaceae, 371 Stemonaceae, 372 Trilliaceae, 373 Smilacaceae (incl. Rhipogonaceae), 374 Petermanniaceae
Asparagales
375 Philesiaceae, 376 *Luzuriagaceae*, 377 Convallariaceae, 378 Asparagaceae, 379 Ruscaceae, 380 Herreriaceae, 381 Dracaenaceae, 382 *Nolinaceae,* 383 Asteliaceae, 384 Hanguanaceae, 385 Dasypogonaceae (incl. Lomandraceae), 386 Calectasiaceae, 387 Blandfordiaceae, 388 Xanthorrhoeaceae, 389 Agavaceae, 390 Hypoxidaceae, 391 Tecophilaeaceae (incl. Lanariaceae), 392 Cyanastraceae, 393 Eriospermaceae, 394 Ixioliriaceae, 395 Phormiaceae, 396 Doryanthaceae, 397 Hemerocallidaceae, 398 Asphodelaceae (incl. Aloaceae), 399 Anthericaceae, 400 Aphyllanthaceae, 401 Funkiaceae [Hostaceae], 402 Hyacinthaceae, 403 Alliaceae, 404 Amaryllidaceae
Melanthiales
405 Melanthiaceae, 406 *Campynemaceae*
Burmanniales
407 Burmanniaceae, 408 *Thismiaceae*, 409 Corsiaceae
Liliales
410 Alstroemeriaceae, 411 Colchicaceae, 412 *Uvulariaceae*, 413 *Calochortaceae*, 414 Liliaceae, 415 *Geosiridaceae*, 416 Iridaceae, 417 *Apostasiaceae*, 418 *Cypripediaceae*, 419 Orchidaceae

ARIFLORAE

Arales
420 Araceae (incl. Acoraceae), 421 Lemnaceae

TRIURIDIFLORAE

Triuridales
422 Triuridaceae

ALISMATIFLORAE

Alismatales
423 Aponogetonaceae, 424 Butomaceae, 425 Limnocharitaceae, 426 Alismataceae, 427 Hydrocharitaceae
Najadales
428 Scheuchzeriaceae, 429 Juncaginaceae (incl. Lilaeaceae), 430 Potamogetonaceae, 431 Posidoniaceae, 432 Zosteraceae, 433 Zannichelliaceae, 474 Cymodoceaceae, 435 *Najadaceae*

BROMELIIFLORAE

Velloziales
436 Velloziaceae
Bromeliales
437 Bromeliaceae
Philydrales
438 Philydraceae
Haemodorales
439 Haemodoraceae
Pontederiales
440 Pontederiaceae
Typhales
441 *Sparganiaceae*, 442 Typhaceae

ZINGIBERIFLORAE

Zingiberales
443 Lowiaceae, 444 Musaceae, 445 Heliconiaceae, 446 Strelitziaceae, 447 Zingiberaceae, 448 Costaceae, 449 Cannaceae, 450 Marantaceae

COMMELINIFLORAE

Commelinales
451 Commelinaceae, 452 Mayacaceae, 453 Xyridaceae, 454 Rapateaceae, 455 Eriocaulaceae
Hydatellales
456 Hydatellaceae
Cyperales
457 Juncaceae, 458 Thurniaceae, 459 Cyperaceae
Poales
460 Flagellariaceae, 461 Joinvilleaceae, 462 Poaceae, 463 Ecdeiocoleaceae, 464 Anarthriaceae, 465 Restionaceae, 466 Centrolepidaceae

CYCLANTHIFLORAE

Cyclanthales
467 Cyclanthaceae

ARECIFLORAE

Arecales
468 Arecaceae

PANDANIFLORAE

Pandanales
469 Pandanaceae

THE YOUNG SYSTEM

Taken from pp. 146–156 of H.G. Bedell & J.L. Reveal, Amended outlines and indices for six recently published systems of angiosperm classification, in *Phytologia* 51:65–156 (1982). This system by D.A. Young was quoted in the above paper as being 'in press', but, as far as we know, has not actually appeared anywhere else. The numbering here is as in Bedell & Reveal (*l.c.*) up to 301 where Grossulariaceae has been inserted from their index and subsequent numbers altered appropriately.

MAGNOLIOPSIDA

MAGNOLIIDAE

MAGNOLIANAE
Magnoliales
> 1 Magnoliaceae, 2 Eupomatiaceae, 3 Winteraceae, 4 Cannellaceae, 5 Annonaceae

Illiciales
> 6 Austrobaileyaceae, 7 Illiciaceae, 8 Schisandraceae, 9 Nelumbonaceae

NYMPHAEANAE
Nymphaeales
> 10 Nymphaeaceae, 11 Cabombaceae

RANUNCULANAE
Degeneriales
> 12 Degeneriaceae

Laurales
> 13 Lactoridaceae, 14 Idiospermaceae, 15 Calycanthaceae, 16 Monimiaceae, 17 Amborellaceae, 18 Chloranthaceae, 19 Trimeniaceae, 20 Gomortegaceae, 21 Lauraceae, 22 Hernandiaceae

Aristolochiales
> 23 Himantandraceae, 24 Myristicaceae, 25 Aristolochiaceae, 26 Piperaceae, 27 Saururaceae

Ranunculales
> 28 Ceratophyllaceae, 29 Lardizabalaceae, 30 *Nandinaceae*, 31 Berberidaceae, 32 Menispermaceae, 33 Sargentodoxaceae, 34 *Podophyllaceae*, 35 Ranunculaceae, 36 Glaucidiaceae, 37 Paeoniaceae, 38 Circaeasteraceae, 39 *Hydrastidaceae*, 40 Papaveraceae, 41 *Fumariaceae*

LILIIDAE

ALISMATANAE
Alismatales
> 42 Alismataceae (incl. Limnocharitaceae), 43 Butomaceae, 44 Hydrocharitaceae

Zosterales
> 45 Aponogetonaceae, 46 Scheuchzeriaceae, 47 Juncaginaceae (incl. Lilaeaceae), 48 Posidoniaceae, 49 Potamogetonaceae, 50 Zannichelliaceae, 51 Cymodoceaceae, 52 Zosteraceae, 53 *Najadaceae*

TRIURIDANAE
Triuridales
> 54 Triuridaceae

ARANAE
Arales
> 55 Araceae (incl. Acoraceae), 56 Lemnaceae

ARECANAE
Arecales
57 Arecaceae
Cyclanthales
58 Cyclanthaceae
Pandanales
59 Pandanaceae

LILIANAE
Dioscoreales
60 Dioscoreaceae, 61 Stemonaceae, 62 Taccaceae
Liliales
63 Liliaceae (apparently incl. Alliaceae, Alstroemeriaceae, Amaryllidaceae, Anthericaceae, Aphyllanthaceae, Asparagaceae, Asphodelaceae, Asteliaceae, Blandfordiaceae, Colchicaceae, Convallariaceae, Eriospermaceae, Hemerocallidaceae, Herreriaceae, Hostaceae, Hyacinthaceae, Hypoxidaceae, Ixioliriaceae, Melanthiaceae, Ruscaceae, Tecophilaeaceae, Trilliaceae), 64 Philydraceae, 65 Pontederiaceae, 66 Haemodoraceae, 67 Cyanastraceae, 68 Iridaceae, 69 *Geosiridaceae*, 70 Velloziaceae, 71 Agavaceae, 72 Aloeaceae, 73 Xanthorrhoeaceae (? incl. Calectasiaceae, Dasypogonaceae, Lomandraceae), 74 Hanguanaceae, 75 Smilacaceae
Typhales
76 Typhaceae
Burmanniales
77 Burmanniaceae, 78 Corsiaceae
Orchidales
79 Orchidaceae

COMMELINANAE
Commelinales
80 Rapateaceae, 81 Xyridaceae, 82 Mayacaceae, 83 Commelinaceae
Eriocaulales
84 Eriocaulaceae
Restionales
85 Flagellariaceae, 86 Joinvilleaceae, 87 Restionaceae, 88 Centrolepidaceae, 89 Ecdeiocoleaceae, 90 Hydatellaceae
Juncales
91 Juncaceae (incl. Thurniaceae)
Cyperales
92 Cyperaceae
Poales
93 Poaceae

ZINGIBERANAE
Bromeliales
94 Bromeliaceae
Zingiberiales
95 Musaceae, 96 Strelitziaceae, 97 Heliconiaceae, 98 Lowiaceae, 99 Zingiberaceae, 100 Costaceae, 101 Cannaceae, 102 Marantaceae

DILLENIIDAE

DILLENIANAE
Dilleniales
103 Dilleniaceae

Theales

104 Actinidiaceae, 105 Stachyuraceae, 106 Theaceae (? incl. Asteropeiaceae, Pellicieraceae, Tetrameristaceae), 107 Icacinaceae, 108 Cardiopteridaceae, 109 Aquifoliaceae, 110 Phellinaceae, 111 Oncothecaceae, 112 Sphenostemonaceae, 113 Paracryphiaceae, 114 Marcgraviaceae, 116 Clethraceae, 117 Cyrillaceae, 118 Pentaphylacaceae

Sarraceniales

119 Sarraceniaceae

Scytopetalales

120 Ochnaceae, 121 Quiinaceae, 122 Scytopetalaceae, 123 Sphaerosepalaceae, 124 Medusagynaceae, 125 Strasburgeriaceae, 126 Dioncophyllaceae

Nepenthales

127 Nepenthaceae

Hypericales

128 Hypericaceae, 129 Elatinaceae

Lecythidales

130 Lecythidaceae

Ericales

131 Ericaceae, 132 Epacridaceae, 133 Empetraceae

Diapensiales

134 Diapensiaceae

Ebenales

135 Ebenaceae, 136 Sapotaceae, 137 Symplocaceae, 138 Lissocarpaceae, 139 Styracaceae

Primulales

140 Myrsinaceae, 141 Theophrastaceae, 142 Primulaceae

Plumbaginales

143 Plumbaginaceae, 144 Polygonaceae

CARYOPHYLLANAE

Caryophyllales

145 Molluginaceae, 146 Aizoaceae, 147 Cactaceae, 148 Caryophyllaceae (incl. Illecebraceae), 149 Portulacaceae, 150 Hectorellaceae

Chenopodiales

151 Barbeuiaceae, 152 Didiereaceae, 153 Stegnospermataceae, 154 Agdestidaceae, 155 *Petiveriaceae*, 156 Gisekiaceae, 157 Nyctaginaceae, 158 Phytolaccaceae, 159 Chenopodiaceae, 160 Amaranthaceae, 161 Achatocarpaceae, 162 Basellaceae, 163 Halophytaceae

MALVANAE

Malvales

164 Sterculiaceae, 165 Huaceae, 166 Elaeocarpaceae, 167 Plagiopteraceae, 168 Tiliaceae, 169 Dipterocarpaceae, 170 Sarcolaenaceae, 171 Bombacaceae, 172 Bixaceae, 173 Cochlospermaceae, 174 Cistaceae, 175 Malvaceae

Urticales

176 Ulmaceae, 177 Cannabaceae, 178 Urticaceae, 179 Cecropiaceae, 180 Moraceae

Rhamnales

181 Rhamnaceae, 182 Elaeagnaceae

Euphorbiales

183 Euphorbiaceae, 184 Thymelaeaceae, 185 Simmondsiaceae, 186 Pandaceae, 187 Aextoxicaceae, 188 Didymelaceae, 189 Dichapetalaceae

VIOLANAE
Violales
190 Flacourtiaceae (incl. Lacistemataceae), 191 Dipentodontaceae, 192 Peridiscaceae, 193 Scyphostegiaceae, 194 Violaceae, 195 Turneraceae, 196 Malesherbiaceae, 197 Passifloraceae, 198 Achariaceae, 199 Caricaceae, 200 Cucurbitaceae, 201 Begoniaceae, 202 Datiscaceae
Loasales
203 Loasaceae
Salicales
204 Salicaceae
Tamaricales
205 Tamaricaceae, 206 Frankeniaceae
Capparales
207 Capparaceae, 208 Tovariaceae, 209 Moringaceae, 210 Bataceae, 211 Gyrostemonaceae, 212 Resedaceae, 213 Brassicaceae
Tropaeolales
214 Tropaeolaceae, 215 Limnanthaceae

ROSIDAE

HAMAMELIDANAE
Trochodendrales
216 Trochodendraceae, 217 Cercidiphyllaceae, 218 Tetracentraceae, 219 Eupteleaceae
Eucommiales
220 Eucommiaceae
Hamamelidales
221 Hamamelidaceae, 222 Platanaceae
Myrothamnales
223 Myrothamnaceae, 224 Geissolomataceae
Casuarinales
225 Casuarinaceae
Fagales
226 Fagaceae, 227 Betulaceae (incl. Corylaceae)

SANTALANAE
Celastrales
228 Medusandraceae, 229 Celastraceae (?incl. Canotiaceae, Goupiaceae), 230 Lophopyxidaceae, 231 Stackhousiaceae, 232 Corynocarpaceae
Vitales
233 Vitaceae (? incl. Leeaceae)
Santales
234 Olacaceae (? incl. Opiliaceae), 235 Santalaceae, 236 Eremolepidaceae, 237 Misodendraceae, 238 Loranthaceae, 239 Viscaceae
Balanophorales
240 Balanophoraceae, 241 Cynomoriaceae
Rafflesiales
243 Rafflesiaceae, 243 Hydnoraceae

RUTANAE
Rutales
244 Rutaceae (Rhabdodendraceae), 245 Cneoraceae, 246 Coriariaceae, 247 Simaroubaceae (? incl. Irvingiaceae), 248 Meliaceae, 249 Ptaeroxylaceae, 250 Burseraceae, 251 Anacardiaceae (incl. Podoaceae), 252 Leitneriaceae
Sapindales
253 Sapindaceae (incl. Emblingiaceae, Stylobasiaceae), 254 Surianaceae, 255 Sabiaceae (incl. Meliosmaceae), 256 Melianthaceae, 257 Akaniaceae, 258 Staphyleaceae, 259 Aceraceae, 260 Hippocastanaceae, 261 Bretschneideraceae

Fabales
262 Fabaceae, 263 *Mimosaceae*, 264 *Caesalpiniaceae*, 265 Connaraceae
Juglandales
266 Rhoipteleaceae, 267 Juglandaceae
Myricales
268 Myricaceae

GERANIANAE
Linales
269 Humiriaceae, 270 Linaceae (? incl. Ctenolophonaceae, Ixonanthaceae), 271 Ancistrocladaceae, 272 Erthroxylaceae, 273 Zygophyllaceae
Geraniales
274 Oxalidaceae (? incl. Lepidobotryaceae), 275 Geraniaceae, 276 Balsaminaceae
Polygalales
277 Malpighiaceae, 278 Polygalaceae, 279 Tremandraceae, 280 Krameriaceae, 281 Trigoniaceae, 282 Vochysiaceae

ROSANAE
Balanopales
283 Balanopaceae
Buxales
284 Buxaceae, 285 Daphniphyllaceae
Cunoniales
286 Cunoniaceae, 287 Brunelliaceae, 288 Eucryphiaceae, 289 Davidsoniaceae
Bruniales
290 Bruniaceae
Rosales
291 Rosaceae (? incl. Neuradaceae), 292 Chrysobalanaceae, 293 Crossosomataceae
Saxifragales
294 Crassulaceae, 295 Cephalotaceae, 296 Droseraceae, 297 Greyiaceae, 298 Pterostemonaceae, 299 *Iteaceae*, 300 *Baueraceae*, 301 Grossulariaceae, 302 Saxifragaceae, 303 *Tetracarpaeaceae*, 304 Penthoraceae, 305 *Brexiaceae*, 306 *Francoaceae*, 307 Parnassiaceae, 308 Vahliaceae
Podostemonales
309 Podostemonaceae
Proteales
310 Proteaceae

MYRTANAE
Myrtales
311 *Sonneratiaceae*, 312 Lythraceae, 313 Trapaceae, 314 Combretaceae, 315 Oliniaceae, 316 Penaeaceae, 317 Myrtaceae, 318 Melastomataceae, 319 Onagraceae
Haloragales
320 Haloragaceae, 321 Gunneraceae
Rhizophorales
322 Rhizophoraceae (? incl. Anisophylleaceae)

CORNANAE
Cornales
323 *Nyssaceae*, 324 *Davidiaceae*, 325 Cornaceae (presumably including several), 326 Alangiaceae, 327 Garryaceae
Hydrangeales
328 Hydrangeaceae, 329 *Philadelphaceae*, 330 Escalloniaceae, 331 Montiniaceae, 332 *Tribelaceae*, 333 Columelliaceae, 334 Alseuosmaceae, 335 *Phyllonomaceae* [*Dulongiaceae*], 336 Eremosynaceae, 337 Stylidiaceae
Pittosporales
338 Pittosporaceae, 339 Byblidaceae, 340 Roridulaceae, 341 Grubbiaceae

Araliales
342 Araliaceae, 343 Apiaceae

GENTIANANAE
Oleales
344 Salvadoraceae, 345 Oleaceae, 346 Barbeyaceae
Gentianales
347 Loganiaceae, 348 Buddlejaceae, 349 Rubiaceae (incl. Theligonaceae), 350 Apocynaceae, 351 Asclepiadaceae, 352 Gentianaceae, 353 Menyanthaceae
Bignoniales
354 Bignoniaceae, 355 Pedaliaceae, 356 *Martyniaceae*, 357 Myoporaceae, 358 Scrophulariaceae, 359 Plantaginaceae, 360 Lentibulariaceae, 361 Acanthaceae, 362 Gesneriaceae

SOLANANAE
Solanales
363 Solanaceae, 364 Convolvulaceae, 365 Polemoniaceae (? incl. Cobaeaceae), 366 Fouquieriaceae
Boraginales
367 Hydrophyllaceae, 368 Boraginaceae, 369 Lennoaceae, 370 Hoplestigmataceae
Campanulales
371 Pentaphragmataceae, 372 Campanulaceae, 373 Goodeniaceae, 374 *Brunoniaceae*

LAMIANAE
Hippuridales
375 Hippuridaceae
Hydrostachyales
376 Hydrostachyaceae
Lamiales
377 Verbenaceae (? incl. Avicenniaceae, Nesogenaceae, Phrymaceae, Stilbaceae), 378 Callitrichaceae, 379 Lamiaceae (? incl. Tetrachondraceae)

DIPSACANAE
Dipsacales
380 Caprifoliaceae, 381 *Sambucaceae*, 382 Adoxaceae, 383 Valerianaceae, 384 Dipsacaceae (? incl. Morinaceae, Triplostegiaceae), 385 Calyceraceae

ASTERANAE
Asterales
386 Asteraceae

THE TAKHTAJAN SYSTEM

Taken from A. Takhtajan, *Systema Magnoliophytorum* [in Russian] (1987). Previous versions of this system appear in (1) pp. 277–288 in A. Takhtajan, *The Origin of Angiospermous Plants* [in Russian] (1954), (2) *Die Evolution der Angiospermen* (1959), (3) *Systema et Phylogenia Magnoliophytorum* [in Russian] (1966), and (4) Outline of the classification of flowering plants (Magnoliophyta), in *Bot. Rev.* 46:225–359 (1980).

MAGNOLIOPSIDA

MAGNOLIIDAE

MAGNOLIANAE
Magnoliales
1 Degeneriaceae, 2 Himantandraceae, 3 Magnoliaceae
Eupomatiales
4 Eupomatiaceae
Annonales
5 Annonaceae, 6 Canellaceae, 7 Myristicaceae
Winterales
8 Winteraceae
Illiciales
9 Illiciaceae, 10 Schisandraceae
Austrobaileyales
11 Austrobaileyaceae
Laurales
12 Amborellaceae, 13 Trimeniaceae, 14 Monimiaceae, 15 *Atherospermataceae*, 16 *Siparunaceae*, 17 Gomortegaceae, 18 Hernandiaceae, 19 Calycanthaceae, 20 Idiospermaceae, 21 Lauraceae, 22 *Gyrocarpaceae*
Lactoridales
23 Lactoridaceae
Chloranthales
24 Chloranthaceae
Piperales
25 Saururaceae, 26 Piperaceae
Aristolochiales
27 Aristolochiaceae

RAFFLESIANAE
Hydnorales
28 Hydnoraceae
Rafflesiales
29 Rafflesiaceae, 30 *Apodanthaceae*, 31 *Mitrastemonaceae*, 32 *Cytinaceae*
Balanophorales
33 *Dactylanthaceae*, 34 *Sarcophytaceae*, 35 *Latraeophilaceae*, 36 *Lophophytaceae*, 37 Balanophoraceae

NEPENTHANAE
Nepenthales
38 Nepenthaceae

NYMPHAEANAE
Nepenthales
39 Cabombaceae, 40 Nymphaeaceae, 41 *Barclayaceae*
Ceratophyllales
42 Ceratophyllaceae

NELUMBONANAE
Nelumbonales
    43 Nelumbonaceae

RANUNCULIDAE

RANUNCULANAE
Ranunculales
  Lardizabalineae
    44 Lardizabalaceae, 45 Sargentodoxaceae
  Menispermineae
    46 Menispermaceae
  Ranunculineae
    47 Ranunculaceae, 48 Circaeasteraceae
  Berberidineae
    49 *Hydrastidaceae*, 50 Berberidaceae, 51 *Nandinaceae*
Glaucidiales
    52 Glaucidiaceae
Paeoniales
    53 Paeoniaceae
Papaverales
    54 Papaveraceae, 55 *Hypecoaceae*, 56 *Fumariaceae*

CARYOPHYLLIDAE

CARYOPHYLLANAE
Caryophyllales
  Caryophyllineae
    57 Phytolaccaceae (incl. Agdestidaceae, Gisekiaceae), 58 Achatocarpaceae, 59
    Barbeuiaceae, 60 Nyctaginaceae, 61 Aizoaceae, 62 *Tetragoniaceae*, 63
    Molluginaceae, 64 Stegnospermataceae, 65 Portulacaceae, 66 Hectorellaceae,
    67 Basellaceae, 68 Halophytaceae, 69 Cactaceae, 70 Didiereaceae, 71 Caryo-
    phyllaceae (incl. Illecebraceae)
  Chenopodiineae
    72 Amaranthaceae, 73 Chenopodiaceae

POLYGONANAE
Polygonales
    74 Polygonaceae

PLUMBAGINANAE
Plumbaginales
    75 Plumbaginaceae

HAMAMELIDIDAE

TROCHODENDRANAE
Trochodendrales
    76 Trochodendraceae, 77 Tetracentraceae
Cercidiphyllales
    78 Cercidiphyllaceae
Eupteleales
    79 Eupteleaceae

EUCOMMIANAE
Eucommiales
    80 Eucommiaceae

HAMAMELIDANAE
Hamamelidales
   81 Hamamelidaceae, 82 *Rhodoleiaceae*, 83 *Altingiaceae*, 84 Platanaceae
Daphniphyllales
   85 Daphniphyllaceae
Balanopales
   86 Balanopaceae
Didymelales
   87 Didymelaceae
Myrothamnales
   88 Myrothamnaceae
Buxales
   89 Buxaceae, 90 *Stylocerataceae*
Simmondsiales
   91 Simmondsiaceae
Casuarinales
   92 Casuarinaceae
Fagales
   93 Fagaceae
Betulales
   94 Betulaceae (incl. Corylaceae)

JUGLANDANAE
Myricales
   95 Myricaceae
Rhoipteleales
   96 Rhoipteleaceae
Juglandales
   97 Juglandaceae

DILLENIIDAE

DILLENIANAE
Dilleniales
   98 Dilleniaceae

THEANAE
Actinidiales
   99 Actinidiaceae
Paracryphiales
   100 Paracryphiaceae
Theales
   101 Stachyuraceae, 102 Theaceae, 103 *Sladeniaceae*, 104 Oncothecaceae, 105 Marcgraviaceae, 106 Pentaphylacaceae, 107 Tetrameristaceae, 108 Symplocaceae, 109 Caryocaraceae, 110 Asteropeiaceae, 111 Pellicieraceae, 112 *Bonnetiaceae*, 113 Clusiaceae
Medusagynales
   114 Medusagynaceae
Ochnales
   115 Ochnaceae, 116 *Lophiraceae*, 117 *Sauvagesiaceae*, 118 Diegodendraceae, 119 Strasburgeriaceae, 120 Quiinaceae, 121 Scytopetalaceae
Ancistrocladales
   122 Ancistrocladaceae
Elatinales
   123 Elatinaceae

Urticineae
185 Moraceae, 186 Cannabaceae, 187 Cecropiaceae, 188 Urticaceae
Barbeyales
189 Barbeyaceae
EUPHORBIANAE
Euphorbiales
190 Euphorbiaceae, 191 Pandaceae, 192 Dichapetalaceae, 193 Aextoxicaceae
Thymelaeales
194 Thymelaeaceae

ROSIDAE

ROSANAE
Cunoniales
195 Cunoniaceae, 196 *Baueraceae*, 197 Davidsoniaceae, 198 Eucryphiaceae, 199 Brunelliaceae
Bruniales
200 Bruniaceae
Geissolomatales
201 Geissolomataceae
Saxifragales
202 Penthoraceae, 203 Crassulaceae, 204 Cephalotaceae, 205 Saxifragaceae, 206 Grossulariaceae, 207 *Iteaceae*, 208 Vahliaceae, 209 Eremosynaceae, 210 *Rousseaceae*, 211 Greyiaceae, 212 *Frankoaceae*, 213 Parnassiaceae, 214 *Lepuropetalaceae*
Droserales
215 Droseraceae
Gunnerales
216 Gunneraceae
Rosales
217 Rosaceae, 218 Chrysobalanaceae, 219 Neuradaceae
Crossosomatales
220 Crossosomataceae
Podostemales
221 Podostemaceae

MYRTANAE
Rhizophorales
223 Anisophylleaceae, 223 Rhizophoraceae
Myrtales
Myrtineae
224 *Psiloxylaceae*, 225 *Heteropyxidaceae*, 226 Myrtaceae, 227 Alzateaceae, 228 Rhynchocalycaceae, 229 Penaeaceae, 230 Oliniaceae, 231 Combretaceae, 232 Crypteroniaceae, 233 Melastomataceae, 234 Lythraceae, 235 *Punicaceae*, 236 *Duabangaceae*, 237 *Sonneratiaceae*
Onagrineae
238 Onagraceae
Trapineae
239 Trapaceae
Haloragales
240 Haloragaceae

FABANAE
Fabales
241 Fabaceae

RUTANAE
Connarales
242 Connaraceae

Sapindales
243 Staphyleaceae, 244 *Tapisceaceae*, 245 Sapindaceae, 246 Aceraceae, 247 Hippocastanaceae, 248 Bretschneideraceae, 249 Melianthaceae, 250 Akaniaceae, 251 Stylobasiaceae, 252 Emblingiaceae, 253 Sabiaceae (incl. Meliosmaceae), 254 Physenaceae

Rutales
255 Rutaceae, 256 Rhabdodendraceae, 257 *Tetradiclidaceae*, 258 Cneoraceae, 259 Simaroubaceae, 260 Irvingiaceae, 261 Surianaceae, 262 Zygophyllaceae, 263 *Nitrariaceae*, 264 Balanitaceae, 265 *Peganaceae*, 266 Meliaceae, 267 *Kirkiaceae*, 268 Ptaeroxylaceae, 269 Tepuianthaceae, 270 Burseraceae, 271 Anacardiaceae (incl. Podoaceae)

Leitneriales
272 Leitneriaceae

Coriariales
273 Coriariaceae

Linales
274 *Hugoniaceae*, 275 Linaceae, 276 Ctenolophonaceae, 277 Ixonanthaceae, 278 Humiriaceae, 279 Erythroxylaceae

Geraniales
280 Oxalidaceae, 281 Lepidobotryaceae, 282 *Hypseocharitaceae*, 283 *Biebersteiniaceae*, 284 Geraniaceae, 285 *Dirachmaceae*, 286 *Ledocarpaceae*, 287 *Rhynchothecaceae*, 288 *Vivianiaceae*

Balsaminales
289 Balsaminacea

Tropaeolales
290 Tropaeolaceae

Limnanthales
291 Limnanthaceae

Polygalales
292 Malpighiaceae, 293 Trigoniaceae (incl. Euphroniaceae), 294 Vochysiaceae, 295 Polygalaceae, 296 Krameriaceae, 297 Tremandraceae

CELASTRANAE
Celastrales
Icacinineae
298 Aquifoliaceae, 299 Phellinaceae, 300 Icacinaceae, 301 Sphenostemonaceae, 302 Cardiopteridaceae, 303 *Brexiaceae*
Celastrineae
304 Celastraceae (incl. Canotiaceae), 305 Goupiaceae, 306 Lophopyxidaceae, 307 Stackhousiaceae, 308 Salvadoraceae, 309 Corynocarpaceae
Santalales
310 Olacaceae, 311 Opiliaceae, 312 *Octoknemataceae*, 313 Medusandraceae, 314 Santalaceae, 315 Misodendraceae, 316 Loranthaceae, 317 Viscaceae, 318 Eremolepidaceae

RHAMNANAE
Rhamnales
319 Rhamnaceae
Elaeagnales
320 Elaeagnaceae

PROTEANAE
Proteales
321 Proteaceae

VITANAE
Vitales
322 Vitaceae, 323 Leeaceae

CORNANAE
Hydrangeales
Tetracarpaeineae
324 *Tetracarpaeaceae*
Escalloniineae
325 Escalloniaceae, 326 *Argophyllaceae*, 327 Griseliniaceae, 328 *Carpodetaceae*, 329 *Polyosmataceae*, 330 Montiniaceae, 331 Melanophyllaceae, 332 Columelliaceae, 333 Alseuosmiaceae, 334 *Dulongiaceae*, 335 *Tribelaceae*
Hydrangeineae
336 Hydrangeaceae, 337 Roridulaceae, 338 *Pottingeriaceae*, 339 Pterostemonaceae
Cornales
340 *Davidiaceae*, 341 *Nyssaceae*, 342 Cornaceae, 343 *Curtisiaceae*, 344 *Mastixiaceae*, 345 Aucubaceae, 346 Garryaceae, 347 Alangiaceae
Aralidiales
348 Aralidiaceae
Torricelliales
349 Torricelliaceae
Apiales
Helwingiinae
350 Helwingiaceae
Apiinae
351 Araliaceae, 352 Apiaceae
Pittosporales
353 Pittosporaceae
Biblydales
354 Biblydaceae
Dipsacales
355 Caprifoliaceae, 356 *Viburnaceae*, 357 *Sambucaceae*, 358 Adoxaceae, 359 Valerianaceae, 360 Triplostegiaceae, 361 Dipsacaceae, 362 Morinaceae
Cynomoriales
363 Cynomoriaceae

LAMIIDAE

GENTIANANAE
Gentianales
364 *Desfontainiaceae*, 365 Loganiaceae, 366 *Spigeliaceae*, 367 Rubiaceae, 368 Theligonaceae, 369 Carlemanniaceae, 370 Dialypetalanthaceae, 371 Gentiananceae, 372 Saccifoliaceae, 373 Menyanthaceae, 374 *Plocospermataceae*, 375 Apocynaceae, 376 Asclepiadaceae
Oleales
377 Oleaceae

LOASANAE
Loasales
378 Loasaceae

SOLANANAE
Solanales
379 Solanaceae, 380 *Nolanaceae*, 381 Duckeodendraceae, 382 *Sclerophylacaceae*, 383 Goetzeaceae
Convolvulales
384 Convolvulaceae, 385 Cuscutaceae
Polemoniales
386 Cobaeaceae, 387 Polemoniaceae

Juncaginales
438 Juncaginaceae, 439 Lilaeaceae, 440 *Maundiaceae*
Potamogetonales
441 Potamogetonaceae, 442 *Ruppiaceae*
Posidoniales
443 Posidoniaceae
Zosterales
444 Zosteraceae
Cymodoceales
445 Zannichelliaceae, 446 Cymodoceaceae
Najadales
447 Najadaceae

TRIURIDIDAE

TRIURIDANAE
Triuridales
448 Triuridaceae

LILIIDAE

LILIANAE
Liliales
449 Melanthiaceae (incl. Colchicaceae), 450 *Calochortaceae*, 451 Iridaceae, 452 *Geosiridaceae*, 453 Tecophilaeaceae, 454 Cyanastraceae, 455 Eriospermaceae, 456 *Medeolaceae*, 457 Liliaceae
Burmanniales
458 Burmanniaceae, 459 Corsiaceae
Amaryllidales
460 Asphodelaceae (incl. Aloaceae, Anthericaceae), 461 Xanthorrhoeaceae, 462 Dasypogonaceae (incl. Calectasiaceae, Lomandraceae), 463 Aphyllanthaceae, 464 Hyacinthaceae, 465 Alliaceae, 466 *Hesperocallidaceae*, 467 Funkiaceae [Hostaceae], 468 Agavaceae, 469 Hemerocallidaceae, 470 Phormiaceae, 471 Blandfordiaceae, 72 Doryanthaceae, 473 Amaryllidaceae, 474 Ixioliriaceae
Asparagales
475 Convallariaceae, 476 Ruscaceae, 477 Asparagaceae, 478 Dracaenaceae, 479 *Nolinaceae*, 480 Herreriaceae, 481 Asteliaceae, 482 Hanguanaceae
Smilacales
483 *Luzuriagaceae*, 484 Philesiaceae, 485 Petermanniaceae, 486 Rhipogonaceae, 487 Smilacaceae
Dioscoreales
488 Trichopodaceae, 489 Stenomeridaceae, 490 Dioscoreaceae, 491 Stemonaceae, 492 Trilliaceae
Taccales
493 Taccaceae
Alstroemeriales
494 Alstroemeriaceae
Haemodorales
495 Haemodoraceae, 496 *Conostylidaceae*, 497 Hypoxidaceae
Orchidales
498 Orchidaceae

PONTEDERIANAE
Pontederiales
499 Pontederiaceae
Phylidrales
500 Phylidraceae

BROMELIANAE
Bromeliales
   501 Bromeliaceae
Velloziales
   502 Velloziaceae

ZINGIBERANAE
Zingiberales
   503 Strelitziaceae, 504 Musaceae, 505 Heliconiaceae, 506 Lowiaceae, 507 Zingiberaceae, 508 Costaceae, 509 Cannaceae, 510 Marantaceae

JUNCANAE
Juncales
   511 Juncaceae, 512 Thurniaceae
Cyperales
   513 Cyperaceae

HYDATELLANAE
Hydatellales
   514 Hydatellaceae

COMMELINANAE
Commelinales
   Commelinineae
     515 Commelinaceae, 516 Mayacaceae
   Xyridineae
     517 Xyridaceae, 518 Rapateaceae
   Eriocaulineae
     519 Eriocaulaceae
Restionales
   Flagellariineae
     520 Flagellariaceae, 521 Joinvilleaceae
   Restionineae
     522 Restionaceae, 523 Anarthriaceae, 524 Ecdeiocoleaceae
   Centrolepidineae
     525 Centrolepidaceae
Poales
   526 Poaceae

ARECIDAE

ARECANAE
Arecales
   527 Arecaceae
Cyclanthales
   528 Cyclanthaceae

PANDANANAE
Pandanales
   529 Pandanaceae

ARANAE
Arales
   530 Araceae (incl. Acoraceae), 531 Lemnaceae

TYPHANAE
Typhales
   532 *Sparganiaceae*, 533 Typhaceae.

THE CRONQUIST SYSTEM

Taken from A. Cronquist, *An Integrated System of Classification of Flowering Plants* (1981), with annotations of a few changes made in *The Evolution and Classification of Flowering Plants*, ed. 2 (1988). An earlier version appeared in ed. 1 of the latter work (1968).

MAGNOLIOPSIDA

MAGNOLIIDAE

Magnoliales
> 1 Winteraceae, 2, Degeneriaceae, 3 Himantandraceae, 4 Eupomatiaceae, 5 Austrobaileyaceae, 6 Magnoliaceae, 7 Lactoridaceae, 8 Annonaceae, 9 Myristacaceae, 10 Canellaceae

Laurales
> 11 Amborellaceae, 12 Trimeniaceae, 13, Monimiaceae, 14 Gomortegaceae, 15 Calycanthaceae, 16 Idiospermaceae, 17 Lauraceae, 18 Hernandiaceae

Piperales
> 19 Chloranthaceae, 20 Saururaceae, 21 Piperaceae

Aristolochiales
> 22 Aristolochiaceae

Illiciales
> 23 Illiciaceae, 24 Schisandraceae

Nymphaeales
> 25 Nelumbonaceae, 26 Nymphaeaceae, 27 *Barclayaceae*, 28 Cabombaceae, 29 Ceratophyllaceae

Ranunculales
> 30 Ranunculaceae (incl. Glaucidiaceae), 31 Circaeasteraceae, 32 Berberidaceae, 33 Sargentodoxaceae, 34 Lardizabalaceae, 35 Menispermaceae, 36 Coriariaceae, 37 Sabiaceae (incl. Meliosmaceae)

Papaverales
> 38 Papaveraceae, 39 *Fumariaceae*

HAMAMELIDIDAE

Trochodendrales
> 40 Tetracentraceae, 41 Trochodendraceae

Hamamelidales
> 42 Cercidiphyllaceae, 43 Eupteleaceae, 44 Platanaceae, 45 Hamamelidaceae, 46 Myrothamnaceae

Daphniphyllales
> 47 Daphniphyllaceae

Didymelales
> 48 Didymelaceae

Eucommiales
> 49 Eucommiaceae

Urticales
> 50 Barbeyaceae, 51 Ulmaceae, 52 Cannabaceae, 53 Moraceae, 54 Cecropiaceae, 55 Urticaceae

Leitneriales
> 56 Leitneriaceae

Juglandales
> 57 Rhoipteleaceae, 58 Juglandaceae

Myricales
> 59 Myricaceae

Fagales
    60 Balanopaceae, 61 Fagaceae, 62 Betulaceae
Casuarinales
    63 Casuarinaceae

CARYOPHYLLIDAE

Caryophyllales
    64 Phytolaccaceae (incl. Agdestidaceae, Barbeuiaceae, Gisekiaceae, Stegnospermataceae), 65 Achatocarpaceae, 66 Nyctaginaceae, 67 Aizoaceae, 68 Didiereaceae, 69 Cactaceae, 70 Chenopodiaceae (incl. Halophytaceae), 71 Amaranthaceae, 72 Portulacaceae (incl. Hectorellaceae), 73 Basellaceae, 74 Molluginaceae, 75 Caryophyllaceae (incl. Illecebraceae)
Polygonales
    76 Polygonaceae
Plumbaginales
    77 Plumbaginaceae

DILLENIIDAE

Dilleniales
    78 Dilleniaceae, 79 Paeoniaceae

Theales
    80 Ochnaceae (incl. Diegodendraceae, Strasburgeriaceae), 81 Sphaerosepalaceae, 82 Sarcolaenaceae, 83 Dipterocarpaceae, 84 Caryocaraceae, 85 Theaceae (incl. Asteropeiaceae), 86 Actinidiaceae, 87 Scytopetalaceae, 88 Pentaphylacaceae, 89 Tetrameristaceae, 90 Pellicieraceae, 91 Oncothecaceae, 92 Marcgraviaceae, 93 Quiinaceae, 94 Elatinaceae, 95 Paracryphiaceae, 96 Medusagynaceae, 97 Clusiaceae
Malvales
    98 Elaeocarpaceae, 99 Tiliaceae, 100 Sterculiaceae, 101 Bombacaceae, 102 Malvaceae
Lecythidales
    103 Lecythidaceae
Nepenthales
    104 Sarraceniaceae, 105 Nepenthaceae, 106 Droseraceae
Violales
    107 Flacourtiaceae (incl. Plagiopteraceae), 108 Peridiscaceae, 109 Bixaceae (incl. Cochlospermaceae), 110 Cistaceae, 111 Huaceae, 112 Lacistemataceae, 113 Scyphostegiaceae, 114 Stachyuraceae, 115 Violaceae, 116 Tamaricaceae, 117 Frankeniaceae, 118 Dioncophyllaceae, 119 Ancistrocladaceae, 120 Turneraceae, 121 Malesherbiaceae, 122 Passifloraceae, 123 Achariaceae, 124 Caricaceae, 125 Fouquieriaceae, 126 Hoplestigmataceae, 127 Cucurbitaceae, 128 Datiscaceae, 129 Begoniaceae, 130 Loasaceae
Salicales
    131 Salicaceae
Capparales
    132 Tovariaceae, 133 Capparaceae, 134 Brassicaceae, 135 Moringaceae, 136 Resedaceae
Batales
    137 Gyrostemonaceae, 138 Bataceae
Ericales
    139 Cyrillaceae, 140 Clethraceae, 141 Grubbiaceae, 142 Empetraceae, 143 Epacridaceae, 144 Ericaceae 145 *Pyrolaceae*, 146 *Monotropaceae*
Diapensiales
    147 Diapensiaceae
Ebenales
    148 Sapotaceae, 149 Ebenaceae, 150 Styracaceae, 151 Lissocarpaceae, 152 Symplocaceae

Primulales

153 Theophrastaceae, 154 Myrsinaceae, 155 Primulaceae

ROSIDAE

Rosales

156 Brunelliaceae, 157 Connaraceae, 158 Eucryphiaceae, 159 Cunoniaceae, 160 Davidsoniaceae, 161 Dialypetalanthaceae, 162 Pittosporaceae, 163 Byblidaceae (incl. Roridulaceae), 164 Hydrangeaceae, 165 Columelliaceae, 166 Grossulariaceae (incl. Escalloniaceae, Montiniaceae, Pterostemonaceae), 167 Greyiaceae, 168 Bruniaceae, 169 Anisophylleaceae, 170 Alseuosmiaceae, 171 Crassulaceae, 172 Cephalotaceae, 173 Saxifragaceae (incl. Eremosynaceae, Parnassiaceae, Penthoraceae, Vahliaceae), 174 Rosaceae, 175 Neuradaceae, 176 Crossosomataceae, 177 Chrysobalanaceae, 178 Surianaceae (incl. Stylobasiaceae), 179 Rhabdodendraceae

Fabales

180 *Mimosaceae*, 181 *Caesalpiniaceae*, 182 Fabaceae

Proteales

183 Elaeagnaceae, 184 Proteaceae

Podostemales

185 Podostemaceae

Haloragales

186 Haloragaceae, 187 Gunneraceae

Myrtales

188 *Sonneratiaceae,* 189 Lythraceae (incl. Rhynchocalycaceae in 1981, not 1988), 190 Penaeaceae, 191 Crypteroniaceae, 192 Thymelaeaceae, 193 Trapaceae, 194 Myrtaceae, 195 *Punicaceae*, 196 Onagraceae, 197 Oliniaceae, 198 Melastomataceae, 199 Combretaceae

Rhizophorales

200 Rhizophoraceae

Cornales

201 Alangiaceae, 202 *Nyssaceae,* 203 Cornaceae (incl. Aralidiaceae, Aucubaceae, Griseliniaceae, Helwingiaceae, Melanophyllaceae, Torricelliaceae), 204 Garryaceae

Santalales

205 Medusandraceae, 206 Dipentodontaceae, 207 Olacaceae, 208 Opiliaceae, 209 Santalaceae, 210 Misodendraceae, 211 Loranthaceae, 212 Viscaceae, 213 Eremolepidaceae, 214 Balanophoraceae (incl. Cynomoriaceae)

Rafflesiales

215 Hydnoraceae, 216 *Mitrastemonaceae*, 217 Rafflesiaceae

Celastrales

218 Geissolomataceae, 219 Celastraceae (incl. Canotiaceae, Goupiaceae, Lophopyxidaceae), 220 *Hippocrateaceae*, 221 Stackhousiaceae, 222 Salvadoraceae, 223 Aquifoliaceae (incl. Phellinaceae, Sphenostemonaceae), 224 Icacinaceae, 225 Aextoxicaceae, 226 Cardiopteridaceae, 227 Corynocarpaceae, 228 Dichapetalaceae

Euphorbiales

229 Buxaceae, 230 Simmondsiaceae, 231 Pandaceae, 232 Euphorbiaceae

Rhamnales

233 Rhamnaceae, 234 Leeaceae, 235 Vitaceae

Linales

236 Erythroxylaceae, 237 Humiriaceae, 238 Ixonanthaceae, 239 *Hugoniaceae*, 240 Linaceae

Polygalales

241 Malpighiaceae, 242 Vochysiaceae (incl. Euphroniaceae), 243 Trigoniaceae, 244 Tremandraceae, 245 Polygalaceae (incl. Emblingiaceae), 246 *Xanthophyllaceae*, 247 Krameriaceae

Sapindales
248 Staphyleaceae, 249 Melianthaceae, 250 Bretschneideraceae, 251 Akaniaceae, 252 Sapindaceae (incl. Ptaeroxylaceae), 253 Hippocastanaceae, 254 Aceraceae, 255 Burseraceae, 256 Anacardiaceae (incl. Podoaceae), 257 *Julianiaceae*, 258 Simaroubaceae (incl. Irvingiaceae), 259 Cneoraceae, 260 Meliaceae, 261 Rutaceae, 262 Zygophyllaceae (incl. Balanitaceae)

Geraniales
263 Oxalidaceae (incl. Lepidobotryaceae), 264 Geraniaceae, 265 Limnanthaceae, 266 Tropaeolaceae, 267 Balsaminaceae

Apiales
268 Araliaceae, 269 Apiaceae

ASTERIDAE

Gentianales
270 Loganiaceae, 271 Retziaceae, 272 Gentianaceae, 273 Saccifoliaceae, 274 Apocynaceae, 275 Asclepiadaceae

Solanales
276 Duckeodendraceae, 277 *Nolanaceae*, 278 Solanaceae, 279 Convolvulaceae, 280 *Cuscutaceae* 281 Menyanthaceae, 282 Polemoniaceae (incl. Cobaeaceae), 283 Hydrophyllaceae

Lamiales
284 Lennoaceae, 285 Boraginaceae, 286 Verbenaceae (incl. Avicenniaceae, Phrymaceae, Stilbaceae), 287 Lamiaceae (incl. Tetrachondraceae)

Callitrichales
288 Hippuridaceae, 289 Callitrichaceae, 290 Hydrostachyaceae

Plantaginales
291 Plantaginaceae

Scrophulariales
292 Buddlejaceae, 293 Oleaceae, 294 Scrophulariaceae, 295 Globulariaceae, 296 Myoporaceae, 297 *Orobanchaceae*, 298 Gesneriaceae, 299 Acanthaceae, 300 Pedaliaceae, 301 Bignoniaceae, 302 *Mendonciaceae*, 303 Lentibulariaceae

Campanulales
304 Pentaphragmataceae, 305 *Sphenocleaceae*, 306 Campanulaceae, 307 Stylidiaceae, 308 *Donatiaceae*, 309 *Brunoniaceae*, 310 Goodeniaceae

Rubiales
311 Rubiaceae, 312 Theligonaceae,

Dipsacales
313 Caprifoliaceae (incl. Carlemanniaceae), 314 Adoxaceae, 315 Valerianaceae (incl. Triplostegiaceae), 316 Dipsacaceae (incl. Morinaceae)

Calycerales
317 Calyceraceae

Asterales
318 Asteraceae

LILIOPSIDA

ALISMATIDAE

Alismatales
319 Butomaceae, 320 Limnocharitaceae, 321 Alismataceae

Hydrocharitales
322 Hydrocharitaceae

Najadales
323 Aponogetonaceae, 324 Scheuchzeriaceae, 325 Juncaginaceae (incl. Lilaeaceae), 326 Potamogetonaceae, 327 *Ruppiaceae*, 328 *Najadaceae*, 329 Zannichelliaceae, 330 Posidoniaceae, 331 Cymodoceaceae, 332 Zosteraceae

Triuridales
333 *Petrosaviaceae*, 334 Triuridaceae

ARECIDAE

Arecales
335 Arecaceae
Cyclanthales
336 Cyclanthales
Pandanales
337 Pandanaceae
Arales
338 Araceae (incl. Acoraceae), 339 Lemnaceae

COMMELINIDAE

Commelinales
340 Rapateaceae, 341 Xyridaceae, 342 Mayacaceae, 343 Commelinaceae
Eriocaulales
344 Eriocaulaceae
Restionales
345 Flagellariaceae, 346 Joinvilleaceae, 347 Restionaceae (incl. Anarthriaceae, Ecdeiocoleaceae), 348 Centrolepidaceae
Juncales
349 Juncaceae, 350 Thurniaceae
Cyperales
351 Cyperaceae, 352 Poaceae
Hydatellales
353 Hydatellaceae
Typhales
354 *Sparganiaceae,* 355 Typhaceae

ZINGIBERIDAE

Bromeliales
356 Bromeliaceae
Zingiberales
357 Strelitziaceae, 358 Heliconiaceae, 359 Musaceae, 360 Lowiaceae, 361 Zingiberaceae, 362 Costaceae, 363 Cannaceae, 364 Marantaceae

LILIIDAE

Liliales
365 Philydraceae, 366 Pontederiaceae, 367 Haemodoraceae, 368 Cyanastraceae, 369 Liliaceae (incl. Alliaceae, Alstroemeriaceae, Amaryllidaceae, Anthericaceae, Aphyllanthaceae, Asparagaceae, Asphodelaceae, Asteliaceae, Blandfordiaceae, Colchicaceae, Convallariaceae, Eriospermaceae, Hemerocallidaceae, Herreriaceae, Hostaceae, Hyacinthaceae, Hypoxidaceae, Ixioliriaceae, Melanthiaceae, Ruscaceae, Tecophilaeaceae, Trilliaceae), 370 Iridaceae, 371 Velloziaceae, 372 Aloeaceae, 373 Agavaceae (incl. Doryanthaceae, Dracaenaceae, Phormiaceae), 374 Xanthorrhoeaceae (incl. Calectasiaceae, Dasypogonaceae, Lomandraceae), 375 Hanguanaceae, 376 Taccaceae, 377 Stemonaceae, 378 Smilacaceae (incl. Petermanniaceae, Philesiaceae, Rhipogonaceae), 379 Dioscoreaceac (incl. Trichopodaceae)
Orchidales
381 *Geosiridaceae,* 381 Burmanniaceae, 382 Corsiaceae, 383 Orchidaceae